OIL INFORMATION

2003
with 2002 data

DONNÉES SUR LE PÉTROLE

INTERNATIONAL ENERGY AGENCY
9, rue de la Fédération,
75739 Paris Cedex 15, France

The International Energy Agency (IEA) is an autonomous body which was established in November 1974 within the framework of the Organisation for Economic Co-operation and Development (OECD) to implement an international energy programme.

It carries out a comprehensive programme of energy co-operation among twenty-six* of the OECD's thirty Member countries. The basic aims of the IEA are:

- to maintain and improve systems for coping with oil supply disruptions;
- to promote rational energy policies in a global context through co-operative relations with non-member countries, industry and international organisations;
- to operate a permanent information system on the international oil market;
- to improve the world's energy supply and demand structure by developing alternative energy sources and increasing the efficiency of energy use;
- to assist in the integration of environmental and energy policies.

* *IEA Member countries: Australia, Austria, Belgium, Canada, the Czech Republic, Denmark, Finland, France, Germany, Greece, Hungary, Ireland, Italy, Japan, the Republic of Korea, Luxembourg, the Netherlands, New Zealand, Norway, Portugal, Spain, Sweden, Switzerland, Turkey, the United Kingdom, the United States. The European Commission also takes part in the work of the IEA.*

ORGANISATION FOR
ECONOMIC CO-OPERATION
AND DEVELOPMENT

Pursuant to Article 1 of the Convention signed in Paris on 14th December 1960, and which came into force on 30th September 1961, the Organisation for Economic Co-operation and Development (OECD) shall promote policies designed:

- to achieve the highest sustainable economic growth and employment and a rising standard of living in Member countries, while maintaining financial stability, and thus to contribute to the development of the world economy;
- to contribute to sound economic expansion in Member as well as non-member countries in the process of economic development; and
- to contribute to the expansion of world trade on a multilateral, non-discriminatory basis in accordance with international obligations.

The original Member countries of the OECD are Austria, Belgium, Canada, Denmark, France, Germany, Greece, Iceland, Ireland, Italy, Luxembourg, the Netherlands, Norway, Portugal, Spain, Sweden, Switzerland, Turkey, the United Kingdom and the United States. The following countries became Members subsequently through accession at the dates indicated hereafter: Japan (28th April 1964), Finland (28th January 1969), Australia (7th June 1971), New Zealand (29th May 1973), Mexico (18th May 1994), the Czech Republic (21st December 1995), Hungary (7th May 1996), Poland (22nd November 1996), the Republic of Korea (12th December 1996) and Slovakia (28th September 2000). The Commission of the European Communities takes part in the work of the OECD (Article 13 of the OECD Convention).

AGENCE INTERNATIONALE DE L'ÉNERGIE
9, rue de la Fédération,
75739 Paris Cedex 15, France

L'agence internationale de l'énergie (AIE) est un organe autonome institué en novembre 1974 dans le cadre de l'Organisation de coopération et de développement économiques (OCDE) afin de mettre en œuvre un programme international de l'énergie.

Elle applique un programme général de coopération dans le domaine de l'énergie entre vingt-six* des trente pays Membres de l'OCDE. Les objectifs fondamentaux de l'AIE sont les suivants :

- tenir à jour et améliorer des systèmes permettant de faire face à des perturbations des approvisionnements pétroliers ;
- œuvrer en faveur de politiques énergétiques rationnelles dans un contexte mondial grâce à des relations de coopération avec les pays non membres, l'industrie et les organisations internationales ;
- gérer un système d'information continue sur le marché international du pétrole ;
- améliorer la structure de l'offre et de la demande mondiales d'énergie en favorisant la mise en valeur de sources d'énergie de substitution et une utilisation plus rationnelle de l'énergie ;
- contribuer à l'intégration des politiques d'énergie et d'environnement.

** Pays Membres de l'AIE : Allemagne, Australie, Autriche, Belgique, Canada, Danemark, Espagne, États-Unis, Finlande, France, Grèce, Hongrie, Irlande, Italie, Japon, Luxembourg, Norvège, Nouvelle-Zélande, Pays-Bas, Portugal, République de Corée, République Tchèque, Royaume-Uni, Suède, Suisse et Turquie. La Commission des Communautés européennes participe également aux travaux de l'AIE.*

ORGANISATION DE COOPÉRATION
ET DE DÉVELOPPEMENT
ÉCONOMIQUES

En vertu de l'article 1er de la Convention signée le 14 décembre 1960, à Paris, et entrée en vigueur le 30 septembre 1961, l'Organisation de Coopération et de Développement Économiques (OCDE) a pour objectif de promouvoir des politiques visant :

- à réaliser la plus forte expansion de l'économie et de l'emploi et une progression du niveau de vie dans les pays Membres, tout en maintenant la stabilité financière, et à contribuer ainsi au développement de l'économie mondiale ;
- à contribuer à une saine expansion économique dans les pays Membres, ainsi que les pays non membres, en voie de développement économique ;
- à contribuer à l'expansion du commerce mondial sur une base multilatérale et non discriminatoire conformément aux obligations internationales.

Les pays Membres originaires de l'OCDE sont : l'Allemagne, l'Autriche, la Belgique, le Canada, le Danemark, l'Espagne, les États-Unis, la France, la Grèce, l'Irlande, l'Islande, l'Italie, le Luxembourg, la Norvège, les Pays-Bas, le Portugal, le Royaume-Uni, la Suède, la Suisse et la Turquie. Les pays suivants sont ultérieurement devenus Membres par adhésion aux dates indiquées ci-après : le Japon (28 avril 1964), la Finlande (28 janvier 1969), l'Australie (7 juin 1971), la Nouvelle-Zélande (29 mai 1973), le Mexique (18 mai 1994), la République Tchèque (21 décembre 1995), la Hongrie (7 mai 1996), la Pologne (22 novembre 1996), la République de Corée (12 décembre 1996) et la République Slovaque (28 septembre 2000). La Commission des Communautés européennes participe aux travaux de l'OCDE (article 13 de la Convention de l'OCDE).

TABLE OF CONTENTS

TABLE DES MATIERES

PARTIE I : SOURCES ET DEFINITIONS

PARTIE II : TENDANCES DU MARCHE MONDIAL DU PETROLE

PARTIE III : DONNEES DETAILLEES SUR LE PETROLE DANS LES PAYS DE L'OCDE

PARTIE IV : SERIES HISTORIQUES DES PAYS DE L'OCDE

PART I

SOURCES AND DEFINITIONS

PREMIERE PARTIE

SOURCES ET DEFINITIONS

1. INTRODUCTION

The *Oil Information 2003* is the latest edition of a publication that has been produced annually since 1989, replacing the publication Annual Oil and Gas Statistics.

Part II, *World Oil Developments,* provides summary tables of world oil market developments, with time series back to 1971.

Part III, *Detailed OECD Oil Data,* provides in tabular form, a more detailed and comprehensive picture of oil supply, demand and end-use consumption for the OECD by region and individual countries.

Part IV, *Historical Series* shows time series of major oil flows for all years beginning in 1973 and ending in 2002, where data are available.

Note that the references to OECD in this publication include the 30 Member countries: Australia, Austria, Belgium, Canada, Czech Republic, Denmark, Finland, France, Germany, Greece, Hungary, Iceland, Ireland, Italy, Japan, Korea, Luxembourg, Mexico, the Netherlands, New Zealand, Norway, Poland, Portugal, Slovak Republic, Spain, Sweden, Switzerland, Turkey, the United Kingdom, and the United States.

Statistics for OECD countries are based on data submissions from national administrations to the Secretariat. Statistics up to 2001 are from the Annual Oil Statistics (AOS) database, while oil data for 2002 are mainly based on the Monthly Oil and Gas Statistics (MOS) database. These monthly oil data, submitted by member governments, are historically lower than the annual statistics as they exclude demand for certain products (e.g. petroleum coke in certain countries). To arrive at a comparable time series between AOS and MOS data, the 2002 monthly data are adjusted upwards for under-reporting. Data for 2002 may therefore differ from similar statistics published in the publication *Oil, Gas, Coal and Electricity Quarterly Statistics* (MOS data). Please see notes related to individual countries for more information on data sources and estimations.

All data for Parts III and IV and selected information from Part II are available on CD-ROM suitable for use on IBM-compatible personal computers. An order form has been provided in the back of this publication.

In addition, a data service is available on the internet. It includes unlimited access through an annual subscription as well as the possibility to obtain data on a pay-per-view basis. Details are available at http:\\www.iea.org.

Annual oil data are collected by the Energy Statistics Division (ESD) of the IEA Secretariat, headed by Mr. Jean-Yves Garnier. Currently Mr. Bruno Castellano is responsible for the annual oil data for OECD countries. Ms. Sharon Michel contributed her desk-top publishing and secretarial support. Ms. Mieke Reece has overall responsibility for this publication.

Enquiries about data or methodology in this publication should be addressed to:

Mrs. Mieke Reece
Energy Statistics Division,
International Energy Agency, OECD
9, rue de la Fédération
75739 PARIS Cedex 15
France

Fax: (33 1) 40.57.66.49
Internet: mieke.reece@iea.org

2. RECENT DATA REVISIONS

2.1 OECD Data

The IEA Secretariat finalised processing the latest submissions of energy statistics from OECD Member countries on 31 March 2003 (covering the period 1960 - 2001). These data were subsequently published in *IEA/OECD Energy Statistics of OECD Countries 2000 - 2001* and in *IEA/OECD Energy Balances of OECD Countries 2000 - 2001*.

Oil supply and demand data for all OECD countries are available for years 1971-2001. For 2002, only supply data are available; end-use consumption data usually lag one to two years.

In addition, supply and demand data are available for years 1960-1970, with the following exceptions:

- Czech Republic 1971 onwards
- Hungary 1965 onwards
- Korea 1970 onwards
- Mexico 1971 onwards
- Slovak Republic 1971 onwards

As a result, regional totals will not be calculated for years 1960-1970, inclusive.

2.2 Non-OECD Data

Data for non-OECD countries are primarily from IEA's database on World Energy Statistics. The database covers supply and demand for all forms of energy for over 100 non-OECD countries for the period 1971 to 2001. The data have been collected from national statistical publications and through collaboration with other international organisations and the energy industry. Non-OECD data for 2001 in Part II of *Oil Information 2003* are provisional. The complete and final results obtained from this work will be published in the annual IEA publications *Energy Statistics of Non-OECD Countries 2000-2001* and *Energy Balances of Non-OECD Countries 2000-2001*.

3. GENERAL DEFINITIONS

3.1. Product Definitions

Inputs to refineries

Crude Oil

Crude oil is a mineral oil of natural origin comprising a mixture of hydrocarbons and associated impurities, such as sulphur. It exists in the liquid phase under normal surface temperature and pressure and its physical characteristics (density, viscosity, etc.) are highly variable. This category includes field or lease condensate recovered from associated and non-associated gas where it is commingled with the commercial crude oil stream.

Natural Gas Liquids (NGL)

NGL are liquid or liquefied hydrocarbons recovered from natural gas in separation facilities or gas processing plants. Natural gas liquids include ethane, propane, butane (normal and iso-), (iso) pentane and pentanes plus (sometimes referred to as natural gasoline or plant condensate).

Refinery Feedstocks

A refinery feedstock is a processed oil destined for further processing (e.g. straight run fuel oil or vacuum gas oil) excluding blending. With further processing, it will be transformed into one or more components and/or finished products. This definition includes returns from the petrochemical industry to the refining industry (e.g. pyrolysis gasoline, C4 fractions, gasoil and fuel oil fractions).

Other Hydrocarbons (than crude oil and NGL)

This category includes synthetic crude oil from tar sands, oil shale, etc., liquids from coal liquefaction, hydrogen and emulsified oils (e.g. Orimulsion). Although not hydrocarbons, additives are also included. Additives are added to or blended with a product to modify fuel properties (octane, cetane, cold properties, etc.) e.g. alcohols (methanol, ethanol), ethers (such as MTBE (methyl tertiary butyl ether), ETBE (ethyl tertiary butyl ether), TAME (tertiary amyl methyl ether) or esters (e.g. rapeseed or dimethylester, etc.) and chemical compounds (such as TML (tetramethyl lead) or TEL (tetraethyl lead)) and detergents.

Outputs from refineries

Refinery Gas (not liquefied)

Refinery gas includes a mixture of non-condensable gases mainly consisting of hydrogen, methane, ethane and olefins obtained during distillation of crude oil or treatment of oil products (e.g. cracking) in refineries. This also includes gases which are returned from the petrochemical industry.

Ethane

A naturally gaseous straight-chain hydrocarbon (C_2H_6), extracted from natural gas and refinery gas streams.

Liquefied Petroleum Gases (LPG)

LPG is light saturated paraffinic hydrocarbons derived from the refinery processes, crude oil stabilisation and natural gas processing plants. They consist mainly of propane (C_3H_8) and butane (C_4H_{10}) or a combination of the two. They are normally liquefied under pressure for transportation and storage.

Naphtha

Naphtha is a feedstock destined for the petrochemical industry (e.g. ethylene manufacture or aromatics production). Naphtha comprises material in the 30°C and 210°C distillation range or part of this range.

Motor Gasoline

Motor gasoline consists of a mixture of light hydrocarbons distilling between 35°C and 215°C. It is used as a fuel for land based spark ignition engines.

Motor gasoline may include additives, oxygenates and octane enhancers, including lead compounds such as TEL (Tetraethyl lead) and TML (tetramethyl lead).

- *Unleaded Motor Gasoline:* motor gasoline where lead compounds have not been added to enhance octane rating. It may contain traces of organic lead.
- *Leaded Motor Gasoline:* motor gaoling with TEL (tetraethyl lead) and/or TML (tetramethyl lead) added to enhance octane rating.

This category includes motor gasoline blending components (excluding additives/oxygenates), e.g. alkylates, isomerate, reformate, cracked gasoline destined for use as finished motor gasoline.

Aviation Gasoline

This is motor spirit prepared especially for aviation piston engines, with an octane number suited to the engine, a freezing point of -60°C and a distillation range usually within the limits of 30°C and 180°C.

Gasoline Type Jet Fuel (Naphtha type Jet Fuel or JP4)

This includes all light hydrocarbon oils for use in aviation turbine power units, distilling between 100°C and 250°C. They are obtained by blending kerosenes and gasoline or naphthas in such a way that the aromatic content does not exceed 25% in volume, and the vapour pressure is between 13.7kPa and 20.6kPa.

Kerosene Type Jet Fuel

This is a distillate used for aviation turbine power units. It has the same distillation characteristics between 150°C and 300°C (generally not above 250°C) and flash point as kerosene. In addition, it has particular specifications (such as freezing point) which are established by the International Air Transport Association (IATA).

Other Kerosene

Kerosene comprises refined petroleum distillate and is used in sectors other than aircraft transport. It distils between 150ºC and 300ºC.

Gas/Diesel Oil (Distillate Fuel Oil)

Gas/diesel oil is primarily a medium distillate distilling between 180ºC and 380ºC. Several grades are available depending on uses: diesel oil for diesel compression ignition (cars, trucks, marine, etc.); light heating oil for industrial and commercial uses; other gas oil including heavy gas oils which distil between 380°C and 540°C and which are used as petrochemical feedstocks.

Fuel Oil

This covers all residual (heavy) fuel oils (including those obtained by blending). Kinematic viscosity is above 10 cSt at 80ºC. The flash point is always above 50°C and density is always more than 0.90 kg/l.

- *Low-sulphur content:* Heavy fuel oil with sulphur content lower than 1%.
- *High-sulphur content:* Heavy fuel oil with sulphur content of 1% or higher.

White Spirit and SBP

White Spirit and SBP are defined as refined distillate intermediates with a distillation in the naphtha/kerosene range. They are sub-divided as:

i. *Industrial Spirit (SBP)*: Light oils distilling between 30° and 200°C. There are 7 or 8 grades of industrial spirit, depending on the position of the cut in the distillation range. The grades are defined according to the temperature difference between the 5% volume and 90% volume distillation points (which is not more than 60°C).

ii. *White Spirit*: Industrial spirit with a flash point above 30°C. The distillation range of white spirit is 135° to 200°C.

Lubricants

Lubricants are hydrocarbons produced from distillate or residue; they are mainly used to reduce friction between bearing surfaces. This category includes all finished grades of lubricating oil, from spindle oil to cylinder oil, and those used in greases, including motor oils and all grades of lubricating oil-based stocks.

Bitumen

Bitumen is a solid, semi-solid or viscous hydrocarbon with a colloidal structure, being brown to black in colour, obtained as a residue in the distillation of crude oil, by vacuum distillation of oil residues from atmospheric distillation. Bitumen is often referred to as asphalt and is primarily used for construction of roads and for roofing material. This category includes fluidized and cut-back bitumen.

Paraffin Waxes

These are saturated aliphatic hydrocarbons. These waxes are residues extracted when dewaxing lubricant oils. They have a crystalline structure which is more-or-less fine according to the grade. Their main characteristics are as follows: they are colourless, odourless and translucent, with a melting point above 45°C.

Petroleum Coke

Petroleum coke is a black solid residue, obtained mainly by cracking and carbonising residue feedstock, tar and pitches in processes such as delayed coking or fluid coking. It consists mainly of carbon (90 to 95%) and has a low ash content. It is used as a feedstock in coke ovens for the steel industry, for heating purposes, for electrode manufacture and for production of chemicals. The two most important qualities are "green coke" and "calcinated coke". This category also includes "catalyst coke" deposited on the catalyst during refining processes; this coke is not recoverable and is usually burned as refinery fuel.

Other Products

All products not specifically mentioned above, for example: tar and sulphur. This category also includes aromatics (e.g. BTX or benzene, toluene and xylene) and olefins (e.g. propylene) produced within refineries.

3.2. Definitions of Supply and Consumption Items

Please note that a major change in the reporting of the petrochemical flows took place a few years ago. Quantities of oil delivered to the petrochemical industry are reported in two sectors; the transformation and industry sectors. The quantities of oil delivered to the petrochemical industry which are returned to refineries are reported in the transformation sector, subdivision *'Petrochemical Industry'*. The amounts reported in the industry sector, *'Chemical (incl. Petrochemical)'* are the net consumption of this industry (i.e. gross deliveries reduced by the amounts of backflow).

Indigenous Production

This includes all production within national boundaries, including offshore production. Production includes only marketable production, and excludes volumes returned to formation. Such production concerns all crude oil, NGL, condensates and other hydrocarbons (including the receipts of additives).

Backflows

Backflows cover cracking residues and other products from the chemical and petrochemical industries, which are returned to refineries and used as feedstocks. Transfers from one refinery to another within the country are not included.

Direct Use

Quantities of crude oil or NGL which are used directly without being processed in refineries, for example crude oil used for electricity generation.

Refinery Intake

This is defined as the total amount of inputs to refinery processes. It covers crude oil, NGL, feedstocks and hydrocarbons of origin other than crude oil and NGL.

Refinery Output (Gross)

This is the production of finished products at a refinery or blending plant. This category excludes refinery losses, but includes refinery fuel.

Transfers

These are finished products which have been reclassified and re-entered for use in a refinery without having been delivered to final consumers. This heading also covers transfers between products which are reclassified after blending, for example transfers between kerosene type jet fuel and kerosene.

Primary Product Receipts

Covers indigenous or imported crude oil and NGL which are used directly without being processed in an oil refinery (for example crude oil used for electricity generation in Japan). See *Direct Use*, above.

Recycled Products

These are finished products which are returned through the refining and/or the marketing network after delivery to final consumers (for example used lubricants which are reprocessed).

International Marine Bunkers

Bunkers cover the quantities of fuels delivered to sea-going ships of all flags, including warships. Consumption by ships engaged in transport in inland and coastal waters is not included (see *Transport Sector*, below).

Imports and Exports:

Amounts are regarded as imported or exported when they have crossed the political boundaries of the country, whether customs clearance has taken place or not. Quantities of crude oil imported or exported under processing agreements (i.e. refining on account) are included. Quantities of oil in transit are excluded. Imports of refinery feedstocks include imported finished products destined for blending in refineries (see refinery feedstocks under product definitions). The import origin of crude oil and NGL is the country of production while for finished products, it is the country of last consignment.

Please note that due to time differences in recording imports and exports (voyage time, delays, re-routing, trans-shipment), the reported exports by one country may not equal the imports reported by importing countries. Additionally, due to differences in product classification, particularly regarding refinery feedstocks, there may be inconsistencies between imports and exports. As a result, total imports by the OECD regions may not equal the OECD regions' exports.

Net Imports

Net imports are calculated to be imports minus exports.

Stock Changes

Stock changes reflect the difference between opening stock level at the first day of the year and closing stock level at the last day of the year of stocks held on national territory. A stock build is shown as a negative number and a stock draw as a positive number.

Statistical Difference

National administrations sometimes obtain the data components of domestic availability from a variety of sources. Owing to differences in concepts, coverage, timing and definitions, observed and calculated consumption are often not identical. This is reflected in statistical differences.

Total Consumption

This is the gross observed inland consumption (including refinery consumption for oil), before the deduction of amounts returned by the petrochemical industry (backflow).

Closing Stocks

Refer to closing stock levels, at the last day of the year, of stocks held on national territory (including government controlled stocks).

Transformation Sector

This sector comprises fuel inputs to both public and private electricity, combined heat and power plants and heat plants. An autoproducer is an industrial establishment which, in addition to its main activities, generates electricity, wholly or partly for its own use. It includes railway's own production of electricity. Heat plants and combined heat and power plants only cover fuel inputs for that part of the heat which is sold to a third party. Transformation sector also comprises fuels used as feedstocks in gas works, coke ovens, blast furnaces and the petrochemical industry.

Public Electricity Plants: Includes inputs of oil for the production of electricity in public electricity plants whose primary purpose is to produce, transmit or distribute electricity.

Autoproducers of Electricity: Includes inputs of oil for the production of electricity by an enterprise which, in addition to its main activities, generates electricity wholly or partly for its own use, e.g. industrial establishments, railways, refineries, etc.

CHP Plants: Includes inputs of oil to combined heat and power plants which generate electricity and useful heat in a single installation. These could be either public CHP plants or autoproducers. Note that for autoproducer's CHP plants, all fuel inputs for electricity production are taken into account, while for heat production, only that part of inputs to heat which is sold to outsiders (e.g. to a network) is shown. Fuel inputs for the production of heat consumed within the autoproducer's establishment are *not* included here. *Information on inputs to and outputs from electricity and heat generation in CHP and Heat plants should be used with caution as a consistent separation of input fuels according to the type of energy output has not always been provided. The IEA Secretariat has undertaken a programme of estimating the missing data where possible.*

Heat Plants: Includes inputs of oil to plants which are designed to produce heat only. Data are available for both autoproducer and public heat plants. Data for autoproducer heat plants represent inputs of fuel to plants which sell heat to a third party under the provisions of a contract.

Petrochemical Industry: Includes quantities of backflows returned from the petrochemical sector, whether returned to refineries for further processing/blending or used directly. It refers to oil products only.

Energy Sector

Fuels used by energy industries for heating purposes and for operation of their equipment are reported under this heading as well as transformation and distribution losses. It includes coal mining and oil and gas extraction (all activities involved in making oil and gas marketable up to the point of shipment from the producing area) as well as transformation industries such as electricity generating plants, gas works, coke ovens, and petroleum refineries (refinery fuels).

Refinery fuel

Includes oil consumed in refineries for the operation of equipment, heating and light.

Total Final Consumption

This is the sum of consumption by the different end-use sectors. Backflows from the petrochemical industry are not included in final consumption.

Transport Sector

This sector covers oil products to all transport activities regardless of sector.

Air: Deliveries of aviation fuels to international civil aviation and to all domestic air transport, commercial, private, agricultural, military, etc. Aviation fuels used for purposes other than flying arc also included, e.g. bench testing of engines, but not airline use of fuel for road vehicles.

Road: All oil products for use in road vehicles, including agricultural highway use. Excludes motor gasoline used in stationary engines, and diesel oil for use in tractors that are not for highway use (included in *Agriculture*).

Rail: All quantities used in rail traffic, including industrial railways.

Inland Waterways: All deliveries to inland waterways and coastal shipping, including small craft, barges and those coasters which are not included in international marine bunkers.

Pipeline Transport: This is all quantities used in support of the operation of pipelines.

Industry Sector

The following industrial categories are defined with reference to their ISIC (International Standard Industrial Classification of All Economic Activities) division or group number. The ISIC numbers refer to series M, N°4/Rev.3, United Nations, New York, 1990.

Iron and Steel: ISIC group no. 271 and class 2731;

Chemical (incl. Petrochem): ISIC division 24. Petrochemical industry includes cracking and reforming processes for the purpose of producing ethylene, propylene, butylene, synthesis gas, aromatics, butadiene and other hydrocarbon based raw materials.

Non-ferrous metals: ISIC group no. 272 and class 2732;

Non-metallic minerals: Glass, ceramic, cement and other building materials industries (ISIC division no. 26);

Transport equipment: ISIC divisions 34 and 35;

Machinery: Fabricated metal products, machinery and equipment other than transport equipment (ISIC divisions 28, 29, 30, 31 and 32);

Mining and quarrying: ISIC divisions 13 and 14;

Food, beverages and tobacco: ISIC divisions 15 and 16;

Pulp, paper and printing: ISIC divisions 21 and 22;

Wood and wood products: ISIC division 20;

Construction: ISIC division 45;

Textiles and leather: ISIC division 17, 18 and 19;

Non-specified: Any manufacturing industry not included in the above categories (ISIC divisions 25, 33, 36 and 37).

Fuels used for production of electricity in the different industry categories (autoproducers) are included in the transformation sector.

Other Sectors

Commerce and Public Services: ISIC divisions 41, 50, 51, 52, 55, 63-67, 70-75, 80, 85, 90-93 and 99.

Residential: All consumption by households (ISIC division 95).

Agriculture: This is all consumption by agriculture, fishing (ocean, coastal and inland fishing), hunting and forestry, including oil consumed by such users for traction (excluding highway use), power and heating. ISIC divisions 01, 02 and 05.

Non-specified: All activities not included elsewhere, e.g. military (except air).

Memo Items

Quantities reported in these categories are already included in the above-mentioned consumption sectors.

Non-energy use: This is the total of *Chem./petchem.feedstocks* and *Other,* below.

Chem./petchem. feedstocks: Concerns feedstocks for non-energy use in the chemical and petrochemical industry.

Other: Petroleum products used for non-energy purposes in all other sectors. This is usually the case for all consumption of lubricants, bitumen, white spirit and paraffin waxes. Petroleum coke is shown as non-energy use only when there is evidence of such use (for example in non-ferrous metals industry sector).

4. SOURCES AND NOTES

4.1 Part II

Sources

The annual historical data in Parts III and IV are from the IEA/OECD energy statistics databases. Please see sections 4.2 and 4.3 for further information. Data for 2001 and 2002 are subject to revisions as final annual statistics become available.

Annual Oil Statistics (AOS): This database covers annual data for crude oil, NGL, refinery feedstock and other hydrocarbons, as well as petroleum products for OECD countries. They include data on detailed supply/demand balances, final consumption by end-use, trade by origin and destination and stock levels and changes, all submitted by national administrations in questionnaires to the Secretariat.

Monthly Oil and Gas Statistics (MOS): The MOS database is compatible with the AOS database with the exception of the detailed breakdown of lubricants, bitumen, paraffin waxes and other products as well as final consumption data by end-use categories which are not available in MOS. Data are collected on a monthly basis from OECD member countries in questionnaires. MOS data form the basis for the estimates of the 2002 oil supply elements of OECD countries.

Statistics up to 2001 are from the Annual Oil Statistics (AOS) database, while oil data for 2002 are mainly based on the Monthly Oil and Gas Statistics (MOS) database. These monthly oil data, submitted by member governments, are historically lower than the annual statistics as they exclude demand for certain products (e.g. petroleum coke in certain countries). To arrive at a comparable time series between AOS and MOS data, the 2002 monthly data are adjusted upwards for the under-reporting. Data for 2002 may therefore differ from similar statistics published in the publication *Oil, Gas, Coal and Electricity Quarterly Statistics* (MOS data).

Energy Balances: These are annual statistics that integrate data from the IEA/OECD coal, electricity, oil and gas databases to provide a summary of energy supply and demand in a common unit (millions of tons of oil equivalent) for each OECD country on an annual basis.

CO_2 Emissions: CO_2 emissions are calculated from the IEA Energy Balances, using the default IPCC methodology and emission factors as described in the *Revised 1996 IPCC Guidelines for National Greenhouse Gas Inventories*.

Energy Statistics and Balances of Non-OECD Countries: The database covers supply and demand balances for all forms of energy for over 100 non-OECD countries on an annual basis.

Energy Prices and Taxes: These are quarterly statistics for OECD countries and certain non-OECD countries which cover import and export prices, spot prices and end-user prices and taxes.

These data are compiled from information contained in *IEA/OECD Energy Prices and Taxes*. This publication provides, on a quarterly basis, detailed country level data from the IEA price database on import and export prices for crude oil, steam coal, coking coal and natural gas; spot prices for crude oil, gasoline, light fuel oil and heavy fuel oil; and end-user prices and taxes by consuming sector for oil products, coal, natural gas and electricity.

More detailed information concerning these prices is available in the Notes on Definitions in *Energy Prices and Taxes* where deviations from the general principles are explained and supplementary information is given.

The *Crude Oil Spot Prices*[1] are monthly average spot prices based on *Platt's* daily quotations. How-

1 Source: Based on Platt's unless otherwise specified. Copyright ©2003, Platt's, a division of the McGraw-Hill Companies, Inc. No reproduction without prior written permission from Platt's. All rights reserved.

ever, prior to July 1987 for Brent and January 1985 for Dubai and WTI, the monthly average prices were calculated from the weekly averages using various published sources. All prices are first month forward supply, except Brent, which is the dated quotation for prompt physical cargoes.

The monthly *Oil Product Spot Prices*[1] are calculated from *Platt's* daily quotations. However, for NW Europe (Rotterdam), the monthly prices were calculated from the weekly price average using various published sources prior to April 1987 for unleaded gasoline, and January 1985 for HSFO. Prices for NW Europe and US (New York Harbour), are the average price between the high and low quote for spot purchases of oil products, barge, FOB. Singapore represents the cargo average price. All prices exclude end-user taxes.

End-Use Prices are collected quarterly from national administrations or other relevant bodies and supplemented with data from national published sources. In general, the end-use prices include 1) transport costs to the consumer; 2) prices actually paid, i.e. net of rebates; and 3) taxes which have to be paid by the consumer as part of the transaction and which are not refundable. This excludes value-added tax paid in many European countries by industry for all goods and services. In these cases value-added tax is refunded to the customer, usually in the form of a tax credit. Therefore, it is not shown as part of prices.

Crude Oil Import Costs are taken from a monthly reporting system of crude oil costs established between national administrations and the IEA Secretariat. The import costs, which are expressed as average unit values (CIF) are recorded at the import stage and are the sum of cost, insurance and freight, excluding import duties.

Crude Oil Tanker Ports Data: The port data provide details of ports capable of accommodating full-laden tankers of 60 000 dwt tonnes size or greater. The data, provided by *SSY Consultancy & Research Ltd* in London, refer to loading and discharge facilities which were in operation as at April 2003, or which were scheduled to enter service by mid-2003.

Ports are categorised as load (L), discharge (D) or as having both (B) facilities. At some ports, several terminals may exist for crude oil tankers; for these, the facility listed is the one at which the largest vessel can be berthed. In selected cases, though, more than one entry has been provided for particular ports to reflect, for example, the presence of two significant refineries at such locations.

Partial discharge of cargo onto a smaller vessel via ship-to-ship transfer (i.e. "lightering") is a common practice in many parts of the world, especially in the United States. This enables ports which lack deep-water facilities to receive cargoes that have been shipped in tankers far larger than would be able to berth directly at the discharge quays available. In some instances, the "mother ship" will be lightered "to extinction", meaning that all cargo is offloaded onto smaller "daughter vessels"; in others, only some of the mother ship's cargo will be offloaded, enabling it then to be accommodated in a part-laden state at the relevant port. Accordingly, *partly-laden* ships larger than the maximum size indicated may be able to berth at a named facility. Therefore, designated lightering areas (e.g. those permitted by the US Coast Guard in the Gulf of Mexico) are included in the information shown.

NB: Apart from lightering areas, the information is not confined to conventional shore-based ports, but also includes single buoy moorings (SBMs), single point moorings (SPMs), offshore platforms, plus floating production, storage and offloading units (FPSOs).

The port tables include terminals for loading, discharging and trans-shipment for crude oil and condensate cargoes, but exclude those which handle only refined products, liquefied natural gas, liquefied petroleum gases and/or chemicals in liquid bulk form.

Figures for vessel length overall LOA, beam (extreme breadth), draft (distance from water-line to lowest point of a vessel's keel) and air draft (distance from water-line to a ship's highest point) are expressed in metres. Data for maximum ship size are in dead-weight (dwt) tonnes (i.e. carrying capacity of a ship, including stores, lubricating oils and water, as well as actual cargo). Information for cargo loading and discharge rates, where available, is reported in tonnes (metric tons) per hour.

Tanker Freight Rates Data: Data for oil tanker freight rates are annual averages in US dollars per tonne on representative trade routes for respective sizes of ship. These correspond to the main vessel size segments within the total oil tanker fleet. Figures are based on reported rates for "single voyage" charters (in which ships are employed on a spot basis, e.g. to carry 260,000 tonnes of crude oil from the Middle East Gulf to Japan). Such spot business forms the basis of most chartering in the tanker market, rather than agreements under which vessels are engaged for a designated time period, such as 12 months. Rates are provided for the carriage of crude oil and refined petroleum products, according to appropriate ship size used to move these cargoes. This information has been supplied by *SSY Consultancy & Research Ltd* (research@ssy.co.uk).

4.2. Part III

Annual Oil Statistics (AOS): All data are extracted from the Annual Oil Statistics (AOS) database (see Sources and Notes for Part II). For definitions see [3.1] Product Definitions and [3.2] Definition of Supply and Consumption Items.

Technical Note on Table 1

Indigenous production includes production of crude oil, NGL and other hydrocarbons of origin other than crude oil and NGL (see product definitions above [3.1]).

+ **Net imports** include imports minus exports (net imports) of crude oil, NGL, refinery feedstocks, other hydrocarbons as well as net imports of refined petroleum products (see product definitions above [3.1]).

- **International marine bunkers** include refined petroleum products used by vessels of all flags (see definitions of supply and consumption items above [3.2]).

+ **Stock changes** include stock changes (positive numbers indicate stock withdrawals, negative numbers indicate additions to stocks) for crude oil, NGL, refinery feedstocks, other hydrocarbons as well as refined petroleum products (see definitions of supply and consumption items above [3.2]).

= **Total supply** = sum of above

Refinery Intake[2] covers inputs of crude oil, NGL, refinery feedstocks and other hydrocarbons to refineries (see definitions of supply and consumption items above [3.2]).

2 Memo items not included in balance.

Refinery Output[2] covers refinery production of petroleum products (see definitions of supply and consumption items above [3.2]).

- **Refinery losses** = refinery intake - refinery output.

+ **Returns** include amounts of backflow returned from the petrochemical sector. This heading also includes recycled products which are refined petroleum products previously sent to consumption and, at a later stage, returned to refineries (see definitions of supply and consumption items above [3.2]).

- **Statistical differences** include statistical differences and transfers of oil products (transfers between products or products returned to refineries after reclassification).

= **Total consumption** = total demand for oil (sum of all sector totals outlined below).

Transformation

Transformation covers oil used in the process of transformation of oil to another energy form, e.g. oil into electricity, heat, etc. (see definitions of supply and consumption items above [3.2]).

Energy

Covers consumption of oil by the energy transformation industries (for heating, light and operation of equipment) as well as oil used in oil and gas extraction, coal mines, pipeline losses and distribution losses (see definitions of supply and consumption items above [3.2]).

Transport

This includes all consumption in transport excluding international marine bunkers, but including lubricants, waxes, etc. used in the transport sector. Also includes consumption in support of the operation of oil and gas pipelines (see definitions of supply and consumption items above [3.2]).

Industry

This includes all industrial consumption of oil, including consumption of petrochemical feedstocks and non-fuel uses, such as consumption of lubricants, waxes, white spirit, bitumen, etc. Industrial consumption of oil for the production of electricity (autoproduction) is excluded (see definitions of transformation, supply and consumption [3.2], above).

Other sectors

Includes residential, commercial and public services, agriculture and other (not specified) use. Consumption of lubricants, waxes, white spirit, etc. are also included (see definitions of supply and consumption items above [3.2] and memo on non-energy use below).

Memo: Non-Energy Use

This is split into chem/petchem. feedstocks and other non-energy use (amounts shown in the memo are already included in transport, industry and other sectors).

Chemical/Petchem. Feedstocks

This covers petroleum products used in the chemical and petrochemical industry in processes such as cracking and reforming, for the purposes of producing ethylene, propylene, butylene, synthesis gas, aromatics butadiene and other hydrocarbon-based raw materials.

Other

This includes non-fuel consumption of petroleum products such as lubricants, paraffin waxes, white spirit, bitumen and that part of petroleum coke which has not been used as a fuel.

Tables 2 to 8B

For explanations on the other tables in Part III, please refer to [3.1] Product Definitions and [3.2] Definitions of Supply and Consumption. Footnotes to tables are available at the end of Part IV.

4.3 Part IV

Annual Oil Statistics (AOS): All data are extracted from the Annual Oil Statistics (AOS) database (see Sources and Notes for Part II). For definitions see [3.1] Product Definitions and [3.2] Definition of Supply and Consumption Items.

Technical note on Tables 8 to 14: Gross consumption includes consumption of fuels by refineries and excludes international marine bunkers.

Table 15: Consumption in the transformation sector includes backflows from the petrochemical industry.

Table 16: Consumption in the energy sector includes refinery fuels.

5. COUNTRY NOTES

Please note that as a result of revisions, data in *Oil Information* may differ from those published earlier in *Energy Statistics of OECD Countries* and *Energy Balances of OECD countries*.

Oil data for 2002 are preliminary, based on *Monthly Oil and Gas Statistics* (MOS) unless otherwise specified. For some countries data in MOS are not fully compatible with *Annual Oil and Gas Statistics* (AOS), therefore MOS 2002 data have been adjusted in order to show consistent time series developments from 2001 to 2002. For general footnotes to the tables in Part III, please consult the fold-out at the end of Part IV.

1. Australia

All data: All data refer to fiscal years which run from 1 July to 30 June (e.g. 2001 = 1 July 2000 to 30 June 2001).

Negative refinery losses are due to differences in treatment of transfers between refineries.

Data after 1973 are based on national surveys. Prior to 1973, there are no detailed data available for autoproducers and for sub-sector industry consumption.

The large drop in industry consumption in 1982-83 is due to the restructuring of the iron and steel industry, the economic recession and one of the worst droughts in history.

Crude Oil: The drop in the production in 1999 is due to a gas explosion at Longford plant.

NGL: Prior to 1992, part of the NGL production is included in crude oil.

Refinery Gas: Prior to 1974, refinery gas is reported net of consumption in refineries.

LPG, Ethane: LPG includes ethane prior to 1991.

Lubricants, White Spirit, Paraffin Waxes: Consumption of lubricants, white spirit/SBP and paraffin waxes is reported as non-specified industry consumption, as no further breakdown is available.

Fuel Oil: Imports of fuel oil are estimated by the Australian administration.

2. Austria

Data were revised by the Austrian Administration back to 1980.

Gas/Diesel Oil: For 1978 and 1979, industry consumption of gas/diesel oil is included with fuel oil.

Other Products: As of 1978, "other products" include paraffin waxes and petroleum coke. From 1991, they include also naphtha.

Lubricants: Before 1990 consumption of lubricants is reported in the road sector as no further breakdown is available.

3. Belgium

Data for non-metallic minerals, machinery, mining/quarrying, food/beverages, tobacco, pulp/paper/printing and construction sectors are not available prior to 1971.

Inputs to electricity and CHP plants have been estimated by the Secretariat from year 1974 based on inputs reported in the Annual Electricity Questionnaire.

Refinery Feedstocks: The decrease of products transferred in 1995 was due to refinery maintenance of one month.

Refinery Gas: Prior to 1975, refinery gas is reported net of consumption in refineries.

LPG, Gas/Diesel Oil: LPG and gas/diesel oil consumption in the commercial/public services sector is included with the residential sector prior to 1976.

Fuel Oil: The break in series for fuel oil between 1975 and 1976 is due to a change in classification between the industrial and "other" sectors. The decrease in industry consumption since 1993 is due to the introduction of an excise tax as well as the increased use of natural gas.

Other Products, Petroleum Coke: The break in series between 1999 and 2000 is due to improvements in the collection system.

4. Canada

Other Hydrocarbons: Indigenous production of other hydrocarbons represents synthetic crude oil produced from tar sands. Prior to 1990, hydrogen used for the upgrading of synthetic crude oil from "other sources" was included in the natural gas supply. From 1990, a different methodology was adopted by the Canadian administration.

Orimulsion imports from Venezuela started in 1994.

International marine bunkers are included with inland waterways prior to 1978.

LPG, Ethane: LPG includes ethane prior to 1990. Ethane is mainly used as petrochemical feedstock.

Kerosene, Gas/Diesel Oil, Fuel Oil: For kerosene, gas/diesel oil and fuel oil, several breaks in series exist for detailed industrial consumption prior to 1980. From 1980, all consumption data are based on one survey. However, from 1988 onwards, data for transport equipment, machinery, food, and textiles are not available as Statistics Canada has ceased the collection of these data. Separate consumption data for agriculture (including forestry) are available from 1983 for kerosene, from 1973 for gas/diesel oil and from 1978 for fuel oil. For previous years, agriculture consumption is included with residential.

Gas/Diesel Oil: As of 1982, gas/diesel oil consumed by fleets in the wholesale/retail service industry is reported in the commerce/public services sector. Prior to 1982, it was included in the road sector.

Gasoline Type Jet Fuel: Prior to 1979, gasoline type jet fuel is included with kerosene type jet fuel.

Naphtha, Other Products: Recycled products for naphtha and "other products" include purchases of feedstock and other additives from non-reporting companies.

The time series from 1990 to 1997 have been revised by the Canadian Administration, which may cause breaks between 1989 and 1990.

5. Czech Republic

Data are available starting in 1971.

Consumption data prior to 1994 are estimated by the Secretariat. The Czech administration submitted official data from 1994 onwards.

Gas/Diesel Oil: From 1999 the Czech Administration has implemented a new updating system. It entails breaks in series for the final consumption.

6. Denmark

From 1990 onwards, Greenland and Faroe Island are not included in oil data.

Other Hydrocarbons: Orimulsion imports and inputs to electricity generation begin in 1995.

Refinery Gas: Prior to 1975, refinery gas is reported net of consumption in refineries.

Gas/Diesel Oil, Fuel Oil: Prior to 1990, gas/diesel oil and fuel oil used for fishing is included in the inland waterways sector, while after this date it is reported in the agriculture sector.

Fuel Oil: From 1974 to 1979, consumption of fuel oil for the CHP production by autoproducers has been estimated. The marked increase in inputs to CHP production in 1994 is due to increased electricity exports to Norway.

All oil consumption in the commercial/public services sector is included in residential prior to 1976 except fuel oil which is shown separately in 1975.

Paraffin Waxes: As of 1987, separate data on paraffin waxes are no longer available.

Gasoline Type Jet Fuel: Consumption of gasoline type jet fuel ceased in 1988. From 1989 onwards, only kerosene type jet fuel is consumed.

Other Products: Information on recycling and end-use consumption begins in 1989, and refers to waste oil.

Data for end-use consumption may vary from year to year due to a detailed survey which is sent to companies in Denmark every other year. For non-survey years, the consumption by end-use is estimated by the Danish Energy Agency.

Due to better survey methods, inputs to electricity and heat generation are reclassified as of 1994. There is therefore a break in series between 1993 and 1994. The oil inputs used in industrial sectors for producing surplus of heat, which is delivered to district heating networks, are allocated to these industrial sectors.

Industry sector detail for 1994 and 1995 is based on a new survey.

7. Finland

Fuel Oil: There is a break in the other sectors between 1998 and 1999 due to a new calculation model.

Petroleum Coke: Data on petroleum coke are only available from 1985 onwards.

In 1995, there is a break in series for petroleum products trade due to the aligning of the National Board of Customs trade data collection system with the European Union's Intrastat system.

8. France

Petroleum Coke: The split of petroleum coke consumption into energy and non-energy use is not available prior to 1982.

LPG, Ethane: LPG includes ethane prior to 1988.

Prior to 1994, the separation of consumption of LPG by the tertiary sector between residential and commerce/public services has been estimated by the Secretariat.

Kerosene Type Jet Fuel: The consumption includes military use as of 1998.

Gas/Diesel Oil: Consumption in the non-metallic minerals sector has been reclassified as non-ferrous metals and vice-versa for years 1982 to 1994. Prior to 1985, residential sector consumption is reported under commerce/public service sector as no separate data were available.

Gas/Diesel Oil, Fuel Oil: Residential and commerce/public services sector consumption for gas/diesel and residual fuel oil has been estimated by the Secretariat from 1985 to 1998 based on information provided by the French Administration.

Additives and oxygenates data are available from 1991.

A different treatment of transfers was adopted from 1998. From 1998, imported petroleum products needing further refinery processing are no longer reported as refinery feedstock imports but as oil product imports and products transferred. And from 1999, due to a reallocation of some products, part of the amounts previously reported in other products are now reported in fuel oil.

9. Germany

Refinery Gas: Prior to 1970, consumption of refinery gas in the chemical industry is included with refineries' own consumption.

Jet Gasoline, Jet Kerosene: Beginning with 1994, there has been a reclassification of jet gasoline to jet kerosene.

Other Kerosene: Due to new information from oil companies, consumption of other kerosene end-use consumption has been modified from 1994 onwards.

Gas/Diesel: In 1989, end-use consumption decreased due to an exceptionally warm winter and a lowering of consumer stocks. In 1995, a break in the industry sectors occurs due to an alignment with the NACE classifications.

Fuel oil: Prior to 1980, consumption of fuel oil in blast furnaces was included in the iron and steel sector. From 2000, part of the product 'Andere Rückstände' is included with fuel oil instead of other products.

Other Products: Prior to 1979, data for "other products" include paraffin waxes, bitumen, white spirit & SBP and lubricants for eastern Germany.

Beginning with 1994, there is a break in series for final consumption by sub-sector due to improved survey methods instituted by Mineralölwirtschafts Verband.

10. Greece

Crude Oil: Crude oil production was stopped on 30th November 1998, and started again in December 1999.

Refinery Feedstocks: Refinery feedstocks are only reported from 1986 onwards.

Refinery Gas: From 1990 onwards, there has been an increased use of refinery gas in electricity generation, replacing fuel oil.

Fuel Oil: From 1993, more information is available on the allocation of fuel oil to specific industry sectors. Fuel oil consumption in the agriculture and residential sectors has been replaced by Gas/Diesel Oil starting in 1993.

Prior to 1987, consumption in the commerce/public services sector is included with residential. Peaks in residential sector consumption in 1978 and 1982 are due to unusually cold winters.

Due to changed reporting methods, more detailed end-use information has become available starting in 1996.

11. Hungary

Refinery Gas, Paraffin Waxes, Lubricants: Data prior to 1993 are partly estimated by the Secretariat.

White Spirit: Prior to 1993, white spirit is included in motor gasoline.

Fuel Oil: The breakdown between low and high sulphur is available as of 1998.

Additives, Aviation Gasoline: The data are available as of 1998.

Beginning with 1993, data were officially submitted to the IEA by questionnaire.

12. Iceland

Consumption data prior to 1980 are estimated by the Secretariat.

Motor Gasoline: Consumption of leaded motor gasoline ceased in 1996.

Fuel Oil: Detailed industry sector consumption is not available prior to 1982.

Bitumen: From 1983 to 1986, end-use consumption increased due to an expansion of public works.

13. Ireland

Gas/Diesel Oil: Gas/diesel oil consumption in the agricultural sector is available from 1986.

Other Kerosene: For the period 1970 to 1977, the split between commercial and public services and agricultural use of kerosene has been estimated by the Secretariat.

Consumption in commercial/public services includes quantities used by state-owned agricultural companies.

LPG, Gas/Diesel Oil, Kerosene, Fuel Oil: End-use data collected for 1993 are based on a detailed survey conducted by a consulting company. Data for historical years back to 1990 were revised by the national administration based on the results of this survey. Due to these revisions, breaks in series will exist for years 1989 and 1990 in the end-use breakdown for LPG, kerosene, gas/diesel oil and fuel oil.

Lubricants: Road consumption is reported in the non-specified transport sector as the detailed breakdown is not available.

The decrease in refinery output in 1996 was due to a partial refinery shutdown.

14. Italy

Aviation Fuels: The aviation fuels show breaks in series in 1988 due to a change in classification.

Gas/Diesel: Non-specified use is included in the commerce/public sector.

Inputs to electricity and heat generation have been estimated by the Secretariat for years 1984 to 1997 based on submissions of the Annual Electricity questionnaire.

All data for the years 1992 to 1997 and the detailed consumption breakdown for other years, have been estimated by the Secretariat based on *Bilancio Energetico Nazionale*. Due to new surveys, breaks appear in the consumption series between 1998 and 1999.

15. Japan

Due to METI's new energy methodology, there are breaks in series between 1999 and 2000. Also all data from 2000 now refer to fiscal years which run from 1 April to 31 March (e.g. 2001 = 1 April 2001 to 31 March 2002).

For 2000 and 2001, some of the data for Japan are estimated by the IEA Secretariat to ensure time-series consistency with the new Japanese energy balance methodology. Therefore, some of the data are different from those that are published by the Japanese Government. A full revised time-series for 1990 to 2001, based on the original data published by the Japanese Government, will be published in next year's edition.

Crude Oil, NGL, Feedstocks, Refinery Gas, LPG, Aviation Gasoline, Motor Gasoline, Jet Gasoline, Jet Kerosene, Other Kerosene: The Japanese Administration has revised supply and demand data for years 1960 to 1969. The classification of fuel use between road transport and internal navigation changed in 1970.

Other Hydrocarbons: Other hydrocarbons comprise sulphuric acid, clay (not white clay) and caustic soda.

Other Kerosene: The decrease in end-use consumption in 1982 was due to an exceptionally warm winter.

Petroleum Coke: Petroleum coke use in blast furnaces series begins in 1990.

Due to improved data collection methods, breaks in series exist for several products between 1981 and 1983. Due to a change in the reporting system to exclude stock levels at service stations and retail stores, a break in series exists between 1981 and 1982.

Orimulsion imports for electricity generation began in 1991.

16. Korea

Data are available starting in 1970.

Stock levels include vessel stocks starting from 1997.

Kerosene Type Jet Fuel: The split between international civil aviation and domestic air transport has been estimated for 1990 to 1995.

Prior to 1994, data are net of backflows to the petrochemical industry.

17. Luxembourg

Fuel Oil: Due to the introduction of an excise tax in 1994, all high-sulphur fuel oil consumption has been replaced by low-sulphur fuel oil. .

The reduction in consumption of fuel oil in the iron and steel industry in the late 1970s is due to substitution by coal.

18. Mexico

Data are available starting in 1971.

Stock levels are available from 1999.

Because of a change in the processing of the data, you may have breaks in series between 1998 and 1999.

Inputs for autoproducer electricity and heat generation are included in the industry sector.

Lubricants, Paraffin Waxes, Petroleum Coke, Other Products: End-use data for these products have been reported as non-specified industrial use as a detailed breakdown is not available.

NGL production reported in the IEA publications may be different from what is reported in the Mexican energy publications as the IEA includes in its oil data liquids produced in conjunction with natural gas.

19. Netherlands

Data were revised by the Dutch Administration back to 1990.

Significant breaks in series between 1983 and 1984 are due to the introduction of a more comprehensive survey on end-use consumption. However, aggregated consumption data for transport, industry and "other" sectors are not affected.

Crude Oil: Prior to 1994, stocks held by stockholding entities were not included.

LPG, Ethane: LPG includes ethane prior to 1994.

Refinery Gas: Refinery gas includes chemical gas from 1978. These quantities are shown as chemical industry consumption. Inputs to public combined heat and power generation begin in 1995.

Motor Gasoline: Motor gasoline includes "other light oils".

Petroleum Coke: Petroleum coke is included with "other products" for all years except 1960 to 1967 and 1978 to 1982.

20. New Zealand

Consumption data for 1960 to 1973 have been estimated by the Secretariat.

Other Hydrocarbons: Other hydrocarbons represent synthetic gasoline production from natural gas. In February 1997 the production of synthetic gasoline ceased.

Gas/Diesel Oil: Consumption of gas/diesel oil in the road and rail sectors is not separately identifiable, and is reported in the non-specified transport sector.

Gas/Diesel Oil, Fuel Oil: The light fuel oil is included in residual fuel oil until 1997. As of 1998, the light fuel oil is included in gas/diesel oil.

From 2000 onwards, Paraffin Waxes is included with Lubricants.

21. Norway

Due to the availability of more detailed information, the Norwegian Administration has revised the reporting of chemical and petrochemical energy and non-energy use beginning with 1993.

The Norwegian Administration revised stock levels for crude oil and some petroleum products. It entails breaks in stocks between 1999 and 2000.

Refinery Feedstocks: Since 1986, imports of refinery feedstocks are reported under the relevant petroleum product imports.

LPG, Ethane: LPG includes ethane prior to 1990.

Lubricants: Consumption of lubricants is reported in the industry sector, as no further breakdown is available.

Beginning in 1991, motor gasoline, gas/diesel and fuel oil end-use consumption data has been revised based on new information.

Gas/Diesel Oil: Quantities of gas/diesel oil for electricity generation from 1970 to 1975 have been estimated by the Secretariat. Starting in 1990, the Norwegian administration has revised the reporting of input of gas/diesel oil for autoproduced electricity on oil and gas platforms, giving rise to a break in series between 1989 and 1990. Data are now reported in the energy sector.

Gas/Diesel Oil, Fuel Oil: The detailed consumption breakdown is available from 1976.

Lubricants: For 1996, some lubricants are included with Other Products.

22. Poland

The stock levels are available as of 1988.

Other Hydrocarbons: From 1997, a hydrocracking complex produces hydrogen from natural gas. These amounts are reported in indigenous production.

Gas/Diesel Oil, Fuel Oil: Prior to 1998, inputs of gas/diesel oil and heavy fuel oil to autoproducer CHP in petroleum refineries have been included in the transformation sector.

23. Portugal

Refinery Gas: As of 1995, there is no longer any production and consumption due to a refinery closure.

Gas/Diesel Oil: Consumption in industry and commercial/public services represents diesel use in the mobile fleets of these sectors.

Fuel Oil: Feedstocks use of fuel oil in the chemical industry beginning in 1984 reflects the opening of a new ammonia plant which uses fuel oil as raw material.

The consumption breakdown for lubricants, bitumen, white spirit/SBP and paraffin waxes has been estimated by the Secretariat in 1986.

24. Slovak Republic

Data are available starting in 1971.

Naphtha: Data are not available and therefore not included in total products before 1989.

Fuel Oil: Fuel oil includes both light and heavy fuel oil.

Data from 1993 were revised by the Slovak Administration for the previous edition. It may entail breaks between 1992 and 1993.

As a result of a new survey more detailed information is available starting in 2001, however there are also many breaks in series between 2000 and 2001.

25. Spain

Detailed consumption data prior to 1981 are partly estimated on the basis of national statistics covering consumption on the Spanish mainland.

Refinery Feedstocks: Data are not available in 1981.

LPG, Fuel Oil: Consumption specifications are less detailed prior to 1973.

Gas/Diesel Oil, Fuel Oil: A reclassification of the "other sector" between 1976 and 1977 has resulted in a break in series for gas/diesel oil and fuel oil.

Petroleum Coke: From 1982 onwards, there are more detailed consumption data available for petroleum coke.

Backflows have been estimated by the Secretariat for 1996.

A change in the reporting system in mid-1996 has resulted in some breaks in series between 1996 and 1997.

26. Sweden

Additives, Ethane: Data are available from 2000 for additives and from 2001 for ethane.

Other Kerosene: Consumption in the road sector is discontinued in 1984 due to product re-classification.

Gas/Diesel Oil: Following a merger between two companies, the new company has revised its stocks levels. It entails a break in stocks between 1999 and 2000.

Fuel Oil: Due to more detailed reporting, there are breaks in consumption series between 1985 and 1986.

Swedish stock data include peacetime crisis stocks. Since these stocks may be held in crude oil instead of oil products, there may be occurrences of negative stock levels for products.

Beginning in 1995, Sweden has changed its standard classification of industry sectors.

27. Switzerland

Refinery Feedstocks: The reduction in refinery intake of refinery feedstocks in 1988 is partly due to a switch to crude oil and partly to a shutdown for maintenance of a refinery.

Naphtha: As of 1993, the refinery gross output figures are net of quantities used for blending into motor gasoline.

Gas/Diesel Oil: Non-specified industry (small manufacturers) consumption of gas/diesel oil is estimated. The method of reporting has been revised from 1987 to 1993. The increase in consumption in 1994 is due to consumer stock-building prior to the introduction of a value-added excise tax on heating fuels as of 1st January 1995.

Since 1st January 1996, the Swiss Administration revised its stock reporting.

The break between 1998 and 1999 in the final consumption of motor gasoline, gas/diesel oil and residual fuel oil is due to a new survey being introduced in 1999.

28. Turkey

Gas/Diesel Oil, Fuel Oil: The end-use classification of gas/diesel oil and fuel oil were changed in the Turkish national statistics between 1977 and 1978. Consequently, breaks in detailed consumption series appear between these two years.

As of 1978, the commercial sector is included with industry, while public services are included with transport sector. Part of non-metallic mineral products industry consumption is included with non-ferrous metals sector.

International marine bunkers are included in exports for the years 1978, 1980, 1981, 1983 and 1984.

Gas/diesel oil and fuel oil consumed to produce electricity are used in both oil and coal-fired plants.

Bitumen: Bitumen consumed in the transformation sector is intended for briquette production.

29. United Kingdom

Crude Oil, NGL: From 1980, NGL includes condensates. For earlier years condensates are included with crude oil.

NGL, Refinery Feedstocks: Stock levels and stock changes of refinery feedstocks and NGL are only reported separately from 1985. Prior to that they were included with crude oil.

Other Hydrocarbons: Until 1994, other sources of other hydrocarbons represents bitumen produced from coal.

LPG, Ethane: LPG includes ethane prior to 1990.

LPG, Fuel Oil: breaks in series for LPG, between 2000 and 2001, and for fuel oil, between 1999 and 2000, due to the re-allocation of some data. And, for fuel oil, because of the introduction of heat production data as well.

Other Products: Imports and exports of "other products" include petroleum coke prior to 1992. Prior to 1979, all petroleum coke data are included with "other products".

Stock levels and stock changes in public utilities are included from 1985.

Prior to 1995, the product breakdown for returns is estimated by the UK Administration. Beginning with 1995, the UK Administration revised their product breakdown for returns and petrochemical reporting methodology.

30. United States

Prior to 1995, a detailed breakdown of oil products consumption in the industry sector is not available. However, data by industry are published for fuel oil for the period 1971 to 1982.

Crude Oil: High statistical differences for crude oil represent *"Unaccounted for crude oil"*, the difference between the supply and disposition of crude oil.

LPG: From 1995, inputs to gas works are included in the industry sector.

LPG, Ethane: LPG includes ethane and pentanes plus prior to 1990.

LPG, NGL: Imports and exports of NGL are reported as LPG trade prior to 1978. From 1992 onwards, the individual components of NGL and LPG have been converted using their respective conversion factor rather than an average factor, resulting in a break in series.

Naphtha: The reduction in naphtha use in the chemical industry beginning in 1978 is due to a reclassification of the product.

Motor Gasoline: Beginning in 1994, consumption in the commerce/public services sector is based on a new model from the Department of Transportation.

Gas/Diesel Oil: Gas/diesel oil consumption in agriculture prior to 1980 is estimated by the Secretariat.

Fuel Oil: International marine bunkers of fuel oil show large increase in 1990 due to a change in the data collection and reporting methodology in the US administration. The consumption breakdown between low and high sulphur fuel oil has been estimated.

Gas/Diesel Oil, Fuel Oil, Petroleum coke: The inputs to autoproducers of electricity are available as of 1997. The inputs to CHP plants are available as of 1999. Before 1999, public CHP plants were included in public electricity plants; autoproducers of CHP were included in autoproducers of electricity and in industry sectors.

The change in the series for transfers in 1990 and 1993 is due to new reporting methods used by the US Administration.

In 1993, the US administration made several adjustments to its collection system for oil statistics in order to accommodate the revisions to the Clean Air Act of 1990. As a result, data for oxygenates (i.e. fuel ethanol, MTBE, etc.) are collected as of 1993.

As a result of the new Manufacturing Energy Consumption Survey (MECS), there are breaks in series between 1999 and 2000 for the industry sector. And again between 2000 and 2001 as the MECS percentages were revised due to revisions in electric cogeneration.

There were significant revisions to residual fuel oil and unfinished oils for 2001 data. Primarily, the changes are a result of importers misclassifying unfinished oils as residual fuel oil.

2002 oil data estimates are based on preliminary figures from *Monthly Oil and Gas Statistics* and will therefore differ from the final figures for 2002 published in *Quarterly Oil, Gas, Coal and Electricity Statistics (2nd Quarter 2003)* and subsequent editions.

6. GEOGRAPHICAL COVERAGE

Australia excludes the overseas territories.

Denmark includes Greenland and the Danish Faroes until 1989.

France includes Monaco and excludes the French overseas territories Guadeloupe, Martinique, Guyane, La Réunion, St.-Pierre et Miquelon, Nouvelle-Calédonie and French Polynesia.

Germany includes the new federal states of Germany from 1970 onwards unless otherwise indicated in a footnote. Supply, consumption and trade data from 1980 onwards have been supplied by the German administration. Data prior to 1980 include an IEA Secretariat estimate of the new federal states in the Germany total.

Japan includes Okinawa.

Italy includes San Marino and the Vatican.

The Netherlands excludes Surinam and the Netherlands Antilles.

Portugal includes the Azores and Madeira.

Spain includes the Canary Islands.

Switzerland includes Liechtenstein.

United States includes Puerto Rico, Guam, the Virgin Islands and the Hawaiian Free Trade Zone.

OECD comprises Australia, Austria, Belgium, Canada, Czech Republic, Denmark, Finland, France, Germany, Greece, Hungary, Iceland, Ireland, Italy, Japan, Korea, Luxembourg, Mexico, Netherlands, New Zealand, Norway, Poland, Portugal, Slovak Republic, Spain, Sweden, Switzerland, Turkey, United Kingdom, United States.

OECD North America comprises Canada, Mexico and the United States.

OECD Pacific comprises Australia, Japan, Korea and New Zealand.

OECD Europe comprises OECD countries except Australia, Canada, Japan, Korea, Mexico, New Zealand and the United States.

IEA comprises OECD countries except Iceland, Mexico, Poland and Slovak Republic.

European Union comprises Austria, Belgium, Denmark, Finland, France, Germany, Greece, Ireland, Italy, Luxembourg, Netherlands, Portugal, Spain, Sweden and United Kingdom. Please note that all these countries are included for all years despite different entry dates into the European Union.

Non-OECD Europe comprises Albania, Bulgaria, Cyprus, Gibraltar, Malta, Romania, and Former Yugoslavia.

Former Yugoslavia comprises Bosnia-Herzegovina, Croatia, Former Yugoslav Republic of Macedonia, Serbia and Montenegro, and Slovenia.

Former USSR comprises Armenia, Azerbaijan, Belarus, Estonia, Georgia, Kazakhstan, Kyrgyzstan, Latvia, Lithuania, Moldova, Russia, Tajikistan, Turkmenistan, Ukraine, and Uzbekistan.

OPEC comprises Algeria, Indonesia, Iran, Iraq, Kuwait, Libya, Nigeria, Qatar, Saudi Arabia, United Arab Emirates and Venezuela.

China includes People's Republic of China and Hong Kong, China.

7. ABBREVIATIONS AND CONVERSION FACTORS

Abbreviations

Mt	million metric tons
Mtoe	million tons of oil equivalent
TPES	Total Primary Energy Supply
kb	Thousand barrels
mb	Million barrels

Conventional Signs

-	Nil
x	Not applicable
..	Not available
0.00	Negligeable (depending on the number of decimals)

Decimal Prefixes

10^{1}	deca (da)	10^{-1}	deci (d)
10^{2}	hecto (h)	10^{-2}	centi (c)
10^{3}	kilo (k)	10^{-3}	milli (m)
10^{6}	mega (M)	10^{-6}	micro (µ)
109	giga (G)	10^{-9}	nano (n)
10^{12}	tera (T)	10^{-12}	pico (p)
10^{15}	peta (P)	10^{-15}	femto (f)
10^{18}	exa (E)	10^{-18}	atto (a)

General Conversion Factors for Energy

To:	TJ	Gcal	Mtoe	MBtu	GWh
From:	multiply by:				
TJ	1	238.8	2.388×10^{-5}	947.8	0.2778
Gcal	4.1868×10^{-3}	1	10^{-7}	3.968	1.163×10^{-3}
Mtoe	4.1868×10^{4}	10^{7}	1	3.968×10^{7}	11630
MBtu	1.0551×10^{-3}	0.252	2.52×10^{-8}	1	2.931×10^{-4}
GWh	3.6	860	8.6×10^{-5}	3412	1

Conversion Factors for Mass

To:	kg	t	lt	st	lb
From:	multiply by:				
kilogramme (kg)	1	0.001	9.84×10^{-4}	1.102×10^{-3}	2.2046
tonne (t)	1000	1	0.984	1.1023	2204.6
long ton (lt)	1016	1.016	1	1.120	2240.0
short ton (st)	907.2	0.9072	0.893	1	2000.0
pound (lb)	0.454	4.54×10^{-4}	4.46×10^{-4}	5.0×10^{-4}	1

Conversion Factors for Volume

To:	gal U.S.	Gal U.K.	bbl	ft^3	l	m^3
From:	multiply by:					
U.S. gallon (gal)	1	0.8327	0.02381	0.1337	3.785	0.0038
U.K. gallon (gal)	1.201	1	0.02859	0.1605	4.546	0.0045
Barrel (bbl)	42.0	34.97	1	5.615	159.0	0.159
Cubic foot (ft^3)	7.48	6.229	0.1781	1	28.3	0.0283
Litre (l)	0.2642	0.220	0.0063	0.0353	1	0.001
Cubic metre (m^3)	264.2	220.0	6.289	35.3147	1000.0	1

Petroleum Products
Average Heat Equivalents of Metric Tons

	Tcal	toe	TJ	Btu (x 10^9)
Refinery gas	0.01150	1.150	0.048148	0.045632
LPG	0.01130	1.130	0.047311	0.044838
Ethane	0.01180	1.180	0.049404	0.046822
Aviation Gasoline	0.01070	1.070	0.044799	0.042458
Motor Gasoline	0.01070	1.070	0.044799	0.042458
Jet Fuel	0.01065	1.065	0.044589	0.042259
Kerosene	0.01045	1.045	0.043752	0.041466
Naphtha	0.01075	1.075	0.045008	0.042656
Gas/Diesel Oil	0.01035	1.035	0.043333	0.041069
Fuel Oil	0.00960	0.960	0.040193	0.038093
Other Products	0.00960	0.960	0.040193	0.038093

Petroleum Products
Average Volume Equivalents of Thousand Metric Tons

	Thousand Barrels	bbl/d	Thousand Kilolitres (1000 m^3)	Thousand US Gallons
LPG	11.60	31.78	1.844	487.20
Aviation Gasoline	8.90	24.38	1.415	373.80
Motor Gasoline	8.47	23.20	1.347	355.73
Gasoline type Jet Fuel	8.25	22.60	1.312	346.50
Naphtha	8.81	24.13	1.400	369.89
White Spirit	8.46	23.18	1.345	355.32
Kerosene type Jet Fuel	7.89	21.61	1.254	331.32
Other Kerosene	7.85	21.49	1.247	329.49
Gas/Diesel Oil	7.46	20.44	1.186	313.32
Fuel Oil	6.50	17.80	1.033	272.85
Lubricants	7.09	19.42	1.127	297.78
Bitumen	6.08	16.66	0.967	255.36
Paraffin Waxes	7.85	21.51	1.248	329.70
Petroleum Coke	5.50	15.07	0.875	231.00

Crude Oil*
Average Volume Equivalent of Thousand Metric Tons

	Thousand Barrels	Thousand Kilolitres
Australia	7.868	1.251
Austria	7.200	1.145
Canada	7.186	1.143
Denmark	7.405	1.177
France	7.332	1.166
Germany	7.330	1.165
Greece	7.231	1.150
Italy	7.300	1.161
Japan	7.357	1.170
Mexico	6.965	1.107
Netherlands	7.239	1.151
New Zealand	7.321	1.164
Norway	7.644	1.215
Spain	7.506	1.193
Sweden	6.623	1.053
Turkey	7.200	1.145
United Kingdom	7.523	1.196
United States	7.400	1.177
Algeria	8.130	1.293
Angola	7.410	1.178
Argentina	7.120	1.132
Brazil	7.056	1.122
China	7.300	1.161
Colombia	7.080	1.126
Egypt	7.260	1.154
India	7.330	1.165
Indonesia	7.360	1.170
Iran	7.350	1.169
Iraq	7.430	1.181
Kuwait	7.250	1.153
Libya	7.580	1.205
Nigeria	7.500	1.192
Oman	7.330	1.165
Qatar	7.500	1.192
Saudi Arabia	7.323	1.164
Syria	7.290	1.159
United Arab Emirates	7.596	1.208
Former Soviet Union	7.300	1.161
Venezuela	6.890	1.095

*Indigenous production, including condensates.

1. INTRODUCTION

Données sur le pétrole 2003 est la dernière parution d'un ouvrage publié tous les ans depuis 1989, remplaçant la publication intitulée *Statistiques annuelles du pétrole et du gaz naturel.*

La deuxième partie, *Tendances du marché mondial du pétrole,* présente des tableaux récapitulatifs de l'évolution des marchés mondiaux du pétrole dont les séries chronologiques remontent jusqu'à 1971.

La troisième partie, *Données détaillées sur le pétrole dans les pays de l'OCDE,* présente sous forme de tableaux une description plus détaillée et exhaustive de l'offre, de la demande et de la consommation finale de pétrole dans la zone de l'OCDE, par région et par pays.

La quatrième partie, *Séries historiques,* présente des séries chronologiques concernant les principaux flux de pétrole pour chaque année allant de 1973 à 2002, lorsque les données sont disponibles.

On notera que les références à l'OCDE dans cette publication renvoient aux 30 pays Membres de l'OCDE : Allemagne, Australie, Autriche, Belgique, Canada, Corée, Danemark, Espagne, Etats-Unis, Finlande, France, Grèce, Hongrie, Irlande, Islande, Italie, Japon, Luxembourg, Mexique, Norvège, Nouvelle-Zélande, Pays-Bas, Pologne, Portugal, République Slovaque, République Tchèque, Royaume-Uni, Suède, Suisse et Turquie.

Les statistiques concernant les pays de l'OCDE sont fondées sur les données communiquées au Secrétariat par les administrations nationales. Jusqu'en 2001, les statistiques sont tirées de la base de données contenant les statistiques annuelles du pétrole (AOS), tandis que les données sur le pétrole pour 2002 proviennent principalement de la base de données contenant les statistiques mensuelles du pétrole et du gaz naturel (MOS). Ces données mensuelles sur le pétrole, communiquées par les administrations nationales des pays Membres, sont plus faibles que toutes les statistiques annuelles précédentes car elles ne tiennent pas compte de la demande de certains produits (par exemple, le coke de pétrole dans certains pays).

Afin d'établir des séries chronologiques comparables entre les données des bases AOS et MOS, les données mensuelles pour 2002 sont corrigées à la hausse pour rectifier des notifications insuffisantes. Les données relatives à l'année 2002 peuvent donc différer des statistiques similaires publiées dans l'ouvrage *Statistiques trimestrielles : électricité, charbon, gaz et pétrole* (données MOS). On trouvera des informations complémentaires sur les sources des données et les estimations dans les notes concernant chaque pays.

Toutes les données figurant dans les troisième et quatrième parties ainsi que certaines informations de la deuxième sont disponibles sur CD-ROM exploitable par des ordinateurs personnels compatibles IBM. Un formulaire de commande est fourni à la fin de cet ouvrage.

En outre, un service de données est disponible sur internet. Ce service comprend une souscription annuelle pour un accès illimité ou bien la possibilité de payer uniquement pour des données sélectionnées.

Les données annuelles sur le pétrole sont collectées par la Division des statistiques de l'énergie (ESD) du Secrétariat de l'AIE, dirigée par M. Jean-Yves Garnier. M. Bruno Castellano est actuellement responsable des données annuelles sur le pétrole concernant les pays de l'OCDE. Mme Sharon Michel a assuré la mise en page et le secrétariat d'édition. Mme Mieke Reece est responsable de l'ensemble de la présente publication.

Toutes les questions concernant les données ou la méthode utilisées dans la présente publication doivent être adressées à :

Mme Mieke Reece
Division des statistiques de l'énergie
Agence internationale de l'énergie
9, rue de la Fédération
75739 Paris Cedex 15, France

Télécopie : (33 1) 40.57.66.49
Internet : mieke.reece@iea.org

2. REVISIONS LES PLUS RECENTES DES DONNEES

2.1 Données des pays de l'OCDE

Le Secrétariat de l'AIE a achevé le traitement des dernières statistiques de l'énergie communiquées par les pays Membres de l'OCDE le 31 mars 2003 (qui couvrent la période 1960 - 2001). Ces données ont été publiées ultérieurement dans les publications de l'AIE/OCDE *Statistiques de l'énergie des pays de l'OCDE 2000 - 2001* et *Bilans énergétiques des pays de l'OCDE 2000 - 2001*.

Les données sur l'offre et la demande de pétrole de tous les pays de l'OCDE sont disponibles pour les années 1971-2001. Pour 2002, on ne dispose que de données sur l'offre ; les données sur la consommation finale sont disponibles, en général, un ou deux ans plus tard.

Il est à noter aussi que l'on dispose de données sur l'offre et la demande pour les années 1960-1970, sauf pour :

• la République tchèque	à partir de 1971
• la Hongrie	à partir de 1965
• la Corée	à partir de 1970
• le Mexique	à partir de 1971
• la République slovaque	à partir de 1971

En conséquence, les totaux régionaux ne seront pas calculés pour la période allant de 1960 à 1970.

2.2 Données des pays non membres de l'OCDE

Les données concernant les pays non membres de l'OCDE proviennent, pour l'essentiel, de la base de données de l'AIE sur les statistiques énergétiques mondiales. Cette base de données englobe l'offre et la demande de toutes les formes d'énergie dans plus de 100 pays non membres de l'OCDE au cours de la période comprise entre 1971 et 2001. Les données ont été collectées à partir de publications de statistiques nationales et grâce à la coopération avec d'autres organisations internationales et avec l'industrie de l'énergie. Les données pour 2001 concernant les pays non membres de l'OCDE présentées dans la première partie de l'ouvrage *Oil Information/Données sur le pétrole 2003* sont provisoires. Les chiffres définitifs obtenus à l'issue de ce travail de collecte seront rendus publics dans la publication annuelle de l'AIE intitulée *Statistiques de l'énergie des pays non membres 2000-2001* et *Bilans énergétiques des pays non membres 2000-2001*.

3. DEFINITIONS GENERALES

3.1. Définition des produits

Entrées en raffineries

Pétrole brut

C'est une huile minérale d'origine naturelle, constituée d'un mélange d'hydrocarbures et d'impuretés associées (soufre, par exemple). Elle existe en phase liquide aux conditions standard de température et de pression et ses caractéristiques physiques (densité, viscosité, etc.) sont extrêmement variables. Cette catégorie comprend aussi les condensats extraits des gaz associés ou non associés sur les gisements et les périmètres d'exploitation lorsque ceux-ci sont mélangés au brut commercial.

Liquides de gaz naturel (LGN)

Les LGN sont des hydrocarbures liquides ou liquéfiés obtenus à partir du gaz naturel dans les installations de séparation ou de traitement du gaz. Les LGN comprennent l'éthane, le propane, le butane (butane normal et isobutane), le pentane et l'isopentane, ainsi que les pentanes plus (essence naturelle ou condensat).

Produits d'alimentation des raffineries

Il s'agit d'un produit ou d'une combinaison de produits dérivé du pétrole brut et destiné à subir un traitement ultérieur autre qu'un mélange dans l'industrie du raffinage (par exemple fuel-oil de distillation directe ou gazole sous vide). Il est transformé en un ou plusieurs constituants et/ou produits finis. Cette définition recouvre également les produits renvoyés par l'industrie pétrochimique aux raffineries (par exemple, essence de pyrolyse, coupes C4, coupes de gazole et de fioul).

Autres hydrocarbures (que le pétrole brut et les LGN)

Cette catégorie comprend le pétrole brut synthétique extrait de schistes bitumineux, de sables asphaltiques, etc. ainsi que les huiles issues de la liquéfaction du charbon, l'hydrogène et les huiles émulsifiées (par exemple l'orimulsion). Bien que les additifs ne soient pas des hydrocarbures, ils sont pris en compte : il s'agit de substances qui sont ajoutées ou mélangées à un produit afin de modifier ses propriétés (indice d'octane ou de cétane, propriétés à froid, etc.) ; ce sont des alcools (méthanol, éthanol), des éthers (comme le MTBE (méthyl tertio-butyl éther)), de l'ETBE (éthyl tertio-butyl éther), du TAME (tertioamylméthyl éther), des esters (par exemple l'ester méthylique de colza, etc.) ainsi que des composés chimiques (tels que le tétraéthylplomb et le tétraméthylplomb) et des détergents.

Production des raffineries

Gaz de raffinerie (non liquéfiés)

Cette catégorie couvre les divers gaz non condensables obtenus dans les raffineries lors de la distillation du pétrole brut ou du traitement des produits pétroliers (par craquage par exemple), essentiellement hydrogène, méthane, éthane et oléfines. Elle comprend également les gaz retournés par l'industrie pétrochimique.

Ethane

L'éthane (C_2H_6) est un hydrocarbure à chaîne droite, gazeux à l'état naturel, que l'on extrait du gaz naturel et des gaz de raffinerie.

Gaz de pétrole liquéfiés (GPL)

Il s'agit des fractions légères d'hydrocarbures paraffiniques saturés qui s'obtiennent lors du raffinage ainsi que dans les installations de stabilisation du pétrole brut et de traitement du gaz naturel. Ce sont principalement le propane (C_3H_8) et le butane (C_4H_{10}) ou un mélange de ces deux hydrocarbures. Ils sont en général liquéfiés sous pression pour le transport et le stockage.

Naphtas

Les naphtas sont un produit d'alimentation des raffineries destiné à l'industrie pétrochimique (par exemple fabrication d'éthylène ou production de composés aromatiques). Les naphtas correspondent aux fractions distillant entre 30°C et 210°C ou sur une partie de cette plage de température.

Essence moteur

C'est un mélange d'hydrocarbures légers distillant entre 35°C et 215°C. Elle est utilisée comme carburant dans les moteurs terrestres à allumage commandé. L'essence moteur peut comprendre des additifs, des composés oxygénés et des pro-octanes, notamment des composés plombés comme le tétraéthylplomb et le tétraméthylplomb.

- *Essence moteur sans plomb :* Essence moteur sans adjonction de composés de plomb pour améliorer l'indice d'octane. Elle peut contenir des traces de plomb organique.
- *Essence moteur au plomb :* Essence moteur additionnée de tétraéthylplomb et/ou de tétraméthylplomb destiné à relever l'indice d'octane.

Cette catégorie comprend les composés pour mélange avec l'essence (à l'exclusion des additifs/composés oxygénés), tels qu'alkylats, isomérats, reformats ou essence de craquage pour utilisation comme essence moteur.

Essence aviation

Il s'agit d'une essence spécialement préparée pour les moteurs alternatifs des avions, avec un indice d'octane adapté au moteur, un point de congélation de -60°C et un intervalle de distillation habituellement compris entre 30°C et 180°C.

Carburéacteur type essence (carburéacteur type naphta ou JP4)

Cette catégorie comprend tous les hydrocarbures légers utilisés dans les turbomoteurs pour avion. Ils distillent entre 100°C et 250ºC et sont obtenus par mélange de kérosène et d'essence ou de naphtas de manière que la teneur en composés aromatiques soit égale ou inférieure à 25 pour cent en volume, et la tension de vapeur se situe entre 13,7kPa et 20,6kPa.

Carburéacteur type kérosène

C'est un distillat utilisé dans les turbomoteurs pour avion, qui répond aux mêmes caractéristiques de distillation (entre 150°C et 300°C mais en général pas au-delà de 250°C) et présente le même point d'éclair que le kérosène. Par ailleurs, il répond à certaines spécifications particulières (concernant par exemple le point de congélation) fixées par l'Association du Transport Aérien International (IATA).

Pétrole lampant

Ce sont des distillats de pétrole raffiné, utilisés dans d'autres secteurs que le transport aérien. Le pétrole lampant distille entre 150°C et 300°C.

Gazole / carburant diesel

Les gazoles/carburants diesel sont essentiellement des huiles intermédiaires qui distillent entre 180°C et 380°C. Plusieurs qualités sont disponibles, suivant l'utilisation: gazole pour moteur diesel à allumage par compression (automobiles, poids lourds, bateaux, etc.) ; fioul pour le chauffage des locaux industriels et commerciaux; autres gazoles, y compris les huiles lourdes distillant entre 380°C et 540°C qui sont utilisées comme produit d'alimentation dans l'industrie pétrochimique.

Fioul

Ce sont tous les fiouls résiduels (lourds), y compris ceux obtenus par mélange, dont la viscosité cinétique est supérieure à 10 cSt à 80°C. Le point d'éclair est toujours supérieur à 50°C et la densité toujours supérieure à 0,90 kg/l.

- *Fioul à faible teneur en soufre :* Fioul lourd dont la teneur en soufre est inférieure à 1%.
- *Fioul à haute teneur en soufre :* Fioul lourd dont la teneur en soufre est de 1% ou plus.

White spirit et SBP (essences spéciales)

Ce sont des distillats intermédiaires raffinés, dont l'intervalle de distillation se situe entre celui des naphtas et celui du kérosène. Ils se subdivisent en :

i. *Essences spéciales (SBP)* : huiles légères distillant entre 30°C et 200°C. On distingue 7 ou 8 qualités d'essences spéciales, selon la position de la coupe dans l'intervalle de distillation. Les qualités sont définies en fonction de la différence de température entre les points de distillation pour 5% et 90% en volume (le point de distillation ne dépasse pas 60°C).

ii. *White spirit :* essence industrielle dont le point d'éclair est supérieur à 30°C. L'intervalle de distillation du white spirit est compris entre 135°C et 200°C.

Lubrifiants

Ce sont des hydrocarbures obtenus à partir de distillats ou de résidus ; ils sont principalement utilisés pour réduire les frottements entre surfaces d'appui. Cette catégorie comprend toutes les qualités d'huiles lubrifiantes, depuis les spindles jusqu'aux huiles à cylindres, y compris les huiles moteur et toutes les qualités d'huiles de base pour lubrifiants.

Bitume

Hydrocarbure solide, semi-solide ou visqueux, à structure colloïdale, de couleur brune à noire ; c'est un résidu de la distillation du pétrole brut, obtenu par distillation sous vide des huiles résiduelles de distillation atmosphérique. Le bitume est aussi souvent appelé asphalte et il est principalement employé pour le revêtement des routes et pour les matériaux de toiture. Cette catégorie comprend le bitume fluidifié et le bitume fluxé.

Paraffines

Hydrocarbures aliphatiques saturés, les paraffines sont des résidus du déparaffinage des huiles lubrifiantes ; elles présentent une structure cristalline, plus ou moins fine, selon le grade. Leurs principales caractéristiques sont les suivantes : incolores, inodores et translucides, avec un point de fusion supérieur à 45°C.

Coke de pétrole

C'est un résidu solide noir, obtenu principalement par craquage et carbonisation de résidus de produits d'alimentation, de goudrons et de poix, dans des procédés tels que la cokéfaction différée ou la cokéfaction fluide. Il se compose essentiellement de carbone (90 à 95%) et brûle en laissant peu de cendres. Il est employé comme produit d'alimentation dans les cokeries des usines sidérurgiques, pour la chauffe, pour la fabrication d'électrodes et pour la production de substances chimiques. Les deux qualités les plus importantes de coke sont le "coke de pétrole" et le "coke de pétrole calciné". Cette catégorie comprend également le "coke de catalyse", qui se dépose sur le catalyseur pendant les opérations de raffinage ; ce coke n'est pas récupérable, et il est en général brûlé comme combustible dans les raffineries.

Autres produits

Il s'agit de tous les produits pétroliers qui ne sont pas expressément cités ci-dessus, comme par exemple le goudron et le soufre. Cette catégorie comprend également les composés aromatiques (p. ex. BTX ou benzène, toluène et xylènes) et les oléfines (p. ex. propylène) produits dans les raffineries.

3.2 Définition des éléments d'information concernant les approvisionnements et la consommation

Il y a lieu de signaler qu'un changement important a été introduit dans la notification des flux de produits pétrochimiques il y a quelques années. Les quantités de pétrole livrées à l'industrie pétrochimique sont indiquées dans deux secteurs : transformation et industrie. Les quantités de pétrole livrées à l'industrie pétrochimique qui sont renvoyées aux raffineries sont indiquées dans le secteur transformation, à la rubrique *Industrie pétrochimique.* Les quantités indiquées dans le secteur industrie, à la rubrique *Chimie (y compris pétrochimie)* représentent la consommation nette de cette industrie (c'est-à-dire, les livraisons brutes, déduction faite des retours aux raffineries).

Production intérieure

La totalité de la production à l'intérieur des frontières nationales, y compris la production en mer. La production ne comprend que la production commercialisable et ne tient pas compte des volumes réinjectés dans les gisements. Cette rubrique comprend la totalité du pétrole brut, des LGN, des condensats et des autres hydrocarbures (y compris les additifs).

Retours aux raffineries

Comprend les résidus de craquage et d'autres produits des industries chimique et pétrochimique qui sont renvoyés aux raffineries et utilisés comme produits d'alimentation. Les transferts d'une raffinerie à l'autre à l'intérieur d'un même pays ne sont pas inclus.

Utilisation directe

Quantités de pétrole brut ou de LGN qui sont utilisées directement sans avoir été traitées en raffinerie, par exemple le pétrole brut utilisé pour la production d'électricité.

Approvisionnements des raffineries

Cette rubrique définit les quantités totales entrées en raffinerie pour les procédés de raffinage. Elle englobe le pétrole brut, les LGN, les produits d'alimentation et les hydrocarbures dont l'origine n'est ni le pétrole brut ni les LGN.

Production des raffineries (brute)

Production de produits finis dans une raffinerie ou une installation de mélange. Cette rubrique ne comprend pas les pertes de raffinage, mais comprend la consommation propre des raffineries.

Transferts

Ils s'appliquent aux produits finis dont la classification a été modifiée et qui ont été renvoyés à la raffinerie pour nouveau traitement sans avoir été livrés aux consommateurs finals. Cette rubrique couvre également les transferts entre produits dont le classement a été modifié après une opération de mélange, par exemple, les transferts entre carburéacteur type kérosène et pétrole lampant.

Produits primaires reçus

Il s'agit du pétrole brut et des LGN d'origine nationale ou importés qui sont utilisés directement sans être passés par une raffinerie de pétrole (par exemple, le pétrole brut utilisé pour la production d'électricité au Japon). Voir ci-dessus la rubrique *Utilisation directe*.

Produits recyclés

Ce sont des produits finis qui sont renvoyés par le biais du raffinage et/ou du réseau de commercialisation après la livraison au consommateur final (par exemple, les lubrifiants usés qui sont retraités).

Soutages maritimes internationaux

Les soutages couvrent les quantités de combustibles livrés aux navires de haute mer, quel que soit leur pavillon, y compris les navires de guerre. La consommation des navires destinés à la navigation fluviale intérieure ou au cabotage n'est pas prise en compte dans cette rubrique (voir *Secteur des transports* ci-dessous).

Importations et exportations

Les quantités sont considérées comme étant importées ou exportées lorsqu'elles ont franchi les frontières politiques du pays, qu'il y ait eu ou non dédouanement. Cette rubrique tient compte des quantités de pétrole brut importées ou exportées, au titre d'accords de traitement (par exemple, raffinage à façon). Les quantités de pétrole en transit en sont exclues. Les importations de produits d'alimentation des raffineries comprennent les produits raffinés importés destinés à des mélanges en raffinerie (voir produits d'alimentation des raffineries au chapitre des définitions des produits). L'origine des importations de pétrole brut et de LGN est le pays de production tandis que pour les produits finis, c'est le dernier pays de provenance qui est pris en compte.

Il est à noter que, en raison des décalages dans le temps d'enregistrements des importations et des exportations (durée du voyage, retards, changements d'itinéraire, transbordements), les exportations notifiées par un pays peuvent ne pas correspondre exactement aux importations notifiées par les pays importateurs. En outre, par suite des différences de classification des produits, en particulier s'agissant des produits d'alimentation des raffineries, il risque d'apparaître des incohérences entre les importations et les exportations. En conséquence, les importations totales des différentes régions de l'OCDE peuvent ne pas être égales aux exportations régionales correspondantes.

Importations nettes

Les importations nettes sont calculées en déduisant les exportations des importations.

Variations des stocks

Les variations des stocks reflètent la différence entre le niveau initial des stocks (le premier jour de l'année) et leur niveau en fin de période (le dernier jour de l'année) ; il s'agit des stocks détenus sur le territoire national. Les chiffres positifs indiquent les prélèvements sur stocks, les chiffres négatifs, les accroissements de stocks.

Ecarts statistiques

Les administrations nationales obtiennent parfois les éléments des données concernant l'approvisionnement intérieur de plusieurs sources. En raison des différences entre les concepts, les couvertures géographiques, les décalages dans le temps et les définitions, il arrive souvent que les chiffres de consommation observée et calculée ne soient pas identiques. Les écarts statistiques correspondent à ces différences.

Consommation totale

Il s'agit de la consommation intérieure brute observée (i.e. comprenant la consommation propre des raffineries), avant d'en déduire les quantités retournées aux raffineries par l'industrie pétrochimique (retour de produits).

Stocks en fin de période

Cette rubrique fait référence au niveau des stocks en fin de période, c'est-à-dire le dernier jour de l'année, s'agissant des stocks détenus sur un territoire national (englobant les stocks contrôlés par les gouvernements).

Secteur transformation

Ce secteur comprend les combustibles utilisés dans des centrales publiques et privées d'électricité, dans des installations de cogénération, et dans des centrales calogènes. Un autoproducteur est un établissement industriel qui, en plus de ses activités principales, produit de l'électricité qu'il consomme en totalité ou en partie. On y compte notamment la production propre d'électricité des chemins de fer. Les données sur les centrales calogènes et les installations de cogénération ne couvrent que les combustibles utilisés pour la production de la chaleur vendue à des tiers. Les combustibles utilisés comme produits d'alimentation dans les usines à gaz, dans les fours à coke, dans les hauts fourneaux ou dans l'industrie pétrochimique entrent également dans la catégorie du secteur transformation.

Centrales publiques d'électricité : Comprend les quantités de pétrole utilisées pour la production d'électricité des centrales publiques, ayant pour activités principales la production, le transport ou la distribution d'électricité.

Autoproducteurs d'électricité : Cette rubrique comprend les quantités de pétrole utilisées pour la production d'électricité par une entreprise qui, en plus de ses activités principales, produit de l'électricité entièrement ou partiellement destinée à sa consommation propre, par exemple, les établissements industriels, les chemins de fer ou les raffineries, notamment.

Centrales de cogénération chaleur/électricité : Cette rubrique comprend les quantités de pétrole utilisées dans des centrales de production combinée de chaleur et d'électricité qui produisent de l'électricité et de la chaleur utile au sein d'une même installation. Il peut s'agir d'installations publiques de cogénération ou d'autoproducteurs. Il convient de noter que, dans le cas des installations de cogénération appartenant à des autoproducteurs, tous les combustibles utilisés pour la production d'électricité sont pris en compte, tandis que pour la production de chaleur on ne tient compte que des combustibles utilisés pour produire celle qui est vendue à l'extérieur (par exemple, à un réseau). Ce décompte *ne comprend pas* les combustibles destinés à la production de chaleur pour la consommation propre des autoproducteurs. *Les informations sur les combustibles utilisés et la production d'électricité et de chaleur dans les installations de cogénération et les centrales calogènes doivent être utilisées avec prudence car les combustibles utilisés selon le type de production d'énergie ne font pas toujours l'objet d'une distinction cohérente. Le Secrétariat de l'AIE a lancé un programme d'estimation des données qui faisaient défaut, dans les cas où cela s'avérait possible.*

Centrales calogènes : Cette rubrique comprend les quantités de pétrole utilisées dans les installations conçues pour produire exclusivement de la chaleur. Des données sont disponibles aussi bien pour les installations calogènes des autoproducteurs que pour les centrales calogènes publiques. Les données concernant les installations calogènes des autoproducteurs représentent les combustibles utilisés dans des installations qui vendent de la chaleur à des tiers au titre des dispositions d'un contrat.

Industrie pétrochimique : Comprend les quantités retournées aux raffineries par le secteur pétrochimique, que ce soit pour un traitement plus poussé ou pour mélange, ou bien pour leur utilisation directe. Il s'agit exclusivement de produits pétroliers.

Secteur énergie

Dans cette catégorie figurent les combustibles utilisés par les entreprises du secteur de l'énergie pour le chauffage ou la chauffe et pour le fonctionnement de leurs équipements, ainsi que les pertes de transformation et de distribution. Ce secteur comprend l'extraction du charbon, du pétrole et du gaz (toutes les activités nécessaires pour rendre commercialisables le pétrole et le gaz depuis la région productrice jusqu'au lieu d'expédition) ainsi que les industries de transformation, qui comprennent notamment les centrales électriques, les usines à gaz, les fours à coke et les raffineries de pétrole (consommation propre des raffineries).

Consommation propre des raffineries

Comprend le pétrole consommé dans les raffineries pour le fonctionnement des équipements, le chauffage et l'éclairage.

Consommation finale totale

La consommation finale totale est la somme des consommations des différents secteurs d'utilisation finale. Les produits renvoyés par l'industrie pétrochimique aux raffineries ne sont pas comptabilisés comme consommation finale.

Secteur transports

Ce secteur englobe les produits pétroliers consommés dans toutes les activités de transport, indépendamment du secteur considéré.

Transport aérien : Livraisons de carburants aviation pour l'aviation civile internationale et pour toutes les activités de transport aérien intérieur, à savoir commerciales, privées, agricoles, militaires, etc. Comprend également les quantités utilisées à des fins autres que le vol proprement dit, par exemple, l'essai de moteurs au banc. Ne tient pas compte du carburant utilisé par les compagnies aériennes pour les véhicules routiers.

Transport routier : La totalité des produits pétroliers utilisés dans les véhicules routiers, y compris le carburant utilisé dans les transports agricoles sur route. Ne tient pas compte du carburant moteur destiné aux moteurs fixes ni du carburant diesel pour les tracteurs utilisés ailleurs que sur route (lequel est comptabilisé dans la rubrique *Agriculture*).

Transport ferroviaire : Toutes les quantités utilisées pour le trafic ferroviaire, y compris les chemins de fer industriels.

Voies navigables intérieures : Ce sont toutes les livraisons effectuées pour la navigation sur les voies navigables intérieures et pour le cabotage, notamment les livraisons aux petites embarcations, péniches et caboteurs qui ne sont pas prises en compte dans les soutages maritimes internationaux.

Transport par conduites : Toutes les quantités utilisées pour l'exploitation des réseaux des conduites.

Secteur industrie

Les catégories industrielles sont définies ci-après par application de leur numéro de catégorie ou de classe CITI (Classification Internationale Type par Industrie de toutes les branches d'activité économique). Les numéros CITI renvoient à la série M, n° 4/Rev.3 Nations Unies, New York, 1990.

Sidérurgie: CITI Groupe 271 et classe n° 2731 ;

Chimie (y compris pétrochimie) : Division CITI 24. L'industrie pétrochimique comprend les procédés de craquage et de reformage mis en œuvre pour produire de l'éthylène, du propylène, du butylène, du gaz de synthèse, des composés aromatiques, du butadiène et d'autres matières premières à base d'hydrocarbures ;

Métaux non ferreux : Groupe 272 et classe CITI n° 2732 ;

Fabrication de produits minéraux non métalliques : Comprend le verre, la céramique, le ciment et les autres matériaux de construction (division CITI n° 26) ;

Matériel de transport : Divisions CITI n° 34 et 35 ;

Machines : Fabrication d'ouvrages en métaux, de machines et de matériel, à l'exclusion du matériel de transport (divisions CITI 28, 29, 30, 31 et 32) ;

Industries extractives : Divisions CITI 13 et 14 ;

Industrie alimentaire, boissons et tabacs : Divisions CITI 15 et 16 ;

Pâtes et papier d'imprimerie : Divisions CITI 21 et 22 ;

Bois et produits dérivés : Division CITI 20 ;

Construction : Division CITI 45 ;

Industries textiles et du cuir : Divisions CITI 17, 18 et 19 ;

Non spécifié : Toute industrie manufacturière n'entrant dans aucune des catégories ci-dessus (divisions CITI 25, 33, 36 et 37).

Les combustibles utilisés pour la production d'électricité dans les différentes catégories industrielles (les autoproducteurs) entrent dans la catégorie transformation.

Autres secteurs

Commerce et services publics : Divisions CITI n° 41, 50, 51, 52, 55, 63-67, 70-75, 80, 85, 90-93 et 99.

Secteur résidentiel : Toutes les quantités consommées par les ménages (division CITI 95).

Agriculture : Toutes les quantités consommées par des utilisateurs classés dans les rubriques agriculture, chasse, sylviculture et pêche (hauturière, côtière et intérieure). Ces quantités comprennent donc les produits pétroliers utilisés par ces utilisateurs, que ce soit pour la traction (à l'exclusion des utilisations routières agricoles), la production d'énergie ou le chauffage. Divisions CITI n° 01, 02 et 05.

Non spécifié : Toutes les activités ne figurant pas ailleurs, par exemple, activités militaires (à l'exception des activités militaires aériennes).

Pour mémoire

Les quantités indiquées dans ces catégories sont déjà prises en compte dans les secteurs de consommation mentionnés ci-dessus.

Utilisations non énergétiques : Il s'agit de la somme des deux catégories ci-après, à savoir, produits d'alimentation de l'industrie chimique/pétrochimique et "autres".

Produits d'alimentation de l'industrie chimique/pétrochimique : Concerne les produits d'alimentation pour les utilisations non énergétiques dans l'industrie chimique/pétrochimique.

Autres : Produits pétroliers utilisés à des fins non énergétiques dans tous les autres secteurs. Cette rubrique concerne généralement toute la consommation de lubrifiants, de bitume, de white spirit et de paraffines. S'agissant du coke de pétrole, il entre dans la catégorie des utilisations non énergétiques uniquement lorsqu'il existe la preuve d'une pareille utilisation (par exemple, dans le secteur industriel des métaux non ferreux).

4. SOURCES ET NOTES

4.1 Deuxième partie

Sources

Les données historiques annuelles figurant dans les troisième et quatrième parties proviennent des bases de données statistiques sur l'énergie de l'AIE/OCDE. Veuillez vous reporter aux sections 3.2 et 3.3 pour un complément d'information. Les données pour 2001 et 2002 feront l'objet de révisions lorsque les statistiques annuelles définitives seront disponibles.

Statistiques annuelles du pétrole (AOS) : Cette base de données couvre les données annuelles concernant le pétrole brut, les LGN, les produits d'alimentation des raffineries et d'autres hydrocarbures, ainsi que les produits pétroliers, pour les pays de l'OCDE. Ces données comprennent notamment des bilans détaillés de l'offre et de la demande, la consommation finale par catégorie d'utilisation finale, les échanges par origine et destination, ainsi que les niveaux et les variations des stocks ; elles sont communiquées au Secrétariat par les administrations nationales en réponse à des questionnaires.

Statistiques mensuelles du pétrole et du gaz naturel (MOS) : Cette base de données est compatible avec les statistiques annuelles, à l'exception de la ventilation détaillée des lubrifiants, du bitume, des paraffines et d'autres produits raffinés, ainsi que des données sur la consommation finale par catégorie d'utilisation finale qui ne figurent pas dans les statistiques mensuelles du pétrole et du gaz naturel. Les données sont recueillies tous les mois au moyen de questionnaires auxquels répondent les pays Membres de l'OCDE. Les estimations des divers éléments de l'approvisionnement pétrolier des pays de l'OCDE pour 2002 sont calculées à partir des statistiques mensuelles du pétrole.

Les statistiques jusqu'en 2001 proviennent de la base de données *Statistiques annuelles du pétrole (AOS)* alors que les données pour 2002 proviennent principalement de la base de données *Statistiques mensuelles du pétrole et du gaz naturel (MOS)*. Les données mensuelles sur le pétrole, communiquées par les administrations nationales des pays Membres, sont plus faibles que toutes les statistiques annuelles précédentes car elles ne tiennent pas compte de la demande de certains produits (par exemple, le coke de pétrole dans certains pays). Afin d'établir des séries chronologiques comparables entre les bases AOS et MOS, les données mensuelles pour 2002 sont corrigées à la hausse pour rectifier des notifications insuffisantes. En conséquence, les données relatives à l'année 2002 peuvent différer des statistiques similaires publiées dans l'ouvrage *Statistiques trimestrielles : électricité, charbon, gaz et pétrole* (données MOS).

Bilans énergétiques : Il s'agit de statistiques annuelles qui regroupent des données tirées des bases de données AIE/OCDE relatives au charbon, à l'électricité, ainsi qu'au pétrole et au gaz, afin de fournir une récapitulation annuelle de l'offre et de la demande d'énergie, en utilisant une unité commune (millions de tonnes d'équivalent pétrole) pour chaque pays de l'OCDE.

Emissions de CO_2 : Les émissions de CO_2 sont calculées à partir des Bilans énergétiques de l'AIE en appliquant la méthodologie par défaut du GIEC et les coefficients d'émission décrits dans la révision de 1996 des *Lignes directrices du GIEC pour les inventaires nationaux de gaz à effet de serre*.

Statistiques et bilans énergétiques des pays non membres : Cette base de données contient les bilans annuels de l'offre et de la demande de toutes les formes d'énergie de plus de 100 pays non membres de l'OCDE.

Energy Prices and Taxes (Enquête trimestrielle de l'AIE sur les prix) : Il s'agit de statistiques trimestrielles concernant les pays de l'OCDE et certains pays non membres de l'OCDE, qui couvrent les prix à l'importation et à l'exportation, les prix au comptant ainsi que les prix et les taxes à la consommation finale.

Ces données sont une compilation d'informations contenues dans *IEA/OECD Energy Prices and Taxes*. Cette publication fournit, tous les trimestres, des données détaillées à l'échelon national, provenant de la base de données de l'AIE, sur les prix à l'importation et à l'exportation du pétrole brut, du charbon vapeur, du charbon à coke et du gaz naturel ; sur les prix au comptant du pétrole brut, de l'essence, du fioul léger et du fioul lourd ; ainsi que sur les prix et les taxes à la consommation finale, par secteur de consommation, qui s'appliquent aux produits pétroliers, au charbon, au gaz naturel et à l'électricité.

Des informations plus détaillées concernant ces prix sont présentées dans le chapitre "Notes on Definitions" de la publication *Energy Prices and Taxes*, où les écarts par rapport aux principes généraux sont expliqués, et des informations complémentaires sont fournies.

Les prix au comptant du pétrole brut[1] sont des moyennes mensuelles basées sur les cotations journalières faites par *Platt's*. Cependant, avant juillet 1987 pour le Brent et avant janvier 1985 pour le Dubai et le WTI, les prix moyens mensuels étaient calculés à partir des moyennes hebdomadaires établies à partir de diverses publications. Tous les prix s'entendent "livraison à 30 jours", sauf pour le Brent pour lequel les prix correspondent au cours du jour pour "livraison immédiate".

Les prix mensuels au comptant des produits pétroliers[1] sont calculés à partir des cotations journalières de *Platt's*. Cependant, pour le nord-ouest de l'Europe (Rotterdam), ils sont calculés à partir des prix hebdomadaires parus dans diverses publications, avant avril 1987 pour l'essence sans plomb, et avant janvier 1985 pour les fiouls HTS. Les prix pour le nord-ouest de l'Europe et les Etats-Unis (port de New York) sont la moyenne des cotations maximales et minimales des achats au comptant de produits pétroliers, sur barge, FAB. Pour Singapour, il s'agit du prix moyen des cargaisons. Tous les prix sont indiqués hors taxes à la consommation finale.

Les prix à la consommation finale sont collectés tous les trimestres et proviennent des administrations nationales ou d'autres organismes compétents. Ils sont complétés par des données provenant de sources nationales publiées. En général, les prix à la consommation finale: 1) comprennent les coûts de transport jusqu'au point de consommation ; 2) indiquent les prix effectivement payés, c'est-à-dire nets de remises ; et 3) comprennent les taxes payables par le consommateur dans le cadre de la transaction et qui ne sont pas remboursables. Ces taxes ne comprennent pas la taxe sur la valeur ajoutée dont s'acquitte l'industrie dans nombre de pays européens pour la totalité des biens et des services. Dans ces cas, la taxe sur la valeur ajoutée est remboursée au consommateur, généralement sous forme de crédit d'impôt. En conséquence, elle n'est pas indiquée comme un élément des prix.

Les coûts du pétrole brut à l'importation sont tirés d'un système de notification mensuelle des coûts du pétrole brut mis en place entre les administrations nationales et le Secrétariat de l'AIE. Les coûts à l'importation, exprimés en valeurs unitaires moyennes (CAF), sont enregistrés au moment de l'importation et représentent la somme du coût, de l'assurance et du fret, mais ne comprennent pas les droits d'importation.

Données sur les ports pétroliers : Les données sur les ports donnent des détails concernant ceux qui sont en en mesure d'accueillir les pétroliers en pleine charge de 60 000 tpl ou davantage. Ces données, fournies par *SSY Consultancy & Research Ltd* de Londres, concernent les installations de chargement et de déchargement en exploitation au mois d'avril 2003 ou dont la mise en service est prévue pour le milieu de l'année 2003.

Les ports sont classés selon qu'il s'agit de ports de chargement (L), de déchargement (D) ou de ports équipés d'installations permettant d'effectuer ces deux opérations (B). Dans certains ports, il peut exister plusieurs terminaux pour navires transporteurs de brut, auquel cas l'installation mentionnée est celle où peut accoster le navire le plus grand. Certains ports, cependant, font l'objet de plus d'une entrée, par exemple afin d'indiquer que deux raffineries importantes y sont implantées.

Il est courant dans de nombreuses parties du monde, notamment aux Etats-Unis, d'effectuer le déchargement partiel de la cargaison par transfert de navire à navire, sur des allèges. Cela permet aux ports qui ne

1 Source: Basé sur Platt's, sauf indication contraire. ©2003, Platt's, qui fait partie du groupe McGraw-Hill Companies, Inc. Reproduction interdite sans autorisation préalable de Platt's. Tous droits réservés.

disposent pas d'installations en eau profonde de recevoir des cargaisons qui ont été transportées sur des pétroliers d'une capacité bien supérieure à celle des navires pouvant accoster le long des terminaux de déchargement disponibles. Dans certains cas, la cargaison du « navire mère » sera totalement déchargée sur des allèges ; dans d'autres, une partie seulement de la cargaison du navire mère sera transférée, de façon qu'il puisse accoster dans le port en charge partielle. Ainsi, il sera possible de recevoir dans l'installation désignée des navires *en charge partielle* dont la taille est supérieure au maximum indiqué. Par conséquent, dans les informations présentées figureront les zones d'aconnage désignées (par exemple, celles qui sont autorisées par la garde côtière des Etats-Unis dans le golfe du Mexique).

NB : Les données fournies ne concernent pas seulement les zones d'aconnage, mais aussi les ports classiques, ainsi que les bouées d'amarrage unique SBM, les systèmes d'amarrage sur plot central, les plateformes au large et les installations flottantes d'extraction, de stockage et de chargement en mer.

Les tableaux sur les ports indiquent les ports de chargement, de déchargement et de transbordement pour les cargaisons de pétrole brut et de condensats, mais ne comprennent pas ceux qui ne traitent que les produits raffinés, les gaz de pétrole liquéfiés et/ou les cargaisons de produits chimiques liquides en vrac.

Les chiffres concernant la longueur hors tout, la largeur maximum, le tirant d'eau (distance entre la ligne de flottaison et le point le plus bas de la quille d'un navire) et la hauteur maximale (distance entre la ligne de flottaison et le point le plus haut d'un navire) sont exprimés en mètres. Les données concernant la taille maximale des navires sont exprimées en tonnes de port en lourd (tpl) (correspondant à la capacité de charge d'un navire, y compris les articles d'avitaillement, les huiles lubrifiantes et l'eau, ainsi que la cargaison proprement dite). Les chiffres sur les débits de chargement et de déchargement, lorsqu'ils sont connus, sont exprimés en tonnes (tonnes métriques) par heure.

Données sur les tarifs de fret des pétroliers : Les données sur les tarifs de fret des pétroliers sont les moyennes annuelles, exprimées en dollars des Etats-Unis par tonne, qui s'appliquent sur les itinéraires commerciaux représentatifs pour les différents tonnages de navires. Ces tonnages correspondent aux principales catégories de navires de la flotte pétrolière, définies en fonction de leur taille. Les chiffres indiqués sont fondés sur les tarifs notifiés d'affrètement au voyage simple (correspondant à l'affrètement au jour le jour, par exemple pour transporter 260 000 tonnes de pétrole brut de la région du Golfe Persique au Japon). Ces transactions au jour le jour constituent l'essentiel de l'affrètement sur le marché des pétroliers, les navires étant rarement engagés pour une période déterminée, de 12 mois par exemple. Les tarifs indiqués concernent le transport de pétrole brut et de produits raffinés, selon le tonnage du navire adapté à l'acheminement de ces cargaisons. Ces informations sont communiquées par *SSY Consultancy & Research Ltd* (research@ssy.co.uk).

4.2. Troisième partie

Statistiques annuelles du pétrole (AOS) : Toutes les données proviennent de la base de données *Statistiques annuelles du pétrole* (AOS) (se reporter aux Sources et Notes de la deuxième partie). On trouvera les définitions aux sections [3.1] Définition des produits et [3.2] Définition des éléments d'information concernant les approvisionnements et la consommation.

Note technique sur le tableau 1

La production nationale comprend la production de pétrole brut, de LGN et des hydrocarbures dont l'origine n'est ni le pétrole brut ni les LGN (voir définitions des produits ci-dessus [3.1]).

+ Les importations nettes représentent les importations moins les exportations de pétrole brut, de LGN, de produits d'alimentation des raffineries, des autres hydrocarbures ainsi que les importations nettes de produits pétroliers raffinés (voir définitions des produits ci-dessus [3.1]).

- Les soutages maritimes internationaux comprennent les produits raffinés utilisés dans les navires quel que soit leur pavillon (voir définitions concernant les approvisionnements et la consommation ci-dessus [3.2]).

+ Les variations des stocks comprennent les variations des stocks (les chiffres positifs indiquent les prélèvements sur stocks, les chiffres négatifs, les accroissements de ces derniers) de pétrole brut, de LGN, de produits d'alimentation des raffineries, des autres hydrocarbures, ainsi que de produits pétroliers raffinés (voir définitions concernant les approvisionnements et la consommation ci-dessus [3.2]).

= **Approvisionnements totaux** = somme des données ci-dessus.

Les approvisionnements des raffineries[2] couvrent les quantités entrées en raffinerie de pétrole brut, de LGN, de produits d'alimentation des raffineries et autres hydrocarbures (voir définitions concernant les approvisionnements et la consommation ci-dessus [3.2]).

La production des raffineries[2] couvre la production de produits pétroliers dans les raffineries (voir définitions des éléments d'information concernant les approvisionnements et la consommation ci-dessus [3.2]).

- **Pertes de raffinage =** entrées en raffineries - production des raffineries.

+ **Les retours** correspondent aux quantités de produits renvoyées par l'industrie pétrochimique. Cette rubrique comprend également les produits recyclés, qui sont des produits pétroliers raffinés précédemment expédiés pour la consommation et qui, ultérieurement, sont renvoyés aux raffineries (voir définitions concernant les approvisionnements et la consommation ci-dessus [3.2]).

- **Les écarts statistiques** comprennent les écarts statistiques et les transferts de produits pétroliers (transferts entre produits ou produits renvoyés aux raffineries après en avoir modifié la classification).

= **Consommation totale =** Demande totale de pétrole (somme des secteurs ci-dessous).

La transformation

Cette catégorie couvre les quantités de pétrole utilisées dans le processus de la transformation du pétrole pour en faire une autre forme d'énergie, par exemple, le pétrole transformé en électricité, en gaz, en chaleur, etc. (voir définitions concernant les approvisionnements et la consommation ci-dessus [3.2]).

L'énergie

Cette catégorie couvre la consommation de pétrole des industries de transformation de l'énergie (pour le chauffage, l'éclairage et le fonctionnement des équipements) ainsi que les quantités de pétrole utilisées lors de l'extraction du pétrole et du gaz, dans les mines de charbon, les pipelines et les pertes de distribution (voir définitions concernant les approvisionnements et la consommation ci-dessus [3.2]).

Les transports

Cette catégorie couvre toute la consommation dans les transports, à l'exclusion des soutages maritimes internationaux, mais comprend les lubrifiants, les paraffines, etc. utilisés dans le secteur des transports. Elle comprend aussi les quantités consommées pour l'exploitation des oléoducs et gazoducs (voir définitions concernant les approvisionnements et la consommation ci-dessus [3.2]).

L'industrie

Cette catégorie couvre la totalité de la consommation industrielle de pétrole, y compris la consommation de produits d'alimentation de l'industrie pétrochimique et les utilisations autres que comme combustibles, par exemple, la consommation de lubrifiants, de paraffines, de white spirit, de bitume, etc. La consommation industrielle de pétrole pour la production d'électricité (autoproduction) n'est pas prise en compte (voir définitions concernant les approvisionnements et la consommation ci-dessus [3.2]).

Les "autres secteurs"

Cette catégorie couvre les secteurs résidentiel, commercial et services publics, l'agriculture et d'autres utilisations (non spécifiées). La consommation de lubrifiants, de paraffines, de white spirit, etc. y est comptabilisée également (voir définitions concernant les approvisionnements et la consommation ci-dessus [3.2] et les éléments d'information pour mémoire concernant les utilisations non énergétiques ci-après).

Pour mémoire: Les utilisations non énergétiques

Se répartissent en produits d'alimentation de l'industrie chimique/pétrochimique et autres utilisations non énergétiques. (Les quantités figurant dans

2 Eléments d'information pour mémoire, non compris dans le bilan.

les éléments d'information pour mémoire sont déjà prises en compte dans les transports, l'industrie et les "autres" secteurs.)

Les produits d'alimentation pour la chimie et la pétrochimie

Cette catégorie couvre les produits pétroliers utilisés dans l'industrie chimique/pétrochimique dans des procédés tels que le craquage et le reformage, pour produire de l'éthylène, du propylène, du butylène, du gaz de synthèse, des composés aromatiques, du butadiène et d'autres matières premières à base d'hydrocarbures.

Autres

Cette catégorie couvre la consommation autre que comme combustible de produits pétroliers tels que lubrifiants, paraffines, white spirit, bitume et la fraction de coke de pétrole qui n'a pas été utilisée comme combustible.

Tableaux 2 à 8B

Pour toute explication concernant les autres tableaux de la partie III, veuillez vous reporter aux sections [3.1] Définition des produits et [3.2] Définition des éléments d'information concernant les approvisionnements et la consommation. De plus, des notes sur les tableaux figurent à la fin de la partie IV.

4.3. Quatrième partie

Statistiques annuelles du pétrole (AOS) : Toutes les données proviennent de la base de données *Statistiques annuelles du pétrole* (AOS) (se reporter aux Sources et Notes de la deuxième partie). On trouvera les définitions aux sections [3.1] Définition des produits et [3.2] Définition des éléments d'information concernant les approvisionnements et la consommation.

Note technique sur les tableaux 8 à 14 : La consommation brute tient compte de la consommation d'énergie des raffineries, mais non des soutages maritimes internationaux.

Tableau 15 : La consommation du secteur transformation comprend les retours aux raffineries provenant de l'industrie pétrochimique.

Tableau 16 : La consommation du secteur énergie comprend la consommation propre des raffineries.

5. NOTES CONCERNANT LES PAYS

On notera que, du fait de révisions, les chiffres dans les *Données sur le pétrole* peuvent différer de ceux publiés antérieurement dans les *Statistiques de l'énergie des pays de l'OCDE* et les *Bilans énergétiques des pays de l'OCDE.*

Les données sur le pétrole pour 2002 sont des chiffres préliminaires fondés sur les *Statistiques mensuelles du pétrole et du gaz naturel* (MOS), sauf indication contraire. Pour certains pays, les données figurant dans les statistiques mensuelles ne sont pas totalement compatibles avec les données des *Statistiques annuelles du pétrole et du gaz naturel* (AOS). Le Secrétariat a donc corrigé pour 2002 les données des statistiques mensuelles du pétrole de ces pays afin de présenter une évolution cohérente des séries chronologiques entre 2001 et 2002. Pour les notes des tableaux figurant dans la troisième partie, veuillez vous référer à la fin de la quatrième partie.

1. Allemagne

Gaz de raffinerie : Avant 1970, la consommation de gaz de raffinerie dans l'industrie chimique est prise en compte dans la consommation propre des raffineries.

Carburéacteur type essence, carburéacteur type kérosène : On a procédé à une nouvelle classification et, à partir de 1994, le carburéacteur type essence entre dans la rubrique carburéacteur type kérosène.

Pétrole lampant : Suite à de nouvelles informations émanant des compagnies pétrolières, la consommation finale a été modifiée à compter de 1994.

Gazole/carburant diesel : En 1989, la consommation finale a diminué en raison d'un hiver exceptionnellement clément et d'un abaissement des stocks des consommateurs. En 1995, il s'est produit une rupture de séries concernant les secteurs industriels suite à un alignement sur les classifications de la NACE.

Fioul : Avant 1980, les quantités de fioul consommées dans les hauts fourneaux étaient comptabilisées avec les données concernant l'industrie sidérurgique. A partir de 2000, une partie du produit "Andere Rückstände" est inclus avec fioul au lieu de autres produits.

Autres produits : Avant 1979, les "autres produits" comprennent les paraffines, le bitume, le white spirit et les essences spéciales (SBP) dans le cas de l'Allemagne orientale.

A partir de 1994, on constate une rupture de série dans les données sur la consommation finale par secteur, parce que le MineralölwirtschaftsVerband a institué de meilleures méthodes d'enquête.

2. Australie

Toutes les données : Toutes les données concernent les exercices budgétaires, qui vont du 1er juillet au 30 juin (par exemple, 2001 = du 1er juillet 2000 au 30 juin 2001).

Les chiffres négatifs concernant les pertes des raffineries s'expliquent par des différences de traitement des transferts entre raffineries.

Après 1973, les données sont fondées sur des enquêtes nationales. On ne dispose pas de données détaillées pour les années antérieures concernant les autoproducteurs et la consommation de l'industrie par sous-secteur.

Le recul important de la consommation de l'industrie observé en 1982-83 est dû à la restructuration de l'industrie sidérurgique, à la récession économique et à l'une des sécheresses les plus graves de l'histoire du pays.

Pétrole brut : La diminution de la production en 1999 est due à une explosion de gaz à l'usine de Longford.

LGN : Avant 1992, une partie de la production de LGN est comptabilisée avec le pétrole brut.

Gaz de raffinerie : Avant 1974, le gaz de raffinerie est indiqué après déduction de la consommation des raffineries.

GPL, éthane : Avant 1991, les données sur l'éthane sont comptabilisées avec le GPL.

Lubrifiants, white spirit, paraffines : La consommation de lubrifiants, de white spirit/SBP et de paraffines est indiquée dans la rubrique consommation de l'industrie non spécifiée car on ne dispose pas d'une ventilation plus détaillée.

Fioul : Les importations de fioul ont été estimées par l'administration australienne.

3. Autriche

Les données ont été révisées par l'administration autrichienne en remontant jusqu'en 1980.

Gazole/carburant diesel : Pour 1978 et 1979, la consommation de gazole/carburant diesel de l'industrie est comptabilisée avec celle de fioul.

Autres produits : A partir de 1978, les paraffines et le coke de pétrole font partie des "autres produits". A partir de 1991, les naphtas sont comptabilisés avec les autres produits.

Lubrifiants : Avant 1990 la consommation de lubrifiants est indiquée dans le secteur des transports car on ne dispose pas d'une ventilation plus détaillée.

4. Belgique

On ne dispose pas des données concernant les secteurs de la fabrication de produits minéraux non métalliques ; de la construction mécanique ; des industries extractives ; de l'industrie alimentaire, des boissons et des tabacs ; de l'industrie des pâtes et papier et de l'imprimerie ainsi que de la construction pour les années antérieures à 1971.

A partir de l'année 1974, les consommations des centrales électriques et des installations de cogénération sont des estimations du Secrétariat établies sur la base des consommations indiquées dans le questionnaire annuel sur l'électricité.

Produits d'alimentation des raffineries : La baisse du volume des transferts de produits observée en 1995 s'explique par des travaux d'entretien des installations qui ont duré un mois.

Gaz de raffinerie : Avant 1975, le gaz de raffinerie est indiqué après déduction de la consommation des raffineries.

GPL, gazole/carburant diesel : Avant 1976, la consommation de GPL et de gazole/carburant diesel dans le secteur commerce/services publics est comptabilisée avec celle du secteur résidentiel.

Fioul : La rupture constatée entre 1975 et 1976 dans la série chronologique concernant le fioul provient d'une modification de classification entre secteurs industriel et "autres". La baisse de la consommation dans l'industrie depuis 1993 est due à l'introduction d'un droit d'accise ainsi qu'à un recours accru au gaz naturel.

Autres produits, coke de pétrole: Les ruptures constatées dans les séries chronologiques entre 1999 et 2000 proviennent d'améliorations apportées au système de collecte des données.

5. Canada

Autres hydrocarbures : La production nationale d'autres hydrocarbures correspond à la production de pétrole brut synthétique à partir de sables asphaltiques. Avant 1990, l'hydrogène utilisé pour la valorisation du pétrole brut synthétique (à partir d'autres sources) était comptabilisé avec les approvisionnements en gaz naturel ; à partir de 1990, l'administration canadienne a adopté une méthode différente.

Les importations d'orimulsion en provenance du Venezuela ont commencé en 1994.

Avant 1978, les soutages maritimes internationaux sont comptabilisés dans la rubrique navigation intérieure.

GPL, éthane : Avant 1990, les données sur l'éthane sont comptabilisées à la rubrique GPL. L'éthane est essentiellement utilisé comme produit d'alimentation pour l'industrie pétrochimique.

Pétrole lampant, gazole/carburant diesel, fioul : Avant 1980, il apparaît plusieurs ruptures de séries dans les données détaillées sur la consommation industrielle de pétrole lampant, de gazole/carburant diesel et de fioul. A partir de 1980, toutes les données relatives à la consommation sont fondées sur une seule enquête. Toutefois, à partir de 1988, les données concernant les matériels de transport, la construction mécanique, les produits alimentaires et les textiles ne sont pas disponibles, car Statistiques Canada a cessé de les collecter. Des chiffres distincts sur la consommation pour l'agriculture (y compris la sylviculture) sont disponibles à partir de 1983 pour le pétrole lampant, de 1973 pour le gazole/carburant diesel et de 1978 pour le fioul. Pour les années antérieures, la consommation du secteur de l'agriculture est comptabilisée avec celle du secteur résidentiel.

Gazole/carburant diesel : A partir de 1982, la consommation de gazole/carburant diesel des parcs automobiles du secteur des services de distribution en gros et au détail est comptabilisée dans le secteur commerce/services publics. Avant 1982, elle entrait dans la catégorie du transport routier.

Carburéacteur type essence : Avant 1979, le carburéacteur type essence est comptabilisé avec le carburéacteur type kérosène.

Naphtas, autres produits : Les achats de produits d'alimentation et d'autres additifs effectués auprès de compagnies non déclarantes sont comptabilisés avec les produits recyclés en naphtas et "autres produits".

Les séries chronologiques allant de 1990 à 1997 ont été révisées par l'administration canadienne, pouvant provoquer des ruptures de séries entre 1989 et 1990.

6. Corée

On dispose de données à partir de 1970.

Les niveaux des stocks tiennent compte des quantités embarquées dans des navires à compter de 1997.

Carburéacteur type kérosène : La répartition de la consommation entre l'aviation civile internationale et le transport aérien intérieur a été estimée par le Secrétariat pour les années 1990 à 1995.

Avant 1994, les retours de l'industrie pétrochimique sont déduits des données.

7. Danemark

A partir de 1990, le Groenland et les îles Féroé danoises ne sont plus inclus dans les statistiques pétrolières.

Autres hydrocarbures : Les importations et la consommation d'orimulsion pour la production d'électricité débutent en 1995.

Gaz de raffinerie : Avant 1975, les quantités de gaz de raffinerie sont indiquées après déduction de la consommation des raffineries.

Gazole/carburant diesel, fioul : Avant 1990, la consommation de gazole/carburant diesel et de fioul pour la pêche entre dans la rubrique navigation intérieure, alors qu'après cette date elle est indiquée dans le secteur de l'agriculture.

Fioul : Entre 1974 et 1979, les données sur la consommation de fioul pour la production de cogénération assurée par des autoproducteurs sont des estimations. La forte augmentation des quantités de fioul utilisées pour la cogénération en 1994 est due à un accroissement des exportations d'électricité à destination de la Norvège.

Avant 1976, la totalité de la consommation de produits pétroliers du secteur commercial/services publics est comptabilisée avec celle du secteur résidentiel; cependant, celle de fioul est indiquée à part en 1975.

Paraffines : A partir de 1987, on ne dispose plus de données distinctes concernant les paraffines.

Carburéacteur type essence : La consommation de carburéacteur type essence a cessé en 1988. Depuis 1989, seul le carburéacteur type kérosène est utilisé.

Autres produits : Les données sur le recyclage et la consommation finale d'huile usagée sont disponibles à partir de 1989.

Les données sur la consommation finale varient d'une année sur l'autre du fait qu'une enquête détaillée est réalisée auprès des entreprises au Danemark tous les deux ans. Les années où l'enquête n'a pas lieu, la consommation par utilisation finale est estimée par l'Agence danoise de l'énergie.

Les méthodes d'enquête ayant été améliorées, les consommations destinées à la cogénération sont classées différemment à partir de 1994. On constate donc une rupture de série entre 1993 et 1994. Le pétrole utilisé dans les secteurs industriels pour produire un excédent de chaleur, qui est fourni aux réseaux de chauffage urbain, est reporté dans ces secteurs industriels.

Les données détaillées pour le secteur industriel en 1994 et 1995 s'appuient sur une nouvelle enquête.

8. Espagne

Avant 1981, les données détaillées relatives à la consommation sont estimées en partie d'après les statistiques nationales qui couvrent la consommation en Espagne continentale.

Produits d'alimentation : On ne dispose pas de données pour l'année 1981.

GPL, fioul : Avant 1973, les spécifications concernant la consommation sont moins détaillées.

Gazole/carburant diesel, fioul : La rupture des séries concernant le gazole/carburant diesel et le fioul résulte de la nouvelle classification des "autres secteurs" adoptée entre 1976 et 1977.

Coke de pétrole : A partir de 1982, on dispose de données plus détaillées sur la consommation de coke de pétrole.

Pour 1996, les retours aux raffineries sont des estimations du Secrétariat.

Le système de notification ayant été modifié au milieu de 1996, il en découle certaines ruptures de séries entre 1996 et 1997.

9. Etats-Unis

Avant 1995, on ne dispose pas de ventilation détaillée de la consommation de produits pétroliers du secteur de l'industrie. Toutefois, des données par branche d'activité ont été publiées pour le fioul pour la période comprise entre 1971 et 1982.

Pétrole brut : Les importants écarts statistiques représentent les quantités "*non comptabilisées*", c'est-à-dire la différence entre les approvisionnements et la demande de pétrole brut.

GPL : A partir de 1995, les apports de combustibles dans les usines à gaz sont comptabilisés avec la consommation du secteur de l'industrie.

GPL, éthane : Les GPL comprennent les pentanes plus et l'éthane avant 1990.

GPL, LGN : Avant 1978, les importations et exportations de LGN sont comptabilisées comme échanges de GPL. A partir de 1992, les données sur les différents produits entrant dans la catégorie des LGN et des GPL ont été converties en utilisant leurs coefficients de conversion respectifs au lieu d'un coefficient moyen, ce qui a entraîné une rupture de série.

Naphtas : La baisse de la consommation de naphtas dans l'industrie chimique à partir de 1978 est due à une nouvelle classification de ce produit.

Essence moteur : A partir de 1994, la consommation du secteur commerce/services publics est calculée sur la base d'un nouveau modèle utilisé par le Ministère des transports.

Gazole/carburant diesel : La consommation de gazole/carburant diesel dans le secteur de l'agriculture avant 1980 est estimée par le Secrétariat.

Fioul : En 1990, l'important accroissement des soutages maritimes internationaux de fioul est dû à des changements dans les méthodes de collecte et de notification des données employées par l'administration américaine. La répartition de la consommation du fioul entre faible et haute teneur en souffre a été estimée.

Gazole/carburant diesel, fioul, coke de pétrole : La quantité de produits pétroliers consommée par les autoproducteurs d'électricité est disponible à partir de 1997. Les données sur les combustibles utilisés dans les installations de cogénération sont disponibles à partir de 1999. Avant 1999, les centrales publiques de cogénération étaient regroupées avec les centrales électriques publiques; les autoproducteurs faisant appel à la cogénération étaient comptabilisés avec les autoproducteurs d'électricité et le secteur de l'industrie.

Les ruptures de série concernant les transferts en 1990 et 1993 sont dues au fait que l'administration américaine emploie de nouvelles méthodes de notification.

En 1993, l'administration américaine a procédé à plusieurs ajustements de son système de collecte de statistiques sur le pétrole afin de le mettre en conformité avec les amendements apportés à la Clean Air Act de 1990. De ce fait, les données sur les composés oxygénés (c'est-à-dire, éthanol utilisé comme carburant, MTBE, etc.) sont recueillies à partir de 1993.

Suite à la nouvelle enquête "Manufacturing Energy Consumption Survey" (MECS), il se produit des ruptures de séries entre 1999 et 2000 dans le secteur de l'industrie, ainsi qu'entre 2000 et 2001 parce que les pourcentages de la MECS ont été modifiés suite à des révisions concernant l'électricité produite par cogénération.

Des révisions importantes des données de l'année 2001 ont été effectuées pour le fioul résiduel et les produits semi-finis. Les changements tiennent, pour l'essentiel, à une erreur de classification des importateurs qui ont inclus les produits semi-finis dans la catégorie du fioul résiduel.

Les données estimatives concernant le pétrole en 2002 sont fondées sur des chiffres préliminaires provenant des *Statistiques mensuelles du pétrole et du gaz naturel* et différeront donc des chiffres définitifs pour 2002 qui seront publiés dans l'édition du 2ème trimestre de 2003 des *Statistiques trimestrielles du pétrole, du gaz, du charbon et de l'électricité*, ainsi que dans les parutions ultérieures.

10. Finlande

Fioul : Un nouveau modèle de calcul ayant été utilisé, il se produit une rupture de série pour la consommation dans les autres secteurs entre 1998 et 1999.

Coke de pétrole : Les données sur le coke de pétrole ne sont disponibles qu'à partir de 1985.

En 1995, il s'est produit une rupture de série pour les échanges de produits pétroliers du fait que le système de collecte de données sur les échanges du Conseil national des douanes ait été aligné sur le système Intrastat de l'Union européenne.

11. France

Coke de pétrole : La répartition de la consommation de coke de pétrole entre utilisations énergétiques et non énergétiques n'est pas disponible avant 1982.

GPL, éthane : Avant 1988, les données sur l'éthane sont comptabilisées dans la rubrique GPL.

Avant 1994, la ventilation de la consommation de GPL du secteur tertiaire entre secteurs résidentiel et commerce/services publics a été estimée par le Secrétariat.

Carburéacteur type kérosène : La consommation comprend les utilisations à des fins militaires à partir de 1998.

Gazole/carburant diesel : Par suite d'un changement de classification, la consommation du secteur de la fabrication de produits minéraux non métalliques figure sous la rubrique des métaux non ferreux, et inversement, pour les années 1982 à 1994. Avant 1985, la consommation du secteur résidentiel est regroupée avec celle du secteur commerce/services publics car on ne disposait pas de données distinctes le concernant.

Gazole/carburant diesel, fioul : La consommation de gazole/carburant diesel et de fioul résiduel des secteurs résidentiel et commerce/services publics est estimée par le Secrétariat à partir d'informations fournies par l'administration française pour les années comprises entre 1985 et 1998.

On dispose de données sur les additifs et les composés oxygénés à partir de 1991.

A partir de 1998, les produits pétroliers raffinés nécessitant un traitement en raffinerie plus poussé ne sont plus comptabilisés avec les importations de produits d'alimentation des raffineries mais avec les importations de produits pétroliers et les transferts de produits. Et à partir de 1999, suite à la reallocation de certains produits, une partie des montants précédemment reportée dans autres produits est maintenant reportée dans fioul.

12. Grèce

Pétrole brut : La production de pétrole brut a cessé le 30 novembre 1998, et a repris en décembre 1999.

Produits d'alimentation des raffineries : Les données sur les produits d'alimentation des raffineries ne sont notifiées qu'à partir de 1986.

Gaz de raffinerie : A compter de 1990, la consommation de gaz de raffinerie pour la production d'électricité a augmenté, en remplacement du fioul.

Fioul : A partir de 1993, la répartition entre les différents secteurs de l'industrie a été améliorée. Dans l'agriculture et le secteur résidentiel, le fuel a été remplacé par le gazole/carburant diesel à partir de 1993.

Avant 1987, la consommation du secteur commerce/services publics est regroupée avec celle du secteur résidentiel. Les pointes de la consommation de ce dernier en 1978 et 1982 sont dues à des hivers exceptionnellement rigoureux.

Les méthodes de notification ayant été modifiées, on dispose de données plus détaillées à partir de 1996.

13. Hongrie

Gaz de raffinerie, paraffines, lubrifiants : Avant 1993, les données sont en partie estimées par le Secrétariat.

White spirit : Avant 1993, le white spirit est comptabilisé avec l'essence moteur.

Fioul : Le détail entre faible et haute teneur en souffre est disponible à partir de 1998.

Additifs, essence aviation : Les données sont disponibles à partir de 1998.

Depuis 1993, des données sont officiellement communiquées à l'AIE dans le cadre du questionnaire.

14. Irlande

Gazole/carburant diesel : On dispose de données sur la consommation de gazole/carburant diesel du secteur agricole à partir de 1986.

Pétrole lampant : Pour la période comprise entre 1970 et 1977, la ventilation de la consommation de pétrole lampant entre le secteur de l'agriculture et le secteur commercial/services publics a été estimée par le Secrétariat.

La consommation du secteur commercial/services publics comprend les quantités utilisées par les entreprises agricoles d'Etat.

GPL, gazole/carburant diesel, kérosène, fioul : Les données sur la consommation finale recueillies pour l'année 1993 sont fondées sur une enquête détaillée réalisée par un cabinet de consultants. L'administration nationale a révisé, sur la base des résultats de cette enquête, les données rétrospectives en remontant jusqu'à l'année 1990. En raison de ces révisions, des ruptures de séries apparaissent en 1989 et 1990, dans la ventilation par utilisation finale de la consommation de GPL, de kérosène, de gazole/carburant diesel et de fioul.

Lubrifiants : La consommation dans le transport routier est indiquée à la rubrique transports non spécifiés car on ne dispose pas de ventilation détaillée.

La baisse de la production des raffineries observée en 1996 était due à une fermeture partielle d'une raffinerie.

15. Islande

Pour la période antérieure à 1980, les données relatives à la consommation ont été estimées par le Secrétariat.

Essence moteur : La consommation d'essence au plomb a cessé en 1996.

Fioul : On ne dispose pas de données détaillées sur la consommation du secteur de l'industrie avant 1982.

Bitume : De 1983 à 1986, la consommation finale a augmenté du fait du développement des travaux publics.

16. Italie

Carburants aviation : En raison d'un changement de classification, on constate certaines ruptures de séries en 1988 dans les données concernant les carburants pour l'aviation.

Gazole/carburant diesel : La consommation non spécifiée est comptabilisée dans le secteur commercial/services publics.

Les consommations pour la production de chaleur et d'électricité ont été estimées par le Secrétariat pour les années 1984 à 1997 à partir des données communiquées en réponse au questionnaire annuel sur l'électricité.

Toutes les données concernant les années 1992 à 1997 et la ventilation détaillée de la consommation des autres années ont été estimées par le Secrétariat à partir du *Bilancio Energetico Nazionale*. De nouvelles enquêtes ayant été réalisées, il apparaît des ruptures dans les séries de données sur la consommation entre 1998 et 1999.

17. Japon

Suite à la nouvelle méthode de calcul des bilans énergétiques du Japon, il y a des ruptures de série entre 1999 et 2000. A partir de 2000 les données concernent les exercices budgétaires, qui vont du 1er avril au 31 mars (par exemple, 2001 = du 1er avril 2001 au 31 mars 2002).

Pour 2000 et 2001, certaines données pour le Japon sont estimées par le Secrétariat de l'AIE de façon à garantir la cohérence des séries temporelles avec la nouvelle méthode de calcul des bilans énergétiques du Japon. Ainsi, certaines données sont différentes de celles publiées par le gouvernement japonais. Une révision complète des séries temporelles de 1990 à 2001, basée sur les données originales publiées par le gouvernement japonais, sera publiée dans l'édition de l' année prochaine.

Pétrole brut, LGN, produits d'alimentation, gaz de raffinerie, essence aviation, essence moteur, carburéacteur type essence, carburéacteur type kérosène, pétrole lampant : L'administration japonaise à révisé les données sur l'offre et la demande pour les années 1960 à 1969. La classification de la consommation de carburants faisant la distinction entre le transport routier et la navigation intérieure a changé en 1970.

Autres hydrocarbures : On compte parmi les autres hydrocarbures l'acide sulfurique, l'argile (kaolin non compris) et la soude caustique.

Pétrole lampant : La baisse de la consommation finale observée en 1982 était due à un hiver exceptionnellement doux.

Coke de pétrole : On dispose d'une série de données sur la consommation de coke de pétrole dans les hauts fourneaux à compter de 1990.

Par suite d'une amélioration des méthodes de collecte de données, on constate des ruptures de séries pour plusieurs produits entre 1981 et 1983. Le système de notification ayant été modifié pour en exclure les niveaux des stocks des stations-service et des points de vente au détail, il y a une rupture de série entre 1981 et 1982.

Les importations d'orimulsion pour la production d'électricité ont commencé en 1991.

18. Luxembourg

Fioul : Par suite de l'adoption d'un droit d'accise, en 1994, il n'a plus été consommé que du fioul à faible teneur en soufre, en lieu et place du fioul à haute teneur en soufre.

La baisse de la consommation de fioul dans l'industrie sidérurgique à la fin des années 70 est due au remplacement de ce combustible par du charbon.

19. Mexique

On dispose de données à partir de l'année 1971.

Les niveaux des stocks sont disponibles à partir de 1999.

Suite à un changement dans le traitement des données, il peut y avoir des ruptures dans les séries entre 1998 et 1999.

La consommation pour l'autoproduction d'électricité et pour la production de chaleur est comptabilisée avec celle du secteur de l'industrie.

Lubrifiants, paraffines, coke de pétrole, autres produits : Les données sur l'utilisation finale de ces produits sont indiquées à la rubrique industrie non spécifiée, car on ne dispose pas d'une ventilation plus détaillée.

La production de LGN indiquée dans les publications de l'AIE peut différer de celle qui figure dans les publications mexicaines sur l'énergie car l'AIE inclut, dans ses données sur le pétrole, les liquides extraits en même temps que le gaz naturel.

20. Norvège

L'administration norvégienne, disposant d'informations plus détaillées, a révisé les données communiquées sur les utilisations énergétiques et non énergétiques dans les industries chimique et pétrochimique à partir de 1993.

L'administration norvégienne a révisé les niveaux des stocks pour le pétrole brut et quelques produits pétroliers. Cela entraine des différences entre les niveaux des stocks fin 1999 et début 2000.

Produits d'alimentation des raffineries : Depuis 1986, les importations de produits d'alimentation des raffineries sont indiquées dans les rubriques des importations de produits pétroliers correspondants.

GPL, éthane : Avant 1990, les données sur l'éthane sont comptabilisées avec le GPL.

Lubrifiants : La consommation de lubrifiants est indiquée dans le secteur de l'industrie, une ventilation plus détaillée n'étant pas disponible.

A partir de 1991, les données sur la consommation finale d'essence moteur, de gazole/carburant diesel et de fioul ont été révisées compte tenu des nouvelles informations concernant ces produits.

Gazole/carburant diesel : Les quantités de gazole/carburant diesel utilisées pour la production d'électricité entre 1970 et 1975 ont été estimées par le Secrétariat. A partir de 1990, l'administration norvégienne a révisé les données communiquées concernant la consommation de gazole/carburant diesel pour l'autoproduction d'électricité sur les plateformes pétrolières et gazières, d'où une rupture de série entre 1989 et 1990. Ces données sont maintenant comptabilisées dans le secteur énergie.

Gazole/carburant diesel, fioul : La ventilation détaillée de la consommation est disponible à partir de 1976.

Lubrifiants : Pour 1996, certains lubrifiants sont comptabilisés avec les "autres produits".

21. Nouvelle-Zélande

Les données sur la consommation des années 1960 à 1973 sont des estimations du Secrétariat.

Autres hydrocarbures : Cette rubrique représente la production d'essence de synthèse à partir de gaz naturel. La production d'essence de synthèse a cessé en février 1997.

Gazole/carburant diesel : Il n'est pas possible de faire la distinction entre la consommation de gazole/carburant diesel pour le transport routier et pour le transport ferroviaire. Cette consommation est prise en compte dans la rubrique non spécifiés du secteur des transports.

Gazole/carburant diesel, fioul : Le fioul léger est regroupé avec le fioul lourd jusqu'en 1997. A compter de 1998, le fioul léger est comptabilisé avec le gazole/carburant diesel.

A partir de 2000, les paraffines sont inclus avec les lubrifiants.

22. Pays-Bas

Les données ont été révisées par l'administration néerlandaise en remontant jusqu'en 1990.

Des ruptures importantes se produisent dans les séries entre 1983 et 1984 par suite de la réalisation d'une enquête plus complète sur la consommation au stade de l'utilisation finale. Toutefois, les données agrégées sur la consommation du secteur des transports, de l'industrie et des "autres" secteurs n'en sont pas modifiées.

Pétrole brut : Avant 1994, les stocks détenus par les entités détentrices n'étaient pas pris en compte.

GPL, éthane : Avant 1994, les données sur l'éthane sont comptabilisées avec le GPL.

Gaz de raffinerie : A partir de 1978, la consommation de gaz de raffinerie comprend le gaz utilisé par l'industrie chimique. Les quantités correspondantes sont indiquées comme consommation de l'industrie chimique. L'utilisation de gaz de raffinerie pour la cogénération publique commence en 1995.

Essence moteur : L'essence moteur comprend les "autres huiles légères".

Coke de pétrole : Le coke de pétrole est comptabilisé avec les "autres produits" pour toutes les années, à l'exception de celles comprises entre 1960 et 1967 et entre 1978 et 1982.

23. Pologne

Les niveaux de stocks sont disponibles à partir de 1988.

Autres hydrocarbures: De l'hydrogène est produit à partir de gaz naturel depuis 1997 dans un complexe doté d'une unité d'hydrocraquage. Ces quantités sont comptabilisées dans la rubrique production.

Gazole/carburant diesel, fioul : Avant 1998, les apports de gazole/carburant diesel et de fioul pour l'autoproduction de cogénération chaleur/électricité dans les raffineries de pétrole sont comptabilisés dans le secteur de transformation.

24. Portugal

Gaz de raffinerie : En 1995, en raison de la fermeture d'une raffinerie, la production et la consommation ont cessé.

Gazole/carburant diesel : La consommation dans l'industrie et le secteur commercial/services publics représente celle des parcs de véhicules de ces secteurs.

Fioul : L'utilisation de fioul comme produit d'alimentation dans l'industrie chimique à partir de 1984 témoigne de l'ouverture d'une nouvelle usine de fabrication d'ammoniac, utilisant du fioul comme matière première.

La ventilation de la consommation de lubrifiants, de bitume, de White spirit/SBP et de paraffines a été estimée par le Secrétariat pour 1986.

25. République slovaque

On dispose de données à partir de l'année 1971.

Naphtas : Les données sont indisponibles et par conséquent non comprises dans le total des produits pétroliers avant 1989.

Fioul : Le données sur le fioul comprennent à la fois le fioul léger et le fioul lourd.

L'administration slovaque a révisé les séries chronologiques jusqu'en 1993 pour la publication précédente. Il peut donc y avoir des ruptures dans les séries entre 1992 et 1993.

Un nouveau système d'enquête appliqué en 2001 entraîne des ruptures de séries importantes concernant la plupart des produits.

26. République tchèque

On dispose de données à partir de l'année 1971.

Avant 1994, les données sur la consommation sont des estimations du Secrétariat. L'administration tchèque a communiqué les données officielles concernant 1994 et les années suivantes.

Gazole/carburant diesel: A partir de 1999, l'administration tchèque a mis en place un nouveau système de mise à jour des données. Cela entraine des ruptures dans les séries pour la consommation finale.

27. Royaume-Uni

Pétrole brut, LGN : A partir de 1980, les LGN comprennent les condensats. Les années précédentes, les condensats étaient comptabilisés avec le pétrole brut.

LGN, produits d'alimentation des raffineries : Les niveaux et les variations des stocks de produits d'alimentation des raffineries et de LGN ne sont indiqués à part que depuis 1985. Avant cette date, ils étaient regroupés avec les données sur le pétrole brut.

Autres hydrocarbures : Jusqu'en 1994 les autres sources d'autres hydrocarbures concernent le bitume produit à partir de charbon.

GPL, éthane : Avant 1990, les données sur l'éthane sont comptabilisées avec le GPL.

GPL, fioul : Il existe des ruptures de séries concernant les GPL entre 2000 et 2001, et pour le fioul entre 1999 et 2000, en raison d'une redistribution des données. De plus on dispose de données sur le fioul utilisé pour la production de chaleur depuis 2000.

Autres produits : Avant 1992, le coke de pétrole est comptabilisé avec les importations et les exportations des "autres produits". Avant 1979, toutes les données sur le coke de pétrole sont regroupées avec celles concernant les "autres produits".

Les niveaux et les variations des stocks des centrales électriques publiques sont pris en compte à partir de 1985.

Avant 1995, la ventilation par produit des retours aux raffineries provenant de l'industrie est estimée par l'administration britannique. A partir de 1995, celle-ci a procédé à une révision dc cette ventilation ainsi que de la méthode de notification concernant l'industrie pétrochimique.

28. Suède

Additifs, éthane : Les données sont disponibles à partir de 2000 pour les additifs, et 2001 pour l'éthane.

Pétrole lampant : La consommation dans le secteur du transport routier s'interrompt en 1984 par suite d'une nouvelle classification des produits.

Gazole/carburant diesel: Suite à une fusion entre deux compagnies, la nouvelle entité a révisé ses niveaux de stocks. Et il y a maintenant des différences entre les niveaux des stocks fin 1999 et début 2000.

Fioul : La notification des données étant plus détaillée, on constate des ruptures de série pour la consommation entre 1985 et 1986.

Les données concernant les stocks suédois comprennent les stocks de crise détenus en temps de paix. Etant donné que ces stocks peuvent être détenus sous forme de pétrole brut au lieu de produits pétroliers, il peut arriver que les niveaux des stocks de produits raffinés soient négatifs.

A partir de 1995, la Suède a modifié sa classification type, par industrie, des branches d'activité économique.

29. Suisse

Produits d'alimentation des raffineries : La baisse des entrées en raffinerie de produits d'alimentation en 1988 est due en partie à une utilisation accrue de pétrole brut et aussi à la fermeture d'une raffinerie pour maintenance.

Naphtas : A partir de 1993, les chiffres de production brute des raffineries sont nets des quantités utilisées pour le mélange d'essence moteur.

Gazole/carburant diesel : La consommation de gazole/carburant diesel indiquée à la rubrique non spécifiés de l'industrie (petites industries) a été estimée. La méthode de notification a été révisée pour les années 1987 à 1993. L'accroissement de la consommation en 1994 est dû au fait que les consommateurs ont augmenté leurs stocks avant l'instauration d'une taxe sur la valeur ajoutée frappant le fioul domestique à partir du 1er janvier 1995.

Depuis le 1er janvier 1996, l'administration suisse a révisé sa méthode de notification des données sur les stocks.

Une nouvelle enquête ayant été introduite en 1999, il apparaît des ruptures dans les séries de données sur la consommation finale d'essence moteur, de gazole/carburant diesel et de fioul entre 1998 et 1999.

30. Turquie

Gazole/carburant diesel, fioul : Entre 1977 et 1978, la classification des utilisations finales du gazole/carburant diesel et du fioul a été modifiée dans les statistiques nationales turques. En conséquence, il y a rupture dans les séries de données détaillées sur la consommation entre ces deux années.

A partir de 1978, le secteur commercial est regroupé avec l'industrie, tandis que les services publics sont regroupés avec le secteur des transports. Une partie de la consommation du secteur de la fabrication de produits minéraux non métalliques est comptabilisée avec celle du secteur des métaux non ferreux.

Les soutages maritimes internationaux sont pris en compte dans les exportations pour les années 1978, 1980, 1981, 1983 et 1984.

Le gazole/carburant diesel et le fioul consommés pour la production d'électricité sont des combustibles utilisés aussi bien dans des centrales alimentées au fioul que dans des centrales à charbon.

Bitume : Le bitume consommé dans le secteur de la transformation est destiné à la production de briquettes.

6. COUVERTURE GEOGRAPHIQUE

Les données concernant l'**Allemagne** tiennent compte des nouveaux Länder à partir de 1970, sauf indication contraire en note de bas de page. Les données sur les approvisionnements, la consommation et les échanges à partir de 1980 sont fournies par l'administration allemande. Dans les données des années antérieures, la fraction concernant les nouveaux Länder dans le total pour l'Allemagne est une estimation du Secrétariat de l'AIE.

L'**Australie** ne comprend pas les territoires d'outre-mer.

Le **Danemark** comprend le Groenland et les îles Féroé danoises jusqu'en 1989.

L'**Espagne** comprend les îles Canaries.

Les **Etats-Unis** comprennent Porto Rico, Guam, les îles Vierges et la zone franche d'Hawaii.

La **France** comprend Monaco mais non les départements et territoires français d'outre-mer, à savoir la Guadeloupe, la Martinique, la Guyane, la Réunion, Saint Pierre et Miquelon, la Nouvelle-Calédonie et la Polynésie française.

L'**Italie** comprend Saint Marin et le Vatican.

Le **Japon** comprend Okinawa.

Les données sur les **Pays-Bas** ne tiennent compte ni du Surinam ni des Antilles néerlandaises.

Le **Portugal** comprend les îles des Açores et de Madère.

La **Suisse** comprend le Liechtenstein.

L'**OCDE** comprend l'Allemagne, l'Australie, l'Autriche, la Belgique, le Canada, la Corée, le Danemark, l'Espagne, les Etats-Unis, la Finlande, la France, la Grèce, la Hongrie, l'Irlande, l'Islande, l'Italie, le Japon, le Luxembourg, le Mexique, la Norvège, la Nouvelle-Zélande, les Pays-Bas, la Pologne, le Portugal, la République Slovaque, la République Tchèque, le Royaume-Uni, la Suède, la Suisse et la Turquie.

La région **OCDE Amérique du Nord** comprend le Canada, le Mexique et les Etats-Unis.

La région **OCDE Pacifique** comprend l'Australie, la Corée, le Japon et la Nouvelle-Zélande.

La région **OCDE Europe** comprend les pays de l'OCDE à l'exception de l'Australie, du Canada, de la Corée, des Etats-Unis, du Japon, du Mexique et de la Nouvelle-Zélande.

L'**AIE** comprend les pays de l'OCDE à l'exception de l'Islande, du Mexique, de la Pologne et de la République Slovaque.

L'**Union européenne** comprend l'Allemagne, l'Autriche, la Belgique, le Danemark, l'Espagne, la Finlande, la France, la Grèce, l'Irlande, l'Italie, le Luxembourg, les Pays-Bas, le Portugal, le Royaume-Uni et la Suède. Il convient de noter que tous ces pays sont pris en compte pour toutes les années bien qu'ils soient entrés à des dates différentes au sein de l'Union européenne.

La région **Europe hors OCDE** comprend l'Albanie, la Bulgarie, la Roumanie, l'ex-Yougoslavie et le groupement Malte-Gibraltar-Chypre.

L'**ex-Yougoslavie** comprend la Bosnie-Herzegovine, la Croatie, l'ex-République yougoslave de Macedoine, la Serbie et Monténégro, et la Slovénie.

L'**ex-URSS** comprend l'Arménie, l'Azerbaïdjan, le Bélarus, l'Estonie, la Géorgie, le Kazakhstan, le Kirghizstan, la Lettonie, la Lituanie, la Moldavie, la Russie, le Tadjikistan, le Turkménistan, l'Ukraine et l'Ouzbékistan.

L'**OPEP** comprend l'Algérie, l'Arabie saoudite, les Emirats arabes unis, l'Indonésie, l'Irak, l'Iran, le Koweït, la Libye, le Nigéria, le Qatar et le Venezuela.

La **Chine** comprend la République populaire de Chine et Hong-Kong, Chine.

7. ABREVIATIONS ET FACTEURS DE CONVERSION

Abréviations

Mt	millions de tonnes
Mtoe	(Mtep) millions de tonnes d'équivalent pétrole
TPES	(ATEP) Approvisionnements totaux en énergie primaire
kb	Milliers de barils
mb	Milliards de barils

Conventions

-	Néant
x	Sans objet
..	Non disponible
0.00	Négligeable (selon le nombre de décimales)

Préfixes décimaux

10^{1}	déca (da)	10^{-1}	déci (d)
10^{2}	hecto (h)	10^{-2}	centi (c)
10^{3}	kilo (k)	10^{-3}	milli (m)
10^{6}	méga (M)	10^{-6}	micro (μ)
10^{9}	giga (G)	10^{-9}	nano (n)
10^{12}	téra (T)	10^{-12}	pico (p)
10^{15}	péta (P)	10^{-15}	femto (f)
10^{18}	exa (E)	10^{-18}	atto (a)

Coefficients de conversion généraux pour l'énergie

Vers :	TJ	Gcal	Mtep	MBtu	GWh
De :	multiplier par :				
TJ	1	238,8	$2{,}388 \times 10^{-5}$	947,8	0,2778
Gcal	$4{,}1868 \times 10^{-3}$	1	10^{-7}	3,968	$1{,}163 \times 10^{-3}$
Mtep	$4{,}1868 \times 10^{4}$	10^{7}	1	$3{,}968 \times 10^{7}$	11630
Mbtu	$1{,}0551 \times 10^{-3}$	0,252	$2{,}52 \times 10^{-8}$	1	$2{,}931 \times 10^{-4}$
GWh	3,6	860	$8{,}6 \times 10^{-5}$	3412	1

Coefficients de conversion pour les mesures de masse

Vers :	kg	t	lt	st	lb
De :	multiplier par :				
Kilogramme (kg)	1	0,001	$9{,}84 \times 10^{-4}$	$1{,}102 \times 10^{-3}$	2,2046
Tonne (t)	1000	1	0,984	1,1023	2204,6
Tonne longue (lt)	1016	1,016	1	1,120	2240,0
Tonne courte (st)	907,2	0,9072	0,893	1	2000,0
livre (lb)	0,454	$4{,}54 \times 10^{-4}$	$4{,}46 \times 10^{-4}$	$5{,}0 \times 10^{-4}$	1

Coefficients de conversion pour les mesures de volume

Vers:	gal U.S.	gal U.K.	bbl	ft^3	l	m^3
De:	multiplier par:					
Gallon U.S. (gal)	1	0,8327	0,02381	0,1337	3,785	0,0038
Gallon U.K. (gal)	1,201	1	0,02859	0,1605	4,546	0,0045
Baril (bbl)	42,0	34,97	1	5,615	159,0	0,159
Pied cube (ft^3)	7,48	6,229	0,1781	1	28,3	0,0283
Litre (l)	0,2642	0,220	0,0063	0,0353	1	0,001
Mètre cube (m^3)	264,2	220,0	6,289	35,3147	1000,0	1

Produits pétroliers
Equivalents thermiques moyens de tonnes

	Tcal	tep	TJ	Btu (x 10^9)
Gaz de raffinerie	0.01150	1.150	0.048148	0.045632
GPL	0.01130	1.130	0.047311	0.044838
Ethane	0.01180	1.180	0.049404	0.046822
Essence aviation	0.01070	1.070	0.044799	0.042458
Essence moteur	0.01070	1.070	0.044799	0.042458
Carburéacteur	0.01065	1.065	0.044589	0.042259
Pétrole lampant	0.01045	1.045	0.043752	0.041466
Naphta	0.01075	1.075	0.045008	0.042656
Gazole/carburant diesel	0.01035	1.035	0.043333	0.041069
Fioul	0.00960	0.960	0.040193	0.038093
Autres produits	0.00960	0.960	0.040193	0.038093

Produits pétroliers
Equivalences moyennes en volume des milliers de tonnes

	Milliers de barils	Barils par jour b/j	Milliers de kilolitres $(1000\ m)^3$	Milliers de gallons américains
GPL	11.60	31.78	1.844	487.20
Essence aviation	8.90	24.38	1.415	373.80
Essence moteur	8.47	23.20	1.347	355.73
Carburéacteur type essence	8.25	22.60	1.312	346.50
Naphta	8.81	24.13	1.400	369.89
White spirit	8.46	23.18	1.345	355.32
Carburéacteur type kérosène	7.89	21.61	1.254	331.32
Autres pétroles lampants	7.85	21.49	1.247	329.49
Gazole/carburant diesel	7.46	20.44	1.186	313.32
Fioul	6.50	17.80	1.033	272.85
Lubrifiants	7.09	19.42	1.127	297.78
Bitume	6.08	16.66	0.967	255.36
Paraffines	7.85	21.51	1.248	329.70
Coke de pétrole	5.50	15.07	0.875	231.00

Pétrole brut*
Equivalences moyennes en volume des milliers de tonnes

	Milliers de barils	Milliers de kilolitres
Allemagne	7.330	1.165
Australie	7.868	1.251
Autriche	7.200	1.145
Canada	7.186	1.143
Danemark	7.405	1.177
Espagne	7.506	1.193
Etats-Unis	7.400	1.177
France	7.332	1.166
Grèce	7.231	1.150
Italie	7.300	1.161
Japon	7.357	1.170
Mexique	6.965	1.107
Norvège	7.644	1.215
Nouvelle-Zélande	7.321	1.164
Pays-Bas	7.239	1.151
Royaume-Uni	7.523	1.196
Suède	6.623	1.053
Turquie	7.200	1.145
Algérie	8.130	1.293
Angola	7.410	1.178
Arabie Saoudite	7.323	1.164
Argentine	7.120	1.132
Brésil	7.056	1.122
Chine	7.300	1.161
Colombie	7.080	1.126
Egypte	7.260	1.154
Emirats Arabes Unis	7.596	1.208
Inde	7.330	1.165
Indonésie	7.360	1.170
Irak	7.430	1.181
Iran	7.350	1.169
Koweït	7.250	1.153
Libye	7.580	1.205
Nigéria	7.500	1.192
Oman	7.330	1.165
Qatar	7.500	1.192
Syrie	7.290	1.159
Ex-URSS	7.300	1.161
Venezuela	6.890	1.095

*Production intérieure, comprend les condensats.

PART II

WORLD OIL DEVELOPMENTS

DEUXIEME PARTIE

TENDANCES DU MARCHE MONDIAL DU PETROLE

Table 1: OECD Energy Consumption
Tableau 1 : Consommation d'énergie au sein de l'OCDE

	1973				1985				2002			
	Oil %	Gas %	Other %	TPES Mtoe	Oil %	Gas %	Other %	TPES Mtoe	Oil %	Gas %	Other %	TPES Mtoe
Canada	50.3	23.2	26.5	161.0	36.3	25.8	37.9	193.4	36.2	27.8	36.0	239.7
Mexico	60.0	19.0	21.0	55.2	66.0	21.4	12.6	111.4	60.0	24.3	15.7	151.8
United States	47.5	29.6	22.9	1736.4	41.4	23.1	35.5	1781.4	39.0	23.4	37.6	2286.7
OECD North America	**48.0**	**28.8**	**23.2**	**1952.6**	**42.2**	**23.3**	**34.5**	**2086.2**	**39.9**	**23.8**	**36.2**	**2678.2**
Austria	56.7	15.3	28.0	21.7	43.0	19.5	37.5	23.3	43.5	22.4	34.1	30.3
Belgium	60.5	15.4	24.1	46.3	39.5	16.5	44.0	44.7	38.2	22.1	39.7	58.8
Czech Republic	19.6	2.2	78.2	45.4	21.0	8.6	70.4	49.4	19.9	18.7	61.3	41.3
Denmark	88.6	.	11.4	19.8	54.9	3.5	41.6	19.9	43.9	23.7	32.5	19.5
Finland	63.6	-	36.4	21.3	39.9	3.1	57.1	26.5	28.6	10.6	60.7	34.7
France	70.4	7.7	21.9	176.6	42.1	12.4	45.5	198.7	34.6	13.9	51.5	266.3
Germany	47.9	8.5	43.6	337.9	34.3	13.5	52.2	361.0	37.6	22.0	40.4	343.1
Greece	77.7	-	22.3	12.4	62.4	0.4	37.2	18.3	56.0	6.0	38.0	29.5
Hungary	38.5	19.6	42.0	21.3	32.8	29.5	37.7	29.9	26.0	43.0	31.1	25.1
Iceland	56.3	-	43.7	1.2	28.4	-	71.6	1.8	24.3	-	75.7	3.4
Ireland	77.1	-	22.9	7.2	47.6	22.4	30.0	8.9	57.4	23.5	19.1	15.5
Italy	77.6	11.03	11.3	128.9	61.0	20.0	19.0	136.2	49.8	34.1	16.1	169.1
Luxembourg	37.1	4.9	58.0	4.5	34.2	9.6	56.2	3.2	63.0	26.1	10.9	4.0
Netherlands	49.5	45.6	4.9	62.4	34.2	52.6	13.2	61.5	38.6	46.0	15.4	77.2
Norway	54.0	0.0	46.0	14.6	41.5	5.8	52.7	20.4	20.8	35.2	44.0	27.7
Poland	11.7	6.72	81.6	93.1	12.7	7.2	80.1	123.3	22.9	11.5	65.6	87.5
Portugal	75.4	0.0	24.6	7.2	75.4	0.0	24.6	11.4	63.4	10.7	25.8	25.4
Slovak Republic	34.6	10.1	55.3	15.5	27.3	19.25	53.4	21.9	17.5	32.71	49.8	18.9
Spain	73.3	1.8	24.9	52.4	55.3	3.0	41.7	71.7	51.2	14.3	34.5	131.3
Sweden	72.2	-	27.8	39.3	36.6	0.2	63.3	46.7	27.1	1.6	71.3	50.3
Switzerland	77.4	0.8	21.9	19.7	54.7	5.5	39.8	23.0	47.8	9.2	43.0	27.1
Turkey	51.3	-	48.7	24.4	45.9	0.1	54.0	39.4	39.8	19.8	40.4	74.5
United Kingdom	50.5	11.4	38.1	220.7	38.7	22.9	38.4	203.9	35.3	37.3	27.4	228.5
OECD Europe	**54.4**	**9.7**	**35.9**	**1394.0**	**39.1**	**14.6**	**46.3**	**1545.2**	**38.1**	**22.9**	**39.0**	**1789.1**
Australia	47.1	5.9	47.0	57.6	37.0	15.2	47.8	73.9	29.1	17.6	53.4	122.7
Japan	77.9	1.6	20.5	323.6	55.8	9.6	34.6	364.9	50.1	12.3	37.6	510.9
Korea	61.9	-	38.1	21.6	48.9	-	51.1	54.3	51.0	10.5	38.5	201.8
New Zealand	53.5	3.4	43.1	8.3	36.4	26.4	37.2	11.5	35.7	27.5	36.8	18.4
OECD Pacific	**72.3**	**2.1**	**25.6**	**411.1**	**51.8**	**9.8**	**38.4**	**504.6**	**47.0**	**12.9**	**40.1**	**853.8**
OECD Total	**53.1**	**18.8**	**28.1**	**3757.7**	**42.2**	**18.4**	**39.4**	**4136.0**	**40.5**	**21.8**	**37.8**	**5321.0**

Source: Energy Balances of OECD Countries. / *Bilans énergétiques des pays de l'OCDE.*

Table 2: OECD Oil Demand and Economic Indicators
Tableau 2 : Demande de pétrole et indicateurs économiques de l'OCDE

	1971	1973	1979	1980	1985	1996	1997	1998	1999	2000	2001	2002
Oil Demand ('000 tonnes)	1759	2031	2119	1975	1741	2111	2138	2149	2179	2188	2182	2171
GDP (billion 1995 $)	11888	13339	15870	16080	18205	24609	25396	25993	26762	27702	27881	28240
IPI (1995=100)	56	65.5	74.3	73.7	80	102.7	108.1	111	113.9	119.6	116.2	115.8
Passenger Cars ('000)[1]	186430	212234	270704	277919	315459	399770	407510	416711	434253	448395	..	..

1. Passenger car figures do not include the Slovak Republic in 1971 and the new federal states of Germany prior to 1992. / *Les données sur les voitures de tourisme ne comprennent pas la République slovaque en 1971 ni les nouveaux états fédéraux d'Allemagne avant 1992.*

Sources: IEA Annual Oil Statistics, OECD Main Economic Indicators, International Road Federation. / *Statistiques annuelles du pétrole de l'AIE, Principaux indicateurs économiques de l'OCDE, Fédération Internationale Routière.*

Table 3: OECD Gross Domestic Product (billion 1995 US$) [1]
Tableau 3 : Production intérieure brute de l'OCDE (milliards de 1995 $US) [1]

	1971	1973	1979	1997	1998	1999	2000	2001	2002
Canada	282.9	319.2	401.4	616.1	641.3	676.3	707.1	717.4	741.4
Mexico	121.9	143.4	204.7	321.3	337.5	349.7	372.9	371.9	377.3
United States	3583.0	4005.1	4783.4	7943.0	8285.9	8629.1	8955.1	8977.8	9186.0
OECD North America	**3987.8**	**4467.7**	**5389.5**	**8880.4**	**9264.6**	**9655.1**	**10035.1**	**10067.0**	**10304.7**
Austria	123.8	137.9	164.2	243.7	253.2	260.2	269.4	271.2	274.0
Belgium	156.1	174.4	200.6	290.0	295.9	305.3	316.7	319.1	321.4
Czech Republic	37.9	40.5	46.9	53.9	53.3	53.5	55.3	57.1	58.1
Denmark	118.6	128.4	140.7	190.3	195.0	199.5	205.5	207.4	210.6
Finland	70.6	81.4	93.7	142.9	150.6	156.7	166.2	167.4	170.0
France	873.1	961.4	1135.7	1600.2	1654.6	1707.8	1772.3	1804.9	1823.6
Germany	1391.4	1514.7	1747.6	2511.6	2560.8	2613.2	2687.8	2703.3	2708.0
Greece	71.0	84.5	102.5	124.7	128.9	133.5	139.2	144.8	150.6
Hungary	28.5	34.0	43.7	47.3	49.6	51.7	54.4	56.5	58.2
Iceland	3.2	3.6	4.9	7.6	8.0	8.3	8.8	9.0	9.0
Ireland	24.1	26.9	35.9	79.7	86.7	96.4	106.0	112.0	116.1
Italy	588.7	647.0	795.8	1131.7	1152.0	1170.3	1203.9	1225.3	1229.8
Luxembourg	7.3	8.4	9.1	20.2	21.7	23.0	25.1	25.3	25.5
Netherlands	233.6	252.6	295.3	443.8	463.1	481.6	497.6	503.9	504.4
Norway	64.3	70.5	92.6	163.8	168.1	171.7	175.8	178.4	181.9
Poland	86.1	99.3	114.4	143.9	150.9	157.0	163.3	164.9	166.8
Portugal	48.0	57.7	68.4	115.4	120.7	124.8	129.3	131.4	131.9
Slovak Republic	13.7	14.7	17.0	21.4	22.3	22.5	23.0	23.8	24.8
Spain	297.5	346.9	397.2	622.5	649.6	676.8	705.2	724.0	737.2
Sweden	161.2	171.4	190.6	257.8	267.2	279.4	291.6	294.0	300.4
Switzerland	231.5	246.2	240.7	313.6	320.9	325.8	336.1	339.1	338.5
Turkey	61.7	68.4	89.2	194.8	200.8	191.4	205.5	190.3	197.1
United Kingdom	673.4	748.4	816.4	1204.8	1240.0	1269.9	1309.1	1334.8	1354.9
OECD Europe	**5365.5**	**5919.3**	**6843.3**	**9925.7**	**10213.9**	**10480.5**	**10847.0**	**10987.8**	**11092.8**
Australia	181.8	197.1	228.4	404.1	425.5	442.6	450.3	468.0	484.6
Japan	2235.8	2618.6	3213.9	5573.9	5512.6	5549.5	5680.6	5647.7	5606.5
Korea	79.1	93.2	152.3	548.5	511.8	567.5	620.4	639.2	677.9
New Zealand	37.8	42.9	43.0	63.8	64.1	67.1	68.9	71.1	73.8
OECD Pacific	**2534.5**	**2951.8**	**3637.6**	**6590.3**	**6514.0**	**6626.7**	**6820.2**	**6826.1**	**6842.8**
OECD Total	**11887.8**	**13338.8**	**15870.4**	**25396.4**	**25992.6**	**26762.3**	**27702.3**	**27880.9**	**28240.3**

1. GDP data for 1971, 1973 and 1979 for the Czech Republic, Hungary, Poland and Slovak Republic are IEA Secretariat estimates. / *Les données du PIB pour 1971, 1973, et 1979 pour la République tchèque, la Hongrie, la Pologne et la République slovaque ont été estimées par le Secrétariat de l'AIE.*

Source: OECD Main Economic Indicators. / *Principaux indicateurs économiques de l'OCDE.*

Table 4: OECD Industrial Production Index
Tableau 4 : Indice de production industrielle de l'OCDE

1995 = 100

	1971	1973	1979	1985	1998	1999	2000	2001	2002
Canada	53.8	65.2	73.8	82.2	110.6	116.3	122.9	118.7	120.7
Mexico	..	..	..	85.0	127.9	133.3	141.3	136.4	136.4
United States	54.4	64.3	73.1	77.1	118.3	123.3	129.1	124.6	123.6
OECD North America	**..**	**..**	**..**	**..**	**..**	**..**	**..**	**..**	**..**
Austria	50.4	57.0	68.3	76.6	116.2	123.2	134.1	135.2	135.4
Belgium	68.0	77.6	84.4	86.8	108.8	109.7	115.6	114.4	115.9
Czech Republic	..	..	..	..	108.3	104.8	110.5	117.7	128.8
Denmark	52.0	58.0	63.0	77.0	110.0	113.0	120.0	122.0	123.0
Finland	45.3	52.8	62.7	78.1	122.1	129.0	144.3	143.1	145.5
France	67.5	77.8	86.3	86.6	110.2	112.3	116.3	117.3	116.1
Germany	74.6	82.4	85.7	88.5	108.8	110.4	117.2	117.8	116.5
Greece	51.8	69.3	91.7	99.2	109.8	114.1	114.7	116.3	116.8
Hungary	..	..	..	128.1	129.1	142.4	168.3	174.3	178.6
Iceland	..	..	..	..	..	..	..	..	..
Ireland	..	..	34.7	43.3	152.1	174.6	201.5	222.1	239.5
Italy	54.5	62.4	73.2	73.9	102.9	102.9	107.8	106.5	105.0
Luxembourg	67.3	78.5	72.6	79.8	105.8	118.0	123.0	125.2	126.4
Netherlands	62.7	71.0	79.9	83.7	104.9	106.3	110.6	112.1	109.8
Norway	31.3	35.3	48.5	61.5	108.4	108.1	111.2	111.0	110.6
Poland	..	..	..	107.3	127.4	133.5	143.5	144.1	146.1
Portugal	35.8	45.2	59.2	71.7	114.1	117.6	115.3	118.8	118.5
Slovak Republic	..	..	..	..	111.2	109.0	118.4	126.7	..
Spain	51.9	67.0	79.6	82.8	111.4	114.2	119.2	117.5	117.6
Sweden	63.1	69.2	69.2	76.3	110.7	114.4	122.2	120.9	120.1
Switzerland	71.0	77.0	73.0	80.0	108.0	112.0	122.0	122.0	116.0
Turkey	..	..	..	57.0	121.4	116.9	124.0	113.2	123.8
United Kingdom	68.9	76.4	82.2	83.0	103.3	104.1	105.9	103.5	100.0
OECD Europe	**..**	**..**	**79.3**	**80.9**	**108.9**	**110.6**	**116.0**	**115.5**	**115.3**
Australia	..	..	68.6	77.3	108.2	110.7	116.8	116.5	120.4
Japan	48.0	59.2	66.8	82.8	99.0	99.8	105.4	97.8	94.4
Korea	..	..	..	36.2	106.2	131.8	154.0	156.0	167.4
New Zealand	..	..	..	..	105.0	109.0	..	..	..
OECD Pacific	**..**	**..**	**..**	**..**	**..**	**..**	**..**	**..**	**..**
OECD Total	**56.0**	**65.5**	**74.3**	**80.0**	**111.0**	**113.9**	**119.6**	**116.2**	**115.8**

Source: OECD Main Economic Indicators. / *Principaux indicateurs économiques de l'OCDE.*

Table 5: OECD Oil Supply per Unit of GDP[1]

Tableau 5 : Approvisionnement en pétrole des pays de l'OCDE par rapport au PIB [1]

	1971	1973	1979	1997	1998	1999	2000	2001	2002
Canada	0.259	0.254	0.229	0.134	0.133	0.129	0.124	0.124	0.117
Mexico	0.213	0.231	0.274	0.275	0.273	0.268	0.246	0.249	0.241
United States	0.203	0.206	0.185	0.107	0.104	0.102	0.100	0.101	0.097
OECD North America	**0.207**	**0.210**	**0.191**	**0.115**	**0.113**	**0.110**	**0.107**	**0.108**	**0.104**
Austria	0.082	0.089	0.079	0.051	0.052	0.048	0.045	0.048	0.048
Belgium	0.161	0.161	0.128	0.083	0.083	0.079	0.075	0.076	0.070
Czech Republic	0.196	0.220	0.244	0.147	0.156	0.154	0.143	0.147	0.142
Denmark	0.149	0.137	0.115	0.048	0.048	0.045	0.043	0.042	0.041
Finland	0.158	0.167	0.141	0.072	0.071	0.067	0.058	0.056	0.059
France	0.118	0.129	0.101	0.054	0.055	0.053	0.049	0.052	0.051
Germany	0.103	0.107	0.092	0.055	0.055	0.052	0.049	0.050	0.048
Greece	0.097	0.114	0.115	0.118	0.118	0.115	0.112	0.111	0.110
Hungary	0.247	0.241	0.261	0.147	0.146	0.135	0.126	0.117	0.112
Iceland	0.178	0.195	0.135	0.104	0.102	0.101	0.096	0.091	0.091
Ireland	0.206	0.206	0.177	0.082	0.084	0.085	0.076	0.076	0.076
Italy	0.150	0.155	0.126	0.081	0.080	0.076	0.073	0.071	0.069
Luxembourg	0.195	0.198	0.144	0.098	0.095	0.096	0.093	0.098	0.100
Netherlands	0.120	0.122	0.107	0.059	0.058	0.058	0.057	0.059	0.059
Norway	0.117	0.112	0.097	0.053	0.051	0.054	0.054	0.045	0.032
Poland	0.099	0.110	0.147	0.130	0.127	0.125	0.123	0.124	0.120
Portugal	0.098	0.095	0.114	0.121	0.131	0.128	0.120	0.121	0.122
Slovak Republic	..	..	..	0.167	0.156	0.139	0.124	0.131	0.133
Spain	0.103	0.111	0.123	0.092	0.095	0.094	0.092	0.093	0.091
Sweden	0.169	0.166	0.138	0.056	0.054	0.051	0.046	0.048	0.046
Switzerland	0.059	0.062	0.056	0.043	0.042	0.041	0.038	0.041	0.038
Turkey	0.148	0.183	0.168	0.154	0.147	0.154	0.151	0.152	0.151
United Kingdom	0.153	0.149	0.116	0.068	0.067	0.066	0.064	0.061	0.060
OECD Europe	**0.124**	**0.128**	**0.111**	**0.069**	**0.068**	**0.066**	**0.063**	**0.063**	**0.062**
Australia	0.136	0.138	0.141	0.088	0.083	0.081	0.081	0.071	0.074
Japan	0.090	0.096	0.081	0.049	0.048	0.048	0.046	0.045	0.046
Korea	0.135	0.144	0.177	0.200	0.180	0.176	0.167	0.158	0.152
New Zealand	0.100	0.103	0.098	0.100	0.100	0.090	0.092	0.088	0.089
OECD Pacific	**0.095**	**0.101**	**0.089**	**0.064**	**0.061**	**0.062**	**0.060**	**0.058**	**0.059**
OECD Total	**0.146**	**0.150**	**0.133**	**0.084**	**0.082**	**0.081**	**0.078**	**0.078**	**0.076**

1. Measured in toe per $1000 of GDP at 1995 prices and exchange rates. / *Mesuré en tep par $1000 de PIB aux prix et aux taux de change de 1995.*

Sources: OECD Main Economic Indicators, IEA Annual Oil Statistics. / *Principaux indicateurs économiques de l'OCDE, Statistiques annuelles du pétrole de l'AIE.*

Table 6: OECD Oil Supply per Capita[1]

Tableau 6 : Approvisionnement en pétrole des pays de l'OCDE par habitant[1]

	1971	1973	1979	1997	1998	1999	2000	2001	2002
Canada	3.338	3.603	3.801	2.750	2.817	2.849	2.854	2.857	2.771
Mexico	0.517	0.618	0.876	0.939	0.962	0.960	0.943	0.934	0.903
United States	3.504	3.888	3.925	3.174	3.196	3.226	3.241	3.161	3.103
OECD North America	**2.955**	**3.257**	**3.292**	**2.606**	**2.628**	**2.646**	**2.657**	**2.608**	**2.549**
Austria	1.361	1.619	1.707	1.532	1.616	1.537	1.502	1.615	1.620
Belgium	2.599	2.881	2.607	2.374	2.408	2.366	2.320	2.360	2.190
Czech Republic	0.753	0.898	1.111	0.767	0.805	0.803	0.769	0.818	0.803
Denmark	3.553	3.498	3.158	1.745	1.757	1.700	1.635	1.627	1.595
Finland	2.420	2.909	2.766	1.999	2.083	2.022	1.860	1.817	1.913
France	1.965	2.333	2.088	1.442	1.510	1.494	1.440	1.540	1.510
Germany	1.830	2.050	2.065	1.697	1.706	1.646	1.603	1.633	1.573
Greece	0.775	1.076	1.236	1.406	1.447	1.456	1.430	1.473	1.506
Hungary	0.681	0.787	1.065	0.675	0.707	0.683	0.672	0.650	0.642
Iceland	2.729	3.293	2.931	2.925	2.993	3.037	2.993	2.874	2.877
Ireland	1.666	1.804	1.883	1.787	1.957	2.180	2.120	2.215	2.280
Italy	1.631	1.829	1.774	1.595	1.592	1.552	1.526	1.494	1.452
Luxembourg	4.164	4.764	3.613	4.688	4.828	5.081	5.327	5.608	5.674
Netherlands	2.119	2.300	2.257	1.690	1.724	1.755	1.793	1.836	1.857
Norway	1.920	1.996	2.209	1.957	1.946	2.067	2.096	1.791	1.269
Poland	0.260	0.326	0.477	0.482	0.497	0.509	0.520	0.531	0.518
Portugal	0.543	0.632	0.807	1.406	1.582	1.605	1.555	1.577	1.600
Slovak Republic	..	..	..	0.663	0.643	0.583	0.531	0.580	0.618
Spain	0.895	1.102	1.319	1.452	1.559	1.609	1.624	1.667	1.661
Sweden	3.363	3.489	3.173	1.618	1.634	1.610	1.505	1.580	1.530
Switzerland	2.169	2.369	2.103	1.889	1.898	1.851	1.772	1.918	1.783
Turkey	0.250	0.325	0.345	0.472	0.456	0.447	0.461	0.421	0.429
United Kingdom	1.844	1.984	1.689	1.404	1.416	1.440	1.427	1.386	1.367
OECD Europe	**1.481**	**1.668**	**1.609**	**1.321**	**1.348**	**1.337**	**1.313**	**1.326**	**1.298**
Australia	1.875	1.994	2.203	1.908	1.864	1.870	1.894	1.706	1.812
Japan	1.918	2.321	2.233	2.157	2.078	2.103	2.061	2.013	2.012
Korea	0.324	0.393	0.717	2.383	1.986	2.139	2.208	2.135	2.157
New Zealand	1.318	1.489	1.338	1.691	1.686	1.577	1.651	1.630	1.684
OECD Pacific	**1.562**	**1.865**	**1.881**	**2.178**	**2.028**	**2.079**	**2.072**	**2.005**	**2.021**
OECD Total	**1.963**	**2.210**	**2.210**	**1.930**	**1.926**	**1.938**	**1.929**	**1.913**	**1.883**

1. Toe per capita. Data for 2002 are based on population estimates. / *Tep par habitant. Les données pour 2002 sont basées sur des estimations de la population.*

Sources: OECD Main Economic Indicators, IEA Annual Oil Statistics. / *Principaux indicateurs économiques de l'OCDE, Statistiques annuelles du pétrole de l'AIE.*

Table 7: Key energy and CO_2 Emissions Data for OECD Countries (2001)
Tableau 7 : Indicateurs principaux et émissions de CO_2 des pays de l'OCDE (2001)

	TPES (Mtoe)	Energy-Related CO_2 emissions (Million tonnes CO_2): from Coal	from Oil	from Gas	from Other*	Total	% Oil in total energy-related CO_2 emissions (%)	Energy-related CO_2/TPES (t CO_2 / toe)	Energy-related CO_2/GDP (kg CO_2/ 1995 US$)	Energy-related CO_2/population (t CO_2 / capita)	Share in total OECD TPES (%)	Share in total OECD energy-related CO_2 emissions (%)
Australia	115.63	218.39	104.29	46.32	0.64	369.64	28.21	3.20	0.79	18.99	2.17	2.95
Austria	30.72	13.95	35.97	15.46	1.24	66.63	53.99	2.17	0.25	8.19	0.58	0.53
Belgium	59.00	28.09	59.47	30.29	1.74	119.60	49.73	2.03	0.37	11.63	1.11	0.96
Canada	248.18	119.49	240.70	159.32	0.00	519.51	46.33	2.09	0.72	16.71	4.65	4.15
Czech Republic	41.40	77.70	21.63	18.30	1.09	118.71	18.22	2.87	2.08	11.57	0.78	0.95
Denmark	19.78	16.25	22.78	10.75	0.66	50.45	45.16	2.55	0.24	9.42	0.37	0.40
Finland	33.82	24.59	26.05	8.62	0.97	60.23	43.25	1.78	0.36	11.61	0.63	0.48
France	265.57	48.00	251.20	85.65	0.00	384.85	65.27	1.45	0.21	6.32	4.98	3.08
Germany	351.09	343.38	332.85	165.45	8.48	850.16	39.15	2.42	0.31	10.33	6.58	6.80
Greece	28.70	38.57	47.54	3.86	0.17	90.15	52.74	3.14	0.62	8.23	0.54	0.72
Hungary	25.34	15.10	17.21	23.98	0.05	56.34	30.55	2.22	1.00	5.53	0.48	0.45
Iceland	3.36	0.37	1.73	0.00	0.00	2.10	82.32	0.63	0.23	7.38	0.06	0.02
Ireland	14.98	10.56	24.64	7.92	0.00	43.11	57.15	2.88	0.38	11.19	0.28	0.34
Italy	172.00	47.07	243.02	134.53	0.65	425.27	57.14	2.47	0.35	7.34	3.23	3.40
Japan	520.73	351.98	621.91	156.97	1.44	1132.31	54.92	2.17	0.20	8.90	9.76	9.05
Korea	194.78	168.73	221.64	42.74	2.67	435.78	50.86	2.24	0.68	9.20	3.65	3.48
Luxembourg	3.83	0.43	6.36	1.62	0.00	8.41	75.63	2.20	0.33	18.99	0.07	0.07
Mexico	152.27	30.70	251.17	77.10	0.00	358.97	69.97	2.36	0.97	3.62	2.86	2.87
Netherlands	77.21	31.72	62.40	82.11	1.26	177.48	35.16	2.30	0.35	11.06	1.45	1.42
New Zealand	18.29	5.01	15.54	10.74	1.57	32.86	47.30	1.80	0.46	8.54	0.34	0.26
Norway	26.61	3.96	21.62	12.29	0.06	37.93	57.00	1.43	0.21	8.41	0.50	0.30
Poland	90.57	215.43	52.77	22.21	2.13	292.53	18.04	3.23	1.77	7.57	1.70	2.34
Portugal	24.73	12.36	41.53	5.15	0.00	59.05	70.33	2.39	0.45	5.87	0.46	0.47
Slovak Republic	18.72	16.62	8.40	14.08	0.27	39.37	21.34	2.10	1.65	7.32	0.35	0.31
Spain	127.38	72.31	175.59	37.70	..	285.60	61.48	2.24	0.39	7.09	2.39	2.28
Sweden	51.05	8.68	37.51	1.74	0.12	48.05	78.06	0.94	0.16	5.40	0.96	0.38
Switzerland	28.02	0.59	34.31	5.86	3.06	43.82	78.30	1.56	0.13	6.06	0.53	0.35
Turkey	72.46	78.59	78.22	31.14	0.06	188.01	41.60	2.59	0.99	2.74	1.36	1.50
United Kingdom	235.16	149.63	190.48	199.85	0.88	540.84	35.22	2.30	0.41	9.20	4.41	4.32
United States	2281.41	2124.18	2333.09	1196.97	19.01	5673.25	41.12	2.49	0.63	19.84	42.78	45.35
European Union	**1495.04**	**845.60**	**1557.39**	**790.70**	**16.19**	**3209.88**	**48.52**	**2.15**	**0.32**	**8.46**	**28.03**	**25.66**
IEA Total	**5067.89**	**4009.32**	**5267.57**	**2495.34**	**45.83**	**11818.05**	**44.57**	**2.33**	**0.43**	**11.88**	**95.03**	**94.46**
OECD Total	**5332.81**	**4272.43**	**5581.64**	**2608.72**	**48.23**	**12511.02**	**44.61**	**2.35**	**0.45**	**10.99**	**100.00**	**100.00**

Note: Energy-Related CO_2 emissions are calculated using the Tier 1 IPCC Sectoral Approach. Emissions from the combustion of biomass-derived fuels are not included in accordance with the IPCC greenhouse gas inventory methodology.

* Other includes industrial wastes and non-renewable municipal wastes.

Sources: IEA, *Energy Balances of OECD Countries, 2000-2001* (Paris: OECD, 2003) and *CO_2 Emissions from Fuel Combustion* (Paris: OECD, 2003).

Table 8: World Oil Demand[1]

Tableau 8 : Demande mondiale de pétrole [1]

000 metric tons

	1971	1973	1978	1998	1999	2000	2001	2002[2]
Canada	74 185	83 889	89 299	88 246	90 329	90 796	91 221	93 349
Mexico	24 268	30 758	45 953	88 503	87 520	90 936	88 526	86 292
United States	727 268	828 784	900 036	854 434	877 933	888 383	888 931	882 865
OECD North America	**825 721**	**943 431**	**1035 288**	**1031 183**	**1055 782**	**1070 115**	**1068 678**	**1062 506**
Austria	9 839	11 605	11 539	12 744	12 077	11 847	12 724	12 899
Belgium	25 490	28 843	27 626	29 233	27 559	28 523	28 843	29 098
Czech Republic	7 511	9 270	11 218	8 184	8 131	7 897	8 294	8 128
Denmark	17 798	17 748	16 222	10 672	10 600	10 292	9 735	9 120
Finland	10 810	13 328	12 715	10 147	10 203	9 903	9 994	10 182
France	102 440	125 963	117 318	96 420	95 751	94 638	96 928	93 592
Germany	143 003	161 998	158 134	136 430	132 412	129 741	131 660	127 339
Greece	7 261	10 089	11 883	19 385	18 931	19 826	20 095	20 228
Hungary	6 828	8 211	12 255	7 494	7 129	6 816	6 558	6 358
Iceland	533	661	642	876	896	912	874	890
Ireland	4 867	5 188	5 987	7 362	8 279	8 345	8 950	8 749
Italy	91 517	103 042	97 181	94 822	91 874	90 396	89 369	90 555
Luxembourg	1 412	1 643	1 403	1 962	2 094	2 227	2 368	2 421
Netherlands	35 401	40 991	38 542	37 940	39 275	40 658	42 644	42 766
Norway	8 015	8 384	9 121	10 039	10 125	9 464	9 688	8 985
Poland	8 777	11 089	16 387	18 894	20 009	19 238	19 075	18 302
Portugal	5 302	6 183	7 296	15 686	16 235	15 967	16 135	16 584
Slovak Republic	4 101	4 657	5 597	3 785	3 431	3 364	3 106	3 231
Spain	31 510	38 386	47 483	65 655	67 652	69 720	72 697	73 417
Sweden	28 060	28 408	26 300	17 140	16 949	16 288	15 481	15 800
Switzerland	13 230	14 728	13 437	13 056	13 108	12 772	12 868	12 272
Turkey	8 763	11 612	16 882	29 331	29 117	30 649	28 582	29 763
United Kingdom	103 760	112 354	93 386	81 875	81 371	80 034	78 344	76 781
OECD Europe	**676 228**	**774 381**	**758 554**	**729 132**	**723 208**	**719 517**	**725 012**	**717 460**
Australia	24 819	26 850	32 100	37 731	38 282	39 473	39 384	39 783
Japan	217 660	267 681	265 257	255 482	258 782	253 283	243 492	244 725
Korea	10 604	14 052	23 980	89 211	96 954	99 413	98 957	100 436
New Zealand	3 722	4 453	4 250	5 850	6 006	6 110	6 072	6 312
OECD Pacific	**256 805**	**313 036**	**325 587**	**388 274**	**400 024**	**398 279**	**387 905**	**391 256**
OECD Total	**1758 754**	**2030 848**	**2119 429**	**2148 589**	**2179 014**	**2187 911**	**2181 595**	**2171 222**
Argentina	23 545	24 224	24 864	24 779	24 295	23 406	20 887	..
Bolivia	650	745	1 152	1 864	1 886	1 752	2 034	..
Brazil	27 944	36 815	53 533	99 056	101 771	101 807	102 037	..
Chile	5 093	5 032	4 939	10 900	11 048	10 454	10 090	..
Colombia	6 381	6 689	6 917	13 891	12 612	12 231	11 526	..
Cuba	6 332	7 840	9 309	9 651	10 035	10 617	10 527	..
Dominican Republic	1 139	1 504	1 892	4 953	5 289	5 744	5 420	..
Ecuador	1 321	1 538	3 137	6 260	5 521	6 089	6 449	..
El Salvador	433	579	704	1 700	1 749	1 750	1 005	..
Guatemala	860	1 006	1 412	2 709	2 745	2 848	3 016	..
Jamaica	1 998	2 888	2 395	3 183	3 261	3 324	3 331	..
Netherlands Antilles	7 228	7 398	5 304	2 829	2 819	2 831	2 819	..
Panama	829	958	963	1 697	1 457	1 402	1 553	..
Peru	4 825	5 170	5 819	7 506	8 234	7 760	7 190	..
Uruguay	1 806	1 859	1 965	2 102	2 455	1 998	1 882	..
Venezuela	12 798	14 892	15 918	22 341	20 952	22 885	25 220	..
Other	7 272	8 074	8 735	10 863	11 302	11 159	11 524	..
Latin America	**110 454**	**127 211**	**148 958**	**226 347**	**227 431**	**228 057**	**227 310**	**222 094**
Bulgaria	9 889	10 849	13 315	4 799	4 453	4 028	4 223	..
Cyprus	632	852	823	2 291	2 445	2 546	2 540	..
Romania	11 232	13 395	18 710	11 976	9 807	10 008	11 260	..
Croatia	..	..	..	4 493	4 432	3 896	4 003	..
Slovenia	..	..	..	2 443	2 460	2 352	2 399	..
Serbia and Montenegro	..	..	..	3 220	2 780	3 018	3 779	..
Other Former Yugoslavia	..	..	..	1 622	1 643	1 844	1 688	..
Former Yugoslavia	8 519	9 896	14 390	11 778	11 315	11 110	11 869	..
Other	1 358	1 276	1 649	2 492	3 028	3 097	3 095	..
Non-OECD Europe	**31 630**	**36 268**	**48 887**	**33 336**	**31 048**	**30 789**	**32 987**	**33 402**

1. Demand refers to net inland deliveries (including refinery fuel and international marine bunkers). / *La demande concerne les livraisons intérieures nettes (comprenant la consommation propre des raffineries et les soutages maritimes internationaux).*
2. 2002 data for non-OECD areas are IEA Secretariat estimates. / *Les données 2002 des régions non-OCDE sont estimées par le Secrétariat de l'AIE.*

Source: Annual Oil Statistics, Energy Statistics of Non-OECD Countries .

Table 8: World Oil Demand[1] (continued)
Tableau 8 : Demande mondiale de pétrole[1] (suite)

000 metric tons

	1971	1973	1978	1998	1999	2000	2001	2002[2]
Azerbaijan	..	..	..	5 942	5 591	6 327	3 819	..
Belarus	..	..	..	8 597	7 553	6 976	7 279	..
Kazakhstan	..	..	..	8 567	7 042	7 397	8 881	..
Latvia	..	..	..	1 628	1 560	1 266	1 505	..
Lithuania	..	..	..	3 763	2 977	2 321	2 563	..
Russia	..	..	..	118 680	120 818	125 280	125 545	..
Turkmenistan	..	..	..	2 617	2 879	2 902	3 782	..
Ukraine	..	..	..	16 436	12 179	11 589	12 671	..
Uzbekistan	..	..	..	7 035	6 894	6 613	6 401	..
Other	..	..	..	5 284	4 991	4 560	4 330	..
Former USSR	**261 000**	**308 298**	**410 840**	**178 549**	**172 484**	**175 231**	**176 776**	**179 657**
Algeria	2 216	2 705	4 521	8 601	9 141	9 711	9 926	..
Angola	860	939	1 053	1 186	1 626	1 771	2 003	..
Côte d'Ivoire	851	962	1 414	1 170	1 103	980	915	..
Egypt	6 085	6 504	10 490	24 394	24 304	26 441	26 541	..
Ghana	721	769	904	1 886	2 038	1 793	1 940	..
Kenya	1 659	1 818	1 898	2 638	2 792	3 050	2 852	..
Libya	629	1 394	3 368	10 481	10 863	11 313	11 788	..
Morocco	1 991	2 532	3 706	6 672	7 515	6 998	7 038	..
Nigeria	1 657	2 569	6 554	9 244	10 094	10 806	12 076	..
Senegal	1 457	1 565	1 225	1 270	1 393	1 452	1 368	..
South Africa	12 761	13 629	13 672	20 994	21 222	19 874	20 020	..
Sudan	1 153	1 594	1 069	1 631	1 791	1 897	2 501	..
Tunisia	1 284	1 373	2 008	3 695	3 842	3 947	3 979	..
Other	7 839	8 474	9 259	13 806	14 835	14 949	15 585	..
Africa	**41 163**	**46 827**	**61 141**	**107 668**	**112 559**	**114 982**	**118 532**	**120 108**
Islamic Republic of Iran	14 304	17 973	29 332	60 465	60 562	62 277	65 228	..
Iraq	3 668	4 264	7 368	21 147	20 628	22 651	23 818	..
Israel	5 230	6 036	7 046	12 056	12 521	14 296	13 573	..
Jordan	501	611	1 279	4 602	4 619	4 870	4 770	..
Kuwait	6 290	7 051	6 356	16 929	15 643	15 202	15 677	..
Lebanon	1 980	2 412	2 055	4 837	4 905	4 475	4 871	..
Oman	1 335	1 326	1 149	2 217	2 775	2 993	3 380	..
Saudi Arabia	16 562	18 795	21 632	56 197	58 706	61 619	64 685	..
Syria	2 263	2 187	4 535	12 799	13 029	12 896	13 200	..
United Arab Emirates	136	280	1 524	16 462	12 041	10 893	13 043	..
Yemen	775	947	1 610	3 056	3 190	3 265	3 296	..
Other	1 683	1 766	1 806	3 445	4 057	4 567	4 423	..
Middle East	**54 727**	**63 648**	**85 692**	**214 212**	**212 676**	**220 004**	**229 964**	**235 621**
Bangladesh	757	876	1 286	3 676	3 284	3 215	3 834	..
India	20 950	23 418	28 894	94 330	102 874	106 141	106 014	..
Indonesia	8 083	10 529	16 817	44 110	46 432	50 840	52 680	..
DPR of Korea	834	1 177	2 077	1 274	834	1 002	1 108	..
Malaysia	4 173	4 273	6 451	20 560	20 784	21 150	22 042	..
Pakistan	3 248	3 442	4 262	17 570	19 135	18 760	18 047	..
Philippines	8 127	9 384	11 291	19 045	18 148	16 731	16 465	..
Singapore	5 199	7 656	9 852	34 764	34 751	34 355	35 436	..
Sri Lanka	1 267	1 604	1 431	2 904	3 149	3 611	3 578	..
Chinese Taipei	6 782	10 333	18 280	37 950	40 673	40 447	41 694	..
Thailand	5 751	7 498	10 202	32 125	33 334	32 395	32 145	..
Vietnam	5 660	5 654	933	6 468	7 559	8 278	8 395	..
Other	3 007	3 234	3 344	4 839	5 006	5 039	4 800	..
Asia	**73 838**	**89 078**	**115 120**	**319 615**	**335 963**	**341 964**	**346 238**	**349 043**
Hong Kong, China	**4 080**	**4 855**	**6 205**	**11 810**	**13 813**	**11 873**	**11 871**	**12 495**
China, People's Republic	**38 957**	**51 392**	**91 885**	**191 361**	**201 198**	**213 420**	**218 147**	**230 887**
Non-OECD	**615 849**	**727 577**	**968 728**	**1282 898**	**1307 172**	**1336 320**	**1361 825**	**1383 307**
World	**2374 603**	**2758 425**	**3088 157**	**3431 487**	**3486 186**	**3524 231**	**3543 420**	**3554 529**

1. Demand refers to net inland deliveries (including refinery fuel and international marine bunkers). / *La demande concerne les livraisons intérieures nettes (comprenant la consommation propre des raffineries et les soutages maritimes internationaux).*
2. 2002 data for non-OECD areas are IEA Secretariat estimates. / *Les données 2002 des régions non-OCDE sont estimées par le Secrétariat de l'AIE.*

Source: Annual Oil Statistics, Energy Statistics of Non-OECD Countries .

Table 9: World Demand by Main Product Groups[1 2]

Tableau 9 : Demande mondiale par principaux groupes de produits raffinés[1 2]

000 metric tons

	1971	1973	1978	1998	1999	2000	2001
WORLD							
LPG/Naphtha	156 245	193 534	194 927	358 090	376 792	394 420	389 901
Motor Gasoline	494 812	559 242	660 057	835 377	845 430	843 953	853 608
Aviation Fuels	106 537	116 137	123 566	196 173	202 227	210 259	204 714
Middle Distillates	584 833	683 611	787 126	1023 972	1052 710	1074 571	1100 411
Heavy Fuel Oil	787 494	910 724	969 711	603 394	586 287	575 199	560 928
Other	244 682	295 177	352 770	414 481	422 740	425 829	433 858
Total Products	**2374 603**	**2758 425**	**3088 157**	**3431 487**	**3486 186**	**3524 231**	**3543 420**
NON-OECD TOTAL							
LPG/Naphtha	12 432	19 006	29 870	118 663	124 091	134 602	142 257
Motor Gasoline	100 563	115 190	154 831	227 607	227 876	228 070	234 338
Aviation Fuels	36 086	39 187	41 505	51 566	52 702	55 149	55 535
Middle Distillates	158 266	186 588	245 526	416 904	436 171	446 233	457 629
Heavy Fuel Oil	242 431	288 726	373 020	324 156	317 532	306 070	301 380
Other	66 071	78 880	123 976	144 002	148 800	166 196	170 686
Total Products	**615 849**	**727 577**	**968 728**	**1282 898**	**1307 172**	**1336 320**	**1361 825**
LATIN AMERICA							
LPG/Naphtha	4 111	5 685	8 003	23 886	24 989	25 672	25 282
Motor Gasoline	23 839	28 122	32 808	55 017	53 643	51 998	50 039
Aviation Fuels	2 812	3 081	4 628	9 274	8 848	8 886	8 600
Middle Distillates	25 808	30 571	39 957	71 896	72 810	72 397	74 271
Heavy Fuel Oil	44 268	49 657	51 189	39 444	39 147	39 768	38 108
Other	9 616	10 095	12 373	26 830	27 994	29 336	31 010
Total Products	**110 454**	**127 211**	**148 958**	**226 347**	**227 431**	**228 057**	**227 310**
NON-OECD EUROPE + FSU							
LPG/Naphtha	4 333	8 085	9 583	7 038	8 192	10 391	9 746
Motor Gasoline	51 243	56 534	76 939	43 787	41 100	39 884	42 756
Aviation Fuels	22 443	23 500	22 909	10 182	10 654	10 751	11 082
Middle Distillates	66 961	78 295	89 925	49 369	47 200	49 023	50 923
Heavy Fuel Oil	111 313	135 444	195 824	66 571	59 710	50 966	47 913
Other	36 337	42 708	64 547	34 938	36 676	45 005	47 343
Total Products	**292 630**	**344 566**	**459 727**	**211 885**	**203 532**	**206 020**	**209 763**
AFRICA							
LPG/Naphtha	712	921	1 625	7 264	7 791	7 998	8 660
Motor Gasoline	7 747	9 138	12 566	22 532	24 020	24 110	24 946
Aviation Fuels	1 950	2 440	3 451	5 743	6 128	6 828	6 838
Middle Distillates	13 949	16 606	22 736	38 677	41 177	42 656	44 321
Heavy Fuel Oil	14 537	15 055	16 588	24 450	24 158	24 100	24 015
Other	2 268	2 667	4 175	9 002	9 285	9 290	9 752
Total Products	**41 163**	**46 827**	**61 141**	**107 668**	**112 559**	**114 982**	**118 532**
MIDDLE EAST							
LPG/Naphtha	1 353	1 584	3 009	19 251	19 063	20 322	22 179
Motor Gasoline	4 106	5 189	10 821	33 925	34 193	35 246	36 980
Aviation Fuels	3 166	3 674	4 774	6 138	5 896	6 554	6 651
Middle Distillates	11 039	13 852	28 386	66 431	66 718	70 509	71 807
Heavy Fuel Oil	32 363	36 253	34 020	62 798	62 045	61 194	65 940
Other	2 700	3 096	4 682	25 669	24 761	26 179	26 407
Total Products	**54 727**	**63 648**	**85 692**	**214 212**	**212 676**	**220 004**	**229 964**
ASIA + CHINA							
LPG/Naphtha	1 923	2 731	7 650	61 224	64 056	70 219	76 390
Motor Gasoline	13 628	16 207	21 697	72 346	74 920	76 832	79 617
Aviation Fuels	5 715	6 492	5 743	20 229	21 176	22 130	22 364
Middle Distillates	40 509	47 264	64 522	190 531	208 266	211 648	216 307
Heavy Fuel Oil	39 950	52 317	75 399	130 893	132 472	130 042	125 404
Other	15 150	20 314	38 199	47 563	50 084	56 386	56 174
Total Products	**116 875**	**145 325**	**213 210**	**522 786**	**550 974**	**567 257**	**576 256**

1. Demand refers to net inland deliveries (including refinery fuel and international marine bunkers). / *La demande concerne les livraisons intérieures nettes (comprenant la consommation propre des raffineries et les soutages maritimes internationaux).*
2. LPG/Naphtha includes ethane. / *LPG/Naphtha comprend l'éthane.*

Source: Annual Oil Statistics, Energy Statistics of Non-OECD Countries .

Table 9: World Demand by Main Product Groups[1 2] (continued)
Tableau 9 : Demande mondiale par principaux groupes de produits raffinés [1 2] (suite)

000 metric tons

	1971	1973	1978	1998	1999	2000	2001
OECD TOTAL							
LPG/Naphtha	143 813	174 528	165 057	239 427	252 701	259 818	247 644
Motor Gasoline	394 249	444 052	505 226	607 770	617 554	615 883	619 270
Aviation Fuels	70 451	76 950	82 061	144 607	149 525	155 110	149 179
Middle Distillates	426 567	497 023	541 600	607 068	616 539	628 338	642 782
Heavy Fuel Oil	545 063	621 998	596 691	279 238	268 755	269 129	259 548
Other	178 611	216 297	228 794	270 479	273 940	259 633	263 172
Total Products	**1758 754**	**2030 848**	**2119 429**	**2148 589**	**2179 014**	**2187 911**	**2181 595**
OECD NORTH AMERICA							
LPG/Naphtha	69 828	81 235	73 640	95 505	104 102	107 825	98 370
Motor Gasoline	284 657	317 497	357 032	407 294	415 010	418 442	423 846
Aviation Fuels	50 517	53 161	54 991	84 016	86 586	89 703	85 081
Middle Distillates	171 719	194 039	217 216	214 007	217 893	227 513	231 712
Heavy Fuel Oil	150 153	182 149	206 243	89 736	85 597	91 618	86 574
Other	98 847	115 350	126 166	140 625	146 594	135 014	143 095
Total Products	**825 721**	**943 431**	**1035 288**	**1031 183**	**1055 782**	**1070 115**	**1068 678**
OECD EUROPE							
LPG/Naphtha	41 766	53 010	48 081	68 927	69 324	71 392	69 578
Motor Gasoline	83 494	95 814	110 576	137 221	137 679	131 336	128 915
Aviation Fuels	16 368	18 928	21 039	43 015	45 384	47 512	45 768
Middle Distillates	210 158	243 923	252 636	271 190	271 484	274 569	286 223
Heavy Fuel Oil	269 358	301 162	266 771	124 713	117 126	112 681	113 030
Other	55 084	61 544	59 451	84 066	82 211	82 027	81 498
Total Products	**676 228**	**774 381**	**758 554**	**729 132**	**723 208**	**719 517**	**725 012**
OECD PACIFIC							
LPG/Naphtha	32 219	40 283	43 336	74 995	79 275	80 601	79 696
Motor Gasoline	26 098	30 741	37 618	63 255	64 865	66 105	66 509
Aviation Fuels	3 566	4 861	6 031	17 576	17 555	17 895	18 330
Middle Distillates	44 690	59 061	71 748	121 871	127 162	126 256	124 847
Heavy Fuel Oil	125 552	138 687	123 677	64 789	66 032	64 830	59 944
Other	24 680	39 403	43 177	45 788	45 135	42 592	38 579
Total Products	**256 805**	**313 036**	**325 587**	**388 274**	**400 024**	**398 279**	**387 905**
AUSTRALIA							
LPG/Naphtha	870	911	837	2 768	2 792	2 822	2 893
Motor Gasoline	7 806	8 894	10 674	13 102	13 306	13 618	13 023
Aviation Fuels	1 122	1 299	1 676	3 928	3 901	4 016	4 444
Middle Distillates	5 610	6 463	7 834	10 792	11 012	11 406	11 500
Heavy Fuel Oil	7 480	7 150	6 675	1 803	1 918	2 120	1 975
Other	1 931	2 133	4 404	5 338	5 353	5 491	5 549
Total Products	**24 819**	**26 850**	**32 100**	**37 731**	**38 282**	**39 473**	**39 384**
AUSTRIA							
LPG/Naphtha	211	256	167	145	151	170	142
Motor Gasoline	1 798	2 218	2 285	2 204	2 047	1 980	1 994
Aviation Fuels	100	112	106	539	528	571	555
Middle Distillates	1 687	2 387	2 831	5 852	5 768	5 867	6 633
Heavy Fuel Oil	4 988	5 487	4 882	1 963	1 608	1 335	1 351
Other	1 055	1 145	1 268	2 041	1 975	1 924	2 049
Total Products	**9 839**	**11 605**	**11 539**	**12 744**	**12 077**	**11 847**	**12 724**
BELGIUM							
LPG/Naphtha	1 878	2 196	1 911	3 361	3 413	3 906	3 975
Motor Gasoline	2 313	2 573	3 108	2 512	2 394	2 245	2 186
Aviation Fuels	399	510	460	1 549	1 513	1 484	1 124
Middle Distillates	8 534	9 815	10 259	11 956	11 499	11 449	12 095
Heavy Fuel Oil	11 090	12 279	10 261	7 450	6 294	6 447	6 629
Other	1 276	1 470	1 627	2 405	2 446	2 992	2 834
Total Products	**25 490**	**28 843**	**27 626**	**29 233**	**27 559**	**28 523**	**28 843**

1. Demand refers to net inland deliveries (including refinery fuel and international marine bunkers). / *La demande concerne les livraisons intérieures nettes (comprenant la consommation propre des raffineries et les soutages maritimes internationaux).*
2. LPG/Naphtha includes ethane. / *LPG/Naphtha comprend l'éthane.*

Source: Annual Oil Statistics, Energy Statistics of Non-OECD Countries .

Table 9: World Demand by Main Product Groups[1 2] (continued)
Tableau 9 : Demande mondiale par principaux groupes de produits raffinés [1 2] (suite)

000 metric tons

	1971	1973	1978	1998	1999	2000	2001
CANADA							
LPG/Naphtha	3 309	3 371	4 970	9 484	11 048	10 342	10 552
Motor Gasoline	20 353	23 570	27 236	27 466	28 168	28 180	28 527
Aviation Fuels	2 326	2 787	3 342	4 855	5 089	5 130	4 685
Middle Distillates	23 324	25 366	26 095	22 916	23 749	24 910	24 038
Heavy Fuel Oil	18 126	20 164	17 367	8 208	7 161	7 410	8 734
Other	6 747	8 631	10 289	15 317	15 114	14 824	14 685
Total Products	**74 185**	**83 889**	**89 299**	**88 246**	**90 329**	**90 796**	**91 221**
CZECH REPUBLIC							
LPG/Naphtha	134	141	340	803	919	804	942
Motor Gasoline	978	1 137	1 241	1 816	1 924	1 858	1 901
Aviation Fuels	228	228	253	189	179	195	208
Middle Distillates	1 976	2 255	3 913	2 714	2 638	2 751	2 937
Heavy Fuel Oil	3 283	4 406	4 512	1 390	1 218	919	915
Other	912	1 103	959	1 272	1 253	1 370	1 391
Total Products	**7 511**	**9 270**	**11 218**	**8 184**	**8 131**	**7 897**	**8 294**
DENMARK							
LPG/Naphtha	297	568	313	83	78	77	71
Motor Gasoline	1 555	1 635	1 767	2 001	1 995	1 968	1 939
Aviation Fuels	632	719	805	779	820	825	823
Middle Distillates	6 453	6 833	6 646	4 455	4 334	4 074	3 924
Heavy Fuel Oil	8 297	7 394	5 985	1 402	1 279	1 279	1 132
Other	564	599	706	1 952	2 094	2 069	1 846
Total Products	**17 798**	**17 748**	**16 222**	**10 672**	**10 600**	**10 292**	**9 735**
FINLAND							
LPG/Naphtha	241	647	736	445	450	458	377
Motor Gasoline	1 086	1 251	1 353	1 858	1 850	1 785	1 823
Aviation Fuels	109	154	203	470	497	495	495
Middle Distillates	4 076	4 569	4 895	4 105	4 175	4 192	4 287
Heavy Fuel Oil	4 408	5 741	4 500	2 005	2 019	1 831	2 014
Other	890	966	1 028	1 264	1 212	1 142	998
Total Products	**10 810**	**13 328**	**12 715**	**10 147**	**10 203**	**9 903**	**9 994**
FRANCE							
LPG/Naphtha	6 048	7 913	7 248	10 089	10 421	10 173	9 984
Motor Gasoline	13 323	15 772	17 587	14 554	14 457	13 896	13 685
Aviation Fuels	1 552	1 860	2 318	5 927	6 307	6 535	6 467
Middle Distillates	39 513	46 407	43 306	45 201	44 728	44 990	47 916
Heavy Fuel Oil	31 700	42 296	36 658	9 436	8 630	7 978	7 179
Other	10 304	11 715	10 201	11 213	11 208	11 066	11 697
Total Products	**102 440**	**125 963**	**117 318**	**96 420**	**95 751**	**94 638**	**96 928**
GERMANY							
LPG/Naphtha	6 832	8 944	8 766	15 497	15 075	15 676	15 536
Motor Gasoline	19 243	20 780	25 829	30 281	30 250	28 806	27 949
Aviation Fuels	2 524	2 825	3 190	6 458	6 827	7 168	6 836
Middle Distillates	60 864	68 235	68 024	62 248	58 895	57 283	60 842
Heavy Fuel Oil	38 643	43 690	36 806	8 802	8 024	8 896	9 542
Other	14 897	17 524	15 519	13 144	13 341	11 912	10 955
Total Products	**143 003**	**161 998**	**158 134**	**136 430**	**132 412**	**129 741**	**131 660**
GREECE							
LPG/Naphtha	266	274	296	509	480	489	488
Motor Gasoline	723	951	1 363	3 156	3 215	3 280	3 385
Aviation Fuels	646	816	1 071	1 169	1 250	1 290	1 160
Middle Distillates	2 177	2 773	3 417	6 995	6 887	7 096	7 294
Heavy Fuel Oil	3 138	4 766	5 420	6 125	5 762	6 077	6 048
Other	311	509	316	1 431	1 337	1 594	1 720
Total Products	**7 261**	**10 089**	**11 883**	**19 385**	**18 931**	**19 826**	**20 095**

1. Demand refers to net inland deliveries (including refinery fuel and international marine bunkers). / *La demande concerne les livraisons intérieures nettes (comprenant la consommation propre des raffineries et les soutages maritimes internationaux).*
2. LPG/Naphtha includes ethane. / *LPG/Naphtha comprend l'éthane.*

Source: Annual Oil Statistics, Energy Statistics of Non-OECD Countries .

Table 9: World Demand by Main Product Groups[1 2] (continued)
Tableau 9 : Demande mondiale par principaux groupes de produits raffinés [1 2] (suite)

000 metric tons

	1971	1973	1978	1998	1999	2000	2001
HUNGARY							
LPG/Naphtha	346	392	1 033	871	838	907	811
Motor Gasoline	752	918	1 398	1 386	1 402	1 336	1 391
Aviation Fuels	50	50	109	190	204	228	214
Middle Distillates	2 399	3 333	4 548	1 865	1 886	2 015	2 069
Heavy Fuel Oil	2 576	2 808	4 189	2 237	2 057	1 519	1 342
Other	705	710	978	945	742	811	731
Total Products	**6 828**	**8 211**	**12 255**	**7 494**	**7 129**	**6 816**	**6 558**
ICELAND							
LPG/Naphtha	1	1	-	2	2	2	2
Motor Gasoline	58	80	91	142	145	143	144
Aviation Fuels	75	108	83	118	125	138	119
Middle Distillates	293	343	323	375	396	403	355
Heavy Fuel Oil	93	106	133	110	79	71	87
Other	13	23	12	129	149	155	167
Total Products	**533**	**661**	**642**	**876**	**896**	**912**	**874**
IRELAND							
LPG/Naphtha	187	215	256	132	136	142	146
Motor Gasoline	666	796	976	1 306	1 414	1 493	1 551
Aviation Fuels	359	280	297	437	516	598	718
Middle Distillates	988	1 271	1 502	3 223	3 642	3 769	3 964
Heavy Fuel Oil	2 445	2 352	2 662	1 821	2 136	1 807	1 896
Other	222	274	294	443	435	536	675
Total Products	**4 867**	**5 188**	**5 987**	**7 362**	**8 279**	**8 345**	**8 950**
ITALY							
LPG/Naphtha	7 822	8 627	5 934	6 497	6 662	6 619	6 206
Motor Gasoline	9 547	11 257	11 008	18 734	17 939	16 877	17 058
Aviation Fuels	1 832	1 917	1 842	3 123	3 485	3 600	3 414
Middle Distillates	17 274	21 616	25 406	26 340	27 060	27 611	28 467
Heavy Fuel Oil	48 904	54 116	47 377	29 433	26 429	25 677	23 705
Other	6 138	5 509	5 614	10 695	10 299	10 012	10 519
Total Products	**91 517**	**103 042**	**97 181**	**94 822**	**91 874**	**90 396**	**89 369**
JAPAN							
LPG/Naphtha	30 849	38 434	40 341	45 959	48 791	47 992	46 669
Motor Gasoline	16 162	19 599	24 371	40 796	41 828	43 011	43 958
Aviation Fuels	2 217	2 975	3 485	10 014	9 512	9 827	9 849
Middle Distillates	35 968	48 821	56 901	84 082	85 736	85 102	84 283
Heavy Fuel Oil	110 409	121 516	102 572	39 263	38 359	35 792	31 253
Other	22 055	36 336	37 587	35 368	34 556	31 559	27 480
Total Products	**217 660**	**267 681**	**265 257**	**255 482**	**258 782**	**253 283**	**243 492**
KOREA							
LPG/Naphtha	479	922	2 144	26 160	27 588	29 673	30 004
Motor Gasoline	689	726	881	7 206	7 548	7 314	7 352
Aviation Fuels	-	368	567	2 818	3 258	3 202	3 152
Middle Distillates	2 350	2 772	5 925	25 068	28 442	27 661	26 980
Heavy Fuel Oil	6 688	8 720	13 674	23 466	25 518	26 647	26 459
Other	398	544	789	4 493	4 600	4 916	5 010
Total Products	**10 604**	**14 052**	**23 980**	**89 211**	**96 954**	**99 413**	**98 957**
LUXEMBOURG							
LPG/Naphtha	27	21	22	24	24	24	28
Motor Gasoline	117	158	244	545	563	582	571
Aviation Fuels	36	51	61	275	323	312	337
Middle Distillates	500	606	560	1 093	1 159	1 291	1 414
Heavy Fuel Oil	691	776	475	7	5	6	7
Other	41	31	41	18	20	12	11
Total Products	**1 412**	**1 643**	**1 403**	**1 962**	**2 094**	**2 227**	**2 368**

1. Demand refers to net inland deliveries (including refinery fuel and international marine bunkers). / *La demande concerne les livraisons intérieures nettes (comprenant la consommation propre des raffineries et les soutages maritimes internationaux).*
2. LPG/Naphtha includes ethane. / *LPG/Naphtha comprend l'éthane.*

Source: Annual Oil Statistics, Energy Statistics of Non-OECD Countries .

Table 9: World Demand by Main Product Groups[1 2] (continued)

Tableau 9 : Demande mondiale par principaux groupes de produits raffinés [1 2] (suite)

000 metric tons

	1971	1973	1978	1998	1999	2000	2001
MEXICO							
LPG/Naphtha	2 751	3 490	4 020	12 546	13 373	14 694	14 457
Motor Gasoline	6 944	8 405	10 743	23 029	22 437	23 118	23 134
Aviation Fuels	440	586	971	2 425	2 543	2 557	2 537
Middle Distillates	6 429	7 752	12 302	15 072	14 063	14 803	14 314
Heavy Fuel Oil	6 222	8 842	15 936	31 237	29 808	30 388	29 055
Other	1 482	1 683	1 981	4 194	5 296	5 376	5 029
Total Products	**24 268**	**30 758**	**45 953**	**88 503**	**87 520**	**90 936**	**88 526**
NETHERLANDS							
LPG/Naphtha	3 162	5 968	5 291	4 409	4 663	5 089	5 257
Motor Gasoline	3 242	3 556	3 954	4 085	4 140	4 031	4 121
Aviation Fuels	778	865	855	3 179	3 304	3 260	3 168
Middle Distillates	8 999	10 108	9 347	8 151	8 461	8 803	8 938
Heavy Fuel Oil	16 056	16 967	15 015	10 892	11 328	11 901	13 193
Other	3 164	3 527	4 080	7 224	7 379	7 574	7 967
Total Products	**35 401**	**40 991**	**38 542**	**37 940**	**39 275**	**40 658**	**42 644**
NEW ZEALAND							
LPG/Naphtha	21	16	14	108	104	114	130
Motor Gasoline	1 441	1 522	1 692	2 151	2 183	2 162	2 176
Aviation Fuels	227	219	303	816	884	850	885
Middle Distillates	762	1 005	1 088	1 929	1 972	2 087	2 084
Heavy Fuel Oil	975	1 301	756	257	237	271	257
Other	296	390	397	589	626	626	540
Total Products	**3 722**	**4 453**	**4 250**	**5 850**	**6 006**	**6 110**	**6 072**
NORWAY							
LPG/Naphtha	258	365	526	1 000	907	1 019	1 362
Motor Gasoline	978	1 095	1 348	1 683	1 671	1 620	1 681
Aviation Fuels	240	272	352	634	735	639	619
Middle Distillates	3 351	3 628	3 959	4 275	4 478	4 000	4 146
Heavy Fuel Oil	2 490	2 225	1 843	764	687	575	621
Other	698	799	1 093	1 683	1 647	1 611	1 259
Total Products	**8 015**	**8 384**	**9 121**	**10 039**	**10 125**	**9 464**	**9 688**
POLAND							
LPG/Naphtha	217	396	439	1 711	1 723	1 858	1 997
Motor Gasoline	1 940	2 422	3 220	4 998	5 755	5 005	4 629
Aviation Fuels	171	180	206	365	360	364	343
Middle Distillates	2 843	3 783	5 656	7 287	7 539	7 363	7 642
Heavy Fuel Oil	2 492	2 970	4 819	2 646	2 654	2 302	2 305
Other	1 114	1 338	2 047	1 887	1 978	2 346	2 159
Total Products	**8 777**	**11 089**	**16 387**	**18 894**	**20 009**	**19 238**	**19 075**
PORTUGAL							
LPG/Naphtha	596	645	719	2 368	2 434	2 421	2 047
Motor Gasoline	553	714	753	2 027	2 079	2 123	1 995
Aviation Fuels	464	580	546	633	724	772	738
Middle Distillates	1 099	1 288	1 648	4 174	4 431	4 974	5 343
Heavy Fuel Oil	2 351	2 677	3 245	4 571	4 551	3 909	3 789
Other	239	279	385	1 913	2 016	1 768	2 223
Total Products	**5 302**	**6 183**	**7 296**	**15 686**	**16 235**	**15 967**	**16 135**
SLOVAK REPUBLIC							
LPG/Naphtha	6	14	17	857	753	781	415
Motor Gasoline	459	532	550	667	643	602	612
Aviation Fuels	-	-	-	29	20	26	31
Middle Distillates	1 056	1 213	1 179	857	845	826	887
Heavy Fuel Oil	2 306	2 503	3 437	657	424	208	401
Other	274	395	414	718	746	921	760
Total Products	**4 101**	**4 657**	**5 597**	**3 785**	**3 431**	**3 364**	**3 106**

1. Demand refers to net inland deliveries (including refinery fuel and international marine bunkers). / *La demande concerne les livraisons intérieures nettes (comprenant la consommation propre des raffineries et les soutages maritimes internationaux).*
2. LPG/Naphtha includes ethane. / *LPG/Naphtha comprend l'éthane.*

Source: Annual Oil Statistics, Energy Statistics of Non-OECD Countries .

Table 9: World Demand by Main Product Groups[1 2] (continued)
Tableau 9 : Demande mondiale par principaux groupes de produits raffinés [1 2] (suite)

000 metric tons

	1971	1973	1978	1998	1999	2000	2001
SPAIN							
LPG/Naphtha	2 786	3 343	5 029	6 782	6 944	7 305	7 302
Motor Gasoline	3 068	4 083	5 454	9 007	8 928	8 524	8 481
Aviation Fuels	1 192	1 703	2 222	3 869	4 097	4 378	4 418
Middle Distillates	6 167	7 322	10 431	22 638	24 316	25 964	27 923
Heavy Fuel Oil	15 974	19 134	21 436	13 483	14 444	13 992	14 536
Other	2 323	2 801	2 911	9 876	8 923	9 557	10 037
Total Products	**31 510**	**38 386**	**47 483**	**65 655**	**67 652**	**69 720**	**72 697**
SWEDEN							
LPG/Naphtha	999	1 242	1 130	1 654	1 250	1 286	1 555
Motor Gasoline	2 824	3 108	3 659	4 017	4 035	3 977	3 961
Aviation Fuels	525	561	566	860	919	909	925
Middle Distillates	9 315	10 062	9 183	5 831	5 738	6 129	4 978
Heavy Fuel Oil	13 211	12 294	10 953	3 932	3 798	3 101	3 206
Other	1 186	1 141	809	846	1 209	886	856
Total Products	**28 060**	**28 408**	**26 300**	**17 140**	**16 949**	**16 288**	**15 481**
SWITZERLAND							
LPG/Naphtha	231	253	173	247	238	237	238
Motor Gasoline	2 355	2 510	2 608	3 851	3 979	3 983	3 873
Aviation Fuels	543	658	765	1 425	1 517	1 582	1 492
Middle Distillates	6 895	7 818	7 584	6 531	6 462	6 139	6 403
Heavy Fuel Oil	2 377	2 641	1 637	386	269	193	224
Other	829	848	670	616	643	638	638
Total Products	**13 230**	**14 728**	**13 437**	**13 056**	**13 108**	**12 772**	**12 868**
TURKEY							
LPG/Naphtha	416	591	1 131	4 775	4 784	5 922	5 276
Motor Gasoline	952	1 341	2 431	4 543	4 257	3 619	3 052
Aviation Fuels	100	142	172	1 521	1 151	1 228	1 208
Middle Distillates	2 605	3 283	5 047	6 837	8 305	9 102	8 775
Heavy Fuel Oil	4 183	5 639	7 199	8 172	7 800	8 030	7 668
Other	507	616	902	3 483	2 820	2 748	2 603
Total Products	**8 763**	**11 612**	**16 882**	**29 331**	**29 117**	**30 649**	**28 582**
UNITED KINGDOM							
LPG/Naphtha	8 805	9 998	6 604	6 666	6 979	6 027	5 421
Motor Gasoline	14 964	16 927	18 349	21 848	22 597	21 603	20 933
Aviation Fuels	3 813	4 337	4 557	9 277	9 983	10 915	10 356
Middle Distillates	21 094	24 975	22 972	28 187	27 842	28 478	28 991
Heavy Fuel Oil	47 662	47 895	33 327	7 029	5 631	4 628	5 240
Other	7 422	8 222	7 577	8 868	8 339	8 383	7 403
Total Products	**103 760**	**112 354**	**93 386**	**81 875**	**81 371**	**80 034**	**78 344**
UNITED STATES							
LPG/Naphtha	63 768	74 374	64 650	73 475	79 681	82 789	73 361
Motor Gasoline	257 360	285 522	319 053	356 799	364 405	367 144	372 185
Aviation Fuels	47 751	49 788	50 678	76 736	78 954	82 016	77 859
Middle Distillates	141 966	160 921	178 819	176 019	180 081	187 800	193 360
Heavy Fuel Oil	125 805	153 143	172 940	50 291	48 628	53 820	48 785
Other	90 618	105 036	113 896	121 114	126 184	114 814	123 381
Total Products	**727 268**	**828 784**	**900 036**	**854 434**	**877 933**	**888 383**	**888 931**

1. Demand refers to net inland deliveries (including refinery fuel and international marine bunkers). / *La demande concerne les livraisons intérieures nettes (comprenant la consommation propre des raffineries et les soutages maritimes internationaux).*
2. LPG/Naphtha includes ethane. / *LPG/Naphtha comprend l'éthane.*

Source: Annual Oil Statistics, Energy Statistics of Non-OECD Countries .

Table 10: World Demand by Main Product Groups[1 2]

Tableau 10 : Demande mondiale par principaux groupes de produits raffinés[1 2]

000 barrels / day

	1971	1973	1978	1998	1999	2000	2001
WORLD							
LPG/Naphtha	4 363	5 378	5 469	10 416	11 040	11 495	11 374
Motor Gasoline	11 542	13 046	15 397	19 470	19 704	19 619	19 902
Aviation Fuels	2 326	2 531	2 698	4 263	4 395	4 555	4 445
Middle Distillates	12 009	14 036	16 149	20 977	21 565	21 952	22 544
Heavy Fuel Oil	14 216	16 438	17 532	10 892	10 585	10 362	10 132
Other	4 696	5 678	6 933	7 831	7 979	8 021	8 220
Total Products	**49 152**	**57 106**	**64 178**	**73 849**	**75 269**	**76 005**	**76 616**
NON-OECD TOTAL							
LPG/Naphtha	356	527	830	3 346	3 503	3 794	4 011
Motor Gasoline	2 348	2 690	3 614	5 300	5 306	5 297	5 461
Aviation Fuels	789	856	905	1 125	1 150	1 200	1 212
Middle Distillates	3 262	3 842	5 051	8 551	8 945	9 126	9 384
Heavy Fuel Oil	4 427	5 267	6 804	5 889	5 768	5 545	5 477
Other	1 335	1 604	2 528	2 804	2 904	3 242	3 353
Total Products	**12 517**	**14 785**	**19 732**	**27 015**	**27 577**	**28 205**	**28 897**
LATIN AMERICA							
LPG/Naphtha	126	167	235	695	724	742	745
Motor Gasoline	555	655	762	1 263	1 231	1 190	1 150
Aviation Fuels	62	68	102	202	192	193	187
Middle Distillates	533	632	825	1 466	1 485	1 473	1 516
Heavy Fuel Oil	811	904	931	699	693	703	676
Other	193	203	249	517	535	548	591
Total Products	**2 281**	**2 628**	**3 104**	**4 841**	**4 860**	**4 848**	**4 866**
NON-OECD EUROPE + FSU							
LPG/Naphtha	122	217	261	209	249	316	296
Motor Gasoline	1 198	1 321	1 798	1 023	961	930	999
Aviation Fuels	488	511	498	221	232	233	241
Middle Distillates	1 377	1 608	1 843	1 010	965	1 000	1 041
Heavy Fuel Oil	2 031	2 471	3 573	1 215	1 090	927	874
Other	752	894	1 358	687	716	889	936
Total Products	**5 968**	**7 022**	**9 331**	**4 365**	**4 211**	**4 294**	**4 387**
AFRICA							
LPG/Naphtha	22	29	51	224	241	247	265
Motor Gasoline	181	214	294	531	565	566	587
Aviation Fuels	43	53	75	125	134	149	149
Middle Distillates	287	342	468	796	847	875	911
Heavy Fuel Oil	265	275	303	444	438	437	436
Other	44	51	81	181	187	188	197
Total Products	**842**	**963**	**1 271**	**2 301**	**2 412**	**2 461**	**2 546**
MIDDLE EAST							
LPG/Naphtha	36	42	82	581	564	597	651
Motor Gasoline	96	121	253	793	799	821	864
Aviation Fuels	69	80	104	134	128	142	145
Middle Distillates	229	287	586	1 366	1 372	1 446	1 476
Heavy Fuel Oil	591	661	621	1 146	1 132	1 114	1 203
Other	54	61	89	503	486	515	520
Total Products	**1 073**	**1 253**	**1 734**	**4 522**	**4 481**	**4 635**	**4 859**
ASIA + CHINA							
LPG/Naphtha	51	72	201	1 638	1 726	1 893	2 055
Motor Gasoline	319	379	508	1 691	1 751	1 791	1 861
Aviation Fuels	127	144	126	443	464	483	490
Middle Distillates	836	974	1 330	3 912	4 276	4 334	4 440
Heavy Fuel Oil	729	955	1 376	2 386	2 415	2 364	2 286
Other	292	395	751	915	980	1 103	1 109
Total Products	**2 354**	**2 919**	**4 291**	**10 986**	**11 612**	**11 967**	**12 240**

1. Demand refers to net inland deliveries (including refinery fuel and international marine bunkers). / *La demande concerne les livraisons intérieures nettes (y compris la consommation propre des raffineries et les soutages maritimes internationaux).*
2. LPG/Naphtha Includes ethane. / *LPG/Naphtha comprend l'éthane.*

Source: Annual Oil Statistics, Energy Statistics of Non-OECD Countries.

Table 10: World Demand by Main Product Groups [1] [2] (continued)
Tableau 10 : Demande mondiale par principaux groupes de produits raffinés [1] [2] *(suite)*

000 barrels / day

	1971	1973	1978	1998	1999	2000	2001
OECD TOTAL							
LPG/Naphtha	4 007	4 851	4 639	7 070	7 537	7 701	7 363
Motor Gasoline	9 194	10 356	11 783	14 170	14 398	14 322	14 441
Aviation Fuels	1 537	1 675	1 793	3 138	3 245	3 355	3 233
Middle Distillates	8 747	10 194	11 098	12 426	12 620	12 826	13 160
Heavy Fuel Oil	9 789	11 171	10 728	5 003	4 817	4 817	4 655
Other	3 361	4 074	4 405	5 027	5 075	4 779	4 867
Total Products	**36 635**	**42 321**	**44 446**	**46 834**	**47 692**	**47 800**	**47 719**
OECD NORTH AMERICA							
LPG/Naphtha	2 060	2 399	2 193	3 174	3 521	3 619	3 330
Motor Gasoline	6 651	7 419	8 342	9 515	9 696	9 750	9 903
Aviation Fuels	1 103	1 160	1 208	1 826	1 882	1 945	1 849
Middle Distillates	3 520	3 975	4 447	4 365	4 445	4 627	4 726
Heavy Fuel Oil	2 737	3 318	3 754	1 616	1 541	1 647	1 560
Other	1 817	2 119	2 388	2 551	2 659	2 440	2 613
Total Products	**17 888**	**20 390**	**22 332**	**23 047**	**23 744**	**24 028**	**23 981**
OECD EUROPE							
LPG/Naphtha	1 116	1 412	1 310	1 913	1 927	1 976	1 942
Motor Gasoline	1 933	2 218	2 561	3 178	3 186	3 033	2 984
Aviation Fuels	356	408	453	931	982	1 023	986
Middle Distillates	4 303	4 996	5 168	5 548	5 550	5 600	5 857
Heavy Fuel Oil	4 761	5 323	4 716	2 205	2 071	1 990	2 001
Other	1 069	1 190	1 156	1 606	1 557	1 547	1 542
Total Products	**13 538**	**15 547**	**15 364**	**15 381**	**15 273**	**15 169**	**15 312**
OECD PACIFIC							
LPG/Naphtha	831	1 040	1 136	1 983	2 089	2 106	2 091
Motor Gasoline	610	719	880	1 477	1 516	1 539	1 554
Aviation Fuels	78	107	132	381	381	387	398
Middle Distillates	924	1 223	1 483	2 513	2 625	2 599	2 577
Heavy Fuel Oil	2 291	2 530	2 258	1 182	1 205	1 180	1 094
Other	475	765	861	870	859	792	712
Total Products	**5 209**	**6 384**	**6 750**	**8 406**	**8 675**	**8 603**	**8 426**
AUSTRALIA							
LPG/Naphtha	23	24	23	95	95	96	98
Motor Gasoline	182	208	249	306	311	317	304
Aviation Fuels	25	29	37	86	85	87	97
Middle Distillates	115	133	161	220	225	233	235
Heavy Fuel Oil	136	130	122	33	35	39	36
Other	40	45	104	108	109	112	112
Total Products	**521**	**569**	**696**	**848**	**860**	**884**	**882**
AUSTRIA							
LPG/Naphtha	6	7	5	5	5	5	5
Motor Gasoline	42	51	53	51	47	46	46
Aviation Fuels	2	2	2	12	11	12	12
Middle Distillates	34	49	58	119	118	119	136
Heavy Fuel Oil	88	97	86	35	28	24	24
Other	18	20	24	42	39	39	41
Total Products	**190**	**226**	**228**	**264**	**248**	**245**	**264**
BELGIUM							
LPG/Naphtha	49	57	50	87	87	99	102
Motor Gasoline	54	60	72	58	55	52	51
Aviation Fuels	8	11	10	33	33	32	24
Middle Distillates	175	201	210	245	235	234	247
Heavy Fuel Oil	196	217	181	132	111	114	117
Other	23	26	31	48	47	57	56
Total Products	**505**	**572**	**554**	**603**	**568**	**588**	**597**

1. Demand refers to net inland deliveries (including refinery fuel and international marine bunkers). / *La demande concerne les livraisons intérieures nettes (y compris la consommation propre des raffineries et les soutages maritimes internationaux).*
2. LPG/Naphtha Includes ethane. / *LPG/Naphtha comprend l'éthane.*

Source: Annual Oil Statistics, Energy Statistics of Non-OECD Countries.

Table 10: World Demand by Main Product Groups [1 2] (continued)

Tableau 10 : Demande mondiale par principaux groupes de produits raffinés [1 2] (suite)

000 barrels / day

	1971	1973	1978	1998	1999	2000	2001
CANADA							
LPG/Naphtha	102	103	145	324	390	367	381
Motor Gasoline	476	551	637	644	661	659	669
Aviation Fuels	51	61	73	104	110	110	101
Middle Distillates	478	520	534	459	476	498	481
Heavy Fuel Oil	331	368	317	148	129	134	158
Other	119	154	181	251	246	244	239
Total Products	**1 557**	**1 757**	**1 887**	**1 930**	**2 012**	**2 012**	**2 029**
CZECH REPUBLIC							
LPG/Naphtha	4	4	9	21	24	21	25
Motor Gasoline	23	26	29	42	45	43	44
Aviation Fuels	5	5	6	4	4	4	4
Middle Distillates	41	46	80	56	54	56	60
Heavy Fuel Oil	58	78	80	25	22	16	16
Other	18	20	19	25	25	26	28
Total Products	**149**	**179**	**223**	**173**	**174**	**166**	**177**
DENMARK							
LPG/Naphtha	9	16	10	3	2	2	2
Motor Gasoline	36	38	41	46	46	45	45
Aviation Fuels	14	15	18	17	18	18	18
Middle Distillates	132	140	136	91	88	83	80
Heavy Fuel Oil	147	131	106	25	23	23	20
Other	9	10	14	34	37	36	30
Total Products	**347**	**350**	**325**	**216**	**214**	**207**	**195**
FINLAND							
LPG/Naphtha	6	17	19	13	12	12	11
Motor Gasoline	25	29	31	43	43	41	42
Aviation Fuels	2	3	4	10	11	11	11
Middle Distillates	84	93	100	84	85	85	88
Heavy Fuel Oil	78	101	80	35	36	32	36
Other	17	18	21	26	24	24	20
Total Products	**212**	**261**	**255**	**211**	**211**	**205**	**208**
FRANCE							
LPG/Naphtha	166	214	200	274	283	276	272
Motor Gasoline	308	365	407	337	335	321	317
Aviation Fuels	33	40	50	128	137	141	140
Middle Distillates	808	949	885	924	914	917	979
Heavy Fuel Oil	560	747	648	167	153	141	127
Other	207	234	198	211	210	207	220
Total Products	**2 082**	**2 549**	**2 388**	**2 041**	**2 032**	**2 003**	**2 055**
GERMANY							
LPG/Naphtha	186	243	237	399	387	402	400
Motor Gasoline	445	481	598	701	700	665	647
Aviation Fuels	55	61	69	140	148	155	148
Middle Distillates	1 244	1 394	1 390	1 273	1 204	1 167	1 244
Heavy Fuel Oil	683	772	650	156	142	157	169
Other	291	342	307	252	255	225	207
Total Products	**2 904**	**3 293**	**3 251**	**2 921**	**2 836**	**2 771**	**2 815**
GREECE							
LPG/Naphtha	8	8	8	16	16	15	15
Motor Gasoline	17	22	32	73	74	76	78
Aviation Fuels	14	17	24	25	27	27	25
Middle Distillates	44	57	70	143	140	144	149
Heavy Fuel Oil	55	84	96	108	102	107	107
Other	5	10	6	27	24	30	31
Total Products	**143**	**198**	**236**	**392**	**383**	**399**	**405**

1. Demand refers to net inland deliveries (including refinery fuel and international marine bunkers). / *La demande concerne les livraisons intérieures nettes (y compris la consommation propre des raffineries et les soutages maritimes internationaux).*
2. LPG/Naphtha Includes ethane. / *LPG/Naphtha comprend l'éthane.*

Source: Annual Oil Statistics, Energy Statistics of Non-OECD Countries.

Table 10: World Demand by Main Product Groups [1 2] (continued)
Tableau 10 : Demande mondiale par principaux groupes de produits raffinés [1 2] (suite)

000 barrels / day

	1971	1973	1978	1998	1999	2000	2001
HUNGARY							
LPG/Naphtha	10	11	27	24	23	25	22
Motor Gasoline	17	21	32	32	32	31	32
Aviation Fuels	1	1	2	4	4	5	5
Middle Distillates	49	68	93	38	38	41	42
Heavy Fuel Oil	46	50	74	40	36	27	24
Other	13	13	17	18	13	15	14
Total Products	**136**	**164**	**245**	**156**	**146**	**144**	**139**
ICELAND							
LPG/Naphtha	-	-	-	-	-	-	-
Motor Gasoline	1	2	2	3	3	3	3
Aviation Fuels	2	2	2	3	3	3	3
Middle Distillates	6	7	7	8	8	8	7
Heavy Fuel Oil	2	2	2	2	1	1	2
Other	-	-	-	1	2	2	3
Total Products	**11**	**13**	**13**	**17**	**17**	**17**	**18**
IRELAND							
LPG/Naphtha	5	6	8	4	4	5	5
Motor Gasoline	15	18	23	30	33	34	36
Aviation Fuels	8	6	6	9	11	13	15
Middle Distillates	20	26	30	66	75	77	82
Heavy Fuel Oil	43	42	47	32	38	32	34
Other	5	5	6	8	7	8	12
Total Products	**96**	**103**	**120**	**149**	**168**	**169**	**184**
ITALY							
LPG/Naphtha	204	225	161	185	193	190	180
Motor Gasoline	221	261	255	434	415	390	395
Aviation Fuels	40	43	40	67	75	77	73
Middle Distillates	355	445	521	539	553	563	582
Heavy Fuel Oil	864	956	837	520	467	453	419
Other	116	103	107	199	187	179	191
Total Products	**1 800**	**2 033**	**1 921**	**1 944**	**1 890**	**1 852**	**1 840**
JAPAN							
LPG/Naphtha	796	993	1 062	1 228	1 294	1 257	1 230
Motor Gasoline	378	458	570	953	978	1 002	1 027
Aviation Fuels	48	65	76	217	206	213	214
Middle Distillates	745	1 012	1 178	1 736	1 771	1 753	1 741
Heavy Fuel Oil	2 015	2 217	1 872	716	700	651	570
Other	420	703	736	669	654	578	497
Total Products	**4 402**	**5 448**	**5 494**	**5 519**	**5 603**	**5 454**	**5 279**
KOREA							
LPG/Naphtha	12	23	51	657	697	749	759
Motor Gasoline	16	17	21	168	176	170	172
Aviation Fuels	-	8	12	61	71	69	68
Middle Distillates	48	57	121	518	589	571	558
Heavy Fuel Oil	122	159	250	428	466	485	483
Other	8	10	14	82	84	90	93
Total Products	**206**	**274**	**469**	**1 914**	**2 083**	**2 134**	**2 133**
LUXEMBOURG							
LPG/Naphtha	1	1	1	1	1	1	1
Motor Gasoline	3	4	6	13	13	13	13
Aviation Fuels	1	1	1	6	7	7	7
Middle Distillates	10	12	11	22	24	26	29
Heavy Fuel Oil	12	14	8	-	-	-	-
Other	-	-	-	-	-	-	-
Total Products	**27**	**32**	**27**	**42**	**45**	**47**	**50**

1. Demand refers to net inland deliveries (including refinery fuel and international marine bunkers). / *La demande concerne les livraisons intérieures nettes (y compris la consommation propre des raffineries et les soutages maritimes internationaux).*
2. LPG/Naphtha Includes ethane. / *LPG/Naphtha comprend l'éthane.*

Source: Annual Oil Statistics, Energy Statistics of Non-OECD Countries.

Table 10: World Demand by Main Product Groups [1 2] (continued)

Tableau 10 : Demande mondiale par principaux groupes de produits raffinés [1 2] (suite)

000 barrels / day

	1971	1973	1978	1998	1999	2000	2001
MEXICO							
LPG/Naphtha	87	113	132	426	467	506	496
Motor Gasoline	161	195	249	533	519	534	536
Aviation Fuels	9	13	21	52	54	55	54
Middle Distillates	132	159	252	306	285	299	291
Heavy Fuel Oil	110	156	281	550	525	534	512
Other	33	36	44	82	106	107	101
Total Products	**532**	**672**	**979**	**1 949**	**1 956**	**2 035**	**1 990**
NETHERLANDS							
LPG/Naphtha	79	149	138	123	129	140	142
Motor Gasoline	75	82	92	95	96	93	95
Aviation Fuels	17	18	18	69	71	70	68
Middle Distillates	186	208	192	167	173	180	183
Heavy Fuel Oil	284	300	265	192	200	210	233
Other	65	73	81	156	159	164	173
Total Products	**706**	**830**	**786**	**802**	**828**	**857**	**894**
NEW ZEALAND							
LPG/Naphtha	-	-	-	3	3	4	4
Motor Gasoline	34	36	40	50	51	50	51
Aviation Fuels	5	5	7	17	19	18	19
Middle Distillates	16	21	23	39	40	42	43
Heavy Fuel Oil	18	24	14	5	4	5	5
Other	7	7	7	11	12	12	10
Total Products	**80**	**93**	**91**	**125**	**129**	**131**	**132**
NORWAY							
LPG/Naphtha	7	9	16	34	31	34	45
Motor Gasoline	23	25	31	39	39	37	39
Aviation Fuels	5	6	7	14	16	14	13
Middle Distillates	69	75	81	87	91	82	85
Heavy Fuel Oil	44	39	33	14	12	10	11
Other	13	14	20	31	31	29	23
Total Products	**161**	**168**	**188**	**219**	**220**	**206**	**216**
POLAND							
LPG/Naphtha	6	10	12	49	50	53	58
Motor Gasoline	45	56	75	116	133	116	107
Aviation Fuels	4	4	5	8	8	8	7
Middle Distillates	58	77	115	149	154	150	156
Heavy Fuel Oil	44	52	85	47	47	41	41
Other	20	24	37	37	39	46	42
Total Products	**177**	**223**	**329**	**406**	**431**	**414**	**411**
PORTUGAL							
LPG/Naphtha	17	19	21	66	68	67	58
Motor Gasoline	13	17	17	47	48	49	46
Aviation Fuels	10	12	11	14	16	17	16
Middle Distillates	23	26	34	86	90	101	109
Heavy Fuel Oil	42	47	57	81	80	69	67
Other	4	5	7	35	37	32	40
Total Products	**109**	**126**	**147**	**329**	**339**	**335**	**336**
SLOVAK REPUBLIC							
LPG/Naphtha	-	-	1	20	18	18	12
Motor Gasoline	11	12	13	16	15	14	14
Aviation Fuels	-	-	-	1	-	1	1
Middle Distillates	21	25	24	17	17	17	18
Heavy Fuel Oil	42	46	63	12	8	4	7
Other	6	8	9	14	14	17	15
Total Products	**80**	**91**	**110**	**80**	**72**	**71**	**67**

1. Demand refers to net inland deliveries (including refinery fuel and international marine bunkers). / *La demande concerne les livraisons intérieures nettes (y compris la consommation propre des raffineries et les soutages maritimes internationaux).*
2. LPG/Naphtha Includes ethane. / *LPG/Naphtha comprend l'éthane.*

Source: Annual Oil Statistics, Energy Statistics of Non-OECD Countries.

Table 10: World Demand by Main Product Groups [1 2] (continued)
Tableau 10 : Demande mondiale par principaux groupes de produits raffinés [1 2] (suite)

000 barrels / day

	1971	1973	1978	1998	1999	2000	2001
SPAIN							
LPG/Naphtha	80	95	141	185	188	196	195
Motor Gasoline	71	95	126	209	207	197	196
Aviation Fuels	26	37	48	84	88	94	95
Middle Distillates	126	150	213	462	497	529	570
Heavy Fuel Oil	282	338	379	238	255	247	257
Other	45	54	56	179	159	170	179
Total Products	**630**	**769**	**963**	**1 357**	**1 394**	**1 433**	**1 492**
SWEDEN							
LPG/Naphtha	26	32	29	46	35	35	43
Motor Gasoline	65	72	85	93	93	92	92
Aviation Fuels	12	12	12	18	20	19	20
Middle Distillates	190	206	187	119	117	125	102
Heavy Fuel Oil	233	217	194	69	67	55	57
Other	21	20	13	15	21	15	14
Total Products	**547**	**559**	**520**	**360**	**353**	**341**	**328**
SWITZERLAND							
LPG/Naphtha	6	7	5	8	7	7	7
Motor Gasoline	55	58	60	89	92	92	90
Aviation Fuels	12	14	16	31	33	34	32
Middle Distillates	141	159	155	133	132	125	131
Heavy Fuel Oil	42	47	29	7	5	3	4
Other	16	15	13	11	11	12	12
Total Products	**272**	**300**	**278**	**279**	**280**	**273**	**276**
TURKEY							
LPG/Naphtha	12	18	33	140	142	177	157
Motor Gasoline	22	31	56	105	99	84	71
Aviation Fuels	2	3	4	33	25	26	26
Middle Distillates	53	68	104	140	170	186	180
Heavy Fuel Oil	74	100	127	144	138	142	136
Other	10	11	16	67	55	53	51
Total Products	**173**	**231**	**340**	**629**	**629**	**668**	**621**
UNITED KINGDOM							
LPG/Naphtha	229	264	179	210	222	196	185
Motor Gasoline	346	392	425	506	523	499	485
Aviation Fuels	83	95	98	201	216	235	223
Middle Distillates	434	515	472	580	573	585	598
Heavy Fuel Oil	842	846	589	124	100	82	93
Other	147	165	154	170	161	161	140
Total Products	**2 081**	**2 277**	**1 917**	**1 791**	**1 795**	**1 758**	**1 724**
UNITED STATES							
LPG/Naphtha	1 871	2 183	1 916	2 424	2 664	2 746	2 453
Motor Gasoline	6 014	6 673	7 456	8 338	8 516	8 557	8 698
Aviation Fuels	1 043	1 086	1 114	1 670	1 718	1 780	1 694
Middle Distillates	2 910	3 296	3 661	3 600	3 684	3 830	3 954
Heavy Fuel Oil	2 296	2 794	3 156	918	887	979	890
Other	1 665	1 929	2 163	2 218	2 307	2 089	2 273
Total Products	**15 799**	**17 961**	**19 466**	**19 168**	**19 776**	**19 981**	**19 962**

1. Demand refers to net inland deliveries (including refinery fuel and international marine bunkers). / *La demande concerne les livraisons intérieures nettes (y compris la consommation propre des raffineries et les soutages maritimes internationaux).*
2. LPG/Naphtha Includes ethane. / *LPG/Naphtha comprend l'éthane.*

Source: Annual Oil Statistics, Energy Statistics of Non-OECD Countries.

Table 11: OECD Consumption of Unleaded Gasoline
Tableau 11 : Consommation d'essence sans plomb au sein de l'OCDE

000 metric tons

	1994	1995	1996	1997	1998	1999	2000	2001	2002
Canada	25 715	25 764	26 089	26 727	27 464	28 167	28 178	28 526	29 095
Mexico	..	10 759	13 340	17 207	21 752	21 339	22 275	22 677	23 278
United States	328 700	336 703	341 071	346 401	356 758	364 363	367 082	372 126	382 005
OECD North America	**354 415**	**373 226**	**380 500**	**390 335**	**405 974**	**413 869**	**417 535**	**423 329**	**434 378**
Austria	2 508	2 398	2 215	2 103	2 204	2 047	1 980	1 994	2 141
Belgium	1 843	1 947	2 027	2 003	2 065	2 394	2 245	2 186	2 082
Czech Republic	..	798	1 006	1 128	1 155	1 341	1 490	1 901	1 926
Denmark	1 837	1 894	1 913	1 963	2 001	1 995	1 968	1 939	1 953
Finland	1 915	1 895	1 834	1 881	1 858	1 850	1 785	1 823	1 840
France	7 578	7 859	8 389	8 906	9 529	10 215	13 894	13 685	13 202
Germany	27 504	28 489	29 486	30 214	30 280	30 250	28 806	27 949	27 194
Greece	743	893	1 107	1 297	1 516	1 737	2 052	2 336	3 543
Hungary	..	..	..	975	1 039	1 374	1 332	1 391	1 434
Iceland	115	118	145	135	142	145	143	144	148
Ireland	482	587	712	870	1 105	1 302	1 460	1 543	1 600
Italy	5 700	7 216	8 175	9 779	11 447	13 502	12 623	14 342	16 640
Luxembourg	410	410	438	475	492	552	582	571	556
Netherlands	3 151	3 373	3 764	4 041	4 085	4 140	4 031	4 121	4 172
Norway	1 554	1 667	1 691	1 666	1 683	1 671	1 620	1 681	1 669
Poland	..	..	..	..	..	4 501	4 591	4 613	4 282
Portugal	548	663	800	904	1 074	1 703	2 123	1 995	2 023
Slovak Republic	..	..	..	..	..	..	..	612	643
Spain	1 508	2 266	3 154	3 645	4 286	4 823	5 415	7 289	8 206
Sweden	4 134	4 245	4 205	4 126	4 017	4 035	3 977	3 961	4 041
Switzerland	2 924	3 006	3 223	3 465	3 590	3 821	3 983	3 873	3 796
Turkey	117	171	255	478	864	1 105	1 359	1 501	1 748
United Kingdom	13 162	13 831	15 231	16 002	17 163	19 926	21 603	20 933	19 918
OECD Europe	**77 733**	**83 726**	**89 770**	**96 056**	**101 595**	**114 429**	**119 062**	**122 383**	**124 757**
Australia	6 245	7 221	7 836	8 438	9 070	9 829	11 457	11 589	12 433
Japan	36 983	37 584	38 791	39 868	40 796	41 828	43 008	43 958	44 617
Korea	6 032	6 989	8 030	8 432	7 206	7 548	7 314	7 352	7 549
New Zealand	913	1 024	1 926	2 130	2 151	2 183	2 162	2 175	2 066
OECD Pacific	**50 173**	**52 818**	**56 583**	**58 868**	**59 223**	**61 388**	**63 941**	**65 074**	**66 665**
OECD Total	**482 321**	**509 770**	**526 853**	**545 259**	**566 792**	**589 686**	**600 538**	**610 786**	**625 800**

Source: IEA Annual Oil Statistics. / *Statistiques annuelles du pétrole de l'AIE.*

Table 12: Percentage Unleaded of Total Motor Gasoline Consumption
Tableau 12 : Part de l'essence sans plomb dans la consommation totale d'essence

per cent

	1994	1995	1996	1997	1998	1999	2000	2001	2002
Canada	100.0	100.0	100.0	100.0	100.0	100.0	100.0	100.0	100.0
Mexico	..	49.1	63.2	80.4	99.0	99.9	100.0	100.0	100.0
United States	100.0	100.0	100.0	100.0	100.0	100.0	100.0	100.0	100.0
OECD North America	**93.9**	**97.1**	**98.0**	**98.9**	**99.9**	**100.0**	**100.0**	**100.0**	**100.0**
Austria	99.8	99.8	99.9	99.9	100.0	100.0	100.0	100.0	100.0
Belgium	64.9	68.7	74.0	79.0	82.2	100.0	100.0	100.0	100.0
Czech Republic	..	48.7	54.6	61.3	63.6	69.7	80.2	100.0	100.0
Denmark	98.0	100.0	100.0	100.0	100.0	100.0	100.0	100.0	100.0
Finland	99.8	99.9	100.0	100.0	100.0	100.0	100.0	100.0	100.0
France	46.0	50.2	55.9	60.9	65.5	70.7	100.0	100.0	100.0
Germany	92.3	94.5	97.4	100.0	100.0	100.0	100.0	100.0	100.0
Greece	27.6	32.2	37.7	42.7	48.0	54.0	62.6	69.0	100.0
Hungary	..	..	..	72.1	75.0	98.0	99.7	100.0	100.0
Iceland	83.9	86.8	92.9	100.0	100.0	100.0	100.0	100.0	100.0
Ireland	49.0	56.6	64.8	74.0	84.6	92.1	97.8	99.5	100.0
Italy	31.1	37.6	42.5	50.7	59.5	74.8	73.3	82.7	100.0
Luxembourg	75.2	79.5	84.1	87.8	90.3	98.0	100.0	100.0	100.0
Netherlands	80.6	83.8	92.3	100.0	100.0	100.0	100.0	100.0	100.0
Norway	92.2	100.0	100.0	100.0	100.0	100.0	100.0	100.0	100.0
Poland	..	..	..	..	..	78.2	91.8	99.7	100.0
Portugal	29.6	35.1	40.7	46.4	53.0	81.9	100.0	100.0	100.0
Slovak Republic	..	..	..	..	..	..	..	100.0	100.0
Spain	16.5	26.6	34.7	40.6	47.6	54.0	63.5	85.9	100.0
Sweden	99.4	99.9	100.0	100.0	100.0	100.0	100.0	100.0	100.0
Switzerland	79.0	83.7	87.5	90.6	93.2	96.0	100.0	100.0	100.0
Turkey	2.9	3.9	5.6	10.4	19.0	26.0	37.6	49.2	55.7
United Kingdom	57.6	63.0	68.0	71.9	78.6	88.2	100.0	100.0	100.0
OECD Europe	**56.6**	**61.3**	**65.1**	**69.8**	**73.8**	**83.0**	**90.4**	**94.7**	**98.9**
Australia	48.9	55.7	60.0	64.6	69.2	73.9	84.1	89.0	92.8
Japan	100.0	100.0	100.0	100.0	100.0	100.0	100.0	100.0	100.0
Korea	99.6	99.7	99.9	100.0	100.0	100.0	100.0	100.0	100.0
New Zealand	44.1	48.9	92.8	100.0	100.0	100.0	100.0	100.0	100.0
OECD Pacific	**86.7**	**88.5**	**91.3**	**92.7**	**93.6**	**94.6**	**96.7**	**97.8**	**98.6**
OECD Total	**84.3**	**87.8**	**89.6**	**91.5**	**93.3**	**95.6**	**97.6**	**98.7**	**99.6**

Source: IEA Annual Oil Statistics. / *Statistiques annuelles du pétrole de l'AIE.*

Table 13: OECD Consumption of Fuel Oil According to Sulphur Content[1]

Tableau 13 : Consommation de fioul selon la teneur en soufre [1]

000 metric tons

	1999			2000			2001		
	Total Fuel Oil	Low [2] Sulphur	High [3] Sulphur	Total Fuel Oil	Low [2] Sulphur	High [3] Sulphur	Total Fuel Oil	Low [2] Sulphur	High [3] Sulphur
Canada	5 453	5 453	..	5 612	5 612	..	7 170	7 170	..
Mexico	27 120	5 296	21 824	27 833	6 590	21 243	26 813	4 846	21 967
United States	27 419	20 217	7 202	30 628	25 251	5 377	34 568	20 856	13 712
OECD North America	**59 992**	..	..	**64 073**	..	..	**68 551**	..	..
Austria	1 431	887	544	1 202	794	408	1 234	799	435
Belgium	1 753	1 608	145	1 147	1 077	70	1 463	1 409	54
Czech Republic	1 165	395	770	868	264	604	853	371	482
Denmark	550	470	80	432	384	48	484	457	27
Finland	1 627	1 353	274	1 347	986	361	1 473	1 242	231
France	4 632	1 949	2 683	4 053	1 692	2 361	3 695	1 648	2 047
Germany	5 079	2 099	2 980	6 260	1 567	4 693	6 928	1 802	5 126
Greece	2 997	506	2 491	2 807	642	2 165	2 757	524	2 233
Hungary	1 802	93	1 709	1 413	56	1 357	1 268	71	1 197
Iceland	77	..	77	63	..	63	85	..	85
Ireland	2 070	..	2 070	1 744	33	1 711	1 834	26	1 808
Italy	23 337	14 682	8 655	21 996	14 393	7 603	19 993	13 000	6 993
Luxembourg	5	5	..	6	6	..	7	7	..
Netherlands	81	81	..	52	52	..	99	99	..
Norway	299	299	..	206	206	..	265	265	..
Poland	1 337	298	1 039	1 150	410	740	1 003	252	751
Portugal	3 898	662	3 236	3 208	546	2 662	3 263	422	2 841
Slovak Republic	424	..	..	208	..	..	398	264	134
Spain	7 442	3 106	4 336	6 735	2 821	3 914	6 765	3 021	3 744
Sweden	2 002	2 002	..	1 395	1 380	15	1 531	1 518	13
Switzerland	228	228	..	146	146	..	177	177	..
Turkey	6 738	..	6 738	6 879	..	6 879	6 508	..	6 508
United Kingdom	4 279	23	4 256	5 119	30	5 089	5 963	50	5 913
OECD Europe	**73 253**	..	..	**68 436**	..	..	**68 046**	..	..
Australia	1 073	66	1 007	1 242	76	1 166	1 186	70	1 116
Japan	30 749	11 026	19 723	30 414	..	30 414	26 477	..	26 477
Korea	17 710	8 180	9 530	19 043	9 967	9 076	19 017	9 782	9 235
New Zealand	31	..	31	94	..	94	84	..	84
OECD Pacific	**49 563**	..	..	**50 793**	..	..	**46 764**	..	..
OECD Total	**182 808**	..	..	**183 302**	..	..	**183 361**	..	..

1. Gross inland consumption, excluding international marine bunkers and refinery fuel. / *Consommation intérieure brute, ne comprenant pas les soutages maritimes internationaux ni la consommation propre des raffineries.*
2. Heavy fuel oil with sulphur content lower than 1%. / *Fioul lourd dont la teneur en soufre est inférieure à 1%.*
3. Heavy fuel oil with sulphur content of 1% or higher. / *Fioul lourd dont la teneur en soufre est de 1% ou plus.*

Source: IEA Annual Oil Statistics. / *Statistiques annuelles du pétrole de l'AIE.*

Table 14: World Crude Oil and NGL Production[1]
Tableau 14 : Production mondiale de pétrole brut et de LGN [1]

000 metric tons

	1971	1973	1978	1998	1999	2000	2001	2002
Canada	70 564	94 128	75 080	124 684	119 858	124 806	126 579	132 691
Mexico	25 368	27 182	69 227	171 909	166 882	169 261	175 498	178 261
United States	527 728	513 296	483 076	373 946	358 436	358 802	356 516	356 379
OECD North America	**623 660**	**634 606**	**627 383**	**670 539**	**645 176**	**652 869**	**658 593**	**667 331**
Austria	2 589	2 596	1 815	1 047	1 063	1 072	1 012	1 003
Czech Republic	35	43	282	433	386	385	366	415
Denmark	..	68	432	11 432	14 465	17 780	16 887	18 183
France	2 497	2 066	1 943	2 092	2 001	1 907	1 897	1 800
Germany	7 565	6 714	5 130	3 825	3 790	4 341	4 304	4 598
Greece	..	..	..	315	16	279	191	189
Hungary	2 000	2 033	2 518	1 834	1 788	1 663	1 549	1 499
Italy	1 320	1 105	1 489	5 819	5 216	4 764	4 223	5 621
Netherlands	1 714	1 556	1 520	2 661	2 535	2 350	2 263	3 073
Norway	293	1 595	17 000	149 826	149 350	160 990	162 536	156 381
Poland	395	392	363	422	519	711	797	711
Slovak Republic	163	130	64	61	66	59	93	96
Spain	125	654	1 211	529	299	227	338	316
Turkey	3 452	3 511	2 736	3 223	2 941	2 762	2 520	2 420
United Kingdom	220	515	54 100	132 498	137 228	126 375	116 805	115 164
Other	..	..	1	95	98	100	94	100
OECD Europe	**22 368**	**22 978**	**90 604**	**316 112**	**321 761**	**325 765**	**315 875**	**311 569**
Australia	14 340	19 182	21 891	29 432	23 657	32 076	33 138	32 453
Japan	763	717	565	640	585	598	585	560
Korea	..	..	..	391	446	670	572	523
New Zealand	..	161	575	2 289	2 047	1 849	1 749	1 633
OECD Pacific	**15 103**	**20 060**	**23 031**	**32 752**	**26 735**	**35 193**	**36 044**	**35 169**
OECD Total	**661 131**	**677 644**	**741 018**	**1019 403**	**993 672**	**1013 827**	**1010 512**	**1014 069**
Argentina	21 918	21 938	23 549	43 793	41 504	40 149	41 236	40 049
Bolivia	1 906	2 487	1 759	2 123	1 853	1 838	2 003	1 887
Brazil	8 313	8 296	8 002	50 895	56 597	63 149	67 334	75 536
Chile	1 723	1 748	1 017	406	386	398	414	414
Colombia	11 376	9 758	6 977	38 673	42 469	35 850	31 645	30 304
Cuba	137	157	330	1 678	2 136	2 695	2 886	3 117
Ecuador	179	10 615	10 346	19 583	19 470	20 887	21 249	20 418
Guatemala	..	..	3	1 389	1 270	1 130	1 149	1 089
Peru	3 135	3 553	7 634	5 797	5 381	5 071	4 960	5 137
Trinidad and Tobago	6 468	8 291	11 446	6 699	6 966	6 812	6 528	7 377
Venezuela	187 399	179 424	115 635	181 399	165 793	173 838	171 776	152 606
Other	..	2	67	426	448	606	611	611
Latin America	**242 554**	**246 269**	**186 765**	**352 861**	**344 273**	**352 423**	**351 791**	**338 545**
Albania	1 657	2 107	2 200	366	331	317	311	267
Bulgaria	305	190	246	33	40	46	35	35
Romania	13 793	14 287	13 724	6 724	6 435	6 339	6 273	6 095
Croatia	..	..	..	1 815	1 385	1 313	1 217	1 143
Serbia and Montenegro	..	..	..	919	718	815	880	821
Other Former Yugoslavia	..	..	..	1	1	1	1	1
Former Yugoslavia	2 961	3 332	4 076	2 735	2 104	2 129	2 098	1 965
Non-OECD Europe	**18 716**	**19 916**	**20 246**	**9 858**	**8 910**	**8 831**	**8 717**	**8 362**
Azerbaijan	..	..	..	11 424	13 807	14 017	14 909	15 306
Belarus	..	..	..	1 830	1 840	1 851	1 852	1 852
Georgia	..	..	..	119	91	110	99	99
Kazakhstan	..	..	..	25 945	30 130	35 317	40 091	47 096
Kyrgyzstan	..	..	..	78	77	77	76	76
Lithuania	..	..	..	277	232	316	471	471
Russia	..	..	..	301 405	303 239	321 692	345 841	378 232
Tajikistan	..	..	..	20	19	18	18	18
Turkmenistan	..	..	..	7 339	7 706	7 729	8 632	9 687
Ukraine	..	..	..	3 903	3 798	3 693	3 703	3 669
Uzbekistan	..	..	..	8 352	8 329	7 707	7 392	7 430
Former USSR	**377 100**	**429 100**	**567 500**	**360 692**	**369 268**	**392 527**	**423 084**	**463 936**

1. Includes additives and other hydrocarbons (other than crude oil and NGL) / *Comprend les additifs et les autres hydrocarbures (autres que le pétrole et les LGN).*

Source: Annual Oil Statistics, Energy Statistics of Non-OECD Countries.

Table 14: World Crude Oil and NGL Production (continued)[1]
Tableau 14 : Production mondiale de pétrole brut et de LGN (suite) [1]

000 metric tons

	1971	1973	1978	1998	1999	2000	2001	2002
Algeria	38 000	51 118	58 021	61 217	63 231	66 787	66 124	67 627
Angola	5 721	8 154	6 469	36 382	36 853	36 825	36 469	44 266
Benin	..	..	..	51	44	37	39	39
Cameroon	..	..	505	7 751	6 849	7 346	6 994	6 267
Congo	15	2 054	2 450	12 776	13 714	13 660	12 686	12 041
Democratic Republic of Congo	..	..	1 145	1 155	1 190	1 207	1 159	1 111
Côte d'Ivoire	..	..	..	521	484	447	410	634
Egypt	14 962	8 504	25 193	42 962	41 367	37 992	37 345	37 052
Gabon	5 785	7 598	10 355	17 792	15 874	13 797	13 441	12 121
Ghana	..	..	..	..	..	..	..	..
Libya	132 535	106 172	97 061	71 916	64 342	67 054	67 355	65 050
Morocco	23	42	24	12	11	12	9	9
Nigeria	75 524	101 412	93 550	109 043	104 563	111 206	116 665	110 667
South Africa	..	..	..	290	341	271	250	225
Sudan	..	..	..	329	3 450	8 857	10 394	11 754
Tunisia	4 109	3 878	4 944	3 978	4 015	3 727	3 386	3 826
Other	..	..	..	4 101	4 822	5 271	5 547	5 547
Africa	**276 674**	**288 932**	**299 717**	**370 276**	**361 150**	**374 496**	**378 273**	**378 236**
Bahrain	3 739	3 411	2 758	2 265	2 251	2 246	2 229	2 209
Islamic Republic of Iran	226 946	293 159	263 734	182 755	176 151	185 988	186 149	175 708
Iraq	83 266	99 542	127 080	104 536	124 684	126 950	116 771	99 776
Israel	5 738	6 005	24	15	12	4	4	4
Jordan	..	..	..	2	2	2	2	2
Kuwait	152 396	150 565	107 890	104 694	95 571	101 894	99 398	92 711
Oman	14 906	14 906	16 021	45 768	47 517	50 277	50 191	47 135
Qatar	20 645	27 502	23 382	32 261	31 686	33 570	35 935	34 553
Saudi Arabia	238 703	380 179	424 553	441 317	409 880	436 917	423 170	408 546
Syria	5 289	5 543	9 924	28 500	28 655	27 252	28 959	27 938
United Arab Emirates	50 892	73 617	87 996	113 214	103 361	109 713	106 661	98 731
Yemen	..	..	..	18 980	19 641	21 396	22 020	22 593
Middle East	**802 520**	**1054 429**	**1063 362**	**1074 307**	**1039 411**	**1096 209**	**1071 489**	**1009 906**
Bangladesh	..	5	6	59	71	95	93	93
Brunei	6 403	11 369	11 796	7 826	9 036	9 628	10 084	9 672
India	7 299	7 189	11 633	34 532	34 431	34 374	33 970	34 918
Indonesia	44 071	66 112	79 272	77 280	74 128	70 319	66 482	61 241
Malaysia	3 276	4 340	11 301	38 825	35 979	34 027	35 927	36 629
Myanmar	872	981	1 353	388	442	420	425	522
Pakistan	415	419	479	2 872	2 909	2 997	3 479	3 303
Philippines	..	..	..	40	46	53	65	156
Chinese Taipei	111	148	218	50	43	33	36	36
Thailand	13	6	11	3 492	3 788	5 095	5 252	5 440
Vietnam	..	..	..	12 400	15 045	15 423	16 732	16 683
Other	1	2	1	3 900	3 998	4 163	4 133	3 789
Asia	**62 461**	**90 571**	**116 070**	**181 664**	**179 916**	**176 627**	**176 678**	**172 482**
China, People's Republic	**39 410**	**53 610**	**104 050**	**161 203**	**160 169**	**163 169**	**164 128**	**168 753**
Non-OECD	**1819 435**	**2182 827**	**2357 710**	**2510 861**	**2463 097**	**2564 282**	**2574 160**	**2540 220**
World	**2480 566**	**2860 471**	**3098 728**	**3530 264**	**3456 769**	**3578 109**	**3584 672**	**3554 289**

1. Includes additives and other hydrocarbons (other than crude oil and NGL) / *Comprend les additifs et les autres hydrocarbures (autres que le pétrole et les LGN).*

Source: Annual Oil Statistics, Energy Statistics of Non-OECD Countries.

Table 15: World Refinery Output
Tableau 15 : Production mondiale des raffineries

000 metric tons

	1971	1973	1978	1998	1999	2000	2001	2002
Canada	68 786	82 322	88 487	90 672	91 583	93 712	96 413	98 193
Mexico	22 209	25 858	40 308	66 816	67 718	64 176	63 886	66 153
United States	598 453	665 486	788 157	808 363	807 518	822 834	819 874	816 056
OECD North America	**689 448**	**773 666**	**916 952**	**965 851**	**966 819**	**980 722**	**980 173**	**980 402**
Austria	7 407	8 817	9 874	9 827	9 413	8 768	9 352	9 340
Belgium	28 475	35 510	33 545	37 944	35 773	38 092	39 863	45 984
Czech Republic	6 115	7 586	9 450	7 090	6 247	6 131	6 309	6 376
Denmark	10 264	9 760	8 020	7 800	8 110	8 231	7 945	7 598
Finland	8 636	9 045	11 104	13 171	12 893	12 777	11 812	12 998
France	106 371	133 792	118 583	93 516	85 283	88 576	88 832	82 927
Germany	123 737	139 508	126 584	117 155	115 017	115 973	114 292	114 085
Greece	5 309	12 423	11 445	21 026	18 894	22 232	21 483	21 319
Hungary	6 509	7 980	10 900	7 992	7 659	7 483	7 452	6 893
Ireland	2 812	2 675	2 236	3 163	2 831	3 286	3 365	3 191
Italy	118 722	130 307	113 861	99 694	94 074	94 771	96 409	97 480
Netherlands	60 850	73 081	60 530	80 051	76 263	80 235	81 306	78 714
Norway	5 695	6 112	8 237	14 799	15 335	15 196	13 984	13 104
Poland	8 328	10 814	16 041	16 682	17 811	18 480	18 411	16 886
Portugal	4 036	4 222	6 471	14 019	13 482	12 308	13 061	12 881
Slovak Republic	5 361	6 043	7 411	5 686	5 638	5 810	6 231	6 297
Spain	34 528	42 423	48 265	61 657	60 841	59 830	57 806	57 931
Sweden	11 500	10 542	15 288	21 001	21 160	21 368	20 446	18 464
Switzerland	5 248	6 126	4 236	5 061	5 100	4 647	4 908	4 904
Turkey	8 589	12 533	13 053	27 346	25 653	23 745	25 374	26 088
United Kingdom	104 425	113 004	95 579	93 564	86 734	86 341	82 063	84 711
OECD Europe	**672 917**	**782 303**	**730 713**	**758 244**	**724 211**	**734 280**	**730 704**	**728 171**
Australia	21 847	25 251	30 065	37 144	36 399	36 708	35 403	34 319
Japan	179 039	222 333	213 895	212 110	208 255	206 162	200 406	198 477
Korea	11 497	14 989	24 214	113 963	120 349	122 001	119 210	111 998
New Zealand	2 910	3 265	2 954	5 145	4 877	5 060	4 981	5 298
OECD Pacific	**215 293**	**265 838**	**271 128**	**368 362**	**369 880**	**369 931**	**360 000**	**350 092**
OECD Total	**1577 658**	**1821 807**	**1918 793**	**2092 457**	**2060 910**	**2084 933**	**2070 877**	**2058 665**
Argentina	23 096	23 491	24 215	27 257	28 232	26 430	25 965	..
Bolivia	668	750	1 276	1 575	1 607	1 549	1 515	..
Brazil	26 309	36 920	52 448	77 502	80 182	81 265	83 887	..
Chile	4 655	4 628	4 669	9 101	9 079	9 138	9 576	..
Colombia	7 679	8 121	7 420	14 481	15 068	15 128	15 841	..
Cuba	4 402	5 436	6 565	1 413	1 153	2 320	2 445	..
Ecuador	1 350	1 570	4 094	6 954	6 339	7 691	7 791	..
Netherlands Antilles	40 927	43 644	26 931	12 842	11 448	11 808	12 916	..
Panama	3 720	3 323	2 355	2 330	2 370	2 131	1 941	..
Peru	4 261	4 729	5 940	8 260	7 463	7 454	7 846	..
Trinidad and Tobago	20 224	19 583	11 765	7 070	7 072	7 425	7 376	..
Uruguay	1 675	1 654	1 822	1 826	1 578	1 906	1 710	..
Venezuela	66 350	69 927	51 715	54 681	52 198	53 472	54 005	..
Other	14 371	19 168	14 229	5 660	5 759	6 084	6 042	..
Latin America	**219 687**	**242 944**	**215 444**	**230 952**	**229 548**	**233 801**	**238 856**	**..**
Albania	1 477	1 646	2 145	356	317	317	311	..
Bulgaria	7 742	9 359	11 851	5 565	5 673	5 181	5 486	..
Romania	16 178	17 986	25 927	13 025	10 458	10 990	12 152	..
Croatia	..	..	..	5 121	5 481	5 214	4 919	..
Former Yugo. Rep. Macedonia	..	..	..	914	729	936	755	..
Serbia and Montenegro	..	..	..	2 426	1 137	1 001	2 395	..
Other Former Yugoslavia	..	..	..	253	284	184	6	..
Former Yugoslavia	8 007	9 029	14 301	8 714	7 631	7 335	8 075	..
Other	..	663	431	1 076	1 176	1 172	1 151	..
Non-OECD Europe	**33 404**	**38 683**	**54 655**	**28 736**	**25 255**	**24 995**	**27 175**	**..**

Source: Annual Oil Statistics, Energy Statistics of Non-OECD Countries.

Table 15: World Refinery Output (continued)
Tableau 15 : Production mondiale des raffineries (suite)

000 metric tons

	1971	1973	1978	1998	1999	2000	2001	2002
Azerbaijan	..	..	..	8 214	7 598	8 367	6 087	..
Belarus	..	..	..	10 907	10 664	11 801	11 513	..
Kazakhstan	..	..	..	8 807	6 819	7 145	9 303	..
Lithuania	..	..	..	6 761	4 527	4 903	6 732	..
Russia	..	..	..	159 749	163 486	173 803	178 073	..
Turkmenistan	..	..	..	5 956	6 092	6 113	6 929	..
Ukraine	..	..	..	13 448	11 237	9 043	16 419	..
Uzbekistan	..	..	..	7 244	7 133	6 826	6 606	..
Other	..	..	..	175	236	178	159	..
Former USSR	**289 600**	**332 100**	**420 690**	**221 261**	**217 792**	**228 179**	**241 821**	**..**
Algeria	2 416	6 127	5 441	18 860	20 390	20 033	20 870	..
Angola	661	725	991	1 743	1 977	1 800	1 869	..
Cameroon	..	..	..	1 635	1 364	1 543	1 490	..
Congo	..	..	..	..	..	379	469	..
Côte d'Ivoire	808	1 154	1 618	3 030	2 652	2 952	3 133	..
Egypt	5 003	6 851	11 753	28 790	28 060	26 233	26 598	..
Gabon	1 002	1 058	1 688	764	791	767	606	..
Ghana	852	971	1 138	725	1 307	1 062	1 103	..
Kenya	2 522	2 638	2 562	1 698	1 677	1 986	1 682	..
Libya	435	1 608	5 217	13 952	15 284	14 848	15 068	..
Morocco	1 485	2 224	2 969	5 870	6 990	6 579	6 682	..
Nigeria	1 949	2 762	2 633	7 314	7 550	4 654	10 484	..
South Africa	12 668	13 106	13 147	23 353	25 162	25 307	25 452	..
Sudan	682	1 145	980	580	417	1 984	2 798	..
Tunisia	1 117	1 036	1 181	1 844	1 906	1 941	1 867	..
Other	4 306	4 786	4 243	3 333	3 057	2 471	2 898	..
Africa	**35 906**	**46 191**	**55 561**	**113 491**	**118 584**	**114 539**	**123 069**	**..**
Bahrain	12 640	12 042	12 540	12 384	13 029	12 720	11 886	..
Islamic Republic of Iran	27 132	28 831	34 491	72 476	76 524	76 138	65 754	..
Iraq	3 578	4 035	9 015	22 220	21 434	21 827	23 576	..
Israel	5 207	6 113	7 392	11 307	10 532	10 566	10 575	..
Jordan	554	671	1 370	3 454	3 462	3 798	3 732	..
Kuwait	18 241	19 252	19 353	46 941	47 128	36 815	39 027	..
Oman	..	..	..	3 902	3 911	3 939	3 285	..
Qatar	21	20	304	2 842	2 803	2 565	2 345	..
Saudi Arabia	27 514	28 244	30 804	77 991	76 493	79 133	82 589	..
Syria	2 112	1 963	4 090	12 496	12 416	12 210	12 677	..
United Arab Emirates	..	..	..	22 110	18 294	19 609	21 438	..
Yemen	3 505	2 852	1 845	4 333	4 375	4 439	4 492	..
Other	2 034	2 420	1 829	..	..	..	..	..
Middle East	**102 538**	**106 443**	**123 033**	**292 456**	**290 401**	**283 759**	**281 376**	**..**
Bangladesh	732	609	1 037	1 271	1 033	1 370	1 354	..
Brunei	..	..	..	474	489	466	476	..
India	19 867	20 724	25 671	68 256	81 066	102 511	107 152	..
Indonesia	8 989	10 130	13 923	44 729	45 739	47 021	49 086	..
Malaysia	3 746	3 858	5 650	16 414	17 345	20 474	22 269	..
Pakistan	3 291	3 360	3 965	6 390	6 611	8 822	9 353	..
Philippines	8 406	8 645	9 443	16 667	16 157	15 195	14 147	..
Singapore	16 489	22 799	27 101	48 340	40 494	35 225	33 378	..
Sri Lanka	1 495	1 711	1 439	2 090	1 807	2 209	1 977	..
Chinese Taipei	5 477	9 004	17 103	34 761	41 933	46 390	50 422	..
Thailand	4 889	6 956	7 906	32 579	33 128	32 625	33 059	..
Other	1 262	974	2 392	1 480	1 269	1 315	1 475	..
Asia	**74 643**	**88 770**	**115 630**	**273 451**	**290 960**	**313 626**	**324 148**	**..**
China, People's Republic	**32 300**	**41 600**	**67 161**	**162 665**	**175 868**	**195 835**	**197 808**	**..**
Non-OECD	**788 078**	**896 731**	**1052 174**	**1323 012**	**1348 408**	**1394 734**	**1434 253**	**..**
World	**2365 736**	**2718 538**	**2970 967**	**3415 469**	**3409 318**	**3479 667**	**3505 130**	**..**

Source: Annual Oil Statistics, Energy Statistics of Non-OECD Countries.

Table 16: Crude Oil Tanker Ports
Tableau 16 : Ports pétroliers

Country/Port		Terminal	Operator(s)	(m) LOA	(m) Beam	(m) Draft	Max Dwt	Load T/Hr	Dis T/Hr	Refinery Name
ALGERIA										
Arzew	L	El Djedid B3 Sea Terminal	Naftal			20.7	250000	5500		Sonatrach
Bejaia	L	Petroleum Port Pier No 3	Naftal	260		12.8	90000	4000		-
Skikda	L	New Port Berth P3	Sonatrach	280		14	100000	5150		Sonatrach
ANGOLA										
Essungo (Quinfuquena)	L	Essungo SPM	Texaco Panama Inc, Angola				150000	700		-
Girassol	L	"Girassol" FPSO	TotalFinaElf			None	150000			-
Kiabo Marine Terminal	L	Kiabo Offshore Field	Sonangol P & P	280.4		None	150000	1900		-
Kiame Marine Terminal	L	"Petroleo Nautipa" FPSO	Ranger Oil Angola Ltd			None	150000	4500		-
Kuito Marine Terminal	L	"Kuito" FPSO	Cabinda Gulf Oil Co Ltd	None	None	None	320000	4750		-
Lombo Terminal	L	"Lombo Este" FPSO	Texaco Panama Inc, Angola	150	None	None	150000	4500		-
Luanda	B	Off-shore Terminal SBM		230	None	12.19	80000	3400		Fina Pet. de Angola
Malongo (Cabinda)	L	Export Berth No 2	Cabinda Gulf Oil Co Ltd	350	None	None	325000	5450		-
Namibe (Mocamedes)	L	Porto Saco				16.45	165000			-
Palanca	L	Palanca SPM	TotalFinaElf Explor. Angola	None	None	None	300000	5150		-
Takula Terminal	L	Takula Export SPM No 2	Cabinda Gulf Oil Co Ltd			None	300000	1700		-
ANTIGUA										
St. John's	D	Sea Buoy CBM				13.1	80000			-
ARGENTINA										
Bahia Blanca	B	Punta Ciguena SPM	Ebytem S.A./Oil Tanking S.A.				96000		2700	-
Buenos Aires	D			None		None	90000			Shell Cia. Argentina
Caleta Cordova	L	TERMAP Monobuoy	Ter. Maritimas Patagonicas S	None	None		150000	4300		-
Caleta Olivia	L	TERMAP Monobuoy	Ter. Maritimas Patagonicas S	None	None		150000	4300		-
Comodoro Rivadavia	L	SBM		None	None	99.99	280000			-
Rio Cullen	L	SPM	Total Austral				130000			-
Rio Gallos (Pta Loyola)	L	Punta Loyola					60000			
ARUBA										
San Nicolas Bay	B	Reef Berth No 2	Coastal Aruba Refining Co	426.7	None	30.48	550000			Coastal, Aruba
AUSTRALIA										
Airlie Island	L	CBM	Novus West Australia Pty Ltd	300	50	13.8	120000	1800		-
Barrow Island Terminal	L	Barrow Island Marine Ter.	Chevron Australia Pty Ltd		47	13.2	80000	1800		-
Botany Bay (Kurnell)	D	Kurnell No 3	Australian Petroleum Pty Ltd	254		11.6	90000			
Brisbane	D	BP Crude Wharf, Lugg. P.	BP Oil Australia				90000	3000		BP Amoco
Brisbane	B	Caltex Crude Berth	Caltex Refineries	260		12.2	90000	3000	3500	Caltex, Lytton
Challis	L	"Challis Venture" FPSO	BHP Petroleum Ltd				150000	4300		-
Elang-Kakatua Field	L	"Modec Venture 1" FPSO	Modec Inc.				150000	3700		-
Fremantle (Kwinawa)	D	BP Ref. Kwinawa Tanker	BP Refin. (Kwinawa) Pty. Ltd	229		12.65	83000		6000	BP Amoco
Geelong	D	Refin. Pier Berth Nos 3&4	Shell Australia Ltd	265		11.6	90000			Shell
Griffin	L	"Griffin Venture" FPSO	BHP Petroleum Pty Ltd				150000	3450		-
Hastings (Westernport)	L	Long Island Pt Jty Bth No 1	Esso (Australia) Ltd	300		13	100000	7250		-
Jabiru	L	"Jabiru Venture" FPSO	Newfield Exploration Austr. Ltd				150000	2900		-
Laminaria-Corallina Ter.	L	"Northern Endeavour"	Woodside Petroleum Ltd				150000	4800		-
Melbourne (Yarraville)	D	Holden Dock Mobil Oil Jty	Mobil Oil Australia	185			81000			
Port Bonython	L		Santos Ltd	265		15.8	110000	5000		-
Port Stanvac	D	Marine Terminal SBM	Mobil Refin. Australia Pty Ltd	345	None		310000		6800	ExxonMobil
Port Walcott	D	Cape Lambert Service Jty	Robe River Iron Associates	180	32	16	120000			-
Saladin (Thevenard Is)	L	Marine Terminal CBM	ChevronTexaco Austr. Pty Ltd		47	16	150000	2850		-
Skua	L		BHP Petroleum Ltd				150000	3000		-
Stag Marine Terminal	L	CALM Buoy & FSO	Apache Energy Ltd				150000	2950		-
Sydney	D	No 1 Gore Bay	Shell Co of Australia Ltd	265		13.7	136000		4000	Shell
Varanus Island	L	Loading Terminal	Apache Energy Ltd	300	46	17	140000	4250		-
Wanaea/Cossack Field	L	"Cossack Pioneer" FPSO	Woodside Pet. (W.A.Oil) Pty				150000	2950		-
Wandoo Marine Ter.	L	Wandoo C.A.L.M.	Wandoo Production Alliance	None	None	50	100000	1700		-
Withnell Bay (Dampier)	L	Withnell Bay LPG/Con. Bth	Woodside Offshore Pet.				95000			-
Yarra Point	D	SBM					120000			-

L = Load, **D** = Discharge, **B** = Both; **NL** = No Limit

Source: SSY Consultancy & Research Ltd

Table 16: Crude Oil Tanker Ports (continued)
Tableau 16 : Ports pétroliers (suite)

Country/Port	Terminal	Operator(s)	(m) LOA	(m) Beam	(m) Draft	Max Dwt	Load T/Hr	Dis T/Hr	Refinery Name
BAHAMAS									
Freeport	B Tanker Berth No 10	Bahamas Oil Refining Corp			27.74	500000			-
South Riding Point	B Berth No 1 ("Outer Berth")	South Riding Point Holding Ltd			30	500000			-
BANGLADESH									
Chittagong	D Kutubdia anchorage					120000			Eastern Refinery
BELGIUM									
Antwerp	D Nafta B 641 North				15.24	110000			Various
Ghent	B Rodenhuizedok jetties	Texaco/Oiltanking Cy.	256	34	12.25	70000		2000	-
BENIN									
Seme Terminal	L CBM	Seme Oilfield Project	270		14	80000	1550		-
BRAZIL									
Angra dos Reis (Gebig)	B Gebig Outer Berth (P-1)	Petroleo Brasiliero S.A.	None	None	25	500000	11050	30900	-
Aratu (Madre de Deus)	B TEMADRE Berth PP4	Petroleo Brasiliero S.A.	275		12.8	85000			Petrobras, Mataripc
Carmopolis (Aracaju)	L DTBASA/SE seaberth	Petroleo Brasiliero S.A.			15	65000	2600		-
Guamare	L DTNEST II/SEGUAM Offs.	Petroleo Brasiliero S.A.			14	116000			-
Rio de Janeiro	L DTSE-GEGUD Terminal	Petroleo Brasiliero S.A.	None		15.8	135000			P'bras, D de Caxias
Santos	D Alemoa Terminal					60000		2000	-
Sao Francisco do Sul	D DTSUL SBM	Petroleo Brasiliero S.A.	None		18	200000		0	P'bras, Araucaria
Sao Sebastiao	D TEBAR S. Pier Bth PP-1	Petroleo Brasiliero S.A.			23	300000		7700	-
Tramandai	D TEDUT SPM No 2	Petroleo Brasiliero S.A.	None		19	200000		6500	Petrobras, Canoas
BRUNEI									
Seria Oil Terminal	L Kuala Belait SBM 1	Brunei Shell Pet. Co Ltd			17.4	210000	3700		Brunei Shell
BULGARIA									
Bourgas	D Druijba Oil Port Berth 2		260		12.65	75000		5150	Neftokhim
CAMEROUN									
Ebome Marine Terminal	L Ebome Marine Terminal	Kelt Cameroun				230000	2700		-
Kole Terminal	L KLB 2	TotalFinaElf E&P Cameroun	None	None	22	250000	4400		-
Limboh Terminal	L Refinery SBM	Ste Nationale de Raffinerie	243.5			90000			Sonara Refinery
Moudi Terminal	L Moudi Terminal SPM	Kelt Cameroun	330	None	27.45	280000	3500		-
CANADA									
Canaport (St John NB)	D Canaport SBM	Canaport Ltd	None	None		350000		8200	Irving Oil
Come-by-Chance	D Oil Refinery Term Berth 1	N. Atlantic Refining Co Ltd	457.2		27.45	325000		3400	North Atlantic Refin
Halifax N.S.	D Imp. Oil South Dartmouth	Imperial Oil Company	282		15.25	135000			Imperial Oil
Hibernia	L Hibernia Plat. Off. Load Sys	Hibernia Management & Dev.				100000			-
Point Tupper	B Statian Terminal Berth 1	Statia Ter.s Pt Tupper Inc	379.45		25.6	350000		90000	Tosco
Quebec (St Romuald)	D South Shore Berth 87	Ultramar Diamond Shamrock	None	None	14.32	160000			Ultramar D. Shamrc
Terra Nova Field	L "Terra Nova" FPSO					130000			-
Vancouver	L Trans. Pipeline Marine Ter.	Trans. Oil Pipeline Co Ltd	250			90000	3550		-
Whiffen Head Marine T.	B Newfoundland Trs.Ltd Bth1	Newfoundland Transship. Ltc	280		17.2	150000	6800	10900	-
CANARY ISLANDS									
Santa Cruz de Tenerife	D Floating Pipeline Terminal	CEPSA				240000			CEPSA
CHILE									
Arica	L Bolivian Export Buoy	Y.P.F. Bolivia			13.1	60000	1200		-
Chanaral	D Barquito Oil Term. Buoys	Codelco	250		15.24	90000		400	-
Gregorio (Punta Arenas)	L		250		14.9	90000	1450		ENAP
Quintero	D Quintero SBM	ENAP	342		25.5	200000		8000	R.P.C., Concon
San Vicente	D Oil Terminal B	Petrox S.A.	250		12.8	85000		1300	Petrox, Talcahuano
Tocopilla	D Central Termoelectr. Pier		234	32.4	13.5	90000			-
CHINA, PEOPLE'S REPUBLIC									
BZ 34 Terminal	L BZ 34 SBM	Japan China Oil Dev. Corp				110000	2575		-
Dalian	B Nianyuwan Oil Terminal		360	None	15	140000	10000		CNPC, Dalian
Huangpu (Whampoa)	D		None	None	13	90000		1800	-
Huizhou	D Huizhou GPC Crude Term.	Huade Petro-Chem. Co Ltd	294		15.2	150000		6850	-
Huizhou Offshore Term.	L "Nan Hai Fa Xian" FPSO	C.A.C.T. Operators Group				160000	4600		-
Jinzhou	D Oil Berth 2	Jinzhou Port Authority				80000		350	-
Liu Hua Terminal	L "Nan Hai Sheng Li" FPSO	Amoco Orient Pet. Co.	265	50	None	140000	4500		-

L = Load, **D** = Discharge, **B** = Both; **NL** = No Limit

Source: SSY Consultancy & Research Ltd

Table 16: Crude Oil Tanker Ports (continued)
Tableau 16 : Ports pétroliers (suite)

Country/Port		Terminal	Operator(s)	(m) LOA	(m) Beam	(m) Draft	Max Dwt	Load T/Hr	Dis T/Hr	Refinery Name
CHINA, PEOPLE'S REPUBLIC ctd										
Lufeng Terminal	L	"Munin" FPSO	JHN Oil Operating Co				100000	3450		-
Ningbo (Beilun)	D	Beilun Term. Berth No 13	Sinochem	326	49.8	20	270000		4500	-
Ningbo (Daxie Island)	D	Daxie Oil Dock No 1	Ningbo Port Tug & Barge Co	485		22.5	250000			-
Qingdao (Tsingtao)	B	Huangdao Oil Storage	Qingdao Harbour Administ.	265	39	20.5	200000	3000	8000	-
Qinhuangdao	L	Oil tanker anchorage		230		12.5	70000	6000		-
Quanzhou	D	Xiaocuo Terminal				16	100000			-
Shanghai	D						85000			Sinochem
Shuidong (Maoming)	D	Shuidong SBM	Sinopec			19.6	250000		3900	Sinopec, Maoming
Tianjin/Xingang	B	Crude Oil Berth No 1				14.93	100000	11000		-
Xijiang	L	"Nan Hai Kai Tuo" FPSO	Phillips Pet. Intern'l. Corp Asia				120000	6150		-
Xingang	L	Crude Oil Berth No 1	Dalian Xingang Port Co	260	50	17	100000			-
Xiuyu (Meizhou)	D	Fujian Term. Oil Jty Ocean	Fujian Refinery		None	16	135000		3300	Fujian
Yantai	D	Transshipment operations					83000		1500	-
Zhangjiagang	D			240	32	12	80000			-
Zhanjiang	D	CNPC Jetty	China National Pet.Corp	240	32	11.5	100000		1750	CNPC, Maoming
Zhoushan (Ao Shan)	B	Ao Shan Oil Ter. Berth No1		346	None	20	250000		9000	-
COLOMBIA										
Covenas Offshore Term.	L	Transfer Loading Unit 2	Ocensa	320	None	17.3	160000	8200		-
Mamonal (Cartagena)	D	Nestor Pineda Terminal	Ecopetrol	260	38.1	10.97	85000	2000		Ecopetrol
Tumaco	L	Offshore loading buoys	Ecopetrol	250		30	100000	3250		-
CONGO										
N'kossa Terminal	L	"N'kossa 1" FPSO	Elf Congo			None	280000	11650		-
Pointe Noire (Djeno)	L	Djeno SBM	Elf Congo	None	None	16	280000	6800		Coraf
Yombo Oil Terminal	L	"Conkouati" FPSO	Walter Internat. Congo, Inc	280.15	53.6	15	155000	5000		-
CONGO, DEMOCRATIC REPUBLIC										
Moanda Term. (Banana)	L	Moanda SBM	Gulf Oil Company			15.2	100000	4100		SOCIR
COTE D'IVOIRE										
Abidjan	D	SBM (Swivel Buoy)	Ste Ivoirienne de Raffinage		None	21	250000		2600	Societe Ivoirienne
Lion Terminal	L	CALM Buoy	United Meridian Inter. Corp	220	35.2		68000	5600		-
CROATIA										
Omisalj	D	Krk Island (Janaf Terminal)		350		27	350000			-
Rijeka	D	INA Pet. Berth, Urinj					200000		4500	Ina Refinery
DENMARK										
Aabenraa (Ensted)	B	Sonderjyllandskajen Bth 1		250	None		90000	4300	4300	-
Fredericia	B	Berth 20		275	None	13	130000			Dansk Shell
Gulfhavn	D	Kuwait Refinery Berth 1	Kuwait Pet. (Danmark) Refin.	280	None	15.5	120000			K.P.C.
Kalundborg	D	Statoil A/S Oil Terminal	Statoil A/S	262		12.5	130000		6400	Dansk Statoil
Siri	L	Siri Offsh. Loading Term.					80000			-
Stignaesvaerkets	D	Oil Pier				15	110000		2500	-
DOMINICAN REPUBLIC										
Palenque (Rio Haina)	D	SBM	Refineria Dominicana S.A.	304.8		13.7	100000		5150	Ref. Dominicana
ECUADOR										
Balao Ter. (Esmeraldas)	L	SBM Buoys X & Y		274.3	39.6	15.2	105000	4350		Petroecuador
EGYPT										
Ain Sukhna	D	SBM Buoy S4	Arab Oil Pipeline Co (SUMED)			99.99	600000		12500	-
East Zeit Terminal	L		Esso Suez Inc	255.4		18.29	85000	3000		-
Marsa Al Hamra	L	WEPCO Monobuoy	WEPCO			11.58	100000	3700		-
Ras Budran	L		Suez Oil Co (SUCO)	344.4		27.4	250000	5000		-
Ras Gharib	L	New Pier No 3	EGPC/General Pet. Co	289.5		16.75	130000	3000		-
Ras Shukheir	L	GUPCO Oil Bth 1 - CBM	Gulf of Suez Pet. Co	350		20.4	260000	3750		-
Sidi Kerir	L	SPMs A1, A2 & A6	Arab Pet.Pipe. Co (SUMED)	None	None	22.85	350000	12000		-
Wadi Feiran	L	Sea Berth No 1	Belayim Pet. Co (Petrobel)	274.3		22.85	105000	2400		-
Zaafarana Terminal	L	"Al Zaafarana" FPSO	Egyptian General Pet. Co	259.07		17.07	150000	3000		-
Zeit Bay Terminal	L	Ras El Behar SBM	Suez Oil Co (SUCO)			25.9	240000	6000		-

L = Load, **D** = Discharge, **B** = Both; **NL** = No Limit

Source: SSY Consultancy & Research Ltd

Table 16: Crude Oil Tanker Ports (continued)
Tableau 16 : Ports pétroliers (suite)

Country/Port	Terminal	Operator(s)	(m) LOA	(m) Beam	(m) Draft	Max Dwt	Load T/Hr	Dis T/Hr	Refinery Name
EL SALVADOR									
Acajutla	D Coastal Acajutla CBM	Coastal Tech. Salvador	231	None	12.8	65000			Esso Carib & C.Am
EQUATORIAL GUINEA									
Ceiba Marine Terminal	L "Sendje Ceiba" FPSO	Triton Equatorial Guinea Inc			None	155000	6800		-
Punta Europa	L Punta Europa Marine Ter.	CMS Oil & Gas Inc	228.6	None	None	80000	2000		-
Zafiro Offshore Terminal	L "Zafiro Producer" FPSO	Mobil Oil Equat. Guinea Inc	None	None	None	310000	4750		-
ESTONIA									
Muuga (Tallinn)	L Berth Nos 9 & 10		280			145000			-
FALKLAND ISLANDS									
Port Stanley	D UK M.O.D. SPM	UK Ministry of Defence				100000			-
FINLAND									
Naantali	D Fortum Oil Harbour Berth 3	Fortum Oil & Gas Oy	250		13	70000			Fortum Oil&Gas Oy
Porvoo (Skoldvik)	D Berth 2	Fortum Oil & Gas Oy	330		15.3	150000		12800	Fortum Oil&Gas Oy
FRANCE									
Antifer	B Antifer West & East		500	63	29	700000		17500	ExxMob, Pt Jerome
Bordeaux	B Ambes No 511					105000			-
Donges (Nantes)	D Berth No 6	Raffineries Elf-France	350		16.1	140000			Elf
Dunkirk (West)	D App. Petrolier des Flandres	Total France	360	60	20.5	265000		23300	TotalFi/BP Amoco
Fos sur Mer	D Berth No 3		415	None	22.25	310000			ExxonMobil
Lavera	D					150000			BP Amoco
Le Havre	B C.I.M. Berth No 10	Cie Industr. Maritime (CIM)				250000		17500	ExxMob, Pt Jerome
GABON									
Cape Lopez	L Elf Gabon Oil Jetty	Elf Gabon	340	None	20.5	250000	10200		SGR
Gamba Terminal	L SBM	Shell Gabon	280	None	17.25	150000	3400		-
Lucina Marine Terminal	L Export SBM	Kelt Energie Gabon			None	130000	1950		-
M'bya Terminal	L CALM buoy				18	150000	2750		-
Oguendjo Terminal	L Oguendo SBM		None	None	18.3	170000	4100		-
Tchatamba Marine Term	L "Madiela" FSO	Marathon Pet. Gabon LDC				135900	3400		-
GEORGIA									
Batumi	L Buoy mooring		265	None	12	70000	1000		
Supsa Marine Terminal	L CALM Buoy	Georgian Pipeline Co (GPC)	290			165000	5000		-
GERMANY									
Brunsbuttel	D Elbehafen Oil Section		400	None	14.8	130000			-
Hamburg	D Kattwykhafen East Pier	Deutsche Shell AG	285	50		120000			-
Rostock	B Oil Harbour Berth No 3	Seehafen Rostock			13	85000	1400	350	PCK Schwedt
Stadersand	D Oil Terminal	Preussen Elektra Aktienge.	230	32	11.5	60000		1200	-
Wilhelmshaven	D N-W Oelleitung Berth No4	Nth-West Oelleitung GmbH	None	None	20	260000		13750	-
GIBRALTAR									
Gibraltar	B Ship-to-ship transfers					600000			-
GREECE									
Agioi Theodoroi (Pir.)	D Berth A	Motor Oil (Hellas) S.A. Ref.	None	None	21.95	400000		15000	Motor Oil Hellas
Aspropyrgos (Piraeus)	D New Pier (1 & 2)	Hellenic Aspropyrgos Ref.	220		11.8	60000			Hellenic
Megara (Piraeus)	D	Hellenic Aspropyrgos Ref.			12.19	75000			Hellenic
Pachi (Piraeus)	D Pachi Oil Terminal	Hellenic Aspropyrgos Ref.	381.9		29	600000		14400	Hellenic
Prinos (Kavala)	L Prinos Offshore Terminal	North Aegean Pet. Corp			22	280000	1700		-
Thessaloniki	D EKO Sea Berth	EKO	258		13.41	95000			EKO,Thessaloniki
GUAM									
Apra	D Foxtrot No 1 Pier	Shell Guam Inc			16.16	135000			-
INDIA									
Chennai (Madras)	D Bharathi Dock berth III		274.3	None	16.15	140000		3500	Chennai Refinery
Cochin (Ambalamugal)	D New Oil Terminal		250	32.9	11.58	63000		4000	Cochin Ref. Ltd
Haldia (Calcutta)	D Oil Jetty Nos 2 & 3		250	44		90000		1500	IOC
Jamnagar	D Reliance Jamnagar SPM N	Reliance Ports & Term.s Ltd	355		22.5	350000		10000	Reliance Oil, Jam.
Mumbai (Bombay)	B Jawahar Dweep Bth No 4	Mumbai Port Trust	300	None	14.33	100000		3000	Bharat+Hindustan
New Mangalore	D Berth No 10 (Tanker J.2)		275	None	14	105000			Mangalore Ref&Pet
Sikka (Bedi)	B Ship-to-ship transfers	Shipping Corp. of India Ltd				150000		12000	

L = Load, **D** = Discharge, **B** = Both; **NL** = No Limit

Source: SSY Consultancy & Research Ltd

Table 16: Crude Oil Tanker Ports (continued)
Tableau 16 : Ports pétroliers (suite)

Country/Port		Terminal	Operator(s)	(m) LOA	(m) Beam	(m) Draft	Max Dwt	Load T/Hr	Dis T/Hr	Refinery Name
INDIA ctd.										
Vadinar (Kandla)	D	Offsh. Oil Term. SBM 1&2	IOC	362.5	60	23	315000		10000	IOC, Koyali & Math.
Visakhapatnam	D	Offhore Tanker Terminal		270	42	15.3	100000			Hindustan Pet.
INDONESIA										
Anoa Marine Terminal	L	"Anoa Natuna" FPSO	Premier Oil Natuna Sea Ltd	293		None	160000	2700		-
Ardjuna Terminal	L	SBM No 4	Atlantic Richfield Indonesia Inc			30.2	200000	3400		-
Arun (Blang Lancang)	L	Arun Terminal SPM	Pertamina PKK				280000	7500		-
Balikpapan	B						130000			Pertamina
Balongan	L	SBM No 1	Pertamina	290			150000	1500		-
Bekapai Terminal	L			295			100000			-
Belida Marine Terminal	L	"Intan" FSO	Conoco Indonesia Inc			None	230000	4350		-
Bima Marine Terminal	L	SBM Nos 8 & 9	Atlantic Richfield Indonesia Inc				250000	2000		-
Camar Marine Terminal	L	Madura Ayu storage tanker	Enterprise Oil (Bawean) Ltd	260			106239	4100		-
Cilacap	D	Pertamina SPM	Pertamina				290000		4000	Pertamina
Cinta Terminal	L	No 1 Export SBM	Maxus Southeast Sumatra Inc.				175000	2350		-
Dumai	L	CPI Oil Wharf 2	Caltex Pacific Oil Co, Jakarta	315		17.4	150000	5400		Pertamina
Kakap Marine Terminal	L	CALM SPM	Marathon Pet. Indonesia Ltd				100000	3250		-
Kasim Marine Terminal	L		Pertamina-Petromer Trend	304.79	45.72	15.24	135000	5300		-
Kuala Beukah	L	5-Point CBM	Asamera Oil (Indonesia) Ltd				90000	800		-
Lalang Marine Terminal	L	Lasmo Riau storage barge	Kondur Petroleum S.A.			16.76	140000	5100		-
Lawe-Lawe Terminal	B	SBM	Union Oil Co of Indonesia	290			135000	2700		-
Madura	L		Kodeco Energy Co Ltd	250			105000	4000		-
Pagerungan Marine Ter.	L	Pagerungan SBM	ARCO Indonesia				125000	900		-
Pangkalan Susu	L	SBM	Pertamina	275			100000	1700		-
Poleng	L	Poleng SBM		None		None	600000	2000		-
Salawati	L		Pertamina			13.11	60000	2700		-
Santan Terminal	L	Sea Terminal SBM No 1	Unocal Indonesia Company	275		18	140000	3700		-
Senipah Terminal	L	Sea Terminal (SBM)	Total Indonesie	275			125000	2850		-
Sungei Pakning	D	"Ladinda" storage barge	Pertamina	220			61000		1000	Pertamina
Tarakan	L	CBM	Tesoro Indonesia Pet. Co	230			55000	550		-
Tuban Marine Terminal	L	"Cilacap" FSO	JOB Pertamina				140000	2700		-
Udang Natuna	L	SBM	Pertamina	260			93000	3400		-
Widuri Terminal	L	SPMs 1 & 2	Maxus Southeast Sumatra Inc			15.24	178000	6000		-
IRAN										
Bandar Abbas	B	Pullad Jetty - Crude Oil Jty		223.4		14	60000		1400	NIOC
Dorood Term. (Darius)	L		National Iranian Oil Co, Offs.				80000	5000		-
Kharg Island	L	Sea Island Berth No 15	National Iranian Oil Co	None	None	31	550000	30000		-
Lavan Island	L	Berth 1 (South Berth)	National Iranian Oil Co			19.5	200000	7100		-
Ras Bahregan (Norwuz)	L	Norwuz Terminal SBM	Ste Irano Italienne de Petroles			23	250000	6000		-
Sirri Island	L		National Iranian Oil Co	365		24	330000	12850		-
Soroosh Terminal	L	"Soreena" FSU	Iran Offshore Oil Co	289.5		14	70000	1600		-
IRAQ										
Mina al Bakr	L	P'frmA Bths 1 & 2; P'frmB E	Iraq National Oil Company	366	None	21	350000	19000		-
IRELAND										
Bantry Bay	B	Whiddy Is. Oil Term. SBM	Bantry Terminals Ltd (INPC)	340		22	320000			-
Whitegate (Cork)	D	Oil Berth No 1		259			95000		3000	INPC
ISRAEL										
Ashkelon	B	Berth Nos 3 & 4	Eilat-Ashkelon Pipeline Co.	350		25	250000	7600		-
Eilat	D	South Oil Jetty		None		27.43	500000			-
Haifa	D	Berth No 3 (Offshore)				15.24	110000		2000	Oil Refineries
ITALY										
Aquila Field	L	"Firenze" FPSO	ENI SpA & AGIP SpA			None	140000			-
Augusta (Priolo)	D	SELM-Priolo Jetty Berth 21	SELM			20.1	260000			Praoil, Priolo
Augusta (Sta Panagia)	D	Santa Panagia Bth No 6	ISAB	365.74		23.77	600000			Isab SpA
Augusta (Siracusa)	D	Esso Pier South Berth 11	Esso Augusta	296.86		16.5	140000			ExxMob, Siracusa
Falconara (Ancona)	D	API Sea Terminal	API Raffineria di Ancona SpA			29.9	400000		1100	A.P.I.

L = Load, **D** = Discharge, **B** = Both; **NL** = No Limit

Source: SSY Consultancy & Research Ltd

Table 16: Crude Oil Tanker Ports (continued)
Tableau 16 : Ports pétroliers (suite)

Country/Port	Terminal	Operator(s)	(m) LOA	(m) Beam	(m) Draft	Max Dwt	Load T/Hr	Dis T/Hr	Refinery Name
ITALY ctd									
Fiumicino	D SPM Berth R1	Raffineria di Roma SpA	None		15.55	150000		4700	Raffineria di Roma
Gaeta (Formia)	B VLCC Sea Berth		349		20.3	260000			-
Gela	D ANIC Sea Line Terminal	ANIC			12.8	80000			Praoil
Genoa	D Multedo Port SPM					500000		11000	-
La Spezia	D Arcola Petrolifera Buoys		230		12.6	80000		1000	Arcola Petrolifera
Livorno (Leghorn)	D Nuova Darsena Bth No 10	AGIP Petroli S.p.a.	300	None	11.4	150000		3000	AGIP
Milazzo	D AGIP Petroli Bth No 5 bis	AGIP Petroli Co				420000			AGIP/KPC
Palermo	D Esso Terminal		None		15	110000		650	-
Porto Torres	D ANIC Terminal Pier C		None		15	110000			-
Quiliano Term. (Savona)	D SARPOM Marine Terminal	Esso Italiana S.p.a.			21.4	260000		11150	SARPOM
Ravenna	D AGIP Rota. P'form - Ter. 4	AGIP	None		21.33	270000			
Sarroch (Cagliari)	D SARAS Island P'form No1	SARAS Refinery	360		20.73	300000		10000	SARAS
Taranto	D Mare Grande CBM	AGIP Raffinazione	350		20	250000			AGIP
Trieste	D S.I.O.T. Terminal Berth 4	Oleodotto Transalpino	328	None	16.5	215000		7700	-
Venice	D San Leonardo		300		13.79	85000			-
JAPAN									
Akita	D Dolphin Berth	Tohoku Electric Co Ltd	239	None	11.7	70000		7500	-
Aokata (Kamigoto)	D Sea Berth					200000			-
Atsumi	D Atsumi	Chubu Electric Power Co Inc	253		12.5	79000		7500	-
Chiba	D Keiyo Sea Berth	Keiyo Sea Berth Co Ltd	348.9	60	19.2	258100		10000	Cosmo, Idem. Kyok
Etajima (Kure)	D C.I. Oil Terminal	C.I. Oil Terminal Ltd	273		14	80000		800	-
Funakawa	D A.O.S. (Sea Berth)	Akita Oil Storage Co Ltd	300	52	16.5	180000			Japan Energy
Hakodate	D Imodco SBM	Cosmo Oil Company Ltd	260		14.5	84000			Cosmo Oil
Heianza (Kin Wan)	B Sea Berth No 1	Okinawa Term. Co Ltd	385	None	26	500000			Various
Himeji	D Sea Berth	Idemitsu Kosan Kaisha Ltd	346	70	19.2	259000		10000	Idemitsu, Hyogo
Iwakuni	D Koa Oil Crude Pier	Koa Oil Company Ltd	None		16.5	150000		6000	Koa (Yamaguchi)
Kanokawa	D MC Terminal Wharf No 3	MC Terminal Company Ltd	270	40	12.6	125000		2600	-
Kashima	D Sea Berth	Kashima Oil Company Ltd	340		19	200000		11000	Kashima Oil
Kawasaki	D Keihin Kawasaki Sea Berth	Showa Shell/Nisseki Mitsubi.	348.5		27	265000		10000	Showa Shell
Kiire	D No 4 Berth	Nisseki Mitsubi. Stag. Ter.	458		30.6	500000		25700	-
Kikuma	D Taiyo Oil Sea Berth	Taiyo Oil Company Ltd	341.5		19.56	280000		8600	Taiyo Oil, Ehime
Kobe	D Shell Sea Berth	Showa Shell Sekiyu Ltd	250		12.6	70000		2000	-
Kudamatsu	D Nippon Oil East Berth	Nippon Pet. Refining Co Ltd	300	None	19	178000		5000	Nippon Petroleum
Mizushima	D Mitsubishi No 6 Pier	Nisseki Mitsubishi	340		16	260000		12000	Nisseki Mitsubishi
Muroran	D Nippon Refining Berth J-1	Nippon Pet. Refining Co Ltd	321.45	58	15	258000		6900	Nippon Petroleum
Nagoya	D Isewan Sea Berth	Isewan Sea Berth Co Ltd	345	60	19.99	280000		15450	Idemitsu+Nipp Min.
Nakagusuku (Okinawa)	D SALM		350	None	21	270000		10200	Nansei Sekiyu
Namikata	D Namikata Terminal	Namikata Terminal Co Ltd	278		25	125000			-
Negishi (Yokohama)	D A-West Jetty (Dolphin)	Nippon Pet. Refining Co Ltd	362		17	160000		7500	Nippon Petroleum
Niigata	D Showa Shell Sealine Buoy	Showa Shell Sekiyu K.K.	270		17	130000		3000	-
Oita	D Kyushu Oil Jetty	Kyushu Oil Company Ltd	350		20.1	273000		8500	Kyushu Oil
Onahama (Iwaki City)	D Onahama Sekiyu S. Bth	Onahama Sekiyu K.K.	270	None	13.5	100000		7650	-
Osaka	D Japan Energy Berth					260000			Koa Oil
Owase	D Toho Oil Sea Berth	Toho Oil Company Ltd	350		19	210000		5950	Toho Oil
Sakai	D Cosmo Oil Berth (South)	Cosmo Oil Company Ltd	333	60	14.5	259999		8000	Cosmo Oil
Sakaide	D Cosmo Oil Co Pier No 1	Cosmo Oil Company Ltd			16.27	130000		8600	Cosmo Oil
Shimotsu (Wakayama)	D Jetty O-1	Tonen Corp., Wakayama Re	349		18.8	245000		13500	Tonen
Shiogama (Sendai)	D Tohoku Oil No 1 Dolphin	Tohoku Oil Company Ltd	339.5	60	15.45	150000			Tohoku Oil
Tokuyama	D Idemitsu Sea Berth	Idemitsu Kosan Co Ltd	340.6	60	19.5	274150		10000	Idemitsu
Tomakomai	D Idemitsu Hokkaido S. Bth	Idemitsu Kosan Co Ltd	None	None	20.5	250000			Idemitsu, Hokkaido
Toyama	D Nihonkai Sekiyu Sea Berth	Nihonkai Oil Company Ltd	340.8		20.78	265000		7500	Nihonkai
Ube	D Seibu Sekiyu Sea Berth	Seibu Sekiyu Co. Ltd	325		19.9	230000		8600	Seibu Oil, Yamagu.
Wakayama (Kainan)	D Kainan Pet. Jetty 9	Wakayama Pet. Ref. Co Ltd	246.9		12.23	77000		5800	Wakayama
Yokkaichi	D Showa Outer SBM	Showa Yokkaichi Sekiyu Co	None		19.95	275000		7700	Showa, Yokkaichi
Yokohama	D Keihin Sea Berth (KYS)	Ohgishima Sekiyu Kichi K.K.	350	60	19.7	200000		8000	Various

L = Load, **D** = Discharge, **B** = Both; **NL** = No Limit

Source: SSY Consultancy & Research Ltd

Table 16: Crude Oil Tanker Ports (continued)
Tableau 16 : Ports pétroliers (suite)

Country/Port		Terminal	Operator(s)	(m) LOA	(m) Beam	(m) Draft	Max Dwt	Load T/Hr	Dis T/Hr	Refinery Name
KENYA										
Mombasa	B	Kipevu Oil Terminal (KOT)		259		13.25	85000			Kenya Petroleum
KOREA, NORTH										
Senbong	D	SBM					250000			North Korean Gov.
KOREA, SOUTH										
Daesan	D	Hyundai Oil Co Ltd SBM	Hyundai Oil Refinery Co Ltd	None		20.5	300000		6000	Hyundai Oil
Inchon	D	Berth No 3		250		12.8	100000			Hanwha Energy
Onsan	D	Ssangyong SBM	Ssangyong Oil Co	None		26	315000			Ssangyong Oil
Ulsan	D	SK Corporation Ter. No 3	SK Corporation	None		22.6	325000			SK Corporation
Yosu (Kwangyang)	D	LG Caltex Crude Wharf 2	LG Caltex Oil Corporation	340		21	280000		10200	LG Caltex Oil
KUWAIT										
Mina Abdulla	L	Sea Island Berth 19		350		16.92	276000	1500		Kuwait Pet. Corp
Mina al Ahmadi	L	SPM CALM Buoys	Kuwait Oil Company (KSC)	None	None	27.43	550000	11000		Kuwait Pet. Corp
Mina Saud	L	Mina Saud Berth 2	Saudi Arabian Texaco Inc	366	None	15.85	150000	8150		Texaco
LATVIA										
Ventspils	L	Berths 32 & 33	Ventbunkers	270	45	15	130000	7000		
LEBANON										
Ras Selaata	D	Sela'ata Terminal	Societe Huiles & Derives S.A.L.			13	70000		200	-
Tripoli	L	Berth No 2	Tripoli Oil Installations Lebanon			18.29	250000	6000		Tripoli Oil
Zahrani Term. (Sidon)	L	Trans Arabian Pipeline Co	Zahrani Oil Installation Sidon Ter.			16.76	100000	7500		Zahrani Oil
LIBYA										
Bouri Terminal	L	"Sloug" FSO	AGIP Oil Ltd	283	44	None	150000	7700		-
Es Sider	L	No 5 Mono Mooring	Waha Oil Co	355.1	56.1	22.25	300000	6800		-
Marsa al Hariga(Tobruk)	L	Oil Jetty Berth No 1	Arabian Gulf Oil Co			17.07	150000	8000		-
Marsa el Brega	L	Crude Berth SPM No 5	Sirte Oil Co	None	None	None	300000	5250		Sirte Oil
Ras Lanuf	L	V.O.O. Berth No 3 (SPM)	Veba Oil Operations B.V.			22.86	300000			Ras Lanuf Oil&Gas
Zawia	D	SALM Berth No 3	National Oil Corp of Libya			19	140000			Azzawiya Oil Refin.
Zuetina	L	Chahett-Monobuoy		None		24.4	275000	4350		-
LITHUANIA										
Butinge Marine Terminal	B	Butinge SPM No 1	Mazeikiai Nafta Butinge Oil Ter.			16	150000	4300		Mazheikiai
MADAGASCAR										
Toamasina (Tamatave)	D	Crude Oil Berth		230	31	12.2	60000			Solima
MALAYSIA										
Bintulu Terminal	L	SBM Berths 1 & 2	Sarawak Shell Berhad	None		15.85	320000	2400		-
Dulang Marine Terminal	L	"Puteri Dulang" FPSO	Petronas Carigali Sdn. Bhd.	242			85000	800		-
Kerteh Terminal	L	TCOT SALM-2	Esso Production Malaysia Inc			20.4	250000	5500		-
Labuan	L	Shell SBM	Sarawak Shell Berhad	290		20.4	320000	5150		-
Malong Marine Terminal	L	"Perintis" FPSO	Trenergy FPSO Sdn Berhad	281			150000			-
Miri (Lutong Terminal)	L	Miri SBM No 5	Sarawak Shell Berhad	None		17.07	150000	2850		Sarawak Shell
Port Dickson	D	Esso/Shell SBM		275	None	15	120000		4500	ExxonMobil + Shell
Sungai Udang	D	Sungai Udang SBM	Sungai Udang Sdn. Bhd.	347.6		22	300000		8600	Petronas, Malacca
Tapis Terminal	L						100000			-
MARTINIQUE										
Fort de France	D			300		17.5	120000			S.A.R.A.
MAYANMAR										
Yetagun Marime Term.	L	"Yetagun" FSO	South East Shipping Co. Ltd				120000	3400		-
MEXICO										
Cayo Arcas Terminal	L	Pemex SBM No 2	Pemex			21	275000	6400	0	-
Coatzacoalcos(Pajaritos	L	Pajaritos Berths 7E/7W	Pemex	250	42	11.89	85000	5450		-
Dos Bocas	L	SBM No 1	Pemex	350	None	21.95	250000	5450		-
Salina Cruz	L	Offshore Terminal SPM 3	Pemex	350		21.3	250000	6800		Pemex
Tuxpan	L	Pemex SBM No 2				13.4	80000			-
MOROCCO										
Mohammedia	D	New Terminal Berth B		290		17	150000		5150	Samir

L = Load, **D** = Discharge, **B** = Both; **NL** = No Limit

Source: SSY Consultancy & Research Ltd

Table 16: Crude Oil Tanker Ports (continued)
Tableau 16 : Ports pétroliers (suite)

Country/Port	Terminal	Operator(s)	(m) LOA	(m) Beam	(m) Draft	Max Dwt	Load T/Hr	Dis T/Hr	Refinery Name
NETHERLANDS									
Amsterdam	D Oiltanking Amsterdam B.V.		325	42	13.7	100000		2000	
Borssele	D					100000			
Flushing (Vlissingen)	D Total Refinery Jetty		280	None	15	100000		3000	Total Fina
Rotterdam (Europoort)	B Maasvlakte Oil Term. 8th		364	60	22.55	350000			
Rotterdam (Europoort)	B Nerefco 6th Pet.havn				22.55	280000			Nerefco
Terneuzen	D Ocean Dock		280		12.34	100000			-
NETHERLANDS ANTILLES									
Bonaire Terminal	B Berth No 1	Bonaire Petroleum Corp N.V.	None	None	33.5	500000			-
Bullen Bay (Curacao)	B COT Tanker Jetty No 4	Refineria Isla Curacao S.A.	411.4	None	28.7	550000	12000	25000	RICSA, Emmastad
Caracas Bay	D Shell Curacao Berths		320		13.71	90000			RICSA, Emmastad
Willemstad/Emmastad	D Refineria Isla Berths 5 & 8	Refineria Isla Curacao S.A.	259	None	12.8	100000			RICSA, Emmastad
NEW ZEALAND									
Maui Field	L "Whakaaropai" FPSO	Shell Todd Oil Services Ltd				100000	2500		-
Whangerei	D Marsden Point Oil Bth No1	New Zealand Refining Co	304		15.24	145000			N Zealand Refining
NICARAGUA									
Puerto Sandino	D Sea Line Terminal		259		12.5	65000		1000	ExxMob, Managua
NIGERIA									
Antan Terminal	L "Knock Taggart" FPSO	Addax Pet. Dev. (Nigeria) Ltd			None	150000			-
Bonny Inshore Terminal	L Inshore Terminal	Shell Pet. Dev. Co of Nigeria	320	None		135000	5000		-
Bonny Offsh Terminal	L Offshore Terminal (SPM)	Shell Pet. Dev. Co of Nigeria	None	None	22.8	300000	6000		-
Brass Terminal	L Brass Term. SBMs 1 & 2	Nigerian AGIP Oil Co. Ltd	366	None	22.9	300000	7000		-
Escravos Terminal	L Escravos SBM Berth 3	Chevron Nigeria Ltd	365	61		300000	4100		-
Escravos Terminal	D Import facility					140000			-
Forcados (Offsh Term.)	L SBM 1	Shell Pet. Dev. Co of Nigeria	None	None	21.94	310000	9000		-
Ima Terminal	L "Ailsa Craig" FSO	Liberty Technical Services Ltd			18	150000	4100		-
Odudu Terminal	L "FSO XV Domy"	Elf Petroleum Nigeria Ltd			None	280000	8600		-
Oso Field	L	Mobil Producing Nigeria				140000			-
Pennington	L Offshore Terminal SBM-B	Texaco Overseas Pet. Co	None	None	24.4	250000	3000		-
Qua Iboe	L SBMs Nos 1, 2 & 3	Mobil Producing Nigeria	None	None	21.95	300000	8850		-
Ukpokiti Terminal	L "Independence" FPSO	Conoco Energy Nigeria Ltd				150000	5450		-
Yoho Terminal	L "FPSO Falcon"	Protrack Shipping Ltd			None	350000	6000		-
NORWAY									
Aasgard	L "Aasgard A" FSO					80000			-
Draugen	L Floating loading buoy	A/S Norske Shell				150000			-
Gullfaks	L Loading buoys	Statoil				140000			-
Kaarsto	L Product Quay No 2	Statoil	275		13	100000	5000		-
Mongstad	L Crude Oil Jetty No 1	Statoil Mongstad	350		23	300000	14500		Statoil
Mongstad	D Crude Oil Jetty No 7	Statoil Mongstad	270		19.7	180000			Statoil
Njord Field	L "Njord B" FSU	Norsk Hydro				90000			-
Norne	L Norne FPSO	Statoil				60000			-
Slagen	D Marine Terminal Berth 1E		350		20.1	190000		10000	ExxonMobil
Statfjord Terminal	L	Statoil/Mobil				150000			-
Stavanger (Sola)	D Shell Refinery	A/S Norsk Shell	335.26		15.85	240000	2200		Shell
Sture	L Crude Oil Term. Jetty No1	Norsk Hydro a.s.			23	300000	12850		-
Varg	L "Petrojarl Varg" FPSO	Golar Nor Offshore				80000			-
Yme Field	L "Polysaga" FSO					100000			-
OMAN									
Mina al Fahal	L SBM No 2	Pet. Dev. Oman L.L.C.	None	None	None	600000			Oman Refinery
PAKISTAN									
Karachi	D Berths OP-1 & OP-3	Karachi Port Trust	259.07	39.62	11.89	75000		1800	Pakistan + Nat Ref.
PANAMA									
Bahia Las Minas	D Refin. Panama SA North	Refineria Panama S.A.	243.83	30.47	11.9	60000			Refin. Panama SA
Charco Azul	B PTP Berth No 1	Petroterminal de Panama	335	None	None	265000	6550	8200	-
Chiriqui Grande	L Petroterminal SBMs 1 & 2	Petroterminal de Panama	294	49	21.9	150000	6540		-
Las Minas Bay	D Colon	Refineria Panama S.A.	245		12.2	60000			Refin. Panama SA

L = Load, **D** = Discharge, **B** = Both; **NL** = No Limit

Source: SSY Consultancy & Research Ltd

Table 16: Crude Oil Tanker Ports (continued)
Tableau 16 : Ports pétroliers (suite)

Country/Port		Terminal	Operator(s)	(m) LOA	(m) Beam	(m) Draft	Max Dwt	Load T/Hr	Dis T/Hr	Refinery Name
PAPUA NEW GUINEA										
Kumul Marine Terminal	L	Kutubu SPM	Chevron Niugini Pty. Ltd			17.1	150000	1000		-
PERU										
La Pampilla	D	Sea Berth No 2		283.45		14.32	70000			Petroperu
Puerto Bayovar	L		Petroperu/ENAPU	400	None	21.95	250000	10200		-
PHILIPPINES										
Bataan (Limay)	D	Bataan SBM	Petron Bataan Refinery	341.38	53.35	22.86	310000			Petron Corp
Batangas (Tabangao)	D	Shell Terminal Jetty No 4	Pilipinas Shell Pet. Corp	348		23	320000			Shell
Malampaya Terminal	L	"Stena Natalita" FPSO					90000			-
Subic Bay	B	POL Pier Southside	Coastal (Subic Bay) Ter. Inc	259.98		12.8	100000			-
POLAND										
Gdansk	B	North Port Liquid Fuel Ter.	Port Polnocny	300		15	150000	4150	4150	Pet. Ref. Gdansk
PORTUGAL										
Leixoes (Oporto)	D	Ter. Oceanico de Leixoes	Petroleos de Portugal E.P.		56.7	17	135000		10000	Petrogal
Lisbon	D	Cabo Ruivo Terminal	Soponata	None	None		65000			-
Sines	D	Berth No 2		350		22.5	350000			Petrogal
PUERTO RICO										
Guayanilla	D	Refinery dock	Caribbean Petroleum	252.97	99.99	12.8	80000			Caribbean Pet.
Yabucoa	D	Sun Refinery	Sun Oil Co			13.11	100000			Sun Oil
QATAR										
Al Rayyan	L	Al Rayyan SPM	ARCO Qatar Inc			21	270000	8200		-
Al Shaheen Terminal	L	Al Shaheen SPM No 1	Maersk Oil Qatar AS	None	None	None	370000	8350		-
Halul Island	L	SBM No 2	Qatar General Pet. Corp	420		29	550000	9500		-
Mesaieed (Umm Said)	L	North Berth SBM	Qatar General Pet. Corp	340	60		260000	7500		National Oil
Ras Laffan	L	Liquid 2B	Qatar General Pet. Corp	345		12.5	60000	4650		-
ROMANIA										
Constantza	B	Berth No 79		None	None	14	85000			-
RUSSIA										
Nakhodka	L	Oil Berth		230	40		60000			-
Novorossiysk	L	Sheskharis Oil Term. Bth1		490		19	250000	10100		Russian Gov.
Primorsk	L	Berth Nos 1 & 2		307	50	15	125000			
Sakhalin	L						90000			
Tuapse	L	Oil Harbour Berths 1 & 2		250		12	85000	2400		Russian Gov.
Vityaz Marine Terminal	L	"Okha" FSO					60000	6850		-
SAUDI ARABIA										
Al Muajjiz Terminal	L	Berth No 102		412	76.2	29.6	500000	16400		-
Juaymah Terminal	L	Crude SPMs	Saudi Aramco			35	750000	17800		-
Rabigh	D	Liquid Cargo Port	Petromin-Petrola Ref. Co	360		23.5	325000			Saudi Aramco
Ras Al Khafji	L	Berth No 4 SBM	Arabian Oil Co Ltd	342		19.5	323000	8500		Arabian Oil
Ras Tanura	L	Sea Islands	Saudi Aramco				450000	38400		Saudi Aramco
Yanbu Crude Oil Term.	L	ARAMCO Berth No 62	Saudi Aramco	525		28.95	500000	17800		Saudi Aramco
SENEGAL										
Mbao Terminal (Dakar)	D	Sea Line	Iransen + Shell			16	130000			Societe Africaine
SINGAPORE										
Jurong	D	Esso/Mobil SBM		366		20.4	330000			ExxonMobil
Pulau Ayer Chawan	D	Berth No 5		305			140000			ExxonMobil
Pulau Bukom	D	Main Strait SBM	Shell Singapore Pte. Ltd	457		23	360000			Shell
Pulau Merlimau	D	P.S.A. SBM (Selat Sinki)	TMS. Co Pte Ltd	500			320000			Sing. Ref. Co Pte Lt
SOUTH AFRICA										
Cape Town	D	No 1 Tanker Basin		256		13.1	85000			Caltex
Durban	D	Reunion SBM					300000			Shell-BP Sapref
Saldanha Bay	B	Ore/Oil Jetty		350		21.5	280000		10000	-
SPAIN										
Algeciras	D	CEPSA Monobuoy	CEPSA	None		None	500000			CEPSA, Cadiz
Bilbao	D	Petronor Refinery No 1 Jty		450		32	500000		16000	Petronor, Somorr.
Cartagena	D	Escombreras Berth E018	Repsol Petroleo S.A.	350		20.4	250000		3300	Repsol

L = Load, **D** = Discharge, **B** = Both; **NL** = No Limit

Source: SSY Consultancy & Research Ltd

Table 16: Crude Oil Tanker Ports (continued)
Tableau 16 : Ports pétroliers (suite)

Country/Port		Terminal	Operator(s)	(m) LOA	(m) Beam	(m) Draft	Max Dwt	Load T/Hr	Dis T/Hr	Refinery Name
SPAIN ctd.										
Castellon de la Plana	D	Sea Berth	BP Oil Espana S.A.	300		22	165000		7000	BP Oil Espana S.A.
Corunna	D	Repsol Jetty No 3	Repsol Petroleo S.A.	300	50	16.5	120000		5150	Repsol
Huelva	D	Rio Tinto Minera S.A.	Rio Tinto Minera S.A.	274.3		21.33	150000		2500	CEPSA
Malaga	D	Sealine Berth	Repsol Petroleo S.A.		None	25	250000		6000	Repsol, Puertollano
Tarragona	D	Repsol Monobuoy	Repsol Petroleo S.A.	350		42.2	250000		11000	Repsol
SRI LANKA										
Colombo	D	Dolphin Tanker Berth		210	32.2	11.8	60000		2500	Ceylon Petroleum
ST. EUSTATIUS ISLAND										
St. Eustatius	B	No 2 (North Berth)	Saudi Aramco			16.75	175000	4100	3400	
ST. LUCIA										
Cul-de-Sac	D	Dock No 2		396.2	None	25	400000			-
SUDAN										
Marsa Bashayer	L	Bashayer Crude Oil Export	Greater Nile Pet. Op. Co Ltd	368	58	20.3	300000	6800		-
SWEDEN										
Brofjorden	D	Crude Oil Jetty		None	None	25	500000			Skandinaviska
Gothenburg	D	Torshamnen Berth 801	Port of Gothenburg A.B.	350	None	19.05	225000			Nynas+Preem+She
Nynashamn	D	Berth No 1	AB Nynas Petroleum	300			130000			Nynas Petroleum
SYRIA										
Banias	B	Berth No 3	Syrian Crude Oil Transp. Co	289.5		16.15	120000			Banias Refining
Tartous	L	North Berth	Syrian Crude Oil Transp. Co	280	40	15.85	100000	4000		-
TAIPEI, CHINESE										
Kaohsiung	D	C.P.C. SBM No 3	Chinese Petroleum Corp				300000		7000	C.P.C., Kaohsiung
Keelung	D	Sha Lun SBM No 2	Chinese Petroleum Corp	None		20	250000		6000	C.P.C., Tao-Yuan
Mailiao	D	Berth W-2		310		19.54	150000			
TANZANIA										
Mjimwema (Dar es Sal.)	D	Monobuoy				16.75	100000			T.I.P.R.
THAILAND										
Benchamas Terminal	L	Benchamas Explorer FSO	Chevron Offsh (Thailand) Ltd			99.99	110000	3250		-
Erawan Terminal	L	Erewan Term. FSO	Unocal Thailand Ltd	243.8			130000	2700		-
Rayong (Map Ta Phut)	D	Star Pet. Refining Co SPM	Star Pet. Refining Co Ltd	345		20.4	280000		8150	Star Pet. + Rayong
Sriracha	D	Thai Oil SBM	Thai Oil Co Ltd				230000		7500	Thai Oil
Tantawan Terminal	L	Tantawan Explorer FPSO	Thaipo Ltd	256.8			110000	3250		-
TOGO										
Lome	D		Societe Togolaise de Stock.	250		13.5	95000		1100	Ref. du Togo
TRINIDAD & TOBAGO										
Galeota Point	L	Galeota SBM	Amoco Trinidad Oil Co				250000	5450		-
Point Fortin	B	Berth No 5 (CBM)	Pet. Co. of Trinidad & Togag	259.07		15.54	70000			Petrotrin
Pointe-a-Pierre	B	Berth No 6 (Sea Island)	Pet. Co. of Trinidad & Togag	289.5		15.54	120000			Petrotrin
TUNISIA										
Ashtart	L	SBM		None			150000	2550		-
Bizerte	D	Stir Ref. Oil Pier A North		240	None		90000			Ste. Tunis. Ind Raff
Didon Terminal	L	"Didon" FPSO	M. P. Zarat Ltd				70000	1350		-
La Skhirra	L	Terminal Platform		300		14.48	150000	7000		-
Tazerka	L		Shell	275	None	99.99	140000	7000		-
Zarzis	L						130000			-
TURKEY										
Aliaga (Izmir)	D	VLCC discharge jetty	TUPRAS AS	360		19.2	250000		15000	TUPRAS, Izmir
Ambarli (Istanbul)	D	TEAS/Petrol Ofisi offsh.	TEAS/Petrol Ofisi	None	None	13	250000			-
Ceyhan Terminal	L	BOTAS Berth Nos 1 & 2	BOTAS Pet. Pipeline Corp	355		23	300000			-
Dortyol	L	Dortyol BOTAS Terminal	BOTAS Pet. Pipeline Corp	250		13	60000	2500		-
Mersin	D	ATAS Refinery Terminal	ATAS	250	None	12	100000		1500	ATAS
Toros Gubre Terminal	L	Jetty No 2	Toros Gubre ve Kimya End. AS				100000			-
Tutunciftlik (Izmit)	D	TUPRAS Refinery Berth 5	TUPRAS	300	None	27	280000		7650	TUPRAS, Izmit

L = Load, **D** = Discharge, **B** = Both; **NL** = No Limit

Source: SSY Consultancy & Research Ltd

Table 16: Crude Oil Tanker Ports (continued)
Tableau 16 : Ports pétroliers (suite)

Country/Port		Terminal	Operator(s)	(m) LOA	(m) Beam	(m) Draft	Max Dwt	Load T/Hr	Dis T/Hr	Refinery Name
UKRAINE										
Odessa	L	Inner Harbour Berth No 2	Ukraine Ministry of Transport	250	40	12.2	90000	2000		Uktatnafta
Theodosia	L	North Oil Berth				12.5	60000	1500		
UNITED ARAB EMIRATES										
Abu al Bukhoosh	L	SPM Buoy		None		22.5	280000	8500		-
Al Ruwais	D	ADNOC Refinery Berth E	Abu Dhabi National Oil Co	360	57.5	17	160000			ADNOC, Ruwais
Arzanah Island	L	SBM	Zakum Development Co			13.1	280000	7500		-
Das Island	L	Tanker Berth 3 (SPM)	Abu Dhabi National Oil Co			24	400000	6000		-
Fateh Terminal	L	SBMs Nos 1 & 2	Dubai Petroleum Co	365.8		30.5	300000	9500		-
Fujairah	B	Vopak (ENOC) Fujairah T.	Vopak ENOC Fujairah Ltd	274	None	18	145000	2575	2575	-
Fujairah Off. Anchorage	B	Ship-to-ship transfers					600000			Metro
Hamriyah	L	SPM	Amoco Sharjah Oil Co			16.76	80000	3100		-
Hulaylah Terminal	L	SPM	Ras al Khaimah Offsh Pet Co				300000	1200		-
Jebel Dhanna	L	Sea Line Load Berth No 4	ADPPOC	377.7	58.8		260000	7750		ADNOC, Ruwais
Khor Fakkan	B	Ship-to-ship transfers					600000			-
Mubarek Terminal	L	"Mubaraka" FPSO	Crescent Petroleum Co				275000	2200		-
Mubarraz Island	L	Mubarraz Terminal SBM	Abu Dhabi Oil Co Ltd (Japan)				235000	4450		-
Ras al Khaimah	L	Salah Field					300000			-
Zirku Island	L	Oil Terminal SPMs 1 & 2	Zakum Development Co	None	67	21	350000			-
UNITED KINGDOM										
Alba Terminal	L	"Alba FSU"	Chevron UK Ltd				130000			-
Banff Field	L	"Ramform Banff" FPSO	Pet. Geo-Services UK Ltd				100000			-
Beryl Terminal	L	SPM	Mobil Oil Co				100000			-
Blenheim Field	L	"Petrojarl 1" FPSO	Pet. Geo-Services UK Ltd				67000			-
Captain	L	"Captain FPSO"	Texaco North Sea UK Ltd				100000			-
Coryton	D	BP Jetty No 4		366		14	100000		14000	-
Curlew Field	L	"Maersk Curlew" FPSO	MAS Prod. Co (Maersk)				100000			-
Fawley	D	Esso Marine Term. Berth 4	Esso Petroleum Co Ltd			14.9	110000		10000	ExxonMobil
Fife Field	L	"Uisge Gorm" FPSO	Amerada Hess Ltd				100000			-
Finnart Terminal	D	BP Oil Berth 3	BP Oil Ltd			22.86	300000			BP Amoco, Grang.
Flotta Terminal	L	SPM Nos 1 & 2	Talisman Energy (UK) Ltd	None	None	21.9	200000	10900		-
Foinaven Field	L	"Petrojarl Foinaven" FPSO	Pet. Geo-Services UK Ltd				100000			-
Gryphon Field	L	"Gryphon A" FPSO	Kerr-McGee				100000			Elf
Hamble	L	BP Terminal	BP Oil UK Ltd	260		13.1	110000	6000		-
Harding	L	Harding Offshore Load. Te	BP				120000			-
Hound Point	L	Berth No 1	BP Oil Ltd			20.24	300000	14000		-
Immingham Oil Terminal	D	Berth Nos 1 & 2	Humber Oil Ts. Trustee Ltd	335.25	None	12.8	140000			Lindsey,Humbersid
Kittiwake	L	Kittiwake Loading Buoy	Shell UK Explor. & Prod.				110000			-
Liverpool (Tranmere)	D	Tranmere Oil Ter. Nth St.	Shell UK Ltd			12.5	95000			Shell, Stanlow
Liverpool Bay	L	Offsh Oil Storage Ins. (OSI	BHP Billiton Petroleum Ltd			28	150000	3850		-
Maureen Field	L	SPM	Phillips Pet. Co UK Ltd				85000			-
Milford Haven	D	Elf Oil Jetty Berth No 1	Elf Oil (GB) Ltd	360			275000			Elf
Nigg Oil Terminal	L		BP Oil Ltd	290	45	15	155000	6800		-
Pembroke	D	Texaco Terminal Berth 6	Texaco Ltd				275000			Texaco
Pierce Field	L	"Haewene Brim" FPSO	Enterprise Oil				100000			-
Ross-Parry Field	L	"Bleo Holm" FPSO	Bluewater Offsh./Talisman	None	None	None	150000	4300		-
Sullom Voe	L	Jetty No 4	BP Explor. Operating Co Ltd	365		22.6	350000	16400		-
Sullom Voe	L	Sullom Voe Jetty No 3	BP Explor. Operating Co Ltd	365		24	270000	7290		-
Teal/West Guillemot	L	"Anasuria" FPSO	Shell UK Explor. & Prod. Ltd				100000			-
Teesport	L	Seal Sands Oil Term. No2	Phillips Pet. Co UK Ltd	305	45.7	17.1	150000			Phillips
Tetney Oil Terminal	D	SPM	Conoco Ltd	None	None	15.24	150000			Conoco
UNITED STATES										
Anacortes (WA)	D	Shell Outer Berth		289.6		13.7	125000			Equilon Enterprises
Atreco (Port Arthur) (TX)	D	Berth 1	Fina Oil & Chemical Co	274.3	42.67	12.19	125000		4750	Fina Oil & Chem.
Barber's Point (HW)	D	Tesoro Hawaii SPM	Tesoro Hawaii Corp	288	45.72	19.8	150000		9000	Tesoro
Barber's Point (Offsh)	D	Chevron Offshore Terminal	Chevron USA Inc	304.8		15.24	150000			Chevron USA Prod.

L = Load, **D** = Discharge, **B** = Both; **NL** = No Limit

Source: SSY Consultancy & Research Ltd

Table 16: Crude Oil Tanker Ports (continued)
Tableau 16 : Ports pétroliers (suite)

Country/Port		Terminal	Operator(s)	(m) LOA	(m) Beam	(m) Draft	Max Dwt	Load T/Hr	Dis T/Hr	Refinery Name
UNITED STATES ctd.										
Baton Rouge (LA)	D	Exxon Refinery Term. Dk 1	Exxon Co USA	246.8	None	13.7	100000			ExxonMobil
Baytown (TX)	D	Exxon Berth No 6	Exxon Company USA	249.92	39.62	12.19	80000			ExxonMobil
Bayway (NJ)	D	Oil Dock No 1	Tosco Refining	246.9		10.97	80000			Bayway Refining
Bridgeport (CT)	D	Shell Terminal		213.35			70000			-
Cherry Pt (Bellingham)	D	Arco Jetty					125000			Arco, Ferndale
Convent (LA)	D	Motiva Star Enterprise Dk N2		274.31		12.19	70000		4100	Star Enterprise
Corpus Christi (TX)	B	Inner Harbour: Pub. Dk 1&8		304.79		13.72	100000			Various
Delaware City (DE)	D	Star Enterprise Bth No 1		281.3	44.2		140000			Star Enterprise
El Segundo (CA)	D	Berth No 4	Chevron USA	305		17	150000			Chevron
Ferndale (WA)	D		Conoco Phillips	304.79		18.3	125000			Conoco Phillips
Freeport (TX)	D	Phillips Term. Dock No 3	Phillips Petroleum	249.9	44.19	13.72	80000			-
Good Hope (LA)	D	Norco (upstream)	Shell Oil Co	274.31			80000			Shell Norco Ref.
Gulfmex No 2	D	Offshore lightering area					500000			-
Houston (Boggy Bayou)	D	Crude Dock	Shell Oil Company	274.31	44.19	12.19	80000			Shell/Pemex, D. Pk
Houston (TX)	D	Oiltanking Deepwater Ter.	Oiltanking of Texas Inc	274.3	44.19	12.19	128000		4100	
Huntington Beach (CA)	D		Golden West Refining Co	304.79	45.72	13.11	150000		3000	Golden West
Jacksonville (FL)	D					15.5	110000			-
Lake Charles (LA)	D	Conoco Westlake Pier 1&2		244	41		90000			Conoco
Long Beach (CA)	D	Berth 121	ARCO Western Pipeline		None	19.8	265000		4350	-
LOOP Terminal (LA)	D	SPM 103	LOOP Inc	427		29.4	700000		13650	-
Los Angeles (CA)	B	Bths 45-47 (Outer Harbour)	Port of Los Angeles	274		15	120000	3400	3400	Various
Los Angeles (Wilm.)(CA	D	Mormon Island Berth 163	Wickland Oil Terminals Inc				100000			Various
Marcus Hook (PA)	D	Sun Oil Co Wharf 3A & 3C	Sun Oil Co	259.6	61		85000			Sun Oil
Martinez (Suisan Bay)	D	Shore Terminals		304.8	None	12.2	80000			Various
Meraux (LA)	D	Murphy Oil Co Dock No 1				12.19	80000		3650	Murphy Oil USA
Mobile (AL)	D	Louisiana Land & Explor.		225.59	33.53	12.19	80000		2700	
Nederland (LA)	D	Sun Marine Ter. Dk N2-5	Sun Co Marketing	304.8	45.72	12.19	70000			-
New Orleans (Chalmet.)	D	Exxon-Mobil Oil Co Dk N1		254.5		12.19	70000			ExxonMobil
New Orleans (St Rose)	D	IMTT Berth No 1	International Matex	259	None	12.2	60000			Shell Chemical
New York (NY)	D						100000			-
Nikiski (Kenai) (AK)	L	Drift River Terminal	Cook Inlet Pipe Line Co	259			80000	6800		-
Norfolk (VA)	D	Sewell's Point	Exxon			12.3	70000			-
Northport (NY)	D	Offshore Oil Discharge	Long Island Lighting Co	247.3			88000			-
Offshore Pascagoula N2	D	Offshore lightering area					500000			Chevron
Ostrica (LA)	B	Gulf Refining Co Terminal	Gulf Refining			12.19	60000	4600		Gulf
Paulsboro (NJ)	D	Citgo Terminal	Citgo Asphalt Refining Co	243.85	38		80000			Citgo Asphalt Ref.
Paulsboro (NJ)	D	Mobil Oil No 1 Berth		274.3	None	11.58	80000			ExxonMobil
Perth Amboy (NJ)	D	Chevron Terminal	Chevron USA				75000			Chevron
Philadelphia (PA)	D	Hog Island Berths 3 & 4	Sun Oil Co	304.8	41.15	11.28	110000			Sun Ref. & Mark.
Port Allen (LA)	D	Placid Refinery Co Dk N1		259.07	35.05	12.19	90000		1000	Placid Refining
Port Arthur (TX)	D	Chevron Refinery Terminal	Chevron USA	236.21	36.57	10.97	40000			Chevron
Portland (ME)	D	Pipeline Pier No 2	Portland Pipeline Corp		30.5	13.72	80000			-
Portland (OR)	D			None	None	11.88	60000			-
Richmond (CA)	D	Long Wharf	Chevron USA	320			190000		7000	Chevron
Riverhead (Long Is.)(NY	D	North Berth, Northville		340.5		18.3	160000			-
Sandwich Oil Term. (MA	D		US Army Corps of Engineers			12.19	70000			-
Santa Catalina Isl (CA)	D						300000			-
Smith's Bluff (Beaumont	D	Sun Marine Berths 2-5	Sun Marine Terminals Inc	304.8	45.72	12.05	80000		6800	ExxonMobil
South Sabine Point (LA)	D	Offshore lightering area					500000			Various
Southtex	D	Offshore lightering area					500000			-
St. James (LA)	D	IMTT St Rose					125000		4100	-
Texas City (TX)	D	BP Amoco - Bth 40 & 41	BP Amoco	320	None	12.04	60000		2000	BP Amoco/Marat.
Valdez Marine Ter.(AK)	L	Berths 4 & 5	Aleyska Pipeline Service Co			22.9	265000	15000		-
Westville (Eagle Pt) (NJ)	D	Eagle Point Berth No 2		277.35	None		140000			Coast. Eagle Pt Oil
Wilmington (DE)	D	West Marine Terminal		None	None	11.6	80000			-

L = Load, **D** = Discharge, **B** = Both; **NL** = No Limit

Source: SSY Consultancy & Research Ltd

Table 16: Crude Oil Tanker Ports (continued)
Tableau 16 : Ports pétroliers (suite)

Country/Port		Terminal	Operator(s)	(m) LOA	(m) Beam	(m) Draft	Max Dwt	Load T/Hr	Dis T/Hr	Refinery Name
UNITED STATES ctd.										
Yorktown (VA)	D	Amoco Dock	Amoco Oil Co	274		11.6	60000			BP Amoco
URUGUAY										
Jose Ignacio Terminal	D	ANCAP SBM	ANCAP			16.3	150000		4250	ANCAP, La Teja
VENEZUELA										
Amuay Bay	L	Jty 1, Bths 1 & 2; Jty4 Bth8	PDVSA	260.9	None	12.8	90000			Lagoven, Judibana
Bajo Grande	L	Berth 3	PDVSA	260		12	85000	2000		PDVSA
El Chaure	L		PDVSA	229		12.19	60000			PDVSA
El Palito	L	Berths 3 & 4	PDVSA	250	None	14.6	85000	3675		PDVSA
Jose Petroterminal	L	Jose Petroter. West Berth	Williams, Enbridge & Co	350		22.85	310000	13600		-
La Salina (Lacustre)	L	Berths 6N, 6S, 7N, 7S	PDVSA	274.5	None		75000	5450		-
Puerto la Cruz	L	Berths No 4 & 5	PDVSA	280.4	49.07	17.68	150000	4600		PDVSA
Puerto Miranda	L	Jetty Nos 2 & 3	PDVSA	277.35	41.15	11.9	60000	6000		-
Punta Cardon	L	Jetty No 3 (South Outer Be	PDVSA	274.31		13.87	90000			PDVSA
Punta Cuchillo	L	Puerto Ordaz	PDVSA			13.1	80000	2000		-
Punta de Palmas	L	Berth No 3	PDVSA	271.3		12.8	80000	4300		-
VIETNAM										
Block 01 (Ruby Field)	L	"Ruby Princess" FPSO	Nortrans Offshore (S) Pte Ltd				150000			-
Dai Hung (Big Bear)	L	SPM No 1		250	42		155000			-
Rang Dong (Daybreak)	L	"Rang Dong I" FPSO	Japan Vietnam Pet. Co Ltd				150000	3400		-
Van Phong Bay	B	Ship-to-ship transfers	Van Phong Bay Petrotech Ltd			20	220000			-
Vung Tau (White Tiger)	L	Bach Ho Crude Oil Ex. Port		250	42	15	110000	3000		-
VIRGIN ISLANDS										
Limetree Bay (St Croix)	D	Dock No 4 - West Basin	HOVENSA, L.L.C.	365.74	None	16.76	300000		12950	HOVENSA, LLC
YEMEN REPUBLIC										
Aden	D	Oil Harbour Berth 4	Aden Refinery Company	286	None	15.5	110000			Aden Refinery Co
Ash Shihr Terminal	L	Petroleum Export Terminal	Canadian Occidental Pet. Ye	None	None	None	300000	7750		-
Ras Isa Terminal	L	Marine Term. FSO "Safer"	Yemen Hunt Oil Co	350.6	56.4	None	307000	13600		-
YUGOSLAVIA, FEDERAL REPUBLIC OF										
Bar	D	New Oil Berth		None	None	12.5	80000			-

L = Load, **D** = Discharge, **B** = Both; **NL** = No Limit

Source: SSY Consultancy & Research Ltd

Table 17: Representative Tanker Single Voyage Freight Rates 1992-2003 [1]

Tableau 17 : Tarifs d'Affrètement des Pétroliers au Voyage 1992-2003 [1]

Rates in US$ / tonne

Cargo size/type:	Caribbean-USAC/US Gulf 30,000 tonnes (clean)	Middle East Gulf-Japan/Korea 55,000 tonnes (clean)	North Africa-Med. 80,000 tonnes (dirty)	West Africa-USAC/US Gulf 130,000 tonnes (dirty)	Middle East Gulf-UK/Cont 280,000 tonnes (dirty)	Middle East Gulf-Japan 260,000 tonnes (dirty)
1st Quarter 1992	9.32	13.33	3.16	5.44	6.51	4.44
2nd Quarter 1992	6.54	11.64	2.90	4.28	6.42	3.99
3rd Quarter 1992	7.02	12.09	2.94	4.23	7.16	4.63
4th Quarter 1992	7.81	12.30	3.70	5.29	8.76	5.66
1st Quarter 1993	9.01	14.61	3.88	6.80	8.22	5.38
2nd Quarter 1993	8.81	12.86	4.09	6.17	7.34	4.78
3rd Quarter 1993	8.36	15.96	3.48	5.69	7.88	5.44
4th Quarter 1993	10.38	18.56	3.70	6.34	6.93	4.68
1st Quarter 1994	9.87	17.34	3.08	5.28	5.59	3.67
2nd Quarter 1994	8.88	15.60	3.52	5.59	5.63	3.64
3rd Quarter 1994	8.22	15.36	3.37	5.67	6.97	4.85
4th Quarter 1994	9.83	18.87	3.90	6.56	7.35	5.21
1st Quarter 1995	12.21	17.96	3.82	6.55	7.64	5.54
2nd Quarter 1995	10.11	16.09	3.88	6.48	7.30	5.12
3rd Quarter 1995	9.59	19.04	4.18	6.75	9.52	7.00
4th Quarter 1995	10.45	21.14	3.61	6.78	8.73	6.15
1st Quarter 1996	13.05	20.29	4.77	7.50	9.92	7.40
2nd Quarter 1996	11.08	17.51	4.60	7.59	8.95	6.62
3rd Quarter 1996	8.61	17.72	4.00	7.16	9.74	7.14
4th Quarter 1996	9.86	20.21	4.05	7.57	8.89	6.64
1st Quarter 1997	13.15	23.22	4.72	8.50	9.86	7.17
2nd Quarter 1997	10.68	20.87	4.49	8.32	9.93	7.29
3rd Quarter 1997	9.89	16.19	4.49	8.02	12.03	9.30
4th Quarter 1997	9.21	19.01	4.25	9.08	12.59	8.89
1st Quarter 1998	7.13	19.42	3.73	8.33	10.27	7.25
2nd Quarter 1998	8.22	15.64	3.73	7.18	12.14	8.30
3rd Quarter 1998	8.03	16.78	3.28	6.85	10.85	7.32
4th Quarter 1998	9.21	16.55	3.31	6.45	9.50	6.23
1st Quarter 1999	8.35	13.32	3.51	6.96	9.66	6.42
2nd Quarter 1999	7.95	13.47	3.42	5.60	7.19	4.69
3rd Quarter 1999	8.32	15.12	2.84	5.24	7.40	4.66
4th Quarter 1999	7.89	14.10	3.37	6.63	7.70	5.11
1st Quarter 2000	8.70	19.80	4.75	8.89	8.82	6.73
2nd Quarter 2000	11.78	23.32	6.18	11.29	14.30	10.66
3rd Quarter 2000	13.45	31.85	7.61	15.01	17.66	13.75
4th Quarter 2000	16.09	34.31	8.28	17.87	21.42	16.79
1st Quarter 2001	17.25	46.20	7.70	15.54	18.00	12.35
2nd Quarter 2001	15.25	31.70	5.97	10.86	12.36	8.14
3rd Quarter 2001	12.76	28.27	4.84	9.93	11.32	7.36
4th Quarter 2001	10.42	22.25	4.19	8.03	8.95	5.86
1st Quarter 2002	8.65	15.94	3.92	6.66	7.39	5.02
2nd Quarter 2002	9.96	16.16	4.41	6.88	7.04	4.80
3rd Quarter 2002	8.74	19.26	3.36	6.80	6.84	4.53
4th Quarter 2002	0.95	21.90	5.84	11.21	13.73	11.27
1st Quarter 2003	13.78	30.47	9.66	16.74	20.11	15.57
1992	7.67	12.34	3.18	4.81	7.21	4.68
1993	9.14	15.50	3.79	6.25	7.59	5.07
1994	9.20	16.79	3.47	5.78	6.38	4.34
1995	10.59	18.56	3.87	6.64	8.30	5.95
1996	10.62	18.93	4.37	7.47	9.37	6.95
1997	10.73	19.82	4.49	8.48	11.11	8.16
1998	8.29	17.10	3.49	7.26	15.88	7.27
1999	8.13	14.01	3.29	6.25	7.99	5.22
2000	12.50	27.32	6.71	13.27	15.55	11.98
2001	13.92	32.11	5.67	11.09	12.66	8.43
2002	9.07	18.33	4.38	7.88	8.75	6.41
2003 [2]	14.11	31.45	8.42	14.87	18.38	13.91

1. Rates for MEG to North-West Europe shipments are calculated on the assumption of a return voyage between the Middle East and North-West Europe via the Cape of Good Hope.
2. 2003 data are average to mid-May.

Source: SSY Consultancy & Research Ltd.

Table 18: World Imports of Crude Oil and NGL [1]

Tableau 18 : Importations mondiales de pétrole brut et LGN [1]

000 metric tons

	1971	1973	1978	1998	1999	2000	2001	2002
Canada	32 997	42 429	31 057	39 433	41 803	46 185	45 755	43 535
Mexico	101	3 552	--	--	462	337	295	392
United States	91 044	171 335	354 157	490 416	493 524	511 411	526 331	517 909
OECD North America	**124 142**	**217 316**	**385 214**	**529 849**	**535 789**	**557 933**	**572 381**	**561 836**
Austria	5 095	6 633	8 158	9 157	8 440	7 949	8 470	8 715
Belgium	30 867	37 650	33 779	37 131	35 276	37 441	35 495	36 636
Czech Republic	5 972	7 193	18 541	6 978	6 064	5 725	6 037	6 189
Denmark	10 544	9 799	7 783	5 975	5 970	5 208	4 429	3 879
Finland	8 945	9 524	10 454	11 931	11 378	11 834	11 095	11 522
France	107 587	134 920	115 635	90 443	82 145	85 698	86 362	81 435
Germany	113 876	126 738	115 771	109 379	103 933	103 684	105 171	104 895
Greece	5 381	13 717	12 630	19 206	17 287	20 383	20 081	20 605
Hungary	4 892	6 555	9 960	6 151	5 786	5 809	5 634	5 038
Ireland	2 967	2 684	2 294	3 148	2 848	2 987	3 371	3 190
Italy	117 582	128 944	111 490	91 594	88 150	90 403	91 250	89 291
Netherlands	60 732	72 161	55 689	62 255	59 251	60 553	60 810	55 236
Norway	5 710	6 076	7 601	2 071	2 085	904	932	642
Poland	7 984	11 140	16 615	15 742	16 627	18 260	17 902	17 785
Portugal	4 067	4 348	6 296	13 919	13 807	12 133	13 158	12 238
Slovak Republic	5 533	6 983	9 157	5 421	5 346	5 329	5 491	5 632
Spain	35 435	42 970	46 758	60 679	58 911	58 782	57 833	57 694
Sweden	11 911	10 606	15 640	20 241	19 693	20 901	20 040	18 280
Switzerland	5 363	6 180	4 246	5 070	5 093	4 649	4 908	4 876
Turkey	5 469	9 306	10 354	23 734	22 985	21 583	23 242	23 662
United Kingdom	107 736	115 472	68 144	47 957	44 891	54 387	53 551	58 254
OECD Europe	**663 648**	**769 599**	**686 995**	**648 182**	**615 966**	**634 602**	**635 262**	**625 694**
Australia	10 228	8 675	10 244	20 372	24 006	21 849	22 709	23 289
Japan	188 992	245 094	230 555	217 559	213 872	216 594	208 806	208 031
Korea	11 591	14 004	22 596	113 736	120 529	123 320	118 896	107 936
New Zealand	2 950	3 176	2 417	4 814	4 637	4 695	4 641	4 945
OECD Pacific	**213 761**	**270 949**	**265 812**	**356 481**	**363 044**	**366 458**	**355 052**	**344 201**
OECD Total	**1001 551**	**1257 864**	**1338 021**	**1534 512**	**1514 799**	**1558 993**	**1562 695**	**1531 731**
Argentina	2 248	3 004	2 192	1 229	962	1 355	1 421	..
Brazil	19 935	34 348	44 957	27 973	23 578	20 023	20 987	..
Chile	3 308	3 141	3 686	8 924	9 219	9 673	9 683	..
Costa Rica	417	420	463	39	--	15	282	..
Cuba	5 438	5 993	7 269	899	801	1 648	1 715	..
Dominican Republic	--	1 253	1 604	2 224	2 017	2 017	1 924	..
El Salvador	480	677	793	948	1 002	1 009	1 041	..
Guatemala	837	946	794	836	890	902	861	..
Jamaica	1 528	1 830	965	712	441	1 059	1 059	..
Netherlands Antilles	39 967	45 293	28 548	13 138	10 829	11 395	12 789	..
Nicaragua	483	598	626	901	850	857	972	..
Panama	4 151	3 848	2 466	2 375	2 572	2 313	2 946	..
Paraguay	180	240	320	138	101	107	97	..
Peru	1 610	1 769	58	4 733	2 911	3 068	3 716	..
Trinidad and Tobago	14 578	14 135	7 750	3 721	4 118	5 051	4 779	..
Uruguay	1 773	1 747	2 132	1 730	1 657	1 958	1 715	..
Other	14 262	17 106	11 524	1 935	1 059	958	1 051	..
Latin America	**111 195**	**136 348**	**116 147**	**72 455**	**63 007**	**63 408**	**67 038**	**..**
Albania	--	--	--	5	--	--	--	..
Bulgaria	7 547	9 652	12 644	5 549	5 613	5 306	5 628	..
Cyprus	--	656	451	1 075	1 186	1 155	1 154	..
Romania	2 858	4 143	12 937	5 977	4 322	4 798	5 544	..
Croatia	..	..	..	3 759	4 529	3 954	3 991	..
Former Yugo. Rep. Macedonia	..	..	..	787	580	801	637	..
Slovenia	..	..	..	239	291	154	6	..
Serbia and Montenegro	..	..	..	1 533	481	161	1 680	..
Former Yugoslavia	4 934	8 317	10 380	6 318	5 881	5 070	6 314	..
Non-OECD Europe	**15 339**	**22 768**	**36 412**	**18 924**	**17 002**	**16 329**	**18 640**	**..**

1. Includes additives, refinery feedstocks and other hydrocarbons (other than crude oil and NGL). / *Comprend les additifs, les produits d'alimentation des raffineries, et les autres hydrocarbures (autres que le pétrole et les LGN).*

Source: Annual Oil Statistics, Energy Statistics of Non-OECD Countries.

Table 18: World Imports of Crude Oil and NGL [1] (continued)
Tableau 18 : Importations mondiales de pétrole brut et LGN [1] (suite)

000 metric tons

	1971	1973	1978	1998	1999	2000	2001	2002
Belarus	..	..	..	10 055	9 900	12 010	11 913	..
Kazakhstan	..	..	..	2 174	715	1 010	2 337	..
Lithuania	..	..	..	6 775	4 645	5 052	6 651	..
Russia	..	..	..	5 575	4 633	5 892	5 118	..
Turkmenistan	..	..	..	900	600	600	600	..
Ukraine	..	..	..	9 893	9 406	6 021	13 531	..
Uzbekistan	..	..	..	7	7	7	7	..
Other	..	..	..	153	78	114	21	..
Former USSR	**5 100**	**13 200**	**8 380**	**35 532**	**29 984**	**30 706**	**40 178**	..
Algeria	..	144	112	232	266	309	209	..
Democratic Republic of Congo	679	732	200	53	55	71	71	..
Côte d'Ivoire	751	1 255	1 551	3 084	3 037	3 125	3 081	..
Ethiopia + Eritrea	665	624	612	..	..	..	..	..
Ghana	900	1 069	1 146	1 200	1 637	1 309	1 576	..
Kenya	2 541	2 696	2 369	2 157	2 139	2 452	1 966	..
Morocco	1 651	2 094	2 815	5 957	7 101	6 696	7 259	..
Senegal	563	666	758	862	852	915	968	..
South Africa	13 220	13 665	14 965	21 043	19 335	18 320	18 981	..
Sudan	970	1 100	1 058	258	..	..	..	..
United Republic of Tanzania	944	879	566	515	488	..	..	..
Tunisia	321	1 030	1 107	953	1 073	1 150	1 089	..
Zambia	..	658	780	626	222	28	444	..
Other	3 139	4 330	1 701	1 394	1 603	1 681	1 769	..
Africa	**26 344**	**30 942**	**29 740**	**38 334**	**37 808**	**36 056**	**37 413**	..
Bahrain	9 022	8 812	9 513	10 526	11 245	10 969	10 113	..
Israel	500	2 700	7 975	12 655	10 328	11 366	11 135	..
Jordan	592	706	1 332	3 556	3 497	3 763	3 879	..
Yemen	3 748	3 124	1 930	..	..	..	..	..
Other	3 364	3 820	4 766	..	..	..	64	..
Middle East	**17 226**	**19 162**	**25 516**	**26 737**	**25 070**	**26 098**	**25 191**	..
Bangladesh	850	719	1 064	1 144	1 190	1 290	1 403	..
India	12 951	13 873	14 657	39 808	57 805	74 097	78 706	..
Indonesia	..	201	2 467	10 602	11 507	10 867	15 337	..
Malaysia	8 751	3 494	4 383	2 221	3 538	7 969	9 460	..
Myanmar	266	..	..	685	675	638	530	..
DPR of Korea	151	412	1 557	506	318	391	581	..
Pakistan	3 057	3 044	3 461	4 582	4 502	7 044	7 075	..
Philippines	9 027	9 266	10 022	15 994	16 131	15 587	15 313	..
Singapore	16 415	22 536	28 220	52 014	44 435	41 832	41 150	..
Sri Lanka	1 536	1 747	1 449	2 155	1 835	2 349	1 955	..
Chinese Taipei	5 805	9 075	17 417	34 320	36 048	38 433	39 483	..
Thailand	5 413	7 643	8 203	32 256	33 599	32 528	34 305	..
Asia	**64 222**	**72 010**	**92 900**	**196 287**	**211 583**	**233 025**	**245 298**	..
China, People's Republic	**50**	..	**690**	**27 320**	**36 614**	**70 265**	**60 260**	..
Non-OECD	**239 476**	**294 430**	**309 785**	**415 589**	**421 068**	**475 887**	**494 018**	..
World	**1241 027**	**1552 294**	**1647 806**	**1950 101**	**1935 867**	**2034 880**	**2056 713**	..

1. Includes additives, refinery feedstocks and other hydrocarbons (other than crude oil and NGL). / *Comprend les additifs, les produits d'alimentation des raffineries, et les autres hydrocarbures (autres que le pétrole et les LGN).*

Source: Annual Oil Statistics, Energy Statistics of Non-OECD Countries.

Table 19: World Imports of Petroleum Products
Tableau 19 : Importations mondiales de produits pétroliers

000 metric tons

	1971	1973	1978	1998	1999	2000	2001	2002
Canada	7 952	5 579	3 046	8 698	8 049	7 557	10 526	9 288
Mexico	1 769	3 279	1 715	13 844	15 551	19 173	16 142	13 498
United States	108 998	140 137	77 157	56 852	60 665	74 404	79 650	69 581
OECD North America	**118 719**	**148 995**	**81 918**	**79 394**	**84 265**	**101 134**	**106 318**	**92 367**
Austria	2 825	3 147	2 340	4 514	4 308	4 247	4 608	4 994
Belgium	7 120	8 024	8 722	13 760	13 811	14 445	15 468	14 921
Czech Republic	1 415	1 697	2 500	2 657	3 023	2 829	3 359	3 104
Denmark	10 258	11 514	10 576	5 362	4 624	4 986	4 966	4 711
Finland	3 121	4 344	3 039	3 346	3 718	3 518	3 712	3 439
France	8 154	7 336	9 303	23 061	26 917	26 715	28 281	31 709
Germany	33 886	41 065	46 249	44 727	40 522	42 171	43 669	36 339
Greece	2 053	2 466	3 824	3 389	4 208	3 255	3 397	4 200
Hungary	661	852	1 773	1 718	1 517	1 170	1 372	2 129
Iceland	544	678	614	878	928	949	871	888
Ireland	2 857	3 213	3 809	5 541	6 558	6 305	6 803	6 987
Italy	3 991	4 989	7 466	21 682	19 957	20 079	15 996	16 037
Luxembourg	1 437	1 678	1 438	2 016	2 069	2 295	2 372	2 452
Netherlands	8 290	9 722	13 615	34 996	37 364	44 675	46 864	51 206
Norway	3 927	4 428	2 898	3 232	3 210	3 626	3 596	3 492
Poland	1 638	1 644	1 890	4 340	3 886	3 185	3 600	3 678
Portugal	1 656	1 955	1 214	4 032	4 302	5 118	4 839	5 187
Slovak Republic	--	--	--	542	580	270	552	743
Spain	1 785	1 502	3 248	15 576	16 991	19 557	21 194	22 642
Sweden	19 023	19 417	13 198	6 395	5 571	5 453	5 984	6 050
Switzerland	8 225	8 720	9 471	8 513	7 569	7 733	8 801	8 219
Turkey	626	171	3 951	5 022	5 593	8 669	5 791	6 505
United Kingdom	19 369	18 299	11 546	11 373	13 896	14 212	16 978	13 475
OECD Europe	**142 861**	**156 861**	**162 684**	**226 672**	**231 122**	**245 462**	**253 073**	**253 107**
Australia	3 161	3 580	3 460	3 069	3 670	4 077	4 246	4 472
Japan	29 286	26 609	28 508	43 109	50 566	50 949	46 747	48 539
Korea	--	--	512	19 447	20 825	23 389	23 654	26 747
New Zealand	830	1 070	1 324	929	1 406	1 175	1 284	1 299
OECD Pacific	**33 277**	**31 259**	**33 804**	**66 554**	**76 467**	**79 590**	**75 931**	**81 057**
OECD Total	**294 857**	**337 115**	**278 406**	**372 620**	**391 854**	**426 186**	**435 322**	**426 531**
Argentina	1 205	914	574	1 533	1 641	1 250	415	..
Brazil	1 515	1 519	817	13 575	13 566	13 849	14 472	..
Chile	485	325	86	1 840	1 840	1 811	1 400	..
Colombia	75	31	981	1 251	243	259	195	..
Costa Rica	70	160	401	1 693	1 887	1 754	1 578	..
Cuba	2 038	2 837	3 556	6 847	6 871	6 199	5 865	..
Dominican Republic	1 139	425	436	3 386	3 702	4 118	4 119	..
Ecuador	18	15	250	1 414	1 454	965	1 289	..
El Salvador	24	23	41	1 104	1 013	1 069	1 102	..
Guatemala	108	86	581	2 019	1 784	1 922	2 143	..
Honduras	11	9	96	1 526	1 522	1 456	1 724	..
Jamaica	589	1 210	1 402	2 418	2 728	2 761	2 570	..
Netherlands Antilles	990	6 110	2 252	1 704	1 706	1 706	1 706	..
Panama	100	264	17	871	463	590	615	..
Paraguay	28	31	152	1 149	1 151	1 014	933	..
Peru	551	458	233	1 146	1 258	1 651	1 182	..
Uruguay	81	107	123	396	963	274	274	..
Other	4 129	3 102	3 759	7 034	7 170	7 452	7 775	..
Latin America	**13 156**	**17 626**	**15 757**	**50 906**	**50 962**	**50 100**	**49 357**	**..**
Cyprus	629	192	402	1 198	1 281	1 417	1 344	..
Gibraltar	217	233	173	1 089	1 133	1 179	1 210	..
Malta	330	343	385	806	952	831	763	..
Romania	2	6	--	2 745	1 592	1 562	2 679	..
Bosnia and Herzegovina	..	..	..	742	735	917	917	..
Former Yugo. Rep. Macedonia	..	..	..	241	364	356	423	..
Slovenia	..	..	..	2 339	2 504	2 449	2 528	..
Other	..	..	..	1 696	2 080	2 251	1 818	..
Former Yugoslavia	932	1 090	1 402	5 018	5 683	5 973	5 686	..
Other	2 188	1 494	1 900	664	934	1 485	1 476	..
Non-OECD Europe	**4 298**	**3 358**	**4 262**	**11 520**	**11 575**	**12 447**	**13 158**	**..**

Source: Annual Oil Statistics, Energy Statistics of Non-OECD Countries.

Table 19: World Imports of Petroleum Products (continued)
Tableau 19 : Importations mondiales de produits pétroliers (suite)

000 metric tons

	1971	1973	1978	1998	1999	2000	2001	2002
Armenia	..	..	..	169	227	554	578	..
Belarus	..	..	..	301	644	881	323	..
Estonia	..	..	..	1 246	1 110	933	922	..
Kazakhstan	..	..	..	799	1 315	1 332	1 488	..
Kyrgyzstan	..	..	..	513	362	304	294	..
Latvia	..	..	..	1 846	1 356	1 117	1 422	..
Lithuania	..	..	..	929	586	325	309	..
Republic of Moldova	..	..	..	629	370	459	491	..
Russia	..	..	..	3 756	796	297	224	..
Tajikistan	..	..	..	1 277	1 308	902	1 138	..
Ukraine	..	..	..	4 834	4 053	4 764	1 938	..
Other	..	..	..	1 537	1 556	1 323	528	..
Former USSR	**1 600**	**1 500**	**700**	**17 836**	**13 683**	**13 191**	**9 655**	**..**
Democratic Rep. Congo	244	270	654	461	406	371	374	..
Côte d'Ivoire	104	100	229	411	361	266	261	..
Egypt	1 382	223	332	2 018	2 338	3 224	3 186	..
Ethiopia + Eritrea	50	89	47	1 279	1 259	1 266	1 374	..
Ghana	40	23	43	1 029	818	883	869	..
Kenya	132	129	292	961	983	981	981	..
Morocco	510	317	824	1 218	1 270	1 455	1 643	..
Mozambique	219	170	46	370	393	436	471	..
Senegal	963	996	664	428	555	585	568	..
South Africa	93	523	525	887	1 114	458	458	..
Sudan	471	464	58	1 066	1 119	617	475	..
Tunisia	240	471	1 219	1 959	2 403	2 210	2 898	..
Zimbabwe	539	679	666	1 463	1 394	1 031	953	..
Other	5 634	5 802	10 284	8 978	11 101	16 708	13 935	..
Africa	**10 621**	**10 256**	**15 883**	**22 528**	**25 514**	**30 491**	**28 446**	**..**
Islamic Republic of Iran	..	30	138	1 014	912	1 421	2 280	..
Israel	..	19	198	2 943	3 953	4 980	5 188	..
Jordan	16	..	..	1 153	1 102	998	967	..
Lebanon	183	221	302	4 746	4 996	4 477	4 850	..
Syria	267	183	1 342	750	683	867	812	..
United Arab Emirates	136	280	1 520	..	..	..	..	..
Other	1 912	2 424	3 629	140	206	425	807	..
Middle East	**2 514**	**3 157**	**7 129**	**10 746**	**11 852**	**13 168**	**14 904**	**..**
Bangladesh	134	169	368	2 026	2 228	1 829	2 481	..
India	1 932	3 734	3 894	23 521	16 608	9 267	7 009	..
Indonesia	2 614	7 511	7 820	7 282	10 618	12 095	11 730	..
DPR of Korea	683	765	580	832	570	669	663	..
Malaysia	832	1 141	1 375	9 190	8 178	6 452	6 738	..
Myanmar	25	77	..	534	749	772	605	..
Nepal	54	70	88	643	697	696	704	..
Pakistan	459	541	1 225	10 847	11 871	10 027	9 032	..
Philippines	198	246	1 058	3 285	2 239	3 070	2 894	..
Singapore	5 478	6 306	3 400	33 357	40 063	39 248	43 147	..
Sri Lanka	72	48	173	909	1 384	1 455	1 577	..
Chinese Taipei	1 723	1 761	1 809	7 887	9 155	9 019	8 874	..
Thailand	1 034	1 072	2 291	1 264	1 847	1 166	387	..
Vietnam	5 660	5 654	933	6 518	7 402	8 125	8 133	..
Other	1 742	2 184	2 173	2 308	2 278	2 286	2 255	..
Asia	**22 640**	**31 279**	**27 187**	**110 403**	**115 887**	**106 176**	**106 229**	**..**
Hong Kong, China	**4 123**	**4 961**	**6 412**	**20 104**	**17 256**	**13 846**	**13 324**	**..**
China, People's Republic	**75**	**140**	**220**	**27 358**	**25 295**	**24 047**	**27 432**	**..**
Non-OECD Total	**59 027**	**72 277**	**77 550**	**271 401**	**272 024**	**263 466**	**262 505**	**..**
World	**353 884**	**409 392**	**355 956**	**644 021**	**663 878**	**689 652**	**697 827**	**..**

Source: Annual Oil Statistics, Energy Statistics of Non-OECD Countries.

Table 20: World Exports of Crude Oil and NGL [1]

Tableau 20 : Exportations mondiales de pétrole brut et LGN [1]

000 metric tons

	1971	1973	1978	1998	1999	2000	2001	2002
Canada	36 853	55 989	16 653	74 116	69 418	76 156	74 495	79 458
Mexico	--	--	19 562	90 886	83 063	91 611	92 740	94 746
United States	68	94	4 446	8 746	8 864	6 300	3 595	3 466
OECD North America	**36 921**	**56 083**	**40 661**	**173 748**	**161 345**	**174 067**	**170 830**	**177 670**
Austria	--	--	--	44	66	137	106	7
Czech Republic	--	--	9 473	106	109	111	119	140
Denmark	--	39	102	8 322	11 025	13 413	11 953	13 635
Germany	116	108	91	1 389	1 741	3 217	1 030	1 321
Greece	--	--	1 010	--	116	--	--	387
Hungary	188	491	1 463	--	2	1	--	--
Italy	--	--	--	408	276	521	605	393
Netherlands	1	--	--	343	307	184	355	894
Norway	238	1 531	15 827	137 093	136 323	146 172	151 426	145 195
Slovak Republic	63	299	314	--	24	15	17	15
United Kingdom	1 569	3 235	25 201	84 612	91 797	92 918	86 746	86 601
Other	135	629	637	4 336	4 881	5 955	4 239	4 127
OECD Europe	**2 247**	**6 033**	**53 804**	**236 653**	**246 643**	**262 629**	**256 579**	**252 700**
Australia	559	18	--	13 437	12 878	18 342	22 309	22 457
New Zealand	--	--	--	1 482	1 355	1 175	1 200	1 030
OECD Pacific	**559**	**18**	**--**	**14 953**	**14 271**	**19 616**	**23 612**	**23 614**
OECD Total	**39 727**	**62 134**	**94 465**	**425 354**	**422 259**	**456 312**	**451 021**	**453 984**
Argentina	27	26	--	16 979	13 817	14 249	14 767	..
Bolivia	1 211	1 698	410	--	--	--	--	..
Brazil	748	838	242	--	29	1 630	5 155	..
Colombia	3 512	1 340	--	22 910	26 573	19 844	15 258	..
Ecuador	48	9 901	6 316	12 257	12 144	12 366	12 846	..
Guatemala	--	--	--	1 204	1 047	1 031	1 060	..
Netherlands Antilles	99	1 852	1 023	--	--	--	--	..
Peru	151	17	1 444	2 118	1 435	673	925	..
Trinidad and Tobago	955	3 222	7 367	4 705	5 327	5 326	5 611	..
Venezuela	120 811	111 350	65 136	122 920	106 650	113 084	109 472	..
Other	--	--	--	1 696	2 190	3 087	2 829	..
Latin America	**127 562**	**130 244**	**81 938**	**184 789**	**169 212**	**171 290**	**167 923**	**..**
Croatia	..	..	..	40	130	35	--	..
Former Yugoslavia	187	91	--	40	130	35	1	..
Other	143	412	--	2	3	--	--	..
Non-OECD Europe	**330**	**503**	**--**	**42**	**133**	**35**	**1**	**..**
Belarus	..	..	..	382	350	351	401	..
Georgia	..	..	..	54	40	75	70	..
Kazakhstan	..	..	..	18 310	23 764	29 170	32 479	..
Kyrgyzstan	..	..	..	--	--	--	--	..
Lithuania	..	..	..	265	228	311	351	..
Russia	..	..	..	137 236	134 544	144 414	162 109	..
Turkmenistan	..	..	..	1 600	1 500	1 500	1 500	..
Uzbekistan	..	..	..	837	573	--	--	..
Other	..	..	..	2 338	5 815	5 917	8 671	..
Former USSR	**74 800**	**85 300**	**121 300**	**161 022**	**166 814**	**181 738**	**205 581**	**..**

1. Includes additives, refinery feedstocks and other hydrocarbons (other than crude oil and NGL). / *Comprend les additifs, les produits d'alimentation des raffineries, et les autres hydrocarbures (autres que le pétrole et les LGN).*

Source: Annual Oil Statistics, Energy Statistics of Non-OECD Countries.

Table 20: World Exports of Crude Oil and NGL (continued) [1]

Tableau 20 : Exportations mondiales de pétrole brut et LGN (suite) [1]

000 metric tons

	1971	1973	1978	1998	1999	2000	2001	2002
Algeria	34 300	45 000	49 200	34 830	34 614	37 332	36 272	..
Angola	4 747	7 323	5 500	34 386	34 448	37 074	34 367	..
Benin	..	..	..	61	44	37	39	..
Cameroon	..	..	505	6 114	5 427	5 717	5 452	..
Congo	21	1 461	2 255	12 776	13 714	13 234	12 213	..
Democratic Republic of Congo	..	..	1 130	1 130	1 174	1 215	1 159	..
Egypt	11 440	4 500	13 251	2 791	2 820	7 077	6 947	..
Gabon	4 842	6 500	8 609	17 009	15 055	13 027	12 832	..
Libya	132 369	106 278	91 617	55 933	47 777	48 418	49 067	..
Nigeria	73 317	97 975	90 183	101 954	95 658	106 326	104 972	..
South Africa	..	..	..	678	1 472	178	184	..
Tunisia	2 943	3 674	4 674	2 842	3 258	2 889	2 736	..
Other	..	..	..	4 602	6 800	12 299	13 578	..
Africa	**263 979**	**272 711**	**266 924**	**275 106**	**262 261**	**284 823**	**279 818**	..
Islamic Republic of Iran	197 994	262 650	186 456	108 649	97 906	107 659	118 903	..
Iraq	79 572	94 991	116 949	79 685	100 637	102 106	89 989	..
Kuwait	137 924	131 524	88 656	60 282	48 031	61 471	61 501	..
Oman	14 767	14 906	16 021	41 993	44 426	47 014	47 690	..
Qatar	20 545	27 485	23 395	28 592	29 180	30 864	32 381	..
Saudi Arabia	209 158	347 986	386 298	339 077	305 169	319 957	302 883	..
Syria	3 465	4 325	8 106	16 673	17 602	16 584	19 613	..
United Arab Emirates	50 892	73 617	87 922	84 986	79 551	84 242	79 427	..
Yemen	..	..	..	14 051	14 629	16 319	16 887	..
Middle East	**714 317**	**957 484**	**913 803**	**773 988**	**737 131**	**786 216**	**769 274**	..
Brunei	5 889	11 597	11 967	7 533	8 670	9 445	9 416	..
India	..	18	..	..	..	..	..	..
Indonesia	32 416	49 438	69 813	38 093	34 090	30 367	29 412	..
Malaysia	7 927	3 827	9 803	19 291	18 995	17 856	18 647	..
Pakistan	..	..	..	434	356	433	239	..
Singapore	106	216	320	1	..	..	..	..
Thailand	..	..	..	1 089	1 013	1 906	1 801	..
Vietnam	..	..	..	12 400	15 045	15 423	16 732	..
Other Asia	..	..	..	3 850	3 947	4 110	4 080	..
Asia	**46 338**	**65 096**	**91 903**	**82 691**	**82 116**	**79 540**	**80 327**	..
China, People's Republic	**263**	**1 834**	**11 313**	**15 600**	**7 167**	**10 306**	**7 550**	..
Non-OECD	**1227 589**	**1513 172**	**1487 181**	**1493 238**	**1424 834**	**1513 948**	**1510 474**	..
World	**1267 316**	**1575 306**	**1581 646**	**1918 592**	**1847 093**	**1970 260**	**1961 495**	..

1. Includes additives, refinery feedstocks and other hydrocarbons (other than crude oil and NGL). / *Comprend les additifs, les produits d'alimentation des raffineries, et les autres hydrocarbures (autres que le pétrole et les LGN).*

Source: Annual Oil Statistics, Energy Statistics of Non-OECD Countries.

Table 21: World Exports of Petroleum Products
Tableau 21 : Exportations mondiales de produits pétroliers

000 metric tons

	1971	1973	1978	1998	1999	2000	2001	2002
Canada	1 900	5 926	6 671	14 112	14 423	14 567	18 815	20 268
Mexico	1 554	1 426	83	4 721	6 664	4 658	4 238	4 377
United States	11 944	12 531	10 680	41 457	40 771	48 957	49 675	50 793
OECD North America	**15 398**	**19 883**	**17 434**	**60 290**	**61 858**	**68 182**	**72 728**	**75 438**
Austria	117	145	177	1 533	1 530	1 304	1 536	1 463
Belgium	9 638	14 507	15 154	18 273	18 852	19 655	19 367	19 977
Czech Republic	33	33	628	1 296	1 196	973	1 156	1 243
Denmark	2 243	2 821	1 976	3 189	4 026	4 590	3 875	3 825
Finland	182	173	1 728	4 655	4 928	4 947	4 815	5 362
France	11 839	13 445	13 156	19 662	17 994	20 676	19 686	18 703
Germany	9 014	9 746	7 423	15 190	16 664	18 601	18 495	18 598
Greece	81	4 892	3 454	3 120	3 540	3 944	3 779	3 783
Hungary	209	427	405	1 819	1 887	1 754	2 260	2 390
Ireland	742	488	51	1 288	1 121	1 346	1 318	1 278
Italy	27 500	28 903	22 809	22 476	20 078	21 050	21 816	21 533
Luxembourg	16	12	30	16	18	17	30	9
Netherlands	34 236	41 684	32 030	60 883	60 063	63 442	65 714	65 429
Norway	1 805	2 134	1 951	9 014	8 437	9 108	8 667	8 548
Poland	947	1 160	1 641	1 637	1 879	1 855	2 272	2 336
Portugal	403	221	166	1 901	1 389	1 424	1 355	1 370
Slovak Republic	1 260	1 386	1 798	2 353	2 764	2 806	3 210	3 489
Spain	4 271	4 216	2 609	8 879	6 926	7 367	6 223	5 926
Sweden	1 652	1 398	2 860	8 835	9 281	10 375	9 435	9 002
Switzerland	142	236	51	475	562	660	575	626
Turkey	93	816	54	2 075	2 459	1 310	2 570	3 052
United Kingdom	17 164	17 404	13 190	24 332	21 655	20 677	19 080	22 107
OECD Europe	**123 587**	**146 247**	**123 341**	**212 901**	**207 249**	**217 881**	**217 234**	**220 049**
Australia	1 823	3 185	3 575	4 081	3 432	3 344	3 725	2 745
Japan	1 893	3 028	718	6 666	5 219	4 328	4 711	3 685
Korea	1 037	999	282	38 687	38 799	39 764	39 007	31 236
New Zealand	18	--	41	214	198	197	139	146
OECD Pacific	**4 771**	**7 212**	**4 616**	**49 648**	**47 648**	**47 633**	**47 582**	**37 812**
OECD Total	**143 756**	**173 342**	**145 391**	**322 839**	**316 755**	**333 696**	**337 544**	**333 299**
Argentina	287	126	772	4 649	5 541	4 872	7 071	..
Brazil	178	2 294	1 458	4 440	3 933	5 308	7 507	..
Colombia	1 535	1 575	1 664	2 905	3 797	3 741	3 900	..
Cuba	--	--	394	--	--	--	--	..
Ecuador	47	110	1 248	2 135	2 088	2 314	2 066	..
El Salvador	5	13	41	232	250	291	318	..
Jamaica	201	84	38	31	68	221	85	..
Netherlands Antilles	34 689	42 356	23 879	11 393	10 011	10 359	11 479	..
Panama	2 957	2 495	1 495	1 394	1 355	1 218	1 000	..
Peru	24	222	347	1 648	980	1 389	1 864	..
Trinidad and Tobago	17 489	16 194	9 888	6 593	6 386	6 878	6 908	..
Venezuela	53 777	59 545	36 748	37 816	37 010	37 076	33 500	..
Other	9 583	10 809	6 951	2 745	3 004	3 668	4 075	..
Latin America	**120 772**	**135 823**	**84 923**	**75 981**	**74 423**	**77 335**	**79 773**	**..**
Bulgaria	41	4	436	1 265	1 662	1 981	2 055	..
Romania	4 948	4 599	7 217	3 232	2 027	2 750	3 096	..
Croatia	..	..	..	1 770	1 817	1 703	1 641	..
Slovenia	..	..	..	168	307	239	168	..
Other	..	..	..	238	116	248	264	..
Former Yugoslavia	425	208	504	2 176	2 240	2 190	2 073	..
Other	666	946	1 055	--	--	--	--	..
Non-OECD Europe	**6 080**	**5 757**	**9 212**	**6 673**	**5 929**	**6 921**	**7 224**	**..**

Source: Annual Oil Statistics, Energy Statistics of Non-OECD Countries.

Table 21: World Exports of Petroleum Products (continued)
Tableau 21 : Exportations mondiales de produits pétroliers (suite)

000 metric tons

	1971	1973	1978	1998	1999	2000	2001	2002
Azerbaijan	..	..	..	2 319	2 431	2 087	2 154	..
Belarus	..	..	..	3 406	4 694	6 845	6 246	..
Estonia	..	..	..	15	12	17	1	..
Kazakhstan	..	..	..	1 039	898	1 029	1 831	..
Latvia	..	..	..	123	184	110	1	..
Lithuania	..	..	..	3 996	2 355	2 790	4 570	..
Russia	..	..	..	50 790	48 417	53 585	56 366	..
Turkmenistan	..	..	..	3 935	3 839	3 839	3 839	..
Ukraine	..	..	..	1 875	3 120	2 228	5 697	..
Uzbekistan	..	..	..	450	441	400	385	..
Other	..	..	..	105	37	26	64	..
Former USSR	**30 300**	**33 000**	**44 100**	**68 053**	**66 428**	**72 956**	**81 154**	**..**
Algeria	378	3 397	1 822	18 663	20 083	21 391	21 898	..
Angola	51	166	188	803	805	673	665	..
Côte d'Ivoire	46	291	478	2 268	1 905	2 234	2 474	..
Egypt	185	460	1 023	4 768	4 327	4 057	4 314	..
Ethiopia + Eritrea	146	164	166	..	..	..	..	..
Gabon	541	520	878	292	350	349	206	..
Ghana	193	199	273	259	463	295	345	..
Kenya	1 016	1 093	763	266	236	235	235	..
Libya	455	953	2 316	6 572	7 209	6 782	7 046	..
Morocco	25	50	25	373	512	1 056	1 212	..
Nigeria	404	654	135	4	391	..	1 346	..
Senegal	77	105	165	76	83	93	120	..
South Africa	..	..	..	3 256	5 066	5 890	5 890	..
Tunisia	159	71	62	703	682	606	597	..
Other	1 084	576	591	856	686	1 611	1 457	..
Africa	**4 760**	**8 699**	**8 885**	**39 159**	**42 798**	**45 272**	**47 805**	**..**
Bahrain	11 045	10 431	11 850	11 993	12 582	12 021	11 532	..
Islamic Republic of Iran	13 138	11 887	7 174	15 419	20 606	16 975	11 602	..
Iraq	126	162	1 958	3 359	3 097	2 159	2 812	..
Israel	101	303	733	2 772	2 539	3 212	2 665	..
Kuwait	11 926	12 177	13 016	36 210	35 969	26 280	28 426	..
Oman	..	..	..	1 807	1 264	1 089	532	..
Qatar	..	..	..	1 907	1 952	1 738	1 283	..
Saudi Arabia	13 226	13 548	20 497	54 724	51 562	52 591	53 944	..
Syria	56	3	948	1 923	1 382	1 433	1 653	..
United Arab Emirates	..	..	139	13 012	13 025	16 164	16 166	..
Yemen	2 936	2 282	651	1 695	1 604	1 648	1 678	..
Other	225	298	89	..	..	..	..	..
Middle East	**52 779**	**51 091**	**57 055**	**144 821**	**145 582**	**135 310**	**132 293**	**..**
Bangladesh	37	..	80	..	..	54	42	..
India	152	162	46	1 003	1 055	8 365	10 065	..
Indonesia	3 677	8 345	6 178	10 827	10 193	9 826	9 674	..
Malaysia	272	614	170	7 016	7 034	8 373	8 734	..
Pakistan	697	502	791	84	126	309	392	..
Philippines	663	420	59	889	1 106	1 854	1 275	..
Singapore	15 355	16 571	19 756	47 015	45 333	40 609	40 134	..
Sri Lanka	300	156	181	150	106	125	82	..
Chinese Taipei	280	560	801	1 710	2 128	2 353	4 515	..
Thailand	194	566	4	4 102	4 858	4 671	5 359	..
Other	85	93	164	138	178	167	165	..
Asia	**21 712**	**27 989**	**28 230**	**72 934**	**72 117**	**76 706**	**80 437**	**..**
Hong Kong, China	**79**	**100**	**225**	**8 437**	**3 584**	**1 826**	**1 418**	**..**
China, People's Republic	**115**	**124**	**1 725**	**6 823**	**8 533**	**10 251**	**11 520**	**..**
Non-OECD	**236 597**	**262 583**	**234 355**	**422 881**	**419 394**	**426 577**	**441 624**	**..**
World	**380 353**	**435 925**	**379 746**	**745 720**	**736 149**	**760 273**	**779 168**	**..**

Source: Annual Oil Statistics, Energy Statistics of Non-OECD Countries.

Table 22: IEA Crude Oil Import Costs by Type of Crude (CIF)
Tableau 22 : Coût du pétrole brut à l'importation au sein de l'AIE par type de brut (CAF)

US dollars / barrel

	Abu Dhabi[1] Murban	Abu Dhabi Up. Zakum	Algeria[1] Sahar. Blend	Canada Heavy	Canada Light Sweet	Iran Heavy	Iran Light	Iraq Basrah Li.
1992	20.23	..	20.92	..	..	17.51	18.56	..
1993	18.29	..	18.46	..	..	15.20	16.18	..
1994	17.38	15.88	16.97	13.64	15.88	14.92	15.59	..
1995	18.62	17.42	18.20	15.50	17.73	16.63	16.91	..
1996	21.50	19.58	22.26	18.64	21.42	19.55	19.89	..
1997	21.69	19.81	20.61	16.10	19.98	18.26	19.02	18.01
1998	14.38	13.32	14.02	10.28	13.52	11.64	12.12	11.59
1999	17.19	17.08	18.80	16.11	18.99	16.01	16.25	17.33
2000	29.54	28.18	29.70	23.78	29.43	26.95	27.19	27.61
2001	26.22	24.62	25.83	17.22	25.05	23.37	23.83	21.05
2002	25.22	24.51	25.41	20.98	25.29	23.69	23.94	22.24
January 1999	11.42	10.83	12.33	9.49	11.74	9.92	10.40	10.22
February 1999	11.97	11.38	12.09	10.03	11.94	9.79	10.57	9.99
March 1999	11.99	11.31	12.68	12.04	13.46	10.38	10.44	11.35
April 1999	14.54	13.63	15.30	13.94	16.27	12.99	12.40	14.18
May 1999	15.97	15.42	15.65	14.70	17.08	14.25	14.89	15.52
June 1999	16.92	16.28	16.45	14.42	16.94	14.65	14.32	15.28
July 1999	17.23	16.65	19.06	16.24	18.88	16.69	17.36	16.93
August 1999	19.83	19.10	20.85	17.79	20.33	18.96	18.71	18.71
September 1999	21.68	20.40	22.70	20.31	22.25	20.29	20.98	20.81
October 1999	24.18	22.96	23.47	20.93	22.90	21.57	21.89	21.30
November 1999	24.17	23.09	25.13	20.02	23.92	22.92	22.89	22.31
December 1999	26.06	24.69	26.61	21.03	25.40	24.81	24.12	23.80
January 2000	26.64	25.19	26.88	22.21	26.83	25.02	25.49	24.35
February 2000	26.97	25.07	29.28	23.76	28.07	25.69	26.40	27.02
March 2000	27.82	26.18	29.09	25.53	28.88	25.61	27.93	28.12
April 2000	27.59	26.74	24.21	23.82	26.10	24.24	23.82	24.45
May 2000	25.82	23.67	28.66	22.88	27.25	24.61	24.95	24.79
June 2000	28.85	27.45	31.07	25.73	29.84	27.38	28.10	27.75
July 2000	29.95	28.78	29.47	26.09	29.52	25.66	27.17	27.20
August 2000	29.33	27.94	30.70	25.70	30.07	27.57	27.47	27.05
September 2000	31.16	28.81	33.09	27.11	32.80	29.55	29.83	29.79
October 2000	33.46	31.79	32.26	22.47	33.51	29.97	30.26	29.52
November 2000	34.41	32.24	33.96	22.47	33.51	29.17	31.40	29.99
December 2000	33.17	31.59	27.93	19.26	28.92	26.49	26.52	27.86
January 2001	25.76	24.47	29.32	16.57	27.50	24.07	25.18	25.64
February 2001	26.71	24.75	28.30	18.20	27.95	24.22	24.92	25.94
March 2001	27.42	25.81	26.87	17.56	26.82	24.52	24.81	22.88
April 2001	26.36	25.20	27.23	17.74	26.77	24.18	24.79	22.09
May 2001	27.38	25.55	28.92	18.26	27.41	25.21	25.52	23.14
June 2001	29.78	26.98	29.17	18.73	26.95	25.60	26.46	22.83
July 2001	28.24	26.77	25.79	19.69	25.95	24.51	24.87	22.58
August 2001	26.57	25.00	26.47	21.81	26.38	23.95	24.51	22.31
September 2001	26.91	25.88	27.13	20.52	25.35	24.26	24.83	21.87
October 2001	25.75	25.01	21.94	15.76	21.98	21.85	22.33	19.91
November 2001	21.58	21.00	20.59	11.44	19.32	19.21	19.28	16.78
December 2001	19.74	19.56	20.05	11.57	18.40	18.02	18.16	15.39
January 2002	20.04	19.22	21.09	13.27	18.39	19.00	18.70	16.20
February 2002	20.74	19.94	20.58	15.66	19.80	19.29	19.28	17.18
March 2002	21.50	20.78	22.96	18.19	22.97	21.12	20.45	20.09
April 2002	25.81	23.99	25.10	21.78	25.28	24.17	24.42	23.17
May 2002	26.68	25.79	25.74	22.60	25.78	24.50	24.61	24.67
June 2002	26.76	25.61	24.40	21.47	25.54	23.99	24.77	22.87
July 2002	26.11	25.44	25.97	22.03	26.51	24.28	24.35	23.95
August 2002	26.69	26.09	27.39	24.11	27.58	24.92	25.75	25.75
September 2002	27.40	26.65	28.84	23.77	28.59	26.08	27.51	26.71
October 2002	28.45	27.93	28.46	22.75	27.91	26.74	27.00	26.66
November 2002	27.81	26.93	24.43	21.64	25.66	25.12	25.10	23.54
December 2002	26.32	25.45	29.36	21.74	26.99	24.72	26.01	24.88
January 2003	29.62	27.36	32.81	25.40	30.53	26.93	28.50	27.47
February 2003	31.33	30.30	34.18	27.52	32.80	30.19	30.25	30.43
March 2003	32.78	31.40	32.40	27.71	32.92	28.59	29.56	30.27

1. Before 1994, Abu Dhabi Murban and Algeria Saharan Blend refer respectively to average Abu Dhabi and average Algeria. / *Avant 1994 Abu Dhabi Murban et Algeria Saharan Blend font référence respectivement à la moyenne Abu Dhabi et à la moyenne Algérie.*

Source: Energy Prices & Taxes, 2nd Quarter 2003.

Table 22: IEA Crude Oil Import Costs by Type of Crude (CIF) (continued)

Tableau 22 : Coût du pétrole brut à l'importation au sein de l'AIE par type de brut (CAF) (suite)

US dollars / barrel

	Iraq Kirkuk	Kuwait Blend	Libya[1] Medium	Libya Light	Mexico Maya	Mexico Olmeca	Nigeria Medium	Nigeria[1] Light
1992	x	17.38	19.18	..	13.60	..	..	20.71
1993	x	14.51	16.92	..	12.13	..	..	18.62
1994	x	14.48	15.77	16.01	12.93	16.50	16.99	17.02
1995	x	16.76	17.36	17.45	14.97	17.78	18.06	18.20
1996	22.84	19.66	21.33	20.99	17.73	21.83	21.70	22.03
1997	18.41	18.28	19.45	19.57	15.56	20.13	20.95	20.94
1998	11.50	11.72	12.96	13.07	9.33	13.65	14.33	14.18
1999	16.42	16.49	17.86	17.98	14.43	17.89	15.98	17.99
2000	27.06	27.15	28.98	28.91	23.82	29.59	30.03	30.31
2001	22.56	22.83	24.78	25.33	17.91	24.62	25.87	26.26
2002	23.20	23.46	24.80	25.27	21.22	24.38	25.29	25.86
January 1999	9.91	9.96	11.04	10.94	7.80	11.30	11.35	11.47
February 1999	9.73	9.81	11.02	11.10	7.68	10.82	11.19	11.46
March 1999	11.57	11.10	13.01	12.37	9.74	12.41	12.49	12.34
April 1999	13.90	13.86	15.17	15.17	12.44	15.70	15.22	15.03
May 1999	14.77	14.71	15.53	15.51	12.80	16.42	16.63	17.08
June 1999	14.46	16.06	16.36	15.84	13.16	16.03	15.41	15.96
July 1999	17.27	17.10	18.35	18.23	15.25	18.76	..	18.65
August 1999	19.07	18.62	20.41	20.60	17.51	20.49	20.55	20.66
September 1999	20.68	20.79	22.41	22.44	19.36	22.71	22.08	22.61
October 1999	21.62	21.23	22.60	22.47	19.12	22.34	23.55	23.58
November 1999	22.72	23.23	24.55	24.55	20.06	24.45	24.18	24.30
December 1999	24.49	23.99	25.59	25.63	20.74	25.79	25.79	26.54
January 2000	25.09	25.45	25.67	26.30	21.58	26.16	25.49	26.50
February 2000	26.95	25.94	28.24	27.75	23.52	28.69	28.59	27.93
March 2000	28.45	25.99	28.02	27.67	24.31	29.52	28.90	29.75
April 2000	20.14	23.93	24.22	25.19	21.25	25.11	23.80	26.46
May 2000	24.96	25.36	27.36	26.68	23.29	28.04	28.00	26.88
June 2000	27.79	29.49	29.94	29.86	25.33	30.61	..	30.56
July 2000	24.59	27.87	30.16	29.45	24.12	29.91	..	31.88
August 2000	25.82	28.05	30.26	30.03	24.13	30.42	27.95	30.42
September 2000	30.08	29.05	31.45	32.32	26.87	34.14	35.11	33.96
October 2000	28.46	30.68	31.58	31.85	25.10	33.13	32.07	32.66
November 2000	30.15	29.25	32.60	32.48	25.29	32.28	34.50	33.61
December 2000	27.29	25.81	27.24	27.57	20.17	28.60	29.98	32.15
January 2001	25.44	22.88	26.77	26.78	19.03	27.19	25.51	26.89
February 2001	25.13	23.78	27.75	28.19	19.57	28.21	28.51	29.09
March 2001	21.59	24.03	25.14	26.06	18.64	25.74	28.16	27.65
April 2001	23.20	22.83	25.67	26.20	16.93	25.28	25.63	26.36
May 2001	25.29	24.52	28.13	28.23	18.77	27.04	28.51	28.63
June 2001	24.99	24.67	27.64	28.43	19.65	26.80	29.69	30.35
July 2001	24.06	23.96	25.09	24.97	18.84	25.17	27.12	27.70
August 2001	24.03	23.13	25.45	25.79	19.83	25.78	26.35	26.22
September 2001	24.96	24.41	25.36	25.00	20.62	25.61	27.26	27.47
October 2001	20.84	22.40	21.45	23.39	16.75	21.34	23.36	24.54
November 2001	18.38	18.89	19.55	19.53	14.26	18.96	21.16	20.76
December 2001	17.32	17.72	18.97	19.18	13.72	18.49	19.89	19.95
January 2002	17.79	17.89	19.73	19.99	14.08	19.33	20.83	20.74
February 2002	17.77	18.66	20.06	20.06	15.18	20.01	21.02	20.83
March 2002	21.28	20.27	23.31	22.70	19.06	23.00	23.07	22.68
April 2002	23.62	23.72	24.78	24.85	22.27	25.87	25.76	26.21
May 2002	23.23	25.07	24.82	24.79	22.84	26.19	26.91	27.06
June 2002	23.58	24.61	23.87	24.51	21.89	24.60	25.60	25.35
July 2002	24.40	24.64	25.29	25.49	22.23	26.37	26.13	26.30
August 2002	25.57	25.48	26.49	26.87	23.47	26.98	27.13	27.04
September 2002	26.39	26.30	28.38	28.35	24.85	28.39	28.63	28.89
October 2002	26.08	26.91	27.86	27.84	23.76	..	27.99	29.62
November 2002	22.59	25.67	25.41	25.36	20.95	..	27.15	26.85
December 2002	25.46	25.09	28.07	29.10	23.14	..	29.00	27.60
January 2003	28.55	26.93	31.76	32.53	27.71	31.99	31.81	32.79
February 2003	30.08	30.03	32.75	33.92	28.22	33.35	34.17	33.67
March 2003	30.40	29.67	31.90	32.48	25.97	32.94	35.65	35.26

1. Before 1994, Libya Medium and Nigeria Light refer respectively to average Libya and Nigeria. / *Avant 1994, Lybia Medium et Nigeria Light font référence respectivement à la moyenne Libye et Nigéria.*

Source: Energy Prices & Taxes, 2nd Quarter 2003.

Table 22: IEA Crude Oil Import Costs by Type of Crude (CIF) (continued)

Tableau 22 : Coût du pétrole brut à l'importation au sein de l'AIE par type de brut (CAF) (suite)

US dollars / barrel

	Norway Ekofisk	Norway Gullfaks	Norway Oseberg	Norway Statfjord	Oman	S. Arabia Heavy	S. Arabia Medium	S. Arabia Light
1992	..	..	..	..	19.13	15.95	17.13	18.52
1993	..	..	..	..	17.10	14.11	15.13	16.28
1994	16.12	16.64	16.42	16.36	16.22	13.98	14.56	15.53
1995	17.62	17.73	17.80	17.57	17.67	16.16	16.53	17.13
1996	21.36	21.10	21.56	21.49	20.49	19.12	19.72	20.37
1997	19.71	20.12	20.20	20.01	20.53	16.77	17.81	18.67
1998	13.25	13.34	13.46	13.36	13.50	10.44	11.49	12.28
1999	17.65	18.03	18.26	18.10	17.21	15.74	16.74	16.96
2000	28.91	29.30	29.18	29.02	30.61	25.94	26.36	27.36
2001	24.35	25.40	25.10	24.94	24.64	21.91	21.42	23.54
2002	25.21	24.99	25.25	25.35	24.60	23.59	23.87	24.13
January 1999	11.30	11.72	11.71	11.71	10.66	8.99	9.53	10.45
February 1999	10.76	11.07	11.13	10.98	11.33	9.16	9.44	10.16
March 1999	12.95	13.02	12.82	12.59	11.13	11.37	11.80	12.18
April 1999	15.52	15.70	15.90	15.52	13.91	13.52	14.14	14.70
May 1999	15.24	15.62	15.58	15.47	16.02	14.47	15.02	15.25
June 1999	15.91	16.25	16.06	15.88	16.30	15.62	15.66	15.58
July 1999	18.54	18.72	19.39	19.45	16.86	16.93	17.39	18.01
August 1999	20.52	20.73	20.53	20.79	19.22	18.26	18.79	19.37
September 1999	22.21	22.28	22.53	22.46	21.51	20.13	20.73	21.40
October 1999	22.27	22.48	22.90	22.58	23.46	20.52	21.22	21.59
November 1999	24.66	24.39	24.93	24.81	23.71	22.18	22.68	23.90
December 1999	26.17	26.33	26.54	26.43	25.44	23.65	23.89	24.73
January 2000	26.27	26.49	26.88	26.46	25.98	24.16	24.52	25.39
February 2000	28.27	28.57	28.73	28.74	25.86	25.62	26.69	27.09
March 2000	28.79	28.22	28.56	29.11	76.57	25.35	25.84	26.95
April 2000	23.82	24.90	24.14	23.47	26.75	22.72	22.80	24.55
May 2000	27.64	27.25	27.94	28.08	24.47	25.04	25.14	26.12
June 2000	30.26	30.44	30.33	30.32	27.92	26.98	27.48	28.07
July 2000	29.53	29.51	29.63	29.43	29.51	26.25	26.44	27.80
August 2000	29.62	30.38	30.02	30.24	28.37	27.29	27.43	28.24
September 2000	32.72	32.77	33.47	32.31	29.41	28.28	28.71	29.11
October 2000	31.25	32.35	31.93	31.85	31.32	28.29	28.84	29.85
November 2000	32.87	33.30	33.66	33.19	32.28	28.52	28.73	29.78
December 2000	26.55	28.35	27.84	25.97	29.45	23.34	23.82	25.82
January 2001	25.47	26.72	26.32	26.35	23.70	22.66	23.20	24.87
February 2001	28.16	28.43	28.16	28.20	24.67	22.84	23.43	25.63
March 2001	25.03	25.35	25.76	24.96	25.94	22.24	22.08	24.75
April 2001	25.86	25.64	26.01	26.22	25.08	22.11	21.94	24.49
May 2001	28.91	28.50	29.00	28.88	26.08	24.20	23.20	25.52
June 2001	28.10	28.01	28.94	28.95	27.21	24.13	22.48	25.23
July 2001	25.08	25.31	24.46	25.17	26.59	22.50	22.41	23.94
August 2001	25.97	27.07	26.95	26.29	25.35	22.95	23.01	24.27
September 2001	26.16	26.12	26.24	26.18	25.85	22.84	22.59	24.37
October 2001	20.69	21.46	21.11	20.94	25.15	19.55	18.96	21.69
November 2001	19.40	20.68	20.05	19.75	20.67	17.46	16.51	18.92
December 2001	18.78	19.59	19.17	19.26	19.09	16.57	16.16	17.87
January 2002	19.73	20.03	20.12	19.98	19.20	16.98	16.82	18.58
February 2002	20.25	20.45	20.90	20.35	20.43	18.47	18.31	19.50
March 2002	23.20	22.45	21.57	23.58	21.34	21.39	22.17	22.10
April 2002	25.35	25.77	25.95	25.43	24.86	23.41	24.00	24.10
May 2002	25.04	25.62	25.09	25.23	26.13	24.07	24.81	24.61
June 2002	24.74	24.13	24.32	24.05	26.47	23.62	24.03	24.36
July 2002	26.06	26.23	26.27	26.33	25.69	24.18	24.58	24.86
August 2002	27.17	27.76	27.15	27.41	26.27	25.13	25.64	25.73
September 2002	28.84	28.64	29.68	28.90	26.74	26.25	26.82	26.99
October 2002	28.20	27.97	28.85	28.24	28.07	26.14	25.94	26.59
November 2002	24.90	25.43	24.57	25.05	27.08	24.43	23.66	25.07
December 2002	28.87	27.58	28.41	29.46	25.77	24.60	25.79	26.18
January 2003	32.73	32.43	33.03	32.75	28.07	27.02	28.33	29.01
February 2003	33.37	33.91	33.28	34.13	30.76	30.76	31.59	31.36
March 2003	32.95	32.45	32.95	32.43	32.54	28.60	28.71	29.90

Source: Energy Prices & Taxes, 2nd Quarter 2003.

Table 22: IEA Crude Oil Import Costs by Type of Crude (CIF) (continued)

Tableau 22 : Coût du pétrole brut à l'importation au sein de l'AIE par type de brut (CAF) (suite)

US dollars / barrel

	S. Arabia Berri (Ex.Li.)	Syria Light	U.K. Brent Blend	U.K. Forties	Urals[1]	Venezuela Extra Heavy	Venezuela Medium	Venezuela Light
1992	..	..	..	..	18.27	..	16.22	17.36
1993	..	..	..	..	15.46	..	14.01	14.89
1994	16.33	16.06	16.54	16.54	15.65	12.32	13.45	14.70
1995	17.75	17.40	17.86	17.40	17.12	14.13	15.13	15.58
1996	21.05	21.12	21.25	21.41	20.20	16.50	18.42	20.16
1997	20.29	19.26	20.90	20.04	18.59	14.53	16.96	18.61
1998	13.27	12.49	13.59	13.42	12.04	8.17	10.92	12.66
1999	17.87	17.73	18.09	18.66	16.74	13.16	15.29	17.33
2000	28.94	27.69	29.23	29.13	26.73	23.02	26.63	29.49
2001	24.91	24.30	26.12	25.83	23.27	17.58	21.00	23.86
2002	24.66	24.93	26.41	25.33	23.77	19.27	21.51	24.73
January 1999	11.10	11.36	11.49	11.75	10.62	6.97	8.87	10.59
February 1999	11.26	10.23	11.28	10.96	9.98	6.08	7.62	10.49
March 1999	12.12	12.21	12.10	12.76	11.83	9.08	11.27	12.59
April 1999	14.52	15.26	14.56	15.74	14.37	11.36	13.91	15.03
May 1999	16.15	15.28	16.09	16.02	14.40	10.68	13.22	16.27
June 1999	16.80	15.60	16.15	16.38	14.73	12.08	13.01	16.16
July 1999	18.12	18.07	18.38	19.13	18.21	13.79	16.16	18.15
August 1999	19.87	20.57	21.36	21.11	19.81	16.03	17.78	19.48
September 1999	21.60	22.52	22.36	22.76	21.64	17.72	20.09	22.10
October 1999	23.00	22.26	22.56	22.49	21.53	18.83	19.11	22.43
November 1999	23.84	25.00	24.77	25.28	23.80	18.37	21.68	24.12
December 1999	25.48	26.13	26.24	26.78	25.17	19.03	23.27	25.35
January 2000	26.38	26.31	26.15	27.12	25.15	20.69	24.37	26.36
February 2000	26.97	28.46	27.97	29.19	27.40	22.78	26.13	27.99
March 2000	27.82	28.01	26.88	28.69	27.05	22.69	26.11	29.81
April 2000	26.20	22.61	24.06	23.62	21.81	21.01	23.56	25.37
May 2000	26.13	27.38	28.09	27.57	26.61	22.42	26.01	28.58
June 2000	28.74	29.65	30.28	30.79	28.10	24.45	28.41	32.04
July 2000	29.32	25.90	32.27	29.96	26.06	22.55	26.13	29.32
August 2000	29.33	27.94	31.36	30.32	26.63	23.60	28.06	30.10
September 2000	30.98	28.65	35.13	33.63	30.76	26.09	30.24	32.16
October 2000	32.19	29.33	31.46	31.99	28.85	24.43	28.94	32.34
November 2000	32.57	32.63	33.80	34.67	31.11	24.31	29.45	32.63
December 2000	28.49	25.97	29.73	26.88	25.27	19.55	24.04	28.09
January 2001	25.68	25.75	26.97	26.89	24.56	19.37	24.86	28.04
February 2001	26.43	27.56	28.89	28.99	25.96	19.00	23.01	28.42
March 2001	26.88	23.06	26.28	25.76	22.01	18.29	22.60	25.27
April 2001	26.09	25.16	26.11	26.86	23.62	18.52	24.85	25.14
May 2001	26.84	27.99	28.85	29.18	26.49	19.44	24.21	26.09
June 2001	26.90	27.38	28.42	28.91	26.38	18.13	21.14	26.20
July 2001	25.43	24.10	25.75	25.37	23.50	17.85	20.39	25.35
August 2001	25.69	25.20	26.68	26.84	24.80	19.17	22.72	25.45
September 2001	25.76	25.25	27.78	26.17	25.30	18.78	20.38	24.75
October 2001	23.37	20.83	22.89	21.92	20.42	16.71	18.03	20.63
November 2001	20.82	19.25	20.34	20.12	18.50	12.57	15.16	17.94
December 2001	19.14	19.34	19.28	19.63	18.34	12.23	14.23	18.37
January 2002	19.40	20.14	20.36	20.02	18.95	13.19	15.03	18.54
February 2002	20.23	20.04	21.07	20.57	19.21	13.17	15.47	19.85
March 2002	21.88	22.87	22.83	23.02	22.60	16.83	21.00	22.74
April 2002	25.09	24.66	24.37	25.90	23.94	17.69	22.88	25.38
May 2002	25.86	25.09	25.58	26.05	24.61	19.87	21.51	25.48
June 2002	25.59	23.54	25.41	24.85	23.06	20.07	22.06	24.72
July 2002	25.75	25.52	26.56	26.58	24.83	22.09	23.55	25.65
August 2002	26.34	26.43	26.89	26.95	25.63	22.02	23.42	26.77
September 2002	27.45	28.21	27.75	28.25	26.94	22.70	25.25	28.36
October 2002	27.85	27.26	28.82	28.66	24.97	21.98	23.88	27.77
November 2002	26.61	23.98	27.86	24.92	23.15	20.39	21.86	25.53
December 2002	26.12	28.40	28.57	29.09	25.10	20.99	..	26.72
January 2003	29.24	30.88	32.41	32.91	30.22	..	..	32.28
February 2003	31.90	32.39	34.69	34.05	30.80	..	32.36	33.89
March 2003	30.79	31.83	33.49	32.92	29.72	24.07	27.54	30.56

1. Before 1994, Urals refers to USSR. / *Avant 1994, Urals fait référence à URSS.*

Source: Energy Prices & Taxes, 2nd Quarter 2003.

Table 23: IEA Crude Oil Import Costs by Country (Average Unit Value, CIF)[1]

Tableau 23 : Coût du pétrole brut à l'importation au sein de l'AIE par pays (Valeur moyenne, CAF)[1]

US dollars / barrel

	1994	1995	1996	1997	1998	1999	2000	2001	2002
Australia	16.76	18.53	21.81	21.78	14.60	18.38	30.79	26.61	25.80
Austria	16.99	18.78	22.06	21.31	14.34	17.54	29.39	25.32	24.64
Belgium	15.41	16.94	20.53	18.65	11.97	17.33	27.87	24.20	24.35
Canada	16.30	17.76	21.26	20.59	13.15	17.85	29.10	24.87	24.97
Denmark	16.12	17.49	21.25	20.15	13.49	17.71	29.66	24.82	24.88
Finland	16.78	17.83	21.69	19.44	12.80	18.31	28.13	23.49	24.51
France	15.76	17.14	20.82	18.99	12.43	17.45	28.18	24.13	24.63
Germany	15.81	17.07	20.68	19.01	12.48	17.51	28.09	24.15	24.40
Greece	15.17	16.54	20.08	18.45	11.66	16.64	26.95	23.22	24.08
Ireland	15.76	17.65	21.19	19.99	13.55	17.14	29.88	25.31	25.52
Italy	15.49	16.90	20.53	18.88	12.21	17.10	27.77	23.87	24.34
Japan	16.48	18.02	20.55	20.55	13.68	17.38	28.72	25.01	24.96
Nethland	15.20	16.92	20.39	18.37	11.98	16.97	27.59	23.48	23.99
New Zealand	16.93	18.73	21.86	21.65	14.63	18.16	29.95	26.14	25.89
Norway	14.44	16.41	21.62	16.71	12.23	17.46	28.91	23.43	24.46
Portugal	15.72	17.22	20.35	18.95	12.21	17.38	28.20	24.02	24.27
Spain	15.24	16.96	20.45	18.34	11.80	16.99	27.16	23.32	23.80
Sweden	15.81	17.23	20.86	18.90	12.61	17.68	28.13	24.03	23.86
Switzerland	16.36	17.69	21.71	20.50	13.38	18.35	29.53	25.04	25.34
Turkey	15.52	16.78	20.25	18.79	11.99	16.07	26.61	22.98	23.57
United Kingdom	15.83	17.29	21.08	19.32	12.64	18.01	28.45	24.45	24.58
United States	15.06	16.74	20.16	18.34	12.02	17.06	27.54	22.07	23.58
Total IEA	**15.66**	**17.19**	**20.52**	**19.11**	**12.52**	**17.26**	**28.00**	**23.65**	**24.18**

1. Excludes Hungary, but includes Czech Republic and Korea since January 2001. / *Ne comprend pas la Hongrie, mais comprend la République tchèque et la Corée depuis janvier 2001.*

Table 24: Crude Oil Spot Prices (FOB)[1]

Tableau 24 : Prix au comptant du pétrole brut (FAB)[1]

US dollars / barrel

API Gravity:	**38.3º**	**39.6º**	**34.2º**	**38.7º**	**31.0º**	**30.2º**	**32.0º**	**35.0º**	**45.5º**
Sulphur Content:	**0.4%**	**0.2%**	**1.3%**	**0.1%**	**2.0%**	**1.8%**	**1.3%**	**0.1%**	**0.1%**
	Brent	**WTI**	**WTS**	**LLS**	**Dubai**	**Iran H.**	**Urals**	**Minas**	**Tapis**
1995	17.04	18.43	17.42	18.60	16.11	16.24	16.87	17.65	18.31
1996	20.66	22.14	20.99	22.40	18.55	18.44	20.36	20.65	21.98
1997	19.10	20.61	18.97	20.76	18.13	18.05	18.44	19.30	20.98
1998	12.74	14.41	12.78	14.14	12.17	11.51	12.02	12.34	13.82
1999	17.87	19.25	17.85	18.93	17.20	16.72	17.34	17.79	18.84
2000	28.39	30.30	28.29	30.45	26.15	26.05	26.97	28.75	29.74
2001	24.46	25.89	23.17	25.95	22.81	22.08	23.12	24.06	25.33
2002	24.98	26.09	24.65	26.14	23.80	23.05	23.76	25.54	25.64
January 2002	19.48	19.68	18.17	20.08	18.48	18.69	18.46	18.86	20.12
February 2002	20.22	20.66	19.29	20.96	19.02	18.58	18.98	18.90	20.17
March 2002	23.73	24.35	22.83	24.48	22.96	21.98	22.47	22.92	23.55
April 2002	25.66	26.26	25.07	26.59	24.52	23.64	24.03	25.81	25.90
May 2002	25.33	27.06	26.03	26.76	24.69	23.40	23.98	25.60	25.97
June 2002	24.13	25.50	24.26	25.44	23.91	22.28	23.15	24.62	24.97
July 2002	25.81	26.92	26.17	27.05	24.67	23.99	24.83	25.28	26.36
August 2002	26.66	28.34	26.89	27.77	25.24	24.72	25.85	25.93	27.40
September 2002	28.38	29.71	27.97	29.44	26.80	26.56	26.84	27.76	28.20
October 2002	27.58	28.87	27.91	29.49	26.32	25.42	26.06	29.75	27.90
November 2002	24.10	26.29	23.90	26.17	23.31	21.75	22.81	28.26	26.89
December 2002	28.67	29.46	27.36	29.43	25.73	25.62	27.64	32.76	30.27
January 2003	31.32	32.99	31.46	33.61	28.02	28.40	29.55	32.37	31.95
February 2003	32.67	35.75	31.10	35.64	30.02	29.21	30.89	32.03	33.96
March 2003	30.54	33.43	28.71	33.45	27.38	26.82	27.84	30.47	31.37
April 2003	24.85	28.26	25.44	27.92	23.45	22.13	22.32	29.76	27.66
May 2003	25.72	28.14	26.68	28.23	24.36	22.58	23.87	28.82	26.76
June 2003*	27.63	30.94	28.53	30.59	25.40	24.10	25.83	27.38	27.13

1. Urals prices are CIF. / *Les prix Urals sont CAF.*

WTI=West Texas Intermediate; **WTS**=West Texas Sour; **LLS**=Louisiana Light Sweet.

* Prices for June 2003 are preliminary. / *Les prix pour juin 2003 sont préliminaires.*

Source: Energy Prices & Taxes, 2nd Quarter 2003.

Table 25: Petroleum Product Spot Prices
Tableau 25 : Prix au comptant des produits pétroliers

US dollars / barrel

	Motor Gasoline			Gasoil			High Sulphur Fuel Oil		
	NW Eur.[1]	USA[2]	Sing.[3]	NW Eur.	USA	Sing.	NW Eur.	USA	Sing.
				0.2%	*#2*	*0.5%*	*3.5% S*	*3% S*	*380 cst*
1990	32.83	30.18	32.31	28.51	28.67	29.30	15.62	16.29	16.31
1991	28.36	27.52	28.59	26.96	25.58	28.27	12.16	11.40	13.26
1992	25.32	24.56	26.59	23.73	24.02	25.15	12.82	12.34	12.84
1993	20.69	21.58	24.02	22.27	22.03	24.01	10.15	11.21	11.04
1994	18.63	20.54	21.08	19.81	20.69	20.85	12.93	12.24	12.65
1995	19.75	22.58	22.12	20.49	20.75	21.64	14.74	14.51	14.66
1996	22.97	25.81	23.58	25.90	26.36	27.07	16.20	16.02	16.27
1997	23.02	25.21	24.56	23.39	23.65	24.24	14.51	15.26	15.43
1998	16.54	17.82	17.19	16.20	16.45	15.46	10.13	10.26	10.34
1999	20.80	22.69	21.01	20.06	20.46	19.13	14.31	13.94	15.47
2000	34.27	35.95	32.51	34.28	36.41	32.44	21.33	20.65	24.12
2001	28.91	30.96	27.50	29.15	29.78	27.29	17.80	17.35	20.37
2002	28.53	30.25	28.00	27.77	28.50	27.45	20.61	20.90	22.88
1st Quarter 2001	30.70	34.08	30.44	30.57	33.14	27.58	17.77	18.53	20.14
2nd Quarter 2001	35.48	36.17	30.76	31.29	32.22	30.23	18.44	17.53	21.91
3rd Quarter 2001	28.63	30.81	26.97	30.52	30.03	28.86	19.11	17.78	21.67
4th Quarter 2001	20.83	22.79	21.84	24.20	23.72	22.50	15.88	15.58	17.76
1st Quarter 2002	22.69	25.03	24.37	23.13	23.95	22.49	16.91	16.03	18.64
2nd Quarter 2002	29.49	30.18	29.47	26.78	27.67	27.67	21.19	21.39	23.31
3rd Quarter 2002	31.46	32.34	28.94	29.90	30.11	28.84	23.18	23.34	25.27
4th Quarter 2002	30.46	33.44	29.22	31.27	32.27	30.81	21.14	22.84	24.29
1st Quarter 2003	37.59	40.46	38.53	40.46	44.01	37.17	26.19	28.85	29.39
2nd Quarter 2003*	33.10	33.20	29.60	30.22	31.99	28.78	20.91	20.59	24.60

1. Premium leaded prior to June 1992, regular unleaded thereafter. / *Supercarburant plombé avant juin 1992, essence ordinaire non-plombée ensuite.*
2. Regular unleaded. / *Essence ordinaire sans plomb.*
3. Regular leaded prior to 1995, premium unleaded from 1995. / *Essence ordinaire plombée avant 1995, supercarburant non plombé à partir de 1995.*

* Prices for the 2nd Quarter 2003 are preliminary. / *Les prix pour le 2ème trimestre 2003 sont préliminaires.*

Table 26: Retail Motor Gasoline Prices
Tableau 26 : Prix au détail de l'essence auto

National currency / litre

	France[1]	Germany[1]	Italy[1]	Spain[2]	UK[1]	Japan[3]	US[3]
	Euro	Euro	Euro	Euro	Pounds	Yen	Dollars
1990	0.795	0.604	0.751	0.497	0.420	121.0	0.307
1991	0.783	0.677	0.762	0.533	0.451	123.0	0.301
1992	0.760	0.716	0.761	0.568	0.461	120.0	0.298
1993	0.778	0.712	0.790	0.609	0.499	118.0	0.293
1994	0.803	0.797	0.815	0.635	0.515	118.5	0.294
1995	0.857	0.793	0.889	0.642	0.538	110.1	0.303
1996	0.909	0.827	0.924	0.667	0.565	104.8	0.325
1997	0.941	0.854	0.944	0.699	0.618	104.4	0.326
1998	0.919	0.814	0.910	0.665	0.648	97.7	0.280
1999	0.954	0.874	0.960	0.700	0.701	98.3	0.308
2000	1.090	1.015	1.081	0.819	0.799	103.9	0.399
2001	1.036	1.024	1.051	0.807	0.758	104.9	0.386
2002	1.014	1.048	1.048	0.814	0.732	101.5	0.360
1st Quarter 2001	1.029	1.038	1.046	0.802	0.764	107.0	0.388
2nd Quarter 2001	1.105	1.087	1.103	0.862	0.777	105.0	0.434
3rd Quarter 2001	1.037	1.019	1.052	0.815	0.770	104.5	0.391
4th Quarter 2001	0.975	0.950	1.009	0.752	0.719	102.9	0.331
1st Quarter 2002	0.966	1.012	1.006	0.776	0.705	100.2	0.309
2nd Quarter 2002	1.029	1.071	1.068	0.834	0.745	101.8	0.373
3rd Quarter 2002	1.026	1.063	1.056	0.825	0.738	102.5	0.375
4th Quarter 2002	1.033	1.046	1.056	0.815	0.741	101.5	0.381
1st Quarter 2003	1.066	1.127	1.085	0.844	0.763	103.0	0.419
2nd Quarter 2003	1.016	..	1.048	0.807	..	..	..

1. Premium leaded prior to 1990, unleaded (95) beginning with 1990. / *Supercarburant plombé avant 1990 et non-plombé (95) à partir de 1990.*
2. Premium leaded prior to 1992, unleaded (95) beginning with 1992. / *Supercarburant plombé avant 1992 et non-plombé (95) à partir de 1992 .*
3. Unleaded regular for all years. / *Le prix de l'essence auto est celui de l'essence ordinaire sans plomb .*

Source: Energy Prices & Taxes, 2nd Quarter 2003.

Table 27: Light Fuel Oil Prices for Households
Tableau 27 : Prix de fuel oil léger destiné aux ménages

National currency / thousand litres

	France Euro	Germany Euro	Italy Euro	Spain Euro	UK Pounds	Japan [1] Yen	US Dollars
1990	333.2	246.0	476.2	233.8	146.4	46 260	287.2
1991	348.9	261.9	569.6	252.3	136.5	51 270	268.9
1992	312.3	229.7	582.0	250.0	124.9	50 530	253.9
1993	318.9	233.5	628.2	285.8	134.2	50 240	258.3
1994	310.8	216.0	633.5	255.0	132.6	47 640	250.5
1995	304.9	207.3	669.2	241.1	138.7	41 454	247.2
1996	334.4	247.8	711.0	273.3	165.3	44 236	279.0
1997	351.0	252.1	725.1	291.5	154.5	49 783	276.0
1998	301.4	207.1	697.1	254.8	124.7	44 738	241.6
1999	326.7	257.6	737.9	287.8	138.9	42 405	252.8
2000	459.6	408.7	855.5	425.2	215.1	47 194	360.6
2001	399.1	373.0	819.5	390.4	191.2	49 783	344.2
2002	364.2	350.7	833.4	369.7	159.3	45 589	312.9
1st Quarter 2001	420.7	367.5	819.3	392.5	204.0	50 614	375.2
2nd Quarter 2001	421.3	395.1	822.9	400.2	200.7	50 245	349.1
3rd Quarter 2001	398.4	397.2	819.7	397.6	196.6	50 010	324.4
4th Quarter 2001	355.9	332.1	816.2	371.3	163.3	48 261	312.9
1st Quarter 2002	339.3	325.0	818.9	349.0	147.1	45 325	309.4
2nd Quarter 2002	359.9	353.3	832.4	375.6	158.3	45 441	307.1
3rd Quarter 2002	367.5	363.7	832.0	368.1	163.0	45 815	299.4
4th Quarter 2002	390.3	360.8	850.3	385.9	168.9	45 773	335.8
1st Quarter 2003	444.6	405.4	900.5	431.4	205.0	47 173	415.3
2nd Quarter 2003	367.7	..	835.1	380.1	..	..	..

1. The price shown is for kerosene, the fuel predominantly used in Japanese households. */ Le prix indiqué concerne le pétrole lampant, principal combustible utilisé par les ménages japonais.*

Table 28: Heavy Fuel Oil Prices for Industry [1]
Tableau 28 : Prix du fuel oil lourd pour l'industrie [1]

National currency / metric ton

	France Euro	Germany Euro	Italy Euro	Spain Euro	UK Pounds	Japan Yen	US Dollars
1990	113.2	118.3	116.3	87.6	76.2	27 190	109.6
1991	100.0	116.7	117.9	86.2	69.9	32 340	83.8
1992	97.0	106.9	112.7	79.7	65.2	26 650	85.8
1993	91.1	100.2	116.5	90.4	66.1	24 120	93.2
1994	103.2	104.3	129.7	109.5	76.2	18 380	98.5
1995	110.4	108.4	146.6	131.0	89.3	17 076	109.8
1996	117.1	116.6	152.8	148.1	98.0	18 793	126.3
1997	116.5	118.1	149.5	150.8	93.6	21 908	118.7
1998	95.0	100.7	134.5	113.9	80.1	20 153	85.5
1999	116.3	116.6	169.4	135.7	93.0	19 358	106.7
2000	194.3	187.9	239.6	202.3	126.7	24 177	168.8
2001	172.4	168.1	220.6	179.8	123.1	26 389	148.1
2002	186.3	183.1	..	195.7	..	25 247	158.8
1st Quarter 2001	174.8	176.9	215.3	176.3	120.7	27 080	167.4
2nd Quarter 2001	175.6	170.8	224.3	183.8	124.8	26 526	148.2
3rd Quarter 2001	182.5	171.5	236.2	188.6	129.9	26 526	144.9
4th Quarter 2001	156.9	153.2	206.8	170.4	117.2	25 421	125.8
1st Quarter 2002	170.7	166.6	218.4	171.7	117.6	23 211	128.8
2nd Quarter 2002	192.6	181.3	242.3	200.4	135.1	25 052	160.6
3rd Quarter 2002	192.3	184.4	..	197.6	141.6	25 846	171.2
4th Quarter 2002	189.7	200.0	..	213.3	x	26 879	174.5
1st Quarter 2003	213.2	213.7	..	..	x	28 113	219.7
2nd Quarter 2003	166.0	..	..	..	x	28 690	..

1. The price shown is for high sulphur fuel oil, except for Germany, for which the price corresponds to low sulphur fuel oil as of 1991. /
Le prix indiqué concerne le fuel oil lourd à haute teneur en souffre, sauf pour l'Allemagne pour laquelle à partir de 1991, le prix correspond au fuel oil lourd à faible teneur en souffre.

Source: Energy Prices and Taxes, 2nd Quarter 2003.

GRAPHS FOR PART II

GRAPHIQUES POUR LA DEUXIEME PARTIE

Graph / *Graphique* 1

Yearly Per Cent Change in Oil Consumption, TPES and GDP*

*Variation annuelle de la consommation de pétrole, de l'ATEP et du PIB**

OECD TOTAL

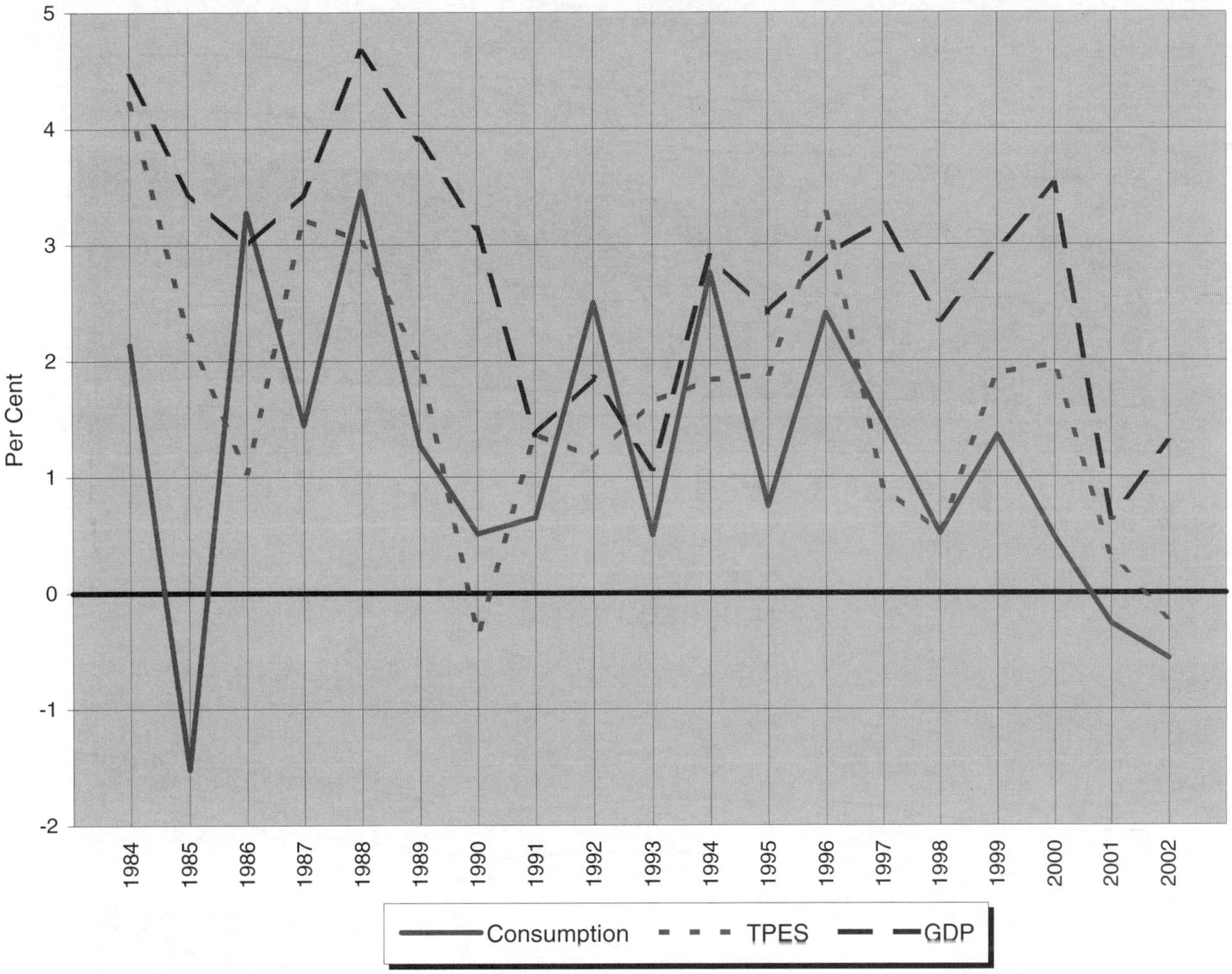

* GDP is calculated in million 1995 US dollars.

** Le PIB est calculé en millions de dollars de 1995.*

Graph / *Graphique* 2

Demand by Main Product Groups *

*Demande des principales catégories de produits raffinés **

OECD TOTAL

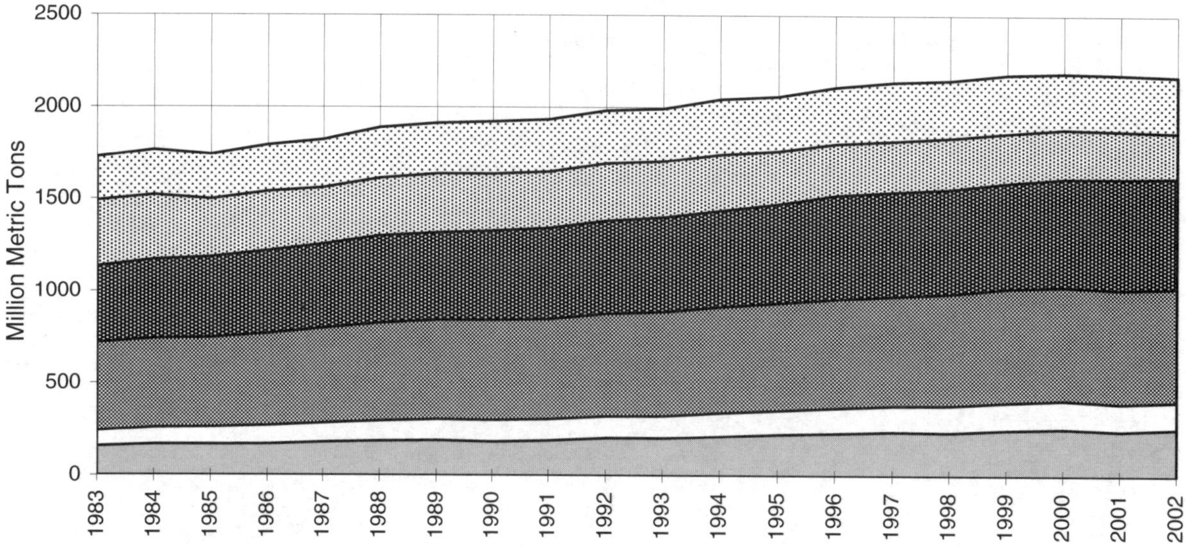

NON-OECD TOTAL

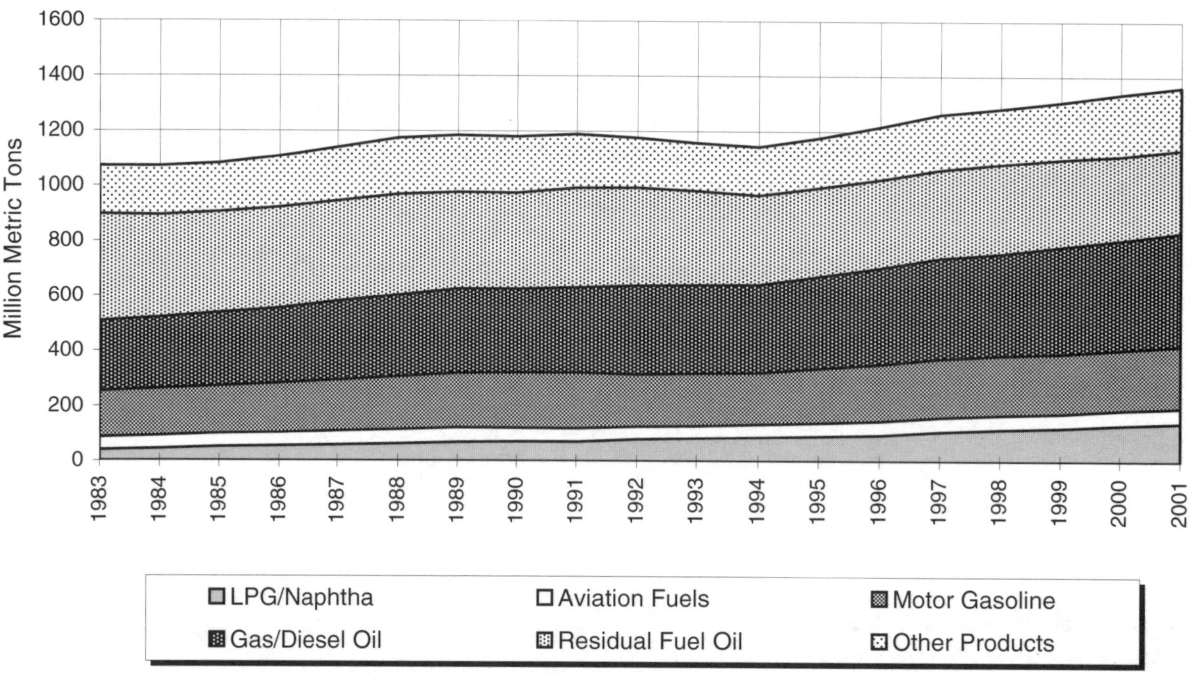

* Demand refers to net inland deliveries (including refinery fuel and international marine bunkers).

* *La demande concerne les livraisons intérieures nettes (comprenant la consommation propre des raffineries et les soutages maritimes internationaux).*

Graph / *Graphique* 3
Unleaded Motor Gasoline Prices and Taxes*, 2002
Les prix et taxes de l'essence auto sans plomb, 2002*

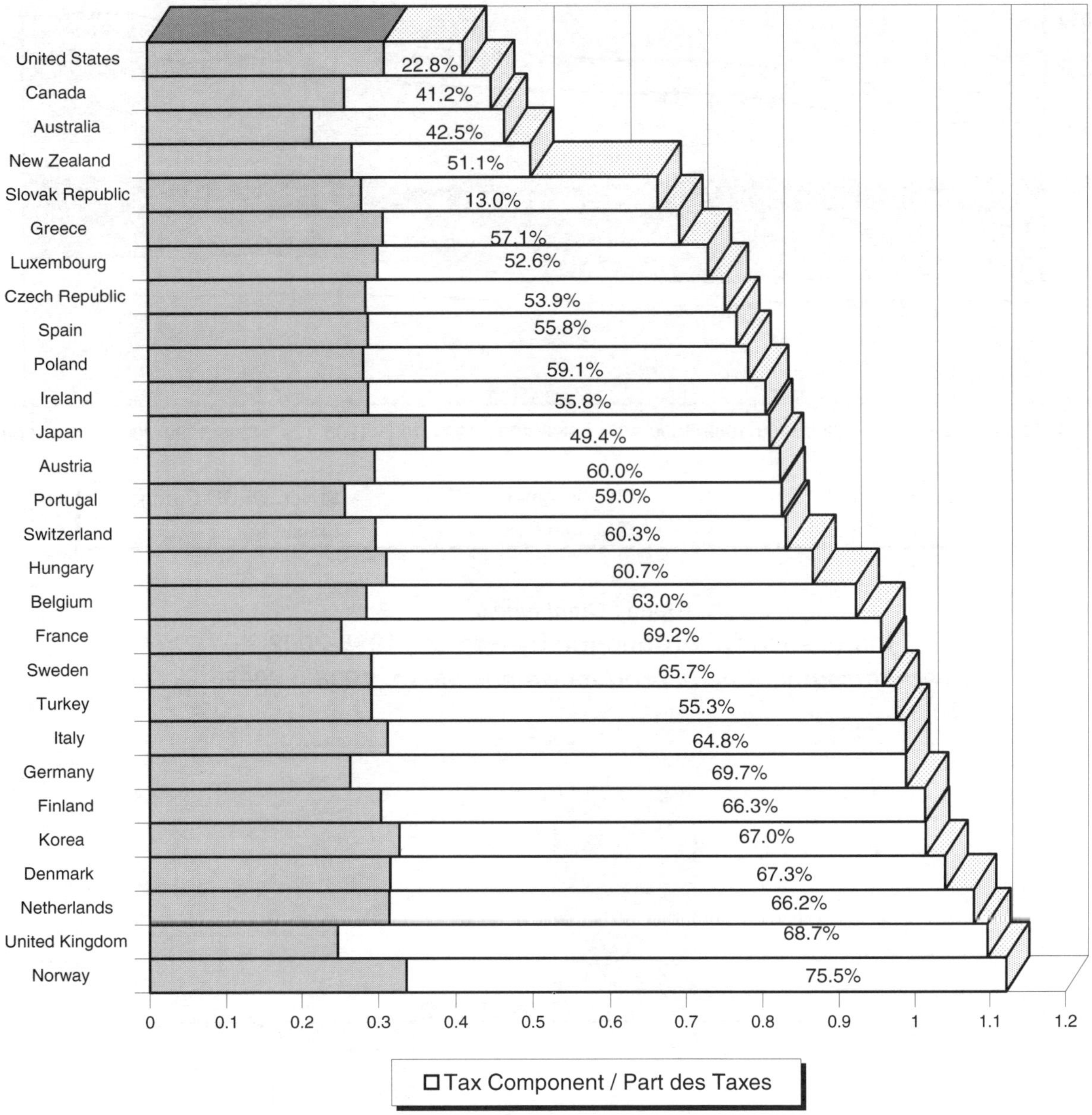

* US Dollars per litre.

** Dollars des Etats-Unis par litre.*

Graph / *Graphique* 4
World Crude Oil and NGL Production by Region
Production mondiale de pétrole brut et de LGN par région

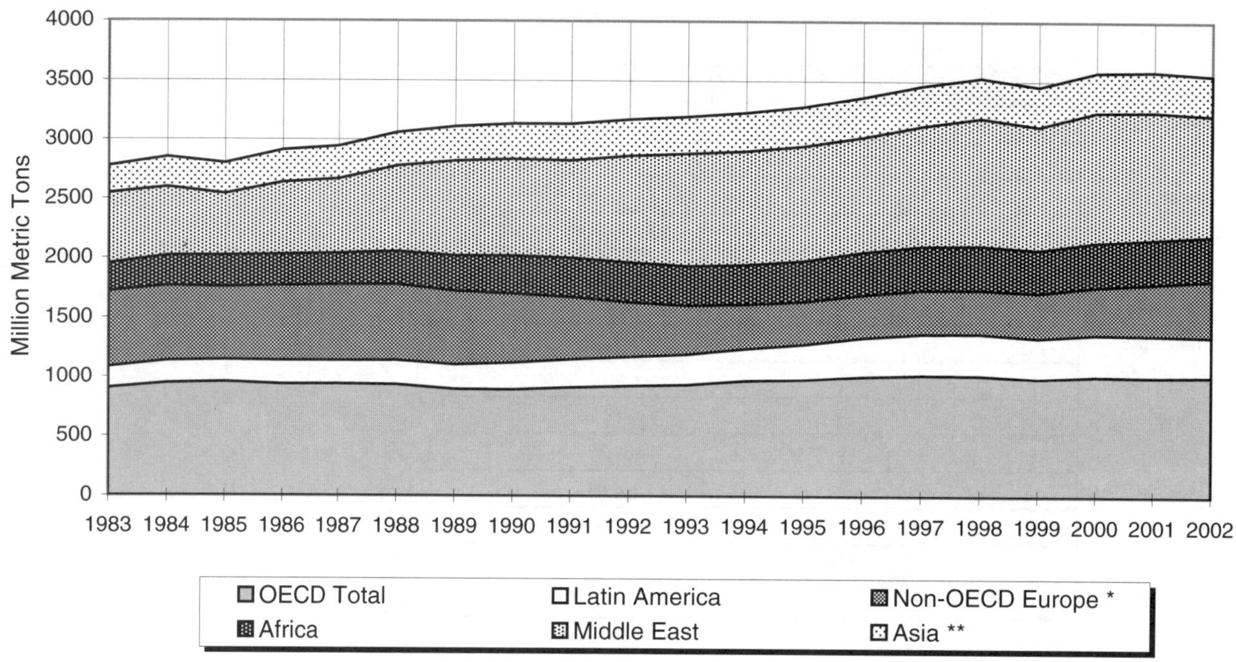

Graph / *Graphique 5*
Increase in Oil Production by Region, 1998-2002
Accroissement de la production pétrolière, 1998 à 2002

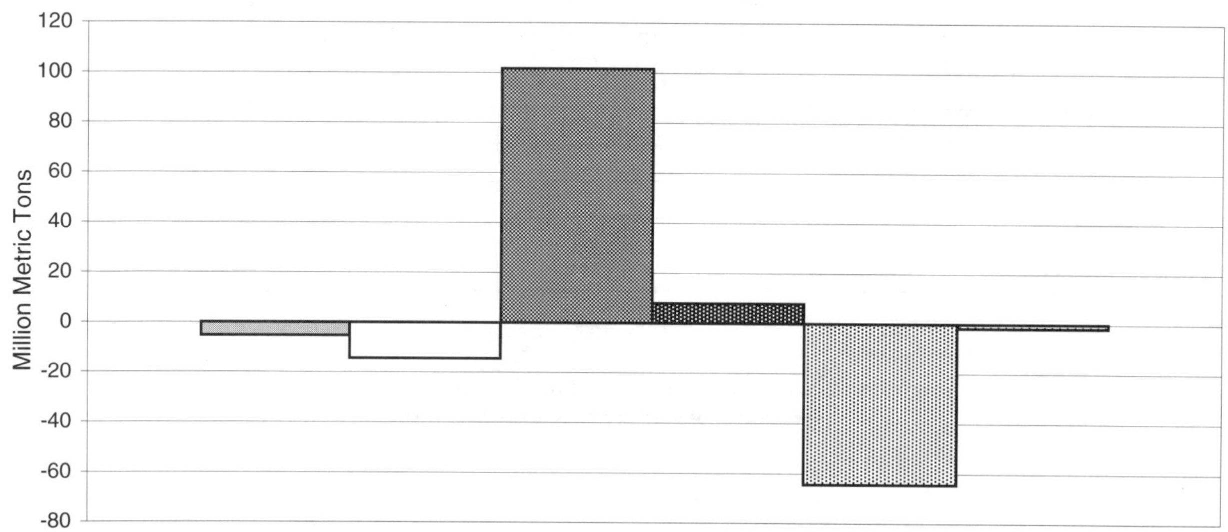

* Non-OECD Europe includes the Former Soviet Union. / *Y compris l'ex-Union Soviétique.*

**Asia includes China. / *Y compris la Chine.*

PART III

DETAILED OECD OIL DATA

TROISIEME PARTIE

DONNEES DETAILLEES SUR LE PETROLE DANS LES PAYS DE L'OCDE

For footnotes to following tables please refer to the fold-out at the end of Part IV.

Pour les notes des tableaux se reporter au dépliant à la fin de la quatrième partie.

OECD TOTAL

Figure 1
FUEL SHARES OF TPES *

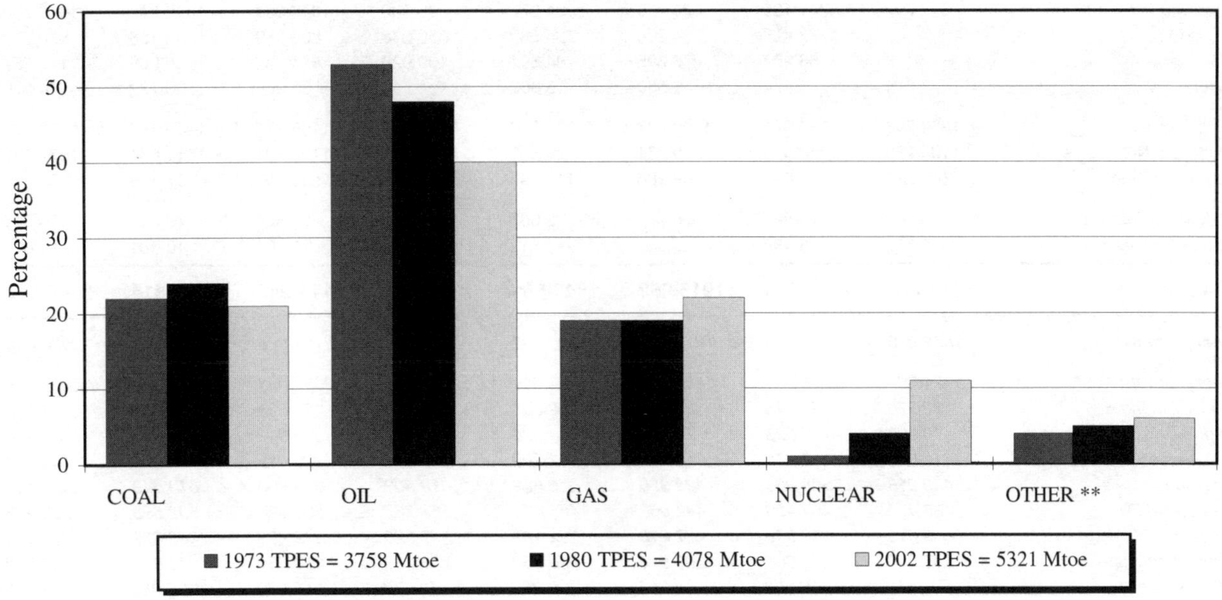

Source: Annual Energy Balances.

* Excluding electricity trade.

** Includes hydro, geothermal, solar, wind, combustible renewables and waste, etc.

Figure 2
OIL PRODUCTION AND CONSUMPTION

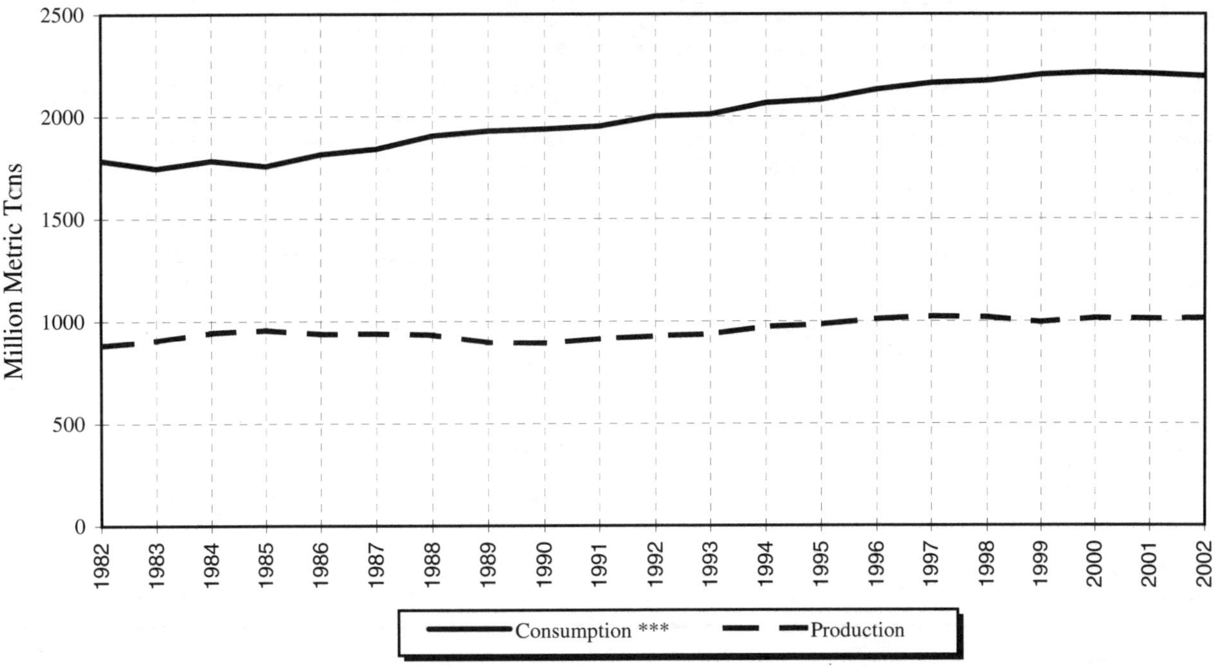

Source: Annual Oil Statistics.

*** Includes refinery fuel and international marine bunkers.

OECD TOTAL
Table 1: Oil Supply and Consumption - Summary Table

000 metric tons

	1973	1978	1980	1990	1999	2000	2001	2002
Indigenous Production	677 751	741 109	820 470	894 077	995 038	1015 394	1012 248	1015 730 e
Crude Oil	..	673 426	734 206	794 683	850 378	868 089	863 158	860 797 e
NGL	..	66 209	76 695	84 391	109 920	111 337	110 968	110 557 e
Other	..	1 474	9 569	15 003	34 740	35 968	38 122	44 376 e
+ Net Imports	1359 503	1376 571	1199 487	1042 812	1167 639	1195 171	1209 452	1170 979 e
Crude Oil, NGL, Other	1195 730	1243 556	1090 221	926 573	1092 540	1102 681	1111 674	1077 747 e
Petroleum Products	163 773	133 015	109 266	116 239	75 099	92 490	97 778	93 232 e
- Intl. Marine Bunkers	73 780	74 828	76 770	75 007	84 768	88 828	79 825	90 914 e
+ Stock Changes	-21 785	16 158	-27 535	-14 313	33 265	-7 491	-10 061	11 073 e
TOTAL SUPPLY	**1941 689**	**2059 010**	**1915 652**	**1847 569**	**2111 174**	**2114 246**	**2131 814**	**2106 868 e**
Refinery Intake	*1822 020*	*1938 250*	*1855 398*	*1803 892*	*2079 586*	*2101 906*	*2091 842*	*2076 674 e*
Refinery Output	*1821 807*	*1918 793*	*1839 032*	*1794 004*	*2060 910*	*2084 933*	*2070 877*	*2058 665 e*
LPG + Ethane	*32 576*	*34 013*	*33 746*	*42 031*	*51 116*	*53 536*	*53 232*	*53 523 e*
Naphtha	*90 597*	*70 523*	*70 800*	*64 521*	*89 675*	*93 349*	*90 050*	*89 420 e*
Kerosenes	*112 629*	*109 951*	*110 774*	*142 399*	*187 261*	*190 075*	*182 018*	*176 797 e*
Gasolines	*439 218*	*508 965*	*492 320*	*546 629*	*613 802*	*613 860*	*613 648*	*624 447 e*
Gas/Diesel Oil	*436 474*	*464 429*	*447 063*	*465 479*	*567 767*	*586 826*	*592 558*	*582 996 e*
Fuel Oil	*515 013*	*518 048*	*477 810*	*292 185*	*277 520*	*272 733*	*266 371*	*251 113 e*
Other Products	*195 300*	*212 864*	*206 519*	*240 760*	*273 769*	*274 554*	*273 000*	*280 369 e*
- Refinery Losses	213	19 457	16 366	9 888	18 676	16 973	20 965	18 009 e
+ Returns	5 557	13 371	20 252	18 421	26 907	28 395	28 963	27 433 e
- Statistical Differences	-15 055	-4 612	1 961	-7 363	1 498	1 253	12 193	12 168 e
TOTAL CONSUMPTION	**1962 088**	**2057 536**	**1917 577**	**1863 465**	**2117 907**	**2124 415**	**2127 619**	**2104 124 e**
Transformation and Energy	**403 264**	**419 173**	**376 314**	**294 991**	**303 937**	**287 054**	**276 580**	..
Crude Oil	21 772	19 669	12 786	19 292	8 632	6 833	4 223	..
LPG + Ethane	3 895	4 828	6 661	7 115	5 519	4 457	4 353	..
Gas/Diesel Oil	11 776	15 965	13 358	9 352	14 492	13 810	13 245	..
Fuel Oil	270 473	286 599	248 431	147 502	139 631	129 763	130 717	..
Petroleum Coke	14 603	15 114	16 014	24 932	34 803	33 804	29 357	..
Other	80 745	76 998	79 064	86 798	100 860	98 387	94 685	..
Transport	**658 831**	**760 357**	**747 139**	**914 894**	**1116 817**	**1135 041**	**1136 710**	..
Motor Gasoline	434 670	497 280	471 564	529 996	610 883	609 821	609 990	..
Gas/Diesel Oil	125 503	160 711	169 451	243 442	328 887	341 842	348 499	..
Subtotal	*560 173*	*657 991*	*641 015*	*773 438*	*939 770*	*951 663*	*958 489*	..
Aviation Fuels	76 140	81 452	84 961	120 062	148 595	154 358	148 604	..
Other	22 518	20 914	21 163	21 394	28 452	29 020	29 617	..
Industry	**513 977**	**509 275**	**486 965**	**405 573**	**446 636**	**441 633**	**445 150**	..
Naphtha	82 529	73 731	69 501	72 248	106 292	113 258	107 524	..
LPG + Ethane	46 653	45 033	55 018	63 495	87 464	85 850	79 935	..
Gas/Diesel Oil	66 097	73 106	66 961	54 634	54 171	49 091	48 681	..
Fuel Oil	208 349	184 142	169 750	76 900	48 814	57 573	56 279	..
Other	110 349	133 263	125 735	138 296	149 895	135 861	152 731	..
Other Sectors	**386 016**	**368 731**	**307 159**	**248 007**	**250 517**	**260 687**	**269 179**	..
Gas/Diesel Oil	241 509	240 553	200 833	161 514	157 939	163 553	173 393	..
Fuel Oil	68 601	53 383	41 761	21 263	8 220	7 668	7 853	..
Other	75 906	74 795	64 565	65 230	84 358	89 466	87 933	..
Memo: Non-Energy Use	*208 479*	*223 870*	*212 164*	*246 638*	*324 379*	*316 403*	*317 260*	..
Chem/Petchem. Feedst.	*114 074*	*129 215*	*123 631*	*146 906*	*204 517*	*209 902*	*203 156*	..
Other	*94 405*	*94 655*	*88 533*	*99 732*	*119 862*	*106 501*	*114 104*	..

Source: Annual Oil Statistics.

OECD TOTAL

Table 2: Supply of Crude Oil, NGL, Feedstocks and Other Hydrocarbons

000 metric tons

	1973	1978	1980	1990	1999	2000	2001	2002
CRUDE OIL								
Indigenous Production	..	673 426	734 206	794 683	850 378	868 089	863 158	860 797 e
+ Imports	..	1320 988	1198 610	1103 690	1449 554	1495 655	1493 153	1459 683 e
- Exports	..	89 660	132 740	236 465	384 100	416 509	413 443	412 587 e
- Direct Use	..	19 815	12 953	19 332	8 661	6 867	4 257	4 429 e
+ Stock Change	..	-3 829	-16 876	-2 571	6 711	2 004	-6 699	-1 768 e
- Statistical Difference	..	3 736	6 169	-9 216	- 440	4 537	13 720	17 270 e
Refinery Intake	**..**	**1877 374**	**1764 078**	**1649 221**	**1914 322**	**1937 835**	**1918 192**	**1884 426 e**
Closing Stock Level	*..*	*133 407*	*162 785*	*219 506*	*240 760*	*235 753*	*242 511*	*244 474 e*
NATURAL GAS LIQUIDS								
Indigenous Production	..	66 209	76 695	84 391	109 920	111 337	110 968	110 557 e
+ Imports	..	6 819	11 663	14 787	16 780	17 472	18 814	19 573 e
- Exports	..	4 191	6 466	11 368	22 714	22 473	21 805	21 818 e
- Direct Use	..	46 524	55 385	64 138	72 247	73 516	71 758	67 489 e
+ Stock Change	..	217	-1 936	- 215	693	- 330	- 551	- 36 e
- Statistical Difference	..	171	-1 138	- 985	3 280	2 755	1 782	1 619 e
Refinery Intake	**..**	**22 359**	**25 709**	**24 442**	**29 152**	**29 735**	**33 886**	**39 168 e**
Closing Stock Level	*..*	*10 493*	*11 433*	*3 369*	*3 476*	*3 806*	*4 355*	*4 392 e*
REFINERY FEEDSTOCKS								
Backflows	..	12 935	19 262	16 654	23 661	25 332	25 849	23 816 e
+ Transfers	..	9 540	14 721	44 566	48 839	51 331	42 659	51 570 e
+ Imports	..	10 207	20 433	64 439	41 566	38 792	44 326	47 499 e
- Exports	..	614	1 290	6 684	9 645	10 711	8 242	7 089 e
+ Stock Change	..	1 206	-1 203	694	1 089	-1 488	1 740	2 248 e
- Statistical Difference	..	-3 777	-4 123	1 799	- 977	-2 338	-4 138	-4 859 e
Refinery Intake	**..**	**37 051**	**56 046**	**117 870**	**106 487**	**105 594**	**110 470**	**122 903 e**
Closing Stock Level	*..*	*24 120*	*31 150*	*33 899*	*30 513*	*32 026*	*30 287*	*28 039 e*
OTHER HYDROCARBONS								
Indigenous Production	..	1 474	9 569	15 003	34 740	35 968	38 122	44 376 e
+ Imports	..	7	11	390	6 899	7 074	6 402	4 976 e
- Exports	..	-	-	2 216	5 800	6 619	7 531	12 490 e
- Direct Use	..	15	15	473	6 526	7 273	6 138	5 719 e
+ Stock Change	..	-	-	- 52	42	199	- 317	159 e
- Statistical Difference	..	-	-	293	- 270	607	1 244	1 125 e
Refinery Intake	**..**	**1 466**	**9 565**	**12 359**	**29 625**	**28 742**	**29 294**	**30 177 e**
Closing Stock Level	*..*	*39*	*39*	*177*	*1 848*	*1 649*	*1 967*	*1 809 e*
CRUDE OIL, NGL, FEEDSTOCKS AND OTHER								
Indigenous Production	677 751	741 109	820 470	894 077	995 038	1015 394	1012 248	1015 730 e
+ Backflows	5 020	12 935	19 262	16 654	23 661	25 332	25 849	23 816 e
+ Transfers	5 472	9 540	14 721	44 566	48 839	51 331	42 659	51 570 e
+ Imports	1257 864	1338 021	1230 717	1183 306	1514 799	1558 993	1562 695	1531 731 e
- Exports	62 134	94 465	140 496	256 733	422 259	456 312	451 021	453 984 e
- Direct Use	64 008	66 354	68 353	83 943	87 434	87 656	82 153	77 637 e
+ Stock Change	-10 779	-2 406	-20 015	-2 144	8 535	385	-5 827	603 e
- Statistical Difference	-12 834	130	908	-8 109	1 593	5 561	12 608	15 155 e
Refinery Intake	**1822 020**	**1938 250**	**1855 398**	**1803 892**	**2079 586**	**2101 906**	**2091 842**	**2076 674 e**
Closing Stock Level	*-*	*168 059*	*205 407*	*256 951*	*276 597*	*273 234*	*279 120*	*278 714 e*

Source: Annual Oil Statistics.

OECD TOTAL
Table 3: Motor Gasoline - Supply

000 metric tons

	1973	1978	1980	1990	1999	2000	2001	2002
MOTOR GASOLINE								
Refinery Output	**435 715**	**495 294**	**477 692**	**534 470**	**611 921**	**612 175**	**612 121**	**623 174 e**
+ Imports	23 846	21 446	24 901	54 032	64 059	69 587	74 223	74 364 e
- Exports	15 579	18 891	21 535	44 624	62 652	65 959	66 968	70 091 e
+ Transfers	1 114	3 096	1 769	-1 330	- 158	2 395	1 940	347 e
+ Stock Change	-1 155	3 444	-4 229	-1 245	5 368	23	-2 138	2 378 e
- Statistical Difference	- 111	- 837	- 398	414	862	1 983	- 402	1 086 e
Total Consumption	**444 052**	**505 226**	**478 996**	**540 889**	**617 676**	**616 238**	**619 580**	**629 086 e**
Closing Stock Level	-	*46 576*	*51 282*	*47 324*	*45 812*	*46 568*	*48 720*	*46 340 e*

Table 4: Aviation Fuels - Supply

000 metric tons

	1973	1978	1980	1990	1999	2000	2001	2002
AVIATION GASOLINE								
Refinery Output	**2 492**	**2 112**	**2 066**	**1 490**	**1 254**	**1 198**	**1 158**	**1 046 e**
+ Imports	438	335	321	215	205	196	171	113 e
- Exports	336	224	400	273	237	278	140	158 e
+ Transfers	-	4	90	-	162	78	126	176 e
+ Stock Change	35	52	- 7	32	27	38	- 9	14 e
- Statistical Difference	39	52	16	21	61	- 8	33	- 61 e
Total Consumption	**2 590**	**2 227**	**2 054**	**1 443**	**1 350**	**1 240**	**1 273**	**1 252 e**
Closing Stock Level	-	*538*	*523*	*341*	*303*	*265*	*274*	*260 e*
GASOLINE TYPE JET FUEL								
Refinery Output	**1 011**	**11 559**	**12 562**	**10 669**	**627**	**487**	**369**	**227 e**
+ Imports	320	623	1 182	305	9	11	9	5 e
- Exports	415	1 038	1 313	728	186	83	62	364 e
+ Transfers	-	- 33	59	- 69	- 212	- 33	- 7	- 8 e
+ Stock Change	6	- 35	21	70	61	- 11	16	9 e
- Statistical Difference	- 16	- 37	97	187	- 93	- 1	- 16	- 372 e
Total Consumption	**938**	**11 113**	**12 414**	**10 060**	**392**	**372**	**341**	**241 e**
Closing Stock Level	-	*1 266*	*1 595*	*896*	*66*	*77*	*61*	*52 e*
KEROSENE TYPE JET FUEL								
Refinery Output	**67 481**	**66 490**	**71 406**	**110 410**	**143 963**	**145 931**	**139 917**	**137 132 e**
+ Imports	15 449	9 340	9 947	16 762	21 551	25 632	28 550	25 416 e
- Exports	7 195	5 970	8 144	14 356	17 121	16 955	17 788	16 139 e
+ Transfers	961	- 157	- 247	- 525	- 556	170	-1 279	-1 308 e
+ Stock Change	- 750	292	- 776	-1 956	391	- 535	- 365	725 e
- Statistical Difference	2 524	1 285	1 658	1 656	447	745	1 470	273 e
Total Consumption	**73 422**	**68 710**	**70 528**	**108 679**	**147 781**	**153 498**	**147 565**	**145 553 e**
Closing Stock Level	-	*6 026*	*8 081*	*9 537*	*10 613*	*11 023*	*11 391*	*10 667 e*

Source: Annual Oil Statistics.

OECD TOTAL
Table 5A: Gas/Diesel Oil - Supply and Consumption

000 metric tons

	1973	1978	1980	1990	1999	2000	2001	2002
Refinery Output	**436 474**	**464 429**	**447 063**	**465 479**	**567 767**	**586 826**	**592 558**	**582 996 e**
+ Recycled Products	-	-	8	30	20	23	30	30 e
+ Imports	78 217	66 294	64 593	89 145	92 576	94 639	107 497	108 033 e
- Exports	49 789	42 489	45 834	64 885	93 978	94 733	95 257	93 994 e
- Intl. Marine Bunkers	10 489	11 662	10 294	16 516	18 648	18 641	18 314	18 001 e
+ Transfers	-1 219	1 619	- 674	-2 077	-5 713	-4 019	-1 361	-3 645 e
+ Stock Changes	-8 338	10 593	-4 400	-5 259	10 343	-1 748	-2 112	2 549 e
- Statistical Differences	- 29	-1 551	- 141	-3 025	-3 122	-5 949	- 777	-2 631 e
Total Consumption	**444 885**	**490 335**	**450 603**	**468 942**	**555 489**	**568 296**	**583 818**	**580 599 e**
Transformation	**10 026**	**13 217**	**10 795**	**7 320**	**12 215**	**11 802**	**11 256**	**..**
Public Electricity	9 568	10 885	7 614	4 706	5 137	6 512	6 084	..
Autoproducer Electricity	-	835	797	1 047	3 153	1 781	2 059	..
Public CHP Plants	57	109	101	55	653	985	895	..
Autoproducers of CHP	-	2	1	26	2 083	1 137	865	..
Public Heat Plants	30	53	42	86	178	161	180	..
Autoproducer Heat Plants	92	319	317	141	5	47	42	..
Petrochemical Industry	221	975	1 878	1 243	997	1 172	1 131	..
Other Transformation	58	39	45	16	9	7	-	..
Energy Sector	**1 750**	**2 748**	**2 563**	**2 032**	**2 277**	**2 008**	**1 989**	**..**
Refinery Fuel	1 094	2 086	1 941	1 255	1 154	1 053	1 052	..
Other Energy Sector	656	662	622	777	1 123	955	937	..
Final Consumption	**433 109**	**474 370**	**437 245**	**459 590**	**540 997**	**554 486**	**570 573**	**..**
Transport	**125 503**	**160 711**	**169 451**	**243 442**	**328 887**	**341 842**	**348 499**	**..**
Air	-	-	-	1	-	1	1	..
Road	88 146	118 635	126 457	213 691	301 737	314 938	322 031	..
Rail	22 159	22 557	21 966	18 016	15 612	16 244	15 699	..
Inland Waterways	10 337	10 025	10 158	8 642	9 170	8 316	7 942	..
Non-Specified	4 861	9 494	10 870	3 092	2 368	2 343	2 826	..
Industry	**66 097**	**73 106**	**66 961**	**54 634**	**54 171**	**49 091**	**48 681**	**..**
Iron and Steel	1 852	1 690	1 397	1 272	1 116	1 716	1 710	..
Chemical (incl. Pet.-Chem.)	2 728	4 400	5 080	5 925	8 879	10 318	9 894	..
Non-Ferrous Metals	820	903	916	1 082	1 928	1 057	898	..
Non-Metallic Minerals	3 764	3 178	2 230	2 191	4 345	3 918	3 949	..
Transport Equipment	807	817	806	496	1 172	2 364	2 566	..
Machinery	2 359	2 352	2 083	2 905	2 762	2 902	2 910	
Mining and Quarrying	1 177	1 827	2 135	2 398	3 407	3 425	3 282	..
Food and Tobacco	2 881	3 464	3 407	3 620	5 522	5 862	5 589	..
Pulp, Paper and Printing	648	615	816	786	2 053	2 067	1 994	..
Wood and Wood Products	548	1 092	1 066	389	3 057	2 129	2 407	..
Construction	4 228	6 622	6 589	8 279	8 155	6 561	6 435	..
Textiles and Leather	944	963	898	1 164	1 863	1 786	1 806	..
Non-Specified	43 341	45 183	39 538	24 127	9 912	4 986	5 241	..
Other Sectors	**241 509**	**240 553**	**200 833**	**161 514**	**157 939**	**163 553**	**173 393**	**..**
Commerce and Public	69 002	71 402	57 480	46 371	44 098	48 975	53 658	..
Residential	127 111	126 095	98 649	71 945	69 447	68 553	73 431	..
Agriculture	26 303	31 047	32 642	41 462	43 675	42 268	42 484	..
Non-Specified	19 093	12 009	12 062	1 736	719	3 757	3 820	..
Memo: Non-Energy Use	*1 403*	*2 810*	*3 202*	*4 129*	*3 911*	*4 327*	*3 426*	..
Chem/Petchem. Feedst.	*1 402*	*2 809*	*3 197*	*4 119*	*3 906*	*4 322*	*3 421*	..
Other	*1*	*1*	*5*	*10*	*5*	*5*	*5*	..
Closing Stock Level	-	*81 731*	*94 554*	*66 043*	*67 759*	*69 846*	*71 954*	*69 353 e*

Source: Annual Oil Statistics.

OECD TOTAL
Table 5B: Fuel Oil (Residual) - Supply and Consumption

000 metric tons

	1973	1978	1980	1990	1999	2000	2001	2002
Refinery Output	**515 013**	**518 048**	**477 810**	**292 185**	**277 520**	**272 733**	**266 371**	**251 113 e**
+ Recycled Products	-	-	1	127	155	184	240	372 e
+ Imports	164 045	116 130	103 767	90 800	64 677	78 634	72 921	66 820 e
- Exports	58 661	37 044	42 861	63 886	66 469	68 888	69 886	64 555 e
- Intl. Marine Bunkers	62 935	62 856	66 129	58 132	65 761	69 793	61 100	72 583 e
+ Transfers	- 160	-2 529	-4 504	-8 451	-6 365	-7 144	-4 913	-6 027 e
+ Stock Changes	816	2 715	2 575	-3 136	4 435	- 25	1 706	1 160 e
- Statistical Differences	-2 131	- 148	- 16	-2 229	2 526	2 256	1 852	893 e
Total Consumption	**560 249**	**534 612**	**470 675**	**251 736**	**205 666**	**203 445**	**203 487**	**175 407 e**
Transformation	**232 652**	**247 008**	**212 885**	**126 086**	**116 302**	**109 323**	**110 157**	..
Public Electricity	217 912	209 591	171 982	107 011	85 827	81 931	80 766	..
Autoproducer Electricity	113	18 649	16 548	7 993	14 317	11 911	12 917	..
Public CHP Plants	7 098	8 821	12 704	2 592	2 437	4 193	4 307	..
Autoproducers of CHP	1 037	5 752	7 374	4 869	7 856	4 015	3 735	..
Public Heat Plants	658	668	852	1 083	1 414	1 043	1 232	..
Autoproducer Heat Plants	158	266	238	197	99	728	704	..
Petrochemical Industry	1 186	777	1 073	1 047	2 672	4 109	5 039	..
Other Transformation	4 490	2 484	2 114	1 294	1 680	1 393	1 457	..
Energy Sector	**37 821**	**39 591**	**35 546**	**21 416**	**23 329**	**20 440**	**20 560**	..
Refinery Fuel	37 292	38 311	35 021	20 995	22 858	20 143	20 126	..
Other Energy Sector	529	1 280	525	421	471	297	434	..
Final Consumption	**289 776**	**248 013**	**222 244**	**104 234**	**66 035**	**73 682**	**72 770**	..
Transport	**12 826**	**10 488**	**10 733**	**6 071**	**9 001**	**8 441**	**8 638**	..
Air	-	-	-	-	-	-	-	..
Road	52	45	67	57	47	1	27	..
Rail	1 965	921	801	54	2	3	1	..
Inland Waterways	10 337	8 864	9 288	5 602	8 815	8 325	8 509	..
Non-Specified	472	658	577	358	137	112	101	..
Industry	**208 349**	**184 142**	**169 750**	**76 900**	**48 814**	**57 573**	**56 279**	..
Iron and Steel	31 301	21 157	16 688	5 720	4 000	4 156	3 592	..
Chemical (incl. Pet.-Chem.)	28 135	29 375	27 340	13 731	9 201	12 705	13 426	..
Non-Ferrous Metals	3 364	6 866	5 717	3 121	2 252	2 553	2 379	..
Non-Metallic Minerals	30 493	26 628	22 687	9 358	8 035	7 849	6 633	..
Transport Equipment	3 324	2 891	2 672	466	1 132	1 038	928	..
Machinery	5 267	4 216	3 565	2 030	1 541	1 164	1 264	..
Mining and Quarrying	2 641	2 191	2 435	1 064	827	853	858	..
Food and Tobacco	16 845	16 501	15 874	7 748	5 031	4 870	4 931	..
Pulp, Paper and Printing	28 518	23 342	22 438	9 481	7 518	10 381	10 184	..
Wood and Wood Products	1 605	1 335	1 241	386	436	391	367	..
Construction	3 993	3 367	2 913	581	426	467	376	..
Textiles and Leather	10 807	9 713	9 200	5 185	4 298	4 555	4 149	..
Non-Specified	42 056	36 560	36 980	18 029	4 117	6 591	7 192	..
Other Sectors	**68 601**	**53 383**	**41 761**	**21 263**	**8 220**	**7 668**	**7 853**	..
Commerce and Public	43 343	35 239	27 436	13 807	4 402	4 336	4 517	..
Residential	18 956	12 733	9 420	5 522	2 495	2 305	2 199	..
Agriculture	2 010	2 111	2 487	1 216	1 218	914	911	..
Non-Specified	4 292	3 300	2 418	718	105	113	226	..
Memo: Non-Energy Use	*628*	*1 641*	*3 024*	*2 864*	*1 663*	*3 787*	*3 804*	..
Chem/Petchem. Feedst.	*510*	*1 569*	*2 966*	*2 570*	*1 484*	*3 493*	*3 589*	..
Other	*118*	*72*	*58*	*294*	*179*	*294*	*215*	..
Closing Stock Level	-	*76 720*	*76 504*	*48 466*	*35 880*	*36 027*	*34 337*	*33 181 e*

Source: Annual Oil Statistics.

OECD TOTAL
Table 5C: Total Products - Supply and Consumption

000 metric tons

	1973	1978	1980	1990	1999	2000	2001	2002
Refinery Output	**1821 807**	**1918 793**	**1839 032**	**1794 004**	**2060 910**	**2084 933**	**2070 877**	**2058 665 e**
+ Recycled Products	537	436	990	1 767	3 246	3 063	3 114	3 617 e
+ Imports	337 115	278 406	271 569	365 297	391 854	426 186	435 322	426 531 e
- Exports	173 342	145 391	162 303	249 058	316 755	333 696	337 544	333 299 e
- Intl. Marine Bunkers	73 780	74 828	76 770	75 007	84 768	88 828	79 825	90 914 e
+ Transfers	58 536	57 635	53 905	39 671	38 595	36 325	39 494	26 067 e
+ Stock Changes	-11 006	18 564	-7 520	-12 169	24 730	-7 876	-4 234	10 470 e
- Statistical Differences	-2 221	-3 921	1 326	1 040	- 95	-4 308	- 415	-2 987 e
Total Consumption	**1962 088**	**2057 536**	**1917 577**	**1863 465**	**2117 907**	**2124 415**	**2127 619**	**2104 124 e**
Transformation	**279 854**	**303 385**	**263 647**	**177 797**	**176 084**	**163 053**	**154 067**	..
Public Electricity	251 572	244 994	195 672	132 867	103 497	98 597	92 265	..
Autoproducer Electricity	290	19 801	17 684	11 110	20 688	17 439	18 040	..
Public CHP Plants	7 155	8 973	12 815	2 676	5 137	6 837	6 741	..
Autoproducers of CHP	1 118	5 864	7 551	5 524	14 339	9 157	5 685	..
Public Heat Plants	688	721	895	1 241	1 664	1 258	1 461	..
Autoproducer Heat Plants	250	585	555	338	105	792	746	..
Petrochemical Industry	5 020	12 935	19 262	16 654	23 661	25 332	25 849	..
Other Transformation	13 761	9 512	9 213	7 387	6 993	3 641	3 280	..
Energy Sector	**123 410**	**115 788**	**112 667**	**117 194**	**127 853**	**124 001**	**122 513**	..
Refinery Fuel	121 130	112 809	110 115	112 628	123 703	120 192	119 004	..
Other Energy Sector	2 280	2 979	2 552	4 566	4 150	3 809	3 509	..
Final Consumption	**1558 824**	**1638 363**	**1541 263**	**1568 474**	**1813 970**	**1837 361**	**1851 039**	..
Transport	**658 831**	**760 357**	**747 139**	**914 894**	**1116 817**	**1135 041**	**1136 710**	..
Air	76 144	81 456	84 965	120 121	148 649	154 409	148 650	..
Road	531 925	625 783	607 719	754 586	927 991	941 210	949 220	..
Rail	24 195	23 509	22 808	18 133	15 721	16 349	15 792	..
Inland Waterways	21 174	19 395	20 149	18 282	21 573	20 316	19 773	..
Non-Specified	5 393	10 214	11 498	3 772	2 883	2 757	3 275	..
Industry	**513 977**	**509 275**	**486 965**	**405 573**	**446 636**	**441 633**	**445 150**	..
Iron and Steel	33 505	23 180	18 887	8 650	6 647	7 781	6 869	..
Chemical (incl. Pet.-Chem.)	146 388	161 855	152 731	163 297	221 930	232 728	227 703	..
Non-Ferrous Metals	4 957	9 404	8 520	7 582	8 312	7 683	7 385	..
Non-Metallic Minerals	34 850	30 552	25 816	18 502	22 378	22 164	21 807	..
Transport Equipment	4 190	3 798	3 602	1 112	2 677	3 764	3 843	..
Machinery	7 967	6 956	6 090	6 689	6 652	6 510	6 705	..
Mining and Quarrying	4 000	4 151	4 686	3 774	4 866	5 070	4 954	..
Food and Tobacco	19 881	20 143	19 644	11 863	11 382	11 496	11 279	..
Pulp, Paper and Printing	29 263	24 075	23 392	10 742	10 311	13 221	12 926	..
Wood and Wood Products	2 202	2 437	2 327	842	3 724	2 706	2 944	..
Construction	69 617	66 753	60 942	68 097	72 314	69 036	68 593	..
Textiles and Leather	11 764	10 711	10 164	6 628	6 607	6 726	6 342	..
Non-Specified	145 393	145 260	150 164	97 795	68 836	52 709	63 800	..
Other Sectors	**386 016**	**368 731**	**307 159**	**248 007**	**250 517**	**260 687**	**269 179**	..
Commerce and Public	129 916	127 731	104 714	76 407	71 404	80 430	85 896	..
Residential	191 089	180 768	141 750	117 338	122 811	123 249	125 586	..
Agriculture	37 015	40 011	41 813	50 662	53 047	50 974	51 736	..
Non-Specified	27 996	20 221	18 882	3 600	3 255	6 034	5 961	..
Memo: Non-Energy Use	*208 479*	*223 870*	*212 164*	*246 638*	*324 379*	*316 403*	*317 260*	..
Chem/Petchem. Feedst.	*114 074*	*129 215*	*123 631*	*146 906*	*204 517*	*209 902*	*203 156*	..
Other	*94 405*	*94 655*	*88 533*	*99 732*	*119 862*	*106 501*	*114 104*	..
Closing Stock Level	-	*250 457*	*270 703*	*212 437*	*201 082*	*208 244*	*212 481*	*201 979 e*

Source: Annual Oil Statistics.

OECD TOTAL
Table 6A: Supply and Consumption

000 metric tons

	1980	2000	2001	2002	1980	2000	2001	2002
	LPG AND ETHANE				**REFINERY GAS**			
Refinery Output	**33 746**	**53 536**	**53 232**	**53 523 e**	**59 424**	**74 280**	**73 567**	**74 105 e**
+ Recycled Products	-	-	-	-	-	-	-	-
+ Imports	16 347	39 823	36 902	36 817 e	-	-	-	-
- Exports	5 044	7 942	7 276	8 014 e	-	-	-	-
- Intl. Marine Bunkers	-	-	-	-	-	-	-	-
+ Transfers	47 243	64 857	63 223	58 778 e	- 127	- 37	- 37	- 118 e
+ Stock Change	- 211	- 524	-3 011	1 780 e	4	3	-	-
- Statistical Difference	-1 216	2 180	1 919	2 531 e	291	65	20	294 e
Total Consumption	**93 297**	**147 570**	**141 151**	**140 353 e**	**59 010**	**74 181**	**73 510**	**73 693 e**
of which: Ethane	*654*	*26 473*	*24 939*	*25 614 e*	-	-	-	-
Transformation	4 672	2 903	2 783	..	1 438	2 389	1 601	..
Energy	1 989	1 554	1 570	..	53 579	68 800	67 770	..
Transport	3 965	11 730	12 410	..	-	-	-	..
Industry	55 018	85 850	79 935	..	3 993	2 992	4 139	..
Other	27 653	45 533	44 453	..	-	-	-	..
Memo: Non-Energy Use	*23 968*	*73 501*	*68 157*	..	*2 848*	*996*	*1 841*	..
Closing Stock Level	*3 327*	*9 998*	*13 008*	*11 231 e*	*4*	-	-	-
	NAPHTHA				**OTHER KEROSENE**			
Refinery Output	**70 800**	**93 349**	**90 050**	**89 420 e**	**39 368**	**44 144**	**42 101**	**39 665 e**
+ Recycled Products	281	216	282	411 e	-	-	-	-
+ Imports	31 744	67 794	65 870	65 175 e	784	5 582	4 481	5 182 e
- Exports	11 544	26 860	26 682	24 710 e	2 725	3 720	3 494	3 611 e
- Intl. Marine Bunkers	-	-	-	-	-	2	-	-
+ Transfers	-2 435	-5 351	-6 525	-3 610 e	-1 085	-1 885	-1 686	-2 127 e
+ Stock Change	- 274	- 653	294	10 e	595	- 848	946	1 121 e
- Statistical Difference	105	-2 871	-1 749	- 373 e	394	570	434	- 44 e
Total Consumption	**88 467**	**131 366**	**125 038**	**127 069 e**	**36 543**	**42 701**	**41 914**	**40 274 e**
Transformation	18 460	17 881	17 105	..	224	438	360	..
Energy	497	225	409	..	211	34	109	..
Transport	-	-	-	..	49	47	52	..
Industry	69 501	113 258	107 524	..	6 834	5 064	5 143	..
Other	9	2	-	..	29 225	37 118	36 250	..
Memo: Non-Energy Use	*68 380*	*112 082*	*106 340*	..	*338*	*701*	*660*	..
Closing Stock Level	*6 871*	*7 232*	*6 910*	*6 923 e*	*9 038*	*5 814*	*4 868*	*3 747 e*
	WHITE SPIRIT + SBP				**LUBRICANTS**			
Refinery Output	**6 686**	**4 305**	**3 855**	**..**	**20 853**	**23 099**	**22 518**	**..**
+ Recycled Products	26	7	7	..	466	759	652	..
+ Imports	934	1 460	1 552	..	3 786	4 589	4 222	..
- Exports	1 233	1 619	1 804	..	5 474	8 055	8 775	..
- Intl. Marine Bunkers	-	-	-	..	330	366	381	..
+ Transfers	102	- 311	- 425	..	- 382	- 868	- 842	..
+ Stock Change	- 103	- 36	- 18	..	- 601	- 305	- 123	..
- Statistical Difference	- 177	- 60	- 160	..	- 153	- 461	- 838	..
Total Consumption	**6 589**	**3 866**	**3 327**	**..**	**18 471**	**19 314**	**18 109**	**..**
Transformation	7	21	23	..	1	13	10	..
Energy	40	-	1	..	11	83	77	..
Transport	-	-	-	..	6 416	8 802	8 517	..
Industry	6 032	3 473	3 019	..	11 190	9 233	8 852	..
Other	510	372	284	..	853	1 183	653	..
Memo: Non-Energy Use	*6 496*	*3 762*	*3 200*	..	*18 457*	*18 119*	*17 493*	..
Closing Stock Level	*1 100*	*533*	*551*	..	*5 949*	*4 344*	*4 469*	..

Source: Annual Oil Statistics.

OECD TOTAL
Table 6B: Supply and Consumption

000 metric tons

	1980	2000	2001	2002	1980	2000	2001	2002
	BITUMEN				**PARAFFIN WAXES**			
Refinery Output	**51 261**	**65 636**	**62 420**	**..**	**1 804**	**2 685**	**2 538**	**..**
+ Recycled Products	-	-	-	..	-	-	-	..
+ Imports	2 127	5 959	6 308	..	438	744	761	..
- Exports	2 416	6 624	6 899	..	521	861	910	..
- Intl. Marine Bunkers	-	-	-	..	-	-	-	..
+ Transfers	- 9	2 523	1 871	..	- 13	- 453	- 368	..
+ Stock Change	- 190	-1 425	764	..	-	7	40	..
- Statistical Difference	45	-1 040	- 226	..	107	122	151	..
Total Consumption	**50 728**	**67 109**	**64 690**	**..**	**1 601**	**2 000**	**1 910**	**..**
Transformation	111	2 810	1 838	..	6	-	-	..
Energy	7	8	11	..	-	13	10	..
Transport	-	-	-	..	-	-	-	..
Industry	50 610	64 153	62 635	..	1 595	1 911	1 838	..
Other	-	138	206	..	-	76	62	..
Memo: Non-Energy Use	*50 610*	*62 370*	*61 108*	..	*1 574*	*1 987*	*1 900*	..
Closing Stock Level	*5 196*	*6 259*	*5 496*	..	*180*	*281*	*241*	..
	PETROLEUM COKE				**OTHER PRODUCTS**			
Refinery Output	**29 252**	**63 455**	**65 833**	**68 173 e**	**37 239**	**41 094**	**42 269**	**138 091 e**
+ Recycled Products	-	-	-	-	208	1 874	1 903	2 804 e
+ Imports	9 188	17 247	17 732	17 379 e	1 510	14 289	14 123	27 227 e
- Exports	9 551	22 613	23 883	24 093 e	3 708	8 506	7 720	27 570 e
- Intl. Marine Bunkers	-	-	-	-	-	23	29	330 e
+ Transfers	-	- 76	- 4	- 59 e	14 118	-13 521	-10 219	-16 332 e
+ Stock Change	187	- 617	- 149	235 e	- 111	-1 220	- 75	489 e
- Statistical Difference	514	399	238	- 426 e	217	-2 235	-2 363	-4 157 e
Total Consumption	**28 562**	**56 997**	**59 291**	**62 061 e**	**49 039**	**36 222**	**42 615**	**128 536 e**
Transformation	968	8 038	3 835	..	14 080	7 077	4 788	..
Energy	15 046	25 766	25 522	..	2 638	4 117	3 940	..
Transport	-	-	-	..	-	-	-	..
Industry	12 548	23 085	29 786	..	32 199	24 091	32 056	..
Other	-	108	148	..	122	937	1 831	..
Memo: Non-Energy Use	*6 807*	*10 399*	*16 764*	..	*25 777*	*22 089*	*30 516*	..
Closing Stock Level	*1 562*	*4 235*	*4 385*	*4 146 e*	*4 937*	*5 742*	*5 816*	*16 079 e*

Source: Annual Oil Statistics.

OECD TOTAL
Table 7A: Imports by Country of Origin

000 metric tons

	2000	2001	2002	2000	2001	2002	2000	2001	2002
	Crude Oil			**Natural Gas Liquids**			**Feedstocks and Other**		
Australia	10 826	9 287	10 809 e	332	352	355 e	204	324	395 e
Belgium	2	2	-	-	-	2 121 e	1 675	2 069	2 184 e
Canada	68 635	68 997	72 490 e	5 593	4 129	4 176 e	1 052	1 066	709 e
Czech Republic	112	124	155 e	-	-	-	8	1	18 e
Denmark	12 078	10 570	10 329 e	-	-	-	50	83	30 e
Finland	-	22	-	-	-	-	119	167	2 e
France	501	402	8 e	-	4	-	887	1 130	729 e
Germany	1 594	280	289 e	11	35	45 e	825	1 611	1 971 e
Greece	-	-	-	-	-	-	42	120	130 e
Hungary	8	-	-	-	-	-	200	154	223 e
Ireland	-	-	-	-	-	-	539	637	390 e
Italy	204	282	905 e	-	-	-	351	406	232 e
Japan	-	-	-	-	-	-	1	33	112 e
Korea	-	-	-	-	-	-	22	55	69 e
Mexico	79 948	82 220	87 147 e	68	182	-	184	41	91 e
Netherlands	354	334	1 276 e	-	-	-	2 372	2 331	2 395 e
New Zealand	1 001	870	860 e	6	-	-	-	19	18 e
Norway	144 636	135 097	131 784 e	1 006	1 267	1 512 e	1 449	1 607	1 408 e
Poland	163	40	206 e	-	-	-	137	76	26 e
Spain	84	-	-	-	7	-	223	201	245 e
Sweden	126	107	80 e	7	49	-	1 532	1 156	1 039 e
Turkey	213	6	-	-	-	-	556	765	427 e
United Kingdom	73 095	75 475	73 598 e	99	209	43 e	839	1 220	745 e
United States	2 135	1 627	1 125 e	324	279	243 e	411	1 024	921 e
Other	284	78	15 e	-	-	-	345	129	125 e
OECD	**395 999**	**385 820**	**391 076 e**	**7 446**	**6 513**	**8 495 e**	**14 023**	**16 425**	**14 634 e**
Argentina	3 471	3 970	4 428 e	-	-	-	37	161	103 e
Brazil	391	2 936	5 778 e	-	-	36 e	177	269	319 e
Colombia	16 307	13 374	12 859 e	-	-	-	129	172	298 e
Ecuador	8 606	7 685	7 500 e	-	-	-	164	43	82 e
Netherlands Antilles	99	238	115 e	-	-	-	2 036	1 766	1 993 e
Peru	575	460	519 e	-	-	-	117	113	92 e
Trinidad and Tobago	2 790	2 558	3 389 e	-	-	14 e	340	208	145 e
Venezuela	88 104	91 978	87 146 e	85	48	97 e	5 589	3 866	4 065 e
Other	1 257	988	1 224 e	-	-	-	51	58	213 e
Latin America	**121 600**	**124 187**	**122 958 e**	**85**	**48**	**147 e**	**8 640**	**6 656**	**7 310 e**
Romania	19	118	373 e	-	-	-	94	15	22 e
Other	-	61	30 e	-	-	-	104	134	154 e
Non-OECD Europe	**19**	**179**	**403 e**	**-**	**-**	**-**	**198**	**149**	**176 e**
FSU	**123 389**	**142 578**	**167 309 e**	**747**	**1 544**	**1 852 e**	**10 205**	**10 209**	**12 753 e**
Iran	75 885	73 120	64 126 e	250	233	1 195 e	30	14	69 e
Iraq	76 065	68 189	47 848 e	-	-	-	-	-	-
Kuwait	53 412	49 378	38 554 e	-	37	-	14	-	-
Oman	18 296	21 489	20 180 e	-	13	24 e	-	-	-
Qatar	27 154	28 123	23 306 e	707	1 335	1 278 e	397	338	261 e
Saudi Arabia	240 136	234 678	220 236 e	433	539	161 e	2 481	1 815	2 190 e
United Arab Emirates	74 397	70 459	69 470 e	565	260	439 e	368	346	-
Other	22 526	26 943	27 038 e	-	-	-	385	1 254	808 e
Middle East	**587 871**	**572 379**	**510 758 e**	**1 955**	**2 417**	**3 097 e**	**3 675**	**3 767**	**3 328 e**
India	-	-	-	-	-	-	66	104	71 e
Indonesia	20 085	20 323	20 899 e	288	281	402 e	355	242	713 e
Singapore	29	41	43 e	-	-	-	388	968	783 e
Other	23 621	23 703	21 721 e	319	166	57 e	687	973	1 042 e
Asia	**43 735**	**44 067**	**42 663 e**	**607**	**447**	**459 e**	**1 496**	**2 287**	**2 609 e**
China	**7 006**	**4 547**	**5 076 e**	**-**	**-**	**-**	**684**	**944**	**775 e**
Algeria	20 498	19 223	19 395 e	6 447	6 917	5 087 e	1 860	5 277	5 289 e
Angola/Cabinada	22 694	25 491	26 579 e	5	-	-	302	121	315 e
Cameroon	3 615	3 918	4 561 e	-	-	-	-	56	21 e
Egypt	6 056	3 667	4 102 e	-	-	-	444	399	180 e
Gabon	11 389	11 582	9 990 e	-	-	-	38	-	-
Libya	50 910	50 528	44 909 e	25	25	75 e	2 086	1 681	1 356 e
Nigeria	74 885	73 816	60 729 e	40	294	139 e	133	664	1 436 e
Tunisia	1 346	1 247	2 192 e	105	70	129 e	239	262	448 e
Other	11 704	14 778	18 615 e	10	46	20 e	837	707	578 e
Africa	**203 097**	**204 250**	**191 072 e**	**6 632**	**7 352**	**5 450 e**	**5 939**	**9 167**	**9 623 e**
Non-spec./Other	**12 939**	**15 146**	**28 368 e**	**-**	**493**	**73 e**	**1 006**	**1 124**	**1 267 e**
Total Imports	**1495 655**	**1493 153**	**1459 683 e**	**17 472**	**18 814**	**19 573 e**	**45 866**	**50 728**	**52 475 e**
Memo: Total OPEC	*801 531*	*779 815*	*696 618 e*	*8 840*	*9 969*	*8 873 e*	*13 313*	*14 243*	*15 379 e*

Source: Annual Oil Statistics.

OECD TOTAL
Table 7B: Imports by Country of Origin

000 metric tons

	2000	2001	2002	2000	2001	2002	2000	2001	2002
	Total Products			**Liquefied Petroleum Gases**			**Naphtha**		
Australia	2 518	2 632	2 925 e	944	1 131	1 453 e	478	386	229 e
Belgium	19 018	19 151	19 249 e	462	436	433 e	1 525	1 126	1 076 e
Canada	12 123	15 081	17 390 e	220	176	283 e	145	154	183 e
Czech Republic	919	995	1 093 e	28	14	36 e	25	24	8 e
Denmark	3 054	2 756	2 617 e	68	65	58 e	96	53	87 e
Finland	4 298	3 649	4 032 e	27	10	25 e	74	78	160 e
France	13 293	12 677	10 622 e	441	372	337 e	1 329	1 288	790 e
Germany	15 279	13 679	13 227 e	649	497	482 e	1 403	981	1 039 e
Greece	1 140	1 058	1 095 e	16	1	120 e	575	425	486 e
Hungary	910	1 176	1 137 e	35	34	33 e	-	-	-
Ireland	240	243	59 e	6	5	7 e	76	16	-
Italy	9 510	11 828	11 286 e	105	90	81 e	338	725	622 e
Japan	1 570	2 200	1 846 e	-	29	177 e	18	12	47 e
Korea	13 811	12 684	11 752 e	133	158	48 e	5 793	5 689	5 370 e
Mexico	2 734	1 989	2 042 e	-	-	9 e	1 593	1 478	1 042 e
Netherlands	44 720	44 187	40 305 e	501	475	523 e	6 448	5 827	5 608 e
New Zealand	439	348	353 e	-	3	15 e	318	334	313 e
Norway	8 838	7 575	8 954 e	1 793	2 128	2 355 e	883	792	564 e
Poland	1 366	1 636	1 648 e	29	7	5 e	75	95	76 e
Spain	4 739	4 049	3 380 e	96	57	57 e	1 200	1 166	1 006 e
Sweden	8 498	7 411	7 161 e	135	76	140 e	151	245	217 e
Turkey	580	1 028	850 e	6	2	-	317	558	358 e
United Kingdom	26 158	25 399	25 768 e	2 550	2 175	2 594 e	2 539	2 160	2 367 e
United States	34 371	31 439	33 202 e	2 245	2 675	2 772 e	1 206	1 066	1 296 e
Other	5 719	5 899	6 299 e	70	108	84 e	158	109	149 e
OECD	**235 845**	**230 769**	**228 292 e**	**10 559**	**10 724**	**12 127 e**	**26 763**	**24 787**	**23 093 e**
Argentina	1 409	2 063	2 357 e	-	-	11 e	12	190	84 e
Brazil	2 847	3 849	2 302 e	-	-	1 e	44	32	30 e
Colombia	1 438	1 899	1 165 e	2	-	-	88	19	173 e
Ecuador	355	735	261 e	-	-	-	19	35	74 e
Netherlands Antilles	3 326	2 918	2 529 e	3	-	-	650	221	209 e
Peru	340	706	375 e	-	-	-	26	96	51 e
Trinidad and Tobago	2 219	2 327	1 948 e	9	8	16 e	147	47	-
Venezuela	18 438	16 263	10 752 e	29	34	81 e	508	480	555 e
Other	1 221	2 003	2 398 e	32	4	21 e	71	150	57 e
Latin America	**31 593**	**32 763**	**24 087 e**	**75**	**46**	**130 e**	**1 565**	**1 270**	**1 233 e**
Romania	275	695	2 050 e	-	-	-	-	1	31 e
Other	1 255	1 582	1 473 e	117	125	133 e	169	158	204 e
Non-OECD Europe	**1 530**	**2 277**	**3 523 e**	**117**	**125**	**133 e**	**169**	**159**	**235 e**
FSU	**39 450**	**40 192**	**47 805 e**	**901**	**812**	**1 122 e**	**2 383**	**2 138**	**2 866 e**
Iran	1 045	1 147	1 086 e	623	862	713 e	359	199	321 e
Iraq	7	97	-	7	97	-	-	-	-
Kuwait	11 290	10 979	9 757 e	2 530	2 434	1 950 e	6 540	6 143	5 916 e
Oman	-	95	34 e	-	-	-	-	13	34 e
Qatar	1 731	2 250	1 955 e	821	904	739 e	849	1 038	1 112 e
Saudi Arabia	18 055	14 666	14 784 e	9 173	6 603	6 288 e	5 773	4 479	5 632 e
United Arab Emirates	12 244	13 728	14 294 e	4 958	4 345	4 517 e	5 792	6 195	7 159 e
Other	4 112	3 174	2 630 e	54	97	117 e	1 592	1 123	1 132 e
Middle East	**48 484**	**46 136**	**44 540 e**	**18 166**	**15 342**	**14 324 e**	**20 905**	**19 190**	**21 306 e**
India	1 825	3 337	3 536 e	-	-	-	1 523	1 993	2 107 e
Indonesia	6 008	6 237	5 272 e	1 055	1 069	711 e	1 127	1 570	1 449 e
Singapore	5 812	4 521	6 165 e	-	23	-	1 822	653	666 e
Other	4 319	4 470	5 235 e	474	262	333 e	2 278	2 156	1 629 e
Asia	**17 964**	**18 565**	**20 208 e**	**1 529**	**1 354**	**1 044 e**	**6 750**	**6 372**	**5 851 e**
China	**2 202**	**2 293**	**2 904 e**	**12**	**3**	**3 e**	**927**	**1 047**	**1 143 e**
Algeria	20 471	17 404	17 904 e	7 121	5 691	5 968 e	2 891	3 746	3 831 e
Angola/Cabinada	189	403	113 e	-	-	-	-	-	22 e
Cameroon	206	114	52 e	-	-	-	-	-	-
Egypt	1 897	2 214	2 890 e	54	214	160 e	1 654	1 290	1 313 e
Gabon	22	-	4 e	-	-	4 e	-	-	-
Libya	5 806	6 236	5 517 e	221	276	124 e	1 497	2 071	1 871 e
Nigeria	1 207	2 355	1 597 e	151	606	559 e	113	147	145 e
Tunisia	436	291	308 e	-	-	-	45	52	114 e
Other	2 116	1 849	2 062 e	36	45	66 e	418	391	373 e
Africa	**32 350**	**30 866**	**30 447 e**	**7 583**	**6 832**	**6 881 e**	**6 618**	**7 697**	**7 669 e**
Non-spec./Other	**16 768**	**31 461**	**24 725 e**	**881**	**1 664**	**1 053 e**	**1 714**	**3 210**	**1 779 e**
Total Imports	**426 186**	**435 322**	**426 531 e**	**39 823**	**36 902**	**36 817 e**	**67 794**	**65 870**	**65 175 e**
Memo: Total OPEC	*96 302*	*91 362*	*82 918 e*	*26 689*	*22 921*	*21 650 e*	*25 449*	*26 068*	*27 991 e*

Source: Annual Oil Statistics.

OECD TOTAL
Table 7C: Imports by Country of Origin

000 metric tons

	2000	2001	2002	2000	2001	2002	2000	2001	2002
	Total Gasoline			**Total Kerosene**			**Gas/Diesel**		
Australia	670	538	624 e	98	184	110 e	31	102	161 e
Belgium	4 574	4 766	4 710 e	1 335	1 427	1 351 e	6 120	7 197	6 174 e
Canada	4 040	6 404	7 525 e	161	189	296 e	5 097	5 191	5 726 e
Czech Republic	103	196	280 e	7	-	-	462	596	618 e
Denmark	1 147	813	758 e	292	324	403 e	835	740	485 e
Finland	2 029	1 919	2 353 e	184	4	43 e	1 794	1 498	1 333 e
France	3 680	3 715	3 478 e	1 080	1 136	742 e	2 174	1 883	1 259 e
Germany	3 807	3 184	2 850 e	387	458	787 e	5 019	4 671	5 223 e
Greece	296	224	265 e	129	88	19 e	68	292	164 e
Hungary	267	375	323 e	-	-	-	505	617	708 e
Ireland	18	44	14 e	8	1	11 e	90	93	19 e
Italy	1 769	2 367	2 141 e	378	414	331 e	4 981	6 871	6 988 e
Japan	162	239	100 e	812	484	452 e	370	944	782 e
Korea	797	1 243	942 e	4 085	2 992	2 805 e	2 534	2 314	1 694 e
Mexico	276	330	282 e	88	120	94 e	-	14	40 e
Netherlands	9 216	8 497	7 075 e	3 941	3 314	2 919 e	18 397	20 734	18 221 e
New Zealand	-	-	-	-	-	-	-	-	8 e
Norway	2 297	1 665	1 797 e	145	47	123 e	1 524	1 274	2 176 e
Poland	262	408	390 e	139	301	157 e	65	113	352 e
Spain	1 465	1 379	905 e	31	5	139 e	320	401	343 e
Sweden	2 119	1 465	1 417 e	32	20	125 e	4 114	3 767	3 290 e
Turkey	-	9	425 e	81	27	-	3	-	-
United Kingdom	5 776	5 907	5 474 e	1 358	1 518	1 538 e	6 382	5 304	6 414 e
United States	6 309	6 657	6 507 e	823	526	1 200 e	3 673	1 905	3 024 e
Other	1 662	2 080	2 116 e	86	21	138 e	1 845	2 050	2 228 e
OECD	**52 741**	**54 424**	**52 751 e**	**15 680**	**13 600**	**13 783 e**	**66 403**	**68 571**	**67 430 e**
Argentina	1 242	1 274	1 562 e	-	-	8 e	14	98	94 e
Brazil	1 539	1 623	1 433 e	-	-	-	-	246	46 e
Colombia	210	460	15 e	119	231	122 e	20	86	101 e
Ecuador	58	21	37 e	-	-	-	-	-	-
Netherlands Antilles	168	276	182 e	753	731	484 e	546	688	685 e
Peru	13	87	-	-	-	-	41	44	-
Trinidad and Tobago	376	376	370 e	125	154	87 e	138	244	98 e
Venezuela	4 508	3 749	2 443 e	1 796	1 536	1 712 e	3 113	2 701	1 645 e
Other	237	328	508 e	101	135	92 e	46	100	434 e
Latin America	**8 351**	**8 194**	**6 550 e**	**2 894**	**2 787**	**2 505 e**	**3 918**	**4 207**	**3 103 e**
Romania	44	123	858 e	-	-	-	207	508	1 020 e
Other	212	508	283 e	45	48	28 e	414	424	366 e
Non-OECD Europe	**256**	**631**	**1 141 e**	**45**	**48**	**28 e**	**621**	**932**	**1 386 e**
FSU	**2 869**	**2 446**	**3 631 e**	**1 051**	**939**	**1 343 e**	**16 977**	**19 792**	**22 021 e**
Iran	-	-	-	-	48	-	31	-	-
Iraq	-	-	-	-	-	-	-	-	-
Kuwait	-	-	-	2 094	2 150	1 673 e	126	-	-
Oman	-	-	-	-	82	-	-	-	-
Qatar	5	19	-	40	40	31 e	14	-	73 e
Saudi Arabia	1 182	1 160	1 446 e	1 308	1 746	1 081 e	370	382	287 e
United Arab Emirates	47	142	241 e	1 289	2 779	2 204 e	27	125	68 e
Other	179	106	339 e	725	823	499 e	149	148	174 e
Middle East	**1 413**	**1 427**	**2 026 e**	**5 456**	**7 668**	**5 488 e**	**717**	**655**	**602 e**
India	144	246	565 e	-	82	56 e	-	640	455 e
Indonesia	5	-	15 e	-	3	-	29	130	-
Singapore	1 266	1 576	1 876 e	562	679	781 e	951	797	1 977 e
Other	125	155	370 e	754	798	1 034 e	557	801	1 200 e
Asia	**1 540**	**1 977**	**2 826 e**	**1 316**	**1 562**	**1 871 e**	**1 537**	**2 368**	**3 632 e**
China	**661**	**635**	**411 e**	**440**	**303**	**423 e**	**19**	**6**	**41 e**
Algeria	268	181	572 e	837	532	501 e	1 835	1 982	1 109 e
Angola/Cabinada	-	-	-	-	-	-	-	101	-
Cameroon	28	-	-	-	-	-	-	73	-
Egypt	63	289	983 e	-	167	301 e	-	-	-
Gabon	-	-	-	-	-	-	-	-	-
Libya	84	37	148 e	575	1 017	515 e	755	487	548 e
Nigeria	25	121	308 e	56	99	-	-	95	62 e
Tunisia	-	-	51 e	74	4	-	-	144	104 e
Other	95	146	53 e	70	30	97 e	63	245	93 e
Africa	**563**	**774**	**2 115 e**	**1 612**	**1 849**	**1 414 e**	**2 653**	**3 127**	**1 916 e**
Non-spec./Other	**1 400**	**3 895**	**3 031 e**	**2 720**	**4 275**	**3 743 e**	**1 794**	**7 839**	**7 902 e**
Total Imports	**69 794**	**74 403**	**74 482 e**	**31 214**	**33 031**	**30 598 e**	**94 639**	**107 497**	**108 033 e**
Memo: Total OPEC	*6 124*	*5 409*	*5 173 e*	*7 995*	*9 950*	*7 717 e*	*6 300*	*5 902*	*3 792 e*

Source: Annual Oil Statistics.

OECD TOTAL
Table 7D: Imports by Country of Origin

000 metric tons

	2000	2001	2002	2000	2001	2002	2000	2001	2002
	Fuel Oil (Residual)			**Petroleum Coke**			**Other Products**		
Australia	-	-	33 e	1	1	-	296	290	315 e
Belgium	3 123	2 428	3 133 e	333	279	398 e	1 546	1 492	1 974 e
Canada	1 425	1 858	1 735 e	80	33	322 e	955	1 076	1 320 e
Czech Republic	117	74	55 e	-	-	-	177	91	96 e
Denmark	560	635	750 e	6	11	6 e	50	115	70 e
Finland	130	32	-	1	-	-	59	108	118 e
France	2 540	2 386	2 415 e	25	13	28 e	2 024	1 884	1 573 e
Germany	2 530	2 271	1 264 e	186	192	273 e	1 298	1 425	1 309 e
Greece	27	25	32 e	-	-	-	29	3	9 e
Hungary	4	89	-	-	2	11 e	99	59	62 e
Ireland	38	84	8 e	-	-	-	4	-	-
Italy	1 208	466	468 e	67	14	144 e	664	881	511 e
Japan	157	377	208 e	25	59	54 e	26	56	26 e
Korea	144	158	625 e	-	-	92 e	325	130	176 e
Mexico	491	-	541 e	-	-	-	286	47	34 e
Netherlands	4 253	3 357	4 103 e	488	637	534 e	1 476	1 346	1 322 e
New Zealand	59	-	-	-	-	-	62	11	17 e
Norway	773	1 017	968 e	34	1	32 e	1 389	651	939 e
Poland	663	585	557 e	-	-	-	133	127	111 e
Spain	799	414	181 e	62	18	89 e	766	609	660 e
Sweden	1 409	1 339	1 448 e	8	21	-	530	478	524 e
Turkey	80	311	40 e	-	-	-	93	121	27 e
United Kingdom	5 768	6 457	5 623 e	335	668	420 e	1 450	1 210	1 338 e
United States	7 680	6 885	3 937 e	9 535	8 270	10 940 e	2 900	3 455	3 526 e
Other	1 461	1 106	1 149 e	16	14	14 e	421	411	421 e
OECD	**35 439**	**32 354**	**29 273 e**	**11 202**	**10 233**	**13 357 e**	**17 058**	**16 076**	**16 478 e**
Argentina	85	239	167 e	32	221	310 e	24	41	121 e
Brazil	1 157	1 860	631 e	-	9	84 e	107	79	77 e
Colombia	960	1 092	741 e	-	-	-	39	11	13 e
Ecuador	220	568	128 e	58	-	-	-	111	22 e
Netherlands Antilles	937	958	562 e	-	-	202 e	269	44	205 e
Peru	260	477	324 e	-	-	-	-	2	-
Trinidad and Tobago	1 279	1 344	1 377 e	-	-	-	145	154	-
Venezuela	5 896	4 917	1 696 e	959	1 442	1 614 e	1 629	1 404	1 006 e
Other	714	1 096	1 085 e	-	115	169 e	20	75	32 e
Latin America	**11 508**	**12 551**	**6 711 e**	**1 049**	**1 787**	**2 379 e**	**2 233**	**1 921**	**1 476 e**
Romania	8	49	106 e	2	-	14 e	14	14	21 e
Other	249	271	393 e	12	5	18 e	37	43	48 e
Non-OECD Europe	**257**	**320**	**499 e**	**14**	**5**	**32 e**	**51**	**57**	**69 e**
FSU	**14 469**	**13 177**	**15 826 e**	**14**	**21**	**28 e**	**786**	**867**	**968 e**
Iran	13	-	21 e	-	-	1 e	19	38	30 e
Iraq	-	-	-	-	-	-	-	-	-
Kuwait	-	69	-	-	183	218 e	-	-	-
Oman	-	-	-	-	-	-	-	-	-
Qatar	-	-	-	-	-	-	2	249	-
Saudi Arabia	9	235	32 e	-	-	-	240	61	18 e
United Arab Emirates	-	42	105 e	-	-	-	131	100	-
Other	1 163	744	337 e	55	-	-	195	133	32 e
Middle East	**1 185**	**1 090**	**495 e**	**55**	**183**	**219 e**	**587**	**581**	**80 e**
India	-	-	-	144	298	200 e	14	78	153 e
Indonesia	3 681	3 218	2 970 e	69	62	57 e	42	185	70 e
Singapore	1 042	719	763 e	-	-	-	169	74	102 e
Other	52	154	467 e	-	78	124 e	79	66	78 e
Asia	**4 775**	**4 091**	**4 200 e**	**213**	**438**	**381 e**	**304**	**403**	**403 e**
China	**48**	**27**	**536 e**	**28**	**169**	**280 e**	**67**	**103**	**67 e**
Algeria	3 043	168	582 e	-	-	-	4 476	5 104	5 341 e
Angola/Cabinada	150	302	62 e	-	-	-	39	-	29 e
Cameroon	171	41	52 e	-	-	-	7	-	-
Egypt	107	247	109 e	-	-	-	19	7	24 e
Gabon	-	-	-	22	-	-	-	-	-
Libya	2 548	2 143	2 257 e	-	-	-	126	205	54 e
Nigeria	862	1 190	511 e	-	-	-	-	97	12 e
Tunisia	317	91	30 e	-	-	-	-	-	9 e
Other	1 169	590	929 e	80	21	21 e	185	381	430 e
Africa	**8 367**	**4 772**	**4 532 e**	**102**	**21**	**21 e**	**4 852**	**5 794**	**5 899 e**
Non-spec./Other	**2 586**	**4 539**	**4 748 e**	**4 570**	**4 875**	**682 e**	**1 103**	**1 164**	**1 787 e**
Total Imports	**78 634**	**72 921**	**66 820 e**	**17 247**	**17 732**	**17 379 e**	**27 041**	**26 966**	**27 227 e**
Memo: Total OPEC	*16 052*	*11 982*	*8 174 e*	*1 028*	*1 687*	*1 890 e*	*6 665*	*7 443*	*6 531 e*

Source: Annual Oil Statistics.

OECD TOTAL
Table 8A: Exports by Country of Destination

000 metric tons

	2000	2001	2002	2000	2001	2002	2000	2001	2002
	Crude Oil			**Natural Gas Liquids**			**Feedstocks and Other**		
Australia	-	-	-	1	42	11 e	-	-	-
Austria	31	62	179 e	-	-	-	-	-	-
Belgium	5 393	4 278	2 571 e	559	749	805 e	410	395	280 e
Canada	19 766	18 757	15 984 e	587	368	286 e	132	237	188 e
Denmark	4 093	2 813	3 239 e	14	23	29 e	27	28	40 e
Finland	5 553	5 980	5 549 e	87	101	26 e	1	4	20 e
France	36 996	36 090	27 677 e	861	782	936 e	624	301	259 e
Germany	21 118	24 632	23 472 e	278	245	192 e	810	460	502 e
Greece	-	-	134 e	-	4	-	-	28	-
Ireland	2 875	3 174	2 387 e	9	15	14 e	7	-	24 e
Italy	4 317	3 554	6 145 e	26	14	38 e	137	56	64 e
Japan	5 124	3 778	2 674 e	1 004	1 253	1 728 e	693	418	473 e
Luxembourg	-	-	-	-	-	-	-	-	-
Mexico	4	-	8 e	1 581	1 014	1 163 e	479	363	375 e
Netherlands	47 279	47 164	44 077 e	1 018	1 368	1 498 e	2 538	1 987	1 477 e
Norway	897	362	723 e	215	41	75 e	50	49	36 e
Portugal	2 546	2 245	2 263 e	201	209	211 e	16	9	31 e
Spain	10 403	10 554	10 062 e	456	362	367 e	241	86	134 e
Sweden	14 746	12 837	10 914 e	579	998	968 e	39	36	168 e
Switzerland	-	-	-	-	-	8 e	28	39	46 e
Turkey	-	-	-	495	493	589 e	52	66	52 e
United Kingdom	34 817	36 635	35 873 e	177	336	233 e	151	149	184 e
United States	174 641	168 652	183 528 e	12 674	11 992	10 922 e	8 315	8 524	13 087 e
Other OECD	6 002	4 206	6 591 e	215	95	198 e	252	323	222 e
OECD	**396 601**	**385 773**	**384 050 e**	**21 037**	**20 504**	**20 297 e**	**15 002**	**13 558**	**17 662 e**
Africa	163	535	501 e	31	51	107 e	175	104	73 e
Latin America	8 425	9 583	7 296 e	496	383	301 e	515	470	344 e
Asia	7 227	9 267	9 407 e	70	63	140 e	538	605	685 e
China	2 132	3 377	4 846 e	384	381	570 e	171	145	7 e
Non-OECD Europe	66	534	542 e	-	-	41 e	105	150	29 e
Former Soviet Union	251	-	1 e	-	-	6 e	7	-	-
Middle East	338	190	259 e	47	9	-	20	4	27 e
Non-specified/Other	1 306	4 184	5 685 e	408	414	356 e	797	737	752 e
Non-OECD	**19 908**	**27 670**	**28 537 e**	**1 436**	**1 301**	**1 521 e**	**2 328**	**2 215**	**1 917 e**
Total Exports	**416 509**	**413 443**	**412 587 e**	**22 473**	**21 805**	**21 818 e**	**17 330**	**15 773**	**19 579 e**
	Total Products			**Liquefied Petroleum Gas**			**Naphtha**		
Australia	1 646	1 171	1 128 e	-	-	-	-	-	3 e
Austria	4 701	4 179	5 377 e	126	114	115 e	43	35	15 e
Belgium	21 062	23 948	21 926 e	805	937	1 117 e	3 196	3 146	3 101 e
Canada	4 465	4 741	5 481 e	4	-	-	-	-	1 e
Denmark	4 830	4 507	4 488 e	19	9	1 e	165	157	146 e
Finland	1 434	1 514	1 676 e	24	36	39 e	152	64	103 e
France	16 378	15 530	14 778 e	183	291	340 e	1 589	1 241	1 359 e
Germany	40 117	39 553	31 548 e	610	432	579 e	6 739	6 545	4 903 e
Greece	1 349	1 301	1 229 e	2	-	2 e	70	50	49 e
Ireland	5 083	4 671	5 310 e	88	51	49 e	-	9	-
Italy	8 844	7 402	7 648 e	304	248	283 e	485	647	602 e
Japan	18 408	17 583	15 422 e	90	60	40 e	5 506	5 951	5 208 e
Luxembourg	2 465	2 499	2 552 e	18	17	15 e	-	-	-
Mexico	15 085	12 237	11 760 e	-	29	15 e	-	-	32 e
Netherlands	16 406	16 374	19 035 e	667	332	418 e	2 356	2 636	3 497 e
Norway	4 888	5 168	5 598 e	129	150	174 e	37	125	52 e
Portugal	3 358	2 233	2 965 e	225	164	188 e	427	248	233 e
Spain	11 160	11 717	13 388 e	152	112	153 e	272	419	294 e
Sweden	5 120	5 112	4 825 e	71	106	195 e	402	295	269 e
Switzerland	7 847	6 825	7 439 e	68	47	75 e	96	67	43 e
Turkey	3 426	1 645	2 147 e	295	96	338 e	142	11	39 e
United Kingdom	14 698	18 856	15 535 e	475	489	446 e	1 432	1 785	1 743 e
United States	41 779	45 408	46 584 e	366	463	568 e	1 348	1 201	1 096 e
Other OECD	8 538	9 406	10 040 e	339	335	383 e	190	273	188 e
OECD	**263 087**	**263 580**	**257 879 e**	**5 060**	**4 518**	**5 533 e**	**24 647**	**24 905**	**22 976 e**
Africa	9 386	8 880	8 992 e	879	792	712 e	81	140	68 e
Latin America	10 316	10 143	10 933 e	296	298	35 e	59	30	43 e
Asia	13 746	16 260	18 516 e	42	45	43 e	292	288	340 e
China	16 441	16 432	9 006 e	621	411	207 e	115	70	134 e
Non-OECD Europe	9 221	8 305	8 043 e	233	200	323 e	290	299	152 e
Former Soviet Union	731	619	435 e	21	-	4 e	15	19	24 e
Middle East	4 318	3 365	2 978 e	257	172	60 e	155	101	97 e
Non-specified/Other	6 450	9 960	16 517 e	533	840	1 097 e	1 206	830	876 e
Non-OECD	**70 609**	**73 964**	**75 420 e**	**2 882**	**2 758**	**2 481 e**	**2 213**	**1 777**	**1 734 e**
Total Exports	**333 696**	**337 544**	**333 299 e**	**7 942**	**7 276**	**8 014 e**	**26 860**	**26 682**	**24 710 e**

Source: Annual Oil Statistics.

OECD TOTAL
Table 8B: Exports by Country of Destination

000 metric tons

	Total Gasoline 2000	Total Gasoline 2001	Total Gasoline 2002	Total Kerosene 2000	Total Kerosene 2001	Total Kerosene 2002	Gas/Diesel Oil 2000	Gas/Diesel Oil 2001	Gas/Diesel Oil 2002
Australia	161	193	85 e	3	17	13 e	686	270	147 e
Austria	743	567	727 e	47	39	89 e	2 964	2 649	3 728 e
Belgium	1 825	1 761	1 462 e	1 046	1 370	895 e	6 652	7 323	5 828 e
Canada	894	1 058	1 211 e	861	537	794 e	470	680	290 e
Denmark	873	851	860 e	510	540	396 e	1 865	1 738	1 648 e
Finland	322	235	297 e	1	56	27 e	318	506	455 e
France	1 731	1 578	1 537 e	241	134	334 e	8 041	7 904	7 002 e
Germany	10 398	8 436	6 605 e	3 225	3 517	3 305 e	14 072	16 194	11 672 e
Greece	290	23	143 e	2	22	6 e	415	637	471 e
Ireland	799	983	1 163 e	641	606	772 e	1 671	1 521	1 788 e
Italy	505	589	579 e	64	42	140 e	347	125	635 e
Japan	568	604	859 e	3 438	3 283	3 111 e	3 767	3 113	1 684 e
Luxembourg	621	567	565 e	313	280	370 e	1 431	1 515	1 521 e
Mexico	5 780	5 067	5 864 e	195	108	238 e	3 737	1 814	989 e
Netherlands	2 971	3 108	3 993 e	191	400	263 e	3 735	3 351	5 189 e
Norway	537	751	830 e	191	316	293 e	1 061	1 217	1 439 e
Portugal	264	30	46 e	42	17	9 e	724	448	513 e
Spain	1 060	909	883 e	195	126	68 e	4 298	4 453	5 691 e
Sweden	1 670	1 650	1 621 e	599	488	381 e	1 334	1 376	1 305 e
Switzerland	2 827	2 382	2 602 e	1 077	990	886 e	3 166	2 819	3 252 e
Turkey	403	128	68 e	60	32	75 e	577	144	188 e
United Kingdom	3 255	5 176	3 227 e	1 288	620	922 e	3 312	4 395	3 400 e
United States	15 281	19 002	20 798 e	2 980	3 501	2 674 e	8 068	7 510	7 991 e
Other OECD	2 709	2 673	3 025 e	312	452	354 e	2 873	3 108	3 740 e
OECD	**56 487**	**58 321**	**59 050 e**	**17 522**	**17 493**	**16 415 e**	**75 584**	**74 810**	**70 566 e**
Africa	2 383	2 038	2 563 e	327	233	249 e	2 675	2 478	2 480 e
Latin America	1 041	1 048	1 668 e	260	186	148 e	2 966	2 417	2 863 e
Asia	1 264	787	1 074 e	563	945	1 028 e	5 919	7 313	8 422 e
China	73	184	12 e	992	1 339	895 e	3 728	3 315	653 e
Non-OECD Europe	1 335	1 357	1 796 e	522	416	515 e	2 321	2 072	2 295 e
Former Soviet Union	267	147	64 e	2	2	5 e	229	176	69 e
Middle East	1 338	771	1 082 e	361	310	64 e	731	196	447 e
Non-specified/Other	2 132	2 517	3 304 e	126	358	431 e	580	2 480	6 199 e
Non-OECD	**9 833**	**8 849**	**11 563 e**	**3 153**	**3 789**	**3 335 e**	**19 149**	**20 447**	**23 428 e**
Total Exports	**66 320**	**67 170**	**70 613 e**	**20 675**	**21 282**	**19 750 e**	**94 733**	**95 257**	**93 994 e**

	Fuel Oil (Residual) 2000	Fuel Oil (Residual) 2001	Fuel Oil (Residual) 2002	Petroleum Coke 2000	Petroleum Coke 2001	Petroleum Coke 2002	Other Products 2000	Other Products 2001	Other Products 2002
Australia	204	49	70 e	563	610	781 e	29	32	29 e
Austria	310	337	256 e	31	30	16 e	437	408	431 e
Belgium	3 555	5 058	4 945 e	871	957	1 282 e	3 112	3 396	3 296 e
Canada	781	853	1 504 e	811	1 085	999 e	644	528	682 e
Denmark	963	707	869 e	149	209	243 e	286	296	325 e
Finland	389	440	572 e	-	-	10 e	228	177	173 e
France	2 561	2 303	1 960 e	723	830	828 e	1 309	1 249	1 418 e
Germany	2 458	1 881	2 049 e	241	230	276 e	2 374	2 318	2 159 e
Greece	144	33	49 e	302	448	371 e	124	88	138 e
Ireland	1 292	1 041	909 e	195	131	291 e	397	329	338 e
Italy	4 692	2 872	2 804 e	2 144	2 103	1 823 e	303	776	702 e
Japan	842	701	508 e	3 371	3 192	3 292 e	826	679	720 e
Luxembourg	12	15	12 e	18	18	16 e	52	87	53 e
Mexico	3 158	2 663	2 299 e	1 808	2 072	1 630 e	407	484	693 e
Netherlands	3 790	3 690	3 425 e	1 590	1 674	1 177 e	1 106	1 183	1 073 e
Norway	2 304	2 005	1 597 e	227	278	345 e	402	326	868 e
Portugal	1 082	956	1 651 e	373	130	77 e	221	240	248 e
Spain	1 760	1 797	2 459 e	2 629	2 944	2 968 e	794	957	872 e
Sweden	547	775	620 e	92	90	97 e	405	332	337 e
Switzerland	245	158	201 e	70	42	57 e	298	320	323 e
Turkey	528	22	130 e	1 148	1 083	1 124 e	273	129	185 e
United Kingdom	3 553	4 976	3 921 e	519	606	672 e	864	809	1 204 e
United States	9 221	9 789	9 430 e	115	149	136 e	4 400	3 793	3 891 e
Other OECD	846	802	632 e	342	452	612 e	927	1 311	1 106 e
OECD	**45 237**	**43 923**	**42 872 e**	**18 332**	**19 363**	**19 123 e**	**20 218**	**20 247**	**21 344 e**
Africa	1 281	1 152	1 094 e	988	1 201	993 e	772	846	833 e
Latin America	2 793	3 732	3 917 e	1 876	1 724	1 403 e	1 025	708	856 e
Asia	4 160	5 570	6 394 e	462	260	89 e	1 044	1 052	1 126 e
China	9 903	9 876	5 076 e	4	86	571 e	1 005	1 151	1 458 e
Non-OECD Europe	3 611	2 867	1 992 e	316	383	444 e	593	711	526 e
Former Soviet Union	30	76	48 e	15	46	71 e	152	153	150 e
Middle East	518	539	408 e	620	812	472 e	338	464	348 e
Non-specified/Other	1 355	2 151	2 754 e	-	8	927 e	518	776	929 e
Non-OECD	**23 651**	**25 963**	**21 683 e**	**4 281**	**4 520**	**4 970 e**	**5 447**	**5 861**	**6 226 e**
Total Exports	**68 888**	**69 886**	**64 555 e**	**22 613**	**23 883**	**24 093 e**	**25 665**	**26 108**	**27 570 e**

Source: Annual Oil Statistics.

OECD NORTH AMERICA

Figure 1
FUEL SHARES OF TPES *

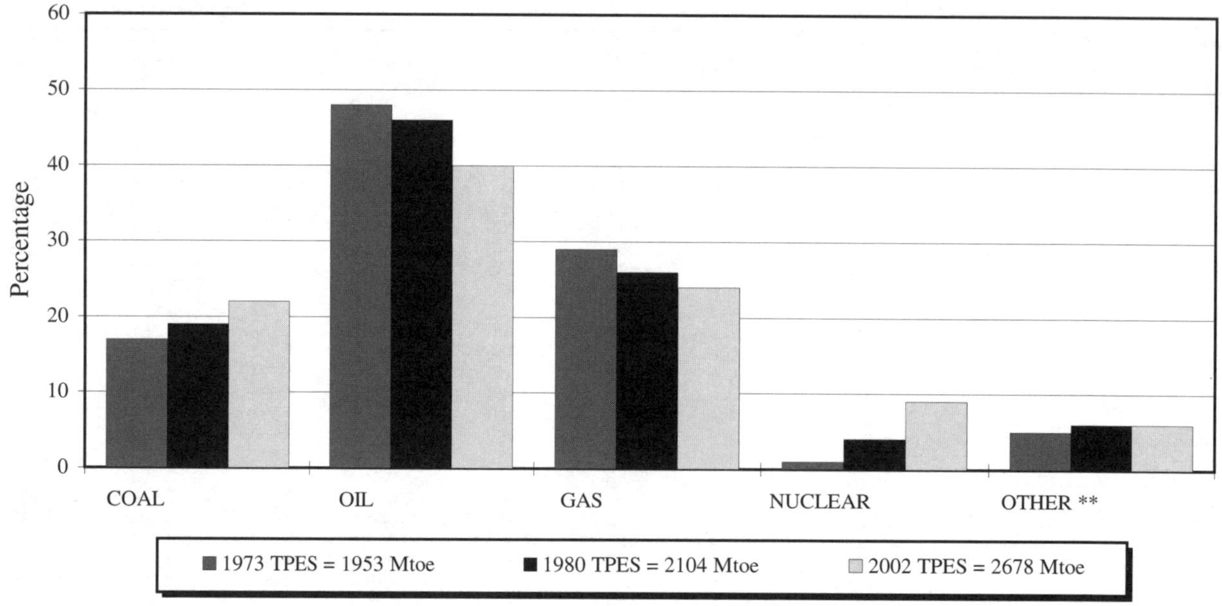

Source: Annual Energy Balances.

* Excluding electricity trade.

** Includes hydro, geothermal, solar, wind, combustible renewables and waste, etc.

Figure 2
OIL PRODUCTION AND CONSUMPTION

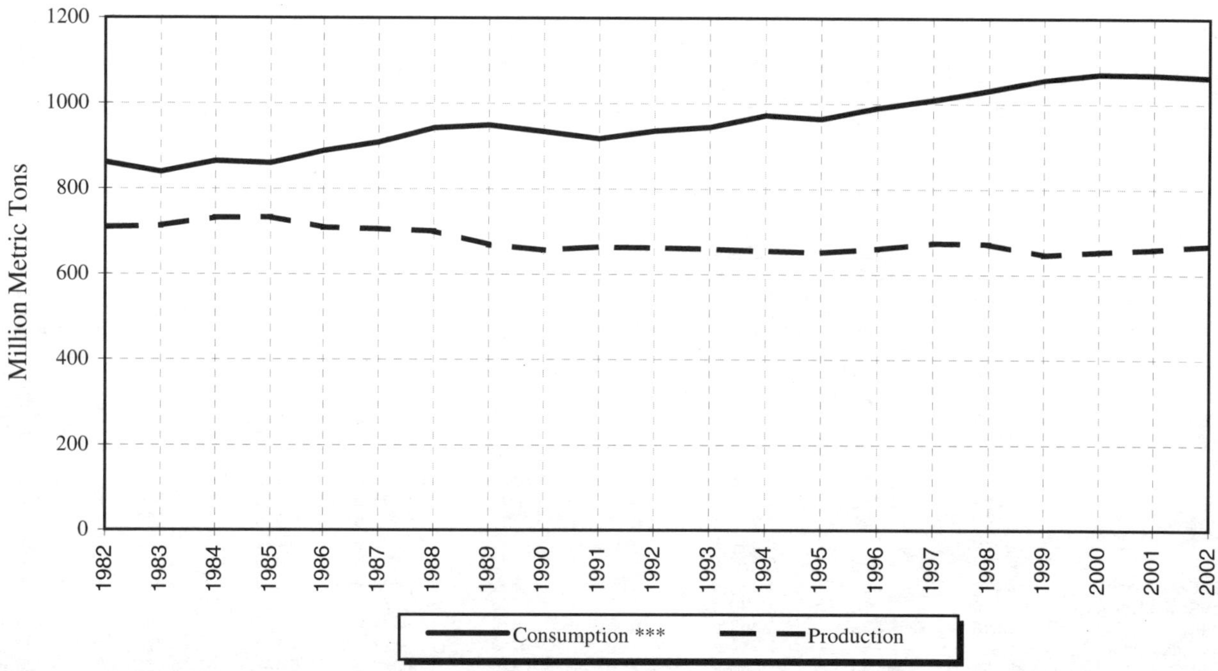

Source: Annual Oil Statistics.

*** Includes refinery fuel and international marine bunkers.

OECD NORTH AMERICA
Table 1: Oil Supply and Consumption - Summary Table

000 metric tons

	1973	1978	1980	1990	1999	2000	2001	2002
Indigenous Production	634 606	627 383	673 365	656 349	645 616	653 312	659 015	667 761 e
Crude Oil	..	564 098	595 298	569 942	526 804	532 273	536 988	541 363 e
NGL	..	62 118	70 343	74 295	87 414	88 972	87 688	85 903 e
Other	..	1 167	7 724	12 112	31 398	32 067	34 339	40 495 e
+ Net Imports	290 345	409 037	290 861	278 223	396 851	416 818	435 141	401 095 e
Crude Oil, NGL, Other	161 233	344 553	256 208	259 704	374 444	383 866	401 551	384 166 e
Petroleum Products	129 112	64 484	34 653	18 519	22 407	32 952	33 590	16 929 e
- Intl. Marine Bunkers	9 629	23 343	30 789	30 934	28 838	31 193	21 997	31 435 e
+ Stock Changes	-6 828	6 440	-8 718	-9 334	18 996	1 357	-13 881	5 321 e
TOTAL SUPPLY	**908 494**	**1019 517**	**924 719**	**894 304**	**1032 625**	**1040 294**	**1058 278**	**1042 742 e**
Refinery Intake	*761 110*	*925 781*	*871 521*	*884 018*	*969 949*	*979 418*	*980 279*	*979 327 e*
Refinery Output	*773 666*	*916 952*	*866 792*	*880 663*	*966 819*	*980 722*	*980 173*	*980 402 e*
LPG + Ethane	*13 492*	*14 082*	*14 038*	*19 264*	*23 261*	*23 848*	*22 763*	*23 091 e*
Naphtha	*23 536*	*14 414*	*17 432*	*12 858*	*13 394*	*13 797*	*11 514*	*15 302 e*
Kerosenes	*56 969*	*53 967*	*55 270*	*73 179*	*85 831*	*87 831*	*83 824*	*81 890 e*
Gasolines	*311 520*	*360 782*	*336 638*	*353 680*	*394 885*	*397 439*	*398 994*	*406 582 e*
Gas/Diesel Oil	*166 147*	*196 404*	*175 978*	*183 674*	*211 284*	*222 324*	*228 368*	*222 606 e*
Fuel Oil	*82 214*	*143 590*	*137 839*	*93 311*	*77 349*	*73 256*	*75 174*	*68 831 e*
Other Products	*119 788*	*133 713*	*129 597*	*144 697*	*160 815*	*162 227*	*159 536*	*162 100 e*
- Refinery Losses	-12 556	8 829	4 729	3 355	3 130	-1 304	106	-1 075 e
+ Returns	344	185	468	1 203	2 527	2 302	2 398	2 951 e
- Statistical Differences	-12 408	-1 072	- 246	-10 533	5 078	4 978	13 889	15 697 e
TOTAL CONSUMPTION	**933 802**	**1011 945**	**920 704**	**902 685**	**1026 944**	**1038 922**	**1046 681**	**1031 071 e**
Transformation and Energy	**167 028**	**169 877**	**136 694**	**116 674**	**130 227**	**125 763**	**122 762**	..
Crude Oil	920	730	697	1 200	303	342	250	..
LPG + Ethane	2 297	2 436	2 702	1 918	765	888	926	..
Gas/Diesel Oil	8 033	9 748	7 200	3 713	7 178	6 235	6 030	..
Fuel Oil	93 988	110 320	79 753	51 677	54 520	51 329	54 155	..
Petroleum Coke	13 054	13 113	13 678	17 622	26 311	25 788	21 952	..
Other	48 736	33 530	32 664	40 544	41 150	41 181	39 449	..
Transport	**428 090**	**484 004**	**460 499**	**530 759**	**634 925**	**647 078**	**646 201**	..
Motor Gasoline	309 337	350 434	321 385	348 335	409 497	413 189	415 908	..
Gas/Diesel Oil	55 856	70 871	74 598	97 145	131 419	136 649	137 712	..
Subtotal	*365 193*	*421 305*	*395 983*	*445 480*	*540 916*	*549 838*	*553 620*	..
Aviation Fuels	52 372	54 418	57 548	77 735	86 586	89 703	85 081	..
Other	10 525	8 281	6 968	7 544	7 423	7 537	7 500	..
Industry	**180 881**	**210 049**	**213 872**	**174 960**	**185 728**	**182 536**	**192 817**	..
Naphtha	23 435	19 670	21 242	13 314	15 926	18 172	15 239	..
LPG + Ethane	35 296	32 494	41 616	46 592	66 361	66 021	60 326	..
Gas/Diesel Oil	25 944	32 360	29 295	20 112	19 814	20 937	22 513	..
Fuel Oil	42 238	46 148	47 064	22 582	8 495	13 328	14 176	..
Other	53 968	79 377	74 655	72 360	75 132	64 078	80 563	..
Other Sectors	**157 803**	**148 015**	**109 639**	**80 292**	**76 064**	**83 545**	**84 901**	..
Gas/Diesel Oil	89 073	89 973	66 655	51 260	47 812	52 168	53 819	..
Fuel Oil	31 719	25 678	17 486	7 318	836	2 863	3 144	..
Other	37 011	32 364	25 498	21 714	27 416	28 514	27 938	..
Memo: Non-Energy Use	*96 492*	*116 321*	*107 740*	*118 771*	*159 495*	*150 126*	*154 895*	..
Chem/Petchem. Feedst.	*44 829*	*61 335*	*59 270*	*63 126*	*87 260*	*87 706*	*83 545*	..
Other	*51 663*	*54 986*	*48 470*	*55 645*	*72 235*	*62 420*	*71 350*	..

Source: Annual Oil Statistics.

OECD NORTH AMERICA
Table 2: Supply of Crude Oil, NGL, Feedstocks and Other Hydrocarbons

000 metric tons

	1973	1978	1980	1990	1999	2000	2001	2002
CRUDE OIL								
Indigenous Production	..	564 098	595 298	569 942	526 804	532 273	536 988	541 363 e
+ Imports	..	380 356	312 560	340 675	506 531	529 209	539 907	530 728 e
- Exports	..	36 812	59 878	97 302	140 904	152 587	150 611	152 749 e
- Direct Use	..	730	697	1 200	303	342	250	353 e
+ Stock Change	..	-4 601	-5 039	286	2 888	2 060	-4 291	-2 499 e
- Statistical Difference	..	5 131	4 390	-9 096	2 675	4 234	13 139	16 094 e
Refinery Intake	**..**	**897 180**	**837 854**	**821 497**	**892 341**	**906 379**	**908 604**	**900 396 e**
Closing Stock Level	..	*44 034*	*55 555*	*116 826*	*112 385*	*110 325*	*114 616*	*117 115 e*
NATURAL GAS LIQUIDS								
Indigenous Production	..	62 118	70 343	74 295	87 414	88 972	87 688	85 903 e
+ Imports	..	4 493	7 309	6 236	6 687	8 049	8 110	6 116 e
- Exports	..	3 849	5 447	8 652	14 681	14 961	12 793	12 549 e
- Direct Use	..	41 163	48 687	55 161	64 397	65 286	63 463	58 578 e
+ Stock Change	..	223	-1 917	- 78	604	- 88	- 486	- 85 e
- Statistical Difference	..	171	-1 120	- 563	- 744	- 177	831	- 29 e
Refinery Intake	**..**	**21 651**	**22 721**	**17 203**	**16 371**	**16 863**	**18 225**	**20 836 e**
Closing Stock Level	..	*10 470*	*11 319*	*2 763*	*2 929*	*3 017*	*3 503*	*3 588 e*
REFINERY FEEDSTOCKS								
Backflows	..	-	-	-	-	-	-	-
+ Transfers	..	-	-	13 943	15 493	14 366	6 325	7 427 e
+ Imports	..	365	1 664	20 954	18 106	16 090	20 637	21 936 e
- Exports	..	-	-	-	-	-	-	-
+ Stock Change	..	523	-1 076	875	648	- 179	340	1 650 e
- Statistical Difference	..	-4 895	-2 634	-	3	-	-	- 4 e
Refinery Intake	**..**	**5 783**	**3 222**	**35 772**	**34 244**	**30 277**	**27 302**	**31 017 e**
Closing Stock Level	..	*14 950*	*17 304*	*14 774*	*12 786*	*12 965*	*12 625*	*10 975 e*
OTHER HYDROCARBONS								
Indigenous Production	..	1 167	7 724	12 112	31 398	32 067	34 339	40 495 e
+ Imports	..	-	-	9	4 465	4 585	3 727	3 056 e
- Exports	..	-	-	2 216	5 760	6 519	7 426	12 372 e
- Direct Use	..	-	-	-	3 387	3 798	3 088	3 029 e
+ Stock Change	..	-	-	- 40	9	174	- 159	80 e
- Statistical Difference	..	-	-	319	- 268	610	1 245	1 152 e
Refinery Intake	**..**	**1 167**	**7 724**	**9 546**	**26 993**	**25 899**	**26 148**	**27 078 e**
Closing Stock Level	..	-	-	*75*	*1 604*	*1 430*	*1 589*	*1 509 e*
CRUDE OIL, NGL, FEEDSTOCKS AND OTHER								
Indigenous Production	634 606	627 383	673 365	656 349	645 616	653 312	659 015	667 761 e
+ Backflows	-	-	-	-	-	-	-	-
+ Transfers	-	-	-	13 943	15 493	14 366	6 325	7 427 e
+ Imports	217 316	385 214	321 533	367 874	535 789	557 933	572 381	561 836 e
- Exports	56 083	40 661	65 325	108 170	161 345	174 067	170 830	177 670 e
- Direct Use	40 745	41 893	49 384	56 361	68 087	69 426	66 801	61 960 e
+ Stock Change	-1 551	-3 855	-8 032	1 043	4 149	1 967	-4 596	- 854 e
- Statistical Difference	-7 567	407	636	-9 340	1 666	4 667	15 215	17 213 e
Refinery Intake	**761 110**	**925 781**	**871 521**	**884 018**	**969 949**	**979 418**	**980 279**	**979 327 e**
Closing Stock Level	-	*69 454*	*84 178*	*134 438*	*129 704*	*127 737*	*132 333*	*133 187 e*

Source: Annual Oil Statistics.

OECD NORTH AMERICA
Table 3: Motor Gasoline - Supply

000 metric tons

	1973	1978	1980	1990	1999	2000	2001	2002
MOTOR GASOLINE								
Refinery Output	**309 516**	**349 933**	**324 751**	**343 877**	**393 661**	**396 276**	**397 887**	**405 574 e**
+ Imports	6 992	3 113	2 232	15 303	27 934	30 310	37 630	39 841 e
- Exports	596	737	609	5 846	12 313	13 947	15 542	15 872 e
+ Transfers	1 060	2 544	4 382	4 304	2 868	5 731	5 223	5 074 e
+ Stock Change	590	2 513	-2 931	- 254	2 847	- 399	-1 808	260 e
- Statistical Difference	65	334	204	244	- 13	- 471	- 456	- 426 e
Total Consumption	**317 497**	**357 032**	**327 621**	**357 140**	**415 010**	**418 442**	**423 846**	**435 303 e**
Closing Stock Level	*-*	*27 437*	*30 268*	*24 379*	*21 881*	*22 280*	*24 088*	*23 828 e*

Table 4: Aviation Fuels - Supply

000 metric tons

	1973	1978	1980	1990	1999	2000	2001	2002
AVIATION GASOLINE								
Refinery Output	**2 004**	**1 750**	**1 551**	**1 053**	**950**	**845**	**837**	**802 e**
+ Imports	13	4	1	24	9	38	67	27 e
- Exports	22	-	-	1	15	17	13	14 e
+ Transfers	-	7	82	- 2	141	92	116	169 e
+ Stock Change	33	32	- 3	50	14	30	-	11 e
- Statistical Difference	3	6	3	1	-	2	-	- 12 e
Total Consumption	**2 025**	**1 787**	**1 628**	**1 123**	**1 099**	**986**	**1 007**	**1 007 e**
Closing Stock Level	*-*	*373*	*362*	*228*	*229*	*199*	*199*	*188 e*
GASOLINE TYPE JET FUEL								
Refinery Output	**-**	**9 099**	**10 336**	**8 750**	**274**	**318**	**270**	**206 e**
+ Imports	-	188	656	72	2	-	-	-
- Exports	-	-	34	338	114	6	30	364 e
+ Transfers	-	-	25	- 49	2	- 49	- 6	- 1 e
+ Stock Change		73	- 130	79	15	- 12	11	7 e
- Statistical Difference	-	-	74	3	- 95	- 2	- 16	- 372 e
Total Consumption	**-**	**9 360**	**10 779**	**8 511**	**274**	**253**	**261**	**220 e**
Closing Stock Level	*-*	*670*	*1 049*	*794*	*52*	*64*	*53*	*46 e*
KEROSENE TYPE JET FUEL								
Refinery Output	**42 426**	**40 971**	**43 322**	**68 266**	**81 032**	**83 139**	**78 654**	**77 542 e**
+ Imports	9 574	3 231	2 368	4 450	5 782	7 233	6 973	5 198 e
- Exports	422	346	193	3 165	1 825	1 803	1 702	967 e
+ Transfers	-	-	- 83	- 50	- 237	- 173	- 188	- 168 e
+ Stock Change	- 420	13	- 350	-1 368	433	- 64	- 18	287 e
- Statistical Difference	22	25	- 77	- 29	- 28	- 132	- 94	- 256 e
Total Consumption	**51 136**	**43 844**	**45 141**	**68 162**	**85 213**	**88 464**	**83 813**	**82 148 e**
Closing Stock Level	*-*	*3 302*	*4 118*	*5 120*	*4 566*	*4 630*	*4 648*	*4 361 e*

Source: Annual Oil Statistics.

OECD NORTH AMERICA
Table 5A: Gas/Diesel Oil - Supply and Consumption

000 metric tons

	1973	1978	1980	1990	1999	2000	2001	2002
Refinery Output	**166 147**	**196 404**	**175 978**	**183 674**	**211 284**	**222 324**	**228 368**	**222 606 e**
+ Recycled Products	-	-	-	-	-	-	-	-
+ Imports	19 294	2 418	1 578	11 494	11 874	14 084	15 247	12 484 e
- Exports	675	1 329	2 015	11 256	14 400	14 182	12 812	12 027 e
- Intl. Marine Bunkers	980	2 300	2 513	7 349	7 725	7 845	7 748	7 531 e
+ Transfers	-	1 435	1 688	- 300	929	781	918	357 e
+ Stock Changes	-6 551	6 073	2 656	-4 486	4 617	880	-4 081	2 345 e
- Statistical Differences	-1 671	- 251	- 376	- 453	356	53	- 182	- 181 e
Total Consumption	**178 906**	**202 952**	**177 748**	**172 230**	**206 223**	**215 989**	**220 074**	**218 415 e**
Transformation	**7 019**	**7 776**	**5 329**	**2 585**	**6 428**	**5 529**	**5 175**	..
Public Electricity	7 019	7 656	5 246	2 559	3 249	3 652	3 398	..
Autoproducer Electricity	-	120	83	26	1 397	764	1 044	..
Public CHP Plants	-	-	-	-	142	88	263	..
Autoproducers of CHP	-	-	-	-	1 640	1 025	470	..
Public Heat Plants	-	-	-	-	-	-	-	..
Autoproducer Heat Plants	-	-	-	-	-	-	-	..
Petrochemical Industry	-	-	-	-	-	-	-	..
Other Transformation	-	-	-	-	-	-	-	..
Energy Sector	**1 014**	**1 972**	**1 871**	**1 128**	**750**	**706**	**855**	..
Refinery Fuel	1 014	1 972	1 871	1 128	750	706	855	..
Other Energy Sector	-	-	-	-	-	-	-	..
Final Consumption	**170 873**	**193 204**	**170 548**	**168 517**	**199 045**	**209 754**	**214 044**	..
Transport	**55 856**	**70 871**	**74 598**	**97 145**	**131 419**	**136 649**	**137 712**	..
Air	-	-	-	-	-	-	-	..
Road	35 191	47 479	50 358	80 887	118 508	122 889	123 913	..
Rail	16 272	16 033	15 711	12 904	11 139	12 154	11 827	..
Inland Waterways	933	867	854	809	837	846	847	..
Non-Specified	3 460	6 492	7 675	2 545	935	760	1 125	..
Industry	**25 944**	**32 360**	**29 295**	**20 112**	**19 814**	**20 937**	**22 513**	..
Iron and Steel	63	24	37	48	127	706	830	..
Chemical (incl. Pet.-Chem.)	59	13	32	47	3 337	4 236	4 917	..
Non-Ferrous Metals	84	18	11	24	1 204	337	317	..
Non-Metallic Minerals	128	29	54	49	2 308	2 049	2 196	..
Transport Equipment	105	-	56	-	665	1 852	1 985	..
Machinery	22	-	26	6	1 137	1 492	1 578	..
Mining and Quarrying	439	741	1 118	962	1 284	1 355	1 203	..
Food and Tobacco	223	-	242	248	1 890	2 350	2 449	..
Pulp, Paper and Printing	112	-	344	247	1 288	1 322	1 359	..
Wood and Wood Products	128	597	640	165	2 861	1 954	2 229	..
Construction	-	1 080	1 047	971	826	878	855	..
Textiles and Leather	-	-	24	-	774	718	849	..
Non-Specified	24 581	29 858	25 664	17 345	2 113	1 688	1 746	..
Other Sectors	**89 073**	**89 973**	**66 655**	**51 260**	**47 812**	**52 168**	**53 819**	..
Commerce and Public	18 717	20 087	15 365	14 560	13 088	15 186	16 008	..
Residential	59 768	56 886	38 894	23 001	21 127	22 435	22 692	..
Agriculture	10 588	13 000	12 396	13 699	13 597	14 547	15 119	..
Non-Specified	-	-	-	-	-	-	-	..
Memo: Non-Energy Use	-	-	-	-	-	-	-	..
Chem/Petchem. Feedst.	-	-	-	-	-	-	-	..
Other	-	-	-	-	-	-	-	..
Closing Stock Level	-	*33 615*	*33 175*	*20 963*	*19 399*	*18 519*	*22 600*	*20 255 e*

Source: Annual Oil Statistics.

OECD NORTH AMERICA
Table 5B: Fuel Oil (Residual) - Supply and Consumption

000 metric tons

	1973	1978	1980	1990	1999	2000	2001	2002
Refinery Output	**82 214**	**143 590**	**137 839**	**93 311**	**77 349**	**73 256**	**75 174**	**68 831 e**
+ Recycled Products	-	-	-	-	-	-	-	-
+ Imports	104 283	64 894	41 787	31 347	18 611	27 925	25 190	13 635 e
- Exports	6 980	4 529	5 237	14 585	11 378	9 566	13 380	16 039 e
- Intl. Marine Bunkers	8 649	21 043	28 276	23 585	21 113	23 348	14 249	23 904 e
+ Transfers	-	105	74	-2 863	206	610	572	937 e
+ Stock Changes	375	1 399	- 143	-3 968	3 203	32	-1 094	2 013 e
- Statistical Differences	-2 257	- 784	- 626	-2 929	2 394	639	- 112	291 e
Total Consumption	**173 500**	**185 200**	**146 670**	**82 586**	**64 484**	**68 270**	**72 325**	**45 182 e**
Transformation	**83 032**	**98 954**	**70 719**	**45 377**	**50 028**	**47 132**	**50 381**	**..**
Public Electricity	83 032	98 738	70 243	45 201	41 891	39 663	41 188	..
Autoproducer Electricity	-	216	193	99	4 272	5 786	7 567	..
Public CHP Plants	-	-	283	77	628	488	419	..
Autoproducers of CHP	-	-	-	-	3 237	1 195	1 207	..
Public Heat Plants	-	-	-	-	-	-	-	..
Autoproducer Heat Plants	-	-	-	-	-	-	-	..
Petrochemical Industry	-	-	-	-	-	-	-	..
Other Transformation	-	-	-	-	-	-	-	..
Energy Sector	**10 956**	**11 366**	**9 034**	**6 300**	**4 492**	**4 197**	**3 774**	**..**
Refinery Fuel	10 956	11 366	9 034	6 300	4 492	4 197	3 774	..
Other Energy Sector	-	-	-	-	-	-	-	..
Final Consumption	**79 512**	**74 880**	**66 917**	**30 909**	**9 964**	**16 941**	**18 170**	**..**
Transport	**5 555**	**3 054**	**2 367**	**1 009**	**633**	**750**	**850**	**..**
Air	-	-	-	-	-	-	-	..
Road	-	-	-	-	-	-	-	..
Rail	359	28	103	-	-	-	-	..
Inland Waterways	5 196	3 026	2 264	1 009	633	750	850	..
Non-Specified	-	-	-	-	-	-	-	..
Industry	**42 238**	**46 148**	**47 064**	**22 582**	**8 495**	**13 328**	**14 176**	**..**
Iron and Steel	5 207	4 144	5 153	957	808	1 080	1 022	..
Chemical (incl. Pet.-Chem.)	1 345	6 155	5 681	1 384	1 591	3 257	3 843	..
Non-Ferrous Metals	489	2 235	1 032	193	101	181	239	..
Non-Metallic Minerals	1 162	317	1 276	2 173	1 869	1 933	1 673	..
Transport Equipment	741	647	826	-	27	87	92	..
Machinery	37	415	387	1	27	87	102	..
Mining and Quarrying	889	743	834	538	415	513	552	..
Food and Tobacco	2 278	2 182	2 438	1 171	1 102	1 146	1 196	..
Pulp, Paper and Printing	14 752	12 145	12 160	3 043	2 046	4 289	4 568	..
Wood and Wood Products	337	299	322	6	14	17	42	..
Construction	-	169	75	31	10	10	11	..
Textiles and Leather	-	-	1 471	-	46	290	314	..
Non-Specified	15 001	16 697	15 409	13 085	439	438	522	..
Other Sectors	**31 719**	**25 678**	**17 486**	**7 318**	**836**	**2 863**	**3 144**	**..**
Commerce and Public	31 229	25 303	17 297	7 270	806	2 817	3 089	..
Residential	490	363	170	35	18	22	11	..
Agriculture	-	12	19	13	12	24	44	..
Non-Specified	-	-	-	-	-	-	-	..
Memo: Non-Energy Use	*245*	-	-	-	-	-	-	..
Chem/Petchem. Feedst.	*245*	-	-	-	-	-	-	..
Other	-	-	-	-	-	-	-	..
Closing Stock Level	-	*31 764*	*34 207*	*19 097*	*11 115*	*11 083*	*12 177*	*10 164 e*

Source: Annual Oil Statistics.

OECD NORTH AMERICA
Table 5C: Total Products - Supply and Consumption

000 metric tons

	1973	1978	1980	1990	1999	2000	2001	2002
Refinery Output	**773 666**	**916 952**	**866 792**	**880 663**	**966 819**	**980 722**	**980 173**	**980 402 e**
+ Recycled Products	344	185	468	1 203	2 527	2 302	2 398	2 951 e
+ Imports	148 995	81 918	56 324	75 179	84 265	101 134	106 318	92 367 e
- Exports	19 883	17 434	21 671	56 660	61 858	68 182	72 728	75 438 e
- Intl. Marine Bunkers	9 629	23 343	30 789	30 934	28 838	31 193	21 997	31 435 e
+ Transfers	40 745	42 714	49 654	42 418	52 594	55 060	60 476	54 533 e
+ Stock Changes	-5 277	10 295	- 686	-10 377	14 847	- 610	-9 285	6 175 e
- Statistical Differences	-4 841	- 658	- 612	-1 193	3 412	311	-1 326	-1 516 e
Total Consumption	**933 802**	**1011 945**	**920 704**	**902 685**	**1026 944**	**1038 922**	**1046 681**	**1031 071 e**
Transformation	**92 032**	**108 681**	**77 430**	**49 187**	**62 778**	**58 334**	**56 559**	**..**
Public Electricity	91 337	107 325	75 651	48 559	47 404	45 216	44 992	..
Autoproducer Electricity	-	336	276	237	6 267	7 027	8 776	..
Public CHP Plants	-	-	283	77	770	614	737	..
Autoproducers of CHP	-	-	-	-	8 337	5 477	2 054	..
Public Heat Plants	-	-	-	-	-	-	-	..
Autoproducer Heat Plants	-	-	-	-	-	-	-	..
Petrochemical Industry	-	-	-	-	-	-	-	..
Other Transformation	695	1 020	1 220	314	-	-	-	..
Energy Sector	**74 996**	**61 196**	**59 264**	**67 487**	**67 449**	**67 429**	**66 203**	**..**
Refinery Fuel	74 076	60 466	58 567	65 054	65 587	65 579	64 454	..
Other Energy Sector	920	730	697	2 433	1 862	1 850	1 749	..
Final Consumption	**766 774**	**842 068**	**784 010**	**786 011**	**896 717**	**913 159**	**923 919**	**..**
Transport	**428 090**	**484 004**	**460 499**	**530 759**	**634 925**	**647 078**	**646 201**	**..**
Air	52 372	54 418	57 548	77 774	86 607	89 722	85 098	..
Road	349 462	403 140	376 344	432 155	531 564	539 566	543 519	..
Rail	16 660	16 061	15 814	12 904	11 139	12 154	11 827	..
Inland Waterways	6 136	3 893	3 118	5 108	4 455	4 675	4 420	..
Non-Specified	3 460	6 492	7 675	2 818	1 160	961	1 337	..
Industry	**180 881**	**210 049**	**213 872**	**174 960**	**185 728**	**182 536**	**192 817**	**..**
Iron and Steel	5 288	4 168	5 190	1 033	1 019	1 807	1 871	..
Chemical (incl. Pet.-Chem.)	46 004	67 503	64 983	64 561	92 217	95 217	92 320	..
Non-Ferrous Metals	1 256	3 239	2 134	1 399	2 754	1 747	1 733	..
Non-Metallic Minerals	1 307	346	1 487	2 466	4 591	4 302	4 208	..
Transport Equipment	863	647	885	-	768	2 020	2 149	..
Machinery	63	415	414	32	1 497	1 841	1 917	..
Mining and Quarrying	1 425	1 544	2 001	1 736	2 177	2 486	2 397	..
Food and Tobacco	2 568	2 182	2 696	1 435	3 193	3 648	3 787	..
Pulp, Paper and Printing	14 898	12 145	12 504	3 295	3 493	5 772	6 055	..
Wood and Wood Products	504	896	968	190	3 049	2 105	2 391	..
Construction	34 651	33 424	28 448	33 887	37 284	35 795	36 106	..
Textiles and Leather	-	-	1 496	-	951	1 066	1 215	..
Non-Specified	72 054	83 540	90 666	64 926	32 735	24 730	36 668	..
Other Sectors	**157 803**	**148 015**	**109 639**	**80 292**	**76 064**	**83 545**	**84 901**	**..**
Commerce and Public	56 414	51 749	38 311	27 003	20 140	24 853	25 881	..
Residential	81 819	75 140	51 255	35 734	38 321	40 291	39 595	..
Agriculture	16 589	17 402	16 637	17 555	17 603	18 401	19 425	..
Non-Specified	2 981	3 724	3 436	-	-	-	-	..
Memo: Non-Energy Use	*96 492*	*116 321*	*107 740*	*118 771*	*159 495*	*150 126*	*154 895*	..
Chem/Petchem. Feedst.	*44 829*	*61 335*	*59 270*	*63 126*	*87 260*	*87 706*	*83 545*	..
Other	*51 663*	*54 986*	*48 470*	*55 645*	*72 235*	*62 420*	*71 350*	..
Closing Stock Level	-	*111 692*	*116 164*	*86 872*	*72 907*	*73 517*	*82 802*	*76 627 e*

Source: Annual Oil Statistics.

OECD NORTH AMERICA
Table 6A: Supply and Consumption

000 metric tons

	1980	2000	2001	2002	1980	2000	2001	2002
	LPG AND ETHANE				**REFINERY GAS**			
Refinery Output	**14 038**	**23 848**	**22 763**	**23 091 e**	**34 047**	**39 291**	**39 562**	**40 026 e**
+ Recycled Products	-	-	-	-	-	-	-	-
+ Imports	576	4 279	3 565	3 776 e	-	-	-	-
- Exports	695	382	418	379 e	-	-	-	-
- Intl. Marine Bunkers	-	-	-	-	-	-	-	-
+ Transfers	44 453	61 658	59 608	54 745 e	16	-	-	-
+ Stock Change	- 79	346	-2 598	1 253 e	2	-	-	-
- Statistical Difference	135	126	- 203	- 237 e	214	-	-	1 e
Total Consumption	**58 158**	**89 623**	**83 123**	**82 723 e**	**33 851**	**39 291**	**39 562**	**40 025 e**
of which: Ethane	*654*	*24 233*	*22 533*	*23 122 e*	-	-	-	-
Transformation	1 220	-	-	..	-	-	-	..
Energy	1 482	888	926	..	31 516	38 898	38 460	..
Transport	576	1 773	1 898	..	-	-	-	..
Industry	41 616	66 021	60 326	..	2 335	393	1 102	..
Other	13 264	20 941	19 973	..	-	-	-	..
Memo: Non-Energy Use	*17 496*	*63 576*	*57 975*	..	*2 335*	*393*	*1 102*	..
Closing Stock Level	*635*	*4 833*	*7 431*	*6 178 e*	*4*	-	-	-
	NAPHTHA				**OTHER KEROSENE**			
Refinery Output	**17 432**	**13 797**	**11 514**	**15 302 e**	**11 948**	**4 692**	**5 170**	**4 348 e**
+ Recycled Products	281	216	282	411 e	-	-	-	-
+ Imports	4 749	4 814	3 951	2 842 e	1	133	228	267 e
- Exports	974	404	351	447 e	32	204	247	979 e
- Intl. Marine Bunkers	-	-	-	-	-	-	-	-
+ Transfers	- 272	- 140	- 230	- 431 e	-1 738	-1 036	-1 071	-1 132 e
+ Stock Change	- 241	- 72	86	93 e	518	85	- 200	32 e
- Statistical Difference	- 380	9	5	49 e	47	- 9	- 10	- 68 e
Total Consumption	**21 355**	**18 202**	**15 247**	**17 721 e**	**10 650**	**3 679**	**3 890**	**2 604 e**
Transformation	-	-	-	..	-	-	-	..
Energy	113	30	8	..	187	1	-	..
Transport	-	-	-	..	-	-	-	..
Industry	21 242	18 172	15 239	..	4 134	343	528	..
Other	-	-	-	..	6 329	3 335	3 362	..
Memo: Non-Energy Use	*21 241*	*18 172*	*15 239*	..	*7*	-	-	..
Closing Stock Level	*694*	*741*	*655*	*502 e*	*2 089*	*618*	*818*	*786 e*
	WHITE SPIRIT + SBP				**LUBRICANTS**			
Refinery Output	**4 616**	**2 676**	**2 272**	**..**	**10 355**	**10 934**	**10 389**	**..**
+ Recycled Products	26	7	7	..	47	242	233	..
+ Imports	366	428	542	..	226	514	494	..
- Exports	222	819	948	..	1 249	1 924	1 924	..
- Intl. Marine Bunkers	-	-	-	..	-	-	-	..
+ Transfers	150	35	- 29	..	8	- 208	- 174	..
+ Stock Change	- 91	30	10	..	- 193	- 34	- 222	..
- Statistical Difference	6	1	1	..	- 13	- 8	- 14	..
Total Consumption	**4 839**	**2 356**	**1 853**	**..**	**9 207**	**9 532**	**8 810**	**..**
Transformation	-	-	-	..	-	-	-	..
Energy	2	-	-	..	3	6	5	..
Transport	-	-	-	..	4 025	5 014	4 752	..
Industry	4 434	2 200	1 755	..	5 179	4 186	3 753	..
Other	403	156	98	..	-	326	300	..
Memo: Non-Energy Use	*4 837*	*2 356*	*1 853*	..	*9 204*	*9 526*	*8 805*	..
Closing Stock Level	*653*	*263*	*253*	..	*2 287*	*2 072*	*2 294*	..

Source: Annual Oil Statistics.

OECD NORTH AMERICA
Table 6B: Supply and Consumption

000 metric tons

	1980	2000	2001	2002	1980	2000	2001	2002
	BITUMEN				PARAFFIN WAXES			
Refinery Output	**27 222**	**36 155**	**33 523**	..	**803**	**907**	**868**	..
+ Recycled Products	-	-	-	..	-	-	-	..
+ Imports	350	1 981	1 880	..	21	109	117	..
- Exports	83	1 622	1 729	..	125	164	166	..
- Intl. Marine Bunkers	-	-	-	..	-	-	-	..
+ Transfers	- 4	744	111	..	-	-	-	..
+ Stock Change	- 139	-1 407	748	..	24	- 9	57	..
- Statistical Difference	26	5	1	..	-	-	1	..
Total Consumption	**27 320**	**35 846**	**34 532**	**..**	**723**	**843**	**875**	**..**
Transformation	-	763	-	..	-	-	-	..
Energy	-	8	11	..	-	-	-	..
Transport	-	-	-	..	-	-	-	..
Industry	27 320	35 010	34 398	..	723	843	875	..
Other	-	65	123	..	-	-	-	..
Memo: Non-Energy Use	*27 320*	*35 075*	*34 521*	..	*723*	*843*	*875*	..
Closing Stock Level	*3 631*	*4 470*	*3 722*	..	*69*	*135*	*78*	..
	PETROLEUM COKE				OTHER PRODUCTS			
Refinery Output	**25 931**	**53 634**	**56 416**	**59 007 e**	**26 623**	**18 630**	**16 506**	**63 067 e**
+ Recycled Products	-	-	-	-	114	1 837	1 876	2 540 e
+ Imports	1 022	989	1 424	2 025 e	391	8 297	9 010	12 272 e
- Exports	8 987	21 260	22 446	22 617 e	1 216	1 882	1 020	5 733 e
- Intl. Marine Bunkers	-	-	-	-	-	-	-	-
+ Transfers	-	-	1	5 e	873	-12 985	-4 375	-5 022 e
+ Stock Change	299	- 109	- 116	73 e	115	93	- 60	- 199 e
- Statistical Difference	41	46	- 229	- 560 e	110	52	- 18	255 e
Total Consumption	**18 224**	**33 208**	**35 508**	**39 053 e**	**26 790**	**13 938**	**21 955**	**66 670 e**
Transformation	162	4 910	1 003	..	-	-	-	..
Energy	13 516	20 878	20 949	..	1 105	910	698	..
Transport	-	-	-	..	-	-	-	..
Industry	4 546	7 420	13 556	..	25 584	12 117	20 207	..
Other	-	-	-	..	101	911	1 050	..
Memo: Non-Energy Use	*4 392*	*7 157*	*13 268*	..	*20 185*	*13 028*	*21 257*	..
Closing Stock Level	*820*	*1 736*	*1 852*	*1 779 e*	*1 903*	*1 874*	*1 934*	*8 480 e*

Source: Annual Oil Statistics.

OECD NORTH AMERICA
Table 7A: Imports by Country of Origin

000 metric tons

	2000	2001	2002	2000	2001	2002	2000	2001	2002
	Crude Oil			**Natural Gas Liquids**			**Feedstocks and Other**		
Australia	2 748	1 658	2 302 e	-	-	-	-	-	-
Belgium	-	-	-	-	-	-	1 255	1 654	1 979 e
Canada	68 555	68 770	72 362 e	5 593	4 129	4 176 e	1 052	1 066	709 e
Czech Republic	-	-	-	-	-	-	-	-	-
Denmark	357	466	170 e	-	-	-	-	60	-
Finland	-	22	-	-	-	-	85	136	-
France	-	30	-	-	4	-	568	745	237 e
Germany	-	-	-	-	-	-	458	1 049	1 333 e
Greece	-	-	-	-	-	-	-	6	-
Hungary	-	-	-	-	-	-	-	-	-
Ireland	-	-	-	-	-	-	226	86	-
Italy	-	32	-	-	-	-	293	349	205 e
Japan	-	-	-	-	-	-	-	-	76 e
Korea	-	-	-	-	-	-	22	17	69 e
Mexico	68 534	72 344	77 149 e	68	182	-	184	41	91 e
Netherlands	73	-	-	-	-	-	252	622	397 e
New Zealand	-	-	-	-	-	-	-	-	-
Norway	28 882	27 096	30 548 e	7	301	156 e	713	710	688 e
Poland	-	-	-	-	-	-	-	2	-
Spain	61	-	-	-	7	-	218	193	78 e
Sweden	37	16	-	7	38	-	848	752	620 e
Turkey	-	6	-	-	-	-	230	210	212 e
United Kingdom	27 256	25 724	30 451 e	51	164	21 e	420	854	399 e
United States	1 217	1 627	955 e	324	279	243 e	337	295	392 e
Other	128	-	-	-	-	-	17	61	52 e
OECD	**197 848**	**197 791**	**213 937 e**	**6 050**	**5 104**	**4 596 e**	**7 178**	**8 908**	**7 537 e**
Argentina	3 156	3 168	3 717 e	-	-	-	37	151	103 e
Brazil	258	850	2 969 e	-	-	36 e	169	259	299 e
Colombia	16 307	13 299	12 404 e	-	-	-	129	172	298 e
Ecuador	6 465	5 859	5 382 e	-	-	-	164	43	82 e
Netherlands Antilles	-	-	-	-	-	-	2 011	1 766	1 993 e
Peru	217	366	519 e	-	-	-	117	113	92 e
Trinidad and Tobago	2 790	2 558	3 389 e	-	-	14 e	340	208	145 e
Venezuela	81 158	82 755	78 106 e	85	48	97 e	3 716	1 791	2 751 e
Other	1 055	988	1 224 e	-	-	-	17	42	213 e
Latin America	**111 406**	**109 843**	**107 710 e**	**85**	**48**	**147 e**	**6 700**	**4 545**	**5 976 e**
Romania	-	-	-	-	-	-	-	-	-
Other	-	-	-	-	-	-	23	-	20 e
Non-OECD Europe	**-**	**-**	**-**	**-**	**-**	**-**	**23**	**-**	**20 e**
FSU	**1 127**	**114**	**4 332 e**	**-**	**-**	**-**	**1 477**	**2 104**	**2 485 e**
Iran	-	-	-	-	-	-	-	-	-
Iraq	34 948	46 543	31 035 e	-	-	-	-	-	-
Kuwait	13 394	12 027	10 745 e	-	37	-	14	-	-
Oman	107	974	826 e	-	-	-	-	-	-
Qatar	-	10	444 e	-	-	-	356	335	261 e
Saudi Arabia	80 700	84 030	79 339 e	108	311	-	1 151	899	856 e
United Arab Emirates	124	969	738 e	-	-	-	348	247	-
Other	1 476	1 321	1 495 e	-	-	-	58	162	333 e
Middle East	**130 749**	**145 874**	**124 622 e**	**108**	**348**	**-**	**1 927**	**1 643**	**1 450 e**
India	-	-	-	-	-	-	66	104	19 e
Indonesia	1 797	1 988	2 458 e	-	-	-	209	82	217 e
Singapore	-	-	-	-	-	-	232	481	153 e
Other	3 071	3 386	2 124 e	4	3	-	508	701	632 e
Asia	**4 868**	**5 374**	**4 582 e**	**4**	**3**	**-**	**1 015**	**1 368**	**1 021 e**
China	**1 656**	**639**	**1 038 e**	**-**	**-**	**-**	**-**	**-**	**11 e**
Algeria	3 074	3 940	6 189 e	1 747	2 302	1 326 e	1 397	4 641	4 621 e
Angola/Cabinada	15 989	16 995	16 314 e	5	-	-	302	121	315 e
Cameroon	545	249	610 e	-	-	-	-	56	21 e
Egypt	235	-	-	-	-	-	237	114	50 e
Gabon	9 959	9 142	8 909 e	-	-	-	38	-	-
Libya	-	-	-	-	-	-	-	-	-
Nigeria	48 549	44 631	35 802 e	40	294	27 e	133	646	1 189 e
Tunisia	-	-	-	-	-	-	-	27	20 e
Other	3 204	5 315	3 931 e	10	11	20 e	248	191	276 e
Africa	**81 555**	**80 272**	**71 755 e**	**1 802**	**2 607**	**1 373 e**	**2 355**	**5 796**	**6 492 e**
Non-spec./Other	**-**	**-**	**2 752 e**	**-**	**-**	**-**	**-**	**-**	**-**
Total Imports	**529 209**	**539 907**	**530 728 e**	**8 049**	**8 110**	**6 116 e**	**20 675**	**24 364**	**24 992 e**
Memo: Total OPEC	*263 744*	*276 893*	*244 856 e*	*1 980*	*2 992*	*1 450 e*	*7 324*	*8 641*	*9 895 e*

Source: Annual Oil Statistics.

OECD NORTH AMERICA
Table 7B: Imports by Country of Origin

000 metric tons

	2000	2001	2002	2000	2001	2002	2000	2001	2002
	Total Products			**Liquefied Petroleum Gases**			**Naphtha**		
Australia	344	414	303 e	-	-	-	-	-	-
Belgium	1 161	1 605	1 586 e	-	-	9 e	-	42	8 e
Canada	11 588	14 544	16 672 e	220	176	283 e	130	150	161 e
Czech Republic	-	-	-	-	-	-	-	-	-
Denmark	243	191	97 e	-	-	-	-	-	-
Finland	778	663	1 099 e	-	-	3 e	-	-	-
France	1 424	1 637	1 144 e	-	-	6 e	116	33	1 e
Germany	933	1 028	913 e	-	-	-	-	-	-
Greece	186	147	98 e	-	-	-	144	60	-
Hungary	2	1	-	-	-	-	-	-	-
Ireland	32	66	-	-	-	-	-	6	-
Italy	1 234	1 917	1 337 e	-	-	2 e	43	-	10 e
Japan	781	578	281 e	-	-	-	4	-	-
Korea	2 790	3 131	2 318 e	-	-	-	28	51	45 e
Mexico	2 524	1 757	1 824 e	-	-	9 e	1 392	1 253	873 e
Netherlands	1 325	1 814	2 358 e	-	-	9 e	57	45	86 e
New Zealand	96	58	-	-	-	-	32	58	-
Norway	1 358	2 122	1 528 e	-	150	71 e	190	218	18 e
Poland	-	-	-	-	-	-	-	-	-
Spain	1 295	1 414	832 e	3	-	-	5	32	-
Sweden	606	481	380 e	-	-	-	11	-	-
Turkey	-	88	383 e	-	-	-	-	23	31 e
United Kingdom	3 452	3 395	3 113 e	-	-	3 e	23	17	14 e
United States	20 489	18 810	17 034 e	2 138	2 445	2 319 e	5	-	4 e
Other	548	451	406 e	-	-	-	-	-	-
OECD	**53 189**	**56 312**	**53 706 e**	**2 361**	**2 771**	**2 714 e**	**2 180**	**1 988**	**1 251 e**
Argentina	1 317	1 941	2 141 e	-	-	-	6	189	64 e
Brazil	2 562	3 784	2 209 e	-	-	1 e	9	12	16 e
Colombia	1 424	1 899	1 047 e	-	-	-	88	19	173 e
Ecuador	297	735	261 e	-	-	-	19	35	74 e
Netherlands Antilles	2 886	2 579	2 350 e	-	-	-	648	180	148 e
Peru	340	680	345 e	-	-	-	26	70	51 e
Trinidad and Tobago	1 407	1 856	1 344 e	-	-	-	147	47	-
Venezuela	14 694	13 949	8 024 e	-	34	-	508	449	500 e
Other	604	1 356	1 572 e	-	-	-	38	41	31 e
Latin America	**25 531**	**28 779**	**19 293 e**	**-**	**34**	**1 e**	**1 489**	**1 042**	**1 057 e**
Romania	-	98	705 e	-	-	-	-	-	-
Other	60	337	58 e	-	-	-	-	-	-
Non-OECD Europe	**60**	**435**	**763 e**	**-**	**-**	**-**	**-**	**-**	**-**
FSU	**4 491**	**3 584**	**4 715 e**	**-**	**-**	**-**	**110**	**64**	**130 e**
Iran	-	-	-	-	-	-	-	-	-
Iraq	-	-	-	-	-	-	-	-	-
Kuwait	380	639	608 e	-	-	-	48	-	-
Oman	-	-	-	-	-	-	-	-	-
Qatar	19	264	-	-	-	-	-	-	-
Saudi Arabia	1 747	1 688	1 286 e	246	218	18 e	142	130	-
United Arab Emirates	211	643	215 e	-	-	-	81	84	-
Other	411	271	267 e	-	-	-	34	37	-
Middle East	**2 768**	**3 505**	**2 376 e**	**246**	**218**	**18 e**	**305**	**251**	**-**
India	222	536	924 e	-	-	-	142	-	34 e
Indonesia	438	540	73 e	-	-	4 e	-	-	-
Singapore	1 167	1 058	595 e	-	-	-	8	9	-
Other	481	873	497 e	-	-	-	-	-	-
Asia	**2 308**	**3 007**	**2 089 e**	**-**	**-**	**4 e**	**150**	**9**	**34 e**
China	**472**	**588**	**275 e**	**-**	**-**	**-**	**-**	**-**	**28 e**
Algeria	9 545	6 390	6 973 e	1 672	542	1 039 e	285	350	164 e
Angola/Cabinada	186	403	113 e	-	-	-	-	-	22 e
Cameroon	143	114	52 e	-	-	-	-	-	-
Egypt	171	219	471 e	-	-	-	124	70	53 e
Gabon	22	-	-	-	-	-	-	-	-
Libya	-	-	-	-	-	-	-	-	-
Nigeria	1 007	1 528	842 e	-	-	-	92	32	49 e
Tunisia	173	213	40 e	-	-	-	-	-	-
Other	1 021	1 241	650 e	-	-	-	79	145	54 e
Africa	**12 268**	**10 108**	**9 141 e**	**1 672**	**542**	**1 039 e**	**580**	**597**	**342 e**
Non-spec./Other	**47**	**-**	**9 e**	**-**	**-**	**-**	**-**	**-**	**-**
Total Imports	**101 134**	**106 318**	**92 367 e**	**4 279**	**3 565**	**3 776 e**	**4 814**	**3 951**	**2 842 e**
Memo: Total OPEC	*28 041*	*25 641*	*18 021 e*	*1 918*	*794*	*1 061 e*	*1 156*	*1 045*	*713 e*

Source: Annual Oil Statistics.

OECD NORTH AMERICA
Table 7C: Imports by Country of Origin

000 metric tons

	2000	2001	2002	2000	2001	2002	2000	2001	2002
	Total Gasoline			**Total Kerosene**			**Gas/Diesel**		
Australia	48	62	-	18	66	-	-	25	-
Belgium	820	1 303	1 371 e	6	13	-	93	13	85 e
Canada	4 039	6 403	7 498 e	161	189	242 e	4 716	4 726	5 391 e
Czech Republic	-	-	-	-	-	-	-	-	-
Denmark	73	1	6 e	-	-	-	-	-	-
Finland	715	537	1 096 e	-	-	-	-	82	-
France	740	1 285	1 070 e	-	-	-	15	138	-
Germany	401	326	651 e	-	-	-	81	132	3 e
Greece	9	62	98 e	-	25	-	33	-	-
Hungary	-	-	-	-	-	-	-	-	-
Ireland	-	1	-	-	-	-	32	44	-
Italy	835	1 258	1 179 e	26	22	-	22	202	92 e
Japan	77	77	-	666	318	193 e	20	109	20 e
Korea	168	655	334 e	2 144	1 837	1 557 e	192	519	227 e
Mexico	276	330	282 e	88	120	94 e	-	14	40 e
Netherlands	713	1 342	1 930 e	19	6	-	131	121	61 e
New Zealand	-	-	-	-	-	-	-	-	-
Norway	386	721	488 e	-	-	10 e	5	-	-
Poland	-	-	-	-	-	-	-	-	-
Spain	932	1 089	707 e	-	-	-	-	34	-
Sweden	103	28	14 e	-	1	-	149	142	28 e
Turkey	-	-	352 e	-	-	-	-	-	-
United Kingdom	1 981	2 425	2 683 e	-	-	1 e	126	180	20 e
United States	6 194	6 549	6 408 e	654	460	788 e	2 882	1 413	2 054 e
Other	307	451	406 e	-	-	-	-	-	-
OECD	**18 817**	**24 905**	**26 573 e**	**3 782**	**3 057**	**2 885 e**	**8 497**	**7 894**	**8 021 e**
Argentina	1 219	1 274	1 562 e	-	-	8 e	7	98	94 e
Brazil	1 539	1 623	1 433 e	-	-	-	-	246	46 e
Colombia	210	460	15 e	107	231	96 e	20	86	101 e
Ecuador	58	21	37 e	-	-	-	-	-	-
Netherlands Antilles	65	268	182 e	630	702	484 e	496	663	667 e
Peru	13	87	-	-	-	-	41	44	-
Trinidad and Tobago	312	269	267 e	28	54	-	28	113	3 e
Venezuela	4 508	3 749	2 443 e	1 431	1 251	909 e	3 108	2 696	1 612 e
Other	71	281	505 e	-	-	-	-	27	299 e
Latin America	**7 995**	**8 032**	**6 444 e**	**2 196**	**2 238**	**1 497 e**	**3 700**	**3 973**	**2 822 e**
Romania	-	33	661 e	-	-	-	-	65	31 e
Other	60	310	58 e	-	27	-	-	-	-
Non-OECD Europe	**60**	**343**	**719 e**	**-**	**27**	**-**	**-**	**65**	**31 e**
FSU	**1 533**	**1 624**	**2 461 e**	**-**	**16**	**-**	**1 284**	**1 595**	**1 219 e**
Iran	-	-	-	-	-	-	-	-	-
Iraq	-	-	-	-	-	-	-	-	-
Kuwait	-	-	-	332	456	390 e	-	-	-
Oman	-	-	-	-	-	-	-	-	-
Qatar	5	19	-	-	-	-	14	-	-
Saudi Arabia	857	826	1 204 e	295	208	58 e	58	126	6 e
United Arab Emirates	-	93	206 e	-	286	9 e	-	80	-
Other	72	28	202 e	34	85	33 e	41	-	32 e
Middle East	**934**	**966**	**1 612 e**	**661**	**1 035**	**490 e**	**113**	**206**	**38 e**
India	80	246	565 e	-	82	-	-	208	174 e
Indonesia	1	-	-	-	-	-	26	15	-
Singapore	332	565	389 e	320	246	68 e	136	80	11 e
Other	11	62	36 e	284	388	418 e	139	415	28 e
Asia	**424**	**873**	**990 e**	**604**	**716**	**486 e**	**301**	**718**	**213 e**
China	**439**	**482**	**163 e**	**-**	**28**	**-**	**-**	**-**	**-**
Algeria	-	95	146 e	95	81	94 e	189	205	85 e
Angola/Cabinada	-	-	-	-	-	-	-	101	-
Cameroon	28	-	-	-	-	-	-	73	-
Egypt	-	149	418 e	-	-	-	-	-	-
Gabon	-	-	-	-	-	-	-	-	-
Libya	-	-	-	-	-	-	-	-	-
Nigeria	25	121	308 e	28	3	-	-	95	55 e
Tunisia	-	-	10 e	-	-	-	-	122	-
Other	93	107	24 e	-	-	13 e	-	200	-
Africa	**146**	**472**	**906 e**	**123**	**84**	**107 e**	**189**	**796**	**140 e**
Non-spec./Other	**-**	**-**	**-**	**-**	**-**	**-**	**-**	**-**	**-**
Total Imports	**30 348**	**37 697**	**39 868 e**	**7 366**	**7 201**	**5 465 e**	**14 084**	**15 247**	**12 484 e**
Memo: Total OPEC	*5 396*	*4 903*	*4 307 e*	*2 181*	*2 285*	*1 460 e*	*3 395*	*3 217*	*1 758 e*

Source: Annual Oil Statistics.

OECD NORTH AMERICA
Table 7D: Imports by Country of Origin

000 metric tons

	2000	2001	2002	2000	2001	2002	2000	2001	2002
	Fuel Oil (Residual)			**Petroleum Coke**			**Other Products**		
Australia	-	-	-	-	-	-	278	261	303 e
Belgium	242	189	-	-	-	-	-	45	113 e
Canada	1 351	1 827	1 683 e	72	13	99 e	899	1 060	1 315 e
Czech Republic	-	-	-	-	-	-	-	-	-
Denmark	170	190	87 e	-	-	-	-	-	4 e
Finland	63	-	-	-	-	-	-	44	-
France	410	40	-	20	8	23 e	123	133	44 e
Germany	445	537	207 e	-	25	22 e	6	8	30 e
Greece	-	-	-	-	-	-	-	-	-
Hungary	-	-	-	-	-	-	2	1	-
Ireland	-	15	-	-	-	-	-	-	-
Italy	187	58	-	-	-	27 e	121	377	27 e
Japan	-	66	62 e	-	-	-	14	8	6 e
Korea	-	-	-	-	-	-	258	69	155 e
Mexico	491	-	494 e	-	-	-	277	40	32 e
Netherlands	315	277	174 e	-	-	-	90	23	98 e
New Zealand	59	-	-	-	-	-	5	-	-
Norway	259	477	181 e	-	-	-	518	556	760 e
Poland	-	-	-	-	-	-	-	-	-
Spain	158	105	-	-	-	-	197	154	125 e
Sweden	332	297	193 e	-	-	-	11	13	145 e
Turkey	-	-	-	-	-	-	-	65	-
United Kingdom	1 312	690	342 e	-	-	-	10	83	50 e
United States	6 787	5 638	2 639 e	875	1 016	1 002 e	954	1 289	1 820 e
Other	241	-	-	-	-	-	-	-	-
OECD	**12 822**	**10 406**	**6 062 e**	**967**	**1 062**	**1 173 e**	**3 763**	**4 229**	**5 027 e**
Argentina	85	239	167 e	-	140	246 e	-	1	-
Brazil	918	1 860	586 e	-	5	75 e	96	38	52 e
Colombia	960	1 092	649 e	-	-	-	39	11	13 e
Ecuador	220	568	128 e	-	-	-	-	111	22 e
Netherlands Antilles	794	729	483 e	-	-	202 e	253	37	184 e
Peru	260	477	294 e	-	-	-	-	2	-
Trinidad and Tobago	749	1 220	1 074 e	-	-	-	143	153	-
Venezuela	3 545	4 417	1 571 e	-	-	43 e	1 594	1 353	946 e
Other	495	993	730 e	-	-	-	-	14	7 e
Latin America	**8 026**	**11 595**	**5 682 e**	**-**	**145**	**566 e**	**2 125**	**1 720**	**1 224 e**
Romania	-	-	-	-	-	13 e	-	-	-
Other	-	-	-	-	-	-	-	-	-
Non-OECD Europe	**-**	**-**	**-**	**-**	**-**	**13 e**	**-**	**-**	**-**
FSU	**1 374**	**249**	**764 e**	**-**	**-**	**-**	**190**	**36**	**141 e**
Iran	-	-	-	-	-	-	-	-	-
Iraq	-	-	-	-	-	-	-	-	-
Kuwait	-	-	-	-	183	218 e	-	-	-
Oman	-	-	-	-	-	-	-	-	-
Qatar	-	-	-	-	-	-	-	245	-
Saudi Arabia	-	150	-	-	-	-	149	30	-
United Arab Emirates	-	-	-	-	-	-	130	100	-
Other	50	-	-	-	-	-	180	121	-
Middle East	**50**	**150**	**-**	**-**	**183**	**218 e**	**459**	**496**	**-**
India	-	-	-	-	-	-	-	-	151 e
Indonesia	411	482	69 e	-	-	-	-	43	-
Singapore	295	158	117 e	-	-	-	76	-	10 e
Other	-	-	-	-	-	-	47	8	15 e
Asia	**706**	**640**	**186 e**	**-**	**-**	**-**	**123**	**51**	**176 e**
China	**-**	**-**	**-**	**-**	**34**	**55 e**	**33**	**44**	**29 e**
Algeria	2 887	61	137 e	-	-	-	4 417	5 056	5 308 e
Angola/Cabinada	150	302	62 e	-	-	-	36	-	29 e
Cameroon	115	41	52 e	-	-	-	-	-	-
Egypt	47	-	-	-	-	-	-	-	-
Gabon	-	-	-	22	-	-	-	-	-
Libya	-	-	-	-	-	-	-	-	-
Nigeria	862	1 180	418 e	-	-	-	-	97	12 e
Tunisia	173	91	30 e	-	-	-	-	-	-
Other	713	475	236 e	-	-	-	136	314	323 e
Africa	**4 947**	**2 150**	**935 e**	**22**	**-**	**-**	**4 589**	**5 467**	**5 672 e**
Non-spec./Other	**-**	**-**	**6 e**	**-**	**-**	**-**	**47**	**-**	**3 e**
Total Imports	**27 925**	**25 190**	**13 635 e**	**989**	**1 424**	**2 025 e**	**11 329**	**12 043**	**12 272 e**
Memo: Total OPEC	*7 705*	*6 290*	*2 195 e*	*-*	*183*	*261 e*	*6 290*	*6 924*	*6 266 e*

Source: Annual Oil Statistics.

OECD NORTH AMERICA
Table 8A: Exports by Country of Destination

000 metric tons

	2000	2001	2002	2000	2001	2002	2000	2001	2002
	Crude Oil			**Natural Gas Liquids**			**Feedstocks and Other**		
Australia	-	-	-	1	10	11 e	-	-	-
Austria	-	-	-	-	-	-	-	-	-
Belgium	-	-	-	2	-	1 e	17	21	18 e
Canada	2 504	2 353	1 589 e	587	368	270 e	132	203	188 e
Denmark	-	-	-	-	-	-	-	-	-
Finland	-	-	-	4	-	-	-	-	-
France	-	-	-	10	4	12 e	2	2	3 e
Germany	-	-	-	28	-	-	-	2	-
Greece	-	-	-	-	-	-	-	-	-
Ireland	-	-	-	-	-	-	4	-	-
Italy	205	183	81 e	-	12	14 e	3	-	-
Japan	2 758	863	510 e	-	9	151 e	63	49	69 e
Luxembourg	-	-	-	-	-	-	-	-	-
Mexico	4	-	8 e	1 379	779	1 021 e	479	363	375 e
Netherlands	73	-	147 e	10	-	-	66	24	8 e
Norway	-	-	-	-	-	-	-	-	-
Portugal	936	811	826 e	-	-	-	-	-	10 e
Spain	7 645	7 709	7 823 e	-	-	7 e	-	-	-
Sweden	-	-	-	-	-	-	-	-	-
Switzerland	-	-	-	-	-	8 e	-	-	-
Turkey	-	-	-	-	-	10 e	-	-	-
United Kingdom	934	764	827 e	2	2	3 e	2	3	2 e
United States	128 319	127 135	132 452 e	12 540	11 393	10 648 e	5 191	6 258	11 263 e
Other OECD	567	-	-	27	-	49 e	23	17	16 e
OECD	**143 945**	**139 818**	**144 263 e**	**14 590**	**12 577**	**12 205 e**	**5 982**	**6 942**	**11 952 e**
Africa	-	344	-	-	5	12 e	1	1	-
Latin America	8 117	9 148	6 254 e	308	195	298 e	491	445	338 e
Asia	266	1 111	1 973 e	6	16	25 e	34	34	71 e
China	-	-	-	57	-	1 e	3	-	2 e
Non-OECD Europe	-	-	-	-	-	-	7	-	-
Former Soviet Union	-	-	-	-	-	-	-	-	-
Middle East	259	190	259 e	-	-	-	1	4	-
Non-specified/Other	-	-	-	-	-	8 e	-	-	9 e
Non-OECD	**8 642**	**10 793**	**8 486 e**	**371**	**216**	**344 e**	**537**	**484**	**420 e**
Total Exports	**152 587**	**150 611**	**152 749 e**	**14 961**	**12 793**	**12 549 e**	**6 519**	**7 426**	**12 372 e**

	2000	2001	2002	2000	2001	2002	2000	2001	2002
	Total Products			**Liquefied Petroleum Gas**			**Naphtha**		
Australia	585	664	792 e	-	-	-	-	-	-
Austria	1	-	-	-	-	-	-	-	-
Belgium	873	996	1 269 e	-	-	-	-	-	-
Canada	3 647	3 998	4 462 e	-	-	-	-	-	-
Denmark	149	205	257 e	-	-	-	-	-	-
Finland	4	190	193 e	-	-	-	-	-	-
France	599	1 455	791 e	-	-	-	-	-	-
Germany	187	149	224 e	-	-	-	-	-	-
Greece	259	395	369 e	-	-	-	-	-	-
Ireland	194	126	338 e	-	-	-	-	-	-
Italy	2 399	2 160	2 011 e	-			-	-	-
Japan	4 195	3 719	3 925 e	-	-	-	-	-	-
Luxembourg	-	-	-	-	-	-	-	-	-
Mexico	14 810	12 111	11 308 e	-	-	-	-	-	-
Netherlands	2 373	2 645	1 593 e	-	-	-	-	-	-
Norway	133	191	223 e	-	-	-	-	-	-
Portugal	347	97	33 e	-	-	-	-	-	-
Spain	2 597	3 370	3 566 e	-	-	-	-	-	-
Sweden	62	198	64 e	-	-	-	-	-	-
Switzerland	188	8	143 e	-	-	-	-	-	-
Turkey	1 150	1 079	1 128 e	-	-	-	-	-	-
United Kingdom	597	820	781 e	-	-	-	-	-	-
United States	18 331	21 656	23 298 e	208	320	366 e	404	351	447 e
Other OECD	536	871	763 e	-	-	-	-	-	-
OECD	**54 216**	**57 103**	**57 531 e**	**208**	**320**	**366 e**	**404**	**351**	**447 e**
Africa	1 086	1 285	1 165 e	-	-	-	-	-	-
Latin America	8 502	8 278	8 287 e	174	98	13 e	-	-	-
Asia	2 755	4 071	4 878 e	-	-	-	-	-	-
China	92	364	853 e	-	-	-	-	-	-
Non-OECD Europe	380	383	510 e	-	-	-	-	-	-
Former Soviet Union	14	51	66 e	-	-	-	-	-	-
Middle East	1 091	1 161	983 e	-	-	-	-	-	-
Non-specified/Other	46	32	1 165 e	-	-	-	-	-	-
Non-OECD	**13 966**	**15 625**	**17 907 e**	**174**	**98**	**13 e**	-	-	-
Total Exports	**68 182**	**72 728**	**75 438 e**	**382**	**418**	**379 e**	**404**	**351**	**447 e**

Source: Annual Oil Statistics.

OECD NORTH AMERICA
Table 8B: Exports by Country of Destination

000 metric tons

	2000	2001	2002	2000	2001	2002	2000	2001	2002
	Total Gasoline			**Total Kerosene**			**Gas/Diesel Oil**		
Australia	1	33	-	-	-	1 e	1	-	-
Austria	-	-	-	-	-	-	-	-	-
Belgium	-	-	1 e	-	-	-	1	39	-
Canada	389	517	429 e	746	524	794 e	466	577	260 e
Denmark	-	-	-	-	-	-	-	-	14 e
Finland	-	-	-	-	44	21 e	2	145	135 e
France	9	32	1 e	3	-	-	109	728	176 e
Germany	-	-	2 e	-	-	-	6	-	2 e
Greece	-	-	-	-	-	-	-	-	-
Ireland	-	-	-	-	-	-	-	-	-
Italy	-	-	-	-	-	-	91	-	2 e
Japan	32	-	133 e	5	-	-	57	15	-
Luxembourg	-	-	-	-	-	-	-	-	-
Mexico	5 705	5 028	5 541 e	195	108	238 e	3 663	1 814	913 e
Netherlands	114	30	20 e	-	205	55 e	601	319	548 e
Norway	-	-	-	-	-	-	-	-	-
Portugal	-	-	-	-	-	-	-	-	-
Spain	-	-	-	-	-	-	48	59	220 e
Sweden	-	-	-	-	-	-	4	130	-
Switzerland	-	-	-	-	-	81 e	101	6	53 e
Turkey	-	-	-	-	-	-	-	-	-
United Kingdom	1	1	18 e	27	68	73 e	44	89	33 e
United States	7 073	9 214	9 006 e	441	476	476 e	5 369	5 784	6 294 e
Other OECD	31	34	9 e	-	-	-	83	25	75 e
OECD	**13 355**	**14 889**	**15 160 e**	**1 417**	**1 425**	**1 739 e**	**10 646**	**9 730**	**8 725 e**
Africa	-	-	20 e	-	28	2 e	1	24	23 e
Latin America	584	530	775 e	238	185	129 e	2 434	1 609	1 606 e
Asia	-	52	1 e	1	9	-	1 076	1 413	1 373 e
China	-	83	-	32	-	-	4	31	218 e
Non-OECD Europe	-	-	-	-	-	-	1	-	50 e
Former Soviet Union	-	-	-	-	-	-	2	1	-
Middle East	31	31	117 e	319	302	64 e	10	-	12 e
Non-specified/Other	-	-	177 e	-	-	12 e	8	4	20 e
Non-OECD	**615**	**696**	**1 090 e**	**590**	**524**	**207 e**	**3 536**	**3 082**	**3 302 e**
Total Exports	**13 970**	**15 585**	**16 250 e**	**2 007**	**1 949**	**1 946 e**	**14 182**	**12 812**	**12 027 e**

	2000	2001	2002	2000	2001	2002	2000	2001	2002
	Fuel Oil (Residual)			**Petroleum Coke**			**Other Products**		
Australia	-	1	-	563	610	781 e	20	20	10 e
Austria	-	-	-	-	-	-	1	-	-
Belgium	1	3	-	847	924	1 247 e	24	30	21 e
Canada	598	775	1 337 e	811	1 085	982 e	637	520	660 e
Denmark	-	-	-	149	205	243 e	-	-	-
Finland	-	-	27 e	-	-	10 e	2	1	-
France	-	1	-	472	683	597 e	6	11	17 e
Germany	-	-	-	168	134	207 e	13	15	13 e
Greece	-	-	-	257	394	368 e	2	1	1 e
Ireland	-	-	50 e	193	126	288 e	1	-	-
Italy	193	85	184 e	2 099	2 058	1 801 e	16	17	24 e
Japan	110	82	102 e	3 301	3 117	3 241 e	690	505	449 e
Luxembourg	-	-	-	-	-	-	-	-	-
Mexico	3 041	2 625	2 299 e	1 808	2 072	1 630 e	398	464	687 e
Netherlands	271	770	168 e	1 376	1 266	795 e	11	55	7 e
Norway	-	-	-	133	190	223 e	-	1	-
Portugal	-	-	-	347	97	33 e	-	-	-
Spain	84	389	529 e	2 464	2 822	2 811 e	1	100	6 e
Sweden	1	-	-	55	65	64 e	2	3	-
Switzerland	31	1	-	54	-	9 e	2	1	-
Turkey	-	-	-	1 146	1 074	1 124 e	4	5	4 e
United Kingdom	-	47	1 e	496	584	643 e	29	31	13 e
United States	1 464	2 552	3 704 e	100	149	136 e	3 272	2 810	2 869 e
Other OECD	22	97	94 e	270	393	537 e	130	322	48 e
OECD	**5 816**	**7 428**	**8 495 e**	**17 109**	**18 048**	**17 770 e**	**5 261**	**4 912**	**4 829 e**
Africa	33	1	62 e	961	1 178	971 e	91	54	87 e
Latin America	2 412	3 597	3 792 e	1 868	1 711	1 387 e	792	548	585 e
Asia	1 090	2 166	3 293 e	439	244	68 e	149	187	143 e
China	1	132	47 e	3	85	571 e	52	33	17 e
Non-OECD Europe	126	49	66 e	252	333	394 e	1	1	-
Former Soviet Union	-	5	-	8	41	65 e	4	4	1 e
Middle East	88	2	274 e	620	806	472 e	23	20	44 e
Non-specified/Other	-	-	10 e	-	-	919 e	38	28	27 e
Non-OECD	**3 750**	**5 952**	**7 544 e**	**4 151**	**4 398**	**4 847 e**	**1 150**	**875**	**904 e**
Total Exports	**9 566**	**13 380**	**16 039 e**	**21 260**	**22 446**	**22 617 e**	**6 411**	**5 787**	**5 733 e**

Source: Annual Oil Statistics.

OECD PACIFIC

Figure 1
FUEL SHARES OF TPES *

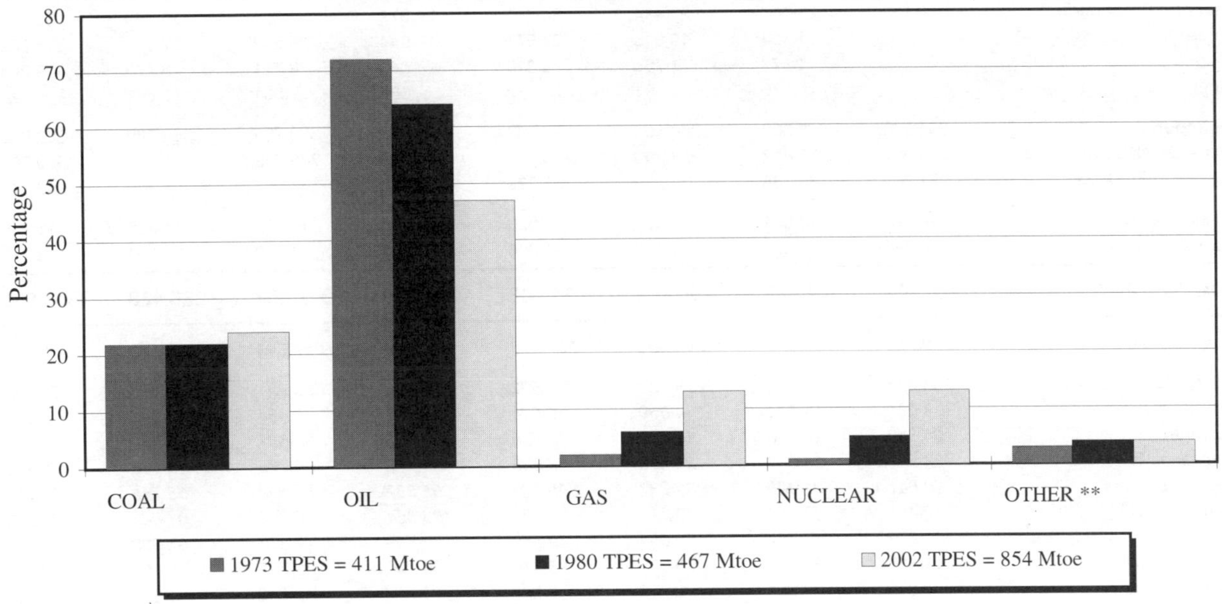

Source: Annual Energy Balances.

* Excluding electricity trade.

** Includes hydro, geothermal, solar, wind, combustible renewables and waste, etc.

Figure 2
OIL PRODUCTION AND CONSUMPTION

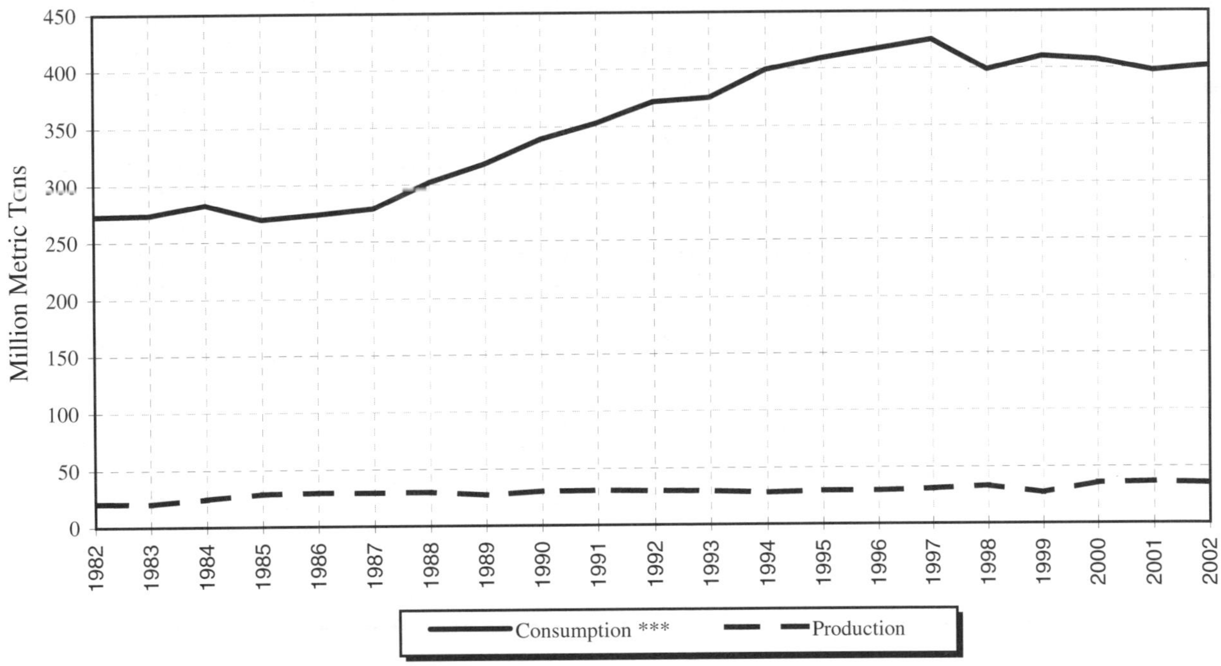

Source: Annual Oil Statistics.

*** Includes refinery fuel and international marine bunkers.

OECD PACIFIC
Table 1: Oil Supply and Consumption - Summary Table

000 metric tons

	1973	1978	1980	1990	1999	2000	2001	2002
Indigenous Production	20 135	23 107	21 444	30 661	26 872	35 324	36 133	35 169 e
Crude Oil	..	21 154	19 618	27 575	17 944	25 812	27 202	25 282 e
NGL	..	1 877	1 735	2 316	8 345	8 711	8 270	9 364 e
Other	..	76	91	770	583	801	661	523 e
+ Net Imports	294 978	295 000	288 602	310 433	377 592	378 799	359 789	363 832 e
Crude Oil, NGL, Other	270 931	265 812	256 303	243 104	348 773	346 842	331 440	320 587 e
Petroleum Products	24 047	29 188	32 299	67 329	28 819	31 957	28 349	43 245 e
- Intl. Marine Bunkers	20 073	14 521	13 577	7 920	13 206	12 382	11 388	11 236 e
+ Stock Changes	-4 796	907	-5 787	-1 966	3 022	-4 059	1 904	2 773 e
TOTAL SUPPLY	**290 244**	**304 493**	**290 682**	**331 208**	**394 280**	**397 682**	**386 438**	**390 538 e**
Refinery Intake	*269 716*	*270 847*	*260 331*	*251 313*	*376 149*	*377 543*	*367 803*	*358 303 e*
Refinery Output	*265 838*	*271 128*	*258 132*	*251 622*	*369 880*	*369 931*	*360 000*	*350 092 e*
LPG + Ethane	*4 931*	*4 931*	*4 859*	*6 004*	*8 557*	*9 434*	*9 841*	*9 227 e*
Naphtha	*24 849*	*21 686*	*19 067*	*12 396*	*32 763*	*32 537*	*32 452*	*32 961 e*
Kerosenes	*24 822*	*28 204*	*26 816*	*28 935*	*53 476*	*55 775*	*54 128*	*52 288 e*
Gasolines	*30 841*	*37 420*	*38 287*	*48 537*	*65 857*	*66 199*	*65 748*	*65 797 e*
Gas/Diesel Oil	*39 451*	*47 104*	*49 218*	*72 097*	*103 564*	*104 330*	*101 863*	*99 298 e*
Fuel Oil	*124 379*	*111 487*	*100 064*	*59 835*	*74 898*	*72 873*	*67 809*	*62 360 e*
Other Products	*16 565*	*20 296*	*19 821*	*23 818*	*30 765*	*28 783*	*28 159*	*28 161 e*
- Refinery Losses	3 878	- 281	2 199	- 309	6 269	7 612	7 803	8 211 e
+ Returns	-	2 479	3 619	3 598	10 331	9 147	9 580	10 561 e
- Statistical Differences	-6 597	-6 292	496	3 775	1 193	4 173	2 118	2 307 e
TOTAL CONSUMPTION	**292 963**	**313 545**	**291 606**	**331 340**	**397 149**	**395 044**	**386 097**	**390 581 e**
Transformation and Energy	**87 762**	**100 981**	**89 742**	**76 416**	**71 524**	**60 913**	**54 961**	..
Crude Oil	20 837	18 895	11 780	17 824	8 265	6 437	3 916	..
LPG + Ethane	350	1 071	1 939	3 714	3 362	1 905	1 833	..
Gas/Diesel Oil	916	2 226	1 370	1 646	2 679	2 376	2 403	..
Fuel Oil	57 411	61 612	57 852	35 471	28 639	25 257	22 545	..
Petroleum Coke	-	-	-	2 401	3 017	2 603	2 340	..
Other	8 248	17 177	16 801	15 360	25 562	22 335	21 924	..
Transport	**56 779**	**71 728**	**75 630**	**107 596**	**143 969**	**148 460**	**150 453**	..
Motor Gasoline	30 609	37 477	38 679	50 265	64 656	66 039	66 448	..
Gas/Diesel Oil	14 970	20 951	23 470	38 026	49 733	52 132	52 641	..
Subtotal	*45 579*	*58 428*	*62 149*	*88 291*	*114 389*	*118 171*	*119 089*	..
Aviation Fuels	4 861	6 031	6 101	12 188	16 679	17 207	17 812	..
Other	6 339	7 269	7 380	7 117	12 901	13 082	13 552	..
Industry	**108 656**	**97 145**	**84 412**	**92 891**	**117 939**	**114 450**	**109 535**	..
Naphtha	24 412	24 260	17 675	24 602	49 794	53 609	52 632	..
LPG + Ethane	5 671	6 708	7 194	7 593	9 696	8 055	7 824	..
Gas/Diesel Oil	11 082	11 824	10 802	17 484	16 925	10 203	8 245	..
Fuel Oil	52 790	39 811	32 829	19 327	16 684	20 969	19 503	..
Other	14 701	14 542	15 912	23 885	24 840	21 614	21 331	..
Other Sectors	**39 766**	**43 691**	**41 822**	**54 437**	**63 717**	**71 221**	**71 148**	..
Gas/Diesel Oil	11 405	12 805	13 651	19 875	22 106	26 654	28 468	..
Fuel Oil	6 039	3 803	3 442	5 538	2 275	979	912	..
Other	22 322	27 083	24 729	29 024	39 336	43 588	41 768	..
Memo: Non-Energy Use	*39 723*	*39 058*	*33 548*	*45 114*	*70 495*	*71 845*	*70 521*	..
Chem/Petchem. Feedst.	*28 932*	*28 497*	*23 034*	*31 627*	*56 197*	*60 797*	*59 407*	..
Other	*10 791*	*10 561*	*10 514*	*13 487*	*14 298*	*11 048*	*11 114*	..

Source: Annual Oil Statistics.

OECD PACIFIC
Table 2: Supply of Crude Oil, NGL, Feedstocks and Other Hydrocarbons

000 metric tons

	1973	1978	1980	1990	1999	2000	2001	2002
CRUDE OIL								
Indigenous Production	..	21 154	19 618	27 575	17 944	25 812	27 202	25 282 e
+ Imports	..	262 136	250 714	240 716	355 224	360 313	348 078	335 548 e
- Exports	..	-	-	6 228	11 261	16 513	20 621	20 529 e
- Direct Use	..	18 968	11 848	17 824	8 282	6 459	3 938	4 016 e
+ Stock Change	..	-3 180	-4 520	602	2 304	- 842	-1 057	- 153 e
- Statistical Difference	..	-3 696	863	1 996	-2 867	-1 221	1 347	1 492 e
Refinery Intake	**..**	**264 838**	**253 101**	**242 845**	**358 796**	**363 532**	**348 317**	**334 640 e**
Closing Stock Level	..	*31 725*	*39 832*	*51 566*	*66 278*	*67 240*	*68 297*	*68 450 e*
NATURAL GAS LIQUIDS								
Indigenous Production	..	1 877	1 735	2 316	8 345	8 711	8 270	9 364 e
+ Imports	..	2 308	4 250	4 939	4 375	2 788	2 933	4 023 e
- Exports	..	-	-	-	1 360	1 555	1 582	1 755 e
- Direct Use	..	4 073	4 516	6 046	3 254	3 642	3 439	4 075 e
+ Stock Change	..	-	-	-	13	- 72	- 18	-
- Statistical Difference	..	-	-	22	4 032	3 219	1 164	2 010 e
Refinery Intake	**..**	**112**	**1 469**	**1 187**	**4 087**	**3 011**	**5 000**	**5 547 e**
Closing Stock Level	..	-	-	-	*180*	*252*	*270*	*270 e*
REFINERY FEEDSTOCKS								
Backflows	..	2 479	3 619	3 598	10 331	9 147	9 580	10 561 e
+ Transfers	..	1 977	-	-	157	112	149	1 801 e
+ Imports	..	1 368	1 339	3 903	2 833	2 696	3 176	4 016 e
- Exports	..	-	-	226	1 612	1 449	1 306	1 212 e
+ Stock Change	..	870	- 755	134	516	- 909	738	891 e
- Statistical Difference	..	873	-1 467	844	- 904	-1 272	-2 060	-2 059 e
Refinery Intake	**..**	**5 821**	**5 670**	**6 565**	**13 129**	**10 869**	**14 397**	**18 116 e**
Closing Stock Level	..	*7 891*	*9 820*	*10 863*	*10 048*	*10 982*	*10 244*	*9 353 e*
OTHER HYDROCARBONS								
Indigenous Production	..	76	91	770	583	801	661	523 e
+ Imports	..	-	-	-	612	661	865	614 e
- Exports	..	-	-	-	38	99	103	118 e
- Direct Use	..	-	-	79	1 031	1 233	1 334	1 006 e
+ Stock Change	..	-	-	- 1	11	1	-	13 e
- Statistical Difference	..	-	-	- 26	-	-	-	-
Refinery Intake	**..**	**76**	**91**	**716**	**137**	**131**	**89**	**-**
Closing Stock Level	..	-	-	*38*	*28*	*27*	*27*	*40 e*
CRUDE OIL, NGL, FEEDSTOCKS AND OTHER								
Indigenous Production	20 135	23 107	21 444	30 661	26 872	35 324	36 133	35 169 e
+ Backflows	-	2 479	3 619	3 598	10 331	9 147	9 580	10 561 e
+ Transfers	-	1 977	-	-	157	112	149	1 801 e
+ Imports	270 949	265 812	256 303	249 558	363 044	366 458	355 052	344 201 e
- Exports	18	-	-	6 454	14 271	19 616	23 612	23 614 e
- Direct Use	22 009	23 041	16 364	23 949	12 567	11 334	8 711	9 097 e
+ Stock Change	-3 646	-2 310	-5 275	735	2 844	-1 822	- 337	725 e
- Statistical Difference	-4 305	-2 823	- 604	2 836	261	726	451	1 443 e
Refinery Intake	**269 716**	**270 847**	**260 331**	**251 313**	**376 149**	**377 543**	**367 803**	**358 303 e**
Closing Stock Level	-	*39 616*	*49 652*	*62 467*	*76 534*	*78 501*	*78 838*	*78 113 e*

Source: Annual Oil Statistics.

OECD PACIFIC
Table 3: Motor Gasoline - Supply

000 metric tons

	1973	1978	1980	1990	1999	2000	2001	2002
MOTOR GASOLINE								
Refinery Output	**30 498**	**36 985**	**37 985**	**47 576**	**65 722**	**66 085**	**65 645**	**65 688 e**
+ Imports	761	892	791	3 317	2 617	2 807	2 880	3 569 e
- Exports	409	337	229	352	2 983	2 964	3 122	2 888 e
+ Transfers	-	-	-	81	418	552	490	258 e
+ Stock Change	- 48	376	54	- 321	- 150	153	455	217 e
- Statistical Difference	61	298	- 214	- 84	759	531	- 160	- 782 e
Total Consumption	**30 741**	**37 618**	**38 815**	**50 385**	**64 865**	**66 102**	**66 508**	**67 626 e**
Closing Stock Level	-	*2 711*	*2 572*	*2 446*	*3 046*	*3 676*	*3 221*	*3 004 e*

Table 4: Aviation Fuels - Supply

000 metric tons

	1973	1978	1980	1990	1999	2000	2001	2002
AVIATION GASOLINE								
Refinery Output	**68**	**59**	**80**	**168**	**135**	**114**	**103**	**109 e**
+ Imports	80	82	64	24	46	17	16	17 e
- Exports	8	16	6	58	52	56	20	51 e
+ Transfers	-	-	-	-	-	-	-	-
+ Stock Change	- 10	8	- 27	- 5	- 3	6	- 1	- 1 e
- Statistical Difference	- 14	18	- 4	12	27	- 9	1	- 17 e
Total Consumption	**144**	**115**	**115**	**117**	**99**	**90**	**97**	**91 e**
Closing Stock Level	-	*44*	*60*	*43*	*35*	*29*	*30*	*31 e*
GASOLINE TYPE JET FUEL								
Refinery Output	**275**	**376**	**222**	**793**	**-**	**-**	**-**	**-**
+ Imports	-	-	-	-	-	-	-	-
- Exports	-	-	-	-	-	-	-	-
+ Transfers	-	-	-	-	-	-	-	-
+ Stock Change	-	-	-	- 17	-	-	-	-
- Statistical Difference	- 4	4	15	- 24	-	-	-	-
Total Consumption	**279**	**372**	**207**	**800**	**-**	**-**	**-**	**-**
Closing Stock Level	-	-	-	*45*	-	-	-	-
KEROSENE TYPE JET FUEL								
Refinery Output	**4 793**	**5 293**	**5 733**	**8 118**	**19 863**	**21 023**	**21 831**	**21 858 e**
+ Imports	793	677	755	3 888	2 655	2 487	2 442	3 109 e
- Exports	266	243	166	538	4 707	5 198	6 110	6 055 e
+ Transfers	-	-	-	- 5	- 105	176	177	- 239 e
+ Stock Change	- 7	4	- 44	- 86	32	- 170	- 80	122 e
- Statistical Difference	875	187	499	92	282	513	27	313 e
Total Consumption	**4 438**	**5 544**	**5 779**	**11 285**	**17 456**	**17 805**	**18 233**	**18 482 e**
Closing Stock Level	-	*550*	*531*	*701*	*1 153*	*1 198*	*1 278*	*1 156 e*

Source: Annual Oil Statistics.

OECD PACIFIC
Table 5A: Gas/Diesel Oil - Supply and Consumption

000 metric tons

	1973	1978	1980	1990	1999	2000	2001	2002
Refinery Output	**39 451**	**47 104**	**49 218**	**72 097**	**103 564**	**104 330**	**101 863**	**99 298 e**
+ Recycled Products	-	-	-	-	-	-	-	-
+ Imports	3 134	2 616	2 265	9 442	3 809	4 080	4 063	4 700 e
- Exports	809	959	707	3 361	14 542	15 532	14 862	12 264 e
- Intl. Marine Bunkers	1 938	1 329	1 022	913	1 568	1 365	1 105	1 108 e
+ Transfers	-	-	-	1	37	237	215	165 e
+ Stock Changes	- 781	621	- 102	- 544	- 153	- 367	1 031	617 e
- Statistical Differences	684	247	359	- 309	- 296	18	- 552	-1 235 e
Total Consumption	**38 373**	**47 806**	**49 293**	**77 031**	**91 443**	**91 365**	**91 757**	**92 643 e**
Transformation	**854**	**2 117**	**1 248**	**1 295**	**2 037**	**1 883**	**1 900**	..
Public Electricity	854	1 745	924	476	482	1 023	1 045	..
Autoproducer Electricity	-	357	302	819	1 549	845	847	..
Public CHP Plants	-	-	-	-	-	12	6	..
Autoproducers of CHP	-	-	-	-	-	-	-	..
Public Heat Plants	-	-	-	-	-	-	-	..
Autoproducer Heat Plants	-	-	-	-	-	-	-	..
Petrochemical Industry	-	-	-	-	6	3	2	..
Other Transformation	-	15	22	-	-	-	-	..
Energy Sector	**62**	**109**	**122**	**351**	**642**	**493**	**503**	..
Refinery Fuel	2	4	4	50	108	12	11	..
Other Energy Sector	60	105	118	301	534	481	492	..
Final Consumption	**37 457**	**45 580**	**47 923**	**75 385**	**88 764**	**88 989**	**89 354**	..
Transport	**14 970**	**20 951**	**23 470**	**38 026**	**49 733**	**52 132**	**52 641**	..
Air	-	-	-	1	-	1	1	..
Road	10 051	13 919	16 351	33 932	43 814	46 730	47 428	..
Rail	1 451	1 517	1 506	1 362	1 358	1 107	1 118	..
Inland Waterways	2 071	2 513	2 424	2 216	3 165	2 737	2 424	..
Non-Specified	1 397	3 002	3 189	515	1 396	1 557	1 670	..
Industry	**11 082**	**11 824**	**10 802**	**17 484**	**16 925**	**10 203**	**8 245**	..
Iron and Steel	878	997	757	833	690	673	479	..
Chemical (incl. Pet.-Chem.)	537	919	926	1 419	1 148	1 343	1 004	..
Non-Ferrous Metals	326	506	531	785	506	478	361	..
Non-Metallic Minerals	523	918	800	1 177	1 047	963	720	..
Transport Equipment	-	13	12	2	69	73	83	..
Machinery	-	23	18	1 083	259	179	123	
Mining and Quarrying	45	512	455	837	1 303	1 225	1 187	..
Food and Tobacco	765	1 200	1 110	1 838	1 768	1 772	1 289	..
Pulp, Paper and Printing	200	243	231	255	427	434	318	..
Wood and Wood Products	-	35	36	22	24	27	28	..
Construction	2 449	3 521	3 690	4 928	4 253	1 941	1 763	..
Textiles and Leather	262	346	316	694	640	639	464	..
Non-Specified	5 097	2 591	1 920	3 611	4 791	456	426	..
Other Sectors	**11 405**	**12 805**	**13 651**	**19 875**	**22 106**	**26 654**	**28 468**	..
Commerce and Public	214	938	1 021	8 885	10 772	15 278	16 948	..
Residential	44	128	101	54	321	119	91	..
Agriculture	2 096	2 458	3 116	10 169	10 564	7 799	7 925	..
Non-Specified	9 051	9 281	9 413	767	449	3 458	3 504	..
Memo: Non-Energy Use	-	*11*	*3*	*4*	*10*	*32*	*28*	..
Chem/Petchem. Feedst.	-	*11*	*2*	-	*6*	*28*	*24*	..
Other	-	-	*1*	*4*	*4*	*4*	*4*	..
Closing Stock Level	-	*3 991*	*4 614*	*4 546*	*5 286*	*5 883*	*4 852*	*4 235 e*

Source: Annual Oil Statistics.

OECD PACIFIC
Table 5B: Fuel Oil (Residual) - Supply and Consumption

000 metric tons

	1973	1978	1980	1990	1999	2000	2001	2002
Refinery Output	**124 379**	**111 487**	**100 064**	**59 835**	**74 898**	**72 873**	**67 809**	**62 360 e**
+ Recycled Products	-	-	-	-	-	-	-	-
+ Imports	14 448	11 559	13 341	15 192	4 748	5 604	5 156	5 942 e
- Exports	3 285	283	350	3 908	13 791	12 340	11 913	6 459 e
- Intl. Marine Bunkers	18 073	13 147	12 499	6 967	11 617	10 988	10 260	10 114 e
+ Transfers	-	-1 977	-	1	13	53	160	-1 024 e
+ Stock Changes	713	1 705	- 483	227	405	- 238	90	53 e
- Statistical Differences	-2 432	-1 186	522	481	54	997	1 240	1 236 e
Total Consumption	**120 614**	**110 530**	**99 551**	**63 899**	**54 602**	**53 967**	**49 802**	**49 522 e**
Transformation	**50 493**	**55 585**	**52 405**	**32 342**	**23 413**	**22 039**	**19 307**	..
Public Electricity	50 493	42 829	41 077	26 107	14 869	15 394	12 992	..
Autoproducer Electricity	-	12 748	11 304	6 203	7 796	5 152	4 726	..
Public CHP Plants	-	-	-	-	69	927	947	..
Autoproducers of CHP	-	-	-	-	-	-	-	..
Public Heat Plants	-	-	-	15	472	418	500	..
Autoproducer Heat Plants	-	-	-	-	-	-	-	..
Petrochemical Industry	-	-	-	-	187	125	118	..
Other Transformation	-	8	24	17	20	23	24	..
Energy Sector	**6 918**	**6 027**	**5 447**	**3 129**	**5 226**	**3 218**	**3 238**	..
Refinery Fuel	6 918	6 023	5 443	2 976	5 039	3 174	3 038	..
Other Energy Sector	-	4	4	153	187	44	200	..
Final Consumption	**63 203**	**48 918**	**41 699**	**28 428**	**25 963**	**28 710**	**27 257**	..
Transport	**4 374**	**5 304**	**5 428**	**3 563**	**7 004**	**6 762**	**6 842**	..
Air	-	-	-	-	-	-	-	..
Road	48	39	62	57	47	1	27	..
Rail	449	375	359	5	1	1	1	..
Inland Waterways	3 705	4 682	4 883	3 393	6 836	6 651	6 713	..
Non-Specified	172	208	124	108	120	109	101	..
Industry	**52 790**	**39 811**	**32 829**	**19 327**	**16 684**	**20 969**	**19 503**	..
Iron and Steel	12 030	5 987	4 182	2 073	1 757	1 786	1 456	..
Chemical (incl. Pet.-Chem.)	8 310	5 381	4 623	3 571	2 410	2 708	2 542	..
Non-Ferrous Metals	1 130	2 280	2 541	1 407	1 011	1 251	1 042	..
Non-Metallic Minerals	9 929	9 042	5 927	2 502	2 342	2 506	2 267	..
Transport Equipment	-	22	10	1	855	772	647	..
Machinery	-	33	23	621	648	372	370	..
Mining and Quarrying	243	404	553	170	119	98	88	..
Food and Tobacco	2 503	2 540	2 379	1 571	913	924	819	..
Pulp, Paper and Printing	3 726	2 539	2 337	3 309	3 151	3 852	3 380	..
Wood and Wood Products	-	30	32	83	78	74	70	..
Construction	463	128	97	161	136	132	125	..
Textiles and Leather	4 761	4 445	3 130	2 820	2 827	2 933	2 636	..
Non-Specified	9 695	6 980	6 995	1 038	437	3 561	4 061	..
Other Sectors	**6 039**	**3 803**	**3 442**	**5 538**	**2 275**	**979**	**912**	..
Commerce and Public	5 827	3 160	2 675	3 526	1 647	396	389	..
Residential	6	317	472	1 754	419	382	342	..
Agriculture	206	326	295	258	192	188	173	..
Non-Specified	-	-	-	-	17	13	8	..
Memo: Non-Energy Use	-	*20*	*2*	*1*	*20*	*463*	*433*	..
Chem/Petchem. Feedst.	-	-	-	-	-	*463*	*431*	..
Other	-	*20*	*2*	*1*	*20*	-	*2*	..
Closing Stock Level	-	*5 889*	*6 286*	*3 279*	*3 527*	*3 808*	*3 718*	*3 665 e*

Source: Annual Oil Statistics.

OECD PACIFIC
Table 5C: Total Products - Supply and Consumption

000 metric tons

	1973	1978	1980	1990	1999	2000	2001	2002
Refinery Output	**265 838**	**271 128**	**258 132**	**251 622**	**369 880**	**369 931**	**360 000**	**350 092 e**
+ Recycled Products	-	-	-	-	-	-	-	-
+ Imports	31 259	33 804	35 944	78 586	76 467	79 590	75 931	81 057 e
- Exports	7 212	4 616	3 645	11 257	47 648	47 633	47 582	37 812 e
- Intl. Marine Bunkers	20 073	14 521	13 577	7 920	13 206	12 382	11 388	11 236 e
+ Transfers	22 009	21 064	16 364	23 949	12 410	11 222	8 562	7 296 e
+ Stock Changes	-1 150	3 217	- 512	-2 701	178	-2 237	2 241	2 048 e
- Statistical Differences	-2 292	-3 469	1 100	939	932	3 447	1 667	864 e
Total Consumption	**292 963**	**313 545**	**291 606**	**331 340**	**397 149**	**395 044**	**386 097**	**390 581 e**
Transformation	**76 085**	**86 424**	**76 162**	**63 082**	**52 226**	**44 535**	**38 820**	..
Public Electricity	73 247	67 977	57 745	46 765	25 537	23 585	18 557	..
Autoproducer Electricity	-	13 105	11 606	8 653	11 495	9 145	8 343	..
Public CHP Plants	-	-	-	-	70	941	955	..
Autoproducers of CHP	-	-	-	-	-	-	-	..
Public Heat Plants	-	-	-	47	499	442	513	..
Autoproducer Heat Plants	-	-	-	-	-	-	-	..
Petrochemical Industry	-	2 479	3 619	3 598	10 331	9 147	9 580	..
Other Transformation	2 838	2 863	3 192	4 019	4 294	1 275	872	..
Energy Sector	**11 677**	**14 557**	**13 580**	**13 334**	**19 298**	**16 378**	**16 141**	..
Refinery Fuel	11 592	14 409	13 419	12 871	18 458	15 729	15 310	..
Other Energy Sector	85	148	161	463	840	649	831	..
Final Consumption	**205 201**	**212 564**	**201 864**	**254 924**	**325 625**	**334 131**	**331 136**	..
Transport	**56 779**	**71 728**	**75 630**	**107 596**	**143 969**	**148 460**	**150 453**	..
Air	4 861	6 031	6 101	12 189	16 674	17 201	17 805	..
Road	42 459	53 235	56 888	87 700	114 206	118 948	120 472	..
Rail	1 925	1 907	1 879	1 376	1 366	1 113	1 123	..
Inland Waterways	5 945	7 333	7 438	5 700	10 140	9 517	9 265	..
Non-Specified	1 589	3 222	3 324	631	1 583	1 681	1 788	..
Industry	**108 656**	**97 145**	**84 412**	**92 891**	**117 939**	**114 450**	**109 535**	..
Iron and Steel	12 922	7 016	5 418	3 959	3 257	3 403	2 807	..
Chemical (incl. Pet.-Chem.)	38 024	35 057	28 761	37 454	61 409	67 162	65 099	..
Non-Ferrous Metals	1 506	2 989	3 290	3 298	2 727	3 053	2 828	..
Non-Metallic Minerals	10 452	9 986	6 750	4 452	5 081	5 364	4 871	..
Transport Equipment	19	68	52	24	1 013	935	815	..
Machinery	-	71	59	2 611	1 829	1 517	1 445	..
Mining and Quarrying	353	967	1 061	1 053	1 504	1 426	1 375	..
Food and Tobacco	3 268	3 766	3 518	3 501	2 834	2 846	2 249	..
Pulp, Paper and Printing	3 926	2 785	2 571	3 845	3 868	4 626	4 030	..
Wood and Wood Products	-	66	70	109	111	104	100	..
Construction	9 446	10 232	10 133	13 213	11 892	10 039	9 681	..
Textiles and Leather	5 023	4 795	3 451	3 700	3 674	3 791	3 316	..
Non-Specified	23 717	19 347	19 278	15 672	18 740	10 154	10 919	..
Other Sectors	**39 766**	**43 691**	**41 822**	**54 437**	**63 717**	**71 221**	**71 148**	..
Commerce and Public	16 620	17 819	16 409	21 964	27 505	34 431	36 056	..
Residential	9 245	11 703	10 475	17 412	20 001	20 723	19 691	..
Agriculture	3 896	4 268	4 872	13 313	13 829	10 896	11 063	..
Non-Specified	10 005	9 901	10 066	1 748	2 382	5 171	4 338	..
Memo: Non-Energy Use	*39 723*	*39 058*	*33 548*	*45 114*	*70 495*	*71 845*	*70 521*	..
Chem/Petchem. Feedst.	*28 932*	*28 497*	*23 034*	*31 627*	*56 197*	*60 797*	*59 407*	..
Other	*10 791*	*10 561*	*10 514*	*13 487*	*14 298*	*11 048*	*11 114*	..
Closing Stock Level	-	*21 582*	*23 290*	*22 473*	*24 505*	*25 809*	*23 568*	*21 520 e*

Source: Annual Oil Statistics.

OECD PACIFIC
Table 6A: Supply and Consumption

000 metric tons

	1980	2000	2001	2002	1980	2000	2001	2002
	LPG AND ETHANE				**REFINERY GAS**			
Refinery Output	**4 859**	**9 434**	**9 841**	**9 227 e**	**8 324**	**12 810**	**12 132**	**11 861 e**
+ Recycled Products	-	-	-	-	-	-	-	-
+ Imports	9 786	20 173	18 719	19 252 e	-	-	-	-
- Exports	1 496	837	577	366 e	-	-	-	-
- Intl. Marine Bunkers	-	-	-	-	-	-	-	-
+ Transfers	1 692	2 030	2 170	2 450 e	-	3	4	-
+ Stock Change	- 72	- 656	- 235	183 e	-	-	-	-
- Statistical Difference	-1 137	2 482	2 096	2 453 e	-	173	166	173 e
Total Consumption	**15 906**	**27 662**	**27 822**	**28 293 e**	**8 324**	**12 640**	**11 970**	**11 688 e**
of which: Ethane	*-*	*410*	*418*	*410 e*	-	-	-	-
Transformation	1 923	1 817	1 735	..	528	926	606	..
Energy	16	88	98	..	7 713	11 624	11 268	..
Transport	1 706	5 645	6 051	..	-	-	-	..
Industry	7 194	8 055	7 824	..	83	90	96	..
Other	5 067	12 057	12 114	..	-	-	-	..
Memo: Non-Energy Use	*3 920*	*3 145*	*3 039*	..	*32*	-	-	..
Closing Stock Level	*1 618*	*3 203*	*3 438*	*3 255 e*	-	-	-	-
	NAPHTHA				**OTHER KEROSENE**			
Refinery Output	**19 067**	**32 537**	**32 452**	**32 961 e**	**21 083**	**34 752**	**32 297**	**30 430 e**
+ Recycled Products	-	-	-	-	-	-	-	-
+ Imports	5 670	35 260	34 818	36 530 e	103	3 410	2 382	2 916 e
- Exports	-	5 855	6 095	5 394 e	23	2 697	2 534	1 524 e
- Intl. Marine Bunkers	-	-	-	-	-	2	-	-
+ Transfers	-	- 82	- 383	-	-	- 348	- 70	- 284 e
+ Stock Change	36	- 284	- 5	- 303 e	77	- 745	950	1 051 e
- Statistical Difference	712	- 287	- 455	- 22 e	429	834	1 031	383 e
Total Consumption	**24 061**	**61 863**	**61 242**	**63 816 e**	**20 811**	**33 536**	**31 994**	**32 206 e**
Transformation	6 131	8 181	8 536	..	-	317	233	..
Energy	247	71	74	..	-	32	54	..
Transport	-	-	-	..	3	14	19	..
Industry	17 675	53 609	52 632	..	1 933	2 998	2 653	..
Other	8	2	-	..	18 875	30 175	29 035	..
Memo: Non-Energy Use	*17 667*	*53 594*	*52 618*	..	*12*	*403*	*381*	..
Closing Stock Level	*1 064*	*2 507*	*2 512*	*2 815 e*	*5 187*	*3 978*	*3 028*	*1 977 e*
	WHITE SPIRIT + SBP				**LUBRICANTS**			
Refinery Output	**286**	**259**	**279**	**..**	**2 635**	**3 902**	**3 887**	**..**
+ Recycled Products	-	-	-	..	-	-	-	..
+ Imports	39	8	4	..	250	119	103	..
- Exports	14	1	-	..	475	923	1 043	..
- Intl. Marine Bunkers	-	-	-	..	56	24	22	..
+ Transfers	-	-	- 9	..	-	56	51	..
+ Stock Change	- 20	7	- 22	..	- 79	10	26	..
- Statistical Difference	- 25	- 61	- 52	..	- 94	115	38	..
Total Consumption	**316**	**334**	**304**	**..**	**2 369**	**3 025**	**2 964**	**..**
Transformation	-	18	18	..	-	13	10	..
Energy	-	-	-	..	-	49	49	..
Transport	-	-	-	..	243	661	640	..
Industry	316	316	286	..	1 474	1 699	2 230	..
Other	-	-	-	..	652	603	35	..
Memo: Non-Energy Use	*271*	*237*	*206*	..	*2 369*	*1 837*	*2 353*	..
Closing Stock Level	*71*	*46*	*68*	..	*729*	*529*	*503*	..

Source: Annual Oil Statistics.

OECD PACIFIC
Table 6B: Supply and Consumption

000 metric tons

	1980	2000	2001	2002	1980	2000	2001	2002
	BITUMEN				**PARAFFIN WAXES**			
Refinery Output	**5 722**	**8 601**	**8 603**	**..**	**178**	**135**	**140**	**..**
+ Recycled Products	-	-	-	..	-	-	-	..
+ Imports	5	94	111	..	5	3	4	..
- Exports	27	746	840	..	80	43	56	..
- Intl. Marine Bunkers	-	-	-	..	-	-	-	..
+ Transfers	-	661	776	..	-	-	-	..
+ Stock Change	- 30	11	13	..	1	- 4	4	..
- Statistical Difference	- 70	-1 187	- 743	..	3	15	19	..
Total Consumption	**5 740**	**9 808**	**9 406**	**..**	**101**	**76**	**73**	**..**
Transformation	-	790	733	..	-	-	-	..
Energy	-	-	-	..	-	9	10	..
Transport	-	-	-	..	-	-	-	..
Industry	5 740	9 018	8 673	..	101	67	63	..
Other	-	-	-	..	-	-	-	..
Memo: Non-Energy Use	*5 740*	*7 247*	*6 997*	..	*101*	*67*	*63*	..
Closing Stock Level	*355*	*388*	*375*	..	*15*	*27*	*23*	..
	PETROLEUM COKE				**OTHER PRODUCTS**			
Refinery Output	**254**	**1 720**	**1 835**	**1 711 e**	**2 422**	**1 356**	**1 283**	**14 589 e**
+ Recycled Products	-	-	-	-	-	-	-	-
+ Imports	2 649	5 365	5 134	4 805 e	221	163	99	217 e
- Exports	27	69	74	51 e	45	372	336	2 760 e
- Intl. Marine Bunkers	-	-	-	-	-	-	-	14 e
+ Transfers	-	-	-	-	14 672	7 884	4 981	5 970 e
+ Stock Change	2	8	7	- 7 e	175	32	8	116 e
- Statistical Difference	1	117	50	51 e	104	- 801	-1 038	-1 689 e
Total Consumption	**2 877**	**6 907**	**6 852**	**6 407 e**	**17 341**	**9 864**	**7 073**	**19 807 e**
Transformation	-	1 911	1 614	..	13 927	6 640	4 128	..
Energy	-	692	726	..	35	102	119	..
Transport	-	-	-	..	-	-	-	..
Industry	2 877	4 270	4 472	..	3 363	3 122	2 825	..
Other	-	34	40	..	16	-	1	..
Memo: Non-Energy Use	*175*	*1 140*	*1 435*	..	*2 613*	*1 465*	*1 507*	..
Closing Stock Level	*4*	*32*	*25*	*32 e*	*184*	*505*	*497*	*1 350 e*

Source: Annual Oil Statistics.

OECD PACIFIC
Table 7A: Imports by Country of Origin

000 metric tons

	2000	2001	2002	2000	2001	2002	2000	2001	2002
	Crude Oil			Natural Gas Liquids			Feedstocks and Other		
Australia	8 078	7 629	8 507 e	332	352	355 e	204	324	395 e
Belgium	-	-	-	-	-	-	-	-	-
Canada	-	227	128 e	-	-	-	-	-	-
Czech Republic	-	-	-	-	-	-	-	-	-
Denmark	-	-	-	-	-	-	-	-	-
Finland	-	-	-	-	-	-	-	-	-
France	-	-	-	-	-	-	-	-	-
Germany	-	-	-	-	-	-	-	-	-
Greece	-	-	-	-	-	-	-	-	-
Hungary	-	-	-	-	-	-	-	-	-
Ireland	-	-	-	-	-	-	-	-	-
Italy	-	-	-	-	-	-	-	-	-
Japan	-	-	-	-	-	-	-	33	36 e
Korea	-	-	-	-	-	-	-	38	-
Mexico	1 644	585	490 e	-	-	-	-	-	-
Netherlands	-	-	-	-	-	-	-	-	-
New Zealand	1 001	870	860 e	6	-	-	-	19	18 e
Norway	849	846	1 232 e	-	-	-	-	-	-
Poland	-	-	-	-	-	-	-	-	-
Spain	-	-	-	-	-	-	-	-	-
Sweden	-	-	-	-	-	-	-	-	-
Turkey	-	-	-	-	-	-	-	-	-
United Kingdom	77	-	162 e	-	-	-	-	-	-
United States	918	-	88 e	-	-	-	-	-	-
Other	-	-	-	-	-	-	-	-	-
OECD	**12 567**	**10 157**	**11 467 e**	**338**	**352**	**355 e**	**204**	**414**	**449 e**
Argentina	184	710	523 e	-	-	-	-	-	-
Brazil	-	674	1 276 e	-	-	-	-	-	-
Colombia	-	-	-	-	-	-	-	-	-
Ecuador	2 141	1 826	2 118 e	-	-	-	-	-	-
Netherlands Antilles	-	-	-	-	-	-	-	-	-
Peru	358	94	-	-	-	-	-	-	-
Trinidad and Tobago	-	-	-	-	-	-	-	-	-
Venezuela	-	200	2 e	-	-	-	659	865	608 e
Other	52	-	-	-	-	-	-	3	-
Latin America	**2 735**	**3 504**	**3 919 e**	**-**	**-**	**-**	**659**	**868**	**608 e**
Romania	-	-	-	-	-	-	-	-	-
Other	-	-	-	-	-	-	-	-	-
Non-OECD Europe	**-**	**-**	**-**	**-**	**-**	**-**	**-**	**-**	**-**
FSU	**1 551**	**2 652**	**3 331 e**	**-**	**68**	**-**	**-**	**-**	**-**
Iran	36 276	36 406	33 222 e	250	233	1 195 e	-	-	39 e
Iraq	5 780	1 194	1 135 e	-	-	-	-	-	-
Kuwait	30 250	29 361	21 893 e	-	-	-	-	-	-
Oman	18 189	20 515	19 170 e	-	13	24 e	-	-	-
Qatar	27 020	28 113	22 834 e	707	1 335	1 278 e	41	3	-
Saudi Arabia	90 567	89 630	83 309 e	325	228	161 e	1 268	803	1 182 e
United Arab Emirates	74 126	69 136	68 732 e	565	260	439 e	20	99	-
Other	5 745	4 473	2 910 e	-	-	-	-	-	-
Middle East	**287 953**	**278 828**	**253 205 e**	**1 847**	**2 069**	**3 097 e**	**1 329**	**905**	**1 221 e**
India	-	-	-	-	-	-	-	-	52 e
Indonesia	18 288	18 335	18 441 e	288	281	402 e	146	160	496 e
Singapore	29	41	43 e	-	-	-	156	487	630 e
Other	20 550	20 317	19 597 e	315	163	57 e	179	272	410 e
Asia	**38 867**	**38 693**	**38 081 e**	**603**	**444**	**459 e**	**481**	**919**	**1 588 e**
China	**5 350**	**3 908**	**4 038 e**	**-**	**-**	**-**	**684**	**935**	**764 e**
Algeria	-	-	-	-	-	-	-	-	-
Angola/Cabinada	2 843	1 839	3 268 e	-	-	-	-	-	-
Cameroon	-	-	-	-	-	-	-	-	-
Egypt	354	142	-	-	-	-	-	-	-
Gabon	868	797	508 e	-	-	-	-	-	-
Libya	-	-	-	-	-	-	-	-	-
Nigeria	2 217	1 984	4 610 e	-	-	112 e	-	-	-
Tunisia	-	-	-	-	-	-	-	-	-
Other	4 811	4 275	8 064 e	-	-	-	-	-	-
Africa	**11 093**	**9 037**	**16 450 e**	**-**	**-**	**112 e**	**-**	**-**	**-**
Non-spec./Other	**197**	**1 299**	**5 057 e**	**-**	**-**	**-**	**-**	**-**	**-**
Total Imports	**360 313**	**348 078**	**335 548 e**	**2 788**	**2 933**	**4 023 e**	**3 357**	**4 041**	**4 630 e**
Memo: Total OPEC	*284 524*	*274 359*	*254 178 e*	*2 135*	*2 337*	*3 587 e*	*2 134*	*1 930*	*2 325 e*

Source: Annual Oil Statistics.

OECD PACIFIC

Table 7B: Imports by Country of Origin

000 metric tons

	2000	2001	2002	2000	2001	2002	2000	2001	2002
	Total Products			**Liquefied Petroleum Gases**			**Naphtha**		
Australia	2 159	2 217	2 597 e	929	1 131	1 453 e	478	386	229 e
Belgium	2	23	-	-	-	-	-	22	-
Canada	1	-	252 e	-	-	-	-	-	-
Czech Republic	-	-	-	-	-	-	-	-	-
Denmark	-	14	7 e	-	-	-	-	14	7 e
Finland	-	-	-	-	-	-	-	-	-
France	5	5	30 e	-	-	-	-	-	25 e
Germany	3	4	-	-	-	-	-	-	-
Greece	38	30	144 e	-	-	71 e	38	30	73 e
Hungary	-	4	-	-	-	-	-	-	-
Ireland	-	-	-	-	-	-	-	-	-
Italy	-	43	-	-	-	-	-	43	-
Japan	704	1 545	1 480 e	-	29	175 e	-	-	19 e
Korea	10 968	9 497	9 211 e	133	158	48 e	5 723	5 589	5 250 e
Mexico	201	120	138 e	-	-	-	201	120	138 e
Netherlands	16	354	505 e	12	-	48 e	-	264	457 e
New Zealand	343	290	353 e	-	3	15 e	286	276	313 e
Norway	45	39	294 e	44	-	294 e	-	39	-
Poland	-	-	-	-	-	-	-	-	-
Spain	1	1	-	-	-	-	-	-	-
Sweden	3	-	-	-	-	-	-	-	-
Turkey	-	104	-	-	-	-	-	104	-
United Kingdom	29	25	154 e	27	21	153 e	-	-	-
United States	1 846	2 018	5 275 e	94	224	408 e	710	743	642 e
Other	20	55	2 e	20	55	-	-	-	-
OECD	**16 384**	**16 388**	**20 442 e**	**1 259**	**1 621**	**2 665 e**	**7 436**	**7 630**	**7 153 e**
Argentina	30	13	-	-	-	-	-	-	-
Brazil	-	-	-	-	-	-	-	-	-
Colombia	-	-	-	-	-	-	-	-	-
Ecuador	-	-	-	-	-	-	-	-	-
Netherlands Antilles	-	37	1 e	-	-	-	-	36	-
Peru	-	26	-	-	-	-	-	26	-
Trinidad and Tobago	1	5	-	-	-	-	-	-	-
Venezuela	7	38	75 e	-	-	44 e	-	31	29 e
Other	-	31	8 e	-	4	8 e	-	27	-
Latin America	**38**	**150**	**84 e**	**-**	**4**	**52 e**	**-**	**120**	**29 e**
Romania	-	-	-	-	-	-	-	-	-
Other	-	1	-	-	-	-	-	-	-
Non-OECD Europe	**-**	**1**	**-**	**-**	**-**	**-**	**-**	**-**	**-**
FSU	**833**	**616**	**845 e**	**-**	**-**	**-**	**590**	**391**	**471 e**
Iran	916	973	1 055 e	594	862	713 e	322	111	321 e
Iraq	7	97	-	7	97	-	-	-	-
Kuwait	8 005	8 178	7 886 e	2 105	2 102	1 908 e	5 626	6 034	5 916 e
Oman	-	13	34 e	-	-	-	-	13	34 e
Qatar	1 690	1 962	1 910 e	821	904	739 e	849	1 038	1 112 e
Saudi Arabia	14 421	10 928	12 025 e	8 619	6 357	6 259 e	5 445	4 241	5 413 e
United Arab Emirates	10 809	10 672	11 850 e	4 958	4 313	4 517 e	5 670	6 111	7 155 e
Other	1 270	902	1 043 e	28	81	93 e	1 242	821	950 e
Middle East	**37 118**	**33 725**	**35 803 e**	**17 132**	**14 716**	**14 229 e**	**19 154**	**18 369**	**20 901 e**
India	1 355	2 068	2 247 e	-	-	-	1 328	1 990	2 063 e
Indonesia	5 416	5 413	4 645 e	1 055	1 069	707 e	1 126	1 570	1 449 e
Singapore	4 578	3 386	4 377 e	-	23	-	1 788	632	663 e
Other	3 603	3 441	4 428 e	438	262	333 e	2 272	2 156	1 614 e
Asia	**14 952**	**14 308**	**15 697 e**	**1 493**	**1 354**	**1 040 e**	**6 514**	**6 348**	**5 789 e**
China	**1 672**	**1 605**	**2 556 e**	**12**	**-**	**3 e**	**924**	**1 047**	**1 115 e**
Algeria	309	704	986 e	138	528	704 e	171	176	282 e
Angola/Cabinada	-	-	-	-	-	-	-	-	-
Cameroon	56	-	-	-	-	-	-	-	-
Egypt	471	661	550 e	-	-	-	471	661	550 e
Gabon	-	-	-	-	-	-	-	-	-
Libya	-	-	56 e	-	-	-	-	-	56 e
Nigeria	139	534	628 e	139	496	559 e	-	38	69 e
Tunisia	-	-	-	-	-	-	-	-	-
Other	49	78	117 e	-	-	-	-	38	115 e
Africa	**1 024**	**1 977**	**2 337 e**	**277**	**1 024**	**1 263 e**	**642**	**913**	**1 072 e**
Non-spec./Other	**7 569**	**7 161**	**3 293 e**	**-**	**-**	**-**	**-**	**-**	**-**
Total Imports	**79 590**	**75 931**	**81 057 e**	**20 173**	**18 719**	**19 252 e**	**35 260**	**34 818**	**36 530 e**
Memo: Total OPEC	*41 719*	*39 499*	*41 116 e*	*18 436*	*16 728*	*16 150 e*	*19 209*	*19 350*	*21 802 e*

Source: Annual Oil Statistics.

OECD PACIFIC
Table 7C: Imports by Country of Origin

000 metric tons

	2000	2001	2002	2000	2001	2002	2000	2001	2002
	Total Gasoline			**Total Kerosene**			**Gas/Diesel**		
Australia	622	476	624 e	80	118	110 e	31	77	161 e
Belgium	-	-	-	-	-	-	-	-	-
Canada	-	-	27 e	-	-	-	-	-	5 e
Czech Republic	-	-	-	-	-	-	-	-	-
Denmark	-	-	-	-	-	-	-	-	-
Finland	-	-	-	-	-	-	-	-	-
France	-	-	-	-	-	-	-	-	-
Germany	-	-	-	-	-	-	-	1	-
Greece	-	-	-	-	-	-	-	-	-
Hungary	-	-	-	-	-	-	-	4	-
Ireland	-	-	-	-	-	-	-	-	-
Italy	-	-	-	-	-	-	-	-	-
Japan	77	162	100 e	117	166	259 e	350	834	762 e
Korea	629	588	608 e	1 941	1 155	1 248 e	2 342	1 795	1 319 e
Mexico	-	-	-	-	-	-	-	-	-
Netherlands	-	-	-	-	-	-	-	-	-
New Zealand	-	-	-	-	-	-	-	-	8 e
Norway	-	-	-	-	-	-	1	-	-
Poland	-	-	-	-	-	-	-	-	-
Spain	-	-	-	-	-	-	-	-	-
Sweden	-	-	-	-	-	-	-	-	-
Turkey	-	-	-	-	-	-	-	-	-
United Kingdom	-	-	-	-	-	-	-	-	-
United States	18	97	58 e	79	-	20 e	-	-	9 e
Other	-	-	-	-	-	-	-	-	2 e
OECD	**1 346**	**1 323**	**1 417 e**	**2 217**	**1 439**	**1 637 e**	**2 724**	**2 711**	**2 266 e**
Argentina	23	-	-	-	-	-	7	-	-
Brazil	-	-	-	-	-	-	-	-	-
Colombia	-	-	-	-	-	-	-	-	-
Ecuador	-	-	-	-	-	-	-	-	-
Netherlands Antilles	-	-	-	-	-	-	-	-	-
Peru	-	-	-	-	-	-	-	-	-
Trinidad and Tobago	-	-	-	-	-	-	-	4	-
Venezuela	-	-	-	-	-	-	-	-	-
Other	-	-	-	-	-	-	-	-	-
Latin America	**23**	**-**	**-**	**-**	**-**	**-**	**7**	**4**	**-**
Romania	-	-	-	-	-	-	-	-	-
Other	-	-	-	-	-	-	-	-	-
Non-OECD Europe	**-**	**-**	**-**	**-**	**-**	**-**	**-**	**-**	**-**
FSU	**1**	**-**	**4 e**	**-**	**-**	**-**	**242**	**225**	**301 e**
Iran	-	-	-	-	-	-	-	-	-
Iraq	-	-	-	-	-	-	-	-	-
Kuwait	-	-	-	274	42	62 e	-	-	-
Oman	-	-	-	-	-	-	-	-	-
Qatar	-	-	-	20	20	31 e	-	-	28 e
Saudi Arabia	154	247	86 e	105	76	223 e	45	7	12 e
United Arab Emirates	-	35	-	179	213	144 e	2	-	34 e
Other	-	-	-	-	-	-	-	-	-
Middle East	**154**	**282**	**86 e**	**578**	**351**	**460 e**	**47**	**7**	**74 e**
India	26	-	-	-	-	19 e	-	48	-
Indonesia	4	-	15 e	-	3	-	3	35	-
Singapore	934	1 011	1 480 e	242	433	473 e	777	684	1 060 e
Other	114	93	334 e	460	410	609 e	261	338	956 e
Asia	**1 078**	**1 104**	**1 829 e**	**702**	**846**	**1 101 e**	**1 041**	**1 105**	**2 016 e**
China	**222**	**153**	**248 e**	**440**	**275**	**423 e**	**19**	**6**	**41 e**
Algeria	-	-	-	-	-	-	-	-	-
Angola/Cabinada	-	-	-	-	-	-	-	-	-
Cameroon	-	-	-	-	-	-	-	-	-
Egypt	-	-	-	-	-	-	-	-	-
Gabon	-	-	-	-	-	-	-	-	-
Libya	-	-	-	-	-	-	-	-	-
Nigeria	-	-	-	-	-	-	-	-	-
Tunisia	-	-	-	-	-	-	-	-	-
Other	-	34	-	-	-	-	-	-	-
Africa	**-**	**34**	**-**	**-**	**-**	**-**	**-**	**-**	**-**
Non-spec./Other	**-**	**-**	**2 e**	**1 960**	**1 913**	**2 404 e**	**-**	**5**	**2 e**
Total Imports	**2 824**	**2 896**	**3 586 e**	**5 897**	**4 824**	**6 025 e**	**4 080**	**4 063**	**4 700 e**
Memo: Total OPEC	*158*	*282*	*101 e*	*578*	*354*	*460 e*	*50*	*42*	*74 e*

Source: Annual Oil Statistics.

OECD PACIFIC
Table 7D: Imports by Country of Origin

000 metric tons

	2000	2001	2002	2000	2001	2002	2000	2001	2002
	Fuel Oil (Residual)			**Petroleum Coke**			**Other Products**		
Australia	-	-	8 e	1	1	-	18	28	12 e
Belgium	-	-	-	-	-	-	2	1	-
Canada	-	-	-	-	-	220 e	1	-	-
Czech Republic	-	-	-	-	-	-	-	-	-
Denmark	-	-	-	-	-	-	-	-	-
Finland	-	-	-	-	-	-	-	-	-
France	-	-	-	-	-	-	5	5	5 e
Germany	-	-	-	-	-	-	3	3	-
Greece	-	-	-	-	-	-	-	-	-
Hungary	-	-	-	-	-	-	-	-	-
Ireland	-	-	-	-	-	-	-	-	-
Italy	-	-	-	-	-	-	-	-	-
Japan	157	311	146 e	-	22	16 e	3	21	3 e
Korea	144	158	625 e	-	-	92 e	56	54	21 e
Mexico	-	-	-	-	-	-	-	-	-
Netherlands	-	87	-	-	-	-	4	3	-
New Zealand	-	-	-	-	-	-	57	11	17 e
Norway	-	-	-	-	-	-	-	-	-
Poland	-	-	-	-	-	-	-	-	-
Spain	-	-	-	-	-	-	1	1	-
Sweden	-	-	-	1	-	-	2	-	-
Turkey	-	-	-	-	-	-	-	-	-
United Kingdom	-	-	-	-	-	-	2	4	1 e
United States	78	90	126 e	801	784	3 953 e	66	80	59 e
Other	-	-	-	-	-	-	-	-	-
OECD	**379**	**646**	**905 e**	**803**	**807**	**4 281 e**	**220**	**211**	**118 e**
Argentina	-	-	-	-	13	-	-	-	-
Brazil	-	-	-	-	-	-	-	-	-
Colombia	-	-	-	-	-	-	-	-	-
Ecuador	-	-	-	-	-	-	-	-	-
Netherlands Antilles	-	-	-	-	-	-	-	1	1 e
Peru	-	-	-	-	-	-	-	-	-
Trinidad and Tobago	-	-	-	-	-	-	1	1	-
Venezuela	-	-	-	-	-	-	7	7	2 e
Other	-	-	-	-	-	-	-	-	-
Latin America	**-**	**-**	**-**	**-**	**13**	**-**	**8**	**9**	**3 e**
Romania	-	-	-	-	-	-	-	-	-
Other	-	-	-	-	-	-	-	1	-
Non-OECD Europe	**-**	**-**	**-**	**-**	**-**	**-**	**-**	**1**	**-**
FSU	**-**	**-**	**69 e**	**-**	**-**	**-**	**-**	**-**	**-**
Iran	-	-	21 e	-	-	-	-	-	-
Iraq	-	-	-	-	-	-	-	-	-
Kuwait	-	-	-	-	-	-	-	-	-
Oman	-	-	-	-	-	-	-	-	-
Qatar	-	-	-	-	-	-	-	-	-
Saudi Arabia			32 e			-	53	-	-
United Arab Emirates	-	-	-	-	-	-	-	-	-
Other	-	-	-	-	-	-	-	-	-
Middle East	**-**	**-**	**53 e**	**-**	**-**	**-**	**53**	**-**	**-**
India	-	-	-	1	30	165 e	-	-	-
Indonesia	3 228	2 736	2 433 e	-	-	41 e	-	-	-
Singapore	745	548	627 e	-	-	-	92	55	74 e
Other	52	154	443 e	-	-	124 e	6	28	15 e
Asia	**4 025**	**3 438**	**3 503 e**	**1**	**30**	**330 e**	**98**	**83**	**89 e**
China	**48**	**27**	**528 e**	**1**	**88**	**194 e**	**6**	**9**	**4 e**
Algeria	-	-	-	-	-	-	-	-	-
Angola/Cabinada	-	-	-	-	-	-	-	-	-
Cameroon	56	-	-	-	-	-	-	-	-
Egypt	-	-	-	-	-	-	-	-	-
Gabon	-	-	-	-	-	-	-	-	-
Libya	-	-	-	-	-	-	-	-	-
Nigeria	-	-	-	-	-	-	-	-	-
Tunisia	-	-	-	-	-	-	-	-	-
Other	44	-	-	3	-	-	2	6	2 e
Africa	**100**	**-**	**-**	**3**	**-**	**-**	**2**	**6**	**2 e**
Non-spec./Other	**1 052**	**1 045**	**884 e**	**4 557**	**4 196**	**-**	**-**	**2**	**1 e**
Total Imports	**5 604**	**5 156**	**5 942 e**	**5 365**	**5 134**	**4 805 e**	**387**	**321**	**217 e**
Memo: Total OPEC	*3 228*	*2 736*	*2 486 e*	*-*	*-*	*41 e*	*60*	*7*	*2 e*

Source: Annual Oil Statistics.

OECD PACIFIC
Table 8A: Exports by Country of Destination

000 metric tons

	2000	2001	2002	2000	2001	2002	2000	2001	2002
	Crude Oil			**Natural Gas Liquids**			**Feedstocks and Other**		
Australia	-	-	-	-	-	-	-	-	-
Austria	-	-	-	-	-	-	-	-	-
Belgium	-	-	-	-	-	-	-	-	-
Canada	-	-	-	-	-	-	-	-	-
Denmark	-	-	-	-	-	-	-	-	-
Finland	-	-	-	-	-	-	-	-	-
France	-	-	-	-	-	-	-	-	-
Germany	-	-	-	-	-	-	-	-	-
Greece	-	-	-	-	-	-	-	-	-
Ireland	-	-	-	-	-	-	-	-	-
Italy	-	-	-	-	-	-	-	-	-
Japan	2 232	2 915	2 154 e	1 004	1 218	1 397 e	630	369	404 e
Luxembourg	-	-	-	-	-	-	-	-	-
Mexico	-	-	-	-	-	-	-	-	-
Netherlands	-	-	-	-	-	-	-	-	-
Norway	-	-	-	-	-	-	-	-	-
Portugal	-	-	-	-	-	-	-	-	-
Spain	-	-	-	-	-	-	-	-	-
Sweden	-	-	-	-	-	-	-	-	-
Switzerland	-	-	-	-	-	-	-	-	-
Turkey	-	-	-	-	-	-	-	-	-
United Kingdom	-	-	-	-	-	-	-	-	-
United States	2 759	2 370	2 250 e	-	-	-	103	86	106 e
Other OECD	3 767	3 753	5 355 e	104	35	79 e	158	238	187 e
OECD	**8 758**	**9 038**	**9 759 e**	**1 108**	**1 253**	**1 476 e**	**891**	**693**	**697 e**
Africa	-	159	237 e	-	-	-	-	-	-
Latin America	84	-	-	9	13	-	-	-	6 e
Asia	5 068	8 074	7 091 e	64	23	35 e	489	571	614 e
China	1 349	730	930 e	327	228	243 e	168	145	5 e
Non-OECD Europe	-	-	-	-	-	-	-	-	-
Former Soviet Union	-	-	-	-	-	-	-	-	-
Middle East	79	-	-	47	-	-	-	-	-
Non-specified/Other	1 175	2 620	2 512 e	-	65	1 e	-	-	8 e
Non-OECD	**7 755**	**11 583**	**10 770 e**	**447**	**329**	**279 e**	**657**	**716**	**633 e**
Total Exports	**16 513**	**20 621**	**20 529 e**	**1 555**	**1 582**	**1 755 e**	**1 548**	**1 409**	**1 330 e**
	Total Products			**Liquefied Petroleum Gas**			**Naphtha**		
Australia	982	491	264 e	-	-	-	-	-	-
Austria	-	-	-	-	-	-	-	-	-
Belgium	-	10	-	-	-	-	-	-	-
Canada	138	-	-	-	-	-	-	-	-
Denmark	712	532	531 e	-	-	-	-	-	-
Finland	-	-	-	-	-	-	-	-	-
France	-	-	1 e	-	-	1 e	-	-	-
Germany	25	3	-	-	-	-	-	-	-
Greece	-	-	-	-	-	-	-	-	-
Ireland	-	-	-	-	-	-	-	-	-
Italy	-	-	-	-	-	-	-	-	-
Japan	14 141	13 776	11 185 e	90	60	37 e	5 469	5 922	5 148 e
Luxembourg	-	-	-	-	-	-	-	-	-
Mexico	39	-	-	-	-	-	-	-	-
Netherlands	-	17	336 e	-	-	-	-	-	-
Norway	-	-	21 e	-	-	-	-	-	-
Portugal	-	-	-	-	-	-	-	-	-
Spain	2	-	-	-	-	-	-	-	-
Sweden	-	-	-	-	-	-	-	-	-
Switzerland	-	-	-	-	-	-	-	-	-
Turkey	-	-	-	-	-	-	-	-	-
United Kingdom	-	18	28 e	-	-	-	-	-	-
United States	3 505	3 803	2 997 e	-	-	-	41	-	-
Other OECD	1 940	2 567	2 443 e	27	70	52 e	-	19	16 e
OECD	**21 484**	**21 217**	**17 806 e**	**117**	**130**	**90 e**	**5 510**	**5 941**	**5 164 e**
Africa	2	12	31 e	-	-	-	-	-	-
Latin America	343	589	669 e	-	-	-	-	-	-
Asia	8 731	9 434	10 516 e	31	41	28 e	260	84	125 e
China	16 161	15 903	7 958 e	602	321	184 e	85	70	105 e
Non-OECD Europe	-	-	-	-	-	-	-	-	-
Former Soviet Union	139	66	11 e	-	-	-	-	-	-
Middle East	516	25	96 e	6	-	-	-	-	-
Non-specified/Other	257	336	725 e	81	85	64 e	-	-	-
Non-OECD	**26 149**	**26 365**	**20 006 e**	**720**	**447**	**276 e**	**345**	**154**	**230 e**
Total Exports	**47 633**	**47 582**	**37 812 e**	**837**	**577**	**366 e**	**5 855**	**6 095**	**5 394 e**

Source: Annual Oil Statistics.

OECD PACIFIC
Table 8B: Exports by Country of Destination

000 metric tons

	2000	2001	2002	2000	2001	2002	2000	2001	2002
	Total Gasoline			**Total Kerosene**			**Gas/Diesel Oil**		
Australia	93	154	66 e	3	16	4 e	681	270	118 e
Austria	-	-	-	-	-	-	-	-	-
Belgium	-	-	-	-	-	-	-	-	-
Canada	22	-	-	115	-	-	-	-	-
Denmark	63	110	50 e	402	401	225 e	247	21	233 e
Finland	-	-	-	-	-	-	-	-	-
France	-	-	-	-	-	-	-	-	-
Germany	-	-	-	-	-	-	-	-	-
Greece	-	-	-	-	-	-	-	-	-
Ireland	-	-	-	-	-	-	-	-	-
Italy	-	-	-	-	-	-	-	-	-
Japan	536	604	519 e	3 433	3 283	3 111 e	3 710	3 055	1 684 e
Luxembourg	-	-	-	-	-	-	-	-	-
Mexico	-	-	-	-	-	-	39	-	-
Netherlands	-	-	-	-	-	-	-	-	318 e
Norway	-	-	-	-	-	-	-	-	-
Portugal	-	-	-	-	-	-	-	-	-
Spain	-	-	-	-	-	-	-	-	-
Sweden	-	-	-	-	-	-	-	-	-
Switzerland	-	-	-	-	-	-	-	-	-
Turkey	-	-	-	-	-	-	-	-	-
United Kingdom	-	-	-	-	-	-	-	-	-
United States	241	732	518 e	2 182	2 284	1 906 e	957	697	441 e
Other OECD	779	653	739 e	201	347	312 e	395	935	964 e
OECD	**1 734**	**2 253**	**1 892 e**	**6 336**	**6 331**	**5 558 e**	**6 029**	**4 978**	**3 758 e**
Africa	-	-	29 e	-	-	-	-	-	2 e
Latin America	-	-	28 e	-	-	-	326	583	639 e
Asia	1 210	733	881 e	559	925	1 028 e	4 776	5 795	6 939 e
China	73	101	12 e	960	1 339	895 e	3 684	3 284	405 e
Non-OECD Europe	-	-	-	-	-	-	-	-	-
Former Soviet Union	-	7	-	-	-	-	129	44	11 e
Middle East	-	25	32 e	35	-	-	440	-	64 e
Non-specified/Other	3	23	65 e	5	49	98 e	148	178	446 e
Non-OECD	**1 286**	**889**	**1 047 e**	**1 559**	**2 313**	**2 021 e**	**9 503**	**9 884**	**8 506 e**
Total Exports	**3 020**	**3 142**	**2 939 e**	**7 895**	**8 644**	**7 579 e**	**15 532**	**14 862**	**12 264 e**
	Fuel Oil (Residual)			**Petroleum Coke**			**Other Products**		
Australia	204	48	70 e	-	-	-	1	3	6 e
Austria	-	-	-	-	-	-	-	-	-
Belgium	-	-	-	-	-	-	-	10	-
Canada	-	-	-	-	-	-	1	-	-
Denmark	-	-	-	-	-	-	-	-	23 e
Finland	-	-	-	-	-	-	-	-	-
France	-	-	-	-	-	-	-	-	-
Germany	-	-	-	-	-	-	25	3	-
Greece	-	-	-	-	-	-	-	-	-
Ireland	-	-	-	-	-	-	-	-	-
Italy	-	-	-	-	-	-	-	-	-
Japan	727	619	385 e	69	74	51 e	107	159	250 e
Luxembourg	-	-	-	-	-	-	-	-	-
Mexico	-	-	-	-	-	-	-	-	-
Netherlands	-	-	-	-	-	-	-	17	18 e
Norway	-	-	-	-	-	-	-	-	21 e
Portugal	-	-	-	-	-	-	-	-	-
Spain	-	-	-	-	-	-	2	-	-
Sweden	-	-	-	-	-	-	-	-	-
Switzerland	-	-	-	-	-	-	-	-	-
Turkey	-	-	-	-	-	-	-	-	-
United Kingdom	-	-	-	-	-	-	-	18	28 e
United States	23	-	86 e	-	-	-	61	90	46 e
Other OECD	292	217	62 e	-	-	-	246	326	298 e
OECD	**1 246**	**884**	**603 e**	**69**	**74**	**51 e**	**443**	**626**	**690 e**
Africa	-	-	-	-	-	-	2	12	-
Latin America	7	-	-	-	-	-	10	6	2 e
Asia	1 188	1 273	856 e	-	-	-	707	583	659 e
China	9 839	9 741	4 950 e	-	-	-	918	1 047	1 407 e
Non-OECD Europe	-	-	-	-	-	-	-	-	-
Former Soviet Union	10	15	-	-	-	-	-	-	-
Middle East	32	-	-	-	-	-	3	-	-
Non-specified/Other	18	-	50 e	-	-	-	2	1	2 e
Non-OECD	**11 094**	**11 029**	**5 856 e**	**-**	**-**	**-**	**1 642**	**1 649**	**2 070 e**
Total Exports	**12 340**	**11 913**	**6 459 e**	**69**	**74**	**51 e**	**2 085**	**2 275**	**2 760 e**

Source: Annual Oil Statistics.

OECD EUROPE

Figure 1
FUEL SHARES OF TPES *

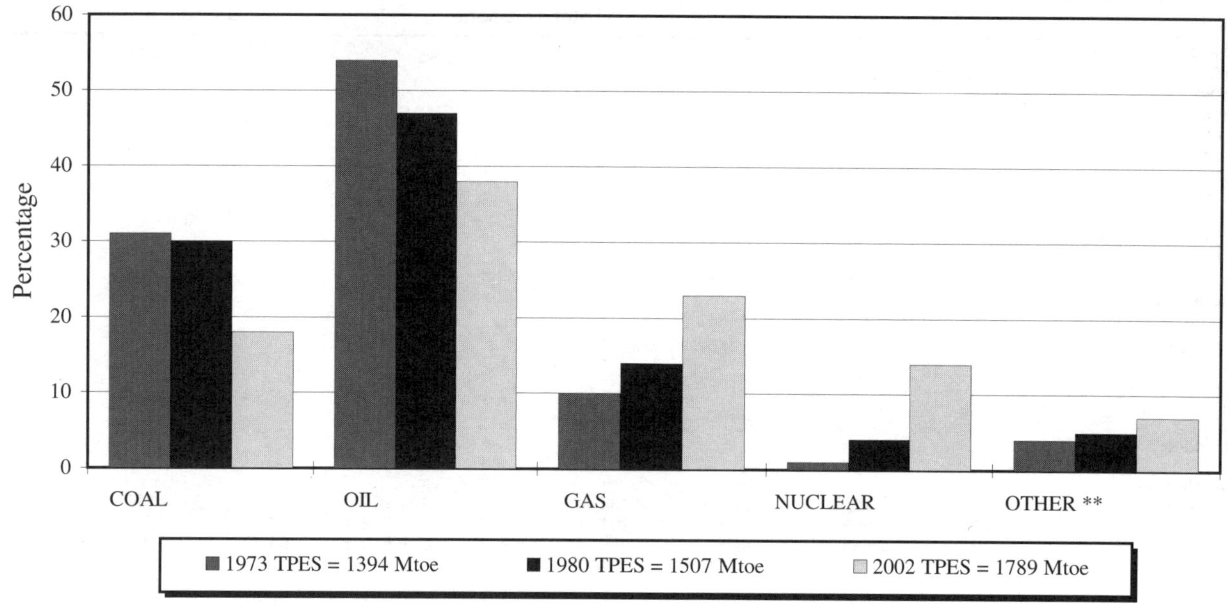

Source: Annual Energy Balances.

* Excluding electricity trade.

** Includes hydro, geothermal, solar, wind, combustible renewables and waste, etc.

Figure 2
OIL PRODUCTION AND CONSUMPTION

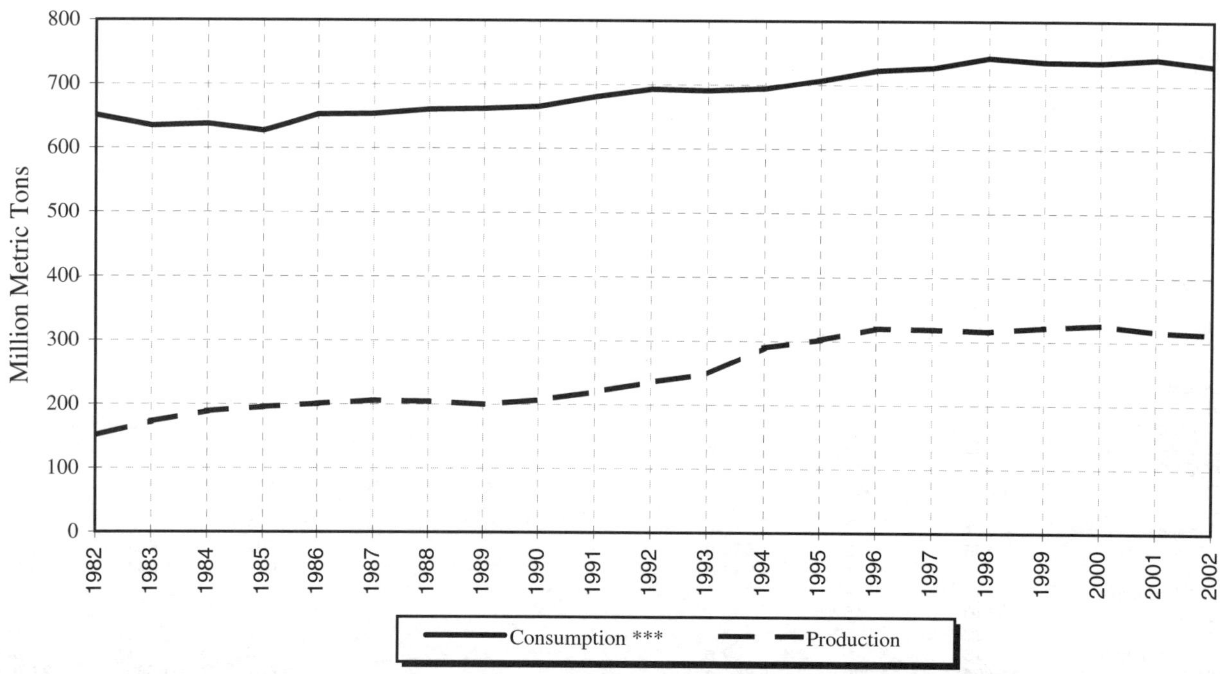

Source: Annual Oil Statistics.

*** Includes refinery fuel and international marine bunkers.

OECD EUROPE

Table 1: Oil Supply and Consumption - Summary Table

000 metric tons

	1973	1978	1980	1990	1999	2000	2001	2002
Indigenous Production	23 010	90 619	125 661	207 067	322 550	326 758	317 100	312 800 e
Crude Oil	..	88 174	119 290	197 166	305 630	310 004	298 968	294 152 e
NGL	..	2 214	4 617	7 780	14 161	13 654	15 010	15 290 e
Other	..	231	1 754	2 121	2 759	3 100	3 122	3 358 e
+ Net Imports	774 180	672 534	620 024	454 156	393 196	399 554	414 522	406 052 e
Crude Oil, NGL, Other	763 566	633 191	577 710	423 765	369 323	371 973	378 683	372 994 e
Petroleum Products	10 614	39 343	42 314	30 391	23 873	27 581	35 839	33 058 e
- Intl. Marine Bunkers	44 078	36 964	32 404	36 153	42 724	45 253	46 440	48 243 e
+ Stock Changes	-10 161	8 811	-13 030	-3 013	11 247	-4 789	1 916	2 979 e
TOTAL SUPPLY	**742 951**	**735 000**	**700 251**	**622 057**	**684 269**	**676 270**	**687 098**	**673 588 e**
Refinery Intake	*791 194*	*741 622*	*723 546*	*668 561*	*733 488*	*744 945*	*743 760*	*739 044 e*
Refinery Output	*782 303*	*730 713*	*714 108*	*661 719*	*724 211*	*734 280*	*730 704*	*728 171 e*
LPG + Ethane	*14 153*	*15 000*	*14 849*	*16 763*	*19 298*	*20 254*	*20 628*	*21 205 e*
Naphtha	*42 212*	*34 423*	*34 301*	*39 267*	*43 518*	*47 015*	*46 084*	*41 157 e*
Kerosenes	*30 838*	*27 780*	*28 688*	*40 285*	*47 954*	*46 469*	*44 066*	*42 619 e*
Gasolines	*96 857*	*110 763*	*117 395*	*144 412*	*153 060*	*150 222*	*148 906*	*152 068 e*
Gas/Diesel Oil	*230 876*	*220 921*	*221 867*	*209 708*	*252 919*	*260 172*	*262 327*	*261 092 e*
Fuel Oil	*308 420*	*262 971*	*239 907*	*139 039*	*125 273*	*126 604*	*123 388*	*119 922 e*
Other Products	*58 947*	*58 855*	*57 101*	*72 245*	*82 189*	*83 544*	*85 305*	*90 108 e*
- Refinery Losses	8 891	10 909	9 438	6 842	9 277	10 665	13 056	10 873 e
+ Returns	5 213	10 707	16 165	13 620	14 049	16 946	16 985	13 921 e
- Statistical Differences	3 950	2 752	1 711	- 605	-4 773	-7 898	-3 814	-5 836 e
TOTAL CONSUMPTION	**735 323**	**732 046**	**705 267**	**629 440**	**693 814**	**690 449**	**694 841**	**682 472 e**
Transformation and Energy	**148 474**	**148 315**	**149 878**	**101 901**	**102 186**	**100 378**	**98 857**	..
Crude Oil	15	44	309	268	64	54	57	..
LPG + Ethane	1 248	1 321	2 020	1 483	1 392	1 664	1 594	..
Gas/Diesel Oil	2 827	3 991	4 788	3 993	4 635	5 199	4 812	..
Fuel Oil	119 074	114 667	110 826	60 354	56 472	53 177	54 017	..
Petroleum Coke	1 549	2 001	2 336	4 909	5 475	5 413	5 065	..
Other	23 761	26 291	29 599	30 894	34 148	34 871	33 312	..
Transport	**173 962**	**204 625**	**211 010**	**276 539**	**337 923**	**339 503**	**340 056**	..
Motor Gasoline	94 724	109 369	111 500	131 396	136 730	130 593	127 634	..
Gas/Diesel Oil	54 677	68 889	71 383	108 271	147 735	153 061	158 146	..
Subtotal	*149 401*	*178 258*	*182 883*	*239 667*	*284 465*	*283 654*	*285 780*	..
Aviation Fuels	18 907	21 003	21 312	30 139	45 330	47 448	45 711	..
Other	5 654	5 364	6 815	6 733	8 128	8 401	8 565	..
Industry	**224 440**	**202 081**	**188 681**	**137 722**	**142 969**	**144 647**	**142 798**	..
Naphtha	34 682	29 801	30 584	34 332	40 572	41 477	39 653	..
LPG + Ethane	5 686	5 831	6 208	9 310	11 407	11 774	11 785	..
Gas/Diesel Oil	29 071	28 922	26 864	17 038	17 432	17 951	17 923	..
Fuel Oil	113 321	98 183	89 857	34 991	23 635	23 276	22 600	..
Other	41 680	39 344	35 168	42 051	49 923	50 169	50 837	..
Other Sectors	**188 447**	**177 025**	**155 698**	**113 278**	**110 736**	**105 921**	**113 130**	..
Gas/Diesel Oil	141 031	137 775	120 527	90 379	88 021	84 731	91 106	..
Fuel Oil	30 843	23 902	20 833	8 407	5 109	3 826	3 797	..
Other	16 573	15 348	14 338	14 492	17 606	17 364	18 227	..
Memo: Non-Energy Use	*72 264*	*68 491*	*70 876*	*82 753*	*94 389*	*94 432*	*91 844*	..
Chem/Petchem. Feedst.	*40 313*	*39 383*	*41 327*	*52 153*	*61 060*	*61 399*	*60 204*	..
Other	*31 951*	*29 108*	*29 549*	*30 600*	*33 329*	*33 033*	*31 640*	..

Source: Annual Oil Statistics.

OECD EUROPE
Table 2: Supply of Crude Oil, NGL, Feedstocks and Other Hydrocarbons

000 metric tons

	1973	1978	1980	1990	1999	2000	2001	2002
CRUDE OIL								
Indigenous Production	..	88 174	119 290	197 166	305 630	310 004	298 968	294 152 e
+ Imports	..	678 496	635 336	522 299	587 799	606 133	605 168	593 407 e
- Exports	..	52 848	72 862	132 935	231 935	247 409	242 211	239 309 e
- Direct Use	..	117	408	308	76	66	69	60 e
+ Stock Change	..	3 952	-7 317	-3 459	1 519	786	-1 351	884 e
- Statistical Difference	..	2 301	916	-2 116	- 248	1 524	- 766	- 316 e
Refinery Intake	**..**	**715 356**	**673 123**	**584 879**	**663 185**	**667 924**	**661 271**	**649 390 e**
Closing Stock Level	..	*57 648*	*67 398*	*51 114*	*62 097*	*58 188*	*59 598*	*58 909 e*
NATURAL GAS LIQUIDS								
Indigenous Production	..	2 214	4 617	7 780	14 161	13 654	15 010	15 290 e
+ Imports	..	18	104	3 612	5 718	6 635	7 771	9 434 e
- Exports	..	342	1 019	2 716	6 673	5 957	7 430	7 514 e
- Direct Use	..	1 288	2 182	2 931	4 596	4 588	4 856	4 836 e
+ Stock Change	..	- 6	- 19	- 137	76	- 170	- 47	49 e
- Statistical Difference	..	-	- 18	- 444	- 8	- 287	- 213	- 362 e
Refinery Intake	**..**	**596**	**1 519**	**6 052**	**8 694**	**9 861**	**10 661**	**12 785 e**
Closing Stock Level	..	*23*	*114*	*606*	*367*	*537*	*582*	*534 e*
REFINERY FEEDSTOCKS								
Backflows	..	10 456	15 643	13 056	13 330	16 185	16 269	13 255 e
+ Transfers	..	7 563	14 721	30 623	33 189	36 853	36 185	42 342 e
+ Imports	..	8 474	17 430	39 582	20 627	20 006	20 513	21 547 e
- Exports	..	614	1 290	6 458	8 033	9 262	6 936	5 877 e
+ Stock Change	..	- 187	628	- 315	- 75	- 400	662	- 293 e
- Statistical Difference	..	245	- 22	955	- 76	-1 066	-2 078	-2 796 e
Refinery Intake	**..**	**25 447**	**47 154**	**75 533**	**59 114**	**64 448**	**68 771**	**73 770 e**
Closing Stock Level	..	*1 279*	*4 026*	*8 262*	*7 679*	*8 079*	*7 418*	*7 711 e*
OTHER HYDROCARBONS								
Indigenous Production	..	231	1 754	2 121	2 759	3 100	3 122	3 358 e
+ Imports	..	7	11	381	1 822	1 828	1 810	1 306 e
- Exports	..	-	-	-	2	1	2	-
- Direct Use	..	15	15	394	2 108	2 242	1 716	1 684 e
+ Stock Change	..	-	-	- 11	22	24	- 158	92 e
- Statistical Difference	..	-	-	-	- 2	- 3	- 1	- 27 e
Refinery Intake	**..**	**223**	**1 750**	**2 097**	**2 495**	**2 712**	**3 057**	**3 099 e**
Closing Stock Level	..	*39*	*39*	*64*	*216*	*192*	*351*	*260 e*
CRUDE OIL, NGL, FEEDSTOCKS AND OTHER								
Indigenous Production	23 010	90 619	125 661	207 067	322 550	326 758	317 100	312 800 e
+ Backflows	5 020	10 456	15 643	13 056	13 330	16 185	16 269	13 255 e
+ Transfers	5 472	7 563	14 721	30 623	33 189	36 853	36 185	42 342 e
+ Imports	769 599	686 995	652 881	565 874	615 966	634 602	635 262	625 694 e
- Exports	6 033	53 804	75 171	142 109	246 643	262 629	256 579	252 700 e
- Direct Use	1 254	1 420	2 605	3 633	6 780	6 896	6 641	6 580 e
+ Stock Change	-5 582	3 759	-6 708	-3 922	1 542	240	- 894	732 e
- Statistical Difference	- 962	2 546	876	-1 605	- 334	168	-3 058	-3 501 e
Refinery Intake	**791 194**	**741 622**	**723 546**	**668 561**	**733 488**	**744 945**	**743 760**	**739 044 e**
Closing Stock Level	-	*58 989*	*71 577*	*60 046*	*70 359*	*66 996*	*67 949*	*67 414 e*

Source: Annual Oil Statistics.

OECD EUROPE
Table 3: Motor Gasoline - Supply

000 metric tons

	1973	1978	1980	1990	1999	2000	2001	2002
MOTOR GASOLINE								
Refinery Output	**95 701**	**108 376**	**114 956**	**143 017**	**152 538**	**149 814**	**148 589**	**151 912 e**
+ Imports	16 093	17 441	21 878	35 412	33 508	36 470	33 713	30 954 e
- Exports	14 574	17 817	20 697	38 426	47 356	49 048	48 304	51 331 e
+ Transfers	54	552	-2 613	-5 715	-3 444	-3 888	-3 773	-4 985 e
+ Stock Change	-1 697	555	-1 352	- 670	2 671	269	- 785	1 901 e
- Statistical Difference	- 237	-1 469	- 388	254	116	1 923	214	2 294 e
Total Consumption	**95 814**	**110 576**	**112 560**	**133 364**	**137 801**	**131 694**	**129 226**	**126 157 e**
Closing Stock Level	-	*16 428*	*18 442*	*20 499*	*20 885*	*20 612*	*21 411*	*19 508 e*

Table 4: Aviation Fuels - Supply

000 metric tons

	1973	1978	1980	1990	1999	2000	2001	2002
AVIATION GASOLINE								
Refinery Output	**420**	**303**	**435**	**269**	**169**	**239**	**218**	**135 e**
+ Imports	345	249	256	167	150	141	88	69 e
- Exports	306	208	394	214	170	205	107	93 e
+ Transfers	-	- 3	8	2	21	- 14	10	7 e
+ Stock Change	12	12	23	- 13	16	2	- 8	4 e
- Statistical Difference	50	28	17	8	34	- 1	32	- 32 e
Total Consumption	**421**	**325**	**311**	**203**	**152**	**164**	**169**	**154 e**
Closing Stock Level	-	*121*	*101*	*70*	*39*	*37*	*45*	*41 e*
GASOLINE TYPE JET FUEL								
Refinery Output	**736**	**2 084**	**2 004**	**1 126**	**353**	**169**	**99**	**21 e**
+ Imports	320	435	526	233	7	11	9	5 e
- Exports	415	1 038	1 279	390	72	77	32	-
+ Transfers	-	- 33	34	- 20	- 214	16	- 1	- 7 e
+ Stock Change	6	- 108	151	8	46	1	5	2 e
- Statistical Difference	- 12	- 41	8	208	2	1	-	-
Total Consumption	**659**	**1 381**	**1 428**	**749**	**118**	**119**	**80**	**21 e**
Closing Stock Level	-	*596*	*546*	*57*	*14*	*13*	*8*	*6 e*
KEROSENE TYPE JET FUEL								
Refinery Output	**20 262**	**20 226**	**22 351**	**34 026**	**43 068**	**41 769**	**39 432**	**37 732 e**
+ Imports	5 082	5 432	6 824	8 424	13 114	15 912	19 135	17 109 e
- Exports	6 507	5 381	7 785	10 653	10 589	9 954	9 976	9 117 e
+ Transfers	961	- 157	- 164	- 470	- 214	167	-1 268	- 901 e
+ Stock Change	- 323	275	- 382	- 502	- 74	- 301	- 267	316 e
- Statistical Difference	1 627	1 073	1 236	1 593	193	364	1 537	216 e
Total Consumption	**17 848**	**19 322**	**19 608**	**29 232**	**45 112**	**47 229**	**45 519**	**44 923 e**
Closing Stock Level	-	*2 174*	*3 432*	*3 716*	*4 894*	*5 195*	*5 465*	*5 150 e*

Source: Annual Oil Statistics.

OECD EUROPE
Table 5A: Gas/Diesel Oil - Supply and Consumption

000 metric tons

	1973	1978	1980	1990	1999	2000	2001	2002
Refinery Output	**230 876**	**220 921**	**221 867**	**209 708**	**252 919**	**260 172**	**262 327**	**261 092 e**
+ Recycled Products	-	-	8	30	20	23	30	30 e
+ Imports	55 789	61 260	60 750	68 209	76 893	76 475	88 187	90 849 e
- Exports	48 305	40 201	43 112	50 268	65 036	65 019	67 583	69 703 e
- Intl. Marine Bunkers	7 571	8 033	6 759	8 254	9 355	9 431	9 461	9 362 e
+ Transfers	-1 219	184	-2 362	-1 778	-6 679	-5 037	-2 494	-4 167 e
+ Stock Changes	-1 006	3 899	-6 954	- 229	5 879	-2 261	938	- 413 e
- Statistical Differences	958	-1 547	- 124	-2 263	-3 182	-6 020	- 43	-1 215 e
Total Consumption	**227 606**	**239 577**	**223 562**	**219 681**	**257 823**	**260 942**	**271 987**	**269 541 e**
Transformation	**2 153**	**3 324**	**4 218**	**3 440**	**3 750**	**4 390**	**4 181**	**..**
Public Electricity	1 695	1 484	1 444	1 671	1 406	1 837	1 641	..
Autoproducer Electricity	-	358	412	202	207	172	168	..
Public CHP Plants	57	109	101	55	511	885	626	..
Autoproducers of CHP	-	2	1	26	443	112	395	..
Public Heat Plants	30	53	42	86	178	161	180	..
Autoproducer Heat Plants	92	319	317	141	5	47	42	..
Petrochemical Industry	221	975	1 878	1 243	991	1 169	1 129	..
Other Transformation	58	24	23	16	9	7	-	..
Energy Sector	**674**	**667**	**570**	**553**	**885**	**809**	**631**	**..**
Refinery Fuel	78	110	66	77	296	335	186	..
Other Energy Sector	596	557	504	476	589	474	445	..
Final Consumption	**224 779**	**235 586**	**218 774**	**215 688**	**253 188**	**255 743**	**267 175**	**..**
Transport	**54 677**	**68 889**	**71 383**	**108 271**	**147 735**	**153 061**	**158 146**	**..**
Air	-	-	-	-	-	-	-	..
Road	42 904	57 237	59 748	98 872	139 415	145 319	150 690	..
Rail	4 436	5 007	4 749	3 750	3 115	2 983	2 754	..
Inland Waterways	7 333	6 645	6 880	5 617	5 168	4 733	4 671	..
Non-Specified	4	-	6	32	37	26	31	..
Industry	**29 071**	**28 922**	**26 864**	**17 038**	**17 432**	**17 951**	**17 923**	**..**
Iron and Steel	911	669	603	391	299	337	401	..
Chemical (incl. Pet.-Chem.)	2 132	3 468	4 122	4 459	4 394	4 739	3 973	..
Non-Ferrous Metals	410	379	374	273	218	242	220	..
Non-Metallic Minerals	3 113	2 231	1 376	965	990	906	1 033	..
Transport Equipment	702	804	738	494	438	439	498	..
Machinery	2 337	2 329	2 039	1 816	1 366	1 231	1 209	..
Mining and Quarrying	693	574	562	599	820	845	892	..
Food and Tobacco	1 893	2 264	2 055	1 534	1 864	1 740	1 851	..
Pulp, Paper and Printing	336	372	241	284	338	311	317	..
Wood and Wood Products	420	460	390	202	172	148	150	..
Construction	1 779	2 021	1 852	2 380	3 076	3 742	3 817	..
Textiles and Leather	682	617	558	470	449	429	493	..
Non-Specified	13 663	12 734	11 954	3 171	3 008	2 842	3 069	..
Other Sectors	**141 031**	**137 775**	**120 527**	**90 379**	**88 021**	**84 731**	**91 106**	**..**
Commerce and Public	50 071	50 377	41 094	22 926	20 238	18 511	20 702	..
Residential	67 299	69 081	59 654	48 890	47 999	45 999	50 648	..
Agriculture	13 619	15 589	17 130	17 594	19 514	19 922	19 440	..
Non-Specified	10 042	2 728	2 649	969	270	299	316	..
Memo: Non-Energy Use	*1 403*	*2 799*	*3 199*	*4 125*	*3 901*	*4 295*	*3 398*	..
Chem/Petchem. Feedst.	*1 402*	*2 798*	*3 195*	*4 119*	*3 900*	*4 294*	*3 397*	..
Other	*1*	*1*	*4*	*6*	*1*	*1*	*1*	..
Closing Stock Level	-	*44 125*	*56 765*	*40 534*	*43 074*	*45 444*	*44 502*	*44 863 e*

Source: Annual Oil Statistics.

OECD EUROPE
Table 5B: Fuel Oil (Residual) - Supply and Consumption

000 metric tons

	1973	1978	1980	1990	1999	2000	2001	2002
Refinery Output	**308 420**	**262 971**	**239 907**	**139 039**	**125 273**	**126 604**	**123 388**	**119 922 e**
+ Recycled Products	-	-	1	127	155	184	240	372 e
+ Imports	45 314	39 677	48 639	44 261	41 318	45 105	42 575	47 243 e
- Exports	48 396	32 232	37 274	45 393	41 300	46 982	44 593	42 057 e
- Intl. Marine Bunkers	36 213	28 666	25 354	27 580	33 031	35 457	36 591	38 565 e
+ Transfers	- 160	- 657	-4 578	-5 589	-6 584	-7 807	-5 645	-5 940 e
+ Stock Changes	- 272	- 389	3 201	605	827	181	2 710	- 906 e
- Statistical Differences	2 558	1 822	88	219	78	620	724	- 634 e
Total Consumption	**266 135**	**238 882**	**224 454**	**105 251**	**86 580**	**81 208**	**81 360**	**80 703 e**
Transformation	**99 127**	**92 469**	**89 761**	**48 367**	**42 861**	**40 152**	**40 469**	..
Public Electricity	84 387	68 024	60 662	35 703	29 067	26 874	26 586	..
Autoproducer Electricity	113	5 685	5 051	1 691	2 249	973	624	..
Public CHP Plants	7 098	8 821	12 421	2 515	1 740	2 778	2 941	..
Autoproducers of CHP	1 037	5 752	7 374	4 869	4 619	2 820	2 528	..
Public Heat Plants	658	668	852	1 068	942	625	732	..
Autoproducer Heat Plants	158	266	238	197	99	728	704	..
Petrochemical Industry	1 186	777	1 073	1 047	2 485	3 984	4 921	..
Other Transformation	4 490	2 476	2 090	1 277	1 660	1 370	1 433	..
Energy Sector	**19 947**	**22 198**	**21 065**	**11 987**	**13 611**	**13 025**	**13 548**	..
Refinery Fuel	19 418	20 922	20 544	11 719	13 327	12 772	13 314	..
Other Energy Sector	529	1 276	521	268	284	253	234	..
Final Consumption	**147 061**	**124 215**	**113 628**	**44 897**	**30 108**	**28 031**	**27 343**	..
Transport	**2 897**	**2 130**	**2 938**	**1 499**	**1 364**	**929**	**946**	..
Air	-	-	-	-	-	-	-	..
Road	4	6	5	-	-	-	-	..
Rail	1 157	518	339	49	1	2	-	..
Inland Waterways	1 436	1 156	2 141	1 200	1 346	924	946	..
Non-Specified	300	450	453	250	17	3	-	..
Industry	**113 321**	**98 183**	**89 857**	**34 991**	**23 635**	**23 276**	**22 600**	..
Iron and Steel	14 064	11 026	7 353	2 690	1 435	1 290	1 114	..
Chemical (incl. Pet.-Chem.)	18 480	17 839	17 036	8 776	5 200	6 740	7 041	..
Non-Ferrous Metals	1 745	2 351	2 144	1 521	1 140	1 121	1 098	..
Non-Metallic Minerals	19 402	17 269	15 484	4 683	3 824	3 410	2 693	..
Transport Equipment	2 583	2 222	1 836	465	250	179	189	..
Machinery	5 230	3 768	3 155	1 408	866	705	792	..
Mining and Quarrying	1 509	1 044	1 048	356	293	242	218	..
Food and Tobacco	12 064	11 779	11 057	5 006	3 016	2 800	2 916	..
Pulp, Paper and Printing	10 040	8 658	7 941	3 129	2 321	2 240	2 236	..
Wood and Wood Products	1 268	1 006	887	297	344	300	255	..
Construction	3 530	3 070	2 741	389	280	325	240	..
Textiles and Leather	6 046	5 268	4 599	2 365	1 425	1 332	1 199	..
Non-Specified	17 360	12 883	14 576	3 906	3 241	2 592	2 609	..
Other Sectors	**30 843**	**23 902**	**20 833**	**8 407**	**5 109**	**3 826**	**3 797**	..
Commerce and Public	6 287	6 776	7 464	3 011	1 949	1 123	1 039	..
Residential	18 460	12 053	8 778	3 733	2 058	1 901	1 846	..
Agriculture	1 804	1 773	2 173	945	1 014	702	694	..
Non-Specified	4 292	3 300	2 418	718	88	100	218	..
Memo: Non-Energy Use	*383*	*1 621*	*3 022*	*2 863*	*1 643*	*3 324*	*3 371*	..
Chem/Petchem. Feedst.	*265*	*1 569*	*2 966*	*2 570*	*1 484*	*3 030*	*3 158*	..
Other	*118*	*52*	*56*	*293*	*159*	*294*	*213*	..
Closing Stock Level	-	*39 067*	*36 011*	*26 090*	*21 238*	*21 136*	*18 442*	*19 352 e*

Source: Annual Oil Statistics.

OECD EUROPE
Table 5C: Total Products - Supply and Consumption

000 metric tons

	1973	1978	1980	1990	1999	2000	2001	2002
Refinery Output	**782 303**	**730 713**	**714 108**	**661 719**	**724 211**	**734 280**	**730 704**	**728 171 e**
+ Recycled Products	193	251	522	564	719	761	716	666 e
+ Imports	156 861	162 684	179 301	211 532	231 122	245 462	253 073	253 107 e
- Exports	146 247	123 341	136 987	181 141	207 249	217 881	217 234	220 049 e
- Intl. Marine Bunkers	44 078	36 964	32 404	36 153	42 724	45 253	46 440	48 243 e
+ Transfers	-4 218	-6 143	-12 113	-26 696	-26 409	-29 957	-29 544	-35 762 e
+ Stock Changes	-4 579	5 052	-6 322	909	9 705	-5 029	2 810	2 247 e
- Statistical Differences	4 912	206	838	1 294	-4 439	-8 066	- 756	-2 335 e
Total Consumption	**735 323**	**732 046**	**705 267**	**629 440**	**693 814**	**690 449**	**694 841**	**682 472 e**
Transformation	**111 737**	**108 280**	**110 055**	**65 528**	**61 080**	**60 184**	**58 688**	..
Public Electricity	86 988	69 692	62 276	37 543	30 556	29 796	28 716	..
Autoproducer Electricity	290	6 360	5 802	2 220	2 926	1 267	921	..
Public CHP Plants	7 155	8 973	12 532	2 599	4 297	5 282	5 049	..
Autoproducers of CHP	1 118	5 864	7 551	5 524	6 002	3 680	3 631	..
Public Heat Plants	688	721	895	1 194	1 165	816	948	..
Autoproducer Heat Plants	250	585	555	338	105	792	746	..
Petrochemical Industry	5 020	10 456	15 643	13 056	13 330	16 185	16 269	..
Other Transformation	10 228	5 629	4 801	3 054	2 699	2 366	2 408	..
Energy Sector	**36 737**	**40 035**	**39 823**	**36 373**	**41 106**	**40 194**	**40 169**	..
Refinery Fuel	35 462	37 934	38 129	34 703	39 658	38 884	39 240	..
Other Energy Sector	1 275	2 101	1 694	1 670	1 448	1 310	929	..
Final Consumption	**586 849**	**583 731**	**555 389**	**527 539**	**591 628**	**590 071**	**595 984**	..
Transport	**173 962**	**204 625**	**211 010**	**276 539**	**337 923**	**339 503**	**340 056**	..
Air	18 911	21 007	21 316	30 158	45 368	47 486	45 747	..
Road	140 004	169 408	174 487	234 731	282 221	282 696	285 229	..
Rail	5 610	5 541	5 115	3 853	3 216	3 082	2 842	..
Inland Waterways	9 093	8 169	9 593	7 474	6 978	6 124	6 088	..
Non-Specified	344	500	499	323	140	115	150	..
Industry	**224 440**	**202 081**	**188 681**	**137 722**	**142 969**	**144 647**	**142 798**	..
Iron and Steel	15 295	11 996	8 279	3 658	2 371	2 571	2 191	..
Chemical (incl. Pet.-Chem.)	62 360	59 295	58 987	61 282	68 304	70 349	70 284	..
Non-Ferrous Metals	2 195	3 176	3 096	2 885	2 831	2 883	2 824	..
Non-Metallic Minerals	23 091	20 220	17 579	11 584	12 706	12 498	12 728	..
Transport Equipment	3 308	3 083	2 665	1 088	896	809	879	..
Machinery	7 904	6 470	5 617	4 046	3 326	3 161	3 343	..
Mining and Quarrying	2 222	1 640	1 624	985	1 185	1 158	1 182	..
Food and Tobacco	14 045	14 195	13 430	6 927	5 355	5 002	5 243	..
Pulp, Paper and Printing	10 439	9 145	8 317	3 602	2 950	2 823	2 841	..
Wood and Wood Products	1 698	1 475	1 289	543	564	497	453	..
Construction	25 520	23 097	22 361	20 997	23 138	23 202	22 806	..
Textiles and Leather	6 741	5 916	5 217	2 928	1 982	1 869	1 811	..
Non-Specified	49 622	42 373	40 220	17 197	17 361	17 825	16 213	..
Other Sectors	**188 447**	**177 025**	**155 698**	**113 278**	**110 736**	**105 921**	**113 130**	..
Commerce and Public	56 882	58 163	49 994	27 440	23 759	21 146	23 959	..
Residential	100 025	93 925	80 020	64 192	64 489	62 235	66 300	..
Agriculture	16 530	18 341	20 304	19 794	21 615	21 677	21 248	..
Non-Specified	15 010	6 596	5 380	1 852	873	863	1 623	..
Memo: Non-Energy Use	*72 264*	*68 491*	*70 876*	*82 753*	*94 389*	*94 432*	*91 844*	..
Chem/Petchem. Feedst.	*40 313*	*39 383*	*41 327*	*52 153*	*61 060*	*61 399*	*60 204*	..
Other	*31 951*	*29 108*	*29 549*	*30 600*	*33 329*	*33 033*	*31 640*	..
Closing Stock Level	-	*117 183*	*131 249*	*103 092*	*103 670*	*108 918*	*106 111*	*103 832 e*

Source: Annual Oil Statistics.

OECD EUROPE
Table 6A: Supply and Consumption

000 metric tons

	1980	2000	2001	2002	1980	2000	2001	2002
	LPG AND ETHANE				**REFINERY GAS**			
Refinery Output	**14 849**	**20 254**	**20 628**	**21 205 e**	**17 053**	**22 179**	**21 873**	**22 218 e**
+ Recycled Products	-	-	-	-	-	-	-	-
+ Imports	5 985	15 371	14 618	13 789 e	-	-	-	-
- Exports	2 853	6 723	6 281	7 269 e	-	-	-	-
- Intl. Marine Bunkers	-	-	-	-	-	-	-	-
+ Transfers	1 098	1 169	1 445	1 583 e	- 143	- 40	- 41	- 118 e
+ Stock Change	- 60	- 214	- 178	344 e	2	3	-	-
- Statistical Difference	- 214	- 428	26	315 e	77	- 108	- 146	120 e
Total Consumption	**19 233**	**30 285**	**30 206**	**29 337 e**	**16 835**	**22 250**	**21 978**	**21 980 e**
of which: Ethane	-	*1 830*	*1 988*	*2 082 e*	-	-	-	-
Transformation	1 529	1 086	1 048	..	910	1 463	995	..
Energy	491	578	546	..	14 350	18 278	18 042	..
Transport	1 683	4 312	4 461	..	-	-	-	..
Industry	6 208	11 774	11 785	..	1 575	2 509	2 941	..
Other	9 322	12 535	12 366	..	-	-	-	..
Memo: Non-Energy Use	*2 552*	*6 780*	*7 143*	..	*481*	*603*	*739*	..
Closing Stock Level	*1 074*	*1 962*	*2 139*	*1 798 e*	-	-	-	-

	1980	2000	2001	2002	1980	2000	2001	2002
	NAPHTHA				**OTHER KEROSENE**			
Refinery Output	**34 301**	**47 015**	**46 084**	**41 157 e**	**6 337**	**4 700**	**4 634**	**4 887 e**
+ Recycled Products	-	-	-	-	-	-	-	-
+ Imports	21 325	27 720	27 101	25 803 e	680	2 039	1 871	1 999 e
- Exports	10 570	20 601	20 236	18 869 e	2 670	819	713	1 108 e
- Intl. Marine Bunkers	-	-	-	-	-	-	-	-
+ Transfers	-2 163	-5 129	-5 912	-3 179 e	653	- 501	- 545	- 711 e
+ Stock Change	- 69	- 297	213	220 e	-	- 188	196	38 e
- Statistical Difference	- 227	-2 593	-1 299	- 400 e	- 82	- 255	- 587	- 359 e
Total Consumption	**43 051**	**51 301**	**48 549**	**45 532 e**	**5 082**	**5 486**	**6 030**	**5 464 e**
Transformation	12 329	9 700	8 569	..	224	121	127	..
Energy	137	124	327	..	24	1	55	..
Transport	-	-	-	..	46	33	33	..
Industry	30 584	41 477	39 653	..	767	1 723	1 962	..
Other	1	-	-	..	4 021	3 608	3 853	..
Memo: Non-Energy Use	*29 472*	*40 316*	*38 483*	..	*319*	*298*	*279*	..
Closing Stock Level	*4 013*	*3 084*	*3 740*	*3 540 e*	*1 782*	*1 218*	*1 022*	*984 e*

	1980	2000	2001	2002	1980	2000	2001	2002
	WHITE SPIRIT + SBP				**LUBRICANTS**			
Refinery Output	**1 784**	**1 370**	**1 304**	**..**	**7 863**	**8 263**	**8 242**	**..**
+ Recycled Products	-	-	-	..	419	517	419	..
+ Imports	529	1 024	1 006	..	3 310	3 956	3 625	..
- Exports	997	799	856	..	3 750	5 208	5 808	..
- Intl. Marine Bunkers	-	-	-	..	274	342	359	..
+ Transfers	- 48	- 346	- 387	..	- 390	- 716	- 719	..
+ Stock Change	8	- 73	- 6	..	- 329	- 281	73	..
- Statistical Difference	- 158	-	- 109	..	- 46	- 568	- 862	..
Total Consumption	**1 434**	**1 176**	**1 170**	**..**	**6 895**	**6 757**	**6 335**	**..**
Transformation	7	3	5	..	1	-	-	..
Energy	38	-	1	..	8	28	23	..
Transport	-	-	-	..	2 148	3 127	3 125	..
Industry	1 282	957	978	..	4 537	3 348	2 869	..
Other	107	216	186	..	201	254	318	..
Memo: Non-Energy Use	*1 388*	*1 169*	*1 141*	..	*6 884*	*6 756*	*6 335*	..
Closing Stock Level	*376*	*224*	*230*	..	*2 933*	*1 743*	*1 672*	..

Source: Annual Oil Statistics.

OECD EUROPE
Table 6B: Supply and Consumption

000 metric tons

	1980	2000	2001	2002	1980	2000	2001	2002
	BITUMEN				PARAFFIN WAXES			
Refinery Output	**18 317**	**20 880**	**20 294**	..	**823**	**1 643**	**1 530**	..
+ Recycled Products	-	-	-	..	-	-	-	..
+ Imports	1 772	3 884	4 317	..	412	632	640	..
- Exports	2 306	4 256	4 330	..	316	654	688	..
- Intl. Marine Bunkers	-	-	-	..	-	-	-	..
+ Transfers	- 5	1 118	984	..	- 13	- 453	- 368	..
+ Stock Change	- 21	- 29	3	..	- 25	20	- 21	..
- Statistical Difference	89	142	516	..	104	107	131	..
Total Consumption	**17 668**	**21 455**	**20 752**	**..**	**777**	**1 081**	**962**	**..**
Transformation	111	1 257	1 105	..	6	-	-	..
Energy	7	-	-	..	-	4	-	..
Transport	-	-	-	..	-	-	-	..
Industry	17 550	20 125	19 564	..	771	1 001	900	..
Other	-	73	83	..	-	76	62	..
Memo: Non-Energy Use	*17 550*	*20 048*	*19 590*	..	*750*	*1 077*	*962*	..
Closing Stock Level	*1 210*	*1 401*	*1 399*	..	*96*	*119*	*140*	..
	PETROLEUM COKE				OTHER PRODUCTS			
Refinery Output	**3 067**	**8 101**	**7 582**	**7 455 e**	**8 194**	**21 108**	**24 480**	**60 435 e**
+ Recycled Products	-	-	-	-	94	37	27	264 e
+ Imports	5 517	10 893	11 174	10 549 e	898	5 829	5 014	14 738 e
- Exports	537	1 284	1 363	1 425 e	2 447	6 252	6 364	19 077 e
- Intl. Marine Bunkers	-	-	-	-	-	23	29	316 e
+ Transfers	-	- 76	- 5	- 64 e	-1 427	-8 420	-10 825	-17 280 e
+ Stock Change	- 114	- 516	- 40	169 e	- 401	-1 345	- 23	572 e
- Statistical Difference	472	236	417	83 e	3	-1 486	-1 307	-2 723 e
Total Consumption	**7 461**	**16 882**	**16 931**	**16 601 e**	**4 908**	**12 420**	**13 587**	**42 059 e**
Transformation	806	1 217	1 218	..	153	437	660	..
Energy	1 530	4 196	3 847	..	1 498	3 105	3 123	..
Transport	-	-	-	..	-	-	-	..
Industry	5 125	11 395	11 758	..	3 252	8 852	9 024	..
Other	-	74	108	..	5	26	780	..
Memo: Non-Energy Use	*2 240*	*2 102*	*2 061*	..	*2 979*	*7 596*	*7 752*	..
Closing Stock Level	*738*	*2 467*	*2 508*	*2 335 e*	*2 850*	*3 363*	*3 385*	*6 249 e*

Source: Annual Oil Statistics.

OECD EUROPE
Table 7A: Imports by Country of Origin

000 metric tons

	2000	2001	2002	2000	2001	2002	2000	2001	2002
	Crude Oil			**Natural Gas Liquids**			**Feedstocks and Other**		
Australia	-	-	-	-	-	-	-	-	-
Belgium	2	2	-	-	-	2 121 e	420	415	205 e
Canada	80	-	-	-	-	-	-	-	-
Czech Republic	112	124	155 e	-	-	-	8	1	18 e
Denmark	11 721	10 104	10 159 e	-	-	-	50	23	30 e
Finland	-	-	-	-	-	-	34	31	2 e
France	501	372	8 e	-	-	-	319	385	492 e
Germany	1 594	280	289 e	11	35	45 e	367	562	638 e
Greece	-	-	-	-	-	-	42	114	130 e
Hungary	8	-	-	-	-	-	200	154	223 e
Ireland	-	-	-	-	-	-	313	551	390 e
Italy	204	250	905 e	-	-	-	58	57	27 e
Japan	-	-	-	-	-	-	1	-	-
Korea	-	-	-	-	-	-	-	-	-
Mexico	9 770	9 291	9 508 e	-	-	-	-	-	-
Netherlands	281	334	1 276 e	-	-	-	2 120	1 709	1 998 e
New Zealand	-	-	-	-	-	-	-	-	-
Norway	114 905	107 155	100 004 e	999	966	1 356 e	736	897	720 e
Poland	163	40	206 e	-	-	-	137	74	26 e
Spain	23	-	-	-	-	-	5	8	167 e
Sweden	89	91	80 e	-	11	-	684	404	419 e
Turkey	213	-	-	-	-	-	326	555	215 e
United Kingdom	45 762	49 751	42 985 e	48	45	22 e	419	366	346 e
United States	-	-	82 e	-	-	-	74	729	529 e
Other	156	78	15 e	-	-	-	328	68	73 e
OECD	**185 584**	**177 872**	**165 672 e**	**1 058**	**1 057**	**3 544 e**	**6 641**	**7 103**	**6 648 e**
Argentina	131	92	188 e	-	-	-	-	10	-
Brazil	133	1 412	1 533 e	-	-	-	8	10	20 e
Colombia	-	75	455 e	-	-	-	-	-	-
Ecuador	-	-	-	-	-	-	-	-	-
Netherlands Antilles	99	238	115 e	-	-	-	25	-	-
Peru	-	-	-	-	-	-	-	-	-
Trinidad and Tobago	-	-	-	-	-	-	-	-	-
Venezuela	6 946	9 023	9 038 e	-	-	-	1 214	1 210	706 e
Other	150	-	-	-	-	-	34	13	-
Latin America	**7 459**	**10 840**	**11 329 e**	**-**	**-**	**-**	**1 281**	**1 243**	**726 e**
Romania	19	118	373 e	-	-	-	94	15	22 e
Other	-	61	30 e	-	-	-	81	134	134 e
Non-OECD Europe	**19**	**179**	**403 e**	**-**	**-**	**-**	**175**	**149**	**156 e**
FSU	**120 711**	**139 812**	**159 646 e**	**747**	**1 476**	**1 852 e**	**8 728**	**8 105**	**10 268 e**
Iran	39 609	36 714	30 904 e	-	-	-	30	14	30 e
Iraq	35 337	20 452	15 678 e	-	-	-	-	-	-
Kuwait	9 768	7 990	5 916 e	-	-	-	-	-	-
Oman	-	-	184 e	-	-	-	-	-	-
Qatar	134	-	28 e	-	-	-	-	-	-
Saudi Arabia	68 869	61 018	57 588 e	-			62	110	152 e
United Arab Emirates	147	354	-	-	-	-	-	-	-
Other	15 305	21 149	22 633 e	-	-	-	327	1 092	475 e
Middle East	**169 169**	**147 677**	**132 931 e**	**-**	**-**	**-**	**419**	**1 219**	**657 e**
India	-	-	-	-	-	-	-	-	-
Indonesia	-	-	-	-	-	-	-	-	-
Singapore	-	-	-	-	-	-	-	-	-
Other	-	-	-	-	-	-	-	-	-
Asia	**-**	**-**	**-**	**-**	**-**	**-**	**-**	**-**	**-**
China	**-**	**-**	**-**	**-**	**-**	**-**	**-**	**9**	**-**
Algeria	17 424	15 283	13 206 e	4 700	4 615	3 761 e	463	636	668 e
Angola/Cabinada	3 862	6 657	6 997 e	-	-	-	-	-	-
Cameroon	3 070	3 669	3 951 e	-	-	-	-	-	-
Egypt	5 467	3 525	4 102 e	-	-	-	207	285	130 e
Gabon	562	1 643	573 e	-	-	-	-	-	-
Libya	50 910	50 528	44 909 e	25	25	75 e	2 086	1 681	1 356 e
Nigeria	24 119	27 201	20 317 e	-	-	-	-	18	247 e
Tunisia	1 346	1 247	2 192 e	105	70	129 e	239	235	428 e
Other	3 689	5 188	6 620 e	-	35	-	589	516	302 e
Africa	**110 449**	**114 941**	**102 867 e**	**4 830**	**4 745**	**3 965 e**	**3 584**	**3 371**	**3 131 e**
Non-spec./Other	**12 742**	**13 847**	**20 559 e**	**-**	**493**	**73 e**	**1 006**	**1 124**	**1 267 e**
Total Imports	**606 133**	**605 168**	**593 407 e**	**6 635**	**7 771**	**9 434 e**	**21 834**	**22 323**	**22 853 e**
Memo: Total OPEC	*253 263*	*228 563*	*197 584 e*	*4 725*	*4 640*	*3 836 e*	*3 855*	*3 672*	*3 159 e*

Source: Annual Oil Statistics.

OECD EUROPE
Table 7B: Imports by Country of Origin

000 metric tons

	2000	2001	2002	2000	2001	2002	2000	2001	2002
	Total Products			**Liquefied Petroleum Gases**			**Naphtha**		
Australia	15	1	25 e	15	-	-	-	-	-
Belgium	17 855	17 523	17 663 e	462	436	424 e	1 525	1 062	1 068 e
Canada	534	537	466 e	-	-	-	15	4	22 e
Czech Republic	919	995	1 093 e	28	14	36 e	25	24	8 e
Denmark	2 811	2 551	2 513 e	68	65	58 e	96	39	80 e
Finland	3 520	2 986	2 933 e	27	10	22 e	74	78	160 e
France	11 864	11 035	9 448 e	441	372	331 e	1 213	1 255	764 e
Germany	14 343	12 647	12 314 e	649	497	482 e	1 403	981	1 039 e
Greece	916	881	853 e	16	1	49 e	393	335	413 e
Hungary	908	1 171	1 137 e	35	34	33 e	-	-	-
Ireland	208	177	59 e	6	5	7 e	76	10	-
Italy	8 276	9 868	9 949 e	105	90	79 e	295	682	612 e
Japan	85	77	85 e	-	-	2 e	14	12	28 e
Korea	53	56	223 e	-	-	-	42	49	75 e
Mexico	9	112	80 e	-	-	-	-	105	31 e
Netherlands	43 379	42 019	37 442 e	489	475	466 e	6 391	5 518	5 065 e
New Zealand	-	-	-	-	-	-	-	-	-
Norway	7 435	5 414	7 132 e	1 749	1 978	1 990 e	693	535	546 e
Poland	1 366	1 636	1 648 e	29	7	5 e	75	95	76 e
Spain	3 443	2 634	2 548 e	93	57	57 e	1 195	1 134	1 006 e
Sweden	7 889	6 930	6 781 e	135	76	140 e	140	245	217 e
Turkey	580	836	467 e	6	2	-	317	431	327 e
United Kingdom	22 677	21 979	22 501 e	2 523	2 154	2 438 e	2 516	2 143	2 353 e
United States	12 036	10 611	10 893 e	13	6	45 e	491	323	650 e
Other	5 151	5 393	5 891 e	50	53	84 e	158	109	149 e
OECD	**166 272**	**158 069**	**154 144 e**	**6 939**	**6 332**	**6 748 e**	**17 147**	**15 169**	**14 689 e**
Argentina	62	109	216 e	-	-	11 e	6	1	20 e
Brazil	285	65	93 e	-	-	-	35	20	14 e
Colombia	14	-	118 e	2	-	-	-	-	-
Ecuador	58	-	-	-	-	-	-	-	-
Netherlands Antilles	440	302	178 e	3	-	-	2	5	61 e
Peru	-	-	30 e	-	-	-	-	-	-
Trinidad and Tobago	811	466	604 e	9	8	16 e	-	-	-
Venezuela	3 737	2 276	2 653 e	29	-	37 e	-	-	26 e
Other	617	616	818 e	32	-	13 e	33	82	26 e
Latin America	**6 024**	**3 834**	**4 710 e**	**75**	**8**	**77 e**	**76**	**108**	**147 e**
Romania	275	597	1 345 e	-	-	-	-	1	31 e
Other	1 195	1 244	1 415 e	117	125	133 e	169	158	204 e
Non-OECD Europe	**1 470**	**1 841**	**2 760 e**	**117**	**125**	**133 e**	**169**	**159**	**235 e**
FSU	**34 126**	**35 992**	**42 245 e**	**901**	**812**	**1 122 e**	**1 683**	**1 683**	**2 265 e**
Iran	129	174	31 e	29	-	-	37	88	-
Iraq	-	-	-	-	-	-	-	-	-
Kuwait	2 905	2 162	1 263 e	425	332	42 e	866	109	-
Oman	-	82	-	-	-	-	-	-	-
Qatar	22	24	45 e	-	-	-	-	-	-
Saudi Arabia	1 887	2 050	1 473 e	308	28	11 e	186	108	219 e
United Arab Emirates	1 224	2 413	2 229 e	-	32	-	41	-	4 e
Other	2 431	2 001	1 320 e	26	16	24 e	316	265	182 e
Middle East	**8 598**	**8 906**	**6 361 e**	**788**	**408**	**77 e**	**1 446**	**570**	**405 e**
India	248	733	365 e	-	-	-	53	3	10 e
Indonesia	154	284	554 e	-	-	-	1	-	-
Singapore	67	77	1 193 e	-	-	-	26	12	3 e
Other	235	156	310 e	36	-	-	6	-	15 e
Asia	**704**	**1 250**	**2 422 e**	**36**	**-**	**-**	**86**	**15**	**28 e**
China	**58**	**100**	**73 e**	**-**	**3**	**-**	**3**	**-**	**-**
Algeria	10 617	10 310	9 945 e	5 311	4 621	4 225 e	2 435	3 220	3 385 e
Angola/Cabinada	3	-	-	-	-	-	-	-	-
Cameroon	7	-	-	-	-	-	-	-	-
Egypt	1 255	1 334	1 869 e	54	214	160 e	1 059	559	710 e
Gabon	-	-	4 e	-	-	4 e	-	-	-
Libya	5 806	6 236	5 461 e	221	276	124 e	1 497	2 071	1 815 e
Nigeria	61	293	127 e	12	110	-	21	77	27 e
Tunisia	263	78	268 e	-	-	-	45	52	114 e
Other	1 046	530	1 295 e	36	45	66 e	339	208	204 e
Africa	**19 058**	**18 781**	**18 969 e**	**5 634**	**5 266**	**4 579 e**	**5 396**	**6 187**	**6 255 e**
Non-spec./Other	**9 152**	**24 300**	**21 423 e**	**881**	**1 664**	**1 053 e**	**1 714**	**3 210**	**1 779 e**
Total Imports	**245 462**	**253 073**	**253 107 e**	**15 371**	**14 618**	**13 789 e**	**27 720**	**27 101**	**25 803 e**
Memo: Total OPEC	*26 542*	*26 222*	*23 781 e*	*6 335*	*5 399*	*4 439 e*	*5 084*	*5 673*	*5 476 e*

Source: Annual Oil Statistics.

OECD EUROPE
Table 7C: Imports by Country of Origin

000 metric tons

	2000	2001	2002	2000	2001	2002	2000	2001	2002
	Total Gasoline			**Total Kerosene**			**Gas/Diesel**		
Australia	-	-	-	-	-	-	-	-	-
Belgium	3 754	3 463	3 339 e	1 329	1 414	1 351 e	6 027	7 184	6 089 e
Canada	1	1	-	-	-	54 e	381	465	330 e
Czech Republic	103	196	280 e	7	-	-	462	596	618 e
Denmark	1 074	812	752 e	292	324	403 e	835	740	485 e
Finland	1 314	1 382	1 257 e	184	4	43 e	1 794	1 416	1 333 e
France	2 940	2 430	2 408 e	1 080	1 136	742 e	2 159	1 745	1 259 e
Germany	3 406	2 858	2 199 e	387	458	787 e	4 938	4 538	5 220 e
Greece	287	162	167 e	129	63	19 e	35	292	164 e
Hungary	267	375	323 e	-	-	-	505	613	708 e
Ireland	18	43	14 e	8	1	11 e	58	49	19 e
Italy	934	1 109	962 e	352	392	331 e	4 959	6 669	6 896 e
Japan	8	-	-	29	-	-	-	1	-
Korea	-	-	-	-	-	-	-	-	148 e
Mexico	-	-	-	-	-	-	-	-	-
Netherlands	8 503	7 155	5 145 e	3 922	3 308	2 919 e	18 266	20 613	18 160 e
New Zealand	-	-	-	-	-	-	-	-	-
Norway	1 911	944	1 309 e	145	47	113 e	1 518	1 274	2 176 e
Poland	262	408	390 e	139	301	157 e	65	113	352 e
Spain	533	290	198 e	31	5	139 e	320	367	343 e
Sweden	2 016	1 437	1 403 e	32	19	125 e	3 965	3 625	3 262 e
Turkey	-	9	73 e	81	27	-	3	-	-
United Kingdom	3 795	3 482	2 791 e	1 358	1 518	1 537 e	6 256	5 124	6 394 e
United States	97	11	41 e	90	66	392 e	791	492	961 e
Other	1 355	1 629	1 710 e	86	21	138 e	1 845	2 050	2 226 e
OECD	**32 578**	**28 196**	**24 761 e**	**9 681**	**9 104**	**9 261 e**	**55 182**	**57 966**	**57 143 e**
Argentina	-	-	-	-	-	-	-	-	-
Brazil	-	-	-	-	-	-	-	-	-
Colombia	-	-	-	12	-	26 e	-	-	-
Ecuador	-	-	-	-	-	-	-	-	-
Netherlands Antilles	103	8	-	123	29	-	50	25	18 e
Peru	-	-	-	-	-	-	-	-	-
Trinidad and Tobago	64	107	103 e	97	100	87 e	110	127	95 e
Venezuela	-	-	-	365	285	803 e	5	5	33 e
Other	166	47	3 e	101	135	92 e	46	73	135 e
Latin America	**333**	**162**	**106 e**	**698**	**549**	**1 008 e**	**211**	**230**	**281 e**
Romania	44	90	197 e	-	-	-	207	443	989 e
Other	152	198	225 e	45	21	28 e	414	424	366 e
Non-OECD Europe	**196**	**288**	**422 e**	**45**	**21**	**28 e**	**621**	**867**	**1 355 e**
FSU	**1 335**	**822**	**1 166 e**	**1 051**	**923**	**1 343 e**	**15 451**	**17 972**	**20 501 e**
Iran	-	-	-	-	48	-	31	-	-
Iraq	-	-	-	-	-	-	-	-	-
Kuwait	-	-	-	1 488	1 652	1 221 e	126	-	-
Oman	-	-	-	-	82	-	-	-	-
Qatar	-	-	-	20	20	-	-	-	45 e
Saudi Arabia	171	87	156 e	908	1 462	800 e	267	240	209 e
United Arab Emirates	47	14	35 e	1 110	2 280	2 051 e	25	45	34 e
Other	107	78	137 e	691	738	466 e	108	148	142 e
Middle East	**325**	**179**	**328 e**	**4 217**	**6 282**	**4 538 e**	**557**	**442**	**490 e**
India	38	-	-	-	-	37 e	-	384	281 e
Indonesia	-	-	-	-	-	-	-	80	-
Singapore	-	-	7 e	-	-	240 e	38	33	906 e
Other	-	-	-	10	-	7 e	157	48	216 e
Asia	**38**	**-**	**7 e**	**10**	**-**	**284 e**	**195**	**545**	**1 403 e**
China	**-**	**-**	**-**	**-**	**-**	**-**	**-**	**-**	**-**
Algeria	268	86	426 e	742	451	407 e	1 646	1 777	1 024 e
Angola/Cabinada	-	-	-	-	-	-	-	-	-
Cameroon	-	-	-	-	-	-	-	-	-
Egypt	63	140	565 e	-	167	301 e	-	-	-
Gabon	-	-	-	-	-	-	-	-	-
Libya	84	37	148 e	575	1 017	515 e	755	487	548 e
Nigeria	-	-	-	28	96	-	-	-	7 e
Tunisia	-	-	41 e	74	4	-	-	22	104 e
Other	2	5	29 e	70	30	84 e	63	45	93 e
Africa	**417**	**268**	**1 209 e**	**1 489**	**1 765**	**1 307 e**	**2 464**	**2 331**	**1 776 e**
Non-spec./Other	**1 400**	**3 895**	**3 029 e**	**760**	**2 362**	**1 339 e**	**1 794**	**7 834**	**7 900 e**
Total Imports	**36 622**	**33 810**	**31 028 e**	**17 951**	**21 006**	**19 108 e**	**76 475**	**88 187**	**90 849 e**
Memo: Total OPEC	*570*	*224*	*765 e*	*5 236*	*7 311*	*5 797 e*	*2 855*	*2 643*	*1 960 e*

Source: Annual Oil Statistics.

OECD EUROPE
Table 7D: Imports by Country of Origin

000 metric tons

	2000	2001	2002	2000	2001	2002	2000	2001	2002
	Fuel Oil (Residual)			**Petroleum Coke**			**Other Products**		
Australia	-	-	25 e	-	-	-	-	1	-
Belgium	2 881	2 239	3 133 e	333	279	398 e	1 544	1 446	1 861 e
Canada	74	31	52 e	8	20	3 e	55	16	5 e
Czech Republic	117	74	55 e	-	-	-	177	91	96 e
Denmark	390	445	663 e	6	11	6 e	50	115	66 e
Finland	67	32	-	1	-	-	59	64	118 e
France	2 130	2 346	2 415 e	5	5	5 e	1 896	1 746	1 524 e
Germany	2 085	1 734	1 057 e	186	167	251 e	1 289	1 414	1 279 e
Greece	27	25	32 e	-	-	-	29	3	9 e
Hungary	4	89	-	-	2	11 e	97	58	62 e
Ireland	38	69	8 e	-	-	-	4	-	-
Italy	1 021	408	468 e	67	14	117 e	543	504	484 e
Japan	-	-	-	25	37	38 e	9	27	17 e
Korea	-	-	-	-	-	-	11	7	-
Mexico	-	-	47 e	-	-	-	9	7	2 e
Netherlands	3 938	2 993	3 929 e	488	637	534 e	1 382	1 320	1 224 e
New Zealand	-	-	-	-	-	-	-	-	-
Norway	514	540	787 e	34	1	32 e	871	95	179 e
Poland	663	585	557 e	-	-	-	133	127	111 e
Spain	641	309	181 e	62	18	89 e	568	454	535 e
Sweden	1 077	1 042	1 255 e	7	21	-	517	465	379 e
Turkey	80	311	40 e	-	-	-	93	56	27 e
United Kingdom	4 456	5 767	5 281 e	335	668	420 e	1 438	1 123	1 287 e
United States	815	1 157	1 172 e	7 859	6 470	5 985 e	1 880	2 086	1 647 e
Other	1 220	1 106	1 149 e	16	14	14 e	421	411	421 e
OECD	**22 238**	**21 302**	**22 306 e**	**9 432**	**8 364**	**7 903 e**	**13 075**	**11 636**	**11 333 e**
Argentina	-	-	-	32	68	64 e	24	40	121 e
Brazil	239	-	45 e	-	4	9 e	11	41	25 e
Colombia	-	-	92 e	-	-	-	-	-	-
Ecuador	-	-	-	58	-	-	-	-	-
Netherlands Antilles	143	229	79 e	-	-	-	16	6	20 e
Peru	-	-	30 e	-	-	-	-	-	-
Trinidad and Tobago	530	124	303 e	-	-	-	1	-	-
Venezuela	2 351	500	125 e	959	1 442	1 571 e	28	44	58 e
Other	219	103	355 e	-	115	169 e	20	61	25 e
Latin America	**3 482**	**956**	**1 029 e**	**1 049**	**1 629**	**1 813 e**	**100**	**192**	**249 e**
Romania	8	49	106 e	2	-	1 e	14	14	21 e
Other	249	271	393 e	12	5	18 e	37	42	48 e
Non-OECD Europe	**257**	**320**	**499 e**	**14**	**5**	**19 e**	**51**	**56**	**69 e**
FSU	**13 095**	**12 928**	**14 993 e**	**14**	**21**	**28 e**	**596**	**831**	**827 e**
Iran	13	-	-	-	-	1 e	19	38	30 e
Iraq	-	-	-	-	-	-	-	-	-
Kuwait	-	69	-	-	-	-	-	-	-
Oman	-	-	-	-	-	-	-	-	-
Qatar	-	-	-	-	-	-	2	4	-
Saudi Arabia	9	85	-	-	-	-	38	31	18 e
United Arab Emirates	-	42	105 e	-	-	-	1	-	-
Other	1 113	744	337 e	55	-	-	15	12	32 e
Middle East	**1 135**	**940**	**442 e**	**55**	**-**	**1 e**	**75**	**85**	**80 e**
India	-	-	-	143	268	35 e	14	78	2 e
Indonesia	42	-	468 e	69	62	16 e	42	142	70 e
Singapore	2	13	19 e	-	-	-	1	19	18 e
Other	-	-	24 e	-	78	-	26	30	48 e
Asia	**44**	**13**	**511 e**	**212**	**408**	**51 e**	**83**	**269**	**138 e**
China	**-**	**-**	**8 e**	**27**	**47**	**31 e**	**28**	**50**	**34 e**
Algeria	156	107	445 e	-	-	-	59	48	33 e
Angola/Cabinada	-	-	-	-	-	-	3	-	-
Cameroon	-	-	-	-	-	-	7	-	-
Egypt	60	247	109 e	-	-	-	19	7	24 e
Gabon	-	-	-	-	-	-	-	-	-
Libya	2 548	2 143	2 257 e	-	-	-	126	205	54 e
Nigeria	-	10	93 e	-	-	-	-	-	-
Tunisia	144	-	-	-	-	-	-	-	9 e
Other	412	115	693 e	77	21	21 e	47	61	105 e
Africa	**3 320**	**2 622**	**3 597 e**	**77**	**21**	**21 e**	**261**	**321**	**225 e**
Non-spec./Other	**1 534**	**3 494**	**3 858 e**	**13**	**679**	**682 e**	**1 056**	**1 162**	**1 783 e**
Total Imports	**45 105**	**42 575**	**47 243 e**	**10 893**	**11 174**	**10 549 e**	**15 325**	**14 602**	**14 738 e**
Memo: Total OPEC	*5 119*	*2 956*	*3 493 e*	*1 028*	*1 504*	*1 588 e*	*315*	*512*	*263 e*

Source: Annual Oil Statistics.

OECD EUROPE
Table 8A: Exports by Country of Destination

000 metric tons

	2000	2001	2002	2000	2001	2002	2000	2001	2002
	Crude Oil			Natural Gas Liquids			Feedstocks and Other		
Australia	-	-	-	-	32	-	-	-	-
Austria	31	62	179 e	-	-	-	-	-	-
Belgium	5 393	4 278	2 571 e	557	749	804 e	393	374	262 e
Canada	17 262	16 404	14 395 e	-	-	16 e	-	34	-
Denmark	4 093	2 813	3 239 e	14	23	29 e	27	28	40 e
Finland	5 553	5 980	5 549 e	83	101	26 e	1	4	20 e
France	36 996	36 090	27 677 e	851	778	924 e	622	299	256 e
Germany	21 118	24 632	23 472 e	250	245	192 e	810	458	502 e
Greece	-	-	134 e	-	4	-	-	28	-
Ireland	2 875	3 174	2 387 e	9	15	14 e	3	-	24 e
Italy	4 112	3 371	6 064 e	26	2	24 e	134	56	64 e
Japan	134	-	10 e	-	26	180 e	-	-	-
Luxembourg	-	-	-	-	-	-	-	-	-
Mexico	-	-	-	202	235	142 e	-	-	-
Netherlands	47 206	47 164	43 930 e	1 008	1 368	1 498 e	2 472	1 963	1 469 e
Norway	897	362	723 e	215	41	75 e	50	49	36 e
Portugal	1 610	1 434	1 437 e	201	209	211 e	16	9	21 e
Spain	2 758	2 845	2 239 e	456	362	360 e	241	86	134 e
Sweden	14 746	12 837	10 914 e	579	998	968 e	39	36	168 e
Switzerland	-	-	-	-	-	-	28	39	46 e
Turkey	-	-	-	495	493	579 e	52	66	52 e
United Kingdom	33 883	35 871	35 046 e	175	334	230 e	149	146	182 e
United States	43 563	39 147	48 826 e	134	599	274 e	3 021	2 180	1 718 e
Other OECD	1 668	453	1 236 e	84	60	70 e	71	68	19 e
OECD	**243 898**	**236 917**	**230 028 e**	**5 339**	**6 674**	**6 616 e**	**8 129**	**5 923**	**5 013 e**
Africa	163	32	264 e	31	46	95 e	174	103	73 e
Latin America	224	435	1 042 e	179	175	3 e	24	25	-
Asia	1 893	82	343 e	-	24	80 e	15	-	-
China	783	2 647	3 916 e	-	153	326 e	-	-	-
Non-OECD Europe	66	534	542 e	-	-	41 e	98	150	29 e
Former Soviet Union	251	-	1 e	-	-	6 e	7	-	-
Middle East	-	-	-	-	9	-	19	-	27 e
Non-specified/Other	131	1 564	3 173 e	408	349	347 e	797	737	735 e
Non-OECD	**3 511**	**5 294**	**9 281 e**	**618**	**756**	**898 e**	**1 134**	**1 015**	**864 e**
Total Exports	**247 409**	**242 211**	**239 309 e**	**5 957**	**7 430**	**7 514 e**	**9 263**	**6 938**	**5 877 e**

	2000	2001	2002	2000	2001	2002	2000	2001	2002
	Total Products			Liquefied Petroleum Gas			Naphtha		
Australia	79	16	72 e	-	-	-	-	-	3 e
Austria	4 700	4 179	5 377 e	126	114	115 e	43	35	15 e
Belgium	20 189	22 942	20 657 e	805	937	1 117 e	3 196	3 146	3 101 e
Canada	680	743	1 019 e	4	-	-	-	-	1 e
Denmark	3 969	3 770	3 700 e	19	9	1 e	165	157	146 e
Finland	1 430	1 324	1 483 e	24	36	39 e	152	64	103 e
France	15 779	14 075	13 986 e	183	291	339 e	1 589	1 241	1 359 e
Germany	39 905	39 401	31 324 e	610	432	579 e	6 739	6 545	4 903 e
Greece	1 090	906	860 e	2	-	2 e	70	50	49 e
Ireland	4 889	4 545	4 972 e	88	51	49 e	-	9	-
Italy	6 445	5 242	5 637 e	304	248	283 e	485	647	602 e
Japan	72	88	312 e	-	-	3 e	37	29	60 e
Luxembourg	2 465	2 499	2 552 e	18	17	15 e	-	-	-
Mexico	236	126	452 e	-	29	15 e	-	-	32 e
Netherlands	14 033	13 712	17 106 e	667	332	418 e	2 356	2 636	3 497 e
Norway	4 755	4 977	5 354 e	129	150	174 e	37	125	52 e
Portugal	3 011	2 136	2 932 e	225	164	188 e	427	248	233 e
Spain	8 561	8 347	9 822 e	152	112	153 e	272	419	294 e
Sweden	5 058	4 914	4 761 e	71	106	195 e	402	295	269 e
Switzerland	7 659	6 817	7 296 e	68	47	75 e	96	67	43 e
Turkey	2 276	566	1 019 e	295	96	338 e	142	11	39 e
United Kingdom	14 101	18 018	14 726 e	475	489	446 e	1 432	1 785	1 743 e
United States	19 943	19 949	20 289 e	158	143	202 e	903	850	649 e
Other OECD	6 062	5 968	6 834 e	312	265	331 e	190	254	172 e
OECD	**187 387**	**185 260**	**182 542 e**	**4 735**	**4 068**	**5 077 e**	**18 733**	**18 613**	**17 365 e**
Africa	8 298	7 583	7 796 e	879	792	712 e	81	140	68 e
Latin America	1 471	1 276	1 977 e	122	200	22 e	59	30	43 e
Asia	2 260	2 755	3 122 e	11	4	15 e	32	204	215 e
China	188	165	195 e	19	90	23 e	30	-	29 e
Non-OECD Europe	8 841	7 922	7 533 e	233	200	323 e	290	299	152 e
Former Soviet Union	578	502	358 e	21	-	4 e	15	19	24 e
Middle East	2 711	2 179	1 899 e	251	172	60 e	155	101	97 e
Non-specified/Other	6 147	9 592	14 627 e	452	755	1 033 e	1 206	830	876 e
Non-OECD	**30 494**	**31 974**	**37 507 e**	**1 988**	**2 213**	**2 192 e**	**1 868**	**1 623**	**1 504 e**
Total Exports	**217 881**	**217 234**	**220 049 e**	**6 723**	**6 281**	**7 269 e**	**20 601**	**20 236**	**18 869 e**

Source: Annual Oil Statistics.

OECD EUROPE
Table 8B: Exports by Country of Destination

000 metric tons

	2000	2001	2002	2000	2001	2002	2000	2001	2002
	Total Gasoline			**Total Kerosene**			**Gas/Diesel Oil**		
Australia	67	6	19 e	-	1	8 e	4	-	29 e
Austria	743	567	727 e	47	39	89 e	2 964	2 649	3 728 e
Belgium	1 825	1 761	1 461 e	1 046	1 370	895 e	6 651	7 284	5 828 e
Canada	483	541	782 e	-	13	-	4	103	30 e
Denmark	810	741	810 e	108	139	171 e	1 618	1 717	1 401 e
Finland	322	235	297 e	1	12	6 e	316	361	320 e
France	1 722	1 546	1 536 e	238	134	334 e	7 932	7 176	6 826 e
Germany	10 398	8 436	6 603 e	3 225	3 517	3 305 e	14 066	16 194	11 670 e
Greece	290	23	143 e	2	22	6 e	415	637	471 e
Ireland	799	983	1 163 e	641	606	772 e	1 671	1 521	1 788 e
Italy	505	589	579 e	64	42	140 e	256	125	633 e
Japan	-	-	207 e	-	-	-	-	43	-
Luxembourg	621	567	565 e	313	280	370 e	1 431	1 515	1 521 e
Mexico	75	39	323 e	-	-	-	35	-	76 e
Netherlands	2 857	3 078	3 973 e	191	195	208 e	3 134	3 032	4 323 e
Norway	537	751	830 e	191	316	293 e	1 061	1 217	1 439 e
Portugal	264	30	46 e	42	17	9 e	724	448	513 e
Spain	1 060	909	883 e	195	126	68 e	4 250	4 394	5 471 e
Sweden	1 670	1 650	1 621 e	599	488	381 e	1 330	1 246	1 305 e
Switzerland	2 827	2 382	2 602 e	1 077	990	805 e	3 065	2 813	3 199 e
Turkey	403	128	68 e	60	32	75 e	577	144	188 e
United Kingdom	3 254	5 175	3 209 e	1 261	552	849 e	3 268	4 306	3 367 e
United States	7 967	9 056	11 274 e	357	741	292 e	1 742	1 029	1 256 e
Other OECD	1 899	1 986	2 277 e	111	105	42 e	2 395	2 148	2 701 e
OECD	**41 398**	**41 179**	**41 998 e**	**9 769**	**9 737**	**9 118 e**	**58 909**	**60 102**	**58 083 e**
Africa	2 383	2 038	2 514 e	327	205	247 e	2 674	2 454	2 455 e
Latin America	457	518	865 e	22	1	19 e	206	225	618 e
Asia	54	2	192 e	3	11	-	67	105	110 e
China	-	-	-	-	-	-	40	-	30 e
Non-OECD Europe	1 335	1 357	1 796 e	522	416	515 e	2 320	2 072	2 245 e
Former Soviet Union	267	140	64 e	2	2	5 e	98	131	58 e
Middle East	1 307	715	933 e	7	8	-	281	196	371 e
Non-specified/Other	2 129	2 494	3 062 e	121	309	321 e	424	2 298	5 733 e
Non-OECD	**7 932**	**7 264**	**9 426 e**	**1 004**	**952**	**1 107 e**	**6 110**	**7 481**	**11 620 e**
Total Exports	**49 330**	**48 443**	**51 424 e**	**10 773**	**10 689**	**10 225 e**	**65 019**	**67 583**	**69 703 e**
	Fuel Oil (Residual)			**Petroleum Coke**			**Other Products**		
Australia	-	-	-	-	-	-	8	9	13 e
Austria	310	337	256 e	31	30	16 e	436	408	431 e
Belgium	3 554	5 055	4 945 e	24	33	35 e	3 088	3 356	3 275 e
Canada	183	78	167 e	-	-	17 e	6	8	22 e
Denmark	963	707	869 e	-	4	-	286	296	302 e
Finland	389	440	545 e	-	-	-	226	176	173 e
France	2 561	2 302	1 960 e	251	147	231 e	1 303	1 238	1 401 e
Germany	2 458	1 881	2 049 e	73	96	69 e	2 336	2 300	2 146 e
Greece	144	33	49 e	45	54	3 e	122	87	137 e
Ireland	1 292	1 041	859 e	2	5	3 e	396	329	338 e
Italy	4 499	2 787	2 620 e	45	45	22 e	287	759	758 e
Japan	5	-	21 e	1	1	-	29	15	21 e
Luxembourg	12	15	12 e	18	18	16 e	52	87	53 e
Mexico	117	38	-	-	-	-	9	20	6 e
Netherlands	3 519	2 920	3 257 e	214	408	382 e	1 095	1 111	1 048 e
Norway	2 304	2 005	1 597 e	94	88	122 e	402	325	847 e
Portugal	1 082	956	1 651 e	26	33	44 e	221	240	248 e
Spain	1 676	1 408	1 930 e	165	122	157 e	791	857	866 e
Sweden	546	775	620 e	37	25	33 e	403	329	337 e
Switzerland	214	157	201 e	16	42	48 e	296	319	323 e
Turkey	528	22	130 e	2	9	-	269	124	181 e
United Kingdom	3 553	4 929	3 920 e	23	22	29 e	835	760	1 163 e
United States	7 734	7 237	5 640 e	15	-	-	1 067	893	976 e
Other OECD	532	488	476 e	72	59	75 e	551	663	760 e
OECD	**38 175**	**35 611**	**33 774 e**	**1 154**	**1 241**	**1 302 e**	**14 514**	**14 709**	**15 825 e**
Africa	1 248	1 151	1 032 e	27	23	22 e	679	780	746 e
Latin America	374	135	125 e	8	13	16 e	223	154	269 e
Asia	1 882	2 131	2 245 e	23	16	21 e	188	282	324 e
China	63	3	79 e	1	1	-	35	71	34 e
Non-OECD Europe	3 485	2 818	1 926 e	64	50	50 e	592	710	526 e
Former Soviet Union	20	56	48 e	7	5	6 e	148	149	149 e
Middle East	398	537	134 e	-	6	-	312	444	304 e
Non-specified/Other	1 337	2 151	2 694 e	-	8	8 e	478	747	900 e
Non-OECD	**8 807**	**8 982**	**8 283 e**	**130**	**122**	**123 e**	**2 655**	**3 337**	**3 252 e**
Total Exports	**46 982**	**44 593**	**42 057 e**	**1 284**	**1 363**	**1 425 e**	**17 169**	**18 046**	**19 077 e**

Source: Annual Oil Statistics.

IEA TOTAL

Figure 1
FUEL SHARES OF TPES *

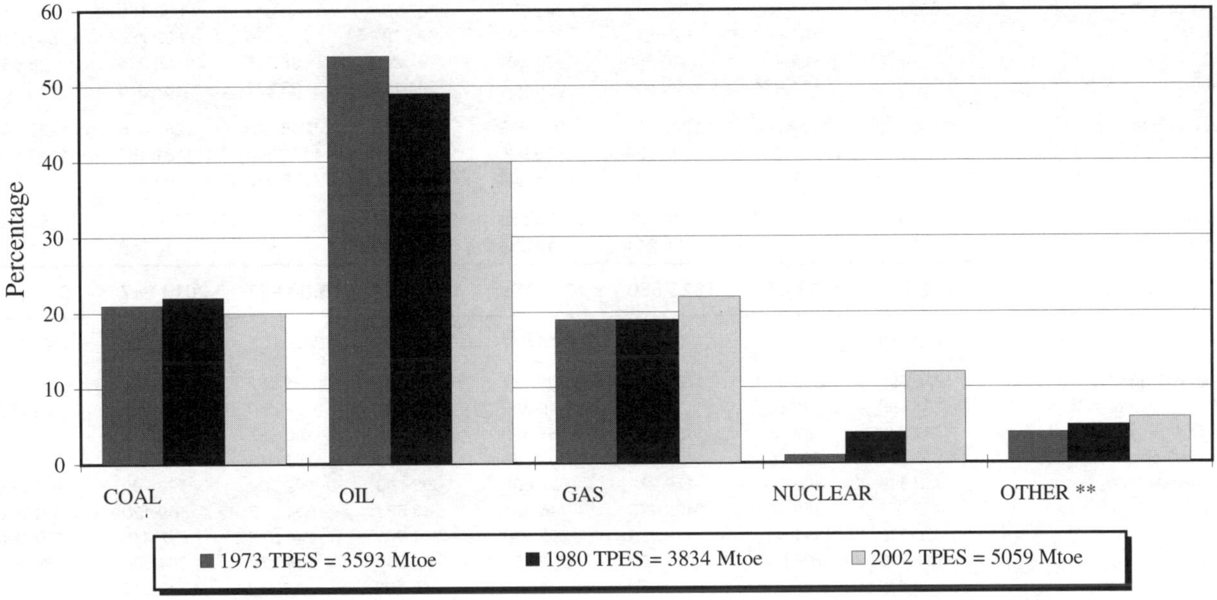

Source: Annual Energy Balances.

* Excluding electricity trade.

** Includes hydro, geothermal, solar, wind, combustible renewables and waste, etc.

Figure 2
OIL PRODUCTION AND CONSUMPTION

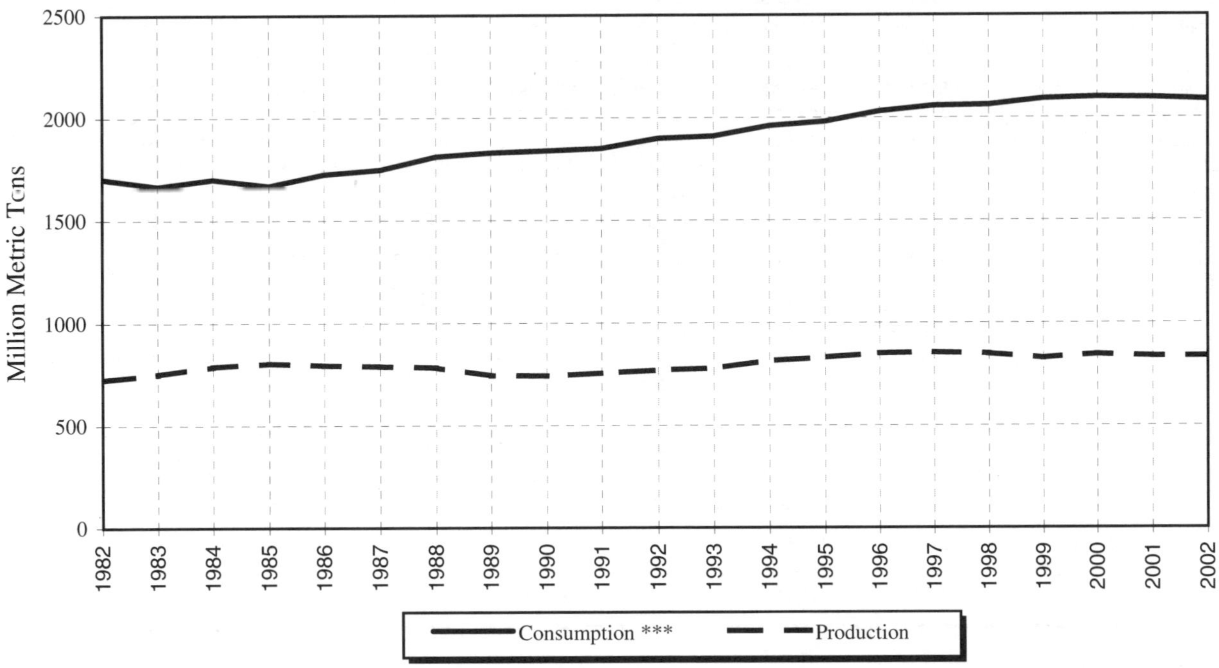

Source: Annual Oil Statistics.

*** Includes refinery fuel and international marine bunkers.

IEA TOTAL
Table 1: Oil Supply and Consumption - Summary Table

000 metric tons

	1973	1978	1980	1990	1999	2000	2001	2002
Indigenous Production	650 047	671 455	706 741	742 770	827 196	844 841	835 100	835 912 e
Crude Oil	..	607 545	626 818	655 794	695 797	710 974	699 786	694 117 e
NGL	..	62 436	70 360	71 991	97 283	98 596	98 236	98 558 e
Other	..	1 474	9 563	14 985	34 116	35 271	37 078	43 237 e
+ Net Imports	1336 498	1369 978	1220 879	1091 909	1218 653	1248 742	1267 516	1233 749 e
Crude Oil, NGL, Other	1174 354	1237 660	1109 694	976 906	1153 192	1170 510	1181 183	1149 122 e
Petroleum Products	162 144	132 318	111 185	115 003	65 461	78 232	86 333	84 627 e
- Intl. Marine Bunkers	73 134	73 779	75 729	73 886	82 997	87 133	78 417	89 732 e
+ Stock Changes	-21 013	15 819	-26 211	-13 234	33 162	-5 513	-10 352	11 017 e
TOTAL SUPPLY	**1892 398**	**1983 473**	**1825 680**	**1747 559**	**1996 014**	**2000 937**	**2013 847**	**1990 946 e**
Refinery Intake	*1775 650*	*1868 849*	*1774 145*	*1715 641*	*1984 935*	*2010 905*	*1999 668*	*1984 307 e*
Refinery Output	*1779 092*	*1855 033*	*1765 075*	*1708 831*	*1969 743*	*1996 467*	*1982 349*	*1969 329 e*
LPG + Ethane	*31 860*	*32 822*	*32 195*	*40 189*	*49 851*	*52 459*	*51 921*	*52 253 e*
Naphtha	*89 942*	*69 754*	*69 705*	*61 399*	*87 395*	*90 957*	*87 912*	*87 320 e*
Kerosenes	*110 234*	*106 880*	*107 128*	*138 848*	*184 268*	*187 016*	*178 687*	*173 541 e*
Gasolines	*430 649*	*495 624*	*476 972*	*527 000*	*592 583*	*592 509*	*592 789*	*603 255 e*
Gas/Diesel Oil	*425 661*	*446 764*	*426 977*	*446 434*	*545 624*	*564 298*	*569 120*	*561 016 e*
Fuel Oil	*499 308*	*494 587*	*449 936*	*259 810*	*244 962*	*244 516*	*237 415*	*221 290 e*
Other Products	*191 438*	*208 602*	*202 162*	*235 151*	*265 060*	*264 712*	*264 505*	*270 654 e*
- Refinery Losses	-3 442	13 816	9 070	6 810	15 192	14 438	17 319	14 978 e
+ Returns	5 476	13 268	20 154	18 189	26 679	28 153	28 398	27 116 e
- Statistical Differences	-14 172	-6 978	-3 194	-6 625	- 261	3 052	7 812	6 760 e
TOTAL CONSUMPTION	**1915 488**	**1989 903**	**1839 958**	**1765 563**	**2007 762**	**2011 600**	**2017 114**	**1996 324 e**
Transformation and Energy	**392 109**	**402 735**	**356 884**	**267 255**	**271 387**	**253 794**	**244 415**	..
Crude Oil	21 772	19 669	12 786	19 283	8 630	6 832	4 223	..
LPG + Ethane	3 191	4 700	6 351	6 451	5 308	4 215	4 051	..
Gas/Diesel Oil	10 394	14 135	10 829	7 913	13 469	12 564	12 093	..
Fuel Oil	262 802	273 503	233 479	125 387	113 489	102 886	104 736	..
Petroleum Coke	14 543	15 114	15 988	24 899	33 291	32 329	27 954	..
Other	79 407	75 614	77 451	83 322	97 200	94 968	91 358	..
Transport	**639 678**	**733 532**	**715 840**	**877 484**	**1071 434**	**1089 477**	**1090 894**	..
Motor Gasoline	423 714	483 052	453 861	506 683	583 038	581 858	581 959	..
Gas/Diesel Oil	118 581	149 657	157 712	231 770	315 835	329 038	335 675	..
Subtotal	*542 295*	*632 709*	*611 573*	*738 453*	*898 873*	*910 896*	*917 634*	..
Aviation Fuels	75 266	80 192	83 348	118 084	145 549	151 275	145 576	..
Other	22 117	20 631	20 919	20 947	27 012	27 306	27 684	..
Industry	**503 665**	**492 974**	**469 938**	**383 853**	**428 487**	**423 240**	**428 169**	..
Naphtha	81 873	72 962	68 405	69 178	104 054	110 901	105 694	..
LPG + Ethane	46 265	44 327	54 166	60 554	83 239	81 486	75 614	..
Gas/Diesel Oil	64 903	71 543	65 092	53 003	52 313	47 211	46 711	..
Fuel Oil	203 225	174 807	161 015	67 415	42 803	52 212	50 959	..
Other	107 399	129 335	121 260	133 703	146 078	131 430	149 191	..
Other Sectors	**380 036**	**360 662**	**297 296**	**236 971**	**236 454**	**245 089**	**253 636**	..
Gas/Diesel Oil	239 890	237 854	197 591	158 207	152 254	157 332	167 210	..
Fuel Oil	67 489	52 268	40 509	20 135	8 060	7 506	7 700	..
Other	72 657	70 540	59 196	58 629	76 140	80 251	78 726	..
Memo: Non-Energy Use	*204 918*	*218 892*	*206 453*	*236 719*	*315 227*	*306 546*	*308 804*	..
Chem/Petchem. Feedst.	*113 121*	*127 846*	*121 858*	*141 144*	*198 731*	*203 263*	*197 116*	..
Other	*91 797*	*91 046*	*84 595*	*95 575*	*116 496*	*103 283*	*111 688*	..

Source: Annual Oil Statistics.

IEA TOTAL

Table 2: Supply of Crude Oil, NGL, Feedstocks and Other Hydrocarbons

000 metric tons

	1973	1978	1980	1990	1999	2000	2001	2002
CRUDE OIL								
Indigenous Production	..	607 545	626 818	655 794	695 797	710 974	699 786	694 117 e
+ Imports	..	1295 216	1172 486	1084 395	1428 186	1472 324	1470 191	1436 445 e
- Exports	..	69 784	87 139	166 834	301 013	324 754	320 246	317 403 e
- Direct Use	..	19 815	12 929	19 311	8 659	6 866	4 257	4 429 e
+ Stock Change	..	-4 052	-15 755	-1 808	7 244	3 291	-6 991	-2 000 e
- Statistical Difference	..	1 031	495	-9 364	- 378	6 135	9 763	12 444 e
Refinery Intake	**..**	**1808 079**	**1682 986**	**1561 600**	**1821 933**	**1848 834**	**1828 720**	**1794 286 e**
Closing Stock Level	..	*133 407*	*162 785*	*218 577*	*237 786*	*231 423*	*238 414*	*240 414 e*
NATURAL GAS LIQUIDS								
Indigenous Production	..	62 436	70 360	71 991	97 283	98 596	98 236	98 558 e
+ Imports	..	6 819	11 663	14 787	16 780	17 472	18 814	19 573 e
- Exports	..	4 191	6 466	11 368	22 714	22 473	21 805	21 818 e
- Direct Use	..	42 754	49 050	51 738	59 610	60 775	59 033	55 499 e
+ Stock Change	..	217	-1 936	- 215	693	- 330	- 549	- 37 e
- Statistical Difference	..	171	-1 138	- 985	3 280	2 755	1 782	1 621 e
Refinery Intake	**..**	**22 356**	**25 709**	**24 442**	**29 152**	**29 735**	**33 881**	**39 156 e**
Closing Stock Level	..	*10 493*	*11 433*	*3 369*	*3 476*	*3 806*	*4 355*	*4 392 e*
REFINERY FEEDSTOCKS								
Backflows	..	12 832	19 164	16 422	23 601	25 272	25 517	23 549 e
+ Transfers	..	9 540	14 668	44 113	48 328	50 760	42 066	51 389 e
+ Imports	..	10 207	20 433	64 439	41 000	38 571	43 958	47 368 e
- Exports	..	614	1 290	6 684	9 645	10 711	8 242	7 089 e
+ Stock Change	..	1 206	-1 203	694	1 089	-1 488	1 739	2 248 e
- Statistical Difference	..	-3 777	-4 123	1 723	- 977	-2 261	-4 138	-4 803 e
Refinery Intake	**..**	**36 948**	**55 895**	**117 261**	**105 350**	**104 665**	**109 176**	**122 268 e**
Closing Stock Level	..	*24 120*	*31 150*	*33 899*	*30 513*	*32 026*	*30 287*	*28 039 e*
OTHER HYDROCARBONS								
Indigenous Production	..	1 474	9 563	14 985	34 116	35 271	37 078	43 237 e
+ Imports	..	7	7	387	6 398	6 700	6 044	4 536 e
- Exports	..	-	-	2 216	5 800	6 619	7 531	12 490 e
- Direct Use	..	15	15	473	6 526	7 273	6 138	5 719 e
+ Stock Change	..			- 52	42	199	- 318	158 e
- Statistical Difference	..	-	-	293	- 270	607	1 244	1 125 e
Refinery Intake	**..**	**1 466**	**9 555**	**12 338**	**28 500**	**27 671**	**27 891**	**28 597 e**
Closing Stock Level	..	*39*	*39*	*177*	*1 848*	*1 649*	*1 967*	*1 809 e*
CRUDE OIL, NGL, FEEDSTOCKS AND OTHER								
Indigenous Production	650 047	671 455	706 741	742 770	827 196	844 841	835 100	835 912 e
+ Backflows	4 939	12 832	19 164	16 422	23 601	25 272	25 517	23 549 e
+ Transfers	5 259	9 540	14 668	44 113	48 328	50 760	42 066	51 389 e
+ Imports	1236 189	1312 249	1204 589	1164 008	1492 364	1535 067	1539 007	1507 922 e
- Exports	61 835	74 589	94 895	187 102	339 172	364 557	357 824	358 800 e
- Direct Use	61 630	62 584	61 994	71 522	74 795	74 914	69 428	65 647 e
+ Stock Change	-10 142	-2 629	-18 894	-1 381	9 068	1 672	-6 119	369 e
- Statistical Difference	-12 823	-2 575	-4 766	-8 333	1 655	7 236	8 651	10 387 e
Refinery Intake	**1775 650**	**1868 849**	**1774 145**	**1715 641**	**1984 935**	**2010 905**	**1999 668**	**1984 307 e**
Closing Stock Level	-	*168 059*	*205 407*	*256 022*	*273 623*	*268 904*	*275 023*	*274 654 e*

Source: Annual Oil Statistics.

IEA TOTAL

Table 3: Motor Gasoline - Supply

000 metric tons

	1973	1978	1980	1990	1999	2000	2001	2002
MOTOR GASOLINE								
Refinery Output	**427 146**	**481 953**	**462 344**	**514 841**	**590 725**	**590 841**	**591 283**	**601 987 e**
+ Imports	21 674	20 971	23 898	51 218	56 521	61 899	65 907	65 958 e
- Exports	15 445	18 010	20 871	44 206	59 216	62 417	62 842	66 398 e
+ Transfers	267	1 478	-1 025	-4 611	-3 715	- 689	-1 219	-2 765 e
+ Stock Change	-1 139	3 385	-4 049	-1 021	5 119	204	-2 034	2 491 e
- Statistical Difference	- 110	- 845	- 387	415	738	2 468	34	1 404 e
Total Consumption	**432 613**	**490 622**	**460 684**	**515 806**	**588 696**	**587 370**	**591 061**	**599 869 e**
Closing Stock Level	-	*46 564*	*51 262*	*46 878*	*44 170*	*44 749*	*46 820*	*44 329 e*

Table 4: Aviation Fuels - Supply

000 metric tons

	1973	1978	1980	1990	1999	2000	2001	2002
AVIATION GASOLINE								
Refinery Output	**2 492**	**2 112**	**2 066**	**1 490**	**1 231**	**1 181**	**1 137**	**1 041 e**
+ Imports	409	301	284	206	200	192	170	111 e
- Exports	336	224	400	273	237	278	137	157 e
+ Transfers	-	4	90	1	162	78	126	176 e
+ Stock Change	35	52	- 7	30	29	35	- 10	14 e
- Statistical Difference	39	51	16	21	59	- 10	33	- 49 e
Total Consumption	**2 561**	**2 194**	**2 017**	**1 433**	**1 326**	**1 218**	**1 253**	**1 234 e**
Closing Stock Level	-	*538*	*523*	*338*	*296*	*261*	*271*	*257 e*
GASOLINE TYPE JET FUEL								
Refinery Output	**1 011**	**11 559**	**12 562**	**10 669**	**627**	**487**	**369**	**227 e**
+ Imports	320	623	1 182	305	9	11	9	5 e
- Exports	415	1 038	1 313	728	186	83	62	364 e
+ Transfers	-	- 33	59	- 69	- 212	- 33	- 7	- 8 e
+ Stock Change	6	- 35	21	70	61	- 11	16	9 e
- Statistical Difference	- 16	- 37	97	187	- 93	- 1	- 16	- 372 e
Total Consumption	**938**	**11 113**	**12 414**	**10 060**	**392**	**372**	**341**	**241 e**
Closing Stock Level	-	*1 266*	*1 595*	*896*	*66*	*77*	*61*	*52 e*
KEROSENE TYPE JET FUEL								
Refinery Output	**66 878**	**65 516**	**70 046**	**107 807**	**141 041**	**142 943**	**136 658**	**133 950 e**
+ Imports	15 190	9 081	9 705	16 645	21 282	25 341	28 309	25 307 e
- Exports	7 178	5 967	8 126	13 626	16 994	16 592	17 276	15 526 e
+ Transfers	961	- 157	- 247	- 514	- 532	182	-1 256	-1 307 e
+ Stock Change	- 750	292	- 776	-1 943	425	- 549	- 330	724 e
- Statistical Difference	2 524	1 282	1 650	1 658	465	890	1 550	516 e
Total Consumption	**72 577**	**67 483**	**68 952**	**106 711**	**144 757**	**150 435**	**144 555**	**142 632 e**
Closing Stock Level	-	*6 026*	*8 081*	*9 487*	*10 425*	*10 849*	*11 179*	*10 455 e*

Source: Annual Oil Statistics.

IEA TOTAL
Table 5A: Gas/Diesel Oil - Supply and Consumption

000 metric tons

	1973	1978	1980	1990	1999	2000	2001	2002
Refinery Output	**425 661**	**446 764**	**426 977**	**446 434**	**545 624**	**564 298**	**569 120**	**561 016 e**
+ Recycled Products	-	-	8	30	20	23	30	30 e
+ Imports	77 224	64 938	62 968	87 318	89 302	91 339	105 207	104 346 e
- Exports	48 799	40 828	43 843	62 462	91 959	93 039	93 080	91 690 e
- Intl. Marine Bunkers	10 199	11 265	9 910	16 219	17 529	17 514	17 350	17 264 e
+ Transfers	-1 219	1 619	- 674	-2 086	-5 487	-3 800	-1 174	-3 592 e
+ Stock Changes	-8 317	10 569	-4 326	-5 164	10 136	-1 272	-2 186	2 583 e
- Statistical Differences	583	-1 392	- 24	-3 042	-3 764	-6 110	-1 122	-3 738 e
Total Consumption	**433 768**	**473 189**	**431 224**	**450 893**	**533 871**	**546 145**	**561 689**	**559 167 e**
Transformation	**9 595**	**12 060**	**9 673**	**6 942**	**11 772**	**11 162**	**10 753**	**..**
Public Electricity	9 139	9 799	6 547	4 345	4 731	5 918	5 629	..
Autoproducer Electricity	-	835	797	1 047	3 153	1 781	2 059	..
Public CHP Plants	57	40	47	52	650	980	890	..
Autoproducers of CHP	-	2	1	23	2 082	1 136	864	..
Public Heat Plants	30	53	42	84	149	132	145	..
Autoproducer Heat Plants	92	319	317	135	1	36	35	..
Petrochemical Industry	219	973	1 877	1 240	997	1 172	1 131	..
Other Transformation	58	39	45	16	9	7	-	..
Energy Sector	**799**	**2 075**	**1 156**	**971**	**1 697**	**1 402**	**1 340**	**..**
Refinery Fuel	160	1 435	572	240	616	489	443	..
Other Energy Sector	639	640	584	731	1 081	913	897	..
Final Consumption	**423 374**	**459 054**	**420 395**	**442 980**	**520 402**	**533 581**	**549 596**	**..**
Transport	**118 581**	**149 657**	**157 712**	**231 770**	**315 835**	**329 038**	**335 675**	**..**
Air	-	-	-	1	-	1	1	..
Road	85 130	114 870	123 047	203 274	289 398	302 842	309 877	..
Rail	21 781	21 846	21 367	16 815	14 927	15 547	15 036	..
Inland Waterways	10 269	9 932	10 077	8 588	9 142	8 306	7 936	..
Non-Specified	1 401	3 009	3 221	3 092	2 368	2 342	2 825	..
Industry	**64 903**	**71 543**	**65 092**	**53 003**	**52 313**	**47 211**	**46 711**	**..**
Iron and Steel	1 842	1 688	1 392	1 249	1 078	1 679	1 686	..
Chemical (incl. Pet.-Chem.)	2 714	4 373	5 044	5 862	8 704	10 138	9 616	..
Non-Ferrous Metals	817	900	914	1 069	1 924	1 051	894	..
Non-Metallic Minerals	3 737	3 155	2 201	2 132	4 262	3 874	3 890	..
Transport Equipment	775	790	777	478	1 151	2 345	2 551	..
Machinery	2 323	2 316	2 034	2 873	2 722	2 861	2 861	
Mining and Quarrying	1 132	1 787	2 088	2 250	3 240	3 251	3 119	..
Food and Tobacco	2 852	3 427	3 352	3 343	5 261	5 614	5 342	..
Pulp, Paper and Printing	645	615	814	772	1 933	2 027	1 961	..
Wood and Wood Products	544	1 089	1 064	385	3 043	2 117	2 395	..
Construction	4 086	6 375	6 336	7 975	7 881	6 263	6 144	..
Textiles and Leather	940	960	895	1 160	1 842	1 770	1 790	..
Non-Specified	42 496	44 068	38 181	23 455	9 272	4 221	4 462	..
Other Sectors	**239 890**	**237 854**	**197 591**	**158 207**	**152 254**	**157 332**	**167 210**	**..**
Commerce and Public	68 947	71 154	57 212	46 340	43 839	48 580	52 992	..
Residential	126 951	126 000	98 589	71 934	69 209	68 145	72 925	..
Agriculture	24 899	28 691	29 728	38 197	38 487	36 850	37 473	..
Non-Specified	19 093	12 009	12 062	1 736	719	3 757	3 820	..
Memo: Non-Energy Use	*1 403*	*2 810*	*3 202*	*4 123*	*3 911*	*4 327*	*3 413*	..
Chem/Petchem. Feedst.	*1 402*	*2 809*	*3 197*	*4 118*	*3 906*	*4 322*	*3 408*	..
Other	*1*	*1*	*5*	*5*	*5*	*5*	*5*	..
Closing Stock Level	-	*81 672*	*94 496*	*65 463*	*66 663*	*68 251*	*70 437*	*67 854 e*

Source: Annual Oil Statistics.

IEA TOTAL

Table 5B: Fuel Oil (Residual) - Supply and Consumption

000 metric tons

	1973	1978	1980	1990	1999	2000	2001	2002
Refinery Output	**499 308**	**494 587**	**449 936**	**259 810**	**244 962**	**244 516**	**237 415**	**221 290 e**
+ Recycled Products	-	-	1	127	155	184	240	372 e
+ Imports	163 237	114 678	103 243	87 439	59 187	71 831	67 783	65 480 e
- Exports	56 525	36 338	39 857	61 729	61 668	67 068	68 096	61 792 e
- Intl. Marine Bunkers	62 579	62 204	65 472	57 308	65 109	69 225	60 656	72 138 e
+ Transfers	- 160	-2 529	-4 504	-8 451	-6 365	-7 144	-4 966	-6 087 e
+ Stock Changes	921	2 643	2 524	-3 072	4 172	- 5	1 623	1 202 e
- Statistical Differences	-1 982	- 102	240	-2 126	1 981	2 045	1 313	588 e
Total Consumption	**546 184**	**510 939**	**445 631**	**218 942**	**173 353**	**171 044**	**172 030**	**147 739 e**
Transformation	**227 327**	**236 662**	**200 640**	**107 757**	**93 530**	**85 602**	**87 336**	..
Public Electricity	214 236	202 078	162 931	90 606	63 655	58 622	58 453	..
Autoproducer Electricity	113	18 649	16 548	7 993	14 317	11 911	12 917	..
Public CHP Plants	6 381	7 604	11 417	2 268	2 256	4 021	4 131	..
Autoproducers of CHP	295	4 422	5 794	3 677	7 774	3 955	3 602	..
Public Heat Plants	597	584	759	823	1 092	873	1 099	..
Autoproducer Heat Plants	29	64	4	49	84	718	691	..
Petrochemical Industry	1 186	777	1 073	1 047	2 672	4 109	4 986	..
Other Transformation	4 490	2 484	2 114	1 294	1 680	1 393	1 457	..
Energy Sector	**35 475**	**36 841**	**32 839**	**17 630**	**19 959**	**17 284**	**17 400**	..
Refinery Fuel	34 949	35 565	32 318	17 210	19 503	16 996	16 968	..
Other Energy Sector	526	1 276	521	420	456	288	432	..
Final Consumption	**283 382**	**237 436**	**212 152**	**93 555**	**59 864**	**68 158**	**67 294**	..
Transport	**12 668**	**10 361**	**10 628**	**6 005**	**9 001**	**8 440**	**8 635**	..
Air	-	-	-	-	-	-	-	..
Road	52	45	67	57	47	1	27	..
Rail	1 895	913	801	54	2	3	1	..
Inland Waterways	10 249	8 745	9 183	5 536	8 815	8 324	8 506	..
Non-Specified	472	658	577	358	137	112	101	..
Industry	**203 225**	**174 807**	**161 015**	**67 415**	**42 803**	**52 212**	**50 959**	..
Iron and Steel	30 640	20 343	15 937	4 753	3 437	3 727	3 256	..
Chemical (incl. Pet.-Chem.)	28 053	29 263	27 161	12 385	7 691	11 423	12 224	..
Non-Ferrous Metals	3 347	6 850	5 706	3 097	2 232	2 533	2 360	..
Non-Metallic Minerals	30 316	26 070	22 288	7 135	6 104	5 936	4 973	..
Transport Equipment	3 308	2 876	2 657	459	1 126	1 031	920	..
Machinery	5 262	4 204	3 553	2 027	1 540	1 163	1 263	..
Mining and Quarrying	2 621	2 132	2 390	855	671	674	649	..
Food and Tobacco	16 825	16 467	15 836	6 486	3 898	3 900	3 914	..
Pulp, Paper and Printing	28 512	23 318	22 423	8 635	7 164	9 982	9 773	..
Wood and Wood Products	1 569	1 299	1 201	358	373	322	309	..
Construction	3 993	3 367	2 913	575	419	460	374	..
Textiles and Leather	10 807	9 713	9 200	5 185	4 297	4 554	4 147	..
Non-Specified	37 972	28 905	29 750	15 465	3 851	6 507	6 797	..
Other Sectors	**67 489**	**52 268**	**40 509**	**20 135**	**8 060**	**7 506**	**7 700**	..
Commerce and Public	42 900	34 733	26 521	12 839	4 395	4 333	4 517	..
Residential	18 946	12 725	9 412	5 519	2 495	2 305	2 199	..
Agriculture	1 831	2 021	2 158	1 059	1 065	755	758	..
Non-Specified	3 812	2 789	2 418	718	105	113	226	..
Memo: Non-Energy Use	*628*	*1 641*	*3 024*	*2 864*	*1 663*	*3 787*	*3 804*	..
Chem/Petchem. Feedst.	*510*	*1 569*	*2 966*	*2 570*	*1 484*	*3 493*	*3 589*	..
Other	*118*	*72*	*58*	*294*	*179*	*294*	*215*	..
Closing Stock Level	-	*76 694*	*76 478*	*48 245*	*35 096*	*35 144*	*33 521*	*32 319 e*

Source: Annual Oil Statistics.

IEA TOTAL
Table 5C: Total Products - Supply and Consumption

000 metric tons

	1973	1978	1980	1990	1999	2000	2001	2002
Refinery Output	**1779 092**	**1855 033**	**1765 075**	**1708 831**	**1969 743**	**1996 467**	**1982 349**	**1969 329 e**
+ Recycled Products	537	436	990	1 767	3 078	2 881	2 881	3 567 e
+ Imports	331 514	274 187	267 014	356 162	370 909	402 609	414 157	407 724 e
- Exports	169 370	141 869	155 829	241 159	305 448	324 377	327 824	323 097 e
- Intl. Marine Bunkers	73 134	73 779	75 729	73 886	82 997	87 133	78 417	89 732 e
+ Transfers	56 371	53 865	47 596	27 409	26 467	24 154	27 362	14 258 e
+ Stock Changes	-10 871	18 448	-7 317	-11 853	24 094	-7 185	-4 233	10 648 e
- Statistical Differences	-1 349	-3 582	1 842	1 708	-1 916	-4 184	- 839	-3 627 e
Total Consumption	**1915 488**	**1989 903**	**1839 958**	**1765 563**	**2007 762**	**2011 600**	**2017 114**	**1996 324 e**
Transformation	**274 002**	**291 762**	**250 159**	**158 803**	**152 800**	**138 624**	**130 453**	..
Public Electricity	247 467	236 395	185 554	116 101	80 919	74 694	69 497	..
Autoproducer Electricity	290	19 801	17 683	11 107	20 688	17 439	18 040	..
Public CHP Plants	6 438	7 687	11 474	2 349	4 953	6 660	6 560	..
Autoproducers of CHP	372	4 528	5 964	4 309	14 256	9 096	5 548	..
Public Heat Plants	627	637	802	979	1 309	1 055	1 289	..
Autoproducer Heat Plants	121	383	321	184	85	771	726	..
Petrochemical Industry	4 939	12 832	19 164	16 422	23 601	25 272	25 517	..
Other Transformation	13 748	9 499	9 197	7 352	6 989	3 637	3 276	..
Energy Sector	**118 107**	**110 973**	**106 725**	**108 452**	**118 587**	**115 170**	**113 962**	..
Refinery Fuel	115 856	108 034	104 232	103 979	114 519	111 442	110 508	..
Other Energy Sector	2 251	2 939	2 493	4 473	4 068	3 728	3 454	..
Final Consumption	**1523 379**	**1587 168**	**1483 074**	**1498 308**	**1736 375**	**1757 806**	**1772 699**	..
Transport	**639 678**	**733 532**	**715 840**	**877 484**	**1071 434**	**1089 477**	**1090 894**	..
Air	75 270	80 196	83 352	118 104	145 582	151 307	145 605	..
Road	517 710	607 634	586 467	720 521	886 394	899 463	907 131	..
Rail	23 747	22 790	22 209	16 926	15 030	15 647	15 124	..
Inland Waterways	21 018	19 183	19 963	18 161	21 545	20 305	19 764	..
Non-Specified	1 933	3 729	3 849	3 772	2 883	2 755	3 270	..
Industry	**503 665**	**492 974**	**469 938**	**383 853**	**428 487**	**423 240**	**428 169**	..
Iron and Steel	32 834	22 364	18 131	7 538	6 019	7 303	6 499	..
Chemical (incl. Pet.-Chem.)	145 337	160 346	150 742	156 115	213 928	224 060	219 534	..
Non-Ferrous Metals	4 937	9 385	8 507	7 493	8 162	7 536	7 228	..
Non-Metallic Minerals	34 646	29 971	25 388	16 041	20 217	20 082	19 983	..
Transport Equipment	4 129	3 744	3 548	1 076	2 637	3 724	3 809	..
Machinery	7 910	6 894	6 016	6 560	6 510	6 400	6 597	..
Mining and Quarrying	3 932	4 050	4 592	3 396	4 443	4 646	4 494	..
Food and Tobacco	19 825	20 066	19 542	10 280	9 940	10 229	9 962	..
Pulp, Paper and Printing	29 253	24 051	23 375	9 875	9 821	12 749	12 472	..
Wood and Wood Products	2 161	2 397	2 284	798	3 640	2 617	2 867	..
Construction	68 582	65 122	59 291	67 325	71 021	67 478	67 524	..
Textiles and Leather	11 759	10 707	10 160	6 614	6 580	6 705	6 320	..
Non-Specified	138 360	133 877	138 362	90 733	65 561	49 675	60 880	..
Other Sectors	**380 036**	**360 662**	**297 296**	**236 971**	**236 454**	**245 089**	**253 636**	..
Commerce and Public	129 418	126 977	103 531	74 953	69 870	78 498	83 461	..
Residential	190 049	179 598	140 518	111 422	116 041	115 626	118 076	..
Agriculture	34 849	36 943	37 808	47 002	47 438	45 135	46 283	..
Non-Specified	25 720	17 144	15 439	3 594	3 105	5 830	5 816	..
Memo: Non-Energy Use	*204 918*	*218 892*	*206 453*	*236 719*	*315 227*	*306 546*	*308 804*	..
Chem/Petchem. Feedst.	*113 121*	*127 846*	*121 858*	*141 144*	*198 731*	*203 263*	*197 116*	..
Other	*91 797*	*91 046*	*84 595*	*95 575*	*116 496*	*103 283*	*111 688*	..
Closing Stock Level	-	*250 360*	*270 599*	*210 981*	*197 028*	*203 409*	*207 683*	*197 035 e*

Source: Annual Oil Statistics.

IEA TOTAL
Table 6A: Supply and Consumption

000 metric tons

	1980	2000	2001	2002	1980	2000	2001	2002
	LPG AND ETHANE				REFINERY GAS			
Refinery Output	**32 195**	**52 459**	**51 921**	**52 253 e**	**58 644**	**71 378**	**70 562**	**71 165 e**
+ Recycled Products	-	-	-	-	-	-	-	-
+ Imports	15 770	34 896	32 526	32 236 e	-	-	-	-
- Exports	4 521	7 721	7 161	7 985 e	-	-	-	-
- Intl. Marine Bunkers	-	-	-	-	-	-	-	-
+ Transfers	43 755	55 264	53 679	49 862 e	- 127	- 37	- 47	- 128 e
+ Stock Change	- 172	- 425	-3 025	1 740 e	4	3	-	-
- Statistical Difference	-1 338	1 886	1 884	2 761 e	295	65	20	294 e
Total Consumption	**88 365**	**132 587**	**126 056**	**125 345 e**	**58 226**	**71 279**	**70 495**	**70 743 e**
of which: Ethane	-	*23 229*	*21 859*	*22 952 e*	-	-	-	-
Transformation	4 643	2 890	2 721	..	1 427	2 379	1 570	..
Energy	1 708	1 325	1 330	..	52 951	66 464	65 251	..
Transport	3 826	10 265	10 766	..	-	-	-	..
Industry	54 166	81 486	75 614	..	3 848	2 436	3 674	..
Other	24 022	36 621	35 625	..	-	-	-	..
Memo: Non-Energy Use	*23 314*	*70 238*	*64 973*	..	*2 848*	*996*	*1 841*	..
Closing Stock Level	*3 327*	*9 792*	*12 821*	*11 081 e*	*4*	-	-	-

	1980	2000	2001	2002	1980	2000	2001	2002
	NAPHTHA				OTHER KEROSENE			
Refinery Output	**69 705**	**90 957**	**87 912**	**87 320 e**	**37 082**	**44 073**	**42 029**	**39 591 e**
+ Recycled Products	281	216	282	411 e	-	-	-	-
+ Imports	31 744	67 794	65 713	65 018 e	782	5 535	4 466	5 147 e
- Exports	11 544	26 860	26 675	24 707 e	2 468	3 712	3 494	3 611 e
- Intl. Marine Bunkers	-	-	-	-	-	2	-	-
+ Transfers	-2 435	-5 351	-6 333	-3 440 e	-1 085	-1 885	-1 686	-2 127 e
+ Stock Change	- 274	- 653	323	- 13 e	600	- 845	939	1 116 e
- Statistical Difference	186	-2 871	-1 795	- 420 e	460	580	445	- 49 e
Total Consumption	**87 291**	**128 974**	**123 017**	**125 009 e**	**34 451**	**42 584**	**41 809**	**40 165 e**
Transformation	18 380	17 846	16 914	..	224	438	360	..
Energy	497	225	409	..	24	33	109	..
Transport	-	-	-	..	49	47	52	..
Industry	68 405	110 901	105 694	..	6 630	5 021	5 090	..
Other	9	2	-	..	27 524	37 045	36 198	..
Memo: Non-Energy Use	*67 285*	*109 725*	*104 510*	..	*326*	*699*	*657*	..
Closing Stock Level	*6 871*	*7 231*	*6 908*	*6 921 e*	*9 038*	*5 799*	*4 860*	*3 744 e*

	1980	2000	2001	2002	1980	2000	2001	2002
	WHITE SPIRIT + SBP				LUBRICANTS			
Refinery Output	**6 675**	**4 218**	**3 748**	**..**	**20 360**	**22 784**	**22 201**	**..**
+ Recycled Products	26	7	7	..	466	577	419	..
+ Imports	933	1 458	1 546	..	3 655	4 299	3 939	..
- Exports	1 233	1 619	1 769	..	5 474	7 885	8 633	..
- Intl. Marine Bunkers	-	-	-	..	330	366	381	..
+ Transfers	102	- 309	- 425	..	- 382	- 679	- 609	..
+ Stock Change	- 103	- 36	- 18	..	- 601	- 321	- 124	..
- Statistical Difference	- 177	- 60	- 160	..	- 153	- 493	- 836	..
Total Consumption	**6 577**	**3 779**	**3 249**	**..**	**17 847**	**18 902**	**17 648**	**..**
Transformation	7	21	23	..	1	13	10	..
Energy	40	-	-	..	11	67	66	..
Transport	-	-	-	..	6 416	8 554	8 231	..
Industry	6 020	3 444	2 994	..	10 572	9 122	8 801	..
Other	510	314	232	..	847	1 146	540	..
Memo: Non-Energy Use	*6 484*	*3 675*	*3 137*	..	*17 833*	*17 708*	*17 032*	..
Closing Stock Level	*1 100*	*527*	*545*	..	*5 949*	*4 323*	*4 447*	..

Source: Annual Oil Statistics.

IEA TOTAL
Table 6B: Supply and Consumption

000 metric tons

	1980	2000	2001	2002	1980	2000	2001	2002
	BITUMEN				PARAFFIN WAXES			
Refinery Output	**50 006**	**64 373**	**61 667**	**..**	**1 780**	**2 590**	**2 466**	**..**
+ Recycled Products	-	-	-	..	-	-	-	..
+ Imports	2 051	5 883	6 164	..	427	723	738	..
- Exports	2 416	6 624	6 863	..	521	856	905	..
- Intl. Marine Bunkers	-	-	-	..	-	-	-	..
+ Transfers	- 9	2 523	1 871	..	- 13	- 437	- 361	..
+ Stock Change	- 190	-1 440	768	..	-	6	38	..
- Statistical Difference	45	-1 040	- 226	..	107	122	151	..
Total Consumption	**49 397**	**65 755**	**63 833**	**..**	**1 566**	**1 904**	**1 825**	**..**
Transformation	111	2 810	1 838	..	6	-	-	..
Energy	7	8	11	..	-	13	10	..
Transport	-	-	-	..	-	-	-	..
Industry	49 279	62 799	61 778	..	1 560	1 890	1 815	..
Other	-	138	206	..	-	1	-	..
Memo: Non-Energy Use	*49 279*	*61 016*	*60 251*	..	*1 539*	*1 891*	*1 815*	..
Closing Stock Level	*5 196*	*6 253*	*5 485*	..	*180*	*277*	*239*	..
	PETROLEUM COKE				OTHER PRODUCTS			
Refinery Output	**29 226**	**61 980**	**64 430**	**65 674 e**	**35 471**	**37 389**	**39 431**	**133 815 e**
+ Recycled Products	-	-	-	-	208	1 874	1 903	2 754 e
+ Imports	9 188	17 131	17 557	17 199 e	1 184	14 277	14 123	26 917 e
- Exports	9 551	22 613	23 882	24 093 e	3 691	7 010	6 949	26 774 e
- Intl. Marine Bunkers	-	-	-	-	-	23	29	330 e
+ Transfers	-	- 76	- 4	- 59 e	14 091	-13 453	-10 227	-16 267 e
+ Stock Change	187	- 617	- 150	239 e	- 155	-1 259	- 63	543 e
- Statistical Difference	514	399	238	- 426 e	328	-2 051	-2 351	-4 136 e
Total Consumption	**28 536**	**55 406**	**57 713**	**59 386 e**	**46 780**	**33 846**	**40 540**	**124 794 e**
Transformation	968	8 038	3 835	..	14 079	7 067	4 782	..
Energy	15 020	24 291	24 119	..	2 366	3 962	3 831	..
Transport	-	-	-	..	-	-	-	..
Industry	12 548	22 969	29 611	..	30 213	21 902	30 174	..
Other	-	108	148	..	122	915	1 753	..
Memo: Non-Energy Use	*6 807*	*10 283*	*16 636*	..	*23 834*	*19 918*	*28 686*	..
Closing Stock Level	*1 562*	*4 235*	*4 385*	*4 146 e*	*4 937*	*5 641*	*5 704*	*15 877 e*

Source: Annual Oil Statistics.

IEA TOTAL
Table 7A: Imports by Country of Origin

000 metric tons

	2000	2001	2002	2000	2001	2002	2000	2001	2002
	Crude Oil			**Natural Gas Liquids**			**Feedstocks and Other**		
Australia	10 826	9 287	10 809 e	332	352	355 e	204	324	395 e
Belgium	2	2	-	-	-	2 121 e	1 675	2 069	2 184 e
Canada	68 635	68 997	72 490 e	5 593	4 129	4 176 e	1 052	1 066	709 e
Czech Republic	31	62	86 e	-	-	-	8	1	2 e
Denmark	12 078	10 570	10 329 e	-	-	-	50	83	30 e
Finland	-	22	-	-	-	-	119	167	2 e
France	501	402	8 e	-	4	-	887	1 130	729 e
Germany	1 594	154	221 e	11	35	45 e	825	1 611	1 969 e
Greece	-	-	-	-	-	-	42	120	130 e
Hungary	-	-	-	-	-	-	200	154	199 e
Ireland	-	-	-	-	-	-	539	637	390 e
Italy	204	282	905 e	-	-	-	351	406	232 e
Japan	-	-	-	-	-	-	1	33	112 e
Korea	-	-	-	-	-	-	22	55	69 e
Mexico	79 948	82 220	87 147 e	68	182	-	184	41	91 e
Netherlands	354	334	1 276 e	-	-	-	2 372	2 331	2 395 e
New Zealand	1 001	870	860 e	6	-	-	-	19	18 e
Norway	144 530	135 097	131 762 e	1 006	1 267	1 512 e	1 449	1 607	1 408 e
Poland	163	40	206 e	-	-	-	137	76	5 e
Spain	84	-	-	-	7	-	223	201	245 e
Sweden	126	107	80 e	7	49	-	1 532	1 156	1 039 e
Turkey	213	6	-	-	-	-	556	765	427 e
United Kingdom	72 727	75 475	73 519 e	99	209	43 e	839	1 220	745 e
United States	2 135	1 627	1 125 e	324	279	243 e	74	729	529 e
Other	269	66	12 e	-	-	-	345	129	113 e
OECD	**395 421**	**385 620**	**390 835 e**	**7 446**	**6 513**	**8 495 e**	**13 686**	**16 130**	**14 167 e**
Argentina	3 471	3 970	4 428 e	-	-	-	37	161	103 e
Brazil	391	2 936	5 778 e	-	-	36 e	177	269	319 e
Colombia	16 307	13 374	12 859 e	-	-	-	129	172	298 e
Ecuador	8 606	7 685	7 500 e	-	-	-	164	43	82 e
Netherlands Antilles	99	238	115 e	-	-	-	2 036	1 766	1 993 e
Peru	575	460	519 e	-	-	-	117	113	92 e
Trinidad and Tobago	2 790	2 558	3 389 e	-	-	14 e	340	208	145 e
Venezuela	88 104	91 978	87 146 e	85	48	97 e	5 589	3 866	4 065 e
Other	1 257	988	1 224 e	-	-	-	51	58	213 e
Latin America	**121 600**	**124 187**	**122 958 e**	**85**	**48**	**147 e**	**8 640**	**6 656**	**7 310 e**
Romania	19	118	373 e	-	-	-	94	15	22 e
Other	-	-	-	-	-	-	104	134	154 e
Non-OECD Europe	**19**	**118**	**373 e**	**-**	**-**	**-**	**198**	**149**	**176 e**
FSU	**105 907**	**119 908**	**144 596 e**	**747**	**1 544**	**1 852 e**	**10 205**	**10 209**	**12 752 e**
Iran	75 885	73 120	64 126 e	250	233	1 195 e	30	14	69 e
Iraq	76 065	68 189	47 848 e	-	-	-	-	-	-
Kuwait	53 412	49 378	38 554 e	-	37	-	14	-	-
Oman	18 296	21 489	20 180 e	-	13	24 e	-	-	-
Qatar	27 154	28 123	23 306 e	707	1 335	1 278 e	397	338	261 e
Saudi Arabia	240 136	234 678	220 236 e	433	539	161 e	2 481	1 815	2 190 e
United Arab Emirates	74 397	70 459	69 470 e	565	260	439 e	368	346	-
Other	22 526	26 943	27 038 e	-	-	-	385	1 254	808 e
Middle East	**587 871**	**572 379**	**510 758 e**	**1 955**	**2 417**	**3 097 e**	**3 675**	**3 767**	**3 328 e**
India	-	-	-	-	-	-	66	104	71 e
Indonesia	20 085	20 323	20 899 e	288	281	402 e	355	242	713 e
Singapore	29	41	43 e	-	-	-	388	968	783 e
Other	23 621	23 703	21 721 e	319	166	57 e	687	973	1 042 e
Asia	**43 735**	**44 067**	**42 663 e**	**607**	**447**	**459 e**	**1 496**	**2 287**	**2 609 e**
China	**7 006**	**4 547**	**5 076 e**	**-**	**-**	**-**	**684**	**944**	**775 e**
Algeria	20 498	19 223	19 395 e	6 447	6 917	5 087 e	1 860	5 277	5 289 e
Angola/Cabinada	22 694	25 491	26 579 e	5	-	-	302	121	315 e
Cameroon	3 615	3 918	4 561 e	-	-	-	-	56	21 e
Egypt	6 056	3 667	4 102 e	-	-	-	444	399	180 e
Gabon	11 389	11 582	9 990 e	-	-	-	38	-	-
Libya	50 910	50 528	44 909 e	25	25	75 e	2 086	1 681	1 356 e
Nigeria	74 885	73 816	60 729 e	40	294	139 e	133	664	1 436 e
Tunisia	1 346	1 247	2 192 e	105	70	129 e	239	262	448 e
Other	11 704	14 778	18 615 e	10	46	20 e	837	707	578 e
Africa	**203 097**	**204 250**	**191 072 e**	**6 632**	**7 352**	**5 450 e**	**5 939**	**9 167**	**9 623 e**
Non-spec./Other	**7 668**	**15 115**	**28 114 e**	**-**	**493**	**73 e**	**748**	**693**	**1 164 e**
Total Imports	**1472 324**	**1470 191**	**1436 445 e**	**17 472**	**18 814**	**19 573 e**	**45 271**	**50 002**	**51 904 e**
Memo: Total OPEC	*801 531*	*779 815*	*696 618 e*	*8 840*	*9 969*	*8 873 e*	*13 313*	*14 243*	*15 379 e*

Source: Annual Oil Statistics.

IEA TOTAL
Table 7B: Imports by Country of Origin

000 metric tons

	2000	2001	2002	2000	2001	2002	2000	2001	2002
	Total Products			Liquefied Petroleum Gases			Naphtha		
Australia	2 518	2 632	2 925 e	944	1 131	1 453 e	478	386	229 e
Belgium	18 919	19 116	19 222 e	459	434	432 e	1 525	1 126	1 076 e
Canada	12 123	15 081	17 390 e	220	176	283 e	145	154	183 e
Czech Republic	683	640	771 e	4	8	13 e	25	24	8 e
Denmark	3 015	2 551	2 545 e	58	59	41 e	96	53	87 e
Finland	4 270	3 646	4 012 e	27	10	11 e	74	78	160 e
France	13 126	12 619	10 580 e	429	367	335 e	1 329	1 288	790 e
Germany	14 525	13 078	12 639 e	527	400	407 e	1 403	981	1 039 e
Greece	1 140	1 058	1 095 e	16	1	120 e	575	425	486 e
Hungary	833	1 065	971 e	35	32	32 e	-	-	-
Ireland	240	243	59 e	6	5	7 e	76	16	-
Italy	9 463	11 827	11 286 e	105	90	81 e	338	725	622 e
Japan	1 570	2 200	1 846 e	-	29	177 e	18	12	47 e
Korea	13 811	12 622	11 752 e	133	158	48 e	5 793	5 689	5 370 e
Mexico	2 734	1 989	2 042 e	-	-	9 e	1 593	1 478	1 042 e
Netherlands	44 427	43 897	40 049 e	498	440	518 e	6 448	5 827	5 608 e
New Zealand	439	348	353 e	-	3	15 e	318	334	313 e
Norway	8 489	7 228	8 695 e	1 718	1 978	2 238 e	883	792	564 e
Poland	1 359	1 608	1 632 e	22	7	4 e	75	95	76 e
Spain	4 739	4 007	3 349 e	96	57	57 e	1 200	1 166	1 006 e
Sweden	8 230	7 175	7 009 e	100	49	113 e	151	245	217 e
Turkey	580	1 028	850 e	6	2	-	317	558	358 e
United Kingdom	25 986	25 314	25 651 e	2 479	2 125	2 545 e	2 539	2 160	2 367 e
United States	18 890	18 476	22 559 e	198	365	673 e	1 206	1 066	1 296 e
Other	5 284	5 173	5 547 e	69	108	84 e	158	109	149 e
OECD	**217 393**	**214 621**	**214 829 e**	**8 149**	**8 034**	**9 696 e**	**26 763**	**24 787**	**23 093 e**
Argentina	1 409	2 063	2 357 e	-	-	11 e	12	190	84 e
Brazil	2 847	3 849	2 302 e	-	-	1 e	44	32	30 e
Colombia	1 205	1 771	1 165 e	2	-	-	88	19	173 e
Ecuador	165	412	261 e	-	-	-	19	35	74 e
Netherlands Antilles	3 233	2 918	2 401 e	3	-	-	650	221	209 e
Peru	193	508	328 e	-	-	-	26	96	51 e
Trinidad and Tobago	2 219	2 327	1 948 e	9	8	16 e	147	47	-
Venezuela	18 438	15 880	10 500 e	29	-	81 e	508	353	419 e
Other	794	1 652	2 079 e	32	4	21 e	71	150	57 e
Latin America	**30 503**	**31 380**	**23 341 e**	**75**	**12**	**130 e**	**1 565**	**1 143**	**1 097 e**
Romania	275	695	2 019 e	-	-	-	-	1	31 e
Other	1 254	1 577	1 439 e	116	121	111 e	169	158	204 e
Non-OECD Europe	**1 529**	**2 272**	**3 458 e**	**116**	**121**	**111 e**	**169**	**159**	**235 e**
FSU	**38 186**	**38 974**	**45 726 e**	**303**	**502**	**365 e**	**2 383**	**2 138**	**2 866 e**
Iran	1 045	1 147	1 086 e	623	862	713 e	359	199	321 e
Iraq	7	97	-	7	97	-	-	-	-
Kuwait	11 290	10 979	9 757 e	2 530	2 434	1 950 e	6 540	6 143	5 916 e
Oman	-	95	34 e	-	-	-	-	13	34 e
Qatar	1 731	2 250	1 955 e	821	904	739 e	849	1 038	1 112 e
Saudi Arabia	17 348	13 963	14 145 e	8 927	6 385	6 270 e	5 773	4 479	5 632 e
United Arab Emirates	12 244	13 728	14 294 e	4 958	4 345	4 517 e	5 792	6 195	7 159 e
Other	4 112	3 174	2 598 e	54	97	117 e	1 592	1 123	1 132 e
Middle East	**47 777**	**45 433**	**43 869 e**	**17 920**	**15 124**	**14 306 e**	**20 905**	**19 190**	**21 306 e**
India	1 825	3 337	3 536 e	-	-	-	1 523	1 993	2 107 e
Indonesia	6 008	6 237	5 272 e	1 055	1 069	711 e	1 127	1 570	1 449 e
Singapore	5 778	4 463	6 165 e	-	23	-	1 822	653	666 e
Other	4 319	4 440	5 235 e	474	262	333 e	2 278	2 156	1 629 e
Asia	**17 930**	**18 477**	**20 208 e**	**1 529**	**1 354**	**1 044 e**	**6 750**	**6 372**	**5 851 e**
China	**2 202**	**2 262**	**2 871 e**	**12**	**3**	**3 e**	**927**	**1 047**	**1 143 e**
Algeria	18 799	16 862	16 865 e	5 449	5 149	4 929 e	2 891	3 746	3 831 e
Angola/Cabinada	189	403	113 e	-	-	-	-	-	22 e
Cameroon	206	114	52 e	-	-	-	-	-	-
Egypt	1 897	2 214	2 861 e	54	214	160 e	1 654	1 290	1 313 e
Gabon	22	-	4 e	-	-	4 e	-	-	-
Libya	5 806	6 231	5 517 e	221	276	124 e	1 497	2 071	1 871 e
Nigeria	1 207	2 355	1 597 e	151	606	559 e	113	147	145 e
Tunisia	436	291	308 e	-	-	-	45	52	114 e
Other	2 116	1 849	2 041 e	36	45	66 e	418	391	352 e
Africa	**30 678**	**30 319**	**29 358 e**	**5 911**	**6 290**	**5 842 e**	**6 618**	**7 697**	**7 648 e**
Non-spec./Other	**16 411**	**30 419**	**24 064 e**	**881**	**1 086**	**739 e**	**1 714**	**3 180**	**1 779 e**
Total Imports	**402 609**	**414 157**	**407 724 e**	**34 896**	**32 526**	**32 236 e**	**67 794**	**65 713**	**65 018 e**
Memo: Total OPEC	*93 923*	*89 729*	*80 988 e*	*24 771*	*22 127*	*20 593 e*	*25 449*	*25 941*	*27 855 e*

Source: Annual Oil Statistics.

IEA TOTAL
Table 7C: Imports by Country of Origin

000 metric tons

	2000	2001	2002	2000	2001	2002	2000	2001	2002
	Total Gasoline			Total Kerosene			Gas/Diesel		
Australia	670	538	624 e	98	184	110 e	31	102	161 e
Belgium	4 539	4 765	4 693 e	1 322	1 412	1 351 e	6 120	7 190	6 174 e
Canada	4 040	6 404	7 525 e	161	189	296 e	5 097	5 191	5 726 e
Czech Republic	28	42	129 e	-	-	-	379	424	485 e
Denmark	1 141	813	758 e	292	307	403 e	835	653	482 e
Finland	2 004	1 919	2 353 e	184	4	42 e	1 791	1 498	1 333 e
France	3 538	3 680	3 446 e	1 080	1 136	742 e	2 174	1 883	1 259 e
Germany	3 505	3 010	2 718 e	386	432	773 e	4 787	4 483	5 003 e
Greece	296	224	265 e	129	88	19 e	68	292	164 e
Hungary	237	286	209 e	-	-	-	459	617	662 e
Ireland	18	44	14 e	8	1	11 e	90	93	19 e
Italy	1 769	2 367	2 141 e	378	414	331 e	4 981	6 871	6 988 e
Japan	162	239	100 e	812	484	452 e	370	944	782 e
Korea	797	1 181	942 e	4 085	2 992	2 805 e	2 534	2 314	1 694 e
Mexico	276	330	282 e	88	120	94 e	-	14	40 e
Netherlands	9 166	8 388	6 910 e	3 827	3 269	2 919 e	18 356	20 725	18 221 e
New Zealand	-	-	-	-	-	-	-	-	8 e
Norway	2 267	1 576	1 785 e	118	45	123 e	1 317	1 169	2 046 e
Poland	262	408	390 e	139	301	157 e	65	113	340 e
Spain	1 465	1 379	874 e	31	5	139 e	320	401	343 e
Sweden	2 051	1 401	1 364 e	32	20	125 e	3 965	3 636	3 250 e
Turkey	-	9	425 e	81	27	-	3	-	-
United Kingdom	5 735	5 905	5 440 e	1 347	1 518	1 538 e	6 362	5 304	6 414 e
United States	407	653	477 e	823	526	1 200 e	1 595	1 086	1 421 e
Other	1 368	1 618	1 692 e	85	21	138 e	1 720	1 827	1 923 e
OECD	**45 741**	**47 179**	**45 556 e**	**15 506**	**13 495**	**13 768 e**	**63 419**	**66 830**	**64 938 e**
Argentina	1 242	1 274	1 562 e	-	-	8 e	14	98	94 e
Brazil	1 539	1 623	1 433 e	-	-	-	-	246	46 e
Colombia	210	460	15 e	119	231	122 e	20	86	101 e
Ecuador	58	21	37 e	-	-	-	-	-	-
Netherlands Antilles	168	276	54 e	753	731	484 e	546	688	685 e
Peru	13	87	-	-	-	-	41	44	-
Trinidad and Tobago	376	376	370 e	125	154	87 e	138	244	98 e
Venezuela	4 508	3 527	2 409 e	1 796	1 536	1 712 e	3 113	2 701	1 563 e
Other	237	305	340 e	101	135	92 e	46	100	328 e
Latin America	**8 351**	**7 949**	**6 220 e**	**2 894**	**2 787**	**2 505 e**	**3 918**	**4 207**	**2 915 e**
Romania	44	123	858 e	-	-	-	207	508	989 e
Other	212	507	283 e	45	48	28 e	414	424	366 e
Non-OECD Europe	**256**	**630**	**1 141 e**	**45**	**48**	**28 e**	**621**	**932**	**1 355 e**
FSU	**2 675**	**2 247**	**3 485 e**	**1 048**	**918**	**1 249 e**	**16 775**	**19 253**	**21 149 e**
Iran	-	-	-	-	48	-	31	-	-
Iraq	-	-	-	-	-	-	-	-	-
Kuwait	-	-	-	2 094	2 150	1 673 e	126	-	-
Oman	-	-	-	-	82	-	-	-	-
Qatar	5	19	-	40	40	31 e	14	-	73 e
Saudi Arabia	721	675	825 e	1 308	1 746	1 081 e	370	382	287 e
United Arab Emirates	47	142	241 e	1 289	2 779	2 204 e	27	125	68 e
Other	179	106	307 e	725	823	499 e	149	148	174 e
Middle East	**952**	**942**	**1 373 e**	**5 456**	**7 668**	**5 488 e**	**717**	**655**	**602 e**
India	144	246	565 e	-	82	56 e	-	640	455 e
Indonesia	5	-	15 e	-	3	-	29	130	-
Singapore	1 232	1 518	1 876 e	562	679	781 e	951	797	1 977 e
Other	125	125	370 e	754	798	1 034 e	557	801	1 200 e
Asia	**1 506**	**1 889**	**2 826 e**	**1 316**	**1 562**	**1 871 e**	**1 537**	**2 368**	**3 632 e**
China	**661**	**604**	**382 e**	**440**	**303**	**423 e**	**19**	**6**	**41 e**
Algeria	268	181	572 e	837	532	501 e	1 835	1 982	1 109 e
Angola/Cabinada	-	-	-	-	-	-	-	101	-
Cameroon	28	-	-	-	-	-	-	73	-
Egypt	63	289	954 e	-	167	301 e	-	-	-
Gabon	-	-	-	-	-	-	-	-	-
Libya	84	37	148 e	575	1 012	515 e	755	487	548 e
Nigeria	25	121	308 e	56	99	-	-	95	62 e
Tunisia	-	-	51 e	74	4	-	-	144	104 e
Other	95	146	53 e	70	30	97 e	63	245	93 e
Africa	**563**	**774**	**2 086 e**	**1 612**	**1 844**	**1 414 e**	**2 653**	**3 127**	**1 916 e**
Non-spec./Other	**1 397**	**3 872**	**3 005 e**	**2 559**	**4 150**	**3 708 e**	**1 680**	**7 829**	**7 798 e**
Total Imports	**62 102**	**66 086**	**66 074 e**	**30 876**	**32 775**	**30 454 e**	**91 339**	**105 207**	**104 346 e**
Memo: Total OPEC	*5 663*	*4 702*	*4 518 e*	*7 995*	*9 945*	*7 717 e*	*6 300*	*5 902*	*3 710 e*

Source: Annual Oil Statistics.

IEA TOTAL
Table 7D: Imports by Country of Origin

000 metric tons

	2000	2001	2002	2000	2001	2002	2000	2001	2002
	Fuel Oil (Residual)			**Petroleum Coke**			**Other Products**		
Australia	-	-	33 e	1	1	-	296	290	315 e
Belgium	3 081	2 428	3 133 e	333	279	398 e	1 540	1 482	1 965 e
Canada	1 425	1 858	1 735 e	80	33	322 e	955	1 076	1 320 e
Czech Republic	70	65	49 e	-	-	-	177	77	87 e
Denmark	558	632	730 e	6	11	6 e	29	23	38 e
Finland	130	32	-	1	-	-	59	105	113 e
France	2 540	2 386	2 415 e	25	12	28 e	2 011	1 867	1 565 e
Germany	2 485	2 222	1 220 e	156	152	191 e	1 276	1 398	1 288 e
Greece	27	25	32 e	-	-	-	29	3	9 e
Hungary	4	70	-	-	2	11 e	98	58	57 e
Ireland	38	84	8 e	-	-	-	4	-	-
Italy	1 162	466	468 e	67	14	144 e	663	880	511 e
Japan	157	377	208 e	25	59	54 e	26	56	26 e
Korea	144	158	625 e	-	-	92 e	325	130	176 e
Mexico	491	-	541 e	-	-	-	286	47	34 e
Netherlands	4 253	3 357	4 103 e	411	552	454 e	1 468	1 339	1 316 e
New Zealand	59	-	-	-	-	-	62	11	17 e
Norway	772	1 016	968 e	25	1	32 e	1 389	651	939 e
Poland	663	567	555 e	-	-	-	133	117	110 e
Spain	799	372	181 e	62	18	89 e	766	609	660 e
Sweden	1 409	1 335	1 431 e	8	21	-	514	468	509 e
Turkey	80	311	40 e	-	-	-	93	121	27 e
United Kingdom	5 768	6 457	5 623 e	335	668	420 e	1 421	1 177	1 304 e
United States	2 226	3 058	3 026 e	9 535	8 270	10 940 e	2 900	3 452	3 526 e
Other	1 460	1 103	1 148 e	16	14	14 e	408	373	399 e
OECD	**29 801**	**28 379**	**28 272 e**	**11 086**	**10 107**	**13 195 e**	**16 928**	**15 810**	**16 311 e**
Argentina	85	239	167 e	32	221	310 e	24	41	121 e
Brazil	1 157	1 860	631 e	-	9	84 e	107	79	77 e
Colombia	727	964	741 e	-	-	-	39	11	13 e
Ecuador	30	245	128 e	58	-	-	-	111	22 e
Netherlands Antilles	844	958	562 e	-	-	202 e	269	44	205 e
Peru	113	279	277 e	-	-	-	-	2	-
Trinidad and Tobago	1 279	1 344	1 377 e	-	-	-	145	154	-
Venezuela	5 896	4 917	1 696 e	959	1 442	1 614 e	1 629	1 404	1 006 e
Other	287	768	1 040 e	-	115	169 e	20	75	32 e
Latin America	**10 418**	**11 574**	**6 619 e**	**1 049**	**1 787**	**2 379 e**	**2 233**	**1 921**	**1 476 e**
Romania	8	49	106 e	2	-	14 e	14	14	21 e
Other	249	271	391 e	12	5	10 e	37	43	46 e
Non-OECD Europe	**257**	**320**	**497 e**	**14**	**5**	**24 e**	**51**	**57**	**67 e**
FSU	**14 394**	**13 112**	**15 702 e**	**14**	**21**	**28 e**	**594**	**783**	**882 e**
Iran	13	-	21 e	-	-	1 e	19	38	30 e
Iraq	-	-	-	-	-	-	-	-	-
Kuwait	-	69	-	-	183	218 e	-	-	-
Oman	-	-	-	-	-	-	-	-	-
Qatar	-	-	-	-	-	-	2	249	-
Saudi Arabia	9	235	32 e	-	-	-	240	61	18 e
United Arab Emirates	-	42	105 e	-	-	-	131	100	-
Other	1 163	744	337 e	55	-	-	195	133	32 e
Middle East	**1 185**	**1 090**	**495 e**	**55**	**183**	**219 e**	**587**	**581**	**80 e**
India	-	-	-	144	298	200 e	14	78	153 e
Indonesia	3 681	3 218	2 970 e	69	62	57 e	42	185	70 e
Singapore	1 042	719	763 e	-	-	-	169	74	102 e
Other	52	154	467 e	-	78	124 e	79	66	78 e
Asia	**4 775**	**4 091**	**4 200 e**	**213**	**438**	**381 e**	**304**	**403**	**403 e**
China	**48**	**27**	**536 e**	**28**	**169**	**280 e**	**67**	**103**	**63 e**
Algeria	3 043	168	582 e	-	-	-	4 476	5 104	5 341 e
Angola/Cabinada	150	302	62 e	-	-	-	39	-	29 e
Cameroon	171	41	52 e	-	-	-	7	-	-
Egypt	107	247	109 e	-	-	-	19	7	24 e
Gabon	-	-	-	22	-	-	-	-	-
Libya	2 548	2 143	2 257 e	-	-	-	126	205	54 e
Nigeria	862	1 190	511 e	-	-	-	-	97	12 e
Tunisia	317	91	30 e	-	-	-	-	-	9 e
Other	1 169	590	929 e	80	21	21 e	185	381	430 e
Africa	**8 367**	**4 772**	**4 532 e**	**102**	**21**	**21 e**	**4 852**	**5 794**	**5 899 e**
Non-spec./Other	**2 586**	**4 418**	**4 627 e**	**4 570**	**4 826**	**672 e**	**1 024**	**1 058**	**1 736 e**
Total Imports	**71 831**	**67 783**	**65 480 e**	**17 131**	**17 557**	**17 199 e**	**26 640**	**26 510**	**26 917 e**
Memo: Total OPEC	*16 052*	*11 982*	*8 174 e*	*1 028*	*1 687*	*1 890 e*	*6 665*	*7 443*	*6 531 e*

Source: Annual Oil Statistics.

IEA TOTAL
Table 8A: Exports by Country of Destination

000 metric tons

	2000	2001	2002	2000	2001	2002	2000	2001	2002
	Crude Oil			Natural Gas Liquids			Feedstocks and Other		
Australia	-	-	-	1	42	11 e	-	-	-
Austria	31	62	167 e	-	-	-	-	-	-
Belgium	5 393	4 278	2 571 e	559	749	805 e	410	395	280 e
Canada	18 286	17 431	14 843 e	587	368	286 e	132	237	188 e
Denmark	4 093	2 813	3 239 e	14	23	29 e	27	28	40 e
Finland	5 553	5 897	5 549 e	87	101	26 e	1	4	20 e
France	36 996	36 090	27 433 e	861	782	936 e	624	301	259 e
Germany	21 118	24 502	23 431 e	278	245	192 e	810	460	502 e
Greece	-	-	134 e	-	4	-	-	28	-
Ireland	2 875	3 174	2 387 e	9	15	14 e	7	-	24 e
Italy	4 112	3 371	6 064 e	26	14	38 e	137	56	64 e
Japan	3 320	2 915	2 164 e	1 004	1 253	1 728 e	693	418	473 e
Luxembourg	-	-	-	-	-	-	-	-	-
Mexico	4	-	8 e	1 581	1 014	1 163 e	479	363	375 e
Netherlands	47 206	47 042	43 930 e	1 018	1 368	1 498 e	2 538	1 987	1 477 e
Norway	897	362	723 e	215	41	75 e	50	49	36 e
Portugal	1 610	1 434	1 437 e	201	209	211 e	16	9	31 e
Spain	2 758	2 845	2 239 e	456	362	367 e	241	86	134 e
Sweden	14 617	12 732	10 914 e	579	998	968 e	39	36	168 e
Switzerland	-	-	-	-	-	8 e	28	39	46 e
Turkey	-	-	-	495	493	589 e	52	66	52 e
United Kingdom	33 883	35 871	34 910 e	177	336	233 e	151	149	184 e
United States	104 747	98 361	108 623 e	12 674	11 992	10 922 e	8 315	8 524	13 087 e
Other OECD	5 987	4 206	6 587 e	215	95	198 e	252	323	222 e
OECD	**313 486**	**303 386**	**297 353 e**	**21 037**	**20 504**	**20 297 e**	**15 002**	**13 558**	**17 662 e**
Africa	163	191	501 e	31	51	107 e	175	104	73 e
Latin America	308	435	1 042 e	496	383	301 e	515	470	344 e
Asia	6 963	8 156	7 434 e	70	63	140 e	538	605	685 e
China	2 132	3 377	4 846 e	384	381	570 e	171	145	7 e
Non-OECD Europe	66	534	542 e	-	-	41 e	105	150	29 e
Former Soviet Union	251	-	-	-	-	6 e	7	-	-
Middle East	79	-	-	47	9	-	20	4	27 e
Non-specified/Other	1 306	4 167	5 685 e	408	414	356 e	797	737	752 e
Non-OECD	**11 268**	**16 860**	**20 050 e**	**1 436**	**1 301**	**1 521 e**	**2 328**	**2 215**	**1 917 e**
Total Exports	**324 754**	**320 246**	**317 403 e**	**22 473**	**21 805**	**21 818 e**	**17 330**	**15 773**	**19 579 e**
	Total Products			Liquefied Petroleum Gas			Naphtha		
Australia	1 646	1 171	1 109 e	-	-	-	-	-	3 e
Austria	4 111	4 172	4 575 e	111	109	115 e	43	35	15 e
Belgium	21 017	23 873	21 868 e	805	937	1 117 e	3 196	3 146	3 101 e
Canada	4 465	4 741	5 454 e	4	-	-	-	-	1 e
Denmark	4 238	4 043	3 961 e	19	9	1 e	165	157	146 e
Finland	1 432	1 503	1 676 e	24	36	39 e	152	64	103 e
France	16 368	15 519	14 776 e	183	291	340 e	1 589	1 241	1 359 e
Germany	39 887	39 454	31 273 e	610	432	579 e	6 739	6 545	4 903 e
Greece	1 347	1 301	1 229 e	2	-	2 e	70	50	49 e
Ireland	5 083	4 671	5 310 e	88	51	49 e	-	9	-
Italy	8 841	7 401	7 636 e	304	248	283 e	485	647	602 e
Japan	18 408	17 583	15 394 e	90	60	40 e	5 506	5 951	5 208 e
Luxembourg	2 464	2 499	2 552 e	18	17	15 e	-	-	-
Mexico	15 085	12 237	11 760 e	-	29	15 e	-	-	32 e
Netherlands	15 955	16 014	18 634 e	667	332	418 e	2 356	2 636	3 497 e
Norway	4 667	4 947	5 374 e	129	150	174 e	37	125	52 e
Portugal	3 358	2 233	2 965 e	225	164	188 e	427	248	233 e
Spain	11 144	11 620	13 243 e	152	112	153 e	272	419	294 e
Sweden	4 944	4 844	4 579 e	71	106	195 e	402	295	269 e
Switzerland	7 846	6 823	7 416 e	68	47	75 e	96	67	43 e
Turkey	3 425	1 643	2 146 e	295	96	338 e	142	11	39 e
United Kingdom	14 602	18 439	15 452 e	475	489	446 e	1 432	1 785	1 743 e
United States	37 808	41 804	42 908 e	366	463	568 e	1 348	1 201	1 096 e
Other OECD	6 825	7 893	7 189 e	307	325	377 e	190	273	185 e
OECD	**254 966**	**256 428**	**248 479 e**	**5 013**	**4 503**	**5 527 e**	**24 647**	**24 905**	**22 973 e**
Africa	9 334	8 879	8 992 e	879	792	712 e	81	140	68 e
Latin America	9 664	9 681	10 570 e	122	200	22 e	59	30	43 e
Asia	13 746	16 260	18 378 e	42	45	43 e	292	288	340 e
China	16 441	16 432	9 006 e	621	411	207 e	115	70	134 e
Non-OECD Europe	9 209	8 296	8 009 e	233	198	322 e	290	299	152 e
Former Soviet Union	708	582	413 e	21	-	4 e	15	19	24 e
Middle East	4 317	3 364	2 978 e	257	172	60 e	155	101	97 e
Non-specified/Other	5 992	7 902	16 272 e	533	840	1 088 e	1 206	823	876 e
Non-OECD	**69 411**	**71 396**	**74 618 e**	**2 708**	**2 658**	**2 458 e**	**2 213**	**1 770**	**1 734 e**
Total Exports	**324 377**	**327 824**	**323 097 e**	**7 721**	**7 161**	**7 985 e**	**26 860**	**26 675**	**24 707 e**

Source: Annual Oil Statistics.

IEA TOTAL
Table 8B: Exports by Country of Destination

000 metric tons

	Total Gasoline 2000	Total Gasoline 2001	Total Gasoline 2002	Total Kerosene 2000	Total Kerosene 2001	Total Kerosene 2002	Gas/Diesel Oil 2000	Gas/Diesel Oil 2001	Gas/Diesel Oil 2002
Australia	161	193	66 e	3	17	13 e	686	270	147 e
Austria	607	567	549 e	47	38	89 e	2 646	2 649	3 203 e
Belgium	1 825	1 761	1 462 e	1 046	1 370	895 e	6 652	7 282	5 812 e
Canada	894	1 058	1 211 e	861	537	767 e	470	680	290 e
Denmark	873	851	860 e	477	497	343 e	1 865	1 738	1 648 e
Finland	322	235	297 e	1	45	27 e	318	506	455 e
France	1 731	1 578	1 537 e	239	129	333 e	8 041	7 904	7 002 e
Germany	10 379	8 436	6 560 e	3 219	3 506	3 300 e	14 004	16 179	11 574 e
Greece	290	23	143 e	2	22	6 e	415	637	471 e
Ireland	799	983	1 163 e	641	606	772 e	1 671	1 521	1 788 e
Italy	505	589	579 e	64	41	140 e	347	125	635 e
Japan	568	604	831 e	3 438	3 283	3 111 e	3 767	3 113	1 684 e
Luxembourg	621	567	565 e	313	280	370 e	1 431	1 515	1 521 e
Mexico	5 780	5 067	5 864 e	195	108	238 e	3 737	1 814	989 e
Netherlands	2 971	3 055	3 823 e	190	397	263 e	3 699	3 326	5 154 e
Norway	537	751	830 e	191	316	282 e	1 061	1 217	1 439 e
Portugal	264	30	46 e	42	17	9 e	724	448	513 e
Spain	1 060	909	883 e	195	126	68 e	4 282	4 453	5 673 e
Sweden	1 659	1 650	1 597 e	478	253	210 e	1 334	1 376	1 305 e
Switzerland	2 827	2 382	2 602 e	1 076	988	885 e	3 166	2 819	3 252 e
Turkey	403	128	68 e	59	30	74 e	577	144	188 e
United Kingdom	3 191	4 909	3 174 e	1 265	577	917 e	3 312	4 364	3 400 e
United States	12 715	16 309	18 768 e	2 817	3 419	2 459 e	7 972	7 351	7 822 e
Other OECD	2 227	2 177	1 921 e	299	422	313 e	1 961	2 305	2 509 e
OECD	**53 209**	**54 812**	**55 399 e**	**17 158**	**17 024**	**15 884 e**	**74 138**	**73 736**	**68 474 e**
Africa	2 383	2 038	2 563 e	326	232	249 e	2 675	2 478	2 480 e
Latin America	1 009	1 001	1 660 e	256	152	102 e	2 842	2 144	2 663 e
Asia	1 264	787	1 074 e	563	945	1 028 e	5 919	7 313	8 414 e
China	73	184	12 e	992	1 339	895 e	3 728	3 315	653 e
Non-OECD Europe	1 335	1 357	1 796 e	522	416	511 e	2 321	2 072	2 295 e
Former Soviet Union	267	147	60 e	1	1	5 e	229	176	69 e
Middle East	1 338	771	1 082 e	360	309	64 e	731	196	447 e
Non-specified/Other	1 900	1 944	3 273 e	126	352	399 e	456	1 650	6 195 e
Non-OECD	**9 569**	**8 229**	**11 520 e**	**3 146**	**3 746**	**3 253 e**	**18 901**	**19 344**	**23 216 e**
Total Exports	**62 778**	**63 041**	**66 919 e**	**20 304**	**20 770**	**19 137 e**	**93 039**	**93 080**	**91 690 e**

	Fuel Oil (Residual) 2000	Fuel Oil (Residual) 2001	Fuel Oil (Residual) 2002	Petroleum Coke 2000	Petroleum Coke 2001	Petroleum Coke 2002	Other Products 2000	Other Products 2001	Other Products 2002
Australia	204	49	70 e	563	610	781 e	29	32	29 e
Austria	192	337	185 e	31	30	16 e	434	407	403 e
Belgium	3 555	5 058	4 945 e	871	957	1 282 e	3 067	3 362	3 254 e
Canada	781	853	1 504 e	811	1 085	999 e	644	528	682 e
Denmark	404	286	421 e	149	209	243 e	286	296	299 e
Finland	389	440	572 e	-	-	10 e	226	177	173 e
France	2 561	2 303	1 960 e	723	830	828 e	1 301	1 243	1 417 e
Germany	2 335	1 823	1 975 e	241	230	276 e	2 360	2 303	2 106 e
Greece	142	33	49 e	302	448	371 e	124	88	138 e
Ireland	1 292	1 041	909 e	195	131	291 e	397	329	338 e
Italy	4 009	2 872	2 803 e	2 144	2 103	1 823 e	303	776	771 e
Japan	842	701	508 e	3 371	3 192	3 292 e	826	679	720 e
Luxembourg	11	15	12 e	18	18	16 e	52	87	53 e
Mexico	3 158	2 663	2 299 e	1 808	2 072	1 630 e	407	484	693 e
Netherlands	3 381	3 411	3 229 e	1 590	1 674	1 177 e	1 101	1 183	1 073 e
Norway	2 083	1 784	1 384 e	227	278	345 e	402	326	868 e
Portugal	1 082	956	1 651 e	373	130	77 e	221	240	248 e
Spain	1 760	1 797	2 332 e	2 629	2 944	2 968 e	794	860	872 e
Sweden	512	751	573 e	92	90	97 e	396	323	333 e
Switzerland	245	158	201 e	70	42	57 e	298	320	301 e
Turkey	528	22	130 e	1 148	1 083	1 124 e	273	129	185 e
United Kingdom	3 550	4 906	3 896 e	519	606	672 e	858	803	1 204 e
United States	9 220	9 639	8 419 e	115	149	136 e	3 255	3 273	3 640 e
Other OECD	636	713	343 e	342	452	612 e	863	1 226	929 e
OECD	**43 552**	**42 611**	**40 370 e**	**18 332**	**19 363**	**19 123 e**	**18 917**	**19 474**	**20 729 e**
Africa	1 276	1 152	1 094 e	988	1 201	993 e	726	846	833 e
Latin America	2 780	3 728	3 822 e	1 876	1 724	1 403 e	720	702	855 e
Asia	4 160	5 570	6 264 e	462	260	89 e	1 044	1 052	1 126 e
China	9 903	9 876	5 076 e	4	86	571 e	1 005	1 151	1 458 e
Non-OECD Europe	3 605	2 867	1 974 e	316	383	444 e	587	704	515 e
Former Soviet Union	16	48	33 e	15	46	71 e	144	145	147 e
Middle East	518	539	408 e	620	812	472 e	338	464	348 e
Non-specified/Other	1 258	1 705	2 751 e	-	7	927 e	513	581	763 e
Non-OECD	**23 516**	**25 485**	**21 422 e**	**4 281**	**4 519**	**4 970 e**	**5 077**	**5 645**	**6 045 e**
Total Exports	**67 068**	**68 096**	**61 792 e**	**22 613**	**23 882**	**24 093 e**	**23 994**	**25 119**	**26 774 e**

Source: Annual Oil Statistics.

EUROPEAN UNION

Figure 1
FUEL SHARES OF TPES *

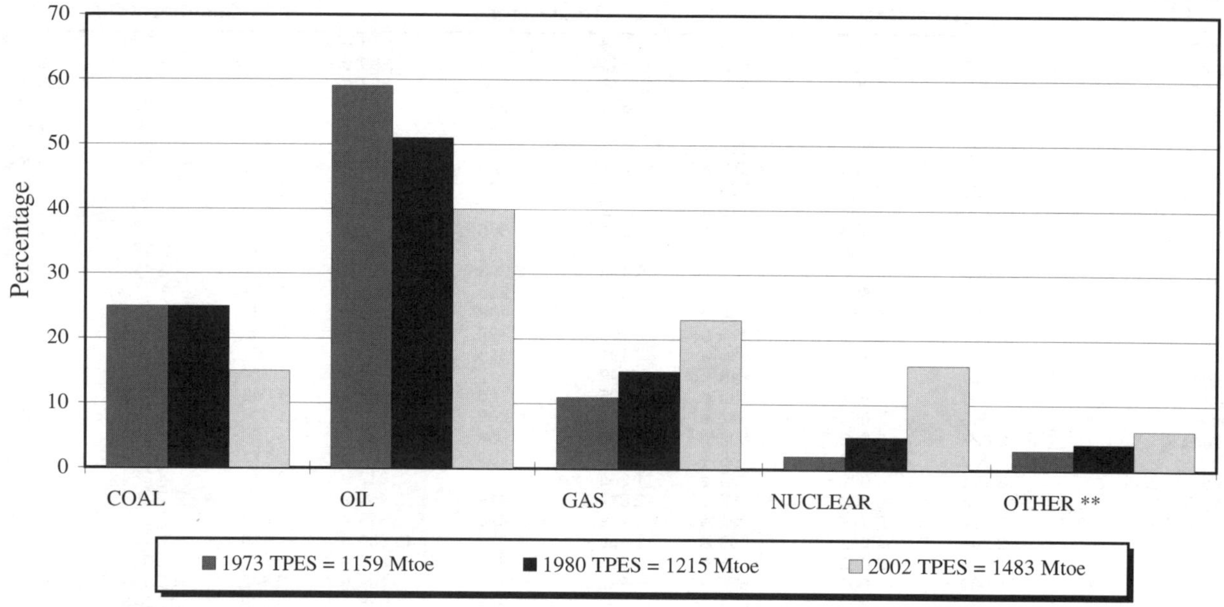

Source: Annual Energy Balances.

* Excluding electricity trade.

** Includes hydro, geothermal, solar, wind, combustible renewables and waste, etc.

Figure 2
OIL PRODUCTION AND CONSUMPTION

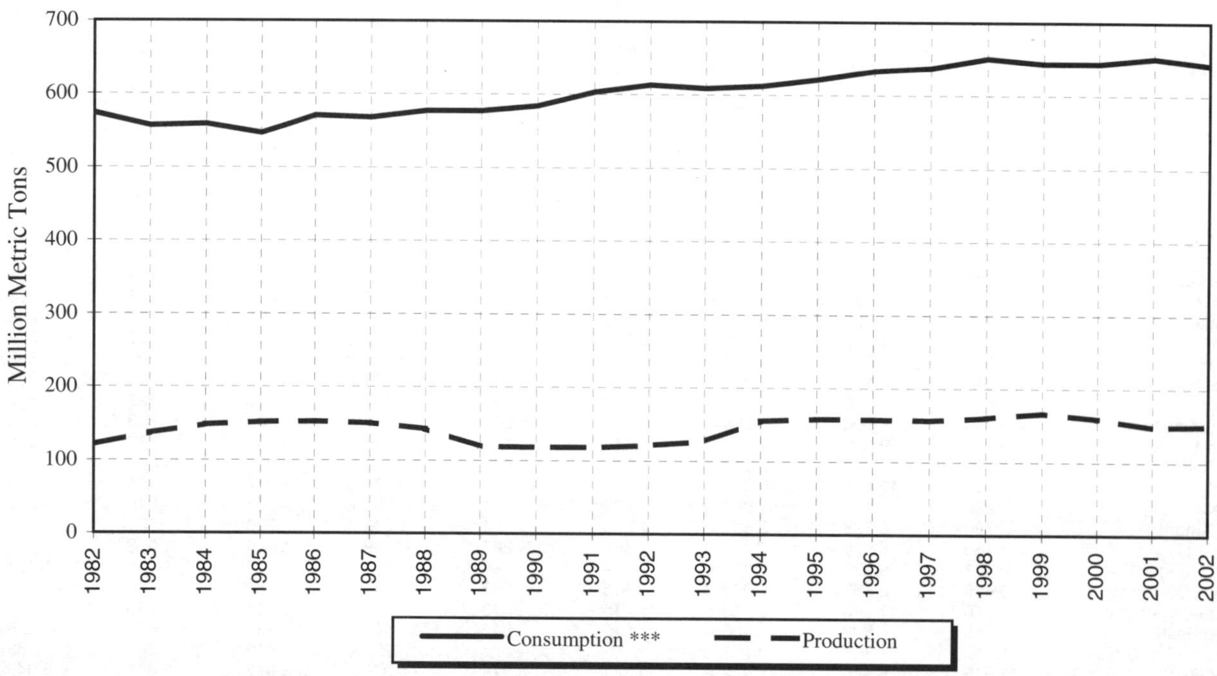

Source: Annual Oil Statistics.

*** Includes refinery fuel and international marine bunkers.

EUROPEAN UNION
Table 1: Oil Supply and Consumption - Summary Table

000 metric tons

	1973	1978	1980	1990	1999	2000	2001	2002
Indigenous Production	15 306	67 656	95 922	118 495	167 125	159 666	148 479	150 528 e
Crude Oil	..	65 750	91 452	111 593	154 920	147 697	136 836	138 584 e
NGL	..	1 891	2 907	4 969	10 120	9 673	9 541	9 583 e
Other	..	15	1 563	1 933	2 085	2 296	2 102	2 361 e
+ Net Imports	711 070	606 568	563 329	457 420	458 546	473 758	495 146	480 927 e
Crude Oil, NGL, Other	712 454	583 794	535 516	434 089	441 795	456 142	466 539	454 943 e
Petroleum Products	-1 384	22 774	27 813	23 331	16 751	17 616	28 607	25 984 e
- Intl. Marine Bunkers	42 808	35 789	31 404	35 101	40 969	43 649	45 066	46 808 e
+ Stock Changes	-9 792	8 807	-11 700	791	10 177	-3 512	- 754	2 476 e
TOTAL SUPPLY	**673 776**	**647 242**	**616 147**	**541 605**	**594 879**	**586 263**	**597 805**	**587 123 e**
Refinery Intake	*732 439*	*668 995*	*652 422*	*592 750*	*649 042*	*661 933*	*659 357*	*656 151 e*
Refinery Output	*725 109*	*661 385*	*645 681*	*587 243*	*640 768*	*652 788*	*648 035*	*647 623 e*
LPG + Ethane	*13 255*	*13 765*	*13 564*	*15 013*	*17 282*	*18 289*	*18 540*	*19 071 e*
Naphtha	*41 052*	*32 100*	*31 151*	*34 207*	*38 189*	*41 286*	*40 508*	*36 250 e*
Kerosenes	*29 187*	*25 993*	*26 957*	*37 376*	*44 803*	*43 314*	*40 740*	*38 939 e*
Gasolines	*89 019*	*100 119*	*107 422*	*131 801*	*137 825*	*135 102*	*133 296*	*136 019 e*
Gas/Diesel Oil	*213 579*	*199 107*	*199 506*	*184 620*	*222 408*	*230 893*	*232 140*	*231 352 e*
Fuel Oil	*284 692*	*237 623*	*215 425*	*119 090*	*106 717*	*109 671*	*106 304*	*105 242 e*
Other Products	*54 325*	*52 678*	*51 656*	*65 136*	*73 544*	*74 233*	*76 507*	*80 750 e*
- Refinery Losses	7 330	7 610	6 741	5 507	8 274	9 145	11 322	8 528 e
+ Returns	5 116	10 417	15 803	12 885	13 488	16 310	16 013	13 294 e
- Statistical Differences	3 678	2 613	1 968	191	-3 137	-7 059	-3 940	-3 519 e
TOTAL CONSUMPTION	**667 884**	**647 436**	**623 241**	**548 792**	**603 230**	**600 487**	**606 436**	**595 408 e**
Transformation and Energy	**140 239**	**135 015**	**138 610**	**92 544**	**92 203**	**90 594**	**88 478**	..
Crude Oil	15	30	300	205	-	-	-	..
LPG + Ethane	1 179	1 268	1 845	1 291	1 224	1 495	1 419	..
Gas/Diesel Oil	2 444	3 019	4 003	3 487	3 967	4 230	4 145	..
Fuel Oil	112 373	103 871	101 889	54 165	49 752	47 002	47 442	..
Petroleum Coke	1 549	2 001	2 336	4 909	5 392	5 318	4 875	..
Other	22 679	24 826	28 237	28 487	31 868	32 549	30 597	..
Transport	**152 815**	**177 802**	**185 385**	**244 227**	**296 238**	**298 815**	**299 596**	..
Motor Gasoline	84 852	96 630	99 210	116 192	117 167	112 639	110 594	..
Gas/Diesel Oil	46 380	57 463	60 647	94 633	131 938	137 393	141 683	..
Subtotal	*131 232*	*154 093*	*159 857*	*210 825*	*249 105*	*250 032*	*252 277*	..
Aviation Fuels	17 260	10 060	19 055	27 405	41 041	43 050	41 479	..
Other	4 314	4 646	6 173	5 997	6 092	5 733	5 840	..
Industry	**202 514**	**175 332**	**160 779**	**115 527**	**121 439**	**122 679**	**122 480**	..
Naphtha	33 656	27 929	28 297	30 196	36 611	37 551	36 183	..
LPG + Ethane	5 573	5 402	5 331	8 088	10 126	9 942	9 491	..
Gas/Diesel Oil	25 588	23 968	21 782	13 267	14 770	15 287	15 279	..
Fuel Oil	100 819	84 456	75 789	27 671	17 254	17 510	17 254	..
Other	36 878	33 577	29 580	36 305	42 678	42 389	44 273	..
Other Sectors	**172 316**	**159 287**	**138 467**	**96 494**	**93 350**	**88 399**	**95 882**	..
Gas/Diesel Oil	129 204	124 663	107 671	78 461	75 637	72 363	78 462	..
Fuel Oil	28 568	21 722	18 724	6 142	4 103	2 891	2 938	..
Other	14 544	12 902	12 072	11 891	13 610	13 145	14 482	..
Memo: Non-Energy Use	*66 017*	*60 201*	*62 296*	*71 456*	*81 816*	*81 277*	*80 088*	..
Chem/Petchem. Feedst.	*38 761*	*36 423*	*37 636*	*46 008*	*54 675*	*54 904*	*53 836*	..
Other	*27 256*	*23 778*	*24 660*	*25 448*	*27 141*	*26 373*	*26 252*	..

Source: Annual Oil Statistics.

EUROPEAN UNION
Table 2: Supply of Crude Oil, NGL, Feedstocks and Other Hydrocarbons

000 metric tons

	1973	1978	1980	1990	1999	2000	2001	2002
CRUDE OIL								
Indigenous Production	..	65 750	91 452	111 593	154 920	147 697	136 836	138 584 e
+ Imports	..	603 171	563 297	458 671	524 482	544 203	541 706	529 883 e
- Exports	..	25 771	42 300	60 164	98 282	103 390	94 117	97 660 e
- Direct Use	..	103	375	233	12	12	12	12 e
+ Stock Change	..	3 831	-6 494	- 312	1 862	1 974	-3 713	932 e
- Statistical Difference	..	2 078	876	-1 970	286	2 061	- 631	1 534 e
Refinery Intake	**..**	**644 800**	**604 704**	**511 525**	**582 684**	**588 411**	**581 331**	**570 193 e**
Closing Stock Level	..	*56 013*	*65 710*	*44 389*	*48 794*	*46 820*	*50 533*	*49 601 e*
NATURAL GAS LIQUIDS								
Indigenous Production	..	1 891	2 907	4 969	10 120	9 673	9 541	9 583 e
+ Imports	..	18	104	3 612	5 718	6 635	7 771	9 434 e
- Exports	..	342	511	1 252	3 870	3 549	3 522	3 390 e
- Direct Use	..	1 288	1 462	1 981	3 867	3 886	3 694	3 500 e
+ Stock Change	..	- 6	- 5	- 99	51	- 105	- 91	- 13 e
- Statistical Difference	..	-	-	- 419	- 8	- 580	- 176	- 163 e
Refinery Intake	**..**	**273**	**1 033**	**5 668**	**8 160**	**9 348**	**10 181**	**12 277 e**
Closing Stock Level	..	*23*	*70*	*529*	*255*	*360*	*451*	*464 e*
REFINERY FEEDSTOCKS								
Backflows	..	10 210	15 328	12 368	12 937	15 731	15 535	12 685 e
+ Transfers	..	7 446	14 334	29 575	31 709	35 452	34 586	41 132 e
+ Imports	..	7 332	16 216	39 391	20 059	19 785	20 015	21 376 e
- Exports	..	614	1 290	6 458	8 033	9 262	6 936	5 877 e
+ Stock Change	..	- 186	589	- 263	- 277	- 363	724	- 401 e
- Statistical Difference	..	266	40	879	- 76	-1 040	-2 078	-2 782 e
Refinery Intake	**..**	**23 922**	**45 137**	**73 734**	**56 471**	**62 383**	**66 002**	**71 697 e**
Closing Stock Level	..	*1 159*	*3 838*	*7 982*	*7 489*	*7 852*	*7 128*	*7 529 e*
OTHER HYDROCARBONS								
Indigenous Production	..	15	1 563	1 933	2 085	2 296	2 102	2 361 e
+ Imports	..	-	-	289	1 721	1 720	1 624	1 177 e
- Exports	..	-	-	-	-	-	2	-
- Direct Use	..	15	15	394	2 108	2 242	1 716	1 676 e
+ Stock Change	..	-	-	- 5	27	14	- 166	96 e
- Statistical Difference	..	-	-	-	- 2	- 3	- 1	- 26 e
Refinery Intake	**..**	**-**	**1 548**	**1 823**	**1 727**	**1 791**	**1 843**	**1 984 e**
Closing Stock Level	..	*39*	*39*	*56*	*192*	*178*	*344*	*248 e*
CRUDE OIL, NGL, FEEDSTOCKS AND OTHER								
Indigenous Production	15 306	67 656	95 922	118 495	167 125	159 666	148 479	150 528 e
+ Backflows	4 923	10 210	15 328	12 368	12 937	15 731	15 535	12 685 e
+ Transfers	5 259	7 446	14 334	29 575	31 709	35 452	34 586	41 132 e
+ Imports	716 166	610 521	579 617	501 963	551 980	572 343	571 116	561 870 e
- Exports	3 712	26 727	44 101	67 874	110 185	116 201	104 577	106 927 e
- Direct Use	1 254	1 406	1 852	2 608	5 987	6 140	5 422	5 188 e
+ Stock Change	-5 366	3 639	-5 910	- 679	1 663	1 520	-3 246	614 e
- Statistical Difference	-1 117	2 344	916	-1 510	200	438	-2 886	-1 437 e
Refinery Intake	**732 439**	**668 995**	**652 422**	**592 750**	**649 042**	**661 933**	**659 357**	**656 151 e**
Closing Stock Level	-	*57 234*	*69 657*	*52 956*	*56 730*	*55 210*	*58 456*	*57 842 e*

Source: Annual Oil Statistics.

EUROPEAN UNION
Table 3: Motor Gasoline - Supply

000 metric tons

	1973	1978	1980	1990	1999	2000	2001	2002
MOTOR GASOLINE								
Refinery Output	**87 878**	**97 898**	**104 996**	**130 431**	**137 307**	**134 701**	**132 990**	**135 872 e**
+ Imports	12 637	13 770	17 769	29 855	26 056	30 039	27 610	25 123 e
- Exports	13 771	16 695	19 507	36 194	44 334	45 641	43 808	45 802 e
+ Transfers	267	552	-2 560	-5 568	-3 677	-4 077	-3 901	-5 177 e
+ Stock Change	-1 661	617	-1 117	- 426	2 704	83	- 526	1 790 e
- Statistical Difference	- 429	-1 547	- 450	84	31	1 577	422	2 688 e
Total Consumption	**85 779**	**97 689**	**100 031**	**118 014**	**118 025**	**113 528**	**111 943**	**109 118 e**
Closing Stock Level	-	*14 833*	*16 715*	*18 391*	*17 116*	*17 033*	*17 596*	*15 806 e*

Table 4: Aviation Fuels - Supply

000 metric tons

	1973	1978	1980	1990	1999	2000	2001	2002
AVIATION GASOLINE								
Refinery Output	**405**	**289**	**422**	**244**	**165**	**232**	**207**	**126 e**
+ Imports	242	143	152	125	138	128	79	64 e
- Exports	306	202	390	214	170	205	102	92 e
+ Transfers	-	- 3	-	3	21	- 14	10	7 e
+ Stock Change	11	13	17	- 14	17	2	- 9	6 e
- Statistical Difference	43	28	17	4	34	- 1	33	- 28 e
Total Consumption	**309**	**212**	**184**	**140**	**137**	**144**	**152**	**139 e**
Closing Stock Level	-	*108*	*95*	*63*	*34*	*32*	*41*	*35 e*
GASOLINE TYPE JET FUEL								
Refinery Output	**736**	**1 932**	**2 004**	**1 126**	**353**	**169**	**99**	**21 e**
+ Imports	320	435	448	233	7	11	9	5 e
- Exports	415	989	1 271	390	72	77	32	-
+ Transfers	-	- 33	42	- 20	- 214	16	- 1	- 7 e
+ Stock Change	6	- 103	145	8	46	1	5	2 e
- Statistical Difference	- 12	- 41	17	208	2	1	-	-
Total Consumption	**659**	**1 283**	**1 351**	**749**	**118**	**119**	**80**	**21 e**
Closing Stock Level	-	*529*	*542*	*57*	*14*	*13*	*8*	*6 e*
KEROSENE TYPE JET FUEL								
Refinery Output	**19 504**	**19 367**	**21 604**	**31 799**	**40 088**	**38 775**	**36 155**	**34 036 e**
+ Imports	4 152	4 492	5 712	7 234	11 440	14 069	17 497	15 671 e
- Exports	6 380	5 353	7 676	10 044	10 295	9 520	9 444	8 707 e
+ Transfers	961	- 157	- 164	- 459	- 186	182	-1 237	- 899 e
+ Stock Change	- 327	255	- 360	- 470	- 49	- 362	- 246	259 e
- Statistical Difference	1 588	1 011	1 261	1 499	162	295	1 423	- 31 e
Total Consumption	**16 322**	**17 593**	**17 855**	**26 561**	**40 836**	**42 849**	**41 302**	**40 391 e**
Closing Stock Level	-	*2 078*	*3 267*	*3 448*	*4 290*	*4 652*	*4 898*	*4 639 e*

Source: Annual Oil Statistics.

EUROPEAN UNION
Table 5A: Gas/Diesel Oil - Supply and Consumption

000 metric tons

	1973	1978	1980	1990	1999	2000	2001	2002
Refinery Output	**213 579**	**199 107**	**199 506**	**184 620**	**222 408**	**230 893**	**232 140**	**231 352 e**
+ Recycled Products	-	-	8	30	20	23	30	30 e
+ Imports	46 773	49 378	49 501	57 758	69 005	67 194	77 747	79 817 e
- Exports	46 739	37 294	39 941	45 071	57 982	58 780	61 075	62 764 e
- Intl. Marine Bunkers	7 074	7 446	6 273	7 764	8 571	8 664	8 820	8 618 e
+ Transfers	-1 219	184	-2 285	-1 722	-6 460	-4 812	-2 313	-4 000 e
+ Stock Changes	- 876	3 768	-6 541	- 169	5 071	-2 321	732	- 375 e
- Statistical Differences	828	-1 416	- 128	-2 166	-2 821	-5 740	-1 128	-1 859 e
Total Consumption	**203 616**	**209 113**	**194 103**	**189 848**	**226 312**	**229 273**	**239 569**	**237 301 e**
Transformation	**1 800**	**2 454**	**3 538**	**3 118**	**3 328**	**3 706**	**3 738**	..
Public Electricity	1 461	1 084	1 187	1 556	1 124	1 284	1 329	..
Autoproducer Electricity	-	336	389	202	207	172	167	..
Public CHP Plants	37	17	44	24	503	877	612	..
Autoproducers of CHP	-	-	-	21	430	102	384	..
Public Heat Plants	26	22	18	61	92	116	123	..
Autoproducer Heat Plants	-	-	-	-	1	36	35	..
Petrochemical Industry	218	972	1 877	1 238	962	1 112	1 088	..
Other Transformation	58	23	23	16	9	7	-	..
Energy Sector	**644**	**565**	**465**	**369**	**639**	**524**	**407**	..
Refinery Fuel	65	102	63	59	277	268	177	..
Other Energy Sector	579	463	402	310	362	256	230	..
Final Consumption	**201 172**	**206 094**	**190 100**	**186 361**	**222 345**	**225 043**	**235 424**	..
Transport	**46 380**	**57 463**	**60 647**	**94 633**	**131 938**	**137 393**	**141 683**	..
Air	-	-	-	-	-	-	-	..
Road	35 888	47 838	50 848	87 131	125 197	131 001	135 574	..
Rail	3 732	3 906	3 738	2 819	2 497	2 412	2 229	..
Inland Waterways	6 756	5 719	6 055	4 651	4 207	3 955	3 850	..
Non-Specified	4	-	6	32	37	25	30	..
Industry	**25 588**	**23 968**	**21 782**	**13 267**	**14 770**	**15 287**	**15 279**	..
Iron and Steel	889	646	570	356	248	285	352	..
Chemical (incl. Pet.-Chem.)	2 033	3 339	4 008	4 067	3 956	4 258	3 460	..
Non-Ferrous Metals	407	328	330	207	182	200	191	..
Non-Metallic Minerals	3 052	2 128	1 287	873	854	772	884	..
Transport Equipment	670	737	668	461	404	407	464	..
Machinery	2 244	2 192	1 898	1 599	1 230	1 097	1 069	..
Mining and Quarrying	638	494	469	531	718	737	790	..
Food and Tobacco	1 798	2 066	1 845	1 320	1 432	1 324	1 448	..
Pulp, Paper and Printing	333	342	220	263	262	246	248	..
Wood and Wood Products	416	428	354	178	142	123	127	..
Construction	1 568	1 429	1 250	1 864	2 587	3 216	3 235	..
Textiles and Leather	662	572	510	405	366	357	418	..
Non-Specified	10 878	9 267	8 373	1 143	2 389	2 265	2 593	..
Other Sectors	**129 204**	**124 663**	**107 671**	**78 461**	**75 637**	**72 363**	**78 462**	..
Commerce and Public	49 817	47 399	38 497	20 560	18 366	16 685	18 491	..
Residential	66 361	63 624	54 348	44 490	44 485	42 588	46 981	..
Agriculture	10 744	11 259	12 475	12 500	12 516	12 791	12 693	..
Non-Specified	2 282	2 381	2 351	911	270	299	297	..
Memo: Non-Energy Use	*1 383*	*2 783*	*3 179*	*3 856*	*3 584*	*3 927*	*3 103*	..
Chem/Petchem. Feedst.	*1 383*	*2 783*	*3 179*	*3 856*	*3 583*	*3 926*	*3 102*	..
Other	-	-	-	-	*1*	*1*	*1*	..
Closing Stock Level	-	*39 479*	*51 436*	*35 148*	*36 953*	*39 360*	*38 628*	*39 003 e*

Source: Annual Oil Statistics.

EUROPEAN UNION
Table 5B: Fuel Oil (Residual) - Supply and Consumption

000 metric tons

	1973	1978	1980	1990	1999	2000	2001	2002
Refinery Output	**284 692**	**237 623**	**215 425**	**119 090**	**106 717**	**109 671**	**106 304**	**105 242 e**
+ Recycled Products	-	-	1	127	155	184	235	365 e
+ Imports	42 873	35 810	44 935	42 612	38 877	42 720	40 493	44 359 e
- Exports	45 247	31 030	35 004	40 687	35 413	42 105	39 932	38 035 e
- Intl. Marine Bunkers	35 449	28 086	24 846	27 018	32 060	34 620	35 858	37 874 e
+ Transfers	- 160	- 657	-4 519	-5 255	-5 618	-6 986	-4 697	-5 059 e
+ Stock Changes	- 401	- 243	3 103	828	487	151	2 585	-1 199 e
- Statistical Differences	2 707	1 727	166	536	785	789	654	- 421 e
Total Consumption	**243 601**	**211 690**	**198 929**	**89 161**	**72 360**	**68 226**	**68 476**	**68 220 e**
Transformation	**93 581**	**83 557**	**82 486**	**43 664**	**38 462**	**36 156**	**36 341**	..
Public Electricity	81 387	63 298	57 699	34 357	26 738	24 517	24 217	..
Autoproducer Electricity	112	5 681	5 037	1 620	2 132	886	523	..
Public CHP Plants	5 643	6 905	10 750	1 998	1 242	2 400	2 573	..
Autoproducers of CHP	230	3 945	5 387	2 857	3 714	1 952	1 555	..
Public Heat Plants	562	487	471	507	492	381	525	..
Autoproducer Heat Plants	-	-	-	1	2	668	648	..
Petrochemical Industry	1 186	774	1 063	1 047	2 482	3 982	4 867	..
Other Transformation	4 461	2 467	2 079	1 277	1 660	1 370	1 433	..
Energy Sector	**18 792**	**20 314**	**19 403**	**10 501**	**11 290**	**10 846**	**11 101**	..
Refinery Fuel	18 272	19 058	18 894	10 240	11 177	10 723	10 987	..
Other Energy Sector	520	1 256	509	261	113	123	114	..
Final Consumption	**131 228**	**107 819**	**97 040**	**34 996**	**22 608**	**21 224**	**21 034**	..
Transport	**1 841**	**1 641**	**2 527**	**1 183**	**1 251**	**823**	**842**	..
Air	-	-	-	-	-	-	-	..
Road	-	-	-	-	-	-	-	..
Rail	772	383	262	5	1	2	-	..
Inland Waterways	769	808	1 812	928	1 233	818	842	..
Non-Specified	300	450	453	250	17	3	-	..
Industry	**100 819**	**84 456**	**75 789**	**27 671**	**17 254**	**17 510**	**17 254**	..
Iron and Steel	13 073	9 604	6 022	1 687	756	636	621	..
Chemical (incl. Pet.-Chem.)	17 831	16 849	16 015	7 782	3 689	5 378	5 582	..
Non-Ferrous Metals	1 609	1 995	1 785	926	819	801	780	..
Non-Metallic Minerals	18 980	16 132	14 534	4 288	3 264	3 012	2 345	..
Transport Equipment	2 509	2 111	1 736	426	244	172	181	..
Machinery	5 134	3 606	3 009	1 311	853	694	783	..
Mining and Quarrying	1 449	906	929	296	258	221	200	..
Food and Tobacco	11 856	11 264	10 573	4 503	2 673	2 462	2 546	..
Pulp, Paper and Printing	10 020	7 989	7 274	2 750	1 854	1 850	1 755	..
Wood and Wood Products	1 209	919	802	249	269	221	186	..
Construction	3 494	3 021	2 691	371	268	310	234	..
Textiles and Leather	5 960	4 905	4 210	1 944	1 107	1 026	882	..
Non-Specified	7 695	5 155	6 209	1 138	1 200	727	1 159	..
Other Sectors	**28 568**	**21 722**	**18 724**	**6 142**	**4 103**	**2 891**	**2 938**	..
Commerce and Public	6 287	6 770	7 149	2 841	1 928	1 104	1 012	..
Residential	17 284	10 905	7 628	2 454	1 315	1 265	1 250	..
Agriculture	1 625	1 641	1 804	772	820	488	498	..
Non-Specified	3 372	2 406	2 143	75	40	34	178	..
Memo: Non-Energy Use	*231*	*1 491*	*2 930*	*2 863*	*1 640*	*3 324*	*3 371*	..
Chem/Petchem. Feedst.	*113*	*1 440*	*2 875*	*2 570*	*1 481*	*3 030*	*3 158*	..
Other	*118*	*51*	*55*	*293*	*159*	*294*	*213*	..
Closing Stock Level	-	*37 992*	*35 041*	*24 768*	*19 390*	*19 239*	*16 654*	*17 853 e*

Source: Annual Oil Statistics.

EUROPEAN UNION
Table 5C: Total Products - Supply and Consumption

000 metric tons

	1973	1978	1980	1990	1999	2000	2001	2002
Refinery Output	**725 109**	**661 385**	**645 681**	**587 243**	**640 768**	**652 788**	**648 035**	**647 623 e**
+ Recycled Products	193	207	475	517	551	579	478	609 e
+ Imports	138 671	139 587	156 684	189 591	204 816	217 031	225 131	224 349 e
- Exports	140 055	116 813	128 871	166 260	188 065	199 415	196 524	198 365 e
- Intl. Marine Bunkers	42 808	35 789	31 404	35 101	40 969	43 649	45 066	46 808 e
+ Transfers	-4 005	-6 040	-12 482	-26 967	-25 722	-29 312	-29 164	-35 944 e
+ Stock Changes	-4 426	5 168	-5 790	1 470	8 514	-5 032	2 492	1 862 e
- Statistical Differences	4 795	269	1 052	1 701	-3 337	-7 497	-1 054	-2 082 e
Total Consumption	**667 884**	**647 436**	**623 241**	**548 792**	**603 230**	**600 487**	**606 436**	**595 408 e**
Transformation	**105 431**	**98 053**	**101 588**	**59 593**	**55 730**	**54 881**	**53 333**	..
Public Electricity	83 735	64 566	59 056	36 082	27 945	26 886	26 035	..
Autoproducer Electricity	289	6 334	5 764	2 146	2 809	1 180	819	..
Public CHP Plants	5 680	6 965	10 804	2 051	3 791	4 896	4 667	..
Autoproducers of CHP	307	4 042	5 546	3 487	4 956	2 608	2 525	..
Public Heat Plants	588	509	490	608	624	523	680	..
Autoproducer Heat Plants	-	-	-	1	3	721	683	..
Petrochemical Industry	4 923	10 210	15 328	12 368	12 937	15 731	15 535	..
Other Transformation	9 909	5 427	4 600	2 850	2 665	2 336	2 389	..
Energy Sector	**34 808**	**36 962**	**37 022**	**32 951**	**36 473**	**35 713**	**35 145**	..
Refinery Fuel	33 568	35 006	35 472	31 557	35 527	34 855	34 637	..
Other Energy Sector	1 240	1 956	1 550	1 394	946	858	508	..
Final Consumption	**527 645**	**512 421**	**484 631**	**456 248**	**511 027**	**509 893**	**517 958**	..
Transport	**152 815**	**177 802**	**185 385**	**244 227**	**296 238**	**298 815**	**299 596**	..
Air	17 273	19 067	19 359	27 424	41 077	43 086	41 513	..
Road	122 911	147 160	153 170	207 518	246 637	247 983	250 565	..
Rail	4 521	4 305	4 027	2 861	2 536	2 451	2 265	..
Inland Waterways	7 766	6 770	8 330	6 101	5 848	5 182	5 108	..
Non-Specified	344	500	499	323	140	113	145	..
Industry	**202 514**	**175 332**	**160 779**	**115 527**	**121 439**	**122 679**	**122 480**	..
Iron and Steel	14 282	10 528	6 886	2 521	1 629	1 843	1 629	..
Chemical (incl. Pet.-Chem.)	60 229	55 359	54 267	53 985	59 733	61 672	61 448	..
Non-Ferrous Metals	2 056	2 616	2 503	1 990	2 169	2 171	2 176	..
Non-Metallic Minerals	22 607	18 977	16 531	10 875	11 794	11 755	12 055	..
Transport Equipment	3 189	2 893	2 484	1 003	838	751	823	..
Machinery	7 698	6 154	5 312	3 666	2 998	2 828	3 042	..
Mining and Quarrying	2 104	1 419	1 408	847	1 042	1 022	1 034	..
Food and Tobacco	13 734	13 474	12 725	6 178	4 543	4 206	4 424	..
Pulp, Paper and Printing	10 415	8 445	7 628	3 198	2 332	2 294	2 221	..
Wood and Wood Products	1 634	1 355	1 167	459	452	384	352	..
Construction	22 428	18 956	18 431	17 868	18 983	18 854	19 039	..
Textiles and Leather	6 634	5 507	4 779	2 432	1 560	1 470	1 389	..
Non-Specified	35 504	29 649	26 658	10 505	13 366	13 429	12 848	..
Other Sectors	**172 316**	**159 287**	**138 467**	**96 494**	**93 350**	**88 399**	**95 882**	..
Commerce and Public	56 554	55 085	46 971	24 796	21 725	19 151	21 382	..
Residential	96 149	85 140	71 554	56 199	56 765	54 558	59 013	..
Agriculture	13 413	13 828	15 199	14 445	14 293	14 199	14 173	..
Non-Specified	6 200	5 234	4 743	1 054	567	491	1 314	..
Memo: Non-Energy Use	*66 017*	*60 201*	*62 296*	*71 456*	*81 816*	*81 277*	*80 088*	..
Chem/Petchem. Feedst.	*38 761*	*36 423*	*37 636*	*46 008*	*54 675*	*54 904*	*53 836*	..
Other	*27 256*	*23 778*	*24 660*	*25 448*	*27 141*	*26 373*	*26 252*	..
Closing Stock Level	-	*109 104*	*122 459*	*93 259*	*90 485*	*95 603*	*93 152*	*91 290 e*

Source: Annual Oil Statistics.

EUROPEAN UNION
Table 6A: Supply and Consumption

000 metric tons

	1980	2000	2001	2002	1980	2000	2001	2002
	LPG AND ETHANE				**REFINERY GAS**			
Refinery Output	**13 564**	**18 289**	**18 540**	**19 071 e**	**15 991**	**19 757**	**19 281**	**19 651 e**
+ Recycled Products	-	-	-	-	-	-	-	-
+ Imports	5 514	10 021	10 056	9 558 e	-	-	-	-
- Exports	2 693	6 123	5 357	6 033 e	-	-	-	-
- Intl. Marine Bunkers	-	-	-	-	-	-	-	-
+ Transfers	378	517	240	204 e	- 143	- 40	- 51	- 128 e
+ Stock Change	- 73	- 130	- 168	357 e	2	3	-	-
- Statistical Difference	- 191	- 186	693	568 e	81	- 108	- 146	120 e
Total Consumption	**16 881**	**22 760**	**22 618**	**22 589 e**	**15 769**	**19 828**	**19 376**	**19 403 e**
of which: Ethane	-	*1 688*	*1 828*	*1 882 e*	-	-	-	-
Transformation	1 422	957	903	..	889	1 453	964	..
Energy	423	538	516	..	13 450	16 475	15 977	..
Transport	1 683	2 513	2 515	..	-	-	-	..
Industry	5 331	9 942	9 491	..	1 430	1 900	2 435	..
Other	8 022	8 810	9 193	..	-	-	-	..
Memo: Non-Energy Use	*1 752*	*5 818*	*5 732*	..	*481*	*603*	*739*	..
Closing Stock Level	*1 069*	*1 795*	*1 967*	*1 610 e*	-	-	-	-

	1980	2000	2001	2002	1980	2000	2001	2002
	NAPHTHA				**OTHER KEROSENE**			
Refinery Output	**31 151**	**41 286**	**40 508**	**36 250 e**	**5 353**	**4 539**	**4 585**	**4 903 e**
+ Recycled Products	-	-	-	-	-	-	-	-
+ Imports	21 085	27 124	26 450	25 193 e	472	1 948	1 824	1 620 e
- Exports	9 911	18 811	18 193	17 149 e	2 411	811	713	1 039 e
- Intl. Marine Bunkers	-	-	-	-	-	-	-	-
+ Transfers	-1 965	-4 926	-5 527	-2 840 e	653	- 501	- 545	- 711 e
+ Stock Change	- 48	- 223	189	207 e	62	- 164	96	38 e
- Statistical Difference	- 186	-2 494	-1 078	- 280 e	- 30	- 255	- 587	- 434 e
Total Consumption	**40 498**	**46 944**	**44 505**	**41 941 e**	**4 159**	**5 266**	**5 834**	**5 245 e**
Transformation	12 063	9 273	7 996	..	224	121	127	..
Energy	137	120	326	..	24	1	55	..
Transport	-	-	-	..	44	32	33	..
Industry	28 297	37 551	36 183	..	736	1 706	1 947	..
Other	1	-	-	..	3 131	3 406	3 672	..
Memo: Non-Energy Use	*27 185*	*36 390*	*35 013*	..	*309*	*290*	*271*	..
Closing Stock Level	*4 794*	*3 731*	*3 542*	*3 335 e*	*1 605*	*1 041*	*945*	*907 e*

	1980	2000	2001	2002	1980	2000	2001	2002
	WHITE SPIRIT + SBP				**LUBRICANTS**			
Refinery Output	**1 611**	**1 139**	**989**	**..**	**6 895**	**7 144**	**7 340**	**..**
+ Recycled Products	-	-	-	..	372	335	186	..
+ Imports	478	839	770	..	2 988	3 211	2 959	..
- Exports	950	744	654	..	3 724	4 850	5 452	..
- Intl. Marine Bunkers	-	-	-	..	268	342	359	..
+ Transfers	- 48	- 344	- 387	..	- 390	- 527	- 485	..
+ Stock Change	3	- 28	2	..	- 341	- 282	62	..
- Statistical Difference	- 171	-	- 108	..	- 58	- 555	- 826	..
Total Consumption	**1 265**	**862**	**828**	**..**	**5 590**	**5 244**	**5 077**	**..**
Transformation	1	-	-	..	1	-	-	..
Energy	-	-	-	..	8	12	12	..
Transport	-	-	-	..	1 919	2 365	2 450	..
Industry	1 157	704	694	..	3 467	2 650	2 410	..
Other	107	158	134	..	195	217	205	..
Memo: Non-Energy Use	*1 264*	*858*	*819*	..	*5 579*	*5 244*	*5 077*	..
Closing Stock Level	*358*	*158*	*156*	..	*2 768*	*1 579*	*1 517*	..

Source: Annual Oil Statistics.

EUROPEAN UNION
Table 6B: Supply and Consumption

000 metric tons

	1980	2000	2001	2002	1980	2000	2001	2002
	BITUMEN				PARAFFIN WAXES			
Refinery Output	**15 417**	**17 392**	**17 674**	**..**	**713**	**1 462**	**1 380**	**..**
+ Recycled Products	-	-	-	..	-	-	-	..
+ Imports	1 341	3 248	3 532	..	372	594	602	..
- Exports	2 204	3 933	4 061	..	315	618	650	..
- Intl. Marine Bunkers	-	-	-	..	-	-	-	..
+ Transfers	- 5	1 118	984	..	- 13	- 437	- 361	..
+ Stock Change	- 31	- 31	- 38	..	- 25	44	- 19	..
- Statistical Difference	119	188	547	..	104	107	131	..
Total Consumption	**14 399**	**17 606**	**17 544**	**..**	**628**	**938**	**821**	**..**
Transformation	-	1 245	1 099	..	6	-	-	..
Energy	7	-	-	..	-	4	-	..
Transport	-	-	-	..	-	-	-	..
Industry	14 392	16 288	16 365	..	622	933	821	..
Other	-	73	80	..	-	1	-	..
Memo: Non-Energy Use	*14 392*	*16 206*	*16 391*	..	*616*	*934*	*821*	..
Closing Stock Level	*1 116*	*1 251*	*1 289*	..	*95*	*86*	*105*	..

	1980	2000	2001	2002	1980	2000	2001	2002
	PETROLEUM COKE				OTHER PRODUCTS			
Refinery Output	**2 967**	**7 838**	**7 201**	**7 082 e**	**8 062**	**19 501**	**22 642**	**54 017 e**
+ Recycled Products	-	-	-	-	94	37	27	214 e
+ Imports	5 100	10 321	10 545	9 873 e	817	5 564	4 958	13 066 e
- Exports	447	1 261	1 352	1 376 e	2 427	5 936	5 699	17 368 e
- Intl. Marine Bunkers	-	-	-	-	-	23	29	316 e
+ Transfers	-	- 76	- 5	- 64 e	-1 463	-8 405	-10 888	-17 270 e
+ Stock Change	- 115	- 513	- 49	173 e	- 471	-1 262	- 124	604 e
- Statistical Difference	517	142	189	- 141 e	1	-1 257	-1 273	-2 264 e
Total Consumption	**6 988**	**16 167**	**16 151**	**15 829 e**	**4 611**	**10 733**	**12 160**	**35 211 e**
Transformation	806	1 205	1 211	..	152	407	643	..
Energy	1 530	4 113	3 664	..	1 484	3 049	3 065	..
Transport	-	-	-	..	-	-	-	..
Industry	4 652	10 775	11 168	..	2 970	7 273	7 750	..
Other	-	74	108	..	5	4	702	..
Memo: Non-Energy Use	*1 834*	*1 510*	*1 542*	..	*2 740*	*6 105*	*6 621*	..
Closing Stock Level	*728*	*2 441*	*2 490*	*2 317 e*	*2 830*	*3 192*	*3 316*	*5 779 e*

Source: Annual Oil Statistics.

EUROPEAN UNION
Table 7A: Imports by Country of Origin

000 metric tons

	2000	2001	2002	2000	2001	2002	2000	2001	2002
	Crude Oil			**Natural Gas Liquids**			**Feedstocks and Other**		
Australia	-	-	-	-	-	-	-	-	-
Belgium	2	2	-	-	-	2 121 e	419	415	205 e
Canada	80	-	-	-	-	-	-	-	-
Czech Republic	31	62	86 e	-	-	-	8	1	1 e
Denmark	11 634	10 104	10 016 e	-	-	-	50	23	30 e
Finland	-	-	-	-	-	-	34	31	2 e
France	501	372	8 e	-	-	-	318	383	490 e
Germany	1 594	154	221 e	11	35	45 e	349	456	575 e
Greece	-	-	-	-	-	-	42	114	130 e
Hungary	-	-	-	-	-	-	198	153	198 e
Ireland	-	-	-	-	-	-	313	551	390 e
Italy	204	250	401 e	-	-	-	58	57	27 e
Japan	-	-	-	-	-	-	1	-	-
Korea	-	-	-	-	-	-	-	-	-
Mexico	9 770	9 291	9 508 e	-	-	-	-	-	-
Netherlands	281	263	1 276 e	-	-	-	2 085	1 654	1 966 e
New Zealand	-	-	-	-	-	-	-	-	-
Norway	114 799	107 155	99 783 e	999	966	1 356 e	736	897	720 e
Poland	163	40	206 e	-	-	-	132	67	-
Spain	23	-	-	-	-	-	5	8	167 e
Sweden	89	91	80 e	-	11	-	684	404	418 e
Turkey	213	-	-	-	-	-	326	555	215 e
United Kingdom	44 816	49 344	42 682 e	48	45	22 e	419	366	346 e
United States	-	-	82 e	-	-	-	74	729	529 e
Other	141	66	12 e	-	-	-	319	41	53 e
OECD	**184 341**	**177 194**	**164 361 e**	**1 058**	**1 057**	**3 544 e**	**6 570**	**6 905**	**6 462 e**
Argentina	131	92	188 e	-	-	-	-	10	-
Brazil	133	1 412	1 533 e	-	-	-	8	10	20 e
Colombia	-	75	455 e	-	-	-	-	-	-
Ecuador	-	-	-	-	-	-	-	-	-
Netherlands Antilles	99	238	115 e	-	-	-	25	-	-
Peru	-	-	-	-	-	-	-	-	-
Trinidad and Tobago	-	-	-	-	-	-	-	-	-
Venezuela	6 946	9 023	9 038 e	-	-	-	1 214	1 210	706 e
Other	150	-	-	-	-	-	34	13	-
Latin America	**7 459**	**10 840**	**11 329 e**	**-**	**-**	**-**	**1 281**	**1 243**	**726 e**
Romania	19	118	373 e	-	-	-	94	15	22 e
Other	-	-	-	-	-	-	81	134	134 e
Non-OECD Europe	**19**	**118**	**373 e**	**-**	**-**	**-**	**175**	**149**	**156 e**
FSU	**89 480**	**101 210**	**122 583 e**	**747**	**1 476**	**1 852 e**	**8 728**	**8 103**	**10 257 e**
Iran	35 475	31 412	25 999 e	-	-	-	30	14	30 e
Iraq	31 209	20 126	15 678 e	-	-	-	-	-	-
Kuwait	9 768	7 990	5 916 e	-	-	-	-	-	-
Oman	-	-	184 e	-	-	-	-	-	-
Qatar	134	-	28 e	-	-	-	-	-	-
Saudi Arabia	65 143	57 493	53 725 e	-	-	-	62	113	152 e
United Arab Emirates	147	354	-	-	-	-	-	-	-
Other	13 087	18 662	19 599 e	-	-	-	327	1 092	475 e
Middle East	**154 963**	**136 037**	**121 129 e**	**-**	**-**	**-**	**419**	**1 219**	**657 e**
India	-	-	-	-	-	-	-	-	-
Indonesia	-	-	-	-	-	-	-	-	-
Singapore	-	-	-	-	-	-	-	-	-
Other	-	-	-	-	-	-	-	-	-
Asia	**-**	**-**	**-**	**-**	**-**	**-**	**-**	**-**	**-**
China	**-**	**-**	**-**	**-**	**-**	**-**	**-**	**9**	**-**
Algeria	16 230	14 507	12 628 e	4 700	4 615	3 761 e	463	583	668 e
Angola/Cabinada	3 862	6 657	6 997 e	-	-	-	-	-	-
Cameroon	3 070	3 669	3 951 e	-	-	-	-	-	-
Egypt	5 467	3 525	3 999 e	-	-	-	207	285	130 e
Gabon	562	1 643	573 e	-	-	-	-	-	-
Libya	45 517	43 138	38 565 e	25	25	75 e	2 086	1 681	1 356 e
Nigeria	22 407	25 723	18 817 e	-	-	-	-	18	247 e
Tunisia	1 346	1 176	1 740 e	105	70	129 e	239	235	428 e
Other	3 677	5 188	6 620 e	-	35	-	589	516	302 e
Africa	**102 138**	**105 226**	**93 890 e**	**4 830**	**4 745**	**3 965 e**	**3 584**	**3 318**	**3 131 e**
Non-spec./Other	**5 803**	**11 081**	**16 218 e**	**-**	**493**	**73 e**	**748**	**693**	**1 164 e**
Total Imports	**544 203**	**541 706**	**529 883 e**	**6 635**	**7 771**	**9 434 e**	**21 505**	**21 639**	**22 553 e**
Memo: Total OPEC	*232 976*	*209 766*	*180 394 e*	*4 725*	*4 640*	*3 836 e*	*3 855*	*3 619*	*3 159 e*

Source: Annual Oil Statistics.

EUROPEAN UNION
Table 7B: Imports by Country of Origin

000 metric tons

	2000	2001	2002	2000	2001	2002	2000	2001	2002
	Total Products			**Liquefied Petroleum Gases**			**Naphtha**		
Australia	15	1	25 e	15	-	-	-	-	-
Belgium	16 513	16 128	16 463 e	458	429	420 e	1 525	1 061	1 054 e
Canada	526	511	463 e	-	-	-	15	4	22 e
Czech Republic	617	523	685 e	2	5	4 e	20	-	-
Denmark	2 378	1 887	1 822 e	58	59	41 e	96	37	75 e
Finland	3 241	2 925	2 827 e	27	10	2 e	74	78	160 e
France	9 532	9 008	7 813 e	383	367	329 e	1 208	1 250	764 e
Germany	10 672	9 247	8 790 e	445	338	354 e	1 254	846	920 e
Greece	535	780	806 e	4	1	49 e	305	310	383 e
Hungary	707	939	938 e	30	26	26 e	-	-	-
Ireland	179	167	59 e	6	5	7 e	76	-	-
Italy	6 516	8 809	8 822 e	63	65	62 e	256	641	594 e
Japan	84	76	85 e	-	-	2 e	14	12	28 e
Korea	53	56	223 e	-	-	-	42	49	75 e
Mexico	9	112	80 e	-	-	-	-	105	31 e
Netherlands	40 679	38 674	34 399 e	480	436	424 e	6 359	5 509	5 064 e
New Zealand	-	-	-	-	-	-	-	-	-
Norway	6 404	4 615	6 120 e	1 030	1 376	1 120 e	693	535	546 e
Poland	927	1 056	927 e	8	2	1 e	3	-	-
Spain	3 358	2 620	2 547 e	93	57	57 e	1 195	1 134	1 006 e
Sweden	6 657	5 515	5 663 e	100	49	113 e	128	240	212 e
Turkey	580	836	467 e	6	2	-	317	431	327 e
United Kingdom	21 689	20 953	21 753 e	2 305	2 019	2 356 e	2 491	2 101	2 348 e
United States	11 885	10 435	10 704 e	4	6	39 e	491	323	650 e
Other	2 343	2 115	2 418 e	43	51	50 e	153	101	144 e
OECD	**146 099**	**137 988**	**134 899 e**	**5 560**	**5 303**	**5 456 e**	**16 715**	**14 767**	**14 403 e**
Argentina	46	86	188 e	-	-	-	6	1	20 e
Brazil	285	65	93 e	-	-	-	35	20	14 e
Colombia	14	-	118 e	2	-	-	-	-	-
Ecuador	58	-	-	-	-	-	-	-	-
Netherlands Antilles	437	302	178 e	2	-	-	2	5	61 e
Peru	-	-	30 e	-	-	-	-	-	-
Trinidad and Tobago	811	466	604 e	9	8	16 e	-	-	-
Venezuela	3 710	2 276	2 607 e	3	-	37 e	-	-	26 e
Other	617	616	792 e	32	-	2 e	33	82	26 e
Latin America	**5 978**	**3 811**	**4 610 e**	**48**	**8**	**55 e**	**76**	**108**	**147 e**
Romania	50	277	1 131 e	-	-	-	-	1	3 e
Other	915	929	916 e	87	74	46 e	92	71	126 e
Non-OECD Europe	**965**	**1 206**	**2 047 e**	**87**	**74**	**46 e**	**92**	**72**	**129 e**
FSU	**30 801**	**32 766**	**36 784 e**	**103**	**163**	**144 e**	**1 621**	**1 634**	**2 229 e**
Iran	107	174	31 e	7	-	-	37	88	-
Iraq	-	-	-	-	-	-	-	-	-
Kuwait	2 479	1 830	1 221 e	-	-	-	866	109	-
Oman	-	82	-	-	-	-	-	-	-
Qatar	22	24	26 e	-	-	-	-	-	-
Saudi Arabia	1 704	2 031	1 470 e	141	11	8 e	186	108	219 e
United Arab Emirates	1 223	2 381	2 192 e	-	-	-	41	-	4 e
Other	2 237	1 954	978 e	19	16	24 e	291	265	87 e
Middle East	**7 772**	**8 476**	**5 918 e**	**167**	**27**	**32 e**	**1 421**	**570**	**310 e**
India	248	733	365 e	-	-	-	53	3	10 e
Indonesia	152	284	554 e	-	-	-	1	-	-
Singapore	67	77	1 193 e	-	-	-	26	12	3 e
Other	235	156	310 e	36	-	-	6	-	15 e
Asia	**702**	**1 250**	**2 422 e**	**36**	**-**	**-**	**86**	**15**	**28 e**
China	**54**	**96**	**65 e**	**-**	**3**	**-**	**3**	**-**	**-**
Algeria	8 385	8 529	8 483 e	3 204	2 943	2 866 e	2 435	3 137	3 324 e
Angola/Cabinada	3	-	-	-	-	-	-	-	-
Cameroon	7	-	-	-	-	-	-	-	-
Egypt	1 250	1 281	1 849 e	49	161	140 e	1 059	559	710 e
Gabon	-	-	4 e	-	-	4 e	-	-	-
Libya	5 689	5 966	5 439 e	154	189	102 e	1 497	2 071	1 815 e
Nigeria	49	293	127 e	-	110	-	21	77	27 e
Tunisia	240	78	242 e	-	-	-	45	52	88 e
Other	1 045	529	1 295 e	36	45	66 e	339	208	204 e
Africa	**16 668**	**16 676**	**17 439 e**	**3 443**	**3 448**	**3 178 e**	**5 396**	**6 104**	**6 168 e**
Non-spec./Other	**7 992**	**22 862**	**20 165 e**	**577**	**1 030**	**647 e**	**1 714**	**3 180**	**1 779 e**
Total Imports	**217 031**	**225 131**	**224 349 e**	**10 021**	**10 056**	**9 558 e**	**27 124**	**26 450**	**25 193 e**
Memo: Total OPEC	*23 520*	*23 788*	*22 150 e*	*3 509*	*3 253*	*3 013 e*	*5 084*	*5 590*	*5 415 e*

Source: Annual Oil Statistics.

EUROPEAN UNION
Table 7C: Imports by Country of Origin

000 metric tons

	2000	2001	2002	2000	2001	2002	2000	2001	2002
	Total Gasoline			**Total Kerosene**			**Gas/Diesel**		
Australia	-	-	-	-	-	-	-	-	-
Belgium	3 600	3 340	3 086 e	1 105	1 152	1 277 e	5 217	6 282	5 331 e
Canada	1	1	-	-	-	54 e	381	450	330 e
Czech Republic	28	30	111 e	-	-	-	365	388	453 e
Denmark	994	757	618 e	149	165	142 e	784	515	372 e
Finland	1 239	1 357	1 204 e	173	4	29 e	1 612	1 395	1 333 e
France	2 103	1 819	1 631 e	494	614	452 e	1 447	998	846 e
Germany	1 862	1 490	928 e	186	140	135 e	4 110	3 663	4 483 e
Greece	75	87	164 e	129	63	19 e	-	291	155 e
Hungary	126	174	201 e	-	-	-	459	613	662 e
Ireland	18	43	14 e	8	1	11 e	58	49	19 e
Italy	617	816	705 e	316	383	312 e	4 112	6 018	6 316 e
Japan	8	-	-	29	-	-	-	-	-
Korea	-	-	-	-	-	-	-	-	148 e
Mexico	-	-	-	-	-	-	-	-	-
Netherlands	7 753	6 384	4 669 e	3 640	3 144	2 823 e	16 826	18 425	15 908 e
New Zealand	-	-	-	-	-	-	-	-	-
Norway	1 881	855	1 297 e	118	45	113 e	1 302	1 169	2 046 e
Poland	75	361	257 e	139	295	123 e	47	55	197 e
Spain	495	290	198 e	31	5	139 e	320	362	343 e
Sweden	1 708	1 117	1 153 e	7	4	14 e	3 465	2 935	2 924 e
Turkey	-	9	73 e	81	27	-	3	-	-
United Kingdom	3 743	3 434	2 781 e	1 335	1 503	1 537 e	6 224	5 110	6 348 e
United States	97	11	41 e	90	66	392 e	791	480	958 e
Other	254	255	299 e	70	15	119 e	571	624	651 e
OECD	**26 677**	**22 630**	**19 430 e**	**8 100**	**7 626**	**7 691 e**	**48 094**	**49 822**	**49 823 e**
Argentina	-	-	-	-	-	-	-	-	-
Brazil	-	-	-	-	-	-	-	-	-
Colombia	-	-	-	12	-	26 e	-	-	-
Ecuador	-	-	-	-	-	-	-	-	-
Netherlands Antilles	103	8	-	123	29	-	50	25	18 e
Peru	-	-	-	-	-	-	-	-	-
Trinidad and Tobago	64	107	103 e	97	100	87 e	110	127	95 e
Venezuela	-	-	-	365	285	803 e	5	5	33 e
Other	166	47	-	101	135	92 e	46	73	123 e
Latin America	**333**	**162**	**103 e**	**698**	**549**	**1 008 e**	**211**	**230**	**269 e**
Romania	6	21	107 e	-	-	-	31	245	952 e
Other	77	101	135 e	45	21	28 e	385	424	279 e
Non-OECD Europe	**83**	**122**	**242 e**	**45**	**21**	**28 e**	**416**	**669**	**1 231 e**
FSU	**1 141**	**619**	**1 016 e**	**1 037**	**896**	**1 217 e**	**13 609**	**15 997**	**17 075 e**
Iran	-	-	-	-	48	-	31	-	-
Iraq	-	-	-	-	-	-	-	-	-
Kuwait	-	-	-	1 488	1 652	1 221 e	125	-	-
Oman	-	-	-	-	82	-	-	-	-
Qatar	-	-	-	20	20	-	-	-	26 e
Saudi Arabia	155	87	156 e	908	1 462	800 e	267	247	260 e
United Arab Emirates	47	14	35 e	1 110	2 280	2 051 e	25	45	34 e
Other	24	31	1 e	626	738	466 e	108	148	115 e
Middle East	**226**	**132**	**192 e**	**4 152**	**6 282**	**4 538 e**	**556**	**440**	**444 e**
India	38	-	-	-	-	37 e	-	384	281 e
Indonesia	-	-	-	-	-	-	-	80	-
Singapore	-	-	7 e	-	-	240 e	38	33	906 e
Other	-	-	-	10	-	7 e	157	48	216 e
Asia	**38**	**-**	**7 e**	**10**	**-**	**284 e**	**195**	**545**	**1 403 e**
China	**-**	**-**	**-**	**-**	**-**	**-**	**-**	**-**	**-**
Algeria	143	66	417 e	742	451	374 e	1 646	1 777	1 024 e
Angola/Cabinada	-	-	-	-	-	-	-	-	-
Cameroon	-	-	-	-	-	-	-	-	-
Egypt	63	140	565 e	-	167	301 e	-	-	-
Gabon	-	-	-	-	-	-	-	-	-
Libya	84	37	148 e	525	1 012	515 e	755	487	548 e
Nigeria	-	-	-	28	96	-	-	-	7 e
Tunisia	-	-	41 e	51	4	-	-	22	104 e
Other	2	5	29 e	70	30	84 e	63	45	93 e
Africa	**292**	**248**	**1 200 e**	**1 416**	**1 760**	**1 274 e**	**2 464**	**2 331**	**1 776 e**
Non-spec./Other	**1 388**	**3 785**	**3 002 e**	**559**	**2 187**	**1 251 e**	**1 649**	**7 713**	**7 796 e**
Total Imports	**30 178**	**27 698**	**25 192 e**	**16 017**	**19 321**	**17 291 e**	**67 194**	**77 747**	**79 817 e**
Memo: Total OPEC	*429*	*204*	*756 e*	*5 186*	*7 306*	*5 764 e*	*2 854*	*2 641*	*1 941 e*

Source: Annual Oil Statistics.

EUROPEAN UNION
Table 7D: Imports by Country of Origin

000 metric tons

	2000	2001	2002	2000	2001	2002	2000	2001	2002
	Fuel Oil (Residual)			**Petroleum Coke**			**Other Products**		
Australia	-	-	25 e	-	-	-	-	1	-
Belgium	2 838	2 239	3 133 e	273	222	352 e	1 497	1 403	1 810 e
Canada	74	31	52 e	-	9	-	55	16	5 e
Czech Republic	26	26	30 e	-	-	-	176	74	87 e
Denmark	273	325	539 e	6	11	6 e	18	18	29 e
Finland	67	32	-	-	-	-	49	49	99 e
France	2 130	2 341	2 380 e	5	4	5 e	1 762	1 615	1 406 e
Germany	1 703	1 583	873 e	105	76	109 e	1 007	1 111	988 e
Greece	-	25	27 e	-	-	-	22	3	9 e
Hungary	-	70	-	-	-	3 e	92	56	46 e
Ireland	9	69	8 e	-	-	-	4	-	-
Italy	607	408	284 e	67	14	113 e	478	464	436 e
Japan	-	-	-	25	37	38 e	8	27	17 e
Korea	-	-	-	-	-	-	11	7	-
Mexico	-	-	47 e	-	-	-	9	7	2 e
Netherlands	3 937	2 986	3 919 e	365	521	411 e	1 319	1 269	1 181 e
New Zealand	-	-	-	-	-	-	-	-	-
Norway	485	539	787 e	25	1	32 e	870	95	179 e
Poland	549	265	281 e	-	-	-	106	78	68 e
Spain	607	301	181 e	62	18	89 e	555	453	534 e
Sweden	972	942	1 028 e	5	19	-	272	209	219 e
Turkey	80	311	40 e	-	-	-	93	56	27 e
United Kingdom	4 006	5 158	4 845 e	217	570	318 e	1 368	1 058	1 220 e
United States	815	1 157	1 172 e	7 726	6 309	5 805 e	1 871	2 083	1 647 e
Other	954	834	863 e	13	5	3 e	285	230	289 e
OECD	**20 132**	**19 642**	**20 514 e**	**8 894**	**7 816**	**7 284 e**	**11 927**	**10 382**	**10 298 e**
Argentina	-	-	-	16	45	47 e	24	40	121 e
Brazil	239	-	45 e	-	4	9 e	11	41	25 e
Colombia	-	-	92 e	-	-	-	-	-	-
Ecuador	-	-	-	58	-	-	-	-	-
Netherlands Antilles	143	229	79 e	-	-	-	14	6	20 e
Peru	-	-	30 e	-	-	-	-	-	-
Trinidad and Tobago	530	124	303 e	-	-	-	1	-	-
Venezuela	2 351	500	79 e	959	1 442	1 571 e	27	44	58 e
Other	219	103	355 e	-	115	169 e	20	61	25 e
Latin America	**3 482**	**956**	**983 e**	**1 033**	**1 606**	**1 796 e**	**97**	**192**	**249 e**
Romania	-	-	48 e	2	-	-	11	10	21 e
Other	192	211	257 e	-	-	-	37	27	45 e
Non-OECD Europe	**192**	**211**	**305 e**	**2**	**-**	**-**	**48**	**37**	**66 e**
FSU	**12 895**	**12 759**	**14 406 e**	**14**	**21**	**21 e**	**381**	**677**	**676 e**
Iran	13	-	-	-	-	1 e	19	38	30 e
Iraq	-	-	-	-	-	-	-	-	-
Kuwait	-	69	-	-	-	-	-	-	-
Oman	-	-	-	-	-	-	-	-	-
Qatar	-	-	-	-	-	-	2	4	-
Saudi Arabia	9	85	-	-	-	-	38	31	18 e
United Arab Emirates	-	42	68 e	-	-	-	-	-	-
Other	1 099	744	257 e	55	-	-	15	12	28 e
Middle East	**1 121**	**940**	**325 e**	**55**	**-**	**1 e**	**74**	**85**	**76 e**
India	-	-	-	143	268	35 e	14	78	2 e
Indonesia	42	-	468 e	67	62	16 e	42	142	70 e
Singapore	2	13	19 e	-	-	-	1	19	18 e
Other	-	-	24 e	-	78	-	26	30	48 e
Asia	**44**	**13**	**511 e**	**210**	**408**	**51 e**	**83**	**269**	**138 e**
China	**-**	**-**	**8 e**	**24**	**44**	**27 e**	**27**	**49**	**30 e**
Algeria	156	107	445 e	-	-	-	59	48	33 e
Angola/Cabinada	-	-	-	-	-	-	3	-	-
Cameroon	-	-	-	-	-	-	7	-	-
Egypt	60	247	109 e	-	-	-	19	7	24 e
Gabon	-	-	-	-	-	-	-	-	-
Libya	2 548	2 143	2 257 e	-	-	-	126	27	54 e
Nigeria	-	10	93 e	-	-	-	-	-	-
Tunisia	144	-	-	-	-	-	-	-	9 e
Other	412	115	693 e	76	20	21 e	47	61	105 e
Africa	**3 320**	**2 622**	**3 597 e**	**76**	**20**	**21 e**	**261**	**143**	**225 e**
Non-spec./Other	**1 534**	**3 350**	**3 710 e**	**13**	**630**	**672 e**	**558**	**987**	**1 308 e**
Total Imports	**42 720**	**40 493**	**44 359 e**	**10 321**	**10 545**	**9 873 e**	**13 456**	**12 821**	**13 066 e**
Memo: Total OPEC	*5 119*	*2 956*	*3 410 e*	*1 026*	*1 504*	*1 588 e*	*313*	*334*	*263 e*

Source: Annual Oil Statistics.

EUROPEAN UNION
Table 8A: Exports by Country of Destination

000 metric tons

	2000	2001	2002	2000	2001	2002	2000	2001	2002
	Crude Oil			**Natural Gas Liquids**			**Feedstocks and Other**		
Australia	-	-	-	-	-	-	-	-	-
Austria	-	-	82 e	-	-	-	-	-	-
Belgium	1 473	626	1 136 e	274	470	538 e	393	374	262 e
Canada	1 605	3 444	3 295 e	-	-	16 e	-	34	-
Denmark	-	112	100 e	13	23	29 e	27	28	40 e
Finland	2 962	3 443	3 528 e	5	26	-	1	4	20 e
France	17 101	16 208	12 172 e	622	402	533 e	622	299	256 e
Germany	12 278	12 727	10 855 e	156	139	98 e	810	458	502 e
Greece	-	-	134 e	-	4	-	-	28	-
Ireland	-	250	250 e	9	12	13 e	3	-	24 e
Italy	1 299	618	839 e	26	2	4 e	134	56	64 e
Japan	-	-	-	-	26	36 e	-	-	-
Luxembourg	-	-	-	-	-	-	-	-	-
Mexico	-	-	-	-	-	-	-	-	-
Netherlands	23 314	23 315	22 202 e	775	1 016	1 155 e	2 472	1 963	1 469 e
Norway	897	362	723 e	215	41	75 e	50	49	36 e
Portugal	825	716	998 e	124	89	77 e	16	9	21 e
Spain	2 594	2 427	1 531 e	302	205	139 e	241	86	134 e
Sweden	5 577	3 908	4 951 e	340	279	330 e	39	36	168 e
Switzerland	-	-	-	-	-	-	28	39	46 e
Turkey	-	-	-	108	29	30 e	52	66	52 e
United Kingdom	1 743	1 601	1 967 e	-	-	-	149	146	182 e
United States	28 108	22 041	27 348 e	68	225	32 e	3 021	2 180	1 718 e
Other OECD	507	1	-	12	22	13 e	71	68	19 e
OECD	**100 283**	**91 799**	**92 111 e**	**3 049**	**3 010**	**3 118 e**	**8 129**	**5 923**	**5 013 e**
Africa	163	-	264 e	16	25	10 e	174	103	73 e
Latin America	84	-	178 e	157	105	-	24	25	-
Asia	1 893	-	-	-	12	73 e	15	-	-
China	519	262	1 634 e	-	110	53 e	-	-	-
Non-OECD Europe	66	534	542 e	-	-	-	97	150	29 e
Former Soviet Union	251	-	-	-	-	-	7	-	-
Middle East	-	-	-	-	-	-	19	-	27 e
Non-specified/Other	131	1 522	2 931 e	327	260	136 e	797	737	735 e
Non-OECD	**3 107**	**2 318**	**5 549 e**	**500**	**512**	**272 e**	**1 133**	**1 015**	**864 e**
Total Exports	**103 390**	**94 117**	**97 660 e**	**3 549**	**3 522**	**3 390 e**	**9 262**	**6 938**	**5 877 e**
	Total Products			**Liquefied Petroleum Gas**			**Naphtha**		
Australia	79	16	53 e	-	-	-	-	-	3 e
Austria	3 188	3 052	3 256 e	98	89	96 e	40	35	15 e
Belgium	19 698	22 237	20 149 e	674	691	915 e	3 065	3 038	3 042 e
Canada	550	623	983 e	4	-	-	-	-	1 e
Denmark	2 502	2 317	2 183 e	19	5	1 e	5	4	12 e
Finland	955	1 114	1 069 e	19	33	14 e	117	64	91 e
France	14 975	13 327	12 907 e	157	218	258 e	1 439	1 221	1 340 e
Germany	38 424	38 185	29 971 e	601	381	554 e	6 647	6 482	4 903 e
Greece	1 065	886	766 e	2	-	2 e	50	50	49 e
Ireland	4 645	4 157	4 458 e	86	51	49 e	-	9	-
Italy	5 466	3 783	0 051 e	280	204	246 e	425	161	189 e
Japan	72	88	309 e	-	-	-	37	29	60 e
Luxembourg	2 464	2 494	2 545 e	18	17	15 e	-	-	-
Mexico	236	97	446 e	-	-	9 e	-	-	32 e
Netherlands	12 007	12 067	15 464 e	615	320	357 e	1 901	2 140	2 930 e
Norway	4 534	4 756	5 130 e	129	150	174 e	37	125	52 e
Portugal	2 938	2 065	2 850 e	221	164	183 e	415	217	233 e
Spain	8 360	8 151	9 599 e	152	95	122 e	252	342	294 e
Sweden	4 604	4 128	4 044 e	49	49	113 e	299	94	85 e
Switzerland	7 602	6 713	7 196 e	68	47	42 e	52	67	31 e
Turkey	2 125	498	754 e	145	30	74 e	142	11	39 e
United Kingdom	12 280	15 916	13 069 e	450	419	337 e	1 307	1 711	1 592 e
United States	18 432	18 368	19 220 e	155	138	182 e	655	679	513 e
Other OECD	3 508	3 421	3 124 e	252	180	212 e	182	189	159 e
OECD	**170 709**	**168 459**	**163 496 e**	**4 194**	**3 281**	**3 955 e**	**17 067**	**16 668**	**15 665 e**
Africa	8 292	7 529	7 581 e	879	789	704 e	81	140	68 e
Latin America	1 430	1 192	1 767 e	122	162	22 e	59	30	43 e
Asia	2 195	2 466	3 122 e	10	4	15 e	32	113	215 e
China	158	118	178 e	19	43	6 e	-	-	29 e
Non-OECD Europe	7 933	7 114	6 457 e	196	165	258 e	282	299	152 e
Former Soviet Union	493	424	276 e	-	-	4 e	15	19	24 e
Middle East	2 612	1 930	1 446 e	251	172	60 e	95	101	97 e
Non-specified/Other	5 593	7 292	14 042 e	452	741	1 009 e	1 180	823	856 e
Non-OECD	**28 706**	**28 065**	**34 869 e**	**1 929**	**2 076**	**2 078 e**	**1 744**	**1 525**	**1 484 e**
Total Exports	**199 415**	**196 524**	**198 365 e**	**6 123**	**5 357**	**6 033 e**	**18 811**	**18 193**	**17 149 e**

Source: Annual Oil Statistics.

EUROPEAN UNION
Table 8B: Exports by Country of Destination

000 metric tons

	2000	2001	2002	2000	2001	2002	2000	2001	2002
	Total Gasoline			**Total Kerosene**			**Gas/Diesel Oil**		
Australia	67	6	-	-	1	8 e	4	-	29 e
Austria	500	447	364 e	47	38	89 e	1 949	1 866	2 230 e
Belgium	1 825	1 761	1 450 e	1 046	1 370	895 e	6 547	7 139	5 770 e
Canada	357	421	776 e	-	13	-	-	103	-
Denmark	708	682	740 e	40	64	71 e	1 153	1 072	795 e
Finland	34	47	10 e	1	1	6 e	245	361	270 e
France	1 712	1 539	1 431 e	224	129	333 e	7 622	6 833	6 222 e
Germany	10 276	8 265	6 401 e	3 219	3 506	3 300 e	13 456	15 583	10 907 e
Greece	290	23	112 e	2	22	6 e	412	617	447 e
Ireland	690	768	923 e	613	574	742 e	1 567	1 382	1 598 e
Italy	505	549	364 e	64	41	130 e	256	122	523 e
Japan	-	-	207 e	-	-	-	-	43	-
Luxembourg	621	567	565 e	313	280	370 e	1 431	1 515	1 521 e
Mexico	75	39	323 e	-	-	-	35	-	76 e
Netherlands	2 771	2 914	3 680 e	178	180	194 e	2 513	2 541	4 171 e
Norway	537	751	830 e	191	316	282 e	1 061	1 217	1 439 e
Portugal	240	30	46 e	42	17	9 e	691	418	513 e
Spain	1 033	879	831 e	195	126	68 e	4 097	4 326	5 468 e
Sweden	1 585	1 581	1 538 e	441	253	210 e	1 307	1 120	1 241 e
Switzerland	2 827	2 382	2 601 e	1 076	988	804 e	3 054	2 813	3 199 e
Turkey	403	128	68 e	59	30	74 e	577	144	188 e
United Kingdom	2 423	4 162	2 422 e	1 168	467	821 e	2 652	3 684	3 015 e
United States	7 737	8 533	10 680 e	357	739	292 e	1 154	802	1 178 e
Other OECD	1 192	1 022	889 e	70	75	1 e	1 143	1 107	1 176 e
OECD	**38 408**	**37 496**	**37 251 e**	**9 346**	**9 230**	**8 705 e**	**52 926**	**54 808**	**51 976 e**
Africa	2 383	2 018	2 416 e	326	204	247 e	2 674	2 424	2 346 e
Latin America	428	518	778 e	22	1	19 e	206	225	618 e
Asia	54	2	192 e	3	11	-	67	105	110 e
China	-	-	-	-	-	-	40	-	30 e
Non-OECD Europe	1 252	1 221	1 554 e	508	407	490 e	2 217	1 891	1 794 e
Former Soviet Union	255	108	31 e	1	1	5 e	69	122	38 e
Middle East	1 307	695	708 e	6	7	-	281	85	188 e
Non-specified/Other	1 836	1 884	2 964 e	119	296	280 e	300	1 415	5 664 e
Non-OECD	**7 515**	**6 446**	**8 643 e**	**985**	**927**	**1 041 e**	**5 854**	**6 267**	**10 788 e**
Total Exports	**45 923**	**43 942**	**45 894 e**	**10 331**	**10 157**	**9 746 e**	**58 780**	**61 075**	**62 764 e**

	2000	2001	2002	2000	2001	2002	2000	2001	2002
	Fuel Oil (Residual)			**Petroleum Coke**			**Other Products**		
Australia	-	-	-	-	-	-	8	9	13 e
Austria	178	241	175 e	31	30	16 e	345	306	271 e
Belgium	3 476	4 898	4 809 e	24	33	35 e	3 041	3 307	3 233 e
Canada	183	78	167 e	-	-	17 e	6	8	22 e
Denmark	300	230	307 e	-	4	-	277	256	257 e
Finland	327	440	505 e	-	-	-	212	168	173 e
France	2 331	2 046	1 740 e	251	147	212 e	1 239	1 194	1 371 e
Germany	2 098	1 699	1 852 e	65	95	68 e	2 062	2 174	1 986 e
Greece	142	33	49 e	45	54	3 e	122	87	98 e
Ireland	1 292	1 041	805 e	2	5	3 e	395	327	338 e
Italy	3 607	2 110	2 040 e	44	45	22 e	285	551	437 e
Japan	5	-	21 e	1	1	-	29	15	21 e
Luxembourg	11	15	12 e	18	18	16 e	52	82	46 e
Mexico	117	38	-	-	-	-	9	20	6 e
Netherlands	2 736	2 550	2 721 e	203	406	368 e	1 090	1 016	1 043 e
Norway	2 083	1 784	1 384 e	94	88	122 e	402	325	847 e
Portugal	1 082	946	1 574 e	26	33	44 e	221	240	248 e
Spain	1 676	1 408	1 800 e	165	119	156 e	790	856	860 e
Sweden	500	695	503 e	36	25	32 e	387	311	322 e
Switzerland	214	56	173 e	16	42	48 e	295	318	298 e
Turkey	528	22	130 e	2	9	-	269	124	181 e
United Kingdom	3 429	4 714	3 703 e	22	20	18 e	829	739	1 161 e
United States	7 506	6 748	5 561 e	15	-	-	853	729	814 e
Other OECD	211	302	151 e	71	58	74 e	387	488	462 e
OECD	**34 032**	**32 094**	**30 182 e**	**1 131**	**1 232**	**1 254 e**	**13 605**	**13 650**	**14 508 e**
Africa	1 243	1 151	1 032 e	27	23	22 e	679	780	746 e
Latin America	362	91	3 e	8	12	15 e	223	153	269 e
Asia	1 818	1 933	2 245 e	23	16	21 e	188	282	324 e
China	63	3	79 e	1	1	-	35	71	34 e
Non-OECD Europe	2 982	2 549	1 791 e	64	50	50 e	432	532	368 e
Former Soviet Union	6	28	25 e	7	5	6 e	140	141	143 e
Middle East	360	499	134 e	-	6	-	312	365	259 e
Non-specified/Other	1 239	1 584	2 544 e	-	7	8 e	467	542	717 e
Non-OECD	**8 073**	**7 838**	**7 853 e**	**130**	**120**	**122 e**	**2 476**	**2 866**	**2 860 e**
Total Exports	**42 105**	**39 932**	**38 035 e**	**1 261**	**1 352**	**1 376 e**	**16 081**	**16 516**	**17 368 e**

Source: Annual Oil Statistics.

AUSTRALIA

Figure 1
FUEL SHARES OF TPES *

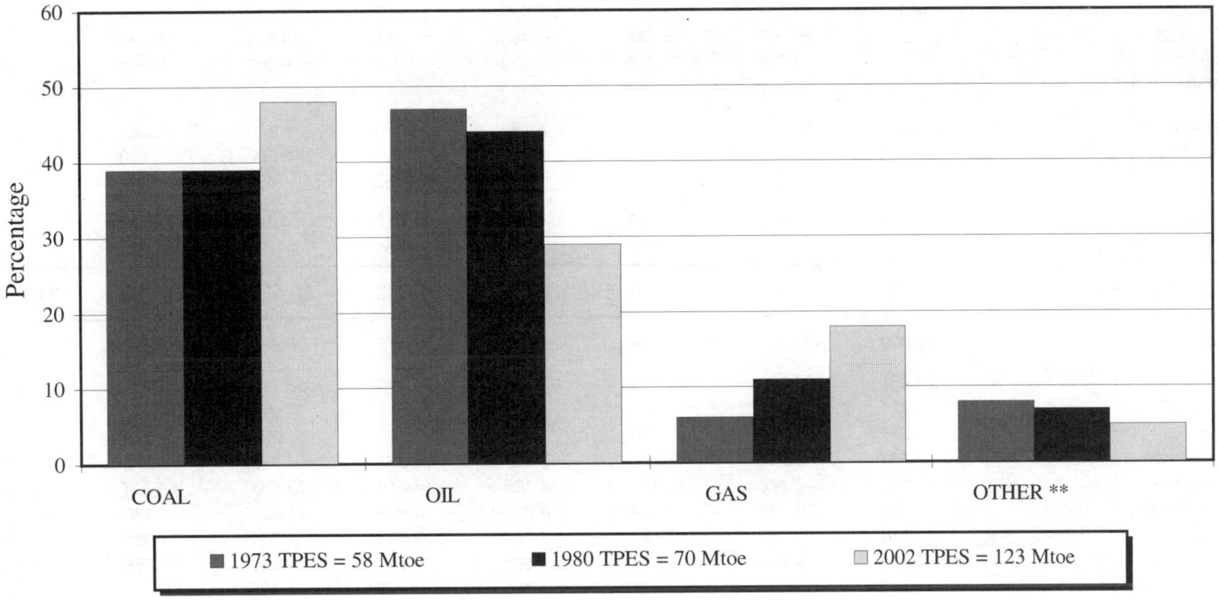

Source: Annual Energy Balances.

* Excluding electricity trade.

** Includes hydro, geothermal, solar, wind, combustible renewables and waste, etc.

Figure 2
OIL PRODUCTION AND CONSUMPTION

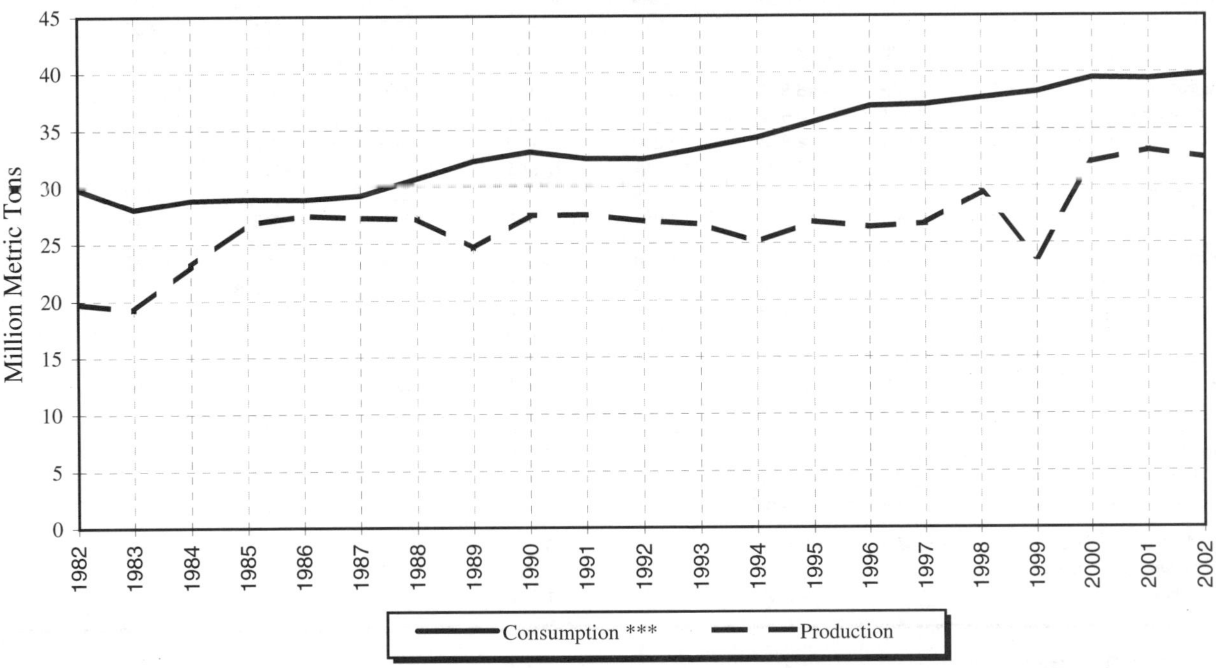

Source: Annual Oil Statistics.

*** Includes refinery fuel and international marine bunkers.

AUSTRALIA
Table 1: Oil Supply and Consumption - Summary Table

000 metric tons

	1973	1978	1980	1990	1999	2000	2001	2002
Indigenous Production	19 182	21 891	20 564	27 493	23 657	32 076	33 138	32 453 e
Crude Oil	..	20 053	18 868	25 465	15 755	23 874	25 331	23 561 e
NGL	..	1 838	1 696	2 028	7 902	8 202	7 807	8 892 e
Other	..	-	-	-	-	-	-	-
+ Net Imports	9 052	10 129	11 136	5 283	11 366	4 240	921	2 559 e
Crude Oil, NGL, Other	8 657	10 244	10 098	4 327	11 128	3 507	400	832 e
Petroleum Products	395	- 115	1 038	956	238	733	521	1 727 e
- Intl. Marine Bunkers	1 828	1 440	1 135	660	815	914 e	793	851 e
+ Stock Changes	- 49	889	- 490	- 785	451	30	- 908	579 e
TOTAL SUPPLY	**26 357**	**31 469**	**30 075**	**31 331**	**34 659**	**35 432 e**	**32 358**	**34 740 e**
Refinery Intake	*26 920*	*29 441*	*29 328*	*29 571*	*36 373*	*36 270*	*34 892*	*33 640 e*
Refinery Output	*25 251*	*30 065*	*28 933*	*30 758*	*36 399*	*36 708*	*35 403*	*34 319 e*
LPG + Ethane	*371*	*349*	*335*	*521*	*929*	*911*	*977*	*943 e*
Naphtha	*550*	*280*	*180*	*217*	*165*	*291*	*376*	*357 e*
Kerosenes	*2 303*	*2 947*	*2 294*	*2 858*	*4 357*	*4 646*	*4 882*	*4 460 e*
Gasolines	*8 815*	*10 242*	*10 861*	*12 128*	*14 215*	*13 863*	*12 406*	*12 482 e*
Gas/Diesel Oil	*5 866*	*6 988*	*7 283*	*8 700*	*10 942*	*11 337*	*11 120*	*11 006 e*
Fuel Oil	*5 010*	*4 726*	*4 150*	*2 440*	*1 592*	*1 769*	*1 886*	*1 629 e*
Other Products	*2 336*	*4 533*	*3 830*	*3 894*	*4 199 e*	*3 891*	*3 756*	*3 442 e*
- Refinery Losses	1 669	- 624	395	-1 187	- 26	- 438	- 511	- 679 e
+ Returns	-	-	-	-	-	-	-	-
- Statistical Differences	- 334	1 433	- 647	124	-2 782	-2 689 e	-5 722	-3 513 e
TOTAL CONSUMPTION	**25 022**	**30 660**	**30 327**	**32 394**	**37 467**	**38 559**	**38 591**	**38 932 e**
Transformation and Energy	**1 079**	**4 312**	**3 619**	**3 154**	**2 980**	**2 956**	**3 049**	..
Crude Oil	-	2	-	7	5	5 e	13	..
LPG + Ethane	50	49	56	32	53	53 e	58	..
Gas/Diesel Oil	170	828	620	834	811	819	872	..
Fuel Oil	634	1 080	986	425	213	210 e	238 e	..
Petroleum Coke	-	-	-	419	709	651	686	..
Other	225	2 353	1 957	1 437	1 189	1 218 e	1 182 e	..
Transport	**12 703**	**15 835**	**16 655**	**21 177**	**25 320**	**26 022**	**26 026**	..
Motor Gasoline	8 894	10 674	10 915	12 943	13 306	13 618	13 023	..
Gas/Diesel Oil	1 620	2 692	3 200	5 043	6 595	6 842	6 917	..
Subtotal	*10 514*	*13 366*	*14 115*	*17 986*	*19 901*	*20 460*	*19 940*	..
Aviation Fuels	1 299	1 676	1 719	2 346	3 901	4 016	4 444	..
Other	890	793	821	845	1 518	1 546	1 642	..
Industry	**7 852**	**8 045**	**7 874**	**6 390**	**7 274**	**7 650**	**7 591**	..
Naphtha	350	328	342	339	26	25 e	26	..
LPG + Ethane	117	216	235	810	948	927	948	..
Gas/Diesel Oil	1 111	1 873	1 676	1 509	2 054	2 124	2 143	..
Fuel Oil	4 136	3 529	3 403	1 061	790	948	798	..
Other	2 138	2 099	2 218	2 671	3 456	3 626	3 676	..
Other Sectors	**3 388**	**2 468**	**2 179**	**1 673**	**1 893**	**1 931**	**1 925**	..
Gas/Diesel Oil	2 284	1 293	1 318	1 165	1 373	1 397	1 396	..
Fuel Oil	110	76	53	21	10	10	10	..
Other	994	1 099	808	487	510	524	519	..
Memo: Non-Energy Use	*2 397*	*2 414*	*2 632*	*3 555*	*4 045*	*4 154*	*4 261*	..
Chem/Petchem. Feedst.	*1 283*	*1 102*	*1 258*	*1 801*	*2 006*	*2 043*	*2 148*	..
Other	*1 114*	*1 312*	*1 374*	*1 754*	*2 039*	*2 111*	*2 113*	..

Source: Annual Oil Statistics.

AUSTRALIA
Table 2: Supply of Crude Oil, NGL, Feedstocks and Other Hydrocarbons

000 metric tons

	1973	1978	1980	1990	1999	2000	2001	2002
CRUDE OIL								
Indigenous Production	..	20 053	18 868	25 465	15 755	23 874	25 331	23 561 e
+ Imports	..	9 612	9 403	6 039	22 421	20 138	21 153	20 910 e
- Exports	..	-	-	5 348	9 906	15 338	19 421	19 499 e
- Direct Use	..	2	-	7	5	5 e	13	10 e
+ Stock Change	..	- 75	- 23	- 529	400	- 54	- 874	293 e
- Statistical Difference	..	781	- 171	- 920	-3 800	-3 637	-2 561	-2 171 e
Refinery Intake	**..**	**28 807**	**28 419**	**26 540**	**32 465**	**32 252**	**28 737**	**27 426 e**
Closing Stock Level	..	*903*	*960*	*1 686*	*1 776*	*1 830*	*2 704*	*2 411 e*
NATURAL GAS LIQUIDS								
Indigenous Production	..	1 838	1 696	2 028	7 902	8 202	7 807	8 892 e
+ Imports	..	-	-	-	-	-	-	-
- Exports	..	-	-	-	1 360	1 555	1 582	1 755 e
- Direct Use	..	1 838	1 696	2 028	1 050	2 005	2 177	2 483 e
+ Stock Change	..	-	-	-	13	- 72	- 18	-
- Statistical Difference	..	-	-	-	3 987	3 174	1 125	1 966 e
Refinery Intake	**..**	**-**	**-**	**-**	**1 518**	**1 396**	**2 905**	**2 688 e**
Closing Stock Level	..	-	-	-	*180*	*252*	*270*	*270 e*
REFINERY FEEDSTOCKS								
Backflows	..	-	-	-	-	-	-	-
+ Transfers	..	-	-	-	-	-	-	-
+ Imports	..	632	695	3 862	1 585	1 711	1 556	2 379 e
- Exports	..	-	-	226	1 612	1 449	1 306	1 203 e
+ Stock Change	..	-	- 33	239	- 25	- 146	- 13	76 e
- Statistical Difference	..	- 2	- 247	844	-2 442	-2 506	-3 013	-2 274 e
Refinery Intake	**..**	**634**	**909**	**3 031**	**2 390**	**2 622**	**3 250**	**3 526 e**
Closing Stock Level	..	-	*425*	*551*	*467*	*613*	*626*	*550 e*
OTHER HYDROCARBONS								
Indigenous Production	..	-	-	-	-	-	-	-
+ Imports	..	-	-	-	-	-	-	-
- Exports	..	-	-	-	-	-	-	-
- Direct Use	..	-	-	-	-	-	-	-
+ Stock Change	..	-	-			-	-	-
- Statistical Difference	..	-	-	-	-	-	-	-
Refinery Intake	**..**	**-**	**-**	**-**	**-**	**-**	**-**	**-**
Closing Stock Level	..	-	-	-	-	-	-	-
CRUDE OIL, NGL, FEEDSTOCKS AND OTHER								
Indigenous Production	19 182	21 891	20 564	27 493	23 657	32 076	33 138	32 453 e
+ Backflows	-	-	-	-	-	-	-	-
+ Transfers	-	-	-	-	-	-	-	-
+ Imports	8 675	10 244	10 098	9 901	24 006	21 849	22 709	23 289 e
- Exports	18	-	-	5 574	12 878	18 342	22 309	22 457 e
- Direct Use	1 042	1 840	1 696	2 035	1 055	2 010	2 190	2 493 e
+ Stock Change	-	- 75	- 56	- 290	388	- 272	- 905	369 e
- Statistical Difference	- 123	779	- 418	- 76	-2 255	-2 969	-4 449	-2 479 e
Refinery Intake	**26 920**	**29 441**	**29 328**	**29 571**	**36 373**	**36 270**	**34 892**	**33 640 e**
Closing Stock Level	-	*903*	*1 385*	*2 237*	*2 423*	*2 695*	*3 600*	*3 231 e*

Source: Annual Oil Statistics.

AUSTRALIA
Table 3: Motor Gasoline - Supply

000 metric tons

	1973	1978	1980	1990	1999	2000	2001	2002
MOTOR GASOLINE								
Refinery Output	**8 787**	**10 203**	**10 795**	**11 969**	**14 092**	**13 752**	**12 309**	**12 379 e**
+ Imports	409	465	359	1 166	651	778	870	1 047 e
- Exports	303	290	229	153	1 121	1 002	941	868 e
+ Transfers	-	-	-	-	-	-	-	-
+ Stock Change	-	553	- 206	- 53	66	92	37	17 e
- Statistical Difference	- 1	257	- 196	- 14	382	2	- 748	- 819 e
Total Consumption	**8 894**	**10 674**	**10 915**	**12 943**	**13 306**	**13 618**	**13 023**	**13 394 e**
Closing Stock Level	-	*710*	*848*	*891*	*769*	*677*	*640*	*623 e*

Table 4: Aviation Fuels - Supply

000 metric tons

	1973	1978	1980	1990	1999	2000	2001	2002
AVIATION GASOLINE								
Refinery Output	**28**	**39**	**66**	**159**	**123**	**111**	**97**	**103 e**
+ Imports	49	50	27	6	31	-	-	4 e
- Exports	8	16	6	58	52	56	20	51 e
+ Transfers	-	-	-	-	-	-	-	-
+ Stock Change	-	13	- 14	- 5	- 1	3	1	- 1 e
- Statistical Difference	- 5	6	- 7	11	28	- 14	1	- 17 e
Total Consumption	**74**	**80**	**80**	**91**	**73**	**72**	**77**	**72 e**
Closing Stock Level	-	*25*	*33*	*36*	*28*	*25*	*24*	*25 e*
GASOLINE TYPE JET FUEL								
Refinery Output	-	-	-	-	-	-	-	-
+ Imports	-	-	-	-	-	-	-	-
- Exports	-	-	-	-	-	-	-	-
+ Transfers	-	-	-	-	-	-	-	-
+ Stock Change	-	-	-	-	-	-	-	-
- Statistical Difference	-	-	-	-	-	-	-	-
Total Consumption	-	-	-	-	-	-	-	-
Closing Stock Level	-	-	-	-	-	-	-	-
KEROSENE TYPE JET FUEL								
Refinery Output	**1 373**	**1 756**	**1 747**	**2 611**	**4 149**	**4 404**	**4 639**	**4 271 e**
+ Imports	79	120	140	185	111	135	307	183 e
- Exports	219	224	166	419	433	459	598	434 e
+ Transfers	-	-	-	-	-	-	-	-
+ Stock Change	-	38	19	- 7	- 72	24	- 28	- 26 e
- Statistical Difference	8	94	101	115	- 73	160	- 47	215 e
Total Consumption	**1 225**	**1 596**	**1 639**	**2 255**	**3 828**	**3 944**	**4 367**	**3 779 e**
Closing Stock Level	-	*151*	*118*	*143*	*256*	*232*	*260*	*286 e*

Source: Annual Oil Statistics.

AUSTRALIA
Table 5A: Gas/Diesel Oil - Supply and Consumption

000 metric tons

	1973	1978	1980	1990	1999	2000	2001	2002
Refinery Output	**5 866**	**6 988**	**7 283**	**8 700**	**10 942**	**11 337**	**11 120**	**11 006 e**
+ Recycled Products	-	-	-	-	-	-	-	-
+ Imports	448	566	566	871	1 211	1 180	952	1 075 e
- Exports	587	751	604	619	1 040	904	1 080	806 e
- Intl. Marine Bunkers	428	209	189	123	89	142 e	104	127 e
+ Transfers	-	-	-	-	-	-	-	-
+ Stock Changes	-	149	- 165	- 116	36	66	31	10 e
- Statistical Differences	114	57	77	162	227	355	- 409	- 614 e
Total Consumption	**5 185**	**6 686**	**6 814**	**8 551**	**10 833**	**11 182**	**11 328**	**11 772 e**
Transformation	**110**	**724**	**502**	**563**	**343**	**350**	**394**	..
Public Electricity	110	352	178	76	106	108 e	130	..
Autoproducer Electricity	-	357	302	487	237	242 e	264	..
Public CHP Plants	-	-	-	-	-	-	-	..
Autoproducers of CHP	-	-	-	-	-	-	-	..
Public Heat Plants	-	-	-	-	-	-	-	..
Autoproducer Heat Plants	-	-	-	-	-	-	-	..
Petrochemical Industry	-	-	-	-	-	-	-	..
Other Transformation	-	15	22	-	-	-	-	..
Energy Sector	**60**	**104**	**118**	**271**	**468**	**469**	**478**	..
Refinery Fuel	-	-	-	-	-	-	-	..
Other Energy Sector	60	104	118	271	468	469	478	..
Final Consumption	**5 015**	**5 858**	**6 194**	**7 717**	**10 022**	**10 363**	**10 456**	..
Transport	**1 620**	**2 692**	**3 200**	**5 043**	**6 595**	**6 842**	**6 917**	..
Air	-	-	-	1	-	-	-	..
Road	915	1 653	2 213	4 385	5 853	6 079	6 144	..
Rail	455	604	615	547	548	572	580	..
Inland Waterways	250	435	372	110	194	191	193	..
Non-Specified	-	-	-	-	-	-	-	..
Industry	**1 111**	**1 873**	**1 676**	**1 509**	**2 054**	**2 124**	**2 143**	..
Iron and Steel	58	207	37	3	4	4	4	..
Chemical (incl. Pet.-Chem.)	-	107	121	5	12	12	12	..
Non-Ferrous Metals	-	76	106	56	47	50	52	..
Non-Metallic Minerals	-	100	72	24	27	28	29	..
Transport Equipment	-	13	12	2	2	2 e	2	..
Machinery	-	23	18	2	2	2	2	..
Mining and Quarrying	-	395	319	487	885	913	932	..
Food and Tobacco	-	127	100	24	20	20	20	..
Pulp, Paper and Printing	-	15	11	1	1	1 e	1	..
Wood and Wood Products	-	35	36	19	23	24	25	..
Construction	-	765	833	884	1 030	1 067	1 063	..
Textiles and Leather	-	10	11	2	1	1	1	..
Non-Specified	1 053	-	-	-	-	-	-	..
Other Sectors	**2 284**	**1 293**	**1 318**	**1 165**	**1 373**	**1 397**	**1 396**	..
Commerce and Public	-	382	339	59	51	50 e	44	..
Residential	-	102	89	52	31	35	35	..
Agriculture	984	809	890	1 054	1 291	1 312	1 317	..
Non-Specified	1 300	-	-	-	-	-	-	..
Memo: Non-Energy Use	-	*11*	*3*	*4*	*10*	*10*	*10*	..
Chem/Petchem. Feedst.	-	*11*	*2*	-	*6*	*6*	*6*	..
Other	-	-	*1*	*4*	*4*	*4*	*4*	..
Closing Stock Level	-	*877*	*666*	*687*	*699*	*633*	*602*	*592 e*

Source: Annual Oil Statistics.

AUSTRALIA
Table 5B: Fuel Oil (Residual) - Supply and Consumption

000 metric tons

	1973	1978	1980	1990	1999	2000	2001	2002
Refinery Output	**5 010**	**4 726**	**4 150**	**2 440**	**1 592**	**1 769**	**1 886**	**1 629 e**
+ Recycled Products	-	-	-	-	-	-	-	-
+ Imports	2 271	1 968	2 498	1 000	565	757	771	884 e
- Exports	620	237	325	561	240	561	685	286 e
- Intl. Marine Bunkers	1 400	1 231	946	537	726	772 e	689	724 e
+ Transfers	-	-	-	-	-	-	-	-
+ Stock Changes	-	193	- 160	- 54	- 16	66	41	65 e
- Statistical Differences	- 489	- 25	31	341	- 17	- 89	38	244 e
Total Consumption	**5 750**	**5 444**	**5 186**	**1 947**	**1 192**	**1 348 e**	**1 286**	**1 324 e**
Transformation	**314**	**594**	**676**	**317**	**93**	**104 e**	**138 e**	..
Public Electricity	314	438	448	284	59	67 e	100 e	..
Autoproducer Electricity	-	148	204	16	14	14 e	14	..
Public CHP Plants	-	-	-	-	-	-	-	..
Autoproducers of CHP	-	-	-	-	-	-	-	..
Public Heat Plants	-	-	-	-	-	-	-	..
Autoproducer Heat Plants	-	-	-	-	-	-	-	..
Petrochemical Industry	-	-	-	-	-	-	-	..
Other Transformation	-	8	24	17	20	23	24	..
Energy Sector	**320**	**486**	**310**	**108**	**120**	**106**	**100**	..
Refinery Fuel	320	484	308	107	119	106	100	..
Other Energy Sector	-	2	2	1	1	-	-	..
Final Consumption	**5 116**	**4 364**	**4 200**	**1 522**	**979**	**1 138**	**1 048**	..
Transport	**870**	**759**	**744**	**440**	**179**	**180**	**240**	..
Air	-	-	-	-	-	-	-	..
Road	-	-	-	-	-	-	-	..
Rail	-	-	-	-	-	-	-	..
Inland Waterways	870	759	744	440	179	180	240	..
Non-Specified	-	-	-	-	-	-	-	..
Industry	**4 136**	**3 529**	**3 403**	**1 061**	**790**	**948**	**798**	..
Iron and Steel	1 020	364	370	5	-	-	-	..
Chemical (incl. Pet.-Chem.)	284	170	115	15	10	10	10	..
Non-Ferrous Metals	-	1 759	1 761	861	696	857	700	..
Non-Metallic Minerals	-	297	154	27	10	10	11	..
Transport Equipment	-	22	10	1	-	-	-	..
Machinery	-	33	23	-	-	-	-	..
Mining and Quarrying	-	226	391	53	22	23	23	..
Food and Tobacco	-	402	323	52	28	30	29	..
Pulp, Paper and Printing	-	156	176	22	19	13	20	..
Wood and Wood Products	-	30	32	18	2	2 e	2	..
Construction	-	2	2	-	1	1	1	..
Textiles and Leather	-	68	46	7	2	2	2	..
Non-Specified	2 832	-	-	-	-	-	-	..
Other Sectors	**110**	**76**	**53**	**21**	**10**	**10**	**10**	..
Commerce and Public	110	76	53	21	10	10	10	..
Residential	-	-	-	-	-	-	-	..
Agriculture	-	-	-	-	-	-	-	..
Non-Specified	-	-	-	-	-	-	-	..
Memo: Non-Energy Use	-	*20*	*2*	*1*	*20*	-	-	..
Chem/Petchem. Feedst.	-	-	-	-	-	-	-	..
Other	-	*20*	*2*	*1*	*20*	-	-	..
Closing Stock Level	-	*1 173*	*1 038*	*425*	*392*	*326*	*285*	*220 e*

Source: Annual Oil Statistics.

AUSTRALIA
Table 5C: Total Products - Supply and Consumption

000 metric tons

	1973	1978	1980	1990	1999	2000	2001	2002
Refinery Output	**25 251**	**30 065**	**28 933**	**30 758**	**36 399**	**36 708**	**35 403**	**34 319 e**
+ Recycled Products	-	-	-	-	-	-	-	-
+ Imports	3 580	3 460	4 165	4 192	3 670	4 077	4 246	4 472 e
- Exports	3 185	3 575	3 127	3 236	3 432	3 344	3 725	2 745 e
- Intl. Marine Bunkers	1 828	1 440	1 135	660	815	914 e	793	851 e
+ Transfers	1 042	1 840	1 696	2 035	1 055	2 010	2 190	2 493 e
+ Stock Changes	- 49	964	- 434	- 495	63	302	- 3	210 e
- Statistical Differences	- 211	654	- 229	200	- 527	280	-1 273	-1 034 e
Total Consumption	**25 022**	**30 660**	**30 327**	**32 394**	**37 467**	**38 559**	**38 591**	**38 932 e**
Transformation	**674**	**1 407**	**1 263**	**911**	**509**	**525**	**612**	..
Public Electricity	424	790	626	360	165	175 e	230	..
Autoproducer Electricity	-	505	506	504	303	307 e	336	..
Public CHP Plants	-	-	-	-	-	-	-	..
Autoproducers of CHP	-	-	-	-	-	-	-	..
Public Heat Plants	-	-	-	-	-	-	-	..
Autoproducer Heat Plants	-	-	-	-	-	-	-	..
Petrochemical Industry	-	-	-	-	-	-	-	..
Other Transformation	250	112	131	47	41	43 e	46	..
Energy Sector	**405**	**2 905**	**2 356**	**2 243**	**2 471**	**2 431**	**2 437**	..
Refinery Fuel	320	2 760	2 197	1 963	1 961	1 920	1 906	..
Other Energy Sector	85	145	159	280	510	511	531	..
Final Consumption	**23 943**	**26 348**	**26 708**	**29 240**	**34 487**	**35 603**	**35 542**	..
Transport	**12 703**	**15 835**	**16 655**	**21 177**	**25 320**	**26 022**	**26 026**	..
Air	1 299	1 676	1 719	2 347	3 901	4 016	4 444	..
Road	9 829	12 361	13 205	17 732	20 498	21 063	20 569	..
Rail	455	604	615	548	548	572	580	..
Inland Waterways	1 120	1 194	1 116	550	373	371	433	..
Non-Specified	-	-	-	-	-	-	-	..
Industry	**7 852**	**8 045**	**7 874**	**6 390**	**7 274**	**7 650**	**7 591**	..
Iron and Steel	1 092	586	419	22	27	27	27	..
Chemical (incl. Pet.-Chem.)	1 673	1 514	1 559	1 844	2 149	2 199	2 272	..
Non-Ferrous Metals	-	1 984	2 034	1 477	1 426	1 589	1 579	..
Non-Metallic Minerals	-	423	249	120	115	118	115	..
Transport Equipment	-	40	27	8	17	17	17	..
Machinery	-	71	59	25	24	21	24	..
Mining and Quarrying	-	629	713	547	913	942	961	..
Food and Tobacco	-	555	452	138	127	131	131	..
Pulp, Paper and Printing	-	174	190	39	31	32	40	..
Wood and Wood Products	-	66	70	41	32	26	27	..
Construction	474	1 340	1 423	1 547	1 777	1 880	1 735	..
Textiles and Leather	-	82	62	21	18	17	17	..
Non-Specified	4 613	581	617	561	618	648	646	..
Other Sectors	**3 388**	**2 468**	**2 179**	**1 673**	**1 893**	**1 931**	**1 925**	..
Commerce and Public	110	541	488	235	274	281	280	..
Residential	769	1 069	754	351	305	318	308	..
Agriculture	1 035	847	921	1 075	1 311	1 332	1 337	..
Non-Specified	1 474	11	16	12	3	-	-	..
Memo: Non-Energy Use	*2 397*	*2 414*	*2 632*	*3 555*	*4 045*	*4 154*	*4 261*	..
Chem/Petchem. Feedst.	*1 283*	*1 102*	*1 258*	*1 801*	*2 006*	*2 043*	*2 148*	..
Other	*1 114*	*1 312*	*1 374*	*1 754*	*2 039*	*2 111*	*2 113*	..
Closing Stock Level	-	*3 873*	*3 231*	*3 243*	*3 023*	*2 721*	*2 724*	*2 514 e*

Source: Annual Oil Statistics.

AUSTRALIA
Table 6A: Supply and Consumption

000 metric tons

	1980	2000	2001	2002	1980	2000	2001	2002
	LPG AND ETHANE				**REFINERY GAS**			
Refinery Output	**335**	**911**	**977**	**943 e**	**1 972**	**1 226**	**1 176**	**1 092 e**
+ Recycled Products	-	-	-	-	-	-	-	-
+ Imports	-	265	322	298 e	-	-	-	-
- Exports	1 490	-	-	-	-	-	-	-
- Intl. Marine Bunkers	-	-	-	-	-	-	-	-
+ Transfers	1 662	2 000	2 172	2 478 e	-	-	-	-
+ Stock Change	2	18	- 83	89 e	-	-	-	-
- Statistical Difference	- 53	397	521	787 e	-	- 77	- 97	- 127 e
Total Consumption	**562**	**2 797**	**2 867**	**3 021 e**	**1 972**	**1 303**	**1 273**	**1 219 e**
of which: Ethane	-	*410*	*418*	*410 e*	-	-	-	-
Transformation	50	20 e	22	..	-	50	57	..
Energy	6	33	36	..	1 889	1 163	1 120	..
Transport	77	1 366	1 402	..	-	-	-	..
Industry	235	927	948	..	83	90	96	..
Other	194	451	459	..	-	-	-	..
Memo: Non-Energy Use	*127*	*662*	*698*	..	*32*	-	-	..
Closing Stock Level	*2*	*113*	*196*	*107 e*	-	-	-	-
	NAPHTHA				**OTHER KEROSENE**			
Refinery Output	**180**	**291**	**376**	**357 e**	**547**	**242**	**243**	**189 e**
+ Recycled Products	-	-	-	-	-	-	-	-
+ Imports	150	-	-	-	-	-	-	8 e
- Exports	-	-	-	22 e	23	8	8	1 e
- Intl. Marine Bunkers	-	-	-	-	-	-	-	-
+ Transfers	-	-	-	-	-	-	-	-
+ Stock Change	11	- 4	3	- 6 e	- 18	- 5	- 6	3 e
- Statistical Difference	- 36	262	353	304 e	- 113	147	161	110 e
Total Consumption	**377**	**25 e**	**26**	**25 e**	**619**	**82**	**68**	**89 e**
Transformation	35	-	-	..	-	-	-	..
Energy	-	-	-	..	-	-	-	..
Transport	-	-	-	..	-	-	-	..
Industry	342	25 e	26	..	21	9	8	..
Other	-	-	-	..	598	73	60	..
Memo: Non-Energy Use	*334*	*14 e*	*15 e*	..	*5*	*6*	*6*	..
Closing Stock Level	*9*	*12*	*9*	*15 e*	*186*	*28*	*34*	*31 e*
	WHITE SPIRIT + SBP				**LUBRICANTS**			
Refinery Output	**177**	**142**	**142**	**..**	**552**	**572**	**564**	**..**
+ Recycled Products	-	-	-	..	-	-	-	..
+ Imports	13	-	-	..	64	73	59	..
- Exports	14	-	-	..	224	221	246	..
- Intl. Marine Bunkers	-	-	-	..	-	-	-	..
+ Transfers	-	-	-	..	-	-	-	..
+ Stock Change	- 21	5	- 6	..	- 18	2	6	..
- Statistical Difference	- 25	- 33	- 10	..	- 51	- 45	- 114	..
Total Consumption	**180**	**180**	**146**	**..**	**425**	**471**	**497**	**..**
Transformation	-	-	-	..	-	-	-	..
Energy	-	-	-	..	-	-	-	..
Transport	-	-	-	..	-	-	-	..
Industry	180	180	146	..	425	471	497	..
Other	-	-	-	..	-	-	-	..
Memo: Non-Energy Use	*180*	*170*	*142*	..	*425*	*471*	*497*	..
Closing Stock Level	*57*	*18*	*24*	..	*150*	*114*	*108*	..

Source: Annual Oil Statistics.

AUSTRALIA
Table 6B: Supply and Consumption

000 metric tons

	1980	2000	2001	2002	1980	2000	2001	2002
		BITUMEN				PARAFFIN WAXES		
Refinery Output	**526**	**725**	**689**	**..**	**12**	**22**	**24**	**..**
+ Recycled Products	-	-	-	..	-	-	-	..
+ Imports	4	56	34	..	-	-	-	..
- Exports	1	-	2	..	-	-	-	..
- Intl. Marine Bunkers	-	-	-	..	-	-	-	..
+ Transfers	-	-	-	..	-	-	-	..
+ Stock Change	- 8	-	4	..	-	- 1	-	..
- Statistical Difference	- 63	- 23	61	..	-	14	17	..
Total Consumption	**584**	**804**	**664**	**..**	**12**	**7**	**7**	**..**
Transformation	-	-	-	..	-	-	-	..
Energy	-	-	-	..	-	-	-	..
Transport	-	-	-	..	-	-	-	..
Industry	584	804	664	..	12	7	7	..
Other	-	-	-	..	-	-	-	..
Memo: Non-Energy Use	*584*	*804*	*664*	..	*12*	*7*	*7*	..
Closing Stock Level	*42*	*62*	*58*	..	-	*4*	*4 e*	..
		PETROLEUM COKE				OTHER PRODUCTS		
Refinery Output	**-**	**651**	**686**	**622 e**	**591**	**553**	**475**	**1 728 e**
+ Recycled Products	-	-	-	-	-	-	-	-
+ Imports	124	671	833	815 e	220	162	98	158 e
- Exports	-	-	-	-	45	133	145	277 e
- Intl. Marine Bunkers	-	-	-	-	-	-	-	-
+ Transfers	-	-	-	-	34	10 e	18	15 e
+ Stock Change	-	-	-	-	144	36	- 3	59 e
- Statistical Difference	-	6	44	49 e	106	- 782	-1 044	-1 166 e
Total Consumption	**124**	**1 316**	**1 475**	**1 388 e**	**838**	**1 410 e**	**1 487**	**2 849 e**
Transformation	-	-	-	..	-	1 e	1	..
Energy	-	651	686	..	33	9 e	17	..
Transport	-	-	-	..	-	-	-	..
Industry	124	665	789	..	789	1 400	1 469	..
Other	-	-	-	..	16	-	-	..
Memo: Non-Energy Use	*124*	*635*	*778*	..	*804*	*1 375*	*1 444*	..
Closing Stock Level	-	-	-	-	*82*	*477*	*480*	*615 e*

Source: Annual Oil Statistics.

AUSTRALIA
Table 7A: Imports by Country of Origin

000 metric tons

	2000	2001	2002	2000	2001	2002	2000	2001	2002
	Crude Oil			Natural Gas Liquids			Feedstocks and Other		
Australia	-	-	-	-	-	-	-	-	-
Belgium	-	-	-	-	-	-	-	-	-
Canada	-	-	-	-	-	-	-	-	-
Czech Republic	-	-	-	-	-	-	-	-	-
Denmark	-	-	-	-	-	-	-	-	-
Finland	-	-	-	-	-	-	-	-	-
France	-	-	-	-	-	-	-	-	-
Germany	-	-	-	-	-	-	-	-	-
Greece	-	-	-	-	-	-	-	-	-
Hungary	-	-	-	-	-	-	-	-	-
Ireland	-	-	-	-	-	-	-	-	-
Italy	-	-	-	-	-	-	-	-	-
Japan	-	-	-	-	-	-	-	-	-
Korea	-	-	-	-	-	-	-	-	-
Mexico	-	-	-	-	-	-	-	-	-
Netherlands	-	-	-	-	-	-	-	-	-
New Zealand	1 001	870	860 e	-	-	-	-	15	9 e
Norway	-	-	-	-	-	-	-	-	-
Poland	-	-	-	-	-	-	-	-	-
Spain	-	-	-	-	-	-	-	-	-
Sweden	-	-	-	-	-	-	-	-	-
Turkey	-	-	-	-	-	-	-	-	-
United Kingdom	-	-	-	-	-	-	-	-	-
United States	-	-	-	-	-	-	-	-	-
Other	-	-	-	-	-	-	-	-	-
OECD	**1 001**	**870**	**860 e**	**-**	**-**	**-**	**-**	**15**	**9 e**
Argentina	-	62	-	-	-	-	-	-	-
Brazil	-	-	-	-	-	-	-	-	-
Colombia	-	-	-	-	-	-	-	-	-
Ecuador	-	-	-	-	-	-	-	-	-
Netherlands Antilles	-	-	-	-	-	-	-	-	-
Peru	-	-	-	-	-	-	-	-	-
Trinidad and Tobago	-	-	-	-	-	-	-	-	-
Venezuela	-	-	-	-	-	-	-	-	-
Other	-	-	-	-	-	-	-	-	-
Latin America	**-**	**62**	**-**	**-**	**-**	**-**	**-**	**-**	**-**
Romania	-	-	-	-	-	-	-	-	-
Other	-	-	-	-	-	-	-	-	-
Non-OECD Europe	**-**	**-**	**-**	**-**	**-**	**-**	**-**	**-**	**-**
FSU	**-**	**-**	**-**	**-**	**-**	**-**	**-**	**-**	**-**
Iran	-	-	-	-	-	-	-	-	39 e
Iraq	296	256	-	-	-	-	-	-	-
Kuwait	52	40	-	-	-	-	-	-	-
Oman	270	66	-	-	-	-	-	-	-
Qatar	341	351	516 e	-	-	-	41	-	-
Saudi Arabia	1 541	2 607	1 400 e	-	-	-	1 268	793	1 164 e
United Arab Emirates	1 297	2 451	1 667 e	-	-	-	20	99	-
Other	170	-	-	-	-	-	-	-	-
Middle East	**3 967**	**5 771**	**3 583 e**	**-**	**-**	**-**	**1 329**	**892**	**1 203 e**
India	-	-	-	-	-	-	-	-	52 e
Indonesia	3 215	2 933	4 995 e	-	-	-	146	160	493 e
Singapore	-	-	43 e	-	-	-	57	267	455 e
Other	10 904	10 053	9 873 e	-	-	-	179	222	167 e
Asia	**14 119**	**12 986**	**14 911 e**	**-**	**-**	**-**	**382**	**649**	**1 167 e**
China	**131**	**165**	**75 e**	**-**	**-**	**-**	**-**	**-**	**-**
Algeria	-	-	-	-	-	-	-	-	-
Angola/Cabinada	-	-	-	-	-	-	-	-	-
Cameroon	-	-	-	-	-	-	-	-	-
Egypt	-	-	-	-	-	-	-	-	-
Gabon	89	-	-	-	-	-	-	-	-
Libya	-	-	-	-	-	-	-	-	-
Nigeria	170	-	-	-	-	-	-	-	-
Tunisia	-	-	-	-	-	-	-	-	-
Other	464	-	-	-	-	-	-	-	-
Africa	**723**	**-**	**-**	**-**	**-**	**-**	**-**	**-**	**-**
Non-spec./Other	**197**	**1 299**	**1 481 e**	**-**	**-**	**-**	**-**	**-**	**-**
Total Imports	**20 138**	**21 153**	**20 910 e**	**-**	**-**	**-**	**1 711**	**1 556**	**2 379 e**
Memo: Total OPEC	*6 912*	*8 638*	*8 578 e*	*-*	*-*	*-*	*1 475*	*1 052*	*1 696 e*

Source: Annual Oil Statistics.

AUSTRALIA
Table 7B: Imports by Country of Origin

000 metric tons

	2000	2001	2002	2000	2001	2002	2000	2001	2002
	Total Products			**Liquefied Petroleum Gases**			**Naphtha**		
Australia	-	-	-	-	-	-	-	-	-
Belgium	2	1	-	-	-	-	-	-	-
Canada	-	-	32 e	-	-	-	-	-	-
Czech Republic	-	-	-	-	-	-	-	-	-
Denmark	-	-	-	-	-	-	-	-	-
Finland	-	-	-	-	-	-	-	-	-
France	2	4	3 e	-	-	-	-	-	-
Germany	3	3	-	-	-	-	-	-	-
Greece	-	-	-	-	-	-	-	-	-
Hungary	-	-	-	-	-	-	-	-	-
Ireland	-	-	-	-	-	-	-	-	-
Italy	-	-	-	-	-	-	-	-	-
Japan	91	245	80 e	-	-	-	-	-	-
Korea	332	119	224 e	-	-	-	-	-	-
Mexico	-	-	-	-	-	-	-	-	-
Netherlands	1	1	-	-	-	-	-	-	-
New Zealand	57	14	40 e	-	3	15 e	-	-	-
Norway	1	-	-	-	-	-	-	-	-
Poland	-	-	-	-	-	-	-	-	-
Spain	1	1	-	-	-	-	-	-	-
Sweden	3	-	-	-	-	-	-	-	-
Turkey	-	-	-	-	-	-	-	-	-
United Kingdom	2	4	1 e	-	-	-	-	-	-
United States	714	743	841 e	-	-	1 e	-	-	-
Other	-	-	2 e	-	-	-	-	-	-
OECD	**1 209**	**1 135**	**1 223 e**	**-**	**3**	**16 e**	**-**	**-**	**-**
Argentina	-	13	-	-	-	-	-	-	-
Brazil	-	-	-	-	-	-	-	-	-
Colombia	-	-	-	-	-	-	-	-	-
Ecuador	-	-	-	-	-	-	-	-	-
Netherlands Antilles	-	1	1 e	-	-	-	-	-	-
Peru	-	-	-	-	-	-	-	-	-
Trinidad and Tobago	-	-	-	-	-	-	-	-	-
Venezuela	-	1	-	-	-	-	-	-	-
Other	-	-	-	-	-	-	-	-	-
Latin America	**-**	**15**	**1 e**	**-**	**-**	**-**	**-**	**-**	**-**
Romania	-	-	-	-	-	-	-	-	-
Other	-	1	-	-	-	-	-	-	-
Non-OECD Europe	**-**	**1**	**-**	**-**	**-**	**-**	**-**	**-**	**-**
FSU	**-**	**-**	**-**	**-**	**-**	**-**	**-**	**-**	**-**
Iran	-	-	-	-	-	-	-	-	-
Iraq	-	-	-	-	-	-	-	-	-
Kuwait	-	2	-	-	2	-	-	-	-
Oman	-	-	-	-	-	-	-	-	-
Qatar	-	2	-	-	2	-	-	-	-
Saudi Arabia	294	448	259 e	204	258	200 e		-	-
United Arab Emirates	22	47	-	20	12	-	-	-	-
Other	-	-	-	-	-	-	-	-	-
Middle East	**316**	**499**	**259 e**	**224**	**274**	**200 e**	**-**	**-**	**-**
India	27	78	-	-	-	-	-	-	-
Indonesia	23	76	44 e	16	38	29 e	-	-	-
Singapore	1 398	1 351	1 713 e	-	-	-	-	-	-
Other	109	59	29 e	13	5	9 e	-	-	-
Asia	**1 557**	**1 564**	**1 786 e**	**29**	**43**	**38 e**	**-**	**-**	**-**
China	**221**	**214**	**269 e**	**-**	**-**	**-**	**-**	**-**	**-**
Algeria	12	2	44 e	12	2	44 e	-	-	-
Angola/Cabinada	-	-	-	-	-	-	-	-	-
Cameroon	-	-	-	-	-	-	-	-	-
Egypt	-	-	-	-	-	-	-	-	-
Gabon	-	-	-	-	-	-	-	-	-
Libya	-	-	-	-	-	-	-	-	-
Nigeria	-	-	-	-	-	-	-	-	-
Tunisia	-	-	-	-	-	-	-	-	-
Other	5	40	2 e	-	-	-	-	-	-
Africa	**17**	**42**	**46 e**	**12**	**2**	**44 e**	**-**	**-**	**-**
Non-spec./Other	**757**	**776**	**888 e**	**-**	**-**	**-**	**-**	**-**	**-**
Total Imports	**4 077**	**4 246**	**4 472 e**	**265**	**322**	**298 e**	**-**	**-**	**-**
Memo: Total OPEC	*351*	*578*	*347 e*	*252*	*314*	*273 e*	-	-	-

Source: Annual Oil Statistics.

AUSTRALIA
Table 7C: Imports by Country of Origin

000 metric tons

	2000	2001	2002	2000	2001	2002	2000	2001	2002
	Total Gasoline			**Total Kerosene**			**Gas/Diesel**		
Australia	-	-	-	-	-	-	-	-	-
Belgium	-	-	-	-	-	-	-	-	-
Canada	-	-	27 e	-	-	-	-	-	5 e
Czech Republic	-	-	-	-	-	-	-	-	-
Denmark	-	-	-	-	-	-	-	-	-
Finland	-	-	-	-	-	-	-	-	-
France	-	-	-	-	-	-	-	-	-
Germany	-	-	-	-	-	-	-	1	-
Greece	-	-	-	-	-	-	-	-	-
Hungary	-	-	-	-	-	-	-	-	-
Ireland	-	-	-	-	-	-	-	-	-
Italy	-	-	-	-	-	-	-	-	-
Japan	36	69	36 e	-	3	-	53	144	27 e
Korea	60	-	121 e	-	3	-	247	93	94 e
Mexico	-	-	-	-	-	-	-	-	-
Netherlands	-	-	-	-	-	-	-	-	-
New Zealand	-	-	-	-	-	-	-	-	8 e
Norway	-	-	-	-	-	-	1	-	-
Poland	-	-	-	-	-	-	-	-	-
Spain	-	-	-	-	-	-	-	-	-
Sweden	-	-	-	-	-	-	-	-	-
Turkey	-	-	-	-	-	-	-	-	-
United Kingdom	-	-	-	-	-	-	-	-	-
United States	-	-	31 e	-	-	-	-	-	-
Other	-	-	-	-	-	-	-	-	2 e
OECD	**96**	**69**	**215 e**	**-**	**6**	**-**	**301**	**238**	**136 e**
Argentina	-	-	-	-	-	-	-	-	-
Brazil	-	-	-	-	-	-	-	-	-
Colombia	-	-	-	-	-	-	-	-	-
Ecuador	-	-	-	-	-	-	-	-	-
Netherlands Antilles	-	-	-	-	-	-	-	-	-
Peru	-	-	-	-	-	-	-	-	-
Trinidad and Tobago	-	-	-	-	-	-	-	-	-
Venezuela	-	-	-	-	-	-	-	-	-
Other	-	-	-	-	-	-	-	-	-
Latin America	**-**	**-**	**-**	**-**	**-**	**-**	**-**	**-**	**-**
Romania	-	-	-	-	-	-	-	-	-
Other	-	-	-	-	-	-	-	-	-
Non-OECD Europe	**-**	**-**	**-**	**-**	**-**	**-**	**-**	**-**	**-**
FSU	**-**	**-**	**-**	**-**	**-**	**-**	**-**	**-**	**-**
Iran	-	-	-	-	-	-	-	-	-
Iraq	-	-	-	-	-	-	-	-	-
Kuwait	-	-	-	-	-	-	-	-	-
Oman	-	-	-	-	-	-	-	-	-
Qatar	-	-	-	-	-	-	-	-	-
Saudi Arabia	31	183	59 e	-	-	-	6	7	-
United Arab Emirates	-	35	-	-	-	-	2	-	-
Other	-	-	-	-	-	-	-	-	-
Middle East	**31**	**218**	**59 e**	**-**	**-**	**-**	**8**	**7**	**-**
India	26	-	-	-	-	-	-	48	-
Indonesia	4	-	15 e	-	3	-	3	35	-
Singapore	412	396	518 e	134	282	188 e	769	619	936 e
Other	-	32	15 e	1	16	3 e	94	-	1 e
Asia	**442**	**428**	**548 e**	**135**	**301**	**191 e**	**866**	**702**	**937 e**
China	**209**	**121**	**227 e**	**-**	**-**	**-**	**5**	**-**	**-**
Algeria	-	-	-	-	-	-	-	-	-
Angola/Cabinada	-	-	-	-	-	-	-	-	-
Cameroon	-	-	-	-	-	-	-	-	-
Egypt	-	-	-	-	-	-	-	-	-
Gabon	-	-	-	-	-	-	-	-	-
Libya	-	-	-	-	-	-	-	-	-
Nigeria	-	-	-	-	-	-	-	-	-
Tunisia	-	-	-	-	-	-	-	-	-
Other	-	34	-	-	-	-	-	-	-
Africa	**-**	**34**	**-**	**-**	**-**	**-**	**-**	**-**	**-**
Non-spec./Other	**-**	**-**	**2 e**	**-**	**-**	**-**	**-**	**5**	**2 e**
Total Imports	**778**	**870**	**1 051 e**	**135**	**307**	**191 e**	**1 180**	**952**	**1 075 e**
Memo: Total OPEC	*35*	*218*	*74 e*	*-*	*3*	*-*	*11*	*42*	*-*

Source: Annual Oil Statistics.

AUSTRALIA
Table 7D: Imports by Country of Origin

000 metric tons

	2000	2001	2002	2000	2001	2002	2000	2001	2002
	Fuel Oil (Residual)			**Petroleum Coke**			**Other Products**		
Australia	-	-	-	-	-	-	-	-	-
Belgium	-	-	-	-	-	-	2	1	-
Canada	-	-	-	-	-	-	-	-	-
Czech Republic	-	-	-	-	-	-	-	-	-
Denmark	-	-	-	-	-	-	-	-	-
Finland	-	-	-	-	-	-	-	-	-
France	-	-	-	-	-	-	2	4	3 e
Germany	-	-	-	-	-	-	3	2	-
Greece	-	-	-	-	-	-	-	-	-
Hungary	-	-	-	-	-	-	-	-	-
Ireland	-	-	-	-	-	-	-	-	-
Italy	-	-	-	-	-	-	-	-	-
Japan	-	-	-	-	22	16 e	2	7	1 e
Korea	-	-	-	-	-	-	25	23	9 e
Mexico	-	-	-	-	-	-	-	-	-
Netherlands	-	-	-	-	-	-	1	1	-
New Zealand	-	-	-	-	-	-	57	11	17 e
Norway	-	-	-	-	-	-	-	-	-
Poland	-	-	-	-	-	-	-	-	-
Spain	-	-	-	-	-	-	1	1	-
Sweden	-	-	-	1	-	-	2	-	-
Turkey	-	-	-	-	-	-	-	-	-
United Kingdom	-	-	-	-	-	-	2	4	1 e
United States	-	-	-	665	680	761 e	49	63	48 e
Other	-	-	-	-	-	-	-	-	-
OECD	**-**	**-**	**-**	**666**	**702**	**777 e**	**146**	**117**	**79 e**
Argentina	-	-	-	-	13	-	-	-	-
Brazil	-	-	-	-	-	-	-	-	-
Colombia	-	-	-	-	-	-	-	-	-
Ecuador	-	-	-	-	-	-	-	-	-
Netherlands Antilles	-	-	-	-	-	-	-	1	1 e
Peru	-	-	-	-	-	-	-	-	-
Trinidad and Tobago	-	-	-	-	-	-	-	-	-
Venezuela	-	-	-	-	-	-	-	1	-
Other	-	-	-	-	-	-	-	-	-
Latin America	**-**	**-**	**-**	**-**	**13**	**-**	**-**	**2**	**1 e**
Romania	-	-	-	-	-	-	-	-	-
Other	-	-	-	-	-	-	-	1	-
Non-OECD Europe	**-**	**-**	**-**	**-**	**-**	**-**	**-**	**1**	**-**
FSU	**-**	**-**	**-**	**-**	**-**	**-**	**-**	**-**	**-**
Iran	-	-	-	-	-	-	-	-	-
Iraq	-	-	-	-	-	-	-	-	-
Kuwait	-	-	-	-	-	-	-	-	-
Oman	-	-	-	-	-	-	-	-	-
Qatar	-	-	-	-	-	-	-	-	-
Saudi Arabia	-	-	-	-	-	-	53		-
United Arab Emirates	-	-	-	-	-	-	-	-	-
Other	-	-	-	-	-	-	-	-	-
Middle East	**-**	**-**	**-**	**-**	**-**	**-**	**53**	**-**	**-**
India	-	-	-	1	30	-	-	-	-
Indonesia	-	-	-	-	-	-	-	-	-
Singapore	-	-	-	-	-	-	83	54	71 e
Other	-	-	-	-	-	-	1	6	1 e
Asia	**-**	**-**	**-**	**1**	**30**	**-**	**84**	**60**	**72 e**
China	**-**	**-**	**-**	**1**	**88**	**38 e**	**6**	**5**	**4 e**
Algeria	-	-	-	-	-	-	-	-	-
Angola/Cabinada	-	-	-	-	-	-	-	-	-
Cameroon	-	-	-	-	-	-	-	-	-
Egypt	-	-	-	-	-	-	-	-	-
Gabon	-	-	-	-	-	-	-	-	-
Libya	-	-	-	-	-	-	-	-	-
Nigeria	-	-	-	-	-	-	-	-	-
Tunisia	-	-	-	-	-	-	-	-	-
Other	-	-	-	3	-	-	2	6	2 e
Africa	**-**	**-**	**-**	**3**	**-**	**-**	**2**	**6**	**2 e**
Non-spec./Other	**757**	**771**	**884 e**	**-**	**-**	**-**	**-**	**-**	**-**
Total Imports	**757**	**771**	**884 e**	**671**	**833**	**815 e**	**291**	**191**	**158 e**
Memo: Total OPEC	-	-	-	-	-	-	*53*	*1*	-

Source: Annual Oil Statistics.

AUSTRALIA
Table 8A: Exports by Country of Destination

000 metric tons

	2000	2001	2002	2000	2001	2002	2000	2001	2002
	Crude Oil			**Natural Gas Liquids**			**Feedstocks and Other**		
Australia	-	-	-	-	-	-	-	-	-
Austria	-	-	-	-	-	-	-	-	-
Belgium	-	-	-	-	-	-	-	-	-
Canada	-	-	-	-	-	-	-	-	-
Denmark	-	-	-	-	-	-	-	-	-
Finland	-	-	-	-	-	-	-	-	-
France	-	-	-	-	-	-	-	-	-
Germany	-	-	-	-	-	-	-	-	-
Greece	-	-	-	-	-	-	-	-	-
Ireland	-	-	-	-	-	-	-	-	-
Italy	-	-	-	-	-	-	-	-	-
Japan	2 232	2 915	2 154 e	1 004	1 218 e	1 397 e	626	361	395 e
Luxembourg	-	-	-	-	-	-	-	-	-
Mexico	-	-	-	-	-	-	-	-	-
Netherlands	-	-	-	-	-	-	-	-	-
Norway	-	-	-	-	-	-	-	-	-
Portugal	-	-	-	-	-	-	-	-	-
Spain	-	-	-	-	-	-	-	-	-
Sweden	-	-	-	-	-	-	-	-	-
Switzerland	-	-	-	-	-	-	-	-	-
Turkey	-	-	-	-	-	-	-	-	-
United Kingdom	-	-	-	-	-	-	-	-	-
United States	2 759	2 370	2 250 e	-	-	-	11	-	-
Other OECD	3 767	3 753	5 355 e	104	35 e	79 e	158	238	187 e
OECD	**8 758**	**9 038**	**9 759 e**	**1 108**	**1 253 e**	**1 476 e**	**795**	**599**	**582 e**
Africa	-	159	237 e	-	-	-	-	-	-
Latin America	84	-	-	9	13 e	-	-	-	-
Asia	5 068	8 074	7 091 e	64	23 e	35 e	487	562	611 e
China	1 349	730	930 e	327	228 e	243 e	167	145	2 e
Non-OECD Europe	-	-	-	-	-	-	-	-	-
Former Soviet Union	-	-	-	-	-	-	-	-	-
Middle East	79	-	-	47	-	-	-	-	-
Non-specified/Other	-	1 420	1 482 e	-	65 e	1 e	-	-	8 e
Non-OECD	**6 580**	**10 383**	**9 740 e**	**447**	**329 e**	**279 e**	**654**	**707**	**621 e**
Total Exports	**15 338**	**19 421**	**19 499 e**	**1 555**	**1 582**	**1 755 e**	**1 449**	**1 306**	**1 203 e**
	Total Products			**Liquefied Petroleum Gas**			**Naphtha**		
Australia	4	2	-	-	-	-	-	-	-
Austria	-	-	-	-	-	-	-	-	-
Belgium	-	-	-	-	-	-	-	-	-
Canada	-	-	-	-	-	-	-	-	-
Denmark	-	-	-	-	-	-	-	-	-
Finland	-	-	-	-	-	-	-	-	-
France	-	-	-	-	-	-	-	-	-
Germany	-	-	-	-	-	-	-	-	-
Greece	-	-	-	-	-	-	-	-	-
Ireland	-	-	-	-	-	-	-	-	-
Italy	-	-	-	-	-	-	-	-	-
Japan	55	16	58 e	-	-	-	-	-	-
Luxembourg	-	-	-	-	-	-	-	-	-
Mexico	-	-	-	-	-	-	-	-	-
Netherlands	-	-	-	-	-	-	-	-	-
Norway	-	-	-	-	-	-	-	-	-
Portugal	-	-	-	-	-	-	-	-	-
Spain	-	-	-	-	-	-	-	-	-
Sweden	-	-	-	-	-	-	-	-	-
Switzerland	-	-	-	-	-	-	-	-	-
Turkey	-	-	-	-	-	-	-	-	-
United Kingdom	-	-	-	-	-	-	-	-	-
United States	198	248	118 e	-	-	-	-	-	-
Other OECD	946	825	957 e	-	-	-	-	-	-
OECD	**1 203**	**1 091**	**1 133 e**	**-**	**-**	**-**	**-**	**-**	**-**
Africa	-	-	31 e	-	-	-	-	-	-
Latin America	7	-	-	-	-	-	-	-	-
Asia	2 083	2 522	1 009 e	-	-	-	-	-	21 e
China	46	84	32 e	-	-	-	-	-	1 e
Non-OECD Europe	-	-	-	-	-	-	-	-	-
Former Soviet Union	-	-	-	-	-	-	-	-	-
Middle East	-	25	64 e	-	-	-	-	-	-
Non-specified/Other	5	3	476 e	-	-	-	-	-	-
Non-OECD	**2 141**	**2 634**	**1 612 e**	**-**	**-**	**-**	**-**	**-**	**22 e**
Total Exports	**3 344**	**3 725**	**2 745 e**	**-**	**-**	**-**	**-**	**-**	**22 e**

Source: Annual Oil Statistics.

AUSTRALIA
Table 8B: Exports by Country of Destination

000 metric tons

	2000	2001	2002	2000	2001	2002	2000	2001	2002
	Total Gasoline			**Total Kerosene**			**Gas/Diesel Oil**		
Australia	-	-	-	-	-	-	-	-	-
Austria	-	-	-	-	-	-	-	-	-
Belgium	-	-	-	-	-	-	-	-	-
Canada	-	-	-	-	-	-	-	-	-
Denmark	-	-	-	-	-	-	-	-	-
Finland	-	-	-	-	-	-	-	-	-
France	-	-	-	-	-	-	-	-	-
Germany	-	-	-	-	-	-	-	-	-
Greece	-	-	-	-	-	-	-	-	-
Ireland	-	-	-	-	-	-	-	-	-
Italy	-	-	-	-	-	-	-	-	-
Japan	-	-	-	29	-	33 e	-	-	1 e
Luxembourg	-	-	-	-	-	-	-	-	-
Mexico	-	-	-	-	-	-	-	-	-
Netherlands	-	-	-	-	-	-	-	-	-
Norway	-	-	-	-	-	-	-	-	-
Portugal	-	-	-	-	-	-	-	-	-
Spain	-	-	-	-	-	-	-	-	-
Sweden	-	-	-	-	-	-	-	-	-
Switzerland	-	-	-	-	-	-	-	-	-
Turkey	-	-	-	-	-	-	-	-	-
United Kingdom	-	-	-	-	-	-	-	-	-
United States	68	133	58 e	38	69	50 e	63	11	-
Other OECD	735	584	645 e	84	88	130 e	45	49	110 e
OECD	**803**	**717**	**703 e**	**151**	**157**	**213 e**	**108**	**60**	**111 e**
Africa	-	-	29 e	-	-	-	-	-	2 e
Latin America	-	-	-	-	-	-	-	-	-
Asia	255	219	92 e	316	449	115 e	792	1 008	398 e
China	-	-	-	-	-	9 e	-	10	-
Non-OECD Europe	-	-	-	-	-	-	-	-	-
Former Soviet Union	-	-	-	-	-	-	-	-	-
Middle East	-	25	32 e	-	-	-	-	-	32 e
Non-specified/Other	-	-	63 e	-	-	98 e	4	2	263 e
Non-OECD	**255**	**244**	**216 e**	**316**	**449**	**222 e**	**796**	**1 020**	**695 e**
Total Exports	**1 058**	**961**	**919 e**	**467**	**606**	**435 e**	**904**	**1 080**	**806 e**
	Fuel Oil (Residual)			**Petroleum Coke**			**Other Products**		
Australia	4	2	-	-	-	-	-	-	-
Austria	-	-	-	-	-	-	-	-	-
Belgium	-	-	-	-	-	-	-	-	-
Canada	-	-	-	-	-	-	-	-	-
Denmark	-	-	-	-	-	-	-	-	-
Finland	-	-	-	-	-	-	-	-	-
France	-	-	-	-	-	-	-	-	-
Germany	-	-	-	-	-	-	-	-	-
Greece	-	-	-	-	-	-	-	-	-
Ireland	-	-	-	-	-	-	-	-	-
Italy	-	-	-	-	-	-	-	-	-
Japan	-	-	-	-	-	-	26	16	24 e
Luxembourg	-	-	-	-	-	-	-	-	-
Mexico	-	-	-	-	-	-	-	-	-
Netherlands	-	-	-	-	-	-	-	-	-
Norway	-	-	-	-	-	-	-	-	-
Portugal	-	-	-	-	-	-	-	-	-
Spain	-	-	-	-	-	-	-	-	-
Sweden	-	-	-	-	-	-	-	-	-
Switzerland	-	-	-	-	-	-	-	-	-
Turkey	-	-	-	-	-	-	-	-	-
United Kingdom	-	-	-	-	-	-	-	-	-
United States	-	-	-	-	-	-	29	35	10 e
Other OECD	6	10	-	-	-	-	76	94	72 e
OECD	**10**	**12**	**-**	**-**	**-**	**-**	**131**	**145**	**106 e**
Africa	-	-	-	-	-	-	-	-	-
Latin America	7	-	-	-	-	-	-	-	-
Asia	544	673	236 e	-	-	-	176	173	147 e
China	-	-	-	-	-	-	46	74	22 e
Non-OECD Europe	-	-	-	-	-	-	-	-	-
Former Soviet Union	-	-	-	-	-	-	-	-	-
Middle East	-	-	-	-	-	-	-	-	-
Non-specified/Other	-	-	50 e	-	-	-	1	1	2 e
Non-OECD	**551**	**673**	**286 e**	**-**	**-**	**-**	**223**	**248**	**171 e**
Total Exports	**561**	**685**	**286 e**	**-**	**-**	**-**	**354**	**393**	**277 e**

Source: Annual Oil Statistics.

AUSTRIA

Figure 1
FUEL SHARES OF TPES *

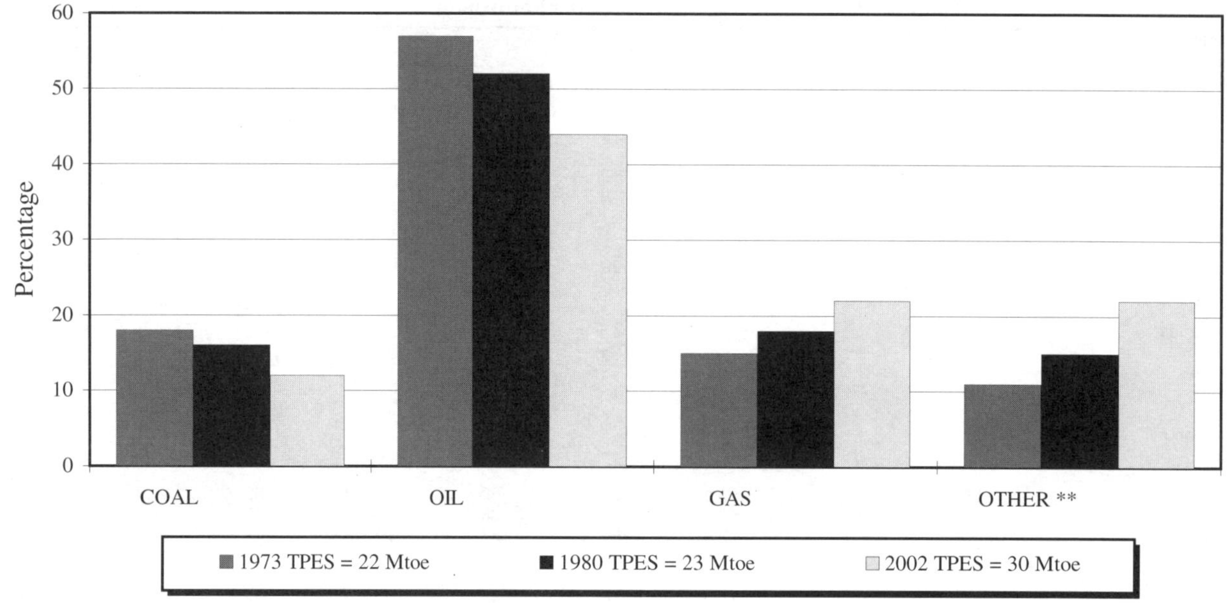

Source: Annual Energy Balances.

* Excluding electricity trade.

** Includes hydro, geothermal, solar, wind, combustible renewables and waste, etc.

Figure 2
OIL PRODUCTION AND CONSUMPTION

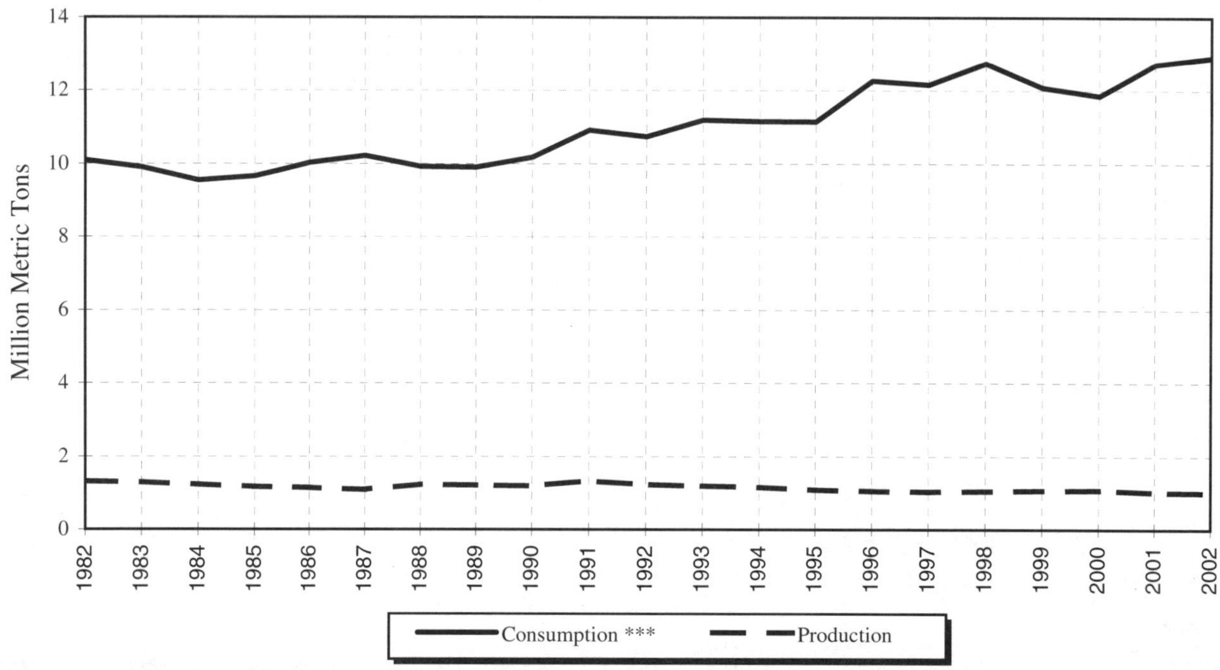

Source: Annual Oil Statistics.

*** Includes refinery fuel and international marine bunkers.

AUSTRIA
Table 1: Oil Supply and Consumption - Summary Table

000 metric tons

	1973	1978	1980	1990	1999	2000	2001	2002
Indigenous Production	2 596	1 815	1 500	1 190	1 063	1 072	1 012	1 003 e
Crude Oil	..	1 791	1 448	1 149	1 002	971	957	947 e
NGL	..	24	52	41	61	101	55	56 e
Other	..	-	-	-	-	-	-	-
+ Net Imports	9 635	10 321	10 894	9 590	11 152	10 755	11 436	12 239 e
Crude Oil, NGL, Other	6 633	8 158	8 868	8 100	8 374	7 812	8 364	8 708 e
Petroleum Products	3 002	2 163	2 026	1 490	2 778	2 943	3 072	3 531 e
- Intl. Marine Bunkers	-	-	-	-	-	-	-	-
+ Stock Changes	- 110	- 646	- 302	- 190	- 2	119	421	- 355 e
TOTAL SUPPLY	**12 121**	**11 490**	**12 092**	**10 590**	**12 213**	**11 946**	**12 869**	**12 887 e**
Refinery Intake	*9 226*	*9 989*	*10 739*	*9 407*	*9 583*	*8 891*	*9 510*	*9 545 e*
Refinery Output	*8 817*	*9 874*	*10 252*	*8 911*	*9 413*	*8 768*	*9 352*	*9 340 e*
LPG + Ethane	*92*	*122*	*199*	*47*	*19*	*34*	-	*27 e*
Naphtha	*134*	*4*	*18*	-	-	-	-	-
Kerosenes	*120*	*128*	*135*	*322*	*509*	*545*	*514*	*484 e*
Gasolines	*1 558*	*1 620*	*1 797*	*2 631*	*2 140*	*1 814*	*1 922*	*1 927 e*
Gas/Diesel Oil	*2 362*	*2 640*	*2 543*	*2 770*	*3 675*	*3 724*	*3 959*	*4 083 e*
Fuel Oil	*3 806*	*4 429*	*4 372*	*1 913*	*1 308*	*979*	*1 047*	*903 e*
Other Products	*745*	*931*	*1 188*	*1 228*	*1 762*	*1 672*	*1 910*	*1 916 e*
- Refinery Losses	409	115	487	496	170	123	158	205 e
+ Returns	-	-	18	86	33	22	10	3 e
- Statistical Differences	107	- 164	- 10	8	- 1	- 2	- 3	- 214 e
TOTAL CONSUMPTION	**11 605**	**11 539**	**11 633**	**10 172**	**12 077**	**11 847**	**12 724**	**12 899 e**
Transformation and Energy	**1 487**	**1 644**	**1 739**	**910**	**1 454**	**1 107**	**1 233**	..
Crude Oil	-	-	-	-	-	-	-	..
LPG + Ethane	36	51	39	9	3	20	-	..
Gas/Diesel Oil	7	4	-	-	9	1	19	..
Fuel Oil	1 289	1 322	1 324	488	817	493	583	..
Petroleum Coke	-	-	-	-	-	-	-	..
Other	155	267	376	413	625	593	631	..
Transport	**3 814**	**4 054**	**4 014**	**4 603**	**5 817**	**6 125**	**6 486**	..
Motor Gasoline	2 218	2 281	2 395	2 506	2 046	1 978	1 992	..
Gas/Diesel Oil	1 193	1 424	1 283	1 725	3 197	3 528	3 891	..
Subtotal	*3 411*	*3 705*	*3 678*	*4 231*	*5 243*	*5 506*	*5 883*	..
Aviation Fuels	112	100	140	301	528	571	555	..
Other	291	243	196	71	46	48	48	..
Industry	**3 192**	**2 737**	**1 975**	**2 052**	**2 226**	**2 288**	**2 316**	..
Naphtha	142	16	35	1	-	-	-	..
LPG + Ethane	37	50	44	68	62	63	55	..
Gas/Diesel Oil	142	-	30	54	341	374	421	..
Fuel Oil	2 140	1 872	914	577	516	564	466	..
Other	731	799	952	1 352	1 307	1 287	1 374	..
Other Sectors	**3 112**	**3 104**	**3 905**	**2 607**	**2 580**	**2 327**	**2 689**	..
Gas/Diesel Oil	1 032	1 391	1 400	1 562	2 205	1 958	2 301	..
Fuel Oil	2 032	1 664	2 415	971	275	278	302	..
Other	48	49	90	74	100	91	86	..
Memo: Non-Energy Use	*1 136*	*1 021*	*1 140*	*1 402*	*1 350*	*1 338*	*1 426*	..
Chem/Petchem. Feedst.	*214*	*208*	*368*	*700*	*687*	*642*	*698*	..
Other	*922*	*813*	*772*	*702*	*663*	*696*	*728*	..

Source: Annual Oil Statistics.

AUSTRIA
Table 2: Supply of Crude Oil, NGL, Feedstocks and Other Hydrocarbons

000 metric tons

	1973	1978	1980	1990	1999	2000	2001	2002
CRUDE OIL								
Indigenous Production	..	1 791	1 448	1 149	1 002	971	957	947 e
+ Imports	..	8 066	8 318	6 797	7 698	7 314	7 940	8 120 e
- Exports	..	-	-	-	51	61	63	-
- Direct Use	..	-	-	-	-	-	-	-
+ Stock Change	..	- 123	- 81	6	- 14	16	- 35	- 128 e
- Statistical Difference	..	- 139	-	-	-	-	-	- 7 e
Refinery Intake	**..**	**9 873**	**9 685**	**7 952**	**8 635**	**8 240**	**8 799**	**8 946 e**
Closing Stock Level	*..*	*381*	*412*	*751*	*793*	*777*	*812*	*940 e*
NATURAL GAS LIQUIDS								
Indigenous Production	..	24	52	41	61	101	55	56 e
+ Imports	..	-	-	-	-	6	-	-
- Exports	..	-	-	-	-	-	-	-
- Direct Use	..	-	-	-	-	-	-	-
+ Stock Change	..	-	-	-	10	-	-	-
- Statistical Difference	..	-	-	-	-	-	-	-
Refinery Intake	**..**	**24**	**52**	**41**	**71**	**107**	**55**	**56 e**
Closing Stock Level	*..*	*-*	*-*	*-*	*-*	*-*	*-*	*-*
REFINERY FEEDSTOCKS								
Backflows	..	-	-	-	-	-	-	-
+ Transfers	..	-	18	86	33	22	18	14 e
+ Imports	..	92	550	1 211	740	627	528	593 e
- Exports	..	-	-	-	15	76	43	7 e
+ Stock Change	..	-	434	30	115	- 33	149	- 148 e
- Statistical Difference	..	-	-	-	-	-	-	- 87 e
Refinery Intake	**..**	**92**	**1 002**	**1 327**	**873**	**540**	**652**	**539 e**
Closing Stock Level	*..*	*-*	*1 740*	*1 247*	*622*	*655*	*506*	*654 e*
OTHER HYDROCARBONS								
Indigenous Production	..	-	-	-	-	-	-	-
+ Imports	..	-	-	92	2	2	2	2 e
- Exports	..	-	-	-	-	-	-	-
- Direct Use	..	-	-	-	-	-	-	-
+ Stock Change	..	-	-	- 5	-	-	-	-
- Statistical Difference	..	-	-	-	- 2	- 2	- 2	- 2 e
Refinery Intake	**..**	**-**	**-**	**87**	**4**	**4**	**4**	**4 e**
Closing Stock Level	*..*	*-*	*-*	*12*	*11*	*11*	*11*	*11 e*
CRUDE OIL, NGL, FEEDSTOCKS AND OTHER								
Indigenous Production	2 596	1 815	1 500	1 190	1 063	1 072	1 012	1 003 e
+ Backflows	-	-	-	-	-	-	-	-
+ Transfers	-	-	18	86	33	22	18	14 e
+ Imports	6 633	8 158	8 868	8 100	8 440	7 949	8 470	8 715 e
- Exports	-	-	-	-	66	137	106	7 e
- Direct Use	-	-	-	-	-	-	-	-
+ Stock Change	-	- 123	353	31	111	- 17	114	- 276 e
- Statistical Difference	3	- 139	-	-	- 2	- 2	- 2	- 96 e
Refinery Intake	**9 226**	**9 989**	**10 739**	**9 407**	**9 583**	**8 891**	**9 510**	**9 545 e**
Closing Stock Level	*-*	*381*	*2 152*	*2 010*	*1 426*	*1 443*	*1 329*	*1 605 e*

Source: Annual Oil Statistics.

AUSTRIA

Table 3: Motor Gasoline - Supply

000 metric tons

	1973	1978	1980	1990	1999	2000	2001	2002
MOTOR GASOLINE								
Refinery Output	**1 558**	**1 620**	**1 797**	**2 631**	**2 140**	**1 814**	**1 922**	**1 927 e**
+ Imports	715	680	741	259	762	670	603	707 e
- Exports	-	-	6	283	824	471	582	495 e
+ Transfers	-	-	-	-	-	-	-	-
+ Stock Change	-	- 20	- 73	- 53	- 31	- 33	51	6 e
- Statistical Difference	55	- 5	2	-	-	-	-	4 e
Total Consumption	**2 218**	**2 285**	**2 457**	**2 554**	**2 047**	**1 980**	**1 994**	**2 141 e**
Closing Stock Level	-	*311*	*420*	*309*	*254*	*287*	*236*	*230 e*

Table 4: Aviation Fuels - Supply

000 metric tons

	1973	1978	1980	1990	1999	2000	2001	2002
AVIATION GASOLINE								
Refinery Output	-	-	-	-	-	-	-	-
+ Imports	-	-	-	-	-	-	-	-
- Exports	-	-	-	-	-	-	-	-
+ Transfers	-	-	-	-	-	-	-	-
+ Stock Change	-	-	-	-	-	-	-	-
- Statistical Difference	-	-	-	-	-	-	-	-
Total Consumption	-	-	-	-	-	-	-	-
Closing Stock Level	-	-	-	-	-	-	-	-
GASOLINE TYPE JET FUEL								
Refinery Output	-	-	-	-	-	-	-	-
+ Imports	3	3	3	3	3	3	4	2 e
- Exports	-	-	-	-	-	1	1	-
+ Transfers	-	-	-	-	-	-	-	-
+ Stock Change	-	-	-	- 1	-	-	- 1	1 o
- Statistical Difference	-	-	-	-	-	-	-	-
Total Consumption	**3**	**3**	**3**	**2**	**3**	**2**	**2**	**3 e**
Closing Stock Level	-	-	-	*1*	*1*	*1*	*2*	*1 e*
KEROSENE TYPE JET FUEL								
Refinery Output	**108**	**102**	**132**	**289**	**508**	**544**	**513**	**483 e**
+ Imports	-	1	-	13	21	35	37	38 e
- Exports	-	-	-	5	6	5	1	-
+ Transfers	-	-	-	-	-	-	-	-
+ Stock Change	-	-	5	2	2	- 5	4	- 3 e
- Statistical Difference	- 1	-	-	-	-	-	-	- 1 e
Total Consumption	**109**	**103**	**137**	**299**	**525**	**569**	**553**	**519 e**
Closing Stock Level	-	*10*	*48*	-	*16*	*21*	*17*	*20 e*

Source: Annual Oil Statistics.

AUSTRIA
Table 5A: Gas/Diesel Oil - Supply and Consumption

000 metric tons

	1973	1978	1980	1990	1999	2000	2001	2002
Refinery Output	**2 362**	**2 640**	**2 543**	**2 770**	**3 675**	**3 724**	**3 959**	**4 083 e**
+ Recycled Products	-	-	-	-	-	-	-	-
+ Imports	54	182	352	576	2 491	2 608	3 059	3 462 e
- Exports	1	-	-	3	459	416	415	520 e
- Intl. Marine Bunkers	-	-	-	-	-	-	-	-
+ Transfers	-	-	-	-	-	-	-	-
+ Stock Changes	-	- 7	- 182	- 2	45	- 55	26	3 e
- Statistical Differences	41	- 4	-	-	-	-	- 3	86 e
Total Consumption	**2 374**	**2 819**	**2 713**	**3 341**	**5 752**	**5 861**	**6 632**	**6 942 e**
Transformation	**7**	**2**	**-**	**-**	**3**	**1**	**19**	**..**
Public Electricity	-	-	-	-	2	-	15	..
Autoproducer Electricity	-	2	-	-	1	1	-	..
Public CHP Plants	-	-	-	-	-	-	4	..
Autoproducers of CHP	-	-	-	-	-	-	-	..
Public Heat Plants	7	-	-	-	-	-	-	..
Autoproducer Heat Plants	-	-	-	-	-	-	-	..
Petrochemical Industry	-	-	-	-	-	-	-	..
Other Transformation	-	-	-	-	-	-	-	..
Energy Sector	**-**	**2**	**-**	**-**	**6**	**-**	**-**	**..**
Refinery Fuel	-	2	-	-	6	-	-	..
Other Energy Sector	-	-	-	-	-	-	-	..
Final Consumption	**2 367**	**2 815**	**2 713**	**3 341**	**5 743**	**5 860**	**6 613**	**..**
Transport	**1 193**	**1 424**	**1 283**	**1 725**	**3 197**	**3 528**	**3 891**	**..**
Air	-	-	-	-	-	-	-	..
Road	1 117	1 364	1 216	1 664	3 148	3 474	3 836	..
Rail	56	60	59	54	42	47	47	..
Inland Waterways	20	-	8	7	7	7	8	..
Non-Specified	-	-	-	-	-	-	-	..
Industry	**142**	**-**	**30**	**54**	**341**	**374**	**421**	**..**
Iron and Steel	4	-	-	-	-	-	-	..
Chemical (incl. Pet.-Chem.)	13	-	-	1	4	4	5	..
Non-Ferrous Metals	-	-	-	-	-	-	-	..
Non-Metallic Minerals	50	-	1	1	7	7	8	..
Transport Equipment	-	-	2	4	23	26	28	..
Machinery	-	-	1	-	1	-	1	..
Mining and Quarrying	-	-	2	4	24	26	29	..
Food and Tobacco	-	-	1	-	2	1	3	..
Pulp, Paper and Printing	-	-	-	-	1	1	1	..
Wood and Wood Products	-	-	-	1	4	5	6	..
Construction	-	-	21	42	268	297	331	..
Textiles and Leather	-	-	-	1	4	4	4	..
Non-Specified	75	-	2	-	3	3	5	..
Other Sectors	**1 032**	**1 391**	**1 400**	**1 562**	**2 205**	**1 958**	**2 301**	**..**
Commerce and Public	-	-	119	32	236	106	246	..
Residential	1 032	1 391	962	1 217	1 648	1 532	1 736	..
Agriculture	-	-	319	313	321	320	319	..
Non-Specified	-	-	-	-	-	-	-	..
Memo: Non-Energy Use	-	-	-	-	-	-	-	..
Chem/Petchem. Feedst.	-	-	-	-	-	-	-	..
Other	-	-	-	-	-	-	-	..
Closing Stock Level	-	*240*	*1 535*	*615*	*356*	*411*	*385*	*382 e*

Source: Annual Oil Statistics.

AUSTRIA
Table 5B: Fuel Oil (Residual) - Supply and Consumption

000 metric tons

	1973	1978	1980	1990	1999	2000	2001	2002
Refinery Output	**3 806**	**4 429**	**4 372**	**1 913**	**1 308**	**979**	**1 047**	**903 e**
+ Recycled Products	-	-	-	-	-	-	-	-
+ Imports	1 801	970	652	401	468	262	316	241 e
- Exports	10	36	-	185	37	152	228	145 e
- Intl. Marine Bunkers	-	-	-	-	-	-	-	-
+ Transfers	-	-	-	-	-	-	-	-
+ Stock Changes	- 110	- 474	- 371	- 93	- 131	246	217	- 95 e
- Statistical Differences	-	7	-	-	-	-	1	- 192 e
Total Consumption	**5 487**	**4 882**	**4 653**	**2 036**	**1 608**	**1 335**	**1 351**	**1 096 e**
Transformation	**1 257**	**1 263**	**1 192**	**359**	**640**	**360**	**466**	**..**
Public Electricity	1 087	788	916 e	28	270	117	196	..
Autoproducer Electricity	-	226	-	-	5	6	4	..
Public CHP Plants	170	249	-	253	281	162	184	..
Autoproducers of CHP	-	-	276 e	32	19	15	21	..
Public Heat Plants	-	-	-	45	63	59	59	..
Autoproducer Heat Plants	-	-	-	1	2	1	2	..
Petrochemical Industry	-	-	-	-	-	-	-	..
Other Transformation	-	-	-	-	-	-	-	..
Energy Sector	**32**	**59**	**132**	**129**	**177**	**133**	**117**	**..**
Refinery Fuel	32	59	132	129	177	133	117	..
Other Energy Sector	-	-	-	-	-	-	-	..
Final Consumption	**4 198**	**3 560**	**3 329**	**1 548**	**791**	**842**	**768**	**..**
Transport	**26**	**24**	**-**	**-**	**-**	**-**	**-**	**..**
Air	-	-	-	-	-	-	-	..
Road	-	-	-	-	-	-	-	..
Rail	26	24	-	-	-	-	-	..
Inland Waterways	-	-	-	-	-	-	-	..
Non-Specified	-	-	-	-	-	-	-	..
Industry	**2 140**	**1 872**	**914**	**577**	**516**	**564**	**466**	**..**
Iron and Steel	451	398	85	88	179	195	161	..
Chemical (incl. Pet.-Chem.)	182	128	38	23	19	19	16	..
Non-Ferrous Metals	-	-	2	4	8	9	8	..
Non-Metallic Minerals	607	481	198	113	89	98	81	..
Transport Equipment	34	23	14	13	2	3	2	..
Machinery	78	91	20	11	10	11	10	..
Mining and Quarrying	116	104	56	6	5	6	5	..
Food and Tobacco	228	227	155	77	39	42	35	..
Pulp, Paper and Printing	188	172	85	123	64	70	58	..
Wood and Wood Products	66	76	23	15	19	21	17	..
Construction	-	-	42	32	20	22	18	..
Textiles and Leather	190	172	76	26	22	24	19	..
Non-Specified	-	-	120	46	40	44	36	..
Other Sectors	**2 032**	**1 664**	**2 415**	**971**	**275**	**278**	**302**	**..**
Commerce and Public	-	-	674	309	32	36	29	..
Residential	2 032	1 664	1 238	471	173	172	194	..
Agriculture	-	-	503	191	70	70	79	..
Non-Specified	-	-	-	-	-	-	-	..
Memo: Non-Energy Use	-	-	-	-	-	-	-	..
Chem/Petchem. Feedst.	-	-	-	-	-	-	-	..
Other	-	-	-	-	-	-	-	..
Closing Stock Level	-	*818*	*644*	*886*	*1 121*	*875*	*658*	*753 e*

Source: Annual Oil Statistics.

AUSTRIA
Table 5C: Total Products - Supply and Consumption

000 metric tons

	1973	1978	1980	1990	1999	2000	2001	2002
Refinery Output	**8 817**	**9 874**	**10 252**	**8 911**	**9 413**	**8 768**	**9 352**	**9 340 e**
+ Recycled Products	-	-	18	86	33	22	10	3 e
+ Imports	3 147	2 340	2 269	2 028	4 308	4 247	4 608	4 994 e
- Exports	145	177	243	538	1 530	1 304	1 536	1 463 e
- Intl. Marine Bunkers	-	-	-	-	-	-	-	-
+ Transfers	-	-	- 18	- 86	- 33	- 22	- 18	- 14 e
+ Stock Changes	- 110	- 523	- 655	- 221	- 113	136	307	- 79 e
- Statistical Differences	104	- 25	- 10	8	1	-	- 1	- 118 e
Total Consumption	**11 605**	**11 539**	**11 633**	**10 172**	**12 077**	**11 847**	**12 724**	**12 899 e**
Transformation	**1 304**	**1 303**	**1 225**	**382**	**644**	**361**	**487**	**..**
Public Electricity	1 087	788	916 e	28	272	117	211	..
Autoproducer Electricity	-	228	-	-	6	7	5	..
Public CHP Plants	170	249	-	253	281	162	188	..
Autoproducers of CHP	-	-	276 e	32	19	15	22	..
Public Heat Plants	7	-	1	46	64	59	59	..
Autoproducer Heat Plants	-	-	-	1	2	1	2	..
Petrochemical Industry	-	-	-	-	-	-	-	..
Other Transformation	40	38	32	22	-	-	-	..
Energy Sector	**183**	**341**	**514**	**528**	**810**	**746**	**746**	**..**
Refinery Fuel	183	341	514	510	799	734	734	..
Other Energy Sector	-	-	-	18	11	12	12	..
Final Consumption	**10 118**	**9 895**	**9 894**	**9 262**	**10 623**	**10 740**	**11 491**	**..**
Transport	**3 814**	**4 054**	**4 014**	**4 603**	**5 817**	**6 125**	**6 486**	**..**
Air	112	106	140	301	528	571	555 e	..
Road	3 600	3 864	3 807	4 240	5 239	5 499	5 875	..
Rail	82	84	59	54	42	47	47	..
Inland Waterways	20	-	8	7	7	7	8	..
Non-Specified	-	-	-	1	1	1	1	..
Industry	**3 192**	**2 737**	**1 975**	**2 052**	**2 226**	**2 288**	**2 316**	**..**
Iron and Steel	463	399	95	108	203	219	183	..
Chemical (incl. Pet.-Chem.)	412	336	406	730	716	671	726	..
Non-Ferrous Metals	-	4	2	14	15	17	15	..
Non-Metallic Minerals	676	500	211	136	118	129	111	..
Transport Equipment	34	23	49	52	35	34	35	..
Machinery	78	95	24	20	18	18	17	..
Mining and Quarrying	116	104	59	14	32	35	37	..
Food and Tobacco	228	228	158	91	50	52	47	..
Pulp, Paper and Printing	188	174	88	133	71	77	65	..
Wood and Wood Products	66	76	23	19	25	28	25	..
Construction	613	607	654	643	858	916	976	..
Textiles and Leather	190	172	78	32	29	31	26	..
Non-Specified	128	19	128	60	56	61	53	..
Other Sectors	**3 112**	**3 104**	**3 905**	**2 607**	**2 580**	**2 327**	**2 689**	**..**
Commerce and Public	13	12	865	397	321	184	307	..
Residential	3 099	3 092	2 216	1 704	1 864	1 748	1 979	..
Agriculture	-	-	824	506	395	395	403	..
Non-Specified	-	-	-	-	-	-	-	..
Memo: Non-Energy Use	*1 136*	*1 021*	*1 140*	*1 402*	*1 350*	*1 338*	*1 426*	..
Chem/Petchem. Feedst.	*214*	*208*	*368*	*700*	*687*	*642*	*698*	..
Other	*922*	*813*	*772*	*702*	*663*	*696*	*728*	..
Closing Stock Level	-	*1 463*	*2 840*	*1 929*	*1 836*	*1 700*	*1 393*	*1 472 e*

Source: Annual Oil Statistics.

AUSTRIA
Table 6A: Supply and Consumption

000 metric tons

	1980	2000	2001	2002	1980	2000	2001	2002
	LPG AND ETHANE				**REFINERY GAS**			
Refinery Output	**199**	**34**	**-**	**27 e**	**344**	**312**	**328**	**293 e**
+ Recycled Products	-	-	-	-	-	-	-	-
+ Imports	35	159	144	155 e	-	-	-	-
- Exports	65	17	4	8 e	-	-	-	-
- Intl. Marine Bunkers	-	-	-	-	-	-	-	-
+ Transfers	-	-	- 4	- 4 e	-	-	-	-
+ Stock Change	- 3	- 6	6	- 1 e	-	-	-	-
- Statistical Difference	-	-	-	- 1 e	-	-	-	- 13 e
Total Consumption	**166**	**170**	**142**	**170 e**	**344**	**312**	**328**	**306 e**
of which: Ethane	-	-	-	-	-	-	-	-
Transformation	1	-	-	..	-	-	2	..
Energy	38	20	-	..	344	310	324	..
Transport	3	4	4	..	-	-	-	..
Industry	44	63	55	..	-	2	2	..
Other	80	83	83	..	-	-	-	..
Memo: Non-Energy Use	-	-	-	..	-	-	-	..
Closing Stock Level	*44*	*17*	*11*	*12 e*	-	-	-	-

	1980	2000	2001	2002	1980	2000	2001	2002
	NAPHTHA				**OTHER KEROSENE**			
Refinery Output	**18**	**-**	**-**	**-**	**3**	**1**	**1**	**1 e**
+ Recycled Products	-	-	-	-	-	-	-	-
+ Imports	17	-	-	-	3	5	4	-
- Exports	1	-	-	-	3	-	-	-
- Intl. Marine Bunkers	-	-	-	-	-	-	-	-
+ Transfers	-	-	-	-	-	-	- 4	-
+ Stock Change	-	-	-	-	6	-	-	-
- Statistical Difference	- 1	-	-	-	-	-	-	- 1 e
Total Consumption	**35**	**-**	**-**	**-**	**9**	**6**	**1**	**2 e**
Transformation	-	-	-	..	-	-	-	..
Energy	-	-	-	..	-	-	-	..
Transport	-	-	-	..	-	-	-	..
Industry	35	-	-	..	1	-	-	..
Other	-	-	-	..	8	6	1	..
Memo: Non-Energy Use	*35*	-	-	..	-	-	-	..
Closing Stock Level	-	-	-	-	*9*	-	-	-

	1980	2000	2001	2002	1980	2000	2001	2002
	WHITE SPIRIT + SBP				**LUBRICANTS**			
Refinery Output	**16**	**-**	**-**	**..**	**119**	**111**	**115**	**..**
+ Recycled Products	-	-	-	..	-	-	-	..
+ Imports	12	7	7	..	140	57	53	..
- Exports	-	-	-	..	77	58	65	..
- Intl. Marine Bunkers	-	-	-	..	-	-	-	..
+ Transfers	-	-	-	..	-	-	-	..
+ Stock Change	-	-	-	..	- 10	- 1	5	..
- Statistical Difference	- 2	-	-	..	- 21	-	-	..
Total Consumption	**30**	**7**	**7**	**..**	**193**	**109**	**108**	**..**
Transformation	-	-	-	..	-	-	-	..
Energy	-	-	-	..	-	12	12	..
Transport	-	-	-	..	193	44	44	..
Industry	30	7	7	..	-	51	50	..
Other	-	-	-	..	-	2	2	..
Memo: Non-Energy Use	*30*	*3*	*3*	..	*193*	*109*	*108*	..
Closing Stock Level	*4*	-	-	..	*80*	*23*	*18*	..

Source: Annual Oil Statistics.

AUSTRIA
Table 6B: Supply and Consumption

000 metric tons

	1980	2000	2001	2002	1980	2000	2001	2002
	BITUMEN				PARAFFIN WAXES			
Refinery Output	**359**	**343**	**402**	**..**	**-**	**-**	**-**	**..**
+ Recycled Products	-	-	-	..	-	-	-	..
+ Imports	244	292	296	..	-	-	-	..
- Exports	11	45	78	..	-	-	-	..
- Intl. Marine Bunkers	-	-	-	..	-	-	-	..
+ Transfers	-	-	-	..	-	-	-	..
+ Stock Change	- 18	- 3	- 1	..	-	-	-	..
- Statistical Difference	- 5	-	1	..	-	-	-	..
Total Consumption	**579**	**587**	**618**	**..**	**-**	**-**	**-**	**..**
Transformation	-	-	-	..	-	-	-	..
Energy	-	-	-	..	-	-	-	..
Transport	-	-	-	..	-	-	-	..
Industry	579	587	618	..	-	-	-	..
Other	-	-	-	..	-	-	-	..
Memo: Non-Energy Use	*579*	*587*	*618*	..	-	-	-	..
Closing Stock Level	*38*	*13*	*14*	..	-	-	-	..
	PETROLEUM COKE				OTHER PRODUCTS			
Refinery Output	**-**	**-**	**-**	**-**	**350**	**906**	**1 065**	**1 623 e**
+ Recycled Products	-	-	-	-	18	22	10	3 e
+ Imports	-	-	-	-	70	149	85	389 e
- Exports	-	-	-	-	80	139	162	295 e
- Intl. Marine Bunkers	-	-	-	-	-	-	-	-
+ Transfers	-	-	-	-	- 18	- 22	- 10	- 10 e
+ Stock Change	-	-	-	-	- 9	- 7	-	10 e
- Statistical Difference	-	-	-	-	17	-	-	-
Total Consumption	**-**	**-**	**-**	**-**	**314**	**909**	**988**	**1 720 e**
Transformation	-	-	-	..	32	-	-	..
Energy	-	-	-	..	-	271	293	..
Transport	-	-	-	..	-	-	-	..
Industry	-	-	-	..	282	638	695	..
Other	-	-	-	..	-	-	-	..
Memo: Non-Energy Use	-	-	-	..	*282*	*638*	*695*	..
Closing Stock Level	-	-	-	-	*18*	*52*	*52*	*74 e*

Source: Annual Oil Statistics.

AUSTRIA
Table 7A: Imports by Country of Origin

000 metric tons

	2000	2001	2002	2000	2001	2002	2000	2001	2002
	Crude Oil			**Natural Gas Liquids**			**Feedstocks and Other**		
Australia	-	-	-	-	-	-	-	-	-
Belgium	-	-	-	-	-	-	24	9	9 e
Canada	-	-	-	-	-	-	-	-	-
Czech Republic	31	62	86 e	-	-	-	8	1	1 e
Denmark	-	-	-	-	-	-	-	-	-
Finland	-	-	-	-	-	-	1	-	-
France	-	-	-	-	-	-	6	6	8 e
Germany	-	-	-	-	-	-	99	183	221 e
Greece	-	-	-	-	-	-	-	-	-
Hungary	-	-	-	-	-	-	198	153	198 e
Ireland	-	-	-	-	-	-	-	-	-
Italy	-	-	-	-	-	-	4	3	-
Japan	-	-	-	-	-	-	-	-	-
Korea	-	-	-	-	-	-	-	-	-
Mexico	-	-	-	-	-	-	-	-	-
Netherlands	-	-	-	-	-	-	65	45	27 e
New Zealand	-	-	-	-	-	-	-	-	-
Norway	-	-	-	-	-	-	-	-	-
Poland	-	-	-	-	-	-	10	4	-
Spain	-	-	-	-	-	-	-	-	-
Sweden	-	-	-	-	-	-	-	1	-
Turkey	-	-	-	-	-	-	-	-	-
United Kingdom	-	-	-	-	-	-	-	-	-
United States	-	-	-	-	-	-	6	-	-
Other	-	3	12 e	-	-	-	71	3	-
OECD	**31**	**65**	**98 e**	**-**	**-**	**-**	**492**	**408**	**464 e**
Argentina	-	-	-	-	-	-	-	-	-
Brazil	-	-	-	-	-	-	-	-	-
Colombia	-	-	-	-	-	-	-	-	-
Ecuador	-	-	-	-	-	-	-	-	-
Netherlands Antilles	-	-	-	-	-	-	-	-	-
Peru	-	-	-	-	-	-	-	-	-
Trinidad and Tobago	-	-	-	-	-	-	-	-	-
Venezuela	-	-	-	-	-	-	-	-	-
Other	-	-	-	-	-	-	-	-	-
Latin America	**-**	**-**	**-**	**-**	**-**	**-**	**-**	**-**	**-**
Romania	-	118	373 e	-	-	-	11	15	3 e
Other	-	-	-	-	-	-	3	1	-
Non-OECD Europe	**-**	**118**	**373 e**	**-**	**-**	**-**	**14**	**16**	**3 e**
FSU	**2 470**	**1 888**	**2 342 e**	**-**	**-**	**-**	**116**	**103**	**116 e**
Iran	-	-	16 e	-	-	-	-	-	-
Iraq	1 506	917	1 061 e	-	-	-	-	-	-
Kuwait	-	-	-	-	-	-	-	-	-
Oman	-	-	-	-	-	-	-	-	-
Qatar	-	-	-	-	-	-	-	-	-
Saudi Arabia	595	641	862 e	-	-	-	-	-	-
United Arab Emirates	-	49	-	-	-	-	-	-	-
Other	272	1 014	930 e	-	-	-	-	-	-
Middle East	**2 373**	**2 621**	**2 869 e**	**-**	**-**	**-**	**-**	**-**	**-**
India	-	-	-	-	-	-	-	-	-
Indonesia	-	-	-	-	-	-	-	-	-
Singapore	-	-	-	-	-	-	-	-	-
Other	-	-	-	-	-	-	-	-	-
Asia	**-**	**-**	**-**	**-**	**-**	**-**	**-**	**-**	**-**
China	**-**	**-**	**-**	**-**	**-**	**-**	**-**	**-**	**-**
Algeria	733	520	259 e	6	-	-	-	-	-
Angola/Cabinada	-	-	-	-	-	-	-	-	-
Cameroon	21	-	14 e	-	-	-	-	-	-
Egypt	-	-	-	-	-	-	-	-	-
Gabon	-	-	-	-	-	-	-	-	-
Libya	700	1 336	678 e	-	-	-	-	-	-
Nigeria	986	1 390	1 481 e	-	-	-	-	-	-
Tunisia	-	-	-	-	-	-	-	-	-
Other	-	-	-	-	-	-	-	-	-
Africa	**2 440**	**3 246**	**2 432 e**	**6**	**-**	**-**	**-**	**-**	**-**
Non-spec./Other	**-**	**2**	**6 e**	**-**	**-**	**-**	**7**	**3**	**12 e**
Total Imports	**7 314**	**7 940**	**8 120 e**	**6**	**-**	**-**	**629**	**530**	**595 e**
Memo: Total OPEC	*4 520*	*4 853*	*4 357 e*	*6*	-	-	-	-	-

Source: Annual Oil Statistics.

AUSTRIA
Table 7B: Imports by Country of Origin

000 metric tons

	2000	2001	2002	2000	2001	2002	2000	2001	2002
	Total Products			**Liquefied Petroleum Gases**			**Naphtha**		
Australia	-	-	-	-	-	-	-	-	-
Belgium	44	26	131 e	2	1	-	-	-	-
Canada	-	-	-	-	-	-	-	-	-
Czech Republic	194	203	261 e	2	5	1 e	-	-	-
Denmark	-	-	-	-	-	-	-	-	-
Finland	-	-	-	-	-	-	-	-	-
France	9	6	1 e	2	-	-	-	-	-
Germany	2 278	2 366	2 102 e	83	86	97 e	-	-	-
Greece	-	-	-	-	-	-	-	-	-
Hungary	443	587	637 e	27	14	14 e	-	-	-
Ireland	-	-	-	-	-	-	-	-	-
Italy	469	639	789 e	-	-	-	-	-	-
Japan	-	-	-	-	-	-	-	-	-
Korea	-	-	-	-	-	-	-	-	-
Mexico	-	-	-	-	-	-	-	-	-
Netherlands	61	30	81 e	9	4	7 e	-	-	-
New Zealand	-	-	-	-	-	-	-	-	-
Norway	-	-	-	-	-	-	-	-	-
Poland	16	8	2 e	6	2	1 e	-	-	-
Spain	1	1	-	-	-	-	-	-	-
Sweden	-	-	-	-	-	-	-	-	-
Turkey	-	-	-	-	-	-	-	-	-
United Kingdom	1	1	-	-	-	-	-	-	-
United States	-	-	-	-	-	-	-	-	-
Other	582	619	739 e	-	-	-	-	-	-
OECD	**4 098**	**4 486**	**4 743 e**	**131**	**112**	**120 e**	**-**	**-**	**-**
Argentina	-	-	-	-	-	-	-	-	-
Brazil	-	-	-	-	-	-	-	-	-
Colombia	-	-	-	-	-	-	-	-	-
Ecuador	-	-	-	-	-	-	-	-	-
Netherlands Antilles	-	-	-	-	-	-	-	-	-
Peru	-	-	-	-	-	-	-	-	-
Trinidad and Tobago	-	-	-	-	-	-	-	-	-
Venezuela	-	-	-	-	-	-	-	-	-
Other	-	-	-	-	-	-	-	-	-
Latin America	**-**	**-**	**-**	**-**	**-**	**-**	**-**	**-**	**-**
Romania	-	-	-	-	-	-	-	-	-
Other	15	22	111 e	6	7	11 e	-	-	-
Non-OECD Europe	**15**	**22**	**111 e**	**6**	**7**	**11 e**	**-**	**-**	**-**
FSU	**23**	**40**	**59 e**	**21**	**25**	**19 e**	**-**	**-**	**-**
Iran	-	-	-	-	-	-	-	-	-
Iraq	-	-	-	-	-	-	-	-	-
Kuwait	-	-	-	-	-	-	-	-	-
Oman	-	-	-	-	-	-	-	-	-
Qatar	-	-	-	-	-	-	-	-	-
Saudi Arabia	-	-	-	-	-	-	-	-	-
United Arab Emirates	-	-	-	-	-	-	-	-	-
Other	-	-	-	-	-	-	-	-	-
Middle East	**-**	**-**	**-**	**-**	**-**	**-**	**-**	**-**	**-**
India	-	-	-	-	-	-	-	-	-
Indonesia	-	-	-	-	-	-	-	-	-
Singapore	-	-	-	-	-	-	-	-	-
Other	-	-	-	-	-	-	-	-	-
Asia	**-**	**-**	**-**	**-**	**-**	**-**	**-**	**-**	**-**
China	**-**	**-**	**-**	**-**	**-**	**-**	**-**	**-**	**-**
Algeria	-	-	-	-	-	-	-	-	-
Angola/Cabinada	-	-	-	-	-	-	-	-	-
Cameroon	-	-	-	-	-	-	-	-	-
Egypt	-	-	-	-	-	-	-	-	-
Gabon	-	-	-	-	-	-	-	-	-
Libya	-	-	-	-	-	-	-	-	-
Nigeria	-	-	-	-	-	-	-	-	-
Tunisia	-	-	-	-	-	-	-	-	-
Other	-	-	-	-	-	-	-	-	-
Africa	**-**	**-**	**-**	**-**	**-**	**-**	**-**	**-**	**-**
Non-spec./Other	**111**	**60**	**81 e**	**1**	**-**	**5 e**	**-**	**-**	**-**
Total Imports	**4 247**	**4 608**	**4 994 e**	**159**	**144**	**155 e**	**-**	**-**	**-**
Memo: Total OPEC	-	-	-	-	-	-	-	-	-

Source: Annual Oil Statistics.

AUSTRIA
Table 7C: Imports by Country of Origin

000 metric tons

	2000	2001	2002	2000	2001	2002	2000	2001	2002
	Total Gasoline			**Total Kerosene**			**Gas/Diesel**		
Australia	-	-	-	-	-	-	-	-	-
Belgium	-	-	-	1	1	-	30	16	116 e
Canada	-	-	-	-	-	-	-	-	-
Czech Republic	5	10	30 e	-	-	-	175	178	218 e
Denmark	-	-	-	-	-	-	-	-	-
Finland	-	-	-	-	-	-	-	-	-
France	-	-	1 e	-	-	-	1	-	-
Germany	390	347	277 e	39	38	38 e	1 438	1 598	1 426 e
Greece	-	-	-	-	-	-	-	-	-
Hungary	105	91	134 e	-	-	-	245	378	464 e
Ireland	-	-	-	-	-	-	-	-	-
Italy	30	31	73 e	-	-	-	360	484	647 e
Japan	-	-	-	-	-	-	-	-	-
Korea	-	-	-	-	-	-	-	-	-
Mexico	-	-	-	-	-	-	-	-	-
Netherlands	7	4	2 e	-	1	-	39	16	65 e
New Zealand	-	-	-	-	-	-	-	-	-
Norway	-	-	-	-	-	-	-	-	-
Poland	-	-	-	-	-	-	-	-	-
Spain	-	-	-	-	-	-	-	-	-
Sweden	-	-	-	-	-	-	-	-	-
Turkey	-	-	-	-	-	-	-	-	-
United Kingdom	-	-	-	-	-	-	-	-	-
United States	-	-	-	-	-	-	-	-	-
Other	136	124	190 e	-	-	-	318	374	483 e
OECD	**673**	**607**	**707 e**	**40**	**40**	**38 e**	**2 606**	**3 044**	**3 419 e**
Argentina	-	-	-	-	-	-	-	-	-
Brazil	-	-	-	-	-	-	-	-	-
Colombia	-	-	-	-	-	-	-	-	-
Ecuador	-	-	-	-	-	-	-	-	-
Netherlands Antilles	-	-	-	-	-	-	-	-	-
Peru	-	-	-	-	-	-	-	-	-
Trinidad and Tobago	-	-	-	-	-	-	-	-	-
Venezuela	-	-	-	-	-	-	-	-	-
Other	-	-	-	-	-	-	-	-	-
Latin America	**-**	**-**	**-**	**-**	**-**	**-**	**-**	**-**	**-**
Romania	-	-	-	-	-	-	-	-	-
Other	-	-	-	-	-	-	-	-	-
Non-OECD Europe	**-**	**-**	**-**	**-**	**-**	**-**	**-**	**-**	**-**
FSU	**-**	**-**	**-**	**-**	**-**	**-**	**2**	**14**	**39 e**
Iran	-	-	-	-	-	-	-	-	-
Iraq	-	-	-	-	-	-	-	-	-
Kuwait	-	-	-	-	-	-	-	-	-
Oman	-	-	-	-	-	-	-	-	-
Qatar	-	-	-	-	-	-	-	-	-
Saudi Arabia	-	-	-	-	-	-	-	-	-
United Arab Emirates	-	-	-	-	-	-	-	-	-
Other	-	-	-	-	-	-	-	-	-
Middle East	**-**	**-**	**-**	**-**	**-**	**-**	**-**	**-**	**-**
India	-	-	-	-	-	-	-	-	-
Indonesia	-	-	-	-	-	-	-	-	-
Singapore	-	-	-	-	-	-	-	-	-
Other	-	-	-	-	-	-	-	-	-
Asia	**-**	**-**	**-**	**-**	**-**	**-**	**-**	**-**	**-**
China	**-**	**-**	**-**	**-**	**-**	**-**	**-**	**-**	**-**
Algeria	-	-	-	-	-	-	-	-	-
Angola/Cabinada	-	-	-	-	-	-	-	-	-
Cameroon	-	-	-	-	-	-	-	-	-
Egypt	-	-	-	-	-	-	-	-	-
Gabon	-	-	-	-	-	-	-	-	-
Libya	-	-	-	-	-	-	-	-	-
Nigeria	-	-	-	-	-	-	-	-	-
Tunisia	-	-	-	-	-	-	-	-	-
Other	-	-	-	-	-	-	-	-	-
Africa	**-**	**-**	**-**	**-**	**-**	**-**	**-**	**-**	**-**
Non-spec./Other	**-**	**-**	**2 e**	**-**	**1**	**-**	**-**	**1**	**4 e**
Total Imports	**673**	**607**	**709 e**	**40**	**41**	**38 e**	**2 608**	**3 059**	**3 462 e**
Memo: Total OPEC	-	-	-	-	-	-	-	-	-

Source: Annual Oil Statistics.

AUSTRIA
Table 7D: Imports by Country of Origin

000 metric tons

	2000	2001	2002	2000	2001	2002	2000	2001	2002
	Fuel Oil (Residual)			Petroleum Coke			Other Products		
Australia	-	-	-	-	-	-	-	-	-
Belgium	-	-	-	-	-	-	11	8	15 e
Canada	-	-	-	-	-	-	-	-	-
Czech Republic	-	-	-	-	-	-	12	10	12 e
Denmark	-	-	-	-	-	-	-	-	-
Finland	-	-	-	-	-	-	-	-	-
France	-	-	-	-	-	-	6	6	-
Germany	144	128	102 e	-	-	-	184	169	162 e
Greece	-	-	-	-	-	-	-	-	-
Hungary	-	70	-	-	-	-	66	34	25 e
Ireland	-	-	-	-	-	-	-	-	-
Italy	-	-	-	-	-	-	79	124	69 e
Japan	-	-	-	-	-	-	-	-	-
Korea	-	-	-	-	-	-	-	-	-
Mexico	-	-	-	-	-	-	-	-	-
Netherlands	-	-	5 e	-	-	-	6	5	2 e
New Zealand	-	-	-	-	-	-	-	-	-
Norway	-	-	-	-	-	-	-	-	-
Poland	-	-	-	-	-	-	10	6	1 e
Spain	-	-	-	-	-	-	1	1	-
Sweden	-	-	-	-	-	-	-	-	-
Turkey	-	-	-	-	-	-	-	-	-
United Kingdom	-	-	-	-	-	-	1	1	-
United States	-	-	-	-	-	-	-	-	-
Other	118	118	61 e	-	-	-	10	3	5 e
OECD	**262**	**316**	**168 e**	**-**	**-**	**-**	**386**	**367**	**291 e**
Argentina	-	-	-	-	-	-	-	-	-
Brazil	-	-	-	-	-	-	-	-	-
Colombia	-	-	-	-	-	-	-	-	-
Ecuador	-	-	-	-	-	-	-	-	-
Netherlands Antilles	-	-	-	-	-	-	-	-	-
Peru	-	-	-	-	-	-	-	-	-
Trinidad and Tobago	-	-	-	-	-	-	-	-	-
Venezuela	-	-	-	-	-	-	-	-	-
Other	-	-	-	-	-	-	-	-	-
Latin America	**-**	**-**	**-**	**-**	**-**	**-**	**-**	**-**	**-**
Romania	-	-	-	-	-	-	-	-	-
Other	-	-	71 e	-	-	-	9	15	29 e
Non-OECD Europe	**-**	**-**	**71 e**	**-**	**-**	**-**	**9**	**15**	**29 e**
FSU	**-**	**-**	**-**	**-**	**-**	**-**	**-**	**1**	**1 e**
Iran	-	-	-	-	-	-	-	-	-
Iraq	-	-	-	-	-	-	-	-	-
Kuwait	-	-	-	-	-	-	-	-	-
Oman	-	-	-	-	-	-	-	-	-
Qatar	-	-	-	-	-	-	-	-	-
Saudi Arabia	-	-	-	-	-	-	-	-	-
United Arab Emirates	-	-	-	-	-	-	-	-	-
Other	-	-	-	-	-	-	-	-	-
Middle East	**-**	**-**	**-**	**-**	**-**	**-**	**-**	**-**	**-**
India	-	-	-	-	-	-	-	-	-
Indonesia	-	-	-	-	-	-	-	-	-
Singapore	-	-	-	-	-	-	-	-	-
Other	-	-	-	-	-	-	-	-	-
Asia	**-**	**-**	**-**	**-**	**-**	**-**	**-**	**-**	**-**
China	**-**	**-**	**-**	**-**	**-**	**-**	**-**	**-**	**-**
Algeria	-	-	-	-	-	-	-	-	-
Angola/Cabinada	-	-	-	-	-	-	-	-	-
Cameroon	-	-	-	-	-	-	-	-	-
Egypt	-	-	-	-	-	-	-	-	-
Gabon	-	-	-	-	-	-	-	-	-
Libya	-	-	-	-	-	-	-	-	-
Nigeria	-	-	-	-	-	-	-	-	-
Tunisia	-	-	-	-	-	-	-	-	-
Other	-	-	-	-	-	-	-	-	-
Africa	**-**	**-**	**-**	**-**	**-**	**-**	**-**	**-**	**-**
Non-spec./Other	**-**	**-**	**2 e**	**-**	**-**	**-**	**110**	**58**	**68 e**
Total Imports	**262**	**316**	**241 e**	**-**	**-**	**-**	**505**	**441**	**389 e**
Memo: Total OPEC	-	-	-	-	-	-	-	-	-

Source: Annual Oil Statistics.

AUSTRIA
Table 8A: Exports by Country of Destination

000 metric tons

	2000	2001	2002	2000	2001	2002	2000	2001	2002
	Crude Oil			**Natural Gas Liquids**			**Feedstocks and Other**		
Australia	-	-	-	-	-	-	-	-	-
Austria	-	-	-	-	-	-	-	-	-
Belgium	-	-	-	-	-	-	10	1	-
Canada	-	-	-	-	-	-	-	-	-
Denmark	-	-	-	-	-	-	-	-	-
Finland	-	-	-	-	-	-	-	-	-
France	-	-	-	-	-	-	-	-	-
Germany	61	63	-	-	-	-	3	18	-
Greece	-	-	-	-	-	-	-	-	-
Ireland	-	-	-	-	-	-	-	-	-
Italy	-	-	-	-	-	-	-	-	-
Japan	-	-	-	-	-	-	-	-	-
Luxembourg	-	-	-	-	-	-	-	-	-
Mexico	-	-	-	-	-	-	-	-	-
Netherlands	-	-	-	-	-	-	46	23	-
Norway	-	-	-	-	-	-	-	-	-
Portugal	-	-	-	-	-	-	-	-	-
Spain	-	-	-	-	-	-	-	-	-
Sweden	-	-	-	-	-	-	-	-	2 e
Switzerland	-	-	-	-	-	-	-	-	-
Turkey	-	-	-	-	-	-	-	-	-
United Kingdom	-	-	-	-	-	-	-	-	-
United States	-	-	-	-	-	-	-	-	-
Other OECD	-	-	-	-	-	-	12	-	2 e
OECD	**61**	**63**	**-**	**-**	**-**	**-**	**71**	**42**	**4 e**
Africa	-	-	-	-	-	-	-	-	-
Latin America	-	-	-	-	-	-	-	-	-
Asia	-	-	-	-	-	-	-	-	-
China	-	-	-	-	-	-	-	-	-
Non-OECD Europe	-	-	-	-	-	-	-	-	-
Former Soviet Union	-	-	-	-	-	-	-	-	-
Middle East	-	-	-	-	-	-	-	-	-
Non-specified/Other	-	-	-	-	-	-	5	1	3 e
Non-OECD	**-**	**-**	**-**	**-**	**-**	**-**	**5**	**1**	**3 e**
Total Exports	**61**	**63**	**-**	**-**	**-**	**-**	**76**	**43**	**7 e**
	Total Products			**Liquefied Petroleum Gas**			**Naphtha**		
Australia	-	-	-	-	-	-	-	-	-
Austria	-	-	-	-	-	-	-	-	-
Belgium	15	14	-	-	-	-	-	-	-
Canada	-	-	-	-	-	-	-	-	-
Denmark	-	-	-	-	-	-	-	-	-
Finland	-	-	-	-	-	-	-	-	-
France	6	8	20 e	-	-	-	-	-	-
Germany	50	83	74 e	-	-	2 e	-	-	-
Greece	2	1	-	-	-	-	-	-	-
Ireland	-	-	-	-	-	-	-	-	-
Italy	26	25	19 e	2	-	-	-	-	-
Japan	-	-	-	-	-	-	-	-	-
Luxembourg	-	-	-	-	-	-	-	-	-
Mexico	-	-	-	-	-	-	-	-	-
Netherlands	31	33	13 e	-	-	-	-	-	-
Norway	-	-	-	-	-	-	-	-	-
Portugal	-	-	-	-	-	-	-	-	-
Spain	-	-	-	-	-	-	-	-	-
Sweden	-	-	2 e	-	-	-	-	-	-
Switzerland	3	3	-	1	-	-	-	-	-
Turkey	-	-	-	-	-	-	-	-	-
United Kingdom	-	-	-	-	-	-	-	-	-
United States	-	-	-	-	-	-	-	-	-
Other OECD	1 029	1 159	1 140 e	14	3	5 e	-	-	-
OECD	**1 162**	**1 326**	**1 268 e**	**17**	**3**	**7 e**	**-**	**-**	**-**
Africa	1	1	-	-	-	-	-	-	-
Latin America	-	-	-	-	-	-	-	-	-
Asia	1	1	-	-	-	-	-	-	-
China	-	-	-	-	-	-	-	-	-
Non-OECD Europe	47	74	32 e	-	-	-	-	-	-
Former Soviet Union	1	1	-	-	-	-	-	-	-
Middle East	-	-	-	-	-	-	-	-	-
Non-specified/Other	92	133	163 e	-	1	1 e	-	-	-
Non-OECD	**142**	**210**	**195 e**	**-**	**1**	**1 e**	**-**	**-**	**-**
Total Exports	**1 304**	**1 536**	**1 463 e**	**17**	**4**	**8 e**	**-**	**-**	**-**

Source: Annual Oil Statistics.

AUSTRIA
Table 8B: Exports by Country of Destination

000 metric tons

	2000	2001	2002	2000	2001	2002	2000	2001	2002
	Total Gasoline			**Total Kerosene**			**Gas/Diesel Oil**		
Australia	-	-	-	-	-	-	-	-	-
Austria	-	-	-	-	-	-	-	-	-
Belgium	-	-	-	-	-	-	-	-	-
Canada	-	-	-	-	-	-	-	-	-
Denmark	-	-	-	-	-	-	-	-	-
Finland	-	-	-	-	-	-	-	-	-
France	-	4	7 e	-	-	-	-	-	-
Germany	1	-	-	-	-	-	5	-	-
Greece	-	-	-	-	-	-	-	-	-
Ireland	-	-	-	-	-	-	-	-	-
Italy	2	-	-	-	-	-	6	-	-
Japan	-	-	-	-	-	-	-	-	-
Luxembourg	-	-	-	-	-	-	-	-	-
Mexico	-	-	-	-	-	-	-	-	-
Netherlands	-	-	-	-	-	-	-	-	-
Norway	-	-	-	-	-	-	-	-	-
Portugal	-	-	-	-	-	-	-	-	-
Spain	-	-	-	-	-	-	-	-	-
Sweden	-	-	-	-	-	-	-	-	-
Switzerland	-	-	-	-	-	-	-	-	-
Turkey	-	-	-	-	-	-	-	-	-
United Kingdom	-	-	-	-	-	-	-	-	-
United States	-	-	-	-	-	-	-	-	-
Other OECD	454	547	455 e	5	1	-	403	404	517 e
OECD	**457**	**551**	**462 e**	**5**	**1**	**-**	**414**	**404**	**517 e**
Africa	-	-	-	-	-	-	-	-	-
Latin America	-	-	-	-	-	-	-	-	-
Asia	-	-	-	-	-	-	-	-	-
China	-	-	-	-	-	-	-	-	-
Non-OECD Europe	14	32	28 e	-	-	-	1	10	-
Former Soviet Union	-	-	-	-	-	-	-	-	-
Middle East	-	-	-	-	-	-	-	-	-
Non-specified/Other	1	-	5 e	-	-	-	1	1	3 e
Non-OECD	**15**	**32**	**33 e**	**-**	**-**	**-**	**2**	**11**	**3 e**
Total Exports	**472**	**583**	**495 e**	**5**	**1**	**-**	**416**	**415**	**520 e**

	2000	2001	2002	2000	2001	2002	2000	2001	2002
	Fuel Oil (Residual)			**Petroleum Coke**			**Other Products**		
Australia	-	-	-	-	-	-	-	-	-
Austria	-	-	-	-	-	-	-	-	-
Belgium	10	14	-	-	-	-	5	-	-
Canada	-	-	-	-	-	-	-	-	-
Denmark	-	-	-	-	-	-	-	-	-
Finland	-	-	-	-	-	-	-	-	-
France	-	-	-	-	-	-	6	4	13 e
Germany	13	69	53 e	-	-	-	31	14	19 e
Greece	-	-	-	-	-	-	2	1	-
Ireland	-	-	-	-	-	-	-	-	-
Italy	-	-	4 e	-	-	-	16	25	15 e
Japan	-	-	-	-	-	-	-	-	-
Luxembourg	-	-	-	-	-	-	-	-	-
Mexico	-	-	-	-	-	-	-	-	-
Netherlands	31	32	13 e	-	-	-	-	1	-
Norway	-	-	-	-	-	-	-	-	-
Portugal	-	-	-	-	-	-	-	-	-
Spain	-	-	-	-	-	-	-	-	-
Sweden	-	-	-	-	-	-	-	-	2 e
Switzerland	-	-	-	-	-	-	2	3	-
Turkey	-	-	-	-	-	-	-	-	-
United Kingdom	-	-	-	-	-	-	-	-	-
United States	-	-	-	-	-	-	-	-	-
Other OECD	76	94	67 e	-	-	-	77	110	96 e
OECD	**130**	**209**	**137 e**	**-**	**-**	**-**	**139**	**158**	**145 e**
Africa	-	-	-	-	-	-	1	1	-
Latin America	-	-	-	-	-	-	-	-	-
Asia	-	-	-	-	-	-	1	1	-
China	-	-	-	-	-	-	-	-	-
Non-OECD Europe	20	19	1 e	-	-	-	12	13	3 e
Former Soviet Union	-	-	-	-	-	-	1	1	-
Middle East	-	-	-	-	-	-	-	-	-
Non-specified/Other	2	-	7 e	-	-	-	88	131	147 e
Non-OECD	**22**	**19**	**8 e**	**-**	**-**	**-**	**103**	**147**	**150 e**
Total Exports	**152**	**228**	**145 e**	**-**	**-**	**-**	**242**	**305**	**295 e**

Source: Annual Oil Statistics.

BELGIUM

Figure 1
FUEL SHARES OF TPES *

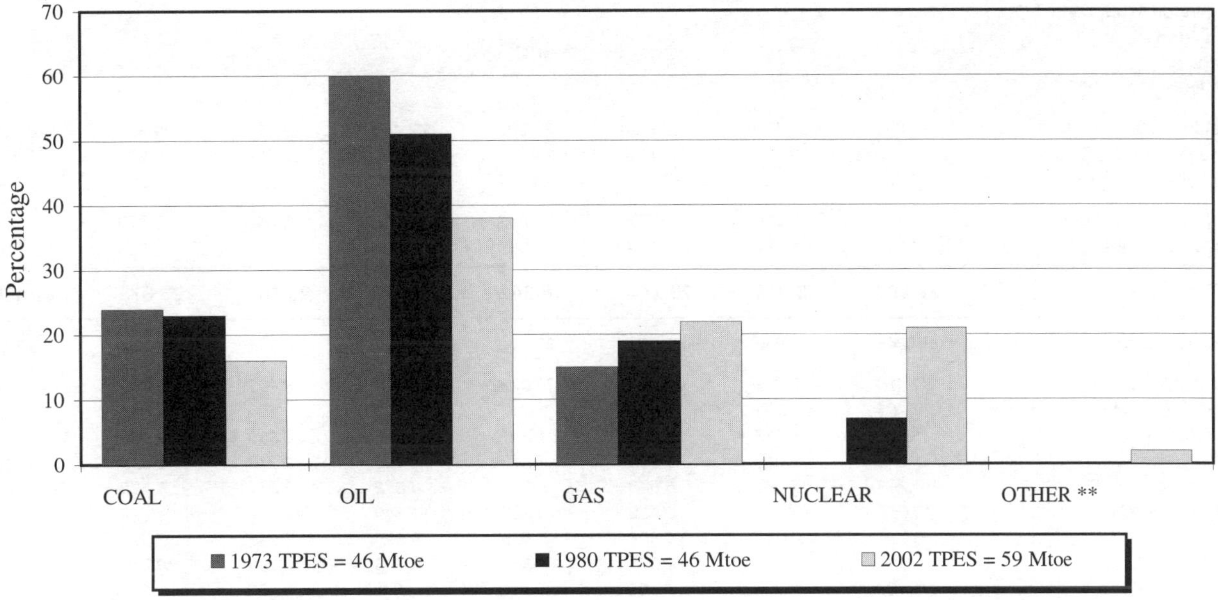

Source: Annual Energy Balances.

* Excluding electricity trade.

** Includes hydro, geothermal, solar, wind, combustible renewables and waste, etc.

Figure 2
OIL PRODUCTION AND CONSUMPTION

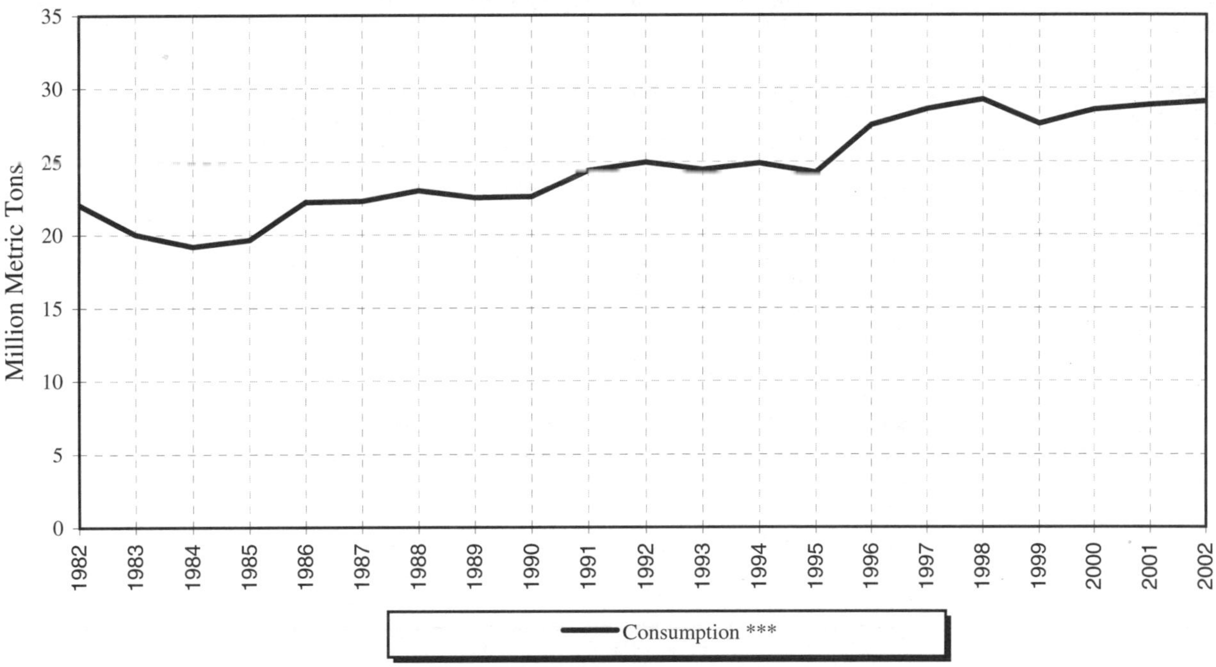

Source: Annual Oil Statistics.

*** Includes refinery fuel and international marine bunkers.

BELGIUM
Table 1: Oil Supply and Consumption - Summary Table

000 metric tons

	1973	1978	1980	1990	1999	2000	2001	2002
Indigenous Production	-	-	-	-	-	-	-	-
Crude Oil	..	-	-	-	-	-	-	-
NGL	..	-	-	-	-	-	-	-
Other	..	-	-	-	-	-	-	-
+ Net Imports	30 837	27 145	25 856	22 227	27 161	28 776	29 335	28 839 e
Crude Oil, NGL, Other	37 320	33 577	33 436	26 708	32 202	33 986	33 234	33 895 e
Petroleum Products	-6 483	-6 432	-7 580	-4 481	-5 041	-5 210	-3 899	-5 056 e
- Intl. Marine Bunkers	3 138	2 820	2 455	4 242	4 535	5 545	5 429	7 273 e
+ Stock Changes	- 211	686	- 257	357	964	- 49	- 233	300 e
TOTAL SUPPLY	**27 488**	**25 011**	**23 144**	**18 342**	**23 590**	**23 182**	**23 673**	**21 866 e**
Refinery Intake	*37 248*	*33 880*	*33 631*	*29 591*	*36 119*	*38 344*	*40 254*	*46 374 e*
Refinery Output	*35 510*	*33 545*	*33 439*	*29 372*	*35 773*	*38 092*	*39 863*	*45 984 e*
LPG + Ethane	*405*	*503*	*516*	*443*	*593*	*764*	*647*	*655 e*
Naphtha	*1 682*	*1 937*	*1 486*	*1 391*	*1 608*	*2 026*	*2 135*	*2 053 e*
Kerosenes	*1 282*	*1 113*	*1 393*	*1 517*	*2 128*	*2 317*	*2 026*	*2 146 e*
Gasolines	*4 768*	*5 288*	*5 843*	*5 271*	*6 032*	*5 311*	*5 493*	*5 776 e*
Gas/Diesel Oil	*12 054*	*10 730*	*10 932*	*10 229*	*11 749*	*12 501*	*12 777*	*12 463 e*
Fuel Oil	*13 715*	*12 053*	*10 898*	*5 636*	*6 895*	*7 990*	*7 015*	*7 604 e*
Other Products	*1 604*	*1 921*	*2 371*	*4 885*	*6 768*	*7 183*	*9 770*	*15 287 e*
- Refinery Losses	1 738	335	192	219	346	252	391	390 e
+ Returns	-	-	-	-	11	-	9	-
- Statistical Differences	45	- 130	- 92	- 236	231	- 48	- 123	- 349 e
TOTAL CONSUMPTION	**25 705**	**24 806**	**23 044**	**18 359**	**23 024**	**22 978**	**23 414**	**21 825 e**
Transformation and Energy	**5 108**	**5 583**	**6 029**	**1 600**	**1 840**	**1 709**	**1 837 e**	..
Crude Oil	-	-	-	-	-	-	-	..
LPG + Ethane	1	33	20	2	-	-	-	..
Gas/Diesel Oil	28	70	28	26	87	53	37	..
Fuel Oil	4 986	4 801	4 955	758	885	728	806 e	..
Petroleum Coke	-	161	187	234	294	284	341	..
Other	93	518	839	580	574	644	653 e	..
Transport	**4 687**	**5 322**	**5 630**	**7 462**	**9 331**	**9 325**	**9 238**	..
Motor Gasoline	2 569	3 103	2 943	2 727	2 394	2 245	2 186	..
Gas/Diesel Oil	1 593	1 725	2 072	3 676	5 071	5 323	5 632	..
Subtotal	*4 162*	*4 828*	*5 015*	*6 403*	*7 465*	*7 568*	*7 818*	..
Aviation Fuels	510	460	475	930	1 513	1 484	1 124	..
Other	15	34	140	129	353	273	296	..
Industry	**7 969**	**6 239**	**4 521**	**4 183**	**5 896**	**6 754**	**7 101**	..
Naphtha	1 646	1 330	1 179	1 345	2 885	3 336	3 390	..
LPG + Ethane	105	69	57	111	272	322	324	..
Gas/Diesel Oil	1 428	1 084	574	273	514	423	761	..
Fuel Oil	3 431	2 832	1 888	1 252	885	669	871	..
Other	1 359	924	823	1 202	1 340	2 004	1 755	..
Other Sectors	**7 941**	**7 662**	**6 864**	**5 114**	**5 957**	**5 190**	**5 238**	..
Gas/Diesel Oil	6 226	6 650	5 858	4 441	5 097	4 742	4 753	..
Fuel Oil	1 247	530	530	228	387	162	190	..
Other	468	482	476	445	473	286	295	..
Memo: Non-Energy Use	*2 512*	*2 191*	*1 933*	*2 639*	*4 365*	*5 107*	*5 454*	..
Chem/Petchem. Feedst.	*1 731*	*1 522*	*1 291*	*2 033*	*3 661*	*4 481*	*4 794*	..
Other	*781*	*669*	*642*	*606*	*704*	*626*	*660*	..

Source: Annual Oil Statistics.

BELGIUM
Table 2: Supply of Crude Oil, NGL, Feedstocks and Other Hydrocarbons

000 metric tons

	1973	1978	1980	1990	1999	2000	2001	2002
CRUDE OIL								
Indigenous Production	..	-	-	-	-	-	-	-
+ Imports	..	33 019	32 000	25 091	32 128	34 002	31 941	33 446 e
- Exports	..	157	37	-	-	-	-	5 e
- Direct Use	..	-	-	-	-	-	-	-
+ Stock Change	..	133	- 12	101	12	- 67	95	74 e
- Statistical Difference	..	- 4	-	- 48	- 8	- 6	- 19	- 22 e
Refinery Intake	**..**	**32 999**	**31 951**	**25 240**	**32 148**	**33 941**	**32 055**	**33 537 e**
Closing Stock Level	..	*1 007*	*1 272*	*535*	*632*	*699*	*604*	*530 e*
NATURAL GAS LIQUIDS								
Indigenous Production	..	-	-	-	-	-	-	-
+ Imports	..	-	-	-	-	-	-	-
- Exports	..	-	-	-	-	-	-	-
- Direct Use	..	-	-	-	-	-	-	-
+ Stock Change	..	-	-	-	-	-	-	-
- Statistical Difference	..	-	-	-	-	-	-	-
Refinery Intake	**..**	**-**	**-**	**-**	**-**	**-**	**-**	**-**
Closing Stock Level	..	-	-	-	-	-	-	-
REFINERY FEEDSTOCKS								
Backflows	..	-	-	-	-	-	-	-
+ Transfers	..	66	108	2 730	3 997	4 355	6 810	12 221 e
+ Imports	..	760	1 474	3 909	3 148	3 439	3 554	3 190 e
- Exports	..	45	1	2 292	3 074	3 455	2 261	2 736 e
+ Stock Change	..	99	80	- 14	- 49	119	34	- 42 e
- Statistical Difference	..	- 1	- 19	- 18	51	55	- 62	- 204 e
Refinery Intake	**..**	**881**	**1 680**	**4 351**	**3 971**	**4 403**	**8 199**	**12 837 e**
Closing Stock Level	..	*466*	*488*	*320*	*349*	*230*	*196*	*238 e*
OTHER HYDROCARBONS								
Indigenous Production	..	-	-	-	-	-	-	-
+ Imports	..	-	-	-	-	-	-	-
- Exports	..	-	-	-	-	-	-	-
- Direct Use	..	-	-	-	-	-	-	-
+ Stock Change	..	-	-	-	-	-	-	-
- Statistical Difference	..	-	-	-	-	-	-	-
Refinery Intake	**..**	**-**	**-**	**-**	**-**	**-**	**-**	**-**
Closing Stock Level	..	-	-	-	-	-	-	-
CRUDE OIL, NGL, FEEDSTOCKS AND OTHER								
Indigenous Production	-	-	-	-	-	-	-	-
+ Backflows	-	-	-	-	-	-	-	-
+ Transfers	-	66	108	2 730	3 997	4 355	6 810	12 221 e
+ Imports	37 650	33 779	33 474	29 000	35 276	37 441	35 495	36 636 e
- Exports	330	202	38	2 292	3 074	3 455	2 261	2 741 e
- Direct Use	-	-	-	-	-	-	-	-
+ Stock Change	- 11	232	68	87	- 37	52	129	32 e
- Statistical Difference	61	- 5	- 19	- 66	43	49	- 81	- 226 e
Refinery Intake	**37 248**	**33 880**	**33 631**	**29 591**	**36 119**	**38 344**	**40 254**	**46 374 e**
Closing Stock Level	-	*1 473*	*1 760*	*855*	*981*	*929*	*800*	*768 e*

Source: Annual Oil Statistics.

BELGIUM
Table 3: Motor Gasoline - Supply

000 metric tons

	1973	1978	1980	1990	1999	2000	2001	2002
MOTOR GASOLINE								
Refinery Output	**4 768**	**4 815**	**5 506**	**5 263**	**6 032**	**5 311**	**5 493**	**5 776 e**
+ Imports	567	813	1 390	1 632	1 572	1 533	1 469	1 215 e
- Exports	2 560	2 617	3 662	4 263	5 257	4 604	4 801	4 934 e
+ Transfers	-	-	-	- 2	- 50	8	20	- 76 e
+ Stock Change	- 174	137	- 294	73	84	- 5	4	61 e
- Statistical Difference	28	40	- 8	- 24	- 13	- 2	- 1	- 40 e
Total Consumption	**2 573**	**3 108**	**2 948**	**2 727**	**2 394**	**2 245**	**2 186**	**2 082 e**
Closing Stock Level	-	*526*	*765*	*442*	*510*	*515*	*511*	*450 e*

Table 4: Aviation Fuels - Supply

000 metric tons

	1973	1978	1980	1990	1999	2000	2001	2002
AVIATION GASOLINE								
Refinery Output	**-**	**-**	**-**	**8**	**-**	**-**	**-**	**-**
+ Imports	10	4	4	3	2	3	3	3 e
- Exports	-	-	-	9	-	-	-	-
+ Transfers	-	-	-	-	-	-	-	-
+ Stock Change	-	-	-	-	-	-	-	-
- Statistical Difference	- 1	-	-	- 1	-	-	-	-
Total Consumption	**11**	**4**	**4**	**3**	**2**	**3**	**3**	**3 e**
Closing Stock Level	-	-	-	*5*	-	-	-	-
GASOLINE TYPE JET FUEL								
Refinery Output	**-**	**473**	**337**	**-**	**-**	**-**	**-**	**-**
+ Imports	-	48	32	-	-	-	-	-
- Exports	-	379	270	-	-	-	-	-
+ Transfers	-	-	-	-	-	-	-	-
+ Stock Change	-	2	2	-	-	-	-	-
- Statistical Difference	-	1	1	-	-	-	-	-
Total Consumption	**-**	**143**	**100**	**-**	**-**	**-**	**-**	**-**
Closing Stock Level	-	*25*	*30*	-	-	-	-	-
KEROSENE TYPE JET FUEL								
Refinery Output	**1 111**	**1 023**	**1 370**	**1 438**	**2 030**	**2 230**	**1 913**	**2 068 e**
+ Imports	73	32	23	461	569	437	466	365 e
- Exports	637	755	1 035	922	1 117	1 166	1 317	1 522 e
+ Transfers	-	-	-	- 72	- 2	- 4	- 5	- 14 e
+ Stock Change	- 49	13	- 22	24	32	- 47	59	8 e
- Statistical Difference	- 1	-	- 35	2	1	- 31	- 5	- 25 e
Total Consumption	**499**	**313**	**371**	**927**	**1 511**	**1 481**	**1 121**	**930 e**
Closing Stock Level	-	*101*	*163*	*135*	*123*	*170*	*111*	*103 e*

Source: Annual Oil Statistics.

BELGIUM
Table 5A: Gas/Diesel Oil - Supply and Consumption

000 metric tons

	1973	1978	1980	1990	1999	2000	2001	2002
Refinery Output	**12 054**	**10 730**	**10 932**	**10 229**	**11 749**	**12 501**	**12 777**	**12 463 e**
+ Recycled Products	-	-	-	-	-	-	-	-
+ Imports	2 488	4 420	4 047	4 754	6 313	6 259	7 281	6 312 e
- Exports	4 817	4 889	5 776	5 645	7 355	7 422	7 589	7 646 e
- Intl. Marine Bunkers	503	702	586	635	570	787	773	805 e
+ Transfers	-	-	-	- 323	- 16	- 17	5	64 e
+ Stock Changes	70	- 63	- 205	26	673	- 4	- 542	205 e
- Statistical Differences	17	- 33	- 120	- 10	25	- 11	- 24	- 79 e
Total Consumption	**9 275**	**9 529**	**8 532**	**8 416**	**10 769**	**10 541**	**11 183**	**10 672 e**
Transformation	**28**	**24**	**5**	**3**	**24**	**14**	**16**	**..**
Public Electricity	28	24 e	5 e	3 e	24	14	16	..
Autoproducer Electricity	-	-	-	-	-	-	-	..
Public CHP Plants	-	-	-	-	-	-	-	..
Autoproducers of CHP	-	-	-	-	-	-	-	..
Public Heat Plants	-	-	-	-	-	-	-	..
Autoproducer Heat Plants	-	-	-	-	-	-	-	..
Petrochemical Industry	-	-	-	-	-	-	-	..
Other Transformation	-	-	-	-	-	-	-	..
Energy Sector	**-**	**46**	**23**	**23**	**63**	**39**	**21**	**..**
Refinery Fuel	-	1	-	14	20	39	21	..
Other Energy Sector	-	45	23	9	43	-	-	..
Final Consumption	**9 247**	**9 459**	**8 504**	**8 390**	**10 682**	**10 488**	**11 146**	**..**
Transport	**1 593**	**1 725**	**2 072**	**3 676**	**5 071**	**5 323**	**5 632**	**..**
Air	-	-	-	-	-	-	-	..
Road	1 251	1 419	1 795	3 479	4 924	5 192	5 515	..
Rail	149	141	146	69	63	59	55	..
Inland Waterways	193	165	131	128	84	72	62	..
Non-Specified	-	-	-	-	-	-	-	..
Industry	**1 428**	**1 084**	**574**	**273**	**514**	**423**	**761**	**..**
Iron and Steel	135	59	38	11	16	20	17	..
Chemical (incl. Pet.-Chem.)	65	52	42	19	14	14	16	..
Non-Ferrous Metals	-	22	13	4	2	13	1	..
Non-Metallic Minerals	196	225	166	7	11	1	8	..
Transport Equipment	-	-	-	-	-	-	17	..
Machinery	125	126	92	37	12	16	17	
Mining and Quarrying	23	25	20	10	-	-	-	..
Food and Tobacco	110	122	90	40	26	23	24	..
Pulp, Paper and Printing	11	18	11	4	2	3	1	..
Wood and Wood Products	-	-	-	-	-	-	-	..
Construction	130 e	205 e	100 e	45	78	97	94	..
Textiles and Leather	-	-	-	6	12	9	12	..
Non-Specified	633	230 e	2 e	90 e	341	228	554	..
Other Sectors	**6 226**	**6 650**	**5 858**	**4 441**	**5 097**	**4 742**	**4 753**	**..**
Commerce and Public	-	1 650	1 232	895	1 009	717	745	..
Residential	5 926	4 650	4 350	3 184	3 578	3 551	3 526	..
Agriculture	300	350	276	362	510	474	482	..
Non-Specified	-	-	-	-	-	-	-	..
Memo: Non-Energy Use	-	-	-	-	-	-	-	..
Chem/Petchem. Feedst.	-	-	-	-	-	-	-	..
Other	-	-	-	-	-	-	-	..
Closing Stock Level	-	*1 653*	*2 110*	*1 079*	*1 026*	*1 030*	*1 572*	*1 367 e*

Source: Annual Oil Statistics.

BELGIUM
Table 5B: Fuel Oil (Residual) - Supply and Consumption

000 metric tons

	1973	1978	1980	1990	1999	2000	2001	2002
Refinery Output	**13 715**	**12 053**	**10 898**	**5 636**	**6 895**	**7 990**	**7 015**	**7 604 e**
+ Recycled Products	-	-	-	-	-	-	-	-
+ Imports	2 882	1 900	2 702	3 201	2 227	2 347	2 886	3 679 e
- Exports	4 352	4 032	4 430	3 249	2 763	3 887	3 413	3 514 e
- Intl. Marine Bunkers	2 615	2 098	1 845	3 565	3 951	4 752	4 641	6 468 e
+ Transfers	-	-	-	- 16	15	21	35	- 16 e
+ Stock Changes	- 23	244	74	170	96	- 2	88	- 28 e
- Statistical Differences	- 57	- 96	26	- 61	176	22	- 18	3 e
Total Consumption	**9 664**	**8 163**	**7 373**	**2 238**	**2 343**	**1 695**	**1 988**	**1 254 e**
Transformation	**4 986**	**3 734**	**3 987**	**254**	**268**	**162**	**261**	..
Public Electricity	4 986	2 989 e	3 352 e	55 e	184	85	180 e	..
Autoproducer Electricity	-	194 e	163 e	79 e	74 e	68	72 e	..
Public CHP Plants	-	551 e	472 e	120 e	10 e	9	9 e	..
Autoproducers of CHP	-	-	-	-	-	-	-	..
Public Heat Plants	-	-	-	-	-	-	-	..
Autoproducer Heat Plants	-	-	-	-	-	-	-	..
Petrochemical Industry	-	-	-	-	-	-	-	..
Other Transformation	-	-	-	-	-	-	-	..
Energy Sector	-	**1 067**	**968**	**504**	**617**	**566**	**545 e**	..
Refinery Fuel	-	800	750	482	590	548	525	..
Other Energy Sector	-	267	218	22	27	18	20 e	..
Final Consumption	**4 678**	**3 362**	**2 418**	**1 480**	**1 458**	**967**	**1 182**	..
Transport	-	-	-	-	**186**	**136**	**121**	..
Air	-	-	-	-	-	-	-	..
Road	-	-	-	-	-	-	-	..
Rail	-	-	-	-	-	-	-	..
Inland Waterways	-	-	-	-	186	136	121	..
Non-Specified	-	-	-	-	-	-	-	..
Industry	**3 431**	**2 832**	**1 888**	**1 252**	**885**	**669**	**871**	..
Iron and Steel	651	375	154	154	9	11	19	..
Chemical (incl. Pet.-Chem.)	733	529	484	428	182	162	213	..
Non-Ferrous Metals	-	51	37	49	22	10	22	..
Non-Metallic Minerals	379	334	192	173	150	146	118	..
Transport Equipment	-	-	-	-	-	-	-	..
Machinery	122	143	116	51	25	-	-	..
Mining and Quarrying	-	-	-	-	-	-	-	..
Food and Tobacco	197	227	194	217	133	139	144	..
Pulp, Paper and Printing	92	71	54	70	40	44	49	..
Wood and Wood Products	-	-	-	-	1	-	-	..
Construction	-	-	23	35	35	36	43	..
Textiles and Leather	-	-	67	67	10	3	4	..
Non-Specified	1 257	1 102 e	567 e	8 e	278	118	259	..
Other Sectors	**1 247**	**530**	**530**	**228**	**387**	**162**	**190**	..
Commerce and Public	435	206	200	110	56	11	19	..
Residential	680	124	120	8	-	-	-	..
Agriculture	132	200	210	110	331	151	171	..
Non-Specified	-	-	-	-	-	-	-	..
Memo: Non-Energy Use	-	-	-	-	*34*	*162*	*213*	..
Chem/Petchem. Feedst.	-	-	-	-	-	-	-	..
Other	-	-	-	-	*34*	*162*	*213*	..
Closing Stock Level	-	*1 827*	*2 378*	*787*	*626*	*628*	*540*	*568 e*

Source: Annual Oil Statistics.

BELGIUM
Table 5C: Total Products - Supply and Consumption

000 metric tons

	1973	1978	1980	1990	1999	2000	2001	2002
Refinery Output	**35 510**	**33 545**	**33 439**	**29 372**	**35 773**	**38 092**	**39 863**	**45 984 e**
+ Recycled Products	-	-	-	-	11	-	9	-
+ Imports	8 024	8 722	10 008	11 971	13 811	14 445	15 468	14 921 e
- Exports	14 507	15 154	17 588	16 452	18 852	19 655	19 367	19 977 e
- Intl. Marine Bunkers	3 138	2 820	2 455	4 242	4 535	5 545	5 429	7 273 e
+ Transfers	-	- 66	- 108	-2 730	-3 997	-4 355	-6 810	-12 221 e
+ Stock Changes	- 200	454	- 325	270	1 001	- 101	- 362	268 e
- Statistical Differences	- 16	- 125	- 73	- 170	188	- 97	- 42	- 123 e
Total Consumption	**25 705**	**24 806**	**23 044**	**18 359**	**23 024**	**22 978**	**23 414**	**21 825 e**
Transformation	**5 108**	**3 783**	**4 023**	**265**	**293**	**178**	**278**	..
Public Electricity	5 107	3 034	3 387	66	209	99	197	..
Autoproducer Electricity	-	194	163	79	74	68	72	..
Public CHP Plants	-	551	472	120	10	9	9	..
Autoproducers of CHP	-	-	-	-	-	-	-	..
Public Heat Plants	-	-	-	-	-	-	-	..
Autoproducer Heat Plants	-	-	-	-	-	-	-	..
Petrochemical Industry	-	-	-	-	-	-	-	..
Other Transformation	1	4	1	-	-	2	-	..
Energy Sector	-	**1 800**	**2 006**	**1 335**	**1 547**	**1 531**	**1 559 e**	..
Refinery Fuel	-	1 488	1 755	1 293	1 447	1 499	1 509	..
Other Energy Sector	-	312	251	42	100	32	50 e	..
Final Consumption	**20 597**	**19 223**	**17 015**	**16 759**	**21 184**	**21 269**	**21 577**	..
Transport	**4 687**	**5 322**	**5 630**	**7 462**	**9 331**	**9 325**	**9 238**	..
Air	510	460	475	930	1 513	1 485	1 124	..
Road	3 835	4 556	4 872	6 322	7 481	7 557	7 864	..
Rail	149	141	150	72	65	61	56	..
Inland Waterways	193	165	133	138	272	222	194	..
Non-Specified	-	-	-	-	-	-	-	..
Industry	**7 969**	**6 239**	**4 521**	**4 183**	**5 896**	**6 754**	**7 101**	..
Iron and Steel	791	436	214	181	48	265	56	..
Chemical (incl. Pet.-Chem.)	2 577	2 125	1 873	2 488	3 923	4 746	5 024	..
Non-Ferrous Metals	-	76	62	60	25	23	24	..
Non-Metallic Minerals	575	565	363	222	254	342	299	..
Transport Equipment	-	-	-	-	-	-	17	..
Machinery	247	281	215	115	45	24	21	
Mining and Quarrying	23	25	20	10	1	-	-	..
Food and Tobacco	307	355	292	266	160	163	169	..
Pulp, Paper and Printing	106	96	78	80	43	49	53	..
Wood and Wood Products	-	-	-	-	1	-	-	..
Construction	668	662	585	384	361	350	373	..
Textiles and Leather	-	-	71	75	22	13	17	..
Non-Specified	2 675	1 618	748	302	1 013	779	1 048	..
Other Sectors	**7 941**	**7 662**	**6 864**	**5 114**	**5 957**	**5 190**	**5 238**	..
Commerce and Public	455	1 973	1 548	1 170	1 172	819	880	..
Residential	7 036	5 119	4 799	3 413	3 767	3 704	3 684	..
Agriculture	450	570	512	489	871	666	673	..
Non-Specified	-	-	5	42	147	1	1	..
Memo: Non-Energy Use	*2 512*	*2 191*	*1 933*	*2 639*	*4 365*	*5 107*	*5 454*	..
Chem/Petchem. Feedst.	*1 731*	*1 522*	*1 291*	*2 033*	*3 661*	*4 481*	*4 794*	..
Other	*781*	*669*	*642*	*606*	*704*	*626*	*660*	..
Closing Stock Level	-	*4 719*	*6 056*	*2 972*	*2 629*	*2 730*	*3 092*	*2 824 e*

Source: Annual Oil Statistics.

BELGIUM
Table 6A: Supply and Consumption

000 metric tons

	1980	2000	2001	2002	1980	2000	2001	2002
	LPG AND ETHANE				REFINERY GAS			
Refinery Output	**516**	**764**	**647**	**655 e**	**616**	**642**	**652**	**684 e**
+ Recycled Products	-	-	-	-	-	-	-	-
+ Imports	267	344	222	157 e	-	-	-	-
- Exports	243	480	277	325 e	-	-	-	-
- Intl. Marine Bunkers	-	-	-	-	-	-	-	-
+ Transfers	-	- 49	- 7	5 e	-	-	-	-
+ Stock Change	4	- 6	6	6 e	-	-	-	-
- Statistical Difference	- 6	3	6	- 5 e	-	-	-	-
Total Consumption	**550**	**570**	**585**	**503 e**	**616**	**642**	**652**	**684 e**
of which: Ethane	-	-	-	-	-	-	-	-
Transformation	1	-	-	..	30	-	-	..
Energy	19	-	-	..	586	642	652	..
Transport	39	89	102	..	-	-	-	..
Industry	57	322	324	..	-	-	-	..
Other	434	159	159	..	-	-	-	..
Memo: Non-Energy Use	*9*	*193*	*286*	..	-	-	-	..
Closing Stock Level	*45*	*30*	*24*	*18 e*	-	-	-	-
	NAPHTHA				OTHER KEROSENE			
Refinery Output	**1 486**	**2 026**	**2 135**	**2 053 e**	**23**	**87**	**113**	**78 e**
+ Recycled Products	-	-	-	-	-	-	-	-
+ Imports	725	1 607	1 569	1 481 e	6	59	69	89 e
- Exports	1 158	291	261	188 e	18	28	31	29 e
- Intl. Marine Bunkers	-	-	-	-	-	-	-	-
+ Transfers	-	- 11	- 41	21 e	-	1	- 7	- 12 e
+ Stock Change	95	- 2	- 16	24 e	20	-	- 5	3 e
- Statistical Difference	- 31	- 7	- 4	39 e	7	- 2	-	- 8 e
Total Consumption	**1 179**	**3 336**	**3 390**	**3 352 e**	**24**	**121**	**139**	**137 e**
Transformation	-	-	-	..	-	-	1	..
Energy	-	-	-	..	-	-	-	..
Transport	-	-	-	..	-	9	8	..
Industry	1 179	3 336	3 390	..	5	6	7	..
Other	-	-	-	..	19	106	123	..
Memo: Non-Energy Use	*1 125*	*3 336*	*3 390*	..	-	-	-	..
Closing Stock Level	*275*	*73*	*89*	*65 e*	*8*	*24*	*29*	*26 e*
	WHITE SPIRIT + SBP				LUBRICANTS			
Refinery Output	**231**	**196**	**112**	**..**	**66**	**-**	**-**	**..**
+ Recycled Products	-	-	-	..	-	-	9	..
+ Imports	13	6	9	..	483	669	555	..
- Exports	201	167	92	..	345	495	457	..
- Intl. Marine Bunkers	-	-	-	..	24	6	15	..
+ Transfers	-	- 11	-	..	-	9	7	..
+ Stock Change	14	-	1	..	- 4	1	- 8	..
- Statistical Difference	-	- 1	- 1	..	- 21	4	- 13	..
Total Consumption	**57**	**25**	**31**	**..**	**197**	**174**	**104**	**..**
Transformation	-	-	-	..	-	-	-	..
Energy	-	-	-	..	4	-	-	..
Transport	-	-	-	..	101	39	65	..
Industry	47	24	30	..	89	126	27	..
Other	10	1	1	..	3	9	12	..
Memo: Non-Energy Use	*57*	*25*	*28*	..	*191*	*174*	*104*	..
Closing Stock Level	*39*	*31*	*30*	..	*158*	*56*	*64*	..

Source: Annual Oil Statistics.

BELGIUM
Table 6B: Supply and Consumption

000 metric tons

	1980	2000	2001	2002	1980	2000	2001	2002
	BITUMEN				PARAFFIN WAXES			
Refinery Output	**757**	**663**	**413**	**..**	**-**	**-**	**-**	**..**
+ Recycled Products	-	-	-	..	-	-	-	..
+ Imports	31	401	607	..	6	12	18	..
- Exports	306	654	693	..	-	-	-	..
- Intl. Marine Bunkers	-	-	-	..	-	-	-	..
+ Transfers	-	2	1	..	-	-	-	..
+ Stock Change	- 8	- 17	27	..	-	-	-	..
- Statistical Difference	15	-	20	..	-	-	-	..
Total Consumption	**459**	**395**	**335**	**..**	**6**	**12**	**18**	**..**
Transformation	-	2	-	..	-	-	-	..
Energy	7	-	-	..	-	-	-	..
Transport	-	-	-	..	-	-	-	..
Industry	452	393	335	..	6	12	18	..
Other	-	-	-	..	-	-	-	..
Memo: Non-Energy Use	*452*	*292*	*335*	..	*6*	*12*	*18*	..
Closing Stock Level	*63*	*91*	*64*	..	-	-	-	..

	1980	2000	2001	2002	1980	2000	2001	2002
	PETROLEUM COKE				OTHER PRODUCTS			
Refinery Output	**181**	**270**	**311**	**327 e**	**520**	**5 412**	**8 282**	**14 276 e**
+ Recycled Products	-	-	-	-	-	-	-	-
+ Imports	144	596	275	371 e	135	172	39	1 249 e
- Exports	2	30	9	3 e	142	431	427	1 816 e
- Intl. Marine Bunkers	-	-	-	-	-	-	-	-
+ Transfers	-	-	-	-	- 108	-4 304	-6 818	-12 193 e
+ Stock Change	-	- 13	23	- 8 e	- 1	- 6	1	- 3 e
- Statistical Difference	-	- 1	1	-	99	- 71	- 3	- 8 e
Total Consumption	**323**	**824**	**599**	**687 e**	**305**	**914**	**1 080**	**1 521 e**
Transformation	-	-	-	..	-	-	-	..
Energy	187	284	341	..	212	-	-	..
Transport	-	-	-	..	-	-	-	..
Industry	136	533	258	..	88	910	1 080	..
Other	-	7	-	..	5	4	-	..
Memo: Non-Energy Use	-	-	-	..	*93*	*913*	*1 080*	..
Closing Stock Level	-	*41*	*18*	*26 e*	*22*	*41*	*40*	*201 e*

Source: Annual Oil Statistics.

BELGIUM
Table 7A: Imports by Country of Origin

000 metric tons

	2000	2001	2002	2000	2001	2002	2000	2001	2002
	Crude Oil			**Natural Gas Liquids**			**Feedstocks and Other**		
Australia	-	-	-	-	-	-	-	-	-
Belgium	-	-	-	-	-	-	-	-	-
Canada	-	-	-	-	-	-	-	-	-
Czech Republic	-	-	-	-	-	-	-	-	-
Denmark	314	397	319 e	-	-	-	50	7	-
Finland	-	-	-	-	-	-	-	-	-
France	-	-	-	-	-	-	29	26	33 e
Germany	-	86	-	-	-	-	3	98	13 e
Greece	-	-	-	-	-	-	-	-	-
Hungary	-	-	-	-	-	-	-	-	-
Ireland	-	-	-	-	-	-	-	-	-
Italy	-	-	-	-	-	-	10	-	-
Japan	-	-	-	-	-	-	-	-	-
Korea	-	-	-	-	-	-	-	-	-
Mexico	126	-	-	-	-	-	-	-	-
Netherlands	27	66	92 e	-	-	-	1 440	1 221	1 335 e
New Zealand	-	-	-	-	-	-	-	-	-
Norway	8 936	8 135	6 783 e	-	-	-	369	315	265 e
Poland	-	-	-	-	-	-	-	-	-
Spain	-	-	-	-	-	-	-	8	-
Sweden	-	-	-	-	-	-	-	15	-
Turkey	-	-	-	-	-	-	-	-	19 e
United Kingdom	6 314	6 107	5 441 e	-	-	-	190	294	225 e
United States	-	-	-	-	-	-	31	721	451 e
Other	-	-	-	-	-	-	107	13	2 e
OECD	**15 717**	**14 791**	**12 635 e**	**-**	**-**	**-**	**2 229**	**2 718**	**2 343 e**
Argentina	-	-	-	-	-	-	-	-	-
Brazil	-	-	-	-	-	-	-	-	-
Colombia	-	-	-	-	-	-	-	-	-
Ecuador	-	-	-	-	-	-	-	-	-
Netherlands Antilles	99	238	115 e	-	-	-	25	-	-
Peru	-	-	-	-	-	-	-	-	-
Trinidad and Tobago	-	-	-	-	-	-	-	-	-
Venezuela	739	1 323	883 e	-	-	-	-	-	-
Other	-	-	-	-	-	-	34	13	-
Latin America	**838**	**1 561**	**998 e**	**-**	**-**	**-**	**59**	**13**	**-**
Romania	19	-	-	-	-	-	61	-	-
Other	-	-	-	-	-	-	-	-	-
Non-OECD Europe	**19**	**-**	**-**	**-**	**-**	**-**	**61**	**-**	**-**
FSU	**5 363**	**5 088**	**9 096 e**	**-**	**-**	**-**	**1 090**	**770**	**811 e**
Iran	2 494	2 895	2 505 e	-	-	-	-	-	-
Iraq	1 814	1 035	686 e	-	-	-	-	-	-
Kuwait	-	87	2 e	-	-	-	-	-	-
Oman	-	-	-	-	-	-	-	-	-
Qatar	33	-	28 e	-	-	-	-	-	-
Saudi Arabia	6 794	5 821	6 338 e	-	-	-	-	-	-
United Arab Emirates	-	50	-	-	-	-	-	-	-
Other	-	56	179 e	-	-	-	-	-	12 e
Middle East	**11 135**	**9 944**	**9 738 e**	**-**	**-**	**-**	**-**	**-**	**12 e**
India	-	-	-	-	-	-	-	-	-
Indonesia	-	-	-	-	-	-	-	-	-
Singapore	-	-	-	-	-	-	-	-	-
Other	-	-	-	-	-	-	-	-	-
Asia	**-**	**-**	**-**	**-**	**-**	**-**	**-**	**-**	**-**
China	**-**	**-**	**-**	**-**	**-**	**-**	**-**	**-**	**-**
Algeria	177	142	631 e	-	-	-	-	53	-
Angola/Cabinada	279	46	2 e	-	-	-	-	-	-
Cameroon	-	53	-	-	-	-	-	-	-
Egypt	25	-	-	-	-	-	-	-	-
Gabon	-	-	-	-	-	-	-	-	-
Libya	160	6	-	-	-	-	-	-	24 e
Nigeria	289	238	186 e	-	-	-	-	-	-
Tunisia	-	-	-	-	-	-	-	-	-
Other	-	70	153 e	-	-	-	-	-	-
Africa	**930**	**555**	**972 e**	**-**	**-**	**-**	**-**	**53**	**24 e**
Non-spec./Other	**-**	**2**	**7 e**	**-**	**-**	**-**	**-**	**-**	**-**
Total Imports	**34 002**	**31 941**	**33 446 e**	**-**	**-**	**-**	**3 439**	**3 554**	**3 190 e**
Memo: Total OPEC	*12 500*	*11 597*	*11 259 e*	-	-	-	-	*53*	*24 e*

Source: Annual Oil Statistics.

BELGIUM
Table 7B: Imports by Country of Origin

000 metric tons

	2000	2001	2002	2000	2001	2002	2000	2001	2002
	Total Products			**Liquefied Petroleum Gases**			**Naphtha**		
Australia	-	-	-	-	-	-	-	-	-
Belgium	-	-	-	-	-	-	-	-	-
Canada	22	12	-	-	-	-	-	-	-
Czech Republic	-	-	-	-	-	-	-	-	-
Denmark	28	8	5 e	-	-	-	26	4	-
Finland	5	16	2 e	-	-	-	-	-	-
France	805	1 138	873 e	1	3	-	55	179	82 e
Germany	298	348	340 e	-	5	10 e	25	-	34 e
Greece	25	29	93 e	-	-	-	25	29	93 e
Hungary	-	-	-	-	-	-	-	-	-
Ireland	60	26	-	-	-	-	58	-	-
Italy	21	44	5 e	-	-	-	-	41	-
Japan	-	-	-	-	-	-	-	-	-
Korea	-	-	-	-	-	-	-	-	-
Mexico	-	-	-	-	-	-	-	-	-
Netherlands	9 062	9 434	9 525 e	93	119	123 e	419	355	306 e
New Zealand	-	-	-	-	-	-	-	-	-
Norway	505	235	117 e	233	57	11 e	164	61	26 e
Poland	42	30	46 e	-	-	-	-	-	-
Spain	119	74	41 e	-	-	-	104	57	38 e
Sweden	79	192	109 e	-	-	-	5	22	-
Turkey	28	-	-	-	-	-	28	-	-
United Kingdom	1 357	1 509	1 338 e	17	37	12 e	530	646	624 e
United States	712	424	456 e	-	-	-	-	33	-
Other	72	170	213 e	-	-	-	-	16	6 e
OECD	**13 240**	**13 689**	**13 163 e**	**344**	**221**	**156 e**	**1 439**	**1 443**	**1 209 e**
Argentina	-	-	-	-	-	-	-	-	-
Brazil	-	-	-	-	-	-	-	-	-
Colombia	-	-	-	-	-	-	-	-	-
Ecuador	-	-	-	-	-	-	-	-	-
Netherlands Antilles	-	-	4 e	-	-	-	-	-	-
Peru	-	-	-	-	-	-	-	-	-
Trinidad and Tobago	-	-	-	-	-	-	-	-	-
Venezuela	-	-	-	-	-	-	-	-	-
Other	33	12	-	-	-	-	33	-	-
Latin America	**33**	**12**	**4 e**	**-**	**-**	**-**	**33**	**-**	**-**
Romania	-	30	-	-	-	-	-	-	-
Other	7	-	4 e	-	-	-	-	-	-
Non-OECD Europe	**7**	**30**	**4 e**	**-**	**-**	**-**	**-**	**-**	**-**
FSU	**1 003**	**1 468**	**1 374 e**	**-**	**-**	**-**	**97**	**40**	**3 e**
Iran	10	4	-	-	-	-	10	-	-
Iraq	-	-	-	-	-	-	-	-	-
Kuwait	-	-	-	-	-	-	-	-	-
Oman	-	-	-	-	-	-	-	-	-
Qatar	-	-	-	-	-	-	-	-	-
Saudi Arabia	-	-	-	-	-	-	-	-	-
United Arab Emirates	-	-	-	-	-	-	-	-	-
Other	-	21	27 e	-	-	-	-	21	27 e
Middle East	**10**	**25**	**27 e**	**-**	**-**	**-**	**10**	**21**	**27 e**
India	-	129	-	-	-	-	-	-	-
Indonesia	-	-	-	-	-	-	-	-	-
Singapore	-	-	-	-	-	-	-	-	-
Other	-	-	-	-	-	-	-	-	-
Asia	**-**	**129**	**-**	**-**	**-**	**-**	**-**	**-**	**-**
China	**-**	**-**	**-**	**-**	**-**	**-**	**-**	**-**	**-**
Algeria	89	27	88 e	-	-	-	27	-	55 e
Angola/Cabinada	-	-	-	-	-	-	-	-	-
Cameroon	-	-	-	-	-	-	-	-	-
Egypt	-	-	27 e	-	-	-	-	-	27 e
Gabon	-	-	-	-	-	-	-	-	-
Libya	58	55	184 e	-	-	-	-	55	159 e
Nigeria	-	10	-	-	-	-	-	10	-
Tunisia	-	-	-	-	-	-	-	-	-
Other	-	19	-	-	-	-	-	-	-
Africa	**147**	**111**	**299 e**	**-**	**-**	**-**	**27**	**65**	**241 e**
Non-spec./Other	**5**	**4**	**50 e**	**-**	**1**	**1 e**	**1**	**-**	**1 e**
Total Imports	**14 445**	**15 468**	**14 921 e**	**344**	**222**	**157 e**	**1 607**	**1 569**	**1 481 e**
Memo: Total OPEC	*157*	*96*	*272 e*	-	-	-	*37*	*65*	*214 e*

Source: Annual Oil Statistics.

BELGIUM
Table 7C: Imports by Country of Origin

000 metric tons

	2000	2001	2002	2000	2001	2002	2000	2001	2002
	Total Gasoline			Total Kerosene			Gas/Diesel		
Australia	-	-	-	-	-	-	-	-	-
Belgium	-	-	-	-	-	-	-	-	-
Canada	-	-	-	-	-	-	22	12	-
Czech Republic	-	-	-	-	-	-	-	-	-
Denmark	-	-	-	-	-	-	-	-	-
Finland	5	-	-	-	-	-	-	16	-
France	37	59	82 e	137	82	50 e	58	77	24 e
Germany	40	120	43 e	-	34	5 e	45	75	85 e
Greece	-	-	-	-	-	-	-	-	-
Hungary	-	-	-	-	-	-	-	-	-
Ireland	2	-	-	-	-	-	-	-	-
Italy	2	-	4 e	-	-	-	12	-	-
Japan	-	-	-	-	-	-	-	-	-
Korea	-	-	-	-	-	-	-	-	-
Mexico	-	-	-	-	-	-	-	-	-
Netherlands	1 307	1 179	1 013 e	358	397	227 e	4 824	5 042	4 740 e
New Zealand	-	-	-	-	-	-	-	-	-
Norway	2	-	-	-	-	-	54	53	27 e
Poland	-	-	-	-	-	-	-	-	15 e
Spain	-	-	-	-	-	-	-	-	-
Sweden	16	-	-	-	-	-	19	101	47 e
Turkey	-	-	-	-	-	-	-	-	-
United Kingdom	109	105	74 e	-	-	48 e	356	379	213 e
United States	-	-	-	-	9	14 e	-	26	-
Other	3	8	-	-	9	109 e	10	6	8 e
OECD	**1 523**	**1 471**	**1 216 e**	**495**	**531**	**453 e**	**5 400**	**5 787**	**5 159 e**
Argentina	-	-	-	-	-	-	-	-	-
Brazil	-	-	-	-	-	-	-	-	-
Colombia	-	-	-	-	-	-	-	-	-
Ecuador	-	-	-	-	-	-	-	-	-
Netherlands Antilles	-	-	-	-	-	-	-	-	-
Peru	-	-	-	-	-	-	-	-	-
Trinidad and Tobago	-	-	-	-	-	-	-	-	-
Venezuela	-	-	-	-	-	-	-	-	-
Other	-	-	-	-	-	-	-	12	-
Latin America	**-**	**-**	**-**	**-**	**-**	**-**	**-**	**12**	**-**
Romania	-	-	-	-	-	-	-	30	-
Other	5	-	-	-	-	-	-	-	-
Non-OECD Europe	**5**	**-**	**-**	**-**	**-**	**-**	**-**	**30**	**-**
FSU	**7**	**-**	**-**	**-**	**-**	**-**	**739**	**1 323**	**1 093 e**
Iran	-	-	-	-	4	-	-	-	-
Iraq	-	-	-	-	-	-	-	-	-
Kuwait	-	-	-	-	-	-	-	-	-
Oman	-	-	-	-	-	-	-	-	-
Qatar	-	-	-	-	-	-	-	-	-
Saudi Arabia	-	-	-	-	-	-	-	-	-
United Arab Emirates	-	-	-	-	-	-	-	-	-
Other	-	-	-	-	-	-	-	-	-
Middle East	**-**	**-**	**-**	**-**	**4**	**-**	**-**	**-**	**-**
India	-	-	-	-	-	-	-	129	-
Indonesia	-	-	-	-	-	-	-	-	-
Singapore	-	-	-	-	-	-	-	-	-
Other	-	-	-	-	-	-	-	-	-
Asia	**-**	**-**	**-**	**-**	**-**	**-**	**-**	**129**	**-**
China	**-**	**-**	**-**	**-**	**-**	**-**	**-**	**-**	**-**
Algeria	-	-	-	-	-	-	62	-	33 e
Angola/Cabinada	-	-	-	-	-	-	-	-	-
Cameroon	-	-	-	-	-	-	-	-	-
Egypt	-	-	-	-	-	-	-	-	-
Gabon	-	-	-	-	-	-	-	-	-
Libya	-	-	-	-	-	-	58	-	25 e
Nigeria	-	-	-	-	-	-	-	-	-
Tunisia	-	-	-	-	-	-	-	-	-
Other	-	-	-	-	-	-	-	-	-
Africa	**-**	**-**	**-**	**-**	**-**	**-**	**120**	**-**	**58 e**
Non-spec./Other	**1**	**1**	**2 e**	**1**	**-**	**1 e**	**-**	**-**	**2 e**
Total Imports	**1 536**	**1 472**	**1 218 e**	**496**	**535**	**454 e**	**6 259**	**7 281**	**6 312 e**
Memo: Total OPEC	*-*	*-*	*-*	*-*	*4*	*-*	*120*	*-*	*58 e*

Source: Annual Oil Statistics.

BELGIUM
Table 7D: Imports by Country of Origin

000 metric tons

	2000	2001	2002	2000	2001	2002	2000	2001	2002
	Fuel Oil (Residual)			**Petroleum Coke**			**Other Products**		
Australia	-	-	-	-	-	-	-	-	-
Belgium	-	-	-	-	-	-	-	-	-
Canada	-	-	-	-	-	-	-	-	-
Czech Republic	-	-	-	-	-	-	-	-	-
Denmark	2	4	3 e	-	-	-	-	-	2 e
Finland	-	-	-	-	-	-	-	-	2 e
France	111	386	307 e	-	-	3 e	406	352	325 e
Germany	137	59	101 e	-	-	-	51	55	62 e
Greece	-	-	-	-	-	-	-	-	-
Hungary	-	-	-	-	-	-	-	-	-
Ireland	-	26	-	-	-	-	-	-	-
Italy	-	-	-	-	-	-	7	3	1 e
Japan	-	-	-	-	-	-	-	-	-
Korea	-	-	-	-	-	-	-	-	-
Mexico	-	-	-	-	-	-	-	-	-
Netherlands	1 642	1 744	2 637 e	16	16	13 e	403	582	466 e
New Zealand	-	-	-	-	-	-	-	-	-
Norway	47	64	51 e	-	-	-	5	-	2 e
Poland	-	-	-	-	-	-	42	30	31 e
Spain	10	15	-	-	-	-	5	2	3 e
Sweden	24	48	27 e	-	-	-	15	21	35 e
Turkey	-	-	-	-	-	-	-	-	-
United Kingdom	171	268	190 e	-	-	-	174	74	177 e
United States	-	-	-	580	258	350 e	132	98	92 e
Other	42	120	83 e	-	-	-	17	11	7 e
OECD	**2 186**	**2 734**	**3 399 e**	**596**	**274**	**366 e**	**1 257**	**1 228**	**1 205 e**
Argentina	-	-	-	-	-	-	-	-	-
Brazil	-	-	-	-	-	-	-	-	-
Colombia	-	-	-	-	-	-	-	-	-
Ecuador	-	-	-	-	-	-	-	-	-
Netherlands Antilles	-	-	-	-	-	-	-	-	4 e
Peru	-	-	-	-	-	-	-	-	-
Trinidad and Tobago	-	-	-	-	-	-	-	-	-
Venezuela	-	-	-	-	-	-	-	-	-
Other	-	-	-	-	-	-	-	-	-
Latin America	**-**	**-**	**-**	**-**	**-**	**-**	**-**	**-**	**4 e**
Romania	-	-	-	-	-	-	-	-	-
Other	2	-	-	-	-	-	-	-	4 e
Non-OECD Europe	**2**	**-**	**-**	**-**	**-**	**-**	**-**	**-**	**4 e**
FSU	**159**	**105**	**277 e**	**-**	**-**	**-**	**1**	**-**	**1 e**
Iran	-	-	-	-	-	-	-	-	-
Iraq	-	-	-	-	-	-	-	-	-
Kuwait	-	-	-	-	-	-	-	-	-
Oman	-	-	-	-	-	-	-	-	-
Qatar	-	-	-	-	-	-	-	-	-
Saudi Arabia	-	-	-	-	-	-	-	-	-
United Arab Emirates	-	-	-	-	-	-	-	-	-
Other	-	-	-	-	-	-	-	-	-
Middle East	**-**	**-**	**-**	**-**	**-**	**-**	**-**	**-**	**-**
India	-	-	-	-	-	-	-	-	-
Indonesia	-	-	-	-	-	-	-	-	-
Singapore	-	-	-	-	-	-	-	-	-
Other	-	-	-	-	-	-	-	-	-
Asia	**-**	**-**	**-**	**-**	**-**	**-**	**-**	**-**	**-**
China	**-**	**-**	**-**	**-**	**-**	**-**	**-**	**-**	**-**
Algeria	-	27	-	-	-	-	-	-	-
Angola/Cabinada	-	-	-	-	-	-	-	-	-
Cameroon	-	-	-	-	-	-	-	-	-
Egypt	-	-	-	-	-	-	-	-	-
Gabon	-	-	-	-	-	-	-	-	-
Libya	-	-	-	-	-	-	-	-	-
Nigeria	-	-	-	-	-	-	-	-	-
Tunisia	-	-	-	-	-	-	-	-	-
Other	-	19	-	-	-	-	-	-	-
Africa	**-**	**46**	**-**	**-**	**-**	**-**	**-**	**-**	**-**
Non-spec./Other	**-**	**1**	**3 e**	**-**	**1**	**5 e**	**2**	**-**	**35 e**
Total Imports	**2 347**	**2 886**	**3 679 e**	**596**	**275**	**371 e**	**1 260**	**1 228**	**1 249 e**
Memo: Total OPEC	*-*	*27*	*-*	*-*	*-*	*-*	*-*	*-*	*-*

Source: Annual Oil Statistics.

BELGIUM
Table 8A: Exports by Country of Destination

000 metric tons

	2000	2001	2002	2000	2001	2002	2000	2001	2002
	Crude Oil			**Natural Gas Liquids**			**Feedstocks and Other**		
Australia	-	-	-	-	-	-	-	-	-
Austria	-	-	-	-	-	-	-	-	-
Belgium	-	-	-	-	-	-	-	-	-
Canada	-	-	-	-	-	-	-	-	-
Denmark	-	-	-	-	-	-	-	-	-
Finland	-	-	-	-	-	-	-	-	-
France	-	-	-	-	-	-	503	169	88 e
Germany	-	-	-	-	-	-	626	289	436 e
Greece	-	-	-	-	-	-	-	-	-
Ireland	-	-	-	-	-	-	-	-	-
Italy	-	-	-	-	-	-	-	-	-
Japan	-	-	-	-	-	-	-	-	-
Luxembourg	-	-	-	-	-	-	-	-	-
Mexico	-	-	-	-	-	-	-	-	-
Netherlands	-	-	5 e	-	-	-	1 085	734	735 e
Norway	-	-	-	-	-	-	15	1	2 e
Portugal	-	-	-	-	-	-	-	-	-
Spain	-	-	-	-	-	-	-	-	-
Sweden	-	-	-	-	-	-	1	-	-
Switzerland	-	-	-	-	-	-	28	39	46 e
Turkey	-	-	-	-	-	-	-	-	-
United Kingdom	-	-	-	-	-	-	72	90	120 e
United States	-	-	-	-	-	-	1 103	899	1 307 e
Other OECD	-	-	-	-	-	-	22	39	-
OECD	**-**	**-**	**5 e**	**-**	**-**	**-**	**3 455**	**2 260**	**2 734 e**
Africa	-	-	-	-	-	-	-	-	-
Latin America	-	-	-	-	-	-	-	-	-
Asia	-	-	-	-	-	-	-	-	-
China	-	-	-	-	-	-	-	-	-
Non-OECD Europe	-	-	-	-	-	-	-	-	-
Former Soviet Union	-	-	-	-	-	-	-	-	-
Middle East	-	-	-	-	-	-	-	-	-
Non-specified/Other	-	-	-	-	-	-	-	1	2 e
Non-OECD	**-**	**-**	**-**	**-**	**-**	**-**	**-**	**1**	**2 e**
Total Exports	**-**	**-**	**5 e**	**-**	**-**	**-**	**3 455**	**2 261**	**2 736 e**
	Total Products			**Liquefied Petroleum Gas**			**Naphtha**		
Australia	1	-	-	-	-	-	-	-	-
Austria	83	77	132 e	9	1	-	-	-	-
Belgium	-	-	-	-	-	-	-	-	-
Canada	-	43	-	-	-	-	-	-	-
Denmark	14	15	63 e	-	-	-	-	-	-
Finland	21	9	15 e	1	2	-	-	-	-
France	2 070	1 591	1 692 e	29	22	7 e	-	-	-
Germany	5 403	5 375	4 601 e	98	85	121 e	191	134	65 e
Greece	6	34	45 e	-	-	-	-	-	-
Ireland	82	106	119 e	-	2	-	-	-	-
Italy	528	190	390 e	-	-	-	-	-	-
Japan	-	2	4 e	-	-	-	-	-	-
Luxembourg	2 016	2 003	2 090 e	17	17	15 e	-	-	-
Mexico	37	-	76 e	-	-	-	-	-	-
Netherlands	3 406	3 745	4 937 e	258	120	128 e	93	112	120 e
Norway	16	26	67 e	2	1	2 e	-	-	-
Portugal	67	46	72 e	34	2	8 e	-	-	-
Spain	377	217	194 e	3	-	2 e	-	-	-
Sweden	106	120	117 e	-	-	-	-	-	-
Switzerland	1 440	1 408	1 259 e	1	1	3 e	-	-	-
Turkey	3	3	4 e	-	-	3 e	-	-	-
United Kingdom	1 059	1 051	779 e	18	22	26 e	5	-	2 e
United States	1 903	2 750	2 476 e	-	-	-	2	15	-
Other OECD	155	160	56 e	4	-	1 e	-	-	-
OECD	**18 793**	**18 971**	**19 188 e**	**474**	**275**	**316 e**	**291**	**261**	**187 e**
Africa	479	147	470 e	6	1	6 e	-	-	-
Latin America	46	16	215 e	-	-	-	-	-	-
Asia	47	7	9 e	-	-	-	-	-	-
China	69	10	8 e	-	-	-	-	-	-
Non-OECD Europe	116	130	3 e	-	-	-	-	-	-
Former Soviet Union	25	18	14 e	-	-	-	-	-	-
Middle East	72	57	17 e	-	1	-	-	-	-
Non-specified/Other	8	11	53 e	-	-	3 e	-	-	1 e
Non-OECD	**862**	**396**	**789 e**	**6**	**2**	**9 e**	**-**	**-**	**1 e**
Total Exports	**19 655**	**19 367**	**19 977 e**	**480**	**277**	**325 e**	**291**	**261**	**188 e**

Source: Annual Oil Statistics.

BELGIUM
Table 8B: Exports by Country of Destination

000 metric tons

	2000	2001	2002	2000	2001	2002	2000	2001	2002
	Total Gasoline			**Total Kerosene**			**Gas/Diesel Oil**		
Australia	-	-	-	-	-	-	-	-	-
Austria	-	-	-	2	-	-	59	61	119 e
Belgium	-	-	-	-	-	-	-	-	-
Canada	-	30	-	-	13	-	-	-	-
Denmark	-	-	20 e	1	2	30 e	-	-	-
Finland	8	-	-	-	-	-	-	-	-
France	223	100	205 e	56	22	19 e	1 085	944	821 e
Germany	1 552	1 166	853 e	569	789	914 e	2 047	2 485	1 688 e
Greece	-	-	-	-	-	-	-	-	38 e
Ireland	-	-	-	23	41	60 e	-	-	23 e
Italy	23	-	-	-	-	-	-	-	-
Japan	-	-	-	-	-	-	-	-	-
Luxembourg	562	500	488 e	113	109	179 e	1 287	1 341	1 369 e
Mexico	37	-	-	-	-	-	-	-	76 e
Netherlands	653	710	1 194 e	107	59	147 e	930	1 328	1 594 e
Norway	-	8	36 e	-	-	-	-	-	3 e
Portugal	-	-	-	-	-	-	30	18	-
Spain	-	-	-	-	-	-	144	91	38 e
Sweden	-	6	-	-	5	-	94	103	81 e
Switzerland	327	237	303 e	223	250	85 e	843	873	795 e
Turkey	-	-	-	-	-	-	-	-	-
United Kingdom	199	547	300 e	100	43	113 e	293	84	192 e
United States	855	1 472	1 509 e	-	-	-	87	64	185 e
Other OECD	15	8	2 e	-	8	-	48	-	-
OECD	**4 454**	**4 784**	**4 910 e**	**1 194**	**1 341**	**1 547 e**	**6 947**	**7 392**	**7 022 e**
Africa	65	-	-	-	-	-	387	128	414 e
Latin America	42	14	9 e	-	-	-	-	-	206 e
Asia	-	-	-	-	-	-	31	-	-
China	-	-	-	-	-	-	-	-	-
Non-OECD Europe	-	-	-	-	-	-	52	35	-
Former Soviet Union	9	3	-	-	-	-	-	-	-
Middle East	34	-	-	-	6	-	-	33	-
Non-specified/Other	-	-	15 e	-	1	4 e	5	1	4 e
Non-OECD	**150**	**17**	**24 e**	**-**	**7**	**4 e**	**475**	**197**	**624 e**
Total Exports	**4 604**	**4 801**	**4 934 e**	**1 194**	**1 348**	**1 551 e**	**7 422**	**7 589**	**7 646 e**

	2000	2001	2002	2000	2001	2002	2000	2001	2002
	Fuel Oil (Residual)			**Petroleum Coke**			**Other Products**		
Australia	-	-	-	-	-	-	1	-	-
Austria	-	-	-	-	-	-	13	15	13 e
Belgium	-	-	-	-	-	-	-	-	-
Canada	-	-	-	-	-	-	-	-	-
Denmark	-	-	-	-	-	-	13	13	13 e
Finland	-	-	-	-	-	-	12	7	15 e
France	253	64	176 e	25	1	-	399	438	464 e
Germany	458	305	598 e	3	3	-	485	408	362 e
Greece	-	30	-	-	-	-	6	4	7 e
Ireland	55	56	33 e	2	5	3 e	2	2	-
Italy	453	161	365 e	-	-	-	52	29	25 e
Japan	-	-	-	-	-	-	-	2	4 e
Luxembourg	11	12	12 e	-	-	-	26	24	27 e
Mexico	-	-	-	-	-	-	-	-	-
Netherlands	958	1 005	1 256 e	-	-	-	407	411	498 e
Norway	-	-	-	-	-	-	14	17	26 e
Portugal	-	22	57 e	-	-	-	3	4	7 e
Spain	160	45	70 e	-	-	-	70	81	84 e
Sweden	-	-	27 e	-	-	-	12	6	9 e
Switzerland	7	29	47 e	-	-	-	39	18	26 e
Turkey	-	-	-	-	-	-	3	3	1 e
United Kingdom	378	292	65 e	-	-	-	66	63	81 e
United States	955	1 194	775 e	-	-	-	4	5	7 e
Other OECD	61	111	-	-	-	-	27	33	53 e
OECD	**3 749**	**3 326**	**3 481 e**	**30**	**9**	**3 e**	**1 654**	**1 583**	**1 722 e**
Africa	-	-	32 e	-	-	-	21	18	18 e
Latin America	-	-	-	-	-	-	4	2	-
Asia	-	-	-	-	-	-	16	7	9 e
China	63	-	-	-	-	-	6	10	8 e
Non-OECD Europe	48	87	-	-	-	-	16	8	3 e
Former Soviet Union	-	-	-	-	-	-	16	15	14 e
Middle East	26	-	-	-	-	-	12	17	17 e
Non-specified/Other	1	-	1 e	-	-	-	2	9	25 e
Non-OECD	**138**	**87**	**33 e**	**-**	**-**	**-**	**93**	**86**	**94 e**
Total Exports	**3 887**	**3 413**	**3 514 e**	**30**	**9**	**3 e**	**1 747**	**1 669**	**1 816 e**

Source: Annual Oil Statistics.

CANADA

Figure 1
FUEL SHARES OF TPES *

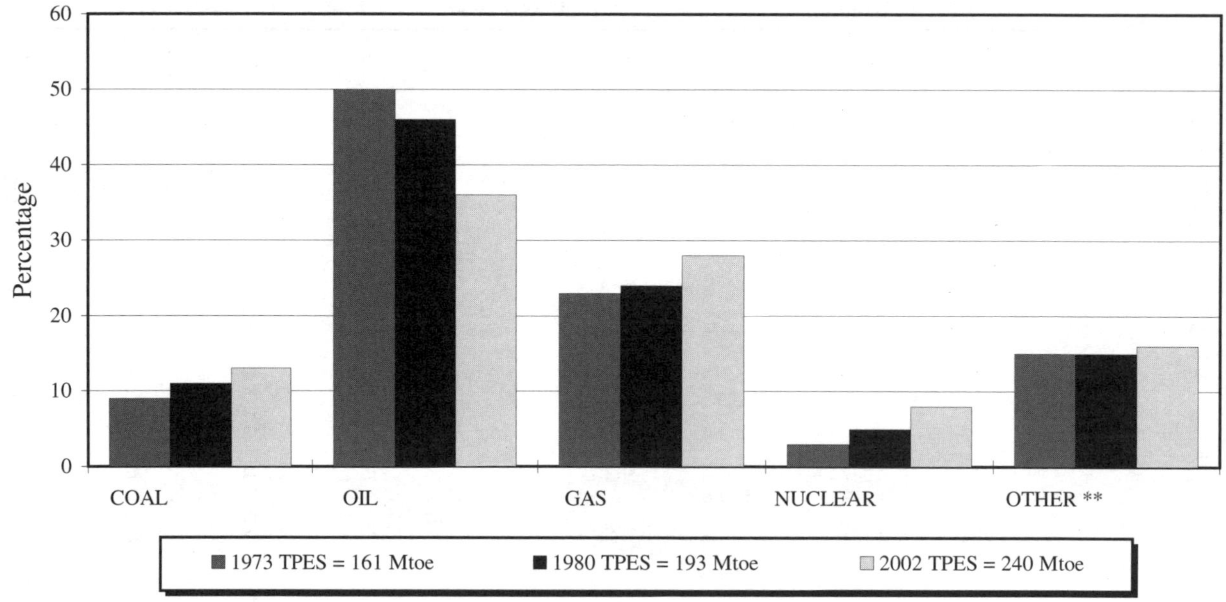

Source: Annual Energy Balances.

* Excluding electricity trade.

** Includes hydro, geothermal, solar, wind, combustible renewables and waste, etc.

Figure 2
OIL PRODUCTION AND CONSUMPTION

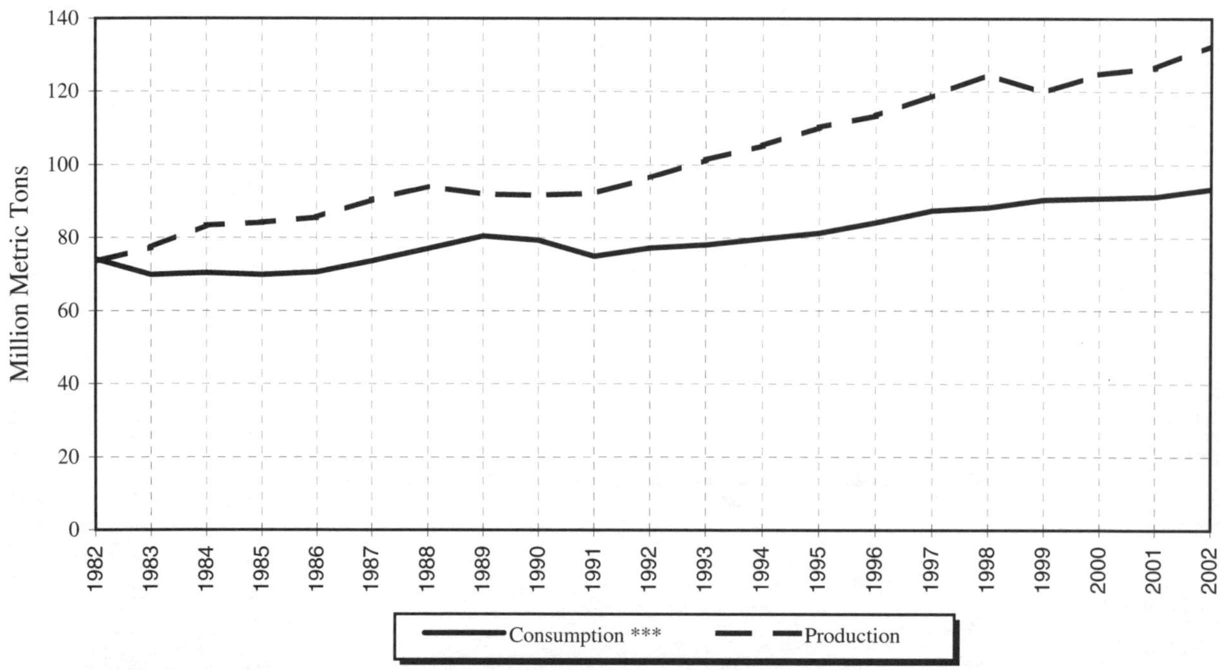

Source: Annual Oil Statistics.

*** Includes refinery fuel and international marine bunkers.

CANADA

Table 1: Oil Supply and Consumption - Summary Table

000 metric tons

	1973	1978	1980	1990	1999	2000	2001	2002
Indigenous Production	94 128	75 080	81 389	91 951	120 298	125 249	127 001	133 121 e
Crude Oil	..	70 194	64 128	68 490	82 624	87 937	88 279	90 993 e
NGL	..	4 886	10 487	12 845	21 286	21 056	21 082	19 781 e
Other	..	-	6 774	10 616	16 388	16 256	17 640	22 347 e
+ Net Imports	-13 907	10 779	8 767	-14 075	-33 989	-36 981	-37 029	-46 903 e
Crude Oil, NGL, Other	-13 560	14 404	12 369	-10 080	-27 615	-29 971	-28 740	-35 923 e
Petroleum Products	- 347	-3 625	-3 602	-3 995	-6 374	-7 010	-8 289	-10 980 e
- Intl. Marine Bunkers	-	1 800	1 526	928	1 065	1 081	1 066	1 151 e
+ Stock Changes	- 981	3 083	-1 528	- 931	719	- 425	-1 332	431 e
TOTAL SUPPLY	**79 240**	**87 142**	**87 102**	**76 017**	**85 963**	**86 762**	**87 574**	**85 498 e**
Refinery Intake	*82 996*	*88 804*	*92 908*	*84 116*	*91 139*	*93 309*	*95 786*	*97 140 e*
Refinery Output	*82 322*	*88 487*	*92 595*	*84 184*	*91 583*	*93 712*	*96 413*	*98 193 e*
LPG + Ethane	*1 325*	*1 641*	*1 932*	*1 716*	*1 882*	*1 874*	*1 904*	*1 952 e*
Naphtha	*1 497*	*3 107*	*3 424*	*3 797*	*3 862*	*3 740*	*3 453*	*3 854 e*
Kerosenes	*5 139*	*6 516*	*5 441*	*5 491*	*6 009*	*5 601*	*5 530*	*5 131 e*
Gasolines	*23 998*	*27 171*	*29 890*	*27 015*	*30 447*	*31 383*	*32 569*	*33 695 e*
Gas/Diesel Oil	*22 731*	*22 961*	*24 962*	*22 670*	*26 685*	*28 198*	*28 722*	*28 919 e*
Fuel Oil	*19 887*	*18 256*	*17 159*	*8 223*	*6 402*	*6 564*	*7 319*	*6 891 e*
Other Products	*7 745*	*8 835*	*9 787*	*15 272*	*16 296*	*16 352*	*16 916*	*17 751 e*
- Refinery Losses	674	317	313	- 68	- 444	- 403	- 627	-1 053 e
+ Returns	344	185	468	1 203	2 527	2 302	2 398	2 951 e
- Statistical Differences	-4 979	- 489	- 818	-1 087	- 330	- 248	444	-2 696 e
TOTAL CONSUMPTION	**83 889**	**87 499**	**88 075**	**78 375**	**89 264**	**89 715**	**90 155**	**92 198 e**
Transformation and Energy	**8 341**	**10 112**	**9 940**	**10 603**	**11 409**	**11 245**	**12 200**	**..**
Crude Oil	120	8	20	5	303	342	250	..
LPG + Ethane	65	95	275	68	147	170	111	..
Gas/Diesel Oil	389	484	460	283	182	172	169	..
Fuel Oil	4 761	4 883	4 604	4 220	2 967	2 815	3 937	..
Petroleum Coke	64	882	895	2 174	3 192	3 316	3 776	..
Other	2 942	3 760	3 686	3 853	4 618	4 430	3 957	..
Transport	**33 024**	**37 794**	**40 622**	**38 904**	**45 781**	**45 804**	**45 492**	**..**
Motor Gasoline	23 545	27 219	28 279	24 948	28 167	28 178	28 526	..
Gas/Diesel Oil	4 532	6 535	7 583	8 199	11 238	11 332	10 970	..
Subtotal	*28 077*	*33 754*	*35 862*	*33 147*	*39 405*	*39 510*	*39 496*	..
Aviation Fuels	2 784	3 340	3 728	4 120	5 089	5 130	4 685	..
Other	2 163	700	1 032	1 637	1 287	1 164	1 311	..
Industry	**21 261**	**20 982**	**20 603**	**17 696**	**20 370**	**19 692**	**19 482**	**..**
Naphtha	1 710	2 883	2 895	2 813	3 610	3 376	3 144	..
LPG + Ethane	390	665	858	3 245	5 871	5 458	5 919	..
Gas/Diesel Oil	4 472	3 637	3 990	2 474	3 080	3 206	3 011	..
Fuel Oil	9 383	8 160	7 323	4 208	2 273	2 549	2 461	..
Other	5 306	5 637	5 537	4 956	5 536	5 103	4 947	..
Other Sectors	**21 263**	**18 611**	**16 910**	**11 172**	**11 704**	**12 974**	**12 981**	**..**
Gas/Diesel Oil	13 700	13 610	12 616	8 836	8 697	9 636	9 414	..
Fuel Oil	3 893	2 094	1 808	307	417	498	666	..
Other	3 670	2 907	2 486	2 029	2 590	2 840	2 901	..
Memo: Non-Energy Use	*7 691*	*9 248*	*9 296*	*11 241*	*15 723*	*14 704*	*14 809*	..
Chem/Petchem. Feedst.	*2 146*	*3 433*	*3 502*	*5 628*	*9 029*	*8 224*	*8 389*	..
Other	*5 545*	*5 815*	*5 794*	*5 613*	*6 694*	*6 480*	*6 420*	..

Source: Annual Oil Statistics.

CANADA
Table 2: Supply of Crude Oil, NGL, Feedstocks and Other Hydrocarbons

000 metric tons

	1973	1978	1980	1990	1999	2000	2001	2002
CRUDE OIL								
Indigenous Production	..	70 194	64 128	68 490	82 624	87 937	88 279	90 993 e
+ Imports	..	31 057	27 308	26 510	40 733	45 098	45 476	43 292 e
- Exports	..	13 373	10 156	27 061	51 870	58 425	56 844	57 547 e
- Direct Use	..	8	20	5	303	342	250	353 e
+ Stock Change	..	964	- 688	- 518	- 314	- 293	- 632	350 e
- Statistical Difference	..	624	327	- 327	299	- 356	- 560	-1 294 e
Refinery Intake	**..**	**88 210**	**80 245**	**67 743**	**70 571**	**74 331**	**76 589**	**78 029 e**
Closing Stock Level	..	*4 193*	*5 228*	*4 799*	*5 048*	*5 341*	*5 973*	*5 623 e*
NATURAL GAS LIQUIDS								
Indigenous Production	..	4 886	10 487	12 845	21 286	21 056	21 082	19 781 e
+ Imports	..	-	-	37	279	324	279	243 e
- Exports	..	3 280	4 783	7 350	13 035	12 540	11 393	10 648 e
- Direct Use	..	550	607	3 557	6 209	5 549	6 113	6 458 e
+ Stock Change	..	- 291	- 328	- 131	107	- 152	- 216	- 35 e
- Statistical Difference	..	171	-1 120	- 565	- 743	- 177	831	- 32 e
Refinery Intake	**..**	**594**	**5 889**	**2 409**	**3 171**	**3 316**	**2 808**	**2 915 e**
Closing Stock Level	..	*1 112*	*1 156*	*989*	*1 482*	*1 634*	*1 850*	*1 885 e*
REFINERY FEEDSTOCKS								
Backflows	..	-	-	-	-	-	-	-
+ Transfers	..	-	-	5 882	5 962	5 040	5 429	5 540 e
+ Imports	..	-	-	-	-	-	-	-
- Exports	..	-	-	-	-	-	-	-
+ Stock Change	..	-	-	-	-	-	-	-
- Statistical Difference	..	-	-	-	-	-	-	-
Refinery Intake	**..**	**-**	**-**	**5 882**	**5 962**	**5 040**	**5 429**	**5 540 e**
Closing Stock Level	..	-	-	-	-	-	-	-
OTHER HYDROCARBONS								
Indigenous Production	..	-	6 774	10 616	16 388	16 256	17 640	22 347 e
+ Imports	..	-	-	-	791	763	-	-
- Exports	..	-	-	2 216	4 513	5 191	6 258	11 263 e
- Direct Use	..	-	-	-	791	763	-	-
+ Stock Change	..	-	-	-	-	-	-	-
- Statistical Difference	..	-	-	318	440	443	422	428 e
Refinery Intake	**..**	**-**	**6 774**	**8 082**	**11 435**	**10 622**	**10 960**	**10 656 e**
Closing Stock Level	..	-	-	-	-	-	-	-
CRUDE OIL, NGL, FEEDSTOCKS AND OTHER								
Indigenous Production	94 128	75 080	81 389	91 951	120 298	125 249	127 001	133 121 e
+ Backflows	-	-	-	-	-	-	-	-
+ Transfers	-	-	-	5 882	5 962	5 040	5 429	5 540 e
+ Imports	42 429	31 057	27 308	26 547	41 803	46 185	45 755	43 535 e
- Exports	55 989	16 653	14 939	36 627	69 418	76 156	74 495	79 458 e
- Direct Use	311	558	627	3 562	7 303	6 654	6 363	6 811 e
+ Stock Change	6	673	-1 016	- 649	- 207	- 445	- 848	315 e
- Statistical Difference	-2 733	795	- 793	- 574	- 4	- 90	693	- 898 e
Refinery Intake	**82 996**	**88 804**	**92 908**	**84 116**	**91 139**	**93 309**	**95 786**	**97 140 e**
Closing Stock Level	-	*5 305*	*6 384*	*5 788*	*6 530*	*6 975*	*7 823*	*7 508 e*

Source: Annual Oil Statistics.

CANADA

Table 3: Motor Gasoline - Supply

000 metric tons

	1973	1978	1980	1990	1999	2000	2001	2002
MOTOR GASOLINE								
Refinery Output	**23 838**	**27 006**	**28 293**	**26 179**	**30 107**	**30 981**	**32 222**	**33 405 e**
+ Imports	16	1	129	720	1 565	993	2 003	1 415 e
- Exports	388	639	494	2 841	4 335	4 489	6 508	6 954 e
+ Transfers	-	926	943	858	725	632	927	1 033 e
+ Stock Change	170	274	- 366	277	15	77	- 137	137 e
- Statistical Difference	66	332	215	243	- 91	14	- 20	- 60 e
Total Consumption	**23 570**	**27 236**	**28 290**	**24 950**	**28 168**	**28 180**	**28 527**	**29 096 e**
Closing Stock Level	-	*3 216*	*3 776*	*2 724*	*1 929*	*1 852*	*1 989*	*1 852 e*

Table 4: Aviation Fuels - Supply

000 metric tons

	1973	1978	1980	1990	1999	2000	2001	2002
AVIATION GASOLINE								
Refinery Output	**160**	**165**	**166**	**94**	**90**	**93**	**88**	**87 e**
+ Imports	13	-	-	15	-	-	-	-
- Exports	-	-	-	1	15	17	13	14 e
+ Transfers	-	7	9	- 2	-	- 1	- 1	-
+ Stock Change	- 2	9	- 3	11	- 1	1	1	1 e
- Statistical Difference	3	6	3	1	- 2	-	-	- 1 e
Total Consumption	**168**	**175**	**169**	**116**	**76**	**76**	**75**	**75 e**
Closing Stock Level	-	*54*	*61*	*27*	*17*	*16*	*15*	*14 e*
GASOLINE TYPE JET FUEL								
Refinery Output	**-**	**-**	**1 431**	**742**	**250**	**309**	**259**	**203 e**
+ Imports	-	-	-	-	2	-	-	-
- Exports	-	-	34	150	-	1	-	8 e
+ Transfers	-	-	25	- 49	2	- 49	- 6	- 1 e
+ Stock Change	-	-	- 10	15	18	- 6	8	4 e
- Statistical Difference	-	-	74	2	- 2	-	-	- 22 e
Total Consumption	**-**	**-**	**1 332**	**556**	**274**	**253**	**261**	**220 e**
Closing Stock Level	-	-	*239*	*161*	*45*	*51*	*43*	*39 e*
KEROSENE TYPE JET FUEL								
Refinery Output	**2 608**	**3 443**	**2 287**	**3 592**	**4 460**	**4 114**	**3 983**	**3 703 e**
+ Imports	275	24	86	633	773	1 013	796	1 205 e
- Exports	211	315	116	651	315	167	248	292 e
+ Transfers	-	-	- 83	- 50	- 237	- 173	- 188	- 168 e
+ Stock Change	- 33	40	- 25	- 103	46	26	- 8	27 e
- Statistical Difference	20	25	- 78	- 27	- 12	12	- 14	- 107 e
Total Consumption	**2 619**	**3 167**	**2 227**	**3 448**	**4 739**	**4 801**	**4 349**	**4 582 e**
Closing Stock Level	-	*251*	*306*	*409*	*286*	*260*	*268*	*241 e*

Source: Annual Oil Statistics.

CANADA
Table 5A: Gas/Diesel Oil - Supply and Consumption

000 metric tons

	1973	1978	1980	1990	1999	2000	2001	2002
Refinery Output	**22 731**	**22 961**	**24 962**	**22 670**	**26 685**	**28 198**	**28 722**	**28 919 e**
+ Recycled Products	-	-	-	-	-	-	-	-
+ Imports	695	72	83	676	871	836	782	454 e
- Exports	240	1 245	1 788	3 955	5 729	5 374	6 288	6 467 e
- Intl. Marine Bunkers	-	270	223	222	177	182	169	157 e
+ Transfers	-	1 435	1 666	1 176	929	781	918	357 e
+ Stock Changes	- 925	1 209	- 226	- 652	468	137	- 299	181 e
- Statistical Differences	- 832	- 104	- 175	- 99	- 150	50	102	- 373 e
Total Consumption	**23 093**	**24 266**	**24 649**	**19 792**	**23 197**	**24 346**	**23 564**	**23 660 e**
Transformation	**309**	**352**	**372**	**256**	**161**	**144**	**149**	..
Public Electricity	309	232	289	230	100	94	101	..
Autoproducer Electricity	-	120	83	26	61	50	48	..
Public CHP Plants	-	-	-	-	-	-	-	..
Autoproducers of CHP	-	-	-	-	-	-	-	..
Public Heat Plants	-	-	-	-	-	-	-	..
Autoproducer Heat Plants	-	-	-	-	-	-	-	..
Petrochemical Industry	-	-	-	-	-	-	-	..
Other Transformation	-	-	-	-	-	-	-	..
Energy Sector	**80**	**132**	**88**	**27**	**21**	**28**	**20**	..
Refinery Fuel	80	132	88	27	21	28	20	..
Other Energy Sector	-	-	-	-	-	-	-	..
Final Consumption	**22 704**	**23 782**	**24 189**	**19 509**	**23 015**	**24 174**	**23 395**	..
Transport	**4 532**	**6 535**	**7 583**	**8 199**	**11 238**	**11 332**	**10 970**	..
Air	-	-	-	-	-	-	-	..
Road	1 642	3 677	4 608	5 388	8 579	8 621	8 298	..
Rail	1 957	1 984	2 095	1 989	1 802	1 847	1 815	..
Inland Waterways	933	867	854	809	837	846	847	..
Non-Specified	-	7	26	13	20	18	10	..
Industry	**4 472**	**3 637**	**3 990**	**2 474**	**3 080**	**3 206**	**3 011**	..
Iron and Steel	63	24	37	29	19	23	13	..
Chemical (incl. Pet.-Chem.)	59	13	32	8	5	5	12	..
Non-Ferrous Metals	84	18	11	12	27	8	3	..
Non-Metallic Minerals	128	29	54	12	5	5	3	..
Transport Equipment	105	-	56	-	-	-	-	..
Machinery	22	-	26	-	-	-	-	..
Mining and Quarrying	439	741	1 118	846	1 173	1 239	1 089	..
Food and Tobacco	223	-	242	-	-	-	-	..
Pulp, Paper and Printing	112	-	344	234	150	143	114	..
Wood and Wood Products	128	597	640	165	322	353	391	..
Construction	-	1 080	1 047	857	697	729	709	..
Textiles and Leather	-	-	24	-	-	-	-	..
Non-Specified	3 109	1 135	359	311	682	701	677	..
Other Sectors	**13 700**	**13 610**	**12 616**	**8 836**	**8 697**	**9 636**	**9 414**	..
Commerce and Public	2 742	3 489	2 788	3 276	3 499	4 366	4 634	..
Residential	10 012	8 905	8 530	3 731	2 615	2 684	2 514	..
Agriculture	946	1 216	1 298	1 829	2 583	2 586	2 266	..
Non-Specified	-	-	-	-	-	-	-	..
Memo: Non-Energy Use	-	-	-	-	-	-	-	..
Chem/Petchem. Feedst.	-	-	-	-	-	-	-	..
Other	-	-	-	-	-	-	-	..
Closing Stock Level	-	*4 902*	*5 543*	*3 152*	*2 242*	*2 105*	*2 404*	*2 223 e*

Source: Annual Oil Statistics.

CANADA
Table 5B: Fuel Oil (Residual) - Supply and Consumption

000 metric tons

	1973	1978	1980	1990	1999	2000	2001	2002
Refinery Output	**19 887**	**18 256**	**17 159**	**8 223**	**6 402**	**6 564**	**7 319**	**6 891 e**
+ Recycled Products	-	-	-	-	-	-	-	-
+ Imports	3 539	1 775	1 060	3 802	1 889	1 906	3 204 e	2 435 e
- Exports	4 752	3 914	2 572	2 299	1 561	1 495	2 402	2 851 e
- Intl. Marine Bunkers	-	1 530	1 303	706	888	899	897 e	994 e
+ Transfers	-	105	74	363	206	610	572	937 e
+ Stock Changes	152	365	320	11	225	- 171	39	- 31 e
- Statistical Differences	-1 338	- 780	- 23	- 8	-	4	- 2	- 657 e
Total Consumption	**20 164**	**15 837**	**14 761**	**9 402**	**6 273**	**6 511**	**7 837**	**7 044 e**
Transformation	**2 306**	**3 091**	**2 918**	**3 306**	**2 147**	**1 916**	**3 270**	..
Public Electricity	2 306	2 875	2 442	3 130	1 957	1 751	3 073	..
Autoproducer Electricity	-	216	193	99	76	71	89	..
Public CHP Plants	-	-	283	77	114	94	108	..
Autoproducers of CHP	-	-	-	-	-	-	-	..
Public Heat Plants	-	-	-	-	-	-	-	..
Autoproducer Heat Plants	-	-	-	-	-	-	-	..
Petrochemical Industry	-	-	-	-	-	-	-	..
Other Transformation	-	-	-	-	-	-	-	..
Energy Sector	**2 455**	**1 792**	**1 686**	**914**	**820**	**899**	**667**	..
Refinery Fuel	2 455	1 792	1 686	914	820	899	667	..
Other Energy Sector	-	-	-	-	-	-	-	..
Final Consumption	**15 403**	**10 954**	**10 157**	**5 182**	**3 306**	**3 696**	**3 900**	..
Transport	**2 127**	**700**	**1 026**	**667**	**616**	**649**	**773**	..
Air	-	-	-	-	-	-	-	..
Road	-	-	-	-	-	-	-	..
Rail	188	-	-	-	-	-	-	..
Inland Waterways	1 939	700	1 026	667	616	649	773	..
Non-Specified	-	-	-	-	-	-	-	..
Industry	**9 383**	**8 160**	**7 323**	**4 208**	**2 273**	**2 549**	**2 461**	..
Iron and Steel	443	961	807	265	189	172	164	..
Chemical (incl. Pet.-Chem.)	1 345	960	530	106	178	88	130	..
Non-Ferrous Metals	489	520	602	193	97	165	223	..
Non-Metallic Minerals	1 162	317	361	54	68	49	56	..
Transport Equipment	141	-	116	-	-	-	-	..
Machinery	37	-	19	-	-	-	-	..
Mining and Quarrying	889	743	834	340	275	343	349	..
Food and Tobacco	518	-	393	-	-	-	-	..
Pulp, Paper and Printing	3 827	3 150	3 072	2 229	1 308	1 295	1 398	..
Wood and Wood Products	62	-	-	6	7	8	15	..
Construction	-	169	75	31	10	10	11	..
Textiles and Leather	-	-	248	-	-	-	-	..
Non-Specified	470	1 340	266	984	141	419	115	..
Other Sectors	**3 893**	**2 094**	**1 808**	**307**	**417**	**498**	**666**	..
Commerce and Public	3 403	1 719	1 619	259	387	452	611	..
Residential	490	363	170	35	18	22	11	..
Agriculture	-	12	19	13	12	24	44	..
Non-Specified	-	-	-	-	-	-	-	..
Memo: Non-Energy Use	*245*	-	-	-	-	-	-	..
Chem/Petchem. Feedst.	*245*	-	-	-	-	-	-	..
Other	-	-	-	-	-	-	-	..
Closing Stock Level	-	*1 982*	*1 736*	*765*	*398*	*569*	*530*	*561 e*

Source: Annual Oil Statistics.

CANADA
Table 5C: Total Products - Supply and Consumption

000 metric tons

	1973	1978	1980	1990	1999	2000	2001	2002
Refinery Output	**82 322**	**88 487**	**92 595**	**84 184**	**91 583**	**93 712**	**96 413**	**98 193 e**
+ Recycled Products	344	185	468	1 203	2 527	2 302	2 398	2 951 e
+ Imports	5 579	3 046	2 612	7 632	8 049	7 557	10 526	9 288 e
- Exports	5 926	6 671	6 214	11 627	14 423	14 567	18 815	20 268 e
- Intl. Marine Bunkers	-	1 800	1 526	928	1 065	1 081	1 066	1 151 e
+ Transfers	311	1 379	897	-2 320	1 341	1 614	934	1 271 e
+ Stock Changes	- 987	2 410	- 512	- 282	926	20	- 484	116 e
- Statistical Differences	-2 246	- 463	245	- 513	- 326	- 158	- 249	-1 798 e
Total Consumption	**83 889**	**87 499**	**88 075**	**78 375**	**89 264**	**89 715**	**90 155**	**92 198 e**
Transformation	**2 615**	**3 443**	**3 290**	**3 689**	**3 251**	**3 003**	**3 655**	**..**
Public Electricity	2 615	3 107	2 731	3 375	2 865	2 655	3 320	..
Autoproducer Electricity	-	336	276	237	272	254	227	..
Public CHP Plants	-	-	283	77	114	94	108	..
Autoproducers of CHP	-	-	-	-	-	-	-	..
Public Heat Plants	-	-	-	-	-	-	-	..
Autoproducer Heat Plants	-	-	-	-	-	-	-	..
Petrochemical Industry	-	-	-	-	-	-	-	..
Other Transformation	-	-	-	-	-	-	-	..
Energy Sector	**5 726**	**6 669**	**6 650**	**6 914**	**8 158**	**8 242**	**8 545**	**..**
Refinery Fuel	5 606	6 661	6 630	5 676	6 297	6 392	6 796	..
Other Energy Sector	120	8	20	1 238	1 861	1 850	1 749	..
Final Consumption	**75 548**	**77 387**	**78 135**	**67 772**	**77 855**	**78 470**	**77 955**	**..**
Transport	**33 024**	**37 794**	**40 622**	**38 904**	**45 781**	**45 804**	**45 492**	**..**
Air	2 784	3 340	3 728	4 120	5 089	5 130	4 685	..
Road	25 187	30 896	32 893	31 033	37 192	37 113	37 150	..
Rail	2 174	1 984	2 095	1 989	1 802	1 847	1 815	..
Inland Waterways	2 879	1 567	1 880	1 476	1 453	1 495	1 620	..
Non-Specified	-	7	26	286	245	219	222	..
Industry	**21 261**	**20 982**	**20 603**	**17 696**	**20 370**	**19 692**	**19 482**	**..**
Iron and Steel	524	985	844	294	208	195	177	..
Chemical (incl. Pet.-Chem.)	3 321	4 406	4 064	5 742	9 212	8 317	8 531	..
Non-Ferrous Metals	1 256	1 524	1 704	1 387	1 504	1 350	1 356	..
Non-Metallic Minerals	1 307	346	572	309	382	317	347	..
Transport Equipment	263	-	175	-	-	-	-	..
Machinery	63	-	46	-	-	-	-	..
Mining and Quarrying	1 425	1 544	2 001	1 409	1 831	2 135	2 018	..
Food and Tobacco	808	-	651	-	-	-	-	..
Pulp, Paper and Printing	3 973	3 150	3 416	2 463	1 458	1 438	1 512	..
Wood and Wood Products	229	597	646	190	355	384	427	..
Construction	3 226	4 486	4 397	3 635	3 711	3 321	3 293	..
Textiles and Leather	-	-	273	-	-	-	-	..
Non-Specified	4 866	3 944	1 814	2 267	1 709	2 235	1 821	..
Other Sectors	**21 263**	**18 611**	**16 910**	**11 172**	**11 704**	**12 974**	**12 981**	**..**
Commerce and Public	6 813	5 817	5 430	4 652	5 859	7 045	7 600	..
Residential	12 308	10 395	10 033	4 527	3 062	3 163	2 912	..
Agriculture	946	1 228	1 447	1 993	2 783	2 766	2 469	..
Non-Specified	1 196	1 171	-	-	-	-	-	..
Memo: Non-Energy Use	*7 691*	*9 248*	*9 296*	*11 241*	*15 723*	*14 704*	*14 809*	..
Chem/Petchem. Feedst.	*2 146*	*3 433*	*3 502*	*5 628*	*9 029*	*8 224*	*8 389*	..
Other	*5 545*	*5 815*	*5 794*	*5 613*	*6 694*	*6 480*	*6 420*	..
Closing Stock Level	-	*13 338*	*14 866*	*9 716*	*7 572*	*7 552*	*8 036*	*7 920 e*

Source: Annual Oil Statistics.

CANADA
Table 6A: Supply and Consumption

000 metric tons

	1980	2000	2001	2002	1980	2000	2001	2002
	LPG AND ETHANE				**REFINERY GAS**			
Refinery Output	**1 932**	**1 874**	**1 904**	**1 952 e**	**3 749**	**3 656**	**3 948**	**4 545 e**
+ Recycled Products	-	-	-	-	-	-	-	-
+ Imports	-	91	135	220 e	-	-	-	-
- Exports	199	208	320	366 e	-	-	-	-
- Intl. Marine Bunkers	-	-	-	-	-	-	-	-
+ Transfers	221	5 019	5 482	5 737 e	16	-	-	-
+ Stock Change	- 6	18	- 37	53 e	2	-	-	-
- Statistical Difference	-	- 167	- 239	- 215 e	214	-	-	-
Total Consumption	**1 948**	**6 961**	**7 403**	**7 811 e**	**3 553**	**3 656**	**3 948**	**4 545 e**
of which: Ethane	-	*4 169*	*4 467*	*5 057 e*	-	-	-	-
Transformation	-	-	-	..	-	-	-	..
Energy	275	170	111	..	3 553	3 656	3 948	..
Transport	6	314	326	..	-	-	-	..
Industry	858	5 458	5 919	..	-	-	-	..
Other	809	1 019	1 047	..	-	-	-	..
Memo: Non-Energy Use	*607*	*4 848*	*5 245*	..	-	-	-	..
Closing Stock Level	*77*	*110*	*147*	*94 e*	*4*	-	-	-
	NAPHTHA				**OTHER KEROSENE**			
Refinery Output	**3 424**	**3 740**	**3 453**	**3 854 e**	**3 154**	**1 487**	**1 547**	**1 428 e**
+ Recycled Products	281	216	282	411 e	-	-	-	-
+ Imports	41	5	-	4 e	1	27	13	23 e
- Exports	750	404	351	447 e	29	111	148	115 e
- Intl. Marine Bunkers	-	-	-	-	-	-	-	-
+ Transfers	- 272	- 140	- 230	- 431 e	-1 743	-1 036	-1 071	-1 132 e
+ Stock Change	- 95	- 28	-	79 e	- 4	17	- 36	38 e
- Statistical Difference	- 379	8	5	44 e	113	2	-	- 71 e
Total Consumption	**3 008**	**3 381**	**3 149**	**3 426 e**	**1 266**	**382**	**305**	**313 e**
Transformation	-	-	-	..	-	-	-	..
Energy	113	5	5	..	-	-	-	..
Transport	-	-	-	..	-	-	-	..
Industry	2 895	3 376	3 144	..	93	19	22	..
Other	-	-	-	..	1 173	363	283	..
Memo: Non-Energy Use	*2 895*	*3 376*	*3 144*	..	-	-	-	..
Closing Stock Level	*350*	*335*	*335*	*256 e*	*600*	*102*	*100*	*100 e*
	WHITE SPIRIT + SBP				**LUBRICANTS**			
Refinery Output	**564**	**112**	**123**	**..**	**844**	**1 209**	**1 140**	**..**
+ Recycled Products	26	7	7	..	47	242	233	..
+ Imports	7	-	-	..	88	261	241	..
- Exports	166	-	-	..	20	555	554	..
- Intl. Marine Bunkers	-	-	-	..	-	-	-	..
+ Transfers	- 15	35	- 29	..	8	- 208	- 174	..
+ Stock Change	- 5	2	- 3	..	- 47	7	4	..
- Statistical Difference	6	-	-	..	- 13	- 9	- 13	..
Total Consumption	**405**	**156**	**98**	**..**	**933**	**965**	**903**	**..**
Transformation	-	-	-	..	-	-	-	..
Energy	2	-	-	..	3	4	3	..
Transport	-	-	-	..	-	201	212	..
Industry	-	-	-	..	930	434	388	..
Other	403	156	98	..	-	326	300	..
Memo: Non-Energy Use	*403*	*156*	*98*	..	*930*	*961*	*900*	..
Closing Stock Level	*75*	*15*	*18*	..	*347*	*343*	*339*	..

Source: Annual Oil Statistics.

CANADA
Table 6B: Supply and Consumption

000 metric tons

	1980	2000	2001	2002	1980	2000	2001	2002
		BITUMEN				PARAFFIN WAXES		
Refinery Output	**3 447**	**4 411**	**4 284**	**..**	**-**	**-**	**-**	**..**
+ Recycled Products	-	-	-	..	-	-	-	..
+ Imports	19	212	254	..	-	-	-	..
- Exports	46	1 268	1 428	..	-	-	-	..
- Intl. Marine Bunkers	-	-	-	..	-	-	-	..
+ Transfers	- 4	744	111	..	-	-	-	..
+ Stock Change	- 121	- 57	22	..	-	-	-	..
- Statistical Difference	26	4	1	..	-	-	-	..
Total Consumption	**3 269**	**4 038**	**3 242**	**..**	**-**	**-**	**-**	**..**
Transformation	-	763	-	..	-	-	-	..
Energy	-	-	-	..	-	-	-	..
Transport	-	-	-	..	-	-	-	..
Industry	3 269	3 210	3 119	..	-	-	-	..
Other	-	65	123	..	-	-	-	..
Memo: Non-Energy Use	*3 269*	*3 275*	*3 242*	..	-	-	-	..
Closing Stock Level	*479*	*338*	*316*	..	-	-	-	..
		PETROLEUM COKE				OTHER PRODUCTS		
Refinery Output	**1 156**	**3 846**	**4 222**	**4 320 e**	**27**	**3 118**	**3 199**	**8 886 e**
+ Recycled Products	-	-	-	-	114	1 837	1 876	2 540 e
+ Imports	1 022	917	1 058	1 151 e	76	1 296	2 040	2 381 e
- Exports	-	100	169	136 e	-	378	386	2 618 e
- Intl. Marine Bunkers	-	-	-	-	-	-	-	-
+ Transfers	-	-	1	5 e	52	-4 600	-5 378	-5 066 e
+ Stock Change	3	10	19	8 e	77	- 13	- 57	- 381 e
- Statistical Difference	41	- 83	- 63	- 268 e	221	7	- 6	- 68 e
Total Consumption	**2 140**	**4 756**	**5 194**	**5 616 e**	**125**	**1 253**	**1 300**	**5 810 e**
Transformation	-	180	236	..	-	-	-	..
Energy	895	3 136	3 540	..	24	342	250	..
Transport	-	-	-	..	-	-	-	..
Industry	1 245	1 440	1 418	..	-	-	-	..
Other	-	-	-	..	101	911	1 050	..
Memo: Non-Energy Use	*1 091*	*1 177*	*1 130*	..	*101*	*911*	*1 050*	..
Closing Stock Level	*5*	*55*	*36*	*28 e*	*1 180*	*1 401*	*1 458*	*2 512 e*

Source: Annual Oil Statistics.

CANADA
Table 7A: Imports by Country of Origin

000 metric tons

	2000	2001	2002	2000	2001	2002	2000	2001	2002
	Crude Oil			Natural Gas Liquids			Feedstocks and Other		
Australia	541	105	-	-	-	-	-	-	-
Belgium	-	-	-	-	-	-	-	-	-
Canada	-	-	-	-	-	-	-	-	-
Czech Republic	-	-	-	-	-	-	-	-	-
Denmark	-	466	-	-	-	-	-	-	-
Finland	-	22	-	-	-	-	-	-	-
France	-	30	-	-	-	-	-	-	-
Germany	-	-	-	-	-	-	-	-	-
Greece	-	-	-	-	-	-	-	-	-
Hungary	-	-	-	-	-	-	-	-	-
Ireland	-	-	-	-	-	-	-	-	-
Italy	-	32	-	-	-	-	-	-	-
Japan	-	-	-	-	-	-	-	-	-
Korea	-	-	-	-	-	-	-	-	-
Mexico	1 373	1 235	1 109 e	-	-	-	-	-	-
Netherlands	40	-	-	-	-	-	-	-	-
New Zealand	-	-	-	-	-	-	-	-	-
Norway	13 010	11 995	12 245 e	-	-	-	-	-	-
Poland	-	-	-	-	-	-	-	-	-
Spain	61	-	-	-	-	-	-	-	-
Sweden	-	16	-	-	-	-	-	-	-
Turkey	-	6	-	-	-	-	-	-	-
United Kingdom	13 101	13 857	10 421 e	-	-	-	-	-	-
United States	1 217	1 627	955 e	324	279	243 e	-	-	-
Other	128	-	-	-	-	-	-	-	-
OECD	**29 471**	**29 391**	**24 730 e**	**324**	**279**	**243 e**	**-**	**-**	**-**
Argentina	-	-	-	-	-	-	-	-	-
Brazil	-	-	-	-	-	-	-	-	-
Colombia	-	117	-	-	-	-	-	-	-
Ecuador	-	-	-	-	-	-	-	-	-
Netherlands Antilles	-	-	-	-	-	-	-	-	-
Peru	-	-	-	-	-	-	-	-	-
Trinidad and Tobago	-	10	-	-	-	-	-	-	-
Venezuela	4 801	4 245	3 154 e	-	-	-	763	-	-
Other	-	-	-	-	-	-	-	-	-
Latin America	**4 801**	**4 372**	**3 154 e**	**-**	**-**	**-**	**763**	**-**	**-**
Romania	-	-	-	-	-	-	-	-	-
Other	-	-	-	-	-	-	-	-	-
Non-OECD Europe	**-**	**-**	**-**	**-**	**-**	**-**	**-**	**-**	**-**
FSU	**777**	**114**	**-**	**-**	**-**	**-**	**-**	**-**	**-**
Iran	-	-	-	-	-	-	-	-	-
Iraq	2 317	3 786	4 245 e	-	-	-	-	-	-
Kuwait	-	-	-	-	-	-	-	-	-
Oman	-	-	-	-	-	-	-	-	-
Qatar	-	-	-	-	-	-	-	-	-
Saudi Arabia	2 908	2 849	2 717 e	-	-	-	-	-	-
United Arab Emirates	-	-	-	-	-	-	-	-	-
Other	114	112	-	-	-	-	-	-	-
Middle East	**5 339**	**6 747**	**6 962 e**	**-**	**-**	**-**	**-**	**-**	**-**
India	-	-	-	-	-	-	-	-	-
Indonesia	-	-	-	-	-	-	-	-	-
Singapore	-	-	-	-	-	-	-	-	-
Other	-	-	-	-	-	-	-	-	-
Asia	**-**	**-**	**-**	**-**	**-**	**-**	**-**	**-**	**-**
China	**-**	**-**	**-**	**-**	**-**	**-**	**-**	**-**	**-**
Algeria	3 047	3 438	4 811 e	-	-	-	-	-	-
Angola/Cabinada	-	176	-	-	-	-	-	-	-
Cameroon	212	36	-	-	-	-	-	-	-
Egypt	-	-	-	-	-	-	-	-	-
Gabon	-	-	-	-	-	-	-	-	-
Libya	-	-	-	-	-	-	-	-	-
Nigeria	1 451	814	883 e	-	-	-	-	-	-
Tunisia	-	-	-	-	-	-	-	-	-
Other	-	388	-	-	-	-	-	-	-
Africa	**4 710**	**4 852**	**5 694 e**	**-**	**-**	**-**	**-**	**-**	**-**
Non-spec./Other	**-**	**-**	**2 752 e**	**-**	**-**	**-**	**-**	**-**	**-**
Total Imports	**45 098**	**45 476**	**43 292 e**	**324**	**279**	**243 e**	**763**	**-**	**-**
Memo: Total OPEC	*14 524*	*15 132*	*15 810 e*	-	-	-	*763*	-	-

Source: Annual Oil Statistics.

CANADA
Table 7B: Imports by Country of Origin

000 metric tons

	2000	2001	2002	2000	2001	2002	2000	2001	2002
	Total Products			Liquefied Petroleum Gases			Naphtha		
Australia	-	-	-	-	-	-	-	-	-
Belgium	6	66	106 e	-	-	-	-	-	-
Canada	-	2	-	-	-	-	-	-	-
Czech Republic	-	-	-	-	-	-	-	-	-
Denmark	-	-	51 e	-	-	-	-	-	-
Finland	102	242	244 e	-	-	-	-	-	-
France	49	249	328 e	-	-	-	-	-	-
Germany	6	32	29 e	-	-	-	-	-	-
Greece	-	-	-	-	-	-	-	-	-
Hungary	-	-	-	-	-	-	-	-	-
Ireland	-	-	-	-	-	-	-	-	-
Italy	66	365	27 e	-	-	-	-	-	-
Japan	88	39	1 e	-	-	-	-	-	-
Korea	161	203	48 e	-	-	-	-	-	-
Mexico	-	1	2 e	-	-	-	-	-	-
Netherlands	210	232	163 e	-	-	-	-	-	-
New Zealand	-	-	-	-	-	-	-	-	-
Norway	37	139	193 e	-	-	-	-	-	-
Poland	-	-	-	-	-	-	-	-	-
Spain	360	169	101 e	-	-	-	-	-	-
Sweden	11	14	145 e	-	-	-	-	-	-
Turkey	-	57	-	-	-	-	-	-	-
United Kingdom	225	437	316 e	-	-	-	-	-	-
United States	5 008	5 850	6 391 e	91	135	220 e	5	-	4 e
Other	-	-	-	-	-	-	-	-	-
OECD	**6 329**	**8 097**	**8 145 e**	**91**	**135**	**220 e**	**5**	**-**	**4 e**
Argentina	-	-	-	-	-	-	-	-	-
Brazil	95	19	25 e	-	-	-	-	-	-
Colombia	24	46	28 e	-	-	-	-	-	-
Ecuador	-	92	-	-	-	-	-	-	-
Netherlands Antilles	-	116	79 e	-	-	-	-	-	-
Peru	-	25	-	-	-	-	-	-	-
Trinidad and Tobago	73	310	-	-	-	-	-	-	-
Venezuela	305	1 121	733 e	-	-	-	-	-	-
Other	-	-	-	-	-	-	-	-	-
Latin America	**497**	**1 729**	**865 e**	**-**	**-**	**-**	**-**	**-**	**-**
Romania	-	-	-	-	-	-	-	-	-
Other	-	62	-	-	-	-	-	-	-
Non-OECD Europe	**-**	**62**	**-**	**-**	**-**	**-**	**-**	**-**	**-**
FSU	**94**	**58**	**51 e**	**-**	**-**	**-**	**-**	**-**	**-**
Iran	-	-	-	-	-	-	-	-	-
Iraq	-	-	-	-	-	-	-	-	-
Kuwait	-	44	17 e	-	-	-	-	-	-
Oman	-	-	-	-	-	-	-	-	-
Qatar	-	-	-	-	-	-	-	-	-
Saudi Arabia	149	34	26 e	-	-	-	-	-	-
United Arab Emirates	-	35	-	-	-	-	-	-	-
Other	138	144	-	-	-	-	-	-	-
Middle East	**287**	**257**	**43 e**	**-**	**-**	**-**	**-**	**-**	**-**
India	-	-	-	-	-	-	-	-	-
Indonesia	-	-	-	-	-	-	-	-	-
Singapore	79	-	-	-	-	-	-	-	-
Other	-	35	17 e	-	-	-	-	-	-
Asia	**79**	**35**	**17 e**	**-**	**-**	**-**	**-**	**-**	**-**
China	**-**	**68**	**9 e**	**-**	**-**	**-**	**-**	**-**	**-**
Algeria	151	72	61 e	-	-	-	-	-	-
Angola/Cabinada	-	-	-	-	-	-	-	-	-
Cameroon	-	-	-	-	-	-	-	-	-
Egypt	-	-	32 e	-	-	-	-	-	-
Gabon	22	-	-	-	-	-	-	-	-
Libya	-	-	-	-	-	-	-	-	-
Nigeria	51	148	58 e	-	-	-	-	-	-
Tunisia	-	-	-	-	-	-	-	-	-
Other	-	-	-	-	-	-	-	-	-
Africa	**224**	**220**	**151 e**	**-**	**-**	**-**	**-**	**-**	**-**
Non-spec./Other	**47**	**-**	**7 e**	**-**	**-**	**-**	**-**	**-**	**-**
Total Imports	**7 557**	**10 526**	**9 288 e**	**91**	**135**	**220 e**	**5**	**-**	**4 e**
Memo: Total OPEC	*656*	*1 454*	*895 e*	-	-	-	-	-	-

Source: Annual Oil Statistics.

CANADA
Table 7C: Imports by Country of Origin

000 metric tons

	2000	2001	2002	2000	2001	2002	2000	2001	2002
	Total Gasoline			**Total Kerosene**			**Gas/Diesel**		
Australia	-	-	-	-	-	-	-	-	-
Belgium	-	42	-	6	13	-	-	-	-
Canada	-	-	-	-	-	-	-	-	-
Czech Republic	-	-	-	-	-	-	-	-	-
Denmark	-	-	-	-	-	-	-	-	-
Finland	102	116	244 e	-	-	-	-	82	-
France	28	200	291 e	-	-	-	-	-	-
Germany	-	-	-	-	-	-	-	-	3 e
Greece	-	-	-	-	-	-	-	-	-
Hungary	-	-	-	-	-	-	-	-	-
Ireland	-	-	-	-	-	-	-	-	-
Italy	-	47	-	-	-	-	-	-	-
Japan	-	38	-	82	-	-	-	-	-
Korea	-	104	-	160	99	48 e	-	-	-
Mexico	-	-	-	-	-	-	-	-	-
Netherlands	70	226	117 e	19	6	-	32	-	-
New Zealand	-	-	-	-	-	-	-	-	-
Norway	37	139	58 e	-	-	-	-	-	-
Poland	-	-	-	-	-	-	-	-	-
Spain	185	33	101 e	-	-	-	-	-	-
Sweden	-	-	-	-	1	-	-	-	-
Turkey	-	-	-	-	-	-	-	-	-
United Kingdom	184	360	151 e	-	-	-	-	-	-
United States	292	545	378 e	654	460	788 e	804	594	451 e
Other	-	-	-	-	-	-	-	-	-
OECD	**898**	**1 850**	**1 340 e**	**921**	**579**	**836 e**	**836**	**676**	**454 e**
Argentina	-	-	-	-	-	-	-	-	-
Brazil	95	19	-	-	-	-	-	-	-
Colombia	-	-	-	-	-	-	-	-	-
Ecuador	-	-	-	-	-	-	-	-	-
Netherlands Antilles	-	-	-	-	-	-	-	-	-
Peru	-	-	-	-	-	-	-	-	-
Trinidad and Tobago	-	-	-	-	-	-	-	-	-
Venezuela	-	-	-	27	50	314 e	-	-	-
Other	-	-	-	-	-	-	-	-	-
Latin America	**95**	**19**	**-**	**27**	**50**	**314 e**	**-**	**-**	**-**
Romania	-	-	-	-	-	-	-	-	-
Other	-	35	-	-	27	-	-	-	-
Non-OECD Europe	**-**	**35**	**-**	**-**	**27**	**-**	**-**	**-**	**-**
FSU	**-**	**40**	**-**	**-**	**16**	**-**	**-**	**-**	**-**
Iran	-	-	-	-	-	-	-	-	-
Iraq	-	-	-	-	-	-	-	-	-
Kuwait	-	-	-	-	44	17 e	-	-	-
Oman	-	-	-	-	-	-	-	-	-
Qatar	-	-	-	-	-	-	-	-	-
Saudi Arabia	-	-	26 e	-	-	-	-	34	-
United Arab Emirates	-	-	-	-	35	-	-	-	-
Other	-	-	-	-	23	-	-	-	-
Middle East	**-**	**-**	**26 e**	**-**	**102**	**17 e**	**-**	**34**	**-**
India	-	-	-	-	-	-	-	-	-
Indonesia	-	-	-	-	-	-	-	-	-
Singapore	-	-	-	33	-	-	-	-	-
Other	-	-	17 e	-	35	-	-	-	-
Asia	**-**	**-**	**17 e**	**33**	**35**	**-**	**-**	**-**	**-**
China	**-**	**59**	**-**	**-**	**-**	**-**	**-**	**-**	**-**
Algeria	-	-	-	59	-	61 e	-	72	-
Angola/Cabinada	-	-	-	-	-	-	-	-	-
Cameroon	-	-	-	-	-	-	-	-	-
Egypt	-	-	32 e	-	-	-	-	-	-
Gabon	-	-	-	-	-	-	-	-	-
Libya	-	-	-	-	-	-	-	-	-
Nigeria	-	-	-	-	-	-	-	-	-
Tunisia	-	-	-	-	-	-	-	-	-
Other	-	-	-	-	-	-	-	-	-
Africa	**-**	**-**	**32 e**	**59**	**-**	**61 e**	**-**	**72**	**-**
Non-spec./Other	**-**	**-**	**-**	**-**	**-**	**-**	**-**	**-**	**-**
Total Imports	**993**	**2 003**	**1 415 e**	**1 040**	**809**	**1 228 e**	**836**	**782**	**454 e**
Memo: Total OPEC	-	-	*26 e*	*86*	*129*	*392 e*	-	*106*	-

Source: Annual Oil Statistics.

CANADA
Table 7D: Imports by Country of Origin

000 metric tons

	2000	2001	2002	2000	2001	2002	2000	2001	2002
	Fuel Oil (Residual)			**Petroleum Coke**			**Other Products**		
Australia	-	-	-	-	-	-	-	-	-
Belgium	-	-	-	-	-	-	-	11	106 e
Canada	-	-	-	-	-	-	-	2	-
Czech Republic	-	-	-	-	-	-	-	-	-
Denmark	-	-	47 e	-	-	-	-	-	4 e
Finland	-	-	-	-	-	-	-	44	-
France	-	-	-	20	8	23 e	1	41	14 e
Germany	-	-	-	-	25	22 e	6	7	4 e
Greece	-	-	-	-	-	-	-	-	-
Hungary	-	-	-	-	-	-	-	-	-
Ireland	-	-	-	-	-	-	-	-	-
Italy	-	-	-	-	-	27 e	66	318	-
Japan	-	-	-	-	-	-	6	1	1 e
Korea	-	-	-	-	-	-	1	-	-
Mexico	-	-	-	-	-	-	-	1	2 e
Netherlands	46	-	-	-	-	-	43	-	46 e
New Zealand	-	-	-	-	-	-	-	-	-
Norway	-	-	-	-	-	-	-	-	135 e
Poland	-	-	-	-	-	-	-	-	-
Spain	131	23	-	-	-	-	44	113	-
Sweden	-	-	-	-	-	-	11	13	145 e
Turkey	-	-	-	-	-	-	-	57	-
United Kingdom	38	-	143 e	-	-	-	3	77	22 e
United States	1 333	1 811 e	1 728 e	875	1 016	1 002 e	954	1 289	1 820 e
Other	-	-	-	-	-	-	-	-	-
OECD	**1 548**	**1 834 e**	**1 918 e**	**895**	**1 049**	**1 074 e**	**1 135**	**1 974**	**2 299 e**
Argentina	-	-	-	-	-	-	-	-	-
Brazil	-	-	-	-	-	25 e	-	-	-
Colombia	24	46	28 e	-	-	-	-	-	-
Ecuador	-	-	-	-	-	-	-	92	-
Netherlands Antilles	-	116	-	-	-	-	-	-	79 e
Peru	-	23	-	-	-	-	-	2	-
Trinidad and Tobago	73	157	-	-	-	-	-	153	-
Venezuela	75	880	376 e	-	-	43 e	203	191	-
Other	-	-	-	-	-	-	-	-	-
Latin America	**172**	**1 222**	**404 e**	**-**	**-**	**68 e**	**203**	**438**	**79 e**
Romania	-	-	-	-	-	-	-	-	-
Other	-	-	-	-	-	-	-	-	-
Non-OECD Europe	**-**	**-**	**-**	**-**	**-**	**-**	**-**	**-**	**-**
FSU	**89**	**-**	**51 e**	**-**	**-**	**-**	**5**	**2**	**-**
Iran	-	-	-	-	-	-	-	-	-
Iraq	-	-	-	-	-	-	-	-	-
Kuwait	-	-	-	-	-	-	-	-	-
Oman	-	-	-	-	-	-	-	-	-
Qatar	-	-	-	-	-	-	-	-	-
Saudi Arabia	-	-	-	-	-	-	149	-	-
United Arab Emirates	-	-	-	-	-	-	-	-	-
Other	-	-	-	-	-	-	138	121	-
Middle East	**-**	**-**	**-**	**-**	**-**	**-**	**287**	**121**	**-**
India	-	-	-	-	-	-	-	-	-
Indonesia	-	-	-	-	-	-	-	-	-
Singapore	46	-	-	-	-	-	-	-	-
Other	-	-	-	-	-	-	-	-	-
Asia	**46**	**-**	**-**	**-**	**-**	**-**	**-**	**-**	**-**
China	**-**	**-**	**-**	**-**	**9**	**9 e**	**-**	**-**	**-**
Algeria	-	-	-	-	-	-	92	-	-
Angola/Cabinada	-	-	-	-	-	-	-	-	-
Cameroon	-	-	-	-	-	-	-	-	-
Egypt	-	-	-	-	-	-	-	-	-
Gabon	-	-	-	22	-	-	-	-	-
Libya	-	-	-	-	-	-	-	-	-
Nigeria	51	148	58 e	-	-	-	-	-	-
Tunisia	-	-	-	-	-	-	-	-	-
Other	-	-	-	-	-	-	-	-	-
Africa	**51**	**148**	**58 e**	**22**	**-**	**-**	**92**	**-**	**-**
Non-spec./Other	**-**	**-**	**4 e**	**-**	**-**	**-**	**47**	**-**	**3 e**
Total Imports	**1 906**	**3 204 e**	**2 435 e**	**917**	**1 058**	**1 151 e**	**1 769**	**2 535**	**2 381 e**
Memo: Total OPEC	*126*	*1 028*	*434 e*	*-*	*-*	*43 e*	*444*	*191*	*-*

Source: Annual Oil Statistics.

CANADA
Table 8A: Exports by Country of Destination

000 metric tons

	2000	2001	2002	2000	2001	2002	2000	2001	2002
	Crude Oil			**Natural Gas Liquids**			**Feedstocks and Other**		
Australia	-	-	-	-	-	-	-	-	-
Austria	-	-	-	-	-	-	-	-	-
Belgium	-	-	-	-	-	-	-	-	-
Canada	-	-	-	-	-	-	-	-	-
Denmark	-	-	-	-	-	-	-	-	-
Finland	-	-	-	-	-	-	-	-	-
France	-	-	-	-	-	-	-	-	-
Germany	-	-	-	-	-	-	-	-	-
Greece	-	-	-	-	-	-	-	-	-
Ireland	-	-	-	-	-	-	-	-	-
Italy	-	-	-	-	-	-	-	-	-
Japan	-	-	-	-	-	-	-	-	-
Luxembourg	-	-	-	-	-	-	-	-	-
Mexico	-	-	-	-	-	-	-	-	-
Netherlands	-	-	-	-	-	-	-	-	-
Norway	-	-	-	-	-	-	-	-	-
Portugal	-	-	-	-	-	-	-	-	-
Spain	-	-	-	-	-	-	-	-	-
Sweden	-	-	-	-	-	-	-	-	-
Switzerland	-	-	-	-	-	-	-	-	-
Turkey	-	-	-	-	-	-	-	-	-
United Kingdom	-	-	-	-	-	-	-	-	-
United States	58 425	56 844	57 547 e	12 540	11 393	10 648 e	5 191	6 258	11 263 e
Other OECD	-	-	-	-	-	-	-	-	-
OECD	**58 425**	**56 844**	**57 547 e**	**12 540**	**11 393**	**10 648 e**	**5 191**	**6 258**	**11 263 e**
Africa	-	-	-	-	-	-	-	-	-
Latin America	-	-	-	-	-	-	-	-	-
Asia	-	-	-	-	-	-	-	-	-
China	-	-	-	-	-	-	-	-	-
Non-OECD Europe	-	-	-	-	-	-	-	-	-
Former Soviet Union	-	-	-	-	-	-	-	-	-
Middle East	-	-	-	-	-	-	-	-	-
Non-specified/Other	-	-	-	-	-	-	-	-	-
Non-OECD	**-**	**-**	**-**	**-**	**-**	**-**	**-**	**-**	**-**
Total Exports	**58 425**	**56 844**	**57 547 e**	**12 540**	**11 393**	**10 648 e**	**5 191**	**6 258**	**11 263 e**
	Total Products			**Liquefied Petroleum Gas**			**Naphtha**		
Australia	3	6	-	-	-	-	-	-	-
Austria	-	-	-	-	-	-	-	-	-
Belgium	-	2	-	-	-	-	-	-	-
Canada	-	-	-	-	-	-	-	-	-
Denmark	-	-	14 e	-	-	-	-	-	-
Finland	-	-	-	-	-	-	-	-	-
France	-	659	67 e	-	-	-	-	-	-
Germany	1	-	-	-	-	-	-	-	-
Greece	-	-	-	-	-	-	-	-	-
Ireland	-	-	-	-	-	-	-	-	-
Italy	-	1	54 e	-	-	-	-	-	-
Japan	3	3	-	-	-	-	-	-	-
Luxembourg	-	-	-	-	-	-	-	-	-
Mexico	1	3	49 e	-	-	-	-	-	-
Netherlands	4	1	211 e	-	-	-	-	-	-
Norway	-	20	-	-	-	-	-	-	-
Portugal	-	-	-	-	-	-	-	-	-
Spain	-	-	-	-	-	-	-	-	-
Sweden	-	1	-	-	-	-	-	-	-
Switzerland	124	-	134 e	-	-	-	-	-	-
Turkey	-	-	-	-	-	-	-	-	-
United Kingdom	3	17	-	-	-	-	-	-	-
United States	14 361	18 054	19 622 e	208	320	366 e	404	351	447 e
Other OECD	1	1	-	-	-	-	-	-	-
OECD	**14 501**	**18 768**	**20 151 e**	**208**	**320**	**366 e**	**404**	**351**	**447 e**
Africa	3	3	-	-	-	-	-	-	-
Latin America	2	4	-	-	-	-	-	-	-
Asia	1	-	86 e	-	-	-	-	-	-
China	14	7	-	-	-	-	-	-	-
Non-OECD Europe	-	-	-	-	-	-	-	-	-
Former Soviet Union	-	-	-	-	-	-	-	-	-
Middle East	-	1	-	-	-	-	-	-	-
Non-specified/Other	46	32	31 e	-	-	-	-	-	-
Non-OECD	**66**	**47**	**117 e**	**-**	**-**	**-**	**-**	**-**	**-**
Total Exports	**14 567**	**18 815**	**20 268 e**	**208**	**320**	**366 e**	**404**	**351**	**447 e**

Source: Annual Oil Statistics.

CANADA
Table 8B: Exports by Country of Destination

000 metric tons

	2000	2001	2002	2000	2001	2002	2000	2001	2002
	Total Gasoline			**Total Kerosene**			**Gas/Diesel Oil**		
Australia	-	-	-	-	-	-	-	-	-
Austria	-	-	-	-	-	-	-	-	-
Belgium	-	-	-	-	-	-	-	-	-
Canada	-	-	-	-	-	-	-	-	-
Denmark	-	-	-	-	-	-	-	-	14 e
Finland	-	-	-	-	-	-	-	-	-
France	-	-	-	-	-	-	-	659	67 e
Germany	-	-	-	-	-	-	-	-	-
Greece	-	-	-	-	-	-	-	-	-
Ireland	-	-	-	-	-	-	-	-	-
Italy	-	-	-	-	-	-	-	-	2 e
Japan	-	-	-	-	-	-	-	-	-
Luxembourg	-	-	-	-	-	-	-	-	-
Mexico	-	-	-	-	-	-	-	-	-
Netherlands	-	-	-	-	-	53 e	-	-	158 e
Norway	-	-	-	-	-	-	-	-	-
Portugal	-	-	-	-	-	-	-	-	-
Spain	-	-	-	-	-	-	-	-	-
Sweden	-	-	-	-	-	-	-	-	-
Switzerland	-	-	-	-	-	81 e	93	-	53 e
Turkey	-	-	-	-	-	-	-	-	-
United Kingdom	-	-	-	-	-	-	-	-	-
United States	4 507	6 521	6 976 e	278	396	261 e	5 273	5 625	6 125 e
Other OECD	-	-	-	-	-	-	-	-	-
OECD	**4 507**	**6 521**	**6 976 e**	**278**	**396**	**395 e**	**5 366**	**6 284**	**6 419 e**
Africa	-	-	-	-	-	-	-	-	-
Latin America	-	-	-	-	-	-	-	-	-
Asia	-	-	-	-	-	-	-	-	29 e
China	-	-	-	-	-	-	-	-	-
Non-OECD Europe	-	-	-	-	-	-	-	-	-
Former Soviet Union	-	-	-	-	-	-	-	-	-
Middle East	-	-	-	-	-	-	-	-	-
Non-specified/Other	-	-	-	-	-	12 e	8	4	19 e
Non-OECD	**-**	**-**	**-**	**-**	**-**	**12 e**	**8**	**4**	**48 e**
Total Exports	**4 507**	**6 521**	**6 976 e**	**278**	**396**	**407 e**	**5 374**	**6 288**	**6 467 e**

	2000	2001	2002	2000	2001	2002	2000	2001	2002
	Fuel Oil (Residual)			**Petroleum Coke**			**Other Products**		
Australia	-	-	-	-	-	-	3	6	-
Austria	-	-	-	-	-	-	-	-	-
Belgium	-	-	-	-	-	-	-	2	-
Canada	-	-	-	-	-	-	-	-	-
Denmark	-	-	-	-	-	-	-	-	-
Finland	-	-	-	-	-	-	-	-	-
France	-	-	-	-	-	-	-	-	-
Germany	-	-	-	-	-	-	1	-	-
Greece	-	-	-	-	-	-	-	-	-
Ireland	-	-	-	-	-	-	-	-	-
Italy	-	-	52 e	-	-	-	-	1	-
Japan	-	-	-	-	-	-	3	3	-
Luxembourg	-	-	-	-	-	-	-	-	-
Mexico	-	-	49 e	-	-	-	1	3	-
Netherlands	-	-	-	-	-	-	4	1	-
Norway	-	-	-	-	20	-	-	-	-
Portugal	-	-	-	-	-	-	-	-	-
Spain	-	-	-	-	-	-	-	-	-
Sweden	-	-	-	-	-	-	-	1	-
Switzerland	31	-	-	-	-	-	-	-	-
Turkey	-	-	-	-	-	-	-	-	-
United Kingdom	-	-	-	-	-	-	3	17	-
United States	1 464	2 402	2 693 e	100	149	136 e	2 127	2 290	2 618 e
Other OECD	-	-	-	-	-	-	1	1	-
OECD	**1 495**	**2 402**	**2 794 e**	**100**	**169**	**136 e**	**2 143**	**2 325**	**2 618 e**
Africa	-	-	-	-	-	-	3	3	-
Latin America	-	-	-	-	-	-	2	4	-
Asia	-	-	57 e	-	-	-	1	-	-
China	-	-	-	-	-	-	14	7	-
Non-OECD Europe	-	-	-	-	-	-	-	-	-
Former Soviet Union	-	-	-	-	-	-	-	-	-
Middle East	-	-	-	-	-	-	-	1	-
Non-specified/Other	-	-	-	-	-	-	38	28	-
Non-OECD	**-**	**-**	**57 e**	**-**	**-**	**-**	**58**	**43**	**-**
Total Exports	**1 495**	**2 402**	**2 851 e**	**100**	**169**	**136 e**	**2 201**	**2 368**	**2 618 e**

Source: Annual Oil Statistics.

CZECH REPUBLIC

Figure 1
FUEL SHARES OF TPES *

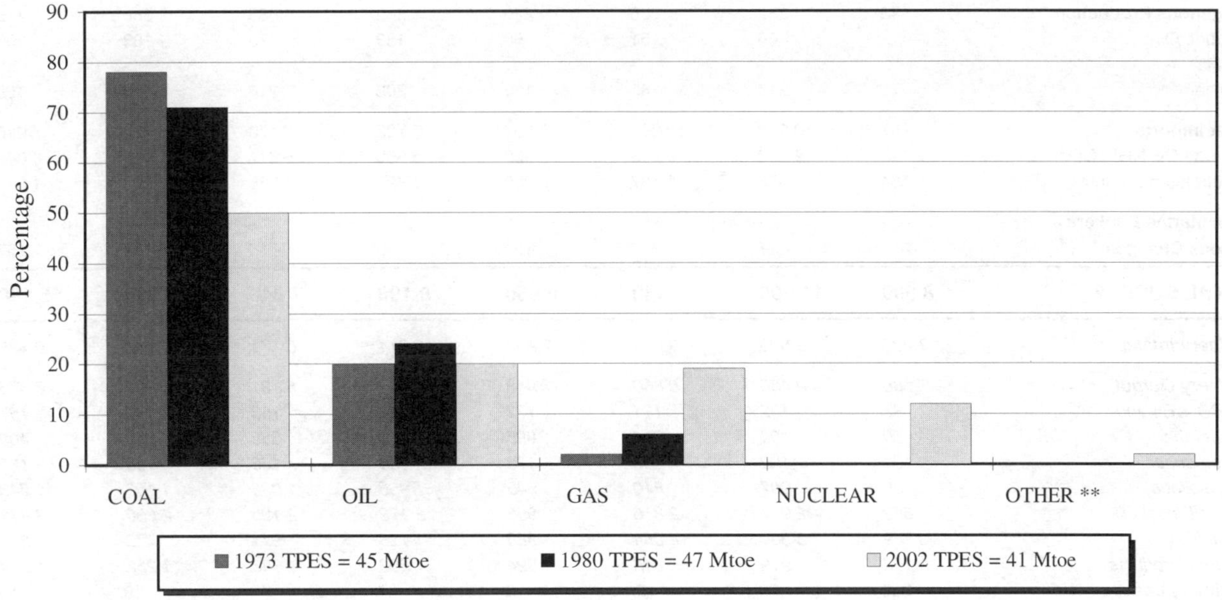

Source: Annual Energy Balances.

* Excluding electricity trade.

** Includes hydro, geothermal, solar, wind, combustible renewables and waste, etc.

Figure 2
OIL PRODUCTION AND CONSUMPTION

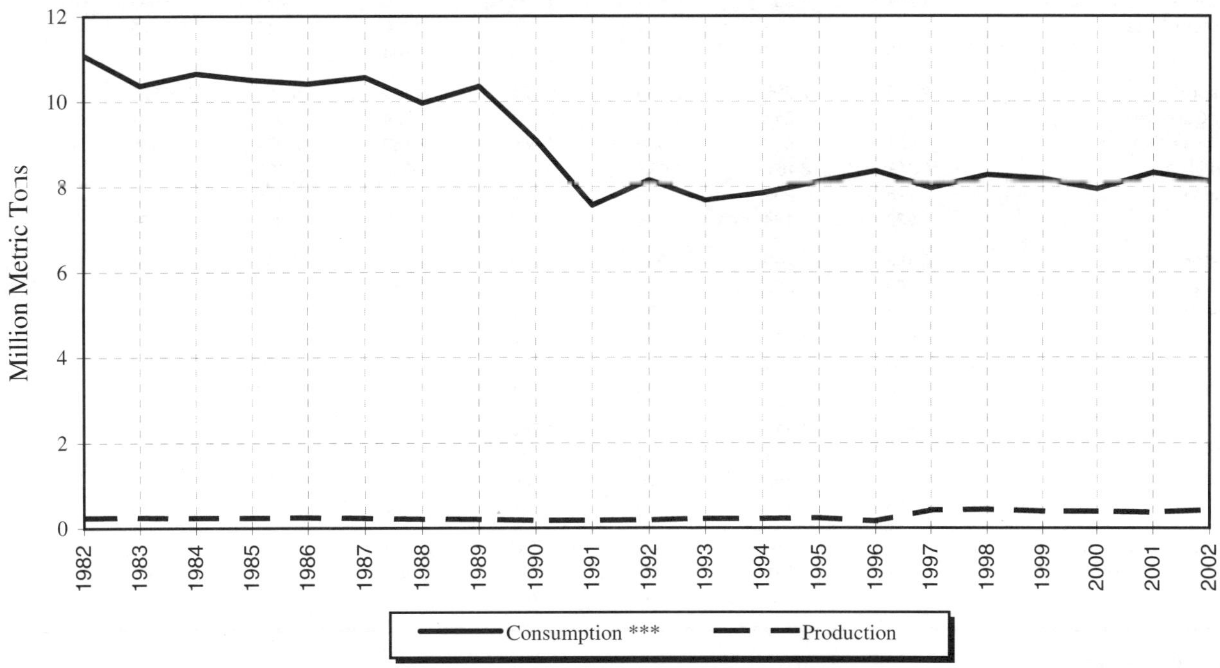

Source: Annual Oil Statistics.

*** Includes refinery fuel and international marine bunkers.

CZECH REPUBLIC
Table 1: Oil Supply and Consumption - Summary Table

000 metric tons

	1973	1978	1980	1990	1999	2000	2001	2002
Indigenous Production	43	282	236	220	386	385	366	415 e
Crude Oil	..	66	51	50	183	175	183	259 e
NGL	..	-	-	-	-	-	-	-
Other	..	216	185	170	203	210	183	156 e
+ Net Imports	8 857	10 940	10 900	8 597	7 782	7 470	8 121	7 910 e
Crude Oil, NGL, Other	7 193	9 068	9 213	7 392	5 955	5 614	5 918	6 049 e
Petroleum Products	1 664	1 872	1 687	1 205	1 827	1 856	2 203	1 861 e
- Intl. Marine Bunkers	-	-	-	-	-	-	-	-
+ Stock Changes	-	- 17	7	152	30	- 15	- 157	- 157 e
TOTAL SUPPLY	**8 900**	**11 205**	**11 143**	**8 969**	**8 198**	**7 840**	**8 330**	**8 168 e**
Refinery Intake	*7 236*	*9 508*	*9 686*	*7 950*	*6 314*	*6 171*	*6 357*	*6 425 e*
Refinery Output	*7 586*	*9 450*	*9 660*	*7 948*	*6 247*	*6 131*	*6 309*	*6 376 e*
LPG + Ethane	*82*	*132*	*131*	*172*	*144*	*153*	*190*	*157 e*
Naphtha	*50*	*494*	*714*	*895*	*413*	*355*	*570*	*409 e*
Kerosenes	*191*	*192*	*257*	*173*	*134*	*163*	*168*	*160 e*
Gasolines	*729*	*907*	*810*	*945*	*956*	*1 036*	*1 106*	*1 248 e*
Gas/Diesel Oil	*1 662*	*2 917*	*2 816*	*2 953*	*2 172*	*2 125*	*2 250*	*2 412 e*
Fuel Oil	*3 769*	*3 889*	*4 084*	*1 886*	*1 128*	*911*	*773*	*376 e*
Other Products	*1 103*	*919*	*848*	*924*	*1 300*	*1 388*	*1 252*	*1 614 e*
- Refinery Losses	- 350	58	26	2	67	40	48	49 e
+ Returns	-	44	114	118	55	52	37	7 e
- Statistical Differences	- 20	- 27	- 63	-	-	- 97	- 7	- 2 e
TOTAL CONSUMPTION	**9 270**	**11 218**	**11 294**	**9 085**	**8 186**	**7 949**	**8 326**	**8 128 e**
Transformation and Energy	**1 226**	**2 059**	**1 774**	**1 198**	**721**	**648**	**710**	..
Crude Oil	-	-	-	54	-	-	-	..
LPG + Ethane	-	-	-	-	2	2	2	..
Gas/Diesel Oil	-	-	-	-	51	56	26	..
Fuel Oil	1 201	1 991	1 634	1 026	556	477	504	..
Petroleum Coke	-	-	-	-	8	15	73	..
Other	25	68	140	118	104	98	105	..
Transport	**2 274**	**2 598**	**2 398**	**2 652**	**4 499**	**4 588**	**4 823**	..
Motor Gasoline	1 137	1 241	1 003	1 161	1 924	1 858	1 901	..
Gas/Diesel Oil	681	920	930	1 080	2 163	2 274	2 485	..
Subtotal	*1 818*	*2 161*	*1 933*	*2 241*	*4 087*	*4 132*	*4 386*	..
Aviation Fuels	228	253	279	214	179	195	208	..
Other	228	184	186	197	233	261	229	..
Industry	**5 192**	**5 249**	**6 030**	**3 969**	**2 833**	**2 535**	**2 638**	..
Naphtha	50	208	414	778	668	538	669	..
LPG + Ethane	48	53	52	52	116	122	110	..
Gas/Diesel Oil	1 039	1 912	2 089	1 726	418	418	420	..
Fuel Oil	3 205	2 369	2 872	722	608	354	359	..
Other	850	707	603	691	1 023	1 103	1 080	..
Other Sectors	**578**	**1 312**	**1 092**	**1 266**	**133**	**178**	**155**	..
Gas/Diesel Oil	445	1 012	874	563	9	7	4	..
Fuel Oil	-	152	128	620	54	88	52	..
Other	133	148	90	83	70	83	99	..
Memo: Non-Energy Use	*1 158*	*1 136*	*1 240*	*1 702*	*2 186*	*2 020*	*2 043*	..
Chem/Petchem. Feedst.	*98*	*261*	*466*	*830*	*1 591*	*1 351*	*1 447*	..
Other	*1 060*	*875*	*774*	*872*	*595*	*669*	*596*	..

Source: Annual Oil Statistics.

CZECH REPUBLIC
Table 2: Supply of Crude Oil, NGL, Feedstocks and Other Hydrocarbons

000 metric tons

	1973	1978	1980	1990	1999	2000	2001	2002
CRUDE OIL								
Indigenous Production	..	66	51	50	183	175	183	259 e
+ Imports	..	18 302	18 934	13 354	6 041	5 699	5 972	6 148 e
- Exports	..	9 473	9 777	6 169	109	111	119	140 e
- Direct Use	..	-	-	54	-	-	-	-
+ Stock Change	..	41	49	116	- 79	103	36	- 44 e
- Statistical Difference	..	-	-	-	-	- 5	-	- 15 e
Refinery Intake	**..**	**8 936**	**9 257**	**7 297**	**6 036**	**5 871**	**6 072**	**6 238 e**
Closing Stock Level	..	*224*	*223*	*133*	*774*	*671*	*635*	*679 e*
NATURAL GAS LIQUIDS								
Indigenous Production	..	-	-	-	-	-	-	-
+ Imports	..	-	-	-	-	-	-	-
- Exports	..	-	-	-	-	-	-	-
- Direct Use	..	-	-	-	-	-	-	-
+ Stock Change	..	-	-	-	-	-	-	-
- Statistical Difference	..	-	-	-	-	-	-	-
Refinery Intake	**..**	**-**	**-**	**-**	**-**	**-**	**-**	**-**
Closing Stock Level	..	-	-	-	-	-	-	-
REFINERY FEEDSTOCKS								
Backflows	..	-	67	71	55	52	32	-
+ Transfers	..	117	121	205	-	-	-	-
+ Imports	..	232	49	191	-	-	-	-
- Exports	..	-	-	-	-	-	-	-
+ Stock Change	..	-	-	-	2	-	-	-
- Statistical Difference	..	-	-	-	-	-	-	-
Refinery Intake	**..**	**349**	**237**	**467**	**57**	**52**	**32**	**-**
Closing Stock Level	..	-	-	-	-	-	-	-
OTHER HYDROCARBONS								
Indigenous Production	..	216	185	170	203	210	183	156 e
+ Imports	..	7	7	16	23	26	65	41 e
- Exports	..	-	-	-	-	-	-	-
- Direct Use	..	-	-	-	-	-	-	8 e
+ Stock Change	..	-	-	-	- 5	12	5	- 2 e
- Statistical Difference	..	-	-	-	-	-	-	-
Refinery Intake	**..**	**223**	**192**	**186**	**221**	**248**	**253**	**187 e**
Closing Stock Level	..	-	-	-	*21*	*9*	*4*	*6 e*
CRUDE OIL, NGL, FEEDSTOCKS AND OTHER								
Indigenous Production	43	282	236	220	386	385	366	415 e
+ Backflows	-	-	67	71	55	52	32	-
+ Transfers	-	117	121	205	-	-	-	-
+ Imports	7 193	18 541	18 990	13 561	6 064	5 725	6 037	6 189 e
- Exports	-	9 473	9 777	6 169	109	111	119	140 e
- Direct Use	-	-	-	54	-	-	-	8 e
+ Stock Change	-	41	49	116	- 82	115	41	- 46 e
- Statistical Difference	-	-	-	-	-	- 5	-	- 15 e
Refinery Intake	**7 236**	**9 508**	**9 686**	**7 950**	**6 314**	**6 171**	**6 357**	**6 425 e**
Closing Stock Level	-	*224*	*223*	*133*	*795*	*680*	*639*	*685 e*

Source: Annual Oil Statistics.

CZECH REPUBLIC
Table 3: Motor Gasoline - Supply

000 metric tons

	1973	1978	1980	1990	1999	2000	2001	2002
MOTOR GASOLINE								
Refinery Output	**714**	**893**	**797**	**920**	**954**	**1 033**	**1 104**	**1 246 e**
+ Imports	423	417	451	249	1 049	928	1 101	988 e
- Exports	-	34	175	8	99	75	209	274 e
+ Transfers	-	-	-	-	-	-	-	3 e
+ Stock Change	-	- 17	- 64	-	20	- 40	- 95	- 37 e
- Statistical Difference	-	18	6	-	-	- 12	-	-
Total Consumption	**1 137**	**1 241**	**1 003**	**1 161**	**1 924**	**1 858**	**1 901**	**1 926 e**
Closing Stock Level	-	*376*	*386*	*243*	*335*	*375*	*470*	*507 e*

Table 4: Aviation Fuels - Supply

000 metric tons

	1973	1978	1980	1990	1999	2000	2001	2002
AVIATION GASOLINE								
Refinery Output	**15**	**14**	**13**	**25**	**2**	**3**	**2**	**2 e**
+ Imports	55	59	59	20	1	1	1	1 e
- Exports	-	6	4	-	-	-	-	-
+ Transfers	-	-	-	-	-	-	-	-
+ Stock Change	-	- 1	8	-	-	- 1	-	-
- Statistical Difference	-	-	-	-	-	-	-	-
Total Consumption	**70**	**66**	**76**	**45**	**3**	**3**	**3**	**3 e**
Closing Stock Level	-	*8*	-	-	-	*1*	*1*	*1 e*
GASOLINE TYPE JET FUEL								
Refinery Output	**-**	**-**	**-**	**-**	**-**	**-**	**-**	**-**
+ Imports	-	-	-	-	-	-	-	-
- Exports	-	-	-	-	-	-	-	-
+ Transfers	-	-	-	-	-	-	-	-
+ Stock Change	-	-	-	-	-	-	-	-
- Statistical Difference	-	-	-	-	-	-	-	-
Total Consumption	**-**	**-**	**-**	**-**	**-**	**-**	**-**	**-**
Closing Stock Level	-	-	-	-	-	-	-	-
KEROSENE TYPE JET FUEL								
Refinery Output	**191**	**187**	**256**	**169**	**134**	**163**	**168**	**160 e**
+ Imports	-	-	-	23	50	33	27	47 e
- Exports	33	-	53	23	-	6	-	-
+ Transfers	-	-	-	-	-	-	- 5	- 1 e
+ Stock Change	-	-	-	-	- 8	- 17	15	- 9 e
- Statistical Difference	-	-	-	-	-	- 19	-	-
Total Consumption	**158**	**187**	**203**	**169**	**176**	**192**	**205**	**197 e**
Closing Stock Level	-	-	-	-	*20*	*37*	*22*	*31 e*

Source: Annual Oil Statistics.

CZECH REPUBLIC
Table 5A: Gas/Diesel Oil - Supply and Consumption

000 metric tons

	1973	1978	1980	1990	1999	2000	2001	2002
Refinery Output	**1 662**	**2 917**	**2 816**	**2 953**	**2 172**	**2 125**	**2 250**	**2 412 e**
+ Recycled Products	-	-	-	-	-	-	-	-
+ Imports	496	1 186	1 203	537	1 040	1 111	1 345	1 255 e
- Exports	-	270	124	167	617	492	603	693 e
- Intl. Marine Bunkers	-	-	-	-	-	-	-	-
+ Transfers	-	-	-	-	-	-	5	2 e
+ Stock Changes	-	- 36	- 34	46	46	- 50	- 62	- 34 e
- Statistical Differences	- 7	- 47	- 32	-	-	- 61	-	-
Total Consumption	**2 165**	**3 844**	**3 893**	**3 369**	**2 641**	**2 755**	**2 935**	**2 942 e**
Transformation	**-**	**-**	**-**	**-**	**11**	**10**	**5**	**..**
Public Electricity	-	-	-	-	1	1	1	..
Autoproducer Electricity	-	-	-	-	-	-	-	..
Public CHP Plants	-	-	-	-	-	-	-	..
Autoproducers of CHP	-	-	-	-	-	-	-	..
Public Heat Plants	-	-	-	-	-	-	-	..
Autoproducer Heat Plants	-	-	-	-	-	-	-	..
Petrochemical Industry	-	-	-	-	10	9	4	..
Other Transformation	-	-	-	-	-	-	-	..
Energy Sector	**-**	**-**	**-**	**-**	**40**	**46**	**21**	**..**
Refinery Fuel	-	-	-	-	17	24	-	..
Other Energy Sector	-	-	-	-	23	22	21	..
Final Consumption	**2 165**	**3 844**	**3 893**	**3 369**	**2 590**	**2 699**	**2 909**	**..**
Transport	**681**	**920**	**930**	**1 080**	**2 163**	**2 274**	**2 485**	**..**
Air	-	-	-	-	-	-	-	..
Road	681	920	930	1 080	2 052	2 173	2 381	..
Rail	-	-	-	-	104	96	98	..
Inland Waterways	-	-	-	-	7	5	6	..
Non-Specified	-	-	-	-	-	-	-	..
Industry	**1 039**	**1 912**	**2 089**	**1 726**	**418**	**418**	**420**	**..**
Iron and Steel	-	-	-	-	10	9	7	..
Chemical (incl. Pet.-Chem.)	-	-	-	-	269	222	198	..
Non-Ferrous Metals	-	-	-	-	1	1	1	..
Non-Metallic Minerals	-	-	-	-	14	16	15	..
Transport Equipment	-	-	-	-	4	5	6	..
Machinery	-	-	-	-	13	18	12	..
Mining and Quarrying	-	-	-	-	12	14	12	..
Food and Tobacco	-	-	-	-	25	31	24	..
Pulp, Paper and Printing	-	-	-	-	4	5	4	..
Wood and Wood Products	-	-	-	-	4	5	3	..
Construction	-	-	-	-	61	91	137	..
Textiles and Leather	-	-	-	-	1	1	1	..
Non-Specified	1 039	1 912	2 089	1 726	-	-	-	..
Other Sectors	**445**	**1 012**	**874**	**563**	**9**	**7**	**4**	**..**
Commerce and Public	-	219	86	50	7	6	3	..
Residential	-	-	-	-	-	-	-	..
Agriculture	445	793	788	513	2	1	1	..
Non-Specified	-	-	-	-	-	-	-	..
Memo: Non-Energy Use	-	-	-	-	*269*	*222*	*198*	..
Chem/Petchem. Feedst.	-	-	-	-	*269*	*222*	*198*	..
Other	-	-	-	-	-	-	-	..
Closing Stock Level	-	*498*	*508*	*395*	*610*	*660*	*722*	*756 e*

Source: Annual Oil Statistics.

CZECH REPUBLIC
Table 5B: Fuel Oil (Residual) - Supply and Consumption

000 metric tons

	1973	1978	1980	1990	1999	2000	2001	2002
Refinery Output	**3 769**	**3 889**	**4 084**	**1 886**	**1 128**	**911**	**773**	**376 e**
+ Recycled Products	-	-	-	-	-	-	5	7 e
+ Imports	625	657	633	558	295	181	271	212 e
- Exports	-	28	168	66	237	160	129	67 e
- Intl. Marine Bunkers	-	-	-	-	-	-	-	-
+ Transfers	-	-	-	-	-	-	3	- 12 e
+ Stock Changes	-	- 4	48	- 10	32	- 13	- 8	17 e
- Statistical Differences	- 12	2	- 37	-	-	-	-	-
Total Consumption	**4 406**	**4 512**	**4 634**	**2 368**	**1 218**	**919**	**915**	**533 e**
Transformation	**1 201**	**1 886**	**1 517**	**1 008**	**348**	**313**	**327**	**..**
Public Electricity	1 201	1 886	1 299	776	13	12	12	..
Autoproducer Electricity	-	-	-	-	30	44	49	..
Public CHP Plants	-	-	-	-	124	80	88	..
Autoproducers of CHP	-	-	-	-	46	75	73	..
Public Heat Plants	-	-	218	232	116	57	62	..
Autoproducer Heat Plants	-	-	-	-	19	45	43	..
Petrochemical Industry	-	-	-	-	-	-	-	..
Other Transformation	-	-	-	-	-	-	-	..
Energy Sector	**-**	**105**	**117**	**18**	**208**	**164**	**177**	**..**
Refinery Fuel	-	105	117	18	53	51	62	..
Other Energy Sector	-	-	-	-	155	113	115	..
Final Consumption	**3 205**	**2 521**	**3 000**	**1 342**	**662**	**442**	**411**	**..**
Transport	**-**	**-**	**-**	**-**	**-**	**-**	**-**	**..**
Air	-	-	-	-	-	-	-	..
Road	-	-	-	-	-	-	-	..
Rail	-	-	-	-	-	-	-	..
Inland Waterways	-	-	-	-	-	-	-	..
Non-Specified	-	-	-	-	-	-	-	..
Industry	**3 205**	**2 369**	**2 872**	**722**	**608**	**354**	**359**	**..**
Iron and Steel	-	-	-	-	162	125	54	..
Chemical (incl. Pet.-Chem.)	-	-	-	-	155	43	172	..
Non-Ferrous Metals	-	-	-	-	-	-	-	..
Non-Metallic Minerals	-	-	-	-	125	92	30	..
Transport Equipment	-	-	-	-	-	-	-	..
Machinery	-	-	-	-	1	2	1	..
Mining and Quarrying	-	-	-	-	15	8	8	..
Food and Tobacco	-	-	-	-	15	13	16	..
Pulp, Paper and Printing	-	-	-	-	20	10	9	..
Wood and Wood Products	-	-	-	-	4	7	2	..
Construction	-	-	-	-	1	5	1	..
Textiles and Leather	-	-	-	-	11	6	4	..
Non-Specified	3 205	2 369	2 872	722	99	43	62	..
Other Sectors	**-**	**152**	**128**	**620**	**54**	**88**	**52**	**..**
Commerce and Public	-	-	-	-	-	-	-	..
Residential	-	-	-	-	-	-	-	..
Agriculture	-	16	7	-	22	37	19	..
Non-Specified	-	136	121	620	32	51	33	..
Memo: Non-Energy Use	-	-	-	-	-	-	-	..
Chem/Petchem. Feedst.	-	-	-	-	-	-	-	..
Other	-	-	-	-	-	-	-	..
Closing Stock Level	-	*46*	*36*	*82*	*148*	*161*	*169*	*152 e*

Source: Annual Oil Statistics.

CZECH REPUBLIC
Table 5C: Total Products - Supply and Consumption

000 metric tons

	1973	1978	1980	1990	1999	2000	2001	2002
Refinery Output	**7 586**	**9 450**	**9 660**	**7 948**	**6 247**	**6 131**	**6 309**	**6 376 e**
+ Recycled Products	-	44	47	47	-	-	5	7 e
+ Imports	1 697	2 500	2 477	1 592	3 023	2 829	3 359	3 104 e
- Exports	33	628	790	387	1 196	973	1 156	1 243 e
- Intl. Marine Bunkers	-	-	-	-	-	-	-	-
+ Transfers	-	- 117	- 121	- 151	-	-	-	8 e
+ Stock Changes	-	- 58	- 42	36	112	- 130	- 198	- 111 e
- Statistical Differences	- 20	- 27	- 63	-	-	- 92	- 7	13 e
Total Consumption	**9 270**	**11 218**	**11 294**	**9 085**	**8 186**	**7 949**	**8 326**	**8 128 e**
Transformation	**1 201**	**1 886**	**1 584**	**1 079**	**404**	**366**	**360**	..
Public Electricity	1 201	1 886	1 299	776	14	13	13	..
Autoproducer Electricity	-	-	-	-	30	44	49	..
Public CHP Plants	-	-	-	-	124	80	88	..
Autoproducers of CHP	-	-	-	-	46	75	73	..
Public Heat Plants	-	-	218	232	116	57	62	..
Autoproducer Heat Plants	-	-	-	-	19	45	43	..
Petrochemical Industry	-	-	67	71	55	52	32	..
Other Transformation	-	-	-	-	-	-	-	..
Energy Sector	**25**	**173**	**190**	**119**	**317**	**282**	**350**	..
Refinery Fuel	25	173	190	65	123	128	200	..
Other Energy Sector	-	-	-	54	194	154	150	..
Final Consumption	**8 044**	**9 159**	**9 520**	**7 887**	**7 465**	**7 301**	**7 616**	..
Transport	**2 274**	**2 598**	**2 398**	**2 652**	**4 499**	**4 588**	**4 823**	..
Air	228	253	279	214	179	195	208	..
Road	2 046	2 345	2 119	2 438	4 178	4 265	4 484	..
Rail	-	-	-	-	135	123	125	..
Inland Waterways	-	-	-	-	7	5	6	..
Non-Specified	-	-	-	-	-	-	-	..
Industry	**5 192**	**5 249**	**6 030**	**3 969**	**2 833**	**2 535**	**2 638**	..
Iron and Steel	-	-	-	-	173	136	63	..
Chemical (incl. Pet.-Chem.)	98	261	466	830	1 782	1 447	1 660	..
Non-Ferrous Metals	-	-	-	-	1	1	1	..
Non-Metallic Minerals	-	-	-	-	143	114	52	..
Transport Equipment	-	-	-	-	4	6	7	..
Machinery	-	-	-	-	19	31	24	..
Mining and Quarrying	-	-	-	-	27	22	20	..
Food and Tobacco	-	-	-	-	42	48	44	..
Pulp, Paper and Printing	-	-	-	-	27	16	14	..
Wood and Wood Products	-	-	-	-	8	13	7	..
Construction	623	512	482	502	430	486	518	..
Textiles and Leather	-	-	-	-	13	8	6	..
Non-Specified	4 471	4 476	5 082	2 637	164	207	222	..
Other Sectors	**578**	**1 312**	**1 092**	**1 266**	**133**	**178**	**155**	..
Commerce and Public	-	219	86	50	7	6	3	..
Residential	43	79	79	79	58	69	78	..
Agriculture	445	809	795	513	29	47	32	..
Non-Specified	90	205	132	624	39	56	42	..
Memo: Non-Energy Use	*1 158*	*1 136*	*1 240*	*1 702*	*2 186*	*2 020*	*2 043*	..
Chem/Petchem. Feedst.	*98*	*261*	*466*	*830*	*1 591*	*1 351*	*1 447*	..
Other	*1 060*	*875*	*774*	*872*	*595*	*669*	*596*	..
Closing Stock Level	-	*928*	*930*	*720*	*1 206*	*1 336*	*1 534*	*1 645 e*

Source: Annual Oil Statistics.

CZECH REPUBLIC
Table 6A: Supply and Consumption

000 metric tons

	1980	2000	2001	2002	1980	2000	2001	2002
	LPG AND ETHANE				REFINERY GAS			
Refinery Output	**131**	**153**	**190**	**157 e**	**35**	**106**	**119**	**112 e**
+ Recycled Products	-	-	-	-	-	-	-	-
+ Imports	-	129	115	120 e	-	-	-	-
- Exports	-	16	32	42 e	-	-	-	-
- Intl. Marine Bunkers	-	-	-	-	-	-	-	-
+ Transfers	-	-	-	5 e	-	-	-	-
+ Stock Change	-	- 2	1	1 e	-	-	-	-
- Statistical Difference	-	-	-	-	-	-	-	-
Total Consumption	**131**	**264**	**274**	**241 e**	**35**	**106**	**119**	**112 e**
of which: Ethane	-	-	-	-	-	-	-	-
Transformation	-	2	2	..	-	-	-	..
Energy	-	-	-	..	35	53	78	..
Transport	-	62	72	..	-	-	-	..
Industry	52	122	110	..	-	53	41	..
Other	79	78	90	..	-	-	-	..
Memo: Non-Energy Use	*52*	*81*	*71*	..	-	-	-	..
Closing Stock Level	-	*9*	*8*	*7 e*	-	-	-	-
	NAPHTHA				OTHER KEROSENE			
Refinery Output	**714**	**355**	**570**	**409 e**	**1**	**-**	**-**	**-**
+ Recycled Products	-	-	-	-	-	-	-	-
+ Imports	121	219	230	200 e	10	5	6	5 e
- Exports	233	8	72	10 e	-	-	-	-
- Intl. Marine Bunkers	-	-	-	-	-	-	-	-
+ Transfers	- 121	-	-	4 e	-	-	-	-
+ Stock Change	-	- 3	- 40	5 e	-	-	-	-
- Statistical Difference	-	-	3	3 e	-	-	-	-
Total Consumption	**481**	**563**	**685**	**605 e**	**11**	**5**	**6**	**5 e**
Transformation	67	21	15	..	-	-	-	..
Energy	-	4	1	..	-	-	-	..
Transport	-	-	-	..	-	-	-	..
Industry	414	538	669	..	-	-	-	..
Other	-	-	-	..	11	5	6	..
Memo: Non-Energy Use	*414*	*538*	*669*	..	-	-	-	..
Closing Stock Level	-	*33*	*73*	*68 e*	-	-	-	-
	WHITE SPIRIT + SBP				LUBRICANTS			
Refinery Output	**118**	**7**	**10**	**..**	**139**	**196**	**126**	**..**
+ Recycled Products	-	-	-	..	47	-	-	..
+ Imports	-	45	28	..	-	70	82	..
- Exports	22	-	-	..	-	28	21	..
- Intl. Marine Bunkers	-	-	-	..	-	-	-	..
+ Transfers	-	-	-	..	-	-	- 1	..
+ Stock Change	-	1	-	..	-	- 10	- 3	..
- Statistical Difference	-	-	-	..	-	-	-	..
Total Consumption	**96**	**53**	**38**	**..**	**186**	**228**	**183**	**..**
Transformation	-	-	-	..	-	-	-	..
Energy	38	-	-	..	-	-	-	..
Transport	-	-	-	..	186	199	157	..
Industry	58	53	38	..	-	29	26	..
Other	-	-	-	..	-	-	-	..
Memo: Non-Energy Use	*58*	*53*	*38*	..	*186*	*228*	*183*	..
Closing Stock Level	-	*1*	*1*	..	-	*26*	*29*	..

Source: Annual Oil Statistics.

CZECH REPUBLIC
Table 6B: Supply and Consumption

000 metric tons

	1980	2000	2001	2002	1980	2000	2001	2002
	BITUMEN				PARAFFIN WAXES			
Refinery Output	**482**	**481**	**325**	**..**	**15**	**9**	**7**	**..**
+ Recycled Products	-	-	-	..	-	-	-	..
+ Imports	-	86	131	..	-	4	6	..
- Exports	-	185	82	..	-	3	8	..
- Intl. Marine Bunkers	-	-	-	..	-	-	-	..
+ Transfers	-	-	-	..	-	-	-	..
+ Stock Change	-	- 4	- 1	..	-	-	-	..
- Statistical Difference	-	-	-	..	-	-	-	..
Total Consumption	**482**	**378**	**373**	**..**	**15**	**10**	**5**	**..**
Transformation	-	-	-	..	-	-	-	..
Energy	-	-	-	..	-	-	-	..
Transport	-	-	-	..	-	-	-	..
Industry	482	378	370	..	15	10	5	..
Other	-	-	3	..	-	-	-	..
Memo: Non-Energy Use	*482*	*378*	*370*	..	-	*10*	*5*	..
Closing Stock Level	-	*12*	*13*	..	-	*2*	*2*	..
	PETROLEUM COKE				OTHER PRODUCTS			
Refinery Output	**-**	**-**	**60**	**58 e**	**59**	**589**	**605**	**1 444 e**
+ Recycled Products	-	-	-	-	-	-	-	-
+ Imports	-	15	13	18 e	-	2	3	258 e
- Exports	-	-	-	-	11	-	-	157 e
- Intl. Marine Bunkers	-	-	-	-	-	-	-	-
+ Transfers	-	-	-	-	-	-	- 2	7 e
+ Stock Change	-	-	-	-	-	9	- 5	- 54 e
- Statistical Difference	-	-	-	-	-	-	- 10	10 e
Total Consumption	**-**	**15**	**73**	**76 e**	**48**	**600**	**611**	**1 488 e**
Transformation	-	-	-	..	-	20	11	..
Energy	-	15	73	..	-	-	-	..
Transport	-	-	-	..	-	-	-	..
Industry	-	-	-	..	48	580	600	..
Other	-	-	-	..	-	-	-	..
Memo: Non-Energy Use	-	-	-	..	*48*	*510*	*509*	..
Closing Stock Level	-	-	-	-	-	*19*	*24*	*123 e*

Source: Annual Oil Statistics.

CZECH REPUBLIC
Table 7A: Imports by Country of Origin

000 metric tons

	Crude Oil			Natural Gas Liquids			Feedstocks and Other		
	2000	2001	2002	2000	2001	2002	2000	2001	2002
Australia	-	-	-	-	-	-	-	-	-
Belgium	-	-	-	-	-	-	1	-	-
Canada	-	-	-	-	-	-	-	-	-
Czech Republic	-	-	-	-	-	-	-	-	-
Denmark	-	-	-	-	-	-	-	-	-
Finland	-	-	-	-	-	-	-	-	-
France	-	-	-	-	-	-	1	2	2 e
Germany	-	-	-	-	-	-	17	28	21 e
Greece	-	-	-	-	-	-	-	-	-
Hungary	-	-	-	-	-	-	2	1	1 e
Ireland	-	-	-	-	-	-	-	-	-
Italy	-	-	-	-	-	-	-	-	-
Japan	-	-	-	-	-	-	-	-	-
Korea	-	-	-	-	-	-	-	-	-
Mexico	-	-	-	-	-	-	-	-	-
Netherlands	-	-	-	-	-	-	-	10	1 e
New Zealand	-	-	-	-	-	-	-	-	-
Norway	-	-	199 e	-	-	-	-	-	-
Poland	-	-	-	-	-	-	5	7	5 e
Spain	-	-	-	-	-	-	-	-	-
Sweden	-	-	-	-	-	-	-	-	1 e
Turkey	-	-	-	-	-	-	-	-	-
United Kingdom	-	-	-	-	-	-	-	-	-
United States	-	-	-	-	-	-	-	-	-
Other	-	-	-	-	-	-	-	15	8 e
OECD	**-**	**-**	**199 e**	**-**	**-**	**-**	**26**	**63**	**39 e**
Argentina	-	-	-	-	-	-	-	-	-
Brazil	-	-	-	-	-	-	-	-	-
Colombia	-	-	-	-	-	-	-	-	-
Ecuador	-	-	-	-	-	-	-	-	-
Netherlands Antilles	-	-	-	-	-	-	-	-	-
Peru	-	-	-	-	-	-	-	-	-
Trinidad and Tobago	-	-	-	-	-	-	-	-	-
Venezuela	-	-	-	-	-	-	-	-	-
Other	-	-	-	-	-	-	-	-	-
Latin America	**-**	**-**	**-**	**-**	**-**	**-**	**-**	**-**	**-**
Romania	-	-	-	-	-	-	-	-	-
Other	-	-	-	-	-	-	-	-	-
Non-OECD Europe	**-**	**-**	**-**	**-**	**-**	**-**	**-**	**-**	**-**
FSU	**5 213**	**5 091**	**4 955 e**	**-**	**-**	**-**	**-**	**2**	**2 e**
Iran	-	-	-	-	-	-	-	-	-
Iraq	41	-	-	-	-	-	-	-	-
Kuwait	-	-	-	-	-	-	-	-	-
Oman	-	-	-	-	-	-	-	-	-
Qatar	-	-	-	-	-	-	-	-	-
Saudi Arabia	-	-	-	-	-	-	-	-	-
United Arab Emirates	-	-	-	-	-	-	-	-	-
Other	-	55	629 e	-	-	-	-	-	-
Middle East	**41**	**55**	**629 e**	**-**	**-**	**-**	**-**	**-**	**-**
India	-	-	-	-	-	-	-	-	-
Indonesia	-	-	-	-	-	-	-	-	-
Singapore	-	-	-	-	-	-	-	-	-
Other	-	-	-	-	-	-	-	-	-
Asia	**-**	**-**	**-**	**-**	**-**	**-**	**-**	**-**	**-**
China	**-**	**-**	**-**	**-**	**-**	**-**	**-**	**-**	**-**
Algeria	445	209	89 e	-	-	-	-	-	-
Angola/Cabinada	-	-	-	-	-	-	-	-	-
Cameroon	-	-	-	-	-	-	-	-	-
Egypt	-	-	-	-	-	-	-	-	-
Gabon	-	-	-	-	-	-	-	-	-
Libya	-	617	276 e	-	-	-	-	-	-
Nigeria	-	-	-	-	-	-	-	-	-
Tunisia	-	-	-	-	-	-	-	-	-
Other	-	-	-	-	-	-	-	-	-
Africa	**445**	**826**	**365 e**	**-**	**-**	**-**	**-**	**-**	**-**
Non-spec./Other	**-**	**-**	**-**	**-**	**-**	**-**	**-**	**-**	**-**
Total Imports	**5 699**	**5 972**	**6 148 e**	**-**	**-**	**-**	**26**	**65**	**41 e**
Memo: Total OPEC	*486*	*826*	*365 e*	-	-	-	-	-	-

Source: Annual Oil Statistics.

CZECH REPUBLIC
Table 7B: Imports by Country of Origin

000 metric tons

	2000	2001	2002	2000	2001	2002	2000	2001	2002
	Total Products			**Liquefied Petroleum Gases**			**Naphtha**		
Australia	-	-	-	-	-	-	-	-	-
Belgium	9	9	13 e	1	-	-	-	-	-
Canada	-	-	-	-	-	-	-	-	-
Czech Republic	-	-	-	-	-	-	-	-	-
Denmark	-	-	-	-	-	-	-	-	-
Finland	-	-	-	-	-	-	-	-	-
France	29	22	16 e	-	-	-	-	-	-
Germany	651	807	589 e	39	29	20 e	142	126	119 e
Greece	-	-	-	-	-	-	-	-	-
Hungary	122	117	25 e	5	6	6 e	-	-	-
Ireland	-	-	-	-	-	-	-	-	-
Italy	3	4	3 e	1	1	-	-	1	-
Japan	-	-	-	-	-	-	-	-	-
Korea	-	-	-	-	-	-	-	-	-
Mexico	-	-	-	-	-	-	-	-	-
Netherlands	9	9	4 e	3	2	1 e	-	-	-
New Zealand	-	-	-	-	-	-	-	-	-
Norway	8	5	9 e	8	5	9 e	-	-	-
Poland	207	334	509 e	12	5	3 e	72	95	76 e
Spain	-	-	1 e	-	-	-	-	-	-
Sweden	-	-	-	-	-	-	-	-	-
Turkey	-	-	-	-	-	-	-	-	-
United Kingdom	-	-	2 e	-	-	1 e	-	-	-
United States	3	-	2 e	-	-	-	-	-	-
Other	1 714	1 983	1 843 e	3	1	2 e	5	8	5 e
OECD	**2 755**	**3 290**	**3 016 e**	**72**	**49**	**42 e**	**219**	**230**	**200 e**
Argentina	-	-	-	-	-	-	-	-	-
Brazil	-	-	-	-	-	-	-	-	-
Colombia	-	-	-	-	-	-	-	-	-
Ecuador	-	-	-	-	-	-	-	-	-
Netherlands Antilles	-	-	-	-	-	-	-	-	-
Peru	-	-	-	-	-	-	-	-	-
Trinidad and Tobago	-	-	-	-	-	-	-	-	-
Venezuela	-	-	-	-	-	-	-	-	-
Other	-	-	-	-	-	-	-	-	-
Latin America	**-**	**-**	**-**	**-**	**-**	**-**	**-**	**-**	**-**
Romania	-	-	1 e	-	-	-	-	-	-
Other	8	18	33 e	8	18	32 e	-	-	-
Non-OECD Europe	**8**	**18**	**34 e**	**8**	**18**	**32 e**	**-**	**-**	**-**
FSU	**65**	**49**	**54 e**	**49**	**48**	**46 e**	**-**	**-**	**-**
Iran	-	-	-	-	-	-	-	-	-
Iraq	-	-	-	-	-	-	-	-	-
Kuwait	-	-	-	-	-	-	-	-	-
Oman	-	-	-	-	-	-	-	-	-
Qatar	-	-	-	-	-	-	-	-	-
Saudi Arabia	-	-	-	-	-	-	-		
United Arab Emirates	-	-	-	-	-	-	-	-	-
Other	-	-	-	-	-	-	-	-	-
Middle East	**-**	**-**	**-**	**-**	**-**	**-**	**-**	**-**	**-**
India	-	-	-	-	-	-	-	-	-
Indonesia	-	-	-	-	-	-	-	-	-
Singapore	-	-	-	-	-	-	-	-	-
Other	-	-	-	-	-	-	-	-	-
Asia	**-**	**-**	**-**	**-**	**-**	**-**	**-**	**-**	**-**
China	**-**	**1**	**-**	**-**	**-**	**-**	**-**	**-**	**-**
Algeria	-	-	-	-	-	-	-	-	-
Angola/Cabinada	-	-	-	-	-	-	-	-	-
Cameroon	-	-	-	-	-	-	-	-	-
Egypt	-	-	-	-	-	-	-	-	-
Gabon	-	-	-	-	-	-	-	-	-
Libya	-	-	-	-	-	-	-	-	-
Nigeria	-	-	-	-	-	-	-	-	-
Tunisia	-	-	-	-	-	-	-	-	-
Other	1	1	-	-	-	-	-	-	-
Africa	**1**	**1**	**-**	**-**	**-**	**-**	**-**	**-**	**-**
Non-spec./Other	**-**	**-**	**-**	**-**	**-**	**-**	**-**	**-**	**-**
Total Imports	**2 829**	**3 359**	**3 104 e**	**129**	**115**	**120 e**	**219**	**230**	**200 e**
Memo: Total OPEC	-	-	-	-	-	-	-	-	-

Source: Annual Oil Statistics.

CZECH REPUBLIC
Table 7C: Imports by Country of Origin

000 metric tons

	2000	2001	2002	2000	2001	2002	2000	2001	2002
	Total Gasoline			**Total Kerosene**			**Gas/Diesel**		
Australia	-	-	-	-	-	-	-	-	-
Belgium	-	-	-	-	-	1 e	-	-	-
Canada	-	-	-	-	-	-	-	-	-
Czech Republic	-	-	-	-	-	-	-	-	-
Denmark	-	-	-	-	-	-	-	-	-
Finland	-	-	-	-	-	-	-	-	-
France	-	-	-	21	15	10 e	-	-	-
Germany	187	197	135 e	2	6	6 e	206	341	182 e
Greece	-	-	-	-	-	-	-	-	-
Hungary	111	108	-	-	-	-	-	-	-
Ireland	-	-	-	-	-	-	-	-	-
Italy	-	-	-	-	-	-	-	-	-
Japan	-	-	-	-	-	-	-	-	-
Korea	-	-	-	-	-	-	-	-	-
Mexico	-	-	-	-	-	-	-	-	-
Netherlands	-	1	-	-	-	-	-	-	-
New Zealand	-	-	-	-	-	-	-	-	-
Norway	-	-	-	-	-	-	-	-	-
Poland	-	42	133 e	-	6	23 e	-	58	143 e
Spain	-	-	-	-	-	-	-	-	-
Sweden	-	-	-	-	-	-	-	-	-
Turkey	-	-	-	-	-	-	-	-	-
United Kingdom	-	-	-	-	-	-	-	-	-
United States	-	-	-	-	-	-	-	-	-
Other	631	754	721 e	15	6	12 e	890	946	923 e
OECD	**929**	**1 102**	**989 e**	**38**	**33**	**52 e**	**1 096**	**1 345**	**1 248 e**
Argentina	-	-	-	-	-	-	-	-	-
Brazil	-	-	-	-	-	-	-	-	-
Colombia	-	-	-	-	-	-	-	-	-
Ecuador	-	-	-	-	-	-	-	-	-
Netherlands Antilles	-	-	-	-	-	-	-	-	-
Peru	-	-	-	-	-	-	-	-	-
Trinidad and Tobago	-	-	-	-	-	-	-	-	-
Venezuela	-	-	-	-	-	-	-	-	-
Other	-	-	-	-	-	-	-	-	-
Latin America	**-**	**-**	**-**	**-**	**-**	**-**	**-**	**-**	**-**
Romania	-	-	-	-	-	-	-	-	-
Other	-	-	-	-	-	-	-	-	-
Non-OECD Europe	**-**	**-**	**-**	**-**	**-**	**-**	**-**	**-**	**-**
FSU	**-**	**-**	**-**	**-**	**-**	**-**	**15**	**-**	**7 e**
Iran	-	-	-	-	-	-	-	-	-
Iraq	-	-	-	-	-	-	-	-	-
Kuwait	-	-	-	-	-	-	-	-	-
Oman	-	-	-	-	-	-	-	-	-
Qatar	-	-	-	-	-	-	-	-	-
Saudi Arabia	-	-	-	-	-	-	-	-	-
United Arab Emirates	-	-	-	-	-	-	-	-	-
Other	-	-	-	-	-	-	-	-	-
Middle East	**-**	**-**	**-**	**-**	**-**	**-**	**-**	**-**	**-**
India	-	-	-	-	-	-	-	-	-
Indonesia	-	-	-	-	-	-	-	-	-
Singapore	-	-	-	-	-	-	-	-	-
Other	-	-	-	-	-	-	-	-	-
Asia	**-**	**-**	**-**	**-**	**-**	**-**	**-**	**-**	**-**
China	**-**	**-**	**-**	**-**	**-**	**-**	**-**	**-**	**-**
Algeria	-	-	-	-	-	-	-	-	-
Angola/Cabinada	-	-	-	-	-	-	-	-	-
Cameroon	-	-	-	-	-	-	-	-	-
Egypt	-	-	-	-	-	-	-	-	-
Gabon	-	-	-	-	-	-	-	-	-
Libya	-	-	-	-	-	-	-	-	-
Nigeria	-	-	-	-	-	-	-	-	-
Tunisia	-	-	-	-	-	-	-	-	-
Other	-	-	-	-	-	-	-	-	-
Africa	**-**	**-**	**-**	**-**	**-**	**-**	**-**	**-**	**-**
Non-spec./Other	**-**	**-**	**-**	**-**	**-**	**-**	**-**	**-**	**-**
Total Imports	**929**	**1 102**	**989 e**	**38**	**33**	**52 e**	**1 111**	**1 345**	**1 255 e**
Memo: Total OPEC	-	-	-	-	-	-	-	-	-

Source: Annual Oil Statistics.

CZECH REPUBLIC
Table 7D: Imports by Country of Origin

000 metric tons

	2000	2001	2002	2000	2001	2002	2000	2001	2002
	Fuel Oil (Residual)			**Petroleum Coke**			**Other Products**		
Australia	-	-	-	-	-	-	-	-	-
Belgium	-	-	-	-	-	1 e	8	9	11 e
Canada	-	-	-	-	-	-	-	-	-
Czech Republic	-	-	-	-	-	-	-	-	-
Denmark	-	-	-	-	-	-	-	-	-
Finland	-	-	-	-	-	-	-	-	-
France	-	-	-	-	-	-	8	7	6 e
Germany	3	16	35 e	5	5	5 e	67	87	87 e
Greece	-	-	-	-	-	-	-	-	-
Hungary	4	-	-	-	2	8 e	2	1	11 e
Ireland	-	-	-	-	-	-	-	-	-
Italy	-	-	-	-	-	-	2	2	3 e
Japan	-	-	-	-	-	-	-	-	-
Korea	-	-	-	-	-	-	-	-	-
Mexico	-	-	-	-	-	-	-	-	-
Netherlands	-	-	-	3	3	-	3	3	3 e
New Zealand	-	-	-	-	-	-	-	-	-
Norway	-	-	-	-	-	-	-	-	-
Poland	100	92	89 e	-	-	-	23	36	42 e
Spain	-	-	-	-	-	-	-	-	1 e
Sweden	-	-	-	-	-	-	-	-	-
Turkey	-	-	-	-	-	-	-	-	-
United Kingdom	-	-	-	-	-	-	-	-	1 e
United States	-	-	-	3	-	2 e	-	-	-
Other	74	163	88 e	3	1	1 e	93	104	91 e
OECD	**181**	**271**	**212 e**	**14**	**11**	**17 e**	**206**	**249**	**256 e**
Argentina	-	-	-	-	-	-	-	-	-
Brazil	-	-	-	-	-	-	-	-	-
Colombia	-	-	-	-	-	-	-	-	-
Ecuador	-	-	-	-	-	-	-	-	-
Netherlands Antilles	-	-	-	-	-	-	-	-	-
Peru	-	-	-	-	-	-	-	-	-
Trinidad and Tobago	-	-	-	-	-	-	-	-	-
Venezuela	-	-	-	-	-	-	-	-	-
Other	-	-	-	-	-	-	-	-	-
Latin America	**-**	**-**	**-**	**-**	**-**	**-**	**-**	**-**	**-**
Romania	-	-	-	-	-	1 e	-	-	-
Other	-	-	-	-	-	-	-	-	1 e
Non-OECD Europe	**-**	**-**	**-**	**-**	**-**	**1 e**	**-**	**-**	**1 e**
FSU	**-**	**-**	**-**	**-**	**-**	**-**	**1**	**1**	**1 e**
Iran	-	-	-	-	-	-	-	-	-
Iraq	-	-	-	-	-	-	-	-	-
Kuwait	-	-	-	-	-	-	-	-	-
Oman	-	-	-	-	-	-	-	-	-
Qatar	-	-	-	-	-	-	-	-	-
Saudi Arabia	-	-	-	-	-	-	-	-	-
United Arab Emirates	-	-	-	-	-	-	-	-	-
Other	-	-	-	-	-	-	-	-	-
Middle East	**-**	**-**	**-**	**-**	**-**	**-**	**-**	**-**	**-**
India	-	-	-	-	-	-	-	-	-
Indonesia	-	-	-	-	-	-	-	-	-
Singapore	-	-	-	-	-	-	-	-	-
Other	-	-	-	-	-	-	-	-	-
Asia	**-**	**-**	**-**	**-**	**-**	**-**	**-**	**-**	**-**
China	**-**	**-**	**-**	**-**	**1**	**-**	**-**	**-**	**-**
Algeria	-	-	-	-	-	-	-	-	-
Angola/Cabinada	-	-	-	-	-	-	-	-	-
Cameroon	-	-	-	-	-	-	-	-	-
Egypt	-	-	-	-	-	-	-	-	-
Gabon	-	-	-	-	-	-	-	-	-
Libya	-	-	-	-	-	-	-	-	-
Nigeria	-	-	-	-	-	-	-	-	-
Tunisia	-	-	-	-	-	-	-	-	-
Other	-	-	-	1	1	-	-	-	-
Africa	**-**	**-**	**-**	**1**	**1**	**-**	**-**	**-**	**-**
Non-spec./Other	**-**	**-**	**-**	**-**	**-**	**-**	**-**	**-**	**-**
Total Imports	**181**	**271**	**212 e**	**15**	**13**	**18 e**	**207**	**250**	**258 e**
Memo: Total OPEC	-	-	-	-	-	-	-	-	-

Source: Annual Oil Statistics.

CZECH REPUBLIC
Table 8A: Exports by Country of Destination

000 metric tons

	2000	2001	2002	2000	2001	2002	2000	2001	2002
	Crude Oil			**Natural Gas Liquids**			**Feedstocks and Other**		
Australia	-	-	-	-	-	-	-	-	-
Austria	31	62	85 e	-	-	-	-	-	-
Belgium	-	-	-	-	-	-	-	-	-
Canada	-	-	-	-	-	-	-	-	-
Denmark	-	-	-	-	-	-	-	-	-
Finland	-	-	-	-	-	-	-	-	-
France	-	-	-	-	-	-	-	-	-
Germany	-	-	-	-	-	-	-	-	-
Greece	-	-	-	-	-	-	-	-	-
Ireland	-	-	-	-	-	-	-	-	-
Italy	-	-	-	-	-	-	-	-	-
Japan	-	-	-	-	-	-	-	-	-
Luxembourg	-	-	-	-	-	-	-	-	-
Mexico	-	-	-	-	-	-	-	-	-
Netherlands	-	-	-	-	-	-	-	-	-
Norway	-	-	-	-	-	-	-	-	-
Portugal	-	-	-	-	-	-	-	-	-
Spain	-	-	-	-	-	-	-	-	-
Sweden	-	-	-	-	-	-	-	-	-
Switzerland	-	-	-	-	-	-	-	-	-
Turkey	-	-	-	-	-	-	-	-	-
United Kingdom	-	-	-	-	-	-	-	-	-
United States	-	-	-	-	-	-	-	-	-
Other OECD	80	57	55 e	-	-	-	-	-	-
OECD	**111**	**119**	**140 e**	**-**	**-**	**-**	**-**	**-**	**-**
Africa	-	-	-	-	-	-	-	-	-
Latin America	-	-	-	-	-	-	-	-	-
Asia	-	-	-	-	-	-	-	-	-
China	-	-	-	-	-	-	-	-	-
Non-OECD Europe	-	-	-	-	-	-	-	-	-
Former Soviet Union	-	-	-	-	-	-	-	-	-
Middle East	-	-	-	-	-	-	-	-	-
Non-specified/Other	-	-	-	-	-	-	-	-	-
Non-OECD	**-**	**-**	**-**	**-**	**-**	**-**	**-**	**-**	**-**
Total Exports	**111**	**119**	**140 e**	**-**	**-**	**-**	**-**	**-**	**-**
	Total Products			**Liquefied Petroleum Gas**			**Naphtha**		
Australia	-	-	-	-	-	-	-	-	-
Austria	242	252	314 e	1	5	4 e	3	-	-
Belgium	-	-	-	-	-	-	-	-	-
Canada	-	-	-	-	-	-	-	-	-
Denmark	-	-	-	-	-	-	-	-	-
Finland	-	-	-	-	-	-	-	-	-
France	-	-	-	-	-	-	-	-	-
Germany	425	316	415 e	-	-	-	-	7	-
Greece	-	-	-	-	-	-	-	-	-
Ireland	-	-	-	-	-	-	-	-	-
Italy	1	2	1 e	-	-	-	-	-	-
Japan	-	-	-	-	-	-	-	-	-
Luxembourg	-	5	7 e	-	-	-	-	-	-
Mexico	-	-	-	-	-	-	-	-	-
Netherlands	-	-	-	-	-	-	-	-	-
Norway	-	-	-	-	-	-	-	-	-
Portugal	-	-	-	-	-	-	-	-	-
Spain	1	1	2 e	-	-	-	-	-	-
Sweden	-	-	-	-	-	-	-	-	-
Switzerland	1	1	1 e	-	-	-	-	-	-
Turkey	-	-	-	-	-	-	-	-	-
United Kingdom	-	-	-	-	-	-	-	-	-
United States	-	-	-	-	-	-	-	-	-
Other OECD	295	573	500 e	13	23	37 e	5	65	10 e
OECD	**965**	**1 150**	**1 240 e**	**14**	**28**	**41 e**	**8**	**72**	**10 e**
Africa	-	-	-	-	-	-	-	-	-
Latin America	-	-	-	-	-	-	-	-	-
Asia	-	-	-	-	-	-	-	-	-
China	-	-	-	-	-	-	-	-	-
Non-OECD Europe	8	6	3 e	2	4	1 e	-	-	-
Former Soviet Union	-	-	-	-	-	-	-	-	-
Middle East	-	-	-	-	-	-	-	-	-
Non-specified/Other	-	-	-	-	-	-	-	-	-
Non-OECD	**8**	**6**	**3 e**	**2**	**4**	**1 e**	**-**	**-**	**-**
Total Exports	**973**	**1 156**	**1 243 e**	**16**	**32**	**42 e**	**8**	**72**	**10 e**

Source: Annual Oil Statistics.

CZECH REPUBLIC
Table 8B: Exports by Country of Destination

000 metric tons

	2000	2001	2002	2000	2001	2002	2000	2001	2002
	Total Gasoline			**Total Kerosene**			**Gas/Diesel Oil**		
Australia	-	-	-	-	-	-	-	-	-
Austria	-	13	33 e	-	-	-	198	206	253 e
Belgium	-	-	-	-	-	-	-	-	-
Canada	-	-	-	-	-	-	-	-	-
Denmark	-	-	-	-	-	-	-	-	-
Finland	-	-	-	-	-	-	-	-	-
France	-	-	-	-	-	-	-	-	-
Germany	-	1	75 e	-	-	-	196	212	239 e
Greece	-	-	-	-	-	-	-	-	-
Ireland	-	-	-	-	-	-	-	-	-
Italy	-	-	-	-	-	-	-	-	-
Japan	-	-	-	-	-	-	-	-	-
Luxembourg	-	-	-	-	-	-	-	-	-
Mexico	-	-	-	-	-	-	-	-	-
Netherlands	-	-	-	-	-	-	-	-	-
Norway	-	-	-	-	-	-	-	-	-
Portugal	-	-	-	-	-	-	-	-	-
Spain	-	-	-	-	-	-	-	-	-
Sweden	-	-	-	-	-	-	-	-	-
Switzerland	-	-	-	-	-	-	-	-	-
Turkey	-	-	-	-	-	-	-	-	-
United Kingdom	-	-	-	-	-	-	-	-	-
United States	-	-	-	-	-	-	-	-	-
Other OECD	75	195	166 e	6	-	-	98	185	201 e
OECD	**75**	**209**	**274 e**	**6**	**-**	**-**	**492**	**603**	**693 e**
Africa	-	-	-	-	-	-	-	-	-
Latin America	-	-	-	-	-	-	-	-	-
Asia	-	-	-	-	-	-	-	-	-
China	-	-	-	-	-	-	-	-	-
Non-OECD Europe	-	-	-	-	-	-	-	-	-
Former Soviet Union	-	-	-	-	-	-	-	-	-
Middle East	-	-	-	-	-	-	-	-	-
Non-specified/Other	-	-	-	-	-	-	-	-	-
Non-OECD	**-**	**-**	**-**	**-**	**-**	**-**	**-**	**-**	**-**
Total Exports	**75**	**209**	**274 e**	**6**	**-**	**-**	**492**	**603**	**693 e**
	Fuel Oil (Residual)			**Petroleum Coke**			**Other Products**		
Australia	-	-	-	-	-	-	-	-	-
Austria	14	17	1 e	-	-	-	26	11	23 e
Belgium	-	-	-	-	-	-	-	-	-
Canada	-	-	-	-	-	-	-	-	-
Denmark	-	-	-	-	-	-	-	-	-
Finland	-	-	-	-	-	-	-	-	-
France	-	-	-	-	-	-	-	-	-
Germany	70	45	30 e	-	-	-	159	51	71 e
Greece	-	-	-	-	-	-	-	-	-
Ireland	-	-	-	-	-	-	-	-	-
Italy	-	-	-	-	-	-	1	2	1 e
Japan	-	-	-	-	-	-	-	-	-
Luxembourg	-	-	-	-	-	-	-	5	7 e
Mexico	-	-	-	-	-	-	-	-	-
Netherlands	-	-	-	-	-	-	-	-	-
Norway	-	-	-	-	-	-	-	-	-
Portugal	-	-	-	-	-	-	-	-	-
Spain	-	-	-	-	-	-	1	1	2 e
Sweden	-	-	-	-	-	-	-	-	-
Switzerland	-	-	-	-	-	-	1	1	1 e
Turkey	-	-	-	-	-	-	-	-	-
United Kingdom	-	-	-	-	-	-	-	-	-
United States	-	-	-	-	-	-	-	-	-
Other OECD	76	67	36 e	-	-	-	22	38	50 e
OECD	**160**	**129**	**67 e**	**-**	**-**	**-**	**210**	**109**	**155 e**
Africa	-	-	-	-	-	-	-	-	-
Latin America	-	-	-	-	-	-	-	-	-
Asia	-	-	-	-	-	-	-	-	-
China	-	-	-	-	-	-	-	-	-
Non-OECD Europe	-	-	-	-	-	-	6	2	2 e
Former Soviet Union	-	-	-	-	-	-	-	-	-
Middle East	-	-	-	-	-	-	-	-	-
Non-specified/Other	-	-	-	-	-	-	-	-	-
Non-OECD	**-**	**-**	**-**	**-**	**-**	**-**	**6**	**2**	**2 e**
Total Exports	**160**	**129**	**67 e**	**-**	**-**	**-**	**216**	**111**	**157 e**

Source: Annual Oil Statistics.

DENMARK

Figure 1
FUEL SHARES OF TPES *

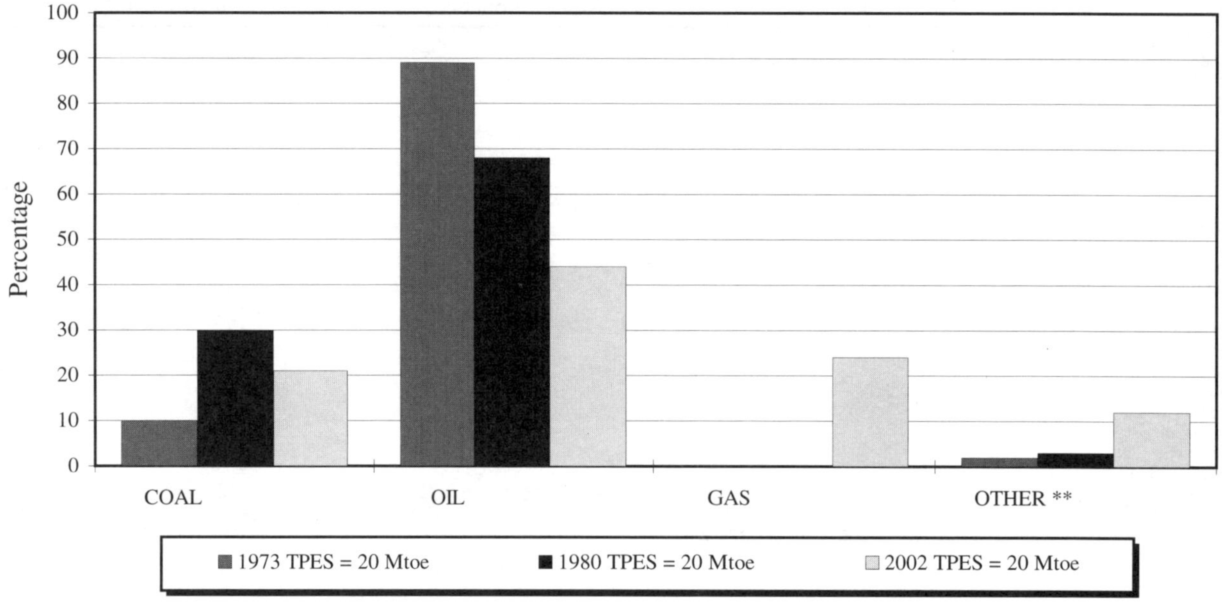

Source: Annual Energy Balances.

* Excluding electricity trade.

** Includes hydro, geothermal, solar, wind, combustible renewables and waste, etc.

Figure 2
OIL PRODUCTION AND CONSUMPTION

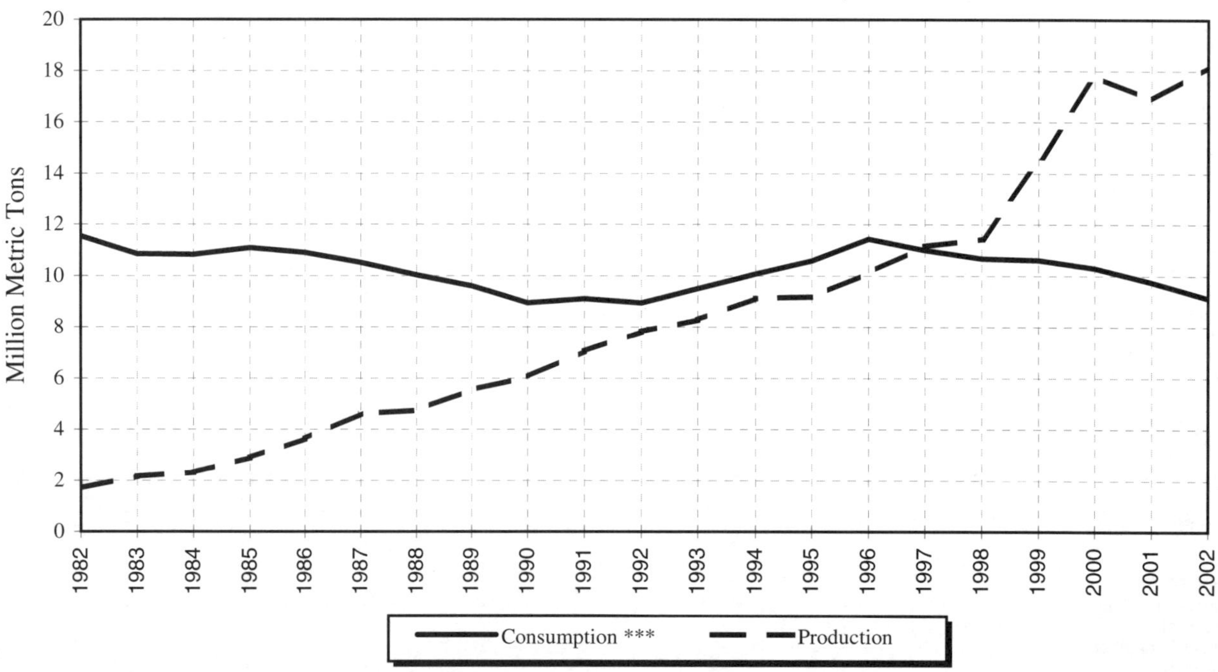

Source: Annual Oil Statistics.

*** Includes refinery fuel and international marine bunkers.

DENMARK
Table 1: Oil Supply and Consumption - Summary Table

000 metric tons

	1973	1978	1980	1990	1999	2000	2001	2002
Indigenous Production	68	432	298	6 037	14 465	17 780	16 887	18 183 e
Crude Oil	..	432	298	6 037	14 465	17 780	16 887	18 183 e
NGL	..	-	-	-	-	-	-	-
Other	..	-	-	-	-	-	-	-
+ Net Imports	18 453	16 281	13 158	2 699	-4 457	-7 809	-6 433	-8 870 e
Crude Oil, NGL, Other	9 760	7 681	6 355	1 988	-5 055	-8 205	-7 524	-9 756 e
Petroleum Products	8 693	8 600	6 803	711	598	396	1 091	886 e
- Intl. Marine Bunkers	702	499	430	981	1 316	1 356	1 122	949 e
+ Stock Changes	- 514	- 26	127	330	581	329	- 422	315 e
TOTAL SUPPLY	**17 305**	**16 188**	**13 153**	**8 085**	**9 273**	**8 944**	**8 910**	**8 679 e**
Refinery Intake	*10 107*	*8 605*	*6 649*	*7 939*	*8 142*	*8 289*	*8 143*	*7 865 e*
Refinery Output	*9 760*	*8 020*	*6 616*	*7 846*	*8 110*	*8 231*	*7 945*	*7 598 e*
LPG + Ethane	*227*	*129*	*107*	*143*	*158*	*175*	*167*	*162 e*
Naphtha	*178*	*85*	*107*	*251*	-	*6*	*5*	-
Kerosenes	*141*	*94*	*36*	*325*	*519*	*496*	*543*	*529 e*
Gasolines	*1 544*	*1 416*	*1 106*	*1 318*	*2 259*	*2 278*	*2 239*	*2 039 e*
Gas/Diesel Oil	*3 551*	*3 304*	*2 789*	*3 290*	*3 273*	*3 260*	*3 220*	*3 190 e*
Fuel Oil	*3 717*	*2 476*	*2 116*	*2 220*	*1 599*	*1 717*	*1 567*	*1 602 e*
Other Products	*402*	*516*	*355*	*299*	*302*	*299*	*204*	*76 e*
- Refinery Losses	347	585	33	93	32	58	198	267 e
+ Returns	-	-	-	18	21	15	17	15 e
- Statistical Differences	- 88	- 120	- 339	54	- 22	- 35	116	256 e
TOTAL CONSUMPTION	**17 046**	**15 723**	**13 459**	**7 956**	**9 284**	**8 936**	**8 613**	**8 171 e**
Transformation and Energy	**3 060**	**2 686**	**1 650**	**635**	**1 849**	**1 750**	**1 601**	..
Crude Oil	-	-	-	-	-	-	-	..
LPG + Ethane	30	16	18	4	-	-	-	..
Gas/Diesel Oil	12	9	39	52	31	31	40	..
Fuel Oil	2 919	2 335	1 333	293	256	170	248	..
Petroleum Coke	-	-	-	-	-	-	-	..
Other	99	326	260	286	1 562	1 549	1 313	..
Transport	**3 314**	**3 614**	**3 434**	**3 914**	**4 663**	**4 620**	**4 415**	..
Motor Gasoline	1 526	1 726	1 505	1 545	1 959	1 931	1 904	..
Gas/Diesel Oil	978	995	1 100	1 566	1 804	1 781	1 606	..
Subtotal	*2 504*	*2 721*	*2 605*	*3 111*	*3 763*	*3 712*	*3 510*	
Aviation Fuels	719	805	723	665	820	825	823	..
Other	91	88	106	138	80	83	82	..
Industry	**3 437**	**2 743**	**2 583**	**1 220**	**1 108**	**1 068**	**1 085**	..
Naphtha	186	-	-	-	-	-	-	..
LPG + Ethane	130	93	100	56	50	43	37	..
Gas/Diesel Oil	837	887	704	371	362	340	385	..
Fuel Oil	1 692	1 270	1 303	421	215	203	184	..
Other	592	493	476	372	481	482	479	..
Other Sectors	**7 235**	**6 680**	**5 792**	**2 187**	**1 664**	**1 498**	**1 512**	..
Gas/Diesel Oil	4 646	4 463	3 744	1 888	1 521	1 381	1 386	..
Fuel Oil	2 222	2 027	1 873	68	71	53	46	..
Other	367	190	175	231	72	64	80	..
Memo: Non-Energy Use	*727*	*478*	*468*	*313*	*310*	*300*	*264*	..
Chem/Petchem. Feedst.	*183*	-	-	-	-	-	-	..
Other	*544*	*478*	*468*	*313*	*310*	*300*	*264*	..

Source: Annual Oil Statistics.

DENMARK
Table 2: Supply of Crude Oil, NGL, Feedstocks and Other Hydrocarbons

000 metric tons

	1973	1978	1980	1990	1999	2000	2001	2002
CRUDE OIL								
Indigenous Production	..	432	298	6 037	14 465	17 780	16 887	18 183 e
+ Imports	..	7 725	5 728	4 112	4 530	3 706	3 042	3 153 e
- Exports	..	102	36	2 789	11 015	13 395	11 934	13 635 e
- Direct Use	..	-	-	-	-	-	-	-
+ Stock Change	..	- 60	37	- 25	- 142	- 52	1	69 e
- Statistical Difference	..	- 12	- 373	55	- 50	- 14	38	49 e
Refinery Intake	**..**	**8 007**	**6 400**	**7 280**	**7 888**	**8 053**	**7 958**	**7 721 e**
Closing Stock Level	..	*531*	*345*	*440*	*503*	*555*	*554*	*485 e*
NATURAL GAS LIQUIDS								
Indigenous Production	..	-	-	-	-	-	-	-
+ Imports	..	-	-	-	-	-	-	-
- Exports	..	-	-	-	-	-	-	-
- Direct Use	..	-	-	-	-	-	-	-
+ Stock Change	..	-	-	-	-	-	-	-
- Statistical Difference	..	-	-	-	-	-	-	-
Refinery Intake	**..**	**-**	**-**	**-**	**-**	**-**	**-**	**-**
Closing Stock Level	..	-	-	-	-	-	-	-
REFINERY FEEDSTOCKS								
Backflows	..	-	-	-	-	-	-	-
+ Transfers	..	-	-	-	-	-	6	-
+ Imports	..	58	1 057	674	241	288	177	20 e
- Exports	..	-	394	9	10	18	19	-
+ Stock Change	..	- 26	83	6	22	- 15	- 5	- 10 e
- Statistical Difference	..	- 566	497	12	- 1	19	- 26	- 134 e
Refinery Intake	**..**	**598**	**249**	**659**	**254**	**236**	**185**	**144 e**
Closing Stock Level	..	*72*	*80*	*115*	*153*	*168*	*173*	*183 e*
OTHER HYDROCARBONS								
Indigenous Production	..	-	-	-	-	-	-	-
+ Imports	..	-	-	-	1 199	1 214	1 210	706 e
- Exports	..	-	-	-	-	-	-	-
- Direct Use	..	-	-	-	1 240	1 236	1 093	857 e
+ Stock Change	..	-	-	-	41	22	- 117	151 e
- Statistical Difference	..	-	-	-	-	-	-	-
Refinery Intake	**..**	**-**	**-**	**-**	**-**	**-**	**-**	**-**
Closing Stock Level	..	-	-	-	*128*	*106*	*223*	*72 e*
CRUDE OIL, NGL, FEEDSTOCKS AND OTHER								
Indigenous Production	68	432	298	6 037	14 465	17 780	16 887	18 183 e
+ Backflows	-	-	-	-	-	-	-	-
+ Transfers	-	-	-	-	-	-	6	-
+ Imports	9 799	7 783	6 785	4 786	5 970	5 208	4 429	3 879 e
- Exports	39	102	430	2 798	11 025	13 413	11 953	13 635 e
- Direct Use	-	-	-	-	1 240	1 236	1 093	857 e
+ Stock Change	- 87	- 86	120	- 19	- 79	- 45	- 121	210 e
- Statistical Difference	- 366	- 578	124	67	- 51	5	12	- 85 e
Refinery Intake	**10 107**	**8 605**	**6 649**	**7 939**	**8 142**	**8 289**	**8 143**	**7 865 e**
Closing Stock Level	-	*603*	*425*	*555*	*784*	*829*	*950*	*740 e*

Source: Annual Oil Statistics.

DENMARK

Table 3: Motor Gasoline - Supply

000 metric tons

	1973	1978	1980	1990	1999	2000	2001	2002
MOTOR GASOLINE								
Refinery Output	**1 544**	**1 416**	**1 106**	**1 318**	**2 259**	**2 278**	**2 239**	**2 039 e**
+ Imports	720	784	735	647	935	1 014	1 031	997 e
- Exports	533	495	346	444	1 160	1 388	1 296	1 114 e
+ Transfers	-	-	-	- 8	- 18	6	- 73	-
+ Stock Change	-	91	28	64	- 11	58	- 14	- 5 e
- Statistical Difference	96	29	- 11	- 5	10	-	- 52	- 36 e
Total Consumption	**1 635**	**1 767**	**1 534**	**1 582**	**1 995**	**1 968**	**1 939**	**1 953 e**
Closing Stock Level	-	*459*	*444*	*478*	*466*	*408*	*422*	*427 e*

Table 4: Aviation Fuels - Supply

000 metric tons

	1973	1978	1980	1990	1999	2000	2001	2002
AVIATION GASOLINE								
Refinery Output	-	-	-	-	-	-	-	-
+ Imports	19	7	4	4	3	2	1	2 e
- Exports	-	-	-	-	-	-	-	-
+ Transfers	-	-	-	-	-	-	-	-
+ Stock Change	3	- 1	1	- 1	-	- 1	-	-
- Statistical Difference	5	1	-	-	1	- 1	- 1	-
Total Consumption	**17**	**5**	**5**	**3**	**2**	**2**	**2**	**2 e**
Closing Stock Level	-	*3*	*2*	*2*	*1*	*2*	*2*	*2 e*
GASOLINE TYPE JET FUEL								
Refinery Output	-	-	-	-	-	-	-	-
+ Imports	-	143	179	-	-	-	-	-
- Exports	-	-	-	-	-	-	-	-
+ Transfers	-	-	-	-	-	-	-	-
+ Stock Change	-	-	-	-	-	-	-	-
- Statistical Difference	-	- 18	22	-	-	-	-	-
Total Consumption	**-**	**161**	**157**	**-**	**-**	**-**	**-**	**-**
Closing Stock Level	-	-	-	-	-	-	-	-
KEROSENE TYPE JET FUEL								
Refinery Output	**13**	**10**	**10**	**249**	**519**	**496**	**543**	**529 e**
+ Imports	719	619	584	516	533	674	800	631 e
- Exports	-	-	-	8	324	348	404	369 e
+ Transfers	-	-	-	- 52	- 4	- 5	- 6	- 6 e
+ Stock Change	- 6	- 10	- 28	- 38	94	7	- 112	- 13 e
- Statistical Difference	24	- 20	5	5	-	1	-	47 e
Total Consumption	**702**	**639**	**561**	**662**	**818**	**823**	**821**	**725 e**
Closing Stock Level	-	*54*	*72*	*67*	*90*	*83*	*195*	*208 e*

Source: Annual Oil Statistics.

DENMARK
Table 5A: Gas/Diesel Oil - Supply and Consumption

000 metric tons

	1973	1978	1980	1990	1999	2000	2001	2002
Refinery Output	**3 551**	**3 304**	**2 789**	**3 290**	**3 273**	**3 260**	**3 220**	**3 190 e**
+ Recycled Products	-	-	-	-	-	-	-	-
+ Imports	4 336	4 119	3 265	1 646	1 945	1 984	1 835	1 833 e
- Exports	852	1 032	436	868	1 177	1 219	888	1 047 e
- Intl. Marine Bunkers	181	186	164	272	611	536	501	522 e
+ Transfers	-	-	-	2	25	- 28	- 25	-
+ Stock Changes	- 62	302	- 114	57	289	68	- 171	93 e
- Statistical Differences	319	153	- 247	- 22	26	- 4	53	248 e
Total Consumption	**6 473**	**6 354**	**5 587**	**3 877**	**3 718**	**3 533**	**3 417**	**3 299 e**
Transformation	**12**	**9**	**39**	**52**	**31**	**31**	**40**	**..**
Public Electricity	-	-	-	1	1	1	1	..
Autoproducer Electricity	-	-	-	-	-	-	-	..
Public CHP Plants	11	8	39	6	9	6	5	..
Autoproducers of CHP	-	-	-	-	1	1	1	..
Public Heat Plants	-	-	-	45	17	20	30	..
Autoproducer Heat Plants	-	-	-	-	1	3	3	..
Petrochemical Industry	-	-	-	-	-	-	-	..
Other Transformation	1	1	-	-	2	-	-	..
Energy Sector	**-**	**-**	**-**	**-**	**-**	**-**	**-**	**..**
Refinery Fuel	-	-	-	-	-	-	-	..
Other Energy Sector	-	-	-	-	-	-	-	..
Final Consumption	**6 461**	**6 345**	**5 548**	**3 825**	**3 687**	**3 502**	**3 377**	**..**
Transport	**978**	**995**	**1 100**	**1 566**	**1 804**	**1 781**	**1 606**	**..**
Air	-	-	-	-	-	-	-	..
Road	434	578	684	1 404	1 628	1 621	1 442	..
Rail	104	97	106	94	73	72	67	..
Inland Waterways	440	320	310	65	80	79	76	..
Non-Specified	-	-	-	3	23	9	21	..
Industry	**837**	**887**	**704**	**371**	**362**	**340**	**385**	**..**
Iron and Steel	59	69	42	4	2	2	3	..
Chemical (incl. Pet.-Chem.)	67	69	61	15	10	10	12	..
Non-Ferrous Metals	-	-	-	3	1	1	1	..
Non-Metallic Minerals	151	145	85	37	20	21	25	..
Transport Equipment	42	38	30	12	12	7	8	..
Machinery	201	138	139	67	53	54	63	..
Mining and Quarrying	25	15	6	15	20	22	25	..
Food and Tobacco	167	199	165	76	48	48	56	..
Pulp, Paper and Printing	25	15	18	11	7	7	8	..
Wood and Wood Products	42	31	24	6	8	7	9	..
Construction	-	122	98	83	154	139	148	..
Textiles and Leather	50	38	30	10	4	3	4	..
Non-Specified	8	8	6	32	23	19	23	..
Other Sectors	**4 646**	**4 463**	**3 744**	**1 888**	**1 521**	**1 381**	**1 386**	**..**
Commerce and Public	-	316	307	276	135	116	111	..
Residential	3 890	3 221	2 634	1 088	835	709	738	..
Agriculture	730	821	743	524	551	556	537	..
Non-Specified	26	105	60	-	-	-	-	..
Memo: Non-Energy Use	-	-	-	-	-	-	-	..
Chem/Petchem. Feedst.	-	-	-	-	-	-	-	..
Other	-	-	-	-	-	-	-	..
Closing Stock Level	-	*1 997*	*2 092*	*1 697*	*862*	*794*	*965*	*872 e*

Source: Annual Oil Statistics.

DENMARK
Table 5B: Fuel Oil (Residual) - Supply and Consumption

000 metric tons

	1973	1978	1980	1990	1999	2000	2001	2002
Refinery Output	**3 717**	**2 476**	**2 116**	**2 220**	**1 599**	**1 717**	**1 567**	**1 602 e**
+ Recycled Products	-	-	-	-	-	-	-	-
+ Imports	4 959	4 228	2 481	219	655	749	736	712 e
- Exports	1 088	180	178	1 092	1 243	1 441	1 133	1 178 e
- Intl. Marine Bunkers	521	307	261	707	702	816	617	427 e
+ Transfers	-	-	-	38	- 10	16	82	-
+ Stock Changes	- 379	- 375	156	233	277	209	- 5	31 e
- Statistical Differences	- 185	164	- 240	41	- 1	- 29	115	89 e
Total Consumption	**6 873**	**5 678**	**4 554**	**870**	**577**	**463**	**515**	**651 e**
Transformation	**2 919**	**2 246**	**1 244**	**261**	**229**	**139**	**217**	..
Public Electricity	-	-	-	44	15	8	5	..
Autoproducer Electricity	-	-	-	-	-	-	-	..
Public CHP Plants	2 919	2 198	1 196	154	164	85	172	..
Autoproducers of CHP	-	48	48	31	25	24	21	..
Public Heat Plants	-	-	-	32	25	22	19	..
Autoproducer Heat Plants	-	-	-	-	-	-	-	..
Petrochemical Industry	-	-	-	-	-	-	-	..
Other Transformation	-	-	-	-	-	-	-	..
Energy Sector	-	**89**	**89**	**32**	**27**	**31**	**31**	..
Refinery Fuel	-	89	89	32	27	31	31	..
Other Energy Sector	-	-	-	-	-	-	-	..
Final Consumption	**3 954**	**3 343**	**3 221**	**577**	**321**	**293**	**267**	..
Transport	**40**	**46**	**45**	**88**	**35**	**37**	**37**	..
Air	-	-	-	-	-	-	-	..
Road	-	-	-	-	-	-	-	..
Rail	-	-	-	-	-	-	-	..
Inland Waterways	40	46	45	88	35	37	37	..
Non-Specified	-	-	-	-	-	-	-	..
Industry	**1 692**	**1 270**	**1 303**	**421**	**215**	**203**	**184**	..
Iron and Steel	102	88	77	1	-	-	-	..
Chemical (incl. Pet.-Chem.)	169	126	142	58	18	17	15	..
Non-Ferrous Metals	-	-	-	1	1	1	-	..
Non-Metallic Minerals	508	290	270	66	40	36	33	..
Transport Equipment	34	25	26	4	-	-	-	..
Machinery	85	76	64	24	3	3	3	..
Mining and Quarrying	17	25	26	9	12	11	10	..
Food and Tobacco	557	504	553	180	124	119	113	..
Pulp, Paper and Printing	118	63	52	25	2	2	1	..
Wood and Wood Products	34	25	26	11	10	10	9	..
Construction	-	10	15	25	2	1	-	..
Textiles and Leather	68	38	52	11	-	-	-	..
Non-Specified	-	-	-	6	3	3	-	..
Other Sectors	**2 222**	**2 027**	**1 873**	**68**	**71**	**53**	**46**	..
Commerce and Public	-	196	193	26	11	8	5	..
Residential	1 939	1 596	1 430	5	1	1	1	..
Agriculture	276	226	242	37	59	44	40	..
Non-Specified	7	9	8	-	-	-	-	..
Memo: Non-Energy Use	-	-	-	-	-	-	-	..
Chem/Petchem. Feedst.	-	-	-	-	-	-	-	..
Other	-	-	-	-	-	-	-	..
Closing Stock Level	-	*3 202*	*2 697*	*1 409*	*617*	*408*	*413*	*382 e*

Source: Annual Oil Statistics.

DENMARK
Table 5C: Total Products - Supply and Consumption

000 metric tons

	1973	1978	1980	1990	1999	2000	2001	2002
Refinery Output	**9 760**	**8 020**	**6 616**	**7 846**	**8 110**	**8 231**	**7 945**	**7 598 e**
+ Recycled Products	-	-	-	18	21	15	17	15 e
+ Imports	11 514	10 576	7 935	3 523	4 624	4 986	4 966	4 711 e
- Exports	2 821	1 976	1 132	2 812	4 026	4 590	3 875	3 825 e
- Intl. Marine Bunkers	702	499	430	981	1 316	1 356	1 122	949 e
+ Transfers	-	-	-	-	1 240	1 236	1 087	857 e
+ Stock Changes	- 427	60	7	349	660	374	- 301	105 e
- Statistical Differences	278	458	- 463	- 13	29	- 40	104	341 e
Total Consumption	**17 046**	**15 723**	**13 459**	**7 956**	**9 284**	**8 936**	**8 613**	**8 171 e**
Transformation	**3 060**	**2 377**	**1 360**	**334**	**1 520**	**1 420**	**1 366**	**..**
Public Electricity	-	-	-	45	16	9	6	..
Autoproducer Electricity	-	-	-	-	-	-	-	..
Public CHP Plants	2 930	2 206	1 235	160	1 413	1 327	1 270	..
Autoproducers of CHP	-	48	48	31	26	25	22	..
Public Heat Plants	-	-	-	94	62	56	65	..
Autoproducer Heat Plants	-	-	-	-	1	3	3	..
Petrochemical Industry	-	-	-	-	-	-	-	..
Other Transformation	130	123	77	4	2	-	-	..
Energy Sector	**-**	**309**	**290**	**301**	**329**	**330**	**235**	**..**
Refinery Fuel	-	309	290	301	329	330	235	..
Other Energy Sector	-	-	-	-	-	-	-	..
Final Consumption	**13 986**	**13 037**	**11 809**	**7 321**	**7 435**	**7 186**	**7 012**	**..**
Transport	**3 314**	**3 614**	**3 434**	**3 914**	**4 663**	**4 620**	**4 415**	**..**
Air	719	805	723	665	822	826	824	..
Road	2 010	2 345	2 250	2 994	3 618	3 587	3 380	..
Rail	104	97	106	97	76	75	70	..
Inland Waterways	481	367	355	155	123	123	120	..
Non-Specified	-	-	-	3	24	9	21	..
Industry	**3 437**	**2 743**	**2 583**	**1 220**	**1 108**	**1 068**	**1 085**	**..**
Iron and Steel	174	165	130	11	6	7	6	..
Chemical (incl. Pet.-Chem.)	459	204	211	78	31	30	30	..
Non-Ferrous Metals	-	-	-	8	3	3	2	..
Non-Metallic Minerals	710	477	432	209	289	297	327	..
Transport Equipment	81	66	59	19	15	10	10	..
Machinery	313	231	225	114	75	75	82	..
Mining and Quarrying	42	40	32	36	36	36	38	..
Food and Tobacco	756	719	729	277	184	178	178	..
Pulp, Paper and Printing	146	81	74	39	12	11	11	..
Wood and Wood Products	81	57	52	23	21	20	21	..
Construction	373	474	424	339	399	369	347	..
Textiles and Leather	120	80	85	25	7	5	6	..
Non-Specified	182	149	130	42	30	27	27	..
Other Sectors	**7 235**	**6 680**	**5 792**	**2 187**	**1 664**	**1 498**	**1 512**	**..**
Commerce and Public	-	516	507	319	151	128	120	..
Residential	6 068	4 950	4 184	1 249	889	757	802	..
Agriculture	1 134	1 097	1 030	619	624	613	590	..
Non-Specified	33	117	71	-	-	-	-	..
Memo: Non-Energy Use	*727*	*478*	*468*	*313*	*310*	*300*	*264*	..
Chem/Petchem. Feedst.	*183*	-	-	-	-	-	-	..
Other	*544*	*478*	*468*	*313*	*310*	*300*	*264*	..
Closing Stock Level	-	*5 843*	*5 496*	*3 839*	*2 208*	*1 834*	*2 135*	*2 030 e*

Source: Annual Oil Statistics.

DENMARK
Table 6A: Supply and Consumption

000 metric tons

	1980	2000	2001	2002	1980	2000	2001	2002
	LPG AND ETHANE				**REFINERY GAS**			
Refinery Output	**107**	**175**	**167**	**162 e**	**203**	**299**	**204**	**76 e**
+ Recycled Products	-	-	-	-	-	-	-	-
+ Imports	143	8	7	1 e	-	-	-	-
- Exports	11	109	102	101 e	-	-	-	-
- Intl. Marine Bunkers	-	-	-	-	-	-	-	-
+ Transfers	-	5	5	-	-	-	-	-
+ Stock Change	- 1	-	- 2	-	-	-	-	-
- Statistical Difference	- 1	2	4	- 6 e	-	-	-	-
Total Consumption	**239**	**77**	**71**	**68 e**	**203**	**299**	**204**	**76 e**
of which: Ethane	-	-	-	-	-	-	-	-
Transformation	18	-	-	..	-	-	-	..
Energy	-	-	-	..	201	299	204	..
Transport	58	12	12	..	-	-	-	..
Industry	100	43	37	..	2	-	-	..
Other	63	22	22	..	-	-	-	..
Memo: Non-Energy Use	-	-	-	..	-	-	-	..
Closing Stock Level	*12*	*6*	*8*	*8 e*	-	-	-	-
	NAPHTHA				**OTHER KEROSENE**			
Refinery Output	**107**	**6**	**5**	**-**	**26**	**-**	**-**	**-**
+ Recycled Products	-	-	-	-	-	-	-	-
+ Imports	36	-	-	-	56	-	-	-
- Exports	61	3	6	-	1	-	-	-
- Intl. Marine Bunkers	-	-	-	-	-	-	-	-
+ Transfers	-	-	-	-	-	4	6	6 e
+ Stock Change	- 3	- 3	2	-	1	-	-	-
- Statistical Difference	20	-	1	-	- 10	- 1	-	-
Total Consumption	**59**	**-**	**-**	**-**	**92**	**5**	**6**	**6 e**
Transformation	59	-	-	..	-	-	-	..
Energy	-	-	-	..	-	-	-	..
Transport	-	-	-	..	3	1	1	..
Industry	-	-	-	..	3	-	-	..
Other	-	-	-	..	86	4	5	..
Memo: Non-Energy Use	-	-	-	..	-	-	-	..
Closing Stock Level	*6*	*3*	*1*	*1 e*	*47*	-	-	-
	WHITE SPIRIT + SBP				**LUBRICANTS**			
Refinery Output	**12**	**-**	**-**	**..**	**-**	**-**	**-**	**..**
+ Recycled Products	-	-	-	..	-	-	-	..
+ Imports	38	27	22	..	117	61	62	..
- Exports	4	2	1	..	18	12	10	..
- Intl. Marine Bunkers	-	-	-	..	5	4	4	..
+ Transfers	-	-	-	..	-	2	1	..
+ Stock Change	3	-	-	..	- 1	1	1	..
- Statistical Difference	17	-	-	..	19	- 4	-	..
Total Consumption	**32**	**25**	**21**	**..**	**74**	**52**	**50**	**..**
Transformation	-	-	-	..	-	-	-	..
Energy	-	-	-	..	-	-	-	..
Transport	-	-	-	..	-	33	32	..
Industry	28	25	21	..	74	15	14	..
Other	4	-	-	..	-	4	4	..
Memo: Non-Energy Use	*32*	*25*	*21*	..	*74*	*52*	*50*	..
Closing Stock Level	*8*	-	-	..	*43*	*8*	*7*	..

Source: Annual Oil Statistics.

DENMARK
Table 6B: Supply and Consumption

000 metric tons

	1980	2000	2001	2002	1980	2000	2001	2002
	BITUMEN				PARAFFIN WAXES			
Refinery Output	**140**	**-**	**-**	**..**	**-**	**-**	**-**	**..**
+ Recycled Products	-	-	-	..	-	-	-	..
+ Imports	205	233	173	..	13	-	-	..
- Exports	73	15	7	..	1	-	-	..
- Intl. Marine Bunkers	-	-	-	..	-	-	-	..
+ Transfers	-	1 236	1 097	..	-	-	-	..
+ Stock Change	7	- 2	5	..	-	-	-	..
- Statistical Difference	- 25	- 7	- 18	..	-	-	-	..
Total Consumption	**304**	**1 459**	**1 286**	**..**	**12**	**-**	**-**	**..**
Transformation	-	1 236	1 093	..	-	-	-	..
Energy	-	-	-	..	-	-	-	..
Transport	-	-	-	..	-	-	-	..
Industry	304	223	193	..	12	-	-	..
Other	-	-	-	..	-	-	-	..
Memo: Non-Energy Use	*304*	*223*	*193*	..	*12*	-	-	..
Closing Stock Level	*31*	*25*	*20*	..	-	-	-	..
	PETROLEUM COKE				OTHER PRODUCTS			
Refinery Output	**-**	**-**	**-**	**-**	**-**	**-**	**-**	**-**
+ Recycled Products	-	-	-	-	-	15	17	15 e
+ Imports	67	234	299	268 e	12	-	-	267 e
- Exports	1	53	28	16 e	2	-	-	-
- Intl. Marine Bunkers	-	-	-	-	-	-	-	-
+ Transfers	-	-	-	-	-	-	-	857 e
+ Stock Change	- 42	37	- 5	3 e	-	-	-	- 4 e
- Statistical Difference	- 12	3	2	5 e	-	-	-	- 6 e
Total Consumption	**36**	**215**	**264**	**250 e**	**10**	**15**	**17**	**1 141 e**
Transformation	-	-	-	..	-	14	16	..
Energy	-	-	-	..	-	-	-	..
Transport	-	-	-	..	-	-	-	..
Industry	36	215	248	..	10	1	1	..
Other	-	-	16	..	-	-	-	..
Memo: Non-Energy Use	*36*	-	-	..	*10*	-	-	..
Closing Stock Level	*42*	*97*	*102*	*99 e*	-	-	-	*31 e*

Source: Annual Oil Statistics.

DENMARK
Table 7A: Imports by Country of Origin

000 metric tons

	2000	2001	2002	2000	2001	2002	2000	2001	2002
	Crude Oil			Natural Gas Liquids			Feedstocks and Other		
Australia	-	-	-	-	-	-	-	-	-
Belgium	-	-	-	-	-	-	-	-	-
Canada	-	-	-	-	-	-	-	-	-
Czech Republic	-	-	-	-	-	-	-	-	-
Denmark	-	-	-	-	-	-	-	-	-
Finland	-	-	-	-	-	-	-	-	-
France	-	-	-	-	-	-	-	-	-
Germany	-	-	-	-	-	-	-	-	-
Greece	-	-	-	-	-	-	-	-	-
Hungary	-	-	-	-	-	-	-	-	-
Ireland	-	-	-	-	-	-	-	-	-
Italy	-	-	-	-	-	-	-	-	-
Japan	-	-	-	-	-	-	-	-	-
Korea	-	-	-	-	-	-	-	-	-
Mexico	-	-	-	-	-	-	-	-	-
Netherlands	-	-	-	-	-	-	-	-	-
New Zealand	-	-	-	-	-	-	-	-	-
Norway	3 596	2 937	3 123 e	-	-	-	147	106	20 e
Poland	13	10	-	-	-	-	122	63	-
Spain	-	-	-	-	-	-	-	-	-
Sweden	-	-	-	-	-	-	-	8	-
Turkey	-	-	-	-	-	-	-	-	-
United Kingdom	-	-	-	-	-	-	2	-	-
United States	-	-	-	-	-	-	-	-	-
Other	-	-	-	-	-	-	-	-	-
OECD	**3 609**	**2 947**	**3 123 e**	**-**	**-**	**-**	**271**	**177**	**20 e**
Argentina	-	-	-	-	-	-	-	-	-
Brazil	-	-	-	-	-	-	-	-	-
Colombia	-	-	-	-	-	-	-	-	-
Ecuador	-	-	-	-	-	-	-	-	-
Netherlands Antilles	-	-	-	-	-	-	-	-	-
Peru	-	-	-	-	-	-	-	-	-
Trinidad and Tobago	-	-	-	-	-	-	-	-	-
Venezuela	-	-	-	-	-	-	1 214	1 210	706 e
Other	-	-	-	-	-	-	-	-	-
Latin America	**-**	**-**	**-**	**-**	**-**	**-**	**1 214**	**1 210**	**706 e**
Romania	-	-	-	-	-	-	-	-	-
Other	-	-	-	-	-	-	-	-	-
Non-OECD Europe	**-**	**-**	**-**	**-**	**-**	**-**	**-**	**-**	**-**
FSU	**97**	**95**	**30 e**	**-**	**-**	**-**	**17**	**-**	**-**
Iran	-	-	-	-	-	-	-	-	-
Iraq	-	-	-	-	-	-	-	-	-
Kuwait	-	-	-	-	-	-	-	-	-
Oman	-	-	-	-	-	-	-	-	-
Qatar	-	-	-	-	-	-	-	-	-
Saudi Arabia	-	-	-	-	-	-	-	-	-
United Arab Emirates	-	-	-	-	-	-	-	-	-
Other	-	-	-	-	-	-	-	-	-
Middle East	**-**	**-**	**-**	**-**	**-**	**-**	**-**	**-**	**-**
India	-	-	-	-	-	-	-	-	-
Indonesia	-	-	-	-	-	-	-	-	-
Singapore	-	-	-	-	-	-	-	-	-
Other	-	-	-	-	-	-	-	-	-
Asia	**-**	**-**	**-**	**-**	**-**	**-**	**-**	**-**	**-**
China	**-**	**-**	**-**	**-**	**-**	**-**	**-**	**-**	**-**
Algeria	-	-	-	-	-	-	-	-	-
Angola/Cabinada	-	-	-	-	-	-	-	-	-
Cameroon	-	-	-	-	-	-	-	-	-
Egypt	-	-	-	-	-	-	-	-	-
Gabon	-	-	-	-	-	-	-	-	-
Libya	-	-	-	-	-	-	-	-	-
Nigeria	-	-	-	-	-	-	-	-	-
Tunisia	-	-	-	-	-	-	-	-	-
Other	-	-	-	-	-	-	-	-	-
Africa	**-**	**-**	**-**	**-**	**-**	**-**	**-**	**-**	**-**
Non-spec./Other	**-**	**-**	**-**	**-**	**-**	**-**	**-**	**-**	**-**
Total Imports	**3 706**	**3 042**	**3 153 e**	**-**	**-**	**-**	**1 502**	**1 387**	**726 e**
Memo: Total OPEC	-	-	-	-	-	-	*1 214*	*1 210*	*706 e*

Source: Annual Oil Statistics.

DENMARK
Table 7B: Imports by Country of Origin

000 metric tons

	Total Products			Liquefied Petroleum Gases			Naphtha		
	2000	2001	2002	2000	2001	2002	2000	2001	2002
Australia	-	-	-	-	-	-	-	-	-
Belgium	24	16	20 e	-	-	-	-	-	-
Canada	64	22	-	-	-	-	-	-	-
Czech Republic	-	-	-	-	-	-	-	-	-
Denmark	-	-	-	-	-	-	-	-	-
Finland	259	185	186 e	-	-	-	-	-	-
France	32	54	15 e	-	-	-	-	-	-
Germany	110	123	169 e	-	-	1 e	-	-	-
Greece	-	-	-	-	-	-	-	-	-
Hungary	-	-	-	-	-	-	-	-	-
Ireland	-	-	-	-	-	-	-	-	-
Italy	-	-	-	-	-	-	-	-	-
Japan	-	-	-	-	-	-	-	-	-
Korea	-	-	-	-	-	-	-	-	-
Mexico	-	-	-	-	-	-	-	-	-
Netherlands	98	71	47 e	-	1	-	-	-	-
New Zealand	-	-	-	-	-	-	-	-	-
Norway	497	493	785 e	2	2	-	-	-	-
Poland	245	188	236 e	1	-	-	-	-	-
Spain	-	-	-	-	-	-	-	-	-
Sweden	1 742	1 630	1 413 e	5	3	-	-	-	-
Turkey	-	-	-	-	-	-	-	-	-
United Kingdom	264	376	217 e	-	1	-	-	-	-
United States	219	220	224 e	-	-	-	-	-	-
Other	-	40	43 e	-	-	-	-	-	-
OECD	**3 554**	**3 418**	**3 355 e**	**8**	**7**	**1 e**	**-**	**-**	**-**
Argentina	-	-	-	-	-	-	-	-	-
Brazil	-	-	-	-	-	-	-	-	-
Colombia	-	-	-	-	-	-	-	-	-
Ecuador	-	-	-	-	-	-	-	-	-
Netherlands Antilles	-	-	-	-	-	-	-	-	-
Peru	-	-	-	-	-	-	-	-	-
Trinidad and Tobago	-	-	-	-	-	-	-	-	-
Venezuela	-	101	298 e	-	-	-	-	-	-
Other	-	-	-	-	-	-	-	-	-
Latin America	**-**	**101**	**298 e**	**-**	**-**	**-**	**-**	**-**	**-**
Romania	-	-	-	-	-	-	-	-	-
Other	-	-	-	-	-	-	-	-	-
Non-OECD Europe	**-**	**-**	**-**	**-**	**-**	**-**	**-**	**-**	**-**
FSU	**917**	**972**	**922 e**	**-**	**-**	**-**	**-**	**-**	**-**
Iran	-	-	-	-	-	-	-	-	-
Iraq	-	-	-	-	-	-	-	-	-
Kuwait	182	157	41 e	-	-	-	-	-	-
Oman	-	-	-	-	-	-	-	-	-
Qatar	-	-	-	-	-	-	-	-	-
Saudi Arabia	137	205	-	-	-	-	-	-	-
United Arab Emirates	-	-	43 e	-	-	-	-	-	-
Other	96	113	52 e	-	-	-	-	-	-
Middle East	**415**	**475**	**136 e**	**-**	**-**	**-**	**-**	**-**	**-**
India	-	-	-	-	-	-	-	-	-
Indonesia	-	-	-	-	-	-	-	-	-
Singapore	-	-	-	-	-	-	-	-	-
Other	68	-	-	-	-	-	-	-	-
Asia	**68**	**-**	**-**	**-**	**-**	**-**	**-**	**-**	**-**
China	**-**	**-**	**-**	**-**	**-**	**-**	**-**	**-**	**-**
Algeria	32	-	-	-	-	-	-	-	-
Angola/Cabinada	-	-	-	-	-	-	-	-	-
Cameroon	-	-	-	-	-	-	-	-	-
Egypt	-	-	-	-	-	-	-	-	-
Gabon	-	-	-	-	-	-	-	-	-
Libya	-	-	-	-	-	-	-	-	-
Nigeria	-	-	-	-	-	-	-	-	-
Tunisia	-	-	-	-	-	-	-	-	-
Other	-	-	-	-	-	-	-	-	-
Africa	**32**	**-**	**-**	**-**	**-**	**-**	**-**	**-**	**-**
Non-spec./Other	**-**	**-**	**-**	**-**	**-**	**-**	**-**	**-**	**-**
Total Imports	**4 986**	**4 966**	**4 711 e**	**8**	**7**	**1 e**	**-**	**-**	**-**
Memo: Total OPEC	*351*	*463*	*382 e*	-	-	-	-	-	-

Source: Annual Oil Statistics.

DENMARK
Table 7C: Imports by Country of Origin

000 metric tons

	2000	2001	2002	2000	2001	2002	2000	2001	2002
	Total Gasoline			**Total Kerosene**			**Gas/Diesel**		
Australia	-	-	-	-	-	-	-	-	-
Belgium	-	-	-	2	-	-	-	-	3 e
Canada	-	-	-	-	-	-	64	22	-
Czech Republic	-	-	-	-	-	-	-	-	-
Denmark	-	-	-	-	-	-	-	-	-
Finland	-	26	30 e	-	-	-	230	137	132 e
France	9	16	-	-	-	-	-	19	-
Germany	4	7	2 e	16	11	-	26	34	81 e
Greece	-	-	-	-	-	-	-	-	-
Hungary	-	-	-	-	-	-	-	-	-
Ireland	-	-	-	-	-	-	-	-	-
Italy	-	-	-	-	-	-	-	-	-
Japan	-	-	-	-	-	-	-	-	-
Korea	-	-	-	-	-	-	-	-	-
Mexico	-	-	-	-	-	-	-	-	-
Netherlands	29	-	31 e	1	-	-	-	23	-
New Zealand	-	-	-	-	-	-	-	-	-
Norway	242	246	250 e	26	2	42 e	208	217	410 e
Poland	-	-	-	32	36	20 e	-	11	67 e
Spain	-	-	-	-	-	-	-	-	-
Sweden	611	543	552 e	6	3	-	762	789	639 e
Turkey	-	-	-	-	-	-	-	-	-
United Kingdom	113	192	134 e	86	101	41 e	62	18	10 e
United States	-	-	-	-	-	-	-	-	-
Other	-	-	-	-	-	-	-	40	43 e
OECD	**1 008**	**1 030**	**999 e**	**169**	**153**	**103 e**	**1 352**	**1 310**	**1 385 e**
Argentina	-	-	-	-	-	-	-	-	-
Brazil	-	-	-	-	-	-	-	-	-
Colombia	-	-	-	-	-	-	-	-	-
Ecuador	-	-	-	-	-	-	-	-	-
Netherlands Antilles	-	-	-	-	-	-	-	-	-
Peru	-	-	-	-	-	-	-	-	-
Trinidad and Tobago	-	-	-	-	-	-	-	-	-
Venezuela	-	-	-	-	101	298 e	-	-	-
Other	-	-	-	-	-	-	-	-	-
Latin America	**-**	**-**	**-**	**-**	**101**	**298 e**	**-**	**-**	**-**
Romania	-	-	-	-	-	-	-	-	-
Other	-	-	-	-	-	-	-	-	-
Non-OECD Europe	**-**	**-**	**-**	**-**	**-**	**-**	**-**	**-**	**-**
FSU	**8**	**2**	**-**	**58**	**71**	**94 e**	**564**	**525**	**448 e**
Iran	-	-	-	-	-	-	-	-	-
Iraq	-	-	-	-	-	-	-	-	-
Kuwait	-	-	-	182	157	41 e	-	-	-
Oman	-	-	-	-	-	-	-	-	-
Qatar	-	-	-	-	-	-	-	-	-
Saudi Arabia	-	-	-	107	205		-	-	-
United Arab Emirates	-	-	-	-	-	43 e	-	-	-
Other	-	-	-	96	113	52 e	-	-	-
Middle East	**-**	**-**	**-**	**415**	**475**	**136 e**	**-**	**-**	**-**
India	-	-	-	-	-	-	-	-	-
Indonesia	-	-	-	-	-	-	-	-	-
Singapore	-	-	-	-	-	-	-	-	-
Other	-	-	-	-	-	-	68	-	-
Asia	**-**	**-**	**-**	**-**	**-**	**-**	**68**	**-**	**-**
China	**-**	**-**	**-**	**-**	**-**	**-**	**-**	**-**	**-**
Algeria	-	-	-	32	-	-	-	-	-
Angola/Cabinada	-	-	-	-	-	-	-	-	-
Cameroon	-	-	-	-	-	-	-	-	-
Egypt	-	-	-	-	-	-	-	-	-
Gabon	-	-	-	-	-	-	-	-	-
Libya	-	-	-	-	-	-	-	-	-
Nigeria	-	-	-	-	-	-	-	-	-
Tunisia	-	-	-	-	-	-	-	-	-
Other	-	-	-	-	-	-	-	-	-
Africa	**-**	**-**	**-**	**32**	**-**	**-**	**-**	**-**	**-**
Non-spec./Other	**-**	**-**	**-**	**-**	**-**	**-**	**-**	**-**	**-**
Total Imports	**1 016**	**1 032**	**999 e**	**674**	**800**	**631 e**	**1 984**	**1 835**	**1 833 e**
Memo: Total OPEC	-	-	-	*351*	*463*	*382 e*	-	-	-

Source: Annual Oil Statistics.

DENMARK
Table 7D: Imports by Country of Origin

000 metric tons

	2000	2001	2002	2000	2001	2002	2000	2001	2002
	Fuel Oil (Residual)			**Petroleum Coke**			**Other Products**		
Australia	-	-	-	-	-	-	-	-	-
Belgium	-	-	-	15	12	16 e	7	4	1 e
Canada	-	-	-	-	-	-	-	-	-
Czech Republic	-	-	-	-	-	-	-	-	-
Denmark	-	-	-	-	-	-	-	-	-
Finland	-	-	-	-	-	-	29	22	24 e
France	11	-	-	-	-	-	12	19	15 e
Germany	-	5	-	-	-	-	64	66	85 e
Greece	-	-	-	-	-	-	-	-	-
Hungary	-	-	-	-	-	-	-	-	-
Ireland	-	-	-	-	-	-	-	-	-
Italy	-	-	-	-	-	-	-	-	-
Japan	-	-	-	-	-	-	-	-	-
Korea	-	-	-	-	-	-	-	-	-
Mexico	-	-	-	-	-	-	-	-	-
Netherlands	-	-	-	-	6	-	68	41	16 e
New Zealand	-	-	-	-	-	-	-	-	-
Norway	-	14	73 e	-	-	-	19	12	10 e
Poland	212	141	149 e	-	-	-	-	-	-
Spain	-	-	-	-	-	-	-	-	-
Sweden	239	202	110 e	-	-	-	119	90	112 e
Turkey	-	-	-	-	-	-	-	-	-
United Kingdom	-	-	-	-	61	28 e	3	3	4 e
United States	-	-	-	219	220	224 e	-	-	-
Other	-	-	-	-	-	-	-	-	-
OECD	**462**	**362**	**332 e**	**234**	**299**	**268 e**	**321**	**257**	**267 e**
Argentina	-	-	-	-	-	-	-	-	-
Brazil	-	-	-	-	-	-	-	-	-
Colombia	-	-	-	-	-	-	-	-	-
Ecuador	-	-	-	-	-	-	-	-	-
Netherlands Antilles	-	-	-	-	-	-	-	-	-
Peru	-	-	-	-	-	-	-	-	-
Trinidad and Tobago	-	-	-	-	-	-	-	-	-
Venezuela	-	-	-	-	-	-	-	-	-
Other	-	-	-	-	-	-	-	-	-
Latin America	**-**	**-**	**-**	**-**	**-**	**-**	**-**	**-**	**-**
Romania	-	-	-	-	-	-	-	-	-
Other	-	-	-	-	-	-	-	-	-
Non-OECD Europe	**-**	**-**	**-**	**-**	**-**	**-**	**-**	**-**	**-**
FSU	**287**	**374**	**380 e**	**-**	**-**	**-**	**-**	**-**	**-**
Iran	-	-	-	-	-	-	-	-	-
Iraq	-	-	-	-	-	-	-	-	-
Kuwait	-	-	-	-	-	-	-	-	-
Oman	-	-	-	-	-	-	-	-	-
Qatar	-	-	-	-	-	-	-	-	-
Saudi Arabia	-	-	-	-	-	-	-	-	-
United Arab Emirates	-	-	-	-	-	-	-	-	-
Other	-	-	-	-	-	-	-	-	-
Middle East	**-**	**-**	**-**	**-**	**-**	**-**	**-**	**-**	**-**
India	-	-	-	-	-	-	-	-	-
Indonesia	-	-	-	-	-	-	-	-	-
Singapore	-	-	-	-	-	-	-	-	-
Other	-	-	-	-	-	-	-	-	-
Asia	**-**	**-**	**-**	**-**	**-**	**-**	**-**	**-**	**-**
China	**-**	**-**	**-**	**-**	**-**	**-**	**-**	**-**	**-**
Algeria	-	-	-	-	-	-	-	-	-
Angola/Cabinada	-	-	-	-	-	-	-	-	-
Cameroon	-	-	-	-	-	-	-	-	-
Egypt	-	-	-	-	-	-	-	-	-
Gabon	-	-	-	-	-	-	-	-	-
Libya	-	-	-	-	-	-	-	-	-
Nigeria	-	-	-	-	-	-	-	-	-
Tunisia	-	-	-	-	-	-	-	-	-
Other	-	-	-	-	-	-	-	-	-
Africa	**-**	**-**	**-**	**-**	**-**	**-**	**-**	**-**	**-**
Non-spec./Other	**-**	**-**	**-**	**-**	**-**	**-**	**-**	**-**	**-**
Total Imports	**749**	**736**	**712 e**	**234**	**299**	**268 e**	**321**	**257**	**267 e**
Memo: Total OPEC	-	-	-	-	-	-	-	-	-

Source: Annual Oil Statistics.

DENMARK
Table 8A: Exports by Country of Destination

000 metric tons

	2000	2001	2002	2000	2001	2002	2000	2001	2002
	Crude Oil			Natural Gas Liquids			Feedstocks and Other		
Australia	-	-	-	-	-	-	-	-	-
Austria	-	-	-	-	-	-	-	-	-
Belgium	-	163	235 e	-	-	-	-	-	-
Canada	-	172	289 e	-	-	-	-	-	-
Denmark	-	-	-	-	-	-	-	-	-
Finland	2 154	2 266	2 875 e	-	-	-	-	-	-
France	702	1 779	842 e	-	-	-	-	-	-
Germany	672	508	944 e	-	-	-	-	-	-
Greece	-	-	-	-	-	-	-	-	-
Ireland	-	169	-	-	-	-	-	-	-
Italy	-	-	-	-	-	-	-	-	-
Japan	-	-	-	-	-	-	-	-	-
Luxembourg	-	-	-	-	-	-	-	-	-
Mexico	-	-	-	-	-	-	-	-	-
Netherlands	3 616	2 229	1 881 e	-	-	-	-	4	-
Norway	400	-	465 e	-	-	-	4	-	-
Portugal	169	85	86 e	-	-	-	-	-	-
Spain	-	-	172 e	-	-	-	-	-	-
Sweden	3 545	2 653	3 258 e	-	-	-	-	-	-
Switzerland	-	-	-	-	-	-	-	-	-
Turkey	-	-	-	-	-	-	-	-	-
United Kingdom	1 717	1 571	1 034 e	-	-	-	6	-	-
United States	420	339	1 554 e	-	-	-	-	15	-
Other OECD	-	-	-	-	-	-	8	-	-
OECD	**13 395**	**11 934**	**13 635 e**	**-**	**-**	**-**	**18**	**19**	**-**
Africa	-	-	-	-	-	-	-	-	-
Latin America	-	-	-	-	-	-	-	-	-
Asia	-	-	-	-	-	-	-	-	-
China	-	-	-	-	-	-	-	-	-
Non-OECD Europe	-	-	-	-	-	-	-	-	-
Former Soviet Union	-	-	-	-	-	-	-	-	-
Middle East	-	-	-	-	-	-	-	-	-
Non-specified/Other	-	-	-	-	-	-	-	-	-
Non-OECD	**-**	**-**	**-**	**-**	**-**	**-**	**-**	**-**	**-**
Total Exports	**13 395**	**11 934**	**13 635 e**	**-**	**-**	**-**	**18**	**19**	**-**

	2000	2001	2002	2000	2001	2002	2000	2001	2002
	Total Products			Liquefied Petroleum Gas			Naphtha		
Australia	-	-	-	-	-	-	-	-	-
Austria	-	-	-	-	-	-	-	-	-
Belgium	15	-	49 e	2	-	1 e	-	-	-
Canada	-	-	-	-	-	-	-	-	-
Denmark	-	-	-	-	-	-	-	-	-
Finland	84	200	186 e	-	-	-	-	-	-
France	121	32	65 e	2	4	1 e	-	-	-
Germany	231	180	126 e	-	2	2 e	-	-	-
Greece	-	-	-	-	-	-	-	-	-
Ireland	1	-	-	1	-	-	-	-	-
Italy	69	27	34 e	-	-	-	-	-	-
Japan	-	-	-	-	-	-	-	-	-
Luxembourg	-	-	-	-	-	-	-	-	-
Mexico	-	-	-	-	-	-	-	-	-
Netherlands	229	72	32 e	15	15	1 e	3	6	-
Norway	2 112	2 018	2 007 e	51	71	76 e	-	-	-
Portugal	-	-	-	-	-	-	-	-	-
Spain	29	12	-	-	-	-	-	-	-
Sweden	1 146	881	930 e	27	1	12 e	-	-	-
Switzerland	-	-	-	-	-	-	-	-	-
Turkey	-	-	-	-	-	-	-	-	-
United Kingdom	346	156	121 e	6	8	-	-	-	-
United States	166	45	38 e	-	-	-	-	-	-
Other OECD	31	93	52 e	5	1	8 e	-	-	-
OECD	**4 580**	**3 716**	**3 640 e**	**109**	**102**	**101 e**	**3**	**6**	**-**
Africa	-	-	-	-	-	-	-	-	-
Latin America	-	-	-	-	-	-	-	-	-
Asia	-	-	-	-	-	-	-	-	-
China	-	-	-	-	-	-	-	-	-
Non-OECD Europe	-	58	80 e	-	-	-	-	-	-
Former Soviet Union	5	46	-	-	-	-	-	-	-
Middle East	-	-	-	-	-	-	-	-	-
Non-specified/Other	5	55	105 e	-	-	-	-	-	-
Non-OECD	**10**	**159**	**185 e**	**-**	**-**	**-**	**-**	**-**	**-**
Total Exports	**4 590**	**3 875**	**3 825 e**	**109**	**102**	**101 e**	**3**	**6**	**-**

Source: Annual Oil Statistics.

DENMARK
Table 8B: Exports by Country of Destination

000 metric tons

	2000	2001	2002	2000	2001	2002	2000	2001	2002
	Total Gasoline			**Total Kerosene**			**Gas/Diesel Oil**		
Australia	-	-	-	-	-	-	-	-	-
Austria	-	-	-	-	-	-	-	-	-
Belgium	13	-	-	-	-	-	-	-	48 e
Canada	-	-	-	-	-	-	-	-	-
Denmark	-	-	-	-	-	-	-	-	-
Finland	-	-	-	-	-	-	17	62	-
France	14	-	5 e	-	-	-	80	11	12 e
Germany	49	19	-	-	-	-	162	136	84 e
Greece	-	-	-	-	-	-	-	-	-
Ireland	-	-	-	-	-	-	-	-	-
Italy	-	-	-	-	-	-	-	-	-
Japan	-	-	-	-	-	-	-	-	-
Luxembourg	-	-	-	-	-	-	-	-	-
Mexico	-	-	-	-	-	-	-	-	-
Netherlands	114	29	10 e	-	-	-	77	22	10 e
Norway	233	395	370 e	176	277	241 e	653	491	638 e
Portugal	-	-	-	-	-	-	-	-	-
Spain	-	-	-	-	-	-	29	12	-
Sweden	870	691	641 e	134	93	128 e	82	13	91 e
Switzerland	-	-	-	-	-	-	-	-	-
Turkey	-	-	-	-	-	-	-	-	-
United Kingdom	78	116	88 e	21	5	-	119	8	15 e
United States	-	-	-	-	-	-	-	-	-
Other OECD	9	-	-	17	19	-	-	73	44 e
OECD	**1 380**	**1 250**	**1 114 e**	**348**	**394**	**369 e**	**1 219**	**828**	**942 e**
Africa	-	-	-	-	-	-	-	-	-
Latin America	-	-	-	-	-	-	-	-	-
Asia	-	-	-	-	-	-	-	-	-
China	-	-	-	-	-	-	-	-	-
Non-OECD Europe	-	-	-	-	-	-	-	21	-
Former Soviet Union	5	46	-	-	-	-	-	-	-
Middle East	-	-	-	-	-	-	-	-	-
Non-specified/Other	3	-	-	-	10	-	-	39	105 e
Non-OECD	**8**	**46**	**-**	**-**	**10**	**-**	**-**	**60**	**105 e**
Total Exports	**1 388**	**1 296**	**1 114 e**	**348**	**404**	**369 e**	**1 219**	**888**	**1 047 e**

	2000	2001	2002	2000	2001	2002	2000	2001	2002
	Fuel Oil (Residual)			**Petroleum Coke**			**Other Products**		
Australia	-	-	-	-	-	-	-	-	-
Austria	-	-	-	-	-	-	-	-	-
Belgium	-	-	-	-	-	-	-	-	-
Canada	-	-	-	-	-	-	-	-	-
Denmark	-	-	-	-	-	-	-	-	-
Finland	67	137	186 e	-	-	-	-	1	-
France	25	17	47 e	-	-	-	-	-	-
Germany	-	-	24 e	20	23	16 e	-	-	-
Greece	-	-	-	-	-	-	-	-	-
Ireland	-	-	-	-	-	-	-	-	-
Italy	69	27	34 e	-	-	-	-	-	-
Japan	-	-	-	-	-	-	-	-	-
Luxembourg	-	-	-	-	-	-	-	-	-
Mexico	-	-	-	-	-	-	-	-	-
Netherlands	20	-	11 e	-	-	-	-	-	-
Norway	951	781	682 e	33	-	-	15	3	-
Portugal	-	-	-	-	-	-	-	-	-
Spain	-	-	-	-	-	-	-	-	-
Sweden	21	75	58 e	-	-	-	12	8	-
Switzerland	-	-	-	-	-	-	-	-	-
Turkey	-	-	-	-	-	-	-	-	-
United Kingdom	122	14	18 e	-	5	-	-	-	-
United States	166	45	38 e	-	-	-	-	-	-
Other OECD	-	-	-	-	-	-	-	-	-
OECD	**1 441**	**1 096**	**1 098 e**	**53**	**28**	**16 e**	**27**	**12**	**-**
Africa	-	-	-	-	-	-	-	-	-
Latin America	-	-	-	-	-	-	-	-	-
Asia	-	-	-	-	-	-	-	-	-
China	-	-	-	-	-	-	-	-	-
Non-OECD Europe	-	37	80 e	-	-	-	-	-	-
Former Soviet Union	-	-	-	-	-	-	-	-	-
Middle East	-	-	-	-	-	-	-	-	-
Non-specified/Other	-	-	-	-	-	-	2	6	-
Non-OECD	**-**	**37**	**80 e**	**-**	**-**	**-**	**2**	**6**	**-**
Total Exports	**1 441**	**1 133**	**1 178 e**	**53**	**28**	**16 e**	**29**	**18**	**-**

Source: Annual Oil Statistics.

FINLAND

Figure 1
FUEL SHARES OF TPES *

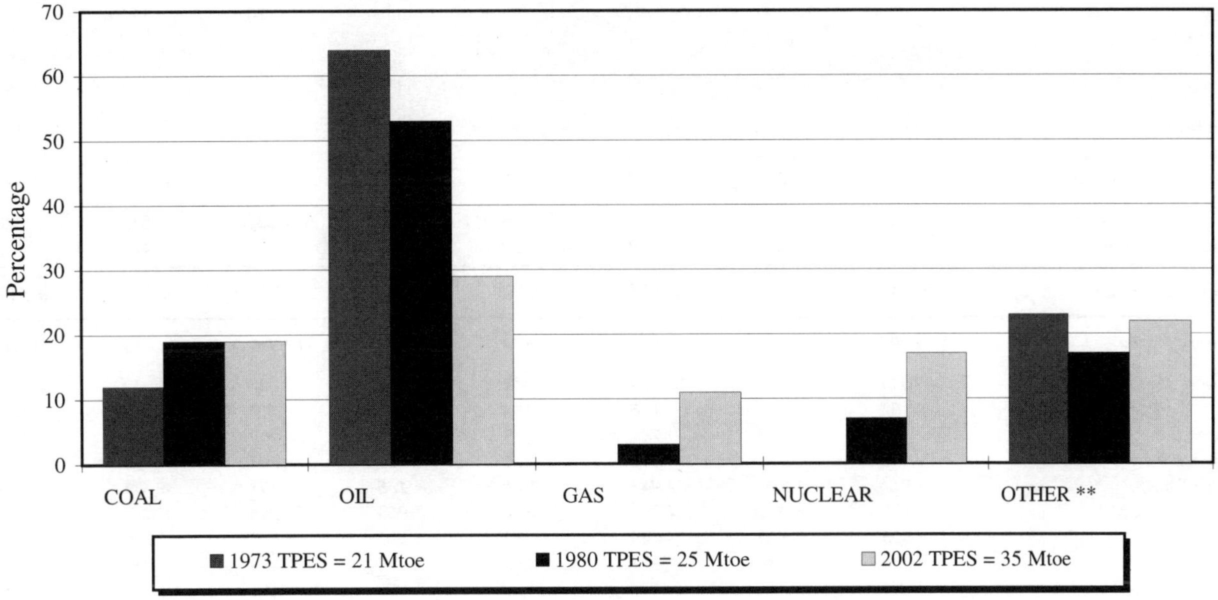

Source: Annual Energy Balances.

* Excluding electricity trade.

** Includes hydro, geothermal, solar, wind, combustible renewables and waste, etc.

Figure 2
OIL PRODUCTION AND CONSUMPTION

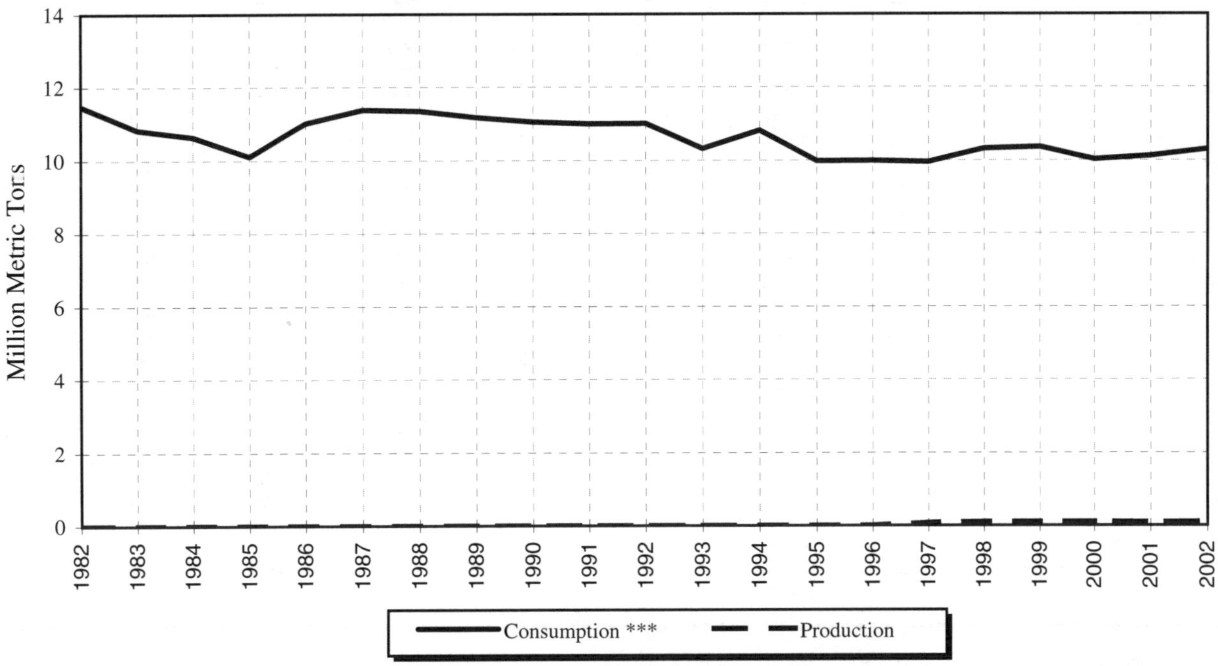

Source: Annual Oil Statistics.

*** Includes refinery fuel and international marine bunkers.

FINLAND
Table 1: Oil Supply and Consumption - Summary Table

000 metric tons

	1973	1978	1980	1990	1999	2000	2001	2002
Indigenous Production	-	-	-	-	114	120	114 e	121 e
Crude Oil	..	-	-	-	-	-	-	-
NGL	..	-	-	-	-	-	-	-
Other	..	-	-	-	114	120	114 e	121 e
+ Net Imports	13 695	11 765	13 743	10 423	10 168	10 405	9 992	9 599 e
Crude Oil, NGL, Other	9 524	10 454	12 876	8 713	11 378	11 834	11 095	11 522 e
Petroleum Products	4 171	1 311	867	1 710	-1 210	-1 429	-1 103	-1 923 e
- Intl. Marine Bunkers	82	261	597	579	567	682	584	620 e
+ Stock Changes	- 171	613	- 328	28	452	- 503 e	- 429	517 e
TOTAL SUPPLY	**13 442**	**12 117**	**12 818**	**9 872**	**10 167**	**9 340 e**	**9 093 e**	**9 617 e**
Refinery Intake	*9 297*	*11 283*	*12 671*	*10 479*	*13 153*	*12 910*	*12 118*	*13 402 e*
Refinery Output	*9 045*	*11 104*	*12 484*	*10 344*	*12 893*	*12 777*	*11 812*	*12 998 e*
LPG + Ethane	*87*	*85*	*116*	*134*	*200*	*217*	*150*	*231 e*
Naphtha	*435*	*613*	*738*	*286*	*270*	*240*	*215*	*204 e*
Kerosenes	*148*	*225*	*248*	*494*	*986*	*765*	*441*	*657 e*
Gasolines	*1 513*	*1 970*	*1 941*	*2 971*	*4 166*	*3 891*	*3 785*	*4 348 e*
Gas/Diesel Oil	*2 572*	*3 618*	*4 281*	*3 663*	*4 965*	*5 248*	*5 058*	*5 176 e*
Fuel Oil	*3 503*	*3 710*	*4 249*	*1 610*	*1 292*	*1 407*	*1 306*	*1 403 e*
Other Products	*787*	*883*	*911*	*1 186*	*1 014*	*1 009*	*857*	*979 e*
- Refinery Losses	252	179	187	135	260	133	306	404 e
+ Returns	156	199	177	89	161	119	127	121 e
- Statistical Differences	- 56	- 516	617	- 648	271	- 14 e	- 623 e	- 349 e
TOTAL CONSUMPTION	**13 402**	**12 653**	**12 191**	**10 474**	**9 797**	**9 340**	**9 537**	**9 683 e**
Transformation and Energy	**2 019**	**2 203**	**2 087**	**1 078**	**1 443**	**1 321**	**1 347**	..
Crude Oil	-	-	-	-	-	-	-	..
LPG + Ethane	9	13	13	16	40	37	60	..
Gas/Diesel Oil	19	24	19	10	14	27	29	..
Fuel Oil	1 392	1 452	1 351	547	771	659	765	..
Petroleum Coke	-	-	-	-	-	-	-	..
Other	599	714	704	505	618	598	493	..
Transport	**2 403**	**2 638**	**2 837**	**4 102**	**4 263**	**4 223**	**4 308**	..
Motor Gasoline	1 229	1 337	1 321	1 971	1 836	1 772	1 809	..
Gas/Diesel Oil	1 002	1 078	1 257	1 636	1 878	1 905	1 966	..
Subtotal	*2 231*	*2 415*	*2 578*	*3 607*	*3 714*	*3 677*	*3 775*	..
Aviation Fuels	154	203	237	451	497	495	495	..
Other	18	20	22	44	52	51	38	..
Industry	**5 138**	**3 885**	**3 791**	**2 608**	**2 028**	**1 797**	**1 705**	..
Naphtha	552	636	607	418	254	256	195	..
LPG + Ethane	19	62	81	129	188	141	123	..
Gas/Diesel Oil	604	549	530	199	248	350	381	..
Fuel Oil	3 463	2 145	2 027	889	638	434	434	..
Other	500	493	546	973	700	616	572	..
Other Sectors	**3 842**	**3 927**	**3 476**	**2 686**	**2 063**	**1 999**	**2 177**	..
Gas/Diesel Oil	2 906	3 123	2 700	2 230	1 875	1 722	1 769	..
Fuel Oil	842	756	724	433	166	203	335	..
Other	94	48	52	23	22	74	73 e	..
Memo: Non-Energy Use	*1 052*	*1 134*	*1 220*	*1 421*	*1 051*	*973*	*841*	..
Chem/Petchem. Feedst.	*559*	*661*	*709*	*720*	*541*	*488*	*396*	..
Other	*493*	*473*	*511*	*701*	*510*	*485*	*445*	..

Source: Annual Oil Statistics.

FINLAND

Table 2: Supply of Crude Oil, NGL, Feedstocks and Other Hydrocarbons

000 metric tons

	1973	1978	1980	1990	1999	2000	2001	2002
CRUDE OIL								
Indigenous Production	..	-	-	-	-	-	-	-
+ Imports	..	10 454	12 876	8 713	10 782	10 874	9 823	10 452 e
- Exports	..	-	-	-	-	-	-	-
- Direct Use	..	-	-	-	-	-	-	-
+ Stock Change	..	526	- 67	548	236	- 177	- 254	370 e
- Statistical Difference	..	- 104	315	- 655	- 88	14	- 334	- 276 e
Refinery Intake	**..**	**11 084**	**12 494**	**9 916**	**11 106**	**10 683**	**9 903**	**11 098 e**
Closing Stock Level	..	*2 200*	*3 037*	*2 284*	*858*	*1 035*	*1 289*	*919 e*
NATURAL GAS LIQUIDS								
Indigenous Production	..	-	-	-	-	-	-	-
+ Imports	..	-	-	-	397	743	1 128	890 e
- Exports	..	-	-	-	-	-	-	-
- Direct Use	..	-	-	-	-	-	-	-
+ Stock Change	..	-	-	-	-	-	-	-
- Statistical Difference	..	-	-	-	- 84	- 23	100	- 202 e
Refinery Intake	**..**	**-**	**-**	**-**	**481**	**766**	**1 028**	**1 092 e**
Closing Stock Level	..	-	-	-	-	-	-	-
REFINERY FEEDSTOCKS								
Backflows	..	199	177	89	161	119	127	121 e
+ Transfers	..	-	-	474	1 181	1 100	811	853 e
+ Imports	..	-	-	-	-	-	-	-
- Exports	..	-	-	-	-	-	-	-
+ Stock Change	..	-	-	-	-	- 152 e	26	- 18 e
- Statistical Difference	..	-	-	-	- 1	- 152	-	-
Refinery Intake	**..**	**199**	**177**	**563**	**1 343**	**1 219**	**964**	**956 e**
Closing Stock Level	..	-	-	-	-	*152 e*	*126*	*144 e*
OTHER HYDROCARBONS								
Indigenous Production	..	-	-	-	114	120	114 e	121 e
+ Imports	..	-	-	-	199	217	144	180 e
- Exports	..	-	-	-	-	-	-	-
- Direct Use	..	-	-	-	90	95	35 e	45 e
+ Stock Change	..	-	-	-	-	-	-	-
- Statistical Difference	..	-	-	-	-	-	-	-
Refinery Intake	**..**	**-**	**-**	**-**	**223**	**242**	**223**	**256 e**
Closing Stock Level	..	-	-	-	-	-	-	-
CRUDE OIL, NGL, FEEDSTOCKS AND OTHER								
Indigenous Production	-	-	-	-	114	120	114 e	121 e
+ Backflows	156	199	177	89	161	119	127	121 e
+ Transfers	-	-	-	474	1 181	1 100	811	853 e
+ Imports	9 524	10 454	12 876	8 713	11 378	11 834	11 095	11 522 e
- Exports	-	-	-	-	-	-	-	-
- Direct Use	-	-	-	-	90	95	35 e	45 e
+ Stock Change	- 320	526	- 67	548	236	- 329	- 228	352 e
- Statistical Difference	63	- 104	315	- 655	- 173	- 161	- 234	- 478 e
Refinery Intake	**9 297**	**11 283**	**12 671**	**10 479**	**13 153**	**12 910**	**12 118**	**13 402 e**
Closing Stock Level	-	*2 200*	*3 037*	*2 284*	*858*	*1 187*	*1 415*	*1 063 e*

Source: Annual Oil Statistics.

FINLAND
Table 3: Motor Gasoline - Supply

000 metric tons

	1973	1978	1980	1990	1999	2000	2001	2002
MOTOR GASOLINE								
Refinery Output	**1 513**	**1 970**	**1 941**	**2 971**	**4 160**	**3 891**	**3 785**	**4 348 e**
+ Imports	92	13	1	-	303	416	363	279 e
- Exports	108	573	590	585	2 501	2 221	2 200	2 751 e
+ Transfers	-	- 2	- 51	- 102	90	95	35 e	45 e
+ Stock Change	- 260	- 17	18	- 110	1	- 137	7	6 e
- Statistical Difference	- 14	38	- 21	188	203	259	167 e	87 e
Total Consumption	**1 251**	**1 353**	**1 340**	**1 986**	**1 850**	**1 785**	**1 823**	**1 840 e**
Closing Stock Level	-	*262*	*470*	*242*	*120*	*257*	*250 e*	*244 e*

Table 4: Aviation Fuels - Supply

000 metric tons

	1973	1978	1980	1990	1999	2000	2001	2002
AVIATION GASOLINE								
Refinery Output	-	-	-	-	**6**	-	-	-
+ Imports	8	7	8	4	2	2	2	3 e
- Exports	-	-	-	-	1	-	-	-
+ Transfers	-	-	-	-	-	-	-	-
+ Stock Change	1	3	1	-	3	-	-	-
- Statistical Difference	-	1	2	-	6	- 1	- 1	-
Total Consumption	**9**	**9**	**7**	**4**	**4**	**3**	**3**	**3 e**
Closing Stock Level	-	*1*	*2*	*1*	-	-	-	-
GASOLINE TYPE JET FUEL								
Refinery Output	-	-	-	-	-	-	-	-
+ Imports	-	-	-	-	-	-	-	-
- Exports	-	-	-	-	-	-	-	-
+ Transfers	-	-	-	-	-	-	-	-
+ Stock Change	-	-	-	-	-	-	-	-
- Statistical Difference	-	-	-	-	-	-	-	-
Total Consumption	-	-	-	-	-	-	-	-
Closing Stock Level	-	-	-	-	-	-	-	-
KEROSENE TYPE JET FUEL								
Refinery Output	**141**	**221**	**242**	**493**	**986**	**765**	**441**	**657 e**
+ Imports	6	2	4	15	-	-	109	-
- Exports	-	20	16	47	451	292	76	64 e
+ Transfers	-	-	-	-	-	-	-	-
+ Stock Change	- 2	- 5	12	- 23	17	20	14	5 e
- Statistical Difference	-	4	12	- 9	59	1	- 4	138 e
Total Consumption	**145**	**194**	**230**	**447**	**493**	**492**	**492**	**460 e**
Closing Stock Level	-	*35*	*24*	*36*	*73*	*53*	*39*	*34 e*

Source: Annual Oil Statistics.

FINLAND
Table 5A: Gas/Diesel Oil - Supply and Consumption

000 metric tons

	1973	1978	1980	1990	1999	2000	2001	2002
Refinery Output	**2 572**	**3 618**	**4 281**	**3 663**	**4 965**	**5 248**	**5 058**	**5 176 e**
+ Recycled Products	-	-	-	-	-	-	-	-
+ Imports	1 783	1 444	1 391	1 098	1 232	1 476	1 670	1 705 e
- Exports	62	534	489	432	1 577	1 947	2 054	2 000 e
- Intl. Marine Bunkers	16	110	168	121	160	188	142	150 e
+ Transfers	-	-	-	- 139	- 390	- 641	- 663	- 670 e
+ Stock Changes	306	277	- 442	- 507	132	- 387	- 295	282 e
- Statistical Differences	52	- 79	67	- 513	187	- 443	- 571	244 e
Total Consumption	**4 531**	**4 774**	**4 506**	**4 075**	**4 015**	**4 004**	**4 145**	**4 099 e**
Transformation	**19**	**24**	**19**	**10**	**14**	**27**	**26**	**..**
Public Electricity	-	2	1	10	14	2	1	..
Autoproducer Electricity	-	-	-	-	-	-	-	..
Public CHP Plants	-	-	-	-	-	2	3	..
Autoproducers of CHP	-	-	-	-	-	1	1	..
Public Heat Plants	19	22	18	-	-	22	20	..
Autoproducer Heat Plants	-	-	-	-	-	-	1	..
Petrochemical Industry	-	-	-	-	-	-	-	..
Other Transformation	-	-	-	-	-	-	-	..
Energy Sector	**-**	**-**	**-**	**-**	**-**	**-**	**3**	**..**
Refinery Fuel	-	-	-	-	-	-	-	..
Other Energy Sector	-	-	-	-	-	-	3	..
Final Consumption	**4 512**	**4 750**	**4 487**	**4 065**	**4 001**	**3 977**	**4 116**	**..**
Transport	**1 002**	**1 078**	**1 257**	**1 636**	**1 878**	**1 905**	**1 966**	**..**
Air	-	-	-	-	-	-	-	..
Road	860	925	1 099	1 544	1 751	1 789	1 824	..
Rail	104	84	88	62	51	47	44	..
Inland Waterways	38	69	70	30	76	69	98	..
Non-Specified	-	-	-	-	-	-	-	..
Industry	**604**	**549**	**530**	**199**	**248**	**350**	**381**	**..**
Iron and Steel	19	35	34	23	32	6	6	..
Chemical (incl. Pet.-Chem.)	44	53	50	24	25	2	2	..
Non-Ferrous Metals	6	8	6	4	1	1	2	..
Non-Metallic Minerals	63	58	51	15	13	15	13	..
Transport Equipment	30	29	26	5	1	3	8	..
Machinery	70	73	73	11	5	3	10	..
Mining and Quarrying	22	21	18	6	4	3	10	..
Food and Tobacco	64	62	56	32	29	3	14	..
Pulp, Paper and Printing	20	20	18	11	13	2	5	..
Wood and Wood Products	25	31	35	6	8	2	5	..
Construction	170	109	113	62	117	303	294	..
Textiles and Leather	15	13	15	-	-	6	8	..
Non-Specified	56	37	35	-	-	1	4	..
Other Sectors	**2 906**	**3 123**	**2 700**	**2 230**	**1 875**	**1 722**	**1 769**	**..**
Commerce and Public	-	-	-	-	278	254	273	..
Residential	2 487	2 617	2 275	1 663	1 145	1 017	928	..
Agriculture	416	500	425	567	452	451	568	..
Non-Specified	3	6	-	-	-	-	-	..
Memo: Non-Energy Use	-	-	-	-	-	-	-	..
Chem/Petchem. Feedst.	-	-	-	-	-	-	-	..
Other	-	-	-	-	-	-	-	..
Closing Stock Level	-	*1 210*	*2 040*	*885*	*872*	*1 259*	*1 554*	*1 272 e*

Source: Annual Oil Statistics.

FINLAND
Table 5B: Fuel Oil (Residual) - Supply and Consumption

000 metric tons

	1973	1978	1980	1990	1999	2000	2001	2002
Refinery Output	**3 503**	**3 710**	**4 249**	**1 610**	**1 292**	**1 407**	**1 306**	**1 403 e**
+ Recycled Products	-	-	-	-	-	-	-	-
+ Imports	2 253	1 377	1 336	1 173	978	674	654	807 e
- Exports	-	519	746	357	-	123	79	88 e
- Intl. Marine Bunkers	66	151	429	458	407	494	442	470 e
+ Transfers	-	-	-	- 268	- 303	- 108	- 37	- 83 e
+ Stock Changes	96	- 225	166	138	24	271	98	- 150 e
- Statistical Differences	81	- 177	457	- 74	- 43	280	- 72	- 212 e
Total Consumption	**5 705**	**4 369**	**4 119**	**1 912**	**1 627**	**1 347**	**1 572**	**1 631 e**
Transformation	**1 392**	**1 452**	**1 351**	**547**	**771**	**659**	**666**	..
Public Electricity	800	300	240	60	30	11	21	..
Autoproducer Electricity	-	30	30	20	2	15	18	..
Public CHP Plants	-	315	355	77	98	62	63	..
Autoproducers of CHP	-	300	240	120	100	41	31	..
Public Heat Plants	562	487	471	266	252	208	296	..
Autoproducer Heat Plants	-	-	-	-	-	6	6	..
Petrochemical Industry	30	20	15	4	15	10	-	..
Other Transformation	-	-	-	-	274	306	231	..
Energy Sector	-	-	-	-	-	-	**99**	..
Refinery Fuel	-	-	-	-	-	-	99	..
Other Energy Sector	-	-	-	-	-	-	-	..
Final Consumption	**4 313**	**2 917**	**2 768**	**1 365**	**856**	**688**	**807**	..
Transport	**8**	**16**	**17**	**43**	**52**	**51**	**38**	..
Air	-	-	-	-	-	-	-	..
Road	-	-	-	-	-	-	-	..
Rail	-	-	-	-	-	-	-	..
Inland Waterways	8	16	17	43	52	51	38	..
Non-Specified	-	-	-	-	-	-	-	..
Industry	**3 463**	**2 145**	**2 027**	**889**	**638**	**434**	**434**	..
Iron and Steel	210	180	358	141	92	23	24	..
Chemical (incl. Pet.-Chem.)	348	440	390	173	152	78	74	..
Non-Ferrous Metals	40	59	50	27	24	8	2	..
Non-Metallic Minerals	150	180	142	85	27	-	16	..
Transport Equipment	28	45	33	36	35	10	5	..
Machinery	66	65	46	88	48	5	4	..
Mining and Quarrying	63	56	61	57	41	19	1	..
Food and Tobacco	360	380	292	139	95	55	28	..
Pulp, Paper and Printing	1 350	570	423	67	52	192	234	..
Wood and Wood Products	354	94	83	41	37	34	31	..
Construction	-	-	-	-	-	-	-	..
Textiles and Leather	70	45	45	34	35	3	7	..
Non-Specified	424	31	104	1	-	7	8	..
Other Sectors	**842**	**756**	**724**	**433**	**166**	**203**	**335**	..
Commerce and Public	-	-	-	-	82	76	85	..
Residential	795	688	646	353	30	81	56	..
Agriculture	47	68	78	80	54	46	55	..
Non-Specified	-	-	-	-	-	-	139	..
Memo: Non-Energy Use	-	-	*49*	*30*	*35*	*30*	*36*	..
Chem/Petchem. Feedst.	-	-	*49*	*30*	*35*	*30*	*36*	..
Other	-	-	-	-	-	-	-	..
Closing Stock Level	-	*1 096*	*979*	*764*	*716*	*445*	*347 e*	*497 e*

Source: Annual Oil Statistics.

FINLAND

Table 5C: Total Products - Supply and Consumption

000 metric tons

	1973	1978	1980	1990	1999	2000	2001	2002
Refinery Output	**9 045**	**11 104**	**12 484**	**10 344**	**12 893**	**12 777**	**11 812**	**12 998 e**
+ Recycled Products	-	-	-	-	-	-	-	-
+ Imports	4 344	3 039	2 963	3 386	3 718	3 518	3 712	3 439 e
- Exports	173	1 728	2 096	1 676	4 928	4 947	4 815	5 362 e
- Intl. Marine Bunkers	82	261	597	579	567	682	584	620 e
+ Transfers	-	-	-	- 474	-1 091	-1 005	- 776 e	- 808 e
+ Stock Changes	149	87	- 261	- 520	216	- 174	- 201	165 e
- Statistical Differences	- 119	- 412	302	7	444	147	- 389	129 e
Total Consumption	**13 402**	**12 653**	**12 191**	**10 474**	**9 797**	**9 340**	**9 537**	**9 683 e**
Transformation	**1 554**	**1 700**	**1 572**	**663**	**941**	**798**	**819**	..
Public Electricity	806	302	241	70	44	13	22	..
Autoproducer Electricity	3	62	60	36	12	15	18	..
Public CHP Plants	-	315	355	77	98	67	66	..
Autoproducers of CHP	-	300	240	120	100	42	32	..
Public Heat Plants	581	509	489	266	252	230	316	..
Autoproducer Heat Plants	-	-	-	-	-	6	7	..
Petrochemical Industry	156	199	177	89	161	119	127	..
Other Transformation	8	13	10	5	274	306	231	..
Energy Sector	**465**	**503**	**515**	**415**	**502**	**523**	**528**	..
Refinery Fuel	465	503	515	415	502	523	525	..
Other Energy Sector	-	-	-	-	-	-	3	..
Final Consumption	**11 383**	**10 450**	**10 104**	**9 396**	**8 354**	**8 019**	**8 190**	..
Transport	**2 403**	**2 638**	**2 837**	**4 102**	**4 263**	**4 223**	**4 308**	..
Air	154	203	237	451	497	495	495	..
Road	2 054	2 266	2 425	3 516	3 587	3 561	3 633	..
Rail	104	84	88	62	51	47	44	..
Inland Waterways	91	85	87	73	128	120	136	..
Non-Specified	-	-	-	-	-	-	-	..
Industry	**5 138**	**3 885**	**3 791**	**2 608**	**2 028**	**1 797**	**1 705**	..
Iron and Steel	232	233	411	176	146	54	54	..
Chemical (incl. Pet.-Chem.)	951	1 154	1 100	887	683	538	436	..
Non-Ferrous Metals	46	69	59	31	25	17	13	..
Non-Metallic Minerals	219	254	208	188	98	36	61	..
Transport Equipment	58	74	59	41	36	13	13	..
Machinery	139	142	127	99	53	9	14	..
Mining and Quarrying	85	77	79	63	45	22	11	..
Food and Tobacco	426	448	351	173	127	61	47	..
Pulp, Paper and Printing	1 390	621	492	122	121	229	291	..
Wood and Wood Products	379	125	118	47	45	36	36	..
Construction	522	427	429	537	433	606	552	..
Textiles and Leather	85	58	60	34	35	9	15	..
Non-Specified	606	203	298	210	181	167	162	..
Other Sectors	**3 842**	**3 927**	**3 476**	**2 686**	**2 063**	**1 999**	**2 177**	..
Commerce and Public	-	-	-	-	360	330	358	..
Residential	3 349	3 337	2 950	2 023	1 183	1 107	993	..
Agriculture	490	584	526	663	520	510	637	..
Non-Specified	3	6	-	-	-	52	189	..
Memo: Non-Energy Use	*1 052*	*1 134*	*1 220*	*1 421*	*1 051*	*973*	*841*	..
Chem/Petchem. Feedst.	*559*	*661*	*709*	*720*	*541*	*488*	*396*	..
Other	*493*	*473*	*511*	*701*	*510*	*485*	*445*	..
Closing Stock Level	-	*2 661*	*3 628*	*2 014*	*1 984*	*2 158*	*2 359*	*2 194 e*

Source: Annual Oil Statistics.

FINLAND
Table 6A: Supply and Consumption

000 metric tons

	1980	2000	2001	2002	1980	2000	2001	2002
	LPG AND ETHANE				**REFINERY GAS**			
Refinery Output	**116**	**217**	**150**	**231 e**	**589**	**568**	**467**	**511 e**
+ Recycled Products	-	-	-	-	-	-	-	-
+ Imports	9	148	217	155 e	-	-	-	-
- Exports	-	-	-	18 e	-	-	-	-
- Intl. Marine Bunkers	-	-	-	-	-	-	-	-
+ Transfers	-	- 98	- 77	- 80 e	-	-	-	-
+ Stock Change	-	- 13	- 19	15 e	-	-	-	-
- Statistical Difference	5	15	29	65 e	-	-	-	-
Total Consumption	**120**	**239**	**242**	**238 e**	**589**	**568**	**467**	**511 e**
of which: Ethane	-	-	-	-	-	-	-	-
Transformation	13	37	60	..	39	30	28	..
Energy	-	-	-	..	515	523	426	..
Transport	-	-	-	..	-	-	-	..
Industry	81	141	123	..	35	15	13	..
Other	26	61	59	..	-	-	-	..
Memo: Non-Energy Use	*18*	*79*	*52*	..	*35*	*15*	*13*	..
Closing Stock Level	*23*	*106*	*125*	*110 e*	-	-	-	-
	NAPHTHA				**OTHER KEROSENE**			
Refinery Output	**738**	**240**	**215**	**204 e**	**6**	**-**	**-**	**-**
+ Recycled Products	-	-	-	-	-	-	-	-
+ Imports	-	158	185	83 e	2	235	73	48 e
- Exports	235	105	68	40 e	-	2	47	66 e
- Intl. Marine Bunkers	-	-	-	-	-	-	-	-
+ Transfers	51	- 20	- 34	- 20 e	-	- 233	-	-
+ Stock Change	- 16	47	- 9	-	-	-	-	-
- Statistical Difference	- 219	19	55	14 e	- 4	-	26	- 18 e
Total Consumption	**757**	**301**	**234**	**213 e**	**12**	**-**	**-**	**-**
Transformation	150	45	39	..	-	-	-	..
Energy	-	-	-	..	-	-	-	..
Transport	-	-	-	..	5	-	-	..
Industry	607	256	195	..	-	-	-	..
Other	-	-	-	..	7	-	-	..
Memo: Non-Energy Use	*607*	*256*	*195*	..	-	-	-	..
Closing Stock Level	*86*	*7*	*16*	*16 e*	*4*	-	-	-
	WHITE SPIRIT + SBP				**LUBRICANTS**			
Refinery Output	**39**	**91**	**50**	**..**	**-**	**71**	**50**	**..**
+ Recycled Products	-	-	-	..	-	-	-	..
+ Imports	6	14	64	..	127	134	125	..
- Exports	8	82	77	..	12	117	159	..
- Intl. Marine Bunkers	-	-	-	..	-	-	-	..
+ Transfers	-	-	-	..	-	-	-	..
+ Stock Change	-	-	-	..	-	-	-	..
- Statistical Difference	-	-	-	..	7	-	- 71	..
Total Consumption	**37**	**23**	**37**	**..**	**108**	**88**	**87**	**..**
Transformation	-	-	-	..	-	-	-	..
Energy	-	-	-	..	-	-	-	..
Transport	-	-	-	..	-	-	-	..
Industry	37	23	37	..	108	88	87	..
Other	-	-	-	..	-	-	-	..
Memo: Non-Energy Use	*37*	*23*	*37*	..	*108*	*88*	*87*	..
Closing Stock Level	-	-	-	..	-	-	-	..

Source: Annual Oil Statistics.

FINLAND
Table 6B: Supply and Consumption

000 metric tons

	1980	2000	2001	2002	1980	2000	2001	2002
	BITUMEN				**PARAFFIN WAXES**			
Refinery Output	**246**	**239**	**244**	**..**	**-**	**-**	**-**	**..**
+ Recycled Products	-	-	-	..	-	-	-	..
+ Imports	66	138	120	..	13	6	6	..
- Exports	-	57	50	..	-	-	-	..
- Intl. Marine Bunkers	-	-	-	..	-	-	-	..
+ Transfers	-	-	-	..	-	-	-	..
+ Stock Change	-	-	-	..	-	-	-	..
- Statistical Difference	- 4	17	56	..	-	-	-	..
Total Consumption	**316**	**303**	**258**	**..**	**13**	**6**	**6**	**..**
Transformation	-	-	-	..	-	-	-	..
Energy	-	-	-	..	-	-	-	..
Transport	-	-	-	..	-	-	-	..
Industry	316	303	258	..	13	6	6	..
Other	-	-	-	..	-	-	-	..
Memo: Non-Energy Use	*316*	*303*	*258*	..	*13*	*6*	*6*	..
Closing Stock Level	-	-	-	..	-	-	-	..
	PETROLEUM COKE				**OTHER PRODUCTS**			
Refinery Output	**-**	**-**	**-**	**-**	**37**	**40**	**46**	**468 e**
+ Recycled Products	-	-	-	-	-	-	-	-
+ Imports	-	8	14	49 e	-	109	110	310 e
- Exports	-	-	-	-	-	1	5	335 e
- Intl. Marine Bunkers	-	-	-	-	-	-	-	-
+ Transfers	-	-	-	-	-	-	-	-
+ Stock Change	-	-	-	-	-	25	3	7 e
- Statistical Difference	-	-	-	-	-	-	- 3	- 189 e
Total Consumption	**-**	**8**	**14**	**49 e**	**37**	**173**	**157**	**639 e**
Transformation	-	-	-	..	-	-	-	..
Energy	-	-	-	..	-	-	-	..
Transport	-	-	-	..	-	-	-	..
Industry	-	8	14	..	37	173	157	..
Other	-	-	-	..	-	-	-	..
Memo: Non-Energy Use	-	-	-	..	*37*	*173*	*157*	..
Closing Stock Level	-	-	-	-	-	*31*	*28*	*21 e*

Source: Annual Oil Statistics.

FINLAND
Table 7A: Imports by Country of Origin

000 metric tons

	2000	2001	2002	2000	2001	2002	2000	2001	2002
	Crude Oil			Natural Gas Liquids			Feedstocks and Other		
Australia	-	-	-	-	-	-	-	-	-
Belgium	-	-	-	-	-	-	-	-	-
Canada	-	-	-	-	-	-	-	-	-
Czech Republic	-	-	-	-	-	-	-	-	-
Denmark	2 375	2 188	2 711 e	-	-	-	-	1	-
Finland	-	-	-	-	-	-	-	-	-
France	-	-	-	-	-	-	-	-	-
Germany	79	-	78 e	-	-	-	-	-	-
Greece	-	-	-	-	-	-	-	-	-
Hungary	-	-	-	-	-	-	-	-	-
Ireland	-	-	-	-	-	-	-	-	-
Italy	-	-	-	-	-	-	-	-	-
Japan	-	-	-	-	-	-	-	-	-
Korea	-	-	-	-	-	-	-	-	-
Mexico	-	-	-	-	-	-	-	-	-
Netherlands	-	-	-	-	-	-	51	27	30 e
New Zealand	-	-	-	-	-	-	-	-	-
Norway	2 510	1 727	1 695 e	77	77	96 e	18	-	10 e
Poland	80	30	-	-	-	-	-	-	-
Spain	-	-	-	-	-	-	-	-	-
Sweden	-	-	-	-	11	-	1	3	5 e
Turkey	-	-	-	-	-	-	-	-	-
United Kingdom	977	1 048	171 e	-	-	-	-	-	-
United States	-	-	-	-	-	-	-	-	-
Other	-	-	-	-	-	-	-	-	-
OECD	**6 021**	**4 993**	**4 655 e**	**77**	**88**	**96 e**	**70**	**31**	**45 e**
Argentina	-	-	-	-	-	-	-	-	-
Brazil	-	-	-	-	-	-	-	-	-
Colombia	-	-	-	-	-	-	-	-	-
Ecuador	-	-	-	-	-	-	-	-	-
Netherlands Antilles	-	-	-	-	-	-	-	-	-
Peru	-	-	-	-	-	-	-	-	-
Trinidad and Tobago	-	-	-	-	-	-	-	-	-
Venezuela	-	-	-	-	-	-	-	-	-
Other	-	-	-	-	-	-	-	-	-
Latin America	**-**	**-**	**-**	**-**	**-**	**-**	**-**	**-**	**-**
Romania	-	-	-	-	-	-	-	-	-
Other	-	-	-	-	-	-	-	-	-
Non-OECD Europe	**-**	**-**	**-**	**-**	**-**	**-**	**-**	**-**	**-**
FSU	**4 853**	**4 830**	**5 797 e**	**666**	**1 040**	**794 e**	**147**	**113**	**135 e**
Iran	-	-	-	-	-	-	-	-	-
Iraq	-	-	-	-	-	-	-	-	-
Kuwait	-	-	-	-	-	-	-	-	-
Oman	-	-	-	-	-	-	-	-	-
Qatar	-	-	-	-	-	-	-	-	-
Saudi Arabia	-	-	-	-	-	-	-	-	-
United Arab Emirates	-	-	-	-	-	-	-	-	-
Other	-	-	-	-	-	-	-	-	-
Middle East	**-**	**-**	**-**	**-**	**-**	**-**	**-**	**-**	**-**
India	-	-	-	-	-	-	-	-	-
Indonesia	-	-	-	-	-	-	-	-	-
Singapore	-	-	-	-	-	-	-	-	-
Other	-	-	-	-	-	-	-	-	-
Asia	**-**	**-**	**-**	**-**	**-**	**-**	**-**	**-**	**-**
China	**-**	**-**	**-**	**-**	**-**	**-**	**-**	**-**	**-**
Algeria	-	-	-	-	-	-	-	-	-
Angola/Cabinada	-	-	-	-	-	-	-	-	-
Cameroon	-	-	-	-	-	-	-	-	-
Egypt	-	-	-	-	-	-	-	-	-
Gabon	-	-	-	-	-	-	-	-	-
Libya	-	-	-	-	-	-	-	-	-
Nigeria	-	-	-	-	-	-	-	-	-
Tunisia	-	-	-	-	-	-	-	-	-
Other	-	-	-	-	-	-	-	-	-
Africa	**-**	**-**	**-**	**-**	**-**	**-**	**-**	**-**	**-**
Non-spec./Other	**-**	**-**	**-**	**-**	**-**	**-**	**-**	**-**	**-**
Total Imports	**10 874**	**9 823**	**10 452 e**	**743**	**1 128**	**890 e**	**217**	**144**	**180 e**
Memo: Total OPEC	-	-	-	-	-	-	-	-	-

Source: Annual Oil Statistics.

FINLAND
Table 7B: Imports by Country of Origin

000 metric tons

	2000	2001	2002	2000	2001	2002	2000	2001	2002
	Total Products			Liquefied Petroleum Gases			Naphtha		
Australia	-	-	-	-	-	-	-	-	-
Belgium	14	6	7 e	-	-	-	-	-	-
Canada	4	12	-	-	-	-	-	-	-
Czech Republic	-	-	-	-	-	-	-	-	-
Denmark	69	142	245 e	-	-	-	-	11	-
Finland	-	-	-	-	-	-	-	-	-
France	18	38	11 e	-	-	-	-	-	-
Germany	29	25	35 e	-	-	-	-	-	-
Greece	-	-	-	-	-	-	-	-	-
Hungary	-	-	-	-	-	-	-	-	-
Ireland	-	-	-	-	-	-	-	-	-
Italy	-	-	-	-	-	-	-	-	-
Japan	-	-	-	-	-	-	-	-	-
Korea	-	-	-	-	-	-	-	-	-
Mexico	-	-	-	-	-	-	-	-	-
Netherlands	115	145	96 e	-	-	-	-	-	-
New Zealand	-	-	-	-	-	-	-	-	-
Norway	491	281	390 e	88	76	47 e	12	-	1 e
Poland	2	11	-	-	-	-	-	-	-
Spain	-	3	3 e	-	-	-	-	-	-
Sweden	639	535	468 e	-	-	3 e	-	-	-
Turkey	-	-	-	-	-	-	-	-	-
United Kingdom	80	113	145 e	-	16	11 e	-	-	-
United States	128	204	220 e	-	-	-	-	-	-
Other	-	-	-	-	-	-	-	-	-
OECD	**1 589**	**1 515**	**1 620 e**	**88**	**92**	**61 e**	**12**	**11**	**1 e**
Argentina	-	-	-	-	-	-	-	-	-
Brazil	-	-	-	-	-	-	-	-	-
Colombia	-	-	-	-	-	-	-	-	-
Ecuador	-	-	-	-	-	-	-	-	-
Netherlands Antilles	-	-	-	-	-	-	-	-	-
Peru	-	-	-	-	-	-	-	-	-
Trinidad and Tobago	-	-	-	-	-	-	-	-	-
Venezuela	-	1	6 e	-	-	-	-	-	-
Other	-	-	-	-	-	-	-	-	-
Latin America	**-**	**1**	**6 e**	**-**	**-**	**-**	**-**	**-**	**-**
Romania	-	-	-	-	-	-	-	-	-
Other	-	-	-	-	-	-	-	-	-
Non-OECD Europe	**-**	**-**	**-**	**-**	**-**	**-**	**-**	**-**	**-**
FSU	**1 919**	**2 095**	**1 809 e**	**60**	**125**	**94 e**	**146**	**174**	**82 e**
Iran	-	-	-	-	-	-	-	-	-
Iraq	-	-	-	-	-	-	-	-	-
Kuwait	-	49	-	-	-	-	-	-	-
Oman	-	-	-	-	-	-	-	-	-
Qatar	-	-	-	-	-	-	-	-	-
Saudi Arabia	-	-	-	-	-	-	-	-	-
United Arab Emirates	-	16	-	-	-	-	-	-	-
Other	-	33	-	-	-	-	-	-	-
Middle East	**-**	**98**	**-**	**-**	**-**	**-**	**-**	**-**	**-**
India	-	-	-	-	-	-	-	-	-
Indonesia	-	-	-	-	-	-	-	-	-
Singapore	-	-	-	-	-	-	-	-	-
Other	-	-	-	-	-	-	-	-	-
Asia	**-**	**-**	**-**	**-**	**-**	**-**	**-**	**-**	**-**
China	**-**	**2**	**4 e**	**-**	**-**	**-**	**-**	**-**	**-**
Algeria	10	-	-	-	-	-	-	-	-
Angola/Cabinada	-	-	-	-	-	-	-	-	-
Cameroon	-	-	-	-	-	-	-	-	-
Egypt	-	1	-	-	-	-	-	-	-
Gabon	-	-	-	-	-	-	-	-	-
Libya	-	-	-	-	-	-	-	-	-
Nigeria	-	-	-	-	-	-	-	-	-
Tunisia	-	-	-	-	-	-	-	-	-
Other	-	-	-	-	-	-	-	-	-
Africa	**10**	**1**	**-**	**-**	**-**	**-**	**-**	**-**	**-**
Non-spec./Other	**-**	**-**	**-**	**-**	**-**	**-**	**-**	**-**	**-**
Total Imports	**3 518**	**3 712**	**3 439 e**	**148**	**217**	**155 e**	**158**	**185**	**83 e**
Memo: Total OPEC	*10*	*66*	*6 e*	-	-	-	-	-	-

Source: Annual Oil Statistics.

FINLAND
Table 7C: Imports by Country of Origin

000 metric tons

	2000	2001	2002	2000	2001	2002	2000	2001	2002
	Total Gasoline			**Total Kerosene**			**Gas/Diesel**		
Australia	-	-	-	-	-	-	-	-	-
Belgium	7	-	-	-	-	-	-	-	-
Canada	-	-	-	-	-	-	3	12	-
Czech Republic	-	-	-	-	-	-	-	-	-
Denmark	5	-	-	-	-	-	6	1	-
Finland	-	-	-	-	-	-	-	-	-
France	-	-	-	-	-	-	-	-	-
Germany	-	-	-	-	-	-	5	-	-
Greece	-	-	-	-	-	-	-	-	-
Hungary	-	-	-	-	-	-	-	-	-
Ireland	-	-	-	-	-	-	-	-	-
Italy	-	-	-	-	-	-	-	-	-
Japan	-	-	-	-	-	-	-	-	-
Korea	-	-	-	-	-	-	-	-	-
Mexico	-	-	-	-	-	-	-	-	-
Netherlands	8	21	2 e	1	-	-	26	42	2 e
New Zealand	-	-	-	-	-	-	-	-	-
Norway	288	192	258 e	-	-	-	22	5	19 e
Poland	-	-	-	-	11	-	-	-	-
Spain	-	-	-	-	-	-	-	-	-
Sweden	24	2	7 e	1	1	-	210	151	192 e
Turkey	-	-	-	-	-	-	-	-	-
United Kingdom	5	19	1 e	-	-	-	61	54	99 e
United States	-	-	-	-	10	22 e	119	179	151 e
Other	-	-	-	-	-	-	-	-	-
OECD	**337**	**234**	**268 e**	**2**	**22**	**22 e**	**452**	**444**	**463 e**
Argentina	-	-	-	-	-	-	-	-	-
Brazil	-	-	-	-	-	-	-	-	-
Colombia	-	-	-	-	-	-	-	-	-
Ecuador	-	-	-	-	-	-	-	-	-
Netherlands Antilles	-	-	-	-	-	-	-	-	-
Peru	-	-	-	-	-	-	-	-	-
Trinidad and Tobago	-	-	-	-	-	-	-	-	-
Venezuela	-	-	-	-	-	-	-	-	-
Other	-	-	-	-	-	-	-	-	-
Latin America	**-**	**-**	**-**	**-**	**-**	**-**	**-**	**-**	**-**
Romania	-	-	-	-	-	-	-	-	-
Other	-	-	-	-	-	-	-	-	-
Non-OECD Europe	**-**	**-**	**-**	**-**	**-**	**-**	**-**	**-**	**-**
FSU	**81**	**131**	**14 e**	**233**	**62**	**26 e**	**1 024**	**1 226**	**1 242 e**
Iran	-	-	-	-	-	-	-	-	-
Iraq	-	-	-	-	-	-	-	-	-
Kuwait	-	-	-	-	49	-	-	-	-
Oman	-	-	-	-	-	-	-	-	-
Qatar	-	-	-	-	-	-	-	-	-
Saudi Arabia	-	-	-	-	-	-	-	-	-
United Arab Emirates	-	-	-	-	16	-	-	-	-
Other	-	-	-	-	33	-	-	-	-
Middle East	**-**	**-**	**-**	**-**	**98**	**-**	**-**	**-**	**-**
India	-	-	-	-	-	-	-	-	-
Indonesia	-	-	-	-	-	-	-	-	-
Singapore	-	-	-	-	-	-	-	-	-
Other	-	-	-	-	-	-	-	-	-
Asia	**-**	**-**	**-**	**-**	**-**	**-**	**-**	**-**	**-**
China	**-**	**-**	**-**	**-**	**-**	**-**	**-**	**-**	**-**
Algeria	-	-	-	-	-	-	-	-	-
Angola/Cabinada	-	-	-	-	-	-	-	-	-
Cameroon	-	-	-	-	-	-	-	-	-
Egypt	-	-	-	-	-	-	-	-	-
Gabon	-	-	-	-	-	-	-	-	-
Libya	-	-	-	-	-	-	-	-	-
Nigeria	-	-	-	-	-	-	-	-	-
Tunisia	-	-	-	-	-	-	-	-	-
Other	-	-	-	-	-	-	-	-	-
Africa	**-**	**-**	**-**	**-**	**-**	**-**	**-**	**-**	**-**
Non-spec./Other	**-**	**-**	**-**	**-**	**-**	**-**	**-**	**-**	**-**
Total Imports	**418**	**365**	**282 e**	**235**	**182**	**48 e**	**1 476**	**1 670**	**1 705 e**
Memo: Total OPEC	-	-	-	-	*65*	-	-	-	-

Source: Annual Oil Statistics.

FINLAND
Table 7D: Imports by Country of Origin

000 metric tons

	2000	2001	2002	2000	2001	2002	2000	2001	2002
	Fuel Oil (Residual)			**Petroleum Coke**			**Other Products**		
Australia	-	-	-	-	-	-	-	-	-
Belgium	-	-	-	-	-	-	7	6	7 e
Canada	-	-	-	-	-	-	1	-	-
Czech Republic	-	-	-	-	-	-	-	-	-
Denmark	58	129	245 e	-	-	-	-	1	-
Finland	-	-	-	-	-	-	-	-	-
France	-	-	-	-	-	-	18	38	11 e
Germany	-	-	11 e	-	-	-	24	25	24 e
Greece	-	-	-	-	-	-	-	-	-
Hungary	-	-	-	-	-	-	-	-	-
Ireland	-	-	-	-	-	-	-	-	-
Italy	-	-	-	-	-	-	-	-	-
Japan	-	-	-	-	-	-	-	-	-
Korea	-	-	-	-	-	-	-	-	-
Mexico	-	-	-	-	-	-	-	-	-
Netherlands	-	-	34 e	-	-	-	80	82	58 e
New Zealand	-	-	-	-	-	-	-	-	-
Norway	81	8	64 e	-	-	-	-	-	1 e
Poland	-	-	-	-	-	-	2	-	-
Spain	-	-	-	-	-	-	-	3	3 e
Sweden	306	332	241 e	1	-	-	97	49	25 e
Turkey	-	-	-	-	-	-	-	-	-
United Kingdom	-	-	34 e	-	-	-	14	24	-
United States	-	-	-	7	14	47 e	2	1	-
Other	-	-	-	-	-	-	-	-	-
OECD	**445**	**469**	**629 e**	**8**	**14**	**47 e**	**245**	**229**	**129 e**
Argentina	-	-	-	-	-	-	-	-	-
Brazil	-	-	-	-	-	-	-	-	-
Colombia	-	-	-	-	-	-	-	-	-
Ecuador	-	-	-	-	-	-	-	-	-
Netherlands Antilles	-	-	-	-	-	-	-	-	-
Peru	-	-	-	-	-	-	-	-	-
Trinidad and Tobago	-	-	-	-	-	-	-	-	-
Venezuela	-	1	-	-	-	-	-	-	6 e
Other	-	-	-	-	-	-	-	-	-
Latin America	**-**	**1**	**-**	**-**	**-**	**-**	**-**	**-**	**6 e**
Romania	-	-	-	-	-	-	-	-	-
Other	-	-	-	-	-	-	-	-	-
Non-OECD Europe	**-**	**-**	**-**	**-**	**-**	**-**	**-**	**-**	**-**
FSU	**229**	**184**	**178 e**	**-**	**-**	**-**	**146**	**193**	**173 e**
Iran	-	-	-	-	-	-	-	-	-
Iraq	-	-	-	-	-	-	-	-	-
Kuwait	-	-	-	-	-	-	-	-	-
Oman	-	-	-	-	-	-	-	-	-
Qatar	-	-	-	-	-	-	-	-	-
Saudi Arabia	-	-	-	-	-	-	-	-	-
United Arab Emirates	-	-	-	-	-	-	-	-	-
Other	-	-	-	-	-	-	-	-	-
Middle East	**-**	**-**	**-**	**-**	**-**	**-**	**-**	**-**	**-**
India	-	-	-	-	-	-	-	-	-
Indonesia	-	-	-	-	-	-	-	-	-
Singapore	-	-	-	-	-	-	-	-	-
Other	-	-	-	-	-	-	-	-	-
Asia	**-**	**-**	**-**	**-**	**-**	**-**	**-**	**-**	**-**
China	**-**	**-**	**-**	**-**	**-**	**2 e**	**-**	**2**	**2 e**
Algeria	-	-	-	-	-	-	10	-	-
Angola/Cabinada	-	-	-	-	-	-	-	-	-
Cameroon	-	-	-	-	-	-	-	-	-
Egypt	-	-	-	-	-	-	-	1	-
Gabon	-	-	-	-	-	-	-	-	-
Libya	-	-	-	-	-	-	-	-	-
Nigeria	-	-	-	-	-	-	-	-	-
Tunisia	-	-	-	-	-	-	-	-	-
Other	-	-	-	-	-	-	-	-	-
Africa	**-**	**-**	**-**	**-**	**-**	**-**	**10**	**1**	**-**
Non-spec./Other	**-**	**-**	**-**	**-**	**-**	**-**	**-**	**-**	**-**
Total Imports	**674**	**654**	**807 e**	**8**	**14**	**49 e**	**401**	**425**	**310 e**
Memo: Total OPEC	*-*	*1*	*-*	*-*	*-*	*-*	*10*	*-*	*6 e*

Source: Annual Oil Statistics.

FINLAND
Table 8A: Exports by Country of Destination

000 metric tons

	2000	2001	2002	2000	2001	2002	2000	2001	2002
	Crude Oil			**Natural Gas Liquids**			**Feedstocks and Other**		
Australia	-	-	-	-	-	-	-	-	-
Austria	-	-	-	-	-	-	-	-	-
Belgium	-	-	-	-	-	-	-	-	-
Canada	-	-	-	-	-	-	-	-	-
Denmark	-	-	-	-	-	-	-	-	-
Finland	-	-	-	-	-	-	-	-	-
France	-	-	-	-	-	-	-	-	-
Germany	-	-	-	-	-	-	-	-	-
Greece	-	-	-	-	-	-	-	-	-
Ireland	-	-	-	-	-	-	-	-	-
Italy	-	-	-	-	-	-	-	-	-
Japan	-	-	-	-	-	-	-	-	-
Luxembourg	-	-	-	-	-	-	-	-	-
Mexico	-	-	-	-	-	-	-	-	-
Netherlands	-	-	-	-	-	-	-	-	-
Norway	-	-	-	-	-	-	-	-	-
Portugal	-	-	-	-	-	-	-	-	-
Spain	-	-	-	-	-	-	-	-	-
Sweden	-	-	-	-	-	-	-	-	-
Switzerland	-	-	-	-	-	-	-	-	-
Turkey	-	-	-	-	-	-	-	-	-
United Kingdom	-	-	-	-	-	-	-	-	-
United States	-	-	-	-	-	-	-	-	-
Other OECD	-	-	-	-	-	-	-	-	-
OECD	**-**	**-**	**-**	**-**	**-**	**-**	**-**	**-**	**-**
Africa	-	-	-	-	-	-	-	-	-
Latin America	-	-	-	-	-	-	-	-	-
Asia	-	-	-	-	-	-	-	-	-
China	-	-	-	-	-	-	-	-	-
Non-OECD Europe	-	-	-	-	-	-	-	-	-
Former Soviet Union	-	-	-	-	-	-	-	-	-
Middle East	-	-	-	-	-	-	-	-	-
Non-specified/Other	-	-	-	-	-	-	-	-	-
Non-OECD	**-**	**-**	**-**	**-**	**-**	**-**	**-**	**-**	**-**
Total Exports	**-**	**-**	**-**	**-**	**-**	**-**	**-**	**-**	**-**
	Total Products			**Liquefied Petroleum Gas**			**Naphtha**		
Australia	-	-	-	-	-	-	-	-	-
Austria	-	-	-	-	-	-	-	-	-
Belgium	85	179	209 e	-	-	-	5	-	-
Canada	62	237	205 e	-	-	-	-	-	-
Denmark	259	206	174 e	-	-	-	-	-	-
Finland	-	-	-	-	-	-	-	-	-
France	557	45	17 e	-	-	-	-	-	-
Germany	174	785	695 e	-	-	-	42	7	30 e
Greece	-	-	-	-	-	-	-	-	-
Ireland	-	-	-	-	-	-	-	-	-
Italy	-	-	-	-	-	-	-	-	-
Japan	-	-	-	-	-	-	-	-	-
Luxembourg	-	-	-	-	-	-	-	-	-
Mexico	-	-	-	-	-	-	-	-	-
Netherlands	142	203	746 e	-	-	-	32	40	5 e
Norway	225	66	75 e	-	-	-	-	-	-
Portugal	56	-	-	-	-	-	-	-	-
Spain	18	16	10 e	-	-	-	-	-	-
Sweden	1 799	1 674	1 787 e	-	-	-	-	-	-
Switzerland	-	-	-	-	-	-	-	-	-
Turkey	-	-	6 e	-	-	6 e	-	-	-
United Kingdom	408	786	175 e	-	-	-	26	21	5 e
United States	917	449	1 155 e	-	-	-	-	-	-
Other OECD	31	10	19 e	-	-	12 e	-	-	-
OECD	**4 733**	**4 656**	**5 273 e**	**-**	**-**	**18 e**	**105**	**68**	**40 e**
Africa	4	-	-	-	-	-	-	-	-
Latin America	-	-	-	-	-	-	-	-	-
Asia	-	-	-	-	-	-	-	-	-
China	2	-	-	-	-	-	-	-	-
Non-OECD Europe	-	-	-	-	-	-	-	-	-
Former Soviet Union	204	158	89 e	-	-	-	-	-	-
Middle East	1	1	-	-	-	-	-	-	-
Non-specified/Other	3	-	-	-	-	-	-	-	-
Non-OECD	**214**	**159**	**89 e**	**-**	**-**	**-**	**-**	**-**	**-**
Total Exports	**4 947**	**4 815**	**5 362 e**	**-**	**-**	**18 e**	**105**	**68**	**40 e**

Source: Annual Oil Statistics.

FINLAND
Table 8B: Exports by Country of Destination

000 metric tons

	Total Gasoline			Total Kerosene			Gas/Diesel Oil		
	2000	**2001**	**2002**	**2000**	**2001**	**2002**	**2000**	**2001**	**2002**
Australia	-	-	-	-	-	-	-	-	-
Austria	-	-	-	-	-	-	-	-	-
Belgium	-	-	-	40	67	67 e	-	47	15 e
Canada	62	151	205 e	-	-	-	-	86	-
Denmark	19	22	40 e	1	-	-	225	167	118 e
Finland	-	-	-	-	-	-	-	-	-
France	321	-	-	-	-	3 e	236	38	11 e
Germany	69	416	391 e	-	-	-	60	330	245 e
Greece	-	-	-	-	-	-	-	-	-
Ireland	-	-	-	-	-	-	-	-	-
Italy	-	-	-	-	-	-	-	-	-
Japan	-	-	-	-	-	-	-	-	-
Luxembourg	-	-	-	-	-	-	-	-	-
Mexico	-	-	-	-	-	-	-	-	-
Netherlands	37	19	71 e	6	-	-	24	123	662 e
Norway	29	25	52 e	11	-	-	167	28	4 e
Portugal	-	-	-	-	-	-	-	-	-
Spain	-	-	-	-	-	-	18	16	-
Sweden	651	800	772 e	236	56	53 e	808	722	857 e
Switzerland	-	-	-	-	-	-	-	-	-
Turkey	-	-	-	-	-	-	-	-	-
United Kingdom	93	358	47 e	-	-	2 e	282	359	80 e
United States	800	396	1 155 e	-	-	-	86	53	-
Other OECD	24	-	-	-	-	-	1	-	-
OECD	**2 105**	**2 187**	**2 733 e**	**294**	**123**	**125 e**	**1 907**	**1 969**	**1 992 e**
Africa	-	-	-	-	-	-	-	-	-
Latin America	-	-	-	-	-	-	-	-	-
Asia	-	-	-	-	-	-	-	-	-
China	-	-	-	-	-	-	-	-	-
Non-OECD Europe	-	-	-	-	-	-	-	-	-
Former Soviet Union	116	13	18 e	-	-	5 e	40	85	8 e
Middle East	-	-	-	-	-	-	-	-	-
Non-specified/Other	-	-	-	-	-	-	-	-	-
Non-OECD	**116**	**13**	**18 e**	**-**	**-**	**5 e**	**40**	**85**	**8 e**
Total Exports	**2 221**	**2 200**	**2 751 e**	**294**	**123**	**130 e**	**1 947**	**2 054**	**2 000 e**

	Fuel Oil (Residual)			Petroleum Coke			Other Products		
	2000	**2001**	**2002**	**2000**	**2001**	**2002**	**2000**	**2001**	**2002**
Australia	-	-	-	-	-	-	-	-	-
Austria	-	-	-	-	-	-	-	-	-
Belgium	-	-	-	-	-	-	40	65	127 e
Canada	-	-	-	-	-	-	-	-	-
Denmark	10	15	15 e	-	-	-	4	2	1 e
Finland	-	-	-	-	-	-	-	-	-
France	-	-	-	-	-	-	-	7	3 e
Germany	-	26	25 e	-	-	-	3	6	4 e
Greece	-	-	-	-	-	-	-	-	-
Ireland	-	-	-	-	-	-	-	-	-
Italy	-	-	-	-	-	-	-	-	-
Japan	-	-	-	-	-	-	-	-	-
Luxembourg	-	-	-	-	-	-	-	-	-
Mexico	-	-	-	-	-	-	-	-	-
Netherlands	29	12	8 e	-	-	-	14	9	-
Norway	-	-	-	-	-	-	18	13	19 e
Portugal	56	-	-	-	-	-	-	-	-
Spain	-	-	10 e	-	-	-	-	-	-
Sweden	-	1	-	-	-	-	104	95	105 e
Switzerland	-	-	-	-	-	-	-	-	-
Turkey	-	-	-	-	-	-	-	-	-
United Kingdom	-	25	30 e	-	-	-	7	23	11 e
United States	28	-	-	-	-	-	3	-	-
Other OECD	-	-	-	-	-	-	6	10	7 e
OECD	**123**	**79**	**88 e**	**-**	**-**	**-**	**199**	**230**	**277 e**
Africa	-	-	-	-	-	-	4	-	-
Latin America	-	-	-	-	-	-	-	-	-
Asia	-	-	-	-	-	-	-	-	-
China	-	-	-	-	-	-	2	-	-
Non-OECD Europe	-	-	-	-	-	-	-	-	-
Former Soviet Union	-	-	-	-	-	-	48	60	58 e
Middle East	-	-	-	-	-	-	1	1	-
Non-specified/Other	-	-	-	-	-	-	3	-	-
Non-OECD	**-**	**-**	**-**	**-**	**-**	**-**	**58**	**61**	**58 e**
Total Exports	**123**	**79**	**88 e**	**-**	**-**	**-**	**257**	**291**	**335 e**

Source: Annual Oil Statistics.

FRANCE

Figure 1
FUEL SHARES OF TPES *

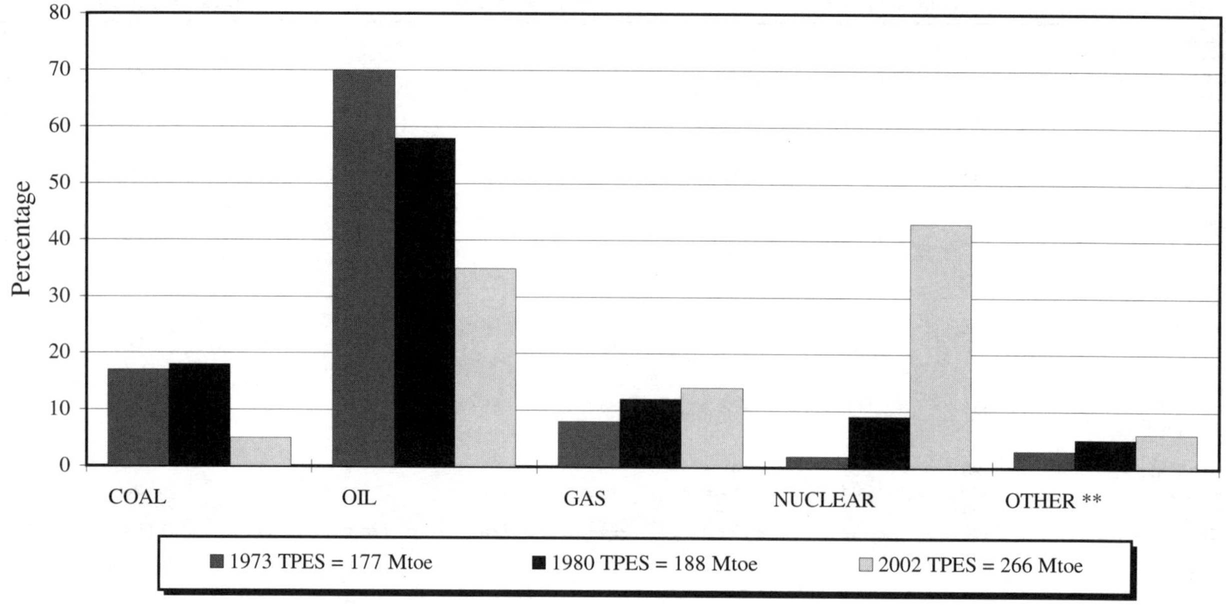

Source: Annual Energy Balances.

* Excluding electricity trade.

** Includes hydro, geothermal, solar, wind, combustible renewables and waste, etc.

Figure 2
OIL PRODUCTION AND CONSUMPTION

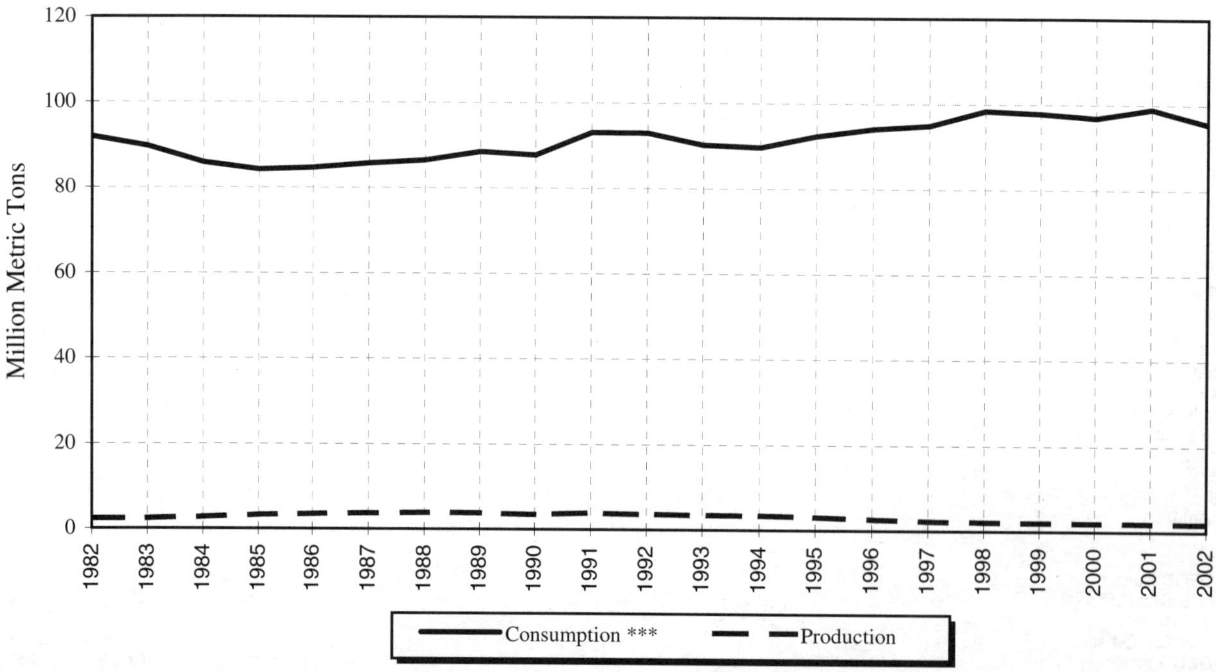

Source: Annual Oil Statistics.

*** Includes refinery fuel and international marine bunkers.

FRANCE
Table 1: Oil Supply and Consumption - Summary Table

000 metric tons

	1973	1978	1980	1990	1999	2000	2001	2002
Indigenous Production	2 066	1 943	2 254	3 470	2 339	2 306	2 294 e	2 195 e
Crude Oil	..	1 117	1 415	3 024	1 539	1 417	1 382	1 313 e
NGL	..	826	839	446	256	249	250	238 e
Other	..	-	-	-	544	640	662 e	644 e
+ Net Imports	128 811	111 782	112 582	85 602	89 724	89 953	93 972 e	93 891 e
Crude Oil, NGL, Other	134 920	115 635	113 557	75 274	80 801	83 914	85 377 e	80 885 e
Petroleum Products	-6 109	-3 853	- 975	10 328	8 923	6 039	8 595	13 006 e
- Intl. Marine Bunkers	5 503	4 889	4 100	2 614	2 987	3 085	2 767	2 645 e
+ Stock Changes	-3 851	5 008	-2 356	243	1 180	-1 572	559	-1 279 e
TOTAL SUPPLY	**121 523**	**113 844**	**108 380**	**86 701**	**90 256**	**87 602**	**94 058 e**	**92 162 e**
Refinery Intake	*134 215*	*119 450*	*116 317*	*78 859*	*86 042*	*89 536*	*91 012*	*84 250 e*
Refinery Output	*133 792*	*118 583*	*115 717*	*78 333*	*85 283*	*88 576*	*88 832*	*82 927 e*
LPG + Ethane	*2 740*	*2 943*	*3 133*	*2 543*	*2 357*	*2 695*	*2 955*	*2 510 e*
Naphtha	*5 034*	*5 716*	*4 349*	*3 480*	*4 642*	*5 776*	*5 695*	*5 374 e*
Kerosenes	*3 353*	*4 093*	*4 739*	*5 057*	*6 510*	*6 112*	*6 162*	*5 255 e*
Gasolines	*16 676*	*17 431*	*18 900*	*17 579*	*17 481*	*17 097*	*16 422*	*15 515 e*
Gas/Diesel Oil	*47 481*	*39 965*	*41 486*	*27 890*	*32 621*	*34 531*	*35 132*	*33 343 e*
Fuel Oil	*46 397*	*38 162*	*33 900*	*12 363*	*10 799*	*11 068*	*11 291*	*10 752 e*
Other Products	*12 111*	*10 273*	*9 210*	*9 421*	*10 873*	*11 297*	*11 175*	*10 178 e*
- Refinery Losses	423	867	600	526	759	960	2 180	1 323 e
+ Returns	-	2 398	2 053	1 777	2 088	2 235	2 047	1 867 e
- Statistical Differences	640	639	2 607	2 870	-3 192	-4 840	-2 206 e	- 28 e
TOTAL CONSUMPTION	**120 460**	**114 736**	**107 226**	**85 082**	**94 777**	**93 717**	**96 131**	**92 734 e**
Transformation and Energy	**23 482**	**21 004**	**18 974**	**8 438**	**9 267**	**8 948**	**8 374**	..
Crude Oil	-	-	-	-	-	-	-	..
LPG + Ethane	361	343	335	151	198	199	182	..
Gas/Diesel Oil	116	184	502	311	388	497	389	..
Fuel Oil	19 497	15 210	13 693	2 794	3 202	2 585	2 199	..
Petroleum Coke	-	-	-	1 064	1 016	1 006	1 062	..
Other	3 508	5 267	4 444	4 118	4 463	4 661	4 542	..
Transport	**25 402**	**30 226**	**30 850**	**40 362**	**48 189**	**49 460**	**49 996**	..
Motor Gasoline	15 772	17 487	17 666	18 193	14 456	13 894	13 685	..
Gas/Diesel Oil	7 274	10 004	10 283	17 908	26 801	28 386	29 179	..
Subtotal	*23 046*	*27 491*	*27 949*	*36 101*	*41 257*	*42 280*	*42 864*	
Aviation Fuels	1 860	2 307	2 458	3 767	6 305	6 535	6 467	..
Other	496	428	443	494	627	645	665	..
Industry	**35 097**	**31 051**	**30 625**	**17 451**	**18 882**	**18 845**	**19 333**	..
Naphtha	4 945	4 259	4 825	4 779	6 652	6 481	6 235	..
LPG + Ethane	605	558	883	1 015	1 202	1 195	1 286	..
Gas/Diesel Oil	6 284	6 269	6 731	2 623	2 475	2 841	2 845	..
Fuel Oil	15 350	13 335	12 418	3 363	1 803	1 818	1 885	..
Other	7 913	6 630	5 768	5 671	6 750	6 510 e	7 082	..
Other Sectors	**36 479**	**32 455**	**26 777**	**18 831**	**18 439**	**16 464**	**18 428**	..
Gas/Diesel Oil	31 969	26 113	21 879	14 906	14 925	13 081	15 141	..
Fuel Oil	2 638	4 039	2 410	1 629	1 082	979	925	..
Other	1 872	2 303	2 488	2 296	2 432	2 404	2 362	..
Memo: Non-Energy Use	*13 047*	*10 522*	*11 106*	*11 147*	*13 873*	*13 476*	*12 278*	..
Chem/Petchem. Feedst.	*8 302*	*6 456*	*7 122*	*6 676*	*9 298*	*9 098*	*7 892*	..
Other	*4 745*	*4 066*	*3 984*	*4 471*	*4 575*	*4 378*	*4 386*	..

Source: Annual Oil Statistics.

FRANCE
Table 2: Supply of Crude Oil, NGL, Feedstocks and Other Hydrocarbons

000 metric tons

	1973	1978	1980	1990	1999	2000	2001	2002
CRUDE OIL								
Indigenous Production	..	1 117	1 415	3 024	1 539	1 417	1 382	1 313 e
+ Imports	..	115 635	109 495	69 566	81 869	85 463	86 156	81 255 e
- Exports	..	-	-	-	-	-	-	-
- Direct Use	..	-	-	-	-	-	-	-
+ Stock Change	..	1 707	- 279	- 308	- 433	- 330	- 452	300 e
- Statistical Difference	..	1 316	509	747	301	1 356	509	2 302 e
Refinery Intake	**..**	**117 143**	**110 122**	**71 535**	**82 674**	**85 194**	**86 577**	**80 566 e**
Closing Stock Level	*..*	*7 896*	*8 935*	*4 621*	*6 679*	*7 009*	*7 461*	*7 161 e*
NATURAL GAS LIQUIDS								
Indigenous Production	..	826	839	446	256	249	250	238 e
+ Imports	..	-	-	-	21	-	86	145 e
- Exports	..	-	-	-	-	-	-	-
- Direct Use	..	826	839	446	221	208	214	220 e
+ Stock Change	..	-	-	-	-	- 5	- 16	16 e
- Statistical Difference	..	-	-	-	34	7	38	- 131 e
Refinery Intake	**..**	**-**	**-**	**-**	**22**	**29**	**68**	**310 e**
Closing Stock Level	*..*	*-*	*-*	*-*	*-*	*5*	*21*	*5 e*
REFINERY FEEDSTOCKS								
Backflows	..	2 307	1 977	1 717	2 013	2 164	1 970	1 787 e
+ Transfers	..	-	-	92	2 753	2 429	1 100	1 003 e
+ Imports	..	-	4 062	6 007	221	163	82	3 e
- Exports	..	-	-	299	1 344	1 784	985 e	550 e
+ Stock Change	..	-	- 331	- 104	134	- 87	126	- 191 e
- Statistical Difference	..	-	- 487	89	431	-1 428	-1 791	-1 065 e
Refinery Intake	**..**	**2 307**	**6 195**	**7 324**	**3 346**	**4 313**	**4 084**	**3 117 e**
Closing Stock Level	*..*	*-*	*539*	*1 984*	*1 565*	*1 652*	*1 526*	*1 717 e*
OTHER HYDROCARBONS								
Indigenous Production	..	-	-	-	544	640	662 e	644 e
+ Imports	..	-	-	-	34	72	38	32 e
- Exports	..	-	-	-	-	-	-	-
- Direct Use	..	-	-	-	570	707	419 e	413 e
+ Stock Change	..	-	-	-	- 8	- 5	2	- 6 e
- Statistical Difference	..	-	-	-	-	-	-	-
Refinery Intake	**..**	**-**	**-**	**-**	**-**	**-**	**283**	**257 e**
Closing Stock Level	*..*	*-*	*-*	*-*	*21*	*26*	*24*	*30 e*
CRUDE OIL, NGL, FEEDSTOCKS AND OTHER								
Indigenous Production	2 066	1 943	2 254	3 470	2 339	2 306	2 294	2 195 e
+ Backflows	-	2 307	1 977	1 717	2 013	2 164	1 970	1 787 e
+ Transfers	-	-	-	92	2 753	2 429	1 100	1 003 e
+ Imports	134 920	115 635	113 557	75 573	82 145	85 698	86 362	81 435 e
- Exports	-	-	-	299	1 344	1 784	985	550 e
- Direct Use	812	826	839	446	791	915	633 e	633 e
+ Stock Change	-1 959	1 707	- 610	- 412	- 307	- 427	- 340	119 e
- Statistical Difference	-	1 316	22	836	766	- 65	-1 244	1 106 e
Refinery Intake	**134 215**	**119 450**	**116 317**	**78 859**	**86 042**	**89 536**	**91 012**	**84 250 e**
Closing Stock Level	*-*	*7 896*	*9 474*	*6 605*	*8 265*	*8 692*	*9 032*	*8 913 e*

Source: Annual Oil Statistics.

FRANCE
Table 3: Motor Gasoline - Supply

000 metric tons

	1973	1978	1980	1990	1999	2000	2001	2002
MOTOR GASOLINE								
Refinery Output	**16 633**	**17 389**	**18 870**	**17 482**	**17 413**	**17 041**	**16 350**	**15 460 e**
+ Imports	431	364	645	4 396	1 957	2 425	1 689	1 541 e
- Exports	1 530	1 752	1 734	2 972	4 992	5 243	4 706	4 255 e
+ Transfers	406	412	393	208	194	190	319 e	320 e
+ Stock Change	- 274	542	131	- 402	168	227	- 105	- 44 e
- Statistical Difference	- 106	- 632	559	481	283	744	- 138 e	- 180 e
Total Consumption	**15 772**	**17 587**	**17 746**	**18 231**	**14 457**	**13 896**	**13 685**	**13 202 e**
Closing Stock Level	-	*3 138*	*3 641*	*3 100*	*2 695*	*2 468*	*2 573*	*2 617 e*

Table 4: Aviation Fuels - Supply

000 metric tons

	1973	1978	1980	1990	1999	2000	2001	2002
AVIATION GASOLINE								
Refinery Output	**43**	**42**	**30**	**97**	**68**	**56**	**72**	**55 e**
+ Imports	29	17	17	8	72	64	26	7 e
- Exports	-	-	1	68	110	71	45	39 e
+ Transfers	-	-	-	-	- 5	- 3	-	-
+ Stock Change	- 3	-	-	- 2	3	4	- 10	6 e
- Statistical Difference	21	23	15	3	1	22	15	3 e
Total Consumption	**48**	**36**	**31**	**32**	**27**	**28**	**28**	**26 e**
Closing Stock Level	-	*5*	*7*	*18*	*14*	*10*	*20*	*14 e*
GASOLINE TYPE JET FUEL								
Refinery Output	-	-	-	-	-	-	-	-
+ Imports	-	-	-	-	-	-	-	-
- Exports	-	-	-	-	-	-	-	-
+ Transfers	-	-	-	-	-	-	-	-
+ Stock Change	-	-	-	-	-	-	-	-
- Statistical Difference	-	-	-	-	-	-	-	-
Total Consumption	-	-	-	-	-	-	-	-
Closing Stock Level	-	-	-	-	-	-	-	-
KEROSENE TYPE JET FUEL								
Refinery Output	**3 319**	**4 002**	**4 603**	**4 997**	**6 426**	**6 030**	**6 051**	**5 192 e**
+ Imports	4	32	-	945	1 519	2 081	2 342	2 702 e
- Exports	626	1 033	1 187	782	1 311	1 184	1 177	936 e
+ Transfers	-	-	-	-	- 102	- 100	-	-
+ Stock Change	- 70	36	- 56	- 103	- 40	- 106	34	- 120 e
- Statistical Difference	815	766	933	1 322	214	214	811	489 e
Total Consumption	**1 812**	**2 271**	**2 427**	**3 735**	**6 278**	**6 507**	**6 439**	**6 349 e**
Closing Stock Level	-	*252*	*479*	*664*	*991*	*1 097*	*1 063*	*1 183 e*

Source: Annual Oil Statistics.

FRANCE
Table 5A: Gas/Diesel Oil - Supply and Consumption

000 metric tons

	1973	1978	1980	1990	1999	2000	2001	2002
Refinery Output	**47 481**	**39 965**	**41 486**	**27 890**	**32 621**	**34 531**	**35 132**	**33 343 e**
+ Recycled Products	-	-	-	-	5	4	7	4 e
+ Imports	2 648	3 446	3 578	11 185	11 668	10 875	14 087	16 202 e
- Exports	4 363	4 118	3 055	3 911	2 230	2 430	1 993	1 968 e
- Intl. Marine Bunkers	720	798	529	311	419	483	561	474 e
+ Transfers	34	37	46	- 49	- 534	- 125	100	93 e
+ Stock Changes	693	2 950	-2 274	146	1 213	- 565	412 e	-1 041 e
- Statistical Differences	130	-1 088	- 143	- 798	-2 265	-2 998	- 370	- 704 e
Total Consumption	**45 643**	**42 570**	**39 395**	**35 748**	**44 589**	**44 805**	**47 554**	**46 863 e**
Transformation	**116**	**163**	**483**	**311**	**384**	**493**	**382**	..
Public Electricity	116	57	29	16	18	30 e	13	..
Autoproducer Electricity	-	-	-	7	23	46 e	72	..
Public CHP Plants	-	-	-	-	-	-	-	..
Autoproducers of CHP	-	-	-	-	-	-	-	..
Public Heat Plants	-	-	-	-	-	-	-	..
Autoproducer Heat Plants	-	-	-	-	-	-	-	..
Petrochemical Industry	-	106	454	288	336	410	297	..
Other Transformation	-	-	-	-	7	7 e	-	..
Energy Sector	**-**	**21**	**19**	**-**	**4**	**4**	**7**	..
Refinery Fuel	-	21	19	-	4	4	7	..
Other Energy Sector	-	-	-	-	-	-	-	..
Final Consumption	**45 527**	**42 386**	**38 893**	**35 437**	**44 201**	**44 308**	**47 165**	..
Transport	**7 274**	**10 004**	**10 283**	**17 908**	**26 801**	**28 386**	**29 179**	..
Air	-	-	-	-	-	-	-	..
Road	5 977	8 225	9 097	17 036	25 948 e	27 539	28 438	..
Rail	646	614	580	380	368	363	241	..
Inland Waterways	651	1 165	606	492	485	484	500	..
Non-Specified	-	-	-	-	-	-	-	..
Industry	**6 284**	**6 269**	**6 731**	**2 623**	**2 475**	**2 841**	**2 845**	..
Iron and Steel	213	130	140	31	35	21 e	97	..
Chemical (incl. Pet.-Chem.)	580	587	1 421	1 033	1 383	1 605	1 527	..
Non-Ferrous Metals	31	10	-	25	15	9 e	10	..
Non-Metallic Minerals	680	239	-	194	122	74 e	128	..
Transport Equipment	-	-	-	71	48	29 e	31	..
Machinery	-	-	-	163	152	93 e	61	..
Mining and Quarrying	-	-	-	31	1	1 e	13	..
Food and Tobacco	-	-	-	109	110	67 e	88	..
Pulp, Paper and Printing	81	72	-	21	14	9 e	11	..
Wood and Wood Products	-	-	-	-	-	-	-	..
Construction	-	-	-	708	409	820	800	..
Textiles and Leather	-	-	-	71	38	23 e	21	..
Non-Specified	4 699	5 231	5 170	166	148	90 e	58	..
Other Sectors	**31 969**	**26 113**	**21 879**	**14 906**	**14 925**	**13 081**	**15 141**	..
Commerce and Public	29 548	23 696	19 479	4 286 e	4 450	3 728	4 580	..
Residential	-	-	-	8 320 e	8 442	7 402	8 590	..
Agriculture	2 421	2 417	2 400	2 300	2 026	1 944	1 900	..
Non-Specified	-	-	-	-	7	7	71	..
Memo: Non-Energy Use	*580*	*587*	*1 421*	*1 033*	*1 383*	*1 605*	*1 527*	..
Chem/Petchem. Feedst.	*580*	*587*	*1 421*	*1 033*	*1 383*	*1 605*	*1 527*	..
Other	-	-	-	-	-	-	-	..
Closing Stock Level	-	*9 186*	*12 362*	*5 662*	*7 226*	*7 791*	*7 379 e*	*8 420 e*

Source: Annual Oil Statistics.

FRANCE
Table 5B: Fuel Oil (Residual) - Supply and Consumption

000 metric tons

	1973	1978	1980	1990	1999	2000	2001	2002
Refinery Output	**46 397**	**38 162**	**33 900**	**12 363**	**10 799**	**11 068**	**11 291**	**10 752 e**
+ Recycled Products	-	-	-	-	32	33	39	39 e
+ Imports	1 902	2 091	3 402	398	3 069 e	4 123	2 845	3 249 e
- Exports	3 760	3 529	5 035	3 108	4 332	5 382	5 699	5 113 e
- Intl. Marine Bunkers	4 760	4 064	3 509	2 262	2 523	2 561	2 170	2 135 e
+ Transfers	-	62	97	- 42	-1 043 e	-1 216	-1 100	-1 003 e
+ Stock Changes	-2 243	- 361	476	424	157	- 533	473	- 254 e
- Statistical Differences	-	- 233	800	- 18	52	115	670	732 e
Total Consumption	**37 536**	**32 594**	**28 531**	**7 791**	**6 107**	**5 417**	**5 009**	**4 803 e**
Transformation	**16 096**	**11 908**	**10 599**	**1 657**	**1 698**	**1 193**	**885**	**..**
Public Electricity	16 089	10 636	9 592	1 173	848	650 e	677	..
Autoproducer Electricity	-	1 269	1 007	484	850	543 e	208	..
Public CHP Plants	-	-	-	-	-	-	-	..
Autoproducers of CHP	-	-	-	-	-	-	-	..
Public Heat Plants	-	-	-	-	-	-	-	..
Autoproducer Heat Plants	-	-	-	-	-	-	-	..
Petrochemical Industry	-	-	-	-	-	-	-	..
Other Transformation	7	3	-	-	-	-	-	..
Energy Sector	**3 401**	**3 302**	**3 094**	**1 137**	**1 504**	**1 392**	**1 314**	**..**
Refinery Fuel	3 401	3 302	3 094	1 043	1 475	1 364	1 314	..
Other Energy Sector	-	-	-	94	29	28	-	..
Final Consumption	**18 039**	**17 384**	**14 838**	**4 997**	**2 905**	**2 832**	**2 810**	**..**
Transport	**51**	**10**	**10**	**5**	**20**	**35**	**-**	**..**
Air	-	-	-	-	-	-	-	..
Road	-	-	-	-	-	-	-	..
Rail	29	7	4	3	1	2	-	..
Inland Waterways	22	3	3	2	8	30	-	..
Non-Specified	-	-	3	-	11	3	-	..
Industry	**15 350**	**13 335**	**12 418**	**3 363**	**1 803**	**1 818**	**1 885**	**..**
Iron and Steel	2 450	1 882	1 190	160	36	30 e	30	..
Chemical (incl. Pet.-Chem.)	2 000	2 262	2 361	614	270	317 e	358	..
Non-Ferrous Metals	540	441	370	136	58	50 e	58	..
Non-Metallic Minerals	741	678	599	625	591	544 e	551	..
Transport Equipment	485	485	297	109	78	70 e	69	..
Machinery	-	-	-	76	20	20 e	20	..
Mining and Quarrying	480	226	237	59	4	4 e	4	..
Food and Tobacco	2 430	2 288	2 429	783	330	338 e	518	..
Pulp, Paper and Printing	1 340	1 116	1 341	416	252	251 e	180	..
Wood and Wood Products	155	120	-	-	-	-	-	..
Construction	3 315	2 880	2 056	148	61	104 e	34	..
Textiles and Leather	1 335	947	498	156	103	90 e	63	..
Non-Specified	79	10	1 040	81	-	-	-	..
Other Sectors	**2 638**	**4 039**	**2 410**	**1 629**	**1 082**	**979**	**925**	**..**
Commerce and Public	2 473	3 713	2 231	546 e	363	300	290	..
Residential	-	-	-	984 e	620	580	540	..
Agriculture	165	326	179	99	80	80	80	..
Non-Specified	-	-	-	-	19	19	15	..
Memo: Non-Energy Use	-	-	-	-	-	-	-	..
Chem/Petchem. Feedst.	-	-	-	-	-	-	-	..
Other	-	-	-	-	-	-	-	..
Closing Stock Level	-	*4 289*	*4 615*	*1 269*	*1 334*	*1 867*	*1 394*	*1 648 e*

Source: Annual Oil Statistics.

FRANCE
Table 5C: Total Products - Supply and Consumption

000 metric tons

	1973	1978	1980	1990	1999	2000	2001	2002
Refinery Output	**133 792**	**118 583**	**115 717**	**78 333**	**85 283**	**88 576**	**88 832**	**82 927 e**
+ Recycled Products	-	91	76	60	75	71	77	80 e
+ Imports	7 336	9 303	12 678	24 446	26 917	26 715	28 281	31 709 e
- Exports	13 445	13 156	13 653	14 118	17 994	20 676	19 686	18 703 e
- Intl. Marine Bunkers	5 503	4 889	4 100	2 614	2 987	3 085	2 767	2 645 e
+ Transfers	812	826	839	354	-1 962	-1 514	- 467 e	- 370 e
+ Stock Changes	-1 892	3 301	-1 746	655	1 487	-1 145	899	-1 398 e
- Statistical Differences	640	- 677	2 585	2 034	-3 958	-4 775	- 962	-1 134 e
Total Consumption	**120 460**	**114 736**	**107 226**	**85 082**	**94 777**	**93 717**	**96 131**	**92 734 e**
Transformation	**16 845**	**14 488**	**12 763**	**3 608**	**3 908**	**3 531**	**3 120**	..
Public Electricity	16 255	10 693	9 621	1 189	866	680 e	690	..
Autoproducer Electricity	-	1 269	1 007	515	880	596 e	340	..
Public CHP Plants	-	-	-	-	-	-	-	..
Autoproducers of CHP	-	-	-	-	-	-	-	..
Public Heat Plants	-	-	-	-	-	-	-	..
Autoproducer Heat Plants	-	-	-	-	-	-	-	..
Petrochemical Industry	-	2 307	1 977	1 717	2 013	2 164	1 970	..
Other Transformation	590	219	158	187	149	91 e	120	..
Energy Sector	**6 637**	**6 516**	**6 211**	**4 830**	**5 359**	**5 417**	**5 254**	..
Refinery Fuel	6 637	6 516	6 211	4 736	5 330	5 389	5 254	..
Other Energy Sector	-	-	-	94	29	28	-	..
Final Consumption	**96 978**	**93 732**	**88 252**	**76 644**	**85 510**	**84 769**	**87 757**	..
Transport	**25 402**	**30 226**	**30 850**	**40 362**	**48 189**	**49 460**	**49 996**	..
Air	1 860	2 307	2 458	3 782	6 319	6 550	6 482	..
Road	22 194	26 124	27 014	35 484	40 752	41 782	42 529	..
Rail	675	621	584	394	373	369	245	..
Inland Waterways	673	1 174	791	702	734	756	740	..
Non-Specified	-	-	3	-	11	3	-	..
Industry	**35 097**	**31 051**	**30 625**	**17 451**	**18 882**	**18 845**	**19 333**	..
Iron and Steel	2 755	2 022	1 330	359	111	121 e	127	..
Chemical (incl. Pet.-Chem.)	10 679	8 825	9 565	7 768	10 434	10 427	10 634	..
Non-Ferrous Metals	571	658	609	566	411	255 e	260	..
Non-Metallic Minerals	1 429	1 033	736	1 443	1 673	1 302 e	1 745	..
Transport Equipment	485	485	297	180	126	101 e	100	..
Machinery	-	-	-	239	172	113 e	126	..
Mining and Quarrying	480	226	237	90	5	5 e	17	..
Food and Tobacco	2 430	2 288	2 574	1 032	595	569 e	756	..
Pulp, Paper and Printing	1 421	1 188	1 341	437	266	260 e	191	..
Wood and Wood Products	155	120	-	-	-	-	-	..
Construction	6 710	5 747	4 829	3 868	3 839	4 245	4 190	..
Textiles and Leather	1 335	947	498	227	141	113 e	84	..
Non-Specified	6 647	7 512	8 609	1 242	1 109	1 334 e	1 103	..
Other Sectors	**36 479**	**32 455**	**26 777**	**18 831**	**18 439**	**16 464**	**18 428**	..
Commerce and Public	32 021	27 409	21 910	5 062	5 106	4 310	5 158	..
Residential	1 665	1 952	1 907	10 835	10 706	9 620	10 724	..
Agriculture	2 793	3 094	2 960	2 934	2 601	2 508	2 460	..
Non-Specified	-	-	-	-	26	26	86	..
Memo: Non-Energy Use	*13 047*	*10 522*	*11 106*	*11 147*	*13 873*	*13 476*	*12 278*	..
Chem/Petchem. Feedst.	*8 302*	*6 456*	*7 122*	*6 676*	*9 298*	*9 098*	*7 892*	..
Other	*4 745*	*4 066*	*3 984*	*4 471*	*4 575*	*4 378*	*4 386*	..
Closing Stock Level	-	*19 142*	*23 302*	*11 745*	*13 350*	*14 495*	*13 596*	*14 994 e*

Source: Annual Oil Statistics.

FRANCE
Table 6A: Supply and Consumption

000 metric tons

	1980	2000	2001	2002	1980	2000	2001	2002
	LPG AND ETHANE				**REFINERY GAS**			
Refinery Output	**3 133**	**2 695**	**2 955**	**2 510 e**	**2 487**	**2 540**	**2 460**	**2 267 e**
+ Recycled Products	-	-	-	-	-	-	-	-
+ Imports	550	1 836	1 888	2 178 e	-	-	-	-
- Exports	660	1 502	1 279	1 248 e	-	-	-	-
- Intl. Marine Bunkers	-	-	-	-	-	-	-	-
+ Transfers	300	67	162	165 e	-	-	-	-
+ Stock Change	64	- 31	44	- 37 e	-	-	-	-
- Statistical Difference	- 152	- 768	- 101	- 124 e	91	-	-	-
Total Consumption	**3 539**	**3 833**	**3 871**	**3 692 e**	**2 396**	**2 540**	**2 460**	**2 267 e**
of which: Ethane	-	*74*	*86*	*98 e*	-	-	-	-
Transformation	221	161	141	..	-	-	-	..
Energy	114	38	41	..	2 314	2 379	2 203	..
Transport	16	217	210	..	-	-	-	..
Industry	883	1 195	1 286	..	82	161	257	..
Other	2 305	2 222	2 193	..	-	-	-	..
Memo: Non-Energy Use	*305*	*555*	*490*	..	-	-	-	..
Closing Stock Level	*97*	*150*	*106*	*143 e*	-	-	-	-
	NAPHTHA				**OTHER KEROSENE**			
Refinery Output	**4 349**	**5 776**	**5 695**	**5 374 e**	**136**	**82**	**111**	**63 e**
+ Recycled Products	-	-	-	-	-	-	-	-
+ Imports	2 412	2 513	2 296	2 704 e	3	147	209	239 e
- Exports	466	1 831	1 956	1 923 e	10	23	54	66 e
- Intl. Marine Bunkers	-	-	-	-	-	-	-	-
+ Transfers	-	- 121	-	-	-	- 7	-	-
+ Stock Change	- 26	- 54	50	36 e	- 6	6	1	1 e
- Statistical Difference	- 16	-1 696	-1 392	- 287 e	58	93	169	166 e
Total Consumption	**6 285**	**7 979**	**7 477**	**6 478 e**	**65**	**112**	**98**	**71 e**
Transformation	1 460	1 498	1 242	..	-	-	-	..
Energy	-	-	-	..	-	-	-	..
Transport	-	-	-	..	6	-	-	..
Industry	4 825	6 481	6 235	..	36	-	-	..
Other	-	-	-	..	23	112	98	..
Memo: Non-Energy Use	*4 746*	*6 481*	*5 363*	..	-	-	-	..
Closing Stock Level	*517*	*431*	*381*	*345 e*	*20*	*9*	*8*	*7 e*
	WHITE SPIRIT + SBP				**LUBRICANTS**			
Refinery Output	**290**	**114**	**98**	**..**	**1 631**	**1 836**	**1 789**	**..**
+ Recycled Products	-	-	-	..	-	34	31	..
+ Imports	6	126	42	..	193	450	384	..
- Exports	111	67	115	..	731	1 306	1 388	..
- Intl. Marine Bunkers	-	-	-	..	45	41	36	..
+ Transfers	-	- 17	-	..	-	- 61	-	..
+ Stock Change	13	2	- 1	..	- 81	- 128	- 25	..
- Statistical Difference	11	50	- 92	..	27	- 105	- 105	..
Total Consumption	**187**	**108**	**116**	**..**	**940**	**889**	**860**	**..**
Transformation	-	-	-	..	-	-	-	..
Energy	-	-	-	..	-	-	-	..
Transport	-	-	-	..	411	393	455	..
Industry	187	108	116	..	449	426	334	..
Other	-	-	-	..	80	70	71	..
Memo: Non-Energy Use	*187*	*108*	*116*	..	*940*	*889*	*860*	..
Closing Stock Level	*32*	*15*	*16*	..	*455*	*303*	*328*	..

Source: Annual Oil Statistics.

FRANCE
Table 6B: Supply and Consumption

000 metric tons

	1980	2000	2001	2002	1980	2000	2001	2002
	BITUMEN				PARAFFIN WAXES			
Refinery Output	**3 210**	**3 733**	**3 712**	..	**80**	**195**	**205**	..
+ Recycled Products	-	-	-	..	-	-	-	..
+ Imports	14	590	662	..	36	96	97	..
- Exports	263	413	494	..	27	167	143	..
- Intl. Marine Bunkers	-	-	-	..	-	-	-	..
+ Transfers	-	- 80	-	..	-	- 13	-	..
+ Stock Change	12	11	17	..	- 1	8	- 1	..
- Statistical Difference	200	466	487	..	4	55	104	..
Total Consumption	**2 773**	**3 375**	**3 410**	..	**84**	**64**	**54**	..
Transformation	-	-	-	..	-	-	-	..
Energy	-	-	-	..	-	4	-	..
Transport	-	-	-	..	-	-	-	..
Industry	2 773	3 375	3 410	..	84	60	54	..
Other	-	-	-	..	-	-	-	..
Memo: Non-Energy Use	*2 773*	*3 321*	*3 356*	..	*84*	*60*	*54*	..
Closing Stock Level	*223*	*234*	*217*	..	*8*	*20*	*21*	..
	PETROLEUM COKE				OTHER PRODUCTS			
Refinery Output	-	**922**	**931**	**862 e**	**1 512**	**1 957**	**1 980**	**7 049 e**
+ Recycled Products	-	-	-	-	76	-	-	37 e
+ Imports	1 722	1 249	1 580	1 246 e	100	140	134	1 641 e
- Exports	-	22	31	40 e	373	1 035	606	3 115 e
- Intl. Marine Bunkers	-	-	-	-	-	-	-	36 e
+ Transfers	-	- 60	-	-	3	32	52	55 e
+ Stock Change	-	-	-	-	2	14	10	55 e
- Statistical Difference	215	- 93	- 22	- 188 e	-	- 874	- 998	-1 041 e
Total Consumption	**1 507**	**2 182**	**2 502**	**2 256 e**	**1 320**	**1 982 e**	**2 568**	**6 727 e**
Transformation	-	71	131	..	-	115 e	339	..
Energy	-	935	931	..	670	663	758	..
Transport	-	-	-	..	-	-	-	..
Industry	1 507	1 176	1 440	..	650	1 204 e	1 471	..
Other	-	-	-	..	-	-	-	..
Memo: Non-Energy Use	-	-	-	..	*650*	*457*	*512*	..
Closing Stock Level	-	-	-	-	*843*	*100*	*90*	*617 e*

Source: Annual Oil Statistics.

FRANCE
Table 7A: Imports by Country of Origin

000 metric tons

	2000	2001	2002	2000	2001	2002	2000	2001	2002
	Crude Oil			**Natural Gas Liquids**			**Feedstocks and Other**		
Australia	-	-	-	-	-	-	-	-	-
Belgium	-	-	-	-	-	-	2	-	-
Canada	-	-	-	-	-	-	-	-	-
Czech Republic	-	-	-	-	-	-	-	-	-
Denmark	853	1 226	735 e	-	-	-	-	-	-
Finland	-	-	-	-	-	-	-	-	-
France	-	-	-	-	-	-	-	-	-
Germany	-	-	23 e	-	-	-	1	-	-
Greece	-	-	-	-	-	-	-	-	-
Hungary	-	-	-	-	-	-	-	-	-
Ireland	-	-	-	-	-	-	-	-	-
Italy	-	25	188 e	-	-	-	-	-	-
Japan	-	-	-	-	-	-	-	-	-
Korea	-	-	-	-	-	-	-	-	-
Mexico	-	37	110 e	-	-	-	-	-	-
Netherlands	-	-	33 e	-	-	-	54	-	-
New Zealand	-	-	-	-	-	-	-	-	-
Norway	21 141	19 371	16 997 e	-	-	-	-	-	-
Poland	-	-	-	-	-	-	-	-	-
Spain	-	-	-	-	-	-	-	-	-
Sweden	-	-	-	-	-	-	-	-	-
Turkey	-	-	-	-	-	-	-	-	-
United Kingdom	9 898	11 347	9 305 e	-	-	-	-	-	-
United States	-	-	-	-	-	-	-	-	-
Other	-	-	-	-	-	-	-	-	-
OECD	**31 892**	**32 006**	**27 391 e**	**-**	**-**	**-**	**57**	**-**	**-**
Argentina	131	-	-	-	-	-	-	-	-
Brazil	103	636	142 e	-	-	-	-	-	-
Colombia	-	-	-	-	-	-	-	-	-
Ecuador	-	-	-	-	-	-	-	-	-
Netherlands Antilles	-	-	-	-	-	-	-	-	-
Peru	-	-	-	-	-	-	-	-	-
Trinidad and Tobago	-	-	-	-	-	-	-	-	-
Venezuela	-	-	-	-	-	-	-	-	-
Other	34	-	-	-	-	-	-	-	-
Latin America	**268**	**636**	**142 e**	**-**	**-**	**-**	**-**	**-**	**-**
Romania	-	-	-	-	-	-	-	-	-
Other	-	-	-	-	-	-	-	-	-
Non-OECD Europe	**-**	**-**	**-**	**-**	**-**	**-**	**-**	**-**	**-**
FSU	**7 897**	**10 106**	**14 000 e**	**-**	**-**	**16 e**	**97**	**-**	**-**
Iran	5 176	3 631	3 797 e	-	-	-	-	-	-
Iraq	7 219	5 366	4 209 e	-	-	-	-	-	-
Kuwait	1 374	1 110	1 154 e	-	-	-	-	-	-
Oman	-	-	-	-	-	-	-	-	-
Qatar	101	-	-	-	-	-	-	-	-
Saudi Arabia	15 226	11 345	11 216 e	-	-	-		20	
United Arab Emirates	6	255	-	-	-	-	-	-	-
Other	2 506	3 129	2 736 e	-	-	-	-	-	-
Middle East	**31 608**	**24 836**	**23 112 e**	**-**	**-**	**-**	**-**	**20**	**-**
India	-	-	-	-	-	-	-	-	-
Indonesia	-	-	-	-	-	-	-	-	-
Singapore	-	-	-	-	-	-	-	-	-
Other	-	-	-	-	-	-	-	-	-
Asia	**-**	**-**	**-**	**-**	**-**	**-**	**-**	**-**	**-**
China	**-**	**-**	**-**	**-**	**-**	**-**	**-**	**-**	**-**
Algeria	3 440	4 233	3 394 e	-	33	-	33	-	-
Angola/Cabinada	1 893	3 699	3 793 e	-	-	-	-	-	-
Cameroon	307	161	653 e	-	-	-	-	-	-
Egypt	-	67	34 e	-	-	-	-	-	-
Gabon	289	1 314	458 e	-	-	-	-	-	-
Libya	2 441	2 988	1 957 e	-	-	-	-	-	-
Nigeria	4 844	5 260	4 423 e	-	-	-	-	-	-
Tunisia	365	110	376 e	-	53	129 e	33	29	-
Other	219	740	1 335 e	-	-	-	-	-	-
Africa	**13 798**	**18 572**	**16 423 e**	**-**	**86**	**129 e**	**66**	**29**	**-**
Non-spec./Other	**-**	**-**	**187 e**	**-**	**-**	**-**	**15**	**71**	**35 e**
Total Imports	**85 463**	**86 156**	**81 255 e**	**-**	**86**	**145 e**	**235**	**120**	**35 e**
Memo: Total OPEC	*39 827*	*34 188*	*30 150 e*	-	*33*	-	*33*	*20*	-

Source: Annual Oil Statistics.

FRANCE
Table 7B: Imports by Country of Origin

000 metric tons

	Total Products			Liquefied Petroleum Gases			Naphtha		
	2000	2001	2002	2000	2001	2002	2000	2001	2002
Australia	-	-	-	-	-	-	-	-	-
Belgium	1 999	1 661	1 701 e	38	27	15 e	136	128	108 e
Canada	122	230	70 e	-	-	-	-	-	-
Czech Republic	6	-	-	-	-	-	-	-	-
Denmark	141	152	176 e	4	7	2 e	-	-	13 e
Finland	610	42	35 e	5	-	-	13	5	-
France	-	-	-	-	-	-	-	-	-
Germany	1 931	1 226	2 127 e	83	66	58 e	248	71	166 e
Greece	18	286	348 e	-	-	39 e	-	141	165 e
Hungary	-	-	-	-	-	-	-	-	-
Ireland	49	13	8 e	-	1	4 e	18	-	-
Italy	1 010	2 272	2 569 e	33	44	48 e	128	274	482 e
Japan	14	21	19 e	-	-	-	7	10	13 e
Korea	1	-	-	-	-	-	-	-	-
Mexico	1	67	32 e	-	-	-	-	67	31 e
Netherlands	2 854	1 993	2 725 e	28	27	36 e	41	49	141 e
New Zealand	-	-	-	-	-	-	-	-	-
Norway	594	523	412 e	179	267	198 e	166	29	23 e
Poland	34	7	5 e	-	-	-	-	-	-
Spain	725	697	885 e	21	2	8 e	83	110	307 e
Sweden	828	483	735 e	2	9	23 e	8	-	6 e
Turkey	94	115	65 e	-	-	-	94	-	8 e
United Kingdom	3 399	2 271	3 214 e	717	508	598 e	208	207	321 e
United States	1 017	1 015	1 178 e	-	-	14 e	37	23	30 e
Other	150	141	267 e	2	3	-	-	-	23 e
OECD	**15 597**	**13 215**	**16 571 e**	**1 112**	**961**	**1 043 e**	**1 187**	**1 114**	**1 837 e**
Argentina	-	9	41 e	-	-	-	-	-	15 e
Brazil	2	4	-	-	-	-	-	-	-
Colombia	-	-	-	-	-	-	-	-	-
Ecuador	-	-	-	-	-	-	-	-	-
Netherlands Antilles	288	17	34 e	2	-	-	-	-	-
Peru	-	-	-	-	-	-	-	-	-
Trinidad and Tobago	364	466	530 e	9	8	10 e	-	-	-
Venezuela	294	114	152 e	-	-	32 e	-	-	26 e
Other	85	133	78 e	32	-	-	-	-	-
Latin America	**1 033**	**743**	**835 e**	**43**	**8**	**42 e**	**-**	**-**	**41 e**
Romania	-	53	407 e	-	-	-	-	-	-
Other	55	113	90 e	-	-	-	14	-	17 e
Non-OECD Europe	**55**	**166**	**497 e**	**-**	**-**	**-**	**14**	**-**	**17 e**
FSU	**4 866**	**5 504**	**5 091 e**	**5**	**1**	**28 e**	**143**	**97**	**165 e**
Iran	16	49	1 e	-	-	-	-	49	-
Iraq	-	-	-	-	-	-	-	-	-
Kuwait	465	439	259 e	-	-	-	-	-	-
Oman	-	-	-	-	-	-	-	-	-
Qatar	20	20	26 e	-	-	-	-	-	-
Saudi Arabia	423	572	462 e	23	7	-	-	28	-
United Arab Emirates	430	599	1 403 e	-	-	-	-	-	-
Other	476	396	259 e	19	16	24 e	103	-	30 e
Middle East	**1 830**	**2 075**	**2 410 e**	**42**	**23**	**24 e**	**103**	**77**	**30 e**
India	42	73	97 e	-	-	-	-	-	-
Indonesia	67	62	16 e	-	-	-	-	-	-
Singapore	-	32	78 e	-	-	-	-	-	-
Other	61	18	21 e	20	-	-	-	-	-
Asia	**170**	**185**	**212 e**	**20**	**-**	**-**	**-**	**-**	**-**
China	**3**	**20**	**27 e**	**-**	**-**	**-**	**-**	**-**	**-**
Algeria	1 528	2 015	1 489 e	575	723	980 e	231	634	261 e
Angola/Cabinada	-	-	-	-	-	-	-	-	-
Cameroon	-	-	-	-	-	-	-	-	-
Egypt	449	235	321 e	-	6	3 e	389	40	161 e
Gabon	-	-	-	-	-	-	-	-	-
Libya	286	334	327 e	4	10	9 e	130	221	155 e
Nigeria	9	110	-	-	110	-	9	-	-
Tunisia	144	-	33 e	-	-	-	-	-	33 e
Other	677	191	345 e	35	45	32 e	307	35	1 e
Africa	**3 093**	**2 885**	**2 515 e**	**614**	**894**	**1 024 e**	**1 066**	**930**	**611 e**
Non-spec./Other	**68**	**3 488**	**3 551 e**	**-**	**1**	**17 e**	**-**	**78**	**3 e**
Total Imports	**26 715**	**28 281**	**31 709 e**	**1 836**	**1 888**	**2 178 e**	**2 513**	**2 296**	**2 704 e**
Memo: Total OPEC	*3 538*	*4 314*	*4 135 e*	*602*	*850*	*1 021 e*	*370*	*932*	*442 e*

Source: Annual Oil Statistics.

FRANCE

Table 7C: Imports by Country of Origin

000 metric tons

	2000	2001	2002	2000	2001	2002	2000	2001	2002
	Total Gasoline			**Total Kerosene**			**Gas/Diesel**		
Australia	-	-	-	-	-	-	-	-	-
Belgium	220	83	159 e	70	75	102 e	981	876	748 e
Canada	-	-	-	-	-	-	119	221	68 e
Czech Republic	-	-	-	-	-	-	-	-	-
Denmark	7	14	-	-	-	-	130	127	146 e
Finland	383	2	-	-	4	-	209	31	29 e
France	-	-	-	-	-	-	-	-	-
Germany	330	247	201 e	2	-	1 e	892	537	1 341 e
Greece	-	8	52 e	18	-	-	-	137	88 e
Hungary	-	-	-	-	-	-	-	-	-
Ireland	-	-	-	-	-	-	23	3	4 e
Italy	48	278	153 e	47	153	103 e	530	1 410	1 510 e
Japan	-	-	-	-	-	-	-	-	-
Korea	-	-	-	-	-	-	-	-	-
Mexico	-	-	-	-	-	-	-	-	-
Netherlands	760	148	174 e	76	17	22 e	1 398	1 474	1 967 e
New Zealand	-	-	-	-	-	-	-	-	-
Norway	11	14	93 e	12	10	-	150	158	67 e
Poland	-	-	-	-	-	-	25	-	-
Spain	158	147	140 e	-	-	29 e	178	211	185 e
Sweden	67	62	19 e	-	-	-	731	373	615 e
Turkey	-	-	17 e	-	-	-	-	-	-
United Kingdom	112	268	211 e	73	60	87 e	1 749	727	1 526 e
United States	-	-	-	5	-	38 e	160	63	278 e
Other	1	23	-	1	-	-	10	6	4 e
OECD	**2 097**	**1 294**	**1 219 e**	**304**	**319**	**382 e**	**7 285**	**6 354**	**8 576 e**
Argentina	-	-	-	-	-	-	-	-	-
Brazil	-	-	-	-	-	-	-	-	-
Colombia	-	-	-	-	-	-	-	-	-
Ecuador	-	-	-	-	-	-	-	-	-
Netherlands Antilles	99	7	-	123	4	-	50	-	18 e
Peru	-	-	-	-	-	-	-	-	-
Trinidad and Tobago	64	107	103 e	97	100	87 e	110	127	95 e
Venezuela	-	-	-	91	-	40 e	5	5	-
Other	-	-	-	7	40	13 e	46	53	61 e
Latin America	**163**	**114**	**103 e**	**318**	**144**	**140 e**	**211**	**185**	**174 e**
Romania	-	-	43 e	-	-	-	-	53	335 e
Other	-	-	15 e	-	-	-	41	113	52 e
Non-OECD Europe	**-**	**-**	**58 e**	**-**	**-**	**-**	**41**	**166**	**387 e**
FSU	**98**	**2**	**9 e**	**62**	**26**	**57 e**	**2 453**	**3 775**	**3 213 e**
Iran	-	-	-	-	-	-	-	-	-
Iraq	-	-	-	-	-	-	-	-	-
Kuwait	-	-	-	421	439	259 e	44	-	-
Oman	-	-	-	-	-	-	-	-	-
Qatar	-	-	-	20	20	-	-	-	26 e
Saudi Arabia	116	72	109 e	124	153	137 e	149	227	216 e
United Arab Emirates	-	14	18 e	427	540	1 314 e	3	3	3 e
Other	15	31	-	279	290	162 e	57	28	6 e
Middle East	**131**	**117**	**127 e**	**1 271**	**1 442**	**1 872 e**	**253**	**258**	**251 e**
India	-	-	-	-	-	35 e	-	23	25 e
Indonesia	-	-	-	-	-	-	-	-	-
Singapore	-	-	7 e	-	-	12 e	-	1	25 e
Other	-	-	-	-	-	-	18	-	-
Asia	**-**	**-**	**7 e**	**-**	**-**	**47 e**	**18**	**24**	**50 e**
China	**-**	**-**	**-**	**-**	**-**	**-**	**-**	**-**	**-**
Algeria	-	-	-	272	141	-	425	517	248 e
Angola/Cabinada	-	-	-	-	-	-	-	-	-
Cameroon	-	-	-	-	-	-	-	-	-
Egypt	-	47	15 e	-	33	85 e	-	-	-
Gabon	-	-	-	-	-	-	-	-	-
Libya	-	-	-	-	-	24 e	152	103	139 e
Nigeria	-	-	-	-	-	-	-	-	-
Tunisia	-	-	-	-	-	-	-	-	-
Other	-	1	-	-	-	-	13	10	49 e
Africa	**-**	**48**	**15 e**	**272**	**174**	**109 e**	**590**	**630**	**436 e**
Non-spec./Other	**-**	**140**	**10 e**	**1**	**446**	**334 e**	**24**	**2 695**	**3 115 e**
Total Imports	**2 489**	**1 715**	**1 548 e**	**2 228**	**2 551**	**2 941 e**	**10 875**	**14 087**	**16 202 e**
Memo: Total OPEC	*116*	*86*	*127 e*	*1 355*	*1 293*	*1 774 e*	*778*	*855*	*632 e*

Source: Annual Oil Statistics.

FRANCE
Table 7D: Imports by Country of Origin

000 metric tons

	2000	2001	2002	2000	2001	2002	2000	2001	2002
	Fuel Oil (Residual)			**Petroleum Coke**			**Other Products**		
Australia	-	-	-	-	-	-	-	-	-
Belgium	169	70	91 e	87	41	57 e	298	361	421 e
Canada	-	-	-	-	9	-	3	-	2 e
Czech Republic	-	-	-	-	-	-	6	-	-
Denmark	-	-	15 e	-	4	-	-	-	-
Finland	-	-	-	-	-	-	-	-	6 e
France	-	-	-	-	-	-	-	-	-
Germany	77	76	29 e	97	70	107 e	202	159	224 e
Greece	-	-	-	-	-	-	-	-	4 e
Hungary	-	-	-	-	-	-	-	-	-
Ireland	5	9	-	-	-	-	3	-	-
Italy	88	27	101 e	31	13	44 e	105	73	128 e
Japan	-	-	-	6	9	6 e	1	2	-
Korea	-	-	-	-	-	-	1	-	-
Mexico	-	-	-	-	-	-	1	-	1 e
Netherlands	321	53	114 e	47	89	114 e	183	136	157 e
New Zealand	-	-	-	-	-	-	-	-	-
Norway	76	45	27 e	-	-	4 e	-	-	-
Poland	-	7	2 e	-	-	-	9	-	3 e
Spain	4	28	50 e	42	-	-	239	199	166 e
Sweden	-	25	68 e	4	-	-	16	14	4 e
Turkey	-	115	40 e	-	-	-	-	-	-
United Kingdom	362	164	176 e	64	262	107 e	114	75	188 e
United States	50	-	53 e	666	858	654 e	99	71	111 e
Other	103	78	185 e	9	1	-	24	30	55 e
OECD	**1 255**	**697**	**951 e**	**1 053**	**1 356**	**1 093 e**	**1 304**	**1 120**	**1 470 e**
Argentina	-	-	-	-	9	26 e	-	-	-
Brazil	-	-	-	-	4	-	2	-	-
Colombia	-	-	-	-	-	-	-	-	-
Ecuador	-	-	-	-	-	-	-	-	-
Netherlands Antilles	-	-	-	-	-	-	14	6	16 e
Peru	-	-	-	-	-	-	-	-	-
Trinidad and Tobago	83	124	235 e	-	-	-	1	-	-
Venezuela	102	58	-	71	37	51 e	25	14	3 e
Other	-	-	-	-	38	-	-	2	4 e
Latin America	**185**	**182**	**235 e**	**71**	**88**	**77 e**	**42**	**22**	**23 e**
Romania	-	-	29 e	-	-	-	-	-	-
Other	-	-	-	-	-	-	-	-	6 e
Non-OECD Europe	**-**	**-**	**29 e**	**-**	**-**	**-**	**-**	**-**	**6 e**
FSU	**2 104**	**1 600**	**1 570 e**	**-**	**3**	**3 e**	**1**	**-**	**46 e**
Iran	13	-	-	-	-	1 e	3	-	-
Iraq	-	-	-	-	-	-	-	-	-
Kuwait	-	-	-	-	-	-	-	-	-
Oman	-	-	-	-	-	-	-	-	-
Qatar	-	-	-	-	-	-	-	-	-
Saudi Arabia	9	85	-	-	-	-	2	-	-
United Arab Emirates	-	42	68 e	-	-	-	-	-	-
Other	-	31	37 e	-	-	-	3	-	-
Middle East	**22**	**158**	**105 e**	**-**	**-**	**1 e**	**8**	**-**	**-**
India	-	-	-	42	50	35 e	-	-	2 e
Indonesia	-	-	-	67	62	16 e	-	-	-
Singapore	-	13	16 e	-	-	-	-	18	18 e
Other	-	-	-	-	-	-	23	18	21 e
Asia	**-**	**13**	**16 e**	**109**	**112**	**51 e**	**23**	**36**	**41 e**
China	**-**	**-**	**8 e**	**3**	**20**	**13 e**	**-**	**-**	**6 e**
Algeria	25	-	-	-	-	-	-	-	-
Angola/Cabinada	-	-	-	-	-	-	-	-	-
Cameroon	-	-	-	-	-	-	-	-	-
Egypt	60	109	57 e	-	-	-	-	-	-
Gabon	-	-	-	-	-	-	-	-	-
Libya	-	-	-	-	-	-	-	-	-
Nigeria	-	-	-	-	-	-	-	-	-
Tunisia	144	-	-	-	-	-	-	-	-
Other	300	84	245 e	-	-	-	22	16	18 e
Africa	**529**	**193**	**302 e**	**-**	**-**	**-**	**22**	**16**	**18 e**
Non-spec./Other	**28**	**2**	**33 e**	**13**	**1**	**8 e**	**2**	**125**	**31 e**
Total Imports	**4 123**	**2 845**	**3 249 e**	**1 249**	**1 580**	**1 246 e**	**1 402**	**1 319**	**1 641 e**
Memo: Total OPEC	*149*	*185*	*68 e*	*138*	*99*	*68 e*	*30*	*14*	*3 e*

Source: Annual Oil Statistics.

FRANCE
Table 8A: Exports by Country of Destination

000 metric tons

	2000	2001	2002	2000	2001	2002	2000	2001	2002
	Crude Oil			**Natural Gas Liquids**			**Feedstocks and Other**		
Australia	-	-	-	-	-	-	-	-	-
Austria	-	-	-	-	-	-	-	-	-
Belgium	-	-	-	-	-	-	40	-	-
Canada	-	-	-	-	-	-	-	-	-
Denmark	-	-	-	-	-	-	-	-	-
Finland	-	-	-	-	-	-	-	-	-
France	-	-	-	-	-	-	-	-	-
Germany	-	-	-	-	-	-	17	1 e	-
Greece	-	-	-	-	-	-	-	-	-
Ireland	-	-	-	-	-	-	-	-	-
Italy	-	-	-	-	-	-	35	-	-
Japan	-	-	-	-	-	-	-	-	-
Luxembourg	-	-	-	-	-	-	-	-	-
Mexico	-	-	-	-	-	-	-	-	-
Netherlands	-	-	-	-	-	-	8	-	-
Norway	-	-	-	-	-	-	-	-	-
Portugal	-	-	-	-	-	-	-	-	-
Spain	-	-	-	-	-	-	40	-	-
Sweden	-	-	-	-	-	-	29	-	-
Switzerland	-	-	-	-	-	-	-	-	-
Turkey	-	-	-	-	-	-	-	-	-
United Kingdom	-	-	-	-	-	-	48	35 e	-
United States	-	-	-	-	-	-	880	231 e	-
Other OECD	-	-	-	-	-	-	25	-	-
OECD	**-**	**-**	**-**	**-**	**-**	**-**	**1 122**	**267 e**	**-**
Africa	-	-	-	-	-	-	-	-	-
Latin America	-	-	-	-	-	-	-	-	-
Asia	-	-	-	-	-	-	-	-	-
China	-	-	-	-	-	-	-	-	-
Non-OECD Europe	-	-	-	-	-	-	-	-	-
Former Soviet Union	-	-	-	-	-	-	-	-	-
Middle East	-	-	-	-	-	-	-	-	-
Non-specified/Other	-	-	-	-	-	-	662	718 e	550 e
Non-OECD	**-**	**-**	**-**	**-**	**-**	**-**	**662**	**718 e**	**550 e**
Total Exports	**-**	**-**	**-**	**-**	**-**	**-**	**1 784**	**985 e**	**550 e**
	Total Products			**Liquefied Petroleum Gas**			**Naphtha**		
Australia	44	6	4 e	-	-	-	-	-	-
Austria	26	13	15 e	1	-	-	-	-	-
Belgium	904	986	1 224 e	43	42	61 e	217	228	248 e
Canada	28	3	273 e	-	-	-	-	-	-
Denmark	33	25	30 e	-	-	-	-	-	-
Finland	32	44	38 e	2	19	2 e	-	-	-
France	-	-	-	-	-	-	-	-	-
Germany	2 369	1 890	1 802 e	110	44	90 e	336	263	314 e
Greece	75	16	53 e	2	-	2 e	-	-	-
Ireland	605	190	226 e	-	-	-	-	-	-
Italy	1 540	1 257	1 524 e	223	179	196 e	19	26	117 e
Japan	17	6	74 e	-	-	-	-	-	60 e
Luxembourg	246	230	164 e	1	-	-	-	-	-
Mexico	41	38	32 e	-	-	-	-	-	32 e
Netherlands	1 668	975	1 470 e	19	16	19 e	427	171	554 e
Norway	47	87	60 e	-	-	6 e	2	25	-
Portugal	517	395	661 e	57	81	82 e	-	-	-
Spain	1 285	1 084	1 380 e	34	34	53 e	33	29	-
Sweden	117	76	69 e	-	-	-	48	-	-
Switzerland	1 578	1 482	1 515 e	1	-	-	-	-	-
Turkey	331	37	154 e	49	-	8 e	1	-	-
United Kingdom	2 820	3 330	2 360 e	256	227	155 e	500	451	247 e
United States	2 492	2 144	1 412 e	45	58	82 e	142	175	85 e
Other OECD	105	75	49 e	12	10	2 e	-	-	-
OECD	**16 920**	**14 389**	**14 589 e**	**855**	**710**	**758 e**	**1 725**	**1 368**	**1 657 e**
Africa	2 025	1 515	1 612 e	437	376	371 e	7	59	1 e
Latin America	264	249	57 e	81	73	17 e	57	16	-
Asia	98	112	159 e	5	-	9 e	-	-	4 e
China	22	9	46 e	7	-	6 e	-	-	-
Non-OECD Europe	430	668	452 e	9	4	4 e	5	29	-
Former Soviet Union	74	10	10 e	-	-	-	10	-	-
Middle East	462	402	349 e	70	23	-	-	-	-
Non-specified/Other	381	2 332	1 429 e	38	93	83 e	27	484	261 e
Non-OECD	**3 756**	**5 297**	**4 114 e**	**647**	**569**	**490 e**	**106**	**588**	**266 e**
Total Exports	**20 676**	**19 686**	**18 703 e**	**1 502**	**1 279**	**1 248 e**	**1 831**	**1 956**	**1 923 e**

Source: Annual Oil Statistics.

FRANCE
Table 8B: Exports by Country of Destination

000 metric tons

	2000	2001	2002	2000	2001	2002	2000	2001	2002
	Total Gasoline			**Total Kerosene**			**Gas/Diesel Oil**		
Australia	39	-	-	-	-	-	-	-	-
Austria	-	-	-	-	-	-	5	-	-
Belgium	4	10	-	65	43	24 e	103	36	40 e
Canada	28	3	213 e	-	-	-	-	-	-
Denmark	1	-	8 e	-	-	14 e	-	-	-
Finland	1	-	-	-	-	-	-	-	-
France	-	-	-	-	-	-	-	-	-
Germany	917	527	443 e	81	127	114 e	431	492	386 e
Greece	24	5	7 e	-	-	-	-	-	11 e
Ireland	-	2	43 e	6	11	-	-	-	-
Italy	307	294	247 e	33	-	-	27	42	72 e
Japan	-	-	-	-	-	-	-	-	-
Luxembourg	-	-	-	199	163	146 e	17	9	-
Mexico	38	-	-	-	-	-	-	-	-
Netherlands	79	97	97 e	-	28	-	213	21	164 e
Norway	6	6	5 e	3	10	11 e	24	31	26 e
Portugal	4	-	-	9	14	-	18	-	14 e
Spain	310	336	278 e	39	9	4 e	454	217	235 e
Sweden	2	-	13 e	-	32	-	-	4	-
Switzerland	558	531	679 e	544	449	377 e	391	396	351 e
Turkey	179	29	28 e	-	-	-	66	-	62 e
United Kingdom	606	482	450 e	196	112	167 e	213	258	119 e
United States	820	947	833 e	-	-	-	15	15	-
Other OECD	39	12	-	-	-	-	-	-	-
OECD	**3 962**	**3 281**	**3 344 e**	**1 175**	**998**	**857 e**	**1 977**	**1 521**	**1 480 e**
Africa	732	240	480 e	26	13	24 e	281	319	216 e
Latin America	47	131	-	-	-	-	15	-	-
Asia	4	-	-	-	-	-	13	17	25 e
China	-	-	-	-	-	-	-	-	30 e
Non-OECD Europe	231	345	156 e	1	2	1 e	42	57	69 e
Former Soviet Union	45	-	1 e	-	1	-	6	-	-
Middle East	224	314	229 e	-	-	-	68	-	68 e
Non-specified/Other	69	440	84 e	5	217	120 e	28	79	80 e
Non-OECD	**1 352**	**1 470**	**950 e**	**32**	**233**	**145 e**	**453**	**472**	**488 e**
Total Exports	**5 314**	**4 751**	**4 294 e**	**1 207**	**1 231**	**1 002 e**	**2 430**	**1 993**	**1 968 e**
	Fuel Oil (Residual)			**Petroleum Coke**			**Other Products**		
Australia	-	-	-	-	-	-	5	6	4 e
Austria	-	-	-	-	-	-	20	13	15 e
Belgium	47	256	308 e	6	11	21 e	419	360	522 e
Canada	-	-	52 e	-	-	-	-	-	8 e
Denmark	11	-	-	-	-	-	21	25	8 e
Finland	-	-	-	-	-	-	29	25	36 e
France	-	-	-	-	-	-	-	-	-
Germany	52	37	48 e	-	-	-	442	400	407 e
Greece	30	-	13 e	-	-	-	19	11	20 e
Ireland	467	87	54 e	-	-	-	132	90	129 e
Italy	816	587	686 e	-	-	-	115	129	206 e
Japan	-	-	-	-	-	-	17	6	14 e
Luxembourg	-	-	-	12	14	16 e	17	44	2 e
Mexico	-	38	-	-	-	-	3	-	-
Netherlands	674	422	408 e	3	-	-	253	220	228 e
Norway	-	11	8 e	-	-	-	12	4	4 e
Portugal	349	222	482 e	-	-	-	80	78	83 e
Spain	200	230	632 e	-	-	-	215	229	178 e
Sweden	29	-	-	-	-	-	38	40	56 e
Switzerland	1	-	-	-	-	-	83	106	108 e
Turkey	-	-	37 e	-	-	-	36	8	19 e
United Kingdom	595	1 613	794 e	-	-	-	454	187	428 e
United States	1 444	939	391 e	-	-	-	26	10	21 e
Other OECD	-	-	-	-	-	-	54	53	47 e
OECD	**4 715**	**4 442**	**3 913 e**	**21**	**25**	**37 e**	**2 490**	**2 044**	**2 543 e**
Africa	323	325	337 e	1	1	-	218	182	183 e
Latin America	-	-	-	-	-	-	64	29	40 e
Asia	-	-	11 e	-	-	-	76	95	110 e
China	-	-	-	-	-	-	15	9	10 e
Non-OECD Europe	129	186	220 e	-	-	-	13	45	2 e
Former Soviet Union	-	-	-	-	-	-	13	9	9 e
Middle East	33	-	-	-	-	-	67	65	52 e
Non-specified/Other	182	746	632 e	-	5	3 e	32	268	166 e
Non-OECD	**667**	**1 257**	**1 200 e**	**1**	**6**	**3 e**	**498**	**702**	**572 e**
Total Exports	**5 382**	**5 699**	**5 113 e**	**22**	**31**	**40 e**	**2 988**	**2 746**	**3 115 e**

Source: Annual Oil Statistics.

GERMANY

Figure 1
FUEL SHARES OF TPES *

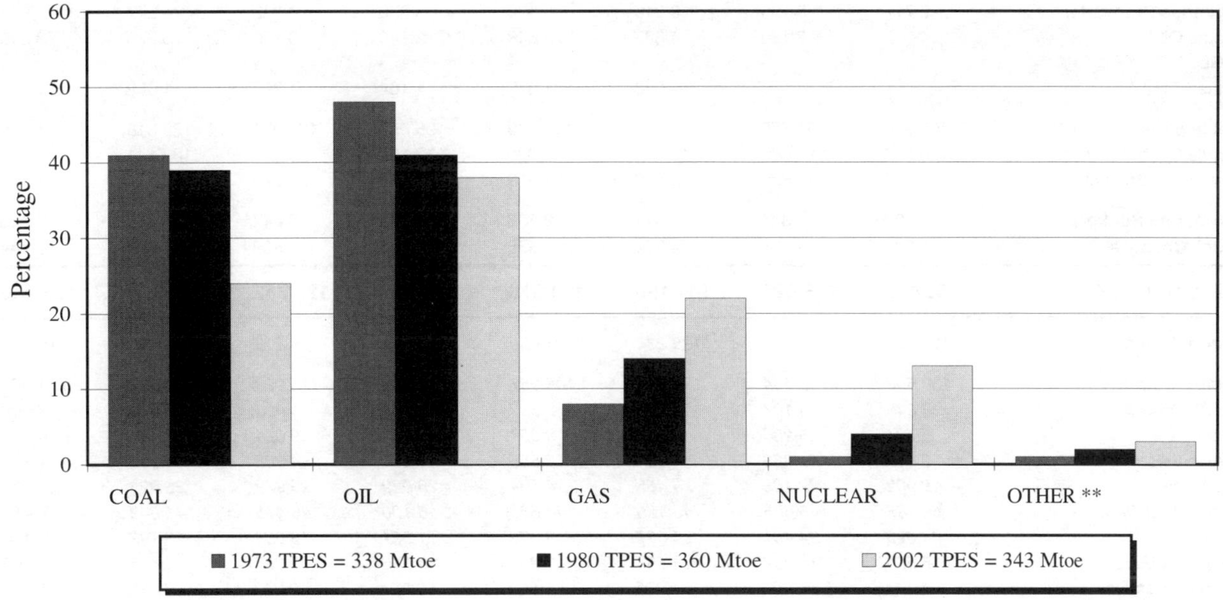

Source: Annual Energy Balances.

* Excluding electricity trade.

** Includes hydro, geothermal, solar, wind, combustible renewables and waste, etc.

Figure 2
OIL PRODUCTION AND CONSUMPTION

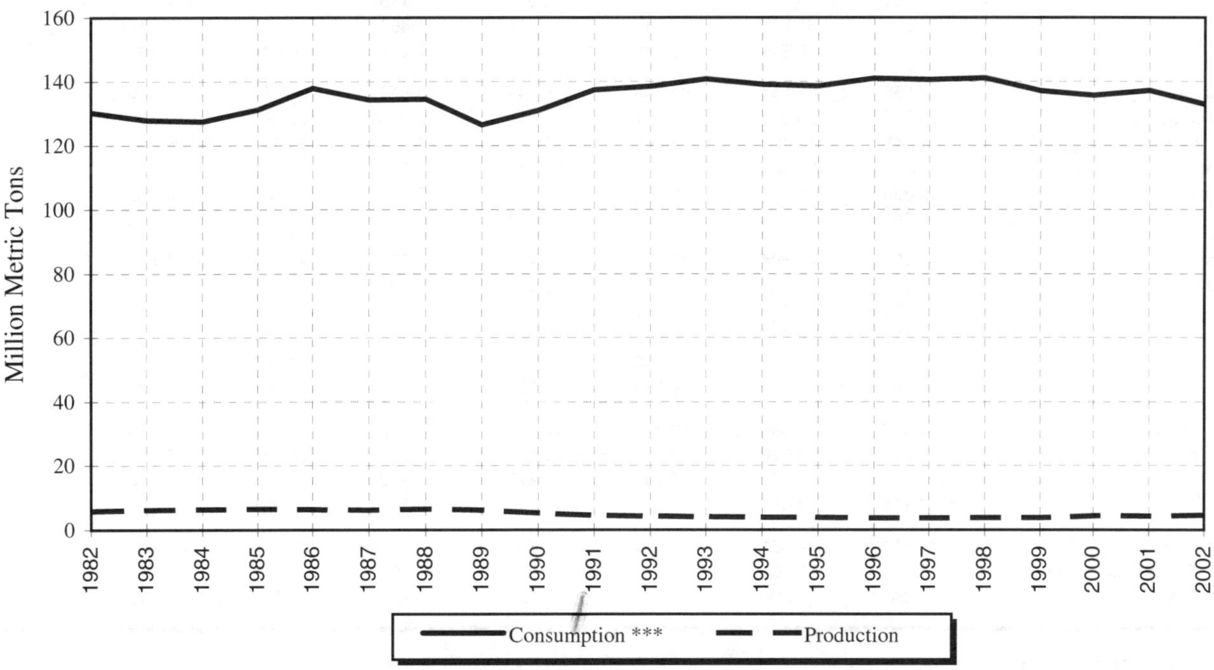

Source: Annual Oil Statistics.

*** Includes refinery fuel and international marine bunkers.

GERMANY
Table 1: Oil Supply and Consumption - Summary Table

000 metric tons

	1973	1978	1980	1990	1999	2000	2001	2002
Indigenous Production	6 714	5 130	6 215	5 520	3 850	4 393	4 347	4 641 e
Crude Oil	..	5 130	4 667	3 648	2 746	3 166	3 278	3 510 e
NGL	..	-	-	-	-	-	-	-
Other	..	-	1 548	1 872	1 104	1 227	1 069	1 131 e
+ Net Imports	157 949	154 506	145 824	119 579	126 050	124 037	129 315	121 315 e
Crude Oil, NGL, Other	126 630	115 680	119 561	88 060	102 192	100 467	104 141	103 574 e
Petroleum Products	31 319	38 826	26 263	31 519	23 858	23 570	25 174	17 741 e
- Intl. Marine Bunkers	4 202	3 374	3 611	2 558	2 139	2 247	2 280	2 445 e
+ Stock Changes	-2 021	- 636	-4 272	1 535	4 161	2 579	- 153	2 444 e
TOTAL SUPPLY	**158 440**	**155 626**	**144 156**	**124 076**	**131 922**	**128 762**	**131 229**	**125 955 e**
Refinery Intake	*138 364*	*126 680*	*138 234*	*107 459*	*117 041*	*118 213*	*116 365*	*116 244 e*
Refinery Output	*139 508*	*126 584*	*136 512*	*105 957*	*115 017*	*115 973*	*114 292*	*114 085 e*
LPG + Ethane	*2 845*	*3 107*	*2 776*	*2 532*	*2 749*	*3 053*	*3 020*	*2 956 e*
Naphtha	*5 794*	*5 208*	*9 371*	*8 554*	*9 663*	*9 024*	*8 495*	*8 662 e*
Kerosenes	*2 143*	*1 363*	*1 521*	*2 656*	*4 259*	*4 328*	*4 206*	*4 166 e*
Gasolines	*18 902*	*22 108*	*25 455*	*24 504*	*26 829*	*26 972*	*26 021*	*25 970 e*
Gas/Diesel Oil	*50 188*	*45 685*	*48 128*	*40 668*	*46 100*	*46 445*	*46 943*	*47 455 e*
Fuel Oil	*44 073*	*34 461*	*34 607*	*13 075*	*11 821*	*13 068*	*13 192*	*12 184 e*
Other Products	*15 563*	*14 652*	*14 654*	*13 968*	*13 596*	*13 083*	*12 415*	*12 692 e*
- Refinery Losses	-1 144	96	1 722	1 502	2 024	2 240	2 073	2 159 e
+ Returns	1 454	1 958	4 352	5 191	4 841	6 136	5 802	5 872 e
- Statistical Differences	1 788	770	- 682	- 714	- 185	- 725	91	- 777 e
TOTAL CONSUMPTION	**159 250**	**156 718**	**147 468**	**128 479**	**134 924**	**133 383**	**134 867**	**130 445 e**
Transformation and Energy	**24 123**	**22 872**	**24 003**	**15 605**	**14 907**	**15 604**	**15 125**	..
Crude Oil	-	-	-	-	-	-	-	..
LPG + Ethane	377	279	372	361	483	480	522	..
Gas/Diesel Oil	71	318	446	1 410	922	831	835	..
Fuel Oil	17 565	15 529	13 997	5 057	4 135	3 844	4 317	..
Petroleum Coke	1 069	1 202	1 045	1 536	1 450	1 510	1 342	..
Other	5 041	5 544	8 143	7 241	7 917	8 939	8 109	..
Transport	**35 535**	**42 556**	**44 508**	**56 149**	**63 843**	**62 856**	**61 360**	..
Motor Gasoline	20 275	25 268	26 284	31 070	30 025	28 581	27 724	..
Gas/Diesel Oil	11 641	13 422	14 089	19 126	26 511	26 646	26 360	..
Subtotal	*31 916*	*38 690*	*40 373*	*50 196*	*56 536*	*55 227*	*54 084*	..
Aviation Fuels	2 825	3 190	3 380	5 446	6 827	7 168	6 836	..
Other	794	676	755	507	480	461	440	..
Industry	**47 083**	**40 604**	**36 452**	**26 269**	**26 831**	**27 069**	**26 553**	..
Naphtha	5 358	5 433	9 651	9 543	12 292	12 823	12 641	..
LPG + Ethane	2 368	1 957	1 185	1 478	1 427	1 475	1 521	..
Gas/Diesel Oil	7 418	6 255	4 862	3 987	3 128	2 784	2 728	..
Fuel Oil	20 343	17 090	14 146	4 752	2 431	4 096	4 342	..
Other	11 596	9 869	6 608	6 509	7 553	5 891	5 321	..
Other Sectors	**52 509**	**50 686**	**42 505**	**30 456**	**29 343**	**27 854**	**31 829**	..
Gas/Diesel Oil	47 973	47 068	39 158	29 132	27 933	26 625	30 588	..
Fuel Oil	3 391	2 182	2 056	279	209	5	6	..
Other	1 145	1 436	1 291	1 045	1 201	1 224	1 235	..
Memo: Non-Energy Use	*13 882*	*14 031*	*18 513*	*19 514*	*21 791*	*22 201*	*21 896*	..
Chem/Petchem. Feedst.	*7 522*	*8 507*	*12 345*	*14 552*	*16 495*	*17 186*	*17 211*	..
Other	*6 360*	*5 524*	*6 168*	*4 962*	*5 296*	*5 015*	*4 685*	..

Source: Annual Oil Statistics.

GERMANY
Table 2: Supply of Crude Oil, NGL, Feedstocks and Other Hydrocarbons

000 metric tons

	1973	1978	1980	1990	1999	2000	2001	2002
CRUDE OIL								
Indigenous Production	..	5 130	4 667	3 648	2 746	3 166	3 278	3 510 e
+ Imports	..	115 771	119 631	88 060	103 933	103 684	105 171	104 895 e
- Exports	..	91	70	-	1 741	3 217	1 030	1 321 e
- Direct Use	..	-	-	-	-	-	-	-
+ Stock Change	..	-1 084	-2 910	186	1 660	3 421	- 750	- 59 e
- Statistical Difference	..	- 33	79	217	- 34	14	- 14	- 6 e
Refinery Intake	**..**	**119 759**	**121 239**	**91 677**	**106 632**	**107 040**	**106 683**	**107 031 e**
Closing Stock Level	..	*17 748*	*21 742*	*18 068*	*18 104*	*14 683*	*15 433*	*15 492 e*
NATURAL GAS LIQUIDS								
Indigenous Production	..	-	-	-	-	-	-	-
+ Imports	..	-	-	-	-	-	-	-
- Exports	..	-	-	-	-	-	-	-
- Direct Use	..	-	-	-	-	-	-	-
+ Stock Change	..	-	-	-	-	-	-	-
- Statistical Difference	..	-	-	-	-	-	-	-
Refinery Intake	**..**	**-**	**-**	**-**	**-**	**-**	**-**	**-**
Closing Stock Level	..	-	-	-	-	-	-	-
REFINERY FEEDSTOCKS								
Backflows	..	1 958	4 081	4 919	4 651	5 889	5 487	5 551 e
+ Transfers	..	5 785	11 366	9 223	4 714	4 109	3 169	2 573 e
+ Imports	..	-	-	-	-	-	-	-
- Exports	..	-	-	-	-	-	-	-
+ Stock Change	..	-	-	-	-	-	-	-
- Statistical Difference	..	822	-	-	-	-	-	- 1 e
Refinery Intake	**..**	**6 921**	**15 447**	**14 142**	**9 365**	**9 998**	**8 656**	**8 125 e**
Closing Stock Level	..	-	-	-	-	-	-	-
OTHER HYDROCARBONS								
Indigenous Production	..	-	1 548	1 872	1 104	1 227	1 069	1 131 e
+ Imports	..	-	-	-	-	-	-	-
- Exports	..	-	-	-	-	-	-	-
- Direct Use	..	-	-	232	60	52	43	43 e
+ Stock Change	..	-	-	-	-	-		
- Statistical Difference	..	-	-	-	-	-	-	-
Refinery Intake	**..**	**-**	**1 548**	**1 640**	**1 044**	**1 175**	**1 026**	**1 088 e**
Closing Stock Level	..	-	-	-	-	-	-	-
CRUDE OIL, NGL, FEEDSTOCKS AND OTHER								
Indigenous Production	6 714	5 130	6 215	5 520	3 850	4 393	4 347	4 641 e
+ Backflows	1 454	1 958	4 081	4 919	4 651	5 889	5 487	5 551 e
+ Transfers	5 259	5 785	11 366	9 223	4 714	4 109	3 169	2 573 e
+ Imports	126 738	115 771	119 631	88 060	103 933	103 684	105 171	104 895 e
- Exports	108	91	70	-	1 741	3 217	1 030	1 321 e
- Direct Use	-	-	-	232	60	52	43	43 e
+ Stock Change	- 808	-1 084	-2 910	186	1 660	3 421	- 750	- 59 e
- Statistical Difference	885	789	79	217	- 34	14	- 14	- 7 e
Refinery Intake	**138 364**	**126 680**	**138 234**	**107 459**	**117 041**	**118 213**	**116 365**	**116 244 e**
Closing Stock Level	-	*17 748*	*21 742*	*18 068*	*18 104*	*14 683*	*15 433*	*15 492 e*

Source: Annual Oil Statistics.

GERMANY

Table 3: Motor Gasoline - Supply

000 metric tons

	1973	1978	1980	1990	1999	2000	2001	2002
MOTOR GASOLINE								
Refinery Output	**18 864**	**21 597**	**25 051**	**24 317**	**26 829**	**26 972**	**26 021**	**25 970 e**
+ Imports	3 669	4 885	5 785	9 910	7 410	8 729	7 813	5 489 e
- Exports	1 442	1 030	1 451	1 841	4 390	4 609	4 659	4 399 e
+ Transfers	- 139	398	-2 622	-1 819	- 957	- 999	- 881	- 458 e
+ Stock Change	- 191	159	- 513	- 9	1 741	- 872	- 93	844 e
- Statistical Difference	- 19	180	- 281	- 716	383	415	252	252 e
Total Consumption	**20 780**	**25 829**	**26 531**	**31 274**	**30 250**	**28 806**	**27 949**	**27 194 e**
Closing Stock Level	-	*3 495*	*3 849*	*5 530*	*4 611*	*5 483*	*5 576*	*4 732 e*

Table 4: Aviation Fuels - Supply

000 metric tons

	1973	1978	1980	1990	1999	2000	2001	2002
AVIATION GASOLINE								
Refinery Output	**38**	**46**	**53**	**27**	**-**	**-**	**-**	**-**
+ Imports	30	29	33	28	26	23	25	19 e
- Exports	29	30	40	24	2	2	2	2 e
+ Transfers	-	-	-	1	-	-	-	-
+ Stock Change	4	1	- 2	2	2	- 1	- 2	1 e
- Statistical Difference	-	-	- 1	5	1	- 6	- 2	- 2 e
Total Consumption	**43**	**46**	**45**	**29**	**25**	**26**	**23**	**20 e**
Closing Stock Level	-	*7*	*8*	*5*	*3*	*4*	*6*	*5 e*
GASOLINE TYPE JET FUEL								
Refinery Output	**-**	**465**	**351**	**160**	**-**	**-**	**-**	**-**
+ Imports	-	9	3	230	-	-	-	-
- Exports	-	89	249	-	-	-	-	-
+ Transfers	-	-	3	10	-	-	-	-
+ Stock Change	-	- 122	165	- 1	-	-	-	-
- Statistical Difference	-	5	- 12	147	-	-	-	-
Total Consumption	**-**	**258**	**285**	**252**	**-**	**-**	**-**	**-**
Closing Stock Level	-	*163*	*120*	*12*	-	-	-	-
KEROSENE TYPE JET FUEL								
Refinery Output	**1 981**	**1 320**	**1 451**	**2 610**	**4 225**	**4 311**	**4 195**	**4 156 e**
+ Imports	1 098	1 730	1 806	2 696	2 836	3 117	3 366	3 170 e
- Exports	187	60	61	83	94	289	513	384 e
+ Transfers	- 83	- 113	- 116	- 146	- 66	- 64	- 64	- 80 e
+ Stock Change	- 8	15	- 6	35	- 53	- 6	- 60	11 e
- Statistical Difference	19	6	24	- 53	46	- 73	111	86 e
Total Consumption	**2 782**	**2 886**	**3 050**	**5 165**	**6 802**	**7 142**	**6 813**	**6 787 e**
Closing Stock Level	-	*235*	*226*	*273*	*656*	*662*	*722*	*711 e*

Source: Annual Oil Statistics.

GERMANY
Table 5A: Gas/Diesel Oil - Supply and Consumption

000 metric tons

	1973	1978	1980	1990	1999	2000	2001	2002
Refinery Output	**50 188**	**45 685**	**48 128**	**40 668**	**46 100**	**46 445**	**46 943**	**47 455 e**
+ Recycled Products	-	-	8	30	15	19	23	26 e
+ Imports	22 084	22 307	15 823	15 836	17 167	16 058	19 545	14 473 e
- Exports	2 226	1 475	2 123	3 518	4 324	5 448	5 576	6 242 e
- Intl. Marine Bunkers	1 063	896	782	548	483	501	458	481 e
+ Transfers	-1 253	474	-1 419	- 505	-1 058	-1 216	- 888	- 829 e
+ Stock Changes	- 611	515	-1 010	951	511	211	558	1 684 e
- Statistical Differences	16	- 453	70	- 741	- 566	-1 318	- 364	-1 169 e
Total Consumption	**67 103**	**67 063**	**58 555**	**53 655**	**58 494**	**56 886**	**60 511**	**57 255 e**
Transformation	**12**	**216**	**375**	**1 339**	**805**	**711**	**731**	..
Public Electricity	12	169	251	776	181	211	269	..
Autoproducer Electricity	-	47	25	108	100	94	58	..
Public CHP Plants	-	-	-	-	409	284	252	..
Autoproducers of CHP	-	-	-	-	-	-	-	..
Public Heat Plants	-	-	-	-	-	-	-	..
Autoproducer Heat Plants	-	-	-	-	-	-	-	..
Petrochemical Industry	-	-	99	453	115	122	152	..
Other Transformation	-	-	-	2	-	-	-	..
Energy Sector	**59**	**102**	**71**	**71**	**117**	**120**	**104**	..
Refinery Fuel	54	29	12	20	78	85	75	..
Other Energy Sector	5	73	59	51	39	35	29	..
Final Consumption	**67 032**	**66 745**	**58 109**	**52 245**	**57 572**	**56 055**	**59 676**	..
Transport	**11 641**	**13 422**	**14 089**	**19 126**	**26 511**	**26 646**	**26 360**	..
Air	-	-	-	-	-	-	-	..
Road	9 502	11 150	11 832	17 564	25 633	25 805	25 548	..
Rail	1 069	1 362	1 402	913	580	565	546	..
Inland Waterways	1 070	910	855	649	298	276	266	..
Non-Specified	-	-	-	-	-	-	-	..
Industry	**7 418**	**6 255**	**4 862**	**3 987**	**3 128**	**2 784**	**2 728**	..
Iron and Steel	139	88	58	139	27	36	25	..
Chemical (incl. Pet.-Chem.)	491	424	289	1 355	447	371	478	..
Non-Ferrous Metals	165	126	119	106	78	75	65	..
Non-Metallic Minerals	1 245	835	547	242	278	225	183	..
Transport Equipment	222	331	305	224	105	102	94	..
Machinery	727	662	558	730	533	472	473	..
Mining and Quarrying	21	24	16	50	90	83	72	..
Food and Tobacco	889	903	782	546	546	486	464	..
Pulp, Paper and Printing	37	21	13	119	108	94	96	..
Wood and Wood Products	257	254	194	116	59	50	43	..
Construction	-	-	-	-	509	490	438	..
Textiles and Leather	331	249	189	179	107	92	87	..
Non-Specified	2 894	2 338	1 792	181	241	208	210	..
Other Sectors	**47 973**	**47 068**	**39 158**	**29 132**	**27 933**	**26 625**	**30 588**	..
Commerce and Public	15 400 e	15 808	12 314	9 705	7 492	6 564	7 877	..
Residential	29 895 e	29 122	23 823	17 100	18 594	18 187	20 872	..
Agriculture	1 300	1 190	2 196	2 161	1 730	1 770	1 735	..
Non-Specified	1 378	948	825	166	117	104	104	..
Memo: Non-Energy Use	*491*	*424*	*289*	*1 355*	*412*	*371*	*478*	..
Chem/Petchem. Feedst.	*491*	*424*	*289*	*1 355*	*412*	*371*	*478*	..
Other	-	-	-	-	-	-	-	..
Closing Stock Level	-	*10 037*	*12 788*	*10 201*	*12 241*	*12 030*	*11 472*	*9 788 e*

Source: Annual Oil Statistics.

GERMANY
Table 5B: Fuel Oil (Residual) - Supply and Consumption

000 metric tons

	1973	1978	1980	1990	1999	2000	2001	2002
Refinery Output	**44 073**	**34 461**	**34 607**	**13 075**	**11 821**	**13 068**	**13 192**	**12 184 e**
+ Recycled Products	-	-	1	127	123	151	196	186 e
+ Imports	4 698	6 105	5 605	4 616	2 443	2 903	2 328	2 924 e
- Exports	2 518	1 937	4 801	2 318	4 012	4 397	4 066	3 611 e
- Intl. Marine Bunkers	3 084	2 412	2 755	1 960	1 613	1 705	1 790	1 927 e
+ Transfers	- 160	- 378	-3 677	-3 504	-2 151	-1 583	-1 250	- 945 e
+ Stock Changes	- 363	- 435	786	314	192	1	111	- 212 e
- Statistical Differences	966	332	- 636	262	28	493	56	81 e
Total Consumption	**41 680**	**35 072**	**30 402**	**10 088**	**6 775**	**7 945**	**8 665**	**8 518 e**
Transformation	**12 634**	**10 735**	**9 326**	**3 650**	**2 429**	**2 150**	**2 571**	..
Public Electricity	10 186	6 473	4 888	1 229	283	139 e	296	..
Autoproducer Electricity	-	2 022	2 065	560	472	206	221	..
Public CHP Plants	1 187	1 393	1 013	632	194	182 e	67	..
Autoproducers of CHP	-	-	-	-	-	-	-	..
Public Heat Plants	-	-	-	-	-	-	-	..
Autoproducer Heat Plants	-	-	-	-	-	-	-	..
Petrochemical Industry	1 074	678	440	503	364	754	913	..
Other Transformation	187	169	920	726	1 116	869	1 074	..
Energy Sector	**4 931**	**4 794**	**4 671**	**1 407**	**1 706**	**1 694**	**1 746**	..
Refinery Fuel	4 455	3 930	4 535	1 393	1 696	1 685	1 737	..
Other Energy Sector	476	864	136	14	10	9	9	..
Final Consumption	**24 115**	**19 543**	**16 405**	**5 031**	**2 640**	**4 101**	**4 348**	..
Transport	**381**	**271**	**203**	-	-	-	-	..
Air	-	-	-	-	-	-	-	..
Road	-	-	-	-	-	-	-	..
Rail	381	264	196	-	-	-	-	..
Inland Waterways	-	7	7	-	-	-	-	..
Non-Specified	-	-	-	-	-	-	-	..
Industry	**20 343**	**17 090**	**14 146**	**4 752**	**2 431**	**4 096**	**4 342**	..
Iron and Steel	4 692	3 028	974	263	29	15	10	..
Chemical (incl. Pet.-Chem.)	3 967 e	4 511 e	3 843	2 229	1 259	3 040	3 282	..
Non-Ferrous Metals	344	324	283	65	27	31	35	..
Non-Metallic Minerals	3 830	2 449	1 979	648	611	491	510	..
Transport Equipment	895	666	668	36	2	1	1	..
Machinery	471	483	347	119	23	22	28	..
Mining and Quarrying	123	58	68	14	15	13	15	..
Food and Tobacco	2 031	2 038	1 895	528	206	314	309	..
Pulp, Paper and Printing	1 417	1 282	1 150	519	104	70	63	..
Wood and Wood Products	286	262	349	82	77	49	37	..
Construction	-	-	418	-	-	-	-	..
Textiles and Leather	788	743	633	202	31	24	24	..
Non-Specified	1 499	1 246	1 539	47	47	26	28	..
Other Sectors	**3 391**	**2 182**	**2 056**	**279**	**209**	**5**	**6**	..
Commerce and Public	117	405	1 939	231 e	204	-	-	..
Residential	2 914	1 433	-	-	-	-	-	..
Agriculture	-	-	-	-	-	-	-	..
Non-Specified	360	344	117	48 e	5	5	6	..
Memo: Non-Energy Use	-	*1 212*	*1 694*	*1 836*	*1 199*	*2 741*	*2 872*	..
Chem/Petchem. Feedst.	-	*1 212*	*1 694*	*1 836*	*1 199*	*2 741*	*2 872*	..
Other	-	-	-	-	-	-	-	..
Closing Stock Level	-	*3 730*	*3 379*	*1 905*	*1 612*	*1 611*	*1 500*	*1 712 e*

Source: Annual Oil Statistics.

GERMANY
Table 5C: Total Products - Supply and Consumption

000 metric tons

	1973	1978	1980	1990	1999	2000	2001	2002
Refinery Output	**139 508**	**126 584**	**136 512**	**105 957**	**115 017**	**115 973**	**114 292**	**114 085 e**
+ Recycled Products	-	-	271	272	190	247	315	321 e
+ Imports	41 065	46 249	37 444	41 672	40 522	42 171	43 669	36 339 e
- Exports	9 746	7 423	11 181	10 153	16 664	18 601	18 495	18 598 e
- Intl. Marine Bunkers	4 202	3 374	3 611	2 558	2 139	2 247	2 280	2 445 e
+ Transfers	-5 259	-5 785	-11 366	-8 991	-4 654	-4 057	-3 126	-2 530 e
+ Stock Changes	-1 213	448	-1 362	1 349	2 501	- 842	597	2 503 e
- Statistical Differences	903	- 19	- 761	- 931	- 151	- 739	105	- 770 e
Total Consumption	**159 250**	**156 718**	**147 468**	**128 479**	**134 924**	**133 383**	**134 867**	**130 445 e**
Transformation	**14 944**	**14 062**	**14 878**	**10 095**	**8 215**	**8 839**	**8 602**	..
Public Electricity	10 398	6 805	5 279	2 008	546	565	709	..
Autoproducer Electricity	-	2 209	2 227	741	659	415	347	..
Public CHP Plants	1 187	1 393	1 015	639	609	468	325	..
Autoproducers of CHP	-	-	-	-	14	10	38	..
Public Heat Plants	-	-	-	-	-	-	-	..
Autoproducer Heat Plants	-	-	-	-	-	-	-	..
Petrochemical Industry	1 454	1 958	4 081	4 919	4 651	5 889	5 487	..
Other Transformation	1 905	1 697	2 276	1 788	1 736	1 492	1 696	..
Energy Sector	**9 179**	**8 810**	**9 125**	**5 510**	**6 692**	**6 765**	**6 523**	..
Refinery Fuel	8 692	7 854	8 914	5 435	6 643	6 721	6 485	..
Other Energy Sector	487	956	211	75	49	44	38	..
Final Consumption	**135 127**	**133 846**	**123 465**	**112 874**	**120 017**	**117 779**	**119 742**	..
Transport	**35 535**	**42 556**	**44 508**	**56 149**	**63 843**	**62 856**	**61 360**	..
Air	2 825	3 190	3 380	5 447	6 828	7 169	6 836	..
Road	30 188	36 822	38 661	49 122	56 069	54 778	53 646	..
Rail	1 452	1 627	1 603	920	584	569	550	..
Inland Waterways	1 070	917	864	660	302	280	268	..
Non-Specified	-	-	-	-	60	60	60	..
Industry	**47 083**	**40 604**	**36 452**	**26 269**	**26 831**	**27 069**	**26 553**	..
Iron and Steel	4 871	3 146	1 073	452	72	64	51	..
Chemical (incl. Pet.-Chem.)	11 660	11 969	14 520	14 954	16 703	17 562	17 657	..
Non-Ferrous Metals	529	480	432	216	137	135	130	..
Non-Metallic Minerals	5 384	3 515	2 739	1 348	1 379	1 096	1 059	..
Transport Equipment	1 117	1 017	993	300	194	181	182	..
Machinery	1 200	1 162	927	929	617	547	556	..
Mining and Quarrying	160	97	89	64	105	96	87	..
Food and Tobacco	2 931	3 003	2 762	1 154	834	875	860	..
Pulp, Paper and Printing	1 469	1 323	1 188	699	265	212	212	..
Wood and Wood Products	543	516	547	203	145	107	89	..
Construction	4 656	3 891	4 279	2 899	3 324	3 131	2 887	..
Textiles and Leather	1 119	992	842	411	142	119	114	..
Non-Specified	11 444	9 493	6 061	2 640	2 914	2 944	2 669	..
Other Sectors	**52 509**	**50 686**	**42 505**	**30 456**	**29 343**	**27 854**	**31 829**	..
Commerce and Public	15 795	16 643	14 785	10 336	7 980	6 879	8 198	..
Residential	33 299	31 230	24 553	17 734	19 495	19 083	21 773	..
Agriculture	1 300	1 190	2 196	2 161	1 743	1 782	1 747	..
Non-Specified	2 115	1 623	971	225	125	110	111	..
Memo: Non-Energy Use	*13 882*	*14 031*	*18 513*	*19 514*	*21 791*	*22 201*	*21 896*	..
Chem/Petchem. Feedst.	*7 522*	*8 507*	*12 345*	*14 552*	*16 495*	*17 186*	*17 211*	..
Other	*6 360*	*5 524*	*6 168*	*4 962*	*5 296*	*5 015*	*4 685*	..
Closing Stock Level	-	*19 968*	*23 138*	*19 612*	*20 715*	*21 557*	*20 960*	*18 457 e*

Source: Annual Oil Statistics.

GERMANY
Table 6A: Supply and Consumption

000 metric tons

	1980	2000	2001	2002	1980	2000	2001	2002
	LPG AND ETHANE				**REFINERY GAS**			
Refinery Output	**2 776**	**3 053**	**3 020**	**2 956 e**	**4 793**	**4 106**	**3 888**	**4 196 e**
+ Recycled Products	-	-	-	-	-	-	-	-
+ Imports	627	652	712	635 e	-	-	-	-
- Exports	539	761	576	601 e	-	-	-	-
- Intl. Marine Bunkers	-	-	-	-	-	-	-	-
+ Transfers	- 356	- 5	- 4	11 e	- 225	15	47	38 e
+ Stock Change	- 14	- 20	12	5 e	-	-	-	-
- Statistical Difference	- 81	- 82	65	130 e	14	- 16	15	18 e
Total Consumption	**2 575**	**3 001**	**3 099**	**2 876 e**	**4 554**	**4 137**	**3 920**	**4 216 e**
of which: Ethane	-	-	-	-	-	-	-	-
Transformation	347	244	283	..	562	149	132	..
Energy	25	236	239	..	3 902	3 585	3 420	..
Transport	3	63	62	..	-	-	-	..
Industry	1 185	1 475	1 521	..	90	403	368	..
Other	1 015	983	994	..	-	-	-	..
Memo: Non-Energy Use	*534*	*831*	*829*	..	*90*	*333*	*353*	..
Closing Stock Level	*81*	*87*	*75*	*70 e*	-	-	-	-
	NAPHTHA				**OTHER KEROSENE**			
Refinery Output	**9 371**	**9 024**	**8 495**	**8 662 e**	**70**	**17**	**11**	**10 e**
+ Recycled Products	-	-	-	-	-	-	-	-
+ Imports	5 474	7 712	7 405	7 013 e	47	13	30	23 e
- Exports	575	1 047	919	954 e	4	10	4	-
- Intl. Marine Bunkers	-	-	-	-	-	-	-	-
+ Transfers	- 881	1 875	1 704	1 717 e	- 45	- 5	- 15	- 8 e
+ Stock Change	- 177	- 13	87	8 e	6	-	-	-
- Statistical Difference	- 118	- 19	36	4 e	8	- 3	- 3	- 1 e
Total Consumption	**13 330**	**17 570**	**16 736**	**16 442 e**	**66**	**18**	**25**	**26 e**
Transformation	3 679	4 746	4 095	..	-	-	-	..
Energy	-	1	-	..	-	-	-	..
Transport	-	-	-	..	3	10	12	..
Industry	9 651	12 823	12 641	..	34	7	12	..
Other	-	-	-	..	29	1	1	..
Memo: Non-Energy Use	*9 651*	*12 823*	*12 641*	..	*55*	*7*	*12*	..
Closing Stock Level	*1 023*	*487*	*400*	*392 e*	*18*	*2*	*2*	*2 e*
	WHITE SPIRIT + SBP				**LUBRICANTS**			
Refinery Output	**443**	**51**	**41**	**..**	**1 696**	**1 508**	**1 525**	**..**
+ Recycled Products	-	-	-	..	262	77	96	..
+ Imports	143	198	188	..	369	643	532	..
- Exports	110	86	81	..	345	455	530	..
- Intl. Marine Bunkers	-	-	-	..	74	41	32	..
+ Transfers	- 33	63	61	..	- 298	- 655	- 589	..
+ Stock Change	6	- 8	2	..	- 49	33	1	..
- Statistical Difference	20	- 11	- 1	..	-	- 12	- 55	..
Total Consumption	**429**	**229**	**212**	**..**	**1 561**	**1 122**	**1 058**	**..**
Transformation	-	-	-	..	-	-	-	..
Energy	-	-	-	..	-	-	-	..
Transport	-	-	-	..	546	388	366	..
Industry	429	229	212	..	1 015	719	677	..
Other	-	-	-	..	-	15	15	..
Memo: Non-Energy Use	*429*	*229*	*210*	..	*1 561*	*1 122*	*1 058*	..
Closing Stock Level	*58*	*16*	*14*	..	*446*	*335*	*334*	..

Source: Annual Oil Statistics.

GERMANY
Table 6B: Supply and Consumption

000 metric tons

	1980	2000	2001	2002	1980	2000	2001	2002
	BITUMEN				PARAFFIN WAXES			
Refinery Output	**4 036**	**3 728**	**3 626**	..	**259**	**225**	**222**	..
+ Recycled Products	-	-	-	..	-	-	-	..
+ Imports	311	516	406	..	198	335	339	..
- Exports	346	723	652	..	164	207	216	..
- Intl. Marine Bunkers	-	-	-	..	-	-	-	..
+ Transfers	- 19	- 171	- 207	..	- 13	- 70	- 86	..
+ Stock Change	- 34	- 2	- 21	..	- 6	1	- 2	..
- Statistical Difference	87	-	17	..	14	2	- 19	..
Total Consumption	**3 861**	**3 348**	**3 135**	**..**	**260**	**282**	**276**	**..**
Transformation	-	7	6	..	-	-	-	..
Energy	-	-	-	..	-	-	-	..
Transport	-	-	-	..	-	-	-	..
Industry	3 861	3 341	3 129	..	260	282	276	..
Other	-	-	-	..	-	-	-	..
Memo: Non-Energy Use	*3 861*	*3 341*	*3 129*	..	*260*	*282*	*276*	..
Closing Stock Level	*320*	*209*	*230*	..	*19*	*19*	*21*	..
	PETROLEUM COKE				OTHER PRODUCTS			
Refinery Output	**1 085**	**1 754**	**1 696**	**1 645 e**	**2 342**	**1 711**	**1 417**	**6 851 e**
+ Recycled Products	-	-	-	-	-	-	-	109 e
+ Imports	1 207	1 104	944	1 032 e	13	168	36	1 561 e
- Exports	348	539	672	682 e	25	28	29	1 723 e
- Intl. Marine Bunkers	-	-	-	-	-	-	-	37 e
+ Transfers	-	-	-	-	-1 665	-1 242	- 954	-1 976 e
+ Stock Change	- 62	- 30	- 27	6 e	- 452	- 136	31	156 e
- Statistical Difference	- 58	- 10	- 3	- 22 e	189	- 99	-	- 147 e
Total Consumption	**1 940**	**2 299**	**1 944**	**2 023 e**	**24**	**572**	**501**	**5 088 e**
Transformation	589	809	747	..	-	23	37	..
Energy	456	701	595	..	-	428	419	..
Transport	-	-	-	..	-	-	-	..
Industry	895	789	602	..	24	121	45	..
Other	-	-	-	..	-	-	-	..
Memo: Non-Energy Use	*87*	-	-	..	*2*	*121*	*38*	..
Closing Stock Level	*105*	*82*	*109*	*103 e*	*698*	*530*	*499*	*942 e*

Source: Annual Oil Statistics.

GERMANY
Table 7A: Imports by Country of Origin

000 metric tons

	2000	2001	2002	2000	2001	2002	2000	2001	2002
	Crude Oil			Natural Gas Liquids			Feedstocks and Other		
Australia	-	-	-	-	-	-	-	-	-
Belgium	-	-	-	-	-	-	-	-	-
Canada	-	-	-	-	-	-	-	-	-
Czech Republic	-	-	-	-	-	-	-	-	-
Denmark	867	1 050	1 640 e	-	-	-	-	-	-
Finland	-	-	-	-	-	-	-	-	-
France	-	-	-	-	-	-	-	-	-
Germany	-	-	-	-	-	-	-	-	-
Greece	-	-	-	-	-	-	-	-	-
Hungary	-	-	-	-	-	-	-	-	-
Ireland	-	-	-	-	-	-	-	-	-
Italy	100	98	128 e	-	-	-	-	-	-
Japan	-	-	-	-	-	-	-	-	-
Korea	-	-	-	-	-	-	-	-	-
Mexico	-	-	-	-	-	-	-	-	-
Netherlands	-	105	515 e	-	-	-	-	-	-
New Zealand	-	-	-	-	-	-	-	-	-
Norway	18 708	20 969	22 281 e	-	-	-	-	-	-
Poland	-	-	118 e	-	-	-	-	-	-
Spain	-	-	-	-	-	-	-	-	-
Sweden	-	-	-	-	-	-	-	-	-
Turkey	-	-	-	-	-	-	-	-	-
United Kingdom	13 041	15 323	11 496 e	-	-	-	-	-	-
United States	-	-	-	-	-	-	-	-	-
Other	57	63	-	-	-	-	-	-	-
OECD	**32 773**	**37 608**	**36 178 e**	**-**	**-**	**-**	**-**	**-**	**-**
Argentina	-	-	-	-	-	-	-	-	-
Brazil	-	126	32 e	-	-	-	-	-	-
Colombia	-	-	-	-	-	-	-	-	-
Ecuador	-	-	-	-	-	-	-	-	-
Netherlands Antilles	-	-	-	-	-	-	-	-	-
Peru	-	-	-	-	-	-	-	-	-
Trinidad and Tobago	-	-	-	-	-	-	-	-	-
Venezuela	1 881	1 670	803 e	-	-	-	-	-	-
Other	-	-	-	-	-	-	-	-	-
Latin America	**1 881**	**1 796**	**835 e**	**-**	**-**	**-**	**-**	**-**	**-**
Romania	-	-	-	-	-	-	-	-	-
Other	-	-	-	-	-	-	-	-	-
Non-OECD Europe	**-**	**-**	**-**	**-**	**-**	**-**	**-**	**-**	**-**
FSU	**34 201**	**35 821**	**38 526 e**	**-**	**-**	**-**	**-**	**-**	**-**
Iran	908	256	64 e	-	-	-	-	-	-
Iraq	220	21	43 e	-	-	-	-	-	-
Kuwait	643	109	237 e	-	-	-	-	-	-
Oman	-	-	-	-	-	-	-	-	-
Qatar	-	-	-	-	-	-	-	-	-
Saudi Arabia	4 568	4 016	3 613 e	-	-	-	-	-	-
United Arab Emirates	-	-	-	-	-	-	-	-	-
Other	7 195	7 312	7 240 e	-	-	-	-	-	-
Middle East	**13 534**	**11 714**	**11 197 e**	**-**	**-**	**-**	**-**	**-**	**-**
India	-	-	-	-	-	-	-	-	-
Indonesia	-	-	-	-	-	-	-	-	-
Singapore	-	-	-	-	-	-	-	-	-
Other	-	-	-	-	-	-	-	-	-
Asia	**-**	**-**	**-**	**-**	**-**	**-**	**-**	**-**	**-**
China	**-**	**-**	**-**	**-**	**-**	**-**	**-**	**-**	**-**
Algeria	6 501	3 958	4 232 e	-	-	-	-	-	-
Angola/Cabinada	446	318	1 185 e	-	-	-	-	-	-
Cameroon	31	-	129 e	-	-	-	-	-	-
Egypt	-	29	-	-	-	-	-	-	-
Gabon	-	-	-	-	-	-	-	-	-
Libya	11 843	10 029	8 732 e	-	-	-	-	-	-
Nigeria	1 983	3 013	2 824 e	-	-	-	-	-	-
Tunisia	154	133	283 e	-	-	-	-	-	-
Other	337	752	774 e	-	-	-	-	-	-
Africa	**21 295**	**18 232**	**18 159 e**	**-**	**-**	**-**	**-**	**-**	**-**
Non-spec./Other	**-**	**-**	**-**	**-**	**-**	**-**	**-**	**-**	**-**
Total Imports	**103 684**	**105 171**	**104 895 e**	**-**	**-**	**-**	**-**	**-**	**-**
Memo: Total OPEC	*28 547*	*23 072*	*20 548 e*	-	-	-	-	-	-

Source: Annual Oil Statistics.

GERMANY
Table 7B: Imports by Country of Origin

000 metric tons

	2000	2001	2002	2000	2001	2002	2000	2001	2002
	Total Products			**Liquefied Petroleum Gases**			**Naphtha**		
Australia	-	1	-	-	-	-	-	-	-
Belgium	5 631	6 194	5 499 e	98	109	148 e	321	288	114 e
Canada	7	25	33 e	-	-	-	-	4	3 e
Czech Republic	380	314	420 e	-	-	3 e	-	-	-
Denmark	392	212	115 e	-	2	1 e	63	6	12 e
Finland	202	831	751 e	3	2	-	4	20	63 e
France	1 815	2 047	1 524 e	18	27	26 e	286	259	128 e
Germany	-	-	-	-	-	-	-	-	-
Greece	-	3	5 e	-	-	-	-	-	-
Hungary	257	335	289 e	-	-	-	-	-	-
Ireland	-	6	2 e	-	1	-	-	-	-
Italy	31	88	139 e	-	-	-	-	-	-
Japan	18	31	27 e	-	-	-	-	-	-
Korea	-	1	1 e	-	-	-	-	-	-
Mexico	-	-	1 e	-	-	-	-	-	-
Netherlands	24 955	24 234	19 355 e	246	242	180 e	5 846	5 105	4 617 e
New Zealand	-	-	-	-	-	-	-	-	-
Norway	477	541	459 e	60	75	113 e	50	106	11 e
Poland	81	40	162 e	-	-	-	-	-	-
Spain	10	5	57 e	-	-	-	-	-	-
Sweden	1 302	1 413	1 088 e	-	16	32 e	4	30	10 e
Turkey	-	-	-	-	-	-	-	-	-
United Kingdom	1 882	2 145	1 572 e	202	229	117 e	64	145	168 e
United States	698	389	345 e	-	-	-	-	-	-
Other	361	312	373 e	5	9	15 e	11	-	3 e
OECD	**38 499**	**39 167**	**32 217 e**	**632**	**712**	**635 e**	**6 649**	**5 963**	**5 129 e**
Argentina	12	27	24 e	-	-	-	1	1	-
Brazil	3	-	12 e	-	-	-	-	-	-
Colombia	-	-	-	-	-	-	-	-	-
Ecuador	-	-	-	-	-	-	-	-	-
Netherlands Antilles	-	-	-	-	-	-	-	-	-
Peru	-	-	-	-	-	-	-	-	-
Trinidad and Tobago	-	-	-	-	-	-	-	-	-
Venezuela	1	-	-	-	-	-	-	-	-
Other	145	-	7 e	-	-	-	-	-	-
Latin America	**161**	**27**	**43 e**	**-**	**-**	**-**	**1**	**1**	**-**
Romania	9	11	1 e	-	-	-	-	-	-
Other	78	4	18 e	19	-	-	-	-	-
Non-OECD Europe	**87**	**15**	**19 e**	**19**	**-**	**-**	**-**	**-**	**-**
FSU	**2 829**	**3 551**	**2 695 e**	**1**	**-**	**-**	**576**	**658**	**714 e**
Iran	-	-	-	-	-	-	-	-	-
Iraq	-	-	-	-	-	-	-	-	-
Kuwait	-	-	-	-	-	-	-	-	-
Oman	-	-	-	-	-	-	-	-	-
Qatar	-	-	-	-	-	-	-	-	-
Saudi Arabia	105	49	192 e	-	-	-	105	49	192 e
United Arab Emirates	-	-	-	-	-	-	-	-	-
Other	2	2	2 e	-	-	-	-	-	-
Middle East	**107**	**51**	**194 e**	**-**	**-**	**-**	**105**	**49**	**192 e**
India	-	-	-	-	-	-	-	-	-
Indonesia	-	-	-	-	-	-	-	-	-
Singapore	-	-	-	-	-	-	-	-	-
Other	-	41	5 e	-	-	-	-	-	-
Asia	**-**	**41**	**5 e**	**-**	**-**	**-**	**-**	**-**	**-**
China	**22**	**30**	**12 e**	**-**	**-**	**-**	**-**	**-**	**-**
Algeria	249	418	494 e	-	-	-	240	413	479 e
Angola/Cabinada	-	-	-	-	-	-	-	-	-
Cameroon	-	-	-	-	-	-	-	-	-
Egypt	105	251	332 e	-	-	-	89	245	308 e
Gabon	-	-	-	-	-	-	-	-	-
Libya	52	52	164 e	-	-	-	52	52	164 e
Nigeria	-	24	27 e	-	-	-	-	24	27 e
Tunisia	-	-	-	-	-	-	-	-	-
Other	57	14	43 e	-	-	-	-	-	-
Africa	**463**	**759**	**1 060 e**	**-**	**-**	**-**	**381**	**734**	**978 e**
Non-spec./Other	**3**	**28**	**94 e**	**-**	**-**	**-**	**-**	**-**	**-**
Total Imports	**42 171**	**43 669**	**36 339 e**	**652**	**712**	**635 e**	**7 712**	**7 405**	**7 013 e**
Memo: Total OPEC	*407*	*543*	*877 e*	-	-	-	*397*	*538*	*862 e*

Source: Annual Oil Statistics.

GERMANY
Table 7C: Imports by Country of Origin

000 metric tons

	2000	2001	2002	2000	2001	2002	2000	2001	2002
	Total Gasoline			Total Kerosene			Gas/Diesel		
Australia	-	-	-	-	-	-	-	-	-
Belgium	1 529	1 200	987 e	516	748	697 e	1 922	2 750	1 770 e
Canada	-	-	-	-	-	-	5	17	27 e
Czech Republic	23	20	81 e	-	-	-	188	210	235 e
Denmark	56	18	-	-	-	-	256	178	96 e
Finland	93	486	438 e	-	-	-	101	320	228 e
France	648	630	484 e	86	122	100 e	439	589	413 e
Germany	-	-	-	-	-	-	-	-	-
Greece	-	-	-	-	-	-	-	-	-
Hungary	21	83	67 e	-	-	-	214	235	198 e
Ireland	-	-	-	-	-	-	-	-	2 e
Italy	-	30	-	-	5	-	-	9	11 e
Japan	-	-	-	-	-	-	-	-	-
Korea	-	-	-	-	-	-	-	-	1 e
Mexico	-	-	-	-	-	-	-	-	-
Netherlands	5 010	4 382	3 053 e	2 509	2 510	2 396 e	9 450	10 891	8 108 e
New Zealand	-	-	-	-	-	-	-	-	-
Norway	210	78	11 e	-	-	-	107	279	293 e
Poland	-	-	-	-	-	-	1	-	59 e
Spain	-	-	-	-	-	-	6	-	-
Sweden	269	261	132 e	-	-	-	1 002	1 073	858 e
Turkey	-	-	-	-	-	-	-	-	-
United Kingdom	637	435	201 e	11	-	-	486	650	539 e
United States	2	-	-	-	-	-	53	20	1 e
Other	54	66	37 e	-	-	-	75	29	73 e
OECD	**8 552**	**7 689**	**5 491 e**	**3 122**	**3 385**	**3 193 e**	**14 305**	**17 250**	**12 912 e**
Argentina	-	-	-	-	-	-	-	-	-
Brazil	-	-	-	-	-	-	-	-	-
Colombia	-	-	-	-	-	-	-	-	-
Ecuador	-	-	-	-	-	-	-	-	-
Netherlands Antilles	-	-	-	-	-	-	-	-	-
Peru	-	-	-	-	-	-	-	-	-
Trinidad and Tobago	-	-	-	-	-	-	-	-	-
Venezuela	-	-	-	-	-	-	-	-	-
Other	145	-	-	-	-	-	-	-	-
Latin America	**145**	**-**	**-**	**-**	**-**	**-**	**-**	**-**	**-**
Romania	-	-	-	-	-	-	2	2	1 e
Other	34	-	4 e	5	-	-	18	1	10 e
Non-OECD Europe	**34**	**-**	**4 e**	**5**	**-**	**-**	**20**	**3**	**11 e**
FSU	**21**	**149**	**13 e**	**-**	**7**	**-**	**1 733**	**2 235**	**1 460 e**
Iran	-	-	-	-	-	-	-	-	-
Iraq	-	-	-	-	-	-	-	-	-
Kuwait	-	-	-	-	-	-	-	-	-
Oman	-	-	-	-	-	-	-	-	-
Qatar	-	-	-	-	-	-	-	-	-
Saudi Arabia	-	-	-	-	-	-	-	-	-
United Arab Emirates	-	-	-	-	-	-	-	-	-
Other	-	-	-	-	-	-	-	-	-
Middle East	**-**	**-**	**-**	**-**	**-**	**-**	**-**	**-**	**-**
India	-	-	-	-	-	-	-	-	-
Indonesia	-	-	-	-	-	-	-	-	-
Singapore	-	-	-	-	-	-	-	-	-
Other	-	-	-	-	-	-	-	34	-
Asia	**-**	**-**	**-**	**-**	**-**	**-**	**-**	**34**	**-**
China	**-**	**-**	**-**	**-**	**-**	**-**	**-**	**-**	**-**
Algeria	-	-	-	-	-	-	-	-	-
Angola/Cabinada	-	-	-	-	-	-	-	-	-
Cameroon	-	-	-	-	-	-	-	-	-
Egypt	-	-	-	-	-	-	-	-	-
Gabon	-	-	-	-	-	-	-	-	-
Libya	-	-	-	-	-	-	-	-	-
Nigeria	-	-	-	-	-	-	-	-	-
Tunisia	-	-	-	-	-	-	-	-	-
Other	-	-	-	-	-	-	-	-	-
Africa	**-**	**-**	**-**	**-**	**-**	**-**	**-**	**-**	**-**
Non-spec./Other	**-**	**-**	**-**	**3**	**4**	**-**	**-**	**23**	**90 e**
Total Imports	**8 752**	**7 838**	**5 508 e**	**3 130**	**3 396**	**3 193 e**	**16 058**	**19 545**	**14 473 e**
Memo: Total OPEC	-	-	-	-	-	-	-	-	-

Source: Annual Oil Statistics.

GERMANY
Table 7D: Imports by Country of Origin

000 metric tons

	2000	2001	2002	2000	2001	2002	2000	2001	2002
	Fuel Oil (Residual)			Petroleum Coke			Other Products		
Australia	-	-	-	-	-	-	-	1	-
Belgium	612	484	1 084 e	150	169	279 e	483	446	420 e
Canada	-	-	-	-	-	-	2	4	3 e
Czech Republic	17	26	30 e	-	-	-	152	58	71 e
Denmark	9	1	-	6	7	6 e	2	-	-
Finland	-	-	-	-	-	-	1	3	22 e
France	90	127	113 e	5	4	2 e	243	289	258 e
Germany	-	-	-	-	-	-	-	-	-
Greece	-	-	-	-	-	-	-	3	5 e
Hungary	-	-	-	-	-	3 e	22	17	21 e
Ireland	-	5	-	-	-	-	-	-	-
Italy	-	3	-	-	1	59 e	31	40	69 e
Japan	-	-	-	18	28	27 e	-	3	-
Korea	-	-	-	-	-	-	-	1	-
Mexico	-	-	-	-	-	-	-	-	1 e
Netherlands	1 220	523	475 e	228	327	203 e	446	254	323 e
New Zealand	-	-	-	-	-	-	-	-	-
Norway	33	-	8 e	17	-	22 e	-	3	1 e
Poland	48	2	76 e	-	-	-	32	38	27 e
Spain	-	-	-	2	3	55 e	2	2	2 e
Sweden	10	-	14 e	-	19	-	17	14	42 e
Turkey	-	-	-	-	-	-	-	-	-
United Kingdom	225	517	427 e	74	61	29 e	183	108	91 e
United States	1	-	8 e	527	275	281 e	115	94	55 e
Other	191	194	209 e	4	-	3 e	21	14	33 e
OECD	**2 456**	**1 882**	**2 444 e**	**1 031**	**894**	**969 e**	**1 752**	**1 392**	**1 444 e**
Argentina	-	-	-	10	26	21 e	1	-	3 e
Brazil	-	-	-	-	-	9 e	3	-	3 e
Colombia	-	-	-	-	-	-	-	-	-
Ecuador	-	-	-	-	-	-	-	-	-
Netherlands Antilles	-	-	-	-	-	-	-	-	-
Peru	-	-	-	-	-	-	-	-	-
Trinidad and Tobago	-	-	-	-	-	-	-	-	-
Venezuela	-	-	-	1	-	-	-	-	-
Other	-	-	7 e	-	-	-	-	-	-
Latin America	**-**	**-**	**7 e**	**11**	**26**	**30 e**	**4**	**-**	**6 e**
Romania	-	-	-	2	-	-	5	9	-
Other	-	-	-	-	-	-	2	3	4 e
Non-OECD Europe	**-**	**-**	**-**	**2**	**-**	**-**	**7**	**12**	**4 e**
FSU	**447**	**446**	**473 e**	**-**	**-**	**-**	**51**	**56**	**35 e**
Iran	-	-	-	-	-	-	-	-	-
Iraq	-	-	-	-	-	-	-	-	-
Kuwait	-	-	-	-	-	-	-	-	-
Oman	-	-	-	-	-	-	-	-	-
Qatar	-	-	-	-	-	-	-	-	-
Saudi Arabia	-	-	-	-	-	-	-	-	-
United Arab Emirates	-	-	-	-	-	-	-	-	-
Other	-	-	-	-	-	-	2	2	2 e
Middle East	**-**	**-**	**-**	**-**	**-**	**-**	**2**	**2**	**2 e**
India	-	-	-	-	-	-	-	-	-
Indonesia	-	-	-	-	-	-	-	-	-
Singapore	-	-	-	-	-	-	-	-	-
Other	-	-	-	-	-	-	-	7	5 e
Asia	**-**	**-**	**-**	**-**	**-**	**-**	**-**	**7**	**5 e**
China	**-**	**-**	**-**	**20**	**24**	**12 e**	**2**	**6**	**-**
Algeria	-	-	-	-	-	-	9	5	15 e
Angola/Cabinada	-	-	-	-	-	-	-	-	-
Cameroon	-	-	-	-	-	-	-	-	-
Egypt	-	-	-	-	-	-	16	6	24 e
Gabon	-	-	-	-	-	-	-	-	-
Libya	-	-	-	-	-	-	-	-	-
Nigeria	-	-	-	-	-	-	-	-	-
Tunisia	-	-	-	-	-	-	-	-	-
Other	-	-	-	40	-	21 e	17	14	22 e
Africa	**-**	**-**	**-**	**40**	**-**	**21 e**	**42**	**25**	**61 e**
Non-spec./Other	**-**	**-**	**-**	**-**	**-**	**-**	**-**	**1**	**4 e**
Total Imports	**2 903**	**2 328**	**2 924 e**	**1 104**	**944**	**1 032 e**	**1 860**	**1 501**	**1 561 e**
Memo: Total OPEC	*-*	*-*	*-*	*1*	*-*	*-*	*9*	*5*	*15 e*

Source: Annual Oil Statistics.

GERMANY
Table 8A: Exports by Country of Destination

000 metric tons

	2000	2001	2002	2000	2001	2002	2000	2001	2002
	Crude Oil			**Natural Gas Liquids**			**Feedstocks and Other**		
Australia	-	-	-	-	-	-	-	-	-
Austria	-	-	82 e	-	-	-	-	-	-
Belgium	-	8	-	-	-	-	-	-	-
Canada	-	-	-	-	-	-	-	-	-
Denmark	-	-	-	-	-	-	-	-	-
Finland	79	-	-	-	-	-	-	-	-
France	590	241	85 e	-	-	-	-	-	-
Germany	-	-	-	-	-	-	-	-	-
Greece	-	-	-	-	-	-	-	-	-
Ireland	-	-	-	-	-	-	-	-	-
Italy	-	-	-	-	-	-	-	-	-
Japan	-	-	-	-	-	-	-	-	-
Luxembourg	-	-	-	-	-	-	-	-	-
Mexico	-	-	-	-	-	-	-	-	-
Netherlands	743	780	322 e	-	-	-	-	-	-
Norway	-	-	-	-	-	-	-	-	-
Portugal	-	-	-	-	-	-	-	-	-
Spain	142	-	-	-	-	-	-	-	-
Sweden	1 311	-	-	-	-	-	-	-	-
Switzerland	-	-	-	-	-	-	-	-	-
Turkey	-	-	-	-	-	-	-	-	-
United Kingdom	26	-	832 e	-	-	-	-	-	-
United States	-	-	-	-	-	-	-	-	-
Other OECD	71	1	-	-	-	-	-	-	-
OECD	**2 962**	**1 030**	**1 321 e**	**-**	**-**	**-**	**-**	**-**	**-**
Africa	-	-	-	-	-	-	-	-	-
Latin America	-	-	-	-	-	-	-	-	-
Asia	255	-	-	-	-	-	-	-	-
China	-	-	-	-	-	-	-	-	-
Non-OECD Europe	-	-	-	-	-	-	-	-	-
Former Soviet Union	-	-	-	-	-	-	-	-	-
Middle East	-	-	-	-	-	-	-	-	-
Non-specified/Other	-	-	-	-	-	-	-	-	-
Non-OECD	**255**	**-**	**-**	**-**	**-**	**-**	**-**	**-**	**-**
Total Exports	**3 217**	**1 030**	**1 321 e**	**-**	**-**	**-**	**-**	**-**	**-**

	2000	2001	2002	2000	2001	2002	2000	2001	2002
	Total Products			**Liquefied Petroleum Gas**			**Naphtha**		
Australia	2	8	-	-	-	-	-	-	-
Austria	2 510	2 633	2 310 e	82	86	95 e	8	17	1 e
Belgium	395	578	704 e	25	15	29 e	31	106	164 e
Canada	-	-	17 e	-	-	-	-	-	-
Denmark	236	233	257 e	-	-	1 e	-	-	-
Finland	29	25	40 e	-	-	-	-	-	-
France	2 235	1 767	1 802 e	67	47	43 e	210	75	183 e
Germany	-	-	-	-	-	-	-	-	-
Greece	11	12	16 e	-	-	-	-	-	-
Ireland	2	6	4 e	1	4	-	-	-	-
Italy	386	163	292 e	21	17	6 e	2	3	-
Japan	5	5	1 e	-	-	-	-	-	-
Luxembourg	84	130	136 e	-	-	-	-	-	-
Mexico	4	3	-	-	-	-	-	-	-
Netherlands	2 595	3 096	3 039 e	167	133	123 e	499	530	392 e
Norway	500	410	389 e	7	13	6 e	-	-	-
Portugal	320	79	95 e	19	8	25 e	-	-	-
Spain	544	465	628 e	1	3	8 e	1	-	-
Sweden	959	753	512 e	22	18	27 e	134	-	16 e
Switzerland	2 091	2 073	1 781 e	64	35	32 e	2	3	1 e
Turkey	20	9	14 e	9	-	3 e	-	-	-
United Kingdom	1 220	2 126	2 011 e	81	39	63 e	10	25	56 e
United States	2 292	2 088	2 636 e	-	-	-	12	18	-
Other OECD	1 542	1 498	1 435 e	163	128	124 e	137	135	141 e
OECD	**17 982**	**18 160**	**18 119 e**	**729**	**546**	**585 e**	**1 046**	**912**	**954 e**
Africa	77	55	67 e	27	24	13 e	-	-	-
Latin America	310	48	206 e	2	-	-	-	1	-
Asia	26	21	22 e	-	-	-	-	-	-
China	48	8	11 e	-	-	-	-	-	-
Non-OECD Europe	95	76	117 e	3	5	3 e	-	-	-
Former Soviet Union	42	37	37 e	-	-	-	-	5	-
Middle East	21	87	10 e	-	-	-	1	1	-
Non-specified/Other	-	3	9 e	-	1	-	-	-	-
Non-OECD	**619**	**335**	**479 e**	**32**	**30**	**16 e**	**1**	**7**	**-**
Total Exports	**18 601**	**18 495**	**18 598 e**	**761**	**576**	**601 e**	**1 047**	**919**	**954 e**

Source: Annual Oil Statistics.

GERMANY
Table 8B: Exports by Country of Destination

000 metric tons

	2000	2001	2002	2000	2001	2002	2000	2001	2002
	Total Gasoline			**Total Kerosene**			**Gas/Diesel Oil**		
Australia	-	6	-	-	1	-	-	-	-
Austria	468	408	286 e	37	37	80 e	1 509	1 660	1 515 e
Belgium	27	84	65 e	-	-	2 e	41	17	50 e
Canada	-	-	-	-	-	-	-	-	-
Denmark	12	8	6 e	20	47	27 e	77	51	50 e
Finland	-	-	-	-	-	6 e	-	-	-
France	312	397	336 e	3	4	-	1 171	973	910 e
Germany	-	-	-	-	-	-	-	-	-
Greece	-	-	-	-	-	-	1	1	-
Ireland	-	-	-	-	-	-	-	-	-
Italy	-	-	-	-	-	-	5	4	199 e
Japan	-	-	-	-	-	-	-	-	-
Luxembourg	25	36	31 e	-	-	-	45	74	88 e
Mexico	-	-	-	-	-	-	-	-	-
Netherlands	853	972	946 e	14	16	11 e	273	235	776 e
Norway	-	17	21 e	-	29	13 e	-	220	139 e
Portugal	18	1	-	-	-	-	281	68	66 e
Spain	-	3	-	-	-	-	138	48	291 e
Sweden	18	32	11 e	21	18	-	252	55	73 e
Switzerland	1 267	1 062	978 e	165	264	244 e	437	530	362 e
Turkey	-	-	-	-	-	-	-	-	-
United Kingdom	388	663	383 e	31	62	-	522	1 074	1 236 e
United States	419	551	777 e	-	-	-	140	29	-
Other OECD	493	380	375 e	8	39	1 e	475	530	486 e
OECD	**4 300**	**4 620**	**4 215 e**	**299**	**517**	**384 e**	**5 367**	**5 569**	**6 241 e**
Africa	1	3	-	-	-	-	23	-	-
Latin America	281	33	184 e	-	-	-	-	1	-
Asia	-	-	-	-	-	-	3	2	-
China	-	-	-	-	-	-	40	-	-
Non-OECD Europe	29	3	2 e	-	-	-	-	4	1 e
Former Soviet Union	-	2	-	-	-	-	15	-	-
Middle East	-	-	-	-	-	-	-	-	-
Non-specified/Other	-	-	-	-	-	-	-	-	-
Non-OECD	**311**	**41**	**186 e**	**-**	**-**	**-**	**81**	**7**	**1 e**
Total Exports	**4 611**	**4 661**	**4 401 e**	**299**	**517**	**384 e**	**5 448**	**5 576**	**6 242 e**

	2000	2001	2002	2000	2001	2002	2000	2001	2002
	Fuel Oil (Residual)			**Petroleum Coke**			**Other Products**		
Australia	-	-	-	-	-	-	2	1	-
Austria	178	238	170 e	18	19	3 e	210	168	160 e
Belgium	150	128	134 e	14	22	13 e	107	206	247 e
Canada	-	-	-	-	-	17 e	-	-	-
Denmark	12	10	12 e	-	4	-	115	113	161 e
Finland	-	-	11 e	-	-	-	29	25	23 e
France	112	78	35 e	184	67	106 e	176	126	189 e
Germany	-	-	-	-	-	-	-	-	-
Greece	-	-	-	-	-	-	10	11	16 e
Ireland	-	-	-	-	-	-	1	2	4 e
Italy	318	07	43 e	8	9	5 e	34	43	39 e
Japan	-	-	-	-	-	-	5	5	1 e
Luxembourg	-	3	-	6	4	-	8	13	17 e
Mexico	-	-	-	-	-	-	4	3	-
Netherlands	441	666	285 e	200	398	364 e	148	146	142 e
Norway	414	86	135 e	1	-	17 e	78	45	58 e
Portugal	-	-	-	-	-	-	2	2	4 e
Spain	372	365	296 e	7	18	-	25	28	33 e
Sweden	437	577	343 e	1	2	-	74	51	42 e
Switzerland	24	2	1 e	16	40	46 e	116	137	117 e
Turkey	-	-	-	2	1	-	9	8	11 e
United Kingdom	151	196	193 e	-	11	16 e	37	56	64 e
United States	1 714	1 480	1 849 e	-	-	-	7	10	10 e
Other OECD	74	74	70 e	52	42	56 e	140	170	182 e
OECD	**4 397**	**3 990**	**3 577 e**	**507**	**637**	**643 e**	**1 337**	**1 369**	**1 520 e**
Africa	-	-	-	-	2	8 e	26	26	46 e
Latin America	-	-	-	-	2	1 e	27	11	21 e
Asia	-	-	-	1	1	-	22	18	22 e
China	-	-	-	-	-	-	8	8	11 e
Non-OECD Europe	-	6	34 e	31	28	30 e	32	30	47 e
Former Soviet Union	-	-	-	-	-	-	27	30	37 e
Middle East	-	70	-	-	-	-	20	16	10 e
Non-specified/Other	-	-	-	-	2	-	-	-	9 e
Non-OECD	**-**	**76**	**34 e**	**32**	**35**	**39 e**	**162**	**139**	**203 e**
Total Exports	**4 397**	**4 066**	**3 611 e**	**539**	**672**	**682 e**	**1 499**	**1 508**	**1 723 e**

Source: Annual Oil Statistics.

GREECE

Figure 1
FUEL SHARES OF TPES *

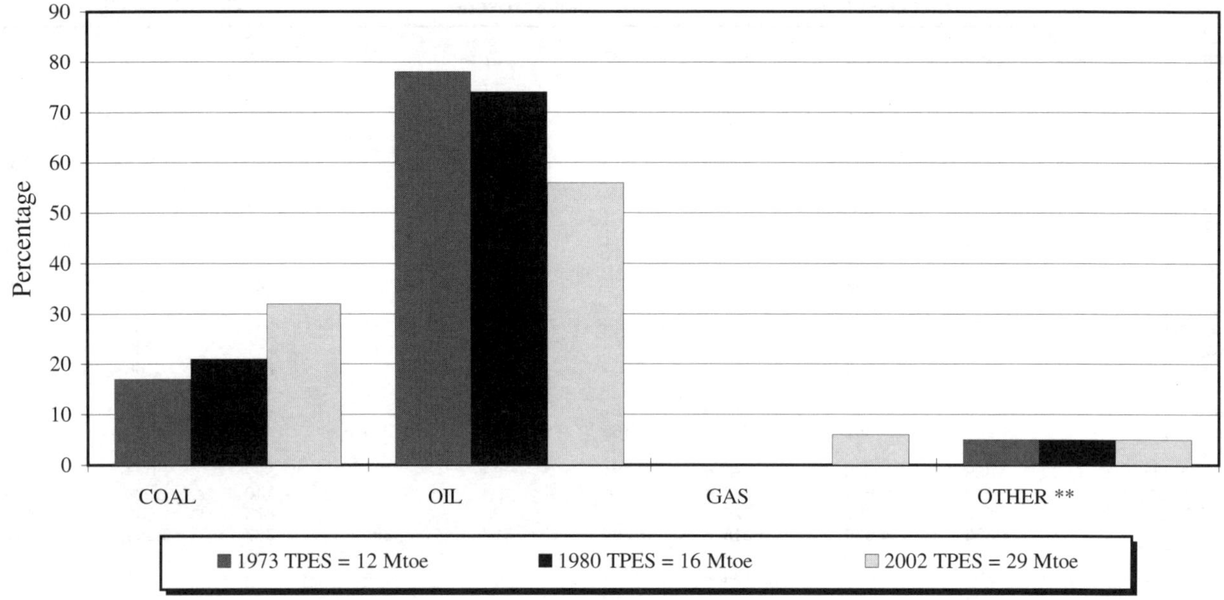

Source: Annual Energy Balances.

* Excluding electricity trade.

** Includes hydro, geothermal, solar, wind, combustible renewables and waste, etc.

Figure 2
OIL PRODUCTION AND CONSUMPTION

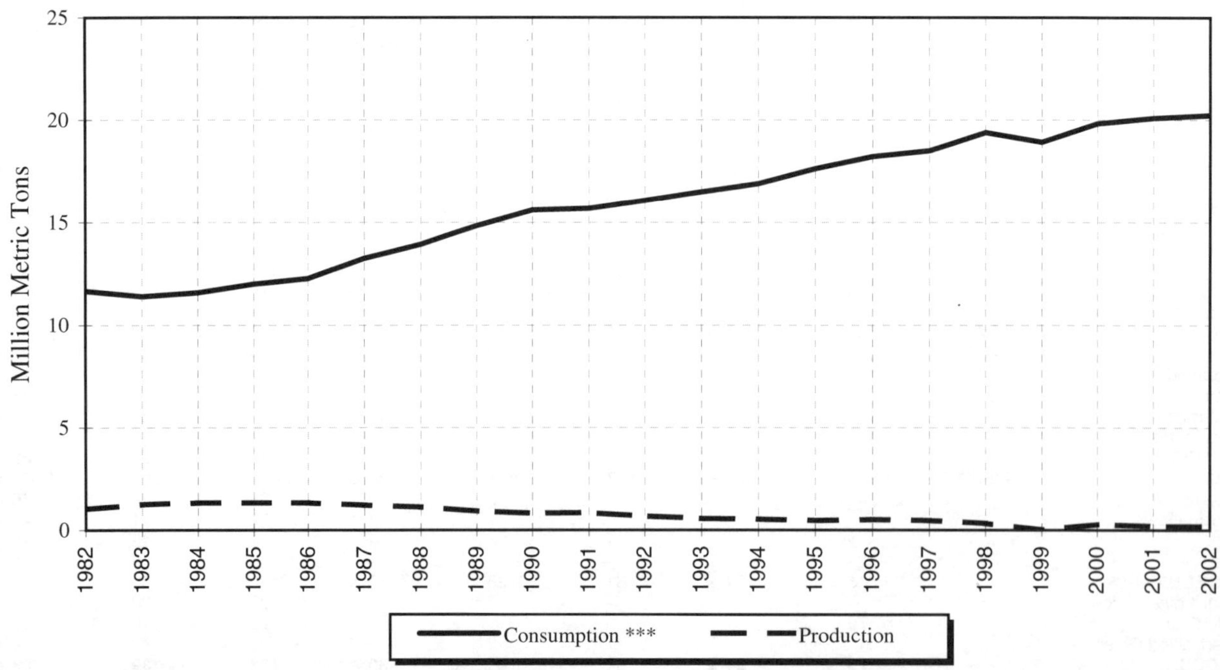

Source: Annual Oil Statistics.

*** Includes refinery fuel and international marine bunkers.

GREECE
Table 1: Oil Supply and Consumption - Summary Table

000 metric tons

	1973	1978	1980	1990	1999	2000	2001	2002
Indigenous Production	-	-	-	830	16	279	191	189 e
Crude Oil	..	-	-	773	15	256	171	165 e
NGL	..	-	-	57	1	23	20	24 e
Other	..	-	-	-	-	-	-	-
+ Net Imports	11 291	11 990	13 191	14 445	17 839	19 694	19 699	20 635 e
Crude Oil, NGL, Other	13 717	11 620	14 534	14 698	17 171	20 383	20 081	20 218 e
Petroleum Products	-2 426	370	-1 343	- 253	668	- 689	- 382	417 e
- Intl. Marine Bunkers	917	1 037	855	2 612	3 190	3 686	3 584	3 222 e
+ Stock Changes	-1 035	183	- 706	202	852	- 346	113	- 816 e
TOTAL SUPPLY	**9 339**	**11 136**	**11 630**	**12 865**	**15 517**	**15 941**	**16 419**	**16 786 e**
Refinery Intake	*12 600*	*11 568*	*14 268*	*16 549*	*18 979*	*22 320*	*21 595*	*21 416 e*
Refinery Output	*12 423*	*11 445*	*14 134*	*16 442*	*18 894*	*22 232*	*21 483*	*21 319 e*
LPG + Ethane	*123*	*146*	*182*	*395*	*553*	*766*	*762*	*694 e*
Naphtha	*1 177*	*513*	*889*	*506*	*880*	*1 036*	*881*	*795 e*
Kerosenes	*576*	*544*	*1 157*	*1 741*	*1 902*	*2 096*	*1 783*	*1 710 e*
Gasolines	*771*	*1 448*	*1 395*	*3 520*	*3 211*	*3 778*	*3 790*	*3 823 e*
Gas/Diesel Oil	*3 457*	*2 801*	*3 649*	*3 663*	*4 866*	*5 647*	*5 452*	*5 624 e*
Fuel Oil	*5 900*	*5 710*	*6 490*	*5 596*	*6 326*	*7 510*	*7 361*	*7 188 e*
Other Products	*419*	*283*	*372*	*1 021*	*1 156*	*1 399*	*1 454*	*1 485 e*
- Refinery Losses	177	123	134	107	85	88	112	97 e
+ Returns	-	-	-	-	-	-	-	140 e
- Statistical Differences	- 10	167	43	- 240	- 309	- 287	- 204	- 177 e
TOTAL CONSUMPTION	**9 172**	**10 846**	**11 453**	**12 998**	**15 741**	**16 140**	**16 511**	**17 006 e**
Transformation and Energy	**2 160**	**2 301**	**2 660**	**2 577**	**2 820**	**3 093**	**3 002**	..
Crude Oil	-	-	-	-	-	-	-	..
LPG + Ethane	-	-	-	26	25	38	38	..
Gas/Diesel Oil	124	90	308	315	336	382	376	..
Fuel Oil	1 952	2 133	2 175	1 721	1 922	2 033	1 916	..
Petroleum Coke	-	-	-	136	131	163	161	..
Other	84	78	177	379	406	477	511	..
Transport	**2 538**	**3 538**	**3 819**	**5 757**	**7 291**	**7 007**	**7 192**	..
Motor Gasoline	899	1 311	1 327	2 373	3 165	3 230	3 336	..
Gas/Diesel Oil	760	1 049	1 250	1 761	2 217	2 193	2 280	..
Subtotal	*1 659*	*2 360*	*2 577*	*4 134*	*5 382*	*5 423*	*5 616*	..
Aviation Fuels	816	1 071	1 082	1 239	1 250	1 290	1 160	..
Other	63	107	160	384	659	294	416	..
Industry	**2 455**	**2 964**	**3 155**	**2 131**	**2 384**	**2 610**	**2 621**	..
Naphtha	109	132	116	59	23	47	38	..
LPG + Ethane	42	41	40	101	315	305	310	..
Gas/Diesel Oil	216	211	245	354	560	504	500	..
Fuel Oil	1 727	2 356	2 462	1 152	769	882	830	..
Other	361	224	292	465	717	872	943	..
Other Sectors	**2 019**	**2 043**	**1 819**	**2 533**	**3 246**	**3 430**	**3 696**	..
Gas/Diesel Oil	1 427	1 724	1 436	2 294	3 055	3 253	3 510	..
Fuel Oil	359	99	148	62	28	28	34	..
Other	233	220	235	177	163	149	152	..
Memo: Non-Energy Use	*458*	*356*	*408*	*514*	*508*	*640*	*685*	..
Chem/Petchem. Feedst.	*109*	*144*	*130*	*118*	*33*	*120*	*131*	..
Other	*349*	*212*	*278*	*396*	*475*	*520*	*554*	..

Source: Annual Oil Statistics.

GREECE
Table 2: Supply of Crude Oil, NGL, Feedstocks and Other Hydrocarbons

000 metric tons

	1973	1978	1980	1990	1999	2000	2001	2002
CRUDE OIL								
Indigenous Production	..	-	-	773	15	256	171	165 e
+ Imports	..	12 574	17 708	14 539	15 944	19 371	18 906	19 116 e
- Exports	..	1 010	3 174	1 233	116	-	-	387 e
- Direct Use	..	12	14	-	-	-	-	-
+ Stock Change	..	- 20	- 247	326	215	- 182	11	- 428 e
- Statistical Difference	..	20	5	- 6	- 9	27	- 4	12 e
Refinery Intake	**..**	**11 512**	**14 268**	**14 411**	**16 067**	**19 418**	**19 092**	**18 454 e**
Closing Stock Level	*..*	*1 236*	*1 739*	*1 114*	*1 101*	*1 283*	*1 272*	*1 700 e*
NATURAL GAS LIQUIDS								
Indigenous Production	..	-	-	57	1	23	20	24 e
+ Imports	..	-	-	-	-	-	-	-
- Exports	..	-	-	-	-	-	-	-
- Direct Use	..	-	-	2	-	-	-	-
+ Stock Change	..	-	-	-	-	-	-	-
- Statistical Difference	..	-	-	1	-	-	-	-
Refinery Intake	**..**	**-**	**-**	**54**	**1**	**23**	**20**	**24 e**
Closing Stock Level	*..*	*-*	*-*	*-*	*-*	*-*	*-*	*-*
REFINERY FEEDSTOCKS								
Backflows	..	-	-	-	-	-	-	-
+ Transfers	..	-	-	733	1 456	1 727	1 449	1 259 e
+ Imports	..	56	-	1 351	1 311	966	1 135	1 432 e
- Exports	..	-	-	-	-	-	-	-
+ Stock Change	..	-	-	- 38	- 30	81	- 162	118 e
- Statistical Difference	..	-	-	3	- 140	- 62	- 20	- 73 e
Refinery Intake	**..**	**56**	**-**	**2 043**	**2 877**	**2 836**	**2 442**	**2 882 e**
Closing Stock Level	*..*	*-*	*-*	*79*	*329*	*248*	*410*	*292 e*
OTHER HYDROCARBONS								
Indigenous Production	..	-	-	-	-	-	-	-
+ Imports	..	-	-	41	32	46	40	57 e
- Exports	..	-	-	-	-	-	-	-
- Direct Use	..	-	-	-	-	-	-	-
+ Stock Change	..	-	-	-	2	- 3	1	- 1 e
- Statistical Difference	..	-	-	-	-	-	-	-
Refinery Intake	**..**	**-**	**-**	**41**	**34**	**43**	**41**	**56 e**
Closing Stock Level	*..*	*-*	*-*	*5*	*1*	*4*	*3*	*4 e*
CRUDE OIL, NGL, FEEDSTOCKS AND OTHER								
Indigenous Production	-	-	-	830	16	279	191	189 e
+ Backflows	-	-	-	-	-	-	-	-
+ Transfers	-	-	-	733	1 456	1 727	1 449	1 259 e
+ Imports	13 717	12 630	17 708	15 931	17 287	20 383	20 081	20 605 e
- Exports	-	1 010	3 174	1 233	116	-	-	387 e
- Direct Use	-	12	14	2	-	-	-	-
+ Stock Change	-1 117	- 20	- 247	288	187	- 104	- 150	- 311 e
- Statistical Difference	-	20	5	- 2	- 149	- 35	- 24	- 61 e
Refinery Intake	**12 600**	**11 568**	**14 268**	**16 549**	**18 979**	**22 320**	**21 595**	**21 416 e**
Closing Stock Level	*-*	*1 236*	*1 739*	*1 198*	*1 431*	*1 535*	*1 685*	*1 996 e*

Source: Annual Oil Statistics.

GREECE

Table 3: Motor Gasoline - Supply

000 metric tons

	1973	1978	1980	1990	1999	2000	2001	2002
MOTOR GASOLINE								
Refinery Output	**771**	**1 250**	**1 139**	**3 379**	**3 205**	**3 758**	**3 770**	**3 802 e**
+ Imports	192	278	898	213	477	415	116	514 e
- Exports	128	166	798	1 097	653	1 011	678	869 e
+ Transfers	-	-	-	-	17	22	167	153 e
+ Stock Change	116	- 6	138	- 45	169	- 27	3	- 122 e
- Statistical Difference	-	- 7	-	27	-	- 123	- 7	- 65 e
Total Consumption	**951**	**1 363**	**1 377**	**2 423**	**3 215**	**3 280**	**3 385**	**3 543 e**
Closing Stock Level	-	*114*	*127*	*319*	*264*	*291*	*288*	*410 e*

Table 4: Aviation Fuels - Supply

000 metric tons

	1973	1978	1980	1990	1999	2000	2001	2002
AVIATION GASOLINE								
Refinery Output	-	-	-	-	-	-	-	-
+ Imports	10	7	6	4	-	-	-	-
- Exports	-	-	-	-	-	-	-	-
+ Transfers	-	-	-	-	-	-	-	-
+ Stock Change	-	2	-	-	-	-	-	-
- Statistical Difference	-	-	-	- 4	-	-	-	-
Total Consumption	**10**	**9**	**6**	**8**	**-**	**-**	**-**	**-**
Closing Stock Level	-	*1*	*1*	-	-	-	-	-
GASOLINE TYPE JET FUEL								
Refinery Output	**-**	**198**	**256**	**141**	**6**	**20**	**20**	**21 e**
+ Imports	-	-	-	-	4	2	5	3 e
- Exports	-	-	-	135	-	-	-	-
+ Transfers	-	-	-	-	-	- 5	- 5	- 7 e
+ Stock Change	-	3	5	0	- 3	-	- 1	1 e
- Statistical Difference	-	- 1	-	- 5	-	-	-	-
Total Consumption	**-**	**202**	**251**	**8**	**7**	**17**	**19**	**18 e**
Closing Stock Level	-	*10*	*17*	*7*	*5*	*5*	*6*	*5 e*
KEROSENE TYPE JET FUEL								
Refinery Output	**456**	**503**	**1 116**	**1 719**	**1 894**	**2 087**	**1 771**	**1 684 e**
+ Imports	576	639	1 291	614	302	95	256	28 e
- Exports	221	308	1 509	1 090	930	755	957	517 e
+ Transfers	-	-	-	-	- 8	- 8	- 119	- 57 e
+ Stock Change	- 5	26	- 74	2	- 6	- 144	145	- 29 e
- Statistical Difference	-	-	- 1	22	9	2	- 45	3 e
Total Consumption	**806**	**860**	**825**	**1 223**	**1 243**	**1 273**	**1 141**	**1 106 e**
Closing Stock Level	-	*65*	*285*	*307*	*195*	*339*	*194*	*223 e*

Source: Annual Oil Statistics.

GREECE
Table 5A: Gas/Diesel Oil - Supply and Consumption

000 metric tons

	1973	1978	1980	1990	1999	2000	2001	2002
Refinery Output	**3 457**	**2 801**	**3 649**	**3 663**	**4 866**	**5 647**	**5 452**	**5 624 e**
+ Recycled Products	-	-	-	-	-	-	-	-
+ Imports	587	1 670	2 448	2 303	2 738	2 013	2 435	2 993 e
- Exports	1 360	817	2 504	556	586	576	794	891 e
- Intl. Marine Bunkers	176	298	201	510	706	750	612	549 e
+ Transfers	-	-	-	-	- 646	- 208	89	36 e
+ Stock Changes	19	- 164	- 149	- 169	419	108	35	- 251 e
- Statistical Differences	-	118	4	7	- 83	- 98	- 61	- 16 e
Total Consumption	**2 527**	**3 074**	**3 239**	**4 724**	**6 168**	**6 332**	**6 666**	**6 978 e**
Transformation	**124**	**90**	**308**	**315**	**336**	**382**	**376**	**..**
Public Electricity	124	90	308	314	336	382	376	..
Autoproducer Electricity	-	-	-	-	-	-	-	..
Public CHP Plants	-	-	-	-	-	-	-	..
Autoproducers of CHP	-	-	-	1	-	-	-	..
Public Heat Plants	-	-	-	-	-	-	-	..
Autoproducer Heat Plants	-	-	-	-	-	-	-	..
Petrochemical Industry	-	-	-	-	-	-	-	..
Other Transformation	-	-	-	-	-	-	-	..
Energy Sector	**-**	**-**	**-**	**-**	**-**	**-**	**-**	**..**
Refinery Fuel	-	-	-	-	-	-	-	..
Other Energy Sector	-	-	-	-	-	-	-	..
Final Consumption	**2 403**	**2 984**	**2 931**	**4 409**	**5 832**	**5 950**	**6 290**	**..**
Transport	**760**	**1 049**	**1 250**	**1 761**	**2 217**	**2 193**	**2 280**	**..**
Air	-	-	-	-	-	-	-	..
Road	508	808	886	1 362	1 888	1 890	1 895	..
Rail	68	46	45	63	40	40	40	..
Inland Waterways	184	195	319	336	289	263	345	..
Non-Specified	-	-	-	-	-	-	-	..
Industry	**216**	**211**	**245**	**354**	**560**	**504**	**500**	**..**
Iron and Steel	50	47	14	41	20	13	13	..
Chemical (incl. Pet.-Chem.)	10	3	2	15	10	9	9	..
Non-Ferrous Metals	-	-	26	-	23	23	23	..
Non-Metallic Minerals	6	14	17	49	53	49	48	..
Transport Equipment	-	-	-	-	17	16	15	..
Machinery	-	-	-	-	-	-	-	..
Mining and Quarrying	24	29	35	43	42	41	40	..
Food and Tobacco	12	14	14	33	53	51	51	..
Pulp, Paper and Printing	4	5	4	12	10	10	10	..
Wood and Wood Products	-	-	-	-	3	2	1	..
Construction	-	-	-	-	126	130	130	..
Textiles and Leather	-	-	-	17	8	10	10	..
Non-Specified	110	99	133	144	195	150	150	..
Other Sectors	**1 427**	**1 724**	**1 436**	**2 294**	**3 055**	**3 253**	**3 510**	**..**
Commerce and Public	-	-	-	145	195	203	270	..
Residential	720	1 087	752	1 292	2 100	2 290	2 470	..
Agriculture	583	593	645	820	760	760	770	..
Non-Specified	124	44	39	37	-	-	-	..
Memo: Non-Energy Use	-	-	-	-	-	-	-	..
Chem/Petchem. Feedst.	-	-	-	-	-	-	-	..
Other	-	-	-	-	-	-	-	..
Closing Stock Level	-	*589*	*844*	*757*	*860*	*752*	*717*	*968 e*

Source: Annual Oil Statistics.

GREECE

Table 5B: Fuel Oil (Residual) - Supply and Consumption

000 metric tons

	1973	1978	1980	1990	1999	2000	2001	2002
Refinery Output	**5 900**	**5 710**	**6 490**	**5 596**	**6 326**	**7 510**	**7 361**	**7 188 e**
+ Recycled Products	-	-	-	-	-	-	-	140 e
+ Imports	828	911	891	2 233	298	174	169	36 e
- Exports	2 135	1 182	1 628	2 026	280	220	255	564 e
- Intl. Marine Bunkers	665	725	654	2 063	2 452	2 898	2 933	2 624 e
+ Transfers	-	-	-	- 733	- 716	-1 392	-1 324	-1 050 e
+ Stock Changes	173	12	- 118	- 80	45	- 32	25	- 66 e
- Statistical Differences	-	31	36	- 245	- 89	- 37	- 72	- 38 e
Total Consumption	**4 101**	**4 695**	**4 945**	**3 172**	**3 310**	**3 179**	**3 115**	**3 098 e**
Transformation	**1 686**	**1 820**	**1 910**	**1 455**	**1 609**	**1 661**	**1 558**	..
Public Electricity	1 686	1 771	1 884	1 421	1 585	1 634	1 539	..
Autoproducer Electricity	-	-	-	-	-	-	-	..
Public CHP Plants	-	-	-	-	-	-	-	..
Autoproducers of CHP	-	49	26	34	24	27	19	..
Public Heat Plants	-	-	-	-	-	-	-	..
Autoproducer Heat Plants	-	-	-	-	-	-	-	..
Petrochemical Industry	-	-	-	-	-	-	-	..
Other Transformation	-	-	-	-	-	-	-	..
Energy Sector	**266**	**313**	**265**	**266**	**313**	**372**	**358**	..
Refinery Fuel	266	313	265	266	313	372	358	..
Other Energy Sector	-	-	-	-	-	-	-	..
Final Consumption	**2 149**	**2 562**	**2 770**	**1 451**	**1 388**	**1 146**	**1 199**	..
Transport	**63**	**107**	**160**	**237**	**591**	**236**	**335**	..
Air	-	-	-	-	-	-	-	..
Road	-	-	-	-	-	-	-	..
Rail	63	1	-	-	-	-	-	..
Inland Waterways	-	106	160	237	591	236	335	..
Non-Specified	-	-	-	-	-	-	-	..
Industry	**1 727**	**2 356**	**2 462**	**1 152**	**769**	**882**	**830**	..
Iron and Steel	210	148	158	101	8	18	19	..
Chemical (incl. Pet.-Chem.)	174	103	79	92	81	87	82	..
Non-Ferrous Metals	180	258	253	185	211	214	216	..
Non-Metallic Minerals	460	934	1 057	159	67	89	86	..
Transport Equipment	-	-	-	-	2	3	3	..
Machinery			-	-	12	15	15	..
Mining and Quarrying	60	70	61	22	42	43	43	..
Food and Tobacco	232	234	164	241	181	208	187	..
Pulp, Paper and Printing	110	100	104	84	66	81	66	..
Wood and Wood Products	-	-	-	-	2	2	2	..
Construction	-	-	-	-	18	30	30	..
Textiles and Leather	-	-	-	111	79	92	81	..
Non-Specified	301	509	586	157	-	-	-	..
Other Sectors	**359**	**99**	**148**	**62**	**28**	**28**	**34**	..
Commerce and Public	-	-	-	13	13	13	16	..
Residential	275	88	44	36	-	-	-	..
Agriculture	2	4	6	-	15	15	18	..
Non-Specified	82	7	98	13	-	-	-	..
Memo: Non-Energy Use	-	-	-	-	-	-	-	..
Chem/Petchem. Feedst.	-	-	-	-	-	-	-	..
Other	-	-	-	-	-	-	-	..
Closing Stock Level	-	*773*	*761*	*876*	*558*	*590*	*565*	*631 e*

Source: Annual Oil Statistics.

GREECE
Table 5C: Total Products - Supply and Consumption

000 metric tons

	1973	1978	1980	1990	1999	2000	2001	2002
Refinery Output	**12 423**	**11 445**	**14 134**	**16 442**	**18 894**	**22 232**	**21 483**	**21 319 e**
+ Recycled Products	-	-	-	-	-	-	-	140 e
+ Imports	2 466	3 824	5 974	5 899	4 208	3 255	3 397	4 200 e
- Exports	4 892	3 454	7 317	6 152	3 540	3 944	3 779	3 783 e
- Intl. Marine Bunkers	917	1 037	855	2 612	3 190	3 686	3 584	3 222 e
+ Transfers	-	12	14	- 731	-1 456	-1 727	-1 449	-1 259 e
+ Stock Changes	82	203	- 459	- 86	665	- 242	263	- 505 e
- Statistical Differences	- 10	147	38	- 238	- 160	- 252	- 180	- 116 e
Total Consumption	**9 172**	**10 846**	**11 453**	**12 998**	**15 741**	**16 140**	**16 511**	**17 006 e**
Transformation	**1 810**	**1 910**	**2 218**	**1 788**	**1 945**	**2 043**	**1 934**	..
Public Electricity	1 810	1 861	2 192	1 735	1 921	2 016	1 915	..
Autoproducer Electricity	-	-	-	-	-	-	-	..
Public CHP Plants	-	-	-	-	-	-	-	..
Autoproducers of CHP	-	49	26	37	24	27	19	..
Public Heat Plants	-	-	-	-	-	-	-	..
Autoproducer Heat Plants	-	-	-	-	-	-	-	..
Petrochemical Industry	-	-	-	-	-	-	-	..
Other Transformation	-	-	-	16	-	-	-	..
Energy Sector	**350**	**391**	**442**	**789**	**875**	**1 050**	**1 068**	..
Refinery Fuel	350	391	442	789	875	1 050	1 068	..
Other Energy Sector	-	-	-	-	-	-	-	..
Final Consumption	**7 012**	**8 545**	**8 793**	**10 421**	**12 921**	**13 047**	**13 509**	..
Transport	**2 538**	**3 538**	**3 819**	**5 757**	**7 291**	**7 007**	**7 192**	..
Air	816	1 071	1 082	1 239	1 250	1 290	1 160	..
Road	1 407	2 119	2 213	3 863	5 105	5 166	5 299	..
Rail	131	47	45	63	41	41	41	..
Inland Waterways	184	301	479	592	895	510	692	..
Non-Specified	-	-	-	-	-	-	-	..
Industry	**2 455**	**2 964**	**3 155**	**2 131**	**2 384**	**2 610**	**2 621**	..
Iron and Steel	260	195	172	142	50	52	53	..
Chemical (incl. Pet.-Chem.)	293	250	211	263	182	275	277	..
Non-Ferrous Metals	180	272	337	242	326	331	334	..
Non-Metallic Minerals	466	948	1 074	310	412	465	506	..
Transport Equipment	-	-	-	-	21	20	19	..
Machinery	-	-	-	-	12	15	15	..
Mining and Quarrying	84	99	96	65	107	108	106	..
Food and Tobacco	244	248	178	274	264	290	272	..
Pulp, Paper and Printing	114	105	108	96	98	113	98	..
Wood and Wood Products	-	-	-	-	6	5	4	..
Construction	194	107	120	204	500	568	576	..
Textiles and Leather	-	-	-	128	88	103	92	..
Non-Specified	620	740	859	407	318	265	269	..
Other Sectors	**2 019**	**2 043**	**1 819**	**2 533**	**3 246**	**3 430**	**3 696**	..
Commerce and Public	-	-	-	161	238	241	312	..
Residential	1 188	1 343	981	1 452	2 182	2 363	2 546	..
Agriculture	625	649	701	870	826	826	838	..
Non-Specified	206	51	137	50	-	-	-	..
Memo: Non-Energy Use	*458*	*356*	*408*	*514*	*508*	*640*	*685*	..
Chem/Petchem. Feedst.	*109*	*144*	*130*	*118*	*33*	*120*	*131*	..
Other	*349*	*212*	*278*	*396*	*475*	*520*	*554*	..
Closing Stock Level	-	*1 807*	*2 523*	*2 483*	*2 042*	*2 284*	*2 021*	*2 526 e*

Source: Annual Oil Statistics.

GREECE

Table 6A: Supply and Consumption

000 metric tons

	1980	2000	2001	2002	1980	2000	2001	2002
	LPG AND ETHANE				**REFINERY GAS**			
Refinery Output	**182**	**766**	**762**	**694 e**	**177**	**488**	**516**	**522 e**
+ Recycled Products	-	-	-	-	-	-	-	-
+ Imports	25	-	1	12 e	-	-	-	-
- Exports	15	309	288	279 e	-	-	-	-
- Intl. Marine Bunkers	-	-	-	-	-	-	-	-
+ Transfers	-	- 8	- 17	- 29 e	-	-	-	-
+ Stock Change	- 3	- 3	- 3	4 e	-	-	-	-
- Statistical Difference	-	4	5	-	-	-	-	-
Total Consumption	**189**	**442**	**450**	**402 e**	**177**	**488**	**516**	**522 e**
of which: Ethane	-	-	-	-	-	-	-	-
Transformation	-	-	-	..	-	-	-	..
Energy	-	38	38	..	177	477	511	..
Transport	-	15	16	..	-	-	-	..
Industry	40	305	310	..	-	11	5	..
Other	149	84	86	..	-	-	-	..
Memo: Non-Energy Use	-	-	-	..	-	-	-	..
Closing Stock Level	*9*	*21*	*24*	*20 e*	-	-	-	-

	1980	2000	2001	2002	1980	2000	2001	2002
	NAPHTHA				**OTHER KEROSENE**			
Refinery Output	**889**	**1 036**	**881**	**795 e**	**41**	**9**	**12**	**26 e**
+ Recycled Products	-	-	-	-	-	-	-	-
+ Imports	260	-	19	57 e	-	-	-	-
- Exports	793	865	615	522 e	-	-	8	6 e
- Intl. Marine Bunkers	-	-	-	-	-	-	-	-
+ Transfers	-	- 132	- 253	- 246 e	-	4	13	- 7 e
+ Stock Change	- 241	8	6	- 64 e	- 5	1	- 1	- 1 e
- Statistical Difference	- 1	-	-	-	-	-	-	-
Total Consumption	**116**	**47**	**38**	**20 e**	**36**	**14**	**16**	**12 e**
Transformation	-	-	-	..	-	-	-	..
Energy	-	-	-	..	-	-	-	..
Transport	-	-	-	..	-	-	-	..
Industry	116	47	38	..	-	1	1	..
Other	-	-	-	..	36	13	15	..
Memo: Non-Energy Use	*116*	*47*	*38*	..	-	-	-	..
Closing Stock Level	*431*	*58*	*52*	*116 e*	*8*	*1*	*2*	*3 e*

	1980	2000	2001	2002	1980	2000	2001	2002
	WHITE SPIRIT + SBP				**LUBRICANTS**			
Refinery Output	**-**	**-**	**-**	**..**	**103**	**160**	**170**	**..**
+ Recycled Products	-	-	-	..	-	-	-	..
+ Imports	-	-	-	..	54	44	30	..
- Exports	-	-	-	..	53	98	100	..
- Intl. Marine Bunkers	-	-	-	..	-	38	39	..
+ Transfers	-	-	-	..	-	- 6	21	..
+ Stock Change	-	-	-	..	- 6	- 6	- 3	..
- Statistical Difference	-	-	-	..	- 2	-	-	..
Total Consumption	**-**	**-**	**-**	**..**	**100**	**56**	**79**	**..**
Transformation	-	-	-	..	-	-	-	..
Energy	-	-	-	..	-	-	-	..
Transport	-	-	-	..	-	43	65	..
Industry	-	-	-	..	100	11	12	..
Other	-	-	-	..	-	2	2	..
Memo: Non-Energy Use	-	-	-	..	*100*	*56*	*79*	..
Closing Stock Level	-	-	-	..	*32*	*33*	*36*	..

Source: Annual Oil Statistics.

GREECE
Table 6B: Supply and Consumption

000 metric tons

	1980	2000	2001	2002	1980	2000	2001	2002
	BITUMEN				PARAFFIN WAXES			
Refinery Output	**75**	**517**	**506**	**..**	**17**	**1**	**1**	**..**
+ Recycled Products	-	-	-	..	-	-	-	..
+ Imports	41	-	-	..	-	-	-	..
- Exports	-	110	84	..	17	-	-	..
- Intl. Marine Bunkers	-	-	-	..	-	-	-	..
+ Transfers	-	5	- 15	..	-	-	-	..
+ Stock Change	4	- 6	7	..	-	-	-	..
- Statistical Difference	-	-	-	..	-	-	-	..
Total Consumption	**120**	**406**	**414**	**..**	**-**	**1**	**1**	**..**
Transformation	-	-	-	..	-	-	-	..
Energy	-	-	-	..	-	-	-	..
Transport	-	-	-	..	-	-	-	..
Industry	120	406	414	..	-	1	1	..
Other	-	-	-	..	-	-	-	..
Memo: Non-Energy Use	*120*	*406*	*414*	*..*	*-*	*1*	*1*	*..*
Closing Stock Level	*8*	*26*	*19*	*..*	*-*	*-*	*-*	*..*
	PETROLEUM COKE				OTHER PRODUCTS			
Refinery Output	**-**	**163**	**161**	**160 e**	**-**	**70**	**100**	**803 e**
+ Recycled Products	-	-	-	-	-	-	-	-
+ Imports	60	512	366	534 e	-	-	-	23 e
- Exports	-	-	-	-	-	-	-	135 e
- Intl. Marine Bunkers	-	-	-	-	-	-	-	49 e
+ Transfers	-	-	-	-	14	1	- 6	- 52 e
+ Stock Change	-	- 142	52	10 e	-	1	- 2	13 e
- Statistical Difference	2	-	-	-	-	-	-	-
Total Consumption	**58**	**533**	**579**	**704 e**	**14**	**72**	**92**	**603 e**
Transformation	-	-	-	..	-	-	-	..
Energy	-	163	161	..	-	-	-	..
Transport	-	-	-	..	-	-	-	..
Industry	58	370	418	..	14	72	92	..
Other	-	-	-	..	-	-	-	..
Memo: Non-Energy Use	*58*	*58*	*61*	*..*	*-*	*72*	*92*	*..*
Closing Stock Level	*-*	*164*	*112*	*102 e*	*-*	*4*	*6*	*48 e*

Source: Annual Oil Statistics.

GREECE
Table 7A: Imports by Country of Origin

000 metric tons

	2000	2001	2002	2000	2001	2002	2000	2001	2002
	Crude Oil			Natural Gas Liquids			Feedstocks and Other		
Australia	-	-	-	-	-	-	-	-	-
Belgium	-	-	-	-	-	-	-	-	-
Canada	-	-	-	-	-	-	-	-	-
Czech Republic	-	-	-	-	-	-	-	-	-
Denmark	-	-	-	-	-	-	-	-	-
Finland	-	-	-	-	-	-	-	-	-
France	-	-	-	-	-	-	-	-	30 e
Germany	-	-	-	-	-	-	-	-	-
Greece	-	-	-	-	-	-	-	-	-
Hungary	-	-	-	-	-	-	-	-	-
Ireland	-	-	-	-	-	-	-	-	-
Italy	-	-	-	-	-	-	-	28	-
Japan	-	-	-	-	-	-	-	-	-
Korea	-	-	-	-	-	-	-	-	-
Mexico	-	-	-	-	-	-	-	-	-
Netherlands	-	-	-	-	-	-	-	-	72 e
New Zealand	-	-	-	-	-	-	-	-	-
Norway	-	-	-	-	-	-	-	-	-
Poland	-	-	-	-	-	-	-	-	-
Spain	-	-	-	-	-	-	-	-	-
Sweden	-	-	-	-	-	-	-	-	-
Turkey	-	-	-	-	-	-	-	44	-
United Kingdom	-	-	-	-	-	-	-	-	-
United States	-	-	-	-	-	-	-	-	-
Other	-	-	-	-	-	-	-	-	-
OECD	**-**	**-**	**-**	**-**	**-**	**-**	**-**	**72**	**102 e**
Argentina	-	-	-	-	-	-	-	-	-
Brazil	-	-	-	-	-	-	-	-	-
Colombia	-	-	-	-	-	-	-	-	-
Ecuador	-	-	-	-	-	-	-	-	-
Netherlands Antilles	-	-	-	-	-	-	-	-	-
Peru	-	-	-	-	-	-	-	-	-
Trinidad and Tobago	-	-	-	-	-	-	-	-	-
Venezuela	-	-	-	-	-	-	-	-	-
Other	-	-	-	-	-	-	-	-	-
Latin America	**-**	**-**	**-**	**-**	**-**	**-**	**-**	**-**	**-**
Romania	-	-	-	-	-	-	9	-	17 e
Other	-	-	-	-	-	-	-	2	10 e
Non-OECD Europe	**-**	**-**	**-**	**-**	**-**	**-**	**9**	**2**	**27 e**
FSU	**4 249**	**5 742**	**9 134 e**	**-**	**-**	**-**	**825**	**874**	**1 212 e**
Iran	6 204	5 382	2 693 e	-	-	-	30	-	-
Iraq	1 841	1 303	927 e	-	-	-	-	-	-
Kuwait	-	-	-	-	-	-	-	-	-
Oman	-	-	-	-	-	-	-	-	-
Qatar	-	-	-	-	-	-	-	-	-
Saudi Arabia	5 821	5 268	5 128 e	-	-	-	40	-	
United Arab Emirates	-	-	-	-	-	-	-	-	-
Other	-	-	-	-	-	-	83	120	30 e
Middle East	**13 866**	**11 953**	**8 748 e**	**-**	**-**	**-**	**159**	**120**	**30 e**
India	-	-	-	-	-	-	-	-	-
Indonesia	-	-	-	-	-	-	-	-	-
Singapore	-	-	-	-	-	-	-	-	-
Other	-	-	-	-	-	-	-	-	-
Asia	**-**	**-**	**-**	**-**	**-**	**-**	**-**	**-**	**-**
China	**-**	**-**	**-**	**-**	**-**	**-**	**-**	**-**	**-**
Algeria	65	-	-	-	-	-	-	1	-
Angola/Cabinada	-	-	-	-	-	-	-	-	-
Cameroon	-	-	-	-	-	-	-	-	-
Egypt	-	-	296 e	-	-	-	-	28	25 e
Gabon	-	-	-	-	-	-	-	-	-
Libya	1 191	1 211	938 e	-	-	-	15	78	93 e
Nigeria	-	-	-	-	-	-	-	-	-
Tunisia	-	-	-	-	-	-	-	-	-
Other	-	-	-	-	-	-	-	-	-
Africa	**1 256**	**1 211**	**1 234 e**	**-**	**-**	**-**	**15**	**107**	**118 e**
Non-spec./Other	**-**	**-**	**-**	**-**	**-**	**-**	**4**	**-**	**-**
Total Imports	**19 371**	**18 906**	**19 116 e**	**-**	**-**	**-**	**1 012**	**1 175**	**1 489 e**
Memo: Total OPEC	*15 122*	*13 164*	*9 686 e*	-	-	-	*91*	*79*	*93 e*

Source: Annual Oil Statistics.

GREECE
Table 7B: Imports by Country of Origin

000 metric tons

	2000	2001	2002	2000	2001	2002	2000	2001	2002
	Total Products			Liquefied Petroleum Gases			Naphtha		
Australia	-	-	-	-	-	-	-	-	-
Belgium	1	-	39 e	-	-	-	-	-	-
Canada	-	-	-	-	-	-	-	-	-
Czech Republic	-	-	-	-	-	-	-	-	-
Denmark	-	-	-	-	-	-	-	-	-
Finland	-	-	-	-	-	-	-	-	-
France	76	9	39 e	-	-	-	-	-	-
Germany	-	2	-	-	-	-	-	-	-
Greece	-	-	-	-	-	-	-	-	-
Hungary	-	-	-	-	-	-	-	-	-
Ireland	-	-	-	-	-	-	-	-	-
Italy	628	598	572 e	-	1	-	-	-	-
Japan	-	-	-	-	-	-	-	-	-
Korea	-	-	-	-	-	-	-	-	-
Mexico	-	-	-	-	-	-	-	-	-
Netherlands	196	19	136 e	-	-	-	-	-	-
New Zealand	-	-	-	-	-	-	-	-	-
Norway	-	-	-	-	-	-	-	-	-
Poland	-	-	-	-	-	-	-	-	-
Spain	117	39	18 e	-	-	-	-	-	-
Sweden	-	-	-	-	-	-	-	-	-
Turkey	3	17	31 e	-	-	-	-	-	-
United Kingdom	17	28	59 e	-	-	-	-	-	-
United States	294	263	279 e	-	-	-	-	-	-
Other	-	-	-	-	-	-	-	-	-
OECD	**1 332**	**975**	**1 173 e**	**-**	**1**	**-**	**-**	**-**	**-**
Argentina	-	-	-	-	-	-	-	-	-
Brazil	-	-	-	-	-	-	-	-	-
Colombia	-	-	-	-	-	-	-	-	-
Ecuador	-	-	-	-	-	-	-	-	-
Netherlands Antilles	-	-	-	-	-	-	-	-	-
Peru	-	-	-	-	-	-	-	-	-
Trinidad and Tobago	-	-	-	-	-	-	-	-	-
Venezuela	118	122	280 e	-	-	-	-	-	-
Other	-	-	39 e	-	-	-	-	-	-
Latin America	**118**	**122**	**319 e**	**-**	**-**	**-**	**-**	**-**	**-**
Romania	35	46	56 e	-	-	-	-	-	-
Other	44	73	227 e	-	-	2 e	-	-	-
Non-OECD Europe	**79**	**119**	**283 e**	**-**	**-**	**2 e**	**-**	**-**	**-**
FSU	**1 039**	**1 600**	**2 045 e**	**-**	**-**	**-**	**-**	**19**	**40 e**
Iran	-	-	-	-	-	-	-	-	-
Iraq	-	-	-	-	-	-	-	-	-
Kuwait	-	-	-	-	-	-	-	-	-
Oman	-	-	-	-	-	-	-	-	-
Qatar	-	-	-	-	-	-	-	-	-
Saudi Arabia	61	106	5 e	-	-	-	-	-	-
United Arab Emirates	29	-	-	-	-	-	-	-	-
Other	143	84	51 e	-	-	-	-	-	-
Middle East	**233**	**190**	**56 e**	**-**	**-**	**-**	**-**	**-**	**-**
India	-	113	165 e	-	-	-	-	-	-
Indonesia	-	-	-	-	-	-	-	-	-
Singapore	-	-	-	-	-	-	-	-	-
Other	-	-	-	-	-	-	-	-	-
Asia	**-**	**113**	**165 e**	**-**	**-**	**-**	**-**	**-**	**-**
China	**-**	**-**	**-**	**-**	**-**	**-**	**-**	**-**	**-**
Algeria	48	20	-	-	-	-	-	-	-
Angola/Cabinada	-	-	-	-	-	-	-	-	-
Cameroon	-	-	-	-	-	-	-	-	-
Egypt	-	-	86 e	-	-	1 e	-	-	-
Gabon	-	-	-	-	-	-	-	-	-
Libya	235	216	73 e	-	-	9 e	-	-	17 e
Nigeria	-	-	-	-	-	-	-	-	-
Tunisia	-	-	-	-	-	-	-	-	-
Other	66	42	-	-	-	-	-	-	-
Africa	**349**	**278**	**159 e**	**-**	**-**	**10 e**	**-**	**-**	**17 e**
Non-spec./Other	**105**	**-**	**-**	**-**	**-**	**-**	**-**	**-**	**-**
Total Imports	**3 255**	**3 397**	**4 200 e**	**-**	**1**	**12 e**	**-**	**19**	**57 e**
Memo: Total OPEC	*491*	*464*	*358 e*	*-*	*-*	*9 e*	*-*	*-*	*17 e*

Source: Annual Oil Statistics.

GREECE
Table 7C: Imports by Country of Origin

000 metric tons

	2000	2001	2002	2000	2001	2002	2000	2001	2002
	Total Gasoline			**Total Kerosene**			**Gas/Diesel**		
Australia	-	-	-	-	-	-	-	-	-
Belgium	-	-	-	-	-	-	-	-	38 e
Canada	-	-	-	-	-	-	-	-	-
Czech Republic	-	-	-	-	-	-	-	-	-
Denmark	-	-	-	-	-	-	-	-	-
Finland	-	-	-	-	-	-	-	-	-
France	49	4	36 e	-	-	-	25	-	-
Germany	-	-	-	-	-	-	-	-	-
Greece	-	-	-	-	-	-	-	-	-
Hungary	-	-	-	-	-	-	-	-	-
Ireland	-	-	-	-	-	-	-	-	-
Italy	57	16	97 e	-	-	-	433	572	461 e
Japan	-	-	-	-	-	-	-	-	-
Korea	-	-	-	-	-	-	-	-	-
Mexico	-	-	-	-	-	-	-	-	-
Netherlands	102	-	24 e	-	-	-	65	-	99 e
New Zealand	-	-	-	-	-	-	-	-	-
Norway	-	-	-	-	-	-	-	-	-
Poland	-	-	-	-	-	-	-	-	-
Spain	117	39	18 e	-	-	-	-	-	-
Sweden	-	-	-	-	-	-	-	-	-
Turkey	-	-	31 e	-	-	-	3	-	-
United Kingdom	11	14	59 e	3	14	-	3	-	-
United States	-	-	-	-	-	-	-	-	41 e
Other	-	-	-	-	-	-	-	-	-
OECD	**336**	**73**	**265 e**	**3**	**14**	**-**	**529**	**572**	**639 e**
Argentina	-	-	-	-	-	-	-	-	-
Brazil	-	-	-	-	-	-	-	-	-
Colombia	-	-	-	-	-	-	-	-	-
Ecuador	-	-	-	-	-	-	-	-	-
Netherlands Antilles	-	-	-	-	-	-	-	-	-
Peru	-	-	-	-	-	-	-	-	-
Trinidad and Tobago	-	-	-	-	-	-	-	-	-
Venezuela	-	-	-	-	-	-	-	-	33 e
Other	-	-	-	-	-	-	-	-	-
Latin America	**-**	**-**	**-**	**-**	**-**	**-**	**-**	**-**	**33 e**
Romania	6	-	27 e	-	-	-	29	46	29 e
Other	7	43	116 e	-	-	28 e	37	30	79 e
Non-OECD Europe	**13**	**43**	**143 e**	**-**	**-**	**28 e**	**66**	**76**	**108 e**
FSU	**-**	**-**	**-**	**-**	**10**	**-**	**1 039**	**1 511**	**1 969 e**
Iran	-	-	-	-	-	-	-	-	-
Iraq	-	-	-	-	-	-	-	-	-
Kuwait	-	-	-	-	-	-	-	-	-
Oman	-	-	-	-	-	-	-	-	-
Qatar	-	-	-	-	-	-	-	-	-
Saudi Arabia	30	5	5 e	22	101		-	-	-
United Arab Emirates	29	-	-	-	-	-	-	-	-
Other	-	-	-	3	-	-	28	28	51 e
Middle East	**68**	**5**	**5 e**	**25**	**101**	**-**	**28**	**28**	**51 e**
India	-	-	-	-	-	-	-	113	165 e
Indonesia	-	-	-	-	-	-	-	-	-
Singapore	-	-	-	-	-	-	-	-	-
Other	-	-	-	-	-	-	-	-	-
Asia	**-**	**-**	**-**	**-**	**-**	**-**	**-**	**113**	**165 e**
China	**-**	**-**	**-**	**-**	**-**	**-**	**-**	**-**	**-**
Algeria	-	-	-	-	-	-	48	20	-
Angola/Cabinada	-	-	-	-	-	-	-	-	-
Cameroon	-	-	-	-	-	-	-	-	-
Egypt	-	-	85 e	-	-	-	-	-	-
Gabon	-	-	-	-	-	-	-	-	-
Libya	-	-	19 e	67	131	-	168	85	28 e
Nigeria	-	-	-	-	-	-	-	-	-
Tunisia	-	-	-	-	-	-	-	-	-
Other	-	-	-	-	-	-	30	30	-
Africa	**-**	**-**	**104 e**	**67**	**131**	**-**	**246**	**135**	**28 e**
Non-spec./Other	**-**	**-**	**-**	**-**	**-**	**-**	**105**	**-**	**-**
Total Imports	**417**	**121**	**517 e**	**95**	**256**	**28 e**	**2 013**	**2 435**	**2 993 e**
Memo: Total OPEC	*68*	*5*	*24 e*	*89*	*232*	*-*	*216*	*105*	*61 e*

Source: Annual Oil Statistics.

GREECE
Table 7D: Imports by Country of Origin

000 metric tons

	2000	2001	2002	2000	2001	2002	2000	2001	2002
	Fuel Oil (Residual)			Petroleum Coke			Other Products		
Australia	-	-	-	-	-	-	-	-	-
Belgium	-	-	-	-	-	-	1	-	1 e
Canada	-	-	-	-	-	-	-	-	-
Czech Republic	-	-	-	-	-	-	-	-	-
Denmark	-	-	-	-	-	-	-	-	-
Finland	-	-	-	-	-	-	-	-	-
France	-	-	-	-	-	-	2	5	3 e
Germany	-	-	-	-	-	-	-	2	-
Greece	-	-	-	-	-	-	-	-	-
Hungary	-	-	-	-	-	-	-	-	-
Ireland	-	-	-	-	-	-	-	-	-
Italy	112	5	-	14	-	10 e	12	4	4 e
Japan	-	-	-	-	-	-	-	-	-
Korea	-	-	-	-	-	-	-	-	-
Mexico	-	-	-	-	-	-	-	-	-
Netherlands	-	-	-	-	-	-	29	19	13 e
New Zealand	-	-	-	-	-	-	-	-	-
Norway	-	-	-	-	-	-	-	-	-
Poland	-	-	-	-	-	-	-	-	-
Spain	-	-	-	-	-	-	-	-	-
Sweden	-	-	-	-	-	-	-	-	-
Turkey	-	17	-	-	-	-	-	-	-
United Kingdom	-	-	-	-	-	-	-	-	-
United States	-	-	-	294	263	238 e	-	-	-
Other	-	-	-	-	-	-	-	-	-
OECD	**112**	**22**	**-**	**308**	**263**	**248 e**	**44**	**30**	**21 e**
Argentina	-	-	-	-	-	-	-	-	-
Brazil	-	-	-	-	-	-	-	-	-
Colombia	-	-	-	-	-	-	-	-	-
Ecuador	-	-	-	-	-	-	-	-	-
Netherlands Antilles	-	-	-	-	-	-	-	-	-
Peru	-	-	-	-	-	-	-	-	-
Trinidad and Tobago	-	-	-	-	-	-	-	-	-
Venezuela	-	31	-	118	91	247 e	-	-	-
Other	-	-	-	-	-	39 e	-	-	-
Latin America	**-**	**31**	**-**	**118**	**91**	**286 e**	**-**	**-**	**-**
Romania	-	-	-	-	-	-	-	-	-
Other	-	-	-	-	-	-	-	-	2 e
Non-OECD Europe	**-**	**-**	**-**	**-**	**-**	**-**	**-**	**-**	**2 e**
FSU	**-**	**60**	**36 e**	**-**	**-**	**-**	**-**	**-**	**-**
Iran	-	-	-	-	-	-	-	-	-
Iraq	-	-	-	-	-	-	-	-	-
Kuwait	-	-	-	-	-	-	-	-	-
Oman	-	-	-	-	-	-	-	-	-
Qatar	-	-	-	-	-	-	-	-	-
Saudi Arabia	-	-	-	-	-	-	-	-	-
United Arab Emirates	-	-	-	-	-	-	-	-	-
Other	62	56	-	50	-	-	-	-	-
Middle East	**62**	**56**	**-**	**50**	**-**	**-**	**-**	**-**	**-**
India	-	-	-	-	-	-	-	-	-
Indonesia	-	-	-	-	-	-	-	-	-
Singapore	-	-	-	-	-	-	-	-	-
Other	-	-	-	-	-	-	-	-	-
Asia	**-**	**-**	**-**	**-**	**-**	**-**	**-**	**-**	**-**
China	**-**	**-**	**-**	**-**	**-**	**-**	**-**	**-**	**-**
Algeria	-	-	-	-	-	-	-	-	-
Angola/Cabinada	-	-	-	-	-	-	-	-	-
Cameroon	-	-	-	-	-	-	-	-	-
Egypt	-	-	-	-	-	-	-	-	-
Gabon	-	-	-	-	-	-	-	-	-
Libya	-	-	-	-	-	-	-	-	-
Nigeria	-	-	-	-	-	-	-	-	-
Tunisia	-	-	-	-	-	-	-	-	-
Other	-	-	-	36	12	-	-	-	-
Africa	**-**	**-**	**-**	**36**	**12**	**-**	**-**	**-**	**-**
Non-spec./Other	**-**	**-**	**-**	**-**	**-**	**-**	**-**	**-**	**-**
Total Imports	**174**	**169**	**36 e**	**512**	**366**	**534 e**	**44**	**30**	**23 e**
Memo: Total OPEC	*-*	*31*	*-*	*118*	*91*	*247 e*	*-*	*-*	*-*

Source: Annual Oil Statistics.

GREECE
Table 8A: Exports by Country of Destination

000 metric tons

	Crude Oil			Natural Gas Liquids			Feedstocks and Other		
	2000	2001	2002	2000	2001	2002	2000	2001	2002
Australia	-	-	-	-	-	-	-	-	-
Austria	-	-	-	-	-	-	-	-	-
Belgium	-	-	-	-	-	-	-	-	-
Canada	-	-	-	-	-	-	-	-	-
Denmark	-	-	-	-	-	-	-	-	-
Finland	-	-	-	-	-	-	-	-	-
France	-	-	-	-	-	-	-	-	-
Germany	-	-	-	-	-	-	-	-	-
Greece	-	-	-	-	-	-	-	-	-
Ireland	-	-	-	-	-	-	-	-	-
Italy	-	-	-	-	-	-	-	-	-
Japan	-	-	-	-	-	-	-	-	-
Luxembourg	-	-	-	-	-	-	-	-	-
Mexico	-	-	-	-	-	-	-	-	-
Netherlands	-	-	-	-	-	-	-	-	-
Norway	-	-	-	-	-	-	-	-	-
Portugal	-	-	-	-	-	-	-	-	-
Spain	-	-	-	-	-	-	-	-	-
Sweden	-	-	-	-	-	-	-	-	-
Switzerland	-	-	-	-	-	-	-	-	-
Turkey	-	-	-	-	-	-	-	-	-
United Kingdom	-	-	-	-	-	-	-	-	-
United States	-	-	-	-	-	-	-	-	-
Other OECD	-	-	-	-	-	-	-	-	-
OECD	**-**	**-**	**-**	**-**	**-**	**-**	**-**	**-**	**-**
Africa	-	-	-	-	-	-	-	-	-
Latin America	-	-	-	-	-	-	-	-	-
Asia	-	-	-	-	-	-	-	-	-
China	-	-	-	-	-	-	-	-	-
Non-OECD Europe	-	-	387 e	-	-	-	-	-	-
Former Soviet Union	-	-	-	-	-	-	-	-	-
Middle East	-	-	-	-	-	-	-	-	-
Non-specified/Other	-	-	-	-	-	-	-	-	-
Non-OECD	**-**	**-**	**387 e**	**-**	**-**	**-**	**-**	**-**	**-**
Total Exports	**-**	**-**	**387 e**	**-**	**-**	**-**	**-**	**-**	**-**

	Total Products			Liquefied Petroleum Gas			Naphtha		
	2000	2001	2002	2000	2001	2002	2000	2001	2002
Australia	-	-	-	-	-	-	-	-	-
Austria	-	-	-	-	-	-	-	-	-
Belgium	119	-	-	-	-	-	90	-	-
Canada	-	-	-	-	-	-	-	-	-
Denmark	-	-	-	-	-	-	-	-	-
Finland	-	-	-	-	-	-	-	-	-
France	39	218	69 e	-	-	-	33	95	33 e
Germany	-	3	3 e	-	-	-	-	-	-
Greece	-	-	-	-	-	-	-	-	-
Ireland	-	-	-	-	-	-	-	-	-
Italy	322	121	101 e				200	91	-
Japan	37	29	-	-	-	-	37	29	-
Luxembourg	-	-	-	-	-	-	-	-	-
Mexico	-	-	-	-	-	-	-	-	-
Netherlands	-	-	63 e	-	-	-	-	-	63 e
Norway	-	-	-	-	-	-	-	-	-
Portugal	-	29	-	-	-	-	-	29	-
Spain	135	269	94 e	-	-	-	56	111	64 e
Sweden	-	-	-	-	-	-	-	-	-
Switzerland	-	-	-	-	-	-	-	-	-
Turkey	558	128	123 e	8	-	5 e	125	9	22 e
United Kingdom	113	-	60 e	-	-	-	-	-	-
United States	550	797	572 e	-	-	-	28	-	3 e
Other OECD	-	-	-	-	-	-	-	-	-
OECD	**1 873**	**1 594**	**1 175 e**	**8**	**-**	**5 e**	**599**	**364**	**185 e**
Africa	347	331	443 e	87	103	64 e	-	-	-
Latin America	10	2	43 e	3	-	-	-	-	28 e
Asia	87	95	405 e	-	-	2 e	-	-	120 e
China	-	-	-	-	-	-	-	-	-
Non-OECD Europe	1 076	1 377	1 491 e	82	82	167 e	181	181	122 e
Former Soviet Union	11	7	12 e	-	-	-	-	-	-
Middle East	540	373	211 e	129	103	41 e	85	70	67 e
Non-specified/Other	-	-	3 e	-	-	-	-	-	-
Non-OECD	**2 071**	**2 185**	**2 608 e**	**301**	**288**	**274 e**	**266**	**251**	**337 e**
Total Exports	**3 944**	**3 779**	**3 783 e**	**309**	**288**	**279 e**	**865**	**615**	**522 e**

Source: Annual Oil Statistics.

GREECE

Table 8B: Exports by Country of Destination

000 metric tons

	Total Gasoline			Total Kerosene			Gas/Diesel Oil		
	2000	2001	2002	2000	2001	2002	2000	2001	2002
Australia	-	-	-	-	-	-	-	-	-
Austria	-	-	-	-	-	-	-	-	-
Belgium	-	-	-	29	-	-	-	-	-
Canada	-	-	-	-	-	-	-	-	-
Denmark	-	-	-	-	-	-	-	-	-
Finland	-	-	-	-	-	-	-	-	-
France	-	-	-	-	33	-	-	83	29 e
Germany	-	-	-	-	-	-	-	-	1 e
Greece	-	-	-	-	-	-	-	-	-
Ireland	-	-	-	-	-	-	-	-	-
Italy	31	25	42 e	-	-	46 e	9	-	100 e
Japan	-	-	-	-	-	-	-	-	-
Luxembourg	-	-	-	-	-	-	-	-	-
Mexico	-	-	-	-	-	-	-	-	-
Netherlands	-	-	-	-	-	-	-	-	-
Norway	-	-	-	-	-	-	-	-	-
Portugal	-	-	-	-	-	-	-	-	-
Spain	-	63	30 e	25	41	-	22	29	-
Sweden	-	-	-	-	-	-	-	-	-
Switzerland	-	-	-	-	-	-	-	-	-
Turkey	214	79	8 e	45	30	64 e	57	6	6 e
United Kingdom	-	-	-	113	-	60 e	-	-	-
United States	84	35	91 e	335	668	238 e	95	94	218 e
Other OECD	-	-	-	-	-	-	-	-	-
OECD	**329**	**202**	**171 e**	**547**	**772**	**408 e**	**183**	**212**	**354 e**
Africa	196	91	191 e	12	25	4 e	9	16	10 e
Latin America	-	-	1 e	-	-	-	-	-	7 e
Asia	-	-	-	-	-	-	20	-	12 e
China	-	-	-	-	-	-	-	-	-
Non-OECD Europe	195	284	412 e	190	167	108 e	359	565	498 e
Former Soviet Union	9	7	12 e	-	-	-	2	-	-
Middle East	282	94	82 e	6	1	-	3	1	10 e
Non-specified/Other	-	-	-	-	-	3 e	-	-	-
Non-OECD	**682**	**476**	**698 e**	**208**	**193**	**115 e**	**393**	**582**	**537 e**
Total Exports	**1 011**	**678**	**869 e**	**755**	**965**	**523 e**	**576**	**794**	**891 e**

	Fuel Oil (Residual)			Petroleum Coke			Other Products		
	2000	2001	2002	2000	2001	2002	2000	2001	2002
Australia	-	-	-	-	-	-	-	-	-
Austria	-	-	-	-	-	-	-	-	-
Belgium	-	-	-	-	-	-	-	-	-
Canada	-	-	-	-	-	-	-	-	-
Denmark	-	-	-	-	-	-	-	-	-
Finland	-	-	-	-	-	-	-	-	-
France	-	-	-	-	-	-	6	7	7 e
Germany	-	-	-	-	-	-	-	3	2 e
Greece	-	-	-	-	-	-	-	-	-
Ireland	-	-	-	-	-	-	-	-	-
Italy	50	-	-	-	-	-	2	5	3 e
Japan	-	-	-	-	-	-	-	-	-
Luxembourg	-	-	-	-	-	-	-	-	-
Mexico	-	-	-	-	-	-	-	-	-
Netherlands	-	-	-	-	-	-	-	-	-
Norway	-	-	-	-	-	-	-	-	-
Portugal	-	-	-	-	-	-	-	-	-
Spain	32	25	-	-	-	-	-	-	-
Sweden	-	-	-	-	-	-	-	-	-
Switzerland	-	-	-	-	-	-	-	-	-
Turkey	42	-	6 e	-	-	-	67	4	12 e
United Kingdom	-	-	-	-	-	-	-	-	-
United States	-	-	-	-	-	-	8	-	22 e
Other OECD	-	-	-	-	-	-	-	-	-
OECD	**124**	**25**	**6 e**	**-**	**-**	**-**	**83**	**19**	**46 e**
Africa	16	80	161 e	-	-	-	27	16	13 e
Latin America	-	-	-	-	-	-	7	2	7 e
Asia	59	71	262 e	-	-	-	8	24	9 e
China	-	-	-	-	-	-	-	-	-
Non-OECD Europe	21	27	135 e	-	-	-	48	71	49 e
Former Soviet Union	-	-	-	-	-	-	-	-	-
Middle East	-	52	-	-	-	-	35	52	11 e
Non-specified/Other	-	-	-	-	-	-	-	-	-
Non-OECD	**96**	**230**	**558 e**	**-**	**-**	**-**	**125**	**165**	**89 e**
Total Exports	**220**	**255**	**564 e**	**-**	**-**	**-**	**208**	**184**	**135 e**

Source: Annual Oil Statistics.

HUNGARY

Figure 1
FUEL SHARES OF TPES *

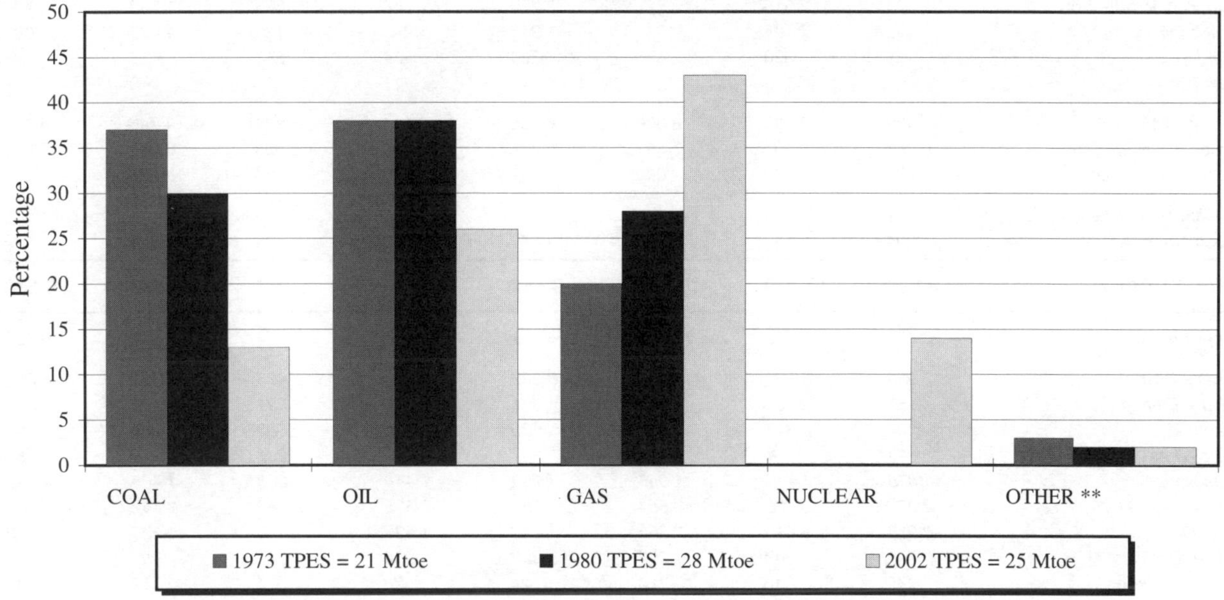

Source: Annual Energy Balances.

* Excluding electricity trade.

** Includes hydro, geothermal, solar, wind, combustible renewables and waste, etc.

Figure 2
OIL PRODUCTION AND CONSUMPTION

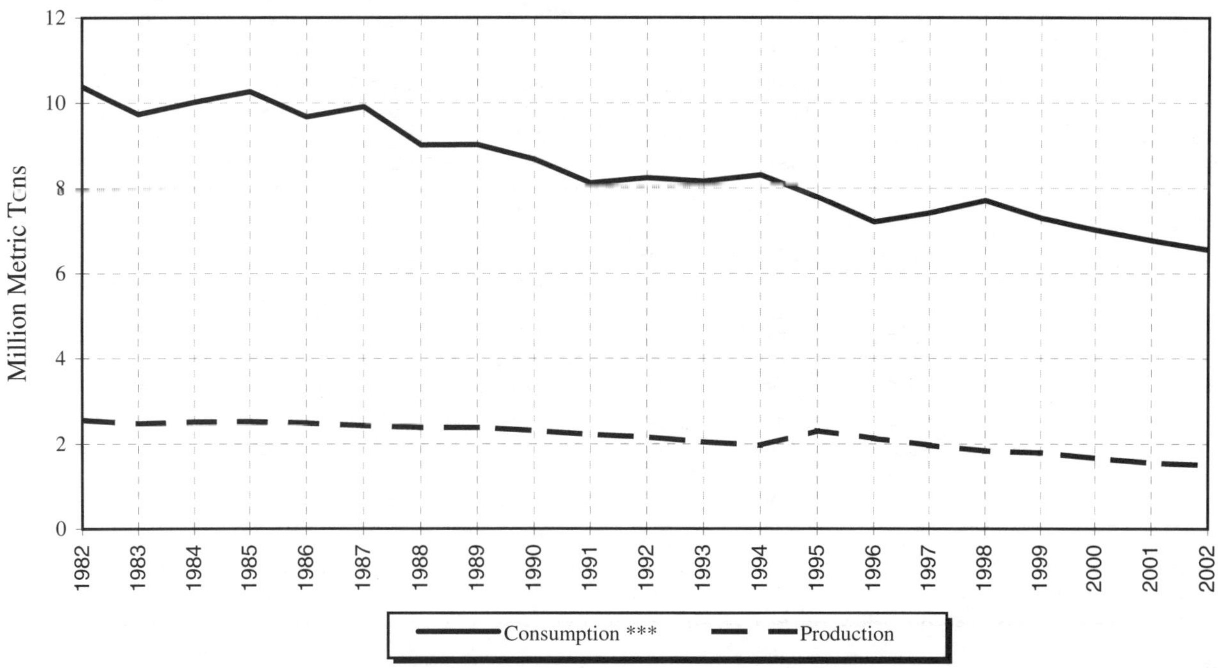

Source: Annual Oil Statistics.

*** Includes refinery fuel and international marine bunkers.

HUNGARY
Table 1: Oil Supply and Consumption - Summary Table

000 metric tons

	1973	1978	1980	1990	1999	2000	2001	2002
Indigenous Production	2 033	2 518	2 517	2 299	1 788	1 663	1 549	1 499 e
Crude Oil	..	2 198	2 031	1 915	1 243	1 136	1 065	991 e
NGL	..	320	486	384	534	513	475	496 e
Other	..	-	-	-	11	14	9	12 e
+ Net Imports	6 489	9 865	8 300	6 555	5 414	5 224	4 746	4 777 e
Crude Oil, NGL, Other	6 064	8 497	7 451	6 418	5 784	5 808	5 634	5 038 e
Petroleum Products	425	1 368	849	137	- 370	- 584	- 888	- 261 e
- Intl. Marine Bunkers	-	-	-	-	-	-	-	-
+ Stock Changes	- 294	- 163	90	- 207	- 45	- 30	324	272 e
TOTAL SUPPLY	**8 228**	**12 220**	**10 907**	**8 647**	**7 157**	**6 857**	**6 619**	**6 548 e**
Refinery Intake	*7 838*	*10 882*	*10 084*	*8 751*	*7 687*	*7 524*	*7 551*	*6 907 e*
Refinery Output	*7 980*	*10 900*	*10 265*	*8 327*	*7 659*	*7 483*	*7 452*	*6 893 e*
LPG + Ethane	*157*	*355*	*317*	*331*	*285*	*291*	*278*	*327 e*
Naphtha	*204*	*817*	*851*	*620*	*734*	*803*	*760*	*695 e*
Kerosenes	-	*1*	*1*	*307*	*196*	*201*	*179*	*182 e*
Gasolines	*1 035*	*1 144*	*1 065*	*1 755*	*1 418*	*1 253*	*1 405*	*1 377 e*
Gas/Diesel Oil	*2 877*	*3 720*	*3 664*	*2 736*	*2 525*	*2 466*	*2 547*	*2 960 e*
Fuel Oil	*2 958*	*3 816*	*3 363*	*1 726*	*1 533*	*1 442*	*1 291*	*419 e*
Other Products	*749 e*	*1 047 e*	*1 004 e*	*852 e*	*968*	*1 027*	*992*	*933 e*
- Refinery Losses	- 142	- 18	- 181	424	28	41	99	14 e
+ Returns	16	41	-	226	156	189	213	189 e
- Statistical Differences	159	- 17	- 57	- 222	-	-	- 38	176 e
TOTAL CONSUMPTION	**8 227**	**12 296**	**11 145**	**8 671**	**7 285**	**7 005**	**6 771**	**6 547 e**
Transformation and Energy	**1 759**	**3 265**	**2 132**	**1 690**	**2 146**	**1 708**	**1 699**	..
Crude Oil	-	3	3	9	62	53	57	..
LPG + Ethane	8	3	8	-	1	1	-	..
Gas/Diesel Oil	100	399	357	180	19	48	105	..
Fuel Oil	1 417	2 584	1 511	932	1 688	1 229	1 106	..
Petroleum Coke	-	-	-	-	75	80	77	..
Other	234	276	253	569	301	297	354	..
Transport	**1 826**	**2 630**	**2 687**	**2 882**	**3 078**	**3 080**	**3 211**	..
Motor Gasoline	918	1 398	1 405	1 790	1 402	1 336	1 391	..
Gas/Diesel Oil	822	1 083	1 125	930	1 470	1 516	1 604	..
Subtotal	*1 740*	*2 481*	*2 530*	*2 720*	*2 872*	*2 852*	*2 995*	..
Aviation Fuels	50	109	119	160	204	228	214	..
Other	36	40	38	2	2	-	2	..
Industry	**2 288**	**3 269**	**3 299**	**1 957**	**1 349**	**1 523**	**1 249**	..
Naphtha	137	748	822	395	507	572	515	..
LPG + Ethane	4	4	5	6	19	23	26	..
Gas/Diesel Oil	284	186	188	410	59	164	107	..
Fuel Oil	1 315	1 558	1 588	745	322	242	181	..
Other	548	773	696	401	442	522	420	..
Other Sectors	**2 354**	**3 132**	**3 027**	**2 142**	**712**	**694**	**612**	..
Gas/Diesel Oil	2 068	2 806	2 675	1 713	350	332	287	..
Fuel Oil	40	7	13	34	47	48	55	..
Other	246	319	339	395	315	314	270	..
Memo: Non-Energy Use	*705*	*1 537*	*1 534*	*1 030*	*1 004*	*1 247*	*1 031*	..
Chem/Petchem. Feedst.	*294*	*970*	*1 016*	*765*	*771*	*1 012*	*780*	..
Other	*411*	*567*	*518*	*265*	*233*	*235*	*251*	..

Source: Annual Oil Statistics.

HUNGARY
Table 2: Supply of Crude Oil, NGL, Feedstocks and Other Hydrocarbons

000 metric tons

	1973	1978	1980	1990	1999	2000	2001	2002
CRUDE OIL								
Indigenous Production	..	2 198	2 031	1 915	1 243	1 136	1 065	991 e
+ Imports	..	9 960	8 336	6 418	5 780	5 800	5 622	5 029 e
- Exports	..	1 463	885	-	-	-	-	-
- Direct Use	..	3	3	9	62	53	57	48 e
+ Stock Change	..	- 192	57	- 183	21	- 82	212	45 e
- Statistical Difference	..	-	-	-	-	-	-	-
Refinery Intake	**..**	**10 500**	**9 536**	**8 141**	**6 982**	**6 801**	**6 842**	**6 017 e**
Closing Stock Level	..	-	-	-	*796*	*878*	*666*	*621 e*
NATURAL GAS LIQUIDS								
Indigenous Production	..	320	486	384	534	513	475	496 e
+ Imports	..	-	-	-	-	-	-	-
- Exports	..	-	-	-	-	-	-	-
- Direct Use	..	-	-	-	-	-	-	-
+ Stock Change	..	-	-	-	-	-	-	-
- Statistical Difference	..	-	-	-	-	-	-	-
Refinery Intake	**..**	**320**	**486**	**384**	**534**	**513**	**475**	**496 e**
Closing Stock Level	..	-	-	-	-	-	-	-
REFINERY FEEDSTOCKS								
Backflows	..	41	-	226	156	189	213	189 e
+ Transfers	..	-	-	-	-	-	-	186 e
+ Imports	..	-	-	-	-	-	-	-
- Exports	..	-	-	-	-	-	-	-
+ Stock Change	..	-	-	-	-	-	-	-
- Statistical Difference	..	- 21	- 62	-	-	-	-	-
Refinery Intake	**..**	**62**	**62**	**226**	**156**	**189**	**213**	**375 e**
Closing Stock Level	..	-	-	-	-	-	-	-
OTHER HYDROCARBONS								
Indigenous Production	..	-	-	-	11	14	9	12 e
+ Imports	..	-	-	-	6	9	12	9 e
- Exports	..	-	-	-	2	1	-	-
- Direct Use	..	-	-	-	-	-	-	-
+ Stock Change	..	-	-	-	-	- 1	-	- 2 e
- Statistical Difference	..	-	-	-	-	-	-	-
Refinery Intake	**..**	**-**	**-**	**-**	**15**	**21**	**21**	**19 e**
Closing Stock Level	..	-	-	-	-	*1*	*1*	*3 e*
CRUDE OIL, NGL, FEEDSTOCKS AND OTHER								
Indigenous Production	2 033	2 518	2 517	2 299	1 788	1 663	1 549	1 499 e
+ Backflows	16	41	-	226	156	189	213	189 e
+ Transfers	-	-	-	-	-	-	-	186 e
+ Imports	6 555	9 960	8 336	6 418	5 786	5 809	5 634	5 038 e
- Exports	491	1 463	885	-	2	1	-	-
- Direct Use	-	3	3	9	62	53	57	48 e
+ Stock Change	- 120	- 192	57	- 183	21	- 83	212	43 e
- Statistical Difference	155	- 21	- 62	-	-	-	-	-
Refinery Intake	**7 838**	**10 882**	**10 084**	**8 751**	**7 687**	**7 524**	**7 551**	**6 907 e**
Closing Stock Level	-	-	-	-	*796*	*879*	*667*	*624 e*

Source: Annual Oil Statistics.

HUNGARY
Table 3: Motor Gasoline - Supply

000 metric tons

	1973	1978	1980	1990	1999	2000	2001	2002
MOTOR GASOLINE								
Refinery Output	**1 035**	**1 144**	**1 065**	**1 755**	**1 416**	**1 249**	**1 401**	**1 372 e**
+ Imports	46	433	604	475	209	178	244	299 e
- Exports	154	128	214	416	415	371 e	513	465 e
+ Transfers	-	-	-	-	183	203	193	189 e
+ Stock Change	- 9	- 51	- 50	- 24	9	77 e	66	38 e
- Statistical Difference	-	-	-	-	-	-	-	- 1 e
Total Consumption	**918**	**1 398**	**1 405**	**1 790**	**1 402**	**1 336**	**1 391**	**1 434 e**
Closing Stock Level	-	-	-	-	*678*	*601 e*	*535*	*497 e*

Table 4: Aviation Fuels - Supply

000 metric tons

	1973	1978	1980	1990	1999	2000	2001	2002
AVIATION GASOLINE								
Refinery Output	**-**	**-**	**-**	**-**	**2**	**4**	**4**	**5 e**
+ Imports	-	-	-	-	-	-	-	-
- Exports	-	-	-	-	-	-	2	-
+ Transfers	-	-	-	-	-	-	-	-
+ Stock Change	-	-	-	-	1	-	-	- 1 e
- Statistical Difference	-	-	-	-	-	-	-	-
Total Consumption	**-**	**-**	**-**	**-**	**3**	**4**	**2**	**4 e**
Closing Stock Level	-	-	-	-	-	-	-	*1 e*
GASOLINE TYPE JET FUEL								
Refinery Output	**-**	**-**	**-**	**-**	**-**	**-**	**-**	**-**
+ Imports	-	-	-	-	-	-	-	-
- Exports	-	-	-	-	-	-	-	-
+ Transfers	-	-	-	-	-	-	-	-
+ Stock Change	-	-	-	-	-	-	-	-
- Statistical Difference	-	-	-	-	-	-	-	-
Total Consumption	**-**	**-**	**-**	**-**	**-**	**-**	**-**	**-**
Closing Stock Level	-	-	-	-	-	-	-	-
KEROSENE TYPE JET FUEL								
Refinery Output	**-**	**-**	**-**	**87**	**194**	**198**	**176**	**179 e**
+ Imports	50	109	119	73	48	40	50	53 e
- Exports	-	-	-	-	39	14	11	21 e
+ Transfers	-	-	-	-	-	-	-	-
+ Stock Change	-	-	-	-	- 2	-	- 3	2 e
- Statistical Difference	-	-	-	-	-	-	-	-
Total Consumption	**50**	**109**	**119**	**160**	**201**	**224**	**212**	**213 e**
Closing Stock Level	-	-	-	-	*2*	*2*	*5*	*3 e*

Source: Annual Oil Statistics.

HUNGARY

Table 5A: Gas/Diesel Oil - Supply and Consumption

000 metric tons

	1973	1978	1980	1990	1999	2000	2001	2002
Refinery Output	**2 877**	**3 720**	**3 664**	**2 736**	**2 525**	**2 466**	**2 547**	**2 960 e**
+ Recycled Products	-	-	-	-	-	-	-	-
+ Imports	513	789	837	825	287	352	530	997 e
- Exports	11	76	84	589	904	788	957	1 293 e
- Intl. Marine Bunkers	-	-	-	-	-	-	-	-
+ Transfers	-	-	-	-	-	-	-	- 116 e
+ Stock Changes	- 101	45	- 67	40	- 10	30	- 17	- 15 e
- Statistical Differences	4	4	5	- 221	-	-	-	172 e
Total Consumption	**3 274**	**4 474**	**4 345**	**3 233**	**1 898**	**2 060**	**2 103**	**2 361 e**
Transformation	**100**	**398**	**357**	**164**	**19**	**48**	**105**	..
Public Electricity	2	42	14	5	1	-	57	..
Autoproducer Electricity	-	-	-	-	-	-	1	..
Public CHP Plants	1	3	1	1	-	-	6	..
Autoproducers of CHP	-	2	1	1	-	-	-	..
Public Heat Plants	4	31	24	21	-	-	4	..
Autoproducer Heat Plants	92	319	317	135	-	-	-	..
Petrochemical Industry	1	1	-	1	18	48	37	..
Other Transformation	-	-	-	-	-	-	-	..
Energy Sector	**-**	**1**	**-**	**16**	**-**	**-**	**-**	..
Refinery Fuel	-	1	-	15	-	-	-	..
Other Energy Sector	-	-	-	1	-	-	-	..
Final Consumption	**3 174**	**4 075**	**3 988**	**3 053**	**1 879**	**2 012**	**1 998**	..
Transport	**822**	**1 083**	**1 125**	**930**	**1 470**	**1 516**	**1 604**	..
Air	-	-	-	-	-	-	-	..
Road	579	807	837	755	1 366	1 434	1 530	..
Rail	201	238	249	166	104	82	73	..
Inland Waterways	42	38	39	9	-	-	1	..
Non-Specified	-	-	-	-	-	-	-	..
Industry	**284**	**186**	**188**	**410**	**59**	**164**	**107**	..
Iron and Steel	12	6	6	8	-	1	1	..
Chemical (incl. Pet.-Chem.)	20	14	14	232	47	145	83	..
Non-Ferrous Metals	-	3	3	1	-	2	1	..
Non-Metallic Minerals	34	20	23	21	-	3	6	..
Transport Equipment	-	-	-	-	-	1	1	..
Machinery	57	30	30	34	-	1	-	..
Mining and Quarrying	10	-	-	-	-	5	3	..
Food and Tobacco	66	38	39	43	5	1	2	..
Pulp, Paper and Printing	-	-	-	3	1	-	-	..
Wood and Wood Products	-	-	-	-	-	-	-	..
Construction	69	60	57	56	-	4	9	..
Textiles and Leather	16	11	12	11	1	-	-	..
Non-Specified	-	4	4	1	5	1	1	..
Other Sectors	**2 068**	**2 806**	**2 675**	**1 713**	**350**	**332**	**287**	..
Commerce and Public	100	322	340	178	-	-	-	..
Residential	778	952	944	837	27	26	40	..
Agriculture	771	1 185	1 093	640	323	306	247	..
Non-Specified	419	347	298	58	-	-	-	..
Memo: Non-Energy Use	*20*	*14*	*14*	*232*	*47*	*145*	*83*	..
Chem/Petchem. Feedst.	*19*	*13*	*14*	*231*	*47*	*145*	*83*	..
Other	*1*	*1*	-	*1*	-	-	-	..
Closing Stock Level	-	-	-	-	*868*	*838*	*855*	*870 e*

Source: Annual Oil Statistics.

HUNGARY
Table 5B: Fuel Oil (Residual) - Supply and Consumption

000 metric tons

	1973	1978	1980	1990	1999	2000	2001	2002
Refinery Output	**2 958**	**3 816**	**3 363**	**1 726**	**1 533**	**1 442**	**1 291**	**419 e**
+ Recycled Products	-	-	-	-	-	-	-	-
+ Imports	120	345	67	187	618	247	158	488 e
- Exports	217	3	420	153	9	172	242	49 e
- Intl. Marine Bunkers	-	-	-	-	-	-	-	-
+ Transfers	-	-	-	-	-	-	-	- 86 e
+ Stock Changes	- 53	31	140	- 48	- 85	2	97	168 e
- Statistical Differences	-	-	-	-	-	-	- 38	84 e
Total Consumption	**2 808**	**4 189**	**3 150**	**1 712**	**2 057**	**1 519**	**1 342**	**856 e**
Transformation	**1 122**	**2 221**	**1 136**	**570**	**1 433**	**1 123**	**1 032**	**..**
Public Electricity	667	1 576	690	234	1 119	962	904	..
Autoproducer Electricity	1	4	14	2	2	-	-	..
Public CHP Plants	325	383	274	139	187	126	104	..
Autoproducers of CHP	65	97	84	78	50	13	12	..
Public Heat Plants	35	97	70	69	12	17	12	..
Autoproducer Heat Plants	29	64	4	48	63	5	-	..
Petrochemical Industry	-	-	-	-	-	-	-	..
Other Transformation	-	-	-	-	-	-	-	..
Energy Sector	**295**	**363**	**375**	**362**	**255**	**106**	**74**	**..**
Refinery Fuel	289	357	371	356	255	106	74	..
Other Energy Sector	6	6	4	6	-	-	-	..
Final Consumption	**1 391**	**1 605**	**1 639**	**780**	**369**	**290**	**236**	**..**
Transport	**36**	**40**	**38**	**1**	**-**	**-**	**-**	**..**
Air	-	-	-	-	-	-	-	..
Road	4	6	5	-	-	-	-	..
Rail	28	16	11	-	-	-	-	..
Inland Waterways	4	18	22	1	-	-	-	..
Non-Specified	-	-	-	-	-	-	-	..
Industry	**1 315**	**1 558**	**1 588**	**745**	**322**	**242**	**181**	**..**
Iron and Steel	239	245	254	126	43	43	-	..
Chemical (incl. Pet.-Chem.)	171	225	253	122	66	42	38	..
Non-Ferrous Metals	119	107	102	46	1	1	1	..
Non-Metallic Minerals	245	283	295	135	145	85	64	..
Transport Equipment	58	71	70	29	-	-	-	..
Machinery	91	107	103	38	8	5	4	..
Mining and Quarrying	40	40	41	29	1	1	-	..
Food and Tobacco	188	274	265	148	37	55	61	..
Pulp, Paper and Printing	14	17	16	6	14	8	10	..
Wood and Wood Products	23	29	27	11	-	-	-	..
Construction	36	49	50	12	-	-	-	..
Textiles and Leather	86	107	102	40	7	2	2	..
Non-Specified	5	4	10	3	-	-	1	..
Other Sectors	**40**	**7**	**13**	**34**	**47**	**48**	**55**	**..**
Commerce and Public	-	-	-	-	14	16	27	..
Residential	-	-	-	-	-	-	-	..
Agriculture	-	-	9	11	17	17	21	..
Non-Specified	40	7	4	23	16	15	7	..
Memo: Non-Energy Use	-	*2*	*1*	-	-	-	-	..
Chem/Petchem. Feedst.	-	*2*	*1*	-	-	-	-	..
Other	-	-	-	-	-	-	-	..
Closing Stock Level	-	-	-	-	*827*	*825*	*728*	*560 e*

Source: Annual Oil Statistics.

HUNGARY
Table 5C: Total Products - Supply and Consumption

000 metric tons

	1973	1978	1980	1990	1999	2000	2001	2002
Refinery Output	**7 980**	**10 900**	**10 265**	**8 327**	**7 659**	**7 483**	**7 452**	**6 893 e**
+ Recycled Products	-	-	-	-	-	-	-	-
+ Imports	852	1 773	1 739	1 612	1 517	1 170	1 372	2 129 e
- Exports	427	405	890	1 475	1 887	1 754	2 260	2 390 e
- Intl. Marine Bunkers	-	-	-	-	-	-	-	-
+ Transfers	-	3	3	9	62	53	57	- 138 e
+ Stock Changes	- 174	29	33	- 24	- 66	53	112	229 e
- Statistical Differences	4	4	5	- 222	-	-	- 38	176 e
Total Consumption	**8 227**	**12 296**	**11 145**	**8 671**	**7 285**	**7 005**	**6 771**	**6 547 e**
Transformation	**1 403**	**2 798**	**1 641**	**1 118**	**1 613**	**1 331**	**1 326**	**..**
Public Electricity	669	1 618	704	239	1 120	962	961	..
Autoproducer Electricity	1	4	14	2	2	-	1	..
Public CHP Plants	326	386	275	140	187	126	110	..
Autoproducers of CHP	65	99	85	79	50	13	12	..
Public Heat Plants	39	128	94	90	13	17	16	..
Autoproducer Heat Plants	121	383	321	183	63	5	-	..
Petrochemical Industry	16	41	-	226	156	189	213	..
Other Transformation	166	139	148	159	22	19	13	..
Energy Sector	**356**	**467**	**491**	**572**	**533**	**377**	**373**	**..**
Refinery Fuel	350	458	484	556	471	323	316	..
Other Energy Sector	6	9	7	16	62	54	57	..
Final Consumption	**6 468**	**9 031**	**9 013**	**6 981**	**5 139**	**5 297**	**5 072**	**..**
Transport	**1 826**	**2 630**	**2 687**	**2 882**	**3 078**	**3 080**	**3 211**	**..**
Air	50	109	119	160	204	228	214	..
Road	1 425	2 158	2 196	2 484	2 770	2 770	2 923	..
Rail	229	254	260	167	104	82	73	..
Inland Waterways	122	109	112	71	-	-	1	..
Non-Specified	-	-	-	-	-	-	-	..
Industry	**2 288**	**3 269**	**3 299**	**1 957**	**1 349**	**1 523**	**1 249**	**..**
Iron and Steel	251	251	260	134	43	44	1	..
Chemical (incl. Pet.-Chem.)	466	1 194	1 268	888	837	1 054	819	..
Non-Ferrous Metals	119	110	105	47	2	3	2	..
Non-Metallic Minerals	280	304	319	157	148	97	72	..
Transport Equipment	58	71	70	29	-	1	1	..
Machinery	149	138	134	73	9	7	5	..
Mining and Quarrying	50	41	42	30	1	6	3	..
Food and Tobacco	255	313	305	192	47	60	68	..
Pulp, Paper and Printing	14	17	16	9	15	8	11	..
Wood and Wood Products	23	29	27	11	-	-	-	..
Construction	476	614	555	303	217	226	249	..
Textiles and Leather	102	118	114	51	8	2	2	..
Non-Specified	45	69	84	33	22	15	16	..
Other Sectors	**2 354**	**3 132**	**3 027**	**2 142**	**712**	**694**	**612**	**..**
Commerce and Public	164	406	435	277	64	65	66	..
Residential	958	1 183	1 186	1 125	282	283	261	..
Agriculture	772	1 188	1 104	658	350	330	277	..
Non-Specified	460	355	302	82	16	16	8	..
Memo: Non-Energy Use	*705*	*1 537*	*1 534*	*1 030*	*1 004*	*1 247*	*1 031*	..
Chem/Petchem. Feedst.	*294*	*970*	*1 016*	*765*	*771*	*1 012*	*780*	..
Other	*411*	*567*	*518*	*265*	*233*	*235*	*251*	..
Closing Stock Level	-	-	-	-	*2 434*	*2 381*	*2 269*	*2 040 e*

Source: Annual Oil Statistics.

HUNGARY
Table 6A: Supply and Consumption

000 metric tons

	1980	2000	2001	2002	1980	2000	2001	2002
	LPG AND ETHANE				**REFINERY GAS**			
Refinery Output	**317**	**291**	**278**	**327 e**	**113 e**	**149**	**172**	**193 e**
+ Recycled Products	-	-	-	-	-	-	-	-
+ Imports	19	124	103	63 e	-	-	-	-
- Exports	72	73	82	110 e	-	-	-	-
- Intl. Marine Bunkers	-	-	-	-	-	-	-	-
+ Transfers	-	-	-	-	-	-	-	-
+ Stock Change	13	- 7	- 3	2 e	-	-	-	-
- Statistical Difference	-	-	-	-	-	-	-	-
Total Consumption	**277**	**335**	**296**	**282 e**	**113 e**	**149**	**172**	**193 e**
of which: Ethane	-	-	-	-	-	-	-	-
Transformation	8	-	-	..	-	-	-	..
Energy	-	1	-	..	113 e	149	172	..
Transport	-	-	2	..	-	-	-	..
Industry	5	23	26	..	-	-	-	..
Other	264	311	268	..	-	-	-	..
Memo: Non-Energy Use	*1*	*8*	*14*	..	-	-	-	..
Closing Stock Level	-	*29*	*32*	*30 e*	-	-	-	-
	NAPHTHA				**OTHER KEROSENE**			
Refinery Output	**851**	**803**	**760**	**695 e**	**1**	**3**	**3**	**3 e**
+ Recycled Products	-	-	-	-	-	-	-	-
+ Imports	-	132	142	125 e	93	-	-	-
- Exports	-	8	-	-	-	-	-	-
- Intl. Marine Bunkers	-	-	-	-	-	-	-	-
+ Transfers	-	- 203	- 193	- 173 e	-	-	-	-
+ Stock Change	-	- 11	- 18	17 e	- 14	-	-	-
- Statistical Difference	-	-	-	-	-	-	-	-
Total Consumption	**851**	**713**	**691**	**664 e**	**80**	**3**	**3**	**3 e**
Transformation	29	141	176	..	-	-	-	..
Energy	-	-	-	..	-	-	-	..
Transport	-	-	-	..	-	-	-	..
Industry	822	572	515	..	5	-	1	..
Other	-	-	-	..	75	3	2	..
Memo: Non-Energy Use	*822*	*572*	*515*	..	*5*	-	-	..
Closing Stock Level	-	*11*	*29*	*12 e*	-	*30*	*30*	*30 e*
	WHITE SPIRIT + SBP				**LUBRICANTS**			
Refinery Output	**-**	**134**	**196**	**..**	**173 e**	**291**	**211**	**..**
+ Recycled Products	-	-	-	..	-	-	-	..
+ Imports	-	-	-	..	-	75	98	..
- Exports	-	55	167	..	-	119	158	..
- Intl. Marine Bunkers	-	-	-	..	-	-	-	..
+ Transfers	-	-	-	..	-	-	-	..
+ Stock Change	-	- 39	- 2	..	-	-	- 10	..
- Statistical Difference	-	-	-	..	-	-	-	..
Total Consumption	**-**	**40**	**27**	**..**	**173**	**247**	**141**	**..**
Transformation	-	-	-	..	-	-	-	..
Energy	-	-	-	..	-	-	-	..
Transport	-	-	-	..	-	-	-	..
Industry	-	40	27	..	173	247	141	..
Other	-	-	-	..	-	-	-	..
Memo: Non-Energy Use	-	*40*	*27*	..	*173*	*247*	*141*	..
Closing Stock Level	-	*39*	*41*	..	-	-	*10*	..

Source: Annual Oil Statistics.

HUNGARY
Table 6B: Supply and Consumption

000 metric tons

	1980	2000	2001	2002	1980	2000	2001	2002
	BITUMEN				PARAFFIN WAXES			
Refinery Output	**648**	**343**	**303**	**..**	**70 e**	**42**	**40**	**..**
+ Recycled Products	-	-	-	..	-	-	-	..
+ Imports	-	10	40	..	-	-	-	..
- Exports	100	126	103	..	-	28	25	..
- Intl. Marine Bunkers	-	-	-	..	-	-	-	..
+ Transfers	-	-	-	..	-	-	-	..
+ Stock Change	11	1	3	..	-	-	- 1	..
- Statistical Difference	-	-	-	..	-	-	-	..
Total Consumption	**559**	**228**	**243**	**..**	**70**	**14**	**14**	**..**
Transformation	111	7	6	..	-	-	-	..
Energy	-	-	-	..	-	-	-	..
Transport	-	-	-	..	-	-	-	..
Industry	448	221	237	..	70	14	14	..
Other	-	-	-	..	-	-	-	..
Memo: Non-Energy Use	*448*	*221*	*237*	..	*70*	*14*	*14*	..
Closing Stock Level	-	*6*	*3*	..	-	-	*1*	..
	PETROLEUM COKE				OTHER PRODUCTS			
Refinery Output	**-**	**68**	**70**	**69 e**	**-**	**-**	**-**	**671 e**
+ Recycled Products	-	-	-	-	-	-	-	-
+ Imports	-	12	7	10 e	-	-	-	94 e
- Exports	-	-	-	-	-	-	-	452 e
- Intl. Marine Bunkers	-	-	-	-	-	-	-	-
+ Transfers	-	-	-	-	3	53	57	48 e
+ Stock Change	-	-	-	- 2 e	-	-	-	20 e
- Statistical Difference	-	-	-	-	-	-	-	- 79 e
Total Consumption	**-**	**80**	**77**	**77 e**	**3**	**53**	**57**	**460 e**
Transformation	-	12	7	..	-	-	-	..
Energy	-	68	70	..	3	53	57	..
Transport	-	-	-	..	-	-	-	..
Industry	-	-	-	..	-	-	-	..
Other	-	-	-	..	-	-	-	..
Memo: Non-Energy Use	-	-	-	..	-	-	-	..
Closing Stock Level	-	-	-	*2 e*	-	-	-	*35 e*

Source: Annual Oil Statistics.

HUNGARY
Table 7A: Imports by Country of Origin

000 metric tons

	2000	2001	2002	2000	2001	2002	2000	2001	2002
	Crude Oil			Natural Gas Liquids			Feedstocks and Other		
Australia	-	-	-	-	-	-	-	-	-
Belgium	-	-	-	-	-	-	-	-	-
Canada	-	-	-	-	-	-	-	-	-
Czech Republic	-	-	-	-	-	-	-	-	1 e
Denmark	-	-	-	-	-	-	-	-	-
Finland	-	-	-	-	-	-	-	-	-
France	-	-	-	-	-	-	-	-	-
Germany	-	-	-	-	-	-	-	-	-
Greece	-	-	-	-	-	-	-	-	-
Hungary	-	-	-	-	-	-	-	-	-
Ireland	-	-	-	-	-	-	-	-	-
Italy	-	-	-	-	-	-	-	-	-
Japan	-	-	-	-	-	-	-	-	-
Korea	-	-	-	-	-	-	-	-	-
Mexico	-	-	-	-	-	-	-	-	-
Netherlands	-	-	-	-	-	-	-	-	-
New Zealand	-	-	-	-	-	-	-	-	-
Norway	-	-	-	-	-	-	-	-	-
Poland	-	-	-	-	-	-	-	-	-
Spain	-	-	-	-	-	-	-	-	-
Sweden	-	-	-	-	-	-	-	-	-
Turkey	-	-	-	-	-	-	-	-	-
United Kingdom	-	-	-	-	-	-	-	-	-
United States	-	-	-	-	-	-	-	-	-
Other	-	-	-	-	-	-	9	12	-
OECD	**-**	**-**	**-**	**-**	**-**	**-**	**9**	**12**	**1 e**
Argentina	-	-	-	-	-	-	-	-	-
Brazil	-	-	-	-	-	-	-	-	-
Colombia	-	-	-	-	-	-	-	-	-
Ecuador	-	-	-	-	-	-	-	-	-
Netherlands Antilles	-	-	-	-	-	-	-	-	-
Peru	-	-	-	-	-	-	-	-	-
Trinidad and Tobago	-	-	-	-	-	-	-	-	-
Venezuela	-	-	-	-	-	-	-	-	-
Other	-	-	-	-	-	-	-	-	-
Latin America	**-**	**-**	**-**	**-**	**-**	**-**	**-**	**-**	**-**
Romania	-	-	-	-	-	-	-	-	-
Other	-	-	-	-	-	-	-	-	-
Non-OECD Europe	**-**	**-**	**-**	**-**	**-**	**-**	**-**	**-**	**-**
FSU	**5 800**	**5 622**	**5 029 e**	**-**	**-**	**-**	**-**	**-**	**8 e**
Iran	-	-	-	-	-	-	-	-	-
Iraq	-	-	-	-	-	-	-	-	-
Kuwait	-	-	-	-	-	-	-	-	-
Oman	-	-	-	-	-	-	-	-	-
Qatar	-	-	-	-	-	-	-	-	-
Saudi Arabia	-	-	-	-	-	-	-	-	-
United Arab Emirates	-	-	-	-	-	-	-	-	-
Other	-	-	-	-	-	-	-	-	-
Middle East	**-**	**-**	**-**	**-**	**-**	**-**	**-**	**-**	**-**
India	-	-	-	-	-	-	-	-	-
Indonesia	-	-	-	-	-	-	-	-	-
Singapore	-	-	-	-	-	-	-	-	-
Other	-	-	-	-	-	-	-	-	-
Asia	**-**	**-**	**-**	**-**	**-**	**-**	**-**	**-**	**-**
China	**-**	**-**	**-**	**-**	**-**	**-**	**-**	**-**	**-**
Algeria	-	-	-	-	-	-	-	-	-
Angola/Cabinada	-	-	-	-	-	-	-	-	-
Cameroon	-	-	-	-	-	-	-	-	-
Egypt	-	-	-	-	-	-	-	-	-
Gabon	-	-	-	-	-	-	-	-	-
Libya	-	-	-	-	-	-	-	-	-
Nigeria	-	-	-	-	-	-	-	-	-
Tunisia	-	-	-	-	-	-	-	-	-
Other	-	-	-	-	-	-	-	-	-
Africa	**-**	**-**	**-**	**-**	**-**	**-**	**-**	**-**	**-**
Non-spec./Other	**-**	**-**	**-**	**-**	**-**	**-**	**-**	**-**	**-**
Total Imports	**5 800**	**5 622**	**5 029 e**	**-**	**-**	**-**	**9**	**12**	**9 e**
Memo: Total OPEC	-	-	-	-	-	-	-	-	-

Source: Annual Oil Statistics.

HUNGARY
Table 7B: Imports by Country of Origin

000 metric tons

	2000	2001	2002	2000	2001	2002	2000	2001	2002
	Total Products			**Liquefied Petroleum Gases**			**Naphtha**		
Australia	-	-	-	-	-	-	-	-	-
Belgium	3	5	1 e	-	-	-	-	-	-
Canada	-	-	-	-	-	-	-	-	-
Czech Republic	66	116	86 e	2	3	9 e	5	24	8 e
Denmark	-	-	-	-	-	-	-	-	-
Finland	-	-	-	-	-	-	-	-	-
France	7	7	-	1	-	-	-	-	-
Germany	31	45	18 e	2	3	-	7	8	-
Greece	-	-	-	-	-	-	-	-	-
Hungary	-	-	-	-	-	-	-	-	-
Ireland	-	-	-	-	-	-	-	-	-
Italy	-	-	1 e	-	-	-	-	-	-
Japan	-	-	-	-	-	-	-	-	-
Korea	-	-	-	-	-	-	-	-	-
Mexico	-	-	-	-	-	-	-	-	-
Netherlands	-	2	2 e	-	-	-	-	-	-
New Zealand	-	-	-	-	-	-	-	-	-
Norway	6	-	1 e	5	-	1 e	-	-	-
Poland	6	3	4 e	2	-	-	-	-	-
Spain	-	1	-	-	-	-	-	-	-
Sweden	-	-	-	-	-	-	-	-	-
Turkey	-	-	-	-	-	-	-	-	-
United Kingdom	-	-	-	-	-	-	-	-	-
United States	-	-	-	-	-	-	-	-	-
Other	657	616	830 e	3	1	1 e	-	-	-
OECD	**776**	**795**	**943 e**	**15**	**7**	**11 e**	**12**	**32**	**8 e**
Argentina	-	-	-	-	-	-	-	-	-
Brazil	-	-	-	-	-	-	-	-	-
Colombia	-	-	-	-	-	-	-	-	-
Ecuador	-	-	-	-	-	-	-	-	-
Netherlands Antilles	3	-	-	1	-	-	-	-	-
Peru	-	-	-	-	-	-	-	-	-
Trinidad and Tobago	-	-	-	-	-	-	-	-	-
Venezuela	1	-	-	1	-	-	-	-	-
Other	-	-	12 e	-	-	-	-	-	-
Latin America	**4**	**-**	**12 e**	**2**	**-**	**-**	**-**	**-**	**-**
Romania	11	12	58 e	-	-	-	-	-	28 e
Other	110	132	148 e	21	29	15 e	77	83	78 e
Non-OECD Europe	**121**	**144**	**206 e**	**21**	**29**	**15 e**	**77**	**83**	**106 e**
FSU	**163**	**382**	**907 e**	**20**	**67**	**37 e**	**43**	**27**	**11 e**
Iran	-	-	-	-	-	-	-	-	-
Iraq	-	-	-	-	-	-	-	-	-
Kuwait	-	-	-	-	-	-	-	-	-
Oman	-	-	-	-	-	-	-	-	-
Qatar	-	-	-	-	-	-	-	-	-
Saudi Arabia	-	-	-	-	-	-	-	-	-
United Arab Emirates	-	-	-	-	-	-	-	-	-
Other	-	-	-	-	-	-	-	-	-
Middle East	**-**	**-**	**-**	**-**	**-**	**-**	**-**	**-**	**-**
India	-	-	-	-	-	-	-	-	-
Indonesia	-	-	-	-	-	-	-	-	-
Singapore	-	-	-	-	-	-	-	-	-
Other	-	-	-	-	-	-	-	-	-
Asia	**-**	**-**	**-**	**-**	**-**	**-**	**-**	**-**	**-**
China	**-**	**-**	**-**	**-**	**-**	**-**	**-**	**-**	**-**
Algeria	-	-	-	-	-	-	-	-	-
Angola/Cabinada	-	-	-	-	-	-	-	-	-
Cameroon	-	-	-	-	-	-	-	-	-
Egypt	-	-	-	-	-	-	-	-	-
Gabon	-	-	-	-	-	-	-	-	-
Libya	-	-	-	-	-	-	-	-	-
Nigeria	-	-	-	-	-	-	-	-	-
Tunisia	-	-	-	-	-	-	-	-	-
Other	-	-	-	-	-	-	-	-	-
Africa	**-**	**-**	**-**	**-**	**-**	**-**	**-**	**-**	**-**
Non-spec./Other	**106**	**51**	**61 e**	**66**	**-**	**-**	**-**	**-**	**-**
Total Imports	**1 170**	**1 372**	**2 129 e**	**124**	**103**	**63 e**	**132**	**142**	**125 e**
Memo: Total OPEC	*1*	-	-	*1*	-	-	-	-	-

Source: Annual Oil Statistics.

HUNGARY
Table 7C: Imports by Country of Origin

000 metric tons

	2000	2001	2002	2000	2001	2002	2000	2001	2002
	Total Gasoline			Total Kerosene			Gas/Diesel		
Australia	-	-	-	-	-	-	-	-	-
Belgium	-	-	-	-	-	-	-	-	1 e
Canada	-	-	-	-	-	-	-	-	-
Czech Republic	-	12	18 e	-	-	-	14	36	32 e
Denmark	-	-	-	-	-	-	-	-	-
Finland	-	-	-	-	-	-	-	-	-
France	-	-	-	-	-	-	-	-	-
Germany	2	17	15 e	-	-	-	4	-	-
Greece	-	-	-	-	-	-	-	-	-
Hungary	-	-	-	-	-	-	-	-	-
Ireland	-	-	-	-	-	-	-	-	-
Italy	-	-	-	-	-	-	-	-	-
Japan	-	-	-	-	-	-	-	-	-
Korea	-	-	-	-	-	-	-	-	-
Mexico	-	-	-	-	-	-	-	-	-
Netherlands	-	-	-	-	-	-	-	-	2 e
New Zealand	-	-	-	-	-	-	-	-	-
Norway	-	-	-	-	-	-	-	-	-
Poland	-	-	-	-	-	-	-	-	-
Spain	-	-	-	-	-	-	-	-	-
Sweden	-	-	-	-	-	-	-	-	-
Turkey	-	-	-	-	-	-	-	-	-
United Kingdom	-	-	-	-	-	-	-	-	-
United States	-	-	-	-	-	-	-	-	-
Other	176	215	266 e	-	-	-	259	257	347 e
OECD	**178**	**244**	**299 e**	**-**	**-**	**-**	**277**	**293**	**382 e**
Argentina	-	-	-	-	-	-	-	-	-
Brazil	-	-	-	-	-	-	-	-	-
Colombia	-	-	-	-	-	-	-	-	-
Ecuador	-	-	-	-	-	-	-	-	-
Netherlands Antilles	-	-	-	-	-	-	-	-	-
Peru	-	-	-	-	-	-	-	-	-
Trinidad and Tobago	-	-	-	-	-	-	-	-	-
Venezuela	-	-	-	-	-	-	-	-	-
Other	-	-	-	-	-	-	-	-	12 e
Latin America	**-**	**-**	**-**	**-**	**-**	**-**	**-**	**-**	**12 e**
Romania	-	-	-	-	-	-	-	3	2 e
Other	-	-	-	-	-	-	-	-	-
Non-OECD Europe	**-**	**-**	**-**	**-**	**-**	**-**	**-**	**3**	**2 e**
FSU	**-**	**-**	**-**	**-**	**-**	**-**	**75**	**234**	**601 e**
Iran	-	-	-	-	-	-	-	-	-
Iraq	-	-	-	-	-	-	-	-	-
Kuwait	-	-	-	-	-	-	-	-	-
Oman	-	-	-	-	-	-	-	-	-
Qatar	-	-	-	-	-	-	-	-	-
Saudi Arabia	-	-	-	-	-	-	-	-	-
United Arab Emirates	-	-	-	-	-	-	-	-	-
Other	-	-	-	-	-	-	-	-	-
Middle East	**-**	**-**	**-**	**-**	**-**	**-**	**-**	**-**	**-**
India	-	-	-	-	-	-	-	-	-
Indonesia	-	-	-	-	-	-	-	-	-
Singapore	-	-	-	-	-	-	-	-	-
Other	-	-	-	-	-	-	-	-	-
Asia	**-**	**-**	**-**	**-**	**-**	**-**	**-**	**-**	**-**
China	**-**	**-**	**-**	**-**	**-**	**-**	**-**	**-**	**-**
Algeria	-	-	-	-	-	-	-	-	-
Angola/Cabinada	-	-	-	-	-	-	-	-	-
Cameroon	-	-	-	-	-	-	-	-	-
Egypt	-	-	-	-	-	-	-	-	-
Gabon	-	-	-	-	-	-	-	-	-
Libya	-	-	-	-	-	-	-	-	-
Nigeria	-	-	-	-	-	-	-	-	-
Tunisia	-	-	-	-	-	-	-	-	-
Other	-	-	-	-	-	-	-	-	-
Africa	**-**	**-**	**-**	**-**	**-**	**-**	**-**	**-**	**-**
Non-spec./Other	**-**	**-**	**-**	**40**	**50**	**53 e**	**-**	**-**	**-**
Total Imports	**178**	**244**	**299 e**	**40**	**50**	**53 e**	**352**	**530**	**997 e**
Memo: Total OPEC	-	-	-	-	-	-	-	-	-

Source: Annual Oil Statistics.

HUNGARY
Table 7D: Imports by Country of Origin

000 metric tons

	2000	2001	2002	2000	2001	2002	2000	2001	2002
	Fuel Oil (Residual)			**Petroleum Coke**			**Other Products**		
Australia	-	-	-	-	-	-	-	-	-
Belgium	1	-	-	-	2	-	2	3	-
Canada	-	-	-	-	-	-	-	-	-
Czech Republic	44	39	19 e	-	-	-	1	2	-
Denmark	-	-	-	-	-	-	-	-	-
Finland	-	-	-	-	-	-	-	-	-
France	-	-	-	-	-	-	6	7	-
Germany	-	-	-	-	-	-	16	17	3 e
Greece	-	-	-	-	-	-	-	-	-
Hungary	-	-	-	-	-	-	-	-	-
Ireland	-	-	-	-	-	-	-	-	-
Italy	-	-	-	-	-	-	-	-	1 e
Japan	-	-	-	-	-	-	-	-	-
Korea	-	-	-	-	-	-	-	-	-
Mexico	-	-	-	-	-	-	-	-	-
Netherlands	-	-	-	-	-	-	-	2	-
New Zealand	-	-	-	-	-	-	-	-	-
Norway	-	-	-	-	-	-	1	-	-
Poland	-	-	4 e	-	-	-	4	3	-
Spain	-	-	-	-	-	-	-	1	-
Sweden	-	-	-	-	-	-	-	-	-
Turkey	-	-	-	-	-	-	-	-	-
United Kingdom	-	-	-	-	-	-	-	-	-
United States	-	-	-	-	-	-	-	-	-
Other	191	106	197 e	-	-	-	28	37	19 e
OECD	**236**	**145**	**220 e**	**-**	**2**	**-**	**58**	**72**	**23 e**
Argentina	-	-	-	-	-	-	-	-	-
Brazil	-	-	-	-	-	-	-	-	-
Colombia	-	-	-	-	-	-	-	-	-
Ecuador	-	-	-	-	-	-	-	-	-
Netherlands Antilles	-	-	-	-	-	-	2	-	-
Peru	-	-	-	-	-	-	-	-	-
Trinidad and Tobago	-	-	-	-	-	-	-	-	-
Venezuela	-	-	-	-	-	-	-	-	-
Other	-	-	-	-	-	-	-	-	-
Latin America	**-**	**-**	**-**	**-**	**-**	**-**	**2**	**-**	**-**
Romania	8	5	28 e	-	-	-	3	4	-
Other	-	-	45 e	12	5	10 e	-	15	-
Non-OECD Europe	**8**	**5**	**73 e**	**12**	**5**	**10 e**	**3**	**19**	**-**
FSU	**3**	**7**	**194 e**	**-**	**-**	**-**	**22**	**47**	**64 e**
Iran	-	-	-	-	-	-	-	-	-
Iraq	-	-	-	-	-	-	-	-	-
Kuwait	-	-	-	-	-	-	-	-	-
Oman	-	-	-	-	-	-	-	-	-
Qatar	-	-	-	-	-	-	-	-	-
Saudi Arabia	-	-	-	-	-	-	-	-	-
United Arab Emirates	-	-	-	-	-	-	-	-	-
Other	-	-	-	-	-	-	-	-	-
Middle East	**-**	**-**	**-**	**-**	**-**	**-**	**-**	**-**	**-**
India	-	-	-	-	-	-	-	-	-
Indonesia	-	-	-	-	-	-	-	-	-
Singapore	-	-	-	-	-	-	-	-	-
Other	-	-	-	-	-	-	-	-	-
Asia	**-**	**-**	**-**	**-**	**-**	**-**	**-**	**-**	**-**
China	**-**	**-**	**-**	**-**	**-**	**-**	**-**	**-**	**-**
Algeria	-	-	-	-	-	-	-	-	-
Angola/Cabinada	-	-	-	-	-	-	-	-	-
Cameroon	-	-	-	-	-	-	-	-	-
Egypt	-	-	-	-	-	-	-	-	-
Gabon	-	-	-	-	-	-	-	-	-
Libya	-	-	-	-	-	-	-	-	-
Nigeria	-	-	-	-	-	-	-	-	-
Tunisia	-	-	-	-	-	-	-	-	-
Other	-	-	-	-	-	-	-	-	-
Africa	**-**	**-**	**-**	**-**	**-**	**-**	**-**	**-**	**-**
Non-spec./Other	**-**	**1**	**1 e**	**-**	**-**	**-**	**-**	**-**	**7 e**
Total Imports	**247**	**158**	**488 e**	**12**	**7**	**10 e**	**85**	**138**	**94 e**
Memo: Total OPEC	-	-	-	-	-	-	-	-	-

Source: Annual Oil Statistics.

HUNGARY
Table 8A: Exports by Country of Destination

000 metric tons

	2000	2001	2002	2000	2001	2002	2000	2001	2002
	Crude Oil			**Natural Gas Liquids**			**Feedstocks and Other**		
Australia	-	-	-	-	-	-	-	-	-
Austria	-	-	-	-	-	-	-	-	-
Belgium	-	-	-	-	-	-	-	-	-
Canada	-	-	-	-	-	-	-	-	-
Denmark	-	-	-	-	-	-	-	-	-
Finland	-	-	-	-	-	-	-	-	-
France	-	-	-	-	-	-	-	-	-
Germany	-	-	-	-	-	-	-	-	-
Greece	-	-	-	-	-	-	-	-	-
Ireland	-	-	-	-	-	-	-	-	-
Italy	-	-	-	-	-	-	-	-	-
Japan	-	-	-	-	-	-	-	-	-
Luxembourg	-	-	-	-	-	-	-	-	-
Mexico	-	-	-	-	-	-	-	-	-
Netherlands	-	-	-	-	-	-	-	-	-
Norway	-	-	-	-	-	-	-	-	-
Portugal	-	-	-	-	-	-	-	-	-
Spain	-	-	-	-	-	-	-	-	-
Sweden	-	-	-	-	-	-	-	-	-
Switzerland	-	-	-	-	-	-	-	-	-
Turkey	-	-	-	-	-	-	-	-	-
United Kingdom	-	-	-	-	-	-	-	-	-
United States	-	-	-	-	-	-	-	-	-
Other OECD	-	-	-	-	-	-	-	-	-
OECD	**-**	**-**	**-**	**-**	**-**	**-**	**-**	**-**	**-**
Africa	-	-	-	-	-	-	-	-	-
Latin America	-	-	-	-	-	-	-	-	-
Asia	-	-	-	-	-	-	-	-	-
China	-	-	-	-	-	-	-	-	-
Non-OECD Europe	-	-	-	-	-	-	1	-	-
Former Soviet Union	-	-	-	-	-	-	-	-	-
Middle East	-	-	-	-	-	-	-	-	-
Non-specified/Other	-	-	-	-	-	-	-	-	-
Non-OECD	**-**	**-**	**-**	**-**	**-**	**-**	**1**	**-**	**-**
Total Exports	**-**	**-**	**-**	**-**	**-**	**-**	**1**	**-**	**-**

	2000	2001	2002	2000	2001	2002	2000	2001	2002
	Total Products			**Liquefied Petroleum Gas**			**Naphtha**		
Australia	-	-	-	-	-	-	-	-	-
Austria	680	862	994 e	12	15	13 e	-	-	-
Belgium	-	6	-	-	2	-	-	-	-
Canada	-	-	-	-	-	-	-	-	-
Denmark	-	13	2 e	-	-	-	-	-	-
Finland	-	-	-	-	-	-	-	-	-
France	14	-	-	-	-	-	-	-	-
Germany	279	337	278 e	-	-	-	-	-	-
Greece	-	-	-	-	-	-	-	-	-
Ireland	-	-	-	-	-	-	-	-	-
Italy	4	109	91 e	3	19	14 e	-	-	-
Japan	-	-	-	-	-	-	-	-	-
Luxembourg	-	-	-	-	-	-	-	-	-
Mexico	-	-	-	-	-	-	-	-	-
Netherlands	-	-	-	-	-	-	-	-	-
Norway	-	-	-	-	-	-	-	-	-
Portugal	-	-	-	-	-	-	-	-	-
Spain	-	-	-	-	-	-	-	-	-
Sweden	-	-	-	-	-	-	-	-	-
Switzerland	-	-	3 e	-	-	-	-	-	-
Turkey	-	-	-	-	-	-	-	-	-
United Kingdom	-	-	-	-	-	-	-	-	-
United States	-	-	-	-	-	-	-	-	-
Other OECD	315	284	254 e	4	18	20 e	-	-	-
OECD	**1 292**	**1 611**	**1 622 e**	**19**	**54**	**47 e**	**-**	**-**	**-**
Africa	-	-	-	-	-	-	-	-	-
Latin America	2	1	-	-	-	-	-	-	-
Asia	1	-	-	1	-	-	-	-	-
China	-	-	-	-	-	-	-	-	-
Non-OECD Europe	370	629	740 e	32	28	63 e	8	-	-
Former Soviet Union	23	-	3 e	21	-	-	-	-	-
Middle East	-	-	-	-	-	-	-	-	-
Non-specified/Other	66	19	25 e	-	-	-	-	-	-
Non-OECD	**462**	**649**	**768 e**	**54**	**28**	**63 e**	**8**	**-**	**-**
Total Exports	**1 754**	**2 260**	**2 390 e**	**73**	**82**	**110 e**	**8**	**-**	**-**

Source: Annual Oil Statistics.

HUNGARY

Table 8B: Exports by Country of Destination

000 metric tons

	Total Gasoline 2000	Total Gasoline 2001	Total Gasoline 2002	Total Kerosene 2000	Total Kerosene 2001	Total Kerosene 2002	Gas/Diesel Oil 2000	Gas/Diesel Oil 2001	Gas/Diesel Oil 2002
Australia	-	-	-	-	-	-	-	-	-
Austria	107	107	152 e	-	-	-	499	577	720 e
Belgium	-	-	-	-	-	-	-	-	-
Canada	-	-	-	-	-	-	-	-	-
Denmark	-	-	-	-	-	-	-	-	-
Finland	-	-	-	-	-	-	-	-	-
France	-	-	-	-	-	-	-	-	-
Germany	23	87	59 e	-	-	-	192	195	191 e
Greece	-	-	-	-	-	-	-	-	-
Ireland	-	-	-	-	-	-	-	-	-
Italy	-	-	-	-	-	-	-	-	-
Japan	-	-	-	-	-	-	-	-	-
Luxembourg	-	-	-	-	-	-	-	-	-
Mexico	-	-	-	-	-	-	-	-	-
Netherlands	-	-	-	-	-	-	-	-	-
Norway	-	-	-	-	-	-	-	-	-
Portugal	-	-	-	-	-	-	-	-	-
Spain	-	-	-	-	-	-	-	-	-
Sweden	-	-	-	-	-	-	-	-	-
Switzerland	-	-	1 e	-	-	-	-	-	-
Turkey	-	-	-	-	-	-	-	-	-
United Kingdom	-	-	-	-	-	-	-	-	-
United States	-	-	-	-	-	-	-	-	-
Other OECD	150	176	117 e	-	-	-	50	9	46 e
OECD	**280**	**370**	**329 e**	**-**	**-**	**-**	**741**	**781**	**957 e**
Africa	-	-	-	-	-	-	-	-	-
Latin America	-	-	-	-	-	-	-	-	-
Asia	-	-	-	-	-	-	-	-	-
China	-	-	-	-	-	-	-	-	-
Non-OECD Europe	30	134	127 e	14	9	21 e	45	176	335 e
Former Soviet Union	-	-	-	-	-	-	2	-	-
Middle East	-	-	-	-	-	-	-	-	-
Non-specified/Other	61 e	11	9 e	-	2	-	-	-	1 e
Non-OECD	**91 e**	**145**	**136 e**	**14**	**11**	**21 e**	**47**	**176**	**336 e**
Total Exports	**371 e**	**515**	**465 e**	**14**	**11**	**21 e**	**788**	**957**	**1 293 e**

	Fuel Oil (Residual) 2000	Fuel Oil (Residual) 2001	Fuel Oil (Residual) 2002	Petroleum Coke 2000	Petroleum Coke 2001	Petroleum Coke 2002	Other Products 2000	Other Products 2001	Other Products 2002
Australia	-	-	-	-	-	-	-	-	-
Austria	-	73	-	-	-	-	62	90	109 e
Belgium	-	-	-	-	-	-	-	4	-
Canada	-	-	-	-	-	-	-	-	-
Denmark	-	-	-	-	-	-	-	13	2 e
Finland	-	-	-	-	-	-	-	-	-
France	-	-	-	-	-	-	14	-	-
Germany	-	-	-	-	-	-	64	55	28 e
Greece	-	-	-	-	-	-	-	-	-
Ireland	-	-	-	-	-	-	-	-	-
Italy	-	4	-	-	-	-	1	00	77 e
Japan	-	-	-	-	-	-	-	-	-
Luxembourg	-	-	-	-	-	-	-	-	-
Mexico	-	-	-	-	-	-	-	-	-
Netherlands	-	-	-	-	-	-	-	-	-
Norway	-	-	-	-	-	-	-	-	-
Portugal	-	-	-	-	-	-	-	-	-
Spain	-	-	-	-	-	-	-	-	-
Sweden	-	-	-	-	-	-	-	-	-
Switzerland	-	-	-	-	-	-	-	-	2 e
Turkey	-	-	-	-	-	-	-	-	-
United Kingdom	-	-	-	-	-	-	-	-	-
United States	-	-	-	-	-	-	-	-	-
Other OECD	35	30	-	-	-	-	76	51	71 e
OECD	**35**	**107**	**-**	**-**	**-**	**-**	**217**	**299**	**289 e**
Africa	-	-	-	-	-	-	-	-	-
Latin America	2	-	-	-	-	-	-	1	-
Asia	-	-	-	-	-	-	-	-	-
China	-	-	-	-	-	-	-	-	-
Non-OECD Europe	135	135	49 e	-	-	-	106	147	145 e
Former Soviet Union	-	-	-	-	-	-	-	-	3 e
Middle East	-	-	-	-	-	-	-	-	-
Non-specified/Other	-	-	-	-	-	-	5	6	15 e
Non-OECD	**137**	**135**	**49 e**	**-**	**-**	**-**	**111**	**154**	**163 e**
Total Exports	**172**	**242**	**49 e**	**-**	**-**	**-**	**328**	**453**	**452 e**

Source: Annual Oil Statistics.

ICELAND

Figure 1
FUEL SHARES OF TPES *

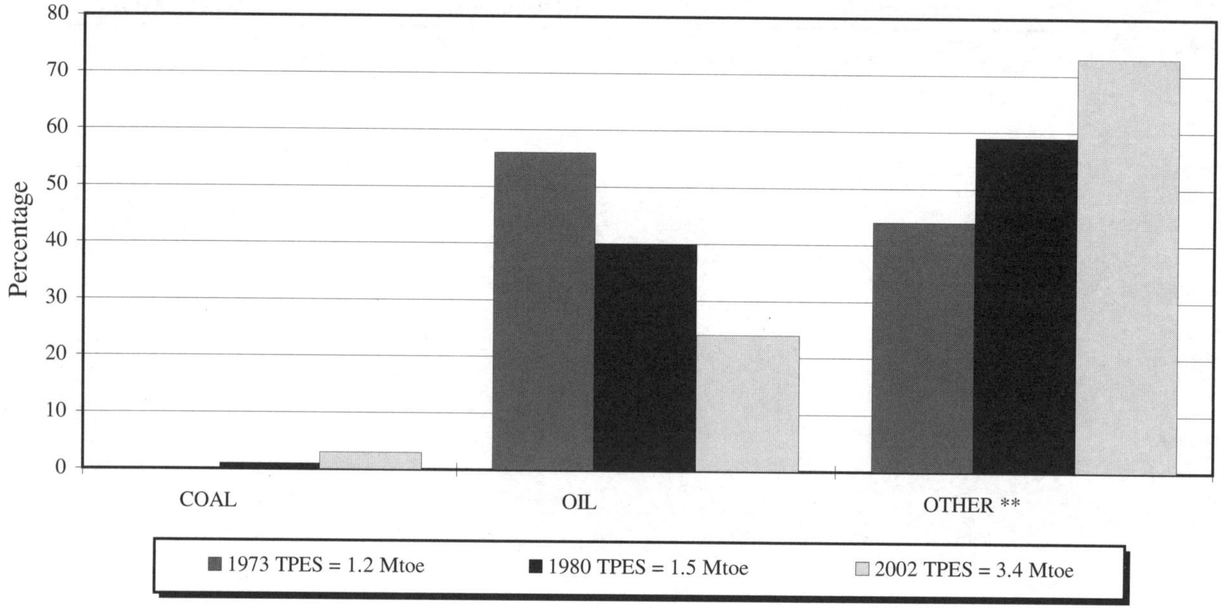

Source: Annual Energy Balances.

* Excluding electricity trade.

** Includes hydro, geothermal, solar, wind, combustible renewables and waste, etc.

Figure 2
OIL PRODUCTION AND CONSUMPTION

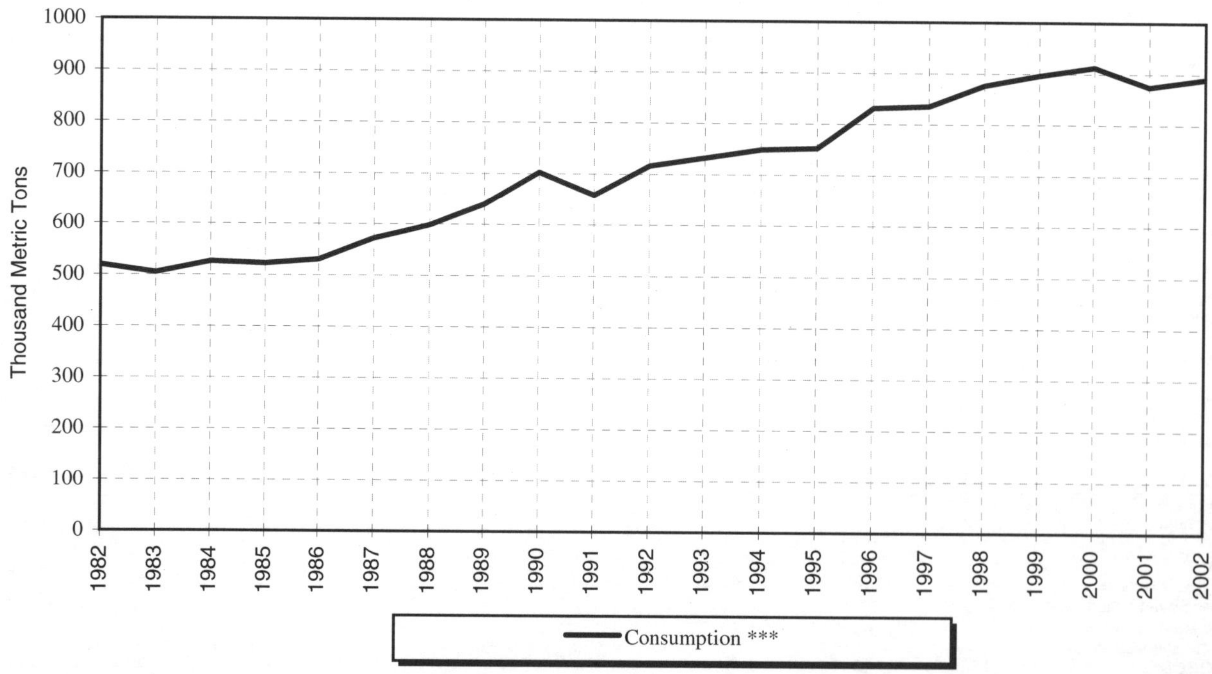

Source: Annual Oil Statistics.

*** Includes refinery fuel and international marine bunkers.

ICELAND
Table 1: Oil Supply and Consumption - Summary Table

000 metric tons

	1973	1978	1980	1990	1999	2000	2001	2002
Indigenous Production	-	-	-	-	-	-	-	-
Crude Oil	..	-	-	-	-	-	-	-
NGL	..	-	-	-	-	-	-	-
Other	..	-	-	-	-	-	-	-
+ Net Imports	678	614	576	729	928	949	871	888 e
Crude Oil, NGL, Other	-	-	-	-	-	-	-	-
Petroleum Products	678	614	576	729	928	949	871	888 e
- Intl. Marine Bunkers	-	-	-	31	51	68	46	55 e
+ Stock Changes	-	15	26	3	- 32	- 37	3	-
TOTAL SUPPLY	**678**	**629**	**602**	**701**	**845**	**844**	**828**	**833 e**
Refinery Intake	-	-	-	-	-	-	-	-
Refinery Output	-	-	-	-	-	-	-	-
LPG + Ethane	-	-	-	-	-	-	-	-
Naphtha	-	-	-	-	-	-	-	-
Kerosenes	-	-	-	-	-	-	-	-
Gasolines	-	-	-	-	-	-	-	-
Gas/Diesel Oil	-	-	-	-	-	-	-	-
Fuel Oil	-	-	-	-	-	-	-	-
Other Products	-	-	-	-	-	-	-	-
- Refinery Losses	-	-	-	-	-	-	-	-
+ Returns	-	-	-	-	-	-	-	-
- Statistical Differences	17	- 13	19	30	-	-	-	- 2 e
TOTAL CONSUMPTION	**661**	**642**	**583**	**671**	**845**	**844**	**828**	**835 e**
Transformation and Energy	**19**	**12**	**10**	**1**	**11**	**1**	**1**	**..**
Crude Oil	-	-	-	-	-	-	-	..
LPG + Ethane	-	-	-	-	-	-	-	..
Gas/Diesel Oil	19	12	10	1	1	1	1	..
Fuel Oil	-	-	-	-	10	-	-	..
Petroleum Coke	-	-	-	-	-	-	-	..
Other	-	-	-	-	-	-	-	..
Transport	**229**	**216**	**180**	**275**	**318**	**333**	**318**	**..**
Motor Gasoline	80	91	90	134	145	143	144	..
Gas/Diesel Oil	38	39	37	52	48	51	52	..
Subtotal	*118*	*130*	*127*	*186*	*193*	*194*	*196*	..
Aviation Fuels	108	83	50	82	125	138	119	..
Other	3	3	3	7	-	1	3	..
Industry	**129**	**150**	**156**	**166**	**257**	**267**	**294**	**..**
Naphtha	-	-	-	-	-	-	-	..
LPG + Ethane	-	-	-	-	1	1	1	..
Gas/Diesel Oil	38	36	34	39	59	72	72	..
Fuel Oil	75	109	110	56	55	46	60	..
Other	16	5	12	71	142	148	161	..
Other Sectors	**284**	**264**	**237**	**229**	**259**	**243**	**215**	**..**
Gas/Diesel Oil	246	234	161	186	239	219	186	..
Fuel Oil	28	21	67	35	12	16	22	..
Other	10	9	9	8	8	8	7	..
Memo: Non-Energy Use	*23*	*12*	*18*	*78*	*149*	*155*	*167*	..
Chem/Petchem. Feedst.	-	-	-	-	-	-	-	..
Other	*23*	*12*	*18*	*78*	*149*	*155*	*167*	..

Source: Annual Oil Statistics.

ICELAND
Table 3: Motor Gasoline - Supply

000 metric tons

	1973	1978	1980	1990	1999	2000	2001	2002
MOTOR GASOLINE								
Refinery Output	-	-	-	-	-	-	-	-
+ Imports	80	95	87	127	148	152	138	148 e
- Exports	-	-	-	-	-	-	-	-
+ Transfers	-	-	-	-	-	-	-	-
+ Stock Change	-	2	2	5	- 3	- 9	6	-
- Statistical Difference	-	6	- 1	- 2	-	-	-	-
Total Consumption	**80**	**91**	**90**	**134**	**145**	**143**	**144**	**148 e**
Closing Stock Level	-	*12*	*20*	*23*	*6*	*15*	*9*	*9 e*

Table 4: Aviation Fuels - Supply

000 metric tons

	1973	1978	1980	1990	1999	2000	2001	2002
AVIATION GASOLINE								
Refinery Output	-	-	-	-	-	-	-	-
+ Imports	2	3	2	2	1	1	1	1 e
- Exports	-	-	-	-	-	-	-	-
+ Transfers	-	-	-	-	-	-	-	-
+ Stock Change	-	-	-	-	-	-	-	-
- Statistical Difference	-	1	-	-	-	-	-	-
Total Consumption	**2**	**2**	**2**	**2**	**1**	**1**	**1**	**1 e**
Closing Stock Level	-	-	-	*1*	*2*	*2*	*2*	*2 e*
GASOLINE TYPE JET FUEL								
Refinery Output	-	-	-	-	-	-	-	-
+ Imports	-	-	-	-	-	-	-	-
- Exports	-	-	-	-	-	-	-	-
+ Transfers	-	-	-	-	-	-	-	-
+ Stock Change	-	-	-	-	-	-	-	-
- Statistical Difference	-	-	-	-	-	-	-	-
Total Consumption	**-**	**-**	**-**	**-**	**-**	**-**	**-**	**-**
Closing Stock Level	-	-	-	-	-	-	-	-
KEROSENE TYPE JET FUEL								
Refinery Output	-	-	-	-	-	-	-	-
+ Imports	106	84	55	79	131	147	117	107 e
- Exports	-	-	-	-	-	-	-	-
+ Transfers	-	-	-	-	-	-	-	-
+ Stock Change	-	-	-	-	- 7	- 10	1	-
- Statistical Difference	-	3	7	- 1	-	-	-	-
Total Consumption	**106**	**81**	**48**	**80**	**124**	**137**	**118**	**107 e**
Closing Stock Level	-	-	-	*11*	*24*	*34*	*33*	*33 e*

Source: Annual Oil Statistics.

ICELAND
Table 5A: Gas/Diesel Oil - Supply and Consumption

000 metric tons

	1973	1978	1980	1990	1999	2000	2001	2002
Refinery Output	-	-	-	-	-	-	-	-
+ Recycled Products	-	-	-	-	-	-	-	-
+ Imports	356	302	231	336	406	428	369	382 e
- Exports	-	-	-	-	-	-	-	-
- Intl. Marine Bunkers	-	-	-	31	49	60	44	53 e
+ Transfers	-	-	-	-	-	-	-	-
+ Stock Changes	-	9	23	- 11	- 10	- 25	- 14	-
- Statistical Differences	15	- 10	12	16	-	-	-	-
Total Consumption	**341**	**321**	**242**	**278**	**347**	**343**	**311**	**329 e**
Transformation	**19**	**12**	**10**	**1**	**1**	**1**	**1**	**..**
Public Electricity	19	12	10	1	1	1	1	..
Autoproducer Electricity	-	-	-	-	-	-	-	..
Public CHP Plants	-	-	-	-	-	-	-	..
Autoproducers of CHP	-	-	-	-	-	-	-	..
Public Heat Plants	-	-	-	-	-	-	-	..
Autoproducer Heat Plants	-	-	-	-	-	-	-	..
Petrochemical Industry	-	-	-	-	-	-	-	..
Other Transformation	-	-	-	-	-	-	-	..
Energy Sector	**-**	**-**	**-**	**-**	**-**	**-**	**-**	**..**
Refinery Fuel	-	-	-	-	-	-	-	..
Other Energy Sector	-	-	-	-	-	-	-	..
Final Consumption	**322**	**309**	**232**	**277**	**346**	**342**	**310**	**..**
Transport	**38**	**39**	**37**	**52**	**48**	**51**	**52**	**..**
Air	-	-	-	-	-	-	-	..
Road	29	26	26	40	42	47	48	..
Rail	-	-	-	-	-	-	-	..
Inland Waterways	9	13	11	12	6	4	4	..
Non-Specified	-	-	-	-	-	-	-	..
Industry	**38**	**36**	**34**	**39**	**59**	**72**	**72**	**..**
Iron and Steel	-	-	-	-	-	-	-	..
Chemical (incl. Pet.-Chem.)	-	-	-	-	-	-	-	..
Non-Ferrous Metals	-	-	-	-	-	-	-	..
Non-Metallic Minerals	-	-	-	-	-	-	-	..
Transport Equipment	-	-	-	-	-	-	-	..
Machinery	-	-	-	-	-	-	-	..
Mining and Quarrying	-	-	-	-	-	-	-	..
Food and Tobacco	-	-	-	-	8	10	11	..
Pulp, Paper and Printing	-	-	-	-	-	-	-	..
Wood and Wood Products	-	-	-	-	-	-	-	..
Construction	-	-	-	34	51	62	61	..
Textiles and Leather	-	-	-	-	-	-	-	..
Non-Specified	38	36	34	5	-	-	-	..
Other Sectors	**246**	**234**	**161**	**186**	**239**	**219**	**186**	**..**
Commerce and Public	-	-	-	-	-	-	-	..
Residential	160	95	60	11	10	8	6	..
Agriculture	86	139	101	175	229	211	180	..
Non-Specified	-	-	-	-	-	-	-	..
Memo: Non-Energy Use	-	-	-	-	-	-	-	..
Chem/Petchem. Feedst.	-	-	-	-	-	-	-	..
Other	-	-	-	-	-	-	-	..
Closing Stock Level	-	*59*	*58*	*53*	*39*	*64*	*78*	*78 e*

Source: Annual Oil Statistics.

ICELAND
Table 5B: Fuel Oil (Residual) - Supply and Consumption

000 metric tons

	1973	1978	1980	1990	1999	2000	2001	2002
Refinery Output	**-**	**-**	**-**	**-**	**-**	**-**	**-**	**-**
+ Recycled Products	-	-	-	-	-	-	-	-
+ Imports	108	115	180	106	91	64	77	93 e
- Exports	-	-	-	-	-	-	-	-
- Intl. Marine Bunkers	-	-	-	-	2	8	2	2 e
+ Transfers	-	-	-	-	-	-	-	-
+ Stock Changes	-	4	1	9	- 12	7	10	-
- Statistical Differences	2	- 14	1	17	-	-	-	- 2 e
Total Consumption	**106**	**133**	**180**	**98**	**77**	**63**	**85**	**93 e**
Transformation	**-**	**-**	**-**	**-**	**10**	**-**	**-**	**..**
Public Electricity	-	-	-	-	-	-	-	..
Autoproducer Electricity	-	-	-	-	-	-	-	..
Public CHP Plants	-	-	-	-	-	-	-	..
Autoproducers of CHP	-	-	-	-	-	-	-	..
Public Heat Plants	-	-	-	-	10	-	-	..
Autoproducer Heat Plants	-	-	-	-	-	-	-	..
Petrochemical Industry	-	-	-	-	-	-	-	..
Other Transformation	-	-	-	-	-	-	-	..
Energy Sector	**-**	**-**	**-**	**-**	**-**	**-**	**-**	**..**
Refinery Fuel	-	-	-	-	-	-	-	..
Other Energy Sector	-	-	-	-	-	-	-	..
Final Consumption	**106**	**133**	**180**	**98**	**67**	**63**	**85**	**..**
Transport	**3**	**3**	**3**	**7**	**-**	**1**	**3**	**..**
Air	-	-	-	-	-	-	-	..
Road	-	-	-	-	-	-	-	..
Rail	-	-	-	-	-	-	-	..
Inland Waterways	3	3	3	7	-	1	3	..
Non-Specified	-	-	-	-	-	-	-	..
Industry	**75**	**109**	**110**	**56**	**55**	**46**	**60**	**..**
Iron and Steel	-	-	-	-	-	-	-	..
Chemical (incl. Pet.-Chem.)	-	-	-	-	-	-	-	..
Non-Ferrous Metals	-	-	-	5	6	6	6	..
Non-Metallic Minerals	-	-	-	-	-	-	-	..
Transport Equipment	-	-	-	-	-	-	-	..
Machinery	-	-	-	-	-	-	-	..
Mining and Quarrying	-	-	-	-	-	-	-	..
Food and Tobacco	-	-	-	50	49	40	54	..
Pulp, Paper and Printing	-	-	-	-	-	-	-	..
Wood and Wood Products	-	-	-	-	-	-	-	..
Construction	-	-	-	-	-	-	-	..
Textiles and Leather	-	-	-	-	-	-	-	..
Non-Specified	75	109	110	1	-	-	-	..
Other Sectors	**28**	**21**	**67**	**35**	**12**	**16**	**22**	**..**
Commerce and Public	-	-	-	-	-	-	-	..
Residential	10	8	8	3	-	-	-	..
Agriculture	18	13	59	32	12	16	22	..
Non-Specified	-	-	-	-	-	-	-	..
Memo: Non-Energy Use	-	-	-	-	-	-	-	..
Chem/Petchem. Feedst.	-	-	-	-	-	-	-	..
Other	-	-	-	-	-	-	-	..
Closing Stock Level	-	*26*	*26*	*18*	*53*	*46*	*36*	*36 e*

Source: Annual Oil Statistics.

ICELAND
Table 5C: Total Products - Supply and Consumption

000 metric tons

	1973	1978	1980	1990	1999	2000	2001	2002
Refinery Output	**-**	**-**	**-**	**-**	**-**	**-**	**-**	**-**
+ Recycled Products	-	-	-	-	-	-	-	-
+ Imports	678	614	576	729	928	949	871	888 e
- Exports	-	-	-	-	-	-	-	-
- Intl. Marine Bunkers	-	-	-	31	51	68	46	55 e
+ Transfers	-	-	-	-	-	-	-	-
+ Stock Changes	-	15	26	3	- 32	- 37	3	-
- Statistical Differences	17	- 13	19	30	-	-	-	- 2 e
Total Consumption	**661**	**642**	**583**	**671**	**845**	**844**	**828**	**835 e**
Transformation	**19**	**12**	**10**	**1**	**11**	**1**	**1**	**..**
Public Electricity	19	12	10	1	1	1	1	..
Autoproducer Electricity	-	-	-	-	-	-	-	..
Public CHP Plants	-	-	-	-	-	-	-	..
Autoproducers of CHP	-	-	-	-	-	-	-	..
Public Heat Plants	-	-	-	-	10	-	-	..
Autoproducer Heat Plants	-	-	-	-	-	-	-	..
Petrochemical Industry	-	-	-	-	-	-	-	..
Other Transformation	-	-	-	-	-	-	-	..
Energy Sector	**-**	**-**	**-**	**-**	**-**	**-**	**-**	**..**
Refinery Fuel	-	-	-	-	-	-	-	..
Other Energy Sector	-	-	-	-	-	-	-	..
Final Consumption	**642**	**630**	**573**	**670**	**834**	**843**	**827**	**..**
Transport	**229**	**216**	**180**	**275**	**318**	**333**	**318**	**..**
Air	108	83	50	82	125	138	119	..
Road	109	117	116	174	187	190	192	..
Rail	-	-	-	-	-	-	-	..
Inland Waterways	12	16	14	19	6	5	7	..
Non-Specified	-	-	-	-	-	-	-	..
Industry	**129**	**150**	**156**	**166**	**257**	**267**	**294**	**..**
Iron and Steel	-	-	-	-	-	-	-	..
Chemical (incl. Pet.-Chem.)	-	-	-	-	-	-	-	..
Non-Ferrous Metals	-	-	-	55	122	123	133	..
Non-Metallic Minerals	-	-	-	-	-	-	-	..
Transport Equipment	-	-	-	-	-	-	-	..
Machinery	-	-	-	-	-	-	-	..
Mining and Quarrying	-	-	-	-	-	-	-	..
Food and Tobacco	-	-	-	50	57	50	65	..
Pulp, Paper and Printing	-	-	-	-	-	-	-	..
Wood and Wood Products	-	-	-	-	-	-	-	..
Construction	15	5	11	53	76	92	93	..
Textiles and Leather	-	-	-	-	-	-	-	..
Non-Specified	114	145	145	8	2	2	3	..
Other Sectors	**284**	**264**	**237**	**229**	**259**	**243**	**215**	**..**
Commerce and Public	-	-	-	-	-	-	-	..
Residential	170	103	68	15	11	9	7	..
Agriculture	111	159	166	214	248	234	208	..
Non-Specified	3	2	3	-	-	-	-	..
Memo: Non-Energy Use	*23*	*12*	*18*	*78*	*149*	*155*	*167*	..
Chem/Petchem. Feedst.	-	-	-	-	-	-	-	..
Other	*23*	*12*	*18*	*78*	*149*	*155*	*167*	..
Closing Stock Level	-	*97*	*104*	*106*	*124*	*161*	*158*	*158 e*

Source: Annual Oil Statistics.

ICELAND
Table 6A: Supply and Consumption

000 metric tons

	1980	2000	2001	2002	1980	2000	2001	2002
	LPG AND ETHANE				**REFINERY GAS**			
Refinery Output	-	-	-	-	-	-	-	-
+ Recycled Products	-	-	-	-	-	-	-	-
+ Imports	1	2	2	1 e	-	-	-	-
- Exports	-	-	-	-	-	-	-	-
- Intl. Marine Bunkers	-	-	-	-	-	-	-	-
+ Transfers	-	-	-	-	-	-	-	-
+ Stock Change	-	-	-	-	-	-	-	-
- Statistical Difference	-	-	-	-	-	-	-	-
Total Consumption	**1**	**2**	**2**	**1 e**	**-**	**-**	**-**	**-**
of which: Ethane	-	-	-	-	-	-	-	-
Transformation	-	-	-	..	-	-	-	..
Energy	-	-	-	..	-	-	-	..
Transport	-	-	-	..	-	-	-	..
Industry	-	1	1	..	-	-	-	..
Other	1	1	1	..	-	-	-	..
Memo: Non-Energy Use	-	-	-	..	-	-	-	..
Closing Stock Level	-	-	-	-	-	-	-	-

	1980	2000	2001	2002	1980	2000	2001	2002
	NAPHTHA				**OTHER KEROSENE**			
Refinery Output	-	-	-	-	-	-	-	-
+ Recycled Products	-	-	-	-	-	-	-	-
+ Imports	-	-	-	-	2	-	-	-
- Exports	-	-	-	-	-	-	-	-
- Intl. Marine Bunkers	-	-	-	-	-	-	-	-
+ Transfers	-	-	-	-	-	-	-	-
+ Stock Change	-	-	-	-	-	-	-	-
- Statistical Difference	-	-	-	-	-	-	-	-
Total Consumption	**-**	**-**	**-**	**-**	**2**	**-**	**-**	**-**
Transformation	-	-	-	..	-	-	-	..
Energy	-	-	-	..	-	-	-	..
Transport	-	-	-	..	-	-	-	..
Industry	-	-	-	..	-	-	-	..
Other	-	-	-	..	2	-	-	..
Memo: Non-Energy Use	-	-	-	..	-	-	-	..
Closing Stock Level	-	-	-	-	-	-	-	-

	1980	2000	2001	2002	1980	2000	2001	2002
	WHITE SPIRIT + SBP				**LUBRICANTS**			
Refinery Output	-	-	-	..	-	-	-	..
+ Recycled Products	-	-	-	..	-	-	-	..
+ Imports	1	2	3	..	6	7	6	..
- Exports	-	-	-	..	-	-	-	..
- Intl. Marine Bunkers	-	-	-	..	-	-	-	..
+ Transfers	-	-	-	..	-	-	-	..
+ Stock Change	-	-	-	..	-	-	-	..
- Statistical Difference	-	-	-	..	-	-	-	..
Total Consumption	**1**	**2**	**3**	**..**	**6**	**7**	**6**	**..**
Transformation	-	-	-	..	-	-	-	..
Energy	-	-	-	..	-	-	-	..
Transport	-	-	-	..	-	-	-	..
Industry	1	2	3	..	-	-	-	..
Other	-	-	-	..	6	7	6	..
Memo: Non-Energy Use	*1*	*2*	*3*	..	*6*	*7*	*6*	..
Closing Stock Level	-	-	-	..	-	-	-	..

Source: Annual Oil Statistics.

ICELAND
Table 6B: Supply and Consumption

000 metric tons

	1980	2000	2001	2002	1980	2000	2001	2002
	BITUMEN				PARAFFIN WAXES			
Refinery Output	-	-	-	..	-	-	-	..
+ Recycled Products	-	-	-	..	-	-	-	..
+ Imports	11	30	32	..	-	-	-	..
- Exports	-	-	-	..	-	-	-	..
- Intl. Marine Bunkers	-	-	-	..	-	-	-	..
+ Transfers	-	-	-	..	-	-	-	..
+ Stock Change	-	-	-	..	-	-	-	..
- Statistical Difference	-	-	-	..	-	-	-	..
Total Consumption	**11**	**30**	**32**	**..**	**-**	**-**	**-**	**..**
Transformation	-	-	-	..	-	-	-	..
Energy	-	-	-	..	-	-	-	..
Transport	-	-	-	..	-	-	-	..
Industry	11	30	32	..	-	-	-	..
Other	-	-	-	..	-	-	-	..
Memo: Non-Energy Use	*11*	*30*	*32*	..	-	-	-	..
Closing Stock Level	-	-	-	..	-	-	-	..
	PETROLEUM COKE				OTHER PRODUCTS			
Refinery Output	-	-	-	-	-	-	-	-
+ Recycled Products	-	-	-	-	-	-	-	-
+ Imports	-	116	126	120 e	-	-	-	36 e
- Exports	-	-	-	-	-	-	-	-
- Intl. Marine Bunkers	-	-	-	-	-	-	-	-
+ Transfers	-	-	-	-	-	-	-	-
+ Stock Change	-	-	-	-	-	-	-	-
- Statistical Difference	-	-	-	-	-	-	-	-
Total Consumption	**-**	**116**	**126**	**120 e**	**-**	**-**	**-**	**36 e**
Transformation	-	-	-	..	-	-	-	..
Energy	-	-	-	..	-	-	-	..
Transport	-	-	-	..	-	-	-	..
Industry	-	116	126	..	-	-	-	..
Other	-	-	-	..	-	-	-	..
Memo: Non-Energy Use	-	*116*	*126*	..	-	-	-	..
Closing Stock Level	-	-	-	-	-	-	-	-

Source: Annual Oil Statistics.

ICELAND
Table 7B: Imports by Country of Origin

000 metric tons

	2000	2001	2002	2000	2001	2002	2000	2001	2002
	Total Products			Liquefied Petroleum Gases			Naphtha		
Australia	-	-	-	-	-	-	-	-	-
Belgium	-	24	17 e	-	-	-	-	-	-
Canada	-	-	-	-	-	-	-	-	-
Czech Republic	-	-	-	-	-	-	-	-	-
Denmark	8	20	20 e	-	-	-	-	-	-
Finland	-	2	2 e	-	-	-	-	-	-
France	34	1	32 e	-	-	-	-	-	-
Germany	117	163	62 e	-	-	-	-	-	-
Greece	-	-	-	-	-	-	-	-	-
Hungary	-	-	-	-	-	-	-	-	-
Ireland	-	-	-	-	-	-	-	-	-
Italy	-	-	-	-	-	-	-	-	-
Japan	-	-	-	-	-	-	-	-	-
Korea	-	-	-	-	-	-	-	-	-
Mexico	-	-	-	-	-	-	-	-	-
Netherlands	271	161	95 e	-	-	-	-	-	-
New Zealand	-	-	-	-	-	-	-	-	-
Norway	269	197	143 e	-	-	1 e	-	-	-
Poland	-	18	14 e	-	-	-	-	-	-
Spain	-	-	-	-	-	-	-	-	-
Sweden	34	6	29 e	-	-	-	-	-	-
Turkey	-	-	-	-	-	-	-	-	-
United Kingdom	90	36	34 e	2	2	-	-	-	-
United States	-	3	-	-	-	-	-	-	-
Other	-	-	6 e	-	-	-	-	-	-
OECD	**823**	**631**	**454 e**	**2**	**2**	**1 e**	**-**	**-**	**-**
Argentina	-	-	-	-	-	-	-	-	-
Brazil	-	-	-	-	-	-	-	-	-
Colombia	-	-	-	-	-	-	-	-	-
Ecuador	-	-	-	-	-	-	-	-	-
Netherlands Antilles	-	-	-	-	-	-	-	-	-
Peru	-	-	-	-	-	-	-	-	-
Trinidad and Tobago	-	-	-	-	-	-	-	-	-
Venezuela	-	-	-	-	-	-	-	-	-
Other	-	-	-	-	-	-	-	-	-
Latin America	**-**	**-**	**-**	**-**	**-**	**-**	**-**	**-**	**-**
Romania	-	-	-	-	-	-	-	-	-
Other	-	-	-	-	-	-	-	-	-
Non-OECD Europe	**-**	**-**	**-**	**-**	**-**	**-**	**-**	**-**	**-**
FSU	**126**	**240**	**434 e**	**-**	**-**	**-**	**-**	**-**	**-**
Iran	-	-	-	-	-	-	-	-	-
Iraq	-	-	-	-	-	-	-	-	-
Kuwait	-	-	-	-	-	-	-	-	-
Oman	-	-	-	-	-	-	-	-	-
Qatar	-	-	-	-	-	-	-	-	-
Saudi Arabia	-	-	-	-	-	-	-	-	-
United Arab Emirates	-	-	-	-	-	-	-	-	-
Other	-	-	-	-	-	-	-	-	-
Middle East	**-**	**-**	**-**	**-**	**-**	**-**	**-**	**-**	**-**
India	-	-	-	-	-	-	-	-	-
Indonesia	-	-	-	-	-	-	-	-	-
Singapore	-	-	-	-	-	-	-	-	-
Other	-	-	-	-	-	-	-	-	-
Asia	**-**	**-**	**-**	**-**	**-**	**-**	**-**	**-**	**-**
China	**-**	**-**	**-**	**-**	**-**	**-**	**-**	**-**	**-**
Algeria	-	-	-	-	-	-	-	-	-
Angola/Cabinada	-	-	-	-	-	-	-	-	-
Cameroon	-	-	-	-	-	-	-	-	-
Egypt	-	-	-	-	-	-	-	-	-
Gabon	-	-	-	-	-	-	-	-	-
Libya	-	-	-	-	-	-	-	-	-
Nigeria	-	-	-	-	-	-	-	-	-
Tunisia	-	-	-	-	-	-	-	-	-
Other	-	-	-	-	-	-	-	-	-
Africa	**-**	**-**	**-**	**-**	**-**	**-**	**-**	**-**	**-**
Non-spec./Other	**-**	**-**	**-**	**-**	**-**	**-**	**-**	**-**	**-**
Total Imports	**949**	**871**	**888 e**	**2**	**2**	**1 e**	**-**	**-**	**-**
Memo: Total OPEC	-	-	-	-	-	-	-	-	-

Source: Annual Oil Statistics.

ICELAND
Table 7C: Imports by Country of Origin

000 metric tons

	2000	2001	2002	2000	2001	2002	2000	2001	2002
	Total Gasoline			**Total Kerosene**			**Gas/Diesel**		
Australia	-	-	-	-	-	-	-	-	-
Belgium	-	1	17 e	-	15	-	-	7	-
Canada	-	-	-	-	-	-	-	-	-
Czech Republic	-	-	-	-	-	-	-	-	-
Denmark	6	-	-	-	14	-	-	3	-
Finland	-	-	-	-	-	1 e	-	-	-
France	34	-	32 e	-	-	-	-	-	-
Germany	-	5	-	-	21	14 e	86	96	8 e
Greece	-	-	-	-	-	-	-	-	-
Hungary	-	-	-	-	-	-	-	-	-
Ireland	-	-	-	-	-	-	-	-	-
Italy	-	-	-	-	-	-	-	-	-
Japan	-	-	-	-	-	-	-	-	-
Korea	-	-	-	-	-	-	-	-	-
Mexico	-	-	-	-	-	-	-	-	-
Netherlands	41	20	14 e	114	45	-	35	9	-
New Zealand	-	-	-	-	-	-	-	-	-
Norway	30	89	12 e	22	2	-	207	105	130 e
Poland	-	-	-	-	-	-	-	-	12 e
Spain	-	-	-	-	-	-	-	-	-
Sweden	13	2	26 e	-	-	-	16	-	-
Turkey	-	-	-	-	-	-	-	-	-
United Kingdom	29	2	-	11	-	-	20	-	-
United States	-	-	-	-	-	-	-	-	-
Other	-	-	6 e	-	-	-	-	-	-
OECD	**153**	**119**	**107 e**	**147**	**97**	**15 e**	**364**	**220**	**150 e**
Argentina	-	-	-	-	-	-	-	-	-
Brazil	-	-	-	-	-	-	-	-	-
Colombia	-	-	-	-	-	-	-	-	-
Ecuador	-	-	-	-	-	-	-	-	-
Netherlands Antilles	-	-	-	-	-	-	-	-	-
Peru	-	-	-	-	-	-	-	-	-
Trinidad and Tobago	-	-	-	-	-	-	-	-	-
Venezuela	-	-	-	-	-	-	-	-	-
Other	-	-	-	-	-	-	-	-	-
Latin America	**-**	**-**	**-**	**-**	**-**	**-**	**-**	**-**	**-**
Romania	-	-	-	-	-	-	-	-	-
Other	-	-	-	-	-	-	-	-	-
Non-OECD Europe	**-**	**-**	**-**	**-**	**-**	**-**	**-**	**-**	**-**
FSU	**-**	**20**	**42 e**	**-**	**20**	**92 e**	**64**	**149**	**232 e**
Iran	-	-	-	-	-	-	-	-	-
Iraq	-	-	-	-	-	-	-	-	-
Kuwait	-	-	-	-	-	-	-	-	-
Oman	-	-	-	-	-	-	-	-	-
Qatar	-	-	-	-	-	-	-	-	-
Saudi Arabia	-	-	-	-	-	-	-	-	-
United Arab Emirates	-	-	-	-	-	-	-	-	-
Other	-	-	-	-	-	-	-	-	-
Middle East	**-**	**-**	**-**	**-**	**-**	**-**	**-**	**-**	**-**
India	-	-	-	-	-	-	-	-	-
Indonesia	-	-	-	-	-	-	-	-	-
Singapore	-	-	-	-	-	-	-	-	-
Other	-	-	-	-	-	-	-	-	-
Asia	**-**	**-**	**-**	**-**	**-**	**-**	**-**	**-**	**-**
China	**-**	**-**	**-**	**-**	**-**	**-**	**-**	**-**	**-**
Algeria	-	-	-	-	-	-	-	-	-
Angola/Cabinada	-	-	-	-	-	-	-	-	-
Cameroon	-	-	-	-	-	-	-	-	-
Egypt	-	-	-	-	-	-	-	-	-
Gabon	-	-	-	-	-	-	-	-	-
Libya	-	-	-	-	-	-	-	-	-
Nigeria	-	-	-	-	-	-	-	-	-
Tunisia	-	-	-	-	-	-	-	-	-
Other	-	-	-	-	-	-	-	-	-
Africa	**-**	**-**	**-**	**-**	**-**	**-**	**-**	**-**	**-**
Non-spec./Other	**-**	**-**	**-**	**-**	**-**	**-**	**-**	**-**	**-**
Total Imports	**153**	**139**	**149 e**	**147**	**117**	**107 e**	**428**	**369**	**382 e**
Memo: Total OPEC	-	-	-	-	-	-	-	-	-

Source: Annual Oil Statistics.

ICELAND
Table 7D: Imports by Country of Origin

000 metric tons

	2000	2001	2002	2000	2001	2002	2000	2001	2002
	Fuel Oil (Residual)			**Petroleum Coke**			**Other Products**		
Australia	-	-	-	-	-	-	-	-	-
Belgium	-	-	-	-	-	-	-	1	-
Canada	-	-	-	-	-	-	-	-	-
Czech Republic	-	-	-	-	-	-	-	-	-
Denmark	1	3	20 e	-	-	-	1	-	-
Finland	-	-	-	-	-	-	-	2	1 e
France	-	-	-	-	1	-	-	-	-
Germany	-	-	-	30	40	40 e	1	1	-
Greece	-	-	-	-	-	-	-	-	-
Hungary	-	-	-	-	-	-	-	-	-
Ireland	-	-	-	-	-	-	-	-	-
Italy	-	-	-	-	-	-	-	-	-
Japan	-	-	-	-	-	-	-	-	-
Korea	-	-	-	-	-	-	-	-	-
Mexico	-	-	-	-	-	-	-	-	-
Netherlands	-	-	-	77	85	80 e	4	2	1 e
New Zealand	-	-	-	-	-	-	-	-	-
Norway	1	1	-	9	-	-	-	-	-
Poland	-	18	2 e	-	-	-	-	-	-
Spain	-	-	-	-	-	-	-	-	-
Sweden	-	4	3 e	-	-	-	5	-	-
Turkey	-	-	-	-	-	-	-	-	-
United Kingdom	-	-	-	-	-	-	28	32	34 e
United States	-	-	-	-	-	-	-	3	-
Other	-	-	-	-	-	-	-	-	-
OECD	**2**	**26**	**25 e**	**116**	**126**	**120 e**	**39**	**41**	**36 e**
Argentina	-	-	-	-	-	-	-	-	-
Brazil	-	-	-	-	-	-	-	-	-
Colombia	-	-	-	-	-	-	-	-	-
Ecuador	-	-	-	-	-	-	-	-	-
Netherlands Antilles	-	-	-	-	-	-	-	-	-
Peru	-	-	-	-	-	-	-	-	-
Trinidad and Tobago	-	-	-	-	-	-	-	-	-
Venezuela	-	-	-	-	-	-	-	-	-
Other	-	-	-	-	-	-	-	-	-
Latin America	**-**	**-**	**-**	**-**	**-**	**-**	**-**	**-**	**-**
Romania	-	-	-	-	-	-	-	-	-
Other	-	-	-	-	-	-	-	-	-
Non-OECD Europe	**-**	**-**	**-**	**-**	**-**	**-**	**-**	**-**	**-**
FSU	**62**	**51**	**68 e**	**-**	**-**	**-**	**-**	**-**	**-**
Iran	-	-	-	-	-	-	-	-	-
Iraq	-	-	-	-	-	-	-	-	-
Kuwait	-	-	-	-	-	-	-	-	-
Oman	-	-	-	-	-	-	-	-	-
Qatar	-	-	-	-	-	-	-	-	-
Saudi Arabia	-	-	-	-	-	-	-	-	-
United Arab Emirates	-	-	-	-	-	-	-	-	-
Other	-	-	-	-	-	-	-	-	-
Middle East	**-**	**-**	**-**	**-**	**-**	**-**	**-**	**-**	**-**
India	-	-	-	-	-	-	-	-	-
Indonesia	-	-	-	-	-	-	-	-	-
Singapore	-	-	-	-	-	-	-	-	-
Other	-	-	-	-	-	-	-	-	-
Asia	**-**	**-**	**-**	**-**	**-**	**-**	**-**	**-**	**-**
China	**-**	**-**	**-**	**-**	**-**	**-**	**-**	**-**	**-**
Algeria	-	-	-	-	-	-	-	-	-
Angola/Cabinada	-	-	-	-	-	-	-	-	-
Cameroon	-	-	-	-	-	-	-	-	-
Egypt	-	-	-	-	-	-	-	-	-
Gabon	-	-	-	-	-	-	-	-	-
Libya	-	-	-	-	-	-	-	-	-
Nigeria	-	-	-	-	-	-	-	-	-
Tunisia	-	-	-	-	-	-	-	-	-
Other	-	-	-	-	-	-	-	-	-
Africa	**-**	**-**	**-**	**-**	**-**	**-**	**-**	**-**	**-**
Non-spec./Other	**-**	**-**	**-**	**-**	**-**	**-**	**-**	**-**	**-**
Total Imports	**64**	**77**	**93 e**	**116**	**126**	**120 e**	**39**	**41**	**36 e**
Memo: Total OPEC	-	-	-	-	-	-	-	-	-

Source: Annual Oil Statistics.

IRELAND

Figure 1
FUEL SHARES OF TPES *

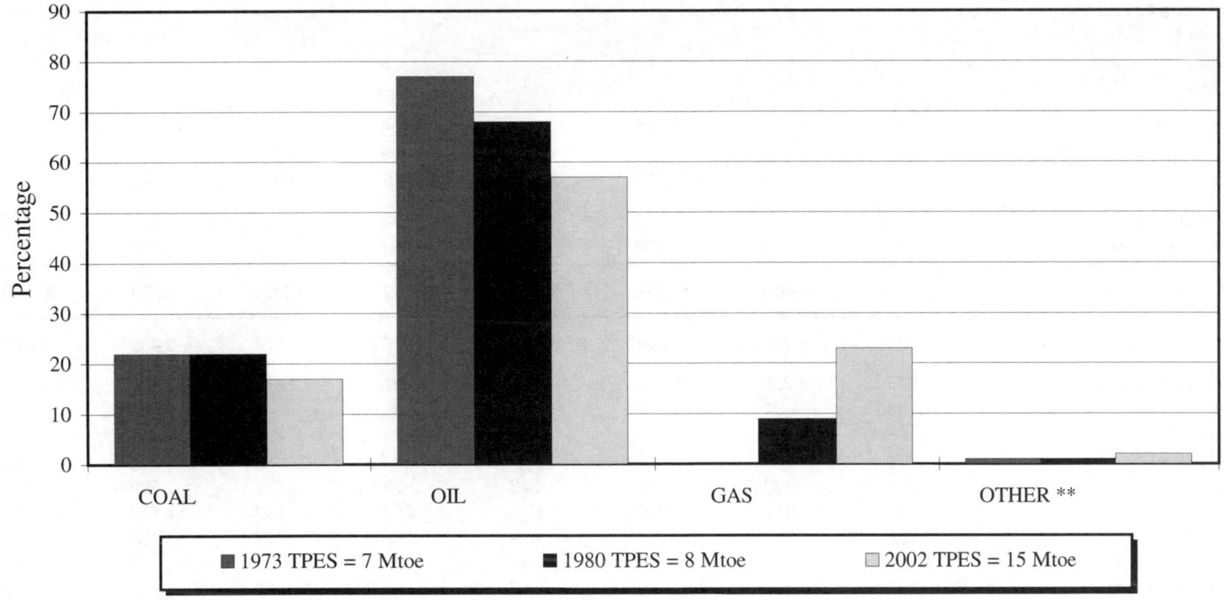

Source: Annual Energy Balances.

* Excluding electricity trade.

** Includes hydro, geothermal, solar, wind, combustible renewables and waste, etc.

Figure 2
OIL PRODUCTION AND CONSUMPTION

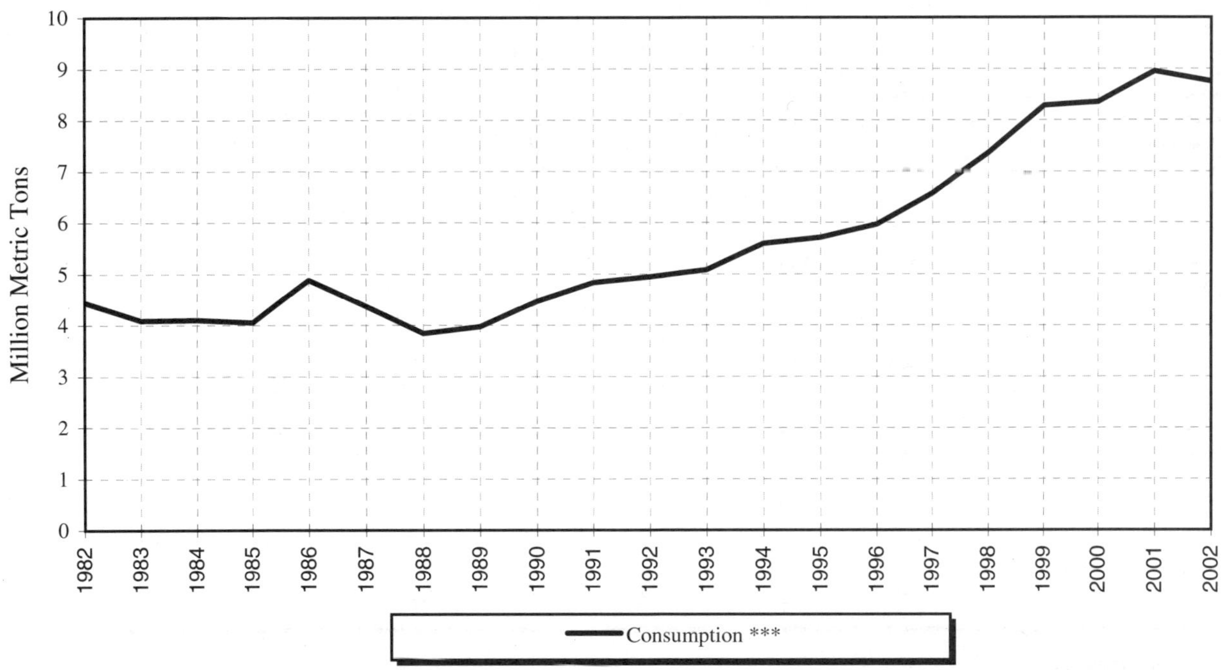

Source: Annual Oil Statistics.

*** Includes refinery fuel and international marine bunkers.

IRELAND
Table 1: Oil Supply and Consumption - Summary Table

000 metric tons

	1973	1978	1980	1990	1999	2000	2001	2002
Indigenous Production	-	-	-	-	-	-	-	-
Crude Oil	..	-	-	-	-	-	-	-
NGL	..	-	-	-	-	-	-	-
Other	..	-	-	-	-	-	-	-
+ Net Imports	5 409	6 052	5 805	4 998	8 285	7 932	8 856	8 899 e
Crude Oil, NGL, Other	2 684	2 294	2 034	2 007	2 848	2 973	3 371	3 190 e
Petroleum Products	2 725	3 758	3 771	2 991	5 437	4 959	5 485	5 709 e
- Intl. Marine Bunkers	92	77	76	18	172	151	161	198 e
+ Stock Changes	149	- 91	- 37	- 216	- 81	111	- 304	- 16 e
TOTAL SUPPLY	**5 466**	**5 884**	**5 692**	**4 764**	**8 032**	**7 892**	**8 391**	**8 685 e**
Refinery Intake	*2 920*	*2 241*	*2 036*	*1 732*	*2 817*	*3 294*	*3 406*	*3 221 e*
Refinery Output	*2 675*	*2 236*	*2 014*	*1 725*	*2 831*	*3 286*	*3 365*	*3 191 e*
LPG + Ethane	*68*	*40*	*30*	*29*	*38*	*47*	*59*	*64 e*
Naphtha	*65*	*6*	*-*	*70*	*114*	*104*	*12*	*12 e*
Kerosenes	*79*	*12*	*19*	*-*	*217*	*249*	*241*	*464 e*
Gasolines	*459*	*530*	*489*	*340*	*435*	*532*	*669*	*680 e*
Gas/Diesel Oil	*694*	*641*	*539*	*618*	*982*	*1 133*	*1 173*	*814 e*
Fuel Oil	*1 253*	*966*	*898*	*629*	*988*	*1 158*	*1 132*	*1 069 e*
Other Products	*57*	*41*	*39*	*39*	*57*	*63*	*79*	*88 e*
- Refinery Losses	245	5	22	7	- 14	8	41	30 e
+ Returns	-	-	-	-	-	-	-	-
- Statistical Differences	125	- 31	- 49	308	- 61	- 310	- 439	104 e
TOTAL CONSUMPTION	**5 096**	**5 910**	**5 719**	**4 449**	**8 107**	**8 194**	**8 789**	**8 551 e**
Transformation and Energy	**1 313**	**1 866**	**1 660**	**411**	**1 525**	**1 176**	**1 388**	..
Crude Oil	-	-	-	-	-	-	-	..
LPG + Ethane	-	15	11	3	3	3	3	..
Gas/Diesel Oil	10	23	17	15	34	36	34	..
Fuel Oil	1 123	1 678	1 498	354	1 404	1 045	1 245	..
Petroleum Coke	-	-	-	-	16	20	27	..
Other	180	150	134	39	68	72	79	..
Transport	**1 363**	**1 811**	**1 714**	**1 962**	**3 598**	**3 917**	**4 177**	..
Motor Gasoline	796	976	1 018	885	1 414	1 493	1 551	..
Gas/Diesel Oil	247	483	421	656	1 614	1 772	1 861	..
Subtotal	*1 043*	*1 459*	*1 439*	*1 541*	*3 028*	*3 265*	*3 412*	..
Aviation Fuels	280	297	223	355	516	598	718	..
Other	40	55	52	66	54	54	47	..
Industry	**1 654**	**1 513**	**1 633**	**867**	**1 086**	**1 208**	**1 241**	..
Naphtha	-	-	-	-	-	-	-	..
LPG + Ethane	30	38	40	55	55	59	60	..
Gas/Diesel Oil	384	420	449	143	179	184	185	..
Fuel Oil	1 033	832	960	428	488	519	435	..
Other	207	223	184	241	364	446	561	..
Other Sectors	**766**	**720**	**712**	**1 209**	**1 898**	**1 893**	**1 983**	..
Gas/Diesel Oil	449	408	411	865	1 040	1 038	1 052	..
Fuel Oil	154	130	119	150	171	181	160	..
Other	163	182	182	194	687	674	771	..
Memo: Non-Energy Use	*217*	*253*	*213*	*204*	*182*	*179*	*209*	..
Chem/Petchem. Feedst.	-	-	-	-	-	-	-	..
Other	*217*	*253*	*213*	*204*	*182*	*179*	*209*	..

Source: Annual Oil Statistics.

IRELAND
Table 2: Supply of Crude Oil, NGL, Feedstocks and Other Hydrocarbons

000 metric tons

	1973	1978	1980	1990	1999	2000	2001	2002
CRUDE OIL								
Indigenous Production	..	-	-	-	-	-	-	-
+ Imports	..	2 196	1 884	1 990	2 826	2 943	3 353	3 177 e
- Exports	..	-	-	-	-	-	-	-
- Direct Use	..	-	-	-	-	-	-	-
+ Stock Change	..	- 37	- 4	- 186	- 31	335	8	- 60 e
- Statistical Difference	..	3	-	89	-	3	- 26	- 88 e
Refinery Intake	**..**	**2 156**	**1 880**	**1 715**	**2 795**	**3 275**	**3 387**	**3 205 e**
Closing Stock Level	..	*234*	*231*	*325*	*450*	*115*	*107*	*167 e*
NATURAL GAS LIQUIDS								
Indigenous Production	..	-	-	-	-	-	-	-
+ Imports	..	-	-	-	-	-	-	-
- Exports	..	-	-	-	-	-	-	-
- Direct Use	..	-	-	-	-	-	-	-
+ Stock Change	..	-	-	-	-	-	-	-
- Statistical Difference	..	-	-	-	-	-	-	-
Refinery Intake	**..**	**-**	**-**	**-**	**-**	**-**	**-**	**-**
Closing Stock Level	..	-	-	-	-	-	-	-
REFINERY FEEDSTOCKS								
Backflows	..	-	-	-	-	-	-	-
+ Transfers	..	-	-	-	-	-	-	-
+ Imports	..	98	150	17	22	44	18	13 e
- Exports	..	-	-	-	-	14	-	-
+ Stock Change	..	- 5	3	- 1	4	- 6	1	3 e
- Statistical Difference	..	8	- 3	- 1	4	5	-	-
Refinery Intake	**..**	**85**	**156**	**17**	**22**	**19**	**19**	**16 e**
Closing Stock Level	..	*12*	*17*	*13*	*7*	*13*	*12*	*9 e*
OTHER HYDROCARBONS								
Indigenous Production	..	-	-	-	-	-	-	-
+ Imports	..	-	-	-	-	-	-	-
- Exports	..	-	-	-	-	-	-	-
- Direct Use	..	-	-	-	-	-	-	-
+ Stock Change	..	-	-	-	-	-	-	-
- Statistical Difference	..	-	-	-	-	-	-	-
Refinery Intake	**..**	**-**	**-**	**-**	**-**	**-**	**-**	**-**
Closing Stock Level	..	-	-	-	-	-	-	-
CRUDE OIL, NGL, FEEDSTOCKS AND OTHER								
Indigenous Production	-	-	-	-	-	-	-	-
+ Backflows	-	-	-	-	-	-	-	-
+ Transfers	-	-	-	-	-	-	-	-
+ Imports	2 684	2 294	2 034	2 007	2 848	2 987	3 371	3 190 e
- Exports	-	-	-	-	-	14	-	-
- Direct Use	-	-	-	-	-	-	-	-
+ Stock Change	124	- 42	- 1	- 187	- 27	329	9	- 57 e
- Statistical Difference	- 112	11	- 3	88	4	8	- 26	- 88 e
Refinery Intake	**2 920**	**2 241**	**2 036**	**1 732**	**2 817**	**3 294**	**3 406**	**3 221 e**
Closing Stock Level	-	*246*	*248*	*338*	*457*	*128*	*119*	*176 e*

Source: Annual Oil Statistics.

IRELAND
Table 3: Motor Gasoline - Supply

000 metric tons

	1973	1978	1980	1990	1999	2000	2001	2002
MOTOR GASOLINE								
Refinery Output	**459**	**530**	**489**	**340**	**435**	**532**	**669**	**680 e**
+ Imports	294	458	532	551	989	1 014	1 001	1 009 e
- Exports	4	-	-	1	7	20	124	106 e
+ Transfers	-	-	-	-	-	- 1	- 1	-
+ Stock Change	-	- 46	- 8	- 10	6	- 100	- 28	32 e
- Statistical Difference	- 47	- 34	- 5	- 5	9	- 68	- 34	15 e
Total Consumption	**796**	**976**	**1 018**	**885**	**1 414**	**1 493**	**1 551**	**1 600 e**
Closing Stock Level	-	*77*	*71*	*99*	*57*	*157*	*185*	*153 e*

Table 4: Aviation Fuels - Supply

000 metric tons

	1973	1978	1980	1990	1999	2000	2001	2002
AVIATION GASOLINE								
Refinery Output	-	-	-	-	-	-	-	-
+ Imports	2	-	1	-	1	1	-	1 e
- Exports	-	-	-	-	-	-	-	-
+ Transfers	-	-	-	-	-	-	1	-
+ Stock Change	-	-	-	1	-	-	-	-
- Statistical Difference	-	- 2	-	1	-	-	-	-
Total Consumption	**2**	**2**	**1**	**-**	**1**	**1**	**1**	**1 e**
Closing Stock Level	-	*1*	*1*	-	-	-	-	-
GASOLINE TYPE JET FUEL								
Refinery Output	-	-	-	-	-	-	-	-
+ Imports	-	-	-	-	-	-	-	-
- Exports	-	-	-	-	-	-	-	-
+ Transfers	-	-	-	-	-	-	-	-
+ Stock Change	-	-	-	-	-	-	-	-
- Statistical Difference	-	-	-	-	-	-	-	-
Total Consumption	**-**	**-**	**-**	**-**	**-**	**-**	**-**	**-**
Closing Stock Level	-	-	-	-	-	-	-	-
KEROSENE TYPE JET FUEL								
Refinery Output	**79**	**12**	**19**	**-**	**40**	**-**	**-**	**149 e**
+ Imports	200	284	201	351	517	687	864	913 e
- Exports	-	-	-	-	-	-	-	-
+ Transfers	-	-	-	-	- 43	- 46	- 130	- 37 e
+ Stock Change	-	4	7	- 14	5	- 15	- 2	- 43 e
- Statistical Difference	1	5	5	- 18	4	29	15	184 e
Total Consumption	**278**	**295**	**222**	**355**	**515**	**597**	**717**	**798 e**
Closing Stock Level	-	*24*	*36*	*35*	*31*	*46*	*48*	*91 e*

Source: Annual Oil Statistics.

IRELAND
Table 5A: Gas/Diesel Oil - Supply and Consumption

000 metric tons

	1973	1978	1980	1990	1999	2000	2001	2002
Refinery Output	**694**	**641**	**539**	**618**	**982**	**1 133**	**1 173**	**814 e**
+ Recycled Products	-	-	-	-	-	-	-	-
+ Imports	696	772	768	1 176	1 948	2 076	2 069	2 569 e
- Exports	30	-	-	4	10	35	47	79 e
- Intl. Marine Bunkers	50	55	35	9	122	114	125	151 e
+ Transfers	-	-	-	-	- 14	- 12	- 8	- 10 e
+ Stock Changes	-	- 57	13	27	- 54	- 198	- 222 e	- 15 e
- Statistical Differences	220	- 33	- 13	129	- 137	- 180	- 292	17 e
Total Consumption	**1 090**	**1 334**	**1 298**	**1 679**	**2 867**	**3 030**	**3 132**	**3 111 e**
Transformation	**10**	**21**	**17**	**11**	**31**	**32**	**29**	**..**
Public Electricity	-	11	2	7	25	28 e	25 e	..
Autoproducer Electricity	-	-	4	4 e	-	-	-	..
Public CHP Plants	-	-	-	-	-	-	-	..
Autoproducers of CHP	-	-	-	-	6	4 e	4 e	..
Public Heat Plants	-	-	-	-	-	-	-	..
Autoproducer Heat Plants	-	-	-	-	-	-	-	..
Petrochemical Industry	-	-	-	-	-	-	-	..
Other Transformation	10	10	11	-	-	-	-	..
Energy Sector	**-**	**2**	**-**	**4**	**3**	**4**	**5**	**..**
Refinery Fuel	-	2	-	4	3	4	5	..
Other Energy Sector	-	-	-	-	-	-	-	..
Final Consumption	**1 080**	**1 311**	**1 281**	**1 664**	**2 833**	**2 994**	**3 098**	**..**
Transport	**247**	**483**	**421**	**656**	**1 614**	**1 772**	**1 861**	**..**
Air	-	-	-	-	-	-	-	..
Road	202	444	380	603	1 482	1 732	1 821	..
Rail	45	36	37	46	114	39	39	..
Inland Waterways	-	3	4	7	18	1	1	..
Non-Specified	-	-	-	-	-	-	-	..
Industry	**384**	**420**	**449**	**143**	**179**	**184**	**185**	**..**
Iron and Steel	-	-	-	1	1	1	1	..
Chemical (incl. Pet.-Chem.)	-	14	12	12	14	14	15	..
Non-Ferrous Metals	-	-	-	-	-	-	-	..
Non-Metallic Minerals	-	-	-	17	21	22	22	..
Transport Equipment	-	17	5	3	4	4	4	..
Machinery	-	20	19	18	22	23	23	..
Mining and Quarrying	-	5	3	19	23	24	24	..
Food and Tobacco	-	85	61	59	72	74	74	..
Pulp, Paper and Printing	-	9	3	3	4	4	4	..
Wood and Wood Products	-	-	-	2	2	2	2	..
Construction	-	-	-	-	-	-	-	..
Textiles and Leather	-	4	4	9	11	11	11	..
Non-Specified	384	266	342	-	5	5	5	..
Other Sectors	**449**	**408**	**411**	**865**	**1 040**	**1 038**	**1 052**	**..**
Commerce and Public	-	198	222	467	562	543	551	..
Residential	449	210	189	190	228	236	239	..
Agriculture	-	-	-	208	250	259	262	..
Non-Specified	-	-	-	-	-	-	-	..
Memo: Non-Energy Use	-	-	-	-	-	-	-	..
Chem/Petchem. Feedst.	-	-	-	-	-	-	-	..
Other	-	-	-	-	-	-	-	..
Closing Stock Level	-	*164*	*183*	*219*	*288*	*486 e*	*708*	*723 e*

Source: Annual Oil Statistics.

IRELAND
Table 5B: Fuel Oil (Residual) - Supply and Consumption

000 metric tons

	1973	1978	1980	1990	1999	2000	2001	2002
Refinery Output	**1 253**	**966**	**898**	**629**	**988**	**1 158**	**1 132**	**1 069 e**
+ Recycled Products	-	-	-	-	-	-	-	-
+ Imports	1 607	1 709	1 911	1 127	2 207	1 582	1 831	1 516 e
- Exports	446	43	179	615	983	1 149	1 096	965 e
- Intl. Marine Bunkers	42	22	41	9	50	37	36	47 e
+ Transfers	-	-	-	-	15	13	8	10 e
+ Stock Changes	-	58	- 43	- 75	-	93	- 58	64 e
- Statistical Differences	62	28	- 31	105	91	- 110	- 79	105 e
Total Consumption	**2 310**	**2 640**	**2 577**	**952**	**2 086**	**1 770**	**1 860**	**1 542 e**
Transformation	**1 123**	**1 660**	**1 490**	**341**	**1 388**	**1 019**	**1 219**	**..**
Public Electricity	1 107	1 617	1 459	339	1 384	1 013	1 213	..
Autoproducer Electricity	-	-	-	-	-	-	-	..
Public CHP Plants	-	-	-	-	-	-	-	..
Autoproducers of CHP	-	39	24	2	4	6 e	6 e	..
Public Heat Plants	-	-	-	-	-	-	-	..
Autoproducer Heat Plants	-	-	-	-	-	-	-	..
Petrochemical Industry	-	-	-	-	-	-	-	..
Other Transformation	16	4	7	-	-	-	-	..
Energy Sector	**-**	**18**	**8**	**13**	**16**	**26**	**26**	**..**
Refinery Fuel	-	18	8	13	16	26	26	..
Other Energy Sector	-	-	-	-	-	-	-	..
Final Consumption	**1 187**	**962**	**1 079**	**598**	**682**	**725**	**615**	**..**
Transport	**-**	**-**	**-**	**20**	**23**	**25**	**20**	**..**
Air	-	-	-	-	-	-	-	..
Road	-	-	-	-	-	-	-	..
Rail	-	-	-	-	-	-	-	..
Inland Waterways	-	-	-	20	23	25	20	..
Non-Specified	-	-	-	-	-	-	-	..
Industry	**1 033**	**832**	**960**	**428**	**488**	**519**	**435**	**..**
Iron and Steel	-	15	15	-	2	-	-	..
Chemical (incl. Pet.-Chem.)	-	130	93	34	39	41	39	..
Non-Ferrous Metals	-	-	-	229	264	280	228	..
Non-Metallic Minerals	-	20	82	17	19	21	17	..
Transport Equipment	-	-	-	3	3	4	3	..
Machinery	-	-	6	24	28	30	24	..
Mining and Quarrying	-	-	-	6	7	7	6	..
Food and Tobacco	-	360	273	79	90	96	84	..
Pulp, Paper and Printing	-	45	16	6	7	7	6	..
Wood and Wood Products	-	-	-	2	2	2	2	..
Construction	-	-	-	-	-	-	-	..
Textiles and Leather	-	-	-	16	18	20	16	..
Non-Specified	1 033	262	475	12	9	11	10	..
Other Sectors	**154**	**130**	**119**	**150**	**171**	**181**	**160**	**..**
Commerce and Public	134	90	95	150	171	181	160	..
Residential	-	-	-	-	-	-	-	..
Agriculture	20	40	24	-	-	-	-	..
Non-Specified	-	-	-	-	-	-	-	..
Memo: Non-Energy Use	-	-	-	-	-	-	-	..
Chem/Petchem. Feedst.	-	-	-	-	-	-	-	..
Other	-	-	-	-	-	-	-	..
Closing Stock Level	-	*604*	*653*	*412*	*516*	*423*	*481*	*417 e*

Source: Annual Oil Statistics.

IRELAND
Table 5C: Total Products - Supply and Consumption

000 metric tons

	1973	1978	1980	1990	1999	2000	2001	2002
Refinery Output	**2 675**	**2 236**	**2 014**	**1 725**	**2 831**	**3 286**	**3 365**	**3 191 e**
+ Recycled Products	-	-	-	-	-	-	-	-
+ Imports	3 213	3 809	3 951	3 689	6 558	6 305	6 803	6 987 e
- Exports	488	51	180	698	1 121	1 346	1 318	1 278 e
- Intl. Marine Bunkers	92	77	76	18	172	151	161	198 e
+ Transfers	-	-	-	-	-	-	-	-
+ Stock Changes	25	- 49	- 36	- 29	- 54	- 218	- 313	41 e
- Statistical Differences	237	- 42	- 46	220	- 65	- 318	- 413	192 e
Total Consumption	**5 096**	**5 910**	**5 719**	**4 449**	**8 107**	**8 194**	**8 789**	**8 551 e**
Transformation	**1 256**	**1 790**	**1 602**	**352**	**1 435**	**1 071**	**1 275**	..
Public Electricity	1 107	1 628	1 461	346	1 409	1 041	1 238	..
Autoproducer Electricity	-	-	4	4	-	-	-	..
Public CHP Plants	-	-	-	-	-	-	-	..
Autoproducers of CHP	-	39	24	2	10	10	10	..
Public Heat Plants	-	-	-	-	-	-	-	..
Autoproducer Heat Plants	-	-	-	-	-	-	-	..
Petrochemical Industry	-	-	-	-	-	-	-	..
Other Transformation	149	123	113	-	16	20	27	..
Energy Sector	**57**	**76**	**58**	**59**	**90**	**105**	**113**	..
Refinery Fuel	57	76	58	59	90	105	113	..
Other Energy Sector	-	-	-	-	-	-	-	..
Final Consumption	**3 783**	**4 044**	**4 059**	**4 038**	**6 582**	**7 018**	**7 401**	..
Transport	**1 363**	**1 811**	**1 714**	**1 962**	**3 598**	**3 917**	**4 177**	..
Air	280	297	223	355	516 e	598 e	718 e	..
Road	998	1 425	1 410	1 494	2 903	3 230	3 377	..
Rail	45	36	37	46	114	39	39	..
Inland Waterways	-	3	4	27	41	26	21	..
Non-Specified	40	50	40	40	24	24	22	..
Industry	**1 654**	**1 513**	**1 633**	**867**	**1 086**	**1 208**	**1 241**	..
Iron and Steel	-	15	15	5	7	6	6	..
Chemical (incl. Pet.-Chem.)	-	149	111	62	128	124	132	..
Non-Ferrous Metals	-	-	-	229	264	280	228	..
Non-Metallic Minerals	-	20	82	102	164	258	326	..
Transport Equipment	-	17	5	10	11	12	11	..
Machinery	4	22	27	68	80	84	80	..
Mining and Quarrying	-	5	3	25	30	31	30	..
Food and Tobacco	-	447	335	140	167	175	164	..
Pulp, Paper and Printing	-	55	20	9	11	11	10	..
Wood and Wood Products	-	-	-	4	4	4	4	..
Construction	115	119	85	135	138	133	163	..
Textiles and Leather	-	4	4	28	33	37	34	..
Non-Specified	1 535	660	946	50	49	53	53	..
Other Sectors	**766**	**720**	**712**	**1 209**	**1 898**	**1 893**	**1 983**	..
Commerce and Public	138	309	349	625	743	734	721	..
Residential	604	366	336	376	904	898	998	..
Agriculture	24	45	27	208	250	259	262	..
Non-Specified	-	-	-	-	1	2	2	..
Memo: Non-Energy Use	*217*	*253*	*213*	*204*	*182*	*179*	*209*	..
Chem/Petchem. Feedst.	-	-	-	-	-	-	-	..
Other	*217*	*253*	*213*	*204*	*182*	*179*	*209*	..
Closing Stock Level	-	*905*	*979*	*781*	*994*	*1 212*	*1 525*	*1 484 e*

Source: Annual Oil Statistics.

IRELAND
Table 6A: Supply and Consumption

000 metric tons

	1980	2000	2001	2002	1980	2000	2001	2002
	LPG AND ETHANE				**REFINERY GAS**			
Refinery Output	**30**	**47**	**59**	**64 e**	**39**	**63**	**79**	**88 e**
+ Recycled Products	-	-	-	-	-	-	-	-
+ Imports	140	112	105	99 e	-	-	-	-
- Exports	-	5	17	22 e	-	-	-	-
- Intl. Marine Bunkers	-	-	-	-	-	-	-	-
+ Transfers	-	-	-	-	-	-	-	-
+ Stock Change	- 1	- 1	- 2	-	-	-	-	-
- Statistical Difference	-	11	- 1	- 5 e	-	-	-	-
Total Consumption	**169**	**142**	**146**	**146 e**	**39**	**63**	**79**	**88 e**
of which: Ethane	-	-	-	-	-	-	-	-
Transformation	-	-	-	..	-	-	-	..
Energy	11	3	3	..	39	63	79	..
Transport	12	5	5	..	-	-	-	..
Industry	40	59	60	..	-	-	-	..
Other	106	75	78	..	-	-	-	..
Memo: Non-Energy Use	-	-	-	..	-	-	-	..
Closing Stock Level	*6*	*8*	*10*	*10 e*	-	-	-	-
	NAPHTHA				**OTHER KEROSENE**			
Refinery Output	**-**	**104**	**12**	**12 e**	**-**	**249**	**241**	**315 e**
+ Recycled Products	-	-	-	-	-	-	-	-
+ Imports	94	-	-	-	86	363	332	352 e
- Exports	-	106	13	11 e	-	25	16	95 e
- Intl. Marine Bunkers	-	-	-	-	-	-	-	-
+ Transfers	-	-	-	-	-	46	130	37 e
+ Stock Change	- 2	2	2	- 1 e	1	- 12	- 3	4 e
- Statistical Difference	- 3	-	1	-	-	- 4	- 23	- 124 e
Total Consumption	**95**	**-**	**-**	**-**	**87**	**625**	**707**	**737 e**
Transformation	95	-	-	..	-	-	-	..
Energy	-	-	-	..	-	-	-	..
Transport	-	-	-	..	-	-	-	..
Industry	-	-	-	..	11	85	96	..
Other	-	-	-	..	76	540	611	..
Memo: Non-Energy Use	-	-	-	..	-	-	-	..
Closing Stock Level	*4*	*7*	*5*	*6 e*	*7*	*45*	*48*	*44 e*
	WHITE SPIRIT + SBP				**LUBRICANTS**			
Refinery Output	**-**	**-**	**-**	**..**	**-**	**-**	**-**	**..**
+ Recycled Products	-	-	-	..	-	-	-	..
+ Imports	3	2	2	..	43	39	38	..
- Exports	-	-	-	..	1	5	5	..
- Intl. Marine Bunkers	-	-	-	..	-	-	-	..
+ Transfers	-	-	-	..	-	-	-	..
+ Stock Change	-	-	-	..	- 1	-	-	..
- Statistical Difference	-	-	-	..	1	-	-	..
Total Consumption	**3**	**2**	**2**	**..**	**40**	**34**	**33**	**..**
Transformation	-	-	-	..	-	-	-	..
Energy	-	-	-	..	-	-	-	..
Transport	-	-	-	..	40	24	22	..
Industry	3	-	-	..	-	10	11	..
Other	-	2	2	..	-	-	-	..
Memo: Non-Energy Use	*3*	*2*	*2*	..	*40*	*34*	*33*	..
Closing Stock Level	-	-	-	..	*5*	-	-	..

Source: Annual Oil Statistics.

IRELAND
Table 6B: Supply and Consumption

000 metric tons

	1980	2000	2001	2002	1980	2000	2001	2002
		BITUMEN				PARAFFIN WAXES		
Refinery Output	-	-	-	..	-	-	-	..
+ Recycled Products	-	-	-	..	-	-	-	..
+ Imports	87	133	163	..	2	11	11	..
- Exports	-	-	-	..	-	1	-	..
- Intl. Marine Bunkers	-	-	-	..	-	-	-	..
+ Transfers	-	-	-	..	-	-	-	..
+ Stock Change	- 2	-	-	..	-	-	-	..
- Statistical Difference	-	-	-	..	-	-	-	..
Total Consumption	**85**	**133**	**163**	**..**	**2**	**10**	**11**	**..**
Transformation	-	-	-	..	-	-	-	..
Energy	-	-	-	..	-	-	-	..
Transport	-	-	-	..	-	-	-	..
Industry	85	133	163	..	2	10	11	..
Other	-	-	-	..	-	-	-	..
Memo: Non-Energy Use	*85*	*133*	*163*	..	*2*	*10*	*11*	..
Closing Stock Level	*11*	-	-	..	-	-	-	..
		PETROLEUM COKE				OTHER PRODUCTS		
Refinery Output	-	-	-	-	-	-	-	-
+ Recycled Products	-	-	-	-	-	-	-	-
+ Imports	-	285	387	328 e	83	-	-	200 e
- Exports	-	-	-	-	-	-	-	-
- Intl. Marine Bunkers	-	-	-	-	-	-	-	-
+ Transfers	-	-	-	-	-	-	-	-
+ Stock Change	-	4	-	-	-	9 e	-	-
- Statistical Difference	-	4	-	-	-	-	-	-
Total Consumption	**-**	**285**	**387**	**328 e**	**83**	**9 e**	**-**	**200 e**
Transformation	-	20	27	..	-	-	-	..
Energy	-	-	-	..	-	9 e	-	..
Transport	-	-	-	..	-	-	-	..
Industry	-	208	280	..	83	-	-	..
Other	-	57	80	..	-	-	-	..
Memo: Non-Energy Use	-	-	-	..	*83*	-	-	..
Closing Stock Level	-	*40*	*40*	*40 e*	*2*	-	-	-

Source: Annual Oil Statistics.

IRELAND
Table 7A: Imports by Country of Origin

000 metric tons

	2000	2001	2002	2000	2001	2002	2000	2001	2002
	Crude Oil			Natural Gas Liquids			Feedstocks and Other		
Australia	-	-	-	-	-	-	-	-	-
Belgium	-	-	-	-	-	-	-	-	-
Canada	-	-	-	-	-	-	-	-	-
Czech Republic	-	-	-	-	-	-	-	-	-
Denmark	-	169	-	-	-	-	-	-	-
Finland	-	-	-	-	-	-	-	-	-
France	-	-	-	-	-	-	-	-	-
Germany	-	-	-	-	-	-	-	-	-
Greece	-	-	-	-	-	-	-	-	-
Hungary	-	-	-	-	-	-	-	-	-
Ireland	-	-	-	-	-	-	-	-	-
Italy	-	-	-	-	-	-	-	-	-
Japan	-	-	-	-	-	-	-	-	-
Korea	-	-	-	-	-	-	-	-	-
Mexico	-	-	-	-	-	-	-	-	-
Netherlands	-	-	-	-	-	-	6	13	3 e
New Zealand	-	-	-	-	-	-	-	-	-
Norway	2 943	2 862	3 011 e	-	-	-	-	-	2 e
Poland	-	-	-	-	-	-	-	-	-
Spain	-	-	-	-	-	-	-	-	-
Sweden	-	-	-	-	-	-	-	-	-
Turkey	-	-	-	-	-	-	-	-	-
United Kingdom	-	322	166 e	-	-	-	38	5	8 e
United States	-	-	-	-	-	-	-	-	-
Other	-	-	-	-	-	-	-	-	-
OECD	**2 943**	**3 353**	**3 177 e**	**-**	**-**	**-**	**44**	**18**	**13 e**
Argentina	-	-	-	-	-	-	-	-	-
Brazil	-	-	-	-	-	-	-	-	-
Colombia	-	-	-	-	-	-	-	-	-
Ecuador	-	-	-	-	-	-	-	-	-
Netherlands Antilles	-	-	-	-	-	-	-	-	-
Peru	-	-	-	-	-	-	-	-	-
Trinidad and Tobago	-	-	-	-	-	-	-	-	-
Venezuela	-	-	-	-	-	-	-	-	-
Other	-	-	-	-	-	-	-	-	-
Latin America	**-**	**-**	**-**	**-**	**-**	**-**	**-**	**-**	**-**
Romania	-	-	-	-	-	-	-	-	-
Other	-	-	-	-	-	-	-	-	-
Non-OECD Europe	**-**	**-**	**-**	**-**	**-**	**-**	**-**	**-**	**-**
FSU	**-**	**-**	**-**	**-**	**-**	**-**	**-**	**-**	**-**
Iran	-	-	-	-	-	-	-	-	-
Iraq	-	-	-	-	-	-	-	-	-
Kuwait	-	-	-	-	-	-	-	-	-
Oman	-	-	-	-	-	-	-	-	-
Qatar	-	-	-	-	-	-	-	-	-
Saudi Arabia	-	-	-	-	-	-	-	-	-
United Arab Emirates	-	-	-	-	-	-	-	-	-
Other	-	-	-	-	-	-	-	-	-
Middle East	**-**	**-**	**-**	**-**	**-**	**-**	**-**	**-**	**-**
India	-	-	-	-	-	-	-	-	-
Indonesia	-	-	-	-	-	-	-	-	-
Singapore	-	-	-	-	-	-	-	-	-
Other	-	-	-	-	-	-	-	-	-
Asia	**-**	**-**	**-**	**-**	**-**	**-**	**-**	**-**	**-**
China	**-**	**-**	**-**	**-**	**-**	**-**	**-**	**-**	**-**
Algeria	-	-	-	-	-	-	-	-	-
Angola/Cabinada	-	-	-	-	-	-	-	-	-
Cameroon	-	-	-	-	-	-	-	-	-
Egypt	-	-	-	-	-	-	-	-	-
Gabon	-	-	-	-	-	-	-	-	-
Libya	-	-	-	-	-	-	-	-	-
Nigeria	-	-	-	-	-	-	-	-	-
Tunisia	-	-	-	-	-	-	-	-	-
Other	-	-	-	-	-	-	-	-	-
Africa	**-**	**-**	**-**	**-**	**-**	**-**	**-**	**-**	**-**
Non-spec./Other	**-**	**-**	**-**	**-**	**-**	**-**	**-**	**-**	**-**
Total Imports	**2 943**	**3 353**	**3 177 e**	**-**	**-**	**-**	**44**	**18**	**13 e**
Memo: Total OPEC	-	-	-	-	-	-	-	-	-

Source: Annual Oil Statistics.

IRELAND
Table 7B: Imports by Country of Origin

000 metric tons

	2000	2001	2002	2000	2001	2002	2000	2001	2002
	Total Products			**Liquefied Petroleum Gases**			**Naphtha**		
Australia	-	-	-	-	-	-	-	-	-
Belgium	37	-	-	-	-	-	-	-	-
Canada	-	-	-	-	-	-	-	-	-
Czech Republic	-	-	-	-	-	-	-	-	-
Denmark	12	-	-	-	-	-	-	-	-
Finland	-	-	-	-	-	-	-	-	-
France	311	176	213 e	-	-	-	-	-	-
Germany	1	-	-	-	-	-	-	-	-
Greece	-	-	-	-	-	-	-	-	-
Hungary	-	-	-	-	-	-	-	-	-
Ireland	-	-	-	-	-	-	-	-	-
Italy	1	1	-	-	-	-	-	-	-
Japan	-	-	-	-	-	-	-	-	-
Korea	-	-	-	-	-	-	-	-	-
Mexico	-	-	-	-	-	-	-	-	-
Netherlands	82	94	80 e	-	-	-	-	-	-
New Zealand	-	-	-	-	-	-	-	-	-
Norway	7	35	63 e	-	-	-	-	-	-
Poland	-	-	-	-	-	-	-	-	-
Spain	-	-	-	-	-	-	-	-	-
Sweden	-	-	-	-	-	-	-	-	-
Turkey	-	-	-	-	-	-	-	-	-
United Kingdom	5 612	6 118	6 383 e	112	105	99 e	-	-	-
United States	113	184	140 e	-	-	-	-	-	-
Other	-	54	-	-	-	-	-	-	-
OECD	**6 176**	**6 662**	**6 879 e**	**112**	**105**	**99 e**	**-**	**-**	**-**
Argentina	-	-	-	-	-	-	-	-	-
Brazil	-	-	-	-	-	-	-	-	-
Colombia	-	-	-	-	-	-	-	-	-
Ecuador	-	-	-	-	-	-	-	-	-
Netherlands Antilles	-	-	-	-	-	-	-	-	-
Peru	-	-	-	-	-	-	-	-	-
Trinidad and Tobago	-	-	-	-	-	-	-	-	-
Venezuela	78	91	90 e	-	-	-	-	-	-
Other	-	-	-	-	-	-	-	-	-
Latin America	**78**	**91**	**90 e**	**-**	**-**	**-**	**-**	**-**	**-**
Romania	-	-	-	-	-	-	-	-	-
Other	-	-	-	-	-	-	-	-	-
Non-OECD Europe	**-**	**-**	**-**	**-**	**-**	**-**	**-**	**-**	**-**
FSU	**49**	**39**	**18 e**	**-**	**-**	**-**	**-**	**-**	**-**
Iran	-	-	-	-	-	-	-	-	-
Iraq	-	-	-	-	-	-	-	-	-
Kuwait	-	-	-	-	-	-	-	-	-
Oman	-	-	-	-	-	-	-	-	-
Qatar	-		-	-	-	-	-	-	-
Saudi Arabia	-	-	-	-	-	-		-	-
United Arab Emirates	-	-	-	-	-	-	-	-	-
Other	-	-	-	-	-	-	-	-	-
Middle East	**-**	**-**	**-**	**-**	**-**	**-**	**-**	**-**	**-**
India	-	-	-	-	-	-	-	-	-
Indonesia	-	-	-	-	-	-	-	-	-
Singapore	2	-	-	-	-	-	-	-	-
Other	-	-	-	-	-	-	-	-	-
Asia	**2**	**-**	**-**	**-**	**-**	**-**	**-**	**-**	**-**
China	**-**	**-**	**-**	**-**	**-**	**-**	**-**	**-**	**-**
Algeria	-	-	-	-	-	-	-	-	-
Angola/Cabinada	-	-	-	-	-	-	-	-	-
Cameroon	-	-	-	-	-	-	-	-	-
Egypt	-	-	-	-	-	-	-	-	-
Gabon	-	-	-	-	-	-	-	-	-
Libya	-	-	-	-	-	-	-	-	-
Nigeria	-	-	-	-	-	-	-	-	-
Tunisia	-	-	-	-	-	-	-	-	-
Other	-	8	-	-	-	-	-	-	-
Africa	**-**	**8**	**-**	**-**	**-**	**-**	**-**	**-**	**-**
Non-spec./Other	**-**	**3**	**-**	**-**	**-**	**-**	**-**	**-**	**-**
Total Imports	**6 305**	**6 803**	**6 987 e**	**112**	**105**	**99 e**	**-**	**-**	**-**
Memo: Total OPEC	*78*	*91*	*90 e*	-	-	-	-	-	-

Source: Annual Oil Statistics.

IRELAND
Table 7C: Imports by Country of Origin

000 metric tons

	2000	2001	2002	2000	2001	2002	2000	2001	2002
	Total Gasoline			**Total Kerosene**			**Gas/Diesel**		
Australia	-	-	-	-	-	-	-	-	-
Belgium	-	-	-	-	-	-	8	-	-
Canada	-	-	-	-	-	-	-	-	-
Czech Republic	-	-	-	-	-	-	-	-	-
Denmark	-	-	-	12	-	-	-	-	-
Finland	-	-	-	-	-	-	-	-	-
France	-	-	-	-	-	-	-	-	-
Germany	-	-	-	-	-	-	-	-	-
Greece	-	-	-	-	-	-	-	-	-
Hungary	-	-	-	-	-	-	-	-	-
Ireland	-	-	-	-	-	-	-	-	-
Italy	-	-	-	-	-	-	-	-	-
Japan	-	-	-	-	-	-	-	-	-
Korea	-	-	-	-	-	-	-	-	-
Mexico	-	-	-	-	-	-	-	-	-
Netherlands	-	-	-	9	11	-	-	-	-
New Zealand	-	-	-	-	-	-	-	-	-
Norway	-	-	12 e	-	-	-	6	-	23 e
Poland	-	-	-	-	-	-	-	-	-
Spain	-	-	-	-	-	-	-	-	-
Sweden	-	-	-	-	-	-	-	-	-
Turkey	-	-	-	-	-	-	-	-	-
United Kingdom	1 015	1 001	998 e	1 029	1 185	1 265 e	2 062	2 069	2 546 e
United States	-	-	-	-	-	-	-	-	-
Other	-	-	-	-	-	-	-	-	-
OECD	**1 015**	**1 001**	**1 010 e**	**1 050**	**1 196**	**1 265 e**	**2 076**	**2 069**	**2 569 e**
Argentina	-	-	-	-	-	-	-	-	-
Brazil	-	-	-	-	-	-	-	-	-
Colombia	-	-	-	-	-	-	-	-	-
Ecuador	-	-	-	-	-	-	-	-	-
Netherlands Antilles	-	-	-	-	-	-	-	-	-
Peru	-	-	-	-	-	-	-	-	-
Trinidad and Tobago	-	-	-	-	-	-	-	-	-
Venezuela	-	-	-	-	-	-	-	-	-
Other	-	-	-	-	-	-	-	-	-
Latin America	**-**	**-**	**-**	**-**	**-**	**-**	**-**	**-**	**-**
Romania	-	-	-	-	-	-	-	-	-
Other	-	-	-	-	-	-	-	-	-
Non-OECD Europe	**-**	**-**	**-**	**-**	**-**	**-**	**-**	**-**	**-**
FSU	**-**	**-**	**-**	**-**	**-**	**-**	**-**	**-**	**-**
Iran	-	-	-	-	-	-	-	-	-
Iraq	-	-	-	-	-	-	-	-	-
Kuwait	-	-	-	-	-	-	-	-	-
Oman	-	-	-	-	-	-	-	-	-
Qatar	-	-	-	-	-	-	-	-	-
Saudi Arabia	-	-	-	-	-	-	-	-	-
United Arab Emirates	-	-	-	-	-	-	-	-	-
Other	-	-	-	-	-	-	-	-	-
Middle East	**-**	**-**	**-**	**-**	**-**	**-**	**-**	**-**	**-**
India	-	-	-	-	-	-	-	-	-
Indonesia	-	-	-	-	-	-	-	-	-
Singapore	-	-	-	-	-	-	-	-	-
Other	-	-	-	-	-	-	-	-	-
Asia	**-**	**-**	**-**	**-**	**-**	**-**	**-**	**-**	**-**
China	**-**	**-**	**-**	**-**	**-**	**-**	**-**	**-**	**-**
Algeria	-	-	-	-	-	-	-	-	-
Angola/Cabinada	-	-	-	-	-	-	-	-	-
Cameroon	-	-	-	-	-	-	-	-	-
Egypt	-	-	-	-	-	-	-	-	-
Gabon	-	-	-	-	-	-	-	-	-
Libya	-	-	-	-	-	-	-	-	-
Nigeria	-	-	-	-	-	-	-	-	-
Tunisia	-	-	-	-	-	-	-	-	-
Other	-	-	-	-	-	-	-	-	-
Africa	**-**	**-**	**-**	**-**	**-**	**-**	**-**	**-**	**-**
Non-spec./Other	**-**	**-**	**-**	**-**	**-**	**-**	**-**	**-**	**-**
Total Imports	**1 015**	**1 001**	**1 010 e**	**1 050**	**1 196**	**1 265 e**	**2 076**	**2 069**	**2 569 e**
Memo: Total OPEC	-	-	-	-	-	-	-	-	-

Source: Annual Oil Statistics.

IRELAND
Table 7D: Imports by Country of Origin

000 metric tons

	2000	2001	2002	2000	2001	2002	2000	2001	2002
	Fuel Oil (Residual)			**Petroleum Coke**			**Other Products**		
Australia	-	-	-	-	-	-	-	-	-
Belgium	28	-	-	-	-	-	1	-	-
Canada	-	-	-	-	-	-	-	-	-
Czech Republic	-	-	-	-	-	-	-	-	-
Denmark	-	-	-	-	-	-	-	-	-
Finland	-	-	-	-	-	-	-	-	-
France	266	117	163 e	-	-	-	45	59	50 e
Germany	-	-	-	-	-	-	1	-	-
Greece	-	-	-	-	-	-	-	-	-
Hungary	-	-	-	-	-	-	-	-	-
Ireland	-	-	-	-	-	-	-	-	-
Italy	-	-	-	-	-	-	1	1	-
Japan	-	-	-	-	-	-	-	-	-
Korea	-	-	-	-	-	-	-	-	-
Mexico	-	-	-	-	-	-	-	-	-
Netherlands	-	-	-	70	81	80 e	3	2	-
New Zealand	-	-	-	-	-	-	-	-	-
Norway	-	34	28 e	-	-	-	1	1	-
Poland	-	-	-	-	-	-	-	-	-
Spain	-	-	-	-	-	-	-	-	-
Sweden	-	-	-	-	-	-	-	-	-
Turkey	-	-	-	-	-	-	-	-	-
United Kingdom	1 251	1 600	1 325 e	10	10	-	133	148	150 e
United States	-	-	-	113	184	140 e	-	-	-
Other	-	54	-	-	-	-	-	-	-
OECD	**1 545**	**1 805**	**1 516 e**	**193**	**275**	**220 e**	**185**	**211**	**200 e**
Argentina	-	-	-	-	-	-	-	-	-
Brazil	-	-	-	-	-	-	-	-	-
Colombia	-	-	-	-	-	-	-	-	-
Ecuador	-	-	-	-	-	-	-	-	-
Netherlands Antilles	-	-	-	-	-	-	-	-	-
Peru	-	-	-	-	-	-	-	-	-
Trinidad and Tobago	-	-	-	-	-	-	-	-	-
Venezuela	-	-	-	78	91	90 e	-	-	-
Other	-	-	-	-	-	-	-	-	-
Latin America	**-**	**-**	**-**	**78**	**91**	**90 e**	**-**	**-**	**-**
Romania	-	-	-	-	-	-	-	-	-
Other	-	-	-	-	-	-	-	-	-
Non-OECD Europe	**-**	**-**	**-**	**-**	**-**	**-**	**-**	**-**	**-**
FSU	**35**	**26**	**-**	**14**	**13**	**18 e**	**-**	**-**	**-**
Iran	-	-	-	-	-	-	-	-	-
Iraq	-	-	-	-	-	-	-	-	-
Kuwait	-	-	-	-	-	-	-	-	-
Oman	-	-	-	-	-	-	-	-	-
Qatar	-	-	-	-	-	-	-	-	-
Saudi Arabia	-	-	-	-	-	-	-	-	-
United Arab Emirates	-	-	-	-	-	-	-	-	-
Other	-	-	-	-	-	-	-	-	-
Middle East	**-**	**-**	**-**	**-**	**-**	**-**	**-**	**-**	**-**
India	-	-	-	-	-	-	-	-	-
Indonesia	-	-	-	-	-	-	-	-	-
Singapore	2	-	-	-	-	-	-	-	-
Other	-	-	-	-	-	-	-	-	-
Asia	**2**	**-**	**-**	**-**	**-**	**-**	**-**	**-**	**-**
China	**-**	**-**	**-**	**-**	**-**	**-**	**-**	**-**	**-**
Algeria	-	-	-	-	-	-	-	-	-
Angola/Cabinada	-	-	-	-	-	-	-	-	-
Cameroon	-	-	-	-	-	-	-	-	-
Egypt	-	-	-	-	-	-	-	-	-
Gabon	-	-	-	-	-	-	-	-	-
Libya	-	-	-	-	-	-	-	-	-
Nigeria	-	-	-	-	-	-	-	-	-
Tunisia	-	-	-	-	-	-	-	-	-
Other	-	-	-	-	8	-	-	-	-
Africa	**-**	**-**	**-**	**-**	**8**	**-**	**-**	**-**	**-**
Non-spec./Other	**-**	**-**	**-**	**-**	**-**	**-**	**-**	**3**	**-**
Total Imports	**1 582**	**1 831**	**1 516 e**	**285**	**387**	**328 e**	**185**	**214**	**200 e**
Memo: Total OPEC	-	-	-	*78*	*91*	*90 e*	-	-	-

Source: Annual Oil Statistics.

IRELAND
Table 8A: Exports by Country of Destination

000 metric tons

	2000	2001	2002	2000	2001	2002	2000	2001	2002
	Crude Oil			**Natural Gas Liquids**			**Feedstocks and Other**		
Australia	-	-	-	-	-	-	-	-	-
Austria	-	-	-	-	-	-	-	-	-
Belgium	-	-	-	-	-	-	-	-	-
Canada	-	-	-	-	-	-	-	-	-
Denmark	-	-	-	-	-	-	-	-	-
Finland	-	-	-	-	-	-	-	-	-
France	-	-	-	-	-	-	-	-	-
Germany	-	-	-	-	-	-	-	-	-
Greece	-	-	-	-	-	-	-	-	-
Ireland	-	-	-	-	-	-	-	-	-
Italy	-	-	-	-	-	-	-	-	-
Japan	-	-	-	-	-	-	-	-	-
Luxembourg	-	-	-	-	-	-	-	-	-
Mexico	-	-	-	-	-	-	-	-	-
Netherlands	-	-	-	-	-	-	11	-	-
Norway	-	-	-	-	-	-	-	-	-
Portugal	-	-	-	-	-	-	-	-	-
Spain	-	-	-	-	-	-	-	-	-
Sweden	-	-	-	-	-	-	-	-	-
Switzerland	-	-	-	-	-	-	-	-	-
Turkey	-	-	-	-	-	-	-	-	-
United Kingdom	-	-	-	-	-	-	3	-	-
United States	-	-	-	-	-	-	-	-	-
Other OECD	-	-	-	-	-	-	-	-	-
OECD	**-**	**-**	**-**	**-**	**-**	**-**	**14**	**-**	**-**
Africa	-	-	-	-	-	-	-	-	-
Latin America	-	-	-	-	-	-	-	-	-
Asia	-	-	-	-	-	-	-	-	-
China	-	-	-	-	-	-	-	-	-
Non-OECD Europe	-	-	-	-	-	-	-	-	-
Former Soviet Union	-	-	-	-	-	-	-	-	-
Middle East	-	-	-	-	-	-	-	-	-
Non-specified/Other	-	-	-	-	-	-	-	-	-
Non-OECD	**-**	**-**	**-**	**-**	**-**	**-**	**-**	**-**	**-**
Total Exports	**-**	**-**	**-**	**-**	**-**	**-**	**14**	**-**	**-**
	Total Products			**Liquefied Petroleum Gas**			**Naphtha**		
Australia	-	-	-	-	-	-	-	-	-
Austria	-	-	-	-	-	-	-	-	-
Belgium	63	29	-	-	-	-	63	3	-
Canada	-	-	-	-	-	-	-	-	-
Denmark	-	-	3 e	-	-	-	-	-	3 e
Finland	-	-	-	-	-	-	-	-	-
France	90	16	4 e	-	1	4 e	18	3	-
Germany	-	-	28 e	-	-	-	-	-	-
Greece	-	-	-	-	-	-	-	-	-
Ireland	-	-	-	-	-	-	-	-	-
Italy	32	-	-	-	-	-	-	-	-
Japan	-	-	-	-	-	-	-	-	-
Luxembourg	-	-	-	-	-	-	-	-	-
Mexico	-	-	-	-	-	-	-	-	-
Netherlands	36	3	2 e	-	-	-	25	3	2 e
Norway	330	289	225 e	-	-	-	-	-	4 e
Portugal	-	-	-	-	-	-	-	-	-
Spain	-	28	59 e	-	-	-	-	-	-
Sweden	-	-	-	-	-	-	-	-	-
Switzerland	-	-	-	-	-	-	-	-	-
Turkey	-	-	-	-	-	-	-	-	-
United Kingdom	462	815	905 e	5	16	18 e	-	4	2 e
United States	296	138	52 e	-	-	-	-	-	-
Other OECD	-	-	-	-	-	-	-	-	-
OECD	**1 309**	**1 318**	**1 278 e**	**5**	**17**	**22 e**	**106**	**13**	**11 e**
Africa	-	-	-	-	-	-	-	-	-
Latin America	-	-	-	-	-	-	-	-	-
Asia	-	-	-	-	-	-	-	-	-
China	-	-	-	-	-	-	-	-	-
Non-OECD Europe	37	-	-	-	-	-	-	-	-
Former Soviet Union	-	-	-	-	-	-	-	-	-
Middle East	-	-	-	-	-	-	-	-	-
Non-specified/Other	-	-	-	-	-	-	-	-	-
Non-OECD	**37**	**-**	**-**	**-**	**-**	**-**	**-**	**-**	**-**
Total Exports	**1 346**	**1 318**	**1 278 e**	**5**	**17**	**22 e**	**106**	**13**	**11 e**

Source: Annual Oil Statistics.

IRELAND

Table 8B: Exports by Country of Destination

000 metric tons

	2000	2001	2002	2000	2001	2002	2000	2001	2002
	Total Gasoline			**Total Kerosene**			**Gas/Diesel Oil**		
Australia	-	-	-	-	-	-	-	-	-
Austria	-	-	-	-	-	-	-	-	-
Belgium	-	-	-	-	-	-	-	-	-
Canada	-	-	-	-	-	-	-	-	-
Denmark	-	-	-	-	-	-	-	-	-
Finland	-	-	-	-	-	-	-	-	-
France	-	-	-	-	-	-	23	12	-
Germany	-	-	-	-	-	-	-	-	-
Greece	-	-	-	-	-	-	-	-	-
Ireland	-	-	-	-	-	-	-	-	-
Italy	-	-	-	-	-	-	-	-	-
Japan	-	-	-	-	-	-	-	-	-
Luxembourg	-	-	-	-	-	-	-	-	-
Mexico	-	-	-	-	-	-	-	-	-
Netherlands	-	-	-	-	-	-	11	-	-
Norway	-	-	-	-	-	-	-	-	-
Portugal	-	-	-	-	-	-	-	-	-
Spain	-	-	-	-	-	-	-	-	26 e
Sweden	-	-	-	-	-	-	-	-	-
Switzerland	-	-	-	-	-	-	-	-	-
Turkey	-	-	-	-	-	-	-	-	-
United Kingdom	20	124	106 e	25	16	95 e	1	35	53 e
United States	-	-	-	-	-	-	-	-	-
Other OECD	-	-	-	-	-	-	-	-	-
OECD	**20**	**124**	**106 e**	**25**	**16**	**95 e**	**35**	**47**	**79 e**
Africa	-	-	-	-	-	-	-	-	-
Latin America	-	-	-	-	-	-	-	-	-
Asia	-	-	-	-	-	-	-	-	-
China	-	-	-	-	-	-	-	-	-
Non-OECD Europe	-	-	-	-	-	-	-	-	-
Former Soviet Union	-	-	-	-	-	-	-	-	-
Middle East	-	-	-	-	-	-	-	-	-
Non-specified/Other	-	-	-	-	-	-	-	-	-
Non-OECD	**-**	**-**	**-**	**-**	**-**	**-**	**-**	**-**	**-**
Total Exports	**20**	**124**	**106 e**	**25**	**16**	**95 e**	**35**	**47**	**79 e**
	Fuel Oil (Residual)			**Petroleum Coke**			**Other Products**		
Australia	-	-	-	-	-	-	-	-	-
Austria	-	-	-	-	-	-	-	-	-
Belgium	-	26	-	-	-	-	-	-	-
Canada	-	-	-	-	-	-	-	-	-
Denmark	-	-	-	-	-	-	-	-	-
Finland	-	-	-	-	-	-	-	-	-
France	49	-	-	-	-	-	-	-	-
Germany	-	-	28 e	-	-	-	-	-	-
Greece	-	-	-	-	-	-	-	-	-
Ireland	-	-	-	-	-	-	-	-	-
Italy	32	-	-	-	-	-	-	-	-
Japan	-	-	-	-	-	-	-	-	-
Luxembourg	-	-	-	-	-	-	-	-	-
Mexico	-	-	-	-	-	-	-	-	-
Netherlands	-	-	-	-	-	-	-	-	-
Norway	330	289	221 e	-	-	-	-	-	-
Portugal	-	-	-	-	-	-	-	-	-
Spain	-	28	33 e	-	-	-	-	-	-
Sweden	-	-	-	-	-	-	-	-	-
Switzerland	-	-	-	-	-	-	-	-	-
Turkey	-	-	-	-	-	-	-	-	-
United Kingdom	405	615	631 e	-	-	-	6	5	-
United States	296	138	52 e	-	-	-	-	-	-
Other OECD	-	-	-	-	-	-	-	-	-
OECD	**1 112**	**1 096**	**965 e**	**-**	**-**	**-**	**6**	**5**	**-**
Africa	-	-	-	-	-	-	-	-	-
Latin America	-	-	-	-	-	-	-	-	-
Asia	-	-	-	-	-	-	-	-	-
China	-	-	-	-	-	-	-	-	-
Non-OECD Europe	37	-	-	-	-	-	-	-	-
Former Soviet Union	-	-	-	-	-	-	-	-	-
Middle East	-	-	-	-	-	-	-	-	-
Non-specified/Other	-	-	-	-	-	-	-	-	-
Non-OECD	**37**	**-**	**-**	**-**	**-**	**-**	**-**	**-**	**-**
Total Exports	**1 149**	**1 096**	**965 e**	**-**	**-**	**-**	**6**	**5**	**-**

Source: Annual Oil Statistics.

ITALY

Figure 1
FUEL SHARES OF TPES *

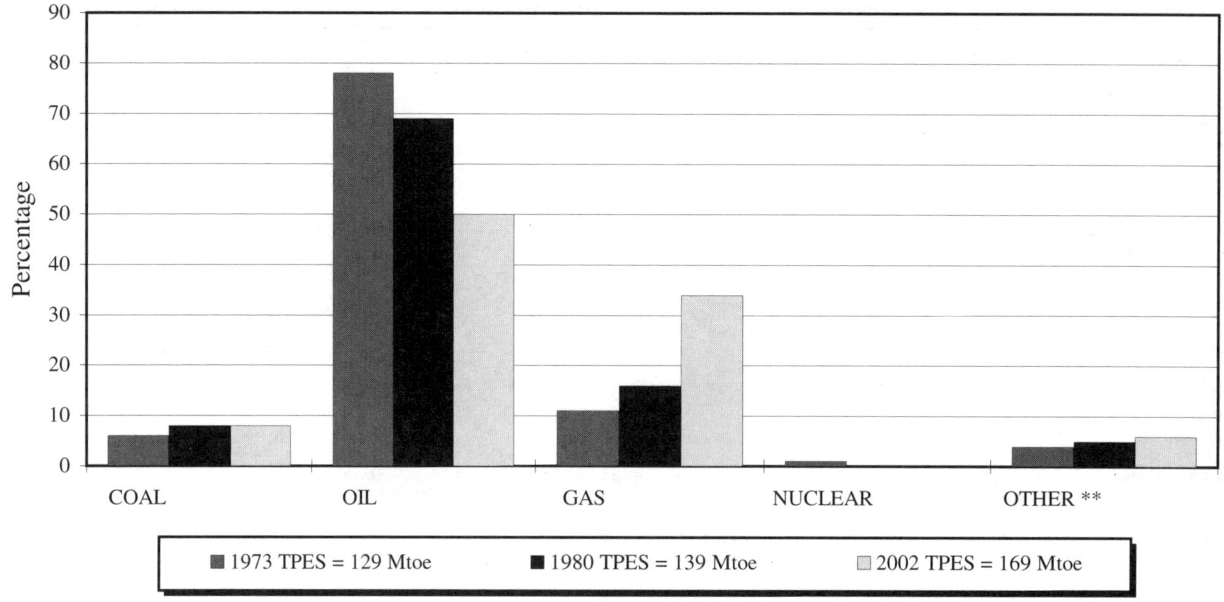

Source: Annual Energy Balances.

* Excluding electricity trade.

** Includes hydro, geothermal, solar, wind, combustible renewables and waste, etc.

Figure 2
OIL PRODUCTION AND CONSUMPTION

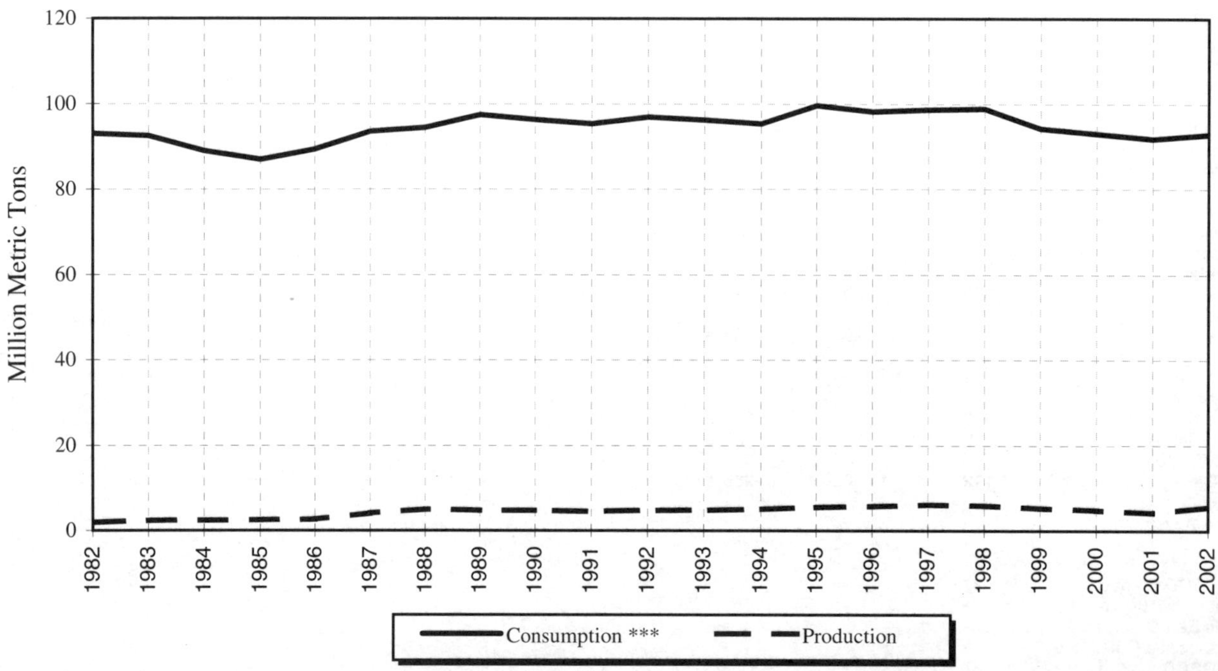

Source: Annual Oil Statistics.

*** Includes refinery fuel and international marine bunkers.

ITALY
Table 1: Oil Supply and Consumption - Summary Table

000 metric tons

	1973	1978	1980	1990	1999	2000	2001	2002
Indigenous Production	1 105	1 489	1 825	4 723	5 216	4 764	4 223	5 643 e
Crude Oil	..	1 422	1 800	4 641	4 998	4 555	4 066	5 273 e
NGL	..	67	25	27	24	30	31	30 e
Other	..	-	-	55	194	179	126	340 e
+ Net Imports	105 030	96 147	98 051	90 361	87 753	88 911	84 825	83 402 e
Crude Oil, NGL, Other	128 944	111 490	94 322	83 713	87 874	89 882	90 645	88 898 e
Petroleum Products	-23 914	-15 343	3 729	6 648	- 121	- 971	-5 820	-5 496 e
- Intl. Marine Bunkers	7 353	5 663	4 283	2 743	2 485	2 785	2 900	3 147 e
+ Stock Changes	- 770	1 645	- 620	-2 112	-	-1 868	1 260	- 609 e
TOTAL SUPPLY	**98 012**	**93 618**	**94 973**	**90 229**	**90 484**	**89 022**	**87 408**	**85 289 e**
Refinery Intake	*132 723*	*116 227*	*99 247*	*90 955*	*95 393*	*96 363*	*97 381*	*97 571 e*
Refinery Output	*130 307*	*113 861*	*97 992*	*90 564*	*94 074*	*94 771*	*96 409*	*97 480 e*
LPG + Ethane	*2 329*	*2 525*	*2 084*	*2 131*	*2 540*	*2 713*	*2 694*	*2 691 e*
Naphtha	*11 192*	*5 958*	*4 184*	*4 129*	*3 483*	*2 998*	*2 203*	*2 178 e*
Kerosenes	*6 907*	*5 058*	*4 263*	*5 161*	*4 847*	*4 477*	*4 071*	*3 514 e*
Gasolines	*15 096*	*16 062*	*15 050*	*18 536*	*20 034*	*19 190*	*20 967*	*21 556 e*
Gas/Diesel Oil	*28 975*	*29 599*	*26 815*	*29 133*	*34 527*	*36 487*	*37 076*	*38 900 e*
Fuel Oil	*60 238*	*49 317*	*39 468*	*23 653*	*19 409*	*19 455*	*19 295*	*19 431 e*
Other Products	*5 570*	*5 342*	*6 128*	*7 821*	*9 234*	*9 451*	*10 103*	*9 210 e*
- Refinery Losses	2 416	2 366	1 255	391	1 319	1 592	972	91 e
+ Returns	3 506	2 445	2 632	3 728	2 535	2 800	2 443	2 324 e
- Statistical Differences	100	- 150	- 89	23	- 49	- 7	- 33	-2 210 e
TOTAL CONSUMPTION	**99 002**	**93 847**	**96 439**	**93 543**	**91 749**	**90 237**	**88 912**	**89 732 e**
Transformation and Energy	**28 030**	**29 533**	**31 255**	**31 122**	**27 274**	**27 188**	**24 840**	..
Crude Oil	15	30	300	205	-	-	-	..
LPG + Ethane	70	146	193	117	108	261	156	..
Gas/Diesel Oil	445	630	613	775	1 094	1 359	984	..
Fuel Oil	21 834	24 683	25 856	23 866	20 662	20 026	18 365	..
Petroleum Coke	417	296	368	920	1 130	1 169	1 146	..
Other	5 249	3 748	3 925	5 239	4 280	4 373	4 189	..
Transport	**18 824**	**21 704**	**24 217**	**32 972**	**39 488**	**39 613**	**40 112**	..
Motor Gasoline	11 036	10 730	11 780	13 073	17 533	16 678	16 348	..
Gas/Diesel Oil	5 133	7 948	9 405	15 819	16 779	17 474	18 521	..
Subtotal	*16 169*	*18 678*	*21 185*	*28 892*	*34 312*	*34 152*	*34 869*	
Aviation Fuels	1 896	1 817	1 842	2 227	3 470	3 567	3 391	..
Other	759	1 209	1 190	1 853	1 706	1 894	1 852	..
Industry	**30 092**	**22 678**	**22 940**	**17 148**	**15 075**	**14 173**	**13 547**	..
Naphtha	6 008	3 593	3 068	3 522	2 481	2 531	2 024	..
LPG + Ethane	379	474	474	298	546	425	422	..
Gas/Diesel Oil	612	1 388	1 133	1 135	1 352	1 423	1 129	..
Fuel Oil	19 314	13 545	13 697	4 607	4 029	3 630	3 176	..
Other	3 779	3 678	4 568	7 586	6 667	6 164	6 796	..
Other Sectors	**22 056**	**19 932**	**18 027**	**12 301**	**9 912**	**9 263**	**10 413**	..
Gas/Diesel Oil	12 455	13 825	13 168	9 630	7 128	6 702	7 197	..
Fuel Oil	6 200	3 950	2 810	550	378	330	350	..
Other	3 401	2 157	2 049	2 121	2 406	2 231	2 866	..
Memo: Non-Energy Use	*8 230*	*6 470*	*6 708*	*8 761*	*8 488*	*6 897*	*7 308*	..
Chem/Petchem. Feedst.	*4 831*	*3 221*	*3 098*	*4 457*	*4 728*	*3 263*	*3 740*	..
Other	*3 399*	*3 249*	*3 610*	*4 304*	*3 760*	*3 634*	*3 568*	..

Source: Annual Oil Statistics.

ITALY
Table 2: Supply of Crude Oil, NGL, Feedstocks and Other Hydrocarbons

000 metric tons

	1973	1978	1980	1990	1999	2000	2001	2002
CRUDE OIL								
Indigenous Production	..	1 422	1 800	4 641	4 998	4 555	4 066	5 273 e
+ Imports	..	108 699	89 058	74 725	80 608	83 653	82 829	80 173 e
- Exports	..	-	-	859	-	-	-	-
- Direct Use	..	30	300	205	-	-	-	-
+ Stock Change	..	1 248	817	- 843	815	- 712	376	169 e
- Statistical Difference	..	-	-	-	-	-	-	- 614 e
Refinery Intake	**..**	**111 339**	**91 375**	**77 459**	**86 421**	**87 496**	**87 271**	**86 229 e**
Closing Stock Level	*..*	*5 220*	*4 795*	*4 832*	*3 684*	*4 396*	*4 020*	*3 851 e*
NATURAL GAS LIQUIDS								
Indigenous Production	..	67	25	27	24	30	31	30 e
+ Imports	..	-	-	-	-	-	-	-
- Exports	..	-	-	-	-	-	-	-
- Direct Use	..	-	-	-	-	-	-	-
+ Stock Change	..	-	-	-	-	-	-	-
- Statistical Difference	..	-	-	-	-	-	-	-
Refinery Intake	**..**	**67**	**25**	**27**	**24**	**30**	**31**	**30 e**
Closing Stock Level	*..*	*-*	*-*	*-*	*-*	*-*	*-*	*-*
REFINERY FEEDSTOCKS								
Backflows	..	2 329	2 522	3 678	2 360	2 626	2 443	2 324 e
+ Transfers	..	-	-	-	-	-	-	-
+ Imports	..	2 791	5 264	11 553	7 323	6 628	8 248	8 957 e
- Exports	..	-	-	1 706	276	521	603	393 e
+ Stock Change	..	- 299	61	- 111	- 712	- 53	- 123	- 81 e
- Statistical Difference	..	-	-	-	-	- 2	-	- 349 e
Refinery Intake	**..**	**4 821**	**7 847**	**13 414**	**8 695**	**8 682**	**9 965**	**11 156 e**
Closing Stock Level	*..*	*243*	*422*	*542*	*1 562*	*1 615*	*1 738*	*1 819 e*
OTHER HYDROCARBONS								
Indigenous Production	..	-	-	55	194	179	126	340 e
+ Imports	..	-	-	-	219	122	173	161 e
- Exports	..	-	-	-	-	-	2	-
- Direct Use	..	-	-	-	148	152	126	318 e
+ Stock Change	..	-	-	-	- 12	6	- 57	- 43 e
- Statistical Difference	..	-	-	-	-	-	-	- 16 e
Refinery Intake	**..**	**-**	**-**	**55**	**253**	**155**	**114**	**156 e**
Closing Stock Level	*..*	*39*	*39*	*39*	*22*	*16*	*73*	*116 e*
CRUDE OIL, NGL, FEEDSTOCKS AND OTHER								
Indigenous Production	1 105	1 489	1 825	4 723	5 216	4 764	4 223	5 643 e
+ Backflows	3 313	2 329	2 522	3 678	2 360	2 626	2 443	2 324 e
+ Transfers	-	-	-	-	-	-	-	-
+ Imports	128 944	111 490	94 322	86 278	88 150	90 403	91 250	89 291 e
- Exports	-	-	-	2 565	276	521	605	393 e
- Direct Use	15	30	300	205	148	152	126	318 e
+ Stock Change	- 638	949	878	- 954	91	- 759	196	45 e
- Statistical Difference	- 14	-	-	-	-	- 2	-	- 979 e
Refinery Intake	**132 723**	**116 227**	**99 247**	**90 955**	**95 393**	**96 363**	**97 381**	**97 571 e**
Closing Stock Level	*-*	*5 502*	*5 256*	*5 413*	*5 268*	*6 027*	*5 831*	*5 786 e*

Source: Annual Oil Statistics.

ITALY

Table 3: Motor Gasoline - Supply

000 metric tons

	1973	1978	1980	1990	1999	2000	2001	2002
MOTOR GASOLINE								
Refinery Output	**14 810**	**15 593**	**14 284**	**18 488**	**20 025**	**19 096**	**20 956**	**21 546 e**
+ Imports	-	281	229	628	726	598	615	571 e
- Exports	3 367	5 083	2 806	4 468	2 729	3 896	4 279	4 805 e
+ Transfers	-	56	691	- 249	148	152	126	318 e
+ Stock Change	- 271	- 146	- 292	114	- 114	1 285	- 49	239 e
- Statistical Difference	- 85	- 307	-	11	- 5	-	-	1 222 e
Total Consumption	**11 257**	**11 008**	**12 106**	**14 502**	**18 061**	**17 235**	**17 369**	**16 647 e**
Closing Stock Level	-	*2 007*	*2 109*	*2 049*	*3 222*	*1 937*	*1 986*	*1 747 e*

Table 4: Aviation Fuels - Supply

000 metric tons

	1973	1978	1980	1990	1999	2000	2001	2002
AVIATION GASOLINE								
Refinery Output	**86**	**34**	**192**	**48**	**9**	**94**	**11**	**10 e**
+ Imports	-	-	-	-	-	-	-	-
- Exports	60	19	180	37	-	89	-	-
+ Transfers	-	-	-	-	-	-	-	-
+ Stock Change	- 4	-	-	-	-	-	-	-
- Statistical Difference	-	-	-	-	-	- 1	-	-
Total Consumption	**22**	**15**	**12**	**11**	**9**	**6**	**11**	**10 e**
Closing Stock Level	-	-	-	-	-	-	-	-
GASOLINE TYPE JET FUEL								
Refinery Output	**200**	**435**	**574**	-	-	-	-	-
+ Imports	-	-	-	-	-	-	-	-
- Exports	-	258	350	-	-	-	-	-
+ Transfers	-	-	30	-	-	-	-	-
+ Stock Change	-	- 5	-			-	-	-
- Statistical Difference	-	-	-	-	-	-	-	-
Total Consumption	**200**	**172**	**254**	**-**	**-**	**-**	**-**	**-**
Closing Stock Level	-	*29*	*29*	-	-	-	-	-
KEROSENE TYPE JET FUEL								
Refinery Output	**2 171**	**2 028**	**1 840**	**2 703**	**4 454**	**4 114**	**3 736**	**3 140 e**
+ Imports	-	-	-	53	36	21	52	19 e
- Exports	1 520	383	229	721	782	630	502	869 e
+ Transfers	1 044	- 5	-	249	-	-	-	100 e
+ Stock Change	-	20	-	- 23	- 232	89	117	- 79 e
- Statistical Difference	-	5	-	-	-	-	-	- 835 e
Total Consumption	**1 695**	**1 655**	**1 611**	**2 261**	**3 476**	**3 594**	**3 403**	**3 146 e**
Closing Stock Level	-	*195*	*195*	*214*	*669*	*580*	*463*	*542 e*

Source: Annual Oil Statistics.

ITALY
Table 5A: Gas/Diesel Oil - Supply and Consumption

000 metric tons

	1973	1978	1980	1990	1999	2000	2001	2002
Refinery Output	**28 975**	**29 599**	**26 815**	**29 133**	**34 527**	**36 487**	**37 076**	**38 900 e**
+ Recycled Products	-	-	-	-	-	-	-	-
+ Imports	483	847	2 947	4 946	1 047	893	748	1 129 e
- Exports	9 394	6 211	2 808	5 398	8 930	8 586	9 881	9 820 e
- Intl. Marine Bunkers	755	810	620	569	794	771	722	668 e
+ Transfers	-	-	- 645	-	-	-	-	-
+ Stock Changes	- 664	366	-1 446	- 744	503	-1 016	597	-1 257 e
- Statistical Differences	-	-	- 76	9	-	49	- 13	- 104 e
Total Consumption	**18 645**	**23 791**	**24 319**	**27 359**	**26 353**	**26 958**	**27 831**	**28 388 e**
Transformation	**380**	**564**	**539**	**722**	**972**	**1 189**	**905**	**..**
Public Electricity	162	167	249	222	226	210 e	150 e	..
Autoproducer Electricity	-	-	-	81	55	-	-	..
Public CHP Plants	-	-	-	-	1	552 e	296 e	..
Autoproducers of CHP	-	-	-	-	278	-	-	..
Public Heat Plants	-	-	-	-	-	-	-	..
Autoproducer Heat Plants	-	-	-	-	-	-	-	..
Petrochemical Industry	218	397	290	419	412	427	459	..
Other Transformation	-	-	-	-	-	-	-	..
Energy Sector	**65**	**66**	**74**	**53**	**122**	**170**	**79**	**..**
Refinery Fuel	5	1	4	1	121	125	64	..
Other Energy Sector	60	65	70	52	1	45	15	..
Final Consumption	**18 200**	**23 161**	**23 706**	**26 584**	**25 259**	**25 599**	**26 847**	**..**
Transport	**5 133**	**7 948**	**9 405**	**15 819**	**16 779**	**17 474**	**18 521**	**..**
Air	-	-	-	-	-	-	-	..
Road	4 663	7 462	8 855	15 427	16 417	17 137	18 149	..
Rail	120	136	170	196	139	137	119	..
Inland Waterways	350	350	380	196	223	200	253	..
Non-Specified	-	-	-	-	-	-	-	..
Industry	**612**	**1 388**	**1 133**	**1 135**	**1 352**	**1 423**	**1 129**	**..**
Iron and Steel	50	38	65	35	8	8	8	..
Chemical (incl. Pet.-Chem.)	333	598	358	761	974	1 035	721	..
Non-Ferrous Metals	10	15	20	13	5	5	5	..
Non-Metallic Minerals	60	140	55	56	28	5	45	..
Transport Equipment	-	-	-	-	-	-	-	..
Machinery	90	270	260	165	154	114	113	..
Mining and Quarrying	-	-	-	-	20	20	18	..
Food and Tobacco	20	80	90	25	45	45	47	..
Pulp, Paper and Printing	10	30	25	14	19	19	16	..
Wood and Wood Products	-	-	-	-	-	-	-	..
Construction	-	-	-	-	53 e	73	52	..
Textiles and Leather	10	70	100	23	46	46	59	..
Non-Specified	29	147	160	43	-	53	45	..
Other Sectors	**12 455**	**13 825**	**13 168**	**9 630**	**7 128**	**6 702**	**7 197**	**..**
Commerce and Public	300	520	400	304	149	285	240	..
Residential	11 000	11 920	11 300	7 048	4 570	4 044	4 480	..
Agriculture	1 155	1 385	1 468	2 278	2 409	2 373	2 477	..
Non-Specified	-	-	-	-	-	-	-	..
Memo: Non-Energy Use	*312*	*558*	*328*	*746*	*944*	*1 005*	*694*	..
Chem/Petchem. Feedst.	*312*	*558*	*328*	*746*	*944*	*1 005*	*694*	..
Other	-	-	-	-	-	-	-	..
Closing Stock Level	-	*3 744*	*5 267*	*5 353*	*2 897*	*3 913*	*3 316*	*4 573 e*

Source: Annual Oil Statistics.

ITALY
Table 5B: Fuel Oil (Residual) - Supply and Consumption

000 metric tons

	1973	1978	1980	1990	1999	2000	2001	2002
Refinery Output	**60 238**	**49 317**	**39 468**	**23 653**	**19 409**	**19 455**	**19 295**	**19 431 e**
+ Recycled Products	-	-	-	-	-	-	-	-
+ Imports	2 547	3 342	8 769	11 821	11 911	11 744	7 690	8 457 e
- Exports	10 168	5 489	1 923	3 442	4 524	4 709	3 516	2 422 e
- Intl. Marine Bunkers	6 550	4 800	3 610	2 129	1 634	1 956	2 120	2 419 e
+ Transfers	-	-	- 211	-	-	-	-	-
+ Stock Changes	1 589	258	364	- 627	- 167	- 555	542	- 37 e
- Statistical Differences	8	-	44	3	- 80	- 7	-	-1 214 e
Total Consumption	**47 648**	**42 628**	**42 813**	**29 273**	**25 075**	**23 986**	**21 891**	**24 224 e**
Transformation	**19 064**	**21 471**	**23 080**	**22 173**	**18 922**	**17 974**	**16 446**	..
Public Electricity	18 974	18 220	19 032	20 342	16 715	16 127 e	14 436 e	..
Autoproducer Electricity	-	-	-	-	689	-	-	..
Public CHP Plants	-	-	-	54	41	1 582 e	1 704 e	..
Autoproducers of CHP	-	3 200	4 000	1 402	1 197	-	-	..
Public Heat Plants	-	-	-	-	-	-	-	..
Autoproducer Heat Plants	-	-	-	-	-	-	-	..
Petrochemical Industry	82	51 e	48	375	280	265	306	..
Other Transformation	8	-	-	-	-	-	-	..
Energy Sector	**2 770**	**3 212**	**2 776**	**1 693**	**1 740**	**2 052**	**1 919**	..
Refinery Fuel	2 730	3 092	2 626	1 588	1 738	1 990	1 898	..
Other Energy Sector	40	120	150	105	2	62	21	..
Final Consumption	**25 814**	**17 945**	**16 957**	**5 407**	**4 413**	**3 960**	**3 526**	..
Transport	**300**	**450**	**450**	**250**	**6**	-	-	..
Air	-	-	-	-	-	-	-	..
Road	-	-	-	-	-	-	-	..
Rail	-	-	-	-	-	-	-	..
Inland Waterways	-	-	-	-	-	-	-	..
Non-Specified	300	450	450	250	6	-	-	..
Industry	**19 314**	**13 545**	**13 697**	**4 607**	**4 029**	**3 630**	**3 176**	..
Iron and Steel	1 050	940	1 000	26	69	65	89	..
Chemical (incl. Pet.-Chem.)	5 214	3 466 e	3 357	1 088	789	810	580	..
Non-Ferrous Metals	150	-	-	-	61	60	62	..
Non-Metallic Minerals	5 500	4 140	4 145	1 169	956	916	396	..
Transport Equipment	-	-	-	-	-	-	-	..
Machinery	2 300	1 095	1 100	464	497	450	510	
Mining and Quarrying	100		-	-	28	27	27	..
Food and Tobacco	1 450	1 165	1 170	416	643	540	580	..
Pulp, Paper and Printing	1 200	735	700	317	155	134	169	..
Wood and Wood Products	-	-	-	-	-	-	-	..
Construction	-	-	-	-	38 e	25 e	25 e	..
Textiles and Leather	1 250	840	1 060	593	343	327	278	..
Non-Specified	1 100	1 164	1 165	534	450	276	460	..
Other Sectors	**6 200**	**3 950**	**2 810**	**550**	**378**	**330**	**350**	..
Commerce and Public	-	-	-	-	-	-	-	..
Residential	5 750	3 650	2 610	435	365	330	350	..
Agriculture	450	300	200	115	13	-	-	..
Non-Specified	-	-	-	-	-	-	-	..
Memo: Non-Energy Use	*118*	*51*	*55*	*288*	*125*	*132*	-	..
Chem/Petchem. Feedst.	-	-	-	-	-	-	-	..
Other	*118*	*51*	*55*	*288*	*125*	*132*	-	..
Closing Stock Level	-	*7 705*	*6 391*	*8 013*	*5 174*	*5 729*	*5 187*	*5 224 e*

Source: Annual Oil Statistics.

ITALY
Table 5C: Total Products - Supply and Consumption

000 metric tons

	1973	1978	1980	1990	1999	2000	2001	2002
Refinery Output	**130 307**	**113 861**	**97 992**	**90 564**	**94 074**	**94 771**	**96 409**	**97 480 e**
+ Recycled Products	193	116	110	50	175	174	-	-
+ Imports	4 989	7 466	15 743	23 613	19 957	20 079	15 996	16 037 e
- Exports	28 903	22 809	12 014	16 965	20 078	21 050	21 816	21 533 e
- Intl. Marine Bunkers	7 353	5 663	4 283	2 743	2 485	2 785	2 900	3 147 e
+ Transfers	15	30	300	205	148	152	126	318 e
+ Stock Changes	- 132	696	-1 498	-1 158	- 91	-1 109	1 064	- 654 e
- Statistical Differences	114	- 150	- 89	23	- 49	- 5	- 33	-1 231 e
Total Consumption	**99 002**	**93 847**	**96 439**	**93 543**	**91 749**	**90 237**	**88 912**	**89 732 e**
Transformation	**23 374**	**24 255**	**26 184**	**26 216**	**22 161**	**22 266**	**19 673**	..
Public Electricity	19 674	18 387	19 281	20 564	16 941	17 207 e	14 930 e	..
Autoproducer Electricity	174	145	171	292	1 110	-	-	..
Public CHP Plants	-	43	8	59	55	2 429 e	2 300 e	..
Autoproducers of CHP	77	3 297	4 159	1 588	1 691	-	-	..
Public Heat Plants	-	-	-	-	-	-	-	..
Autoproducer Heat Plants	-	-	-	-	-	-	-	..
Petrochemical Industry	3 313	2 329	2 522	3 678	2 360	2 626	2 443	..
Other Transformation	136	54	43	35	4	4	-	..
Energy Sector	**4 656**	**5 278**	**5 071**	**4 906**	**5 113**	**4 922**	**5 167**	..
Refinery Fuel	4 421	4 948	4 426	4 422	5 108	4 774	5 120	..
Other Energy Sector	235	330	645	484	5	148	47	..
Final Consumption	**70 972**	**64 314**	**65 184**	**62 421**	**64 475**	**63 049**	**64 072**	..
Transport	**18 824**	**21 704**	**24 217**	**32 972**	**39 488**	**39 613**	**40 112**	..
Air	1 896	1 817	1 842	2 227	3 470	3 567	3 391	..
Road	16 066	18 797	21 187	29 921	35 645	35 705	36 341	..
Rail	120	136	170	196	142	139	121	..
Inland Waterways	442	504	568	378	225	202	259	..
Non-Specified	300	450	450	250	6	-	-	..
Industry	**30 092**	**22 678**	**22 940**	**17 148**	**15 075**	**14 173**	**13 547**	..
Iron and Steel	1 120	1 008	1 098	86	97	95	117	..
Chemical (incl. Pet.-Chem.)	12 076	8 211	7 833	7 051	5 638	4 964	4 670	..
Non-Ferrous Metals	170	25	37	19	89	92	93	..
Non-Metallic Minerals	5 595	4 355	4 252	3 499	3 321	3 488	2 948	..
Transport Equipment	-	-	-	-	-	-	-	..
Machinery	2 643	1 590	1 620	811	792	713	771	..
Mining and Quarrying	100	-	-	-	51	50	48	..
Food and Tobacco	1 480	1 255	1 272	457	731	629	674	..
Pulp, Paper and Printing	1 220	795	737	338	181	160	192	..
Wood and Wood Products	-	-	-	-	-	-	-	..
Construction	2 057	1 742	2 016	2 431	2 613	2 526	2 641	..
Textiles and Leather	1 263	925	1 178	629	430	414	379	..
Non-Specified	2 368	2 772	2 897	1 827	1 132	1 042	1 014	..
Other Sectors	**22 056**	**19 932**	**18 027**	**12 301**	**9 912**	**9 263**	**10 413**	..
Commerce and Public	300	520	400	310	188	314	267	..
Residential	19 954	17 485	15 692	9 256	7 119	6 422	6 842	..
Agriculture	1 802	1 927	1 935	2 722	2 591	2 515	2 602	..
Non-Specified	-	-	-	13	14	12	702	..
Memo: Non-Energy Use	*8 230*	*6 470*	*6 708*	*8 761*	*8 488*	*6 897*	*7 308*	..
Chem/Petchem. Feedst.	*4 831*	*3 221*	*3 098*	*4 457*	*4 728*	*3 263*	*3 740*	..
Other	*3 399*	*3 249*	*3 610*	*4 304*	*3 760*	*3 634*	*3 568*	..
Closing Stock Level	-	*16 611*	*16 922*	*17 767*	*14 672*	*15 781*	*14 717*	*15 371 e*

Source: Annual Oil Statistics.

ITALY
Table 6A: Supply and Consumption

000 metric tons

	1980	2000	2001	2002	1980	2000	2001	2002
	LPG AND ETHANE				**REFINERY GAS**			
Refinery Output	**2 084**	**2 713**	**2 694**	**2 691 e**	**1 979**	**2 611**	**2 741**	**2 816 e**
+ Recycled Products	-	-	-	-	-	-	-	-
+ Imports	626	1 978	1 843	1 378 e	-	-	-	-
- Exports	230	408	425	452 e	-	-	-	-
- Intl. Marine Bunkers	-	-	-	-	-	-	-	-
+ Transfers	-	-	-	-	-	-	-	-
+ Stock Change	- 49	- 102	- 111	195 e	-	-	-	-
- Statistical Difference	- 57	-	4	- 106 e	-	-	-	171 e
Total Consumption	**2 488**	**4 181**	**3 997**	**3 918 e**	**1 979**	**2 611**	**2 741**	**2 645 e**
of which: Ethane	-	-	-	-	-	-	-	-
Transformation	120	199	111	..	171	913	424	..
Energy	73	62	45	..	1 516	1 609	1 891	..
Transport	736	1 422	1 390	..	-	-	-	..
Industry	474	425	422	..	292	89	426	..
Other	1 085	2 073	2 029	..	-	-	-	..
Memo: Non-Energy Use	*128*	*222*	*71*	..	*292*	*89*	*110*	..
Closing Stock Level	*160*	*343*	*454*	*259 e*	-	-	-	-

	1980	2000	2001	2002	1980	2000	2001	2002
	NAPHTHA				**OTHER KEROSENE**			
Refinery Output	**4 184**	**2 998**	**2 203**	**2 178 e**	**2 423**	**363**	**335**	**374 e**
+ Recycled Products	-	-	-	-	-	-	-	-
+ Imports	1 985	1 893	2 115	1 762 e	43	296	286	179 e
- Exports	525	923	942	689 e	1 912	200	136	203 e
- Intl. Marine Bunkers	-	-	-	-	-	-	-	-
+ Transfers	- 697	-	-	-	817	-	-	- 100 e
+ Stock Change	- 13	- 131	- 21	19 e	- 68	- 29	14	61 e
- Statistical Difference	- 34	-	-	219 e	4	-	-	-
Total Consumption	**4 968**	**3 837**	**3 355**	**3 051 e**	**1 299**	**430**	**499**	**311 e**
Transformation	1 891	1 225	1 054	..	224	121	126	..
Energy	9	81	277	..	24	1	55	..
Transport	-	-	-	..	4	-	-	..
Industry	3 068	2 531	2 024	..	269	261	266	..
Other	-	-	-	..	778	47	52	..
Memo: Non-Energy Use	*2 096*	*1 684 e*	*2 024*	..	*254*	*251*	*256*	..
Closing Stock Level	*880*	*1 108*	*1 129*	*1 110 e*	*658*	*321*	*307*	*246 e*

	1980	2000	2001	2002	1980	2000	2001	2002
	WHITE SPIRIT + SBP				**LUBRICANTS**			
Refinery Output	**117**	**12**	**14**	**..**	**1 028**	**1 264**	**1 577**	**..**
+ Recycled Products	-	-	-	..	110	174	-	..
+ Imports	5	-	-	..	199	123	98	..
- Exports	56	-	-	..	584	669	958	..
- Intl. Marine Bunkers	-	-	-	..	53	58	58	..
+ Transfers	-	-	-	..	-	-	-	..
+ Stock Change	-	-	-	..	- 43	- 62	81	..
- Statistical Difference	-	-	-	..	-	- 54	- 24	..
Total Consumption	**66**	**12**	**14**	**..**	**657**	**826**	**764**	**..**
Transformation	-	-	-	..	-	-	-	..
Energy	-	-	-	..	-	-	-	..
Transport	-	-	-	..	-	472	462	..
Industry	66	12	14	..	657	322	286	..
Other	-	-	-	..	-	32	16	..
Memo: Non-Energy Use	*66*	*12*	*14*	..	*657*	*826*	*764*	..
Closing Stock Level	*9*	-	-	..	*417*	*367*	*286*	..

Source: Annual Oil Statistics.

ITALY
Table 6B: Supply and Consumption

000 metric tons

	1980	2000	2001	2002	1980	2000	2001	2002
	BITUMEN				**PARAFFIN WAXES**			
Refinery Output	**2 065**	**2 705**	**2 924**	**..**	**32**	**40**	**42**	**..**
+ Recycled Products	-	-	-	..	-	-	-	..
+ Imports	38	46	2	..	34	-	-	..
- Exports	162	315	387	..	1	-	-	..
- Intl. Marine Bunkers	-	-	-	..	-	-	-	..
+ Transfers	-	-	-	..	-	-	-	..
+ Stock Change	49	-	25	..	-	-	-	..
- Statistical Difference	-	8	-	..	-	-	-	..
Total Consumption	**1 990**	**2 428**	**2 564**	**..**	**65**	**40**	**42**	**..**
Transformation	-	-	-	..	-	-	-	..
Energy	-	-	-	..	-	-	-	..
Transport	-	-	-	..	-	-	-	..
Industry	1 990	2 428	2 564	..	65	40	42	..
Other	-	-	-	..	-	-	-	..
Memo: Non-Energy Use	*1 990*	*2 428*	*2 564*	..	*65*	*40*	*42*	..
Closing Stock Level	*77*	*224*	*199*	..	*7*	-	-	..
	PETROLEUM COKE				**OTHER PRODUCTS**			
Refinery Output	**542**	**1 909**	**1 679**	**1 528 e**	**365**	**910**	**1 126**	**4 866 e**
+ Recycled Products	-	-	-	-	-	-	-	-
+ Imports	821	2 055	1 875	1 583 e	47	432	672	959 e
- Exports	24	45	54	3 e	224	580	736	2 270 e
- Intl. Marine Bunkers	-	-	-	-	-	-	-	60 e
+ Transfers	-	-	-	-	315	-	-	-
+ Stock Change	-	- 321	17	114 e	-	- 267	- 148	91 e
- Statistical Difference	-	-	-	- 214 e	30	-	-	- 370 e
Total Consumption	**1 339**	**3 598**	**3 517**	**3 436 e**	**473**	**495**	**914**	**3 956 e**
Transformation	159	274	281	..	-	13	15	..
Energy	209	895	865	..	300	23	14	..
Transport	-	-	-	..	-	-	-	..
Industry	971	2 429	2 371	..	173	459	183	..
Other	-	-	-	..	-	-	702	..
Memo: Non-Energy Use	*604*	-	-	..	*173*	*142*	*183*	..
Closing Stock Level	*502*	*808*	*791*	*677 e*	*221*	*451*	*599*	*993 e*

Source: Annual Oil Statistics.

ITALY
Table 7A: Imports by Country of Origin

000 metric tons

	2000	2001	2002	2000	2001	2002	2000	2001	2002
	Crude Oil			Natural Gas Liquids			Feedstocks and Other		
Australia	-	-	-	-	-	-	-	-	-
Belgium	-	-	-	-	-	-	-	65	28 e
Canada	-	-	-	-	-	-	-	-	-
Czech Republic	-	-	-	-	-	-	-	-	-
Denmark	-	-	-	-	-	-	-	-	30 e
Finland	-	-	-	-	-	-	-	-	-
France	-	-	-	-	-	-	93	70	213 e
Germany	-	-	-	-	-	-	28	9	34 e
Greece	-	-	-	-	-	-	9	114	130 e
Hungary	-	-	-	-	-	-	-	-	-
Ireland	-	-	-	-	-	-	48	-	-
Italy	-	-	-	-	-	-	-	-	-
Japan	-	-	-	-	-	-	-	-	-
Korea	-	-	-	-	-	-	-	-	-
Mexico	285	87	81 e	-	-	-	-	-	-
Netherlands	-	-	-	-	-	-	-	-	3 e
New Zealand	-	-	-	-	-	-	-	-	-
Norway	3 650	2 927	5 183 e	-	-	-	-	-	-
Poland	-	-	-	-	-	-	-	-	-
Spain	-	-	-	-	-	-	-	-	151 e
Sweden	-	-	-	-	-	-	-	-	26 e
Turkey	-	-	-	-	-	-	326	490	196 e
United Kingdom	489	343	258 e	-	-	-	55	33	-
United States	-	-	-	-	-	-	-	7	78 e
Other	-	-	-	-	-	-	75	4	-
OECD	**4 424**	**3 357**	**5 522 e**	**-**	**-**	**-**	**634**	**792**	**889 e**
Argentina	-	-	-	-	-	-	-	-	-
Brazil	-	150	89 e	-	-	-	-	-	-
Colombia	-	-	-	-	-	-	-	-	-
Ecuador	-	-	-	-	-	-	-	-	-
Netherlands Antilles	-	-	-	-	-	-	-	-	-
Peru	-	-	-	-	-	-	-	-	-
Trinidad and Tobago	-	-	-	-	-	-	-	-	-
Venezuela	131	139	31 e	-	-	-	-	-	-
Other	116	-	-	-	-	-	-	-	-
Latin America	**247**	**289**	**120 e**	**-**	**-**	**-**	**-**	**-**	**-**
Romania	-	-	-	-	-	-	13	-	2 e
Other	-	-	-	-	-	-	78	105	124 e
Non-OECD Europe	**-**	**-**	**-**	**-**	**-**	**-**	**91**	**105**	**126 e**
FSU	**16 137**	**19 501**	**18 672 e**	**-**	**-**	**-**	**3 442**	**4 429**	**5 749 e**
Iran	10 386	10 443	9 411 e	-	-	-	-	-	30 e
Iraq	8 285	3 925	2 654 e	-	-	-	-	-	-
Kuwait	1 574	653	55 e	-	-	-	-	-	-
Oman	-	-	-	-	-	-	-	-	-
Qatar	-	-	-	-	-	-	-	-	-
Saudi Arabia	8 407	8 933	8 337 e				16	93	3 e
United Arab Emirates	-	-	-	-	-	-	-	-	-
Other	2 004	5 063	4 868 e	-	-	-	244	972	423 e
Middle East	**30 656**	**29 017**	**25 325 e**	**-**	**-**	**-**	**260**	**1 065**	**456 e**
India	-	-	-	-	-	-	-	-	-
Indonesia	-	-	-	-	-	-	-	-	-
Singapore	-	-	-	-	-	-	-	-	-
Other	-	-	-	-	-	-	-	-	-
Asia	**-**	**-**	**-**	**-**	**-**	**-**	**-**	**-**	**-**
China	**-**	**-**	**-**	**-**	**-**	**-**	**-**	**-**	**-**
Algeria	3 190	2 765	2 049 e	-	-	-	412	494	639 e
Angola/Cabinada	82	314	1 311 e	-	-	-	-	-	-
Cameroon	2 330	2 284	1 389 e	-	-	-	-	-	-
Egypt	3 320	2 868	3 526 e	-	-	-	155	125	78 e
Gabon	-	-	-	-	-	-	-	-	-
Libya	21 940	20 284	19 620 e	-	-	-	1 050	689	375 e
Nigeria	1 142	1 752	1 571 e	-	-	-	-	-	149 e
Tunisia	55	222	385 e	-	-	-	177	206	355 e
Other	130	176	683 e	-	-	-	529	516	302 e
Africa	**32 189**	**30 665**	**30 534 e**	**-**	**-**	**-**	**2 323**	**2 030**	**1 898 e**
Non-spec./Other	**-**	**-**	**-**	**-**	**-**	**-**	**-**	**-**	**-**
Total Imports	**83 653**	**82 829**	**80 173 e**	**-**	**-**	**-**	**6 750**	**8 421**	**9 118 e**
Memo: Total OPEC	*55 055*	*48 894*	*43 728 e*	-	-	-	*1 478*	*1 276*	*1 196 e*

Source: Annual Oil Statistics.

ITALY
Table 7B: Imports by Country of Origin

000 metric tons

	2000	2001	2002	2000	2001	2002	2000	2001	2002
	Total Products			**Liquefied Petroleum Gases**			**Naphtha**		
Australia	15	-	-	15	-	-	-	-	-
Belgium	545	227	271 e	-	-	6 e	-	-	-
Canada	5	-	-	-	-	-	-	-	-
Czech Republic	-	-	-	-	-	-	-	-	-
Denmark	168	70	162 e	-	-	-	-	-	-
Finland	2	-	-	-	-	-	-	-	-
France	1 368	1 565	1 144 e	190	195	184 e	-	-	13 e
Germany	447	163	69 e	19	17	6 e	124	7	-
Greece	173	127	98 e	4	1	10 e	138	58	15 e
Hungary	3	12	12 e	3	12	12 e	-	-	-
Ireland	-	-	-	-	-	-	-	-	-
Italy	-	-	-	-	-	-	-	-	-
Japan	-	-	-	-	-	-	-	-	-
Korea	-	-	-	-	-	-	-	-	-
Mexico	-	6	-	-	-	-	-	6	-
Netherlands	372	261	358 e	31	12	22 e	-	-	-
New Zealand	-	-	-	-	-	-	-	-	-
Norway	167	246	186 e	10	-	3 e	-	-	-
Poland	-	-	-	-	-	-	-	-	-
Spain	287	208	26 e	-	-	-	76	-	17 e
Sweden	216	129	46 e	31	-	-	-	-	-
Turkey	222	578	321 e	3	2	-	127	377	314 e
United Kingdom	1 696	1 452	1 355 e	30	-	19 e	85	-	-
United States	2 159	1 617	1 721 e	-	-	16 e	-	-	-
Other	385	215	196 e	23	26	17 e	41	-	-
OECD	**8 230**	**6 876**	**5 965 e**	**359**	**265**	**295 e**	**591**	**448**	**359 e**
Argentina	-	-	-	-	-	-	-	-	-
Brazil	239	6	2 e	-	-	-	-	-	-
Colombia	-	-	92 e	-	-	-	-	-	-
Ecuador	-	-	-	-	-	-	-	-	-
Netherlands Antilles	-	-	-	-	-	-	-	-	-
Peru	-	-	-	-	-	-	-	-	-
Trinidad and Tobago	392	-	68 e	-	-	-	-	-	-
Venezuela	2 170	437	369 e	-	-	-	-	-	-
Other	105	54	219 e	-	-	2 e	-	-	-
Latin America	**2 906**	**497**	**750 e**	**-**	**-**	**2 e**	**-**	**-**	**-**
Romania	-	50	229 e	-	-	-	-	-	-
Other	582	539	428 e	62	67	33 e	72	68	106 e
Non-OECD Europe	**582**	**589**	**657 e**	**62**	**67**	**33 e**	**72**	**68**	**106 e**
FSU	**1 139**	**1 321**	**1 617 e**	**2**	**2**	**3 e**	**54**	**78**	**249 e**
Iran	7	23	-	7	-	-	-	23	-
Iraq	-	-	-	-	-	-	-	-	-
Kuwait	-	-	-	-	-	-	-	-	-
Oman	-	-	-	-	-	-	-	-	-
Qatar	-	-	-	-	-	-	-	-	-
Saudi Arabia	132	4	8 e	118	4	8 e	-	-	-
United Arab Emirates	-	-	-	-	-	-	-	-	-
Other	1 106	704	270 e	-	-	-	87	29	19 e
Middle East	**1 245**	**731**	**278 e**	**125**	**4**	**8 e**	**87**	**52**	**19 e**
India	107	58	-	-	-	-	-	-	-
Indonesia	42	-	468 e	-	-	-	-	-	-
Singapore	-	-	-	-	-	-	-	-	-
Other	2	-	24 e	-	-	-	-	-	-
Asia	**151**	**58**	**492 e**	**-**	**-**	**-**	**-**	**-**	**-**
China	**-**	**-**	**-**	**-**	**-**	**-**	**-**	**-**	**-**
Algeria	1 745	1 554	1 516 e	1 259	1 187	811 e	63	36	34 e
Angola/Cabinada	-	-	-	-	-	-	-	-	-
Cameroon	-	-	-	-	-	-	-	-	-
Egypt	186	318	237 e	43	155	133 e	140	50	52 e
Gabon	-	-	4 e	-	-	4 e	-	-	-
Libya	3 838	3 928	3 362 e	128	163	84 e	862	1 294	819 e
Nigeria	-	-	100 e	-	-	-	-	-	-
Tunisia	24	22	77 e	-	-	-	24	-	-
Other	33	102	520 e	-	-	5 e	-	89	124 e
Africa	**5 826**	**5 924**	**5 816 e**	**1 430**	**1 505**	**1 037 e**	**1 089**	**1 469**	**1 029 e**
Non-spec./Other	**-**	**-**	**462 e**	**-**	**-**	**-**	**-**	**-**	**-**
Total Imports	**20 079**	**15 996**	**16 037 e**	**1 978**	**1 843**	**1 378 e**	**1 893**	**2 115**	**1 762 e**
Memo: Total OPEC	*7 934*	*5 946*	*5 823 e*	*1 512*	*1 354*	*903 e*	*925*	*1 353*	*853 e*

Source: Annual Oil Statistics.

ITALY
Table 7C: Imports by Country of Origin

000 metric tons

	2000	2001	2002	2000	2001	2002	2000	2001	2002
	Total Gasoline			**Total Kerosene**			**Gas/Diesel**		
Australia	-	-	-	-	-	-	-	-	-
Belgium	49	-	-	-	-	-	4	9	-
Canada	-	-	-	-	-	-	5	-	-
Czech Republic	-	-	-	-	-	-	-	-	-
Denmark	-	-	-	-	-	-	20	-	-
Finland	-	-	-	-	-	-	-	-	-
France	298	345	381 e	9	-	-	46	42	54 e
Germany	-	47	-	-	-	-	-	-	-
Greece	31	-	27 e	-	-	19 e	-	68	27 e
Hungary	-	-	-	-	-	-	-	-	-
Ireland	-	-	-	-	-	-	-	-	-
Italy	-	-	-	-	-	-	-	-	-
Japan	-	-	-	-	-	-	-	-	-
Korea	-	-	-	-	-	-	-	-	-
Mexico	-	-	-	-	-	-	-	-	-
Netherlands	57	68	88 e	-	-	-	19	25	62 e
New Zealand	-	-	-	-	-	-	-	-	-
Norway	21	-	-	-	-	-	-	-	-
Poland	-	-	-	-	-	-	-	-	-
Spain	20	29	-	-	-	-	-	-	-
Sweden	-	-	-	-	-	-	-	-	-
Turkey	-	9	-	-	-	-	-	-	-
United Kingdom	57	-	54 e	-	-	-	25	-	1 e
United States	-	-	-	-	-	-	37	-	107 e
Other	12	14	11 e	-	-	-	-	-	-
OECD	**545**	**512**	**561 e**	**9**	**-**	**19 e**	**156**	**144**	**251 e**
Argentina	-	-	-	-	-	-	-	-	-
Brazil	-	-	-	-	-	-	-	-	-
Colombia	-	-	-	-	-	-	-	-	-
Ecuador	-	-	-	-	-	-	-	-	-
Netherlands Antilles	-	-	-	-	-	-	-	-	-
Peru	-	-	-	-	-	-	-	-	-
Trinidad and Tobago	-	-	-	-	-	-	-	-	-
Venezuela	-	-	-	-	18	-	-	-	-
Other	-	-	-	-	-	-	-	-	-
Latin America	**-**	**-**	**-**	**-**	**18**	**-**	**-**	**-**	**-**
Romania	-	21	-	-	-	-	-	29	210 e
Other	20	58	-	-	16	-	215	138	103 e
Non-OECD Europe	**20**	**79**	**-**	**-**	**16**	**-**	**215**	**167**	**313 e**
FSU	**-**	**-**	**-**	**51**	**-**	**23 e**	**214**	**159**	**128 e**
Iran	-	-	-	-	-	-	-	-	-
Iraq	-	-	-	-	-	-	-	-	-
Kuwait	-	-	-	-	-	-	-	-	-
Oman	-	-	-	-	-	-	-	-	-
Qatar	-	-	-	-	-	-	-	-	-
Saudi Arabia	-	-	-	12	-	-	-	-	-
United Arab Emirates	-	-	-	-	-	-	-	-	-
Other	-	-	-	-	18	-	-	-	58 e
Middle East	**-**	**-**	**-**	**12**	**18**	**-**	**-**	**-**	**58 e**
India	-	-	-	-	-	-	-	-	-
Indonesia	-	-	-	-	-	-	-	-	-
Singapore	-	-	-	-	-	-	-	-	-
Other	-	-	-	-	-	-	-	-	-
Asia	**-**	**-**	**-**	**-**	**-**	**-**	**-**	**-**	**-**
China	**-**	**-**	**-**	**-**	**-**	**-**	**-**	**-**	**-**
Algeria	25	-	-	-	-	28 e	288	236	195 e
Angola/Cabinada	-	-	-	-	-	-	-	-	-
Cameroon	-	-	-	-	-	-	-	-	-
Egypt	-	-	-	-	-	-	-	-	-
Gabon	-	-	-	-	-	-	-	-	-
Libya	8	24	10 e	245	286	128 e	20	16	100 e
Nigeria	-	-	-	-	-	-	-	-	7 e
Tunisia	-	-	-	-	-	-	-	22	77 e
Other	-	-	-	-	-	-	-	4	-
Africa	**33**	**24**	**10 e**	**245**	**286**	**156 e**	**308**	**278**	**379 e**
Non-spec./Other	**-**	**-**	**-**	**-**	**-**	**-**	**-**	**-**	**-**
Total Imports	**598**	**615**	**571 e**	**317**	**338**	**198 e**	**893**	**748**	**1 129 e**
Memo: Total OPEC	*33*	*24*	*10 e*	*257*	*304*	*156 e*	*308*	*252*	*302 e*

Source: Annual Oil Statistics.

ITALY
Table 7D: Imports by Country of Origin

000 metric tons

	2000	2001	2002	2000	2001	2002	2000	2001	2002
	Fuel Oil (Residual)			**Petroleum Coke**			**Other Products**		
Australia	-	-	-	-	-	-	-	-	-
Belgium	460	199	252 e	-	-	-	32	19	13 e
Canada	-	-	-	-	-	-	-	-	-
Czech Republic	-	-	-	-	-	-	-	-	-
Denmark	148	70	162 e	-	-	-	-	-	-
Finland	-	-	-	-	-	-	2	-	-
France	740	894	426 e	-	-	-	85	89	86 e
Germany	287	75	43 e	-	-	-	17	17	20 e
Greece	-	-	-	-	-	-	-	-	-
Hungary	-	-	-	-	-	-	-	-	-
Ireland	-	-	-	-	-	-	-	-	-
Italy	-	-	-	-	-	-	-	-	-
Japan	-	-	-	-	-	-	-	-	-
Korea	-	-	-	-	-	-	-	-	-
Mexico	-	-	-	-	-	-	-	-	-
Netherlands	247	135	176 e	-	-	-	18	21	10 e
New Zealand	-	-	-	-	-	-	-	-	-
Norway	136	246	183 e	-	-	-	-	-	-
Poland	-	-	-	-	-	-	-	-	-
Spain	133	161	-	6	-	5 e	52	18	4 e
Sweden	184	117	46 e	-	-	-	1	12	-
Turkey	80	179	-	-	-	-	12	11	7 e
United Kingdom	1 339	1 313	1 183 e	-	-	-	160	139	98 e
United States	294	86	220 e	1 704	1 439	1 196 e	124	92	182 e
Other	286	151	160 e	-	-	-	23	24	8 e
OECD	**4 334**	**3 626**	**2 851 e**	**1 710**	**1 439**	**1 201 e**	**526**	**442**	**428 e**
Argentina	-	-	-	-	-	-	-	-	-
Brazil	239	-	-	-	-	-	-	6	2 e
Colombia	-	-	92 e	-	-	-	-	-	-
Ecuador	-	-	-	-	-	-	-	-	-
Netherlands Antilles	-	-	-	-	-	-	-	-	-
Peru	-	-	-	-	-	-	-	-	-
Trinidad and Tobago	392	-	68 e	-	-	-	-	-	-
Venezuela	1 931	46	-	239	373	369 e	-	-	-
Other	105	54	191 e	-	-	13 e	-	-	13 e
Latin America	**2 667**	**100**	**351 e**	**239**	**373**	**382 e**	**-**	**6**	**15 e**
Romania	-	-	19 e	-	-	-	-	-	-
Other	190	184	186 e	-	-	-	23	8	-
Non-OECD Europe	**190**	**184**	**205 e**	**-**	**-**	**-**	**23**	**8**	**-**
FSU	**818**	**787**	**1 214 e**	**-**	**5**	**-**	**-**	**290**	**-**
Iran	-	-	-	-	-	-	-	-	-
Iraq	-	-	-	-	-	-	-	-	-
Kuwait	-	-	-	-	-	-	-	-	-
Oman	-	-	-	-	-	-	-	-	-
Qatar	-	-	-	-	-	-	-	-	-
Saudi Arabia	-	-	-	-	-	-	2	-	-
United Arab Emirates	-	-	-	-	-	-	-	-	-
Other	1 008	657	193 e	5	-	-	6	-	-
Middle East	**1 008**	**657**	**193 e**	**5**	**-**	**-**	**8**	**-**	**-**
India	-	-	-	101	58	-	6	-	-
Indonesia	42	-	468 e	-	-	-	-	-	-
Singapore	-	-	-	-	-	-	-	-	-
Other	-	-	24 e	-	-	-	2	-	-
Asia	**42**	**-**	**492 e**	**101**	**58**	**-**	**8**	**-**	**-**
China	**-**	**-**	**-**	**-**	**-**	**-**	**-**	**-**	**-**
Algeria	104	80	445 e	-	-	-	6	15	3 e
Angola/Cabinada	-	-	-	-	-	-	-	-	-
Cameroon	-	-	-	-	-	-	-	-	-
Egypt	-	113	52 e	-	-	-	3	-	-
Gabon	-	-	-	-	-	-	-	-	-
Libya	2 548	2 143	2 194 e	-	-	-	27	2	27 e
Nigeria	-	-	93 e	-	-	-	-	-	-
Tunisia	-	-	-	-	-	-	-	-	-
Other	33	-	367 e	-	-	-	-	9	24 e
Africa	**2 685**	**2 336**	**3 151 e**	**-**	**-**	**-**	**36**	**26**	**54 e**
Non-spec./Other	**-**	**-**	**-**	**-**	**-**	**-**	**-**	**-**	**462 e**
Total Imports	**11 744**	**7 690**	**8 457 e**	**2 055**	**1 875**	**1 583 e**	**601**	**772**	**959 e**
Memo: Total OPEC	*4 625*	*2 269*	*3 200 e*	*239*	*373*	*369 e*	*35*	*17*	*30 e*

Source: Annual Oil Statistics.

ITALY
Table 8A: Exports by Country of Destination

000 metric tons

	Crude Oil			Natural Gas Liquids			Feedstocks and Other		
	2000	2001	2002	2000	2001	2002	2000	2001	2002
Australia	-	-	-	-	-	-	-	-	-
Austria	-	-	-	-	-	-	-	-	-
Belgium	-	-	-	-	-	-	-	-	-
Canada	-	-	-	-	-	-	-	34	-
Denmark	-	-	-	-	-	-	-	-	-
Finland	-	-	-	-	-	-	-	-	-
France	-	-	-	-	-	-	32	-	-
Germany	-	-	-	-	-	-	-	-	-
Greece	-	-	-	-	-	-	-	28	-
Ireland	-	-	-	-	-	-	-	-	-
Italy	-	-	-	-	-	-	-	-	-
Japan	-	-	-	-	-	-	-	-	-
Luxembourg	-	-	-	-	-	-	-	-	-
Mexico	-	-	-	-	-	-	-	-	-
Netherlands	-	-	-	-	-	-	10	-	-
Norway	-	-	-	-	-	-	-	-	-
Portugal	-	-	-	-	-	-	-	-	-
Spain	-	-	-	-	-	-	37	-	68 e
Sweden	-	-	-	-	-	-	-	-	-
Switzerland	-	-	-	-	-	-	-	-	-
Turkey	-	-	-	-	-	-	-	-	-
United Kingdom	-	-	-	-	-	-	-	-	9 e
United States	-	-	-	-	-	-	323	453	201 e
Other OECD	-	-	-	-	-	-	-	-	-
OECD	**-**	**-**	**-**	**-**	**-**	**-**	**402**	**515**	**278 e**
Africa	-	-	-	-	-	-	21	55	73 e
Latin America	-	-	-	-	-	-	-	-	-
Asia	-	-	-	-	-	-	15	-	-
China	-	-	-	-	-	-	-	-	-
Non-OECD Europe	-	-	-	-	-	-	64	35	15 e
Former Soviet Union	-	-	-	-	-	-	-	-	-
Middle East	-	-	-	-	-	-	19	-	27 e
Non-specified/Other	-	-	-	-	-	-	-	-	-
Non-OECD	**-**	**-**	**-**	**-**	**-**	**-**	**119**	**90**	**115 e**
Total Exports	**-**	**-**	**-**	**-**	**-**	**-**	**521**	**605**	**393 e**

	Total Products			Liquefied Petroleum Gas			Naphtha		
	2000	2001	2002	2000	2001	2002	2000	2001	2002
Australia	28	1	9 e	-	-	-	-	-	-
Austria	427	242	738 e	-	-	-	-	-	1 e
Belgium	116	55	73 e	3	-	19 e	-	-	23 e
Canada	34	17	4 e	-	-	-	-	-	-
Denmark	3	2	1 e	-	-	-	-	-	-
Finland	8	-	-	-	-	-	-	-	-
France	1 653	2 133	1 849 e	19	35	73 e	612	432	347 e
Germany	93	121	187 e	13	6	14 e	-	-	47 e
Greece	690	718	480 e	-	-	-	-	-	-
Ireland	-	-	-	-	-	-	-	-	-
Italy	-	-	-	-	-	-	-	-	-
Japan	2	-	-	-	-	-	-	-	-
Luxembourg	-	-	-	-	-	-	-	-	-
Mexico	2	17	40 e	-	-	6 e	-	-	-
Netherlands	218	298	140 e	2	9	20 e	106	178	34 e
Norway	-	22	2 e	-	-	-	-	22	-
Portugal	193	203	160 e	1	-	2 e	-	-	-
Spain	3 990	4 196	4 716 e	14	23	25 e	40	74	82 e
Sweden	7	5	-	-	-	-	-	-	-
Switzerland	1 118	928	1 218 e	-	-	2 e	43	37	15 e
Turkey	1 009	215	247 e	43	20	28 e	-	-	-
United Kingdom	566	1 152	1 309 e	-	4	15 e	25	159	89 e
United States	1 699	2 457	2 667 e	42	18	16 e	9	-	-
Other OECD	5	6	1 e	-	1	-	-	-	-
OECD	**11 861**	**12 788**	**13 841 e**	**137**	**116**	**220 e**	**835**	**902**	**638 e**
Africa	3 273	4 043	3 021 e	149	197	140 e	56	21	21 e
Latin America	413	320	237 e	-	-	3 e	-	-	-
Asia	640	479	261 e	5	4	3 e	12	-	-
China	-	-	5 e	-	-	-	-	-	-
Non-OECD Europe	3 941	3 555	3 144 e	71	63	67 e	20	19	30 e
Former Soviet Union	20	33	22 e	-	-	-	-	-	-
Middle East	902	598	444 e	46	45	19 e	-	-	-
Non-specified/Other	-	-	558 e	-	-	-	-	-	-
Non-OECD	**9 189**	**9 028**	**7 692 e**	**271**	**309**	**232 e**	**88**	**40**	**51 e**
Total Exports	**21 050**	**21 816**	**21 533 e**	**408**	**425**	**452 e**	**923**	**942**	**689 e**

Source: Annual Oil Statistics.

ITALY
Table 8B: Exports by Country of Destination

000 metric tons

	2000	2001	2002	2000	2001	2002	2000	2001	2002
	Total Gasoline			**Total Kerosene**			**Gas/Diesel Oil**		
Australia	28	-	-	-	-	-	-	-	-
Austria	20	34	76 e	-	-	9 e	330	132	581 e
Belgium	-	-	-	-	-	-	-	3	-
Canada	30	-	-	-	-	-	-	17	-
Denmark	-	-	-	-	-	-	-	-	-
Finland	-	-	-	-	-	-	-	-	-
France	59	242	92 e	-	9	59 e	867	1 305	1 151 e
Germany	-	30	-	-	-	-	-	6	19 e
Greece	127	18	86 e	-	-	-	366	616	353 e
Ireland	-	-	-	-	-	-	-	-	-
Italy	-	-	-	-	-	-	-	-	-
Japan	-	-	-	-	-	-	-	-	-
Luxembourg	-	-	-	-	-	-	-	-	-
Mexico	-	-	28 e	-	-	-	-	-	-
Netherlands	46	22	33 e	27	-	-	-	44	-
Norway	-	-	-	-	-	-	-	-	-
Portugal	29	-	-	25	-	-	85	190	149 e
Spain	589	407	383 e	96	76	64 e	2 549	3 166	3 888 e
Sweden	-	-	-	-	-	-	-	-	-
Switzerland	374	262	357 e	40	5	10 e	494	554	677 e
Turkey	-	-	8 e	14	-	-	440	96	10 e
United Kingdom	31	65	91 e	114	99	186 e	10	399	199 e
United States	1 103	1 378	1 871 e	5	67	39 e	244	307	187 e
Other OECD	-	-	-	-	-	-	-	-	-
OECD	**2 436**	**2 458**	**3 025 e**	**321**	**256**	**367 e**	**5 385**	**6 835**	**7 214 e**
Africa	712	1 209	989 e	225	153	213 e	1 398	1 691	1 046 e
Latin America	-	30	-	-	-	15 e	158	178	139 e
Asia	49	-	20 e	-	-	-	-	30	33 e
China	-	-	-	-	-	-	-	-	-
Non-OECD Europe	229	355	557 e	284	229	375 e	1 473	1 063	925 e
Former Soviet Union	20	-	-	-	-	-	-	33	-
Middle East	539	227	214 e	-	-	-	172	51	110 e
Non-specified/Other	-	-	-	-	-	102 e	-	-	353 e
Non-OECD	**1 549**	**1 821**	**1 780 e**	**509**	**382**	**705 e**	**3 201**	**3 046**	**2 606 e**
Total Exports	**3 985**	**4 279**	**4 805 e**	**830**	**638**	**1 072 e**	**8 586**	**9 881**	**9 820 e**

	2000	2001	2002	2000	2001	2002	2000	2001	2002
	Fuel Oil (Residual)			**Petroleum Coke**			**Other Products**		
Australia	-	-	-	-	-	-	-	1	9 e
Austria	-	-	5 e	-	-	-	77	76	66 e
Belgium	-	-	-	-	-	-	113	52	31 e
Canada	-	-	-	-	-	-	4	-	4 e
Denmark	-	-	-	-	-	-	3	2	1 e
Finland	-	-	-	-	-	-	8	-	-
France	13	-	20 e	-	-	-	83	110	107 e
Germany	-	-	-	-	-	-	80	79	107 e
Greece	112	-	-	45	54	3 e	40	30	38 e
Ireland	-	-	-	-	-	-	-	-	-
Italy	-	-	-	-	-	-	-	-	-
Japan	-	-	-	-	-	-	2	-	-
Luxembourg	-	-	-	-	-	-	-	-	-
Mexico	-	-	-	-	-	-	2	17	6 e
Netherlands	-	20	25 e	-	-	-	37	25	28 e
Norway	-	-	-	-	-	-	-	-	2 e
Portugal	37	-	-	-	-	-	16	13	9 e
Spain	482	225	22 e	-	-	-	220	225	252 e
Sweden	-	-	-	-	-	-	7	5	-
Switzerland	133	25	120 e	-	-	-	34	45	37 e
Turkey	407	-	87 e	-	-	-	105	99	114 e
United Kingdom	295	356	486 e	-	-	-	91	70	243 e
United States	216	366	185 e	-	-	-	80	321	369 e
Other OECD	-	-	-	-	-	-	5	5	1 e
OECD	**1 695**	**992**	**950 e**	**45**	**54**	**3 e**	**1 007**	**1 175**	**1 424 e**
Africa	591	552	385 e	-	-	-	142	220	227 e
Latin America	213	51	-	-	-	-	42	61	80 e
Asia	536	336	81 e	-	-	-	38	109	124 e
China	-	-	-	-	-	-	-	-	5 e
Non-OECD Europe	1 625	1 482	960 e	-	-	-	239	344	230 e
Former Soviet Union	-	-	22 e	-	-	-	-	-	-
Middle East	49	103	24 e	-	-	-	96	172	77 e
Non-specified/Other	-	-	-	-	-	-	-	-	103 e
Non-OECD	**3 014**	**2 524**	**1 472 e**	**-**	**-**	**-**	**557**	**906**	**846 e**
Total Exports	**4 709**	**3 516**	**2 422 e**	**45**	**54**	**3 e**	**1 564**	**2 081**	**2 270 e**

Source: Annual Oil Statistics.

JAPAN

Figure 1
FUEL SHARES OF TPES *

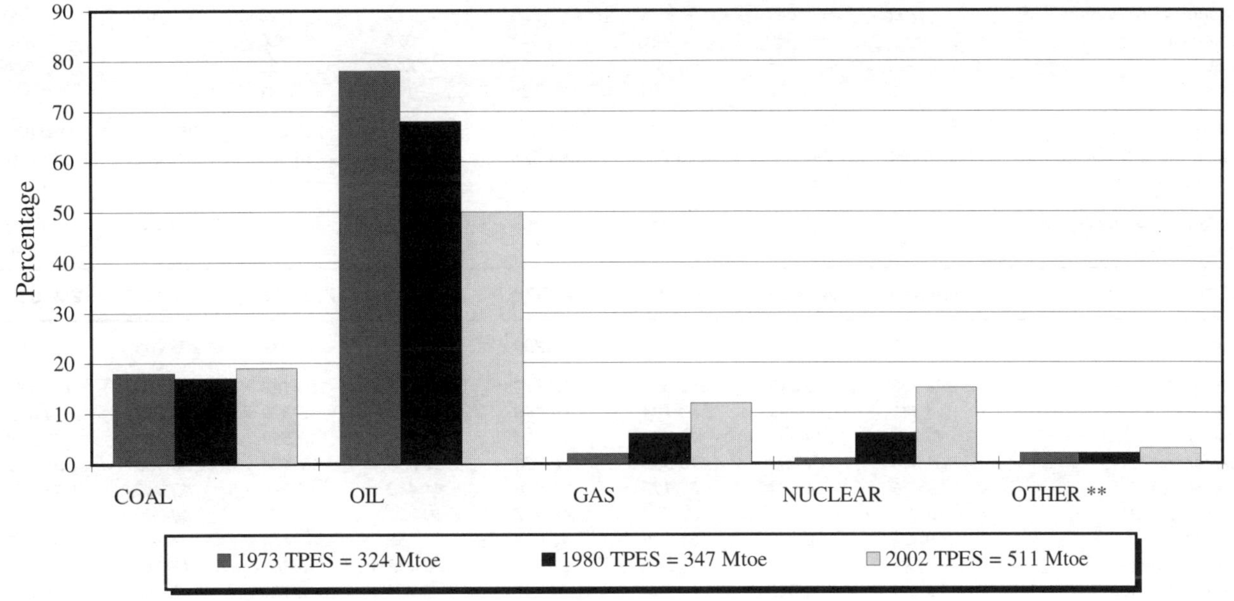

Source: Annual Energy Balances.

* Excluding electricity trade.

** Includes hydro, geothermal, solar, wind, combustible renewables and waste, etc.

Figure 2
OIL PRODUCTION AND CONSUMPTION

Source: Annual Oil Statistics.

*** Includes refinery fuel and international marine bunkers.

JAPAN
Table 1: Oil Supply and Consumption - Summary Table

000 metric tons

	1973	1978	1980	1990	1999	2000	2001	2002
Indigenous Production	792	641	542	635	722	729	674	560 e
Crude Oil	..	539	430	343	378	330	374	326 e
NGL	..	26	21	164	207	268	211	234 e
Other	..	76	91	128	137	131	89	-
+ Net Imports	268 675	258 345	246 681	252 374	259 219	263 215	250 842	252 885 e
Crude Oil, NGL, Other	245 094	230 555	218 760	194 623	213 872	216 594	208 806	208 031 e
Petroleum Products	23 581	27 790	27 921	57 751	45 347	46 621	42 036	44 854 e
- Intl. Marine Bunkers	17 350	12 657	11 981	5 308	5 416	5 003	4 227	4 305 e
+ Stock Changes	-5 473	- 723	-4 986	-1 173	2 776	-3 504	2 698	496 e
TOTAL SUPPLY	**246 644**	**245 606**	**230 256**	**246 528**	**257 301**	**255 437**	**249 987**	**249 636 e**
Refinery Intake	*224 920*	*215 755*	*203 677*	*175 290*	*211 996*	*210 912*	*205 708*	*204 259 e*
Refinery Output	*222 333*	*213 895*	*200 694*	*173 817*	*208 255*	*206 162*	*200 406*	*198 477 e*
LPG + Ethane	*4 443*	*4 299*	*4 194*	*4 449*	*4 873*	*5 339*	*5 298*	*4 608 e*
Naphtha	*23 035*	*18 933*	*16 441*	*8 004*	*13 251*	*12 598*	*12 938*	*13 286 e*
Kerosenes	*21 954*	*24 098*	*23 002*	*22 298*	*29 896*	*31 434*	*30 450*	*30 253 e*
Gasolines	*19 853*	*24 624*	*25 163*	*31 076*	*41 460*	*41 974*	*42 873*	*42 771 e*
Gas/Diesel Oil	*30 321*	*34 481*	*35 774*	*48 997*	*60 904*	*59 878*	*58 980*	*57 838 e*
Fuel Oil	*109 328*	*92 645*	*81 175*	*41 023*	*37 432*	*36 774*	*32 243*	*32 489 e*
Other Products	*13 399*	*14 815*	*14 945*	*17 970*	*20 439*	*18 165*	*17 624*	*17 232 e*
- Refinery Losses	2 587	1 860	2 983	1 473	3 741	4 750	5 302	5 782 e
+ Returns	-	2 479	3 619	3 598	5 210	4 310	4 606	5 187 e
- Statistical Differences	-6 274	-8 854	- 74	2 805	194	2 407 e	5 420 e	3 434 e
TOTAL CONSUMPTION	**250 331**	**255 079**	**230 966**	**245 848**	**258 576**	**252 590 e**	**243 871 e**	**245 607 e**
Transformation and Energy	**82 406**	**88 827**	**78 042**	**66 980**	**54 593**	**42 042 e**	**35 327 e**	..
Crude Oil	20 837	18 893	11 780	17 817	8 260	6 432	3 903	..
LPG + Ethane	300	1 022	1 883	3 682	3 092	1 644	1 599	..
Gas/Diesel Oil	688	739	635	564	1 698	1 476	1 476	..
Fuel Oil	52 926	53 851	49 442	29 914	22 038	16 537	13 145	..
Petroleum Coke	-	-	-	1 982	2 279	1 911	1 614	..
Other	7 655	14 322	14 302	13 021	17 226	14 042 e	13 590 e	..
Transport	**39 577**	**48 995**	**51 774**	**68 962**	**87 302**	**88 775**	**89 785**	..
Motor Gasoline	19 599	24 371	25 432	32 735	41 828	43 008	43 958	..
Gas/Diesel Oil	11 939	15 237	16 999	25 200	30 773	31 377	31 306	..
Subtotal	*31 538*	*39 608*	*42 431*	*57 935*	*72 601*	*74 385*	*75 264*	..
Aviation Fuels	2 975	3 485	3 599	7 077	9 512	9 827	9 849	..
Other	5 064	5 902	5 744	3 950	5 189	4 563	4 672	..
Industry	**93 518**	**79 040**	**65 831**	**68 615**	**74 371**	**69 867 e**	**65 743 e**	..
Naphtha	23 225	22 001	14 812	18 694	28 654	31 054	29 754	..
LPG + Ethane	5 542	6 433	6 844	6 574	7 192	5 644 e	5 584 e	..
Gas/Diesel Oil	8 938	8 883	8 157	14 098	13 693	6 774 e	4 890 e	..
Fuel Oil	44 013	30 104	23 020	9 742	6 707	11 679 e	10 983 e	..
Other	11 800	11 619	12 998	19 507	18 125	14 716 e	14 532 e	..
Other Sectors	**34 830**	**38 217**	**35 319**	**41 291**	**42 310**	**51 906**	**53 016**	..
Gas/Diesel Oil	8 409	10 308	10 261	14 218	15 215	20 776	22 666	..
Fuel Oil	5 496	2 941	2 440	3 367	1 262	18	18	..
Other	20 925	24 968	22 618	23 706	25 833	31 112	30 332	..
Memo: Non-Energy Use	*36 000*	*34 078*	*27 749*	*34 663*	*42 357*	*42 245*	*40 588*	..
Chem/Petchem. Feedst.	*26 800*	*25 408*	*19 134*	*24 120*	*32 138*	*35 344*	*33 659*	..
Other	*9 200*	*8 670*	*8 615*	*10 543*	*10 219*	*6 901*	*6 929*	..

Source: Annual Oil Statistics.

JAPAN
Table 2: Supply of Crude Oil, NGL, Feedstocks and Other Hydrocarbons

000 metric tons

	1973	1978	1980	1990	1999	2000	2001	2002
CRUDE OIL								
Indigenous Production	..	539	430	343	378	330	374	326 e
+ Imports	..	228 247	214 510	189 684	208 885	213 147	205 008	203 400 e
- Exports	..	-	-	-	-	-	-	-
- Direct Use	..	18 966	11 848	17 817	8 277	6 454	3 925	4 006 e
+ Stock Change	..	-3 187	-4 076	496	1 751	-1 187	- 77	-1 362 e
- Statistical Difference	..	-4 478	1 038	2 224	- 483	238	3 125	2 923 e
Refinery Intake	**..**	**211 111**	**197 978**	**170 482**	**203 220**	**205 598**	**198 255**	**195 435 e**
Closing Stock Level	..	*30 528*	*37 889*	*49 193*	*59 729*	*61 036*	*61 113*	*62 475 e*
NATURAL GAS LIQUIDS								
Indigenous Production	..	26	21	164	207	268	211	234 e
+ Imports	..	2 308	4 250	4 939	4 375	2 788	2 933	4 023 e
- Exports	..	-	-	-	-	-	-	-
- Direct Use	..	2 222	2 802	3 916	2 013	1 442	1 049	1 398 e
+ Stock Change	..	-	-	-	-	-	-	-
- Statistical Difference	..	-	-	-	-	- 1	-	-
Refinery Intake	**..**	**112**	**1 469**	**1 187**	**2 569**	**1 615**	**2 095**	**2 859 e**
Closing Stock Level	..	-	-	-	-	-	-	-
REFINERY FEEDSTOCKS								
Backflows	..	2 479	3 619	3 598	5 210	4 310	4 606	5 187 e
+ Transfers	..	1 977	-	-	-	-	-	-
+ Imports	..	-	-	-	-	-	-	-
- Exports	..	-	-	-	-	-	-	-
+ Stock Change	..	874	- 701	- 105	860	- 742	663	778 e
- Statistical Difference	..	874	-1 221	-	-	-	-	-
Refinery Intake	**..**	**4 456**	**4 139**	**3 493**	**6 070**	**3 568**	**5 269**	**5 965 e**
Closing Stock Level	..	*7 738*	*9 264*	*10 312*	*8 893*	*9 660*	*8 997*	*8 219 e*
OTHER HYDROCARBONS								
Indigenous Production	..	76	91	128	137	131	89	-
+ Imports	..	-	-	-	612	659	865	608 e
- Exports	..	-	-	-	-	-	-	-
- Direct Use	..	-	-	-	612	659	865	608 e
+ Stock Change	..	-	-	-	-	-	-	-
- Statistical Difference	..	-	-	-	-	-	-	-
Refinery Intake	**..**	**76**	**91**	**128**	**137**	**131**	**89**	**-**
Closing Stock Level	..	-	-	-	-	-	-	-
CRUDE OIL, NGL, FEEDSTOCKS AND OTHER								
Indigenous Production	792	641	542	635	722	729	674	560 e
+ Backflows	-	2 479	3 619	3 598	5 210	4 310	4 606	5 187 e
+ Transfers	-	1 977	-	-	-	-	-	-
+ Imports	245 094	230 555	218 760	194 623	213 872	216 594	208 806	208 031 e
- Exports	-	-	-	-	-	-	-	-
- Direct Use	20 966	21 188	14 650	21 733	10 902	8 555	5 839	6 012 e
+ Stock Change	-3 522	-2 313	-4 777	391	2 611	-1 929	586	- 584 e
- Statistical Difference	-3 522	-3 604	- 183	2 224	- 483	237	3 125	2 923 e
Refinery Intake	**224 920**	**215 755**	**203 677**	**175 290**	**211 996**	**210 912**	**205 708**	**204 259 e**
Closing Stock Level	-	*38 266*	*47 153*	*59 505*	*68 622*	*70 696*	*70 110*	*70 694 e*

Source: Annual Oil Statistics.

JAPAN
Table 3: Motor Gasoline - Supply

000 metric tons

	1973	1978	1980	1990	1999	2000	2001	2002
MOTOR GASOLINE								
Refinery Output	**19 813**	**24 604**	**25 149**	**31 067**	**41 448**	**41 971**	**42 867**	**42 765 e**
+ Imports	9	-	-	1 870	1 020	1 200	896	1 221 e
- Exports	65	-	-	77	583	257	286	202 e
+ Transfers	-	-	-	-	-	- 21	34	-
+ Stock Change	27	- 183	233	- 190	- 65	59	428	246 e
- Statistical Difference	185	50	- 50	- 65	- 8	- 56	- 19	- 587 e
Total Consumption	**19 599**	**24 371**	**25 432**	**32 735**	**41 828**	**43 008**	**43 958**	**44 617 e**
Closing Stock Level	-	*1 846*	*1 579*	*1 153*	*1 542*	*2 266*	*1 838*	*1 592 e*

Table 4: Aviation Fuels - Supply

000 metric tons

	1973	1978	1980	1990	1999	2000	2001	2002
AVIATION GASOLINE								
Refinery Output	**40**	**20**	**14**	**9**	**12**	**3**	**6**	**6 e**
+ Imports	-	-	-	-	-	-	-	-
- Exports	-	-	-	-	-	-	-	-
+ Transfers	-	-	-	-	-	-	-	-
+ Stock Change	-	-	- 1	-	-	-	-	-
- Statistical Difference	-	12	-	-	-	-	-	-
Total Consumption	**40**	**8**	**13**	**9**	**12**	**3**	**6**	**6 e**
Closing Stock Level	-	*3*	*2*	-	-	-	-	-
GASOLINE TYPE JET FUEL								
Refinery Output	-	-	-	-	-	-	-	-
+ Imports	-	-	-	-	-	-	-	-
- Exports	-	-	-	-	-	-	-	-
+ Transfers	-	-	-	-	-	-	-	-
+ Stock Change	-	-	-	-	-	-	-	-
- Statistical Difference	-	-	-	-	-	-	-	-
Total Consumption	-	-	-	-	-	-	-	-
Closing Stock Level	-	-	-	-	-	-	-	-
KEROSENE TYPE JET FUEL								
Refinery Output	**3 244**	**3 249**	**3 597**	**3 479**	**8 187**	**8 269**	**8 149**	**8 177 e**
+ Imports	525	277	310	3 681	1 991	2 036	1 974	2 465 e
- Exports	-	-	-	15	431	600	531	343 e
+ Transfers	-	-	-	-	-	151	227	-
+ Stock Change	- 4	- 38	- 9	- 55	48	- 52	6	93 e
- Statistical Difference	830	11	312	22	295	- 20	- 18	17 e
Total Consumption	**2 935**	**3 477**	**3 586**	**7 068**	**9 500**	**9 824**	**9 843**	**10 375 e**
Closing Stock Level	-	*354*	*358*	*408*	*616*	*543*	*537*	*444 e*

Source: Annual Oil Statistics.

JAPAN
Table 5A: Gas/Diesel Oil - Supply and Consumption

000 metric tons

	1973	1978	1980	1990	1999	2000	2001	2002
Refinery Output	**30 321**	**34 481**	**35 774**	**48 997**	**60 904**	**59 878**	**58 980**	**57 838 e**
+ Recycled Products	-	-	-	-	-	-	-	-
+ Imports	2 367	1 537	1 275	6 741	2 037	2 365	1 921	1 562 e
- Exports	31	197	103	1 106	1 689	1 694	1 659	1 234 e
- Intl. Marine Bunkers	1 369	966	720	396	248	298	119	103 e
+ Transfers	-	-	-	-	-	215	171	345 e
+ Stock Changes	- 804	513	106	- 484	- 108	- 364	826	379 e
- Statistical Differences	510	201	280	- 328	- 483	- 301	- 218	- 695 e
Total Consumption	**29 974**	**35 167**	**36 052**	**54 080**	**61 379**	**60 403**	**60 338**	**59 482 e**
Transformation	**688**	**739**	**635**	**494**	**1 542**	**1 464**	**1 465**	..
Public Electricity	688	739	635	162	230	861	882	..
Autoproducer Electricity	-	-	-	332	1 312	603	583	..
Public CHP Plants	-	-	-	-	-	-	-	..
Autoproducers of CHP	-	-	-	-	-	-	-	..
Public Heat Plants	-	-	-	-	-	-	-	..
Autoproducer Heat Plants	-	-	-	-	-	-	-	..
Petrochemical Industry	-	-	-	-	-	-	-	..
Other Transformation	-	-	-	-	-	-	-	..
Energy Sector	-	-	-	**70**	**156**	**12**	**11**	..
Refinery Fuel	-	-	-	40	108	12	11	..
Other Energy Sector	-	-	-	30	48	-	-	..
Final Consumption	**29 286**	**34 428**	**35 417**	**53 516**	**59 681**	**58 927**	**58 862**	..
Transport	**11 939**	**15 237**	**16 999**	**25 200**	**30 773**	**31 377**	**31 306**	..
Air	-	-	-	-	-	-	-	..
Road	9 136	12 266	14 138	23 108	28 379	29 521	29 551	..
Rail	996	913	891	540	511	227	227	..
Inland Waterways	1 807	2 058	1 970	1 552	1 883	1 629	1 528	..
Non-Specified	-	-	-	-	-	-	-	..
Industry	**8 938**	**8 883**	**8 157**	**14 098**	**13 693**	**6 774 e**	**4 890 e**	..
Iron and Steel	820	785	715	724	642	638 e	460 e	..
Chemical (incl. Pet.-Chem.)	537	812	805	1 255	1 045	1 242 e	897 e	..
Non-Ferrous Metals	326	430	425	701	445	418 e	302 e	..
Non-Metallic Minerals	523	818	728	914	930	867 e	626 e	..
Transport Equipment	-	-	-	-	-	-	-	..
Machinery	-	-	-	932	188	110 e	79 e	..
Mining and Quarrying	45	71	89	257	348	228 e	165 e	..
Food and Tobacco	765	1 039	981	1 690	1 680	1 697 e	1 225 e	..
Pulp, Paper and Printing	200	228	220	248	407	407 e	294 e	..
Wood and Wood Products	-	-	-	-	-	-	-	..
Construction	2 449	2 677	2 790	3 649	2 996	549 e	396 e	..
Textiles and Leather	262	336	305	657	619	618 e	446 e	..
Non-Specified	3 011	1 687	1 099	3 071	4 393	-	-	..
Other Sectors	**8 409**	**10 308**	**10 261**	**14 218**	**15 215**	**20 776**	**22 666**	..
Commerce and Public	-	-	-	6 502	8 398	13 852	15 742	..
Residential	-	-	-	-	-	-	-	..
Agriculture	998	1 541	1 721	7 716	6 817	3 839	3 839	..
Non-Specified	7 411	8 767	8 540	-	-	3 085	3 085	..
Memo: Non-Energy Use	-	-	-	-	-	*22*	*18*	..
Chem/Petchem. Feedst.	-	-	-	-	-	*22*	*18*	..
Other	-	-	-	-	-	-	-	..
Closing Stock Level	-	*2 874*	*3 665*	*3 263*	*2 896*	*3 490*	*2 664*	*2 285 e*

Source: Annual Oil Statistics.

JAPAN
Table 5B: Fuel Oil (Residual) - Supply and Consumption

000 metric tons

	1973	1978	1980	1990	1999	2000	2001	2002
Refinery Output	**109 328**	**92 645**	**81 175**	**41 023**	**37 432**	**36 774**	**32 243**	**32 489 e**
+ Recycled Products	-	-	-	-	-	-	-	-
+ Imports	12 177	9 265	8 994	10 039	1 655	1 086	976	1 336 e
- Exports	2 455	5	5	1 516	1 112	885	1 112	842 e
- Intl. Marine Bunkers	15 921	11 646	11 205	4 874	5 147	4 676	4 086	4 188 e
+ Transfers	-	-1 977	-	-	-	- 17	76	- 158 e
+ Stock Changes	- 180	381	- 633	112	410	- 119	176	- 135 e
- Statistical Differences	-2 646	-2 263	- 604	- 105	26	1 047	1 106	745 e
Total Consumption	**105 595**	**90 926**	**78 930**	**44 889**	**33 212**	**31 116**	**27 167**	**27 757 e**
Transformation	**46 838**	**49 086**	**45 121**	**27 826**	**19 435**	**15 835**	**12 455**	..
Public Electricity	46 838	36 486	34 021	21 624	11 634	10 677	7 722	..
Autoproducer Electricity	-	12 600	11 100	6 187	7 782	5 138	4 712	..
Public CHP Plants	-	-	-	-	-	-	-	..
Autoproducers of CHP	-	-	-	-	-	-	-	..
Public Heat Plants	-	-	-	15	19	20	21	..
Autoproducer Heat Plants	-	-	-	-	-	-	-	..
Petrochemical Industry	-	-	-	-	-	-	-	..
Other Transformation	-	-	-	-	-	-	-	..
Energy Sector	**6 088**	**4 765**	**4 321**	**2 088**	**2 603**	**702**	**690**	..
Refinery Fuel	6 088	4 765	4 321	1 936	2 463	702	690	..
Other Energy Sector	-	-	-	152	140	-	-	..
Final Consumption	**52 669**	**37 075**	**29 488**	**14 975**	**11 174**	**14 579 e**	**14 022 e**	..
Transport	**3 160**	**4 030**	**4 028**	**1 866**	**3 205**	**2 882**	**3 021**	..
Air	-	-	-	-	-	-	-	..
Road	48	39	62	47	45	-	-	..
Rail	449	375	359	-	-	-	-	..
Inland Waterways	2 556	3 468	3 539	1 785	3 075	2 882	3 021	..
Non-Specified	107	148	68	34	85	-	-	..
Industry	**44 013**	**30 104**	**23 020**	**9 742**	**6 707**	**11 679 e**	**10 983 e**	..
Iron and Steel	11 010	5 621	3 811	1 290	893	1 061 e	908 e	..
Chemical (incl. Pet.-Chem.)	8 026	5 211	4 508	1 129	532	979 e	838 e	..
Non-Ferrous Metals	1 130	521	780	442	222	280 e	239 e	..
Non-Metallic Minerals	9 929	8 745	5 773	1 629	1 631	1 769 e	1 514 e	..
Transport Equipment	-	-	-	-	-	-	-	..
Machinery	-	-	-	37	6	9 e	8 e	..
Mining and Quarrying	243	170	158	64	80	52 e	44 e	..
Food and Tobacco	2 503	1 989	1 938	634	146	168 e	144 e	..
Pulp, Paper and Printing	3 726	2 383	2 161	2 490	1 917	2 729 e	2 335 e	..
Wood and Wood Products	-	-	-	-	-	-	-	..
Construction	463	122	93	30	15	23 e	20 e	..
Textiles and Leather	4 761	4 377	3 084	1 325	1 235	1 415 e	1 210 e	..
Non-Specified	2 222	965	714	672	30	3 194	3 723	..
Other Sectors	**5 496**	**2 941**	**2 440**	**3 367**	**1 262**	**18**	**18**	..
Commerce and Public	5 300	2 621	2 151	3 183	1 239	-	-	..
Residential	-	-	-	-	-	-	-	..
Agriculture	196	320	289	184	23	14	14	..
Non-Specified	-	-	-	-	-	4	4	..
Memo: Non-Energy Use	-	-	-	-	-	*463*	*433*	..
Chem/Petchem. Feedst.	-	-	-	-	-	*463*	*431*	..
Other	-	-	-	-	-	-	*2*	..
Closing Stock Level	-	*4 442*	*4 972*	*2 287*	*1 466*	*1 628*	*1 452*	*1 587 e*

Source: Annual Oil Statistics.

JAPAN

Table 5C: Total Products - Supply and Consumption

000 metric tons

	1973	1978	1980	1990	1999	2000	2001	2002
Refinery Output	**222 333**	**213 895**	**200 694**	**173 817**	**208 255**	**206 162**	**200 406**	**198 477 e**
+ Recycled Products	-	-	-	-	-	-	-	-
+ Imports	26 609	28 508	28 408	61 575	50 566	50 949	46 747	48 539 e
- Exports	3 028	718	487	3 824	5 219	4 328	4 711	3 685 e
- Intl. Marine Bunkers	17 350	12 657	11 981	5 308	5 416	5 003	4 227	4 305 e
+ Transfers	20 966	19 211	14 650	21 733	10 902	8 555	5 839	6 012 e
+ Stock Changes	-1 951	1 590	- 209	-1 564	165	-1 575	2 112	1 080 e
- Statistical Differences	-2 752	-5 250	109	581	677	2 170	2 295	511 e
Total Consumption	**250 331**	**255 079**	**230 966**	**245 848**	**258 576**	**252 590**	**243 871**	**245 607 e**
Transformation	**72 001**	**78 447**	**68 172**	**57 734**	**42 608**	**33 009**	**26 533**	..
Public Electricity	69 426	60 628	50 400	41 968	21 917	18 593	13 061	..
Autoproducer Electricity	-	12 600	11 100	8 149	11 192	8 838	8 007	..
Public CHP Plants	-	-	-	-	-	-	-	..
Autoproducers of CHP	-	-	-	-	-	-	-	..
Public Heat Plants	-	-	-	47	36	36	33	..
Autoproducer Heat Plants	-	-	-	-	-	-	-	..
Petrochemical Industry	-	2 479	3 619	3 598	5 210	4 310	4 606	..
Other Transformation	2 575	2 740	3 053	3 972	4 253	1 232	826	..
Energy Sector	**10 405**	**10 380**	**9 870**	**9 246**	**11 985**	**9 033**	**8 794**	..
Refinery Fuel	10 405	10 380	9 870	9 063	11 797	9 033	8 794	..
Other Energy Sector	-	-	-	183	188	-	-	..
Final Consumption	**167 925**	**166 252**	**152 924**	**178 868**	**203 983**	**210 548 e**	**208 544 e**	..
Transport	**39 577**	**48 995**	**51 774**	**68 962**	**87 302**	**88 775**	**89 785**	..
Air	2 975	3 485	3 599	7 077	9 512	9 827	9 849	..
Road	30 473	38 383	41 192	57 867	72 045	74 079	75 029	..
Rail	1 470	1 303	1 264	548	516	231	231	..
Inland Waterways	4 532	5 664	5 640	3 428	5 095	4 638	4 676	..
Non-Specified	127	160	79	42	134	-	-	..
Industry	**93 518**	**79 040**	**65 831**	**68 615**	**74 371**	**69 867 e**	**65 743 e**	..
Iron and Steel	11 830	6 423	4 984	3 027	2 179	2 504 e	2 138 e	..
Chemical (incl. Pet.-Chem.)	35 456	31 504	24 515	27 264	34 809	39 328 e	36 950 e	..
Non-Ferrous Metals	1 456	951	1 205	1 566	1 034	1 155 e	991 e	..
Non-Metallic Minerals	10 452	9 563	6 501	3 170	3 905	4 193 e	3 730 e	..
Transport Equipment	-	-	-	-	-	-	-	..
Machinery	-	-	-	1 780	995	985 e	932 e	..
Mining and Quarrying	353	284	297	356	501	375 e	300 e	..
Food and Tobacco	3 268	3 028	2 919	2 324	1 826	1 865 e	1 369 e	..
Pulp, Paper and Printing	3 926	2 611	2 381	2 998	2 576	3 437 e	2 928 e	..
Wood and Wood Products	-	-	-	-	-	-	-	..
Construction	8 630	8 363	8 242	10 110	7 861	5 821 e	5 544 e	..
Textiles and Leather	5 023	4 713	3 389	2 143	2 019	2 204 e	1 828 e	..
Non-Specified	13 124	11 600	11 398	13 877	16 666	8 000 e	9 033 e	..
Other Sectors	**34 830**	**38 217**	**35 319**	**41 291**	**42 310**	**51 906**	**53 016**	..
Commerce and Public	15 688	15 406	13 628	14 851	15 515	25 246	28 462	..
Residential	8 330	10 251	9 202	15 170	16 851	16 587	15 049	..
Agriculture	2 621	3 192	3 327	10 547	9 345	6 284	6 284	..
Non-Specified	8 191	9 368	9 162	723	599	3 789	3 221	..
Memo: Non-Energy Use	*36 000*	*34 078*	*27 749*	*34 663*	*42 357*	*42 245*	*40 588*	..
Chem/Petchem. Feedst.	*26 800*	*25 408*	*19 134*	*24 120*	*32 138*	*35 344*	*33 659*	..
Other	*9 200*	*8 670*	*8 615*	*10 543*	*10 219*	*6 901*	*6 929*	..
Closing Stock Level	-	*16 906*	*19 184*	*16 182*	*14 586*	*15 228*	*13 116*	*12 036 e*

Source: Annual Oil Statistics.

JAPAN
Table 6A: Supply and Consumption

000 metric tons

	1980	2000	2001	2002	1980	2000	2001	2002
	LPG AND ETHANE				**REFINERY GAS**			
Refinery Output	**4 194**	**5 339**	**5 298**	**4 608 e**	**5 824**	**9 303**	**8 761**	**8 684 e**
+ Recycled Products	-	-	-	-	-	-	-	-
+ Imports	9 725	14 777	14 336	14 048 e	-	-	-	-
- Exports	6	55	76	56 e	-	-	-	-
- Intl. Marine Bunkers	-	-	-	-	-	-	-	-
+ Transfers	16	- 164	- 213	- 177 e	-	3	4	-
+ Stock Change	- 72	- 424	- 323	288 e	-	-	-	-
- Statistical Difference	-1 085	2 031	1 478	1 449 e	-	250	263	300 e
Total Consumption	**14 942**	**17 442**	**17 544**	**17 262 e**	**5 824**	**9 056 e**	**8 502 e**	**8 384 e**
of which: Ethane	-	-	-	-	-	-	-	-
Transformation	1 873	1 598	1 545	..	528	876	549	..
Energy	10	46	54	..	5 296	8 180 e	7 953 e	..
Transport	1 473	1 492	1 462	..	-	-	-	..
Industry	6 844	5 644 e	5 584 e	..	-	-	-	..
Other	4 742	8 662	8 899	..	-	-	-	..
Memo: Non-Energy Use	*3 679*	*1 633 e*	*1 616 e*	..	-	-	-	..
Closing Stock Level	*1 616*	*2 501*	*2 824*	*2 536 e*	-	-	-	-
	NAPHTHA				**OTHER KEROSENE**			
Refinery Output	**16 441**	**12 598**	**12 938**	**13 286 e**	**19 405**	**23 165**	**22 301**	**22 076 e**
+ Recycled Products	-	-	-	-	-	-	-	-
+ Imports	5 427	22 228	20 731	22 018 e	61	2 634	1 653	1 962 e
- Exports	-	107	68	84 e	-	125	200	210 e
- Intl. Marine Bunkers	-	-	-	-	-	2	-	-
+ Transfers	-	- 82	- 377	-	-	- 145	79	-
+ Stock Change	34	- 202	202	- 453 e	167	- 483	723	610 e
- Statistical Difference	755	- 425	- 305	- 227 e	544	645	730	147 e
Total Consumption	**21 147**	**34 860**	**33 731**	**34 994 e**	**19 089**	**24 399**	**23 826**	**24 291 e**
Transformation	6 092	3 766	3 942	..	-	185	160	..
Energy	243	40	35	..	-	4	2	..
Transport	-	-	-	..	-	-	-	..
Industry	14 812	31 054	29 754	..	1 865	2 363 e	2 266 e	..
Other	-	-	-	..	17 224	21 847	21 398	..
Memo: Non-Energy Use	*14 812*	*31 050*	*29 751*	..	-	*367 e*	*352 e*	..
Closing Stock Level	*1 055*	*1 627*	*1 425*	*1 878 e*	*4 982*	*2 520*	*1 797*	*1 187 e*
	WHITE SPIRIT + SBP				**LUBRICANTS**			
Refinery Output	**65**	**27**	**32**	**..**	**2 083**	**2 380**	**2 378**	**..**
+ Recycled Products	-	-	-	..	-	-	-	..
+ Imports	-	-	-	..	138	32	27	..
- Exports	-	-	-	..	251	340	439	..
- Intl. Marine Bunkers	-	-	-	..	56	24	22	..
+ Transfers	-	-	-	..	-	56	51	..
+ Stock Change	1	-	- 9	..	- 61	10	58	..
- Statistical Difference	1	- 32	- 37	..	- 43	112	104	..
Total Consumption	**65**	**59**	**60**	**..**	**1 896**	**2 002**	**1 949**	**..**
Transformation	-	-	-	..	-	-	-	..
Energy	-	-	-	..	-	49	49	..
Transport	-	-	-	..	243	189	189	..
Industry	65	59	60	..	1 001	1 161	1 676	..
Other	-	-	-	..	652	603	35	..
Memo: Non-Energy Use	*65*	*59*	*60*	..	*1 896*	*1 349*	*1 842*	..
Closing Stock Level	*14*	*10*	*19*	..	*547*	*355*	*297*	..

Source: Annual Oil Statistics.

JAPAN
Table 6B: Supply and Consumption

000 metric tons

	1980	2000	2001	2002	1980	2000	2001	2002
	BITUMEN				PARAFFIN WAXES			
Refinery Output	**4 777**	**5 510**	**5 424**	**..**	**166**	**103**	**106**	**..**
+ Recycled Products	-	-	-	..	-	-	-	..
+ Imports	1	30	32	..	2	3	4	..
- Exports	15	222	284	..	80	43	56	..
- Intl. Marine Bunkers	-	-	-	..	-	-	-	..
+ Transfers	-	661	813	..	-	-	-	..
+ Stock Change	- 7	- 8	18	..	1	- 2	4	..
- Statistical Difference	- 3	-1 187	- 800	..	3	1	2	..
Total Consumption	**4 759**	**7 158**	**6 803**	**..**	**86**	**60**	**56**	**..**
Transformation	-	790	733	..	-	-	-	..
Energy	-	-	-	..	-	-	-	..
Transport	-	-	-	..	-	-	-	..
Industry	4 759	6 368 e	6 070 e	..	86	60	56	..
Other	-	-	-	..	-	-	-	..
Memo: Non-Energy Use	*4 759*	*4 597*	*4 394*	..	*86*	*60*	*56*	..
Closing Stock Level	*273*	*232*	*214*	..	*15*	*22*	*18*	..
	PETROLEUM COKE				OTHER PRODUCTS			
Refinery Output	**254**	**786**	**872**	**814 e**	**1 776**	**56**	**51**	**7 734 e**
+ Recycled Products	-	-	-	-	-	-	-	-
+ Imports	2 474	4 557	4 196	3 890 e	1	1	1	37 e
- Exports	27	-	-	-	-	-	-	714 e
- Intl. Marine Bunkers	-	-	-	-	-	-	-	14 e
+ Transfers	-	-	-	-	14 634	7 898	4 974	6 002 e
+ Stock Change	2	9	-	1 e	30	1	3	51 e
- Statistical Difference	1	110	7	-	- 2	- 2	2	- 638 e
Total Consumption	**2 702**	**5 242**	**5 061**	**4 705 e**	**16 443**	**7 958**	**5 027**	**13 734 e**
Transformation	-	1 911	1 614	..	13 923	6 584	4 070	..
Energy	-	-	-	..	-	-	-	..
Transport	-	-	-	..	-	-	-	..
Industry	2 702	3 331 e	3 447 e	..	2 520	1 374	957	..
Other	-	-	-	..	-	-	-	..
Memo: Non-Energy Use	-	*368*	*552*	..	*1 809*	*62*	*53*	..
Closing Stock Level	*4*	*25*	*25*	*24 e*	*102*	*9*	*6*	*503 e*

Source: Annual Oil Statistics.

JAPAN
Table 7A: Imports by Country of Origin

000 metric tons

	2000	2001	2002	2000	2001	2002	2000	2001	2002
	Crude Oil			**Natural Gas Liquids**			**Feedstocks and Other**		
Australia	2 751	2 316	2 362 e	332	352	355 e	-	-	-
Belgium	-	-	-	-	-	-	-	-	-
Canada	-	-	-	-	-	-	-	-	-
Czech Republic	-	-	-	-	-	-	-	-	-
Denmark	-	-	-	-	-	-	-	-	-
Finland	-	-	-	-	-	-	-	-	-
France	-	-	-	-	-	-	-	-	-
Germany	-	-	-	-	-	-	-	-	-
Greece	-	-	-	-	-	-	-	-	-
Hungary	-	-	-	-	-	-	-	-	-
Ireland	-	-	-	-	-	-	-	-	-
Italy	-	-	-	-	-	-	-	-	-
Japan	-	-	-	-	-	-	-	-	-
Korea	-	-	-	-	-	-	-	-	-
Mexico	1 644	585	490 e	-	-	-	-	-	-
Netherlands	-	-	-	-	-	-	-	-	-
New Zealand	-	-	-	6	-	-	-	-	-
Norway	137	159	280 e	-	-	-	-	-	-
Poland	-	-	-	-	-	-	-	-	-
Spain	-	-	-	-	-	-	-	-	-
Sweden	-	-	-	-	-	-	-	-	-
Turkey	-	-	-	-	-	-	-	-	-
United Kingdom	-	-	-	-	-	-	-	-	-
United States	234	-	-	-	-	-	-	-	-
Other	-	-	-	-	-	-	-	-	-
OECD	**4 766**	**3 060**	**3 132 e**	**338**	**352**	**355 e**	**-**	**-**	**-**
Argentina	-	-	-	-	-	-	-	-	-
Brazil	-	-	139 e	-	-	-	-	-	-
Colombia	-	-	-	-	-	-	-	-	-
Ecuador	275	268	188 e	-	-	-	-	-	-
Netherlands Antilles	-	-	-	-	-	-	-	-	-
Peru	-	-	-	-	-	-	-	-	-
Trinidad and Tobago	-	-	-	-	-	-	-	-	-
Venezuela	-	-	-	-	-	-	659	865	608 e
Other	-	-	-	-	-	-	-	-	-
Latin America	**275**	**268**	**327 e**	**-**	**-**	**-**	**659**	**865**	**608 e**
Romania	-	-	-	-	-	-	-	-	-
Other	-	-	-	-	-	-	-	-	-
Non-OECD Europe	**-**	**-**	**-**	**-**	**-**	**-**	**-**	**-**	**-**
FSU	**-**	**282**	**753 e**	**-**	**68**	**-**	**-**	**-**	**-**
Iran	24 692	25 542	25 935 e	250	233	1 195 e	-	-	-
Iraq	3 066	504	625 e	-	-	-	-	-	-
Kuwait	21 423	20 038	14 472 e	-	-	-	-	-	-
Oman	9 807	12 314	11 507 e	-	13	24 e	-	-	-
Qatar	20 141	21 883	17 398 e	707	1 335	1 278 e	-	-	-
Saudi Arabia	51 864	50 942	49 676 e	325	228	161 e	-	-	-
United Arab Emirates	54 995	48 906	52 153 e	565	260	439 e	-	-	-
Other	281	137	292 e	-	-	-	-	-	-
Middle East	**186 269**	**180 266**	**172 058 e**	**1 847**	**2 069**	**3 097 e**	**-**	**-**	**-**
India	-	-	-	-	-	-	-	-	-
Indonesia	10 134	8 884	7 672 e	288	281	402 e	-	-	-
Singapore	-	-	-	-	-	-	-	-	-
Other	5 492	5 329	4 464 e	315	163	57 e	-	-	-
Asia	**15 626**	**14 213**	**12 136 e**	**603**	**444**	**459 e**	**-**	**-**	**-**
China	**4 692**	**3 460**	**3 409 e**	**-**	**-**	**-**	**-**	**-**	**-**
Algeria	-	-	-	-	-	-	-	-	-
Angola/Cabinada	-	657	2 095 e	-	-	-	-	-	-
Cameroon	-	-	-	-	-	-	-	-	-
Egypt	234	-	-	-	-	-	-	-	-
Gabon	-	-	369 e	-	-	-	-	-	-
Libya	-	-	-	-	-	-	-	-	-
Nigeria	517	1 453	4 224 e	-	-	112 e	-	-	-
Tunisia	-	-	-	-	-	-	-	-	-
Other	768	1 349	4 897 e	-	-	-	-	-	-
Africa	**1 519**	**3 459**	**11 585 e**	**-**	**-**	**112 e**	**-**	**-**	**-**
Non-spec./Other	**-**	**-**	**-**	**-**	**-**	**-**	**-**	**-**	**-**
Total Imports	**213 147**	**205 008**	**203 400 e**	**2 788**	**2 933**	**4 023 e**	**659**	**865**	**608 e**
Memo: Total OPEC	*186 832*	*178 152*	*172 155 e*	*2 135*	*2 337*	*3 587 e*	*659*	*865*	*608 e*

Source: Annual Oil Statistics.

JAPAN
Table 7B: Imports by Country of Origin

000 metric tons

	2000	2001	2002	2000	2001	2002	2000	2001	2002
	Total Products			Liquefied Petroleum Gases			Naphtha		
Australia	1 181	1 329	1 227 e	773	948	1 011 e	384	369	216 e
Belgium	-	-	-	-	-	-	-	-	-
Canada	1	-	220 e	-	-	-	-	-	-
Czech Republic	-	-	-	-	-	-	-	-	-
Denmark	-	-	7 e	-	-	-	-	-	7 e
Finland	-	-	-	-	-	-	-	-	-
France	3	1	2 e	-	-	-	-	-	-
Germany	-	1	-	-	-	-	-	-	-
Greece	38	-	59 e	-	-	-	38	-	59 e
Hungary	-	4	-	-	-	-	-	-	-
Ireland	-	-	-	-	-	-	-	-	-
Italy	-	-	-	-	-	-	-	-	-
Japan	-	-	-	-	-	-	-	-	-
Korea	10 636	9 338	8 953 e	133	158	48 e	5 723	5 589	5 250 e
Mexico	115	120	138 e	-	-	-	115	120	138 e
Netherlands	15	266	357 e	12	-	-	-	264	357 e
New Zealand	252	244	282 e	-	-	-	252	244	282 e
Norway	-	-	255 e	-	-	255 e	-	-	-
Poland	-	-	-	-	-	-	-	-	-
Spain	-	-	-	-	-	-	-	-	-
Sweden	-	-	-	-	-	-	-	-	-
Turkey	-	-	-	-	-	-	-	-	-
United Kingdom	27	21	14 e	27	21	14 e	-	-	-
United States	487	700	3 911 e	-	192	289 e	454	491	499 e
Other	20	55	-	20	55	-	-	-	-
OECD	**12 775**	**12 079**	**15 425 e**	**965**	**1 374**	**1 617 e**	**6 966**	**7 077**	**6 808 e**
Argentina	-	-	-	-	-	-	-	-	-
Brazil	-	-	-	-	-	-	-	-	-
Colombia	-	-	-	-	-	-	-	-	-
Ecuador	-	-	-	-	-	-	-	-	-
Netherlands Antilles	-	36	-	-	-	-	-	36	-
Peru	-	-	-	-	-	-	-	-	-
Trinidad and Tobago	1	5	-	-	-	-	-	-	-
Venezuela	7	6	2 e	-	-	-	-	-	-
Other	-	-	-	-	-	-	-	-	-
Latin America	**8**	**47**	**2 e**	**-**	**-**	**-**	**-**	**36**	**-**
Romania	-	-	-	-	-	-	-	-	-
Other	-	-	-	-	-	-	-	-	-
Non-OECD Europe	**-**	**-**	**-**	**-**	**-**	**-**	**-**	**-**	**-**
FSU	**307**	**308**	**485 e**	**-**	**-**	**-**	**70**	**83**	**184 e**
Iran	302	696	559 e	302	696	501 e	-	-	58 e
Iraq	7	48	-	7	48	-	-	-	-
Kuwait	4 885	5 269	4 871 e	1 447	1 681	1 469 e	3 360	3 588	3 340 e
Oman	-	13	32 e	-	-	-	-	13	32 e
Qatar	1 008	1 199	1 421 e	513	467	631 e	495	732	759 e
Saudi Arabia	8 144	6 550	7 820 e	5 999	4 609	4 763 e	2 129	1 907	2 957 e
United Arab Emirates	7 660	6 024	6 798 e	4 021	3 704	3 776 e	3 609	2 267	2 991 e
Other	515	502	564 e	14	48	53 e	501	454	511 e
Middle East	**22 521**	**20 301**	**22 065 e**	**12 303**	**11 253**	**11 193 e**	**10 094**	**8 961**	**10 648 e**
India	923	757	734 e	-	-	-	923	757	550 e
Indonesia	2 551	2 853	2 584 e	1 018	969	634 e	911	1 371	1 270 e
Singapore	1 659	839	1 164 e	-	-	-	1 079	379	520 e
Other	2 254	1 819	1 882 e	322	233	110 e	1 425	1 309	1 070 e
Asia	**7 387**	**6 268**	**6 364 e**	**1 340**	**1 202**	**744 e**	**4 338**	**3 816**	**3 410 e**
China	**703**	**494**	**867 e**	**12**	**-**	**-**	**481**	**398**	**535 e**
Algeria	76	244	190 e	18	154	85 e	58	90	105 e
Angola/Cabinada	-	-	-	-	-	-	-	-	-
Cameroon	-	-	-	-	-	-	-	-	-
Egypt	221	194	156 e	-	-	-	221	194	156 e
Gabon	-	-	-	-	-	-	-	-	-
Libya	-	-	-	-	-	-	-	-	-
Nigeria	139	391	478 e	139	353	409 e	-	38	69 e
Tunisia	-	-	-	-	-	-	-	-	-
Other	-	38	103 e	-	-	-	-	38	103 e
Africa	**436**	**867**	**927 e**	**157**	**507**	**494 e**	**279**	**360**	**433 e**
Non-spec./Other	**6 812**	**6 383**	**2 404 e**	**-**	**-**	**-**	**-**	**-**	**-**
Total Imports	**50 949**	**46 747**	**48 539 e**	**14 777**	**14 336**	**14 048 e**	**22 228**	**20 731**	**22 018 e**
Memo: Total OPEC	*24 779*	*23 280*	*24 723 e*	*13 464*	*12 681*	*12 268 e*	*10 562*	*9 993*	*11 549 e*

Source: Annual Oil Statistics.

JAPAN
Table 7C: Imports by Country of Origin

000 metric tons

	2000	2001	2002	2000	2001	2002	2000	2001	2002
	Total Gasoline			**Total Kerosene**			**Gas/Diesel**		
Australia	24	-	-	-	12	-	-	-	-
Belgium	-	-	-	-	-	-	-	-	-
Canada	-	-	-	-	-	-	-	-	-
Czech Republic	-	-	-	-	-	-	-	-	-
Denmark	-	-	-	-	-	-	-	-	-
Finland	-	-	-	-	-	-	-	-	-
France	-	-	-	-	-	-	-	-	-
Germany	-	-	-	-	-	-	-	-	-
Greece	-	-	-	-	-	-	-	-	-
Hungary	-	-	-	-	-	-	-	4	-
Ireland	-	-	-	-	-	-	-	-	-
Italy	-	-	-	-	-	-	-	-	-
Japan	-	-	-	-	-	-	-	-	-
Korea	569	582	487 e	1 941	1 152	1 248 e	2 095	1 668	1 191 e
Mexico	-	-	-	-	-	-	-	-	-
Netherlands	-	-	-	-	-	-	-	-	-
New Zealand	-	-	-	-	-	-	-	-	-
Norway	-	-	-	-	-	-	-	-	-
Poland	-	-	-	-	-	-	-	-	-
Spain	-	-	-	-	-	-	-	-	-
Sweden	-	-	-	-	-	-	-	-	-
Turkey	-	-	-	-	-	-	-	-	-
United Kingdom	-	-	-	-	-	-	-	-	-
United States	18	-	-	-	-	20 e	-	-	-
Other	-	-	-	-	-	-	-	-	-
OECD	**611**	**582**	**487 e**	**1 941**	**1 164**	**1 268 e**	**2 095**	**1 672**	**1 191 e**
Argentina	-	-	-	-	-	-	-	-	-
Brazil	-	-	-	-	-	-	-	-	-
Colombia	-	-	-	-	-	-	-	-	-
Ecuador	-	-	-	-	-	-	-	-	-
Netherlands Antilles	-	-	-	-	-	-	-	-	-
Peru	-	-	-	-	-	-	-	-	-
Trinidad and Tobago	-	-	-	-	-	-	-	4	-
Venezuela	-	-	-	-	-	-	-	-	-
Other	-	-	-	-	-	-	-	-	-
Latin America	**-**	**-**	**-**	**-**	**-**	**-**	**-**	**4**	**-**
Romania	-	-	-	-	-	-	-	-	-
Other	-	-	-	-	-	-	-	-	-
Non-OECD Europe	**-**	**-**	**-**	**-**	**-**	**-**	**-**	**-**	**-**
FSU	**-**	**-**	**-**	**-**	**-**	**-**	**237**	**225**	**301 e**
Iran	-	-	-	-	-	-	-	-	-
Iraq	-	-	-	-	-	-	-	-	-
Kuwait	-	-	-	78	-	62 e	-	-	-
Oman	-	-	-	-	-	-	-	-	-
Qatar	-	-	-	-	-	31 e	-	-	-
Saudi Arabia	-	-	-	16	34	68 e	-	-	-
United Arab Emirates	-	-	-	30	53	31 e	-	-	-
Other	-	-	-	-	-	-	-	-	-
Middle East	**-**	**-**	**-**	**124**	**87**	**192 e**	**-**	**-**	**-**
India	-	-	-	-	-	19 e	-	-	-
Indonesia	-	-	-	-	-	-	-	-	-
Singapore	480	289	573 e	97	139	68 e	-	-	-
Other	109	25	161 e	338	232	300 e	33	17	70 e
Asia	**589**	**314**	**734 e**	**435**	**371**	**387 e**	**33**	**17**	**70 e**
China	**-**	**-**	**-**	**210**	**92**	**176 e**	**-**	**3**	**-**
Algeria	-	-	-	-	-	-	-	-	-
Angola/Cabinada	-	-	-	-	-	-	-	-	-
Cameroon	-	-	-	-	-	-	-	-	-
Egypt	-	-	-	-	-	-	-	-	-
Gabon	-	-	-	-	-	-	-	-	-
Libya	-	-	-	-	-	-	-	-	-
Nigeria	-	-	-	-	-	-	-	-	-
Tunisia	-	-	-	-	-	-	-	-	-
Other	-	-	-	-	-	-	-	-	-
Africa	**-**	**-**	**-**	**-**	**-**	**-**	**-**	**-**	**-**
Non-spec./Other	**-**	**-**	**-**	**1 960**	**1 913**	**2 404 e**	**-**	**-**	**-**
Total Imports	**1 200**	**896**	**1 221 e**	**4 670**	**3 627**	**4 427 e**	**2 365**	**1 921**	**1 562 e**
Memo: Total OPEC	*-*	*-*	*-*	*124*	*87*	*192 e*	*-*	*-*	*-*

Source: Annual Oil Statistics.

JAPAN
Table 7D: Imports by Country of Origin

000 metric tons

	2000	2001	2002	2000	2001	2002	2000	2001	2002
	Fuel Oil (Residual)			**Petroleum Coke**			**Other Products**		
Australia	-	-	-	-	-	-	-	-	-
Belgium	-	-	-	-	-	-	-	-	-
Canada	-	-	-	-	-	220 e	1	-	-
Czech Republic	-	-	-	-	-	-	-	-	-
Denmark	-	-	-	-	-	-	-	-	-
Finland	-	-	-	-	-	-	-	-	-
France	-	-	-	-	-	-	3	1	2 e
Germany	-	-	-	-	-	-	-	1	-
Greece	-	-	-	-	-	-	-	-	-
Hungary	-	-	-	-	-	-	-	-	-
Ireland	-	-	-	-	-	-	-	-	-
Italy	-	-	-	-	-	-	-	-	-
Japan	-	-	-	-	-	-	-	-	-
Korea	144	158	625 e	-	-	92 e	31	31	12 e
Mexico	-	-	-	-	-	-	-	-	-
Netherlands	-	-	-	-	-	-	3	2	-
New Zealand	-	-	-	-	-	-	-	-	-
Norway	-	-	-	-	-	-	-	-	-
Poland	-	-	-	-	-	-	-	-	-
Spain	-	-	-	-	-	-	-	-	-
Sweden	-	-	-	-	-	-	-	-	-
Turkey	-	-	-	-	-	-	-	-	-
United Kingdom	-	-	-	-	-	-	-	-	-
United States	-	-	-	-	-	3 092 e	15	17	11 e
Other	-	-	-	-	-	-	-	-	-
OECD	**144**	**158**	**625 e**	**-**	**-**	**3 404 e**	**53**	**52**	**25 e**
Argentina	-	-	-	-	-	-	-	-	-
Brazil	-	-	-	-	-	-	-	-	-
Colombia	-	-	-	-	-	-	-	-	-
Ecuador	-	-	-	-	-	-	-	-	-
Netherlands Antilles	-	-	-	-	-	-	-	-	-
Peru	-	-	-	-	-	-	-	-	-
Trinidad and Tobago	-	-	-	-	-	-	1	1	-
Venezuela	-	-	-	-	-	-	7	6	2 e
Other	-	-	-	-	-	-	-	-	-
Latin America	**-**	**-**	**-**	**-**	**-**	**-**	**8**	**7**	**2 e**
Romania	-	-	-	-	-	-	-	-	-
Other	-	-	-	-	-	-	-	-	-
Non-OECD Europe	**-**	**-**	**-**	**-**	**-**	**-**	**-**	**-**	**-**
FSU	**-**	**-**	**-**	**-**	**-**	**-**	**-**	**-**	**-**
Iran	-	-	-	-	-	-	-	-	-
Iraq	-	-	-	-	-	-	-	-	-
Kuwait	-	-	-	-	-	-	-	-	-
Oman	-	-	-	-	-	-	-	-	-
Qatar	-	-	-	-	-	-	-	-	-
Saudi Arabia	-	-	32 e	-	-	-	-	-	-
United Arab Emirates	-	-	-	-	-	-	-	-	-
Other	-	-	-	-	-	-	-	-	-
Middle East	**-**	**-**	**32 e**	**-**	**-**	**-**	**-**	**-**	**-**
India	-	-	-	-	-	165 e	-	-	-
Indonesia	622	513	639 e	-	-	41 e	-	-	-
Singapore	-	31	-	-	-	-	3	1	3 e
Other	25	-	40 e	-	-	124 e	2	3	7 e
Asia	**647**	**544**	**679 e**	**-**	**-**	**330 e**	**5**	**4**	**10 e**
China	**-**	**-**	**-**	**-**	**-**	**156 e**	**-**	**1**	**-**
Algeria	-	-	-	-	-	-	-	-	-
Angola/Cabinada	-	-	-	-	-	-	-	-	-
Cameroon	-	-	-	-	-	-	-	-	-
Egypt	-	-	-	-	-	-	-	-	-
Gabon	-	-	-	-	-	-	-	-	-
Libya	-	-	-	-	-	-	-	-	-
Nigeria	-	-	-	-	-	-	-	-	-
Tunisia	-	-	-	-	-	-	-	-	-
Other	-	-	-	-	-	-	-	-	-
Africa	**-**	**-**	**-**	**-**	**-**	**-**	**-**	**-**	**-**
Non-spec./Other	**295**	**274**	**-**	**4 557**	**4 196**	**-**	**-**	**-**	**-**
Total Imports	**1 086**	**976**	**1 336 e**	**4 557**	**4 196**	**3 890 e**	**66**	**64**	**37 e**
Memo: Total OPEC	*622*	*513*	*671 e*	*-*	*-*	*41 e*	*7*	*6*	*2 e*

Source: Annual Oil Statistics.

JAPAN
Table 8A: Exports by Country of Destination

000 metric tons

	2000	2001	2002	2000	2001	2002	2000	2001	2002
	Crude Oil			**Natural Gas Liquids**			**Feedstocks and Other**		
Australia	-	-	-	-	-	-	-	-	-
Austria	-	-	-	-	-	-	-	-	-
Belgium	-	-	-	-	-	-	-	-	-
Canada	-	-	-	-	-	-	-	-	-
Denmark	-	-	-	-	-	-	-	-	-
Finland	-	-	-	-	-	-	-	-	-
France	-	-	-	-	-	-	-	-	-
Germany	-	-	-	-	-	-	-	-	-
Greece	-	-	-	-	-	-	-	-	-
Ireland	-	-	-	-	-	-	-	-	-
Italy	-	-	-	-	-	-	-	-	-
Japan	-	-	-	-	-	-	-	-	-
Luxembourg	-	-	-	-	-	-	-	-	-
Mexico	-	-	-	-	-	-	-	-	-
Netherlands	-	-	-	-	-	-	-	-	-
Norway	-	-	-	-	-	-	-	-	-
Portugal	-	-	-	-	-	-	-	-	-
Spain	-	-	-	-	-	-	-	-	-
Sweden	-	-	-	-	-	-	-	-	-
Switzerland	-	-	-	-	-	-	-	-	-
Turkey	-	-	-	-	-	-	-	-	-
United Kingdom	-	-	-	-	-	-	-	-	-
United States	-	-	-	-	-	-	-	-	-
Other OECD	-	-	-	-	-	-	-	-	-
OECD	**-**	**-**	**-**	**-**	**-**	**-**	**-**	**-**	**-**
Africa	-	-	-	-	-	-	-	-	-
Latin America	-	-	-	-	-	-	-	-	-
Asia	-	-	-	-	-	-	-	-	-
China	-	-	-	-	-	-	-	-	-
Non-OECD Europe	-	-	-	-	-	-	-	-	-
Former Soviet Union	-	-	-	-	-	-	-	-	-
Middle East	-	-	-	-	-	-	-	-	-
Non-specified/Other	-	-	-	-	-	-	-	-	-
Non-OECD	**-**	**-**	**-**	**-**	**-**	**-**	**-**	**-**	**-**
Total Exports	**-**	**-**	**-**	**-**	**-**	**-**	**-**	**-**	**-**
	Total Products			**Liquefied Petroleum Gas**			**Naphtha**		
Australia	478	209	5 e	-	-	-	-	-	-
Austria	-	-	-	-	-	-	-	-	-
Belgium	-	-	-	-	-	-	-	-	-
Canada	1	-	-	-	-	-	-	-	-
Denmark	-	-	-	-	-	-	-	-	-
Finland	-	-	-	-	-	-	-	-	-
France	-	-	-	-	-	-	-	-	-
Germany	-	3	-	-	-	-	-	-	-
Greece	-	-	-	-	-	-	-	-	-
Ireland	-	-	-	-	-	-	-	-	-
Italy	-	-	-	-	-	-	-	-	-
Japan	-	-	-	-	-	-	-	-	-
Luxembourg	-	-	-	-	-	-	-	-	-
Mexico	39	-	-	-	-	-	-	-	-
Netherlands	-	-	-	-	-	-	-	-	-
Norway	-	-	-	-	-	-	-	-	-
Portugal	-	-	-	-	-	-	-	-	-
Spain	-	-	-	-	-	-	-	-	-
Sweden	-	-	-	-	-	-	-	-	-
Switzerland	-	-	-	-	-	-	-	-	-
Turkey	-	-	-	-	-	-	-	-	-
United Kingdom	-	1	7 e	-	-	-	-	-	-
United States	701	580	362 e	-	-	-	-	-	-
Other OECD	994	1 669	1 372 e	27	70	52 e	-	19	16 e
OECD	**2 213**	**2 462**	**1 746 e**	**27**	**70**	**52 e**	**-**	**19**	**16 e**
Africa	2	1	-	-	-	-	-	-	-
Latin America	10	6	2 e	-	-	-	-	-	-
Asia	574	255	511 e	21	-	1 e	22	-	8 e
China	1 353	1 739	1 241 e	1	6	3 e	85	49	60 e
Non-OECD Europe	-	-	-	-	-	-	-	-	-
Former Soviet Union	-	-	-	-	-	-	-	-	-
Middle East	6	-	-	6	-	-	-	-	-
Non-specified/Other	170	248	185 e	-	-	-	-	-	-
Non-OECD	**2 115**	**2 249**	**1 939 e**	**28**	**6**	**4 e**	**107**	**49**	**68 e**
Total Exports	**4 328**	**4 711**	**3 685 e**	**55**	**76**	**56 e**	**107**	**68**	**84 e**

Source: Annual Oil Statistics.

JAPAN
Table 8B: Exports by Country of Destination

000 metric tons

	2000	2001	2002	2000	2001	2002	2000	2001	2002
	Total Gasoline			**Total Kerosene**			**Gas/Diesel Oil**		
Australia	61	25	-	-	-	-	416	181	-
Austria	-	-	-	-	-	-	-	-	-
Belgium	-	-	-	-	-	-	-	-	-
Canada	-	-	-	-	-	-	-	-	-
Denmark	-	-	-	-	-	-	-	-	-
Finland	-	-	-	-	-	-	-	-	-
France	-	-	-	-	-	-	-	-	-
Germany	-	-	-	-	-	-	-	-	-
Greece	-	-	-	-	-	-	-	-	-
Ireland	-	-	-	-	-	-	-	-	-
Italy	-	-	-	-	-	-	-	-	-
Japan	-	-	-	-	-	-	-	-	-
Luxembourg	-	-	-	-	-	-	-	-	-
Mexico	-	-	-	-	-	-	39	-	-
Netherlands	-	-	-	-	-	-	-	-	-
Norway	-	-	-	-	-	-	-	-	-
Portugal	-	-	-	-	-	-	-	-	-
Spain	-	-	-	-	-	-	-	-	-
Sweden	-	-	-	-	-	-	-	-	-
Switzerland	-	-	-	-	-	-	-	-	-
Turkey	-	-	-	-	-	-	-	-	-
United Kingdom	-	-	-	-	-	-	-	-	-
United States	19	150	35 e	554	338	151 e	83	56	59 e
Other OECD	44	69	80 e	117	259	182 e	350	813	754 e
OECD	**124**	**244**	**115 e**	**671**	**597**	**333 e**	**888**	**1 050**	**813 e**
Africa	-	-	-	-	-	-	-	-	-
Latin America	-	-	-	-	-	-	-	-	-
Asia	91	13	85 e	-	2	104 e	197	69	178 e
China	39	6	-	49	83	116 e	465	364	60 e
Non-OECD Europe	-	-	-	-	-	-	-	-	-
Former Soviet Union	-	-	-	-	-	-	-	-	-
Middle East	-	-	-	-	-	-	-	-	-
Non-specified/Other	3	23	2 e	5	49	-	144	176	183 e
Non-OECD	**133**	**42**	**87 e**	**54**	**134**	**220 e**	**806**	**609**	**421 e**
Total Exports	**257**	**286**	**202 e**	**725**	**731**	**553 e**	**1 694**	**1 659**	**1 234 e**

	2000	2001	2002	2000	2001	2002	2000	2001	2002
	Fuel Oil (Residual)			**Petroleum Coke**			**Other Products**		
Australia	-	-	-	-	-	-	1	3	5 e
Austria	-	-	-	-	-	-	-	-	-
Belgium	-	-	-	-	-	-	-	-	-
Canada	-	-	-	-	-	-	1	-	-
Denmark	-	-	-	-	-	-	-	-	-
Finland	-	-	-	-	-	-	-	-	-
France	-	-	-	-	-	-	-	-	-
Germany	-	-	-	-	-	-	-	3	-
Greece	-	-	-	-	-	-	-	-	-
Ireland	-	-	-	-	-	-	-	-	-
Italy	-	-	-	-	-	-	-	-	-
Japan	-	-	-	-	-	-	-	-	-
Luxembourg	-	-	-	-	-	-	-	-	-
Mexico	-	-	-	-	-	-	-	-	-
Netherlands	-	-	-	-	-	-	-	-	-
Norway	-	-	-	-	-	-	-	-	-
Portugal	-	-	-	-	-	-	-	-	-
Spain	-	-	-	-	-	-	-	-	-
Sweden	-	-	-	-	-	-	-	-	-
Switzerland	-	-	-	-	-	-	-	-	-
Turkey	-	-	-	-	-	-	-	-	-
United Kingdom	-	-	-	-	-	-	-	1	7 e
United States	22	-	86 e	-	-	-	23	36	31 e
Other OECD	286	207	62 e	-	-	-	170	232	226 e
OECD	**308**	**207**	**148 e**	**-**	**-**	**-**	**195**	**275**	**269 e**
Africa	-	-	-	-	-	-	2	1	-
Latin America	-	-	-	-	-	-	10	6	2 e
Asia	116	61	-	-	-	-	127	110	135 e
China	443	844	694 e	-	-	-	271	387	308 e
Non-OECD Europe	-	-	-	-	-	-	-	-	-
Former Soviet Union	-	-	-	-	-	-	-	-	-
Middle East	-	-	-	-	-	-	-	-	-
Non-specified/Other	18	-	-	-	-	-	-	-	-
Non-OECD	**577**	**905**	**694 e**	**-**	**-**	**-**	**410**	**504**	**445 e**
Total Exports	**885**	**1 112**	**842 e**	**-**	**-**	**-**	**605**	**779**	**714 e**

Source: Annual Oil Statistics.

KOREA

Figure 1
FUEL SHARES OF TPES *

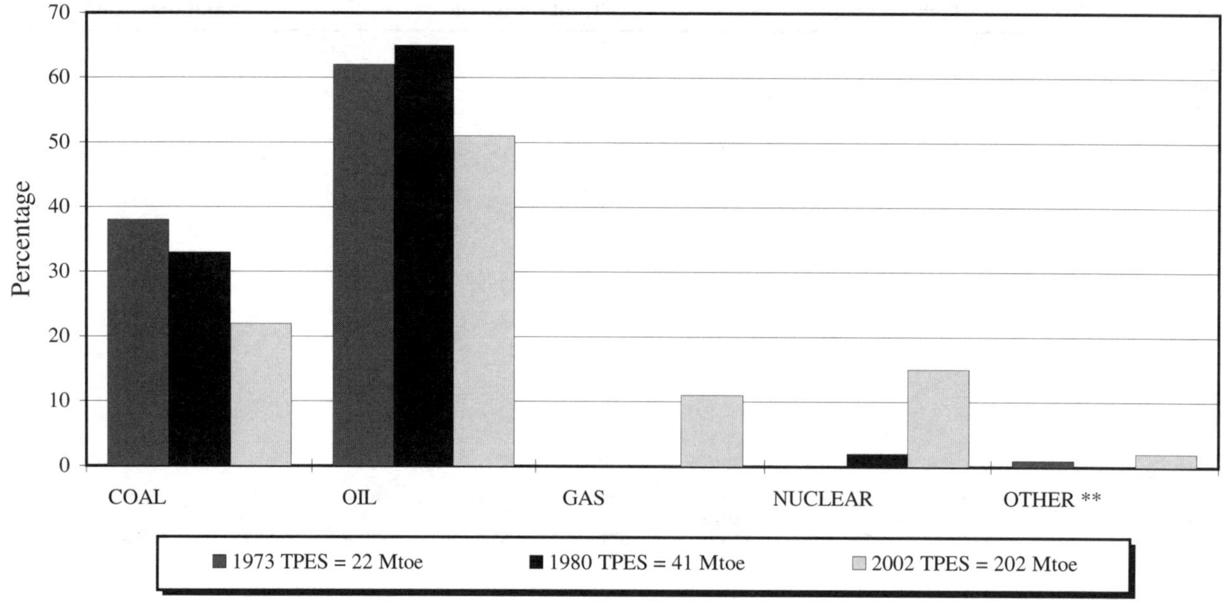

Source: Annual Energy Balances.

* Excluding electricity trade.

** Includes hydro, geothermal, solar, wind, combustible renewables and waste, etc.

Figure 2
OIL PRODUCTION AND CONSUMPTION

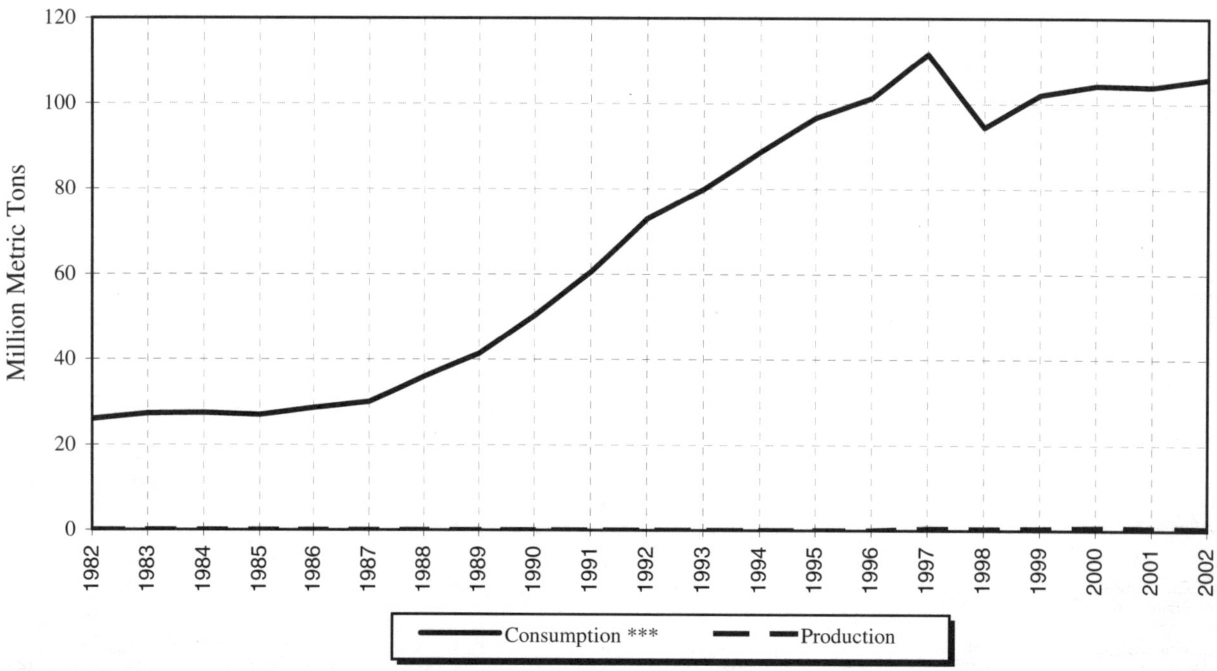

Source: Annual Oil Statistics.

*** Includes refinery fuel and international marine bunkers.

KOREA
Table 1: Oil Supply and Consumption - Summary Table

000 metric tons

	1973	1978	1980	1990	1999	2000	2001	2002
Indigenous Production	-	-	-	-	446	670	572	523 e
Crude Oil	..	-	-	-	-	-	-	-
NGL	..	-	-	-	-	-	-	-
Other	..	-	-	-	446	670	572	523 e
+ Net Imports	13 005	22 826	26 805	50 428	102 517	106 846	103 440	103 320 e
Crude Oil, NGL, Other	14 004	22 596	24 812	41 841	120 491	123 221	118 793	107 809 e
Petroleum Products	- 999	230	1 993	8 587	-17 974	-16 375	-15 353	-4 489 e
- Intl. Marine Bunkers	576	74	96	1 626	6 689	6 232	6 120	5 826 e
+ Stock Changes	758	668	- 251	- 26	117	- 705	27	1 678 e
TOTAL SUPPLY	**13 187**	**23 420**	**26 458**	**48 776**	**96 391**	**100 579**	**97 919**	**99 695 e**
Refinery Intake	*14 565*	*22 662*	*24 413*	*41 693*	*122 838*	*125 265*	*122 148*	*115 178 e*
Refinery Output	*14 989*	*24 214*	*25 601*	*42 365*	*120 349*	*122 001*	*119 210*	*111 998 e*
LPG + Ethane	*117 e*	*283 e*	*330 e*	*1 034 e*	*2 755*	*3 184*	*3 566*	*3 676 e*
Naphtha	*1 264 e*	*2 473 e*	*2 446 e*	*4 175 e*	*19 347*	*19 648*	*19 138*	*19 318 e*
Kerosenes	*565 e*	*1 159 e*	*1 520 e*	*2 991 e*	*18 384*	*18 855*	*17 978*	*16 702 e*
Gasolines	*1 040 e*	*1 312 e*	*1 053 e*	*3 599 e*	*8 811*	*8 933*	*8 969*	*9 027 e*
Gas/Diesel Oil	*2 663 e*	*5 006 e*	*5 507 e*	*13 049 e*	*29 831*	*31 132*	*29 854*	*28 406 e*
Fuel Oil	*8 742 e*	*13 265 e*	*13 954 e*	*15 982 e*	*35 542*	*33 967*	*33 349*	*27 814 e*
Other Products	*598 e*	*716 e*	*791 e*	*1 535 e*	*5 679*	*6 282*	*6 356*	*7 055 e*
- Refinery Losses	- 424	-1 552	-1 188	- 672	2 489	3 264	2 938	3 180 e
+ Returns	-	-	-	-	5 121	4 837	4 974	5 374 e
- Statistical Differences	135	1 066	1 179	866	3 637	4 134	2 144	1 905 e
TOTAL CONSUMPTION	**13 476**	**23 906**	**26 467**	**48 582**	**95 386**	**98 018**	**97 811**	**99 984 e**
Transformation and Energy	**3 709**	**7 640**	**7 905**	**6 014**	**13 672**	**15 633**	**16 302**	..
Crude Oil	-	-	-	-	-	-	-	..
LPG + Ethane	-	-	-	-	217	208	176	..
Gas/Diesel Oil	19 e	627 e	112 e	246 e	170	81	55	..
Fuel Oil	3 465 e	6 652 e	7 410 e	5 131 e	6 388	8 510	9 162	..
Petroleum Coke	-	-	-	-	29	41	40	..
Other	225 e	361 e	383 e	637 e	6 868	6 793	6 869	..
Transport	**2 473**	**4 580**	**4 859**	**14 178**	**26 818**	**29 007**	**29 934**	..
Motor Gasoline	707	853	798	2 762	7 375	7 284	7 323	..
Gas/Diesel Oil	1 080	2 647	2 843	7 163	10 920	12 347	12 850	..
Subtotal	*1 787*	*3 500*	*3 641*	*9 925*	*18 205*	*10 631*	*20 170*	..
Aviation Fuels	368	567	469	2 077	2 382	2 514	2 634	..
Other	318	513	749	2 176	6 141	6 862	7 127	..
Industry	**6 297**	**9 179**	**9 875**	**17 271**	**35 721**	**36 303**	**35 665**	..
Naphtha	837	1 931	2 521	5 569	21 114	22 530	22 852	..
LPG + Ethane	12	56	114	193	1 533	1 461	1 265	..
Gas/Diesel Oil	806	855	739	1 696	993	1 076	991	..
Fuel Oil	4 142	5 782	6 043	8 474	9 176	8 316	7 698	..
Other	500	555	458	1 339	2 905	2 920	2 859	..
Other Sectors	**997**	**2 507**	**3 828**	**11 119**	**19 175**	**17 075**	**15 910**	..
Gas/Diesel Oil	442	903	1 745	4 288	5 261	4 250	4 189	..
Fuel Oil	309	734	912	2 105	1 000	951	884	..
Other	246	870	1 171	4 726	12 914	11 874	10 837	..
Memo: Non-Energy Use	*1 068*	*2 304*	*2 935*	*6 547*	*23 746*	*25 102*	*25 415*	..
Chem/Petchem. Feedst.	*849*	*1 987*	*2 642*	*5 706*	*22 053*	*23 410*	*23 600*	..
Other	*219*	*317*	*293*	*841*	*1 693*	*1 692*	*1 815*	..

Source: Annual Oil Statistics.

KOREA
Table 2: Supply of Crude Oil, NGL, Feedstocks and Other Hydrocarbons

000 metric tons

	1973	1978	1980	1990	1999	2000	2001	2002
CRUDE OIL								
Indigenous Production	..	-	-	-	-	-	-	-
+ Imports	..	22 596	24 812	41 841	119 533	122 546	117 697	106 712 e
- Exports	..	-	-	-	-	-	-	-
- Direct Use	..	-	-	-	-	-	-	-
+ Stock Change	..	69	- 401	556	224	393	- 225	850 e
- Statistical Difference	..	3	- 2	704	1 339	2 138	767	594 e
Refinery Intake	**..**	**22 662**	**24 413**	**41 693**	**118 418**	**120 801**	**116 705**	**106 968 e**
Closing Stock Level	..	*103*	*859*	*457*	*4 292*	*3 899*	*4 124*	*3 274 e*
NATURAL GAS LIQUIDS								
Indigenous Production	..	-	-	-	-	-	-	-
+ Imports	..	-	-	-	-	-	-	-
- Exports	..	-	-	-	-	-	-	-
- Direct Use	..	-	-	-	-	-	-	-
+ Stock Change	..	-	-	-	-	-	-	-
- Statistical Difference	..	-	-	-	-	-	-	-
Refinery Intake	**..**	**-**	**-**	**-**	**-**	**-**	**-**	**-**
Closing Stock Level	..	-	-	-	-	-	-	-
REFINERY FEEDSTOCKS								
Backflows	..	-	-	-	5 121	4 837	4 974	5 374 e
+ Transfers	..	-	-	-	157	112	149	1 801 e
+ Imports	..	-	-	-	996	772	1 199	1 218 e
- Exports	..	-	-	-	-	-	-	9 e
+ Stock Change	..	-	-	-	- 317	- 22	92	47 e
- Statistical Difference	..	-	-	-	1 537	1 235	971	221 e
Refinery Intake	**..**	**-**	**-**	**-**	**4 420**	**4 464**	**5 443**	**8 210 e**
Closing Stock Level	..	-	-	-	*670*	*692*	*600*	*553 e*
OTHER HYDROCARBONS								
Indigenous Production	..	-	-	-	446	670	572	523 e
+ Imports	..	-	-	-	-	2	-	6 e
- Exports	..	-	-	-	38	99	103	118 e
- Direct Use	..	-	-	-	419	574	469	398 e
+ Stock Change	..	-	-	-	11	1	-	- 13 e
- Statistical Difference	..	-	-	-	-	-	-	-
Refinery Intake	**..**	**-**	**-**	**-**	**-**	**-**	**-**	**-**
Closing Stock Level	..	-	-	-	*28*	*27*	*27*	*40 e*
CRUDE OIL, NGL, FEEDSTOCKS AND OTHER								
Indigenous Production	-	-	-	-	446	670	572	523 e
+ Backflows	-	-	-	-	5 121	4 837	4 974	5 374 e
+ Transfers	-	-	-	-	157	112	149	1 801 e
+ Imports	14 004	22 596	24 812	41 841	120 529	123 320	118 896	107 936 e
- Exports	-	-	-	-	38	99	103	127 e
- Direct Use	-	-	-	-	419	574	469	398 e
+ Stock Change	-	69	- 401	556	- 82	372	- 133	884 e
- Statistical Difference	- 561	3	- 2	704	2 876	3 373	1 738	815 e
Refinery Intake	**14 565**	**22 662**	**24 413**	**41 693**	**122 838**	**125 265**	**122 148**	**115 178 e**
Closing Stock Level	-	*103*	*859*	*457*	*4 990*	*4 618*	*4 751*	*3 867 e*

Source: Annual Oil Statistics.

KOREA
Table 3: Motor Gasoline - Supply

000 metric tons

	1973	1978	1980	1990	1999	2000	2001	2002
MOTOR GASOLINE								
Refinery Output	**765 e**	**936 e**	**831 e**	**2 806 e**	**8 811**	**8 933**	**8 969**	**9 027 e**
+ Imports	-	-	-	-	51	85	337	512 e
- Exports	41	47	-	4	1 264	1 692	1 895	1 817 e
+ Transfers	-	-	-	-	417	572	453	257 e
+ Stock Change	- 121	- 9	- 8	- 28	- 41	- 63	- 21	22 e
- Statistical Difference	- 123 e	- 1 e	-	- 4 e	426	521	491	452 e
Total Consumption	**726**	**881**	**823**	**2 778**	**7 548**	**7 314**	**7 352**	**7 549 e**
Closing Stock Level	-	-	-	*133*	*387*	*450*	*471*	*449 e*

Table 4: Aviation Fuels - Supply

000 metric tons

	1973	1978	1980	1990	1999	2000	2001	2002
AVIATION GASOLINE								
Refinery Output	-	-	-	-	-	-	-	-
+ Imports	-	-	-	1	-	-	-	1 e
- Exports	-	-	-	-	-	-	-	-
+ Transfers	-	-	-	-	-	-	-	-
+ Stock Change	-	-	-	-	-	-	-	- 1 e
- Statistical Difference	-	-	-	-	-	-	-	-
Total Consumption	**-**	**-**	**-**	**1**	**-**	**-**	**-**	**-**
Closing Stock Level	-	-	-	-	-	-	-	*1 e*
GASOLINE TYPE JET FUEL								
Refinery Output	**275 e**	**376 e**	**222 e**	**793 e**	-	-	-	-
+ Imports	-	-	-	-	-	-	-	-
- Exports	-	-	-	-	-	-	-	-
+ Transfers	-	-	-	-	-	-	-	-
+ Stock Change	-	-	-	- 17	-	-		
- Statistical Difference	- 4 e	4 e	15 e	- 24 e	-	-	-	-
Total Consumption	**279**	**372**	**207**	**800**	**-**	**-**	**-**	**-**
Closing Stock Level	-	-	-	*45*	-	-	-	-
KEROSENE TYPE JET FUEL								
Refinery Output	**176 e**	**288 e**	**389 e**	**1 242 e**	**6 691**	**7 512**	**8 228**	**8 540 e**
+ Imports	-	-	1	-	453	214	43	351 e
- Exports	47	19	-	-	3 840	4 139	4 981	5 274 e
+ Transfers	-	-	-	-	- 103	27	- 49	- 237 e
+ Stock Change	- 5	- 3	- 49	- 25	68	- 152	- 64	118 e
- Statistical Difference	35 e	71 e	79 e	- 73 e	11	260	25	30 e
Total Consumption	**89**	**195**	**262**	**1 290**	**3 258**	**3 202**	**3 152**	**3 468 e**
Closing Stock Level	-	-	-	*78*	*186*	*338*	*402*	*284 e*

Source: Annual Oil Statistics.

KOREA
Table 5A: Gas/Diesel Oil - Supply and Consumption

000 metric tons

	1973	1978	1980	1990	1999	2000	2001	2002
Refinery Output	**2 663 e**	**5 006 e**	**5 507 e**	**13 049 e**	**29 831**	**31 132**	**29 854**	**28 406 e**
+ Recycled Products	-	-	-	-	-	-	-	-
+ Imports	-	132	-	1 821	326	386	952	1 796 e
- Exports	191	11	-	1 401	11 793	12 929	12 123	10 217 e
- Intl. Marine Bunkers	51	23	13	264	1 151	869	808	798 e
+ Transfers	-	-	-	-	37	24	36	- 181 e
+ Stock Changes	- 14	- 64	- 41	87	61	- 97	162	208 e
- Statistical Differences	60 e	8 e	14 e	- 101 e	- 33	- 107	- 12	103 e
Total Consumption	**2 347**	**5 032**	**5 439**	**13 393**	**17 344**	**17 754**	**18 085**	**19 111 e**
Transformation	**17**	**623**	**108**	**236**	**152**	**69**	**41**	**..**
Public Electricity	17	623	108	236	146	54	33	..
Autoproducer Electricity	-	-	-	-	-	-	-	..
Public CHP Plants	-	-	-	-	-	12	6	..
Autoproducers of CHP	-	-	-	-	-	-	-	..
Public Heat Plants	-	-	-	-	-	-	-	..
Autoproducer Heat Plants	-	-	-	-	-	-	-	..
Petrochemical Industry	-	-	-	-	6	3	2	..
Other Transformation	-	-	-	-	-	-	-	..
Energy Sector	**2**	**4**	**4**	**10**	**18**	**12**	**14**	**..**
Refinery Fuel	2	4	4	10	-	-	-	..
Other Energy Sector	-	-	-	-	18	12	14	..
Final Consumption	**2 328**	**4 405**	**5 327**	**13 147**	**17 174**	**17 673**	**18 030**	**..**
Transport	**1 080**	**2 647**	**2 843**	**7 163**	**10 920**	**12 347**	**12 850**	**..**
Air	-	-	-	-	-	1	1	..
Road	-	-	-	6 439	9 582	11 130	11 733	..
Rail	-	-	-	275	299	308	311	..
Inland Waterways	-	-	-	449	966	802	602	..
Non-Specified	1 080	2 647	2 843	-	73	106	203	..
Industry	**806**	**855**	**739**	**1 696**	**993**	**1 076**	**991**	**..**
Iron and Steel	-	-	-	100	44	31	15	..
Chemical (incl. Pet.-Chem.)	-	-	-	159	91	89	95	..
Non-Ferrous Metals	-	-	-	28	14	10	7	..
Non-Metallic Minerals	-	-	-	239	90	68	65	..
Transport Equipment	-	-	-	-	67	71	81	..
Machinery	-	-	-	149	69	67	42	..
Mining and Quarrying	-	-	-	55	15	20	24	..
Food and Tobacco	-	-	-	106	45	44	38	..
Pulp, Paper and Printing	-	-	-	6	19	26	23	..
Wood and Wood Products	-	-	-	3	1	3	3	..
Construction	-	-	-	331	174	278	258	..
Textiles and Leather	-	-	-	35	20	20	17	..
Non-Specified	806	855	739	485	344	349	323	..
Other Sectors	**442**	**903**	**1 745**	**4 288**	**5 261**	**4 250**	**4 189**	**..**
Commerce and Public	102	389	472	2 221	2 251	1 301	1 105	..
Residential	-	-	-	-	288	83	56	..
Agriculture	-	-	400	1 300	2 273	2 493	2 609	..
Non-Specified	340	514	873	767	449	373	419	..
Memo: Non-Energy Use	-	-	-	-	-	-	-	..
Chem/Petchem. Feedst.	-	-	-	-	-	-	-	..
Other	-	-	-	-	-	-	-	..
Closing Stock Level	-	-	-	*463*	*1 344*	*1 441*	*1 279*	*1 071 e*

Source: Annual Oil Statistics.

KOREA
Table 5B: Fuel Oil (Residual) - Supply and Consumption

000 metric tons

	1973	1978	1980	1990	1999	2000	2001	2002
Refinery Output	**8 742 e**	**13 265 e**	**13 954 e**	**15 982 e**	**35 542**	**33 967**	**33 349**	**27 814 e**
+ Recycled Products	-	-	-	-	-	-	-	-
+ Imports	-	326	1 839	4 153	2 528	3 761	3 409	3 714 e
- Exports	210	-	-	1 815	12 367	10 796	10 062	5 261 e
- Intl. Marine Bunkers	525	51	83	1 362	5 538	5 363	5 312	5 028 e
+ Transfers	-	-	-	-	13	65	84	- 866 e
+ Stock Changes	891	1 101	339	149	11	- 171	- 80	71 e
- Statistical Differences	703 e	1 018 e	1 084 e	214 e	22	54	123	150 e
Total Consumption	**8 195**	**13 623**	**14 965**	**16 893**	**20 167**	**21 409**	**21 265**	**20 294 e**
Transformation	**2 955**	**5 878**	**6 596**	**4 198**	**3 885**	**6 100**	**6 714**	..
Public Electricity	2 955	5 878	6 596	4 198	3 176	4 650	5 170	..
Autoproducer Electricity	-	-	-	-	-	-	-	..
Public CHP Plants	-	-	-	-	69	927	947	..
Autoproducers of CHP	-	-	-	-	-	-	-	..
Public Heat Plants	-	-	-	-	453	398	479	..
Autoproducer Heat Plants	-	-	-	-	-	-	-	..
Petrochemical Industry	-	-	-	-	187	125	118	..
Other Transformation	-	-	-	-	-	-	-	..
Energy Sector	**510**	**774**	**814**	**933**	**2 503**	**2 410**	**2 448**	..
Refinery Fuel	510	774	814	933	2 457	2 366	2 248	..
Other Energy Sector	-	-	-	-	46	44	200	..
Final Consumption	**4 730**	**6 971**	**7 555**	**11 762**	**13 779**	**12 899**	**12 103**	..
Transport	**279**	**455**	**600**	**1 183**	**3 603**	**3 632**	**3 521**	..
Air	-	-	-	-	-	-	-	..
Road	-	-	-	10	2	1	27	..
Rail	-	-	-	5	1	1	1	..
Inland Waterways	279	455	600	1 168	3 582	3 589	3 452	..
Non-Specified	-	-	-	-	18	41	41	..
Industry	**4 142**	**5 782**	**6 043**	**8 474**	**9 176**	**8 316**	**7 698**	..
Iron and Steel	-	-	-	776	864	725	548	..
Chemical (incl. Pet.-Chem.)	-	-	-	2 427	1 868	1 719	1 694	..
Non-Ferrous Metals	-	-	-	104	93	114	103	..
Non-Metallic Minerals	-	-	-	846	701	727	742	..
Transport Equipment	-	-	-	-	855	772	647	..
Machinery				584	642	363	362	..
Mining and Quarrying	-	-	-	53	17	23	21	..
Food and Tobacco	-	-	-	873	737	726	646	..
Pulp, Paper and Printing	-	-	-	797	1 215	1 110	1 025	..
Wood and Wood Products	-	-	-	65	76	72	68	..
Construction	-	-	-	130	118	108	104	..
Textiles and Leather	-	-	-	1 488	1 590	1 516	1 424	..
Non-Specified	4 142	5 782	6 043	331	400	341	314	..
Other Sectors	**309**	**734**	**912**	**2 105**	**1 000**	**951**	**884**	..
Commerce and Public	303	417	440	301	397	386	379	..
Residential	6	317	472	1 754	419	382	342	..
Agriculture	-	-	-	50	167	174	159	..
Non-Specified	-	-	-	-	17	9	4	..
Memo: Non-Energy Use	-	-	-	-	-	-	-	..
Chem/Petchem. Feedst.	-	-	-	-	-	-	-	..
Other	-	-	-	-	-	-	-	..
Closing Stock Level	-	-	-	*523*	*1 589*	*1 760*	*1 840*	*1 769 e*

Source: Annual Oil Statistics.

KOREA
Table 5C: Total Products - Supply and Consumption

000 metric tons

	1973	1978	1980	1990	1999	2000	2001	2002
Refinery Output	**14 989**	**24 214**	**25 601**	**42 365**	**120 349**	**122 001**	**119 210**	**111 998 e**
+ Recycled Products	-	-	-	-	-	-	-	-
+ Imports	-	512	2 004	12 300	20 825	23 389	23 654	26 747 e
- Exports	999	282	11	3 713	38 799	39 764	39 007	31 236 e
- Intl. Marine Bunkers	576	74	96	1 626	6 689	6 232	6 120	5 826 e
+ Transfers	-	-	-	-	262	462	320	-1 403 e
+ Stock Changes	758	599	150	- 582	199	-1 077	160	794 e
- Statistical Differences	696	1 063	1 181	162	761	761	406	1 090 e
Total Consumption	**13 476**	**23 906**	**26 467**	**48 582**	**95 386**	**98 018**	**97 811**	**99 984 e**
Transformation	**2 972**	**6 501**	**6 704**	**4 434**	**9 109**	**11 001**	**11 675**	**..**
Public Electricity	2 972	6 501	6 704	4 434	3 455	4 817	5 266	..
Autoproducer Electricity	-	-	-	-	-	-	-	..
Public CHP Plants	-	-	-	-	70	941	955	..
Autoproducers of CHP	-	-	-	-	-	-	-	..
Public Heat Plants	-	-	-	-	463	406	480	..
Autoproducer Heat Plants	-	-	-	-	-	-	-	..
Petrochemical Industry	-	-	-	-	5 121	4 837	4 974	..
Other Transformation	-	-	-	-	-	-	-	..
Energy Sector	**737**	**1 139**	**1 201**	**1 580**	**4 563**	**4 632**	**4 627**	**..**
Refinery Fuel	737	1 139	1 201	1 580	4 421	4 494	4 327	..
Other Energy Sector	-	-	-	-	142	138	300	..
Final Consumption	**9 767**	**16 266**	**18 562**	**42 568**	**81 714**	**82 385**	**81 509**	**..**
Transport	**2 473**	**4 580**	**4 859**	**14 178**	**26 818**	**29 007**	**29 934**	**..**
Air	368	567	469	2 077	2 377	2 508	2 627	..
Road	746	911	947	10 204	19 480	21 634	22 679	..
Rail	-	-	-	280	302	310	312	..
Inland Waterways	279	455	600	1 617	4 550	4 393	4 055	..
Non-Specified	1 080	2 647	2 843	-	109	162	261	..
Industry	**6 297**	**9 179**	**9 875**	**17 271**	**35 721**	**36 303**	**35 665**	**..**
Iron and Steel	-	-	9	901	1 051	872	642	..
Chemical (incl. Pet.-Chem.)	895	2 039	2 687	8 346	24 451	25 635	25 877	..
Non-Ferrous Metals	-	-	-	150	156	172	153	..
Non-Metallic Minerals	-	-	-	1 162	1 061	1 053	1 026	..
Transport Equipment	19	28	25	16	996	918	798	..
Machinery	-	-	-	806	810	538	489	..
Mining and Quarrying	-	-	-	111	35	45	48	..
Food and Tobacco	-	-	-	1 007	856	839	743	..
Pulp, Paper and Printing	-	-	-	808	1 261	1 157	1 062	..
Wood and Wood Products	-	-	-	68	79	78	73	..
Construction	219	317	293	1 342	2 027	2 133	2 229	..
Textiles and Leather	-	-	-	1 536	1 637	1 570	1 471	..
Non-Specified	5 164	6 795	6 861	1 018	1 301	1 293	1 054	..
Other Sectors	**997**	**2 507**	**3 828**	**11 119**	**19 175**	**17 075**	**15 910**	**..**
Commerce and Public	588	1 650	2 045	6 722	11 626	8 812	7 238	..
Residential	69	335	495	1 881	2 810	3 783	4 298	..
Agriculture	-	-	400	1 503	2 959	3 098	3 257	..
Non-Specified	340	522	888	1 013	1 780	1 382	1 117	..
Memo: Non-Energy Use	*1 068*	*2 304*	*2 935*	*6 547*	*23 746*	*25 102*	*25 415*	..
Chem/Petchem. Feedst.	*849*	*1 987*	*2 642*	*5 706*	*22 053*	*23 410*	*23 600*	..
Other	*219*	*317*	*293*	*841*	*1 693*	*1 692*	*1 815*	..
Closing Stock Level	-	-	-	*2 468*	*5 928*	*7 005*	*6 845*	*6 051 e*

Source: Annual Oil Statistics.

KOREA
Table 6A: Supply and Consumption

000 metric tons

	1980	2000	2001	2002	1980	2000	2001	2002
	LPG AND ETHANE				**REFINERY GAS**			
Refinery Output	**330 e**	**3 184**	**3 566**	**3 676 e**	**377 e**	**2 052**	**1 969**	**1 841 e**
+ Recycled Products	-	-	-	-	-	-	-	-
+ Imports	61	5 131	4 061	4 906 e	-	-	-	-
- Exports	-	701	416	246 e	-	-	-	-
- Intl. Marine Bunkers	-	-	-	-	-	-	-	-
+ Transfers	-	- 1	- 2	- 45 e	-	-	-	-
+ Stock Change	- 2	- 250	172	- 194 e	-	-	-	-
- Statistical Difference	1 e	54	100	216 e	-	-	-	-
Total Consumption	**388**	**7 309**	**7 281**	**7 881 e**	**377 e**	**2 052**	**1 969**	**1 841 e**
of which: Ethane	-	-	-	-	-	-	-	-
Transformation	-	199	168	..	-	-	-	..
Energy	-	9	8	..	377 e	2 052	1 969	..
Transport	147	2 745	3 137	..	-	-	-	..
Industry	114	1 461	1 265	..	-	-	-	..
Other	127	2 895	2 703	..	-	-	-	..
Memo: Non-Energy Use	*114*	*850*	*725*	..	-	-	-	..
Closing Stock Level	-	*588*	*416*	*610 e*	-	-	-	-
	NAPHTHA				**OTHER KEROSENE**			
Refinery Output	**2 446 e**	**19 648**	**19 138**	**19 318 e**	**1 131 e**	**11 343**	**9 750**	**8 162 e**
+ Recycled Products	-	-	-	-	-	-	-	-
+ Imports	91	13 032	14 087	14 512 e	12	776	729	946 e
- Exports	-	5 748	6 027	5 288 e	-	2 564	2 326	1 313 e
- Intl. Marine Bunkers	-	-	-	-	-	-	-	-
+ Transfers	-	-	- 6	-	-	- 204	- 150	- 284 e
+ Stock Change	- 13	- 78	- 210	156 e	- 66	- 257	233	438 e
- Statistical Difference	- 9 e	- 124	- 503	- 99 e	- 2 e	44	140	127 e
Total Consumption	**2 533 e**	**26 978**	**27 485**	**28 797 e**	**1 079**	**9 050**	**8 096**	**7 822 e**
Transformation	-	4 415	4 594	..	-	132	73	..
Energy	4 e	31	39	..	-	28	52	..
Transport	-	-	-	..	2	13	18	..
Industry	2 521	22 530	22 852	..	41	622	376	..
Other	8	2	-	..	1 036	8 255	7 577	..
Memo: Non-Energy Use	*2 521*	*22 530*	*22 852*	..	*7*	*30*	*23*	..
Closing Stock Level	-	*808*	*1 078*	*922 e*	-	*1 430*	*1 197*	*759 e*
	WHITE SPIRIT + SBP				**LUBRICANTS**			
Refinery Output	**44 e**	**90**	**105**	**..**	**-**	**950**	**945**	**..**
+ Recycled Products	-	-	-	..	-	-	-	..
+ Imports	-	-	-	..	-	-	3	..
- Exports	-	1	-	..	-	362	358	..
- Intl. Marine Bunkers	-	-	-	..	-	-	-	..
+ Transfers	-	-	- 9	..	-	-	-	..
+ Stock Change	-	2	- 7	..	-	- 5	- 38	..
- Statistical Difference	- 1 e	4	- 5	..	-	48	48	..
Total Consumption	**45**	**87**	**94**	**..**	**-**	**535**	**504**	**..**
Transformation	-	18	18	..	-	13	10	..
Energy	-	-	-	..	-	-	-	..
Transport	-	-	-	..	-	472	451	..
Industry	45	69	76	..	-	50	43	..
Other	-	-	-	..	-	-	-	..
Memo: Non-Energy Use	-	-	-	..	-	-	-	..
Closing Stock Level	-	*18*	*25*	..	-	*48*	*86*	..

Source: Annual Oil Statistics.

KOREA
Table 6B: Supply and Consumption

000 metric tons

	1980	2000	2001	2002	1980	2000	2001	2002
	BITUMEN				PARAFFIN WAXES			
Refinery Output	**315 e**	**2 234**	**2 370**	**..**	**-**	**10**	**10**	**..**
+ Recycled Products	-	-	-	..	-	-	-	..
+ Imports	-	4	33	..	-	-	-	..
- Exports	11	524	554	..	-	-	-	..
- Intl. Marine Bunkers	-	-	-	..	-	-	-	..
+ Transfers	-	-	- 37	..	-	-	-	..
+ Stock Change	- 11	1	- 1	..	-	- 1	-	..
- Statistical Difference	-	23	- 4	..	-	-	-	..
Total Consumption	**293**	**1 692**	**1 815**	**..**	**-**	**9**	**10**	**..**
Transformation	-	-	-	..	-	-	-	..
Energy	-	-	-	..	-	9	10	..
Transport	-	-	-	..	-	-	-	..
Industry	293	1 692	1 815	..	-	-	-	..
Other	-	-	-	..	-	-	-	..
Memo: Non-Energy Use	*293*	*1 692*	*1 815*	..	-	-	-	..
Closing Stock Level	-	*39*	*40*	..	-	*1*	*1*	..
	PETROLEUM COKE				OTHER PRODUCTS			
Refinery Output	**-**	**283**	**277**	**275 e**	**55 e**	**663**	**680**	**4 939 e**
+ Recycled Products	-	-	-	-	-	-	-	-
+ Imports	-	-	-	-	-	-	-	9 e
- Exports	-	69	74	51 e	-	239	191	1 769 e
- Intl. Marine Bunkers	-	-	-	-	-	-	-	-
+ Transfers	-	-	-	-	-	- 21	-	- 47 e
+ Stock Change	-	- 1	7	- 8 e	1	- 5	7	- 16 e
- Statistical Difference	-	1	- 1	2 e	-	- 17	4	109 e
Total Consumption	**-**	**212**	**211**	**214 e**	**56 e**	**415**	**492**	**3 007 e**
Transformation	-	-	-	..	-	55	57	..
Energy	-	41	40	..	2 e	40	45	..
Transport	-	-	-	..	-	-	-	..
Industry	-	137	131	..	54	320	389	..
Other	-	34	40	..	-	-	1	..
Memo: Non-Energy Use	-	-	-	..	-	-	-	..
Closing Stock Level	-	*7*	-	*8 e*	-	*17*	*10*	*178 e*

Source: Annual Oil Statistics.

KOREA
Table 7A: Imports by Country of Origin

000 metric tons

	2000	2001	2002	2000	2001	2002	2000	2001	2002
	Crude Oil			Natural Gas Liquids			Feedstocks and Other		
Australia	4 711	4 421	5 232 e	-	-	-	-	-	-
Belgium	-	-	-	-	-	-	-	-	-
Canada	-	227	128 e	-	-	-	-	-	-
Czech Republic	-	-	-	-	-	-	-	-	-
Denmark	-	-	-	-	-	-	-	-	-
Finland	-	-	-	-	-	-	-	-	-
France	-	-	-	-	-	-	-	-	-
Germany	-	-	-	-	-	-	-	-	-
Greece	-	-	-	-	-	-	-	-	-
Hungary	-	-	-	-	-	-	-	-	-
Ireland	-	-	-	-	-	-	-	-	-
Italy	-	-	-	-	-	-	-	-	-
Japan	-	-	-	-	-	-	-	6	36 e
Korea	-	-	-	-	-	-	-	-	-
Mexico	-	-	-	-	-	-	-	-	-
Netherlands	-	-	-	-	-	-	-	-	-
New Zealand	-	-	-	-	-	-	-	4	9 e
Norway	712	687	952 e	-	-	-	-	-	-
Poland	-	-	-	-	-	-	-	-	-
Spain	-	-	-	-	-	-	-	-	-
Sweden	-	-	-	-	-	-	-	-	-
Turkey	-	-	-	-	-	-	-	-	-
United Kingdom	-	-	162 e	-	-	-	-	-	-
United States	684	-	-	-	-	-	-	-	-
Other	-	-	-	-	-	-	-	-	-
OECD	**6 107**	**5 335**	**6 474 e**	**-**	**-**	**-**	**-**	**10**	**45 e**
Argentina	-	648	523 e	-	-	-	-	-	-
Brazil	-	674	1 137 e	-	-	-	-	-	-
Colombia	-	-	-	-	-	-	-	-	-
Ecuador	1 866	1 558	1 930 e	-	-	-	-	-	-
Netherlands Antilles	-	-	-	-	-	-	-	-	-
Peru	358	94	-	-	-	-	-	-	-
Trinidad and Tobago	-	-	-	-	-	-	-	-	-
Venezuela	-	200	2 e	-	-	-	-	-	-
Other	-	-	-	-	-	-	-	3	-
Latin America	**2 224**	**3 174**	**3 592 e**	**-**	**-**	**-**	**-**	**3**	**-**
Romania	-	-	-	-	-	-	-	-	-
Other	-	-	-	-	-	-	-	-	-
Non-OECD Europe	**-**	**-**	**-**	**-**	**-**	**-**	**-**	**-**	**-**
FSU	**1 551**	**2 370**	**2 578 e**	**-**	**-**	**-**	**-**	**-**	**-**
Iran	11 471	10 784	7 106 e	-	-	-	-	-	-
Iraq	2 418	434	510 e	-	-	-	-	-	-
Kuwait	8 669	9 003	7 381 e	-	-	-	-	-	-
Oman	7 437	7 317	6 773 e	-	-	-	-	-	-
Qatar	6 287	5 654	4 682 e	-	-	-	-	3	-
Saudi Arabia	36 185	35 439	31 527 e	-	-		-	10	18 e
United Arab Emirates	17 618	17 425	14 840 e	-	-	-	-	-	-
Other	4 876	4 336	2 386 e	-	-	-	-	-	-
Middle East	**94 961**	**90 392**	**75 205 e**	**-**	**-**	**-**	**-**	**13**	**18 e**
India	-	-	-	-	-	-	-	-	-
Indonesia	4 827	6 337	5 571 e	-	-	-	-	-	3 e
Singapore	-	-	-	-	-	-	90	188	151 e
Other	3 623	4 228	4 425 e	-	-	-	-	50	243 e
Asia	**8 450**	**10 565**	**9 996 e**	**-**	**-**	**-**	**90**	**238**	**397 e**
China	**527**	**283**	**554 e**	**-**	**-**	**-**	**684**	**935**	**764 e**
Algeria	-	-	-	-	-	-	-	-	-
Angola/Cabinada	2 843	1 182	1 173 e	-	-	-	-	-	-
Cameroon	-	-	-	-	-	-	-	-	-
Egypt	120	142	-	-	-	-	-	-	-
Gabon	779	797	139 e	-	-	-	-	-	-
Libya	-	-	-	-	-	-	-	-	-
Nigeria	1 405	531	258 e	-	-	-	-	-	-
Tunisia	-	-	-	-	-	-	-	-	-
Other	3 579	2 926	3 167 e	-	-	-	-	-	-
Africa	**8 726**	**5 578**	**4 737 e**	**-**	**-**	**-**	**-**	**-**	**-**
Non-spec./Other	**-**	**-**	**3 576 e**	**-**	**-**	**-**	**-**	**-**	**-**
Total Imports	**122 546**	**117 697**	**106 712 e**	**-**	**-**	**-**	**774**	**1 199**	**1 224 e**
Memo: Total OPEC	*88 880*	*85 807*	*71 877 e*	-	-	-	-	*13*	*21 e*

Source: Annual Oil Statistics.

KOREA
Table 7B: Imports by Country of Origin

000 metric tons

	2000	2001	2002	2000	2001	2002	2000	2001	2002
	Total Products			Liquefied Petroleum Gases			Naphtha		
Australia	250	200	456 e	156	183	442 e	94	17	13 e
Belgium	-	22	-	-	-	-	-	22	-
Canada	-	-	-	-	-	-	-	-	-
Czech Republic	-	-	-	-	-	-	-	-	-
Denmark	-	14	-	-	-	-	-	14	-
Finland	-	-	-	-	-	-	-	-	-
France	-	-	25 e	-	-	-	-	-	25 e
Germany	-	-	-	-	-	-	-	-	-
Greece	-	30	85 e	-	-	71 e	-	30	14 e
Hungary	-	-	-	-	-	-	-	-	-
Ireland	-	-	-	-	-	-	-	-	-
Italy	-	43	-	-	-	-	-	43	-
Japan	613	1 297	1 371 e	-	29	175 e	-	-	19 e
Korea	-	-	-	-	-	-	-	-	-
Mexico	86	-	-	-	-	-	86	-	-
Netherlands	-	87	148 e	-	-	48 e	-	-	100 e
New Zealand	34	32	31 e	-	-	-	34	32	31 e
Norway	44	39	39 e	44	-	39 e	-	39	-
Poland	-	-	-	-	-	-	-	-	-
Spain	-	-	-	-	-	-	-	-	-
Sweden	-	-	-	-	-	-	-	-	-
Turkey	-	104	-	-	-	-	-	104	-
United Kingdom	-	-	139 e	-	-	139 e	-	-	-
United States	507	374	387 e	94	32	118 e	256	252	143 e
Other	-	-	-	-	-	-	-	-	-
OECD	**1 534**	**2 242**	**2 681 e**	**294**	**244**	**1 032 e**	**470**	**553**	**345 e**
Argentina	-	-	-	-	-	-	-	-	-
Brazil	-	-	-	-	-	-	-	-	-
Colombia	-	-	-	-	-	-	-	-	-
Ecuador	-	-	-	-	-	-	-	-	-
Netherlands Antilles	-	-	-	-	-	-	-	-	-
Peru	-	26	-	-	-	-	-	26	-
Trinidad and Tobago	-	-	-	-	-	-	-	-	-
Venezuela	-	31	73 e	-	-	44 e	-	31	29 e
Other	-	31	8 e	-	4	8 e	-	27	-
Latin America	**-**	**88**	**81 e**	**-**	**4**	**52 e**	**-**	**84**	**29 e**
Romania	-	-	-	-	-	-	-	-	-
Other	-	-	-	-	-	-	-	-	-
Non-OECD Europe	**-**	**-**	**-**	**-**	**-**	**-**	**-**	**-**	**-**
FSU	**526**	**308**	**360 e**	**-**	**-**	**-**	**520**	**308**	**287 e**
Iran	614	277	496 e	292	166	212 e	322	111	263 e
Iraq	-	49	-	-	49	-	-	-	-
Kuwait	3 120	2 907	3 015 e	658	419	439 e	2 266	2 446	2 576 e
Oman	-	-	2 e	-	-	-	-	-	2 e
Qatar	682	761	489 e	308	435	108 e	354	306	353 e
Saudi Arabia	5 810	3 866	3 907 e	2 416	1 490	1 296 e	3 316	2 334	2 456 e
United Arab Emirates	3 127	4 601	5 052 e	917	597	741 e	2 061	3 844	4 164 e
Other	755	400	479 e	14	33	40 e	741	367	439 e
Middle East	**14 108**	**12 861**	**13 440 e**	**4 605**	**3 189**	**2 836 e**	**9 060**	**9 408**	**10 253 e**
India	405	1 233	1 513 e	-	-	-	405	1 233	1 513 e
Indonesia	2 842	2 484	2 017 e	21	62	44 e	215	199	179 e
Singapore	1 487	972	1 390 e	-	23	-	709	253	143 e
Other	1 168	1 501	2 481 e	103	24	214 e	847	847	544 e
Asia	**5 902**	**6 190**	**7 401 e**	**124**	**109**	**258 e**	**2 176**	**2 532**	**2 379 e**
China	**748**	**897**	**1 420 e**	**-**	**-**	**3 e**	**443**	**649**	**580 e**
Algeria	221	458	752 e	108	372	575 e	113	86	177 e
Angola/Cabinada	-	-	-	-	-	-	-	-	-
Cameroon	56	-	-	-	-	-	-	-	-
Egypt	250	467	394 e	-	-	-	250	467	394 e
Gabon	-	-	-	-	-	-	-	-	-
Libya	-	-	56 e	-	-	-	-	-	56 e
Nigeria	-	143	150 e	-	143	150 e	-	-	-
Tunisia	-	-	-	-	-	-	-	-	-
Other	44	-	12 e	-	-	-	-	-	12 e
Africa	**571**	**1 068**	**1 364 e**	**108**	**515**	**725 e**	**363**	**553**	**639 e**
Non-spec./Other	**-**	**-**	**-**	**-**	**-**	**-**	**-**	**-**	**-**
Total Imports	**23 389**	**23 654**	**26 747 e**	**5 131**	**4 061**	**4 906 e**	**13 032**	**14 087**	**14 512 e**
Memo: Total OPEC	*16 416*	*15 577*	*16 007 e*	*4 720*	*3 733*	*3 609 e*	*8 647*	*9 357*	*10 253 e*

Source: Annual Oil Statistics.

KOREA
Table 7C: Imports by Country of Origin

000 metric tons

	2000	2001	2002	2000	2001	2002	2000	2001	2002
	Total Gasoline			**Total Kerosene**			**Gas/Diesel**		
Australia	-	-	1 e	-	-	-	-	-	-
Belgium	-	-	-	-	-	-	-	-	-
Canada	-	-	-	-	-	-	-	-	-
Czech Republic	-	-	-	-	-	-	-	-	-
Denmark	-	-	-	-	-	-	-	-	-
Finland	-	-	-	-	-	-	-	-	-
France	-	-	-	-	-	-	-	-	-
Germany	-	-	-	-	-	-	-	-	-
Greece	-	-	-	-	-	-	-	-	-
Hungary	-	-	-	-	-	-	-	-	-
Ireland	-	-	-	-	-	-	-	-	-
Italy	-	-	-	-	-	-	-	-	-
Japan	41	90	64 e	117	163	259 e	297	690	706 e
Korea	-	-	-	-	-	-	-	-	-
Mexico	-	-	-	-	-	-	-	-	-
Netherlands	-	-	-	-	-	-	-	-	-
New Zealand	-	-	-	-	-	-	-	-	-
Norway	-	-	-	-	-	-	-	-	-
Poland	-	-	-	-	-	-	-	-	-
Spain	-	-	-	-	-	-	-	-	-
Sweden	-	-	-	-	-	-	-	-	-
Turkey	-	-	-	-	-	-	-	-	-
United Kingdom	-	-	-	-	-	-	-	-	-
United States	-	-	-	79	-	-	-	-	-
Other	-	-	-	-	-	-	-	-	-
OECD	**41**	**90**	**65 e**	**196**	**163**	**259 e**	**297**	**690**	**706 e**
Argentina	-	-	-	-	-	-	-	-	-
Brazil	-	-	-	-	-	-	-	-	-
Colombia	-	-	-	-	-	-	-	-	-
Ecuador	-	-	-	-	-	-	-	-	-
Netherlands Antilles	-	-	-	-	-	-	-	-	-
Peru	-	-	-	-	-	-	-	-	-
Trinidad and Tobago	-	-	-	-	-	-	-	-	-
Venezuela	-	-	-	-	-	-	-	-	-
Other	-	-	-	-	-	-	-	-	-
Latin America	**-**	**-**	**-**	**-**	**-**	**-**	**-**	**-**	**-**
Romania	-	-	-	-	-	-	-	-	-
Other	-	-	-	-	-	-	-	-	-
Non-OECD Europe	**-**	**-**	**-**	**-**	**-**	**-**	**-**	**-**	**-**
FSU	**1**	**-**	**4 e**	**-**	**-**	**-**	**5**	**-**	**-**
Iran	-	-	-	-	-	-	-	-	-
Iraq	-	-	-	-	-	-	-	-	-
Kuwait	-	-	-	196	42	-	-	-	-
Oman	-	-	-	-	-	-	-	-	-
Qatar	-	-	-	20	20	-	-	-	28 e
Saudi Arabia	-	-	-	78	42	155 e	-	-	-
United Arab Emirates	-	-	-	149	160	113 e	-	-	34 e
Other	-	-	-	-	-	-	-	-	-
Middle East	**-**	**-**	**-**	**443**	**264**	**268 e**	**-**	**-**	**62 e**
India	-	-	-	-	-	-	-	-	-
Indonesia	-	-	-	-	-	-	-	-	-
Singapore	25	179	295 e	-	-	217 e	8	-	108 e
Other	5	36	128 e	121	162	306 e	62	259	879 e
Asia	**30**	**215**	**423 e**	**121**	**162**	**523 e**	**70**	**259**	**987 e**
China	**13**	**32**	**21 e**	**230**	**183**	**247 e**	**14**	**3**	**41 e**
Algeria	-	-	-	-	-	-	-	-	-
Angola/Cabinada	-	-	-	-	-	-	-	-	-
Cameroon	-	-	-	-	-	-	-	-	-
Egypt	-	-	-	-	-	-	-	-	-
Gabon	-	-	-	-	-	-	-	-	-
Libya	-	-	-	-	-	-	-	-	-
Nigeria	-	-	-	-	-	-	-	-	-
Tunisia	-	-	-	-	-	-	-	-	-
Other	-	-	-	-	-	-	-	-	-
Africa	**-**	**-**	**-**	**-**	**-**	**-**	**-**	**-**	**-**
Non-spec./Other	**-**	**-**	**-**	**-**	**-**	**-**	**-**	**-**	**-**
Total Imports	**85**	**337**	**513 e**	**990**	**772**	**1 297 e**	**386**	**952**	**1 796 e**
Memo: Total OPEC	-	-	-	*443*	*264*	*268 e*	-	-	*62 e*

Source: Annual Oil Statistics.

KOREA
Table 7D: Imports by Country of Origin

000 metric tons

	2000	2001	2002	2000	2001	2002	2000	2001	2002
	Fuel Oil (Residual)			**Petroleum Coke**			**Other Products**		
Australia	-	-	-	-	-	-	-	-	-
Belgium	-	-	-	-	-	-	-	-	-
Canada	-	-	-	-	-	-	-	-	-
Czech Republic	-	-	-	-	-	-	-	-	-
Denmark	-	-	-	-	-	-	-	-	-
Finland	-	-	-	-	-	-	-	-	-
France	-	-	-	-	-	-	-	-	-
Germany	-	-	-	-	-	-	-	-	-
Greece	-	-	-	-	-	-	-	-	-
Hungary	-	-	-	-	-	-	-	-	-
Ireland	-	-	-	-	-	-	-	-	-
Italy	-	-	-	-	-	-	-	-	-
Japan	157	311	146 e	-	-	-	1	14	2 e
Korea	-	-	-	-	-	-	-	-	-
Mexico	-	-	-	-	-	-	-	-	-
Netherlands	-	87	-	-	-	-	-	-	-
New Zealand	-	-	-	-	-	-	-	-	-
Norway	-	-	-	-	-	-	-	-	-
Poland	-	-	-	-	-	-	-	-	-
Spain	-	-	-	-	-	-	-	-	-
Sweden	-	-	-	-	-	-	-	-	-
Turkey	-	-	-	-	-	-	-	-	-
United Kingdom	-	-	-	-	-	-	-	-	-
United States	78	90	126 e	-	-	-	-	-	-
Other	-	-	-	-	-	-	-	-	-
OECD	**235**	**488**	**272 e**	**-**	**-**	**-**	**1**	**14**	**2 e**
Argentina	-	-	-	-	-	-	-	-	-
Brazil	-	-	-	-	-	-	-	-	-
Colombia	-	-	-	-	-	-	-	-	-
Ecuador	-	-	-	-	-	-	-	-	-
Netherlands Antilles	-	-	-	-	-	-	-	-	-
Peru	-	-	-	-	-	-	-	-	-
Trinidad and Tobago	-	-	-	-	-	-	-	-	-
Venezuela	-	-	-	-	-	-	-	-	-
Other	-	-	-	-	-	-	-	-	-
Latin America	**-**	**-**	**-**	**-**	**-**	**-**	**-**	**-**	**-**
Romania	-	-	-	-	-	-	-	-	-
Other	-	-	-	-	-	-	-	-	-
Non-OECD Europe	**-**	**-**	**-**	**-**	**-**	**-**	**-**	**-**	**-**
FSU	**-**	**-**	**69 e**	**-**	**-**	**-**	**-**	**-**	**-**
Iran	-	-	21 e	-	-	-	-	-	-
Iraq	-	-	-	-	-	-	-	-	-
Kuwait	-	-	-	-	-	-	-	-	-
Oman	-	-	-	-	-	-	-	-	-
Qatar	-	-	-	-	-	-	-	-	-
Saudi Arabia	-	-	-	-	-	-	-	-	-
United Arab Emirates	-	-	-	-	-	-	-	-	-
Other	-	-	-	-	-	-	-	-	-
Middle East	**-**	**-**	**21 e**	**-**	**-**	**-**	**-**	**-**	**-**
India	-	-	-	-	-	-	-	-	-
Indonesia	2 606	2 223	1 794 e	-	-	-	-	-	-
Singapore	745	517	627 e	-	-	-	-	-	-
Other	27	154	403 e	-	-	-	3	19	7 e
Asia	**3 378**	**2 894**	**2 824 e**	**-**	**-**	**-**	**3**	**19**	**7 e**
China	**48**	**27**	**528 e**	**-**	**-**	**-**	**-**	**3**	**-**
Algeria	-	-	-	-	-	-	-	-	-
Angola/Cabinada	-	-	-	-	-	-	-	-	-
Cameroon	56	-	-	-	-	-	-	-	-
Egypt	-	-	-	-	-	-	-	-	-
Gabon	-	-	-	-	-	-	-	-	-
Libya	-	-	-	-	-	-	-	-	-
Nigeria	-	-	-	-	-	-	-	-	-
Tunisia	-	-	-	-	-	-	-	-	-
Other	44	-	-	-	-	-	-	-	-
Africa	**100**	**-**	**-**	**-**	**-**	**-**	**-**	**-**	**-**
Non-spec./Other	**-**	**-**	**-**	**-**	**-**	**-**	**-**	**-**	**-**
Total Imports	**3 761**	**3 409**	**3 714 e**	**-**	**-**	**-**	**4**	**36**	**9 e**
Memo: Total OPEC	*2 606*	*2 223*	*1 815 e*	*-*	*-*	*-*	*-*	*-*	*-*

Source: Annual Oil Statistics.

KOREA
Table 8A: Exports by Country of Destination

000 metric tons

	2000	2001	2002	2000	2001	2002	2000	2001	2002
	Crude Oil			**Natural Gas Liquids**			**Feedstocks and Other**		
Australia	-	-	-	-	-	-	-	-	-
Austria	-	-	-	-	-	-	-	-	-
Belgium	-	-	-	-	-	-	-	-	-
Canada	-	-	-	-	-	-	-	-	-
Denmark	-	-	-	-	-	-	-	-	-
Finland	-	-	-	-	-	-	-	-	-
France	-	-	-	-	-	-	-	-	-
Germany	-	-	-	-	-	-	-	-	-
Greece	-	-	-	-	-	-	-	-	-
Ireland	-	-	-	-	-	-	-	-	-
Italy	-	-	-	-	-	-	-	-	-
Japan	-	-	-	-	-	-	4	8	9 e
Luxembourg	-	-	-	-	-	-	-	-	-
Mexico	-	-	-	-	-	-	-	-	-
Netherlands	-	-	-	-	-	-	-	-	-
Norway	-	-	-	-	-	-	-	-	-
Portugal	-	-	-	-	-	-	-	-	-
Spain	-	-	-	-	-	-	-	-	-
Sweden	-	-	-	-	-	-	-	-	-
Switzerland	-	-	-	-	-	-	-	-	-
Turkey	-	-	-	-	-	-	-	-	-
United Kingdom	-	-	-	-	-	-	-	-	-
United States	-	-	-	-	-	-	92	86	106 e
Other OECD	-	-	-	-	-	-	-	-	-
OECD	**-**	**-**	**-**	**-**	**-**	**-**	**96**	**94**	**115 e**
Africa	-	-	-	-	-	-	-	-	-
Latin America	-	-	-	-	-	-	-	-	6 e
Asia	-	-	-	-	-	-	2	9	3 e
China	-	-	-	-	-	-	1	-	3 e
Non-OECD Europe	-	-	-	-	-	-	-	-	-
Former Soviet Union	-	-	-	-	-	-	-	-	-
Middle East	-	-	-	-	-	-	-	-	-
Non-specified/Other	-	-	-	-	-	-	-	-	-
Non-OECD	**-**	**-**	**-**	**-**	**-**	**-**	**3**	**9**	**12 e**
Total Exports	**-**	**-**	**-**	**-**	**-**	**-**	**99**	**103**	**127 e**

	2000	2001	2002	2000	2001	2002	2000	2001	2002
	Total Products			**Liquefied Petroleum Gas**			**Naphtha**		
Australia	405	234	177 e	-	-	-	-	-	-
Austria	-	-	-	-	-	-	-	-	-
Belgium	-	10	-	-	-	-	-	-	-
Canada	137	-	-	-	-	-	-	-	-
Denmark	712	532	531 e	-	-	-	-	-	-
Finland	-	-	-	-	-	-	-	-	-
France	-	-	1 e	-	-	1 e	-	-	-
Germany	25	-	-	-	-	-	-	-	-
Greece	-	-	-	-	-	-	-	-	-
Ireland	-	-	-	-	-	-	-	-	-
Italy	-	-	-	-	-	-	-	-	-
Japan	14 086	13 760	11 127 e	90	60	37 e	5 469	5 922	5 148 e
Luxembourg	-	-	-	-	-	-	-	-	-
Mexico	-	-	-	-	-	-	-	-	-
Netherlands	-	17	336 e	-	-	-	-	-	-
Norway	-	-	21 e	-	-	-	-	-	-
Portugal	-	-	-	-	-	-	-	-	-
Spain	2	-	-	-	-	-	-	-	-
Sweden	-	-	-	-	-	-	-	-	-
Switzerland	-	-	-	-	-	-	-	-	-
Turkey	-	-	-	-	-	-	-	-	-
United Kingdom	-	17	21 e	-	-	-	-	-	-
United States	2 606	2 975	2 517 e	-	-	-	41	-	-
Other OECD	-	73	114 e	-	-	-	-	-	-
OECD	**17 973**	**17 618**	**14 845 e**	**90**	**60**	**38 e**	**5 510**	**5 922**	**5 148 e**
Africa	-	11	-	-	-	-	-	-	-
Latin America	326	583	667 e	-	-	-	-	-	-
Asia	6 053	6 649	8 996 e	10	41	27 e	238	84	96 e
China	14 762	14 080	6 685 e	601	315	181 e	-	21	44 e
Non-OECD Europe	-	-	-	-	-	-	-	-	-
Former Soviet Union	139	66	11 e	-	-	-	-	-	-
Middle East	510	-	32 e	-	-	-	-	-	-
Non-specified/Other	1	-	-	-	-	-	-	-	-
Non-OECD	**21 791**	**21 389**	**16 391 e**	**611**	**356**	**208 e**	**238**	**105**	**140 e**
Total Exports	**39 764**	**39 007**	**31 236 e**	**701**	**416**	**246 e**	**5 748**	**6 027**	**5 288 e**

Source: Annual Oil Statistics.

KOREA
Table 8B: Exports by Country of Destination

000 metric tons

	2000	2001	2002	2000	2001	2002	2000	2001	2002
	Total Gasoline			Total Kerosene			Gas/Diesel Oil		
Australia	19	129	65 e	3	16	-	260	89	111 e
Austria	-	-	-	-	-	-	-	-	-
Belgium	-	-	-	-	-	-	-	-	-
Canada	22	-	-	115	-	-	-	-	-
Denmark	63	110	50 e	402	401	225 e	247	21	233 e
Finland	-	-	-	-	-	-	-	-	-
France	-	-	-	-	-	-	-	-	-
Germany	-	-	-	-	-	-	-	-	-
Greece	-	-	-	-	-	-	-	-	-
Ireland	-	-	-	-	-	-	-	-	-
Italy	-	-	-	-	-	-	-	-	-
Japan	536	604	519 e	3 404	3 283	3 078 e	3 710	3 055	1 683 e
Luxembourg	-	-	-	-	-	-	-	-	-
Mexico	-	-	-	-	-	-	-	-	-
Netherlands	-	-	-	-	-	-	-	-	318 e
Norway	-	-	-	-	-	-	-	-	-
Portugal	-	-	-	-	-	-	-	-	-
Spain	-	-	-	-	-	-	-	-	-
Sweden	-	-	-	-	-	-	-	-	-
Switzerland	-	-	-	-	-	-	-	-	-
Turkey	-	-	-	-	-	-	-	-	-
United Kingdom	-	-	-	-	-	-	-	-	-
United States	154	449	425 e	1 590	1 877	1 705 e	811	630	382 e
Other OECD	-	-	14 e	-	-	-	-	73	100 e
OECD	**794**	**1 292**	**1 073 e**	**5 514**	**5 577**	**5 008 e**	**5 028**	**3 868**	**2 827 e**
Africa	-	-	-	-	-	-	-	-	-
Latin America	-	-	28 e	-	-	-	326	583	639 e
Asia	864	501	704 e	243	474	809 e	3 787	4 718	6 363 e
China	34	95	12 e	911	1 256	770 e	3 219	2 910	345 e
Non-OECD Europe	-	-	-	-	-	-	-	-	-
Former Soviet Union	-	7	-	-	-	-	129	44	11 e
Middle East	-	-	-	35	-	-	440	-	32 e
Non-specified/Other	-	-	-	-	-	-	-	-	-
Non-OECD	**898**	**603**	**744 e**	**1 189**	**1 730**	**1 579 e**	**7 901**	**8 255**	**7 390 e**
Total Exports	**1 692**	**1 895**	**1 817 e**	**6 703**	**7 307**	**6 587 e**	**12 929**	**12 123**	**10 217 e**

	2000	2001	2002	2000	2001	2002	2000	2001	2002
	Fuel Oil (Residual)			Petroleum Coke			Other Products		
Australia	123	-	-	-	-	-	-	-	1 e
Austria	-	-	-	-	-	-	-	-	-
Belgium	-	-	-	-	-	-	-	10	-
Canada	-	-	-	-	-	-	-	-	-
Denmark	-	-	-	-	-	-	-	-	23 e
Finland	-	-	-	-	-	-	-	-	-
France	-	-	-	-	-	-	-	-	-
Germany	-	-	-	-	-	-	25	-	-
Greece	-	-	-	-	-	-	-	-	-
Ireland	-	-	-	-	-	-	-	-	-
Italy	-	-	-	-	-	-	-	-	-
Japan	727	619	385 e	69	74	51 e	81	143	226 e
Luxembourg	-	-	-	-	-	-	-	-	-
Mexico	-	-	-	-	-	-	-	-	-
Netherlands	-	-	-	-	-	-	-	17	18 e
Norway	-	-	-	-	-	-	-	-	21 e
Portugal	-	-	-	-	-	-	-	-	-
Spain	-	-	-	-	-	-	2	-	-
Sweden	-	-	-	-	-	-	-	-	-
Switzerland	-	-	-	-	-	-	-	-	-
Turkey	-	-	-	-	-	-	-	-	-
United Kingdom	-	-	-	-	-	-	-	17	21 e
United States	1	-	-	-	-	-	9	19	5 e
Other OECD	-	-	-	-	-	-	-	-	-
OECD	**851**	**619**	**385 e**	**69**	**74**	**51 e**	**117**	**206**	**315 e**
Africa	-	-	-	-	-	-	-	11	-
Latin America	-	-	-	-	-	-	-	-	-
Asia	507	531	620 e	-	-	-	404	300	377 e
China	9 396	8 897	4 256 e	-	-	-	601	586	1 077 e
Non-OECD Europe	-	-	-	-	-	-	-	-	-
Former Soviet Union	10	15	-	-	-	-	-	-	-
Middle East	32	-	-	-	-	-	3	-	-
Non-specified/Other	-	-	-	-	-	-	1	-	-
Non-OECD	**9 945**	**9 443**	**4 876 e**	**-**	**-**	**-**	**1 009**	**897**	**1 454 e**
Total Exports	**10 796**	**10 062**	**5 261 e**	**69**	**74**	**51 e**	**1 126**	**1 103**	**1 769 e**

Source: Annual Oil Statistics.

LUXEMBOURG

Figure 1
FUEL SHARES OF TPES *

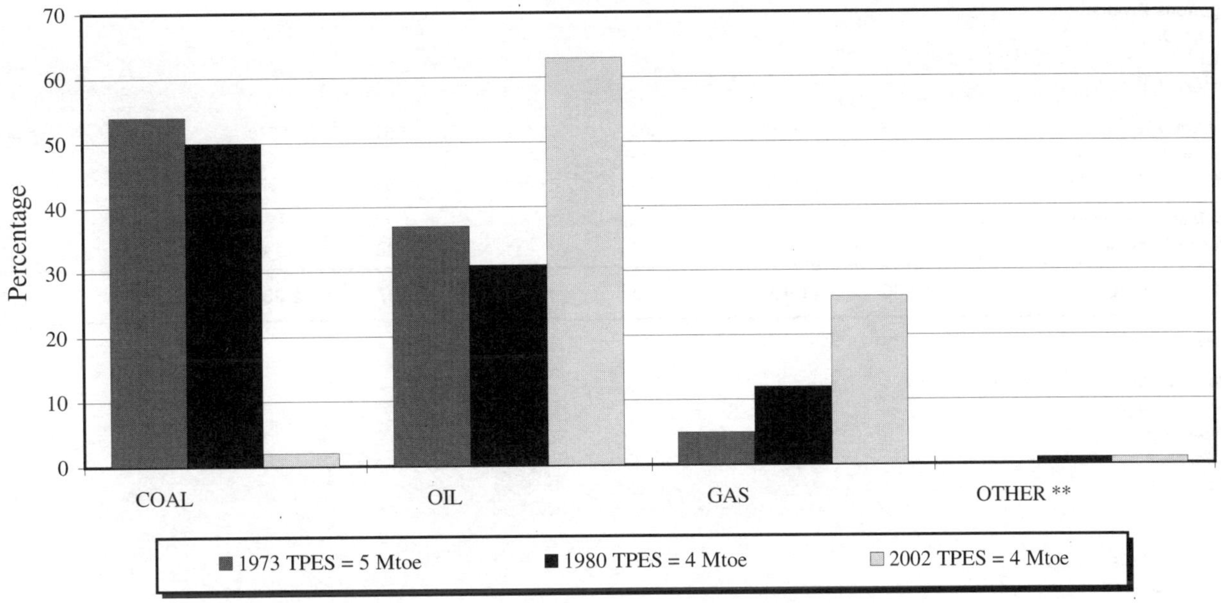

Source: Annual Energy Balances.

* Excluding electricity trade.

** Includes hydro, geothermal, solar, wind, combustible renewables and waste, etc.

Figure 2
OIL PRODUCTION AND CONSUMPTION

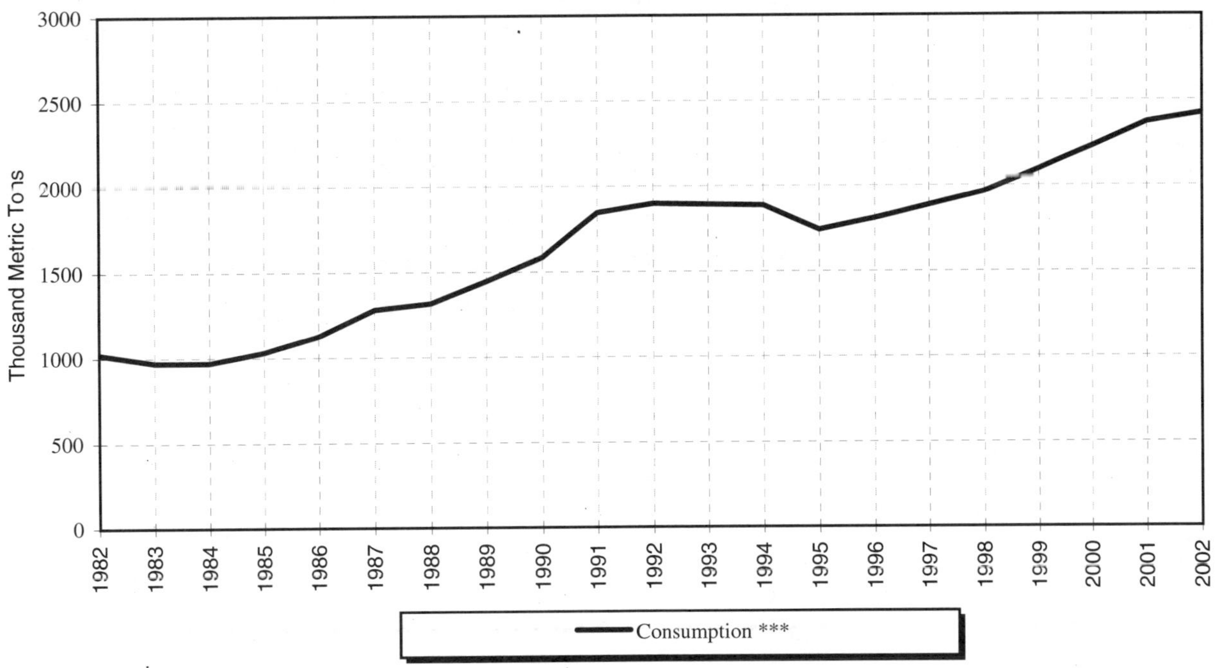

Source: Annual Oil Statistics.

*** Includes refinery fuel and international marine bunkers.

LUXEMBOURG
Table 1: Oil Supply and Consumption - Summary Table

000 metric tons

	1973	1978	1980	1990	1999	2000	2001	2002
Indigenous Production	-	-	-	-	-	-	-	-
Crude Oil	..	-	-	-	-	-	-	-
NGL	..	-	-	-	-	-	-	-
Other	..	-	-	-	-	-	-	-
+ Net Imports	1 666	1 408	1 083	1 596	2 051	2 278	2 342	2 443 e
Crude Oil, NGL, Other	-	-	-	-	-	-	-	-
Petroleum Products	1 666	1 408	1 083	1 596	2 051	2 278	2 342	2 443 e
- Intl. Marine Bunkers	-	-	-	-	-	-	-	-
+ Stock Changes	-	- 2	- 1	- 11	46	- 48	28	- 17 e
TOTAL SUPPLY	**1 666**	**1 406**	**1 082**	**1 585**	**2 097**	**2 230**	**2 370**	**2 426 e**
Refinery Intake	-	-	-	-	-	-	-	-
Refinery Output	-	-	-	-	-	-	-	-
LPG + Ethane	-	-	-	-	-	-	-	-
Naphtha	-	-	-	-	-	-	-	-
Kerosenes	-	-	-	-	-	-	-	-
Gasolines	-	-	-	-	-	-	-	-
Gas/Diesel Oil	-	-	-	-	-	-	-	-
Fuel Oil	-	-	-	-	-	-	-	-
Other Products	-	-	-	-	-	-	-	-
- Refinery Losses	-	-	-	-	-	-	-	-
+ Returns	-	-	-	-	-	-	-	-
- Statistical Differences	23	3	7	2	3	3	2	5 e
TOTAL CONSUMPTION	**1 643**	**1 403**	**1 075**	**1 583**	**2 094**	**2 227**	**2 368**	**2 421 e**
Transformation and Energy	**118**	**55**	**24**	**3**	**-**	**-**	**-**	**..**
Crude Oil	-	-	-	-	-	-	-	..
LPG + Ethane	6	-	-	-	-	-	-	..
Gas/Diesel Oil	-	-	-	-	-	-	-	..
Fuel Oil	112	55	24	3	-	-	-	..
Petroleum Coke	-	-	-	-	-	-	-	..
Other	-	-	-	-	-	-	-	..
Transport	**273**	**409**	**479**	**977**	**1 664**	**1 823**	**1 929**	**..**
Motor Gasoline	144	243	284	412	560	579	569	..
Gas/Diesel Oil	68	93	122	429	778	929	1 019	..
Subtotal	*212*	*336*	*406*	*841*	*1 338*	*1 508*	*1 588*	..
Aviation Fuels	51	61	63	129	323	312	337	..
Other	10	12	10	7	3	3	4	..
Industry	**828**	**511**	**207**	**301**	**96**	**83**	**77**	**..**
Naphtha	-	-	-	-	-	-	-	..
LPG + Ethane	3	8	11	5	13	13	12	..
Gas/Diesel Oil	138	70	63	36	57	51	46	..
Fuel Oil	645	394	93	242	5	6	7	..
Other	42	39	40	18	21	13	12	..
Other Sectors	**424**	**428**	**365**	**302**	**334**	**321**	**362**	**..**
Gas/Diesel Oil	399	397	343	291	323	310	348	..
Fuel Oil	12	20	11	-	-	-	-	..
Other	13	11	11	11	11	11	14	..
Memo: Non-Energy Use	*31*	*41*	*42*	*22*	*22*	*14*	*12*	..
Chem/Petchem. Feedst.	*1*	*3*	*3*	*5*	*1*	*1*	*1*	..
Other	*30*	*38*	*39*	*17*	*21*	*13*	*11*	..

Source: Annual Oil Statistics.

LUXEMBOURG

Table 3: Motor Gasoline - Supply

000 metric tons

	1973	1978	1980	1990	1999	2000	2001	2002
MOTOR GASOLINE								
Refinery Output	-	-	-	-	-	-	-	-
+ Imports	162	259	293	419	555	595	574	552 e
- Exports	1	5	2	1	4	3	3	-
+ Transfers	-	-	-	-	-	-	-	-
+ Stock Change	-	- 6	- 5	- 4	12	- 12	-	6 e
- Statistical Difference	3	4	-	2	-	- 2	-	2 e
Total Consumption	**158**	**244**	**286**	**412**	**563**	**582**	**571**	**556 e**
Closing Stock Level	-	*14*	*26*	*42*	*26*	*38*	*38*	*32 e*

Table 4: Aviation Fuels - Supply

000 metric tons

	1973	1978	1980	1990	1999	2000	2001	2002
AVIATION GASOLINE								
Refinery Output	-	-	-	-	-	-	-	-
+ Imports	1	-	-	1	-	-	-	-
- Exports	-	-	-	-	-	-	-	-
+ Transfers	-	-	-	-	-	-	-	-
+ Stock Change	-	-	-	-	-	-	-	-
- Statistical Difference	-	-	-	-	-	-	-	-
Total Consumption	**1**	**-**	**-**	**1**	**-**	**-**	**-**	**-**
Closing Stock Level	-	-	-	-	-	-	-	-
GASOLINE TYPE JET FUEL								
Refinery Output	-	-	-	-	-	-	-	-
+ Imports	-	-	-	-	-	-	-	-
- Exports	-	-	-	-	-	-	-	-
+ Transfers	-	-	-	-	-	-	-	-
+ Stock Change	-	-	-	-	-	-	-	-
- Statistical Difference	-	-	-	-	-	-	-	-
Total Consumption	**-**	**-**	**-**	**-**	**-**	**-**	**-**	**-**
Closing Stock Level	-	-	-	-	-	-	-	-
KEROSENE TYPE JET FUEL								
Refinery Output	-	-	-	-	-	-	-	-
+ Imports	52	61	63	129	322	318	330	369 e
- Exports	-	-	-	-	-	-	-	-
+ Transfers	-	-	-	-	-	-	-	-
+ Stock Change	-	-	-	- 1	1	- 6	7	- 5 e
- Statistical Difference	2	-	-	-	-	-	-	-
Total Consumption	**50**	**61**	**63**	**128**	**323**	**312**	**337**	**364 e**
Closing Stock Level	-	*2*	*2*	*4*	*6*	*12*	*5*	*10 e*

Source: Annual Oil Statistics.

LUXEMBOURG
Table 5A: Gas/Diesel Oil - Supply and Consumption

000 metric tons

	1973	1978	1980	1990	1999	2000	2001	2002
Refinery Output	-	-	-	-	-	-	-	-
+ Recycled Products	-	-	-	-	-	-	-	-
+ Imports	620	566	549	773	1 134	1 333	1 416	1 490 e
- Exports	9	7	16	12	6	8	21	5 e
- Intl. Marine Bunkers	-	-	-	-	-	-	-	-
+ Transfers	-	-	-	-	-	-	-	-
+ Stock Changes	-	-	- 4	- 6	32	- 30	21	- 18 e
- Statistical Differences	6	- 1	1	- 1	2	5	3	3 e
Total Consumption	**605**	**560**	**528**	**756**	**1 158**	**1 290**	**1 413**	**1 464 e**
Transformation	-	-	-	-	-	-	-	..
Public Electricity	-	-	-	-	-	-	-	..
Autoproducer Electricity	-	-	-	-	-	-	-	..
Public CHP Plants	-	-	-	-	-	-	-	..
Autoproducers of CHP	-	-	-	-	-	-	-	..
Public Heat Plants	-	-	-	-	-	-	-	..
Autoproducer Heat Plants	-	-	-	-	-	-	-	..
Petrochemical Industry	-	-	-	-	-	-	-	..
Other Transformation	-	-	-	-	-	-	-	..
Energy Sector	-	-	-	-	-	-	-	..
Refinery Fuel	-	-	-	-	-	-	-	..
Other Energy Sector	-	-	-	-	-	-	-	..
Final Consumption	**605**	**560**	**528**	**756**	**1 158**	**1 290**	**1 413**	..
Transport	**68**	**93**	**122**	**429**	**778**	**929**	**1 019**	..
Air	-	-	-	-	-	-	-	..
Road	57	82	114	429	756	906	1 007	..
Rail	11	9	8	-	8	7	3	..
Inland Waterways	-	2	-	-	-	-	-	..
Non-Specified	-	-	-	-	14	16	9	..
Industry	**138**	**70**	**63**	**36**	**57**	**51**	**46**	..
Iron and Steel	58	25	20	7	17	7	4	..
Chemical (incl. Pet.-Chem.)	36	9	2	6	3	2	2	..
Non-Ferrous Metals	-	-	-	-	-	-	-	..
Non-Metallic Minerals	-	-	-	-	10	8	8	..
Transport Equipment	-	-	-	-	-	-	-	..
Machinery	-	-	-	-	-	-	-	..
Mining and Quarrying	-	-	-	-	2	5	4	..
Food and Tobacco	-	-	6	2	5	4	3	..
Pulp, Paper and Printing	-	-	-	-	-	-	-	..
Wood and Wood Products	-	-	-	-	-	-	-	..
Construction	44	36	35	13	12	15	14	..
Textiles and Leather	-	-	-	-	-	-	-	..
Non-Specified	-	-	-	8	8	10	11	..
Other Sectors	**399**	**397**	**343**	**291**	**323**	**310**	**348**	..
Commerce and Public	-	-	-	-	-	-	-	..
Residential	389	393	340	286	304	290	328	..
Agriculture	10	4	3	5	9	7	7	..
Non-Specified	-	-	-	-	10	13	13	..
Memo: Non-Energy Use	-	-	-	-	*1*	*1*	*1*	..
Chem/Petchem. Feedst.	-	-	-	-	-	-	-	..
Other	-	-	-	-	*1*	*1*	*1*	..
Closing Stock Level	-	*37*	*75*	*77*	*56*	*86*	*65*	*83 e*

Source: Annual Oil Statistics.

LUXEMBOURG
Table 5B: Fuel Oil (Residual) - Supply and Consumption

000 metric tons

	1973	1978	1980	1990	1999	2000	2001	2002
Refinery Output	-	-	-	-	-	-	-	-
+ Recycled Products	-	-	-	-	-	-	-	-
+ Imports	789	482	156	247	5	6	7	5 e
- Exports	1	9	28	1	-	-	-	-
- Intl. Marine Bunkers	-	-	-	-	-	-	-	-
+ Transfers	-	-	-	-	-	-	-	-
+ Stock Changes	-	3	8	-	1	-	-	-
- Statistical Differences	12	1	5	1	1	-	-	-
Total Consumption	**776**	**475**	**131**	**245**	**5**	**6**	**7**	**5 e**
Transformation	**112**	**55**	**24**	**3**	-	-	-	..
Public Electricity	-	-	-	-	-	-	-	..
Autoproducer Electricity	112	55	24	3	-	-	-	..
Public CHP Plants	-	-	-	-	-	-	-	..
Autoproducers of CHP	-	-	-	-	-	-	-	..
Public Heat Plants	-	-	-	-	-	-	-	..
Autoproducer Heat Plants	-	-	-	-	-	-	-	..
Petrochemical Industry	-	-	-	-	-	-	-	..
Other Transformation	-	-	-	-	-	-	-	..
Energy Sector	-	-	-	-	-	-	-	..
Refinery Fuel	-	-	-	-	-	-	-	..
Other Energy Sector	-	-	-	-	-	-	-	..
Final Consumption	**664**	**420**	**107**	**242**	**5**	**6**	**7**	..
Transport	**7**	**6**	**3**	-	-	-	-	..
Air	-	-	-	-	-	-	-	..
Road	-	-	-	-	-	-	-	..
Rail	7	-	-	-	-	-	-	..
Inland Waterways	-	6	3	-	-	-	-	..
Non-Specified	-	-	-	-	-	-	-	..
Industry	**645**	**394**	**93**	**242**	**5**	**6**	**7**	..
Iron and Steel	588	320	34	208	4	2	2	..
Chemical (incl. Pet.-Chem.)	30	60	44	31	1	-	-	..
Non-Ferrous Metals	-	-	-	-	-	-	-	..
Non-Metallic Minerals	-	-	-	-	-	-	-	..
Transport Equipment	-	-	-	-	-	-	-	..
Machinery	-	-	-	-	-	-	-	..
Mining and Quarrying	-	-	-	-	-	-	-	..
Food and Tobacco	-	-	2	2	-	-	-	..
Pulp, Paper and Printing	-	-	-	-	-	-	-	..
Wood and Wood Products	-	-	-	-	-	-	-	..
Construction	-	-	-	-	-	-	-	..
Textiles and Leather	-	-	-	-	-	-	-	..
Non-Specified	27	14	13	1	-	4	5	..
Other Sectors	**12**	**20**	**11**	-	-	-	-	..
Commerce and Public	-	-	-	-	-	-	-	..
Residential	12	5	-	-	-	-	-	..
Agriculture	-	11	10	-	-	-	-	..
Non-Specified	-	4	1	-	-	-	-	..
Memo: Non-Energy Use	-	-	-	-	-	-	-	..
Chem/Petchem. Feedst.	-	-	-	-	-	-	-	..
Other	-	-	-	-	-	-	-	..
Closing Stock Level	-	*23*	*12*	*12*	-	-	-	-

Source: Annual Oil Statistics.

LUXEMBOURG
Table 5C: Total Products - Supply and Consumption

000 metric tons

	1973	1978	1980	1990	1999	2000	2001	2002
Refinery Output	-	-	-	-	-	-	-	-
+ Recycled Products	-	-	-	-	-	-	-	-
+ Imports	1 678	1 438	1 132	1 610	2 069	2 295	2 372	2 452 e
- Exports	12	30	49	14	18	17	30	9 e
- Intl. Marine Bunkers	-	-	-	-	-	-	-	-
+ Transfers	-	-	-	-	-	-	-	-
+ Stock Changes	-	- 2	- 1	- 11	46	- 48	28	- 17 e
- Statistical Differences	23	3	7	2	3	3	2	5 e
Total Consumption	**1 643**	**1 403**	**1 075**	**1 583**	**2 094**	**2 227**	**2 368**	**2 421 e**
Transformation	**118**	**55**	**24**	**3**	-	-	-	..
Public Electricity	-	-	-	-	-	-	-	..
Autoproducer Electricity	112	55	24	3	-	-	-	..
Public CHP Plants	-	-	-	-	-	-	-	..
Autoproducers of CHP	-	-	-	-	-	-	-	..
Public Heat Plants	-	-	-	-	-	-	-	..
Autoproducer Heat Plants	-	-	-	-	-	-	-	..
Petrochemical Industry	-	-	-	-	-	-	-	..
Other Transformation	6	-	-	-	-	-	-	..
Energy Sector	-	-	-	-	-	-	-	..
Refinery Fuel	-	-	-	-	-	-	-	..
Other Energy Sector	-	-	-	-	-	-	-	..
Final Consumption	**1 525**	**1 348**	**1 051**	**1 580**	**2 094**	**2 227**	**2 368**	..
Transport	**273**	**409**	**479**	**977**	**1 664**	**1 823**	**1 929**	..
Air	51	61	63	129	323	312	337	..
Road	204	331	405	848	1 319	1 488	1 580	..
Rail	18	9	8	-	8	7	3	..
Inland Waterways	-	8	3	-	-	-	-	..
Non-Specified	-	-	-	-	14	16	9	..
Industry	**828**	**511**	**207**	**301**	**96**	**83**	**77**	..
Iron and Steel	651	350	60	220	24	12	8	..
Chemical (incl. Pet.-Chem.)	68	77	53	42	8	6	6	..
Non-Ferrous Metals	-	-	-	-	-	-	-	..
Non-Metallic Minerals	-	-	-	-	10	8	8	..
Transport Equipment	-	-	-	-	-	-	-	..
Machinery	-	-	-	-	-	-	-	..
Mining and Quarrying	-	-	-	-	2	5	4	..
Food and Tobacco	-	-	8	4	5	4	3	..
Pulp, Paper and Printing	-	-	-	-	-	-	-	..
Wood and Wood Products	-	-	-	-	-	-	-	..
Construction	60	65	64	26	25	21	19	..
Textiles and Leather	-	-	-	-	-	-	-	..
Non-Specified	49	19	22	9	22	27	29	..
Other Sectors	**424**	**428**	**365**	**302**	**334**	**321**	**362**	..
Commerce and Public	-	-	-	-	-	-	-	..
Residential	410	409	350	297	310	296	334	..
Agriculture	13	15	13	5	9	7	7	..
Non-Specified	1	4	2	-	15	18	21	..
Memo: Non-Energy Use	*31*	*41*	*42*	*22*	*22*	*14*	*12*	..
Chem/Petchem. Feedst.	*1*	*3*	*3*	*5*	*1*	*1*	*1*	..
Other	*30*	*38*	*39*	*17*	*21*	*13*	*11*	..
Closing Stock Level	-	*77*	*116*	*137*	*88*	*136*	*108*	*125 e*

Source: Annual Oil Statistics.

LUXEMBOURG
Table 6A: Supply and Consumption

000 metric tons

	1980	2000	2001	2002	1980	2000	2001	2002
	LPG AND ETHANE				REFINERY GAS			
Refinery Output	-	-	-	-	-	-	-	-
+ Recycled Products	-	-	-	-	-	-	-	-
+ Imports	28	30	33	24 e	-	-	-	-
- Exports	3	6	6	4 e	-	-	-	-
- Intl. Marine Bunkers	-	-	-	-	-	-	-	-
+ Transfers	-	-	-	-	-	-	-	-
+ Stock Change	-	-	-	-	-	-	-	-
- Statistical Difference	-	-	- 1	-	-	-	-	-
Total Consumption	**25**	**24**	**28**	**20 e**	**-**	**-**	**-**	**-**
of which: Ethane	-	-	-	-	-	-	-	-
Transformation	-	-	-	..	-	-	-	..
Energy	-	-	-	..	-	-	-	..
Transport	4	2	3	..	-	-	-	..
Industry	11	13	12	..	-	-	-	..
Other	10	9	13	..	-	-	-	..
Memo: Non-Energy Use	-	-	-	..	-	-	-	..
Closing Stock Level	*1*	-	-	-	-	-	-	-
	NAPHTHA				OTHER KEROSENE			
Refinery Output	-	-	-	-	-	-	-	-
+ Recycled Products	-	-	-	-	-	-	-	-
+ Imports	-	-	-	-	-	1	1	1 e
- Exports	-	-	-	-	-	-	-	-
- Intl. Marine Bunkers	-	-	-	-	-	-	-	-
+ Transfers	-	-	-	-	-	-	-	-
+ Stock Change	-	-	-	-	-	-	-	-
- Statistical Difference	-	-	-	-	-	-	-	-
Total Consumption	**-**	**-**	**-**	**-**	**-**	**1**	**1**	**1 e**
Transformation	-	-	-	..	-	-	-	..
Energy	-	-	-	..	-	-	-	..
Transport	-	-	-	..	-	-	-	..
Industry	-	-	-	..	-	-	-	..
Other	-	-	-	..	-	1	1	..
Memo: Non-Energy Use	-	-	-	..	-	-	-	..
Closing Stock Level	-	-	-	-	-	-	-	-
	WHITE SPIRIT + SBP				LUBRICANTS			
Refinery Output	-	-	-	..	-	-	-	..
+ Recycled Products	-	-	-	..	-	-	-	..
+ Imports	3	-	-	..	10	7	7	..
- Exports	-	-	-	..	-	-	-	..
- Intl. Marine Bunkers	-	-	-	..	-	-	-	..
+ Transfers	-	-	-	..	-	-	-	..
+ Stock Change	-	-	-	..	-	-	-	..
- Statistical Difference	-	-	-	..	-	-	-	..
Total Consumption	**3**	**-**	**-**	**..**	**10**	**7**	**7**	**..**
Transformation	-	-	-	..	-	-	-	..
Energy	-	-	-	..	-	-	-	..
Transport	-	-	-	..	3	1	1	..
Industry	2	-	-	..	7	6	6	..
Other	1	-	-	..	-	-	-	..
Memo: Non-Energy Use	*3*	-	-	..	*10*	*7*	*7*	..
Closing Stock Level	-	-	-	..	-	-	-	..

Source: Annual Oil Statistics.

LUXEMBOURG
Table 6B: Supply and Consumption

000 metric tons

	1980	2000	2001	2002	1980	2000	2001	2002
	BITUMEN				PARAFFIN WAXES			
Refinery Output	-	-	-	..	-	-	-	..
+ Recycled Products	-	-	-	..	-	-	-	..
+ Imports	29	5	4	..	-	-	-	..
- Exports	-	-	-	..	-	-	-	..
- Intl. Marine Bunkers	-	-	-	..	-	-	-	..
+ Transfers	-	-	-	..	-	-	-	..
+ Stock Change	-	-	-	..	-	-	-	..
- Statistical Difference	-	-	-	..	-	-	-	..
Total Consumption	**29**	**5**	**4**	**..**	**-**	**-**	**-**	**..**
Transformation	-	-	-	..	-	-	-	..
Energy	-	-	-	..	-	-	-	..
Transport	-	-	-	..	-	-	-	..
Industry	29	5	4	..	-	-	-	..
Other	-	-	-	..	-	-	-	..
Memo: Non-Energy Use	*29*	*5*	*4*	..	-	-	-	..
Closing Stock Level	-	-	-	..	-	-	-	..
	PETROLEUM COKE				OTHER PRODUCTS			
Refinery Output	-	-	-	-	-	-	-	-
+ Recycled Products	-	-	-	-	-	-	-	-
+ Imports	-	-	-	-	1	-	-	11 e
- Exports	-	-	-	-	-	-	-	-
- Intl. Marine Bunkers	-	-	-	-	-	-	-	-
+ Transfers	-	-	-	-	-	-	-	-
+ Stock Change	-	-	-	-	-	-	-	-
- Statistical Difference	-	-	-	-	1	-	-	-
Total Consumption	**-**	**-**	**-**	**-**	**-**	**-**	**-**	**11 e**
Transformation	-	-	-	..	-	-	-	..
Energy	-	-	-	..	-	-	-	..
Transport	-	-	-	..	-	-	-	..
Industry	-	-	-	..	-	-	-	..
Other	-	-	-	..	-	-	-	..
Memo: Non-Energy Use	-	-	-	..	-	-	-	..
Closing Stock Level	-	-	-	-	-	-	-	-

Source: Annual Oil Statistics.

LUXEMBOURG
Table 7B: Imports by Country of Origin

000 metric tons

	2000	2001	2002	2000	2001	2002	2000	2001	2002
	Total Products			**Liquefied Petroleum Gases**			**Naphtha**		
Australia	-	-	-	-	-	-	-	-	-
Belgium	2 057	2 025	2 096 e	29	32	23 e	-	-	-
Canada	-	-	-	-	-	-	-	-	-
Czech Republic	-	-	-	-	-	-	-	-	-
Denmark	-	-	-	-	-	-	-	-	-
Finland	-	-	-	-	-	-	-	-	-
France	98	167	154 e	1	1	1 e	-	-	-
Germany	68	109	122 e	-	-	-	-	-	-
Greece	-	-	-	-	-	-	-	-	-
Hungary	-	-	-	-	-	-	-	-	-
Ireland	-	-	-	-	-	-	-	-	-
Italy	-	-	-	-	-	-	-	-	-
Japan	-	-	-	-	-	-	-	-	-
Korea	-	-	-	-	-	-	-	-	-
Mexico	-	-	-	-	-	-	-	-	-
Netherlands	72	71	80 e	-	-	-	-	-	-
New Zealand	-	-	-	-	-	-	-	-	-
Norway	-	-	-	-	-	-	-	-	-
Poland	-	-	-	-	-	-	-	-	-
Spain	-	-	-	-	-	-	-	-	-
Sweden	-	-	-	-	-	-	-	-	-
Turkey	-	-	-	-	-	-	-	-	-
United Kingdom	-	-	-	-	-	-	-	-	-
United States	-	-	-	-	-	-	-	-	-
Other	-	-	-	-	-	-	-	-	-
OECD	**2 295**	**2 372**	**2 452 e**	**30**	**33**	**24 e**	**-**	**-**	**-**
Argentina	-	-	-	-	-	-	-	-	-
Brazil	-	-	-	-	-	-	-	-	-
Colombia	-	-	-	-	-	-	-	-	-
Ecuador	-	-	-	-	-	-	-	-	-
Netherlands Antilles	-	-	-	-	-	-	-	-	-
Peru	-	-	-	-	-	-	-	-	-
Trinidad and Tobago	-	-	-	-	-	-	-	-	-
Venezuela	-	-	-	-	-	-	-	-	-
Other	-	-	-	-	-	-	-	-	-
Latin America	**-**	**-**	**-**	**-**	**-**	**-**	**-**	**-**	**-**
Romania	-	-	-	-	-	-	-	-	-
Other	-	-	-	-	-	-	-	-	-
Non-OECD Europe	**-**	**-**	**-**	**-**	**-**	**-**	**-**	**-**	**-**
FSU	**-**	**-**	**-**	**-**	**-**	**-**	**-**	**-**	**-**
Iran	-	-	-	-	-	-	-	-	-
Iraq	-	-	-	-	-	-	-	-	-
Kuwait	-	-	-	-	-	-	-	-	-
Oman	-	-	-	-	-	-	-	-	-
Qatar	-	-	-	-	-	-	-	-	-
Saudi Arabia	-	-	-	-	-	-	-	-	-
United Arab Emirates	-	-	-	-	-	-	-	-	-
Other	-	-	-	-	-	-	-	-	-
Middle East	**-**	**-**	**-**	**-**	**-**	**-**	**-**	**-**	**-**
India	-	-	-	-	-	-	-	-	-
Indonesia	-	-	-	-	-	-	-	-	-
Singapore	-	-	-	-	-	-	-	-	-
Other	-	-	-	-	-	-	-	-	-
Asia	**-**	**-**	**-**	**-**	**-**	**-**	**-**	**-**	**-**
China	**-**	**-**	**-**	**-**	**-**	**-**	**-**	**-**	**-**
Algeria	-	-	-	-	-	-	-	-	-
Angola/Cabinada	-	-	-	-	-	-	-	-	-
Cameroon	-	-	-	-	-	-	-	-	-
Egypt	-	-	-	-	-	-	-	-	-
Gabon	-	-	-	-	-	-	-	-	-
Libya	-	-	-	-	-	-	-	-	-
Nigeria	-	-	-	-	-	-	-	-	-
Tunisia	-	-	-	-	-	-	-	-	-
Other	-	-	-	-	-	-	-	-	-
Africa	**-**	**-**	**-**	**-**	**-**	**-**	**-**	**-**	**-**
Non-spec./Other	**-**	**-**	**-**	**-**	**-**	**-**	**-**	**-**	**-**
Total Imports	**2 295**	**2 372**	**2 452 e**	**30**	**33**	**24 e**	**-**	**-**	**-**
Memo: Total OPEC	-	-	-	-	-	-	-	-	-

Source: Annual Oil Statistics.

LUXEMBOURG
Table 7C: Imports by Country of Origin

000 metric tons

	2000	2001	2002	2000	2001	2002	2000	2001	2002
	Total Gasoline			**Total Kerosene**			**Gas/Diesel**		
Australia	-	-	-	-	-	-	-	-	-
Belgium	545	507	494 e	213	146	183 e	1 260	1 329	1 389 e
Canada	-	-	-	-	-	-	-	-	-
Czech Republic	-	-	-	-	-	-	-	-	-
Denmark	-	-	-	-	-	-	-	-	-
Finland	-	-	-	-	-	-	-	-	-
France	-	-	-	92	163	150 e	-	-	-
Germany	25	35	30 e	-	-	-	42	73	92 e
Greece	-	-	-	-	-	-	-	-	-
Hungary	-	-	-	-	-	-	-	-	-
Ireland	-	-	-	-	-	-	-	-	-
Italy	-	-	-	-	-	-	-	-	-
Japan	-	-	-	-	-	-	-	-	-
Korea	-	-	-	-	-	-	-	-	-
Mexico	-	-	-	-	-	-	-	-	-
Netherlands	25	32	28 e	14	22	37 e	31	14	9 e
New Zealand	-	-	-	-	-	-	-	-	-
Norway	-	-	-	-	-	-	-	-	-
Poland	-	-	-	-	-	-	-	-	-
Spain	-	-	-	-	-	-	-	-	-
Sweden	-	-	-	-	-	-	-	-	-
Turkey	-	-	-	-	-	-	-	-	-
United Kingdom	-	-	-	-	-	-	-	-	-
United States	-	-	-	-	-	-	-	-	-
Other	-	-	-	-	-	-	-	-	-
OECD	**595**	**574**	**552 e**	**319**	**331**	**370 e**	**1 333**	**1 416**	**1 490 e**
Argentina	-	-	-	-	-	-	-	-	-
Brazil	-	-	-	-	-	-	-	-	-
Colombia	-	-	-	-	-	-	-	-	-
Ecuador	-	-	-	-	-	-	-	-	-
Netherlands Antilles	-	-	-	-	-	-	-	-	-
Peru	-	-	-	-	-	-	-	-	-
Trinidad and Tobago	-	-	-	-	-	-	-	-	-
Venezuela	-	-	-	-	-	-	-	-	-
Other	-	-	-	-	-	-	-	-	-
Latin America	**-**	**-**	**-**	**-**	**-**	**-**	**-**	**-**	**-**
Romania	-	-	-	-	-	-	-	-	-
Other	-	-	-	-	-	-	-	-	-
Non-OECD Europe	**-**	**-**	**-**	**-**	**-**	**-**	**-**	**-**	**-**
FSU	**-**	**-**	**-**	**-**	**-**	**-**	**-**	**-**	**-**
Iran	-	-	-	-	-	-	-	-	-
Iraq	-	-	-	-	-	-	-	-	-
Kuwait	-	-	-	-	-	-	-	-	-
Oman	-	-	-	-	-	-	-	-	-
Qatar	-	-	-	-	-	-	-	-	-
Saudi Arabia	-	-	-	-	-	-	-	-	-
United Arab Emirates	-	-	-	-	-	-	-	-	-
Other	-	-	-	-	-	-	-	-	-
Middle East	**-**	**-**	**-**	**-**	**-**	**-**	**-**	**-**	**-**
India	-	-	-	-	-	-	-	-	-
Indonesia	-	-	-	-	-	-	-	-	-
Singapore	-	-	-	-	-	-	-	-	-
Other	-	-	-	-	-	-	-	-	-
Asia	**-**	**-**	**-**	**-**	**-**	**-**	**-**	**-**	**-**
China	**-**	**-**	**-**	**-**	**-**	**-**	**-**	**-**	**-**
Algeria	-	-	-	-	-	-	-	-	-
Angola/Cabinada	-	-	-	-	-	-	-	-	-
Cameroon	-	-	-	-	-	-	-	-	-
Egypt	-	-	-	-	-	-	-	-	-
Gabon	-	-	-	-	-	-	-	-	-
Libya	-	-	-	-	-	-	-	-	-
Nigeria	-	-	-	-	-	-	-	-	-
Tunisia	-	-	-	-	-	-	-	-	-
Other	-	-	-	-	-	-	-	-	-
Africa	**-**	**-**	**-**	**-**	**-**	**-**	**-**	**-**	**-**
Non-spec./Other	**-**	**-**	**-**	**-**	**-**	**-**	**-**	**-**	**-**
Total Imports	**595**	**574**	**552 e**	**319**	**331**	**370 e**	**1 333**	**1 416**	**1 490 e**
Memo: Total OPEC	-	-	-	-	-	-	-	-	-

Source: Annual Oil Statistics.

LUXEMBOURG
Table 7D: Imports by Country of Origin

000 metric tons

	2000	2001	2002	2000	2001	2002	2000	2001	2002
	Fuel Oil (Residual)			**Petroleum Coke**			**Other Products**		
Australia	-	-	-	-	-	-	-	-	-
Belgium	6	7	5 e	-	-	-	4	4	2 e
Canada	-	-	-	-	-	-	-	-	-
Czech Republic	-	-	-	-	-	-	-	-	-
Denmark	-	-	-	-	-	-	-	-	-
Finland	-	-	-	-	-	-	-	-	-
France	-	-	-	-	-	-	5	3	3 e
Germany	-	-	-	-	-	-	1	1	-
Greece	-	-	-	-	-	-	-	-	-
Hungary	-	-	-	-	-	-	-	-	-
Ireland	-	-	-	-	-	-	-	-	-
Italy	-	-	-	-	-	-	-	-	-
Japan	-	-	-	-	-	-	-	-	-
Korea	-	-	-	-	-	-	-	-	-
Mexico	-	-	-	-	-	-	-	-	-
Netherlands	-	-	-	-	-	-	2	3	6 e
New Zealand	-	-	-	-	-	-	-	-	-
Norway	-	-	-	-	-	-	-	-	-
Poland	-	-	-	-	-	-	-	-	-
Spain	-	-	-	-	-	-	-	-	-
Sweden	-	-	-	-	-	-	-	-	-
Turkey	-	-	-	-	-	-	-	-	-
United Kingdom	-	-	-	-	-	-	-	-	-
United States	-	-	-	-	-	-	-	-	-
Other	-	-	-	-	-	-	-	-	-
OECD	**6**	**7**	**5 e**	**-**	**-**	**-**	**12**	**11**	**11 e**
Argentina	-	-	-	-	-	-	-	-	-
Brazil	-	-	-	-	-	-	-	-	-
Colombia	-	-	-	-	-	-	-	-	-
Ecuador	-	-	-	-	-	-	-	-	-
Netherlands Antilles	-	-	-	-	-	-	-	-	-
Peru	-	-	-	-	-	-	-	-	-
Trinidad and Tobago	-	-	-	-	-	-	-	-	-
Venezuela	-	-	-	-	-	-	-	-	-
Other	-	-	-	-	-	-	-	-	-
Latin America	**-**	**-**	**-**	**-**	**-**	**-**	**-**	**-**	**-**
Romania	-	-	-	-	-	-	-	-	-
Other	-	-	-	-	-	-	-	-	-
Non-OECD Europe	**-**	**-**	**-**	**-**	**-**	**-**	**-**	**-**	**-**
FSU	**-**	**-**	**-**	**-**	**-**	**-**	**-**	**-**	**-**
Iran	-	-	-	-	-	-	-	-	-
Iraq	-	-	-	-	-	-	-	-	-
Kuwait	-	-	-	-	-	-	-	-	-
Oman	-	-	-	-	-	-	-	-	-
Qatar	-	-	-	-	-	-	-	-	-
Saudi Arabia	-	-	-	-	-	-	-	-	-
United Arab Emirates	-	-	-	-	-	-	-	-	-
Other	-	-	-	-	-	-	-	-	-
Middle East	**-**	**-**	**-**	**-**	**-**	**-**	**-**	**-**	**-**
India	-	-	-	-	-	-	-	-	-
Indonesia	-	-	-	-	-	-	-	-	-
Singapore	-	-	-	-	-	-	-	-	-
Other	-	-	-	-	-	-	-	-	-
Asia	**-**	**-**	**-**	**-**	**-**	**-**	**-**	**-**	**-**
China	**-**	**-**	**-**	**-**	**-**	**-**	**-**	**-**	**-**
Algeria	-	-	-	-	-	-	-	-	-
Angola/Cabinada	-	-	-	-	-	-	-	-	-
Cameroon	-	-	-	-	-	-	-	-	-
Egypt	-	-	-	-	-	-	-	-	-
Gabon	-	-	-	-	-	-	-	-	-
Libya	-	-	-	-	-	-	-	-	-
Nigeria	-	-	-	-	-	-	-	-	-
Tunisia	-	-	-	-	-	-	-	-	-
Other	-	-	-	-	-	-	-	-	-
Africa	**-**	**-**	**-**	**-**	**-**	**-**	**-**	**-**	**-**
Non-spec./Other	**-**	**-**	**-**	**-**	**-**	**-**	**-**	**-**	**-**
Total Imports	**6**	**7**	**5 e**	**-**	**-**	**-**	**12**	**11**	**11 e**
Memo: Total OPEC	-	-	-	-	-	-	-	-	-

Source: Annual Oil Statistics.

LUXEMBOURG
Table 8A: Exports by Country of Destination

000 metric tons

	Crude Oil 2000	Crude Oil 2001	Crude Oil 2002	Natural Gas Liquids 2000	Natural Gas Liquids 2001	Natural Gas Liquids 2002	Feedstocks and Other 2000	Feedstocks and Other 2001	Feedstocks and Other 2002
Australia	-	-	-	-	-	-	-	-	-
Austria	-	-	-	-	-	-	-	-	-
Belgium	-	-	-	-	-	-	-	-	-
Canada	-	-	-	-	-	-	-	-	-
Denmark	-	-	-	-	-	-	-	-	-
Finland	-	-	-	-	-	-	-	-	-
France	-	-	-	-	-	-	-	-	-
Germany	-	-	-	-	-	-	-	-	-
Greece	-	-	-	-	-	-	-	-	-
Ireland	-	-	-	-	-	-	-	-	-
Italy	-	-	-	-	-	-	-	-	-
Japan	-	-	-	-	-	-	-	-	-
Luxembourg	-	-	-	-	-	-	-	-	-
Mexico	-	-	-	-	-	-	-	-	-
Netherlands	-	-	-	-	-	-	-	-	-
Norway	-	-	-	-	-	-	-	-	-
Portugal	-	-	-	-	-	-	-	-	-
Spain	-	-	-	-	-	-	-	-	-
Sweden	-	-	-	-	-	-	-	-	-
Switzerland	-	-	-	-	-	-	-	-	-
Turkey	-	-	-	-	-	-	-	-	-
United Kingdom	-	-	-	-	-	-	-	-	-
United States	-	-	-	-	-	-	-	-	-
Other OECD	-	-	-	-	-	-	-	-	-
OECD	**-**	**-**	**-**	**-**	**-**	**-**	**-**	**-**	**-**
Africa	-	-	-	-	-	-	-	-	-
Latin America	-	-	-	-	-	-	-	-	-
Asia	-	-	-	-	-	-	-	-	-
China	-	-	-	-	-	-	-	-	-
Non-OECD Europe	-	-	-	-	-	-	-	-	-
Former Soviet Union	-	-	-	-	-	-	-	-	-
Middle East	-	-	-	-	-	-	-	-	-
Non-specified/Other	-	-	-	-	-	-	-	-	-
Non-OECD	**-**	**-**	**-**	**-**	**-**	**-**	**-**	**-**	**-**
Total Exports	**-**	**-**	**-**	**-**	**-**	**-**	**-**	**-**	**-**

	Total Products 2000	Total Products 2001	Total Products 2002	Liquefied Petroleum Gas 2000	Liquefied Petroleum Gas 2001	Liquefied Petroleum Gas 2002	Naphtha 2000	Naphtha 2001	Naphtha 2002
Australia	-	-	-	-	-	-	-	-	-
Austria	-	-	-	-	-	-	-	-	-
Belgium	2	4	-	-	-	-	-	-	-
Canada	-	-	-	-	-	-	-	-	-
Denmark	-	-	-	-	-	-	-	-	-
Finland	-	-	-	-	-	-	-	-	-
France	-	-	-	-	-	-	-	-	-
Germany	15	26	9 e	6	6	4 e	-	-	-
Greece	-	-	-	-	-	-	-	-	-
Ireland	-	-	-	-	-	-	-	-	-
Italy	-	-	-	-	-	-	-	-	-
Japan	-	-	-	-	-	-	-	-	-
Luxembourg	-	-	-	-	-	-	-	-	-
Mexico	-	-	-	-	-	-	-	-	-
Netherlands	-	-	-	-	-	-	-	-	-
Norway	-	-	-	-	-	-	-	-	-
Portugal	-	-	-	-	-	-	-	-	-
Spain	-	-	-	-	-	-	-	-	-
Sweden	-	-	-	-	-	-	-	-	-
Switzerland	-	-	-	-	-	-	-	-	-
Turkey	-	-	-	-	-	-	-	-	-
United Kingdom	-	-	-	-	-	-	-	-	-
United States	-	-	-	-	-	-	-	-	-
Other OECD	-	-	-	-	-	-	-	-	-
OECD	**17**	**30**	**9 e**	**6**	**6**	**4 e**	**-**	**-**	**-**
Africa	-	-	-	-	-	-	-	-	-
Latin America	-	-	-	-	-	-	-	-	-
Asia	-	-	-	-	-	-	-	-	-
China	-	-	-	-	-	-	-	-	-
Non-OECD Europe	-	-	-	-	-	-	-	-	-
Former Soviet Union	-	-	-	-	-	-	-	-	-
Middle East	-	-	-	-	-	-	-	-	-
Non-specified/Other	-	-	-	-	-	-	-	-	-
Non-OECD	**-**	**-**	**-**	**-**	**-**	**-**	**-**	**-**	**-**
Total Exports	**17**	**30**	**9 e**	**6**	**6**	**4 e**	**-**	**-**	**-**

Source: Annual Oil Statistics.

LUXEMBOURG
Table 8B: Exports by Country of Destination

000 metric tons

	2000	2001	2002	2000	2001	2002	2000	2001	2002
	Total Gasoline			**Total Kerosene**			**Gas/Diesel Oil**		
Australia	-	-	-	-	-	-	-	-	-
Austria	-	-	-	-	-	-	-	-	-
Belgium	1	1	-	-	-	-	1	3	-
Canada	-	-	-	-	-	-	-	-	-
Denmark	-	-	-	-	-	-	-	-	-
Finland	-	-	-	-	-	-	-	-	-
France	-	-	-	-	-	-	-	-	-
Germany	2	2	-	-	-	-	7	18	5 e
Greece	-	-	-	-	-	-	-	-	-
Ireland	-	-	-	-	-	-	-	-	-
Italy	-	-	-	-	-	-	-	-	-
Japan	-	-	-	-	-	-	-	-	-
Luxembourg	-	-	-	-	-	-	-	-	-
Mexico	-	-	-	-	-	-	-	-	-
Netherlands	-	-	-	-	-	-	-	-	-
Norway	-	-	-	-	-	-	-	-	-
Portugal	-	-	-	-	-	-	-	-	-
Spain	-	-	-	-	-	-	-	-	-
Sweden	-	-	-	-	-	-	-	-	-
Switzerland	-	-	-	-	-	-	-	-	-
Turkey	-	-	-	-	-	-	-	-	-
United Kingdom	-	-	-	-	-	-	-	-	-
United States	-	-	-	-	-	-	-	-	-
Other OECD	-	-	-	-	-	-	-	-	-
OECD	**3**	**3**	**-**	**-**	**-**	**-**	**8**	**21**	**5 e**
Africa	-	-	-	-	-	-	-	-	-
Latin America	-	-	-	-	-	-	-	-	-
Asia	-	-	-	-	-	-	-	-	-
China	-	-	-	-	-	-	-	-	-
Non-OECD Europe	-	-	-	-	-	-	-	-	-
Former Soviet Union	-	-	-	-	-	-	-	-	-
Middle East	-	-	-	-	-	-	-	-	-
Non-specified/Other	-	-	-	-	-	-	-	-	-
Non-OECD	**-**	**-**	**-**	**-**	**-**	**-**	**-**	**-**	**-**
Total Exports	**3**	**3**	**-**	**-**	**-**	**-**	**8**	**21**	**5 e**
	Fuel Oil (Residual)			**Petroleum Coke**			**Other Products**		
Australia	-	-	-	-	-	-	-	-	-
Austria	-	-	-	-	-	-	-	-	-
Belgium	-	-	-	-	-	-	-	-	-
Canada	-	-	-	-	-	-	-	-	-
Denmark	-	-	-	-	-	-	-	-	-
Finland	-	-	-	-	-	-	-	-	-
France	-	-	-	-	-	-	-	-	-
Germany	-	-	-	-	-	-	-	-	-
Greece	-	-	-	-	-	-	-	-	-
Ireland	-	-	-	-	-	-	-	-	-
Italy	-	-	-	-	-	-	-	-	-
Japan	-	-	-	-	-	-	-	-	-
Luxembourg	-	-	-	-	-	-	-	-	-
Mexico	-	-	-	-	-	-	-	-	-
Netherlands	-	-	-	-	-	-	-	-	-
Norway	-	-	-	-	-	-	-	-	-
Portugal	-	-	-	-	-	-	-	-	-
Spain	-	-	-	-	-	-	-	-	-
Sweden	-	-	-	-	-	-	-	-	-
Switzerland	-	-	-	-	-	-	-	-	-
Turkey	-	-	-	-	-	-	-	-	-
United Kingdom	-	-	-	-	-	-	-	-	-
United States	-	-	-	-	-	-	-	-	-
Other OECD	-	-	-	-	-	-	-	-	-
OECD	**-**	**-**	**-**	**-**	**-**	**-**	**-**	**-**	**-**
Africa	-	-	-	-	-	-	-	-	-
Latin America	-	-	-	-	-	-	-	-	-
Asia	-	-	-	-	-	-	-	-	-
China	-	-	-	-	-	-	-	-	-
Non-OECD Europe	-	-	-	-	-	-	-	-	-
Former Soviet Union	-	-	-	-	-	-	-	-	-
Middle East	-	-	-	-	-	-	-	-	-
Non-specified/Other	-	-	-	-	-	-	-	-	-
Non-OECD	**-**	**-**	**-**	**-**	**-**	**-**	**-**	**-**	**-**
Total Exports	**-**	**-**	**-**	**-**	**-**	**-**	**-**	**-**	**-**

Source: Annual Oil Statistics.

MEXICO

Figure 1
FUEL SHARES OF TPES *

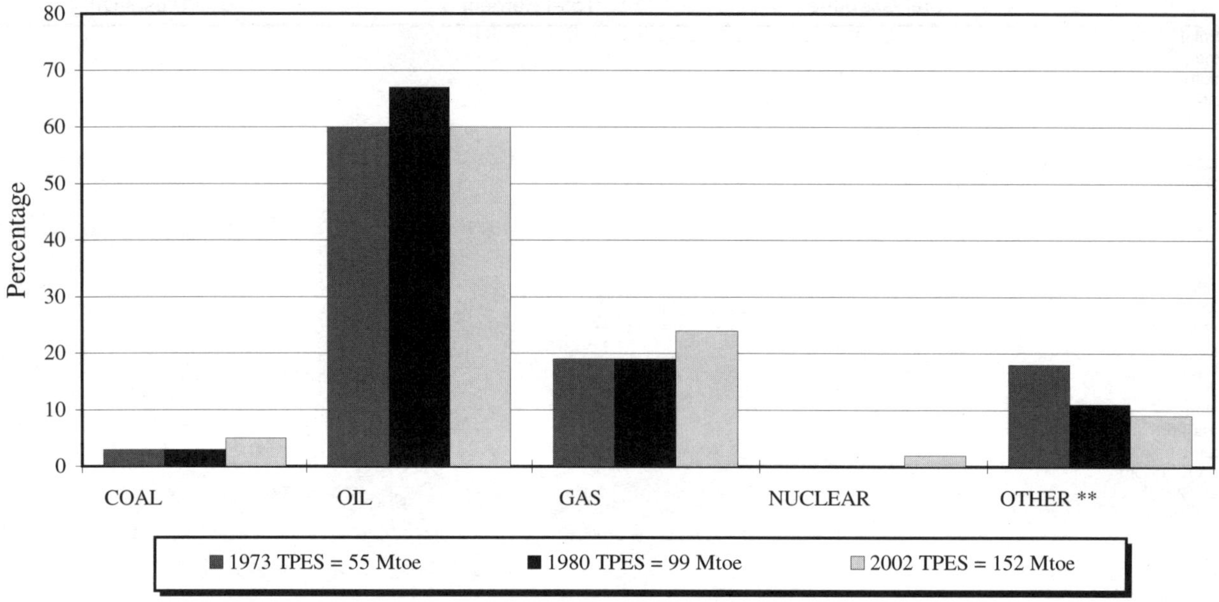

Source: Annual Energy Balances.

* Excluding electricity trade.

** Includes hydro, geothermal, solar, wind, combustible renewables and waste, etc.

Figure 2
OIL PRODUCTION AND CONSUMPTION

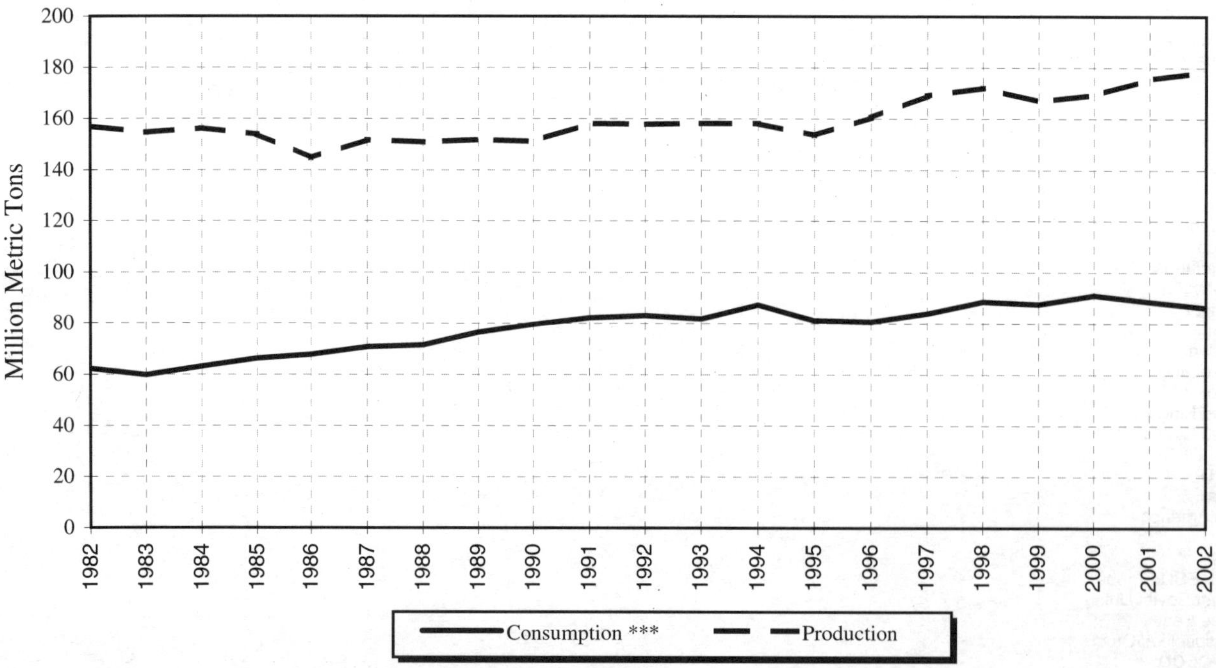

Source: Annual Oil Statistics.

*** Includes refinery fuel and international marine bunkers.

MEXICO
Table 1: Oil Supply and Consumption - Summary Table

000 metric tons

	1973	1978	1980	1990	1999	2000	2001	2002
Indigenous Production	27 182	69 227	113 350	151 054	166 882	169 261	175 498	178 261 e
Crude Oil	..	65 457	107 017	138 656	154 081	156 403	162 559	165 963 e
NGL	..	3 770	6 333	12 398	12 637	12 741	12 723	11 988 e
Other	..	-	-	-	164	117	216	310 e
+ Net Imports	5 405	-17 930	-47 117	-68 918	-73 714	-76 759	-80 541	-85 233 e
Crude Oil, NGL, Other	3 552	-19 562	-45 601	-69 631	-82 601	-91 274	-92 445	-94 354 e
Petroleum Products	1 853	1 632	-1 516	713	8 887	14 515	11 904	9 121 e
- Intl. Marine Bunkers	113	304	325	657	1 169	1 335	1 095	861 e
+ Stock Changes	- 7	252	- 841	- 175	- 16	-1 342	46	13 e
TOTAL SUPPLY	**32 467**	**51 245**	**65 067**	**81 304**	**91 983**	**89 825**	**93 908**	**92 180 e**
Refinery Intake	*28 348*	*43 394*	*55 212*	*68 467*	*70 658*	*65 371*	*66 529*	*67 360 e*
Refinery Output	*25 858*	*40 308*	*50 432*	*65 927*	*67 718*	*64 176*	*63 886*	*66 153 e*
LPG + Ethane	*572*	*947*	*1 321*	*1 679*	*955*	*782*	*870*	*870 e*
Naphtha	*390*	*504*	*519*	*1 620*	*854*	*850*	*833*	*874 e*
Kerosenes	*2 249*	*2 838*	*3 371*	*3 207*	*2 690*	*2 584*	*2 654*	*2 628 e*
Gasolines	*6 024*	*9 190*	*11 615*	*16 776*	*16 115*	*15 717*	*15 174*	*15 616 e*
Gas/Diesel Oil	*5 434*	*10 289*	*12 693*	*13 285*	*13 303*	*13 249*	*13 891*	*13 353 e*
Fuel Oil	*9 084*	*14 762*	*18 735*	*26 394*	*27 734*	*24 116*	*24 814*	*26 156 e*
Other Products	*2 105*	*1 778*	*2 178*	*2 966*	*6 067*	*6 878*	*5 650*	*6 656 e*
- Refinery Losses	2 490	3 086	4 780	2 540	2 940	1 195	2 643	1 207 e
+ Returns	-	-	-	-	-	-	-	-
- Statistical Differences	- 668	2 510	5 494	- 79	2 692	- 971	3 834	5 542 e
TOTAL CONSUMPTION	**30 645**	**45 649**	**54 793**	**78 843**	**86 351**	**89 601**	**87 431**	**85 431 e**
Transformation and Energy	**8 770**	**12 436**	**15 352**	**24 876**	**30 613**	**31 482**	**29 576**	..
Crude Oil	-	-	-	9	-	-	-	..
LPG + Ethane	675	95	274	575	194	222	240	..
Gas/Diesel Oil	1 344	1 725	2 426	1 374	941	1 125	1 054	..
Fuel Oil	5 620	9 573	11 317	19 884	24 560	25 511	24 353	..
Petroleum Coke	60	-	26	33	1 512	1 475	1 363	..
Other	1 071	1 043	1 309	3 001	3 406	3 149	2 566	..
Transport	**12 309**	**18 074**	**22 983**	**29 549**	**33 592**	**35 208**	**35 403**	..
Motor Gasoline	8 020	10 462	13 854	19 728	21 364	22 275	22 677	..
Gas/Diesel Oil	3 460	6 485	7 649	7 785	8 937	9 336	9 111	..
Subtotal	*11 480*	*16 947*	*21 503*	*27 513*	*30 301*	*31 611*	*31 788*	..
Aviation Fuels	586	971	1 341	1 697	2 543	2 557	2 537	..
Other	243	156	139	339	748	1 040	1 078	..
Industry	**5 385**	**9 215**	**9 163**	**15 623**	**12 757**	**12 560**	**12 318**	..
Naphtha	391	504	520	1 620	854	850	960	..
LPG + Ethane	385	698	845	2 848	4 143	4 222	3 975	..
Gas/Diesel Oil	646	851	1 014	1 040	1 256	1 292	1 272	..
Fuel Oil	2 666	5 553	4 777	7 808	5 038	4 565	4 505	..
Other	1 297	1 609	2 007	2 307	1 466	1 631	1 606	..
Other Sectors	**4 181**	**5 924**	**7 295**	**8 795**	**9 389**	**10 351**	**10 134**	..
Gas/Diesel Oil	631	1 346	1 535	1 492	1 917	1 959	1 893	..
Fuel Oil	443	506	611	807	-	-	-	..
Other	3 107	4 072	5 149	6 496	7 472	8 392	8 241	..
Memo: Non-Energy Use	*1 727*	*2 471*	*2 996*	*6 382*	*5 642*	*5 709*	*5 600*	..
Chem/Petchem. Feedst.	*612*	*994*	*1 103*	*4 058*	*4 221*	*4 873*	*4 754*	..
Other	*1 115*	*1 477*	*1 893*	*2 324*	*1 421*	*836*	*846*	..

Source: Annual Oil Statistics.

MEXICO
Table 2: Supply of Crude Oil, NGL, Feedstocks and Other Hydrocarbons

000 metric tons

	1973	1978	1980	1990	1999	2000	2001	2002
CRUDE OIL								
Indigenous Production	..	65 457	107 017	138 656	154 081	156 403	162 559	165 963 e
+ Imports	..	-	-	-	-	-	-	-
- Exports	..	19 562	45 601	69 631	83 063	91 611	92 740	94 746 e
- Direct Use	..	-	-	9	-	-	-	-
+ Stock Change	..	204	- 530	- 302	- 527	- 911	6	267 e
- Statistical Difference	..	2 705	5 674	247	459	-1 036	3 807	4 826 e
Refinery Intake	**..**	**43 394**	**55 212**	**68 467**	**70 032**	**64 917**	**66 018**	**66 658 e**
Closing Stock Level	..	-	-	-	*1 976*	*2 887*	*2 881*	*2 614 e*
NATURAL GAS LIQUIDS								
Indigenous Production	..	3 770	6 333	12 398	12 637	12 741	12 723	11 988 e
+ Imports	..	-	-	-	-	-	-	-
- Exports	..	-	-	-	-	-	-	-
- Direct Use	..	3 770	6 333	12 398	12 637	12 741	12 723	11 988 e
+ Stock Change	..	-	-	-	-	-	-	-
- Statistical Difference	..	-	-	-	-	-	-	-
Refinery Intake	**..**	**-**	**-**	**-**	**-**	**-**	**-**	**-**
Closing Stock Level	..	-	-	-	-	-	-	-
REFINERY FEEDSTOCKS								
Backflows	..	-	-	-	-	-	-	-
+ Transfers	..	-	-	-	-	-	-	-
+ Imports	..	-	-	-	-	-	-	-
- Exports	..	-	-	-	-	-	-	-
+ Stock Change	..	-	-	-	-	-	-	-
- Statistical Difference	..	-	-	-	-	-	-	-
Refinery Intake	**..**	**-**	**-**	**-**	**-**	**-**	**-**	**-**
Closing Stock Level	..	-	-	-	-	-	-	-
OTHER HYDROCARBONS								
Indigenous Production	..	-	-	-	164	117	216	310 e
+ Imports	..	-	-	-	462	337	295	392 e
- Exports	..	-	-	-	-	-	-	-
- Direct Use	..	-	-	-	-	-	-	-
+ Stock Change	..	-	-	-	-	-	-	-
- Statistical Difference	..	-	-	-	-	-	-	-
Refinery Intake	**..**	**-**	**-**	**-**	**626**	**454**	**511**	**702 e**
Closing Stock Level	..	-	-	-	-	-	-	-
CRUDE OIL, NGL, FEEDSTOCKS AND OTHER								
Indigenous Production	27 182	69 227	113 350	151 054	166 882	169 261	175 498	178 261 e
+ Backflows	-	-	-	-	-	-	-	-
+ Transfers	-	-	-	-	-	-	-	-
+ Imports	3 552	-	-	-	462	337	295	392 e
- Exports	-	19 562	45 601	69 631	83 063	91 611	92 740	94 746 e
- Direct Use	2 378	3 770	6 333	12 407	12 637	12 741	12 723	11 988 e
+ Stock Change	- 19	204	- 530	- 302	- 527	- 911	6	267 e
- Statistical Difference	- 11	2 705	5 674	247	459	-1 036	3 807	4 826 e
Refinery Intake	**28 348**	**43 394**	**55 212**	**68 467**	**70 658**	**65 371**	**66 529**	**67 360 e**
Closing Stock Level	-	-	-	-	*1 976*	*2 887*	*2 881*	*2 614 e*

Source: Annual Oil Statistics.

MEXICO
Table 3: Motor Gasoline - Supply

000 metric tons

	1973	1978	1980	1990	1999	2000	2001	2002
MOTOR GASOLINE								
Refinery Output	**6 024**	**9 190**	**11 615**	**16 776**	**16 092**	**15 700**	**15 158**	**15 613 e**
+ Imports	1 336	9	27	1 614	5 465	6 540	7 096	7 283 e
- Exports	36	74	58	375	2 810	2 598	2 740	2 066 e
+ Transfers	1 060	1 618	2 847	3 446	3 507	3 098	3 224	3 112 e
+ Stock Change	20	2	- 163	- 68	261	- 107	- 40	- 164 e
- Statistical Difference	- 1	2	- 11	1	78	- 485	- 436	- 366 e
Total Consumption	**8 405**	**10 743**	**14 279**	**21 392**	**22 437**	**23 118**	**23 134**	**24 144 e**
Closing Stock Level	-	-	-	-	*1 250*	*1 357*	*1 397*	*1 561 e*

Table 4: Aviation Fuels - Supply

000 metric tons

	1973	1978	1980	1990	1999	2000	2001	2002
AVIATION GASOLINE								
Refinery Output	-	-	-	-	**23**	**17**	**16**	**3 e**
+ Imports	-	-	-	-	-	-	-	1 e
- Exports	-	-	-	-	-	-	-	-
+ Transfers	-	-	-	-	-	-	-	-
+ Stock Change	-	-	-	-	- 1	3	-	-
- Statistical Difference	-	-	-	-	2	2	-	- 12 e
Total Consumption	-	-	-	-	**20**	**18**	**16**	**16 e**
Closing Stock Level	-	-	-	-	*4*	*1*	*1*	*1 e*
GASOLINE TYPE JET FUEL								
Refinery Output	-	-	-	-	-	-	-	-
+ Imports	-	-	-	-	-	-	-	-
- Exports	-	-	-	-	-	-	-	-
+ Transfers	-	-	-	-	-	-	-	-
+ Stock Change	-	-	-	-	-	-		
- Statistical Difference	-	-	-	-	-	-	-	-
Total Consumption	-	-	-	-	-	-	-	-
Closing Stock Level	-	-	-	-	-	-	-	-
KEROSENE TYPE JET FUEL								
Refinery Output	**603**	**974**	**1 360**	**2 402**	**2 635**	**2 530**	**2 586**	**2 564 e**
+ Imports	-	-	-	-	-	-	-	-
- Exports	17	3	18	706	107	167	114	288 e
+ Transfers	-	-	-	-	-	-	-	-
+ Stock Change	-	-	-	-	- 23	31	- 31	- 2 e
- Statistical Difference	-	-	1	- 1	- 18	- 145	- 80	- 156 e
Total Consumption	**586**	**971**	**1 341**	**1 697**	**2 523**	**2 539**	**2 521**	**2 430 e**
Closing Stock Level	-	-	-	-	*149*	*118*	*149*	*151 e*

Source: Annual Oil Statistics.

MEXICO
Table 5A: Gas/Diesel Oil - Supply and Consumption

000 metric tons

	1973	1978	1980	1990	1999	2000	2001	2002
Refinery Output	**5 434**	**10 289**	**12 693**	**13 285**	**13 303**	**13 249**	**13 891**	**13 353 e**
+ Recycled Products	-	-	-	-	-	-	-	-
+ Imports	-	-	-	-	1 747	2 078	819	1 822 e
- Exports	1	6	63	1 600	671	220	442	395 e
- Intl. Marine Bunkers	-	-	-	115	959	1 023	898	644 e
+ Transfers	-	-	-	-	-	-	-	-
+ Stock Changes	23	- 23	- 134	122	165	- 229	34	28 e
- Statistical Differences	- 625	- 147	- 128	1	534	143	74	924 e
Total Consumption	**6 081**	**10 407**	**12 624**	**11 691**	**13 051**	**13 712**	**13 330**	**13 240 e**
Transformation	**410**	**1 074**	**1 057**	**360**	**405**	**593**	**454**	**..**
Public Electricity	410	1 074	1 057	360	405	593	454	..
Autoproducer Electricity	-	-	-	-	-	-	-	..
Public CHP Plants	-	-	-	-	-	-	-	..
Autoproducers of CHP	-	-	-	-	-	-	-	..
Public Heat Plants	-	-	-	-	-	-	-	..
Autoproducer Heat Plants	-	-	-	-	-	-	-	..
Petrochemical Industry	-	-	-	-	-	-	-	..
Other Transformation	-	-	-	-	-	-	-	..
Energy Sector	**934**	**651**	**1 369**	**1 014**	**536**	**532**	**600**	**..**
Refinery Fuel	934	651	1 369	1 014	536	532	600	..
Other Energy Sector	-	-	-	-	-	-	-	..
Final Consumption	**4 737**	**8 682**	**10 198**	**10 317**	**12 110**	**12 587**	**12 276**	**..**
Transport	**3 460**	**6 485**	**7 649**	**7 785**	**8 937**	**9 336**	**9 111**	**..**
Air	-	-	-	-	-	-	-	..
Road	-	-	-	7 138	8 432	8 804	8 609	..
Rail	-	-	-	647	505	532	502	..
Inland Waterways	-	-	-	-	-	-	-	..
Non-Specified	3 460	6 485	7 649	-	-	-	-	..
Industry	**646**	**851**	**1 014**	**1 040**	**1 256**	**1 292**	**1 272**	**..**
Iron and Steel	-	-	-	19	30	25	20	..
Chemical (incl. Pet.-Chem.)	-	-	-	39	144	156	140	..
Non-Ferrous Metals	-	-	-	12	-	-	-	..
Non-Metallic Minerals	-	-	-	37	43	6	10	..
Transport Equipment	-	-	-	-	-	-	-	..
Machinery	-	-	-	6	7	7	7	..
Mining and Quarrying	-	-	-	116	111	116	114	..
Food and Tobacco	-	-	-	248	79	67	64	..
Pulp, Paper and Printing	-	-	-	13	101	26	21	..
Wood and Wood Products	-	-	-	-	-	-	-	..
Construction	-	-	-	114	129	149	146	..
Textiles and Leather	-	-	-	-	-	-	-	..
Non-Specified	646	851	1 014	436	612	740	750	..
Other Sectors	**631**	**1 346**	**1 535**	**1 492**	**1 917**	**1 959**	**1 893**	**..**
Commerce and Public	55	248	268	31	81	84	82	..
Residential	-	-	-	-	-	-	-	..
Agriculture	576	1 098	1 267	1 461	1 836	1 875	1 811	..
Non-Specified	-	-	-	-	-	-	-	..
Memo: Non-Energy Use	-	-	-	-	-	-	-	..
Chem/Petchem. Feedst.	-	-	-	-	-	-	-	..
Other	-	-	-	-	-	-	-	..
Closing Stock Level	-	-	-	-	*683*	*912*	*878*	*850 e*

Source: Annual Oil Statistics.

MEXICO
Table 5B: Fuel Oil (Residual) - Supply and Consumption

000 metric tons

	1973	1978	1980	1990	1999	2000	2001	2002
Refinery Output	**9 084**	**14 762**	**18 735**	**26 394**	**27 734**	**24 116**	**24 814**	**26 156 e**
+ Recycled Products	-	-	-	-	-	-	-	-
+ Imports	638	1 078	-	3 141	5 296	6 590	4 846	1 005 e
- Exports	842	-	1 752	465	2 413	3	221	1 363 e
- Intl. Marine Bunkers	113	304	325	542	210	312	197	217 e
+ Transfers	-	-	-	-	-	-	-	-
+ Stock Changes	- 38	92	48	- 30	133	- 94	73	- 128 e
- Statistical Differences	-	- 4	1	- 1	942	221	457	362 e
Total Consumption	**8 729**	**15 632**	**16 705**	**28 499**	**29 598**	**30 076**	**28 858**	**25 091 e**
Transformation	**3 676**	**7 513**	**9 051**	**16 405**	**22 082**	**23 268**	**22 308**	**..**
Public Electricity	3 676	7 513	9 051	16 405	22 082	23 268	22 308	..
Autoproducer Electricity	-	-	-	-	-	-	-	..
Public CHP Plants	-	-	-	-	-	-	-	..
Autoproducers of CHP	-	-	-	-	-	-	-	..
Public Heat Plants	-	-	-	-	-	-	-	..
Autoproducer Heat Plants	-	-	-	-	-	-	-	..
Petrochemical Industry	-	-	-	-	-	-	-	..
Other Transformation	-	-	-	-	-	-	-	..
Energy Sector	**1 944**	**2 060**	**2 266**	**3 479**	**2 478**	**2 243**	**2 045**	**..**
Refinery Fuel	1 944	2 060	2 266	3 479	2 478	2 243	2 045	..
Other Energy Sector	-	-	-	-	-	-	-	..
Final Consumption	**3 109**	**6 059**	**5 388**	**8 615**	**5 038**	**4 565**	**4 505**	**..**
Transport	**-**	**-**	**-**	**-**	**-**	**-**	**-**	**..**
Air	-	-	-	-	-	-	-	..
Road	-	-	-	-	-	-	-	..
Rail	-	-	-	-	-	-	-	..
Inland Waterways	-	-	-	-	-	-	-	..
Non-Specified	-	-	-	-	-	-	-	..
Industry	**2 666**	**5 553**	**4 777**	**7 808**	**5 038**	**4 565**	**4 505**	**..**
Iron and Steel	-	-	-	692	518	403	320	..
Chemical (incl. Pet.-Chem.)	-	-	-	1 278	1 118	944	797	..
Non-Ferrous Metals	-	-	-	-	-	-	-	..
Non-Metallic Minerals	-	-	-	2 119	1 783	1 816	1 539	..
Transport Equipment	-	-	-	-	-	-	-	..
Machinery		-	-	1	-	-	-	..
Mining and Quarrying	-	-	-	198	140	170	203	..
Food and Tobacco	-	-	-	1 171	1 027	867	891	..
Pulp, Paper and Printing	-	-	-	814	321	365	360	..
Wood and Wood Products	-	-	-	-	-	-	-	..
Construction	-	-	-	-	-	-	-	..
Textiles and Leather	-	-	-	-	-	-	-	..
Non-Specified	2 666	5 553	4 777	1 535	131	-	395	..
Other Sectors	**443**	**506**	**611**	**807**	**-**	**-**	**-**	**..**
Commerce and Public	443	506	611	807	-	-	-	..
Residential	-	-	-	-	-	-	-	..
Agriculture	-	-	-	-	-	-	-	..
Non-Specified	-	-	-	-	-	-	-	..
Memo: Non-Energy Use	-	-	-	-	-	-	-	..
Chem/Petchem. Feedst.	-	-	-	-	-	-	-	..
Other	-	-	-	-	-	-	-	..
Closing Stock Level	-	-	-	-	*608*	*702*	*629*	*757 e*

Source: Annual Oil Statistics.

MEXICO
Table 5C: Total Products - Supply and Consumption

000 metric tons

	1973	1978	1980	1990	1999	2000	2001	2002
Refinery Output	**25 858**	**40 308**	**50 432**	**65 927**	**67 718**	**64 176**	**63 886**	**66 153 e**
+ Recycled Products	-	-	-	-	-	-	-	-
+ Imports	3 279	1 715	888	5 462	15 551	19 173	16 142	13 498 e
- Exports	1 426	83	2 404	4 749	6 664	4 658	4 238	4 377 e
- Intl. Marine Bunkers	113	304	325	657	1 169	1 335	1 095	861 e
+ Transfers	2 378	3 770	6 333	12 407	12 637	12 741	12 723	11 988 e
+ Stock Changes	12	48	- 311	127	511	- 431	40	- 254 e
- Statistical Differences	- 657	- 195	- 180	- 326	2 233	65	27	716 e
Total Consumption	**30 645**	**45 649**	**54 793**	**78 843**	**86 351**	**89 601**	**87 431**	**85 431 e**
Transformation	**4 086**	**8 587**	**10 108**	**16 765**	**22 487**	**23 861**	**22 762**	**..**
Public Electricity	4 086	8 587	10 108	16 765	22 487	23 861	22 762	..
Autoproducer Electricity	-	-	-	-	-	-	-	..
Public CHP Plants	-	-	-	-	-	-	-	..
Autoproducers of CHP	-	-	-	-	-	-	-	..
Public Heat Plants	-	-	-	-	-	-	-	..
Autoproducer Heat Plants	-	-	-	-	-	-	-	..
Petrochemical Industry	-	-	-	-	-	-	-	..
Other Transformation	-	-	-	-	-	-	-	..
Energy Sector	**4 684**	**3 849**	**5 244**	**8 111**	**8 126**	**7 621**	**6 814**	**..**
Refinery Fuel	4 684	3 849	5 244	8 102	8 125	7 621	6 814	..
Other Energy Sector	-	-	-	9	1	-	-	..
Final Consumption	**21 875**	**33 213**	**39 441**	**53 967**	**55 738**	**58 119**	**57 855**	**..**
Transport	**12 309**	**18 074**	**22 983**	**29 549**	**33 592**	**35 208**	**35 403**	**..**
Air	586	971	1 341	1 736	2 564	2 576	2 554	..
Road	8 263	10 618	13 993	27 166	30 523	32 100	32 347	..
Rail	-	-	-	647	505	532	502	..
Inland Waterways	-	-	-	-	-	-	-	..
Non-Specified	3 460	6 485	7 649	-	-	-	-	..
Industry	**5 385**	**9 215**	**9 163**	**15 623**	**12 757**	**12 560**	**12 318**	**..**
Iron and Steel	-	-	-	739	567	428	340	..
Chemical (incl. Pet.-Chem.)	612	994	1 103	5 379	5 512	5 991	5 706	..
Non-Ferrous Metals	-	-	-	12	7	1	1	..
Non-Metallic Minerals	-	-	-	2 157	1 830	1 824	1 551	..
Transport Equipment	-	-	-	-	-	-	-	..
Machinery	-	-	-	32	73	44	42	..
Mining and Quarrying	-	-	-	327	346	351	379	..
Food and Tobacco	-	-	-	1 435	1 138	961	985	..
Pulp, Paper and Printing	-	-	-	832	433	417	388	..
Wood and Wood Products	-	-	-	-	-	-	-	..
Construction	-	-	-	114	129	149	146	..
Textiles and Leather	-	-	-	-	-	-	-	..
Non-Specified	4 773	8 221	8 060	4 596	2 722	2 394	2 780	..
Other Sectors	**4 181**	**5 924**	**7 295**	**8 795**	**9 389**	**10 351**	**10 134**	**..**
Commerce and Public	498	754	879	1 288	1 281	1 540	1 575	..
Residential	746	904	989	5 849	6 090	6 762	6 541	..
Agriculture	1 152	1 713	1 991	1 658	2 018	2 049	2 018	..
Non-Specified	1 785	2 553	3 436	-	-	-	-	..
Memo: Non-Energy Use	*1 727*	*2 471*	*2 996*	*6 382*	*5 642*	*5 709*	*5 600*	..
Chem/Petchem. Feedst.	*612*	*994*	*1 103*	*4 058*	*4 221*	*4 873*	*4 754*	..
Other	*1 115*	*1 477*	*1 893*	*2 324*	*1 421*	*836*	*846*	..
Closing Stock Level	-	-	-	-	*2 932*	*3 363*	*3 323*	*3 577 e*

Source: Annual Oil Statistics.

MEXICO
Table 6A: Supply and Consumption

000 metric tons

	1980	2000	2001	2002	1980	2000	2001	2002
	LPG AND ETHANE				**REFINERY GAS**			
Refinery Output	**1 321**	**782**	**870**	**870 e**	**431**	**2 153**	**2 001**	**1 918 e**
+ Recycled Products	-	-	-	-	-	-	-	-
+ Imports	576	3 965	3 254	3 230 e	-	-	-	-
- Exports	496	174	98	13 e	-	-	-	-
- Intl. Marine Bunkers	-	-	-	-	-	-	-	-
+ Transfers	3 486	9 643	9 499	8 876 e	-	-	-	-
+ Stock Change	- 39	- 78	7	36 e	-	-	-	-
- Statistical Difference	135	294	35	- 337 e	-	-	-	-
Total Consumption	**4 713**	**13 844**	**13 497**	**13 336 e**	**431**	**2 153**	**2 001**	**1 918 e**
of which: Ethane	*654*	*3 244*	*3 058*	*2 640 e*	-	-	-	-
Transformation	-	-	-	..	-	-	-	..
Energy	274	222	240	..	431	2 153	2 001	..
Transport	139	1 040	1 078	..	-	-	-	..
Industry	845	4 222	3 975	..	-	-	-	..
Other	3 455	8 360	8 204	..	-	-	-	..
Memo: Non-Energy Use	*654*	*3 263*	*3 083*	..	-	-	-	..
Closing Stock Level	-	*167*	*160*	*124 e*	-	-	-	-
	NAPHTHA				**OTHER KEROSENE**			
Refinery Output	**519**	**850**	**833**	**874 e**	**2 011**	**54**	**68**	**64 e**
+ Recycled Products	-	-	-	-	-	-	-	-
+ Imports	-	-	127	157 e	-	-	-	-
- Exports	-	-	-	-	-	-	-	-
- Intl. Marine Bunkers	-	-	-	-	-	-	-	-
+ Transfers	-	-	-	-	-	-	-	-
+ Stock Change	-	-	-	-	- 6	4	7	5 e
- Statistical Difference	- 1	-	-	-	- 66	- 10	- 11	4 e
Total Consumption	**520**	**850**	**960**	**1 031 e**	**2 071**	**68**	**86**	**65 e**
Transformation	-	-	-	..	-	-	-	..
Energy	-	-	-	..	187	1	-	..
Transport	-	-	-	..	-	-	-	..
Industry	520	850	960	..	191	35	49	..
Other	-	-	-	..	1 693	32	37	..
Memo: Non-Energy Use	*519*	*850*	*960*	..	*7*	-	-	..
Closing Stock Level	-	-	-	-	-	*13*	*6*	*1 e*
	WHITE SPIRIT + SBP				**LUBRICANTS**			
Refinery Output	**-**	**-**	**-**	**..**	**-**	**-**	**-**	**..**
+ Recycled Products	-	-	-	..	-	-	-	..
+ Imports	-	-	-	..	-	-	-	..
- Exports	-	-	-	..	-	-	-	..
- Intl. Marine Bunkers	-	-	-	..	-	-	-	..
+ Transfers	-	-	-	..	-	-	-	..
+ Stock Change	-	-	-	..	-	-	-	..
- Statistical Difference	-	-	-	..	-	-	-	..
Total Consumption	**-**	**-**	**-**	**..**	**-**	**-**	**-**	**..**
Transformation	-	-	-	..	-	-	-	..
Energy	-	-	-	..	-	-	-	..
Transport	-	-	-	..	-	-	-	..
Industry	-	-	-	..	-	-	-	..
Other	-	-	-	..	-	-	-	..
Memo: Non-Energy Use	-	-	-	..	-	-	-	..
Closing Stock Level	-	-	-	..	-	-	-	..

Source: Annual Oil Statistics.

MEXICO
Table 6B: Supply and Consumption

000 metric tons

	1980	2000	2001	2002	1980	2000	2001	2002
	BITUMEN				PARAFFIN WAXES			
Refinery Output	-	-	-	..	-	-	-	..
+ Recycled Products	-	-	-	..	-	-	-	..
+ Imports	-	-	-	..	-	-	-	..
- Exports	-	-	-	..	-	-	-	..
- Intl. Marine Bunkers	-	-	-	..	-	-	-	..
+ Transfers	-	-	-	..	-	-	-	..
+ Stock Change	-	-	-	..	-	-	-	..
- Statistical Difference	-	-	-	..	-	-	-	..
Total Consumption	-	-	-	..	-	-	-	..
Transformation	-	-	-	..	-	-	-	..
Energy	-	-	-	..	-	-	-	..
Transport	-	-	-	..	-	-	-	..
Industry	-	-	-	..	-	-	-	..
Other	-	-	-	..	-	-	-	..
Memo: Non-Energy Use	-	-	-	..	-	-	-	..
Closing Stock Level	-	-	-	..	-	-	-	..
	PETROLEUM COKE				OTHER PRODUCTS			
Refinery Output	**26**	**1 475**	**1 363**	**2 453 e**	**1 721**	**3 250**	**2 286**	**2 285 e**
+ Recycled Products	-	-	-	-	-	-	-	-
+ Imports	-	-	-	-	285	-	-	-
- Exports	-	-	-	-	17	1 496	623	252 e
- Intl. Marine Bunkers	-	-	-	-	-	-	-	-
+ Transfers	-	-	-	-	-	-	-	-
+ Stock Change	-	-	-	-	- 17	39	- 10	- 29 e
- Statistical Difference	-	-	-	-	- 111	45	- 12	297 e
Total Consumption	**26**	**1 475**	**1 363**	**2 453 e**	**2 083**	**1 748**	**1 665**	**1 707 e**
Transformation	-	-	-	..	-	-	-	..
Energy	26	1 475	1 363	..	267	152	108	..
Transport	-	-	-	..	-	-	-	..
Industry	-	-	-	..	1 816	1 596	1 557	..
Other	-	-	-	..	-	-	-	..
Memo: Non-Energy Use	-	-	-	..	*1 816*	*1 596*	*1 557*	..
Closing Stock Level	-	-	-	-	-	*93*	*103*	*132 e*

Source: Annual Oil Statistics.

MEXICO
Table 7B: Imports by Country of Origin

000 metric tons

	2000	2001	2002	2000	2001	2002	2000	2001	2002
	Total Products			**Liquefied Petroleum Gases**			**Naphtha**		
Australia	-	-	-	-	-	-	-	-	-
Belgium	35	-	-	-	-	-	-	-	-
Canada	-	-	-	-	-	-	-	-	-
Czech Republic	-	-	-	-	-	-	-	-	-
Denmark	-	-	-	-	-	-	-	-	-
Finland	-	-	-	-	-	-	-	-	-
France	108	35	-	-	-	-	-	-	-
Germany	-	-	-	-	-	-	-	-	-
Greece	-	-	-	-	-	-	-	-	-
Hungary	-	-	-	-	-	-	-	-	-
Ireland	-	-	-	-	-	-	-	-	-
Italy	46	-	-	-	-	-	-	-	-
Japan	-	-	-	-	-	-	-	-	-
Korea	-	62	-	-	-	-	-	-	-
Mexico	-	-	-	-	-	-	-	-	-
Netherlands	-	89	151 e	-	-	-	-	-	-
New Zealand	-	-	-	-	-	-	-	-	-
Norway	-	150	71 e	-	150	71 e	-	-	-
Poland	-	-	-	-	-	-	-	-	-
Spain	-	42	31 e	-	-	-	-	-	-
Sweden	-	-	-	-	-	-	-	-	-
Turkey	-	-	-	-	-	-	-	-	-
United Kingdom	-	-	37 e	-	-	3 e	-	-	-
United States	15 481	12 960	10 643 e	2 047	2 310	2 099 e	-	-	-
Other	-	57	-	-	-	-	-	-	-
OECD	**15 670**	**13 395**	**10 933 e**	**2 047**	**2 460**	**2 173 e**	**-**	**-**	**-**
Argentina	-	-	-	-	-	-	-	-	-
Brazil	-	-	-	-	-	-	-	-	-
Colombia	233	128	-	-	-	-	-	-	-
Ecuador	190	323	-	-	-	-	-	-	-
Netherlands Antilles	93	-	128 e	-	-	-	-	-	-
Peru	147	198	47 e	-	-	-	-	-	-
Trinidad and Tobago	-	-	-	-	-	-	-	-	-
Venezuela	-	383	252 e	-	34	-	-	127	136 e
Other	427	351	316 e	-	-	-	-	-	-
Latin America	**1 090**	**1 383**	**743 e**	**-**	**34**	**-**	**-**	**127**	**136 e**
Romania	-	-	31 e	-	-	-	-	-	-
Other	-	-	-	-	-	-	-	-	-
Non-OECD Europe	**-**	**-**	**31 e**	**-**	**-**	**-**	**-**	**-**	**-**
FSU	**-**	**-**	**-**	**-**	**-**	**-**	**-**	**-**	**-**
Iran	-	-	-	-	-	-	-	-	-
Iraq	-	-	-	-	-	-	-	-	-
Kuwait	-	-	-	-	-	-	-	-	-
Oman	-	-	-	-	-	-	-	-	-
Qatar	-	-	-	-	-	-	-	-	-
Saudi Arabia	707	703	639 e	246	218	18 e	-	-	
United Arab Emirates	-	-	-	-	-	-	-	-	-
Other	-	-	32 e	-	-	-	-	-	-
Middle East	**707**	**703**	**671 e**	**246**	**218**	**18 e**	**-**	**-**	**-**
India	-	-	-	-	-	-	-	-	-
Indonesia	-	-	-	-	-	-	-	-	-
Singapore	34	58	-	-	-	-	-	-	-
Other	-	30	-	-	-	-	-	-	-
Asia	**34**	**88**	**-**	**-**	**-**	**-**	**-**	**-**	**-**
China	**-**	**31**	**29 e**	**-**	**-**	**-**	**-**	**-**	**-**
Algeria	1 672	542	1 039 e	1 672	542	1 039 e	-	-	-
Angola/Cabinada	-	-	-	-	-	-	-	-	-
Cameroon	-	-	-	-	-	-	-	-	-
Egypt	-	-	29 e	-	-	-	-	-	-
Gabon	-	-	-	-	-	-	-	-	-
Libya	-	-	-	-	-	-	-	-	-
Nigeria	-	-	-	-	-	-	-	-	-
Tunisia	-	-	-	-	-	-	-	-	-
Other	-	-	21 e	-	-	-	-	-	21 e
Africa	**1 672**	**542**	**1 089 e**	**1 672**	**542**	**1 039 e**	**-**	**-**	**21 e**
Non-spec./Other	**-**	**-**	**2 e**	**-**	**-**	**-**	**-**	**-**	**-**
Total Imports	**19 173**	**16 142**	**13 498 e**	**3 965**	**3 254**	**3 230 e**	**-**	**127**	**157 e**
Memo: Total OPEC	*2 379*	*1 628*	*1 930 e*	*1 918*	*794*	*1 057 e*	*-*	*127*	*136 e*

Source: Annual Oil Statistics.

MEXICO
Table 7C: Imports by Country of Origin

000 metric tons

	2000	2001	2002	2000	2001	2002	2000	2001	2002
	Total Gasoline			Total Kerosene			Gas/Diesel		
Australia	-	-	-	-	-	-	-	-	-
Belgium	35	-	-	-	-	-	-	-	-
Canada	-	-	-	-	-	-	-	-	-
Czech Republic	-	-	-	-	-	-	-	-	-
Denmark	-	-	-	-	-	-	-	-	-
Finland	-	-	-	-	-	-	-	-	-
France	108	35	-	-	-	-	-	-	-
Germany	-	-	-	-	-	-	-	-	-
Greece	-	-	-	-	-	-	-	-	-
Hungary	-	-	-	-	-	-	-	-	-
Ireland	-	-	-	-	-	-	-	-	-
Italy	-	-	-	-	-	-	-	-	-
Japan	-	-	-	-	-	-	-	-	-
Korea	-	62	-	-	-	-	-	-	-
Mexico	-	-	-	-	-	-	-	-	-
Netherlands	-	89	151 e	-	-	-	-	-	-
New Zealand	-	-	-	-	-	-	-	-	-
Norway	-	-	-	-	-	-	-	-	-
Poland	-	-	-	-	-	-	-	-	-
Spain	-	-	31 e	-	-	-	-	-	-
Sweden	-	-	-	-	-	-	-	-	-
Turkey	-	-	-	-	-	-	-	-	-
United Kingdom	-	-	34 e	-	-	-	-	-	-
United States	5 902	6 004	6 030 e	-	-	-	2 078	819	1 603 e
Other	-	57	-	-	-	-	-	-	-
OECD	**6 045**	**6 247**	**6 246 e**	**-**	**-**	**-**	**2 078**	**819**	**1 603 e**
Argentina	-	-	-	-	-	-	-	-	-
Brazil	-	-	-	-	-	-	-	-	-
Colombia	-	-	-	-	-	-	-	-	-
Ecuador	-	-	-	-	-	-	-	-	-
Netherlands Antilles	-	-	128 e	-	-	-	-	-	-
Peru	-	-	-	-	-	-	-	-	-
Trinidad and Tobago	-	-	-	-	-	-	-	-	-
Venezuela	-	222	34 e	-	-	-	-	-	82 e
Other	-	23	165 e	-	-	-	-	-	106 e
Latin America	**-**	**245**	**327 e**	**-**	**-**	**-**	**-**	**-**	**188 e**
Romania	-	-	-	-	-	-	-	-	31 e
Other	-	-	-	-	-	-	-	-	-
Non-OECD Europe	**-**	**-**	**-**	**-**	**-**	**-**	**-**	**-**	**31 e**
FSU	**-**	**-**	**-**	**-**	**-**	**-**	**-**	**-**	**-**
Iran	-	-	-	-	-	-	-	-	-
Iraq	-	-	-	-	-	-	-	-	-
Kuwait	-	-	-	-	-	-	-	-	-
Oman	-	-	-	-	-	-	-	-	-
Qatar	-	-	-	-	-	-	-	-	-
Saudi Arabia	461	485	621 e	-	-	-	-	-	-
United Arab Emirates	-	-	-	-	-	-	-	-	-
Other	-	-	32 e	-	-	-	-	-	-
Middle East	**461**	**485**	**653 e**	**-**	**-**	**-**	**-**	**-**	**-**
India	-	-	-	-	-	-	-	-	-
Indonesia	-	-	-	-	-	-	-	-	-
Singapore	34	58	-	-	-	-	-	-	-
Other	-	30	-	-	-	-	-	-	-
Asia	**34**	**88**	**-**	**-**	**-**	**-**	**-**	**-**	**-**
China	**-**	**31**	**29 e**	**-**	**-**	**-**	**-**	**-**	**-**
Algeria	-	-	-	-	-	-	-	-	-
Angola/Cabinada	-	-	-	-	-	-	-	-	-
Cameroon	-	-	-	-	-	-	-	-	-
Egypt	-	-	29 e	-	-	-	-	-	-
Gabon	-	-	-	-	-	-	-	-	-
Libya	-	-	-	-	-	-	-	-	-
Nigeria	-	-	-	-	-	-	-	-	-
Tunisia	-	-	-	-	-	-	-	-	-
Other	-	-	-	-	-	-	-	-	-
Africa	**-**	**-**	**29 e**	**-**	**-**	**-**	**-**	**-**	**-**
Non-spec./Other	**-**	**-**	**-**	**-**	**-**	**-**	**-**	**-**	**-**
Total Imports	**6 540**	**7 096**	**7 284 e**	**-**	**-**	**-**	**2 078**	**819**	**1 822 e**
Memo: Total OPEC	*461*	*707*	*655 e*	*-*	*-*	*-*	*-*	*-*	*82 e*

Source: Annual Oil Statistics.

MEXICO
Table 7D: Imports by Country of Origin

000 metric tons

	2000	2001	2002	2000	2001	2002	2000	2001	2002
	Fuel Oil (Residual)			**Petroleum Coke**			**Other Products**		
Australia	-	-	-	-	-	-	-	-	-
Belgium	-	-	-	-	-	-	-	-	-
Canada	-	-	-	-	-	-	-	-	-
Czech Republic	-	-	-	-	-	-	-	-	-
Denmark	-	-	-	-	-	-	-	-	-
Finland	-	-	-	-	-	-	-	-	-
France	-	-	-	-	-	-	-	-	-
Germany	-	-	-	-	-	-	-	-	-
Greece	-	-	-	-	-	-	-	-	-
Hungary	-	-	-	-	-	-	-	-	-
Ireland	-	-	-	-	-	-	-	-	-
Italy	46	-	-	-	-	-	-	-	-
Japan	-	-	-	-	-	-	-	-	-
Korea	-	-	-	-	-	-	-	-	-
Mexico	-	-	-	-	-	-	-	-	-
Netherlands	-	-	-	-	-	-	-	-	-
New Zealand	-	-	-	-	-	-	-	-	-
Norway	-	-	-	-	-	-	-	-	-
Poland	-	-	-	-	-	-	-	-	-
Spain	-	42	-	-	-	-	-	-	-
Sweden	-	-	-	-	-	-	-	-	-
Turkey	-	-	-	-	-	-	-	-	-
United Kingdom	-	-	-	-	-	-	-	-	-
United States	5 454	3 827	911 e	-	-	-	-	-	-
Other	-	-	-	-	-	-	-	-	-
OECD	**5 500**	**3 869**	**911 e**	**-**	**-**	**-**	**-**	**-**	**-**
Argentina	-	-	-	-	-	-	-	-	-
Brazil	-	-	-	-	-	-	-	-	-
Colombia	233	128	-	-	-	-	-	-	-
Ecuador	190	323	-	-	-	-	-	-	-
Netherlands Antilles	93	-	-	-	-	-	-	-	-
Peru	147	198	47 e	-	-	-	-	-	-
Trinidad and Tobago	-	-	-	-	-	-	-	-	-
Venezuela	-	-	-	-	-	-	-	-	-
Other	427	328	45 e	-	-	-	-	-	-
Latin America	**1 090**	**977**	**92 e**	**-**	**-**	**-**	**-**	**-**	**-**
Romania	-	-	-	-	-	-	-	-	-
Other	-	-	-	-	-	-	-	-	-
Non-OECD Europe	**-**	**-**	**-**	**-**	**-**	**-**	**-**	**-**	**-**
FSU	**-**	**-**	**-**	**-**	**-**	**-**	**-**	**-**	**-**
Iran	-	-	-	-	-	-	-	-	-
Iraq	-	-	-	-	-	-	-	-	-
Kuwait	-	-	-	-	-	-	-	-	-
Oman	-	-	-	-	-	-	-	-	-
Qatar	-	-	-	-	-	-	-	-	-
Saudi Arabia	-	-	-	-	-	-	-	-	-
United Arab Emirates	-	-	-	-	-	-	-	-	-
Other	-	-	-	-	-	-	-	-	-
Middle East	**-**	**-**	**-**	**-**	**-**	**-**	**-**	**-**	**-**
India	-	-	-	-	-	-	-	-	-
Indonesia	-	-	-	-	-	-	-	-	-
Singapore	-	-	-	-	-	-	-	-	-
Other	-	-	-	-	-	-	-	-	-
Asia	**-**	**-**	**-**	**-**	**-**	**-**	**-**	**-**	**-**
China	**-**	**-**	**-**	**-**	**-**	**-**	**-**	**-**	**-**
Algeria	-	-	-	-	-	-	-	-	-
Angola/Cabinada	-	-	-	-	-	-	-	-	-
Cameroon	-	-	-	-	-	-	-	-	-
Egypt	-	-	-	-	-	-	-	-	-
Gabon	-	-	-	-	-	-	-	-	-
Libya	-	-	-	-	-	-	-	-	-
Nigeria	-	-	-	-	-	-	-	-	-
Tunisia	-	-	-	-	-	-	-	-	-
Other	-	-	-	-	-	-	-	-	-
Africa	**-**	**-**	**-**	**-**	**-**	**-**	**-**	**-**	**-**
Non-spec./Other	**-**	**-**	**2 e**	**-**	**-**	**-**	**-**	**-**	**-**
Total Imports	**6 590**	**4 846**	**1 005 e**	**-**	**-**	**-**	**-**	**-**	**-**
Memo: Total OPEC	-	-	-	-	-	-	-	-	-

Source: Annual Oil Statistics.

MEXICO
Table 8A: Exports by Country of Destination

000 metric tons

	Crude Oil 2000	Crude Oil 2001	Crude Oil 2002	Natural Gas Liquids 2000	Natural Gas Liquids 2001	Natural Gas Liquids 2002	Feedstocks and Other 2000	Feedstocks and Other 2001	Feedstocks and Other 2002
Australia	-	-	-	-	-	-	-	-	-
Austria	-	-	-	-	-	-	-	-	-
Belgium	-	-	-	-	-	-	-	-	-
Canada	1 480	1 326	1 141 e	-	-	-	-	-	-
Denmark	-	-	-	-	-	-	-	-	-
Finland	-	-	-	-	-	-	-	-	-
France	-	-	-	-	-	-	-	-	-
Germany	-	-	-	-	-	-	-	-	-
Greece	-	-	-	-	-	-	-	-	-
Ireland	-	-	-	-	-	-	-	-	-
Italy	205	183	81 e	-	-	-	-	-	-
Japan	1 804	863	510 e	-	-	-	-	-	-
Luxembourg	-	-	-	-	-	-	-	-	-
Mexico	-	-	-	-	-	-	-	-	-
Netherlands	73	-	147 e	-	-	-	-	-	-
Norway	-	-	-	-	-	-	-	-	-
Portugal	936	811	826 e	-	-	-	-	-	-
Spain	7 645	7 709	7 823 e	-	-	-	-	-	-
Sweden	-	-	-	-	-	-	-	-	-
Switzerland	-	-	-	-	-	-	-	-	-
Turkey	-	-	-	-	-	-	-	-	-
United Kingdom	934	764	827 e	-	-	-	-	-	-
United States	69 894	70 291	74 905 e	-	-	-	-	-	-
Other OECD	-	-	-	-	-	-	-	-	-
OECD	**82 971**	**81 947**	**86 260 e**	**-**	**-**	**-**	**-**	**-**	**-**
Africa	-	344	-	-	-	-	-	-	-
Latin America	8 117	9 148	6 254 e	-	-	-	-	-	-
Asia	264	1 111	1 973 e	-	-	-	-	-	-
China	-	-	-	-	-	-	-	-	-
Non-OECD Europe	-	-	-	-	-	-	-	-	-
Former Soviet Union	-	-	-	-	-	-	-	-	-
Middle East	259	190	259 e	-	-	-	-	-	-
Non-specified/Other	-	-	-	-	-	-	-	-	-
Non-OECD	**8 640**	**10 793**	**8 486 e**	**-**	**-**	**-**	**-**	**-**	**-**
Total Exports	**91 611**	**92 740**	**94 746 e**	**-**	**-**	**-**	**-**	**-**	**-**

	Total Products 2000	Total Products 2001	Total Products 2002	Liquefied Petroleum Gas 2000	Liquefied Petroleum Gas 2001	Liquefied Petroleum Gas 2002	Naphtha 2000	Naphtha 2001	Naphtha 2002
Australia	-	-	-	-	-	-	-	-	-
Austria	-	-	-	-	-	-	-	-	-
Belgium	-	-	-	-	-	-	-	-	-
Canada	-	-	27 e	-	-	-	-	-	-
Denmark	-	-	-	-	-	-	-	-	-
Finland	-	-	-	-	-	-	-	-	-
France	-	-	-	-	-	-	-	-	-
Germany	-	-	-	-	-	-	-	-	-
Greece	-	-	-	-	-	-	-	-	-
Ireland	-	-	-	-	-	-	-	-	-
Italy	-	-	-	-	-	-	-	-	-
Japan	-	-	28 e	-	-	-	-	-	-
Luxembourg	-	-	-	-	-	-	-	-	-
Mexico	-	-	-	-	-	-	-	-	-
Netherlands	-	77	-	-	-	-	-	-	-
Norway	-	-	-	-	-	-	-	-	-
Portugal	-	-	-	-	-	-	-	-	-
Spain	-	97	145 e	-	-	-	-	-	-
Sweden	-	-	-	-	-	-	-	-	-
Switzerland	-	-	-	-	-	-	-	-	-
Turkey	-	-	-	-	-	-	-	-	-
United Kingdom	-	-	-	-	-	-	-	-	-
United States	3 970	3 602	3 676 e	-	-	-	-	-	-
Other OECD	-	-	-	-	-	-	-	-	-
OECD	**3 970**	**3 776**	**3 876 e**	**-**	**-**	**-**	**-**	**-**	**-**
Africa	46	-	-	-	-	-	-	-	-
Latin America	642	462	363 e	174	98	13 e	-	-	-
Asia	-	-	138 e	-	-	-	-	-	-
China	-	-	-	-	-	-	-	-	-
Non-OECD Europe	-	-	-	-	-	-	-	-	-
Former Soviet Union	-	-	-	-	-	-	-	-	-
Middle East	-	-	-	-	-	-	-	-	-
Non-specified/Other	-	-	-	-	-	-	-	-	-
Non-OECD	**688**	**462**	**501 e**	**174**	**98**	**13 e**	**-**	**-**	**-**
Total Exports	**4 658**	**4 238**	**4 377 e**	**174**	**98**	**13 e**	**-**	**-**	**-**

Source: Annual Oil Statistics.

MEXICO
Table 8B: Exports by Country of Destination

000 metric tons

	Total Gasoline			Total Kerosene			Gas/Diesel Oil		
	2000	**2001**	**2002**	**2000**	**2001**	**2002**	**2000**	**2001**	**2002**
Australia	-	-	-	-	-	-	-	-	-
Austria	-	-	-	-	-	-	-	-	-
Belgium	-	-	-	-	-	-	-	-	-
Canada	-	-	-	-	-	27 e	-	-	-
Denmark	-	-	-	-	-	-	-	-	-
Finland	-	-	-	-	-	-	-	-	-
France	-	-	-	-	-	-	-	-	-
Germany	-	-	-	-	-	-	-	-	-
Greece	-	-	-	-	-	-	-	-	-
Ireland	-	-	-	-	-	-	-	-	-
Italy	-	-	-	-	-	-	-	-	-
Japan	-	-	28 e	-	-	-	-	-	-
Luxembourg	-	-	-	-	-	-	-	-	-
Mexico	-	-	-	-	-	-	-	-	-
Netherlands	-	-	-	-	-	-	-	10	-
Norway	-	-	-	-	-	-	-	-	-
Portugal	-	-	-	-	-	-	-	-	-
Spain	-	-	-	-	-	-	-	-	18 e
Sweden	-	-	-	-	-	-	-	-	-
Switzerland	-	-	-	-	-	-	-	-	-
Turkey	-	-	-	-	-	-	-	-	-
United Kingdom	-	-	-	-	-	-	-	-	-
United States	2 566	2 693	2 030 e	163	80	215 e	96	159	169 e
Other OECD	-	-	-	-	-	-	-	-	-
OECD	**2 566**	**2 693**	**2 058 e**	**163**	**80**	**242 e**	**96**	**169**	**187 e**
Africa	-	-	-	-	-	-	-	-	-
Latin America	32	47	8 e	4	34	46 e	124	273	200 e
Asia	-	-	-	-	-	-	-	-	8 e
China	-	-	-	-	-	-	-	-	-
Non-OECD Europe	-	-	-	-	-	-	-	-	-
Former Soviet Union	-	-	-	-	-	-	-	-	-
Middle East	-	-	-	-	-	-	-	-	-
Non-specified/Other	-	-	-	-	-	-	-	-	-
Non-OECD	**32**	**47**	**8 e**	**4**	**34**	**46 e**	**124**	**273**	**208 e**
Total Exports	**2 598**	**2 740**	**2 066 e**	**167**	**114**	**288 e**	**220**	**442**	**395 e**

	Fuel Oil (Residual)			Petroleum Coke			Other Products		
	2000	**2001**	**2002**	**2000**	**2001**	**2002**	**2000**	**2001**	**2002**
Australia	-	-	-	-	-	-	-	-	-
Austria	-	-	-	-	-	-	-	-	-
Belgium	-	-	-	-	-	-	-	-	-
Canada	-	-	-	-	-	-	-	-	-
Denmark	-	-	-	-	-	-	-	-	-
Finland	-	-	-	-	-	-	-	-	-
France	-	-	-	-	-	-	-	-	-
Germany	-	-	-	-	-	-	-	-	-
Greece	-	-	-	-	-	-	-	-	-
Ireland	-	-	-	-	-	-	-	-	-
Italy	-	-	-	-	-	-	-	-	-
Japan	-	-	-	-	-	-	-	-	-
Luxembourg	-	-	-	-	-	-	-	-	-
Mexico	-	-	-	-	-	-	-	-	-
Netherlands	-	67	-	-	-	-	-	-	-
Norway	-	-	-	-	-	-	-	-	-
Portugal	-	-	-	-	-	-	-	-	-
Spain	-	-	127 e	-	-	-	-	97	-
Sweden	-	-	-	-	-	-	-	-	-
Switzerland	-	-	-	-	-	-	-	-	-
Turkey	-	-	-	-	-	-	-	-	-
United Kingdom	-	-	-	-	-	-	-	-	-
United States	-	150	1 011 e	-	-	-	1 145	520	251 e
Other OECD	-	-	-	-	-	-	-	-	-
OECD	**-**	**217**	**1 138 e**	**-**	**-**	**-**	**1 145**	**617**	**251 e**
Africa	-	-	-	-	-	-	46	-	-
Latin America	3	4	95 e	-	-	-	305	6	1 e
Asia	-	-	130 e	-	-	-	-	-	-
China	-	-	-	-	-	-	-	-	-
Non-OECD Europe	-	-	-	-	-	-	-	-	-
Former Soviet Union	-	-	-	-	-	-	-	-	-
Middle East	-	-	-	-	-	-	-	-	-
Non-specified/Other	-	-	-	-	-	-	-	-	-
Non-OECD	**3**	**4**	**225 e**	**-**	**-**	**-**	**351**	**6**	**1 e**
Total Exports	**3**	**221**	**1 363 e**	**-**	**-**	**-**	**1 496**	**623**	**252 e**

Source: Annual Oil Statistics.

NETHERLANDS

Figure 1
FUEL SHARES OF TPES *

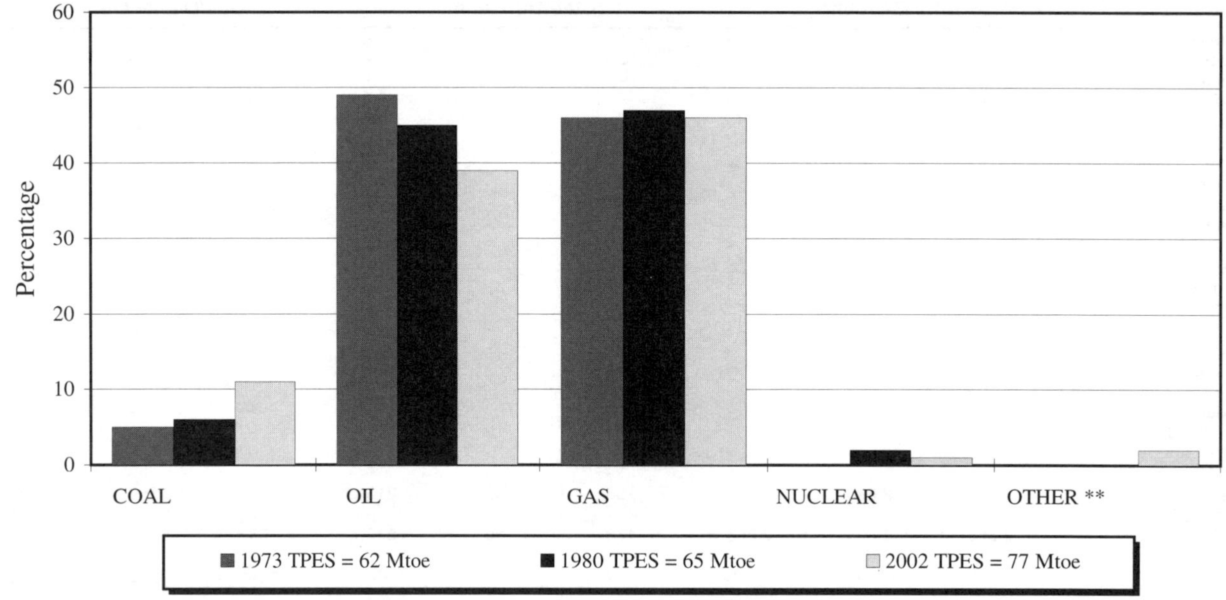

Source: Annual Energy Balances.

* Excluding electricity trade.

** Includes hydro, geothermal, solar, wind, combustible renewables and waste, etc.

Figure 2
OIL PRODUCTION AND CONSUMPTION

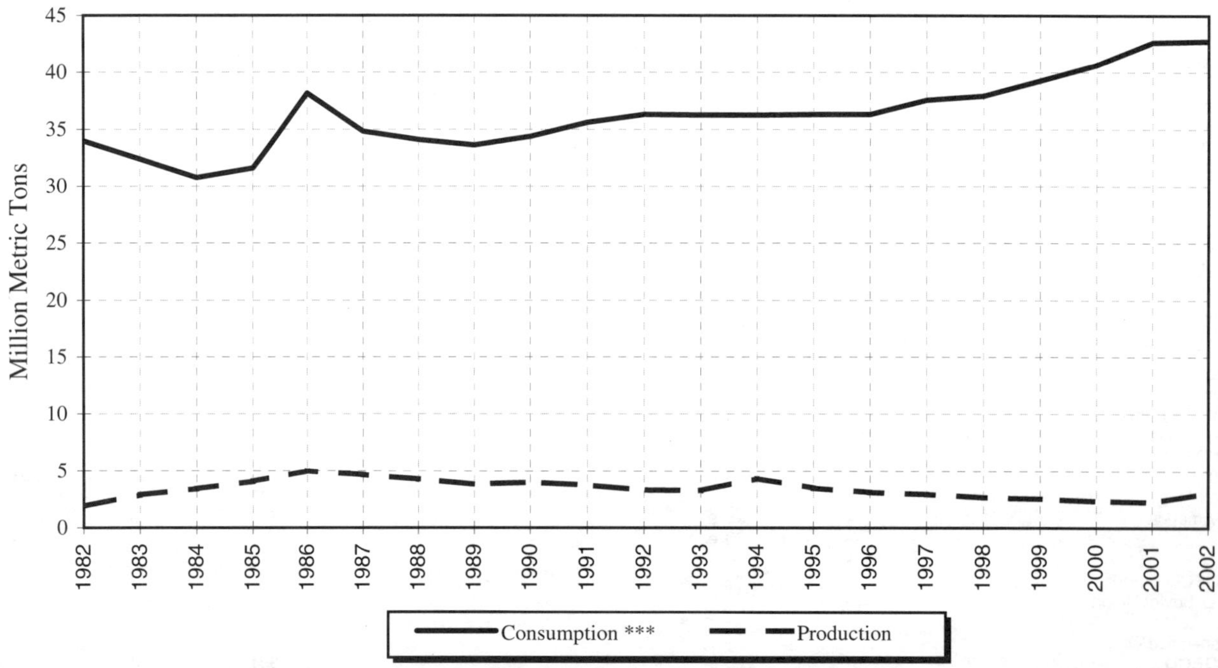

Source: Annual Oil Statistics.

*** Includes refinery fuel and international marine bunkers.

NETHERLANDS
Table 1: Oil Supply and Consumption - Summary Table

000 metric tons

	1973	1978	1980	1990	1999	2000	2001	2002
Indigenous Production	1 556	1 520	1 568	3 976	2 535	2 350	2 263	3 073 e
Crude Oil	..	1 402	1 280	3 533	1 594	1 443	1 370	2 232 e
NGL	..	118	288	443	941	907	893	841 e
Other	..	-	-	-	-	-	-	-
+ Net Imports	40 199	37 274	37 525	30 436	36 245	41 602	41 605	40 119 e
Crude Oil, NGL, Other	72 161	55 689	49 891	47 694	58 944	60 369	60 455	54 342 e
Petroleum Products	-31 962	-18 415	-12 366	-17 258	-22 699	-18 767	-18 850	-14 223 e
- Intl. Marine Bunkers	11 972	11 427	9 590	11 202	13 063	13 788	15 140	14 957 e
+ Stock Changes	- 511	151	- 720	301	1 359	-2 285	18	638 e
TOTAL SUPPLY	**29 272**	**27 518**	**28 783**	**23 511**	**27 076**	**27 879**	**28 746**	**28 873 e**
Refinery Intake	*73 505*	*60 803*	*57 537*	*68 287*	*77 173*	*81 294*	*82 598*	*79 828 e*
Refinery Output	*73 081*	*60 530*	*57 238*	*67 914*	*76 263*	*80 235*	*81 306*	*78 714 e*
LPG + Ethane	*966*	*1 231*	*1 632*	*2 640*	*3 871*	*3 684*	*3 974*	*4 663 e*
Naphtha	*6 594*	*4 074*	*3 646*	*9 794*	*9 122*	*11 303*	*12 116*	*8 655 e*
Kerosenes	*4 477*	*2 943*	*3 306*	*5 168*	*7 162*	*7 487*	*7 068*	*6 659 e*
Gasolines	*6 173*	*7 732*	*9 694*	*12 907*	*13 251*	*14 314*	*14 476*	*15 809 e*
Gas/Diesel Oil	*20 559*	*18 807*	*17 511*	*15 524*	*21 682*	*21 994*	*22 167*	*19 582 e*
Fuel Oil	*29 165*	*20 659*	*17 046*	*12 704*	*11 145*	*11 215*	*10 780*	*12 707 e*
Other Products	*5 147*	*5 084*	*4 403*	*9 177*	*10 030*	*10 238*	*10 725*	*10 639 e*
- Refinery Losses	424	273	299	373	910	1 059	1 292	1 114 e
+ Returns	-	2 304	4 435	31	46	50	50	50 e
- Statistical Differences	- 171	130	- 75	12	-	-	-	-
TOTAL CONSUMPTION	**29 019**	**29 419**	**32 994**	**23 157**	**26 212**	**26 870**	**27 504**	**27 809 e**
Transformation and Energy	**5 268**	**7 481**	**12 836**	**4 368**	**4 451**	**3 667**	**3 805**	..
Crude Oil	-	-	-	-	-	-	-	..
LPG + Ethane	-	110	461	20	1	3	1	..
Gas/Diesel Oil	26	480	1 042	25	17	11	10	..
Fuel Oil	3 602	4 023	7 075	1 077	339	276	347	..
Petroleum Coke	-	250	227	-	-	-	-	..
Other	1 640	2 618	4 031	3 246	4 094	3 377	3 447	..
Transport	**6 985**	**8 020**	**8 180**	**9 918**	**13 567**	**13 704**	**13 763**	..
Motor Gasoline	3 556	3 954	3 854	3 445	4 140	4 031	4 120	..
Gas/Diesel Oil	2 457	2 854	2 600	3 933	5 410	5 779	5 881	..
Subtotal	*6 013*	*6 808*	*6 454*	*7 378*	*9 550*	*9 810*	*10 001*	..
Aviation Fuels	865	855	943	1 567	3 304	3 260	3 168	..
Other	107	357	783	973	713	634	594	..
Industry	**9 925**	**9 989**	**9 008**	**7 575**	**7 193**	**8 596**	**9 049**	..
Naphtha	5 493	4 169	3 225	3 153	2 625	2 984	3 337	..
LPG + Ethane	201	378	772	1 766	1 323	1 465	1 316	..
Gas/Diesel Oil	417	1 406	1 634	386	217	193	168	..
Fuel Oil	2 110	1 248	1 016	178	48	41	44	..
Other	1 704	2 788	2 361	2 092	2 980	3 913	4 184	..
Other Sectors	**6 841**	**3 929**	**2 970**	**1 296**	**1 001**	**903**	**887**	..
Gas/Diesel Oil	4 429	2 712	2 147	937	720	683	608	..
Fuel Oil	992	381	304	32	15	2	27	..
Other	1 420	836	519	327	266	218	252	..
Memo: Non-Energy Use	*7 202*	*7 673*	*6 814 e*	*7 031*	*6 637*	*7 360*	*7 806*	..
Chem/Petchem. Feedst.	*5 494*	*6 314*	*5 526 e*	*5 959*	*5 351*	*6 046*	*6 353*	..
Other	*1 708*	*1 359*	*1 288 e*	*1 072*	*1 286*	*1 314*	*1 453*	..

Source: Annual Oil Statistics.

NETHERLANDS
Table 2: Supply of Crude Oil, NGL, Feedstocks and Other Hydrocarbons

000 metric tons

	1973	1978	1980	1990	1999	2000	2001	2002
CRUDE OIL								
Indigenous Production	..	1 402	1 280	3 533	1 594	1 443	1 370	2 232 e
+ Imports	..	54 737	49 748	45 398	53 951	54 667	54 253	46 837 e
- Exports	..	-	-	1 261	307	184	355	894 e
- Direct Use	..	-	-	-	-	-	-	-
+ Stock Change	..	342	- 924	125	- 211	- 848	- 526	129 e
- Statistical Difference	..	54	- 107	-	-	-	-	-
Refinery Intake	**..**	**56 427**	**50 211**	**47 795**	**55 027**	**55 078**	**54 742**	**48 304 e**
Closing Stock Level	..	*5 463*	*5 936*	*2 523*	*3 951*	*4 799*	*5 325*	*5 196 e*
NATURAL GAS LIQUIDS								
Indigenous Production	..	118	288	443	941	907	893	841 e
+ Imports	..	18	104	3 612	5 300	5 886	6 557	8 399 e
- Exports	..	-	23	55	-	-	-	-
- Direct Use	..	-	-	-	-	-	-	-
+ Stock Change	..	- 6	- 5	- 30	24	- 83	- 58	- 31 e
- Statistical Difference	..	-	-	-	-	-	-	-
Refinery Intake	**..**	**130**	**364**	**3 970**	**6 265**	**6 710**	**7 392**	**9 209 e**
Closing Stock Level	..	*23*	*70*	*380*	*174*	*257*	*315*	*346 e*
REFINERY FEEDSTOCKS								
Backflows	..	2 304	4 435	-	-	-	-	-
+ Transfers	..	959	2 465	16 522	15 881	19 506	20 464	22 315 e
+ Imports	..	934	62	-	-	-	-	-
- Exports	..	-	-	-	-	-	-	-
+ Stock Change	..	-	-	-	-	-	-	-
- Statistical Difference	..	- 49	-	-	-	-	-	-
Refinery Intake	**..**	**4 246**	**6 962**	**16 522**	**15 881**	**19 506**	**20 464**	**22 315 e**
Closing Stock Level	..	-	-	-	-	-	-	-
OTHER HYDROCARBONS								
Indigenous Production	..	-	-	-	-	-	-	-
+ Imports	..	-	-	-	-	-	-	-
- Exports	..	-	-	-	-	-	-	-
- Direct Use	..	-	-	-	-	-	-	-
+ Stock Change	..	-	-	-	-	-	-	-
- Statistical Difference	..	-	-	-	-	-	-	-
Refinery Intake	**..**	**-**	**-**	**-**	**-**	**-**	**-**	**-**
Closing Stock Level	..	-	-	-	-	-	-	-
CRUDE OIL, NGL, FEEDSTOCKS AND OTHER								
Indigenous Production	1 556	1 520	1 568	3 976	2 535	2 350	2 263	3 073 e
+ Backflows	-	2 304	4 435	-	-	-	-	-
+ Transfers	-	959	2 465	16 522	15 881	19 506	20 464	22 315 e
+ Imports	72 161	55 689	49 914	49 010	59 251	60 553	60 810	55 236 e
- Exports	-	-	23	1 316	307	184	355	894 e
- Direct Use	-	-	-	-	-	-	-	-
+ Stock Change	- 331	336	- 929	95	- 187	- 931	- 584	98 e
- Statistical Difference	- 119	5	- 107	-	-	-	-	-
Refinery Intake	**73 505**	**60 803**	**57 537**	**68 287**	**77 173**	**81 294**	**82 598**	**79 828 e**
Closing Stock Level	-	*5 486*	*6 006*	*2 903*	*4 125*	*5 056*	*5 640*	*5 542 e*

Source: Annual Oil Statistics.

NETHERLANDS

Table 3: Motor Gasoline - Supply

000 metric tons

	1973	1978	1980	1990	1999	2000	2001	2002
MOTOR GASOLINE								
Refinery Output	**5 776**	**7 331**	**9 140**	**12 843**	**13 190**	**14 263**	**14 423**	**15 767 e**
+ Imports	352	499	2 144	7 311	5 394	7 516	6 175	7 262 e
- Exports	2 453	3 491	6 550	13 326	11 226	12 703	11 402	12 281 e
+ Transfers	-	- 317	- 972	-3 735	-3 491	-4 502	-4 924	-7 051 e
+ Stock Change	- 119	- 43	111	352	273	- 543	- 151	475 e
- Statistical Difference	-	25	18	-	-	-	-	-
Total Consumption	**3 556**	**3 954**	**3 855**	**3 445**	**4 140**	**4 031**	**4 121**	**4 172 e**
Closing Stock Level	*-*	*650*	*767*	*1 366*	*986*	*1 529*	*1 680*	*1 205 e*

Table 4: Aviation Fuels - Supply

000 metric tons

	1973	1978	1980	1990	1999	2000	2001	2002
AVIATION GASOLINE								
Refinery Output	**175**	**130**	**87**	**64**	**61**	**51**	**53**	**42 e**
+ Imports	1	-	26	17	-	-	1	9 e
- Exports	174	117	120	75	56	43	49	45 e
+ Transfers	-	-	-	2	- 1	- 1	- 1	- 3 e
+ Stock Change	7	- 7	13	- 2	1	- 2	-	1 e
- Statistical Difference	-	- 1	- 1	-	-	-	-	-
Total Consumption	**9**	**7**	**7**	**6**	**5**	**5**	**4**	**4 e**
Closing Stock Level	*-*	*30*	*21*	*13*	*4*	*6*	*6*	*5 e*
GASOLINE TYPE JET FUEL								
Refinery Output	**222**	**271**	**467**	**-**	**-**	**-**	**-**	**-**
+ Imports	2	-	-	-	-	-	-	-
- Exports	113	169	393	-	-	-	-	-
+ Transfers	-	-	-	-	-	-	-	-
+ Stock Change	- 11	- 10	1		-	-	-	-
- Statistical Difference	-	-	- 7	-	-	-	-	-
Total Consumption	**100**	**92**	**77**	**-**	**-**	**-**	**-**	**-**
Closing Stock Level	*-*	*29*	*42*	*-*	*-*	*-*	*-*	*-*
KEROSENE TYPE JET FUEL								
Refinery Output	**3 258**	**2 445**	**2 834**	**4 943**	**6 967**	**7 111**	**6 655**	**6 175 e**
+ Imports	96	272	423	579	721	558	941	695 e
- Exports	2 527	1 942	2 335	3 657	4 123	4 147	3 929	3 218 e
+ Transfers	-	-	17	- 267	- 276	- 207	- 571	- 375 e
+ Stock Change	- 75	- 15	- 87	- 37	10	- 60	68	25 e
- Statistical Difference	- 4	4	- 7	-	-	-	-	-
Total Consumption	**756**	**756**	**859**	**1 561**	**3 299**	**3 255**	**3 164**	**3 302 e**
Closing Stock Level	*-*	*204*	*399*	*349*	*292*	*352*	*284*	*259 e*

Source: Annual Oil Statistics.

NETHERLANDS
Table 5A: Gas/Diesel Oil - Supply and Consumption

000 metric tons

	1973	1978	1980	1990	1999	2000	2001	2002
Refinery Output	**20 559**	**18 807**	**17 511**	**15 524**	**21 682**	**21 994**	**22 167**	**19 582 e**
+ Recycled Products	-	-	-	-	-	-	-	-
+ Imports	2 335	2 962	8 041	9 190	7 332	8 007	9 239	12 509 e
- Exports	14 055	11 961	16 442	16 408	20 202	19 418	22 097	21 626 e
- Intl. Marine Bunkers	1 653	2 000	1 729	2 083	2 009	2 055	2 193	2 081 e
+ Transfers	-	- 94	- 137	- 815	-1 843	-1 620	- 355	-1 352 e
+ Stock Changes	138	- 273	212	- 127	1 404	- 242	- 94	- 152 e
- Statistical Differences	- 5	- 11	33	-	-	-	-	-
Total Consumption	**7 329**	**7 452**	**7 423**	**5 281**	**6 364**	**6 666**	**6 667**	**6 880 e**
Transformation	**26**	**478**	**1 039**	**20**	**10**	**7**	**7**	**..**
Public Electricity	-	-	-	-	-	-	-	..
Autoproducer Electricity	-	-	-	-	-	-	-	..
Public CHP Plants	26	9	5	6	4	3	5	..
Autoproducers of CHP	-	-	-	13	6	4	2	..
Public Heat Plants	-	-	-	-	-	-	-	..
Autoproducer Heat Plants	-	-	-	-	-	-	-	..
Petrochemical Industry	-	469	1 034	-	-	-	-	..
Other Transformation	-	-	-	1	-	-	-	..
Energy Sector	**-**	**2**	**3**	**5**	**7**	**4**	**3**	**..**
Refinery Fuel	-	2	3	5	5	2	1	..
Other Energy Sector	-	-	-	-	2	2	2	..
Final Consumption	**7 303**	**6 972**	**6 381**	**5 256**	**6 347**	**6 655**	**6 657**	**..**
Transport	**2 457**	**2 854**	**2 600**	**3 933**	**5 410**	**5 779**	**5 881**	**..**
Air	-	-	-	-	-	-	-	..
Road	1 254	2 120	1 986	3 725	5 122	5 416	5 530	..
Rail	54	39	47	37	29	43	41	..
Inland Waterways	1 149	695	567	171	259	320	310	..
Non-Specified	-	-	-	-	-	-	-	..
Industry	**417**	**1 406**	**1 634**	**386**	**217**	**193**	**168**	**..**
Iron and Steel	44	55	56	2	2	3	3	..
Chemical (incl. Pet.-Chem.)	21	1 214	1 398	184	5	3	3	..
Non-Ferrous Metals	-	-	-	1	-	-	-	..
Non-Metallic Minerals	13	10	10	11	7	6	8	..
Transport Equipment	-	-	-	9	5	5	5	..
Machinery	-	-	-	11	8	10	11	..
Mining and Quarrying	-	-	-	16	33	24	5	..
Food and Tobacco	7	19	19	22	26	12	13	..
Pulp, Paper and Printing	1	5	6	5	-	3	1	..
Wood and Wood Products	-	-	-	4	2	-	-	..
Construction	-	-	-	120	124	112	117	..
Textiles and Leather	-	-	-	-	1	2	1	..
Non-Specified	331	103	145	1	4	13	1	..
Other Sectors	**4 429**	**2 712**	**2 147**	**937**	**720**	**683**	**608**	**..**
Commerce and Public	1 032	1 129	677	497	221	199	96	..
Residential	3 153	1 288	1 095	165	22	22	23	..
Agriculture	244	295	375	275	477	462	489	..
Non-Specified	-	-	-	-	-	-	-	..
Memo: Non-Energy Use	-	*1 214*	*1 141*	*161*	-	-	-	..
Chem/Petchem. Feedst.	-	*1 214*	*1 141*	*161*	-	-	-	..
Other	-	-	-	-	-	-	-	..
Closing Stock Level	-	*2 202*	*3 864*	*2 784*	*2 281*	*2 523*	*2 617*	*2 769 e*

Source: Annual Oil Statistics.

NETHERLANDS
Table 5B: Fuel Oil (Residual) - Supply and Consumption

000 metric tons

	1973	1978	1980	1990	1999	2000	2001	2002
Refinery Output	**29 165**	**20 659**	**17 046**	**12 704**	**11 145**	**11 215**	**10 780**	**12 707 e**
+ Recycled Products	-	-	-	-	-	-	-	-
+ Imports	1 668	2 706	7 508	8 684	9 456	13 059	14 695	15 565 e
- Exports	13 854	7 998	7 492	9 301	7 995	10 210	10 894	12 113 e
- Intl. Marine Bunkers	10 263	9 363	7 799	9 019	10 926	11 582	12 775	12 757 e
+ Transfers	-	- 195	- 689	-1 740	-1 091	-1 746	-1 956	-3 264 e
+ Stock Changes	- 32	- 168	- 174	- 41	- 187	- 417	568	139 e
- Statistical Differences	- 20	- 11	5	-	-	-	-	-
Total Consumption	**6 704**	**5 652**	**8 395**	**1 287**	**402**	**319**	**418**	**277 e**
Transformation	**1 597**	**2 174**	**5 379**	**276**	**18**	**9**	**28**	..
Public Electricity	-	-	-	-	-	-	-	..
Autoproducer Electricity	-	-	-	-	-	-	-	..
Public CHP Plants	1 367	1 940	5 185	42	10	3	21	..
Autoproducers of CHP	230	234	194	234	8	6	7	..
Public Heat Plants	-	-	-	-	-	-	-	..
Autoproducer Heat Plants	-	-	-	-	-	-	-	..
Petrochemical Industry	-	-	-	-	-	-	-	..
Other Transformation	-	-	-	-	-	-	-	..
Energy Sector	**2 005**	**1 849**	**1 696**	**801**	**321**	**267**	**319**	..
Refinery Fuel	2 005	1 849	1 696	801	321	267	319	..
Other Energy Sector	-	-	-	-	-	-	-	..
Final Consumption	**3 102**	**1 629**	**1 320**	**210**	**63**	**43**	**71**	..
Transport	-	-	-	-	-	-	-	..
Air	-	-	-	-	-	-	-	..
Road	-	-	-	-	-	-	-	..
Rail	-	-	-	-	-	-	-	..
Inland Waterways	-	-	-	-	-	-	-	..
Non-Specified	-	-	-	-	-	-	-	..
Industry	**2 110**	**1 248**	**1 016**	**178**	**48**	**41**	**44**	..
Iron and Steel	484	375	295	2	-	-	-	..
Chemical (incl. Pet.-Chem.)	302	397	345	27	6	7	13	..
Non-Ferrous Metals	-	-	-	-	-	-	-	..
Non-Metallic Minerals	113	111	79	97	34	26	21	..
Transport Equipment	-	-	-	-	-	-	-	..
Machinery	-	-		8	3	3	3	..
Mining and Quarrying	-	-	-	-	-	-	-	..
Food and Tobacco	270	211	169	34	5	4	7	..
Pulp, Paper and Printing	60	31	9	1	-	-	-	..
Wood and Wood Products	-	-	-	-	-	-	-	..
Construction	-	-	-	-	-	-	-	..
Textiles and Leather	-	-	-	2	-	-	-	..
Non-Specified	881	123	119	7	-	1	-	..
Other Sectors	**992**	**381**	**304**	**32**	**15**	**2**	**27**	..
Commerce and Public	-	-	-	32	15	2	27	..
Residential	271	166	64	-	-	-	-	..
Agriculture	32	2	-	-	-	-	-	..
Non-Specified	689	213	240	-	-	-	-	..
Memo: Non-Energy Use	-	*192*	*158*	*12*	-	-	-	..
Chem/Petchem. Feedst.	-	*192*	*158 e*	*7*	-	-	-	..
Other	-	-	-	*5*	-	-	-	..
Closing Stock Level	-	*2 126*	*2 930*	*1 836*	*1 614*	*2 031*	*1 463*	*1 324 e*

Source: Annual Oil Statistics.

NETHERLANDS
Table 5C: Total Products - Supply and Consumption

000 metric tons

	1973	1978	1980	1990	1999	2000	2001	2002
Refinery Output	**73 081**	**60 530**	**57 238**	**67 914**	**76 263**	**80 235**	**81 306**	**78 714 e**
+ Recycled Products	-	-	-	31	46	50	50	50 e
+ Imports	9 722	13 615	28 548	39 520	37 364	44 675	46 864	51 206 e
- Exports	41 684	32 030	40 914	56 778	60 063	63 442	65 714	65 429 e
- Intl. Marine Bunkers	11 972	11 427	9 590	11 202	13 063	13 788	15 140	14 957 e
+ Transfers	-	- 959	-2 465	-16 522	-15 881	-19 506	-20 464	-22 315 e
+ Stock Changes	- 180	- 185	209	206	1 546	-1 354	602	540 e
- Statistical Differences	- 52	125	32	12	-	-	-	-
Total Consumption	**29 019**	**29 419**	**32 994**	**23 157**	**26 212**	**26 870**	**27 504**	**27 809 e**
Transformation	**1 623**	**4 487**	**9 835**	**696**	**1 370**	**616**	**647**	**..**
Public Electricity	-	-	-	-	-	-	-	..
Autoproducer Electricity	-	-	-	-	-	-	-	..
Public CHP Plants	1 393	1 949	5 190	48	781	76	93	..
Autoproducers of CHP	230	234	194	646	589	540	554	..
Public Heat Plants	-	-	-	-	-	-	-	..
Autoproducer Heat Plants	-	-	-	-	-	-	-	..
Petrochemical Industry	-	2 304	4 435	-	-	-	-	..
Other Transformation	-	-	16	2	-	-	-	..
Energy Sector	**3 645**	**2 994**	**3 001**	**3 672**	**3 081**	**3 051**	**3 158**	**..**
Refinery Fuel	3 645	2 994	3 001	3 672	3 079	3 049	3 156	..
Other Energy Sector	-	-	-	-	2	2	2	..
Final Consumption	**23 751**	**21 938**	**20 158**	**18 789**	**21 761**	**23 203**	**23 699**	**..**
Transport	**6 985**	**8 020**	**8 180**	**9 918**	**13 567**	**13 704**	**13 763**	**..**
Air	865	855	943	1 567	3 321	3 276	3 183	..
Road	4 907	6 427	6 607	8 131	9 936	10 042	10 198	..
Rail	54	39	47	37	38	52	50	..
Inland Waterways	1 159	699	583	183	272	334	332	..
Non-Specified	-	-	-	-	-	-	-	..
Industry	**9 925**	**9 989**	**9 008**	**7 575**	**7 193**	**8 596**	**9 049**	**..**
Iron and Steel	537	441	364	5	3	4	4	..
Chemical (incl. Pet.-Chem.)	6 078	7 579	6 960	6 257	5 826	7 220	7 560	..
Non-Ferrous Metals	-	-	-	82	97	99	100	..
Non-Metallic Minerals	130	135	103	156	63	56	48	..
Transport Equipment	-	-	-	21	6	7	7	..
Machinery	-	-	-	265	481	578	693	..
Mining and Quarrying	-	-	-	16	33	24	5	..
Food and Tobacco	277	230	188	61	35	16	20	..
Pulp, Paper and Printing	61	36	15	8	1	4	2	..
Wood and Wood Products	-	-	-	4	2	-	-	..
Construction	765	662	586	594	482	399	498	..
Textiles and Leather	-	-	-	2	1	4	2	..
Non-Specified	2 077	906	792	104	163	185	110	..
Other Sectors	**6 841**	**3 929**	**2 970**	**1 296**	**1 001**	**903**	**887**	**..**
Commerce and Public	1 054	1 185	733	747	438	353	313	..
Residential	4 652	2 060	1 382	235	56	56	59	..
Agriculture	316	372	409	314	507	494	515	..
Non-Specified	819	312	446	-	-	-	-	..
Memo: Non-Energy Use	*7 202*	*7 673*	*6 814*	*7 031*	*6 637*	*7 360*	*7 806*	..
Chem/Petchem. Feedst.	*5 494*	*6 314*	*5 526 e*	*5 959*	*5 351*	*6 046*	*6 353*	..
Other	*1 708*	*1 359*	*1 288 e*	*1 072*	*1 286*	*1 314*	*1 453*	..
Closing Stock Level	-	*6 470*	*9 700*	*8 207*	*6 800*	*8 154*	*7 552*	*7 012 e*

Source: Annual Oil Statistics.

NETHERLANDS
Table 6A: Supply and Consumption

000 metric tons

	1980	2000	2001	2002	1980	2000	2001	2002
	LPG AND ETHANE				**REFINERY GAS**			
Refinery Output	**1 632**	**3 684**	**3 974**	**4 663 e**	**1 913**	**3 684**	**3 860**	**4 139 e**
+ Recycled Products	-	-	-	-	-	-	-	-
+ Imports	995	1 708	1 963	1 826 e	-	-	-	-
- Exports	545	1 416	1 669	2 103 e	-	-	-	-
- Intl. Marine Bunkers	-	-	-	-	-	-	-	-
+ Transfers	- 1	-1 906	-2 357	-2 341 e	- 13	- 84	- 88	- 78 e
+ Stock Change	- 3	35	9	3 e	-	-	-	-
- Statistical Difference	7	-	-	-	-	-	-	-
Total Consumption	**2 071**	**2 105**	**1 920**	**2 048 e**	**1 900**	**3 600**	**3 772**	**4 061 e**
of which: Ethane	-	*24*	*17*	*19 e*	-	-	-	-
Transformation	455	-	-	..	2	361	378	..
Energy	6	3	1	..	1 049	2 186	2 293	..
Transport	696	567	521	..	-	-	-	..
Industry	772	1 465	1 316	..	849	1 053	1 101	..
Other	142	70	82	..	-	-	-	..
Memo: Non-Energy Use	*698*	*1 384*	*1 306*	..	*4*	-	-	..
Closing Stock Level	*79*	*176*	*167*	*164 e*	-	-	-	-
	NAPHTHA				**OTHER KEROSENE**			
Refinery Output	**3 646**	**11 303**	**12 116**	**8 655 e**	**472**	**376**	**413**	**484 e**
+ Recycled Products	-	-	-	-	-	-	-	-
+ Imports	7 835	9 595	9 197	8 981 e	153	739	624	500 e
- Exports	5 254	10 965	10 586	9 130 e	300	293	246	205 e
- Intl. Marine Bunkers	-	-	-	-	-	-	-	-
+ Transfers	- 416	-6 904	-7 546	-5 281 e	- 103	- 673	- 748	- 713 e
+ Stock Change	227	- 45	156	42 e	22	- 67	35	4 e
- Statistical Difference	- 4	-	-	-	-	-	-	-
Total Consumption	**6 042**	**2 984**	**3 337**	**3 267 e**	**244**	**82**	**78**	**70 e**
Transformation	2 816	-	-	..	-	-	-	..
Energy	1	-	-	..	-	-	-	..
Transport	-	-	-	..	-	-	-	..
Industry	3 225	2 984	3 337	..	4	34	4	..
Other	-	-	-	..	240	48	74	..
Memo: Non-Energy Use	*3 225*	*2 984*	*3 337*	..	-	*32*	*3*	..
Closing Stock Level	*000*	*981*	*825*	*783 e*	*212*	*109*	*74*	*70 e*
	WHITE SPIRIT + SBP				**LUBRICANTS**			
Refinery Output	**217**	**345**	**386**	**..**	**566**	**647**	**574**	**..**
+ Recycled Products	-	-	-	..	-	50	50	..
+ Imports	64	371	370	..	441	538	597	..
- Exports	180	209	197	..	674	841	820	..
- Intl. Marine Bunkers	-	-	-	..	62	151	172	..
+ Transfers	- 1	- 385	- 461	..	- 15	- 44	- 51	..
+ Stock Change	- 15	- 12	10	..	- 60	- 15	- 2	..
- Statistical Difference	- 1	-	-	..	5	-	-	..
Total Consumption	**86**	**110**	**108**	**..**	**191**	**184**	**176**	**..**
Transformation	1	-	-	..	1	-	-	..
Energy	-	-	-	..	1	-	-	..
Transport	-	-	-	..	87	67	73	..
Industry	1	45	54	..	49	82	61	..
Other	84	65	54	..	53	35	42	..
Memo: Non-Energy Use	*85*	*110*	*108*	..	*189*	*184*	*176*	..
Closing Stock Level	*90*	*53*	*43*	..	*203*	*141*	*143*	..

Source: Annual Oil Statistics.

NETHERLANDS
Table 6B: Supply and Consumption

000 metric tons

	1980	2000	2001	2002	1980	2000	2001	2002
	BITUMEN				**PARAFFIN WAXES**			
Refinery Output	**879**	**740**	**714**	**..**	**95**	**118**	**103**	**..**
+ Recycled Products	-	-	-	..	-	-	-	..
+ Imports	172	157	281	..	54	97	103	..
- Exports	455	464	484	..	80	125	158	..
- Intl. Marine Bunkers	-	-	-	..	-	-	-	..
+ Transfers	-	- 178	- 131	..	-	6	5	..
+ Stock Change	- 8	32	1	..	- 15	1	- 6	..
- Statistical Difference	2	-	-	..	-	-	-	..
Total Consumption	**586**	**287**	**381**	**..**	**54**	**97**	**47**	**..**
Transformation	-	-	-	..	6	-	-	..
Energy	-	-	-	..	-	-	-	..
Transport	-	-	-	..	-	-	-	..
Industry	586	287	381	..	48	97	47	..
Other	-	-	-	..	-	-	-	..
Memo: Non-Energy Use	*586*	*287*	*381*	..	*42*	*97*	*47*	..
Closing Stock Level	*63*	*25*	*24*	..	*25*	*18*	*24*	..
	PETROLEUM COKE				**OTHER PRODUCTS**			
Refinery Output	**211**	**-**	**-**	**-**	**522**	**4 704**	**5 088**	**6 500 e**
+ Recycled Products	-	-	-	-	-	-	-	50 e
+ Imports	543	-	-	-	149	2 330	2 678	3 859 e
- Exports	72	-	-	-	22	2 608	3 183	4 708 e
- Intl. Marine Bunkers	-	-	-	-	-	-	-	119 e
+ Transfers	-	-	-	-	- 135	-1 262	-1 280	-1 857 e
+ Stock Change	-	-	-	-	- 10	- 19	8	3 e
- Statistical Difference	-	-	-	-	- 18	-	-	-
Total Consumption	**682**	**-**	**-**	**-**	**522**	**3 145**	**3 311**	**3 728 e**
Transformation	16	-	-	..	120	239	234	..
Energy	211	-	-	..	33	591	542	..
Transport	-	-	-	..	-	-	-	..
Industry	455	-	-	..	369	2 315	2 535	..
Other	-	-	-	..	-	-	-	..
Memo: Non-Energy Use	*455*	-	-	..	*231*	*2 282*	*2 448*	..
Closing Stock Level	-	-	-	-	*125*	*210*	*202*	*433 e*

Source: Annual Oil Statistics.

NETHERLANDS
Table 7A: Imports by Country of Origin

000 metric tons

	Crude Oil 2000	Crude Oil 2001	Crude Oil 2002	Natural Gas Liquids 2000	Natural Gas Liquids 2001	Natural Gas Liquids 2002	Feedstocks and Other 2000	Feedstocks and Other 2001	Feedstocks and Other 2002
Australia	-	-	-	-	-	-	-	-	-
Belgium	-	-	-	-	-	2 121 e	-	-	-
Canada	-	-	-	-	-	-	-	-	-
Czech Republic	-	-	-	-	-	-	-	-	-
Denmark	2 146	1 462	1 207 e	-	-	-	-	-	-
Finland	-	-	-	-	-	-	-	-	-
France	-	-	-	-	-	-	-	-	-
Germany	-	-	48 e	11	35	45 e	-	-	-
Greece	-	-	-	-	-	-	-	-	-
Hungary	-	-	-	-	-	-	-	-	-
Ireland	-	-	-	-	-	-	-	-	-
Italy	-	-	-	-	-	-	-	-	-
Japan	-	-	-	-	-	-	-	-	-
Korea	-	-	-	-	-	-	-	-	-
Mexico	66	-	85 e	-	-	-	-	-	-
Netherlands	-	-	-	-	-	-	-	-	-
New Zealand	-	-	-	-	-	-	-	-	-
Norway	10 418	9 893	6 848 e	922	889	1 260 e	-	-	-
Poland	-	-	-	-	-	-	-	-	-
Spain	-	-	-	-	-	-	-	-	-
Sweden	-	-	-	-	-	-	-	-	-
Turkey	-	-	-	-	-	-	-	-	-
United Kingdom	10 963	10 738	11 212 e	48	45	22 e	-	-	-
United States	-	-	-	-	-	-	-	-	-
Other	-	-	-	-	-	-	-	-	-
OECD	**23 593**	**22 093**	**19 400 e**	**981**	**969**	**3 448 e**	**-**	**-**	**-**
Argentina	-	-	-	-	-	-	-	-	-
Brazil	-	-	11 e	-	-	-	-	-	-
Colombia	-	75	123 e	-	-	-	-	-	-
Ecuador	-	-	-	-	-	-	-	-	-
Netherlands Antilles	-	-	-	-	-	-	-	-	-
Peru	-	-	-	-	-	-	-	-	-
Trinidad and Tobago	-	-	-	-	-	-	-	-	-
Venezuela	511	459	232 e	-	-	-	-	-	-
Other	-	-	-	-	-	-	-	-	-
Latin America	**511**	**534**	**366 e**	**-**	**-**	**-**	**-**	**-**	**-**
Romania	-	-	-	-	-	-	-	-	-
Other	-	-	-	-	-	-	-	-	-
Non-OECD Europe	**-**	**-**	**-**	**-**	**-**	**-**	**-**	**-**	**-**
FSU	**4 450**	**6 767**	**7 677 e**	**81**	**436**	**1 042 e**	**-**	**-**	**-**
Iran	3 304	1 157	1 508 e	-	-	-	-	-	-
Iraq	3 198	4 718	3 384 e	-	-	-	-	-	-
Kuwait	6 177	6 031	4 468 e	-	-	-	-	-	-
Oman	-	-	184 e	-	-	-	-	-	-
Qatar	-	-	-	-	-	-	-	-	-
Saudi Arabia	12 265	10 536	7 568 e	-	-	-	-	-	-
United Arab Emirates	141	-	-	-	-	-	-	-	-
Other	-	225	1 388 e	-	-	-	-	-	-
Middle East	**25 085**	**22 667**	**18 500 e**	**-**	**-**	**-**	**-**	**-**	**-**
India	-	-	-	-	-	-	-	-	-
Indonesia	-	-	-	-	-	-	-	-	-
Singapore	-	-	-	-	-	-	-	-	-
Other	-	-	-	-	-	-	-	-	-
Asia	**-**	**-**	**-**	**-**	**-**	**-**	**-**	**-**	**-**
China	**-**	**-**	**-**	**-**	**-**	**-**	**-**	**-**	**-**
Algeria	84	84	75 e	4 694	4 582	3 761 e	-	-	-
Angola/Cabinada	243	721	183 e	-	-	-	-	-	-
Cameroon	-	-	164 e	-	-	-	-	-	-
Egypt	-	-	75 e	-	-	-	-	-	-
Gabon	-	-	-	-	-	-	-	-	-
Libya	-	-	-	25	25	75 e	-	-	-
Nigeria	455	1 387	346 e	-	-	-	-	-	-
Tunisia	-	-	-	105	17	-	-	-	-
Other	-	-	51 e	-	35	-	-	-	-
Africa	**782**	**2 192**	**894 e**	**4 824**	**4 659**	**3 836 e**	**-**	**-**	**-**
Non-spec./Other	**246**	**-**	**-**	**-**	**493**	**73 e**	**-**	**-**	**-**
Total Imports	**54 667**	**54 253**	**46 837 e**	**5 886**	**6 557**	**8 399 e**	**-**	**-**	**-**
Memo: Total OPEC	*26 135*	*24 372*	*17 581 e*	*4 719*	*4 607*	*3 836 e*	-	-	-

Source: Annual Oil Statistics.

NETHERLANDS
Table 7B: Imports by Country of Origin

000 metric tons

	2000	2001	2002	2000	2001	2002	2000	2001	2002
	Total Products			**Liquefied Petroleum Gases**			**Naphtha**		
Australia	-	-	-	-	-	-	-	-	-
Belgium	5 309	5 109	6 264 e	244	208	197 e	1 068	645	832 e
Canada	216	142	192 e	-	-	-	15	-	19 e
Czech Republic	35	6	4 e	-	-	-	20	-	-
Denmark	80	86	71 e	3	-	2 e	-	5	11 e
Finland	128	134	500 e	5	-	2 e	57	53	71 e
France	2 218	1 374	1 265 e	36	14	5 e	749	626	541 e
Germany	3 018	2 392	2 096 e	188	134	121 e	721	756	700 e
Greece	86	27	134 e	-	-	-	86	-	80 e
Hungary	4	5	-	-	-	-	-	-	-
Ireland	27	3	16 e	-	-	-	-	-	-
Italy	428	375	210 e	29	20	14 e	117	158	72 e
Japan	13	24	32 e	-	-	-	7	2	15 e
Korea	51	55	222 e	-	-	-	42	49	75 e
Mexico	-	32	-	-	-	-	-	32	-
Netherlands	-	-	-	-	-	-	-	-	-
New Zealand	-	-	-	-	-	-	-	-	-
Norway	874	539	1 275 e	99	229	196 e	180	137	281 e
Poland	289	129	239 e	-	-	-	3	-	-
Spain	696	715	596 e	-	-	2 e	505	589	458 e
Sweden	613	492	930 e	13	14	37 e	111	188	196 e
Turkey	8	21	5 e	3	-	-	5	15	5 e
United Kingdom	5 203	4 527	4 429 e	430	487	665 e	1 497	1 086	1 235 e
United States	2 554	2 671	2 515 e	4	2	9 e	454	267	620 e
Other	316	182	250 e	-	3	2 e	101	68	93 e
OECD	**22 166**	**19 040**	**21 245 e**	**1 054**	**1 111**	**1 252 e**	**5 738**	**4 676**	**5 304 e**
Argentina	26	40	123 e	-	-	-	5	-	5 e
Brazil	41	46	74 e	-	-	-	35	20	14 e
Colombia	-	-	-	-	-	-	-	-	-
Ecuador	-	-	-	-	-	-	-	-	-
Netherlands Antilles	149	235	140 e	-	-	-	2	5	61 e
Peru	-	-	-	-	-	-	-	-	-
Trinidad and Tobago	-	-	-	-	-	-	-	-	-
Venezuela	391	459	163 e	3	-	5 e	-	-	-
Other	51	129	26 e	-	-	-	-	82	26 e
Latin America	**658**	**909**	**526 e**	**3**	**-**	**5 e**	**42**	**107**	**106 e**
Romania	-	59	3 e	-	-	-	-	1	3 e
Other	29	31	3 e	-	-	-	6	3	3 e
Non-OECD Europe	**29**	**90**	**6 e**	**-**	**-**	**-**	**6**	**4**	**6 e**
FSU	**12 090**	**11 134**	**15 241 e**	**10**	**-**	**-**	**351**	**286**	**647 e**
Iran	27	16	-	-	-	-	27	16	-
Iraq	-	-	-	-	-	-	-	-	-
Kuwait	921	240	161 e	-	-	-	866	109	-
Oman	-	-	-	-	-	-	-	-	-
Qatar	-	-	-	-	-	-	-	-	-
Saudi Arabia	93	47	29 e	-	-	-	81	31	27 e
United Arab Emirates	61	18	4 e	-	-	-	41	-	4 e
Other	141	158	12 e	-	-	-	21	135	11 e
Middle East	**1 243**	**479**	**206 e**	**-**	**-**	**-**	**1 036**	**291**	**42 e**
India	99	200	101 e	-	-	-	53	3	10 e
Indonesia	43	192	70 e	-	-	-	1	-	-
Singapore	26	45	887 e	-	-	-	26	12	3 e
Other	73	19	253 e	16	-	-	6	-	15 e
Asia	**241**	**456**	**1 311 e**	**16**	**-**	**-**	**86**	**15**	**28 e**
China	**28**	**41**	**22 e**	**-**	**-**	**-**	**3**	**-**	**-**
Algeria	427	655	1 291 e	27	52	139 e	248	440	695 e
Angola/Cabinada	-	-	-	-	-	-	-	-	-
Cameroon	-	-	-	-	-	-	-	-	-
Egypt	295	174	579 e	-	-	-	232	174	138 e
Gabon	-	-	-	-	-	-	-	-	-
Libya	168	124	302 e	22	16	-	75	53	159 e
Nigeria	12	43	-	-	-	-	12	43	-
Tunisia	21	-	27 e	-	-	-	21	-	27 e
Other	40	41	139 e	-	-	-	32	20	79 e
Africa	**963**	**1 037**	**2 338 e**	**49**	**68**	**139 e**	**620**	**730**	**1 098 e**
Non-spec./Other	**7 257**	**13 678**	**10 311 e**	**576**	**784**	**430 e**	**1 713**	**3 088**	**1 750 e**
Total Imports	**44 675**	**46 864**	**51 206 e**	**1 708**	**1 963**	**1 826 e**	**9 595**	**9 197**	**8 981 e**
Memo: Total OPEC	*2 143*	*1 794*	*2 020 e*	*52*	*68*	*144 e*	*1 351*	*692*	*885 e*

Source: Annual Oil Statistics.

NETHERLANDS
Table 7C: Imports by Country of Origin

000 metric tons

	2000	2001	2002	2000	2001	2002	2000	2001	2002
	Total Gasoline			**Total Kerosene**			**Gas/Diesel**		
Australia	-	-	-	-	-	-	-	-	-
Belgium	1 126	1 262	1 389 e	151	100	249 e	660	1 005	1 066 e
Canada	1	1	-	-	-	-	145	98	173 e
Czech Republic	-	-	-	-	-	-	-	-	-
Denmark	30	11	-	-	12	-	31	44	21 e
Finland	53	51	59 e	1	-	8 e	4	4	359 e
France	462	302	239 e	12	5	42 e	223	62	54 e
Germany	821	411	336 e	51	21	91 e	370	210	282 e
Greece	-	-	54 e	-	-	-	-	27	-
Hungary	-	-	-	-	-	-	-	-	-
Ireland	5	3	14 e	-	-	-	22	-	2 e
Italy	154	111	89 e	11	-	-	38	46	-
Japan	-	-	-	-	-	-	-	-	-
Korea	-	-	-	-	-	-	-	-	147 e
Mexico	-	-	-	-	-	-	-	-	-
Netherlands	-	-	-	-	-	-	-	-	-
New Zealand	-	-	-	-	-	-	-	-	-
Norway	146	47	295 e	-	2	-	220	47	344 e
Poland	-	65	172 e	-	-	-	-	-	14 e
Spain	129	65	27 e	25	-	23 e	-	52	76 e
Sweden	289	129	362 e	-	-	14 e	146	106	214 e
Turkey	-	-	-	-	-	-	-	-	-
United Kingdom	1 619	1 379	844 e	67	63	70 e	925	676	855 e
United States	95	4	31 e	13	9	16 e	330	98	338 e
Other	48	18	57 e	20	6	-	2	41	-
OECD	**4 978**	**3 859**	**3 968 e**	**351**	**218**	**513 e**	**3 116**	**2 516**	**3 945 e**
Argentina	-	-	-	-	-	-	-	-	-
Brazil	-	-	-	-	-	-	-	-	-
Colombia	-	-	-	-	-	-	-	-	-
Ecuador	-	-	-	-	-	-	-	-	-
Netherlands Antilles	4	1	-	-	-	-	-	-	-
Peru	-	-	-	-	-	-	-	-	-
Trinidad and Tobago	-	-	-	-	-	-	-	-	-
Venezuela	-	-	-	70	65	30 e	-	-	-
Other	21	47	-	-	-	-	-	-	-
Latin America	**25**	**48**	**-**	**70**	**65**	**30 e**	**-**	**-**	**-**
Romania	-	-	-	-	-	-	-	58	-
Other	11	-	-	-	-	-	11	27	-
Non-OECD Europe	**11**	**-**	**-**	**-**	**-**	**-**	**11**	**85**	**-**
FSU	**828**	**335**	**980 e**	**198**	**63**	**354 e**	**3 269**	**2 349**	**3 799 e**
Iran	-	-	-	-	-	-	-	-	-
Iraq	-	-	-	-	-	-	-	-	-
Kuwait	-	-	-	51	62	161 e	4	-	-
Oman	-	-	-	-	-	-	-	-	-
Qatar	-	-	-	-	-	-	-	-	-
Saudi Arabia	-	-	-	10	15	2 e	2		
United Arab Emirates	18	-	-	2	18	-	-	-	-
Other	9	-	1 e	69	23	-	13	-	-
Middle East	**27**	**-**	**1 e**	**132**	**118**	**163 e**	**19**	**-**	**-**
India	38	-	-	-	-	-	-	119	91 e
Indonesia	-	-	-	-	-	-	-	50	-
Singapore	-	-	-	-	-	-	-	32	881 e
Other	-	-	-	10	-	-	40	14	216 e
Asia	**38**	**-**	**-**	**10**	**-**	**-**	**40**	**215**	**1 188 e**
China	**-**	**-**	**-**	**-**	**-**	**-**	**-**	**-**	**-**
Algeria	109	66	417 e	-	-	32 e	34	92	-
Angola/Cabinada	-	-	-	-	-	-	-	-	-
Cameroon	-	-	-	-	-	-	-	-	-
Egypt	63	-	414 e	-	-	27 e	-	-	-
Gabon	-	-	-	-	-	-	-	-	-
Libya	50	-	119 e	-	-	-	21	55	24 e
Nigeria	-	-	-	-	-	-	-	-	-
Tunisia	-	-	-	-	-	-	-	-	-
Other	-	4	26 e	1	1	-	2	1	27 e
Africa	**222**	**70**	**976 e**	**1**	**1**	**59 e**	**57**	**148**	**51 e**
Non-spec./Other	**1 387**	**1 864**	**1 346 e**	**535**	**1 100**	**76 e**	**1 495**	**3 926**	**3 526 e**
Total Imports	**7 516**	**6 176**	**7 271 e**	**1 297**	**1 565**	**1 195 e**	**8 007**	**9 239**	**12 509 e**
Memo: Total OPEC	*177*	*66*	*536 e*	*133*	*160*	*225 e*	*61*	*197*	*24 e*

Source: Annual Oil Statistics.

NETHERLANDS
Table 7D: Imports by Country of Origin

000 metric tons

	2000	2001	2002	2000	2001	2002	2000	2001	2002
	Fuel Oil (Residual)			**Petroleum Coke**			**Other Products**		
Australia	-	-	-	-	-	-	-	-	-
Belgium	1 439	1 375	1 633 e	-	-	-	621	514	898 e
Canada	49	31	-	-	-	-	6	12	-
Czech Republic	9	-	-	-	-	-	6	6	4 e
Denmark	-	3	11 e	-	-	-	16	11	26 e
Finland	-	26	-	-	-	-	8	-	1 e
France	510	160	205 e	-	-	-	226	205	179 e
Germany	444	272	171 e	-	-	-	423	588	395 e
Greece	-	-	-	-	-	-	-	-	-
Hungary	-	-	-	-	-	-	4	5	-
Ireland	-	-	-	-	-	-	-	-	-
Italy	24	-	-	-	-	-	55	40	35 e
Japan	-	-	-	-	-	-	6	22	17 e
Korea	-	-	-	-	-	-	9	6	-
Mexico	-	-	-	-	-	-	-	-	-
Netherlands	-	-	-	-	-	-	-	-	-
New Zealand	-	-	-	-	-	-	-	-	-
Norway	52	-	3 e	-	-	-	177	77	156 e
Poland	275	60	49 e	-	-	-	11	4	4 e
Spain	28	-	-	-	-	-	9	9	10 e
Sweden	48	46	106 e	-	-	-	6	9	1 e
Turkey	-	-	-	-	-	-	-	6	-
United Kingdom	330	517	385 e	-	-	-	335	319	375 e
United States	306	567	295 e	-	-	-	1 352	1 724	1 206 e
Other	129	39	81 e	-	-	-	16	7	17 e
OECD	**3 643**	**3 096**	**2 939 e**	**-**	**-**	**-**	**3 286**	**3 564**	**3 324 e**
Argentina	-	-	-	-	-	-	21	40	118 e
Brazil	-	-	45 e	-	-	-	6	26	15 e
Colombia	-	-	-	-	-	-	-	-	-
Ecuador	-	-	-	-	-	-	-	-	-
Netherlands Antilles	143	229	79 e	-	-	-	-	-	-
Peru	-	-	-	-	-	-	-	-	-
Trinidad and Tobago	-	-	-	-	-	-	-	-	-
Venezuela	318	364	79 e	-	-	-	-	30	49 e
Other	30	-	-	-	-	-	-	-	-
Latin America	**491**	**593**	**203 e**	**-**	**-**	**-**	**27**	**96**	**182 e**
Romania	-	-	-	-	-	-	-	-	-
Other	-	-	-	-	-	-	1	1	-
Non-OECD Europe	**-**	**-**	**-**	**-**	**-**	**-**	**1**	**1**	**-**
FSU	**7 390**	**8 083**	**9 435 e**	**-**	**-**	**-**	**44**	**18**	**26 e**
Iran	-	-	-	-	-	-	-	-	-
Iraq	-	-	-	-	-	-	-	-	-
Kuwait	-	69	-	-	-	-	-	-	-
Oman	-	-	-	-	-	-	-	-	-
Qatar	-	-	-	-	-	-	-	-	-
Saudi Arabia	-	-	-	-	-	-	-	1	-
United Arab Emirates	-	-	-	-	-	-	-	-	-
Other	29	-	-	-	-	-	-	-	-
Middle East	**29**	**69**	**-**	**-**	**-**	**-**	**-**	**1**	**-**
India	-	-	-	-	-	-	8	78	-
Indonesia	-	-	-	-	-	-	42	142	70 e
Singapore	-	-	3 e	-	-	-	-	1	-
Other	-	-	-	-	-	-	1	5	22 e
Asia	**-**	**-**	**3 e**	**-**	**-**	**-**	**51**	**226**	**92 e**
China	**-**	**-**	**-**	**-**	**-**	**-**	**25**	**41**	**22 e**
Algeria	-	-	-	-	-	-	9	5	8 e
Angola/Cabinada	-	-	-	-	-	-	-	-	-
Cameroon	-	-	-	-	-	-	-	-	-
Egypt	-	-	-	-	-	-	-	-	-
Gabon	-	-	-	-	-	-	-	-	-
Libya	-	-	-	-	-	-	-	-	-
Nigeria	-	-	-	-	-	-	-	-	-
Tunisia	-	-	-	-	-	-	-	-	-
Other	-	12	-	-	-	-	5	3	7 e
Africa	**-**	**12**	**-**	**-**	**-**	**-**	**14**	**8**	**15 e**
Non-spec./Other	**1 506**	**2 842**	**2 985 e**	**-**	**-**	**-**	**45**	**74**	**198 e**
Total Imports	**13 059**	**14 695**	**15 565 e**	**-**	**-**	**-**	**3 493**	**4 029**	**3 859 e**
Memo: Total OPEC	*318*	*433*	*79 e*	*-*	*-*	*-*	*51*	*178*	*127 e*

Source: Annual Oil Statistics.

NETHERLANDS

Table 8A: Exports by Country of Destination

000 metric tons

	Crude Oil			Natural Gas Liquids			Feedstocks and Other		
	2000	2001	2002	2000	2001	2002	2000	2001	2002
Australia	-	-	-	-	-	-	-	-	-
Austria	-	-	-	-	-	-	-	-	-
Belgium	184	206	194 e	-	-	-	-	-	-
Canada	-	-	-	-	-	-	-	-	-
Denmark	-	33	36 e	-	-	-	-	-	-
Finland	-	-	-	-	-	-	-	-	-
France	-	-	33 e	-	-	-	-	-	-
Germany	-	52	426 e	-	-	-	-	-	-
Greece	-	-	-	-	-	-	-	-	-
Ireland	-	-	-	-	-	-	-	-	-
Italy	-	-	-	-	-	-	-	-	-
Japan	-	-	-	-	-	-	-	-	-
Luxembourg	-	-	-	-	-	-	-	-	-
Mexico	-	-	-	-	-	-	-	-	-
Netherlands	-	-	-	-	-	-	-	-	-
Norway	-	33	33 e	-	-	-	-	-	-
Portugal	-	-	-	-	-	-	-	-	-
Spain	-	-	-	-	-	-	-	-	-
Sweden	-	-	69 e	-	-	-	-	-	-
Switzerland	-	-	-	-	-	-	-	-	-
Turkey	-	-	-	-	-	-	-	-	-
United Kingdom	-	30	101 e	-	-	-	-	-	-
United States	-	-	-	-	-	-	-	-	-
Other OECD	-	-	-	-	-	-	-	-	-
OECD	**184**	**354**	**892 e**	**-**	**-**	**-**	**-**	**-**	**-**
Africa	-	-	-	-	-	-	-	-	-
Latin America	-	-	-	-	-	-	-	-	-
Asia	-	-	-	-	-	-	-	-	-
China	-	-	-	-	-	-	-	-	-
Non-OECD Europe	-	-	-	-	-	-	-	-	-
Former Soviet Union	-	-	-	-	-	-	-	-	-
Middle East	-	-	-	-	-	-	-	-	-
Non-specified/Other	-	1	2 e	-	-	-	-	-	-
Non-OECD	**-**	**1**	**2 e**	**-**	**-**	**-**	**-**	**-**	**-**
Total Exports	**184**	**355**	**894 e**	**-**	**-**	**-**	**-**	**-**	**-**

	Total Products			Liquefied Petroleum Gas			Naphtha		
	2000	2001	2002	2000	2001	2002	2000	2001	2002
Australia	4	1	3 e	-	-	-	-	-	3 e
Austria	129	72	48 e	6	2	1 e	32	18	13 e
Belgium	16 597	18 899	16 298 e	471	555	691 e	2 038	2 184	1 980 e
Canada	125	103	216 e	1	-	-	-	-	1 e
Denmark	115	178	96 e	-	2	-	5	4	3 e
Finland	235	256	174 e	4	8	-	117	64	91 e
France	4 032	4 030	3 752 e	20	15	38 e	308	301	304 e
Germany	27 395	26 921	20 147 e	362	219	294 e	6 065	6 078	4 447 e
Greece	184	84	138 e	-	-	-	50	50	49 e
Ireland	85	179	87 e	3	8	-	-	6	-
Italy	701	945	245 e	1	-	5 e	51	90	27 e
Japan	2	45	230 e	-	-	-	-	-	-
Luxembourg	118	131	155 e	-	-	-	-	-	-
Mexico	-	39	202 e	-	-	-	-	-	-
Netherlands	-	-	-	-	-	-	-	-	-
Norway	92	76	146 e	-	1	1 e	35	34	48 e
Portugal	457	350	403 e	21	17	6 e	22	11	18 e
Spain	476	607	637 e	-	9	-	122	128	148 e
Sweden	209	169	194 e	-	-	15 e	39	46	38 e
Switzerland	1 323	807	1 419 e	1	2	4 e	7	27	15 e
Turkey	74	14	51 e	29	10	21 e	16	2	17 e
United Kingdom	2 603	4 067	3 584 e	52	72	39 e	378	654	855 e
United States	1 581	1 138	2 954 e	15	30	60 e	426	390	406 e
Other OECD	154	86	119 e	8	2	3 e	45	30	18 e
OECD	**56 721**	**58 597**	**51 298 e**	**994**	**952**	**1 178 e**	**9 756**	**10 057**	**8 481 e**
Africa	452	502	679 e	-	9	2 e	18	50	12 e
Latin America	81	138	583 e	5	13	-	2	13	9 e
Asia	1 063	1 566	2 216 e	-	-	1 e	20	113	91 e
China	4	48	108 e	-	43	-	-	-	29 e
Non-OECD Europe	294	112	308 e	3	9	-	2	6	-
Former Soviet Union	58	41	17 e	-	-	-	5	14	-
Middle East	161	42	46 e	-	-	-	9	19	30 e
Non-specified/Other	4 608	4 668	10 174 e	414	643	922 e	1 153	314	478 e
Non-OECD	**6 721**	**7 117**	**14 131 e**	**422**	**717**	**925 e**	**1 209**	**529**	**649 e**
Total Exports	**63 442**	**65 714**	**65 429 e**	**1 416**	**1 669**	**2 103 e**	**10 965**	**10 586**	**9 130 e**

Source: Annual Oil Statistics.

NETHERLANDS
Table 8B: Exports by Country of Destination

000 metric tons

	2000	2001	2002	2000	2001	2002	2000	2001	2002
	Total Gasoline			**Total Kerosene**			**Gas/Diesel Oil**		
Australia	-	-	-	-	-	-	4	-	-
Austria	12	5	2 e	8	1	-	46	13	15 e
Belgium	1 753	1 603	1 299 e	901	1 260	802 e	6 028	6 604	5 239 e
Canada	25	20	126 e	-	-	-	-	-	-
Denmark	11	16	39 e	12	15	-	10	59	-
Finland	6	40	3 e	1	1	-	38	84	-
France	545	315	377 e	65	25	37 e	1 214	1 389	1 259 e
Germany	6 860	5 568	4 461 e	2 568	2 590	2 272 e	9 383	10 455	6 942 e
Greece	55	-	4 e	2	2	-	35	-	34 e
Ireland	4	31	10 e	1	38	12 e	35	50	-
Italy	22	68	58 e	31	1	-	78	1	77 e
Japan	-	-	207 e	-	-	-	-	43	-
Luxembourg	34	31	46 e	1	8	45 e	82	91	64 e
Mexico	-	39	202 e	-	-	-	-	-	-
Netherlands	-	-	-	-	-	-	-	-	-
Norway	10	7	16 e	1	-	-	-	3	1 e
Portugal	156	27	13 e	-	-	-	106	53	89 e
Spain	88	-	87 e	19	-	-	86	266	324 e
Sweden	32	35	30 e	45	12	-	8	10	-
Switzerland	300	289	284 e	104	20	88 e	889	460	1 013 e
Turkey	-	-	-	-	-	-	-	-	-
United Kingdom	501	1 215	737 e	523	126	154 e	794	987	796 e
United States	355	344	1 121 e	-	4	-	179	33	135 e
Other OECD	28	12	26 e	40	8	-	11	17	46 e
OECD	**10 797**	**9 665**	**9 148 e**	**4 322**	**4 111**	**3 410 e**	**19 026**	**20 618**	**16 034 e**
Africa	98	231	219 e	1	-	5 e	129	110	362 e
Latin America	28	82	246 e	2	1	4 e	33	14	266 e
Asia	1	2	172 e	3	11	-	-	56	40 e
China	-	-	-	-	-	-	-	-	-
Non-OECD Europe	32	19	39 e	-	-	-	22	-	191 e
Former Soviet Union	36	10	-	1	-	-	1	4	9 e
Middle East	25	-	7 e	-	-	-	-	-	-
Non-specified/Other	1 729	1 442	2 495 e	111	52	4 e	207	1 295	4 724 e
Non-OECD	**1 949**	**1 786**	**3 178 e**	**118**	**64**	**13 e**	**392**	**1 479**	**5 592 e**
Total Exports	**12 746**	**11 451**	**12 326 e**	**4 440**	**4 175**	**3 423 e**	**19 418**	**22 097**	**21 626 e**
	Fuel Oil (Residual)			**Petroleum Coke**			**Other Products**		
Australia	-	-	-	-	-	-	-	1	-
Austria	-	3	-	-	-	-	25	30	17 e
Belgium	3 205	4 303	4 249 e	-	-	-	2 201	2 390	2 038 e
Canada	97	78	79 e	-	-	-	2	5	10 e
Denmark	2	18	32 e	-	-	-	75	64	22 e
Finland	-	-	44 e	-	-	-	69	59	36 e
France	1 624	1 709	1 363 e	-	-	-	256	276	374 e
Germany	1 257	863	712 e	-	-	-	900	1 148	1 019 e
Greece	-	3	36 e	-	-	-	42	29	15 e
Ireland	10	4	44 e	-	-	-	32	42	21 e
Italy	511	199	51 e	-	-	-	37	46	27 e
Japan	-	-	21 e	-	-	-	2	2	2 e
Luxembourg	-	-	-	-	-	-	1	1	-
Mexico	-	-	-	-	-	-	-	-	-
Netherlands	-	-	-	-	-	-	-	-	-
Norway	-	5	4 e	-	-	-	46	26	76 e
Portugal	151	237	277 e	-	-	-	1	5	-
Spain	90	144	19 e	-	-	-	71	60	59 e
Sweden	13	8	48 e	-	-	-	72	58	63 e
Switzerland	7	-	5 e	-	-	-	15	9	10 e
Turkey	21	-	-	-	-	-	8	2	13 e
United Kingdom	243	750	730 e	-	-	-	112	263	273 e
United States	586	258	960 e	-	-	-	20	79	272 e
Other OECD	-	-	14 e	-	-	-	22	17	12 e
OECD	**7 817**	**8 582**	**8 688 e**	**-**	**-**	**-**	**4 009**	**4 612**	**4 359 e**
Africa	153	69	48 e	-	-	-	53	33	31 e
Latin America	-	-	3 e	-	-	-	11	15	55 e
Asia	1 022	1 370	1 891 e	-	-	-	17	14	21 e
China	-	3	79 e	-	-	-	4	2	-
Non-OECD Europe	223	67	67 e	-	-	-	12	11	11 e
Former Soviet Union	6	3	3 e	-	-	-	9	10	5 e
Middle East	99	-	-	-	-	-	28	23	9 e
Non-specified/Other	890	800	1 334 e	-	-	-	104	122	217 e
Non-OECD	**2 393**	**2 312**	**3 425 e**	**-**	**-**	**-**	**238**	**230**	**349 e**
Total Exports	**10 210**	**10 894**	**12 113 e**	**-**	**-**	**-**	**4 247**	**4 842**	**4 708 e**

Source: Annual Oil Statistics.

NEW ZEALAND

Figure 1
FUEL SHARES OF TPES *

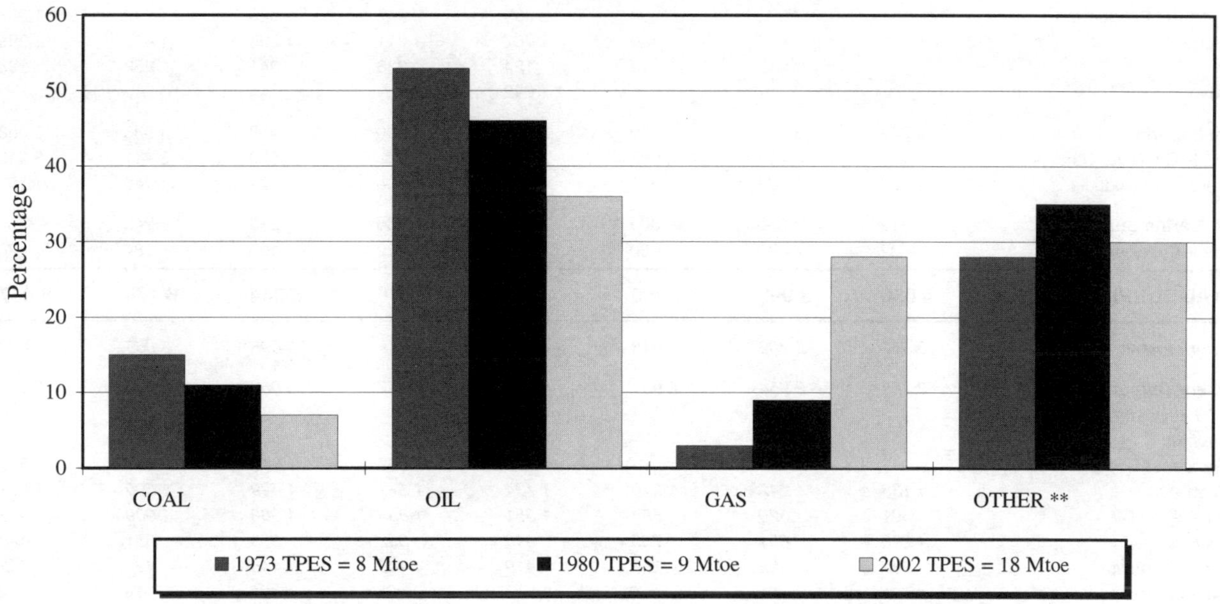

Source: Annual Energy Balances.

* Excluding electricity trade.

** Includes hydro, geothermal, solar, wind, combustible renewables and waste, etc.

Figure 2
OIL PRODUCTION AND CONSUMPTION

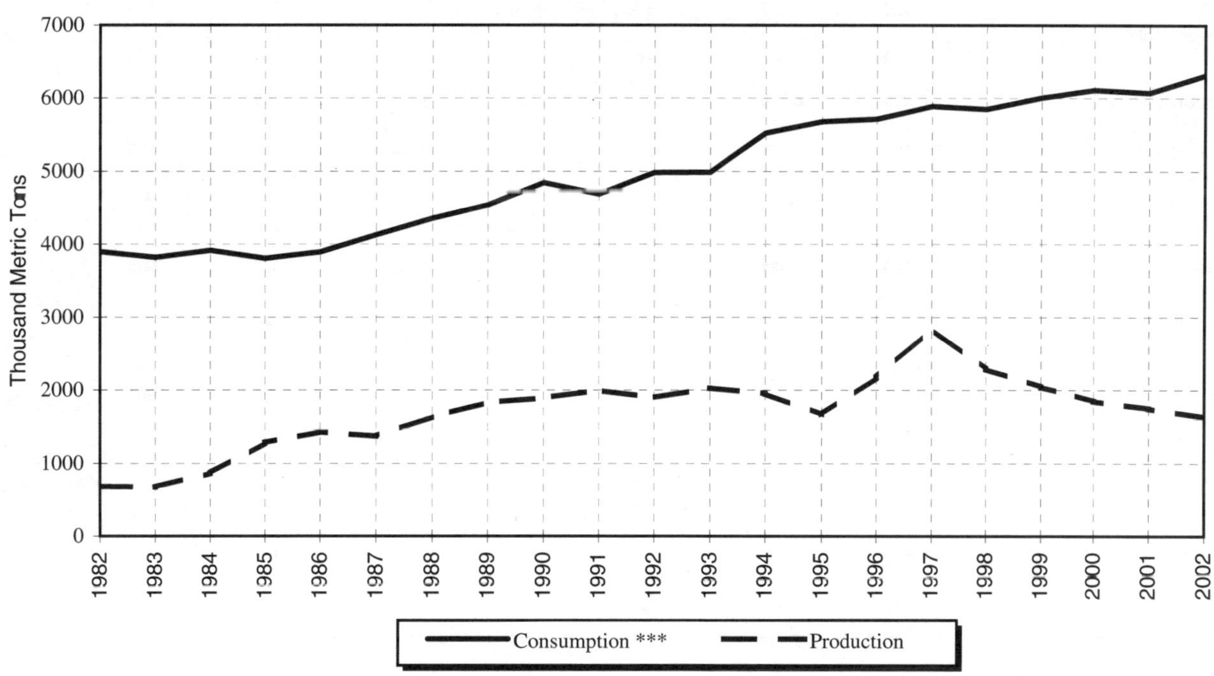

Source: Annual Oil Statistics.

*** Includes refinery fuel and international marine bunkers.

NEW ZEALAND
Table 1: Oil Supply and Consumption - Summary Table

000 metric tons

	1973	1978	1980	1990	1999	2000	2001	2002
Indigenous Production	161	575	338	2 533	2 047	1 849	1 749	1 633 e
Crude Oil	..	562	320	1 767	1 811	1 608	1 497	1 395 e
NGL	..	13	18	124	236	241	252	238 e
Other	..	-	-	642	-	-	-	-
+ Net Imports	4 246	3 700	3 980	2 348	4 490	4 498	4 586	5 068 e
Crude Oil, NGL, Other	3 176	2 417	2 633	2 313	3 282	3 520	3 441	3 915 e
Petroleum Products	1 070	1 283	1 347	35	1 208	978	1 145	1 153 e
- Intl. Marine Bunkers	319	350	365	326	286	233	248	254 e
+ Stock Changes	- 32	73	- 60	18	- 322	120	87	20 e
TOTAL SUPPLY	**4 056**	**3 998**	**3 893**	**4 573**	**5 929**	**6 234**	**6 174**	**6 467 e**
Refinery Intake	*3 311*	*2 989*	*2 913*	*4 759*	*4 942*	*5 096*	*5 055*	*5 226 e*
Refinery Output	*3 265*	*2 954*	*2 904*	*4 682*	*4 877*	*5 060*	*4 981*	*5 298 e*
LPG + Ethane	-	-	-	-	-	-	-	-
Naphtha	-	-	-	-	-	-	-	-
Kerosenes	-	-	-	*788*	*839*	*840*	*818*	*873 e*
Gasolines	*1 133 e*	*1 242*	*1 210*	*1 734*	*1 371*	*1 429*	*1 500*	*1 517 e*
Gas/Diesel Oil	*601 e*	*629*	*654*	*1 351*	*1 887*	*1 983*	*1 909*	*2 048 e*
Fuel Oil	*1 299 e*	*851*	*785*	*390*	*332*	*363*	*331*	*428 e*
Other Products	*232 e*	*232*	*255*	*419*	*448*	*445*	*423*	*432 e*
- Refinery Losses	46	35	9	77	65	36	74	- 72 e
+ Returns	-	-	-	-	-	-	-	-
- Statistical Differences	- 124	63	38	- 20	144	321	276	481 e
TOTAL CONSUMPTION	**4 134**	**3 900**	**3 846**	**4 516**	**5 720**	**5 877**	**5 824**	**6 058 e**
Transformation and Energy	**568**	**202**	**176**	**268**	**279**	**282**	**283**	..
Crude Oil	-	-	-	-	-	-	-	..
LPG + Ethane	-	-	-	-	-	-	-	..
Gas/Diesel Oil	39	32	3	2	-	-	-	..
Fuel Oil	386	29	14	1	-	-	-	..
Petroleum Coke	-	-	-	-	-	-	-	..
Other	143	141	159	265	279	282	283	..
Transport	**2 026**	**2 318**	**2 342**	**3 279**	**4 529**	**4 656**	**4 708**	..
Motor Gasoline	1 409	1 579	1 534	1 825	2 147	2 129	2 144	..
Gas/Diesel Oil	331	375	428	620	1 445	1 566	1 568	..
Subtotal	*1 740*	*1 954*	*1 962*	*2 445*	*3 592*	*3 695*	*3 712*	..
Aviation Fuels	219	303	314	688	884	850	885	..
Other	67	61	66	146	53	111	111	..
Industry	**989**	**881**	**832**	**615**	**573**	**630**	**536**	..
Naphtha	-	-	-	-	-	-	-	..
LPG + Ethane	-	3	1	16	23	23	27	..
Gas/Diesel Oil	227	213	230	181	185	229	221	..
Fuel Oil	499	396	363	50	11	26	24	..
Other	263	269	238	368	354	352	264	..
Other Sectors	**551**	**499**	**496**	**354**	**339**	**309**	**297**	..
Gas/Diesel Oil	270	301	327	204	257	231	217	..
Fuel Oil	124	52	37	45	3	-	-	..
Other	157	146	132	105	79	78	80	..
Memo: Non-Energy Use	*258*	*262*	*232*	*349*	*347*	*344*	*257*	..
Chem/Petchem. Feedst.	-	-	-	-	-	-	-	..
Other	*258*	*262*	*232*	*349*	*347*	*344*	*257*	..

Source: Annual Oil Statistics.

NEW ZEALAND
Table 2: Supply of Crude Oil, NGL, Feedstocks and Other Hydrocarbons

000 metric tons

	1973	1978	1980	1990	1999	2000	2001	2002
CRUDE OIL								
Indigenous Production	..	562	320	1 767	1 811	1 608	1 497	1 395 e
+ Imports	..	1 681	1 989	3 152	4 385	4 482	4 220	4 526 e
- Exports	..	-	-	880	1 355	1 175	1 200	1 030 e
- Direct Use	..	-	-	-	-	-	-	-
+ Stock Change	..	13	- 20	79	- 71	6	119	66 e
- Statistical Difference	..	- 2	- 2	- 12	77	40	16	146 e
Refinery Intake	**..**	**2 258**	**2 291**	**4 130**	**4 693**	**4 881**	**4 620**	**4 811 e**
Closing Stock Level	..	*191*	*124*	*230*	*481*	*475*	*356*	*290 e*
NATURAL GAS LIQUIDS								
Indigenous Production	..	13	18	124	236	241	252	238 e
+ Imports	..	-	-	-	-	-	-	-
- Exports	..	-	-	-	-	-	-	-
- Direct Use	..	13	18	102	191	195	213	194 e
+ Stock Change	..	-	-	-	-	-	-	-
- Statistical Difference	..	-	-	22	45	46	39	44 e
Refinery Intake	**..**	**-**	**-**	**-**	**-**	**-**	**-**	**-**
Closing Stock Level	..	-	-	-	-	-	-	-
REFINERY FEEDSTOCKS								
Backflows	..	-	-	-	-	-	-	-
+ Transfers	..	-	-	-	-	-	-	-
+ Imports	..	736	644	41	252	213	421	419 e
- Exports	..	-	-	-	-	-	-	-
+ Stock Change	..	- 4	- 21	-	- 2	1	- 4	- 10 e
- Statistical Difference	..	1	1	-	1	- 1	- 18	- 6 e
Refinery Intake	**..**	**731**	**622**	**41**	**249**	**215**	**435**	**415 e**
Closing Stock Level	..	*153*	*131*	-	*18*	*17*	*21*	*31 e*
OTHER HYDROCARBONS								
Indigenous Production	..	-	-	642	-	-	-	-
+ Imports	..	-	-	-	-	-	-	-
- Exports	..	-	-	-	-	-	-	-
- Direct Use	..	-	-	79	-	-	-	-
+ Stock Change	..	-	-	- 1	-	-	-	-
- Statistical Difference	..	-	-	- 26	-	-	-	-
Refinery Intake	**..**	**-**	**-**	**588**	**-**	**-**	**-**	**-**
Closing Stock Level	..	-	-	*38*	-	-	-	-
CRUDE OIL, NGL, FEEDSTOCKS AND OTHER								
Indigenous Production	161	575	338	2 533	2 047	1 849	1 749	1 633 e
+ Backflows	-	-	-	-	-	-	-	-
+ Transfers	-	-	-	-	-	-	-	-
+ Imports	3 176	2 417	2 633	3 193	4 637	4 695	4 641	4 945 e
- Exports	-	-	-	880	1 355	1 175	1 200	1 030 e
- Direct Use	1	13	18	181	191	195	213	194 e
+ Stock Change	- 124	9	- 41	78	- 73	7	115	56 e
- Statistical Difference	- 99	- 1	- 1	- 16	123	85	37	184 e
Refinery Intake	**3 311**	**2 989**	**2 913**	**4 759**	**4 942**	**5 096**	**5 055**	**5 226 e**
Closing Stock Level	-	*344*	*255*	*268*	*499*	*492*	*377*	*321 e*

Source: Annual Oil Statistics.

NEW ZEALAND
Table 3: Motor Gasoline - Supply

000 metric tons

	1973	1978	1980	1990	1999	2000	2001	2002
MOTOR GASOLINE								
Refinery Output	**1 133 e**	**1 242**	**1 210**	**1 734**	**1 371**	**1 429**	**1 500**	**1 517 e**
+ Imports	343 e	427	432	281	895	744	777	789 e
- Exports	-	-	-	118	15	13	-	1 e
+ Transfers	-	-	-	81	1	1	3	1 e
+ Stock Change	46 e	15	35	- 50	- 110	65	11	- 68 e
- Statistical Difference	-	- 8	32	- 1	- 41	64	116	172 e
Total Consumption	**1 522**	**1 692**	**1 645**	**1 929**	**2 183**	**2 162**	**2 175**	**2 066 e**
Closing Stock Level	-	*155*	*145*	*269*	*348*	*283*	*272*	*340 e*

Table 4: Aviation Fuels - Supply

000 metric tons

	1973	1978	1980	1990	1999	2000	2001	2002
AVIATION GASOLINE								
Refinery Output	-	-	-	-	-	-	-	-
+ Imports	31 e	32	37	17	15	17	16	12 e
- Exports	-	-	-	-	-	-	-	-
+ Transfers	-	-	-	-	-	-	-	-
+ Stock Change	- 10 e	- 5	- 12	-	- 2	3	- 2	1 e
- Statistical Difference	- 9 e	-	3	1	- 1	5	-	-
Total Consumption	**30**	**27**	**22**	**16**	**14**	**15**	**14**	**13 e**
Closing Stock Level	-	*16*	*25*	*7*	*7*	*4*	*6*	*5 e*
GASOLINE TYPE JET FUEL								
Refinery Output	-	-	-	-	-	-	-	-
+ Imports	-	-	-	-	-	-	-	-
- Exports	-	-	-	-	-	-	-	-
+ Transfers	-	-	-	-	-	-	-	-
+ Stock Change	-	-	-	-	-	-	-	-
- Statistical Difference	-	-	-	-	-	-	-	-
Total Consumption	**-**	**-**	**-**	**-**	**-**	**-**	**-**	**-**
Closing Stock Level	-	-	-	-	-	-	-	-
KEROSENE TYPE JET FUEL								
Refinery Output	-	-	-	**786**	**836**	**838**	**815**	**870 e**
+ Imports	189 e	280	304	22	100	102	118	110 e
- Exports	-	-	-	104	3	-	-	4 e
+ Transfers	-	-	-	- 5	- 2	- 2	- 1	- 2 e
+ Stock Change	2 e	7	- 5	1	- 12	10	6	- 63 e
- Statistical Difference	2 e	11	7	28	49	113	67	51 e
Total Consumption	**189**	**276**	**292**	**672**	**870**	**835**	**871**	**860 e**
Closing Stock Level	-	*45*	*55*	*72*	*95*	*85*	*79*	*142 e*

Source: Annual Oil Statistics.

NEW ZEALAND
Table 5A: Gas/Diesel Oil - Supply and Consumption

000 metric tons

	1973	1978	1980	1990	1999	2000	2001	2002
Refinery Output	**601 e**	**629**	**654**	**1 351**	**1 887**	**1 983**	**1 909**	**2 048 e**
+ Recycled Products	-	-	-	-	-	-	-	-
+ Imports	319 e	381	424	9	235	149	238	267 e
- Exports	-	-	-	235	20	5	-	7 e
- Intl. Marine Bunkers	90 e	131	100	130	80	56	74	80 e
+ Transfers	-	-	-	1	-	- 2	8	1 e
+ Stock Changes	37 e	23	- 2	- 31	- 142	28	12	20 e
- Statistical Differences	-	- 19	- 12	- 42	- 7	71	87	- 29 e
Total Consumption	**867**	**921**	**988**	**1 007**	**1 887**	**2 026**	**2 006**	**2 278 e**
Transformation	**39**	**31**	**3**	**2**	**-**	**-**	**-**	**..**
Public Electricity	39 e	31	3	2	-	-	-	..
Autoproducer Electricity	-	-	-	-	-	-	-	..
Public CHP Plants	-	-	-	-	-	-	-	..
Autoproducers of CHP	-	-	-	-	-	-	-	..
Public Heat Plants	-	-	-	-	-	-	-	..
Autoproducer Heat Plants	-	-	-	-	-	-	-	..
Petrochemical Industry	-	-	-	-	-	-	-	..
Other Transformation	-	-	-	-	-	-	-	..
Energy Sector	**-**	**1**	**-**	**-**	**-**	**-**	**-**	**..**
Refinery Fuel	-	-	-	-	-	-	-	..
Other Energy Sector	-	1	-	-	-	-	-	..
Final Consumption	**828**	**889**	**985**	**1 005**	**1 887**	**2 026**	**2 006**	**..**
Transport	**331**	**375**	**428**	**620**	**1 445**	**1 566**	**1 568**	**..**
Air	-	-	-	-	-	-	-	..
Road	-	-	-	-	-	-	-	..
Rail	-	-	-	-	-	-	-	..
Inland Waterways	14 e	20	82	105	122	115	101	..
Non-Specified	317 e	355	346	515	1 323	1 451	1 467	..
Industry	**227**	**213**	**230**	**181**	**185**	**229**	**221**	**..**
Iron and Steel	-	5	5	6	-	-	-	..
Chemical (incl. Pet.-Chem.)	-	-	-	-	-	-	-	..
Non-Ferrous Metals	-	-	-	-	-	-	-	..
Non-Metallic Minerals	-	-	-	-	-	-	-	..
Transport Equipment	-	-	-	-	-	-	-	..
Machinery	-	-						..
Mining and Quarrying	-	46	47	38	55	64	66	..
Food and Tobacco	-	34	29	18	23	11	6	..
Pulp, Paper and Printing	-	-	-	-	-	-	-	..
Wood and Wood Products	-	-	-	-	-	-	-	..
Construction	-	79	67	64	53	47	46	..
Textiles and Leather	-	-	-	-	-	-	-	..
Non-Specified	227 e	49	82	55	54	107	103	..
Other Sectors	**270**	**301**	**327**	**204**	**257**	**231**	**217**	**..**
Commerce and Public	112 e	167	210	103	72	75	57	..
Residential	44 e	26	12	2	2	1	-	..
Agriculture	114 e	108	105	99	183	155	160	..
Non-Specified	-	-	-	-	-	-	-	..
Memo: Non-Energy Use	-	-	-	-	-	-	-	..
Chem/Petchem. Feedst.	-	-	-	-	-	-	-	..
Other	-	-	-	-	-	-	-	..
Closing Stock Level	-	*240*	*283*	*133*	*347*	*319*	*307*	*287 e*

Source: Annual Oil Statistics.

NEW ZEALAND
Table 5B: Fuel Oil (Residual) - Supply and Consumption

000 metric tons

	1973	1978	1980	1990	1999	2000	2001	2002
Refinery Output	**1 299 e**	**851**	**785**	**390**	**332**	**363**	**331**	**428 e**
+ Recycled Products	-	-	-	-	-	-	-	-
+ Imports	-	-	10	-	-	-	-	8 e
- Exports	-	41	20	16	72	98	54	70 e
- Intl. Marine Bunkers	227 e	219	265	194	206	177	173	174 e
+ Transfers	-	-	-	1	-	5	-	-
+ Stock Changes	2 e	30	- 29	20	-	- 14	- 47	52 e
- Statistical Differences	-	84	11	31	23	- 15	- 27	97 e
Total Consumption	**1 074**	**537**	**470**	**170**	**31**	**94**	**84**	**147 e**
Transformation	**386**	**27**	**12**	**1**	**-**	**-**	**-**	**..**
Public Electricity	386 e	27	12	1	-	-	-	..
Autoproducer Electricity	-	-	-	-	-	-	-	..
Public CHP Plants	-	-	-	-	-	-	-	..
Autoproducers of CHP	-	-	-	-	-	-	-	..
Public Heat Plants	-	-	-	-	-	-	-	..
Autoproducer Heat Plants	-	-	-	-	-	-	-	..
Petrochemical Industry	-	-	-	-	-	-	-	..
Other Transformation	-	-	-	-	-	-	-	..
Energy Sector	**-**	**2**	**2**	**-**	**-**	**-**	**-**	**..**
Refinery Fuel	-	-	-	-	-	-	-	..
Other Energy Sector	-	2	2	-	-	-	-	..
Final Consumption	**688**	**508**	**456**	**169**	**31**	**94**	**84**	**..**
Transport	**65**	**60**	**56**	**74**	**17**	**68**	**60**	**..**
Air	-	-	-	-	-	-	-	..
Road	-	-	-	-	-	-	-	..
Rail	-	-	-	-	-	-	-	..
Inland Waterways	-	-	-	-	-	-	-	..
Non-Specified	65 e	60	56	74	17	68	60	..
Industry	**499**	**396**	**363**	**50**	**11**	**26**	**24**	**..**
Iron and Steel	-	2	1	2	-	-	-	..
Chemical (incl. Pet.-Chem.)	-	-	-	-	-	-	-	..
Non-Ferrous Metals	-	-	-	-	-	-	-	..
Non-Metallic Minerals	-	-	-	-	-	-	-	..
Transport Equipment	-	-	-	-	-	-	-	..
Machinery	-	-	-	-	-	-	-	..
Mining and Quarrying	-	8	4	-	-	-	-	..
Food and Tobacco	-	149	118	12	2	-	-	..
Pulp, Paper and Printing	-	-	-	-	-	-	-	..
Wood and Wood Products	-	-	-	-	-	-	-	..
Construction	-	4	2	1	2	-	-	..
Textiles and Leather	-	-	-	-	-	-	-	..
Non-Specified	499 e	233	238	35	7	26	24	..
Other Sectors	**124**	**52**	**37**	**45**	**3**	**-**	**-**	**..**
Commerce and Public	114 e	46	31	21	1	-	-	..
Residential	-	-	-	-	-	-	-	..
Agriculture	10 e	6	6	24	2	-	-	..
Non-Specified	-	-	-	-	-	-	-	..
Memo: Non-Energy Use	-	-	-	-	-	-	-	..
Chem/Petchem. Feedst.	-	-	-	-	-	-	-	..
Other	-	-	-	-	-	-	-	..
Closing Stock Level	-	*274*	*276*	*44*	*80*	*94*	*141*	*89 e*

Source: Annual Oil Statistics.

NEW ZEALAND
Table 5C: Total Products - Supply and Consumption

000 metric tons

	1973	1978	1980	1990	1999	2000	2001	2002
Refinery Output	**3 265**	**2 954**	**2 904**	**4 682**	**4 877**	**5 060**	**4 981**	**5 298 e**
+ Recycled Products	-	-	-	-	-	-	-	-
+ Imports	1 070	1 324	1 367	519	1 406	1 175	1 284	1 299 e
- Exports	-	41	20	484	198	197	139	146 e
- Intl. Marine Bunkers	319	350	365	326	286	233	248	254 e
+ Transfers	1	13	18	181	191	195	213	194 e
+ Stock Changes	92	64	- 19	- 60	- 249	113	- 28	- 36 e
- Statistical Differences	- 25	64	39	- 4	21	236	239	297 e
Total Consumption	**4 134**	**3 900**	**3 846**	**4 516**	**5 720**	**5 877**	**5 824**	**6 058 e**
Transformation	**438**	**69**	**23**	**3**	**-**	**-**	**-**	**..**
Public Electricity	425	58	15	3	-	-	-	..
Autoproducer Electricity	-	-	-	-	-	-	-	..
Public CHP Plants	-	-	-	-	-	-	-	..
Autoproducers of CHP	-	-	-	-	-	-	-	..
Public Heat Plants	-	-	-	-	-	-	-	..
Autoproducer Heat Plants	-	-	-	-	-	-	-	..
Petrochemical Industry	-	-	-	-	-	-	-	..
Other Transformation	13	11	8	-	-	-	-	..
Energy Sector	**130**	**133**	**153**	**265**	**279**	**282**	**283**	**..**
Refinery Fuel	130	130	151	265	279	282	283	..
Other Energy Sector	-	3	2	-	-	-	-	..
Final Consumption	**3 566**	**3 698**	**3 670**	**4 248**	**5 441**	**5 595**	**5 541**	**..**
Transport	**2 026**	**2 318**	**2 342**	**3 279**	**4 529**	**4 656**	**4 708**	**..**
Air	219	303	314	688	884	850	885	..
Road	1 411	1 580	1 544	1 897	2 183	2 172	2 195	..
Rail	-	-	-	-	-	-	-	..
Inland Waterways	14	20	82	105	122	115	101	..
Non-Specified	382	415	402	589	1 340	1 519	1 527	..
Industry	**989**	**881**	**832**	**615**	**573**	**630**	**536**	**..**
Iron and Steel	-	7	6	9	-	-	-	..
Chemical (incl. Pet.-Chem.)	-	-	-	-	-	-	-	..
Non-Ferrous Metals	50	54	51	105	111	137	105	..
Non-Metallic Minerals	-	-	-	-	-	-	-	..
Transport Equipment	-	-	-	-	-	-	-	..
Machinery	-	-	-	-	-	-	-	..
Mining and Quarrying	-	54	51	39	55	64	66	..
Food and Tobacco	-	183	147	32	25	11	6	..
Pulp, Paper and Printing	-	-	-	-	-	-	-	..
Wood and Wood Products	-	-	-	-	-	-	-	..
Construction	123	212	175	214	227	205	173	..
Textiles and Leather	-	-	-	-	-	-	-	..
Non-Specified	816	371	402	216	155	213	186	..
Other Sectors	**551**	**499**	**496**	**354**	**339**	**309**	**297**	**..**
Commerce and Public	234	222	248	156	90	92	76	..
Residential	77	48	24	10	35	35	36	..
Agriculture	240	229	224	188	214	182	185	..
Non-Specified	-	-	-	-	-	-	-	..
Memo: Non-Energy Use	*258*	*262*	*232*	*349*	*347*	*344*	*257*	..
Chem/Petchem. Feedst.	-	-	-	-	-	-	-	..
Other	*258*	*262*	*232*	*349*	*347*	*344*	*257*	..
Closing Stock Level	-	*803*	*875*	*580*	*968*	*855*	*883*	*919 e*

Source: Annual Oil Statistics.

NEW ZEALAND
Table 6A: Supply and Consumption

000 metric tons

	1980	2000	2001	2002	1980	2000	2001	2002
	LPG AND ETHANE				**REFINERY GAS**			
Refinery Output	**-**	**-**	**-**	**-**	**151**	**229**	**226**	**244 e**
+ Recycled Products	-	-	-	-	-	-	-	-
+ Imports	-	-	-	-	-	-	-	-
- Exports	-	81	85	64 e	-	-	-	-
- Intl. Marine Bunkers	-	-	-	-	-	-	-	-
+ Transfers	14	195	213	194 e	-	-	-	-
+ Stock Change	-	-	- 1	-	-	-	-	-
- Statistical Difference	-	-	- 3	1 e	-	-	-	-
Total Consumption	**14**	**114**	**130**	**129 e**	**151**	**229**	**226**	**244 e**
of which: Ethane	-	-	-	-	-	-	-	-
Transformation	-	-	-	..	-	-	-	..
Energy	-	-	-	..	151	229	226	..
Transport	9	42	50	..	-	-	-	..
Industry	1	23	27	..	-	-	-	..
Other	4	49	53	..	-	-	-	..
Memo: Non-Energy Use	-	-	-	..	-	-	-	..
Closing Stock Level	-	*1*	*2*	*2 e*	-	-	-	-
	NAPHTHA				**OTHER KEROSENE**			
Refinery Output	**-**	**-**	**-**	**-**	**-**	**2**	**3**	**3 e**
+ Recycled Products	-	-	-	-	-	-	-	-
+ Imports	2	-	-	-	30	-	-	-
- Exports	-	-	-	-	-	-	-	-
- Intl. Marine Bunkers	-	-	-	-	-	-	-	-
+ Transfers	-	-	-	-	-	1	1	-
+ Stock Change	4	-	-	-	- 6	-	-	-
- Statistical Difference	2	-	-	-	-	- 2	-	- 1 e
Total Consumption	**4**	**-**	**-**	**-**	**24**	**5**	**4**	**4 e**
Transformation	4	-	-	..	-	-	-	..
Energy	-	-	-	..	-	-	-	..
Transport	-	-	-	..	1	1	1	..
Industry	-	-	-	..	6	4	3	..
Other	-	-	-	..	17	-	-	..
Memo: Non-Energy Use	-	-	-	..	-	-	-	..
Closing Stock Level	-	-	-	-	*19*	-	-	-
	WHITE SPIRIT + SBP				**LUBRICANTS**			
Refinery Output	**-**	**-**	**-**	**..**	**-**	**-**	**-**	**..**
+ Recycled Products	-	-	-	..	-	-	-	..
+ Imports	26	8	4	..	48	14	14	..
- Exports	-	-	-	..	-	-	-	..
- Intl. Marine Bunkers	-	-	-	..	-	-	-	..
+ Transfers	-	-	-	..	-	-	-	..
+ Stock Change	-	-	-	..	-	3	-	..
- Statistical Difference	-	-	-	..	-	-	-	..
Total Consumption	**26**	**8**	**4**	**..**	**48**	**17**	**14**	**..**
Transformation	-	-	-	..	-	-	-	..
Energy	-	-	-	..	-	-	-	..
Transport	-	-	-	..	-	-	-	..
Industry	26	8	4	..	48	17	14	..
Other	-	-	-	..	-	-	-	..
Memo: Non-Energy Use	*26*	*8*	*4*	..	*48*	*17*	*14*	..
Closing Stock Level	-	-	-	..	*32*	*12*	*12*	..

Source: Annual Oil Statistics.

NEW ZEALAND
Table 6B: Supply and Consumption

000 metric tons

	1980	2000	2001	2002	1980	2000	2001	2002
	BITUMEN				PARAFFIN WAXES			
Refinery Output	**104**	**132**	**120**	**..**	**-**	**-**	**-**	**..**
+ Recycled Products	-	-	-	..	-	-	-	..
+ Imports	-	4	12	..	3	-	-	..
- Exports	-	-	-	..	-	-	-	..
- Intl. Marine Bunkers	-	-	-	..	-	-	-	..
+ Transfers	-	-	-	..	-	-	-	..
+ Stock Change	- 4	18	- 8	..	-	-	-	..
- Statistical Difference	- 4	-	-	..	-	-	-	..
Total Consumption	**104**	**154**	**124**	**..**	**3**	**-**	**-**	**..**
Transformation	-	-	-	..	-	-	-	..
Energy	-	-	-	..	-	-	-	..
Transport	-	-	-	..	-	-	-	..
Industry	104	154	124	..	3	-	-	..
Other	-	-	-	..	-	-	-	..
Memo: Non-Energy Use	*104*	*154*	*124*	..	*3*	-	-	..
Closing Stock Level	*40*	*55*	*63*	..	-	-	-	..

	1980	2000	2001	2002	1980	2000	2001	2002
	PETROLEUM COKE				OTHER PRODUCTS			
Refinery Output	**-**	**-**	**-**	**-**	**-**	**84**	**77**	**188 e**
+ Recycled Products	-	-	-	-	-	-	-	-
+ Imports	51	137	105	100 e	-	-	-	13 e
- Exports	-	-	-	-	-	-	-	-
- Intl. Marine Bunkers	-	-	-	-	-	-	-	-
+ Transfers	-	-	-	-	4	- 3	- 11	-
+ Stock Change	-	-	-	-	-	-	1	22 e
- Statistical Difference	-	-	-	-	-	-	-	6 e
Total Consumption	**51**	**137**	**105**	**100 e**	**4**	**81**	**67**	**217 e**
Transformation	-	-	-	..	4	-	-	..
Energy	-	-	-	..	-	53	57	..
Transport	-	-	-	..	-	-	-	..
Industry	51	137	105	..	-	28	10	..
Other	-	-	-	..	-	-	-	..
Memo: Non-Energy Use	*51*	*137*	*105*	..	-	*28*	*10*	..
Closing Stock Level	-	-	-	-	-	*2*	*1*	*54 e*

Source: Annual Oil Statistics.

NEW ZEALAND
Table 7A: Imports by Country of Origin

000 metric tons

	2000	2001	2002	2000	2001	2002	2000	2001	2002
	Crude Oil			Natural Gas Liquids			Feedstocks and Other		
Australia	616	892	913 e	-	-	-	204	324	395 e
Belgium	-	-	-	-	-	-	-	-	-
Canada	-	-	-	-	-	-	-	-	-
Czech Republic	-	-	-	-	-	-	-	-	-
Denmark	-	-	-	-	-	-	-	-	-
Finland	-	-	-	-	-	-	-	-	-
France	-	-	-	-	-	-	-	-	-
Germany	-	-	-	-	-	-	-	-	-
Greece	-	-	-	-	-	-	-	-	-
Hungary	-	-	-	-	-	-	-	-	-
Ireland	-	-	-	-	-	-	-	-	-
Italy	-	-	-	-	-	-	-	-	-
Japan	-	-	-	-	-	-	-	27	-
Korea	-	-	-	-	-	-	-	38	-
Mexico	-	-	-	-	-	-	-	-	-
Netherlands	-	-	-	-	-	-	-	-	-
New Zealand	-	-	-	-	-	-	-	-	-
Norway	-	-	-	-	-	-	-	-	-
Poland	-	-	-	-	-	-	-	-	-
Spain	-	-	-	-	-	-	-	-	-
Sweden	-	-	-	-	-	-	-	-	-
Turkey	-	-	-	-	-	-	-	-	-
United Kingdom	77	-	-	-	-	-	-	-	-
United States	-	-	88 e	-	-	-	-	-	-
Other	-	-	-	-	-	-	-	-	-
OECD	**693**	**892**	**1 001 e**	**-**	**-**	**-**	**204**	**389**	**395 e**
Argentina	184	-	-	-	-	-	-	-	-
Brazil	-	-	-	-	-	-	-	-	-
Colombia	-	-	-	-	-	-	-	-	-
Ecuador	-	-	-	-	-	-	-	-	-
Netherlands Antilles	-	-	-	-	-	-	-	-	-
Peru	-	-	-	-	-	-	-	-	-
Trinidad and Tobago	-	-	-	-	-	-	-	-	-
Venezuela	-	-	-	-	-	-	-	-	-
Other	52	-	-	-	-	-	-	-	-
Latin America	**236**	**-**	**-**	**-**	**-**	**-**	**-**	**-**	**-**
Romania	-	-	-	-	-	-	-	-	-
Other	-	-	-	-	-	-	-	-	-
Non-OECD Europe	**-**	**-**	**-**	**-**	**-**	**-**	**-**	**-**	**-**
FSU	**-**	**-**	**-**	**-**	**-**	**-**	**-**	**-**	**-**
Iran	113	80	181 e	-	-	-	-	-	-
Iraq	-	-	-	-	-	-	-	-	-
Kuwait	106	280	40 e	-	-	-	-	-	-
Oman	675	818	890 e	-	-	-	-	-	-
Qatar	251	225	238 e	-	-	-	-	-	-
Saudi Arabia	977	642	706 e	-	-	-	-	-	-
United Arab Emirates	216	354	72 e	-	-	-	-	-	-
Other	418	-	232 e	-	-	-	-	-	-
Middle East	**2 756**	**2 399**	**2 359 e**	**-**	**-**	**-**	**-**	**-**	**-**
India	-	-	-	-	-	-	-	-	-
Indonesia	112	181	203 e	-	-	-	-	-	-
Singapore	29	41	-	-	-	-	9	32	24 e
Other	531	707	835 e	-	-	-	-	-	-
Asia	**672**	**929**	**1 038 e**	**-**	**-**	**-**	**9**	**32**	**24 e**
China	**-**	**-**	**-**	**-**	**-**	**-**	**-**	**-**	**-**
Algeria	-	-	-	-	-	-	-	-	-
Angola/Cabinada	-	-	-	-	-	-	-	-	-
Cameroon	-	-	-	-	-	-	-	-	-
Egypt	-	-	-	-	-	-	-	-	-
Gabon	-	-	-	-	-	-	-	-	-
Libya	-	-	-	-	-	-	-	-	-
Nigeria	125	-	128 e	-	-	-	-	-	-
Tunisia	-	-	-	-	-	-	-	-	-
Other	-	-	-	-	-	-	-	-	-
Africa	**125**	**-**	**128 e**	**-**	**-**	**-**	**-**	**-**	**-**
Non-spec./Other	**-**	**-**	**-**	**-**	**-**	**-**	**-**	**-**	**-**
Total Imports	**4 482**	**4 220**	**4 526 e**	**-**	**-**	**-**	**213**	**421**	**419 e**
Memo: Total OPEC	*1 900*	*1 762*	*1 568 e*	-	-	-	-	-	-

Source: Annual Oil Statistics.

NEW ZEALAND
Table 7B: Imports by Country of Origin

000 metric tons

	2000	2001	2002	2000	2001	2002	2000	2001	2002
	Total Products			**Liquefied Petroleum Gases**			**Naphtha**		
Australia	728	688	914 e	-	-	-	-	-	-
Belgium	-	-	-	-	-	-	-	-	-
Canada	-	-	-	-	-	-	-	-	-
Czech Republic	-	-	-	-	-	-	-	-	-
Denmark	-	-	-	-	-	-	-	-	-
Finland	-	-	-	-	-	-	-	-	-
France	-	-	-	-	-	-	-	-	-
Germany	-	-	-	-	-	-	-	-	-
Greece	-	-	-	-	-	-	-	-	-
Hungary	-	-	-	-	-	-	-	-	-
Ireland	-	-	-	-	-	-	-	-	-
Italy	-	-	-	-	-	-	-	-	-
Japan	-	3	29 e	-	-	-	-	-	-
Korea	-	40	34 e	-	-	-	-	-	-
Mexico	-	-	-	-	-	-	-	-	-
Netherlands	-	-	-	-	-	-	-	-	-
New Zealand	-	-	-	-	-	-	-	-	-
Norway	-	-	-	-	-	-	-	-	-
Poland	-	-	-	-	-	-	-	-	-
Spain	-	-	-	-	-	-	-	-	-
Sweden	-	-	-	-	-	-	-	-	-
Turkey	-	-	-	-	-	-	-	-	-
United Kingdom	-	-	-	-	-	-	-	-	-
United States	138	201	136 e	-	-	-	-	-	-
Other	-	-	-	-	-	-	-	-	-
OECD	**866**	**932**	**1 113 e**	**-**	**-**	**-**	**-**	**-**	**-**
Argentina	30	-	-	-	-	-	-	-	-
Brazil	-	-	-	-	-	-	-	-	-
Colombia	-	-	-	-	-	-	-	-	-
Ecuador	-	-	-	-	-	-	-	-	-
Netherlands Antilles	-	-	-	-	-	-	-	-	-
Peru	-	-	-	-	-	-	-	-	-
Trinidad and Tobago	-	-	-	-	-	-	-	-	-
Venezuela	-	-	-	-	-	-	-	-	-
Other	-	-	-	-	-	-	-	-	-
Latin America	**30**	**-**	**-**	**-**	**-**	**-**	**-**	**-**	**-**
Romania	-	-	-	-	-	-	-	-	-
Other	-	-	-	-	-	-	-	-	-
Non-OECD Europe	**-**	**-**	**-**	**-**	**-**	**-**	**-**	**-**	**-**
FSU	**-**	**-**	**-**	**-**	**-**	**-**	**-**	**-**	**-**
Iran	-	-	-	-	-	-	-	-	-
Iraq	-	-	-	-	-	-	-	-	-
Kuwait	-	-	-	-	-	-	-	-	-
Oman	-	-	-	-	-	-	-	-	-
Qatar	-	-	-	-	-	-	-	-	-
Saudi Arabia	173	64	39 e	-	-	-			
United Arab Emirates	-	-	-	-	-	-	-	-	-
Other	-	-	-	-	-	-	-	-	-
Middle East	**173**	**64**	**39 e**	**-**	**-**	**-**	**-**	**-**	**-**
India	-	-	-	-	-	-	-	-	-
Indonesia	-	-	-	-	-	-	-	-	-
Singapore	34	224	110 e	-	-	-	-	-	-
Other	72	62	36 e	-	-	-	-	-	-
Asia	**106**	**286**	**146 e**	**-**	**-**	**-**	**-**	**-**	**-**
China	**-**	**-**	**-**	**-**	**-**	**-**	**-**	**-**	**-**
Algeria	-	-	-	-	-	-	-	-	-
Angola/Cabinada	-	-	-	-	-	-	-	-	-
Cameroon	-	-	-	-	-	-	-	-	-
Egypt	-	-	-	-	-	-	-	-	-
Gabon	-	-	-	-	-	-	-	-	-
Libya	-	-	-	-	-	-	-	-	-
Nigeria	-	-	-	-	-	-	-	-	-
Tunisia	-	-	-	-	-	-	-	-	-
Other	-	-	-	-	-	-	-	-	-
Africa	**-**	**-**	**-**	**-**	**-**	**-**	**-**	**-**	**-**
Non-spec./Other	**-**	**2**	**1 e**	**-**	**-**	**-**	**-**	**-**	**-**
Total Imports	**1 175**	**1 284**	**1 299 e**	**-**	**-**	**-**	**-**	**-**	**-**
Memo: Total OPEC	*173*	*64*	*39 e*	-	-	-	-	-	-

Source: Annual Oil Statistics.

NEW ZEALAND
Table 7C: Imports by Country of Origin

000 metric tons

	2000	2001	2002	2000	2001	2002	2000	2001	2002
	Total Gasoline			Total Kerosene			Gas/Diesel		
Australia	598	476	623 e	80	106	110 e	31	77	161 e
Belgium	-	-	-	-	-	-	-	-	-
Canada	-	-	-	-	-	-	-	-	-
Czech Republic	-	-	-	-	-	-	-	-	-
Denmark	-	-	-	-	-	-	-	-	-
Finland	-	-	-	-	-	-	-	-	-
France	-	-	-	-	-	-	-	-	-
Germany	-	-	-	-	-	-	-	-	-
Greece	-	-	-	-	-	-	-	-	-
Hungary	-	-	-	-	-	-	-	-	-
Ireland	-	-	-	-	-	-	-	-	-
Italy	-	-	-	-	-	-	-	-	-
Japan	-	3	-	-	-	-	-	-	29 e
Korea	-	6	-	-	-	-	-	34	34 e
Mexico	-	-	-	-	-	-	-	-	-
Netherlands	-	-	-	-	-	-	-	-	-
New Zealand	-	-	-	-	-	-	-	-	-
Norway	-	-	-	-	-	-	-	-	-
Poland	-	-	-	-	-	-	-	-	-
Spain	-	-	-	-	-	-	-	-	-
Sweden	-	-	-	-	-	-	-	-	-
Turkey	-	-	-	-	-	-	-	-	-
United Kingdom	-	-	-	-	-	-	-	-	-
United States	-	97	27 e	-	-	-	-	-	9 e
Other	-	-	-	-	-	-	-	-	-
OECD	**598**	**582**	**650 e**	**80**	**106**	**110 e**	**31**	**111**	**233 e**
Argentina	23	-	-	-	-	-	7	-	-
Brazil	-	-	-	-	-	-	-	-	-
Colombia	-	-	-	-	-	-	-	-	-
Ecuador	-	-	-	-	-	-	-	-	-
Netherlands Antilles	-	-	-	-	-	-	-	-	-
Peru	-	-	-	-	-	-	-	-	-
Trinidad and Tobago	-	-	-	-	-	-	-	-	-
Venezuela	-	-	-	-	-	-	-	-	-
Other	-	-	-	-	-	-	-	-	-
Latin America	**23**	**-**	**-**	**-**	**-**	**-**	**7**	**-**	**-**
Romania	-	-	-	-	-	-	-	-	-
Other	-	-	-	-	-	-	-	-	-
Non-OECD Europe	**-**	**-**	**-**	**-**	**-**	**-**	**-**	**-**	**-**
FSU	**-**	**-**	**-**	**-**	**-**	**-**	**-**	**-**	**-**
Iran	-	-	-	-	-	-	-	-	-
Iraq	-	-	-	-	-	-	-	-	-
Kuwait	-	-	-	-	-	-	-	-	-
Oman	-	-	-	-	-	-	-	-	-
Qatar	-	-	-	-	-	-	-	-	-
Saudi Arabia	123	64	27 e	11	-	-	39	-	12 e
United Arab Emirates	-	-	-	-	-	-	-	-	-
Other	-	-	-	-	-	-	-	-	-
Middle East	**123**	**64**	**27 e**	**11**	**-**	**-**	**39**	**-**	**12 e**
India	-	-	-	-	-	-	-	-	-
Indonesia	-	-	-	-	-	-	-	-	-
Singapore	17	147	94 e	11	12	-	-	65	16 e
Other	-	-	30 e	-	-	-	72	62	6 e
Asia	**17**	**147**	**124 e**	**11**	**12**	**-**	**72**	**127**	**22 e**
China	**-**	**-**	**-**	**-**	**-**	**-**	**-**	**-**	**-**
Algeria	-	-	-	-	-	-	-	-	-
Angola/Cabinada	-	-	-	-	-	-	-	-	-
Cameroon	-	-	-	-	-	-	-	-	-
Egypt	-	-	-	-	-	-	-	-	-
Gabon	-	-	-	-	-	-	-	-	-
Libya	-	-	-	-	-	-	-	-	-
Nigeria	-	-	-	-	-	-	-	-	-
Tunisia	-	-	-	-	-	-	-	-	-
Other	-	-	-	-	-	-	-	-	-
Africa	**-**	**-**	**-**	**-**	**-**	**-**	**-**	**-**	**-**
Non-spec./Other	**-**	**-**	**-**	**-**	**-**	**-**	**-**	**-**	**-**
Total Imports	**761**	**793**	**801 e**	**102**	**118**	**110 e**	**149**	**238**	**267 e**
Memo: Total OPEC	*123*	*64*	*27 e*	*11*	-	-	*39*	-	*12 e*

Source: Annual Oil Statistics.

NEW ZEALAND
Table 7D: Imports by Country of Origin

000 metric tons

	2000	2001	2002	2000	2001	2002	2000	2001	2002
	Fuel Oil (Residual)			**Petroleum Coke**			**Other Products**		
Australia	-	-	8 e	1	1	-	18	28	12 e
Belgium	-	-	-	-	-	-	-	-	-
Canada	-	-	-	-	-	-	-	-	-
Czech Republic	-	-	-	-	-	-	-	-	-
Denmark	-	-	-	-	-	-	-	-	-
Finland	-	-	-	-	-	-	-	-	-
France	-	-	-	-	-	-	-	-	-
Germany	-	-	-	-	-	-	-	-	-
Greece	-	-	-	-	-	-	-	-	-
Hungary	-	-	-	-	-	-	-	-	-
Ireland	-	-	-	-	-	-	-	-	-
Italy	-	-	-	-	-	-	-	-	-
Japan	-	-	-	-	-	-	-	-	-
Korea	-	-	-	-	-	-	-	-	-
Mexico	-	-	-	-	-	-	-	-	-
Netherlands	-	-	-	-	-	-	-	-	-
New Zealand	-	-	-	-	-	-	-	-	-
Norway	-	-	-	-	-	-	-	-	-
Poland	-	-	-	-	-	-	-	-	-
Spain	-	-	-	-	-	-	-	-	-
Sweden	-	-	-	-	-	-	-	-	-
Turkey	-	-	-	-	-	-	-	-	-
United Kingdom	-	-	-	-	-	-	-	-	-
United States	-	-	-	136	104	100 e	2	-	-
Other	-	-	-	-	-	-	-	-	-
OECD	**-**	**-**	**8 e**	**137**	**105**	**100 e**	**20**	**28**	**12 e**
Argentina	-	-	-	-	-	-	-	-	-
Brazil	-	-	-	-	-	-	-	-	-
Colombia	-	-	-	-	-	-	-	-	-
Ecuador	-	-	-	-	-	-	-	-	-
Netherlands Antilles	-	-	-	-	-	-	-	-	-
Peru	-	-	-	-	-	-	-	-	-
Trinidad and Tobago	-	-	-	-	-	-	-	-	-
Venezuela	-	-	-	-	-	-	-	-	-
Other	-	-	-	-	-	-	-	-	-
Latin America	**-**	**-**	**-**	**-**	**-**	**-**	**-**	**-**	**-**
Romania	-	-	-	-	-	-	-	-	-
Other	-	-	-	-	-	-	-	-	-
Non-OECD Europe	**-**	**-**	**-**	**-**	**-**	**-**	**-**	**-**	**-**
FSU	**-**	**-**	**-**	**-**	**-**	**-**	**-**	**-**	**-**
Iran	-	-	-	-	-	-	-	-	-
Iraq	-	-	-	-	-	-	-	-	-
Kuwait	-	-	-	-	-	-	-	-	-
Oman	-	-	-	-	-	-	-	-	-
Qatar	-	-	-	-	-	-	-	-	-
Saudi Arabia	-	-	-	-	-	-	-	-	-
United Arab Emirates	-	-	-	-	-	-	-	-	-
Other	-	-	-	-	-	-	-	-	-
Middle East	**-**	**-**	**-**	**-**	**-**	**-**	**-**	**-**	**-**
India	-	-	-	-	-	-	-	-	-
Indonesia	-	-	-	-	-	-	-	-	-
Singapore	-	-	-	-	-	-	6	-	-
Other	-	-	-	-	-	-	-	-	-
Asia	**-**	**-**	**-**	**-**	**-**	**-**	**6**	**-**	**-**
China	**-**	**-**	**-**	**-**	**-**	**-**	**-**	**-**	**-**
Algeria	-	-	-	-	-	-	-	-	-
Angola/Cabinada	-	-	-	-	-	-	-	-	-
Cameroon	-	-	-	-	-	-	-	-	-
Egypt	-	-	-	-	-	-	-	-	-
Gabon	-	-	-	-	-	-	-	-	-
Libya	-	-	-	-	-	-	-	-	-
Nigeria	-	-	-	-	-	-	-	-	-
Tunisia	-	-	-	-	-	-	-	-	-
Other	-	-	-	-	-	-	-	-	-
Africa	**-**	**-**	**-**	**-**	**-**	**-**	**-**	**-**	**-**
Non-spec./Other	**-**	**-**	**-**	**-**	**-**	**-**	**-**	**2**	**1 e**
Total Imports	**-**	**-**	**8 e**	**137**	**105**	**100 e**	**26**	**30**	**13 e**
Memo: Total OPEC	-	-	-	-	-	-	-	-	-

Source: Annual Oil Statistics.

NEW ZEALAND
Table 8A: Exports by Country of Destination

000 metric tons

	Crude Oil 2000	Crude Oil 2001	Crude Oil 2002	Natural Gas Liquids 2000	Natural Gas Liquids 2001	Natural Gas Liquids 2002	Feedstocks and Other 2000	Feedstocks and Other 2001	Feedstocks and Other 2002
Australia	-	-	-	-	-	-	-	-	-
Austria	-	-	-	-	-	-	-	-	-
Belgium	-	-	-	-	-	-	-	-	-
Canada	-	-	-	-	-	-	-	-	-
Denmark	-	-	-	-	-	-	-	-	-
Finland	-	-	-	-	-	-	-	-	-
France	-	-	-	-	-	-	-	-	-
Germany	-	-	-	-	-	-	-	-	-
Greece	-	-	-	-	-	-	-	-	-
Ireland	-	-	-	-	-	-	-	-	-
Italy	-	-	-	-	-	-	-	-	-
Japan	-	-	-	-	-	-	-	-	-
Luxembourg	-	-	-	-	-	-	-	-	-
Mexico	-	-	-	-	-	-	-	-	-
Netherlands	-	-	-	-	-	-	-	-	-
Norway	-	-	-	-	-	-	-	-	-
Portugal	-	-	-	-	-	-	-	-	-
Spain	-	-	-	-	-	-	-	-	-
Sweden	-	-	-	-	-	-	-	-	-
Switzerland	-	-	-	-	-	-	-	-	-
Turkey	-	-	-	-	-	-	-	-	-
United Kingdom	-	-	-	-	-	-	-	-	-
United States	-	-	-	-	-	-	-	-	-
Other OECD	-	-	-	-	-	-	-	-	-
OECD	**-**	**-**	**-**	**-**	**-**	**-**	**-**	**-**	**-**
Africa	-	-	-	-	-	-	-	-	-
Latin America	-	-	-	-	-	-	-	-	-
Asia	-	-	-	-	-	-	-	-	-
China	-	-	-	-	-	-	-	-	-
Non-OECD Europe	-	-	-	-	-	-	-	-	-
Former Soviet Union	-	-	-	-	-	-	-	-	-
Middle East	-	-	-	-	-	-	-	-	-
Non-specified/Other	1 175	1 200	1 030 e	-	-	-	-	-	-
Non-OECD	**1 175**	**1 200**	**1 030 e**	**-**	**-**	**-**	**-**	**-**	**-**
Total Exports	**1 175**	**1 200**	**1 030 e**	**-**	**-**	**-**	**-**	**-**	**-**

	Total Products 2000	Total Products 2001	Total Products 2002	Liquefied Petroleum Gas 2000	Liquefied Petroleum Gas 2001	Liquefied Petroleum Gas 2002	Naphtha 2000	Naphtha 2001	Naphtha 2002
Australia	95	46	82 e	-	-	-	-	-	-
Austria	-	-	-	-	-	-	-	-	-
Belgium	-	-	-	-	-	-	-	-	-
Canada	-	-	-	-	-	-	-	-	-
Denmark	-	-	-	-	-	-	-	-	-
Finland	-	-	-	-	-	-	-	-	-
France	-	-	-	-	-	-	-	-	-
Germany	-	-	-	-	-	-	-	-	-
Greece	-	-	-	-	-	-	-	-	-
Ireland	-	-	-	-	-	-	-	-	-
Italy	-	-	-	-	-	-	-	-	-
Japan	-	-	-	-	-	-	-	-	-
Luxembourg	-	-	-	-	-	-	-	-	-
Mexico	-	-	-	-	-	-	-	-	-
Netherlands	-	-	-	-	-	-	-	-	-
Norway	-	-	-	-	-	-	-	-	-
Portugal	-	-	-	-	-	-	-	-	-
Spain	-	-	-	-	-	-	-	-	-
Sweden	-	-	-	-	-	-	-	-	-
Switzerland	-	-	-	-	-	-	-	-	-
Turkey	-	-	-	-	-	-	-	-	-
United Kingdom	-	-	-	-	-	-	-	-	-
United States	-	-	-	-	-	-	-	-	-
Other OECD	-	-	-	-	-	-	-	-	-
OECD	**95**	**46**	**82 e**	**-**	**-**	**-**	**-**	**-**	**-**
Africa	-	-	-	-	-	-	-	-	-
Latin America	-	-	-	-	-	-	-	-	-
Asia	21	8	-	-	-	-	-	-	-
China	-	-	-	-	-	-	-	-	-
Non-OECD Europe	-	-	-	-	-	-	-	-	-
Former Soviet Union	-	-	-	-	-	-	-	-	-
Middle East	-	-	-	-	-	-	-	-	-
Non-specified/Other	81	85	64 e	81	85	64 e	-	-	-
Non-OECD	**102**	**93**	**64 e**	**81**	**85**	**64 e**	**-**	**-**	**-**
Total Exports	**197**	**139**	**146 e**	**81**	**85**	**64 e**	**-**	**-**	**-**

Source: Annual Oil Statistics.

NEW ZEALAND

Table 8B: Exports by Country of Destination

000 metric tons

	Total Gasoline 2000	Total Gasoline 2001	Total Gasoline 2002	Total Kerosene 2000	Total Kerosene 2001	Total Kerosene 2002	Gas/Diesel Oil 2000	Gas/Diesel Oil 2001	Gas/Diesel Oil 2002
Australia	13	-	1 e	-	-	4 e	5	-	7 e
Austria	-	-	-	-	-	-	-	-	-
Belgium	-	-	-	-	-	-	-	-	-
Canada	-	-	-	-	-	-	-	-	-
Denmark	-	-	-	-	-	-	-	-	-
Finland	-	-	-	-	-	-	-	-	-
France	-	-	-	-	-	-	-	-	-
Germany	-	-	-	-	-	-	-	-	-
Greece	-	-	-	-	-	-	-	-	-
Ireland	-	-	-	-	-	-	-	-	-
Italy	-	-	-	-	-	-	-	-	-
Japan	-	-	-	-	-	-	-	-	-
Luxembourg	-	-	-	-	-	-	-	-	-
Mexico	-	-	-	-	-	-	-	-	-
Netherlands	-	-	-	-	-	-	-	-	-
Norway	-	-	-	-	-	-	-	-	-
Portugal	-	-	-	-	-	-	-	-	-
Spain	-	-	-	-	-	-	-	-	-
Sweden	-	-	-	-	-	-	-	-	-
Switzerland	-	-	-	-	-	-	-	-	-
Turkey	-	-	-	-	-	-	-	-	-
United Kingdom	-	-	-	-	-	-	-	-	-
United States	-	-	-	-	-	-	-	-	-
Other OECD	-	-	-	-	-	-	-	-	-
OECD	**13**	**-**	**1 e**	**-**	**-**	**4 e**	**5**	**-**	**7 e**
Africa	-	-	-	-	-	-	-	-	-
Latin America	-	-	-	-	-	-	-	-	-
Asia	-	-	-	-	-	-	-	-	-
China	-	-	-	-	-	-	-	-	-
Non-OECD Europe	-	-	-	-	-	-	-	-	-
Former Soviet Union	-	-	-	-	-	-	-	-	-
Middle East	-	-	-	-	-	-	-	-	-
Non-specified/Other	-	-	-	-	-	-	-	-	-
Non-OECD	**-**	**-**	**-**	**-**	**-**	**-**	**-**	**-**	**-**
Total Exports	**13**	**-**	**1 e**	**-**	**-**	**4 e**	**5**	**-**	**7 e**

	Fuel Oil (Residual) 2000	Fuel Oil (Residual) 2001	Fuel Oil (Residual) 2002	Petroleum Coke 2000	Petroleum Coke 2001	Petroleum Coke 2002	Other Products 2000	Other Products 2001	Other Products 2002
Australia	77	46	70 e	-	-	-	-	-	-
Austria	-	-	-	-	-	-	-	-	-
Belgium	-	-	-	-	-	-	-	-	-
Canada	-	-	-	-	-	-	-	-	-
Denmark	-	-	-	-	-	-	-	-	-
Finland	-	-	-	-	-	-	-	-	-
France	-	-	-	-	-	-	-	-	-
Germany	-	-	-	-	-	-	-	-	-
Greece	-	-	-	-	-	-	-	-	-
Ireland	-	-	-	-	-	-	-	-	-
Italy	-	-	-	-	-	-	-	-	-
Japan	-	-	-	-	-	-	-	-	-
Luxembourg	-	-	-	-	-	-	-	-	-
Mexico	-	-	-	-	-	-	-	-	-
Netherlands	-	-	-	-	-	-	-	-	-
Norway	-	-	-	-	-	-	-	-	-
Portugal	-	-	-	-	-	-	-	-	-
Spain	-	-	-	-	-	-	-	-	-
Sweden	-	-	-	-	-	-	-	-	-
Switzerland	-	-	-	-	-	-	-	-	-
Turkey	-	-	-	-	-	-	-	-	-
United Kingdom	-	-	-	-	-	-	-	-	-
United States	-	-	-	-	-	-	-	-	-
Other OECD	-	-	-	-	-	-	-	-	-
OECD	**77**	**46**	**70 e**	**-**	**-**	**-**	**-**	**-**	**-**
Africa	-	-	-	-	-	-	-	-	-
Latin America	-	-	-	-	-	-	-	-	-
Asia	21	8	-	-	-	-	-	-	-
China	-	-	-	-	-	-	-	-	-
Non-OECD Europe	-	-	-	-	-	-	-	-	-
Former Soviet Union	-	-	-	-	-	-	-	-	-
Middle East	-	-	-	-	-	-	-	-	-
Non-specified/Other	-	-	-	-	-	-	-	-	-
Non-OECD	**21**	**8**	**-**	**-**	**-**	**-**	**-**	**-**	**-**
Total Exports	**98**	**54**	**70 e**	**-**	**-**	**-**	**-**	**-**	**-**

Source: Annual Oil Statistics.

NORWAY

Figure 1
FUEL SHARES OF TPES *

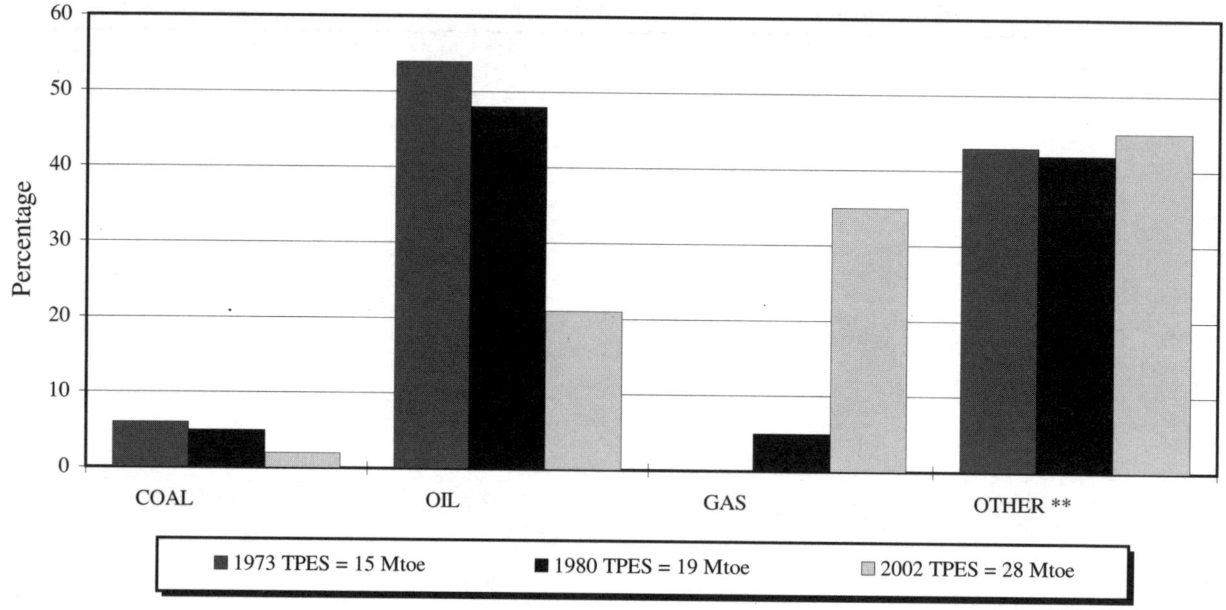

Source: Annual Energy Balances.

* Excluding electricity trade.

** Includes hydro, geothermal, solar, wind, combustible renewables and waste, etc.

Figure 2
OIL PRODUCTION AND CONSUMPTION

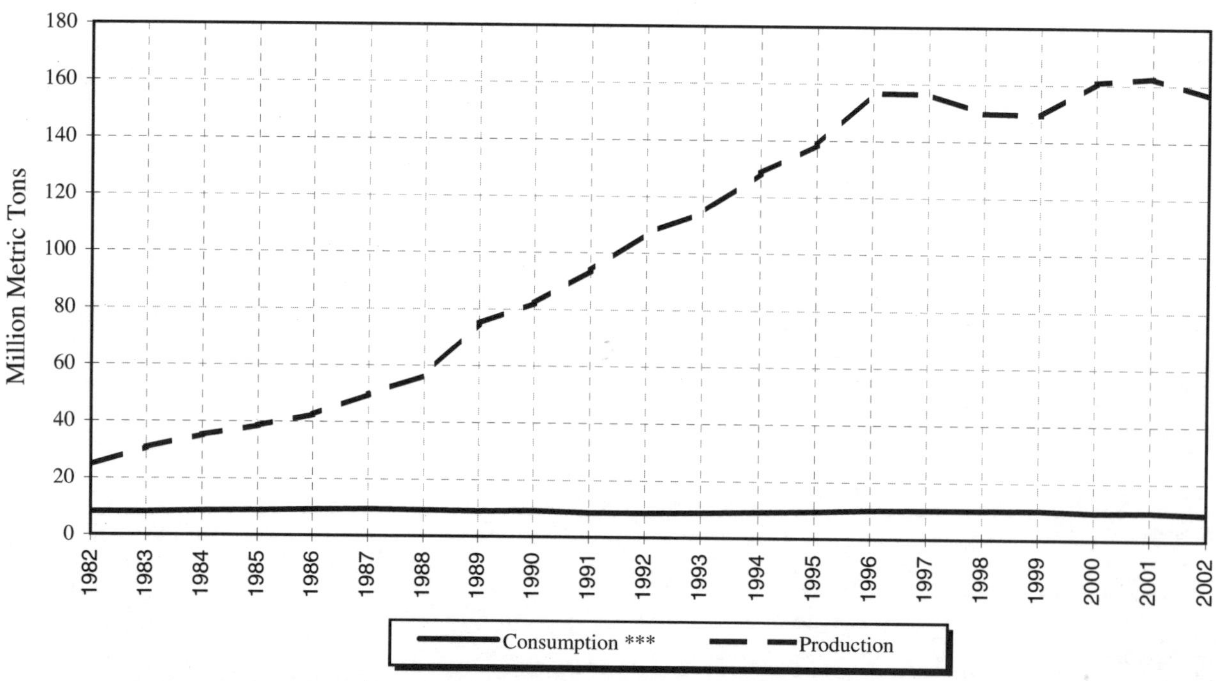

Source: Annual Oil Statistics.

*** Includes refinery fuel and international marine bunkers.

NORWAY
Table 1: Oil Supply and Consumption - Summary Table

000 metric tons

	1973	1978	1980	1990	1999	2000	2001	2002
Indigenous Production	1 595	17 000	24 277	82 088	149 350	160 990	162 536	156 381 e
Crude Oil	..	17 000	23 055	79 663	145 843	157 522	157 551	151 181 e
NGL	..	-	1 222	2 425	3 507	3 468	4 985	5 200 e
Other	..	-	-	-	-	-	-	-
+ Net Imports	6 839	-7 279	-14 620	-71 352	-139 465	-150 750	-155 565	-149 609 e
Crude Oil, NGL, Other	4 545	-8 226	-15 692	-66 622	-134 238	-145 268	-150 494	-144 553 e
Petroleum Products	2 294	947	1 072	-4 730	-5 227	-5 482	-5 071	-5 056 e
- Intl. Marine Bunkers	647	430	284	450	858	828	812	693 e
+ Stock Changes	409	316	- 397	-1 808	- 23	- 392	2 198	- 40 e
TOTAL SUPPLY	**8 196**	**9 607**	**8 976**	**8 478**	**9 004**	**9 020**	**8 357**	**6 039 e**
Refinery Intake	*6 497*	*8 739*	*7 884*	*13 229*	*15 368*	*15 076*	*14 367*	*13 321 e*
Refinery Output	*6 112*	*8 237*	*7 770*	*13 024*	*15 335*	*15 196*	*13 984*	*13 104 e*
LPG + Ethane	*48*	*28*	*119*	*245*	*318*	*330*	*284*	*289 e*
Naphtha	*269*	*330*	*485*	*564*	*886*	*1 161*	*920*	*1 125 e*
Kerosenes	*379*	*406*	*347*	*1 066*	*876*	*812*	*665*	*652 e*
Gasolines	*704*	*1 339*	*1 186*	*3 203*	*3 202*	*3 385*	*3 287*	*2 920 e*
Gas/Diesel Oil	*1 948*	*3 316*	*3 364*	*5 906*	*7 253*	*6 916*	*6 478*	*6 009 e*
Fuel Oil	*2 558*	*2 226*	*1 830*	*1 129*	*1 897*	*1 703*	*1 647*	*1 503 e*
Other Products	*206*	*592*	*439*	*911*	*903*	*889*	*703*	*606 e*
- Refinery Losses	385	502	114	205	33	- 120	383	217 e
+ Returns	-	15	77	97	60	51	53	41 e
- Statistical Differences	74	414	199	- 238	- 296	504	- 902	-2 470 e
TOTAL CONSUMPTION	**7 737**	**8 706**	**8 740**	**8 608**	**9 327**	**8 687**	**8 929**	**8 333 e**
Transformation and Energy	**130**	**484**	**480**	**898**	**1 017**	**970**	**754**	..
Crude Oil	-	-	-	-	-	-	-	..
LPG + Ethane	-	7	120	96	77	80	82	..
Gas/Diesel Oil	33	96	92	125	220	181	171	..
Fuel Oil	-	45	57	4	4	13	7	..
Petroleum Coke	-	-	-	-	-	-	-	..
Other	97	336	211	673	716	696	494	..
Transport	**2 442**	**2 927**	**2 964**	**3 940**	**4 582**	**4 226**	**4 296**	..
Motor Gasoline	1 078	1 330	1 371	1 771	1 648	1 598	1 658	..
Gas/Diesel Oil	1 073	1 137	1 101	1 562	2 187	1 974	2 013	..
Subtotal	*2 151*	*2 467*	*2 472*	*3 333*	*3 835*	*3 572*	*3 671*	
Aviation Fuels	272	352	387	492	735	639	619	..
Other	19	108	105	115	12	15	6	..
Industry	**3 130**	**3 523**	**3 616**	**2 788**	**2 487**	**2 451**	**2 703**	..
Naphtha	324	187	63	-	-	-	-	..
LPG + Ethane	25	317	758	952	879	979	1 323	..
Gas/Diesel Oil	428	822	763	565	387	369	354	..
Fuel Oil	1 642	1 418	1 280	343	281	180	252	..
Other	711	779	752	928	940	923	774	..
Other Sectors	**2 035**	**1 772**	**1 680**	**982**	**1 241**	**1 040**	**1 176**	..
Gas/Diesel Oil	1 445	1 296	1 243	793	1 064	896	1 016	..
Fuel Oil	135	45	55	15	2	1	3	..
Other	455	431	382	174	175	143	157	..
Memo: Non-Energy Use	*1 191*	*1 398*	*1 606*	*1 847*	*1 712*	*1 698*	*1 906*	..
Chem/Petchem. Feedst.	*513*	*832*	*1 054*	*1 186*	*925*	*928*	*1 278*	..
Other	*678*	*566*	*552*	*661*	*787*	*770*	*628*	..

Source: Annual Oil Statistics.

NORWAY
Table 2: Supply of Crude Oil, NGL, Feedstocks and Other Hydrocarbons

000 metric tons

	1973	1978	1980	1990	1999	2000	2001	2002
CRUDE OIL								
Indigenous Production	..	17 000	23 055	79 663	145 843	157 522	157 551	151 181 e
+ Imports	..	7 064	4 323	1 444	2 085	904	932	642 e
- Exports	..	15 827	19 900	66 602	133 520	143 764	147 518	141 071 e
- Direct Use	..	-	-	-	-	-	-	-
+ Stock Change	..	173	- 288	-1 786	- 80	- 566	1 958	- 62 e
- Statistical Difference	..	223	40	- 23	- 11	- 150	- 403	-1 837 e
Refinery Intake	**..**	**8 187**	**7 150**	**12 742**	**14 339**	**14 246**	**13 326**	**12 527 e**
Closing Stock Level	*..*	*874*	*1 048*	*3 331*	*6 068*	*3 442*	*1 484*	*1 546 e*
NATURAL GAS LIQUIDS								
Indigenous Production	..	-	1 222	2 425	3 507	3 468	4 985	5 200 e
+ Imports	..	-	-	-	-	-	-	-
- Exports	..	-	508	1 464	2 803	2 408	3 908	4 124 e
- Direct Use	..	-	718	948	729	702	1 160	1 334 e
+ Stock Change	..	-	- 14	- 38	25	- 65	46	61 e
- Statistical Difference	..	-	- 18	- 25	-	293	- 37	- 197 e
Refinery Intake	**..**	**-**	**-**	**-**	**-**	**-**	**-**	**-**
Closing Stock Level	*..*	*-*	*44*	*77*	*112*	*177*	*131*	*70 e*
REFINERY FEEDSTOCKS								
Backflows	..	15	77	97	60	51	53	41 e
+ Transfers	..	-	213	390	969	830	988	794 e
+ Imports	..	537	393	-	-	-	-	-
- Exports	..	-	-	-	-	-	-	-
+ Stock Change	..	-	51	-	-	-	-	-
- Statistical Difference	..	-	-	-	-	51	-	41 e
Refinery Intake	**..**	**552**	**734**	**487**	**1 029**	**830**	**1 041**	**794 e**
Closing Stock Level	*..*	*-*	*23*	*-*	*-*	*-*	*-*	*-*
OTHER HYDROCARBONS								
Indigenous Production	..	-	-	-	-	-	-	-
+ Imports	..	-	-	-	-	-	-	-
- Exports	..	-	-	-	-	-	-	-
- Direct Use	..	-	-	-	-	-	-	-
+ Stock Change	..	-	-	-	-	-	-	-
- Statistical Difference	..	-	-	-	-	-	-	-
Refinery Intake	**..**	**-**	**-**	**-**	**-**	**-**	**-**	**-**
Closing Stock Level	*..*	*-*	*-*	*-*	*-*	*-*	*-*	*-*
CRUDE OIL, NGL, FEEDSTOCKS AND OTHER								
Indigenous Production	1 595	17 000	24 277	82 088	149 350	160 990	162 536	156 381 e
+ Backflows	-	15	77	97	60	51	53	41 e
+ Transfers	-	-	213	390	969	830	988	794 e
+ Imports	6 076	7 601	4 716	1 444	2 085	904	932	642 e
- Exports	1 531	15 827	20 408	68 066	136 323	146 172	151 426	145 195 e
- Direct Use	-	-	718	948	729	702	1 160	1 334 e
+ Stock Change	357	173	- 251	-1 824	- 55	- 631	2 004	- 1 e
- Statistical Difference	-	223	22	- 48	- 11	194	- 440	-1 993 e
Refinery Intake	**6 497**	**8 739**	**7 884**	**13 229**	**15 368**	**15 076**	**14 367**	**13 321 e**
Closing Stock Level	*-*	*874*	*1 115*	*3 408*	*6 180*	*3 619*	*1 615*	*1 616 e*

Source: Annual Oil Statistics.

NORWAY
Table 3: Motor Gasoline - Supply

000 metric tons

	1973	1978	1980	1990	1999	2000	2001	2002
MOTOR GASOLINE								
Refinery Output	**704**	**1 187**	**1 186**	**3 203**	**3 202**	**3 385**	**3 287**	**2 920 e**
+ Imports	553	333	393	494	340	619	405	414 e
- Exports	182	129	114	1 717	1 840	2 016	2 214	2 050 e
+ Transfers	-	-	-	-	-	-	-	-
+ Stock Change	20	11	- 17	- 20	- 14	21	- 6	- 37 e
- Statistical Difference	-	54	57	172	17	389	- 209	- 422 e
Total Consumption	**1 095**	**1 348**	**1 391**	**1 788**	**1 671**	**1 620**	**1 681**	**1 669 e**
Closing Stock Level	-	*141*	*200*	*209*	*196*	*175*	*181*	*218 e*

Table 4: Aviation Fuels - Supply

000 metric tons

	1973	1978	1980	1990	1999	2000	2001	2002
AVIATION GASOLINE								
Refinery Output	-	-	-	-	-	-	-	-
+ Imports	14	5	-	7	1	3	2	1 e
- Exports	-	-	-	-	-	-	-	-
+ Transfers	-	-	8	-	-	-	-	-
+ Stock Change	-	-	- 1	-	- 1	-	1	- 1 e
- Statistical Difference	7	- 1	-	4	-	-	-	- 1 e
Total Consumption	**7**	**6**	**7**	**3**	**-**	**3**	**3**	**1 e**
Closing Stock Level	-	*4*	*4*	*2*	*1*	*1*	-	*1 e*
GASOLINE TYPE JET FUEL								
Refinery Output	-	**152**	-	-	-	-	-	-
+ Imports	-	-	78	-	-	-	-	-
- Exports	-	49	8	-	-	-	-	-
+ Transfers	-	-	- 8	-	-	-	-	-
+ Stock Change	-	- 5	0	-	-	-	-	-
- Statistical Difference	-	-	- 9	-	-	-	-	-
Total Consumption	**-**	**98**	**77**	**-**	**-**	**-**	**-**	**-**
Closing Stock Level	-	*67*	*4*	-	-	-	-	-
KEROSENE TYPE JET FUEL								
Refinery Output	**202**	**299**	**120**	**918**	**768**	**697**	**654**	**726 e**
+ Imports	96	37	189	86	216	174	212	264 e
- Exports	-	23	49	412	200	216	118	55 e
+ Transfers	-	-	-	-	-	-	-	-
+ Stock Change	1	- 6	10	- 8	- 18	72	- 21	15 e
- Statistical Difference	34	59	- 33	95	31	91	111	340 e
Total Consumption	**265**	**248**	**303**	**489**	**735**	**636**	**616**	**610 e**
Closing Stock Level	-	*57*	*74*	*48*	*117*	*45*	*66*	*51 e*

Source: Annual Oil Statistics.

NORWAY
Table 5A: Gas/Diesel Oil - Supply and Consumption

000 metric tons

	1973	1978	1980	1990	1999	2000	2001	2002
Refinery Output	**1 948**	**3 316**	**3 364**	**5 906**	**7 253**	**6 916**	**6 478**	**6 009 e**
+ Recycled Products	-	-	-	-	-	-	-	-
+ Imports	1 702	917	1 160	827	389	649	917	540 e
- Exports	377	877	1 002	3 609	3 484	3 459	3 005	2 462 e
- Intl. Marine Bunkers	207	190	102	218	467	460	458	420 e
+ Transfers	-	-	- 77	- 56	-	-	-	-
+ Stock Changes	- 87	109	- 124	- 4	153	83	44	43 e
- Statistical Differences	-	- 76	20	- 199	- 14	309	422	152 e
Total Consumption	**2 979**	**3 351**	**3 199**	**3 045**	**3 858**	**3 420**	**3 554**	**3 558 e**
Transformation	**33**	**23**	**27**	**4**	**58**	**16**	**18**	**..**
Public Electricity	33	1	4	2	-	-	-	..
Autoproducer Electricity	-	22	23	-	-	-	-	..
Public CHP Plants	-	-	-	-	-	-	-	..
Autoproducers of CHP	-	-	-	-	-	-	-	..
Public Heat Plants	-	-	-	2	57	16	18	..
Autoproducer Heat Plants	-	-	-	-	-	-	-	..
Petrochemical Industry	-	-	-	-	1	-	-	..
Other Transformation	-	-	-	-	-	-	-	..
Energy Sector	**-**	**73**	**65**	**121**	**162**	**165**	**153**	**..**
Refinery Fuel	-	1	1	2	-	11	-	..
Other Energy Sector	-	72	64	119	162	154	153	..
Final Consumption	**2 946**	**3 255**	**3 107**	**2 920**	**3 638**	**3 239**	**3 383**	**..**
Transport	**1 073**	**1 137**	**1 101**	**1 562**	**2 187**	**1 974**	**2 013**	**..**
Air	-	-	-	-	-	-	-	..
Road	627	399	440	725	1 355	1 308	1 352	..
Rail	24	28	28	32	20	17	17	..
Inland Waterways	422	710	633	805	812	649	644	..
Non-Specified	-	-	-	-	-	-	-	..
Industry	**428**	**822**	**763**	**565**	**387**	**369**	**354**	**..**
Iron and Steel	-	11	14	9	9	11	14	..
Chemical (incl. Pet.-Chem.)	65	87	62	38	17	22	20	..
Non-Ferrous Metals	-	30	24	51	24	26	16	..
Non-Metallic Minerals	-	60	37	30	33	30	29	..
Transport Equipment	-	40	41	15	9	7	12	..
Machinery	-	69	61	33	20	16	16	..
Mining and Quarrying	-	40	46	36	34	31	38	..
Food and Tobacco	-	123	116	109	97	89	71	..
Pulp, Paper and Printing	-	30	19	9	5	3	6	..
Wood and Wood Products	-	29	34	20	12	8	8	..
Construction	-	285	292	207	113	112	113	..
Textiles and Leather	-	16	15	7	6	5	4	..
Non-Specified	363	2	2	1	8	9	7	..
Other Sectors	**1 445**	**1 296**	**1 243**	**793**	**1 064**	**896**	**1 016**	**..**
Commerce and Public	-	691	650	326	312	231	312	..
Residential	-	428	430	285	149	109	122	..
Agriculture	-	177	163	182	603	556	582	..
Non-Specified	1 445	-	-	-	-	-	-	..
Memo: Non-Energy Use	-	*1*	*4*	-	*1*	*1*	*1*	..
Chem/Petchem. Feedst.	-	*1*	-	-	*1*	*1*	*1*	..
Other	-	-	*4*	-	-	-	-	..
Closing Stock Level	-	*622*	*749*	*438*	*563*	*480*	*436*	*393 e*

Source: Annual Oil Statistics.

NORWAY
Table 5B: Fuel Oil (Residual) - Supply and Consumption

000 metric tons

	1973	1978	1980	1990	1999	2000	2001	2002
Refinery Output	**2 558**	**2 226**	**1 830**	**1 129**	**1 897**	**1 703**	**1 647**	**1 503 e**
+ Recycled Products	-	-	-	-	-	-	-	-
+ Imports	776	164	514	592	1 164	1 039	1 135	1 067 e
- Exports	1 335	442	409	972	1 638	1 544	1 211	1 509 e
- Intl. Marine Bunkers	431	232	176	232	391	368	354	273 e
+ Transfers	-	-	- 59	- 334	- 969	- 830	- 988	- 794 e
+ Stock Changes	239	33	9	39	- 26	60	50	- 55 e
- Statistical Differences	13	135	214	- 255	- 262	- 149	11	- 249 e
Total Consumption	**1 794**	**1 614**	**1 495**	**477**	**299**	**209**	**268**	**188 e**
Transformation	**-**	**3**	**10**	**-**	**3**	**2**	**1**	**..**
Public Electricity	-	-	-	-	-	-	-	..
Autoproducer Electricity	-	-	-	-	-	-	-	..
Public CHP Plants	-	-	-	-	-	-	-	..
Autoproducers of CHP	-	-	-	-	-	-	-	..
Public Heat Plants	-	-	-	-	-	-	-	..
Autoproducer Heat Plants	-	-	-	-	-	-	-	..
Petrochemical Industry	-	3	10	-	3	2	1	..
Other Transformation	-	-	-	-	-	-	-	..
Energy Sector	**-**	**42**	**47**	**4**	**1**	**11**	**6**	**..**
Refinery Fuel	-	32	43	4	-	3	3	..
Other Energy Sector	-	10	4	-	1	8	3	..
Final Consumption	**1 794**	**1 569**	**1 438**	**473**	**295**	**196**	**261**	**..**
Transport	**17**	**106**	**103**	**115**	**12**	**15**	**6**	**..**
Air	-	-	-	-	-	-	-	..
Road	-	-	-	-	-	-	-	..
Rail	-	-	-	-	-	-	-	..
Inland Waterways	17	106	103	115	12	15	6	..
Non-Specified	-	-	-	-	-	-	-	..
Industry	**1 642**	**1 418**	**1 280**	**343**	**281**	**180**	**252**	**..**
Iron and Steel	-	12	18	19	-	-	-	..
Chemical (incl. Pet.-Chem.)	215	287	236	74	68	43	46	..
Non-Ferrous Metals	-	95	91	25	23	22	22	..
Non-Metallic Minerals	-	296	256	32	-	-	-	..
Transport Equipment	-	25	15	3	-	-	-	..
Machinery	-	24	21	4	2	2	2	..
Mining and Quarrying	-	39	33	20	3	3	4	..
Food and Tobacco	-	207	181	87	48	19	27	..
Pulp, Paper and Printing	-	395	393	65	126	86	140	..
Wood and Wood Products	-	22	18	9	8	3	9	..
Construction	-	-	-	-	-	-	-	..
Textiles and Leather	-	16	17	5	2	2	2	..
Non-Specified	1 427	-	1	-	1	-	-	..
Other Sectors	**135**	**45**	**55**	**15**	**2**	**1**	**3**	**..**
Commerce and Public	-	6	11	9	-	-	-	..
Residential	-	13	20	1	-	-	-	..
Agriculture	-	26	24	5	2	1	3	..
Non-Specified	135	-	-	-	-	-	-	..
Memo: Non-Energy Use	*152*	*128*	*91*	-	*3*	-	-	..
Chem/Petchem. Feedst.	*152*	*127*	*90*	-	*3*	-	-	..
Other	-	*1*	*1*	-	-	-	-	..
Closing Stock Level	-	*294*	*185*	*224*	*215*	*155*	*105*	*160 e*

Source: Annual Oil Statistics.

NORWAY
Table 5C: Total Products - Supply and Consumption

000 metric tons

	1973	1978	1980	1990	1999	2000	2001	2002
Refinery Output	**6 112**	**8 237**	**7 770**	**13 024**	**15 335**	**15 196**	**13 984**	**13 104 e**
+ Recycled Products	-	-	-	-	-	-	-	-
+ Imports	4 428	2 898	3 162	3 022	3 210	3 626	3 596	3 492 e
- Exports	2 134	1 951	2 090	7 752	8 437	9 108	8 667	8 548 e
- Intl. Marine Bunkers	647	430	284	450	858	828	812	693 e
+ Transfers	-	-	505	558	- 240	- 128	172	540 e
+ Stock Changes	52	143	- 146	16	32	239	194	- 39 e
- Statistical Differences	74	191	177	- 190	- 285	310	- 462	- 477 e
Total Consumption	**7 737**	**8 706**	**8 740**	**8 608**	**9 327**	**8 687**	**8 929**	**8 333 e**
Transformation	**46**	**55**	**110**	**101**	**117**	**67**	**71**	**..**
Public Electricity	33	1	4	2	-	-	-	..
Autoproducer Electricity	-	22	23	-	-	-	-	..
Public CHP Plants	-	-	-	-	-	-	-	..
Autoproducers of CHP	-	-	-	-	-	-	-	..
Public Heat Plants	-	-	-	2	57	16	18	..
Autoproducer Heat Plants	-	-	-	-	-	-	-	..
Petrochemical Industry	-	15	77	97	60	51	53	..
Other Transformation	13	17	6	-	-	-	-	..
Energy Sector	**84**	**429**	**370**	**797**	**900**	**903**	**683**	**..**
Refinery Fuel	84	344	298	675	735	740	525	..
Other Energy Sector	-	85	72	122	165	163	158	..
Final Consumption	**7 607**	**8 222**	**8 260**	**7 710**	**8 310**	**7 717**	**8 175**	**..**
Transport	**2 442**	**2 927**	**2 964**	**3 940**	**4 582**	**4 226**	**4 296**	**..**
Air	272	352	387	492	737	641	621	..
Road	1 700	1 724	1 806	2 493	2 962	2 865	2 969	..
Rail	24	28	28	32	20	17	17	..
Inland Waterways	446	823	743	923	863	703	689	..
Non-Specified	-	-	-	-	-	-	-	..
Industry	**3 130**	**3 523**	**3 616**	**2 788**	**2 487**	**2 451**	**2 703**	**..**
Iron and Steel	-	45	60	33	12	19	22	..
Chemical (incl. Pet.-Chem.)	641	1 078	1 262	1 298	1 008	1 081	1 431	..
Non-Ferrous Metals	-	278	305	232	209	254	182	..
Non-Metallic Minerals	-	358	301	77	86	90	80	..
Transport Equipment	-	65	57	20	14	11	14	..
Machinery	-	95	86	42	158	150	125	..
Mining and Quarrying	-	79	80	57	38	35	44	..
Food and Tobacco	-	331	298	199	154	115	107	..
Pulp, Paper and Printing	-	426	413	76	138	97	151	..
Wood and Wood Products	-	51	52	29	20	11	17	..
Construction	264	608	591	600	540	473	437	..
Textiles and Leather	-	32	32	12	8	8	6	..
Non-Specified	2 225	77	79	113	102	107	87	..
Other Sectors	**2 035**	**1 772**	**1 680**	**982**	**1 241**	**1 040**	**1 176**	**..**
Commerce and Public	10	707	677	339	315	234	316	..
Residential	427	846	799	441	313	241	267	..
Agriculture	18	219	204	202	613	565	593	..
Non-Specified	1 580	-	-	-	-	-	-	..
Memo: Non-Energy Use	*1 191*	*1 398*	*1 606*	*1 847*	*1 712*	*1 698*	*1 906*	..
Chem/Petchem. Feedst.	*513*	*832*	*1 054*	*1 186*	*925*	*928*	*1 278*	..
Other	*678*	*566*	*552*	*661*	*787*	*770*	*628*	..
Closing Stock Level	-	*1 457*	*1 502*	*1 110*	*1 292*	*1 096*	*902*	*941 e*

Source: Annual Oil Statistics.

NORWAY
Table 6A: Supply and Consumption

000 metric tons

	1980	2000	2001	2002	1980	2000	2001	2002
	LPG AND ETHANE				REFINERY GAS			
Refinery Output	**119**	**330**	**284**	**289 e**	**193**	**694**	**492**	**406 e**
+ Recycled Products	-	-	-	-	-	-	-	-
+ Imports	92	226	54	92 e	-	-	-	-
- Exports	53	440	751	1 019 e	-	-	-	-
- Intl. Marine Bunkers	-	-	-	-	-	-	-	-
+ Transfers	718	702	1 160	1 334 e	-	-	-	-
+ Stock Change	- 1	7	-	- 2 e	-	-	-	-
- Statistical Difference	- 10	- 242	- 667	- 360 e	-	-	-	-
Total Consumption	**885**	**1 067**	**1 414**	**1 054 e**	**193**	**694**	**492**	**406 e**
of which: Ethane	-	*142*	*138*	*178 e*	-	-	-	-
Transformation	59	48	52	..	-	-	-	..
Energy	61	32	30	..	193	694	492	..
Transport	-	-	-	..	-	-	-	..
Industry	758	979	1 323	..	-	-	-	..
Other	7	8	9	..	-	-	-	..
Memo: Non-Energy Use	*704*	*791*	*1 145*	..	-	-	-	..
Closing Stock Level	*2*	*7*	*7*	*9 e*	-	-	-	-
	NAPHTHA				OTHER KEROSENE			
Refinery Output	**485**	**1 161**	**920**	**1 125 e**	**227**	**115**	**11**	**- 74 e**
+ Recycled Products	-	-	-	-	-	-	-	-
+ Imports	52	73	71	30 e	98	33	21	335 e
- Exports	326	1 343	1 284	1 306 e	1	-	-	59 e
- Intl. Marine Bunkers	-	-	-	-	-	-	-	-
+ Transfers	- 77	-	-	-	-	-	-	-
+ Stock Change	- 20	18	24	- 15 e	- 8	- 28	102	9 e
- Statistical Difference	39	- 91	- 269	- 166 e	- 52	-	-	77 e
Total Consumption	**75**	**-**	**-**	**-**	**368**	**120**	**134**	**134 e**
Transformation	12	-	-	..	-	-	-	..
Energy	-	-	-	..	-	-	-	..
Transport	-	-	-	..	2	-	-	..
Industry	63	-	-	..	7	6	7	..
Other	-	-	-	..	359	114	127	..
Memo: Non-Energy Use	*63*	-	-	..	-	*5*	*5*	..
Closing Stock Level	*80*	*52*	*28*	*43 e*	*98*	*121*	*19*	*10 e*
	WHITE SPIRIT + SBP				LUBRICANTS			
Refinery Output	**28**	**-**	**-**	**..**	**5**	**-**	**-**	**..**
+ Recycled Products	-	-	-	..	-	-	-	..
+ Imports	28	15	17	..	78	75	67	..
- Exports	25	-	-	..	1	37	29	..
- Intl. Marine Bunkers	-	-	-	..	6	-	-	..
+ Transfers	-	-	-	..	-	-	-	..
+ Stock Change	2	1	- 6	..	- 2	6	-	..
- Statistical Difference	13	-	-	..	12	- 45	- 34	..
Total Consumption	**20**	**16**	**11**	**..**	**62**	**89**	**72**	**..**
Transformation	2	1	-	..	-	-	-	..
Energy	-	-	-	..	-	-	-	..
Transport	-	-	-	..	-	-	-	..
Industry	18	15	11	..	62	89	72	..
Other	-	-	-	..	-	-	-	..
Memo: Non-Energy Use	*17*	*15*	*11*	..	*62*	*89*	*72*	..
Closing Stock Level	*9*	-	*6*	..	*27*	*10*	*10*	..

Source: Annual Oil Statistics.

NORWAY
Table 6B: Supply and Consumption

000 metric tons

	1980	2000	2001	2002	1980	2000	2001	2002
	BITUMEN				**PARAFFIN WAXES**			
Refinery Output	**92**	**-**	**-**	**..**	**1**	**-**	**-**	**..**
+ Recycled Products	-	-	-	..	-	-	-	..
+ Imports	147	287	272	..	11	8	6	..
- Exports	2	10	4	..	1	-	-	..
- Intl. Marine Bunkers	-	-	-	..	-	-	-	..
+ Transfers	-	-	-	..	-	-	-	..
+ Stock Change	- 11	2	- 2	..	-	-	-	..
- Statistical Difference	- 31	- 46	- 31	..	-	-	-	..
Total Consumption	**257**	**325**	**297**	**..**	**11**	**8**	**6**	**..**
Transformation	-	-	-	..	-	-	-	..
Energy	-	-	-	..	-	-	-	..
Transport	-	-	-	..	-	-	-	..
Industry	257	325	297	..	11	8	6	..
Other	-	-	-	..	-	-	-	..
Memo: Non-Energy Use	*257*	*325*	*297*	..	*11*	*8*	*6*	..
Closing Stock Level	*54*	*24*	*26*	..	-	-	-	..
	PETROLEUM COKE				**OTHER PRODUCTS**			
Refinery Output	**100**	**195**	**211**	**200 e**	**20**	**-**	**-**	**-**
+ Recycled Products	-	-	-	-	-	-	-	-
+ Imports	300	383	388	415 e	22	42	29	334 e
- Exports	90	23	10	49 e	9	20	41	39 e
- Intl. Marine Bunkers	-	-	-	-	-	-	-	-
+ Transfers	-	-	-	-	-	-	-	-
+ Stock Change	1	- 3	8	2 e	10	-	-	2 e
- Statistical Difference	- 45	94	228	224 e	2	-	- 24	- 72 e
Total Consumption	**356**	**458**	**369**	**344 e**	**41**	**22**	**12**	**369 e**
Transformation	-	-	-	..	-	-	-	..
Energy	-	-	-	..	-	-	-	..
Transport	-	-	-	..	-	-	-	..
Industry	356	458	369	..	41	22	12	..
Other	-	-	-	..	-	-	-	..
Memo: Non-Energy Use	*356*	*442*	*357*	..	*41*	*22*	*12*	..
Closing Stock Level	-	*26*	*18*	*16 e*	*16*	-	-	*40 e*

Source: Annual Oil Statistics.

NORWAY
Table 7A: Imports by Country of Origin

000 metric tons

	2000	2001	2002	2000	2001	2002	2000	2001	2002
	Crude Oil			**Natural Gas Liquids**			**Feedstocks and Other**		
Australia	-	-	-	-	-	-	-	-	-
Belgium	-	-	-	-	-	-	-	-	-
Canada	-	-	-	-	-	-	-	-	-
Czech Republic	-	-	-	-	-	-	-	-	-
Denmark	87	-	143 e	-	-	-	-	-	-
Finland	-	-	-	-	-	-	-	-	-
France	-	-	-	-	-	-	-	-	-
Germany	-	-	-	-	-	-	-	-	-
Greece	-	-	-	-	-	-	-	-	-
Hungary	-	-	-	-	-	-	-	-	-
Ireland	-	-	-	-	-	-	-	-	-
Italy	-	-	-	-	-	-	-	-	-
Japan	-	-	-	-	-	-	-	-	-
Korea	-	-	-	-	-	-	-	-	-
Mexico	-	-	-	-	-	-	-	-	-
Netherlands	-	71	-	-	-	-	-	-	-
New Zealand	-	-	-	-	-	-	-	-	-
Norway	-	-	-	-	-	-	-	-	-
Poland	-	-	-	-	-	-	-	-	-
Spain	-	-	-	-	-	-	-	-	-
Sweden	-	-	-	-	-	-	-	-	-
Turkey	-	-	-	-	-	-	-	-	-
United Kingdom	578	407	224 e	-	-	-	-	-	-
United States	-	-	-	-	-	-	-	-	-
Other	-	-	-	-	-	-	-	-	-
OECD	**665**	**478**	**367 e**	**-**	**-**	**-**	**-**	**-**	**-**
Argentina	-	-	-	-	-	-	-	-	-
Brazil	-	-	-	-	-	-	-	-	-
Colombia	-	-	-	-	-	-	-	-	-
Ecuador	-	-	-	-	-	-	-	-	-
Netherlands Antilles	-	-	-	-	-	-	-	-	-
Peru	-	-	-	-	-	-	-	-	-
Trinidad and Tobago	-	-	-	-	-	-	-	-	-
Venezuela	-	-	-	-	-	-	-	-	-
Other	-	-	-	-	-	-	-	-	-
Latin America	**-**	**-**	**-**	**-**	**-**	**-**	**-**	**-**	**-**
Romania	-	-	-	-	-	-	-	-	-
Other	-	-	-	-	-	-	-	-	-
Non-OECD Europe	**-**	**-**	**-**	**-**	**-**	**-**	**-**	**-**	**-**
FSU	**239**	**454**	**275 e**	**-**	**-**	**-**	**-**	**-**	**-**
Iran	-	-	-	-	-	-	-	-	-
Iraq	-	-	-	-	-	-	-	-	-
Kuwait	-	-	-	-	-	-	-	-	-
Oman	-	-	-	-	-	-	-	-	-
Qatar	-	-	-	-	-	-	-	-	-
Saudi Arabia	-	-	-	-	-	-	-	-	-
United Arab Emirates	-	-	-	-	-	-	-	-	-
Other	-	-	-	-	-	-	-	-	-
Middle East	**-**	**-**	**-**	**-**	**-**	**-**	**-**	**-**	**-**
India	-	-	-	-	-	-	-	-	-
Indonesia	-	-	-	-	-	-	-	-	-
Singapore	-	-	-	-	-	-	-	-	-
Other	-	-	-	-	-	-	-	-	-
Asia	**-**	**-**	**-**	**-**	**-**	**-**	**-**	**-**	**-**
China	**-**	**-**	**-**	**-**	**-**	**-**	**-**	**-**	**-**
Algeria	-	-	-	-	-	-	-	-	-
Angola/Cabinada	-	-	-	-	-	-	-	-	-
Cameroon	-	-	-	-	-	-	-	-	-
Egypt	-	-	-	-	-	-	-	-	-
Gabon	-	-	-	-	-	-	-	-	-
Libya	-	-	-	-	-	-	-	-	-
Nigeria	-	-	-	-	-	-	-	-	-
Tunisia	-	-	-	-	-	-	-	-	-
Other	-	-	-	-	-	-	-	-	-
Africa	**-**	**-**	**-**	**-**	**-**	**-**	**-**	**-**	**-**
Non-spec./Other	**-**	**-**	**-**	**-**	**-**	**-**	**-**	**-**	**-**
Total Imports	**904**	**932**	**642 e**	**-**	**-**	**-**	**-**	**-**	**-**
Memo: Total OPEC	-	-	-	-	-	-	-	-	-

Source: Annual Oil Statistics.

NORWAY
Table 7B: Imports by Country of Origin

000 metric tons

	2000	2001	2002	2000	2001	2002	2000	2001	2002
	Total Products			**Liquefied Petroleum Gases**			**Naphtha**		
Australia	-	-	-	-	-	-	-	-	-
Belgium	76	73	91 e	-	-	-	-	-	14 e
Canada	8	26	3 e	-	-	-	-	-	-
Czech Republic	-	-	-	-	-	-	-	-	-
Denmark	394	459	619 e	-	-	-	-	2	5 e
Finland	251	58	79 e	-	-	-	-	-	-
France	36	69	58 e	-	-	-	5	5	-
Germany	450	247	348 e	-	-	-	-	-	-
Greece	-	-	-	-	-	-	-	-	-
Hungary	-	-	-	-	-	-	-	-	-
Ireland	29	10	-	-	-	-	-	10	-
Italy	-	-	3 e	-	-	-	-	-	-
Japan	1	1	-	-	-	-	-	-	-
Korea	-	-	-	-	-	-	-	-	-
Mexico	-	-	-	-	-	-	-	-	-
Netherlands	136	91	88 e	-	-	-	31	7	1 e
New Zealand	-	-	-	-	-	-	-	-	-
Norway	-	-	-	-	-	-	-	-	-
Poland	219	215	192 e	-	-	-	-	-	-
Spain	-	5	-	-	-	-	-	-	-
Sweden	963	1 178	966 e	-	-	-	12	5	5 e
Turkey	-	-	-	-	-	-	-	-	-
United Kingdom	663	856	614 e	-	-	-	25	42	5 e
United States	133	169	175 e	-	-	-	-	-	-
Other	-	8	17 e	-	-	-	-	-	-
OECD	**3 359**	**3 465**	**3 253 e**	**-**	**-**	**-**	**73**	**71**	**30 e**
Argentina	14	23	17 e	-	-	-	-	-	-
Brazil	-	-	-	-	-	-	-	-	-
Colombia	-	-	-	-	-	-	-	-	-
Ecuador	-	-	-	-	-	-	-	-	-
Netherlands Antilles	-	-	-	-	-	-	-	-	-
Peru	-	-	-	-	-	-	-	-	-
Trinidad and Tobago	-	-	-	-	-	-	-	-	-
Venezuela	1	-	-	-	-	-	-	-	-
Other	-	-	-	-	-	-	-	-	-
Latin America	**15**	**23**	**17 e**	**-**	**-**	**-**	**-**	**-**	**-**
Romania	-	-	-	-	-	-	-	-	-
Other	-	-	-	-	-	-	-	-	-
Non-OECD Europe	**-**	**-**	**-**	**-**	**-**	**-**	**-**	**-**	**-**
FSU	**22**	**51**	**93 e**	**-**	**-**	**-**	**-**	**-**	**-**
Iran	-	-	-	-	-	-	-	-	-
Iraq	-	-	-	-	-	-	-	-	-
Kuwait	-	-	-	-	-	-	-	-	-
Oman	-	-	-	-	-	-	-	-	-
Qatar	-	-	-	-	-	-	-	-	-
Saudi Arabia	-	-	-	-	-	-	-	-	-
United Arab Emirates	-	-	-	-	-	-	-	-	-
Other	-	-	-	-	-	-	-	-	-
Middle East	**-**	**-**	**-**	**-**	**-**	**-**	**-**	**-**	**-**
India	-	-	-	-	-	-	-	-	-
Indonesia	-	-	-	-	-	-	-	-	-
Singapore	-	-	-	-	-	-	-	-	-
Other	-	-	-	-	-	-	-	-	-
Asia	**-**	**-**	**-**	**-**	**-**	**-**	**-**	**-**	**-**
China	**4**	**3**	**4 e**	**-**	**-**	**-**	**-**	**-**	**-**
Algeria	-	-	33 e	-	-	-	-	-	-
Angola/Cabinada	-	-	-	-	-	-	-	-	-
Cameroon	-	-	-	-	-	-	-	-	-
Egypt	-	-	-	-	-	-	-	-	-
Gabon	-	-	-	-	-	-	-	-	-
Libya	-	-	-	-	-	-	-	-	-
Nigeria	-	-	-	-	-	-	-	-	-
Tunisia	-	-	-	-	-	-	-	-	-
Other	-	-	-	-	-	-	-	-	-
Africa	**-**	**-**	**33 e**	**-**	**-**	**-**	**-**	**-**	**-**
Non-spec./Other	**226**	**54**	**92 e**	**226**	**54**	**92 e**	**-**	**-**	**-**
Total Imports	**3 626**	**3 596**	**3 492 e**	**226**	**54**	**92 e**	**73**	**71**	**30 e**
Memo: Total OPEC	*1*	*-*	*33 e*	*-*	*-*	*-*	*-*	*-*	*-*

Source: Annual Oil Statistics.

NORWAY
Table 7C: Imports by Country of Origin

000 metric tons

	2000	2001	2002	2000	2001	2002	2000	2001	2002
	Total Gasoline			**Total Kerosene**			**Gas/Diesel**		
Australia	-	-	-	-	-	-	-	-	-
Belgium	-	-	-	3	5	-	-	3	13 e
Canada	-	-	-	-	-	-	-	15	-
Czech Republic	-	-	-	-	-	-	-	-	-
Denmark	74	55	134 e	143	142	261 e	51	138	110 e
Finland	50	25	52 e	11	-	13 e	179	21	-
France	-	-	5 e	-	20	11 e	7	29	-
Germany	55	5	16 e	1	29	119 e	1	78	40 e
Greece	-	-	-	-	-	-	-	-	-
Hungary	-	-	-	-	-	-	-	-	-
Ireland	-	-	-	-	-	-	-	-	-
Italy	-	-	-	-	-	-	-	-	-
Japan	-	-	-	-	-	-	-	1	-
Korea	-	-	-	-	-	-	-	-	-
Mexico	-	-	-	-	-	-	-	-	-
Netherlands	5	15	1 e	1	1	1 e	19	2	6 e
New Zealand	-	-	-	-	-	-	-	-	-
Norway	-	-	-	-	-	-	-	-	-
Poland	187	5	-	-	-	11 e	18	-	-
Spain	-	-	-	-	-	-	-	5	-
Sweden	240	256	197 e	25	15	111 e	351	559	298 e
Turkey	-	-	-	-	-	-	-	-	-
United Kingdom	11	46	10 e	12	15	-	12	14	46 e
United States	-	-	-	-	-	-	-	12	-
Other	-	-	-	-	-	7 e	-	-	-
OECD	**622**	**407**	**415 e**	**196**	**227**	**534 e**	**638**	**877**	**513 e**
Argentina	-	-	-	-	-	-	-	-	-
Brazil	-	-	-	-	-	-	-	-	-
Colombia	-	-	-	-	-	-	-	-	-
Ecuador	-	-	-	-	-	-	-	-	-
Netherlands Antilles	-	-	-	-	-	-	-	-	-
Peru	-	-	-	-	-	-	-	-	-
Trinidad and Tobago	-	-	-	-	-	-	-	-	-
Venezuela	-	-	-	-	-	-	-	-	-
Other	-	-	-	-	-	-	-	-	-
Latin America	**-**	**-**	**-**	**-**	**-**	**-**	**-**	**-**	**-**
Romania	-	-	-	-	-	-	-	-	-
Other	-	-	-	-	-	-	-	-	-
Non-OECD Europe	**-**	**-**	**-**	**-**	**-**	**-**	**-**	**-**	**-**
FSU	**-**	**-**	**-**	**11**	**6**	**32 e**	**11**	**40**	**27 e**
Iran	-	-	-	-	-	-	-	-	-
Iraq	-	-	-	-	-	-	-	-	-
Kuwait	-	-	-	-	-	-	-	-	-
Oman	-	-	-	-	-	-	-	-	-
Qatar	-	-	-	-	-	-	-	-	-
Saudi Arabia	-	-	-	-	-	-	-	-	-
United Arab Emirates	-	-	-	-	-	-	-	-	-
Other	-	-	-	-	-	-	-	-	-
Middle East	**-**	**-**	**-**	**-**	**-**	**-**	**-**	**-**	**-**
India	-	-	-	-	-	-	-	-	-
Indonesia	-	-	-	-	-	-	-	-	-
Singapore	-	-	-	-	-	-	-	-	-
Other	-	-	-	-	-	-	-	-	-
Asia	**-**	**-**	**-**	**-**	**-**	**-**	**-**	**-**	**-**
China	**-**	**-**	**-**	**-**	**-**	**-**	**-**	**-**	**-**
Algeria	-	-	-	-	-	33 e	-	-	-
Angola/Cabinada	-	-	-	-	-	-	-	-	-
Cameroon	-	-	-	-	-	-	-	-	-
Egypt	-	-	-	-	-	-	-	-	-
Gabon	-	-	-	-	-	-	-	-	-
Libya	-	-	-	-	-	-	-	-	-
Nigeria	-	-	-	-	-	-	-	-	-
Tunisia	-	-	-	-	-	-	-	-	-
Other	-	-	-	-	-	-	-	-	-
Africa	**-**	**-**	**-**	**-**	**-**	**33 e**	**-**	**-**	**-**
Non-spec./Other	**-**	**-**	**-**	**-**	**-**	**-**	**-**	**-**	**-**
Total Imports	**622**	**407**	**415 e**	**207**	**233**	**599 e**	**649**	**917**	**540 e**
Memo: Total OPEC	-	-	-	-	-	*33 e*	-	-	-

Source: Annual Oil Statistics.

NORWAY
Table 7D: Imports by Country of Origin

000 metric tons

	2000	2001	2002	2000	2001	2002	2000	2001	2002
	Fuel Oil (Residual)			**Petroleum Coke**			**Other Products**		
Australia	-	-	-	-	-	-	-	-	-
Belgium	-	-	-	60	55	45 e	13	10	19 e
Canada	-	-	-	8	11	3 e	-	-	-
Czech Republic	-	-	-	-	-	-	-	-	-
Denmark	115	117	104 e	-	-	-	11	5	5 e
Finland	-	-	-	1	-	-	10	12	14 e
France	-	5	11 e	-	-	-	24	10	31 e
Germany	331	86	105 e	4	4	9 e	58	45	59 e
Greece	-	-	-	-	-	-	-	-	-
Hungary	-	-	-	-	-	-	-	-	-
Ireland	29	-	-	-	-	-	-	-	-
Italy	-	-	-	-	-	-	-	-	3 e
Japan	-	-	-	-	-	-	1	-	-
Korea	-	-	-	-	-	-	-	-	-
Mexico	-	-	-	-	-	-	-	-	-
Netherlands	-	7	10 e	43	28	43 e	37	31	26 e
New Zealand	-	-	-	-	-	-	-	-	-
Norway	-	-	-	-	-	-	-	-	-
Poland	14	210	181 e	-	-	-	-	-	-
Spain	-	-	-	-	-	-	-	-	-
Sweden	105	96	210 e	2	2	-	228	245	145 e
Turkey	-	-	-	-	-	-	-	-	-
United Kingdom	445	609	419 e	118	98	102 e	40	32	32 e
United States	-	-	-	130	157	175 e	3	-	-
Other	-	-	-	-	8	10 e	-	-	-
OECD	**1 039**	**1 130**	**1 040 e**	**366**	**363**	**387 e**	**425**	**390**	**334 e**
Argentina	-	-	-	14	23	17 e	-	-	-
Brazil	-	-	-	-	-	-	-	-	-
Colombia	-	-	-	-	-	-	-	-	-
Ecuador	-	-	-	-	-	-	-	-	-
Netherlands Antilles	-	-	-	-	-	-	-	-	-
Peru	-	-	-	-	-	-	-	-	-
Trinidad and Tobago	-	-	-	-	-	-	-	-	-
Venezuela	-	-	-	-	-	-	1	-	-
Other	-	-	-	-	-	-	-	-	-
Latin America	**-**	**-**	**-**	**14**	**23**	**17 e**	**1**	**-**	**-**
Romania	-	-	-	-	-	-	-	-	-
Other	-	-	-	-	-	-	-	-	-
Non-OECD Europe	**-**	**-**	**-**	**-**	**-**	**-**	**-**	**-**	**-**
FSU	**-**	**5**	**27 e**	**-**	**-**	**7 e**	**-**	**-**	**-**
Iran	-	-	-	-	-	-	-	-	-
Iraq	-	-	-	-	-	-	-	-	-
Kuwait	-	-	-	-	-	-	-	-	-
Oman	-	-	-	-	-	-	-	-	-
Qatar	-	-	-	-	-	-	-	-	-
Saudi Arabia	-	-	-	-	-	-	-	-	-
United Arab Emirates	-	-	-	-	-	-	-	-	-
Other	-	-	-	-	-	-	-	-	-
Middle East	**-**	**-**	**-**	**-**	**-**	**-**	**-**	**-**	**-**
India	-	-	-	-	-	-	-	-	-
Indonesia	-	-	-	-	-	-	-	-	-
Singapore	-	-	-	-	-	-	-	-	-
Other	-	-	-	-	-	-	-	-	-
Asia	**-**	**-**	**-**	**-**	**-**	**-**	**-**	**-**	**-**
China	**-**	**-**	**-**	**3**	**2**	**4 e**	**1**	**1**	**-**
Algeria	-	-	-	-	-	-	-	-	-
Angola/Cabinada	-	-	-	-	-	-	-	-	-
Cameroon	-	-	-	-	-	-	-	-	-
Egypt	-	-	-	-	-	-	-	-	-
Gabon	-	-	-	-	-	-	-	-	-
Libya	-	-	-	-	-	-	-	-	-
Nigeria	-	-	-	-	-	-	-	-	-
Tunisia	-	-	-	-	-	-	-	-	-
Other	-	-	-	-	-	-	-	-	-
Africa	**-**	**-**	**-**	**-**	**-**	**-**	**-**	**-**	**-**
Non-spec./Other	**-**	**-**	**-**	**-**	**-**	**-**	**-**	**-**	**-**
Total Imports	**1 039**	**1 135**	**1 067 e**	**383**	**388**	**415 e**	**427**	**391**	**334 e**
Memo: Total OPEC	-	-	-	-	-	-	*1*	-	-

Source: Annual Oil Statistics.

NORWAY
Table 8A: Exports by Country of Destination

000 metric tons

	2000	2001	2002	2000	2001	2002	2000	2001	2002
	Crude Oil			**Natural Gas Liquids**			**Feedstocks and Other**		
Australia	-	-	-	-	32	-	-	-	-
Austria	-	-	-	-	-	-	-	-	-
Belgium	3 920	3 652	1 435 e	283	279	266 e	-	-	-
Canada	15 657	12 960	11 100 e	-	-	-	-	-	-
Denmark	4 093	2 701	3 139 e	1	-	-	-	-	-
Finland	2 591	2 454	2 021 e	78	75	26 e	-	-	-
France	19 895	19 882	15 261 e	229	376	391 e	-	-	-
Germany	8 840	11 775	12 576 e	94	106	94 e	-	-	-
Greece	-	-	-	-	-	-	-	-	-
Ireland	2 875	2 924	2 137 e	-	3	1 e	-	-	-
Italy	2 813	2 753	5 225 e	-	-	20 e	-	-	-
Japan	134	-	10 e	-	-	144 e	-	-	-
Luxembourg	-	-	-	-	-	-	-	-	-
Mexico	-	-	-	202	235	142 e	-	-	-
Netherlands	23 892	23 727	21 728 e	233	352	343 e	-	-	-
Norway	-	-	-	-	-	-	-	-	-
Portugal	785	718	439 e	77	120	134 e	-	-	-
Spain	164	418	708 e	154	157	221 e	-	-	-
Sweden	9 040	8 824	5 963 e	239	719	638 e	-	-	-
Switzerland	-	-	-	-	-	-	-	-	-
Turkey	-	-	-	387	464	549 e	-	-	-
United Kingdom	32 140	34 270	32 943 e	175	334	230 e	-	-	-
United States	15 455	17 106	21 478 e	66	374	242 e	-	-	-
Other OECD	1 066	395	1 177 e	72	38	57 e	-	-	-
OECD	**143 360**	**144 559**	**137 340 e**	**2 290**	**3 664**	**3 498 e**	**-**	**-**	**-**
Africa	-	32	-	15	21	85 e	-	-	-
Latin America	140	435	864 e	22	70	3 e	-	-	-
Asia	-	82	343 e	-	12	7 e	-	-	-
China	264	2 385	2 282 e	-	43	273 e	-	-	-
Non-OECD Europe	-	-	-	-	-	41 e	-	-	-
Former Soviet Union	-	-	-	-	-	6 e	-	-	-
Middle East	-	-	-	-	9	-	-	-	-
Non-specified/Other	-	25	242 e	81	89	211 e	-	-	-
Non-OECD	**404**	**2 959**	**3 731 e**	**118**	**244**	**626 e**	**-**	**-**	**-**
Total Exports	**143 764**	**147 518**	**141 071 e**	**2 408**	**3 908**	**4 124 e**	**-**	**-**	**-**
	Total Products			**Liquefied Petroleum Gas**			**Naphtha**		
Australia	-	-	-	-	-	-	-	-	-
Austria	-	-	2 e	-	-	2 e	-	-	-
Belgium	403	517	382 e	131	244	202 e	131	108	59 e
Canada	130	120	36 e	-	-	-	-	-	-
Denmark	875	976	988 e	-	4	-	160	153	134 e
Finland	473	199	414 e	5	3	25 e	35	-	12 e
France	583	478	844 e	26	73	81 e	102	13	11 e
Germany	365	384	283 e	7	49	15 e	52	56	-
Greece	-	20	-	-	-	-	-	-	-
Ireland	244	388	514 e	2	-	-	-	-	-
Italy	162	3	322 e	-	-	-	-	-	60 e
Japan	-	-	3 e	-	-	3 e	-	-	-
Luxembourg	-	-	-	-	-	-	-	-	-
Mexico	-	29	6 e	-	29	6 e	-	-	-
Netherlands	1 434	1 206	1 149 e	52	12	61 e	389	496	567 e
Norway	-	-	-	-	-	-	-	-	-
Portugal	73	42	82 e	4	-	5 e	12	12	-
Spain	164	118	189 e	-	17	31 e	-	-	-
Sweden	277	517	471 e	22	57	82 e	103	201	184 e
Switzerland	16	-	45 e	-	-	33 e	5	-	12 e
Turkey	150	66	264 e	150	66	264 e	-	-	-
United Kingdom	1 725	1 685	1 574 e	25	70	109 e	125	74	151 e
United States	1 244	1 349	586 e	3	5	20 e	196	171	116 e
Other OECD	229	176	104 e	10	34	55 e	3	-	-
OECD	**8 547**	**8 273**	**8 258 e**	**437**	**663**	**994 e**	**1 313**	**1 284**	**1 306 e**
Africa	-	3	8 e	-	3	8 e	-	-	-
Latin America	29	83	143 e	-	38	-	-	-	-
Asia	-	115	-	-	-	-	-	-	-
China	30	47	17 e	-	47	17 e	30	-	-
Non-OECD Europe	448	114	68 e	3	-	-	-	-	-
Former Soviet Union	16	32	46 e	-	-	-	-	-	-
Middle East	38	-	8 e	-	-	-	-	-	-
Non-specified/Other	-	-	-	-	-	-	-	-	-
Non-OECD	**561**	**394**	**290 e**	**3**	**88**	**25 e**	**30**	**-**	**-**
Total Exports	**9 108**	**8 667**	**8 548 e**	**440**	**751**	**1 019 e**	**1 343**	**1 284**	**1 306 e**

Source: Annual Oil Statistics.

NORWAY
Table 8B: Exports by Country of Destination

000 metric tons

	2000	2001	2002	2000	2001	2002	2000	2001	2002
	Total Gasoline			**Total Kerosene**			**Gas/Diesel Oil**		
Australia	-	-	-	-	-	-	-	-	-
Austria	-	-	-	-	-	-	-	-	-
Belgium	-	-	11 e	-	-	-	104	104	42 e
Canada	126	120	6 e	-	-	-	4	-	30 e
Denmark	102	59	70 e	35	32	47 e	465	645	606 e
Finland	288	188	287 e	-	-	-	71	-	50 e
France	9	7	105 e	12	-	-	310	343	604 e
Germany	80	83	23 e	-	-	-	154	189	235 e
Greece	-	-	-	-	-	-	-	20	-
Ireland	109	215	240 e	28	32	30 e	104	139	190 e
Italy	-	-	100 e	-	-	-	-	3	27 e
Japan	-	-	-	-	-	-	-	-	-
Luxembourg	-	-	-	-	-	-	-	-	-
Mexico	-	-	-	-	-	-	-	-	-
Netherlands	86	111	44 e	12	12	14 e	585	476	117 e
Norway	-	-	-	-	-	-	-	-	-
Portugal	24	-	-	-	-	-	33	30	-
Spain	27	30	20 e	-	-	-	137	68	3 e
Sweden	74	69	59 e	37	-	-	23	126	64 e
Switzerland	-	-	-	-	-	-	11	-	-
Turkey	-	-	-	-	-	-	-	-	-
United Kingdom	767	746	734 e	70	42	23 e	616	591	352 e
United States	230	457	293 e	-	-	-	588	227	78 e
Other OECD	-	97	1 e	22	-	-	192	44	47 e
OECD	**1 922**	**2 182**	**1 993 e**	**216**	**118**	**114 e**	**3 397**	**3 005**	**2 445 e**
Africa	-	-	-	-	-	-	-	-	-
Latin America	29	-	20 e	-	-	-	-	-	-
Asia	-	-	-	-	-	-	-	-	-
China	-	-	-	-	-	-	-	-	-
Non-OECD Europe	53	-	-	-	-	-	58	-	-
Former Soviet Union	12	32	29 e	-	-	-	4	-	17 e
Middle East	-	-	8 e	-	-	-	-	-	-
Non-specified/Other	-	-	-	-	-	-	-	-	-
Non-OECD	**94**	**32**	**57 e**	**-**	**-**	**-**	**62**	**-**	**17 e**
Total Exports	**2 016**	**2 214**	**2 050 e**	**216**	**118**	**114 e**	**3 459**	**3 005**	**2 462 e**

	2000	2001	2002	2000	2001	2002	2000	2001	2002
	Fuel Oil (Residual)			**Petroleum Coke**			**Other Products**		
Australia	-	-	-	-	-	-	-	-	-
Austria	-	-	-	-	-	-	-	-	-
Belgium	35	50	68 e	-	-	-	2	11	-
Canada	-	-	-	-	-	-	-	-	-
Denmark	104	56	114 e	-	-	-	9	27	17 e
Finland	62	-	40 e	-	-	-	12	8	-
France	124	42	24 e	-	-	19 e	-	-	-
Germany	28	3	9 e	8	1	1 e	36	3	-
Greece	-	-	-	-	-	-	-	-	-
Ireland	-	-	54 e	-	-	-	1	2	-
Italy	161	-	135 e	1	-	-	-	-	-
Japan	-	-	-	-	-	-	-	-	-
Luxembourg	-	-	-	-	-	-	-	-	-
Mexico	-	-	-	-	-	-	-	-	-
Netherlands	299	97	327 e	11	2	14 e	-	-	5 e
Norway	-	-	-	-	-	-	-	-	-
Portugal	-	-	77 e	-	-	-	-	-	-
Spain	-	-	130 e	-	3	1 e	-	-	4 e
Sweden	11	56	70 e	1	-	1 e	6	8	11 e
Switzerland	-	-	-	-	-	-	-	-	-
Turkey	-	-	-	-	-	-	-	-	-
United Kingdom	121	145	192 e	1	2	11 e	-	15	2 e
United States	227	489	79 e	-	-	-	-	-	-
Other OECD	-	-	-	1	1	1 e	1	-	-
OECD	**1 172**	**938**	**1 319 e**	**23**	**9**	**48 e**	**67**	**74**	**39 e**
Africa	-	-	-	-	-	-	-	-	-
Latin America	-	44	122 e	-	1	1 e	-	-	-
Asia	-	115	-	-	-	-	-	-	-
China	-	-	-	-	-	-	-	-	-
Non-OECD Europe	334	114	68 e	-	-	-	-	-	-
Former Soviet Union	-	-	-	-	-	-	-	-	-
Middle East	38	-	-	-	-	-	-	-	-
Non-specified/Other	-	-	-	-	-	-	-	-	-
Non-OECD	**372**	**273**	**190 e**	**-**	**1**	**1 e**	**-**	**-**	**-**
Total Exports	**1 544**	**1 211**	**1 509 e**	**23**	**10**	**49 e**	**67**	**74**	**39 e**

Source: Annual Oil Statistics.

POLAND

Figure 1
FUEL SHARES OF TPES *

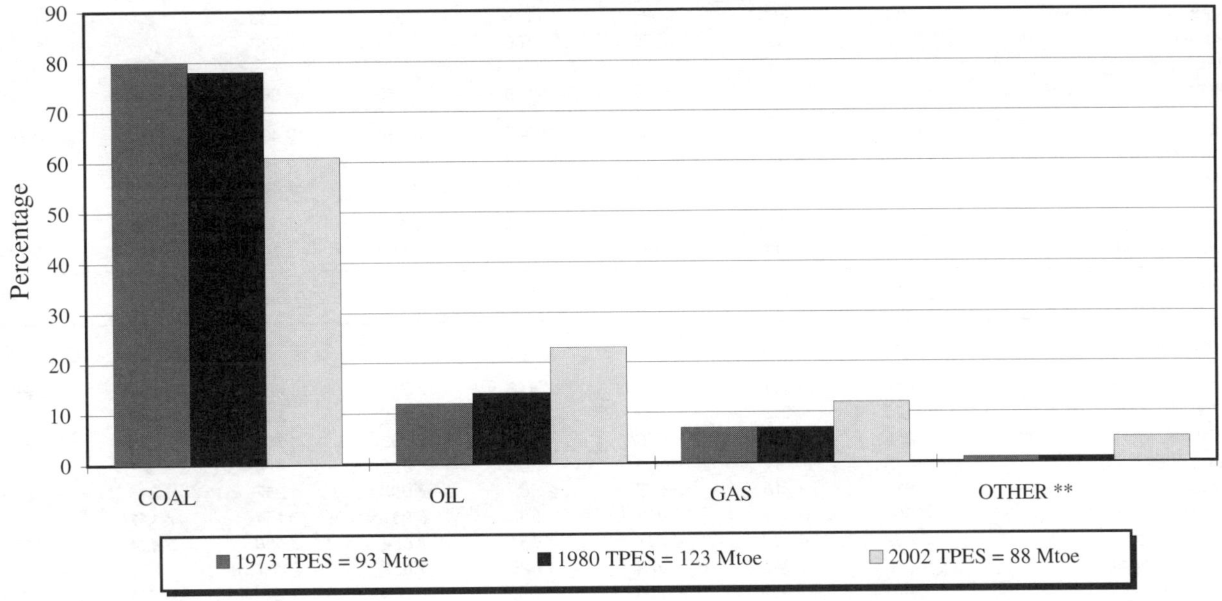

Source: Annual Energy Balances.

* Excluding electricity trade.

** Includes hydro, geothermal, solar, wind, combustible renewables and waste, etc.

Figure 2
OIL PRODUCTION AND CONSUMPTION

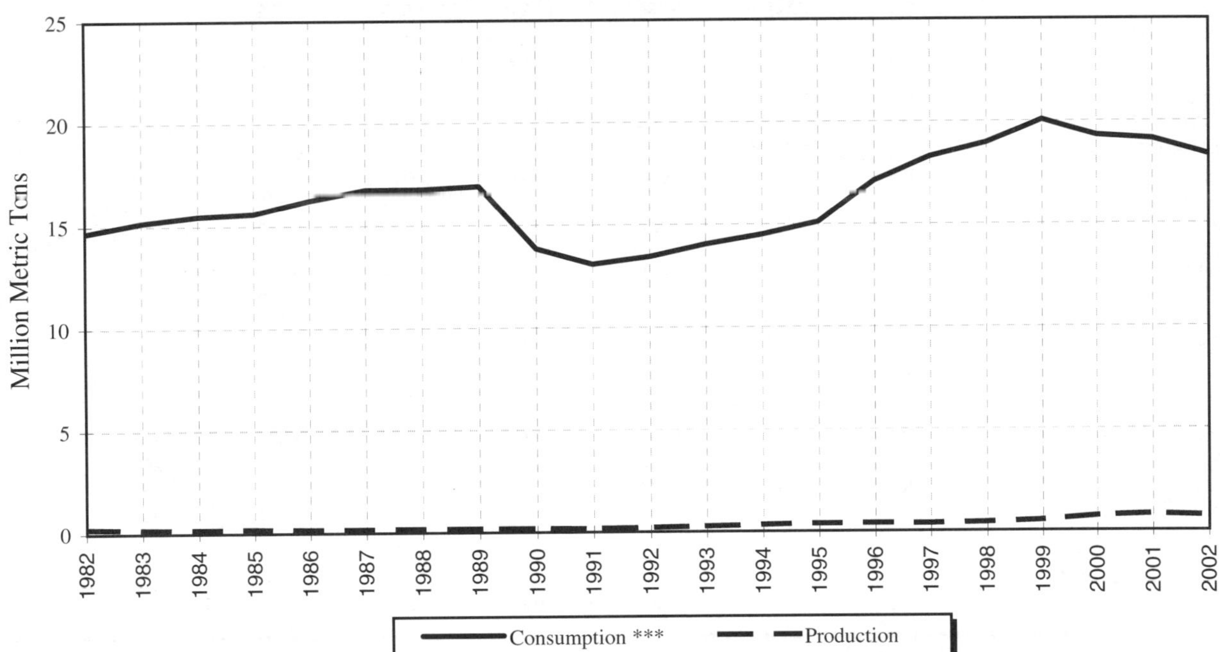

Source: Annual Oil Statistics.

*** Includes refinery fuel and international marine bunkers.

POLAND
Table 1: Oil Supply and Consumption - Summary Table

000 metric tons

	1973	1978	1980	1990	1999	2000	2001	2002
Indigenous Production	392	363	335	178	894	1 233	1 297	1 211 e
Crude Oil	..	363	329	160	434	653	767	672 e
NGL	..	-	-	-	-	-	-	-
Other	..	-	6	18	460	580	530	539 e
+ Net Imports	11 624	16 864	17 668	14 347	18 634	19 461	18 790	18 704 e
Crude Oil, NGL, Other	11 140	16 615	16 351	13 129	16 627	18 131	17 462	17 362 e
Petroleum Products	484	249	1 317	1 218	2 007	1 330	1 328	1 342 e
- Intl. Marine Bunkers	533	745	716	433	551	292	267	266 e
+ Stock Changes	- 752	88	- 491	- 817	133	- 758	221	- 122 e
TOTAL SUPPLY	**10 731**	**16 570**	**16 796**	**13 275**	**19 110**	**19 644**	**20 041**	**19 527 e**
Refinery Intake	*11 221*	*17 100*	*16 315*	*13 198*	*18 355*	*19 820*	*19 330*	*18 525 e*
Refinery Output	*10 814*	*16 041*	*15 426*	*12 812*	*17 811*	*18 480*	*18 411*	*16 886 e*
LPG + Ethane	*122*	*215*	*203*	*141*	*290*	*291*	*254*	*253 e*
Naphtha	*265*	*265*	*576*	*985*	*705*	*791*	*745*	*745 e*
Kerosenes	*5*	*11*	*5*	*202*	*262*	*428*	*627*	*585 e*
Gasolines	*1 916*	*3 216*	*2 927*	*2 365*	*4 087*	*4 257*	*4 246*	*4 019 e*
Gas/Diesel Oil	*3 306*	*5 355*	*5 093*	*3 963*	*6 951*	*7 154*	*7 141*	*6 175 e*
Fuel Oil	*3 862*	*4 932*	*4 647*	*3 631*	*3 654*	*3 558*	*3 424*	*2 970 e*
Other Products	*1 338*	*2 047*	*1 975*	*1 525*	*1 862*	*2 001*	*1 974*	*2 139 e*
- Refinery Losses	407	1 059	889	386	544	1 340	919	1 639 e
+ Returns	81	103	98	232	228	242	289	104 e
- Statistical Differences	- 232	- 131	- 355	- 296	- 724	- 460	547	- 98 e
TOTAL CONSUMPTION	**10 637**	**15 745**	**16 360**	**13 417**	**19 518**	**19 006**	**18 864**	**18 090 e**
Transformation and Energy	**1 586**	**2 937**	**3 265**	**2 433**	**1 648**	**1 653**	**1 809**	..
Crude Oil	-	-	-	-	2	1	-	..
LPG + Ethane	29	33	36	89	17	20	11	..
Gas/Diesel Oil	19	93	93	64	81	120	94	..
Fuel Oil	1 371	2 590	2 894	1 921	1 299	1 252	1 426	..
Petroleum Coke	-	-	-	-	-	-	-	..
Other	167	221	242	359	249	260	278	..
Transport	**5 039**	**7 045**	**6 966**	**6 653**	**10 135**	**8 729**	**8 728**	..
Motor Gasoline	2 324	3 125	3 163	3 032	5 710	4 960	4 602	..
Gas/Diesel Oil	2 380	3 590	3 479	3 321	3 375	2 734	2 934	..
Subtotal	*4 704*	*6 715*	*6 642*	*6 353*	*9 085*	*7 694*	*7 536*	..
Aviation Fuels	180	206	222	199	358	362	341	..
Other	155	124	102	101	692	673	851	..
Industry	**2 999**	**4 421**	**4 729**	**2 958**	**3 526**	**3 796**	**3 379**	..
Naphtha	265	265	576	933	663	756	669	..
LPG + Ethane	3	8	7	93	79	132	163	..
Gas/Diesel Oil	349	448	512	324	543	516	589	..
Fuel Oil	1 040	1 680	1 505	593	778	662	503	..
Other	1 342	2 020	2 129	1 015	1 463	1 730	1 455	..
Other Sectors	**1 013**	**1 342**	**1 400**	**1 373**	**4 209**	**4 828**	**4 948**	..
Gas/Diesel Oil	742	1 119	1 100	1 247	3 376	3 901	3 984	..
Fuel Oil	161	77	143	69	137	140	131	..
Other	110	146	157	57	696	787	833	..
Memo: Non-Energy Use	*1 516*	*2 201*	*2 613*	*2 091*	*1 975*	*2 406*	*2 090*	..
Chem/Petchem. Feedst.	*341*	*375*	*670*	*1 187*	*844*	*1 015*	*870*	..
Other	*1 175*	*1 826*	*1 943*	*904*	*1 131*	*1 391*	*1 220*	..

Source: Annual Oil Statistics.

POLAND

Table 2: Supply of Crude Oil, NGL, Feedstocks and Other Hydrocarbons

000 metric tons

	1973	1978	1980	1990	1999	2000	2001	2002
CRUDE OIL								
Indigenous Production	..	363	329	160	434	653	767	672 e
+ Imports	..	16 615	16 347	13 126	16 022	18 002	17 558	17 682 e
- Exports	..	-	-	-	-	129	440	423 e
- Direct Use	..	-	-	-	2	1	-	-
+ Stock Change	..	19	- 522	- 440	- 47	- 445	227	- 230 e
- Statistical Difference	..	-	-	-	- 312	- 194	150	-
Refinery Intake	**..**	**16 997**	**16 154**	**12 846**	**16 719**	**18 274**	**17 962**	**17 701 e**
Closing Stock Level	..	-	-	*929*	*998*	*1 443*	*1 216*	*1 446 e*
NATURAL GAS LIQUIDS								
Indigenous Production	..	-	-	-	-	-	-	-
+ Imports	..	-	-	-	-	-	-	-
- Exports	..	-	-	-	-	-	-	-
- Direct Use	..	-	-	-	-	-	-	-
+ Stock Change	..	-	-	-	-	-	-	-
- Statistical Difference	..	-	-	-	-	-	-	-
Refinery Intake	**..**	**-**	**-**	**-**	**-**	**-**	**-**	**-**
Closing Stock Level	..	-	-	-	-	-	-	-
REFINERY FEEDSTOCKS								
Backflows	..	103	98	232	60	60	56	54 e
+ Transfers	..	-	53	175	511	571	438	70 e
+ Imports	..	-	-	-	566	221	331	90 e
- Exports	..	-	-	-	-	-	-	-
+ Stock Change	..	-	-	-	-	-	-	-
- Statistical Difference	..	-	-	76	-	- 77	-	- 56 e
Refinery Intake	**..**	**103**	**151**	**331**	**1 137**	**929**	**825**	**270 e**
Closing Stock Level	..	-	-	-	-	-	-	-
OTHER HYDROCARBONS								
Indigenous Production	..	-	6	18	460	580	530	539 e
+ Imports	..	-	4	3	39	37	13	13 e
- Exports	..	-	-	-	-	-	-	-
- Direct Use	..	-	-	-	-	-	-	-
+ Stock Change		-	-			-	-	-
- Statistical Difference	..	-	-	-	-	-	-	- 2 e
Refinery Intake	**..**	**-**	**10**	**21**	**499**	**617**	**543**	**554 e**
Closing Stock Level	..	-	-	-	-	-	-	-
CRUDE OIL, NGL, FEEDSTOCKS AND OTHER								
Indigenous Production	392	363	335	178	894	1 233	1 297	1 211 e
+ Backflows	81	103	98	232	60	60	56	54 e
+ Transfers	213	-	53	175	511	571	438	70 e
+ Imports	11 140	16 615	16 351	13 129	16 627	18 260	17 902	17 785 e
- Exports	-	-	-	-	-	129	440	423 e
- Direct Use	-	-	-	-	2	1	-	-
+ Stock Change	- 605	19	- 522	- 440	- 47	- 445	227	- 230 e
- Statistical Difference	-	-	-	76	- 312	- 271	150	- 58 e
Refinery Intake	**11 221**	**17 100**	**16 315**	**13 198**	**18 355**	**19 820**	**19 330**	**18 525 e**
Closing Stock Level	-	-	-	*929*	*998*	*1 443*	*1 216*	*1 446 e*

Source: Annual Oil Statistics.

POLAND
Table 3: Motor Gasoline - Supply

000 metric tons

	1973	1978	1980	1990	1999	2000	2001	2002
MOTOR GASOLINE								
Refinery Output	**1 916**	**3 216**	**2 927**	**2 365**	**4 087**	**4 257**	**4 242**	**4 018 e**
+ Imports	756	371	677	1 073	1 670	898	861	649 e
- Exports	1	422	244	6	-	75	362	389 e
+ Transfers	- 213	-	- 53	- 165	50	- 14	- 65	-
+ Stock Change	- 36	55	- 38	- 179	- 6	- 61	- 47	53 e
- Statistical Difference	-	-	-	-	46	-	-	49 e
Total Consumption	**2 422**	**3 220**	**3 269**	**3 088**	**5 755**	**5 005**	**4 629**	**4 282 e**
Closing Stock Level	-	-	-	*423*	*386*	*447*	*494*	*441 e*

Table 4: Aviation Fuels - Supply

000 metric tons

	1973	1978	1980	1990	1999	2000	2001	2002
AVIATION GASOLINE								
Refinery Output	-	-	-	-	-	-	**4**	**1 e**
+ Imports	27	31	35	7	4	3	-	-
- Exports	-	-	-	-	-	-	2	-
+ Transfers	-	-	-	- 1	-	-	-	-
+ Stock Change	-	-	-	2	- 1	-	1	-
- Statistical Difference	-	-	-	-	-	-	-	-
Total Consumption	**27**	**31**	**35**	**8**	**3**	**3**	**3**	**1 e**
Closing Stock Level	-	-	-	*2*	*1*	*1*	-	-
GASOLINE TYPE JET FUEL								
Refinery Output	-	-	-	-	-	-	-	-
+ Imports	-	-	-	-	-	-	-	-
- Exports	-	-	-	-	-	-	-	-
+ Transfers	-	-	-	-	-	-	-	-
+ Stock Change	-	-	-	-	-	-	-	-
- Statistical Difference	-	-	-	-	-	-	-	-
Total Consumption	-	-	-	-	-	-	-	-
Closing Stock Level	-	-	-	-	-	-	-	-
KEROSENE TYPE JET FUEL								
Refinery Output	-	-	-	**201**	**259**	**427**	**623**	**575 e**
+ Imports	153	175	187	38	138	144	121	1 e
- Exports	-	-	-	24	12	191	392	316 e
+ Transfers	-	-	-	- 11	- 24	- 12	- 4	- 1 e
+ Stock Change	-	-	-	- 13	- 4	- 7	- 8	2 e
- Statistical Difference	-	-	-	-	-	-	-	- 88 e
Total Consumption	**153**	**175**	**187**	**191**	**357**	**361**	**340**	**349 e**
Closing Stock Level	-	-	-	*39*	*15*	*22*	*30*	*28 e*

Source: Annual Oil Statistics.

POLAND
Table 5A: Gas/Diesel Oil - Supply and Consumption

000 metric tons

	1973	1978	1980	1990	1999	2000	2001	2002
Refinery Output	**3 306**	**5 355**	**5 093**	**3 963**	**6 951**	**7 154**	**7 141**	**6 175 e**
+ Recycled Products	-	-	-	-	-	-	-	-
+ Imports	637	1 054	1 393	1 491	888	695	984	1 214 e
- Exports	121	802	1 022	185	62	52	166	270 e
- Intl. Marine Bunkers	290	397	384	151	111	44	22	40 e
+ Transfers	-	-	-	9	- 226	- 219	- 123	- 3 e
+ Stock Changes	- 44	38	103	- 171	43	- 245	58	- 10 e
- Statistical Differences	- 2	- 2	- 1	-	108	18	271	187 e
Total Consumption	**3 490**	**5 250**	**5 184**	**4 956**	**7 375**	**7 271**	**7 601**	**6 879 e**
Transformation	**2**	**71**	**55**	**17**	**37**	**46**	**48**	..
Public Electricity	-	-	-	-	-	-	-	..
Autoproducer Electricity	-	-	-	-	-	-	-	..
Public CHP Plants	-	69	54	3	3	5	5	..
Autoproducers of CHP	-	-	-	3	1	1	1	..
Public Heat Plants	-	-	-	2	29	29	35	..
Autoproducer Heat Plants	-	-	-	6	4	11	7	..
Petrochemical Industry	2	2	1	3	-	-	-	..
Other Transformation	-	-	-	-	-	-	-	..
Energy Sector	**17**	**22**	**38**	**47**	**44**	**74**	**46**	..
Refinery Fuel	-	-	-	1	2	32	9	..
Other Energy Sector	17	22	38	46	42	42	37	..
Final Consumption	**3 471**	**5 157**	**5 091**	**4 892**	**7 294**	**7 151**	**7 507**	..
Transport	**2 380**	**3 590**	**3 479**	**3 321**	**3 375**	**2 734**	**2 934**	..
Air	-	-	-	-	-	-	-	..
Road	1 943	2 799	2 810	2 725	3 173	2 562	2 770	..
Rail	378	711	599	554	180	165	161	..
Inland Waterways	59	80	70	42	22	6	2	..
Non-Specified	-	-	-	-	-	1	1	..
Industry	**349**	**448**	**512**	**324**	**543**	**516**	**589**	..
Iron and Steel	10	2	5	4	8	12	4	..
Chemical (incl. Pet.-Chem.)	14	27	36	24	31	24	125	..
Non-Ferrous Metals	3	3	2	1	4	6	4	..
Non-Metallic Minerals	27	23	29	22	40	38	46	..
Transport Equipment	32	27	29	18	21	19	14	..
Machinery	36	36	49	26	33	34	41	..
Mining and Quarrying	45	40	47	32	56	58	47	..
Food and Tobacco	29	37	55	29	174	171	169	..
Pulp, Paper and Printing	3	-	2	1	19	14	11	..
Wood and Wood Products	4	3	2	4	14	12	11	..
Construction	142	247	253	156	94	87	74	..
Textiles and Leather	4	3	3	4	21	16	16	..
Non-Specified	-	-	-	3	28	25	27	..
Other Sectors	**742**	**1 119**	**1 100**	**1 247**	**3 376**	**3 901**	**3 984**	..
Commerce and Public	-	-	-	-	178	311	539	..
Residential	-	-	-	-	228	400	500	..
Agriculture	742	1 119	1 100	1 247	2 970	3 190	2 945	..
Non-Specified	-	-	-	-	-	-	-	..
Memo: Non-Energy Use	-	-	-	*6*	-	-	-	..
Chem/Petchem. Feedst.	-	-	-	*1*	-	-	-	..
Other	-	-	-	*5*	-	-	-	..
Closing Stock Level	-	-	-	*527*	*374*	*619*	*561*	*571 e*

Source: Annual Oil Statistics.

POLAND
Table 5B: Fuel Oil (Residual) - Supply and Consumption

000 metric tons

	1973	1978	1980	1990	1999	2000	2001	2002
Refinery Output	**3 862**	**4 932**	**4 647**	**3 631**	**3 654**	**3 558**	**3 424**	**2 970 e**
+ Recycled Products	-	-	-	-	-	-	-	-
+ Imports	62	259	343	101	32	113	185	218 e
- Exports	1 038	376	237	927	1 566	1 367	1 206	1 126 e
- Intl. Marine Bunkers	243	348	332	282	440	248	245	226 e
+ Transfers	-	-	-	-	-	-	-	-
+ Stock Changes	- 67	- 24	- 34	-	137	- 12	- 16	82 e
- Statistical Differences	- 151	- 28	- 257	- 119	- 397	- 10	82	- 23 e
Total Consumption	**2 727**	**4 471**	**4 644**	**2 642**	**2 214**	**2 054**	**2 060**	**1 941 e**
Transformation	**1 208**	**2 226**	**2 523**	**1 634**	**418**	**342**	**367**	**..**
Public Electricity	-	-	-	-	-	-	-	..
Autoproducer Electricity	-	-	-	-	-	-	-	..
Public CHP Plants	486	899	936	175	181	172	174	..
Autoproducers of CHP	593	1 125	1 353	1 087	82	60	50	..
Public Heat Plants	-	-	-	224	140	100	130	..
Autoproducer Heat Plants	129	202	234	148	15	10 e	13 e	..
Petrochemical Industry	-	-	-	-	-	-	-	..
Other Transformation	-	-	-	-	-	-	-	..
Energy Sector	**163**	**364**	**371**	**287**	**881**	**910**	**1 059**	**..**
Refinery Fuel	160	360	367	286	877	904	1 057	..
Other Energy Sector	3	4	4	1	4	6	2	..
Final Consumption	**1 356**	**1 881**	**1 750**	**721**	**915**	**802**	**634**	**..**
Transport	**155**	**124**	**102**	**59**	**-**	**-**	**-**	**..**
Air	-	-	-	-	-	-	-	..
Road	-	-	-	-	-	-	-	..
Rail	70	8	-	-	-	-	-	..
Inland Waterways	85	116	102	59	-	-	-	..
Non-Specified	-	-	-	-	-	-	-	..
Industry	**1 040**	**1 680**	**1 505**	**593**	**778**	**662**	**503**	**..**
Iron and Steel	661	814	751	275	45	26	16	..
Chemical (incl. Pet.-Chem.)	82	112	179	68	392	338	184	..
Non-Ferrous Metals	17	16	11	19	14	14	13	..
Non-Metallic Minerals	177	558	399	104	148	97	108	..
Transport Equipment	16	15	15	7	6	7	8	..
Machinery	5	12	12	2	1	1	1	..
Mining and Quarrying	20	59	45	11	16	9	6	..
Food and Tobacco	20	34	38	41	57	63	68	..
Pulp, Paper and Printing	6	24	15	32	33	34	37	..
Wood and Wood Products	36	36	40	28	63	69	58	..
Construction	-	-	-	6	2	3	2	..
Textiles and Leather	-	-	-	-	1	1	2	..
Non-Specified	-	-	-	-	-	-	-	..
Other Sectors	**161**	**77**	**143**	**69**	**137**	**140**	**131**	**..**
Commerce and Public	-	-	-	-	-	-	-	..
Residential	-	-	-	-	-	-	-	..
Agriculture	161	77	143	69	137	140	131	..
Non-Specified	-	-	-	-	-	-	-	..
Memo: Non-Energy Use	-	-	-	-	-	-	-	..
Chem/Petchem. Feedst.	-	-	-	-	-	-	-	..
Other	-	-	-	-	-	-	-	..
Closing Stock Level	-	-	-	*203*	*123*	*135*	*151*	*69 e*

Source: Annual Oil Statistics.

POLAND
Table 5C: Total Products - Supply and Consumption

000 metric tons

	1973	1978	1980	1990	1999	2000	2001	2002
Refinery Output	**10 814**	**16 041**	**15 426**	**12 812**	**17 811**	**18 480**	**18 411**	**16 886 e**
+ Recycled Products	-	-	-	-	168	182	233	50 e
+ Imports	1 644	1 890	2 836	2 766	3 886	3 185	3 600	3 678 e
- Exports	1 160	1 641	1 519	1 548	1 879	1 855	2 272	2 336 e
- Intl. Marine Bunkers	533	745	716	433	551	292	267	266 e
+ Transfers	- 213	-	- 53	- 175	- 509	- 570	- 438	- 70 e
+ Stock Changes	- 147	69	31	- 377	180	- 313	- 6	108 e
- Statistical Differences	- 232	- 131	- 355	- 372	- 412	- 189	397	- 40 e
Total Consumption	**10 637**	**15 745**	**16 360**	**13 417**	**19 518**	**19 006**	**18 864**	**18 090 e**
Transformation	**1 306**	**2 417**	**2 698**	**1 911**	**524**	**456**	**479**	..
Public Electricity	-	-	-	-	-	-	-	..
Autoproducer Electricity	-	-	-	-	-	-	-	..
Public CHP Plants	486	968	990	178	184	177	179	..
Autoproducers of CHP	597	1 131	1 360	1 110	83	61	51	..
Public Heat Plants	-	-	-	226	173	133	169	..
Autoproducer Heat Plants	129	202	234	154	20	21 e	20 e	..
Petrochemical Industry	81	103	98	232	60	60	56	..
Other Transformation	13	13	16	11	4	4	4	..
Energy Sector	**280**	**520**	**567**	**522**	**1 124**	**1 197**	**1 330**	..
Refinery Fuel	251	480	508	438	1 059	1 129	1 279	..
Other Energy Sector	29	40	59	84	65	68	51	..
Final Consumption	**9 051**	**12 808**	**13 095**	**10 984**	**17 870**	**17 353**	**17 055**	..
Transport	**5 039**	**7 045**	**6 966**	**6 653**	**10 135**	**8 729**	**8 728**	..
Air	180	206	222	199	358	362	341	..
Road	4 267	5 924	5 973	5 792	9 569	8 189	8 215	..
Rail	448	719	599	560	186	170	166	..
Inland Waterways	144	196	172	102	22	6	2	..
Non-Specified	-	-	-	-	-	2	4	..
Industry	**2 999**	**4 421**	**4 729**	**2 958**	**3 526**	**3 796**	**3 379**	..
Iron and Steel	671	816	756	373	61	50	30	..
Chemical (incl. Pet.-Chem.)	439	515	886	1 286	1 769	1 926	1 719	..
Non-Ferrous Metals	20	19	13	22	21	23	21	..
Non-Metallic Minerals	204	581	428	304	331	258	257	..
Transport Equipment	61	54	54	36	40	39	33	..
Machinery	57	62	74	88	61	69	64	..
Mining and Quarrying	68	101	94	51	77	73	69	..
Food and Tobacco	56	77	102	98	247	256	260	..
Pulp, Paper and Printing	10	24	17	35	57	55	51	..
Wood and Wood Products	41	40	43	44	84	89	76	..
Construction	1 020	1 626	1 640	604	690	875	717	..
Textiles and Leather	5	4	4	14	27	21	22	..
Non-Specified	347	502	618	3	61	62	60	..
Other Sectors	**1 013**	**1 342**	**1 400**	**1 373**	**4 209**	**4 828**	**4 948**	..
Commerce and Public	-	-	-	5	239	383	705	..
Residential	110	146	157	37	648	840	950	..
Agriculture	903	1 196	1 243	1 330	3 172	3 401	3 148	..
Non-Specified	-	-	-	1	150	204	145	..
Memo: Non-Energy Use	*1 516*	*2 201*	*2 613*	*2 091*	*1 975*	*2 406*	*2 090*	..
Chem/Petchem. Feedst.	*341*	*375*	*670*	*1 187*	*844*	*1 015*	*870*	..
Other	*1 175*	*1 826*	*1 943*	*904*	*1 131*	*1 391*	*1 220*	..
Closing Stock Level	-	-	-	*1 350*	*998*	*1 311*	*1 317*	*1 209 e*

Source: Annual Oil Statistics.

POLAND
Table 6A: Supply and Consumption

000 metric tons

	1980	2000	2001	2002	1980	2000	2001	2002
	LPG AND ETHANE				**REFINERY GAS**			
Refinery Output	**203**	**291**	**254**	**253 e**	**149**	**664**	**666**	**666 e**
+ Recycled Products	-	-	-	-	-	-	-	-
+ Imports	-	933	1 088	1 317 e	-	-	-	-
- Exports	16	47	17	15 e	-	-	-	-
- Intl. Marine Bunkers	-	-	-	-	-	-	-	-
+ Transfers	-	- 50	- 6	-	-	-	-	-
+ Stock Change	-	- 20	12	1 e	-	-	-	-
- Statistical Difference	- 13	-	-	107 e	- 4	-	-	-
Total Consumption	**200**	**1 107**	**1 331**	**1 449 e**	**153**	**664**	**666**	**666 e**
of which: Ethane	-	-	-	-	-	-	-	-
Transformation	29	13	11	..	11	10	18	..
Energy	7	7	-	..	141	183	211	..
Transport	-	425	566	..	-	-	-	..
Industry	7	132	163	..	1	471	437	..
Other	157	530	591	..	-	-	-	..
Memo: Non-Energy Use	-	-	-	..	-	-	-	..
Closing Stock Level	-	*39*	*27*	*26 e*	-	-	-	-
	NAPHTHA				**OTHER KEROSENE**			
Refinery Output	**576**	**791**	**745**	**745 e**	**5**	**1**	**4**	**10 e**
+ Recycled Products	-	-	-	-	-	-	-	-
+ Imports	-	-	-	-	-	47	15	35 e
- Exports	-	-	-	-	-	-	-	-
- Intl. Marine Bunkers	-	-	-	-	-	-	-	-
+ Transfers	-	-	-	-	-	-	-	-
+ Stock Change	-	-	- 1	-	-	-	-	-
- Statistical Difference	- 80	-	46	47 e	-	-	-	1 e
Total Consumption	**656**	**791**	**698**	**698 e**	**5**	**48**	**19**	**44 e**
Transformation	80	35	29	..	-	-	-	..
Energy	-	-	-	..	-	-	-	..
Transport	-	-	-	..	-	-	-	..
Industry	576	756	669	..	5	7	4	..
Other	-	-	-	..	-	41	15	..
Memo: Non-Energy Use	*576*	*756*	*669*	..	*5*	*2*	*3*	..
Closing Stock Level	-	*1*	*2*	*2 e*	-	*2*	*2*	*2 e*
	WHITE SPIRIT + SBP				**LUBRICANTS**			
Refinery Output	**11**	**87**	**61**	**..**	**493**	**208**	**224**	**..**
+ Recycled Products	-	-	-	..	-	182	233	..
+ Imports	-	-	-	..	125	283	253	..
- Exports	-	-	-	..	-	118	101	..
- Intl. Marine Bunkers	-	-	-	..	-	-	-	..
+ Transfers	-	- 2	-	..	-	- 189	- 233	..
+ Stock Change	-	-	-	..	-	16	- 1	..
- Statistical Difference	-	-	-	..	-	32	- 2	..
Total Consumption	**11**	**85**	**61**	**..**	**618**	**350**	**377**	**..**
Transformation	-	-	-	..	-	-	-	..
Energy	-	-	1	..	-	16	11	..
Transport	-	-	-	..	-	248	285	..
Industry	11	27	22	..	618	56	50	..
Other	-	58	38	..	-	30	31	..
Memo: Non-Energy Use	*11*	*85*	*60*	..	*618*	*349*	*377*	..
Closing Stock Level	-	*6*	*6*	..	-	*21*	*22*	..

Source: Annual Oil Statistics.

POLAND
Table 6B: Supply and Consumption

000 metric tons

	1980	2000	2001	2002	1980	2000	2001	2002
	BITUMEN				**PARAFFIN WAXES**			
Refinery Output	**1 255**	**831**	**678**	**..**	**24**	**95**	**68**	**..**
+ Recycled Products	-	-	-	..	-	-	-	..
+ Imports	65	46	70	..	11	21	23	..
- Exports	-	-	18	..	-	5	5	..
- Intl. Marine Bunkers	-	-	-	..	-	-	-	..
+ Transfers	-	-	-	..	-	- 16	- 6	..
+ Stock Change	-	15	- 5	..	-	1	2	..
- Statistical Difference	-	-	-	..	-	-	-	..
Total Consumption	**1 320**	**892**	**725**	**..**	**35**	**96**	**82**	**..**
Transformation	-	-	-	..	-	-	-	..
Energy	-	-	-	..	-	-	-	..
Transport	-	-	-	..	-	-	-	..
Industry	1 320	892	725	..	35	21	20	..
Other	-	-	-	..	-	75	62	..
Memo: Non-Energy Use	*1 320*	*892*	*725*	..	*35*	*96*	*82*	..
Closing Stock Level	-	*6*	*11*	..	-	*4*	*2*	..
	PETROLEUM COKE				**OTHER PRODUCTS**			
Refinery Output	**-**	**-**	**-**	**-**	**43**	**116**	**277**	**1 473 e**
+ Recycled Products	-	-	-	-	-	-	-	50 e
+ Imports	-	-	-	-	-	2	-	244 e
- Exports	-	-	-	-	-	-	3	220 e
- Intl. Marine Bunkers	-	-	-	-	-	-	-	-
+ Transfers	-	-	-	-	-	- 68	- 1	- 66 e
+ Stock Change	-	-	-	-	-	-	- 1	- 20 e
- Statistical Difference	-	-	-	-	-	- 229	-	- 320 e
Total Consumption	**-**	**-**	**-**	**-**	**43**	**279**	**272**	**1 781 e**
Transformation	-	-	-	..	-	10	6	..
Energy	-	-	-	..	-	3	1	..
Transport	-	-	-	..	-	-	-	..
Industry	-	-	-	..	43	244	190	..
Other	-	-	-	..	-	22	75	..
Memo: Non-Energy Use	-	-	-	..	*43*	*226*	*174*	..
Closing Stock Level	-	-	-	-	-	*8*	*9*	*70 e*

Source: Annual Oil Statistics.

POLAND
Table 7A: Imports by Country of Origin

000 metric tons

	2000	2001	2002	2000	2001	2002	2000	2001	2002
	Crude Oil			**Natural Gas Liquids**			**Feedstocks and Other**		
Australia	-	-	-	-	-	-	-	-	-
Belgium	-	-	-	-	-	-	-	-	-
Canada	-	-	-	-	-	-	-	-	-
Czech Republic	23	12	31 e	-	-	-	-	-	-
Denmark	-	-	-	-	-	-	-	-	-
Finland	-	-	-	-	-	-	-	-	-
France	-	-	-	-	-	-	-	-	-
Germany	-	126	68 e	-	-	-	-	-	-
Greece	-	-	-	-	-	-	-	-	-
Hungary	8	-	-	-	-	-	-	-	-
Ireland	-	-	-	-	-	-	-	-	-
Italy	-	-	-	-	-	-	-	-	-
Japan	-	-	-	-	-	-	-	-	-
Korea	-	-	-	-	-	-	-	-	-
Mexico	-	-	-	-	-	-	-	-	-
Netherlands	-	-	-	-	-	-	-	-	-
New Zealand	-	-	-	-	-	-	-	-	-
Norway	106	-	22 e	-	-	-	-	-	-
Poland	-	-	-	-	-	-	-	-	-
Spain	-	-	-	-	-	-	-	-	-
Sweden	-	-	-	-	-	-	-	-	-
Turkey	-	-	-	-	-	-	-	-	-
United Kingdom	368	-	79 e	-	-	-	-	-	-
United States	-	-	-	-	-	-	-	-	-
Other	15	12	3 e	-	-	-	-	-	-
OECD	**520**	**150**	**203 e**	**-**	**-**	**-**	**-**	**-**	**-**
Argentina	-	-	-	-	-	-	-	-	-
Brazil	-	-	-	-	-	-	-	-	-
Colombia	-	-	-	-	-	-	-	-	-
Ecuador	-	-	-	-	-	-	-	-	-
Netherlands Antilles	-	-	-	-	-	-	-	-	-
Peru	-	-	-	-	-	-	-	-	-
Trinidad and Tobago	-	-	-	-	-	-	-	-	-
Venezuela	-	-	-	-	-	-	-	-	-
Other	-	-	-	-	-	-	-	-	-
Latin America	**-**	**-**	**-**	**-**	**-**	**-**	**-**	**-**	**-**
Romania	-	-	-	-	-	-	-	-	-
Other	-	61	30 e	-	-	-	-	-	-
Non-OECD Europe	**-**	**61**	**30 e**	**-**	**-**	**-**	**-**	**-**	**-**
FSU	**17 482**	**17 316**	**17 195 e**	**-**	**-**	**-**	**-**	**-**	**-**
Iran	-	-	-	-	-	-	-	-	-
Iraq	-	-	-	-	-	-	-	-	-
Kuwait	-	-	-	-	-	-	-	-	-
Oman	-	-	-	-	-	-	-	-	-
Qatar	-	-	-	-	-	-	-	-	-
Saudi Arabia	-	-	-	-	-	-	-	-	-
United Arab Emirates	-	-	-	-	-	-	-	-	-
Other	-	-	-	-	-	-	-	-	-
Middle East	**-**	**-**	**-**	**-**	**-**	**-**	**-**	**-**	**-**
India	-	-	-	-	-	-	-	-	-
Indonesia	-	-	-	-	-	-	-	-	-
Singapore	-	-	-	-	-	-	-	-	-
Other	-	-	-	-	-	-	-	-	-
Asia	**-**	**-**	**-**	**-**	**-**	**-**	**-**	**-**	**-**
China	**-**	**-**	**-**	**-**	**-**	**-**	**-**	**-**	**-**
Algeria	-	-	-	-	-	-	-	-	-
Angola/Cabinada	-	-	-	-	-	-	-	-	-
Cameroon	-	-	-	-	-	-	-	-	-
Egypt	-	-	-	-	-	-	-	-	-
Gabon	-	-	-	-	-	-	-	-	-
Libya	-	-	-	-	-	-	-	-	-
Nigeria	-	-	-	-	-	-	-	-	-
Tunisia	-	-	-	-	-	-	-	-	-
Other	-	-	-	-	-	-	-	-	-
Africa	**-**	**-**	**-**	**-**	**-**	**-**	**-**	**-**	**-**
Non-spec./Other	**-**	**31**	**254 e**	**-**	**-**	**-**	**258**	**344**	**103 e**
Total Imports	**18 002**	**17 558**	**17 682 e**	**-**	**-**	**-**	**258**	**344**	**103 e**
Memo: Total OPEC	-	-	-	-	-	-	-	-	-

Source: Annual Oil Statistics.

POLAND
Table 7B: Imports by Country of Origin

000 metric tons

	2000	2001	2002	2000	2001	2002	2000	2001	2002
	Total Products			**Liquefied Petroleum Gases**			**Naphtha**		
Australia	-	-	-	-	-	-	-	-	-
Belgium	64	9	10 e	3	2	1 e	-	-	-
Canada	-	-	-	-	-	-	-	-	-
Czech Republic	155	301	159 e	4	4	22 e	-	-	-
Denmark	31	185	52 e	10	6	17 e	-	-	-
Finland	28	1	18 e	-	-	14 e	-	-	-
France	25	21	10 e	12	5	2 e	-	-	-
Germany	637	431	473 e	122	96	75 e	-	-	-
Greece	-	-	-	-	-	-	-	-	-
Hungary	4	7	1 e	-	2	1 e	-	-	-
Ireland	-	-	-	-	-	-	-	-	-
Italy	1	1	-	-	-	-	-	-	-
Japan	-	-	-	-	-	-	-	-	-
Korea	-	-	-	-	-	-	-	-	-
Mexico	-	-	-	-	-	-	-	-	-
Netherlands	22	40	10 e	3	35	5 e	-	-	-
New Zealand	-	-	-	-	-	-	-	-	-
Norway	80	-	45 e	75	-	45 e	-	-	-
Poland	-	-	-	-	-	-	-	-	-
Spain	-	-	-	-	-	-	-	-	-
Sweden	234	230	123 e	35	27	27 e	-	-	-
Turkey	-	-	-	-	-	-	-	-	-
United Kingdom	82	49	46 e	69	48	46 e	-	-	-
United States	-	-	-	-	-	-	-	-	-
Other	336	461	458 e	1	-	-	-	-	-
OECD	**1 699**	**1 736**	**1 405 e**	**334**	**225**	**255 e**	**-**	**-**	**-**
Argentina	-	-	-	-	-	-	-	-	-
Brazil	-	-	-	-	-	-	-	-	-
Colombia	-	-	-	-	-	-	-	-	-
Ecuador	-	-	-	-	-	-	-	-	-
Netherlands Antilles	-	-	-	-	-	-	-	-	-
Peru	-	-	-	-	-	-	-	-	-
Trinidad and Tobago	-	-	-	-	-	-	-	-	-
Venezuela	-	-	-	-	-	-	-	-	-
Other	-	-	-	-	-	-	-	-	-
Latin America	**-**	**-**	**-**	**-**	**-**	**-**	**-**	**-**	**-**
Romania	-	-	-	-	-	-	-	-	-
Other	1	1	17 e	1	-	15 e	-	-	-
Non-OECD Europe	**1**	**1**	**17 e**	**1**	**-**	**15 e**	**-**	**-**	**-**
FSU	**1 138**	**978**	**1 603 e**	**598**	**310**	**733 e**	**-**	**-**	**-**
Iran	-	-	-	-	-	-	-	-	-
Iraq	-	-	-	-	-	-	-	-	-
Kuwait	-	-	-	-	-	-	-	-	-
Oman	-	-	-	-	-	-	-	-	-
Qatar	-	-	-	-	-	-	-	-	-
Saudi Arabia	-	-	-	-	-	-			
United Arab Emirates	-	-	-	-	-	-	-	-	-
Other	-	-	-	-	-	-	-	-	-
Middle East	**-**	**-**	**-**	**-**	**-**	**-**	**-**	**-**	**-**
India	-	-	-	-	-	-	-	-	-
Indonesia	-	-	-	-	-	-	-	-	-
Singapore	-	-	-	-	-	-	-	-	-
Other	-	-	-	-	-	-	-	-	-
Asia	**-**	**-**	**-**	**-**	**-**	**-**	**-**	**-**	**-**
China	**-**	**-**	**4 e**	**-**	**-**	**-**	**-**	**-**	**-**
Algeria	-	-	-	-	-	-	-	-	-
Angola/Cabinada	-	-	-	-	-	-	-	-	-
Cameroon	-	-	-	-	-	-	-	-	-
Egypt	-	-	-	-	-	-	-	-	-
Gabon	-	-	-	-	-	-	-	-	-
Libya	-	5	-	-	-	-	-	-	-
Nigeria	-	-	-	-	-	-	-	-	-
Tunisia	-	-	-	-	-	-	-	-	-
Other	-	-	-	-	-	-	-	-	-
Africa	**-**	**5**	**-**	**-**	**-**	**-**	**-**	**-**	**-**
Non-spec./Other	**347**	**880**	**649 e**	**-**	**553**	**314 e**	**-**	**-**	**-**
Total Imports	**3 185**	**3 600**	**3 678 e**	**933**	**1 088**	**1 317 e**	**-**	**-**	**-**
Memo: Total OPEC	-	*5*	-	-	-	-	-	-	-

Source: Annual Oil Statistics.

POLAND
Table 7C: Imports by Country of Origin

000 metric tons

	2000	2001	2002	2000	2001	2002	2000	2001	2002
	Total Gasoline			**Total Kerosene**			**Gas/Diesel**		
Australia	-	-	-	-	-	-	-	-	-
Belgium	-	-	-	13	-	-	-	-	-
Canada	-	-	-	-	-	-	-	-	-
Czech Republic	69	154	87 e	7	-	-	64	134	48 e
Denmark	-	-	-	-	3	-	-	84	3 e
Finland	25	-	-	-	-	-	3	-	-
France	-	-	-	-	-	-	-	-	-
Germany	302	169	132 e	1	5	-	146	92	212 e
Greece	-	-	-	-	-	-	-	-	-
Hungary	3	5	-	-	-	-	-	-	-
Ireland	-	-	-	-	-	-	-	-	-
Italy	-	-	-	-	-	-	-	-	-
Japan	-	-	-	-	-	-	-	-	-
Korea	-	-	-	-	-	-	-	-	-
Mexico	-	-	-	-	-	-	-	-	-
Netherlands	9	-	-	-	-	-	6	-	-
New Zealand	-	-	-	-	-	-	-	-	-
Norway	-	-	-	5	-	-	-	-	-
Poland	-	-	-	-	-	-	-	-	-
Spain	-	-	-	-	-	-	-	-	-
Sweden	55	62	27 e	-	-	-	133	131	40 e
Turkey	-	-	-	-	-	-	-	-	-
United Kingdom	12	-	-	-	-	-	-	-	-
United States	-	-	-	-	-	-	-	-	-
Other	229	291	276 e	1	-	-	91	151	174 e
OECD	**704**	**681**	**522 e**	**27**	**8**	**-**	**443**	**592**	**477 e**
Argentina	-	-	-	-	-	-	-	-	-
Brazil	-	-	-	-	-	-	-	-	-
Colombia	-	-	-	-	-	-	-	-	-
Ecuador	-	-	-	-	-	-	-	-	-
Netherlands Antilles	-	-	-	-	-	-	-	-	-
Peru	-	-	-	-	-	-	-	-	-
Trinidad and Tobago	-	-	-	-	-	-	-	-	-
Venezuela	-	-	-	-	-	-	-	-	-
Other	-	-	-	-	-	-	-	-	-
Latin America	**-**	**-**	**-**	**-**	**-**	**-**	**-**	**-**	**-**
Romania	-	-	-	-	-	-	-	-	-
Other	-	1	-	-	-	-	-	-	-
Non-OECD Europe	**-**	**1**	**-**	**-**	**-**	**-**	**-**	**-**	**-**
FSU	**194**	**179**	**101 e**	**3**	**1**	**1 e**	**138**	**390**	**633 e**
Iran	-	-	-	-	-	-	-	-	-
Iraq	-	-	-	-	-	-	-	-	-
Kuwait	-	-	-	-	-	-	-	-	-
Oman	-	-	-	-	-	-	-	-	-
Qatar	-	-	-	-	-	-	-	-	-
Saudi Arabia	-	-	-	-	-	-	-	-	-
United Arab Emirates	-	-	-	-	-	-	-	-	-
Other	-	-	-	-	-	-	-	-	-
Middle East	**-**	**-**	**-**	**-**	**-**	**-**	**-**	**-**	**-**
India	-	-	-	-	-	-	-	-	-
Indonesia	-	-	-	-	-	-	-	-	-
Singapore	-	-	-	-	-	-	-	-	-
Other	-	-	-	-	-	-	-	-	-
Asia	**-**	**-**	**-**	**-**	**-**	**-**	**-**	**-**	**-**
China	**-**	**-**	**-**	**-**	**-**	**-**	**-**	**-**	**-**
Algeria	-	-	-	-	-	-	-	-	-
Angola/Cabinada	-	-	-	-	-	-	-	-	-
Cameroon	-	-	-	-	-	-	-	-	-
Egypt	-	-	-	-	-	-	-	-	-
Gabon	-	-	-	-	-	-	-	-	-
Libya	-	-	-	-	5	-	-	-	-
Nigeria	-	-	-	-	-	-	-	-	-
Tunisia	-	-	-	-	-	-	-	-	-
Other	-	-	-	-	-	-	-	-	-
Africa	**-**	**-**	**-**	**-**	**5**	**-**	**-**	**-**	**-**
Non-spec./Other	**3**	**-**	**26 e**	**161**	**122**	**35 e**	**114**	**2**	**104 e**
Total Imports	**901**	**861**	**649 e**	**191**	**136**	**36 e**	**695**	**984**	**1 214 e**
Memo: Total OPEC	-	-	-	-	*5*	-	-	-	-

Source: Annual Oil Statistics.

POLAND

Table 7D: Imports by Country of Origin

000 metric tons

	2000	2001	2002	2000	2001	2002	2000	2001	2002
	Fuel Oil (Residual)			**Petroleum Coke**			**Other Products**		
Australia	-	-	-	-	-	-	-	-	-
Belgium	42	-	-	-	-	-	6	7	9 e
Canada	-	-	-	-	-	-	-	-	-
Czech Republic	11	9	1 e	-	-	-	-	-	1 e
Denmark	1	-	-	-	-	-	20	92	32 e
Finland	-	-	-	-	-	-	-	1	4 e
France	-	-	-	-	-	-	13	16	8 e
Germany	45	49	34 e	-	-	-	21	20	20 e
Greece	-	-	-	-	-	-	-	-	-
Hungary	-	-	-	-	-	-	1	-	-
Ireland	-	-	-	-	-	-	-	-	-
Italy	-	-	-	-	-	-	1	1	-
Japan	-	-	-	-	-	-	-	-	-
Korea	-	-	-	-	-	-	-	-	-
Mexico	-	-	-	-	-	-	-	-	-
Netherlands	-	-	-	-	-	-	4	5	5 e
New Zealand	-	-	-	-	-	-	-	-	-
Norway	-	-	-	-	-	-	-	-	-
Poland	-	-	-	-	-	-	-	-	-
Spain	-	-	-	-	-	-	-	-	-
Sweden	-	-	14 e	-	-	-	11	10	15 e
Turkey	-	-	-	-	-	-	-	-	-
United Kingdom	-	-	-	-	-	-	1	1	-
United States	-	-	-	-	-	-	-	-	-
Other	1	3	-	-	-	-	13	16	8 e
OECD	**100**	**61**	**49 e**	**-**	**-**	**-**	**91**	**169**	**102 e**
Argentina	-	-	-	-	-	-	-	-	-
Brazil	-	-	-	-	-	-	-	-	-
Colombia	-	-	-	-	-	-	-	-	-
Ecuador	-	-	-	-	-	-	-	-	-
Netherlands Antilles	-	-	-	-	-	-	-	-	-
Peru	-	-	-	-	-	-	-	-	-
Trinidad and Tobago	-	-	-	-	-	-	-	-	-
Venezuela	-	-	-	-	-	-	-	-	-
Other	-	-	-	-	-	-	-	-	-
Latin America	**-**	**-**	**-**	**-**	**-**	**-**	**-**	**-**	**-**
Romania	-	-	-	-	-	-	-	-	-
Other	-	-	-	-	-	-	-	-	2 e
Non-OECD Europe	**-**	**-**	**-**	**-**	**-**	**-**	**-**	**-**	**2 e**
FSU	**13**	**14**	**50 e**	**-**	**-**	**-**	**192**	**84**	**85 e**
Iran	-	-	-	-	-	-	-	-	-
Iraq	-	-	-	-	-	-	-	-	-
Kuwait	-	-	-	-	-	-	-	-	-
Oman	-	-	-	-	-	-	-	-	-
Qatar	-	-	-	-	-	-	-	-	-
Saudi Arabia	-	-	-	-	-	-	-	-	-
United Arab Emirates	-	-	-	-	-	-	-	-	-
Other	-	-	-	-	-	-	-	-	-
Middle East	**-**	**-**	**-**	**-**	**-**	**-**	**-**	**-**	**-**
India	-	-	-	-	-	-	-	-	-
Indonesia	-	-	-	-	-	-	-	-	-
Singapore	-	-	-	-	-	-	-	-	-
Other	-	-	-	-	-	-	-	-	-
Asia	**-**	**-**	**-**	**-**	**-**	**-**	**-**	**-**	**-**
China	**-**	**-**	**-**	**-**	**-**	**-**	**-**	**-**	**4 e**
Algeria	-	-	-	-	-	-	-	-	-
Angola/Cabinada	-	-	-	-	-	-	-	-	-
Cameroon	-	-	-	-	-	-	-	-	-
Egypt	-	-	-	-	-	-	-	-	-
Gabon	-	-	-	-	-	-	-	-	-
Libya	-	-	-	-	-	-	-	-	-
Nigeria	-	-	-	-	-	-	-	-	-
Tunisia	-	-	-	-	-	-	-	-	-
Other	-	-	-	-	-	-	-	-	-
Africa	**-**	**-**	**-**	**-**	**-**	**-**	**-**	**-**	**-**
Non-spec./Other	**-**	**110**	**119 e**	**-**	**-**	**-**	**69**	**93**	**51 e**
Total Imports	**113**	**185**	**218 e**	**-**	**-**	**-**	**352**	**346**	**244 e**
Memo: Total OPEC	-	-	-	-	-	-	-	-	-

Source: Annual Oil Statistics.

POLAND
Table 8A: Exports by Country of Destination

000 metric tons

	Crude Oil 2000	Crude Oil 2001	Crude Oil 2002	Natural Gas Liquids 2000	Natural Gas Liquids 2001	Natural Gas Liquids 2002	Feedstocks and Other 2000	Feedstocks and Other 2001	Feedstocks and Other 2002
Australia	-	-	-	-	-	-	-	-	-
Austria	-	-	-	-	-	-	-	-	-
Belgium	-	-	-	-	-	-	-	-	-
Canada	-	-	-	-	-	-	-	-	-
Denmark	-	-	-	-	-	-	-	-	-
Finland	-	83	-	-	-	-	-	-	-
France	-	-	244 e	-	-	-	-	-	-
Germany	-	130	41 e	-	-	-	-	-	-
Greece	-	-	-	-	-	-	-	-	-
Ireland	-	-	-	-	-	-	-	-	-
Italy	-	-	-	-	-	-	-	-	-
Japan	-	-	-	-	-	-	-	-	-
Luxembourg	-	-	-	-	-	-	-	-	-
Mexico	-	-	-	-	-	-	-	-	-
Netherlands	-	122	-	-	-	-	-	-	-
Norway	-	-	-	-	-	-	-	-	-
Portugal	-	-	-	-	-	-	-	-	-
Spain	-	-	-	-	-	-	-	-	-
Sweden	129	105	-	-	-	-	-	-	-
Switzerland	-	-	-	-	-	-	-	-	-
Turkey	-	-	-	-	-	-	-	-	-
United Kingdom	-	-	136 e	-	-	-	-	-	-
United States	-	-	-	-	-	-	-	-	-
Other OECD	-	-	1 e	-	-	-	-	-	-
OECD	**129**	**440**	**422 e**	**-**	**-**	**-**	**-**	**-**	**-**
Africa	-	-	-	-	-	-	-	-	-
Latin America	-	-	-	-	-	-	-	-	-
Asia	-	-	-	-	-	-	-	-	-
China	-	-	-	-	-	-	-	-	-
Non-OECD Europe	-	-	-	-	-	-	-	-	-
Former Soviet Union	-	-	1 e	-	-	-	-	-	-
Middle East	-	-	-	-	-	-	-	-	-
Non-specified/Other	-	-	-	-	-	-	-	-	-
Non-OECD	**-**	**-**	**1 e**	**-**	**-**	**-**	**-**	**-**	**-**
Total Exports	**129**	**440**	**423 e**	**-**	**-**	**-**	**-**	**-**	**-**

	Total Products 2000	Total Products 2001	Total Products 2002	Liquefied Petroleum Gas 2000	Liquefied Petroleum Gas 2001	Liquefied Petroleum Gas 2002	Naphtha 2000	Naphtha 2001	Naphtha 2002
Australia	-	-	-	-	-	-	-	-	-
Austria	18	7	-	15	5	-	-	-	-
Belgium	45	75	58 e	-	-	-	-	-	-
Canada	-	-	-	-	-	-	-	-	-
Denmark	592	464	527 e	-	-	-	-	-	-
Finland	2	11	-	-	-	-	-	-	-
France	10	11	2 e	-	-	-	-	-	-
Germany	116	99	162 e	-	-	-	-	-	-
Greece	2	-	-	-	-	-	-	-	-
Ireland	-	-	-	-	-	-	-	-	-
Italy	3	1	-	-	-	-	-	-	-
Japan	-	-	-	-	-	-	-	-	-
Luxembourg	1	-	-	-	-	-	-	-	-
Mexico	-	-	-	-	-	-	-	-	-
Netherlands	430	283	401 e	-	-	-	-	-	-
Norway	221	221	224 e	-	-	-	-	-	-
Portugal	-	-	-	-	-	-	-	-	-
Spain	16	-	-	-	-	-	-	-	-
Sweden	176	268	246 e	-	-	-	-	-	-
Switzerland	1	2	1 e	-	-	-	-	-	-
Turkey	1	2	1 e	-	-	-	-	-	-
United Kingdom	96	417	83 e	-	-	-	-	-	-
United States	1	2	-	-	-	-	-	-	-
Other OECD	67	254	469 e	32	10	5 e	-	-	-
OECD	**1 798**	**2 117**	**2 174 e**	**47**	**15**	**5 e**	**-**	**-**	**-**
Africa	6	1	-	-	-	-	-	-	-
Latin America	10	-	-	-	-	-	-	-	-
Asia	-	-	-	-	-	-	-	-	-
China	-	-	-	-	-	-	-	-	-
Non-OECD Europe	12	8	27 e	-	2	1 e	-	-	-
Former Soviet Union	23	35	21 e	-	-	-	-	-	-
Middle East	1	1	-	-	-	-	-	-	-
Non-specified/Other	5	110	114 e	-	-	9 e	-	-	-
Non-OECD	**57**	**155**	**162 e**	**-**	**2**	**10 e**	**-**	**-**	**-**
Total Exports	**1 855**	**2 272**	**2 336 e**	**47**	**17**	**15 e**	**-**	**-**	**-**

Source: Annual Oil Statistics.

POLAND
Table 8B: Exports by Country of Destination

000 metric tons

	Total Gasoline 2000	Total Gasoline 2001	Total Gasoline 2002	Total Kerosene 2000	Total Kerosene 2001	Total Kerosene 2002	Gas/Diesel Oil 2000	Gas/Diesel Oil 2001	Gas/Diesel Oil 2002
Australia	-	-	-	-	-	-	-	-	-
Austria	-	-	-	-	1	-	-	-	-
Belgium	-	-	-	-	-	-	-	41	16 e
Canada	-	-	-	-	-	-	-	-	-
Denmark	-	-	-	33	43	53 e	-	-	-
Finland	-	-	-	-	11	-	-	-	-
France	-	-	-	2	5	1 e	-	-	-
Germany	-	-	-	6	11	5 e	-	15	53 e
Greece	-	-	-	-	-	-	-	-	-
Ireland	-	-	-	-	-	-	-	-	-
Italy	-	-	-	-	1	-	-	-	-
Japan	-	-	-	-	-	-	-	-	-
Luxembourg	-	-	-	-	-	-	-	-	-
Mexico	-	-	-	-	-	-	-	-	-
Netherlands	-	53	170 e	1	3	-	36	15	35 e
Norway	-	-	-	-	-	11 e	-	-	-
Portugal	-	-	-	-	-	-	-	-	-
Spain	-	-	-	-	-	-	16	-	-
Sweden	11	-	24 e	121	235	171 e	-	-	-
Switzerland	-	-	-	1	2	1 e	-	-	-
Turkey	-	-	-	1	2	1 e	-	-	-
United Kingdom	64	267	53 e	23	43	5 e	-	31	-
United States	-	-	-	-	2	-	-	-	-
Other OECD	-	44	138 e	-	30	32 e	-	64	162 e
OECD	**75**	**364**	**385 e**	**188**	**389**	**280 e**	**52**	**166**	**266 e**
Africa	-	-	-	1	1	-	-	-	-
Latin America	-	-	-	-	-	-	-	-	-
Asia	-	-	-	-	-	-	-	-	-
China	-	-	-	-	-	-	-	-	-
Non-OECD Europe	-	-	-	-	-	4 e	-	-	-
Former Soviet Union	-	-	4 e	1	1	-	-	-	-
Middle East	-	-	-	1	1	-	-	-	-
Non-specified/Other	-	-	-	-	-	32 e	-	-	4 e
Non-OECD	**-**	**-**	**4 e**	**3**	**3**	**36 e**	**-**	**-**	**4 e**
Total Exports	**75**	**364**	**389 e**	**191**	**392**	**316 e**	**52**	**166**	**270 e**

	Fuel Oil (Residual) 2000	Fuel Oil (Residual) 2001	Fuel Oil (Residual) 2002	Petroleum Coke 2000	Petroleum Coke 2001	Petroleum Coke 2002	Other Products 2000	Other Products 2001	Other Products 2002
Australia	-	-	-	-	-	-	-	-	-
Austria	-	-	-	-	-	-	3	1	-
Belgium	-	-	-	-	-	-	45	34	42 e
Canada	-	-	-	-	-	-	-	-	-
Denmark	559	421	448 e	-	-	-	-	-	26 e
Finland	-	-	-	-	-	-	2	-	-
France	-	-	-	-	-	-	8	6	1 e
Germany	96	58	74 e	-	-	-	14	15	30 e
Greece	2	-	-	-	-	-	-	-	-
Ireland	-	-	-	-	-	-	-	-	-
Italy	3	-	-	-	-	-	-	-	-
Japan	-	-	-	-	-	-	-	-	-
Luxembourg	1	-	-	-	-	-	-	-	-
Mexico	-	-	-	-	-	-	-	-	-
Netherlands	388	212	196 e	-	-	-	5	-	-
Norway	221	221	213 e	-	-	-	-	-	-
Portugal	-	-	-	-	-	-	-	-	-
Spain	-	-	-	-	-	-	-	-	-
Sweden	35	24	47 e	-	-	-	9	9	4 e
Switzerland	-	-	-	-	-	-	-	-	-
Turkey	-	-	-	-	-	-	-	-	-
United Kingdom	3	70	25 e	-	-	-	6	6	-
United States	1	-	-	-	-	-	-	-	-
Other OECD	23	89	88 e	-	-	-	12	17	44 e
OECD	**1 332**	**1 095**	**1 091 e**	**-**	**-**	**-**	**104**	**88**	**147 e**
Africa	5	-	-	-	-	-	-	-	-
Latin America	10	-	-	-	-	-	-	-	-
Asia	-	-	-	-	-	-	-	-	-
China	-	-	-	-	-	-	-	-	-
Non-OECD Europe	6	-	18 e	-	-	-	6	6	4 e
Former Soviet Union	14	28	14 e	-	-	-	8	6	3 e
Middle East	-	-	-	-	-	-	-	-	-
Non-specified/Other	-	83	3 e	-	-	-	5	27	66 e
Non-OECD	**35**	**111**	**35 e**	**-**	**-**	**-**	**19**	**39**	**73 e**
Total Exports	**1 367**	**1 206**	**1 126 e**	**-**	**-**	**-**	**123**	**127**	**220 e**

Source: Annual Oil Statistics.

PORTUGAL

Figure 1
FUEL SHARES OF TPES *

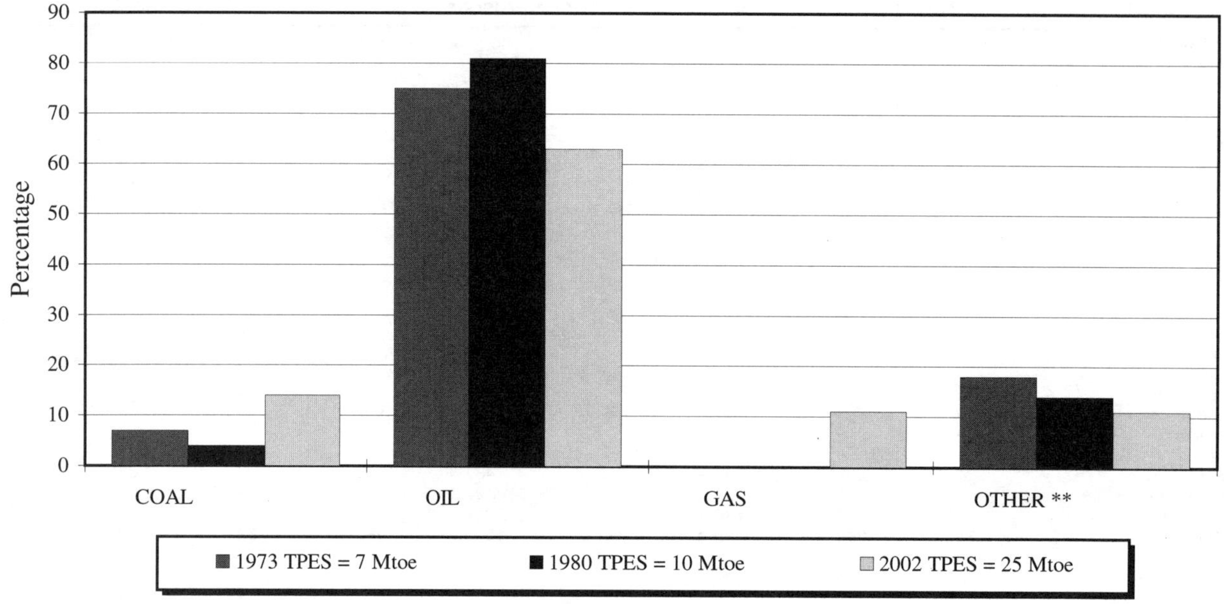

Source: Annual Energy Balances.

* Excluding electricity trade.

** Includes hydro, geothermal, solar, wind, combustible renewables and waste, etc.

Figure 2
OIL PRODUCTION AND CONSUMPTION

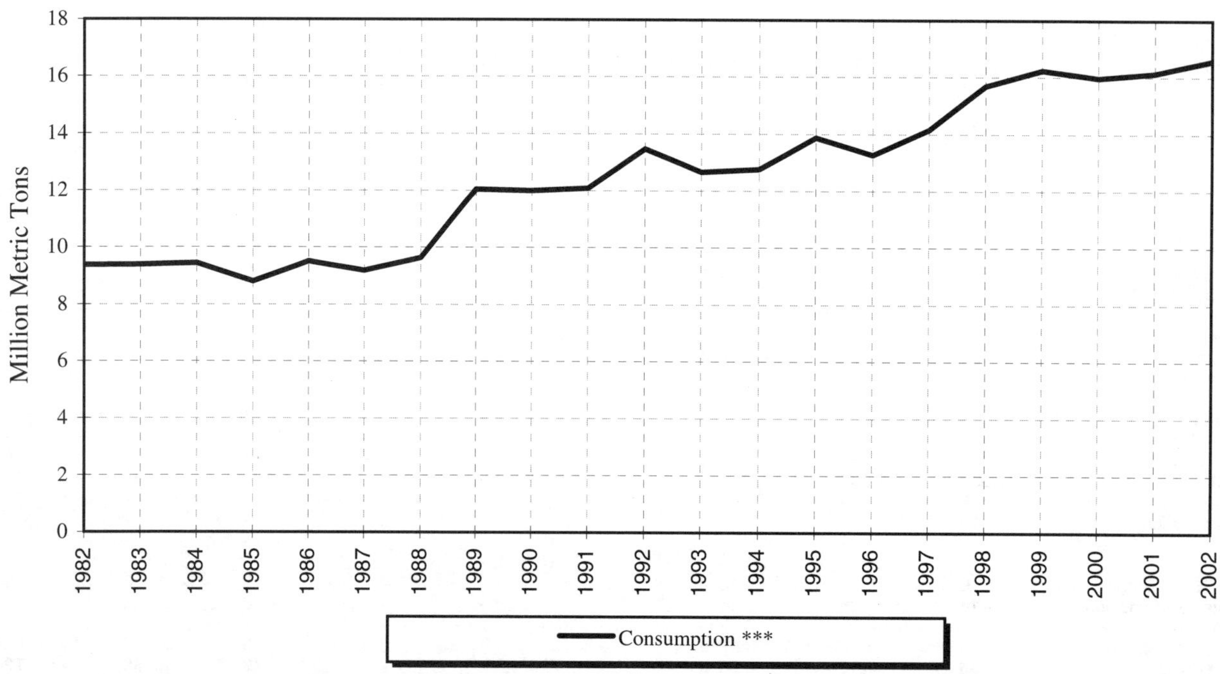

Source: Annual Oil Statistics.

*** Includes refinery fuel and international marine bunkers.

PORTUGAL
Table 1: Oil Supply and Consumption - Summary Table

000 metric tons

	1973	1978	1980	1990	1999	2000	2001	2002
Indigenous Production	-	-	-	-	-	-	-	-
Crude Oil	..	-	-	-	-	-	-	-
NGL	..	-	-	-	-	-	-	-
Other	..	-	-	-	-	-	-	-
+ Net Imports	6 082	7 344	9 272	12 197	16 720	15 827	16 642	16 055 e
Crude Oil, NGL, Other	4 348	6 296	8 316	11 189	13 807	12 133	13 158	12 238 e
Petroleum Products	1 734	1 048	956	1 008	2 913	3 694	3 484	3 817 e
- Intl. Marine Bunkers	820	583	435	619	599	676	483	495 e
+ Stock Changes	40	60	- 702	- 101	- 251	133	- 456	404 e
TOTAL SUPPLY	**5 302**	**6 821**	**8 135**	**11 477**	**15 870**	**15 284**	**15 703**	**15 964 e**
Refinery Intake	*4 274*	*6 581*	*7 685*	*10 984*	*13 594*	*12 362*	*13 079*	*12 733 e*
Refinery Output	*4 222*	*6 471*	*7 609*	*11 428*	*13 482*	*12 308*	*13 061*	*12 881 e*
LPG + Ethane	*92*	*132*	*220*	*357*	*341*	*277*	*371*	*334 e*
Naphtha	*193*	*296*	*203*	*811*	*1 007*	*996*	*986*	*1 078 e*
Kerosenes	*393*	*423*	*492*	*911*	*965*	*799*	*621*	*452 e*
Gasolines	*548*	*790*	*988*	*1 710*	*2 654*	*2 296*	*2 615*	*2 465 e*
Gas/Diesel Oil	*921*	*1 307*	*1 866*	*2 984*	*4 279*	*3 794*	*4 533*	*4 670 e*
Fuel Oil	*1 852*	*2 843*	*3 473*	*3 816*	*3 141*	*3 000*	*2 703*	*2 591 e*
Other Products	*223*	*680*	*367*	*839 e*	*1 095*	*1 146*	*1 232*	*1 291 e*
- Refinery Losses	52	110	76	- 444	112	54	18	- 148 e
+ Returns	-	-	-	-	-	-	-	-
- Statistical Differences	- 113	- 2	32	539	122	- 61	33	23 e
TOTAL CONSUMPTION	**5 363**	**6 713**	**8 027**	**11 382**	**15 636**	**15 291**	**15 652**	**16 089 e**
Transformation and Energy	**898**	**1 209**	**2 003**	**2 752**	**3 281**	**2 599**	**2 920**	..
Crude Oil	-	-	-	-	-	-	-	..
LPG + Ethane	2	3	2	6	10	7	16	..
Gas/Diesel Oil	25	30	67	25	17	26	35	..
Fuel Oil	692	1 021	1 799	2 403	2 551	1 949	2 167	..
Petroleum Coke	-	-	-	-	-	-	-	..
Other	179	155	135	318	703	617	702	..
Transport	**1 847**	**2 202**	**2 523**	**3 669**	**5 918**	**6 381**	**6 414**	..
Motor Gasoline	714	753	751	1 369	2 079	2 123	1 995	..
Gas/Diesel Oil	479	854	1 198	1 669	3 028	3 404	3 599	..
Subtotal	*1 193*	*1 607*	*1 949*	*3 038*	*5 107*	*5 527*	*5 594*	..
Aviation Fuels	580	546	516	567	724	772	738	..
Other	74	49	58	64	87	82	82	..
Industry	**1 786**	**2 377**	**2 609**	**3 791**	**4 809**	**4 633**	**4 490**	..
Naphtha	192	231	223	1 351	1 340	1 369	1 024	..
LPG + Ethane	71	89	106	249	356	340	340	..
Gas/Diesel Oil	101	113	131	245	383	431	469	..
Fuel Oil	1 291	1 752	1 926	1 601	1 584	1 489	1 292	..
Other	131	192	223	345	1 146	1 004	1 365	..
Other Sectors	**832**	**925**	**892**	**1 170**	**1 628**	**1 678**	**1 828**	..
Gas/Diesel Oil	430	403	346	503	800	899	1 080	..
Fuel Oil	33	62	72	66	-	-	-	..
Other	369	460	474	601	828	779	748	..
Memo: Non-Energy Use	*372*	*483*	*519*	*2 061*	*2 276*	*2 283*	*2 148*	..
Chem/Petchem. Feedst.	*203*	*252*	*245*	*1 669*	*1 587*	*1 626*	*1 252*	..
Other	*169*	*231*	*274*	*392*	*689*	*657*	*896*	..

Source: Annual Oil Statistics.

PORTUGAL

Table 2: Supply of Crude Oil, NGL, Feedstocks and Other Hydrocarbons

000 metric tons

	1973	1978	1980	1990	1999	2000	2001	2002
CRUDE OIL								
Indigenous Production	..	-	-	-	-	-	-	-
+ Imports	..	6 270	8 260	10 607	13 302	11 498	12 617	11 466 e
- Exports	..	-	-	-	-	-	-	-
- Direct Use	..	-	-	-	-	-	-	-
+ Stock Change	..	238	- 704	- 130	- 375	120	- 181	331 e
- Statistical Difference	..	13	9	4	-	1	-	-
Refinery Intake	**..**	**6 495**	**7 547**	**10 473**	**12 927**	**11 617**	**12 436**	**11 797 e**
Closing Stock Level	..	*659*	*1 546*	*792*	*1 126*	*1 006*	*1 187*	*856 e*
NATURAL GAS LIQUIDS								
Indigenous Production	..	-	-	-	-	-	-	-
+ Imports	..	-	-	-	-	-	-	-
- Exports	..	-	-	-	-	-	-	-
- Direct Use	..	-	-	-	-	-	-	-
+ Stock Change	..	-	-	-	-	-	-	-
- Statistical Difference	..	-	-	-	-	-	-	-
Refinery Intake	**..**	**-**	**-**	**-**	**-**	**-**	**-**	**-**
Closing Stock Level	..	-	-	-	-	-	-	-
REFINERY FEEDSTOCKS								
Backflows	..	-	-	-	-	-	-	-
+ Transfers	..	-	75	33	154	102	127	126 e
+ Imports	..	26	56	582	505	635	541	772 e
- Exports	..	-	-	-	-	-	-	-
+ Stock Change	..	- 26	11	- 104	29	18	- 13	38 e
- Statistical Difference	..	- 86	4	-	21	10	12	-
Refinery Intake	**..**	**86**	**138**	**511**	**667**	**745**	**643**	**936 e**
Closing Stock Level	..	*50*	*44*	*131*	*73*	*55*	*68*	*30 e*
OTHER HYDROCARBONS								
Indigenous Production	..	-	-	-	-	-	-	-
+ Imports	..	-	-	-	-	-	-	-
- Exports	..	-	-	-	-	-	-	-
- Direct Use	..	-	-	-	-	-	-	-
+ Stock Change	..	-	-	-	-	-	-	-
- Statistical Difference	..	-	-	-	-	-	-	-
Refinery Intake	**..**	**-**	**-**	**-**	**-**	**-**	**-**	**-**
Closing Stock Level	..	-	-	-	-	-	-	-
CRUDE OIL, NGL, FEEDSTOCKS AND OTHER								
Indigenous Production	-	-	-	-	-	-	-	-
+ Backflows	-	-	-	-	-	-	-	-
+ Transfers	-	-	75	33	154	102	127	126 e
+ Imports	4 348	6 296	8 316	11 189	13 807	12 133	13 158	12 238 e
- Exports	-	-	-	-	-	-	-	-
- Direct Use	-	-	-	-	-	-	-	-
+ Stock Change	- 147	212	- 693	- 234	- 346	138	- 194	369 e
- Statistical Difference	- 73	- 73	13	4	21	11	12	-
Refinery Intake	**4 274**	**6 581**	**7 685**	**10 984**	**13 594**	**12 362**	**13 079**	**12 733 e**
Closing Stock Level	-	*709*	*1 590*	*923*	*1 199*	*1 061*	*1 255*	*886 e*

Source: Annual Oil Statistics.

PORTUGAL
Table 3: Motor Gasoline - Supply

000 metric tons

	1973	1978	1980	1990	1999	2000	2001	2002
MOTOR GASOLINE								
Refinery Output	**548**	**776**	**975**	**1 704**	**2 654**	**2 296**	**2 615**	**2 465 e**
+ Imports	153	6	-	-	11	154	10	34 e
- Exports	10	22	185	294	588	397	635	541 e
+ Transfers	-	-	-	4	-	2	-	-
+ Stock Change	17	- 4	- 34	- 33	4	27	5	35 e
- Statistical Difference	- 6	3	5	11	2	- 41	-	- 30 e
Total Consumption	**714**	**753**	**751**	**1 370**	**2 079**	**2 123**	**1 995**	**2 023 e**
Closing Stock Level	-	*140*	*240*	*204*	*193*	*166*	*198*	*163 e*

Table 4: Aviation Fuels - Supply

000 metric tons

	1973	1978	1980	1990	1999	2000	2001	2002
AVIATION GASOLINE								
Refinery Output	-	-	-	-	-	-	-	-
+ Imports	11	4	1	2	3	3	2	2 e
- Exports	2	1	-	-	-	-	-	-
+ Transfers	-	-	-	-	-	-	-	-
+ Stock Change	-	-	1	-	-	-	-	-
- Statistical Difference	1	-	-	-	-	-	- 1	-
Total Consumption	**8**	**3**	**2**	**2**	**3**	**3**	**3**	**2 e**
Closing Stock Level	-	*2*	*1*	*1*	*1*	*1*	*1*	*1 e*
GASOLINE TYPE JET FUEL								
Refinery Output	-	**14**	**13**	**6**	-	-	-	-
+ Imports	-	-	8	-	-	-	-	-
- Exports	-	-	-	-	-	-	-	-
+ Transfers	-	-	-	-	-	-	-	-
+ Stock Change	-	3	- 6	1	-	-	-	-
- Statistical Difference	-	-	- 2	1	-	-	-	-
Total Consumption	-	**17**	**17**	**6**	-	-	-	-
Closing Stock Level	-	*1*	*13*	*2*	-	-	-	-
KEROSENE TYPE JET FUEL								
Refinery Output	**252**	**340**	**419**	**885**	**965**	**798**	**618**	**448 e**
+ Imports	354	167	70	67	-	73	151	190 e
- Exports	16	7	-	369	221	99	13	-
+ Transfers	-	-	-	26	- 18	- 5	-	-
+ Stock Change	- 26	11	16	- 26	- 8	9	- 15	64 e
- Statistical Difference	- 8	- 15	8	24	- 3	7	6	- 17 e
Total Consumption	**572**	**526**	**497**	**559**	**721**	**769**	**735**	**719 e**
Closing Stock Level	-	*37*	*91*	*98*	*103*	*94*	*109*	*45 e*

Source: Annual Oil Statistics.

PORTUGAL
Table 5A: Gas/Diesel Oil - Supply and Consumption

000 metric tons

	1973	1978	1980	1990	1999	2000	2001	2002
Refinery Output	**921**	**1 307**	**1 866**	**2 984**	**4 279**	**3 794**	**4 533**	**4 670 e**
+ Recycled Products	-	-	-	-	-	-	-	-
+ Imports	289	227	150	64	268	1 271	943	580 e
- Exports	66	24	49	344	78	152	90	6 e
- Intl. Marine Bunkers	184	166	168	212	183	205	153	128 e
+ Transfers	-	-	- 9	- 30	- 15	- 20	25	-
+ Stock Changes	96	82	- 18	- 18	- 17	-	- 111	25 e
- Statistical Differences	21	26	30	2	26	- 72	- 36	- 80 e
Total Consumption	**1 035**	**1 400**	**1 742**	**2 442**	**4 228**	**4 760**	**5 183**	**5 221 e**
Transformation	**19**	**25**	**56**	**20**	**17**	**25**	**34**	**..**
Public Electricity	19	22	54	19	15	23	32	..
Autoproducer Electricity	-	3	2	1	2	1	1	..
Public CHP Plants	-	-	-	-	-	-	-	..
Autoproducers of CHP	-	-	-	-	-	1	1	..
Public Heat Plants	-	-	-	-	-	-	-	..
Autoproducer Heat Plants	-	-	-	-	-	-	-	..
Petrochemical Industry	-	-	-	-	-	-	-	..
Other Transformation	-	-	-	-	-	-	-	..
Energy Sector	**6**	**5**	**11**	**5**	**-**	**1**	**1**	**..**
Refinery Fuel	6	3	6	-	-	1	-	..
Other Energy Sector	-	2	5	5	-	-	1	..
Final Consumption	**1 010**	**1 370**	**1 675**	**2 417**	**4 211**	**4 734**	**5 148**	**..**
Transport	**479**	**854**	**1 198**	**1 669**	**3 028**	**3 404**	**3 599**	**..**
Air	-	-	-	-	-	-	-	..
Road	393	760	1 101	1 571	2 940	3 305	3 508	..
Rail	46	52	50	55	50	56	42	..
Inland Waterways	40	42	47	43	38	43	49	..
Non-Specified	-	-	-	-	-	-	-	..
Industry	**101**	**113**	**131**	**245**	**383**	**431**	**469**	**..**
Iron and Steel	1	2	2	2	3	3	3	..
Chemical (incl. Pet.-Chem.)	5	4	4	6	8	9	10	..
Non-Ferrous Metals	-	-	-	-	3	3	3	..
Non-Metallic Minerals	10	11	15	21	35	39	43	..
Transport Equipment	4	6	15	4	3	3	3	..
Machinery	-	-	1	2	2	2	2	..
Mining and Quarrying	5	-	-	21	47	53	58	..
Food and Tobacco	11	14	12	22	45	51	56	..
Pulp, Paper and Printing	-	3	4	4	6	7	8	..
Wood and Wood Products	-	5	6	13	13	15	16	..
Construction	48	51	61	138	211	237	258	..
Textiles and Leather	-	-	1	3	4	5	6	..
Non-Specified	17	17	10	9	3	4	3	..
Other Sectors	**430**	**403**	**346**	**503**	**800**	**899**	**1 080**	**..**
Commerce and Public	116	98	45	85	240	270	426	..
Residential	3	2	1	9	-	-	-	..
Agriculture	311	303	300	409	560	629	654	..
Non-Specified	-	-	-	-	-	-	-	..
Memo: Non-Energy Use	-	-	-	-	-	-	-	..
Chem/Petchem. Feedst.	-	-	-	-	-	-	-	..
Other	-	-	-	-	-	-	-	..
Closing Stock Level	-	*118*	*294*	*357*	*376*	*376*	*487*	*462 e*

Source: Annual Oil Statistics.

PORTUGAL
Table 5B: Fuel Oil (Residual) - Supply and Consumption

000 metric tons

	1973	1978	1980	1990	1999	2000	2001	2002
Refinery Output	**1 852**	**2 843**	**3 473**	**3 816**	**3 141**	**3 000**	**2 703**	**2 591 e**
+ Recycled Products	-	-	-	-	-	-	-	-
+ Imports	694	336	692	1 830	1 795	1 226	1 313	2 097 e
- Exports	-	72	139	1 137	164	483	195	318 e
- Intl. Marine Bunkers	629	410	262	407	416	471	330	367 e
+ Transfers	-	-	-	55	4	21	9	-
+ Stock Changes	78	153	40	- 106	- 136	220	4	- 99 e
- Statistical Differences	- 53	15	7	- 19	89	75	45	109 e
Total Consumption	**2 048**	**2 835**	**3 797**	**4 070**	**4 135**	**3 438**	**3 459**	**3 795 e**
Transformation	**424**	**757**	**1 511**	**2 185**	**2 314**	**1 719**	**1 971**	..
Public Electricity	424	682	1 271	1 957	1 629	1 173	1 499	..
Autoproducer Electricity	-	-	-	-	-	-	-	..
Public CHP Plants	-	-	136	94	196	169	164	..
Autoproducers of CHP	-	75	104	129	489	377	308	..
Public Heat Plants	-	-	-	-	-	-	-	..
Autoproducer Heat Plants	-	-	-	-	-	-	-	..
Petrochemical Industry	-	-	-	-	-	-	-	..
Other Transformation	-	-	-	5	-	-	-	..
Energy Sector	**268**	**264**	**288**	**218**	**237**	**230**	**196**	..
Refinery Fuel	268	262	286	213	237	230	196	..
Other Energy Sector	-	2	2	5	-	-	-	..
Final Consumption	**1 356**	**1 814**	**1 998**	**1 667**	**1 584**	**1 489**	**1 292**	..
Transport	**32**	-	-	-	-	-	-	..
Air	-	-	-	-	-	-	-	..
Road	-	-	-	-	-	-	-	..
Rail	31	-	-	-	-	-	-	..
Inland Waterways	1	-	-	-	-	-	-	..
Non-Specified	-	-	-	-	-	-	-	..
Industry	**1 291**	**1 752**	**1 926**	**1 601**	**1 584**	**1 489**	**1 292**	..
Iron and Steel	25	31	26	45	19	19	14	..
Chemical (incl. Pet.-Chem.)	140	174	158	476	272	284	400	..
Non-Ferrous Metals	-	5	10	8	9	9	6	..
Non-Metallic Minerals	581	800	870	192	346	310	234	..
Transport Equipment	-	3	6	7	7	7	5	..
Machinery	-	-	10	18	7	7	5	..
Mining and Quarrying	12	-	12	3	3	3	2	..
Food and Tobacco	164	207	195	199	177	177	130	..
Pulp, Paper and Printing	137	218	254	225	337	311	233	..
Wood and Wood Products	10	24	30	30	69	58	44	..
Construction	4	4	5	17	50	50	37	..
Textiles and Leather	161	236	272	298	270	254	179	..
Non-Specified	57	50	78	83	18	-	3	..
Other Sectors	**33**	**62**	**72**	**66**	-	-	-	..
Commerce and Public	25	46	44	53	-	-	-	..
Residential	4	10	21	-	-	-	-	..
Agriculture	4	6	7	13	-	-	-	..
Non-Specified	-	-	-	-	-	-	-	..
Memo: Non-Energy Use	-	-	-	*262*	*209*	*223*	*191*	..
Chem/Petchem. Feedst.	-	-	-	*262*	*209*	*223*	*191*	..
Other	-	-	-	-	-	-	-	..
Closing Stock Level	-	*73*	*310*	*676*	*706*	*486*	*482*	*581 e*

Source: Annual Oil Statistics.

PORTUGAL
Table 5C: Total Products - Supply and Consumption

000 metric tons

	1973	1978	1980	1990	1999	2000	2001	2002
Refinery Output	**4 222**	**6 471**	**7 609**	**11 428**	**13 482**	**12 308**	**13 061**	**12 881 e**
+ Recycled Products	-	-	-	-	-	-	-	-
+ Imports	1 955	1 214	1 343	3 480	4 302	5 118	4 839	5 187 e
- Exports	221	166	387	2 472	1 389	1 424	1 355	1 370 e
- Intl. Marine Bunkers	820	583	435	619	599	676	483	495 e
+ Transfers	-	-	- 75	- 33	- 154	- 102	- 127	- 126 e
+ Stock Changes	187	- 152	- 9	133	95	- 5	- 262	35 e
- Statistical Differences	- 40	71	19	535	101	- 72	21	23 e
Total Consumption	**5 363**	**6 713**	**8 027**	**11 382**	**15 636**	**15 291**	**15 652**	**16 089 e**
Transformation	**557**	**837**	**1 625**	**2 268**	**2 338**	**1 758**	**2 007**	..
Public Electricity	443	704	1 325	1 976	1 644	1 196	1 531	..
Autoproducer Electricity	-	3	2	1	2	1	1	..
Public CHP Plants	-	-	136	94	196	169	164	..
Autoproducers of CHP	-	75	104	129	496	392	311	..
Public Heat Plants	-	-	-	-	-	-	-	..
Autoproducer Heat Plants	-	-	-	-	-	-	-	..
Petrochemical Industry	-	-	-	-	-	-	-	..
Other Transformation	114	55	58	68	-	-	-	..
Energy Sector	**341**	**372**	**378**	**484**	**943**	**841**	**913**	..
Refinery Fuel	341	368	371	472	943	841	911	..
Other Energy Sector	-	4	7	12	-	-	2	..
Final Consumption	**4 465**	**5 504**	**6 024**	**8 630**	**12 355**	**12 692**	**12 732**	..
Transport	**1 847**	**2 202**	**2 523**	**3 669**	**5 918**	**6 381**	**6 414**	..
Air	580	546	516	567	724	772	738	..
Road	1 148	1 561	1 908	2 998	5 101	5 508	5 583	..
Rail	78	53	52	56	50	56	42	..
Inland Waterways	41	42	47	48	43	45	51	..
Non-Specified	-	-	-	-	-	-	-	..
Industry	**1 786**	**2 377**	**2 609**	**3 791**	**4 809**	**4 633**	**4 490**	..
Iron and Steel	34	37	38	62	38	37	32	..
Chemical (incl. Pet.-Chem.)	352	431	409	1 896	1 764	1 784	1 561	..
Non-Ferrous Metals	-	6	12	11	17	17	14	..
Non-Metallic Minerals	627	856	908	371	1 098	959	1 002	..
Transport Equipment	9	15	27	24	29	26	24	..
Machinery	4	2	17	28	17	17	15	..
Mining and Quarrying	18	1	13	27	76	81	64	..
Food and Tobacco	190	244	225	245	253	257	215	..
Pulp, Paper and Printing	139	223	261	233	351	326	249	..
Wood and Wood Products	13	32	37	47	101	91	78	..
Construction	144	195	231	364	700	711	1 029	..
Textiles and Leather	163	239	276	311	291	275	201	..
Non-Specified	93	96	155	172	74	52	6	..
Other Sectors	**832**	**925**	**892**	**1 170**	**1 628**	**1 678**	**1 828**	..
Commerce and Public	154	174	213	199	403	424	564	..
Residential	319	433	366	533	654	615	600	..
Agriculture	359	318	313	438	571	639	664	..
Non-Specified	-	-	-	-	-	-	-	..
Memo: Non-Energy Use	*372*	*483*	*519*	*2 061*	*2 276*	*2 283*	*2 148*	..
Chem/Petchem. Feedst.	*203*	*252*	*245*	*1 669*	*1 587*	*1 626*	*1 252*	..
Other	*169*	*231*	*274*	*392*	*689*	*657*	*896*	..
Closing Stock Level	-	*949*	*1 608*	*1 776*	*1 898*	*1 903*	*2 202*	*2 167 e*

Source: Annual Oil Statistics.

PORTUGAL
Table 6A: Supply and Consumption

000 metric tons

	1980	2000	2001	2002	1980	2000	2001	2002
	LPG AND ETHANE				**REFINERY GAS**			
Refinery Output	**220**	**277**	**371**	**334 e**	**28**	**-**	**-**	**-**
+ Recycled Products	-	-	-	-	-	-	-	-
+ Imports	280	767	704	706 e	-	-	-	-
- Exports	1	23	21	28 e	-	-	-	-
- Intl. Marine Bunkers	-	-	-	-	-	-	-	-
+ Transfers	-	-	-	-	-	-	-	-
+ Stock Change	5	- 9	- 26	-	-	-	-	-
- Statistical Difference	4	- 29	5	10 e	- 1	-	-	-
Total Consumption	**500**	**1 041**	**1 023**	**1 002 e**	**29**	**-**	**-**	**-**
of which: Ethane	-	-	-	-	-	-	-	-
Transformation	-	-	-	..	28	-	-	..
Energy	2	7	16	..	1	-	-	..
Transport	-	20	20	..	-	-	-	..
Industry	106	340	340	..	-	-	-	..
Other	392	674	647	..	-	-	-	..
Memo: Non-Energy Use	-	-	-	..	-	-	-	..
Closing Stock Level	*35*	*41*	*67*	*67 e*	-	-	-	-

	1980	2000	2001	2002	1980	2000	2001	2002
	NAPHTHA				**OTHER KEROSENE**			
Refinery Output	**203**	**996**	**986**	**1 078 e**	**73**	**1**	**3**	**4 e**
+ Recycled Products	-	-	-	-	-	-	-	-
+ Imports	47	387	179	151 e	-	4	5	2 e
- Exports	-	-	117	121 e	1	-	-	-
- Intl. Marine Bunkers	-	-	-	-	-	-	-	-
+ Transfers	-	-	-	-	-	4	-	-
+ Stock Change	5	- 5	- 25	29 e	5	1	- 1	-
- Statistical Difference	2	- 2	- 1	4 e	10	1	-	2 e
Total Consumption	**253**	**1 380**	**1 024**	**1 133 e**	**67**	**9**	**7**	**4 e**
Transformation	30	11	-	..	-	-	-	..
Energy	-	-	-	..	-	-	-	..
Transport	-	-	-	..	-	-	-	..
Industry	223	1 369	1 024	..	11	-	-	..
Other	-	-	-	..	56	9	7	..
Memo: Non-Energy Use	*223*	*1 369*	*1 024*	..	-	-	-	..
Closing Stock Level	*28*	*45*	*70*	*41 e*	*15*	*1*	*2*	*2 e*

	1980	2000	2001	2002	1980	2000	2001	2002
	WHITE SPIRIT + SBP				**LUBRICANTS**			
Refinery Output	**9**	**44**	**24**	**..**	**73**	**139**	**117**	**..**
+ Recycled Products	-	-	-	..	-	-	-	..
+ Imports	27	2	-	..	46	76	66	..
- Exports	-	34	13	..	10	93	92	..
- Intl. Marine Bunkers	-	-	-	..	5	-	-	..
+ Transfers	-	-	-	..	-	- 2	- 1	..
+ Stock Change	- 2	2	- 2	..	- 3	- 19	12	..
- Statistical Difference	1	- 2	- 4	..	1	- 10	- 2	..
Total Consumption	**33**	**16**	**13**	**..**	**100**	**111**	**104**	**..**
Transformation	-	-	-	..	-	-	-	..
Energy	-	-	-	..	-	-	-	..
Transport	-	-	-	..	58	62	62	..
Industry	25	11	11	..	24	32	30	..
Other	8	5	2	..	18	17	12	..
Memo: Non-Energy Use	*33*	*16*	*13*	..	*100*	*111*	*104*	..
Closing Stock Level	*10*	*3*	*5*	..	*53*	*48*	*36*	..

Source: Annual Oil Statistics.

PORTUGAL
Table 6B: Supply and Consumption

000 metric tons

	1980	2000	2001	2002	1980	2000	2001	2002
	BITUMEN				PARAFFIN WAXES			
Refinery Output	**147**	**267**	**424**	**..**	**4**	**15**	**18**	**..**
+ Recycled Products	-	-	-	..	-	-	-	..
+ Imports	21	365	464	..	1	-	3	..
- Exports	-	70	90	..	2	6	7	..
- Intl. Marine Bunkers	-	-	-	..	-	-	-	..
+ Transfers	-	-	11	..	-	-	2	..
+ Stock Change	- 16	- 6	4	..	2	-	- 3	..
- Statistical Difference	- 6	4	13	..	-	- 3	- 3	..
Total Consumption	**158**	**552**	**800**	**..**	**5**	**12**	**16**	**..**
Transformation	-	-	-	..	-	-	-	..
Energy	-	-	-	..	-	-	-	..
Transport	-	-	-	..	-	-	-	..
Industry	158	479	720	..	5	11	16	..
Other	-	73	80	..	-	1	-	..
Memo: Non-Energy Use	*158*	*552*	*800*	..	*5*	*12*	*16*	..
Closing Stock Level	*31*	*38*	*34*	..	*2*	*2*	*5*	..
	PETROLEUM COKE				OTHER PRODUCTS			
Refinery Output	**-**	**-**	**-**	**-**	**106**	**681**	**649**	**1 291 e**
+ Recycled Products	-	-	-	-	-	-	-	-
+ Imports	-	469	594	632 e	-	321	405	793 e
- Exports	-	-	-	-	-	67	82	356 e
- Intl. Marine Bunkers	-	-	-	-	-	-	-	-
+ Transfers	-	-	-	-	- 66	- 102	- 173	- 126 e
+ Stock Change	-	- 59	- 69	19 e	- 4	- 166	- 35	- 38 e
- Statistical Difference	-	-	-	- 1 e	- 40	-	- 1	26 e
Total Consumption	**-**	**410**	**525**	**652 e**	**76**	**667**	**765**	**1 538 e**
Transformation	-	-	-	..	-	3	2	..
Energy	-	-	-	..	76	603	700	..
Transport	-	-	-	..	-	-	-	..
Industry	-	410	525	..	-	61	63	..
Other	-	-	-	..	-	-	-	..
Memo: Non-Energy Use	-	-	-	..	-	-	-	..
Closing Stock Level	-	*123*	*192*	*173 e*	*485*	*479*	*514*	*632 e*

Source: Annual Oil Statistics.

PORTUGAL
Table 7A: Imports by Country of Origin

000 metric tons

	2000	2001	2002	2000	2001	2002	2000	2001	2002
	Crude Oil			**Natural Gas Liquids**			**Feedstocks and Other**		
Australia	-	-	-	-	-	-	-	-	-
Belgium	-	-	-	-	-	-	-	-	-
Canada	-	-	-	-	-	-	-	-	-
Czech Republic	-	-	-	-	-	-	-	-	-
Denmark	338	84	86 e	-	-	-	-	-	-
Finland	-	-	-	-	-	-	-	-	-
France	-	-	-	-	-	-	-	-	-
Germany	-	-	-	-	-	-	-	-	-
Greece	-	-	-	-	-	-	-	-	-
Hungary	-	-	-	-	-	-	-	-	-
Ireland	-	-	-	-	-	-	-	-	-
Italy	-	-	-	-	-	-	-	-	-
Japan	-	-	-	-	-	-	-	-	-
Korea	-	-	-	-	-	-	-	-	-
Mexico	942	944	818 e	-	-	-	-	-	-
Netherlands	-	-	-	-	-	-	-	-	-
New Zealand	-	-	-	-	-	-	-	-	-
Norway	868	799	344 e	-	-	-	-	-	-
Poland	-	-	-	-	-	-	-	-	-
Spain	-	-	-	-	-	-	-	-	-
Sweden	-	-	-	-	-	-	-	-	-
Turkey	-	-	-	-	-	-	-	-	-
United Kingdom	585	479	1 170 e	-	-	-	-	-	-
United States	-	-	-	-	-	-	-	-	-
Other	-	-	-	-	-	-	-	-	-
OECD	**2 733**	**2 306**	**2 418 e**	**-**	**-**	**-**	**-**	**-**	**-**
Argentina	-	-	-	-	-	-	-	-	-
Brazil	-	291	1 008 e	-	-	-	-	-	-
Colombia	-	-	-	-	-	-	-	-	-
Ecuador	-	-	-	-	-	-	-	-	-
Netherlands Antilles	-	-	-	-	-	-	-	-	-
Peru	-	-	-	-	-	-	-	-	-
Trinidad and Tobago	-	-	-	-	-	-	-	-	-
Venezuela	-	852	424 e	-	-	-	-	-	-
Other	-	-	-	-	-	-	-	-	-
Latin America	**-**	**1 143**	**1 432 e**	**-**	**-**	**-**	**-**	**-**	**-**
Romania	-	-	-	-	-	-	-	-	-
Other	-	-	-	-	-	-	-	-	-
Non-OECD Europe	**-**	**-**	**-**	**-**	**-**	**-**	**-**	**-**	**-**
FSU	**285**	**1 268**	**578 e**	**-**	**-**	**-**	**-**	**-**	**-**
Iran	1 682	469	777 e	-	-	-	-	-	-
Iraq	808	273	362 e	-	-	-	-	-	-
Kuwait	-	-	-	-	-	-	-	-	-
Oman	-	-	-	-	-	-	-	-	-
Qatar	-	-	-	-	-	-	-	-	-
Saudi Arabia	1 954	1 965	1 872 e	-	-	-	-	-	-
United Arab Emirates	-	-	-	-	-	-	-	-	-
Other	78	81	159 e	-	-	-	-	-	-
Middle East	**4 522**	**2 788**	**3 170 e**	**-**	**-**	**-**	**-**	**-**	**-**
India	-	-	-	-	-	-	-	-	-
Indonesia	-	-	-	-	-	-	-	-	-
Singapore	-	-	-	-	-	-	-	-	-
Other	-	-	-	-	-	-	-	-	-
Asia	**-**	**-**	**-**	**-**	**-**	**-**	**-**	**-**	**-**
China	**-**	**-**	**-**	**-**	**-**	**-**	**-**	**-**	**-**
Algeria	-	319	554 e	-	-	-	-	-	-
Angola/Cabinada	275	642	392 e	-	-	-	-	-	-
Cameroon	-	-	-	-	-	-	-	-	-
Egypt	406	292	68 e	-	-	-	-	-	-
Gabon	-	143	47 e	-	-	-	-	-	-
Libya	-	79	-	-	-	-	-	-	-
Nigeria	3 277	3 637	2 580 e	-	-	-	-	-	-
Tunisia	-	-	-	-	-	-	-	-	-
Other	-	-	227 e	-	-	-	-	-	-
Africa	**3 958**	**5 112**	**3 868 e**	**-**	**-**	**-**	**-**	**-**	**-**
Non-spec./Other	**-**	**-**	**-**	**-**	**-**	**-**	**635**	**541**	**772 e**
Total Imports	**11 498**	**12 617**	**11 466 e**	**-**	**-**	**-**	**635**	**541**	**772 e**
Memo: Total OPEC	*7 721*	*7 594*	*6 569 e*	-	-	-	-	-	-

Source: Annual Oil Statistics.

PORTUGAL
Table 7B: Imports by Country of Origin

000 metric tons

	2000	2001	2002	2000	2001	2002	2000	2001	2002
	Total Products			Liquefied Petroleum Gases			Naphtha		
Australia	-	-	-	-	-	-	-	-	-
Belgium	62	70	64 e	25	13	5 e	-	-	-
Canada	-	-	-	-	-	-	-	-	-
Czech Republic	-	-	-	-	-	-	-	-	-
Denmark	4	31	8 e	4	3	8 e	-	-	-
Finland	56	20	-	-	-	-	-	-	-
France	572	595	827 e	102	98	93 e	-	-	-
Germany	383	84	90 e	14	8	19 e	-	-	-
Greece	-	30	-	-	-	-	-	-	-
Hungary	-	-	-	-	-	-	-	-	-
Ireland	-	1	-	-	1	-	-	-	-
Italy	119	178	200 e	1	-	-	-	-	-
Japan	-	-	-	-	-	-	-	-	-
Korea	-	-	-	-	-	-	-	-	-
Mexico	-	-	-	-	-	-	-	-	-
Netherlands	597	451	437 e	31	11	15 e	-	-	-
New Zealand	-	-	-	-	-	-	-	-	-
Norway	63	71	111 e	12	21	35 e	-	-	-
Poland	-	-	-	-	-	-	-	-	-
Spain	1 233	676	801 e	63	55	47 e	387	179	151 e
Sweden	59	119	258 e	23	5	2 e	-	-	-
Turkey	-	-	-	-	-	-	-	-	-
United Kingdom	650	792	1 046 e	468	347	443 e	-	-	-
United States	515	198	32 e	-	-	-	-	-	-
Other	-	-	-	-	-	-	-	-	-
OECD	**4 313**	**3 316**	**3 874 e**	**743**	**562**	**667 e**	**387**	**179**	**151 e**
Argentina	-	-	-	-	-	-	-	-	-
Brazil	-	-	-	-	-	-	-	-	-
Colombia	-	-	-	-	-	-	-	-	-
Ecuador	-	-	-	-	-	-	-	-	-
Netherlands Antilles	-	-	-	-	-	-	-	-	-
Peru	-	-	-	-	-	-	-	-	-
Trinidad and Tobago	-	-	-	-	-	-	-	-	-
Venezuela	35	316	633 e	-	-	-	-	-	-
Other	-	-	-	-	-	-	-	-	-
Latin America	**35**	**316**	**633 e**	**-**	**-**	**-**	**-**	**-**	**-**
Romania	-	-	-	-	-	-	-	-	-
Other	-	-	-	-	-	-	-	-	-
Non-OECD Europe	**-**	**-**	**-**	**-**	**-**	**-**	**-**	**-**	**-**
FSU	**279**	**109**	**163 e**	**-**	**-**	**-**	**-**	**-**	**-**
Iran	-	-	-	-	-	-	-	-	-
Iraq	-	-	-	-	-	-	-	-	-
Kuwait	-	-	-	-	-	-	-	-	-
Oman	-	-	-	-	-	-	-	-	-
Qatar	-	-	-	-	-	-	-	-	-
Saudi Arabia	30	-	-	-	-	-	-	-	-
United Arab Emirates	-	120	96 e	-	-	-	-	-	-
Other	-	-	4 e	-	-	-	-	-	-
Middle East	**30**	**120**	**100 e**	**-**	**-**	**-**	**-**	**-**	**-**
India	-	-	-	-	-	-	-	-	-
Indonesia	-	-	-	-	-	-	-	-	-
Singapore	-	-	-	-	-	-	-	-	-
Other	31	68	-	-	-	-	-	-	-
Asia	**31**	**68**	**-**	**-**	**-**	**-**	**-**	**-**	**-**
China	**-**	**-**	**-**	**-**	**-**	**-**	**-**	**-**	**-**
Algeria	78	128	60 e	18	57	28 e	-	-	-
Angola/Cabinada	-	-	-	-	-	-	-	-	-
Cameroon	-	-	-	-	-	-	-	-	-
Egypt	6	-	-	6	-	-	-	-	-
Gabon	-	-	-	-	-	-	-	-	-
Libya	-	-	-	-	-	-	-	-	-
Nigeria	-	-	-	-	-	-	-	-	-
Tunisia	-	-	-	-	-	-	-	-	-
Other	-	-	11 e	-	-	11 e	-	-	-
Africa	**84**	**128**	**71 e**	**24**	**57**	**39 e**	**-**	**-**	**-**
Non-spec./Other	**346**	**782**	**346 e**	**-**	**85**	**-**	**-**	**-**	**-**
Total Imports	**5 118**	**4 839**	**5 187 e**	**767**	**704**	**706 e**	**387**	**179**	**151 e**
Memo: Total OPEC	*143*	*564*	*789 e*	*18*	*57*	*28 e*	*-*	*-*	*-*

Source: Annual Oil Statistics.

PORTUGAL
Table 7C: Imports by Country of Origin

000 metric tons

	Total Gasoline			Total Kerosene			Gas/Diesel		
	2000	**2001**	**2002**	**2000**	**2001**	**2002**	**2000**	**2001**	**2002**
Australia	-	-	-	-	-	-	-	-	-
Belgium	-	-	-	-	14	-	30	18	-
Canada	-	-	-	-	-	-	-	-	-
Czech Republic	-	-	-	-	-	-	-	-	-
Denmark	-	-	-	-	-	-	-	28	-
Finland	-	-	-	-	-	-	-	20	-
France	5	2	-	3	-	-	32	-	-
Germany	-	-	-	-	-	-	357	67	67 e
Greece	-	-	-	-	-	-	-	30	-
Hungary	-	-	-	-	-	-	-	-	-
Ireland	-	-	-	-	-	-	-	-	-
Italy	-	-	-	38	-	-	76	174	142 e
Japan	-	-	-	-	-	-	-	-	-
Korea	-	-	-	-	-	-	-	-	-
Mexico	-	-	-	-	-	-	-	-	-
Netherlands	106	-	14 e	-	-	-	202	78	53 e
New Zealand	-	-	-	-	-	-	-	-	-
Norway	29	-	-	-	-	-	22	20	-
Poland	-	-	-	-	-	-	-	-	-
Spain	2	10	13 e	6	5	2 e	136	99	82 e
Sweden	-	-	-	-	-	-	36	66	56 e
Turkey	-	-	-	-	-	-	-	-	-
United Kingdom	15	-	9 e	-	-	-	59	96	17 e
United States	-	-	-	-	-	-	-	-	-
Other	-	-	-	-	-	-	-	-	-
OECD	**157**	**12**	**36 e**	**47**	**19**	**2 e**	**950**	**696**	**417 e**
Argentina	-	-	-	-	-	-	-	-	-
Brazil	-	-	-	-	-	-	-	-	-
Colombia	-	-	-	-	-	-	-	-	-
Ecuador	-	-	-	-	-	-	-	-	-
Netherlands Antilles	-	-	-	-	-	-	-	-	-
Peru	-	-	-	-	-	-	-	-	-
Trinidad and Tobago	-	-	-	-	-	-	-	-	-
Venezuela	-	-	-	-	-	62 e	-	-	-
Other	-	-	-	-	-	-	-	-	-
Latin America	**-**	**-**	**-**	**-**	**-**	**62 e**	**-**	**-**	**-**
Romania	-	-	-	-	-	-	-	-	-
Other	-	-	-	-	-	-	-	-	-
Non-OECD Europe	**-**	**-**	**-**	**-**	**-**	**-**	**-**	**-**	**-**
FSU	**-**	**-**	**-**	**-**	**-**	**-**	**205**	**84**	**163 e**
Iran	-	-	-	-	-	-	-	-	-
Iraq	-	-	-	-	-	-	-	-	-
Kuwait	-	-	-	-	-	-	-	-	-
Oman	-	-	-	-	-	-	-	-	-
Qatar	-	-	-	-	-	-	-	-	-
Saudi Arabia	-	-	-	30	-	-	-	-	-
United Arab Emirates	-	-	-	-	120	96 e	-	-	-
Other	-	-	-	-	-	-	-	-	-
Middle East	**-**	**-**	**-**	**30**	**120**	**96 e**	**-**	**-**	**-**
India	-	-	-	-	-	-	-	-	-
Indonesia	-	-	-	-	-	-	-	-	-
Singapore	-	-	-	-	-	-	-	-	-
Other	-	-	-	-	-	-	31	-	-
Asia	**-**	**-**	**-**	**-**	**-**	**-**	**31**	**-**	**-**
China	**-**	**-**	**-**	**-**	**-**	**-**	**-**	**-**	**-**
Algeria	-	-	-	-	17	32 e	60	54	-
Angola/Cabinada	-	-	-	-	-	-	-	-	-
Cameroon	-	-	-	-	-	-	-	-	-
Egypt	-	-	-	-	-	-	-	-	-
Gabon	-	-	-	-	-	-	-	-	-
Libya	-	-	-	-	-	-	-	-	-
Nigeria	-	-	-	-	-	-	-	-	-
Tunisia	-	-	-	-	-	-	-	-	-
Other	-	-	-	-	-	-	-	-	-
Africa	**-**	**-**	**-**	**-**	**17**	**32 e**	**60**	**54**	**-**
Non-spec./Other	**-**	**-**	**-**	**-**	**-**	**-**	**25**	**109**	**-**
Total Imports	**157**	**12**	**36 e**	**77**	**156**	**192 e**	**1 271**	**943**	**580 e**
Memo: Total OPEC	-	-	-	*30*	*137*	*190 e*	*60*	*54*	-

Source: Annual Oil Statistics.

PORTUGAL
Table 7D: Imports by Country of Origin

000 metric tons

	2000	2001	2002	2000	2001	2002	2000	2001	2002
	Fuel Oil (Residual)			**Petroleum Coke**			**Other Products**		
Australia	-	-	-	-	-	-	-	-	-
Belgium	-	21	57 e	-	-	-	7	4	2 e
Canada	-	-	-	-	-	-	-	-	-
Czech Republic	-	-	-	-	-	-	-	-	-
Denmark	-	-	-	-	-	-	-	-	-
Finland	56	-	-	-	-	-	-	-	-
France	257	389	650 e	-	-	-	173	106	84 e
Germany	-	8	-	-	-	-	12	1	4 e
Greece	-	-	-	-	-	-	-	-	-
Hungary	-	-	-	-	-	-	-	-	-
Ireland	-	-	-	-	-	-	-	-	-
Italy	-	-	51 e	-	-	-	4	4	7 e
Japan	-	-	-	-	-	-	-	-	-
Korea	-	-	-	-	-	-	-	-	-
Mexico	-	-	-	-	-	-	-	-	-
Netherlands	256	353	355 e	-	-	-	2	9	-
New Zealand	-	-	-	-	-	-	-	-	-
Norway	-	30	76 e	-	-	-	-	-	-
Poland	-	-	-	-	-	-	-	-	-
Spain	393	97	131 e	11	12	29 e	235	219	346 e
Sweden	-	48	200 e	-	-	-	-	-	-
Turkey	-	-	-	-	-	-	-	-	-
United Kingdom	98	342	577 e	-	-	-	10	7	-
United States	92	-	-	423	198	32 e	-	-	-
Other	-	-	-	-	-	-	-	-	-
OECD	**1 152**	**1 288**	**2 097 e**	**434**	**210**	**61 e**	**443**	**350**	**443 e**
Argentina	-	-	-	-	-	-	-	-	-
Brazil	-	-	-	-	-	-	-	-	-
Colombia	-	-	-	-	-	-	-	-	-
Ecuador	-	-	-	-	-	-	-	-	-
Netherlands Antilles	-	-	-	-	-	-	-	-	-
Peru	-	-	-	-	-	-	-	-	-
Trinidad and Tobago	-	-	-	-	-	-	-	-	-
Venezuela	-	-	-	35	316	571 e	-	-	-
Other	-	-	-	-	-	-	-	-	-
Latin America	**-**	**-**	**-**	**35**	**316**	**571 e**	**-**	**-**	**-**
Romania	-	-	-	-	-	-	-	-	-
Other	-	-	-	-	-	-	-	-	-
Non-OECD Europe	**-**	**-**	**-**	**-**	**-**	**-**	**-**	**-**	**-**
FSU	**74**	**25**	**-**	**-**	**-**	**-**	**-**	**-**	**-**
Iran	-	-	-	-	-	-	-	-	-
Iraq	-	-	-	-	-	-	-	-	-
Kuwait	-	-	-	-	-	-	-	-	-
Oman	-	-	-	-	-	-	-	-	-
Qatar	-	-	-	-	-	-	-	-	-
Saudi Arabia	-	-	-	-	-	-	-	-	-
United Arab Emirates	-	-	-	-	-	-	-	-	-
Other	-	-	-	-	-	-	-	-	4 e
Middle East	**-**	**-**	**-**	**-**	**-**	**-**	**-**	**-**	**4 e**
India	-	-	-	-	-	-	-	-	-
Indonesia	-	-	-	-	-	-	-	-	-
Singapore	-	-	-	-	-	-	-	-	-
Other	-	-	-	-	68	-	-	-	-
Asia	**-**	**-**	**-**	**-**	**68**	**-**	**-**	**-**	**-**
China	**-**	**-**	**-**	**-**	**-**	**-**	**-**	**-**	**-**
Algeria	-	-	-	-	-	-	-	-	-
Angola/Cabinada	-	-	-	-	-	-	-	-	-
Cameroon	-	-	-	-	-	-	-	-	-
Egypt	-	-	-	-	-	-	-	-	-
Gabon	-	-	-	-	-	-	-	-	-
Libya	-	-	-	-	-	-	-	-	-
Nigeria	-	-	-	-	-	-	-	-	-
Tunisia	-	-	-	-	-	-	-	-	-
Other	-	-	-	-	-	-	-	-	-
Africa	**-**	**-**	**-**	**-**	**-**	**-**	**-**	**-**	**-**
Non-spec./Other	**-**	**-**	**-**	**-**	**-**	**-**	**321**	**588**	**346 e**
Total Imports	**1 226**	**1 313**	**2 097 e**	**469**	**594**	**632 e**	**764**	**938**	**793 e**
Memo: Total OPEC	-	-	-	*35*	*316*	*571 e*	-	-	-

Source: Annual Oil Statistics.

PORTUGAL

Table 8A: Exports by Country of Destination

000 metric tons

	2000	2001	2002	2000	2001	2002	2000	2001	2002
	Crude Oil			**Natural Gas Liquids**			**Feedstocks and Other**		
Australia	-	-	-	-	-	-	-	-	-
Austria	-	-	-	-	-	-	-	-	-
Belgium	-	-	-	-	-	-	-	-	-
Canada	-	-	-	-	-	-	-	-	-
Denmark	-	-	-	-	-	-	-	-	-
Finland	-	-	-	-	-	-	-	-	-
France	-	-	-	-	-	-	-	-	-
Germany	-	-	-	-	-	-	-	-	-
Greece	-	-	-	-	-	-	-	-	-
Ireland	-	-	-	-	-	-	-	-	-
Italy	-	-	-	-	-	-	-	-	-
Japan	-	-	-	-	-	-	-	-	-
Luxembourg	-	-	-	-	-	-	-	-	-
Mexico	-	-	-	-	-	-	-	-	-
Netherlands	-	-	-	-	-	-	-	-	-
Norway	-	-	-	-	-	-	-	-	-
Portugal	-	-	-	-	-	-	-	-	-
Spain	-	-	-	-	-	-	-	-	-
Sweden	-	-	-	-	-	-	-	-	-
Switzerland	-	-	-	-	-	-	-	-	-
Turkey	-	-	-	-	-	-	-	-	-
United Kingdom	-	-	-	-	-	-	-	-	-
United States	-	-	-	-	-	-	-	-	-
Other OECD	-	-	-	-	-	-	-	-	-
OECD	**-**	**-**	**-**	**-**	**-**	**-**	**-**	**-**	**-**
Africa	-	-	-	-	-	-	-	-	-
Latin America	-	-	-	-	-	-	-	-	-
Asia	-	-	-	-	-	-	-	-	-
China	-	-	-	-	-	-	-	-	-
Non-OECD Europe	-	-	-	-	-	-	-	-	-
Former Soviet Union	-	-	-	-	-	-	-	-	-
Middle East	-	-	-	-	-	-	-	-	-
Non-specified/Other	-	-	-	-	-	-	-	-	-
Non-OECD	**-**	**-**	**-**	**-**	**-**	**-**	**-**	**-**	**-**
Total Exports	**-**	**-**	**-**	**-**	**-**	**-**	**-**	**-**	**-**
	Total Products			**Liquefied Petroleum Gas**			**Naphtha**		
Australia	-	-	-	-	-	-	-	-	-
Austria	-	-	-	-	-	-	-	-	-
Belgium	7	61	83 e	-	-	-	-	49	16 e
Canada	-	-	-	-	-	-	-	-	-
Denmark	-	-	-	-	-	-	-	-	-
Finland	-	-	-	-	-	-	-	-	-
France	26	67	53 e	-	-	-	-	18	-
Germany	5	5	-	-	-	-	-	-	-
Greece	-	-	2 e	-	-	-	-	-	-
Ireland	-	27	-	-	-	-	-	-	-
Italy	-	-	32 e	-	-	-	-	-	-
Japan	-	-	-	-	-	-	-	-	-
Luxembourg	-	-	-	-	-	-	-	-	-
Mexico	-	-	-	-	-	-	-	-	-
Netherlands	67	173	229 e	-	-	-	-	50	87 e
Norway	39	-	2 e	-	-	-	-	-	-
Portugal	-	-	-	-	-	-	-	-	-
Spain	380	445	347 e	18	20	21 e	-	-	-
Sweden	-	-	18 e	-	-	-	-	-	18 e
Switzerland	2	-	1 e	-	-	1 e	-	-	-
Turkey	-	-	-	-	-	-	-	-	-
United Kingdom	128	28	38 e	-	-	-	-	-	-
United States	562	459	474 e	-	-	-	-	-	-
Other OECD	-	1	-	-	-	-	-	-	-
OECD	**1 216**	**1 266**	**1 279 e**	**18**	**20**	**22 e**	**-**	**117**	**121 e**
Africa	178	66	45 e	4	1	6 e	-	-	-
Latin America	28	23	42 e	1	-	-	-	-	-
Asia	-	-	2 e	-	-	-	-	-	-
China	-	-	-	-	-	-	-	-	-
Non-OECD Europe	-	-	-	-	-	-	-	-	-
Former Soviet Union	-	-	-	-	-	-	-	-	-
Middle East	2	-	2 e	-	-	-	-	-	-
Non-specified/Other	-	-	-	-	-	-	-	-	-
Non-OECD	**208**	**89**	**91 e**	**5**	**1**	**6 e**	**-**	**-**	**-**
Total Exports	**1 424**	**1 355**	**1 370 e**	**23**	**21**	**28 e**	**-**	**117**	**121 e**

Source: Annual Oil Statistics.

PORTUGAL
Table 8B: Exports by Country of Destination

000 metric tons

	Total Gasoline 2000	Total Gasoline 2001	Total Gasoline 2002	Total Kerosene 2000	Total Kerosene 2001	Total Kerosene 2002	Gas/Diesel Oil 2000	Gas/Diesel Oil 2001	Gas/Diesel Oil 2002
Australia	-	-	-	-	-	-	-	-	-
Austria	-	-	-	-	-	-	-	-	-
Belgium	-	-	37 e	-	-	-	-	-	-
Canada	-	-	-	-	-	-	-	-	-
Denmark	-	-	-	-	-	-	-	-	-
Finland	-	-	-	-	-	-	-	-	-
France	-	23	-	-	-	-	-	-	-
Germany	-	-	-	-	-	-	-	-	-
Greece	-	-	-	-	-	-	-	-	-
Ireland	-	-	-	-	-	-	-	-	-
Italy	-	-	-	-	-	-	-	-	-
Japan	-	-	-	-	-	-	-	-	-
Luxembourg	-	-	-	-	-	-	-	-	-
Mexico	-	-	-	-	-	-	-	-	-
Netherlands	6	94	107 e	-	-	-	-	-	-
Norway	-	-	-	-	-	-	-	-	-
Portugal	-	-	-	-	-	-	-	-	-
Spain	-	70	-	16	-	-	66	82	5 e
Sweden	-	-	-	-	-	-	-	-	-
Switzerland	-	-	-	-	-	-	-	-	-
Turkey	-	-	-	-	-	-	-	-	-
United Kingdom	-	22	-	21	-	-	-	-	-
United States	383	410	397 e	-	-	-	5	-	-
Other OECD	-	-	-	-	-	-	-	-	-
OECD	**389**	**619**	**541 e**	**37**	**-**	**-**	**71**	**82**	**5 e**
Africa	8	5	-	62	13	-	81	8	1 e
Latin America	-	11	-	-	-	-	-	-	-
Asia	-	-	-	-	-	-	-	-	-
China	-	-	-	-	-	-	-	-	-
Non-OECD Europe	-	-	-	-	-	-	-	-	-
Former Soviet Union	-	-	-	-	-	-	-	-	-
Middle East	-	-	-	-	-	-	-	-	-
Non-specified/Other	-	-	-	-	-	-	-	-	-
Non-OECD	**8**	**16**	**-**	**62**	**13**	**-**	**81**	**8**	**1 e**
Total Exports	**397**	**635**	**541 e**	**99**	**13**	**-**	**152**	**90**	**6 e**

	Fuel Oil (Residual) 2000	Fuel Oil (Residual) 2001	Fuel Oil (Residual) 2002	Petroleum Coke 2000	Petroleum Coke 2001	Petroleum Coke 2002	Other Products 2000	Other Products 2001	Other Products 2002
Australia	-	-	-	-	-	-	-	-	-
Austria	-	-	-	-	-	-	-	-	-
Belgium	-	-	28 e	-	-	-	7	12	2 e
Canada	-	-	-	-	-	-	-	-	-
Denmark	-	-	-	-	-	-	-	-	-
Finland	-	-	-	-	-	-	-	-	-
France	-	-	35 e	-	-	-	26	26	18 e
Germany	-	-	-	-	-	-	5	5	-
Greece	-	-	-	-	-	-	-	-	2 e
Ireland	-	27	-	-	-	-	-	-	-
Italy	-	-	32 e	-	-	-	-	-	-
Japan	-	-	-	-	-	-	-	-	-
Luxembourg	-	-	-	-	-	-	-	-	-
Mexico	-	-	-	-	-	-	-	-	-
Netherlands	59	29	35 e	-	-	-	2	-	-
Norway	39	-	-	-	-	-	-	-	2 e
Portugal	-	-	-	-	-	-	-	-	-
Spain	108	90	95 e	-	-	-	172	183	226 e
Sweden	-	-	-	-	-	-	-	-	-
Switzerland	-	-	-	-	-	-	2	-	-
Turkey	-	-	-	-	-	-	-	-	-
United Kingdom	78	-	16 e	-	-	-	29	6	22 e
United States	174	49	77 e	-	-	-	-	-	-
Other OECD	-	-	-	-	-	-	-	1	-
OECD	**458**	**195**	**318 e**	**-**	**-**	**-**	**243**	**233**	**272 e**
Africa	-	-	-	-	-	-	23	39	38 e
Latin America	25	-	-	-	-	-	2	12	42 e
Asia	-	-	-	-	-	-	-	-	2 e
China	-	-	-	-	-	-	-	-	-
Non-OECD Europe	-	-	-	-	-	-	-	-	-
Former Soviet Union	-	-	-	-	-	-	-	-	-
Middle East	-	-	-	-	-	-	2	-	2 e
Non-specified/Other	-	-	-	-	-	-	-	-	-
Non-OECD	**25**	**-**	**-**	**-**	**-**	**-**	**27**	**51**	**84 e**
Total Exports	**483**	**195**	**318 e**	**-**	**-**	**-**	**270**	**284**	**356 e**

Source: Annual Oil Statistics.

SLOVAK REPUBLIC

Figure 1
FUEL SHARES OF TPES *

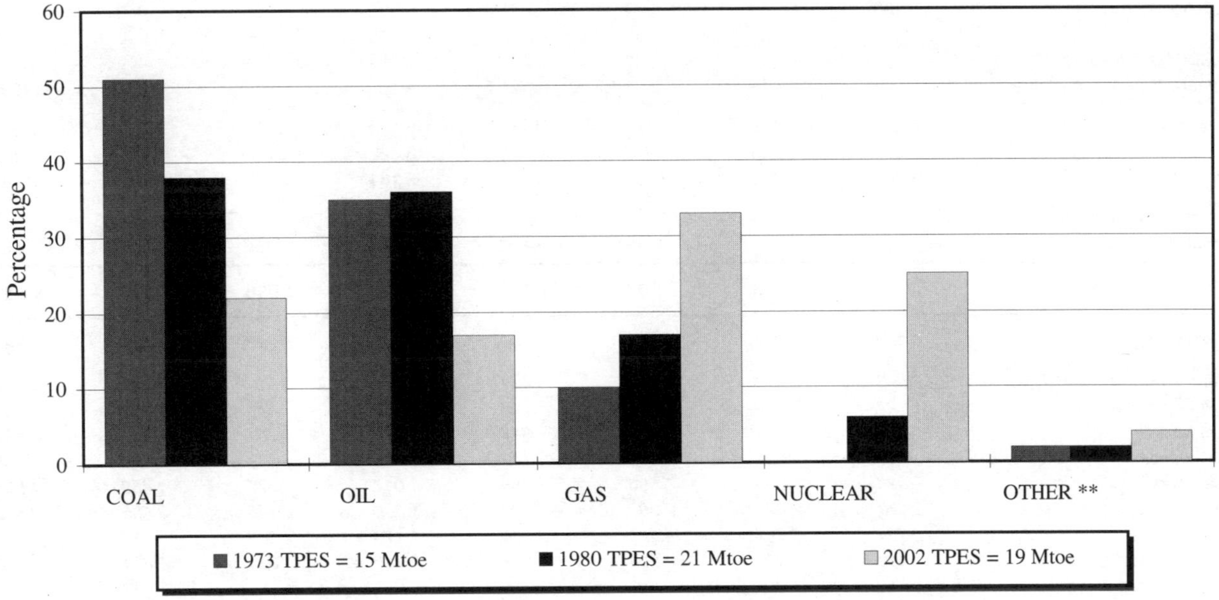

Source: Annual Energy Balances.

* Excluding electricity trade.

** Includes hydro, geothermal, solar, wind, combustible renewables and waste, etc.

Figure 2
OIL PRODUCTION AND CONSUMPTION

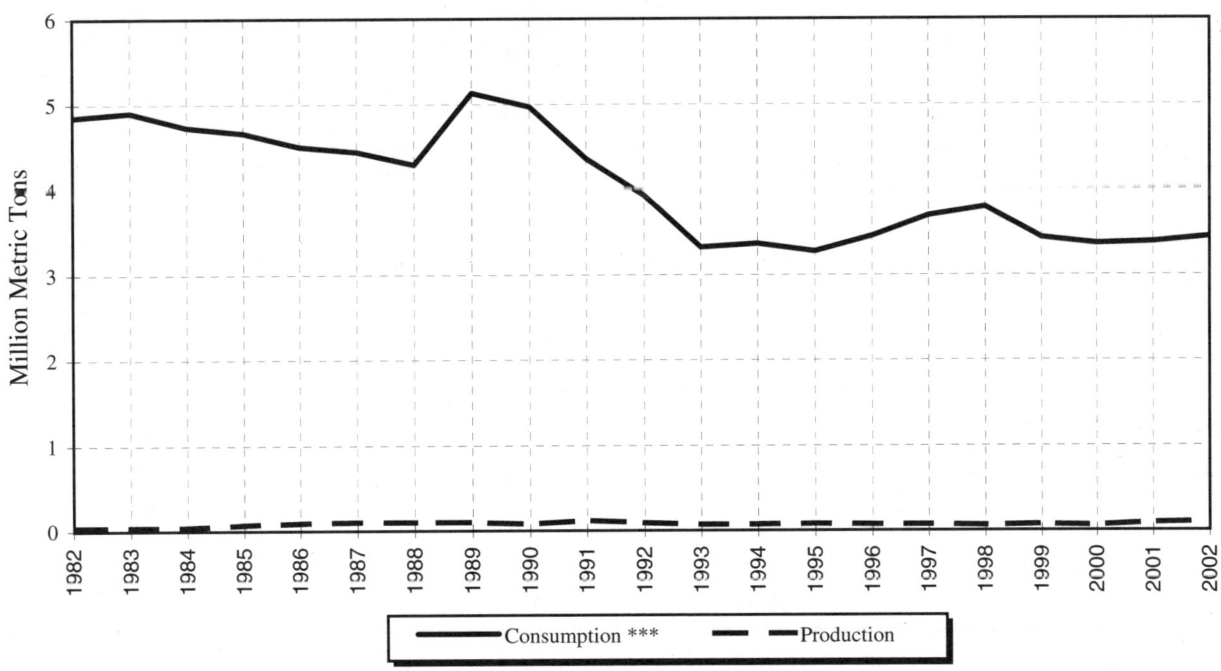

Source: Annual Oil Statistics.

*** Includes refinery fuel and international marine bunkers.

SLOVAK REPUBLIC
Table 1: Oil Supply and Consumption - Summary Table

000 metric tons

	1973	1978	1980	1990	1999	2000	2001	2002
Indigenous Production	130	64	44	75	66	59	353	346 e
Crude Oil	..	61	42	73	66	59	46	45 e
NGL	..	3	2	2	-	-	9	11 e
Other	..	-	-	-	-	-	298	290 e
+ Net Imports	5 298	7 045	7 481	4 745	3 138	2 778	2 816	2 871 e
Crude Oil, NGL, Other	6 684	8 843	9 777	6 169	5 322	5 314	5 474	5 617 e
Petroleum Products	-1 386	-1 798	-2 296	-1 424	-2 184	-2 536	-2 658	-2 746 e
- Intl. Marine Bunkers	-	-	-	-	-	-	-	-
+ Stock Changes	- 13	- 16	- 18	- 90	18	159	21	165 e
TOTAL SUPPLY	**5 415**	**7 093**	**7 507**	**4 730**	**3 222**	**2 996**	**3 190**	**3 382 e**
Refinery Intake	*6 801*	*8 907*	*9 726*	*6 586*	*5 638*	*5 810*	*6 315*	*6 482 e*
Refinery Output	*6 043*	*7 411*	*8 099*	*6 434*	*5 638*	*5 810*	*6 231*	*6 297 e*
LPG + Ethane	*22*	*29*	*27*	*22*	*20*	*4*	*187*	*147 e*
Naphtha	-	-	-	*517*	*721 e*	*751 e*	*560*	*481 e*
Kerosenes	*141*	*222*	*270*	*142*	*41 e*	*47 e*	*50*	*43 e*
Gasolines	*629*	*935*	*806*	*488*	*1 017*	*1 377*	*1 439*	*1 557 e*
Gas/Diesel Oil	*2 073*	*2 021*	*2 300*	*1 797*	*1 889*	*2 125*	*2 406*	*2 452 e*
Fuel Oil	*2 759*	*3 767*	*4 492*	*2 350*	*1 170*	*543*	*718*	*697 e*
Other Products	*419*	*437*	*204*	*1 118*	*780 e*	*963 e*	*871*	*920 e*
- Refinery Losses	758	1 496	1 627	152	-	-	84	185 e
+ Returns	-	-	-	-	-	-	276	213 e
- Statistical Differences	-	-	- 3	- 393	- 209	- 368	-	- 34 e
TOTAL CONSUMPTION	**4 657**	**5 597**	**5 883**	**4 971**	**3 431**	**3 364**	**3 382**	**3 444 e**
Transformation and Energy	**780**	**1 053**	**803**	**426**	**278**	**124**	**779**	..
Crude Oil	-	-	-	-	-	-	-	..
LPG + Ethane	-	-	-	-	-	-	51	..
Gas/Diesel Oil	-	-	-	-	-	-	3	..
Fuel Oil	680	933	741	310	273	114	202	..
Petroleum Coke	-	-	-	-	-	-	40	..
Other	100	120	62	116	5	10	483	..
Transport	**1 576**	**1 490**	**1 170**	**933**	**1 338**	**1 294**	**1 367**	..
Motor Gasoline	532	550	596	419	626	585	608	..
Gas/Diesel Oil	1 044	940	574	514	692	683	727	..
Subtotal	*1 576*	*1 490*	*1 170*	*933*	*1 318*	*1 268*	*1 335*	..
Aviation Fuels	-	-	-	-	20	26	31	..
Other	-	-	-	-	-	-	1	..
Industry	**1 799**	**2 515**	**2 979**	**2 973**	**1 609**	**1 770**	**990**	..
Naphtha	-	-	-	517	721	751	201	..
LPG + Ethane	-	-	-	-	2	9	182	..
Gas/Diesel Oil	161	228	309	228	-	-	37	..
Fuel Oil	1 343	1 993	2 343	1 028	140	88	252	..
Other	295	294	327	1 200	746	922	318	..
Other Sectors	**502**	**539**	**931**	**639**	**206**	**176**	**246**	..
Gas/Diesel Oil	-	-	446	382	153	142	120	..
Fuel Oil	480	511	431	217	11	6	-	..
Other	22	28	54	40	42	28	126	..
Memo: Non-Energy Use	*295*	*294*	*84*	*1 368*	*1 386*	*1 587*	*599*	..
Chem/Petchem. Feedst.	-	-	-	*517*	*721*	*751*	*416*	..
Other	*295*	*294*	*84*	*851*	*665*	*836*	*183*	..

Source: Annual Oil Statistics.

SLOVAK REPUBLIC

Table 2: Supply of Crude Oil, NGL, Feedstocks and Other Hydrocarbons

000 metric tons

	1973	1978	1980	1990	1999	2000	2001	2002
CRUDE OIL								
Indigenous Production	..	61	42	73	66	59	46	45 e
+ Imports	..	9 157	9 777	6 169	5 346	5 329	5 404	5 556 e
- Exports	..	314	-	-	24	15	17	15 e
- Direct Use	..	-	24	12	-	-	-	-
+ Stock Change	..	-	- 69	- 21	41	69	59	195 e
- Statistical Difference	..	-	-	- 99	- 209	- 368	-	-
Refinery Intake	**..**	**8 904**	**9 726**	**6 308**	**5 638**	**5 810**	**5 492**	**5 781 e**
Closing Stock Level	..	-	-	-	-	-	-	-
NATURAL GAS LIQUIDS								
Indigenous Production	..	3	2	2	-	-	9	11 e
+ Imports	..	-	-	-	-	-	-	-
- Exports	..	-	-	-	-	-	-	-
- Direct Use	..	-	2	2	-	-	2	2 e
+ Stock Change	..	-	-	-	-	-	- 2	1 e
- Statistical Difference	..	-	-	-	-	-	-	- 2 e
Refinery Intake	**..**	**3**	**-**	**-**	**-**	**-**	**5**	**12 e**
Closing Stock Level	..	-	-	-	-	-	-	-
REFINERY FEEDSTOCKS								
Backflows	..	-	-	-	-	-	276	213 e
+ Transfers	..	-	-	278	-	-	155	111 e
+ Imports	..	-	-	-	-	-	37	41 e
- Exports	..	-	-	-	-	-	-	-
+ Stock Change	..	-	-	-	-	-	1	-
- Statistical Difference	..	-	-	-	-	-	-	-
Refinery Intake	**..**	**-**	**-**	**278**	**-**	**-**	**469**	**365 e**
Closing Stock Level	..	-	-	-	-	-	-	-
OTHER HYDROCARBONS								
Indigenous Production	..	-	-	-	-	-	298	290 e
+ Imports	..	-	-	-	-	-	50	35 e
- Exports	..	-	-	-	-	-	-	-
- Direct Use	..	-	-	-	-	-	-	-
+ Stock Change	..	-	-	-	-	-	1	1 e
- Statistical Difference	..	-	-	-	-	-	-	2 e
Refinery Intake	**..**	**-**	**-**	**-**	**-**	**-**	**349**	**324 e**
Closing Stock Level	..	-	-	-	-	-	-	-
CRUDE OIL, NGL, FEEDSTOCKS AND OTHER								
Indigenous Production	130	64	44	75	66	59	353	346 e
+ Backflows	-	-	-	-	-	-	276	213 e
+ Transfers	-	-	-	278	-	-	155	111 e
+ Imports	6 983	9 157	9 777	6 169	5 346	5 329	5 491	5 632 e
- Exports	299	314	-	-	24	15	17	15 e
- Direct Use	-	-	26	14	-	-	2	2 e
+ Stock Change	- 13	-	- 69	- 21	41	69	59	197 e
- Statistical Difference	-	-	-	- 99	- 209	- 368	-	-
Refinery Intake	**6 801**	**8 907**	**9 726**	**6 586**	**5 638**	**5 810**	**6 315**	**6 482 e**
Closing Stock Level	-	-	-	-	-	-	-	-

Source: Annual Oil Statistics.

SLOVAK REPUBLIC
Table 3: Motor Gasoline - Supply

000 metric tons

	1973	1978	1980	1990	1999	2000	2001	2002
MOTOR GASOLINE								
Refinery Output	**629**	**935**	**806**	**488**	**1 017**	**1 377**	**1 438**	**1 556 e**
+ Imports	-	-	212	-	255	98	221	326 e
- Exports	97	385	362	37	626	869	1 024	1 238 e
+ Transfers	-	-	-	-	-	-	-	-
+ Stock Change	-	-	19	18	- 3	- 4	- 23	- 2 e
- Statistical Difference	-	-	1	-	-	-	-	- 1 e
Total Consumption	**532**	**550**	**674**	**469**	**643**	**602**	**612**	**643 e**
Closing Stock Level	-	-	-	-	-	-	-	-

Table 4: Aviation Fuels - Supply

000 metric tons

	1973	1978	1980	1990	1999	2000	2001	2002
AVIATION GASOLINE								
Refinery Output	-	-	-	-	-	-	**1**	**1 e**
+ Imports	-	-	-	-	-	-	-	-
- Exports	-	-	-	-	-	-	1	1 e
+ Transfers	-	-	-	-	-	-	-	-
+ Stock Change	-	-	-	-	-	-	-	-
- Statistical Difference	-	-	-	-	-	-	-	-
Total Consumption	-	-	-	-	-	-	-	-
Closing Stock Level	-	-	-	-	-	-	-	-
GASOLINE TYPE JET FUEL								
Refinery Output	-	-	-	-	-	-	-	-
+ Imports	-	-	-	-	-	-	-	-
- Exports	-	-	-	-	-	-	-	-
+ Transfers	-	-	-	-	-	-	-	-
+ Stock Change	-	-	-	-	-	-	-	-
- Statistical Difference	-	-	-	-	-	-	-	-
Total Consumption	-	-	-	-	-	-	-	-
Closing Stock Level	-	-	-	-	-	-	-	-
KEROSENE TYPE JET FUEL								
Refinery Output	-	-	-	-	**28 e**	**31 e**	**50**	**43 e**
+ Imports	-	-	-	-	-	-	3	1 e
- Exports	-	-	-	-	8 e	5 e	6	9 e
+ Transfers	-	-	-	-	-	-	- 19	-
+ Stock Change	-	-	-	-	-	-	3	1 e
- Statistical Difference	-	-	-	-	-	-	-	1 e
Total Consumption	-	-	-	-	**20**	**26**	**31**	**35 e**
Closing Stock Level	-	-	-	-	-	-	-	-

Source: Annual Oil Statistics.

SLOVAK REPUBLIC
Table 5A: Gas/Diesel Oil - Supply and Consumption

000 metric tons

	1973	1978	1980	1990	1999	2000	2001	2002
Refinery Output	**2 073**	**2 021**	**2 300**	**1 797**	**1 889**	**2 125**	**2 406**	**2 452 e**
+ Recycled Products	-	-	-	-	-	-	-	-
+ Imports	-	-	1	-	233	99	118	269 e
- Exports	868	853	906	638	1 286	1 422	1 569	1 639 e
- Intl. Marine Bunkers	-	-	-	-	-	-	-	-
+ Transfers	-	-	-	-	-	-	- 64	- 50 e
+ Stock Changes	-	-	- 66	- 35	9	23	- 4	- 52 e
- Statistical Differences	-	-	-	-	-	-	-	- 4 e
Total Consumption	**1 205**	**1 168**	**1 329**	**1 124**	**845**	**825**	**887**	**984 e**
Transformation	**-**	**-**	**-**	**-**	**-**	**-**	**-**	**..**
Public Electricity	-	-	-	-	-	-	-	..
Autoproducer Electricity	-	-	-	-	-	-	-	..
Public CHP Plants	-	-	-	-	-	-	-	..
Autoproducers of CHP	-	-	-	-	-	-	-	..
Public Heat Plants	-	-	-	-	-	-	-	..
Autoproducer Heat Plants	-	-	-	-	-	-	-	..
Petrochemical Industry	-	-	-	-	-	-	-	..
Other Transformation	-	-	-	-	-	-	-	..
Energy Sector	**-**	**-**	**-**	**-**	**-**	**-**	**3**	**..**
Refinery Fuel	-	-	-	-	-	-	-	..
Other Energy Sector	-	-	-	-	-	-	3	..
Final Consumption	**1 205**	**1 168**	**1 329**	**1 124**	**845**	**825**	**884**	**..**
Transport	**1 044**	**940**	**574**	**514**	**692**	**683**	**727**	**..**
Air	-	-	-	-	-	-	-	..
Road	1 044	940	574	514	692	683	727	..
Rail	-	-	-	-	-	-	-	..
Inland Waterways	-	-	-	-	-	-	-	..
Non-Specified	-	-	-	-	-	-	-	..
Industry	**161**	**228**	**309**	**228**	**-**	**-**	**37**	**..**
Iron and Steel	-	-	-	-	-	-	-	..
Chemical (incl. Pet.-Chem.)	-	-	-	-	-	-	13	..
Non-Ferrous Metals	-	-	-	-	-	-	-	..
Non-Metallic Minerals	-	-	-	-	-	-	3	..
Transport Equipment	-	-	-	-	-	-	1	..
Machinery	-	-	-	-	-	-	1	..
Mining and Quarrying	-	-	-	-	-	-	2	..
Food and Tobacco	-	-	-	-	-	-	3	..
Pulp, Paper and Printing	-	-	-	-	-	-	1	..
Wood and Wood Products	-	-	-	-	-	-	1	..
Construction	-	-	-	-	-	-	10	..
Textiles and Leather	-	-	-	-	-	-	-	..
Non-Specified	161	228	309	228	-	-	2	..
Other Sectors	**-**	**-**	**446**	**382**	**153**	**142**	**120**	**..**
Commerce and Public	-	-	-	-	-	-	45	..
Residential	-	-	-	-	-	-	-	..
Agriculture	-	-	446	382	153	142	75	..
Non-Specified	-	-	-	-	-	-	-	..
Memo: Non-Energy Use	-	-	-	-	-	-	*13*	..
Chem/Petchem. Feedst.	-	-	-	-	-	-	*13*	..
Other	-	-	-	-	-	-	-	..
Closing Stock Level	-	-	-	-	-	-	-	-

Source: Annual Oil Statistics.

SLOVAK REPUBLIC
Table 5B: Fuel Oil (Residual) - Supply and Consumption

000 metric tons

	1973	1978	1980	1990	1999	2000	2001	2002
Refinery Output	**2 759**	**3 767**	**4 492**	**2 350**	**1 170**	**543**	**718**	**697 e**
+ Recycled Products	-	-	-	-	-	-	-	-
+ Imports	-	-	1	13	71	36	30	24 e
- Exports	256	330	1 015	765	822	450	363	274 e
- Intl. Marine Bunkers	-	-	-	-	-	-	-	-
+ Transfers	-	-	-	-	-	-	53	60 e
+ Stock Changes	-	-	36	- 43	5	79	16	4 e
- Statistical Differences	-	-	- 1	-	-	-	-	- 32 e
Total Consumption	**2 503**	**3 437**	**3 515**	**1 555**	**424**	**208**	**454**	**543 e**
Transformation	**441**	**607**	**671**	**290**	**262**	**111**	**146**	..
Public Electricity	-	-	-	-	90	41	5	..
Autoproducer Electricity	-	-	-	-	-	-	-	..
Public CHP Plants	231	318	351	149	-	-	2	..
Autoproducers of CHP	149	205	227	105	-	-	83	..
Public Heat Plants	61	84	93	36	172	70	3	..
Autoproducer Heat Plants	-	-	-	-	-	-	-	..
Petrochemical Industry	-	-	-	-	-	-	53	..
Other Transformation	-	-	-	-	-	-	-	..
Energy Sector	**239**	**326**	**70**	**20**	**11**	**3**	**56**	..
Refinery Fuel	239	326	70	20	-	-	56	..
Other Energy Sector	-	-	-	-	11	3	-	..
Final Consumption	**1 823**	**2 504**	**2 774**	**1 245**	**151**	**94**	**252**	..
Transport	-	-	-	-	-	-	-	..
Air	-	-	-	-	-	-	-	..
Road	-	-	-	-	-	-	-	..
Rail	-	-	-	-	-	-	-	..
Inland Waterways	-	-	-	-	-	-	-	..
Non-Specified	-	-	-	-	-	-	-	..
Industry	**1 343**	**1 993**	**2 343**	**1 028**	**140**	**88**	**252**	..
Iron and Steel	-	-	-	-	-	-	-	..
Chemical (incl. Pet.-Chem.)	-	-	-	-	-	-	221	..
Non-Ferrous Metals	-	-	-	-	-	-	-	..
Non-Metallic Minerals	-	-	-	-	-	-	13	..
Transport Equipment	-	-	-	-	-	-	-	..
Machinery	-	-	-	-	-	-	-	..
Mining and Quarrying	-	-	-	-	-	-	-	..
Food and Tobacco	-	-	-	-	-	-	4	..
Pulp, Paper and Printing	-	-	-	-	-	-	14	..
Wood and Wood Products	-	-	-	-	-	-	-	..
Construction	-	-	-	-	5	4	-	..
Textiles and Leather	-	-	-	-	-	-	-	..
Non-Specified	1 343	1 993	2 343	1 028	135	84	-	..
Other Sectors	**480**	**511**	**431**	**217**	**11**	**6**	-	..
Commerce and Public	-	-	304	161	7	3	-	..
Residential	-	-	-	-	-	-	-	..
Agriculture	-	-	127	56	4	3	-	..
Non-Specified	480	511	-	-	-	-	-	..
Memo: Non-Energy Use	-	-	-	-	-	-	-	..
Chem/Petchem. Feedst.	-	-	-	-	-	-	-	..
Other	-	-	-	-	-	-	-	..
Closing Stock Level	-	-	-	-	-	-	-	-

Source: Annual Oil Statistics.

SLOVAK REPUBLIC
Table 5C: Total Products - Supply and Consumption

000 metric tons

	1973	1978	1980	1990	1999	2000	2001	2002
Refinery Output	**6 043**	**7 411**	**8 099**	**6 434**	**5 638**	**5 810**	**6 231**	**6 297 e**
+ Recycled Products	-	-	-	-	-	-	-	-
+ Imports	-	-	255	178	580	270	552	743 e
- Exports	1 386	1 798	2 551	1 602	2 764	2 806	3 210	3 489 e
- Intl. Marine Bunkers	-	-	-	-	-	-	-	-
+ Transfers	-	-	29	30	-	-	- 153	- 109 e
+ Stock Changes	-	- 16	51	- 69	- 23	90	- 38	- 32 e
- Statistical Differences	-	-	-	-	-	-	-	- 34 e
Total Consumption	**4 657**	**5 597**	**5 883**	**4 971**	**3 431**	**3 364**	**3 382**	**3 444 e**
Transformation	**441**	**607**	**672**	**317**	**262**	**111**	**372**	**..**
Public Electricity	-	-	-	-	90	41	5	..
Autoproducer Electricity	-	-	1	3	-	-	-	..
Public CHP Plants	231	318	351	149	-	-	2	..
Autoproducers of CHP	149	205	227	105	-	-	86	..
Public Heat Plants	61	84	93	36	172	70	3	..
Autoproducer Heat Plants	-	-	-	-	-	-	-	..
Petrochemical Industry	-	-	-	-	-	-	276	..
Other Transformation	-	-	-	24	-	-	-	..
Energy Sector	**339**	**446**	**131**	**109**	**16**	**13**	**407**	**..**
Refinery Fuel	339	446	131	109	-	-	403	..
Other Energy Sector	-	-	-	-	16	13	4	..
Final Consumption	**3 877**	**4 544**	**5 080**	**4 545**	**3 153**	**3 240**	**2 603**	**..**
Transport	**1 576**	**1 490**	**1 170**	**933**	**1 338**	**1 294**	**1 367**	**..**
Air	-	-	-	-	20	26	31	..
Road	1 576	1 490	1 170	933	1 318	1 268	1 335	..
Rail	-	-	-	-	-	-	-	..
Inland Waterways	-	-	-	-	-	-	-	..
Non-Specified	-	-	-	-	-	-	1	..
Industry	**1 799**	**2 515**	**2 979**	**2 973**	**1 609**	**1 770**	**990**	**..**
Iron and Steel	-	-	-	-	-	-	-	..
Chemical (incl. Pet.-Chem.)	-	-	-	517	721	751	744	..
Non-Ferrous Metals	-	-	-	-	-	-	2	..
Non-Metallic Minerals	-	-	-	-	-	-	16	..
Transport Equipment	-	-	-	-	-	1	1	..
Machinery	-	-	-	-	-	-	2	..
Mining and Quarrying	-	-	-	-	-	-	12	..
Food and Tobacco	-	-	-	-	-	-	7	..
Pulp, Paper and Printing	-	-	-	-	-	-	15	..
Wood and Wood Products	-	-	-	-	-	-	1	..
Construction	-	-	-	1	398	442	113	..
Textiles and Leather	-	-	-	-	-	-	-	..
Non-Specified	1 799	2 515	2 979	2 455	490	576	77	..
Other Sectors	**502**	**539**	**931**	**639**	**206**	**176**	**246**	**..**
Commerce and Public	-	-	304	161	14	9	155	..
Residential	14	17	18	15	21	12	12	..
Agriculture	-	-	605	458	171	155	79	..
Non-Specified	488	522	4	5	-	-	-	..
Memo: Non-Energy Use	*295*	*294*	*84*	*1 368*	*1 386*	*1 587*	*599*	..
Chem/Petchem. Feedst.	-	-	-	*517*	*721*	*751*	*416*	..
Other	*295*	*294*	*84*	*851*	*665*	*836*	*183*	..
Closing Stock Level	-	-	-	-	-	-	-	-

Source: Annual Oil Statistics.

SLOVAK REPUBLIC
Table 6A: Supply and Consumption

000 metric tons

	1980	2000	2001	2002	1980	2000	2001	2002
	LPG AND ETHANE				**REFINERY GAS**			
Refinery Output	**27**	**4**	**187**	**147 e**	**200**	**85 e**	**338**	**356 e**
+ Recycled Products	-	-	-	-	-	-	-	-
+ Imports	-	27	32	33 e	-	-	-	-
- Exports	11	-	-	1 e	-	-	-	-
- Intl. Marine Bunkers	-	-	-	-	-	-	-	-
+ Transfers	2	-	51	40 e	-	-	10	10 e
+ Stock Change	-	- 1	- 5	3 e	-	-	-	-
- Statistical Difference	-	-	-	-	-	-	-	-
Total Consumption	**18**	**30**	**265**	**222 e**	**200**	**85**	**348**	**366 e**
of which: Ethane	-	-	*22*	*22 e*	-	-	-	-
Transformation	-	-	51	..	-	-	13	..
Energy	-	-	-	..	56	-	307	..
Transport	-	-	-	..	-	-	-	..
Industry	-	9	182	..	144	85	28	..
Other	18	21	32	..	-	-	-	..
Memo: Non-Energy Use	-	-	*101*	..	-	-	-	..
Closing Stock Level	-	-	-	-	-	-	-	-

	1980	2000	2001	2002	1980	2000	2001	2002
	NAPHTHA				**OTHER KEROSENE**			
Refinery Output	**-**	**751 e**	**560**	**481 e**	**270**	**16 e**	**-**	**-**
+ Recycled Products	-	-	-	-	-	-	-	-
+ Imports	-	-	30	-	-	-	-	-
- Exports	-	-	7	3 e	257	8 e	-	-
- Intl. Marine Bunkers	-	-	-	-	-	-	-	-
+ Transfers	-	-	- 192	- 170 e	-	-	-	-
+ Stock Change	-	-	- 28	23 e	1	- 7	-	-
- Statistical Difference	-	-	-	-	-	-	-	-
Total Consumption	**-**	**751**	**363**	**331 e**	**14**	**1**	**-**	**-**
Transformation	-	-	162	..	-	-	-	..
Energy	-	-	-	..	-	-	-	..
Transport	-	-	-	..	-	-	-	..
Industry	-	751	201	..	8	1	-	..
Other	-	-	-	..	6	-	-	..
Memo: Non-Energy Use	-	*751*	*201*	..	-	-	-	..
Closing Stock Level	-	-	-	-	-	-	-	-

	1980	2000	2001	2002	1980	2000	2001	2002
	WHITE SPIRIT + SBP				**LUBRICANTS**			
Refinery Output	**-**	**-**	**46**	**..**	**-**	**107 e**	**93**	**..**
+ Recycled Products	-	-	-	..	-	-	-	..
+ Imports	-	-	3	..	-	-	24	..
- Exports	-	-	35	..	-	52 e	41	..
- Intl. Marine Bunkers	-	-	-	..	-	-	-	..
+ Transfers	-	-	-	..	-	-	-	..
+ Stock Change	-	-	-	..	-	-	2	..
- Statistical Difference	-	-	-	..	-	-	-	..
Total Consumption	**-**	**-**	**14**	**..**	**-**	**55**	**78**	**..**
Transformation	-	-	-	..	-	-	-	..
Energy	-	-	-	..	-	-	-	..
Transport	-	-	-	..	-	-	1	..
Industry	-	-	-	..	-	55	1	..
Other	-	-	14	..	-	-	76	..
Memo: Non-Energy Use	-	-	-	..	-	*55*	*78*	..
Closing Stock Level	-	-	-	..	-	-	-	..

Source: Annual Oil Statistics.

SLOVAK REPUBLIC
Table 6B: Supply and Consumption

000 metric tons

	1980	2000	2001	2002	1980	2000	2001	2002
	BITUMEN				PARAFFIN WAXES			
Refinery Output	**-**	**432 e**	**75**	**..**	**-**	**-**	**4**	**..**
+ Recycled Products	-	-	-	..	-	-	-	..
+ Imports	-	-	42	..	-	-	-	..
- Exports	-	-	18	..	-	-	-	..
- Intl. Marine Bunkers	-	-	-	..	-	-	-	..
+ Transfers	-	-	-	..	-	-	- 1	..
+ Stock Change	-	-	1	..	-	-	-	..
- Statistical Difference	-	-	-	..	-	-	-	..
Total Consumption	**-**	**432**	**100**	**..**	**-**	**-**	**3**	**..**
Transformation	-	-	-	..	-	-	-	..
Energy	-	-	-	..	-	-	-	..
Transport	-	-	-	..	-	-	-	..
Industry	-	432	100	..	-	-	3	..
Other	-	-	-	..	-	-	-	..
Memo: Non-Energy Use	*-*	*432*	*100*	*..*	*-*	*-*	*3*	*..*
Closing Stock Level	-	-	-	..	-	-	-	..
	PETROLEUM COKE				OTHER PRODUCTS			
Refinery Output	**-**	**-**	**40**	**46 e**	**4**	**339 e**	**275**	**518 e**
+ Recycled Products	-	-	-	-	-	-	-	-
+ Imports	-	-	49	60 e	41	10	-	30 e
- Exports	-	-	1	-	-	-	145	324 e
- Intl. Marine Bunkers	-	-	-	-	-	-	-	-
+ Transfers	-	-	-	-	27	-	9	1 e
+ Stock Change	-	-	1	- 4 e	61	-	- 1	- 5 e
- Statistical Difference	-	-	-	-	-	-	-	2 e
Total Consumption	**-**	**-**	**89**	**102 e**	**133**	**349**	**138**	**218 e**
Transformation	-	-	-	..	1	-	-	..
Energy	-	-	40	..	5	-	-	..
Transport	-	-	-	..	-	-	-	..
Industry	-	-	49	..	127	349	135	..
Other	-	-	-	..	-	-	3	..
Memo: Non-Energy Use	*-*	*-*	*2*	*..*	*84*	*349*	*99*	*..*
Closing Stock Level	-	-	-	-	-	-	-	-

Source: Annual Oil Statistics.

SLOVAK REPUBLIC
Table 7A: Imports by Country of Origin

000 metric tons

	2000	2001	2002	2000	2001	2002	2000	2001	2002
	Crude Oil			Natural Gas Liquids			Feedstocks and Other		
Australia	-	-	-	-	-	-	-	-	-
Belgium	-	-	-	-	-	-	-	-	-
Canada	-	-	-	-	-	-	-	-	-
Czech Republic	58 e	50	38 e	-	-	-	-	-	16 e
Denmark	-	-	-	-	-	-	-	-	-
Finland	-	-	-	-	-	-	-	-	-
France	-	-	-	-	-	-	-	-	-
Germany	-	-	-	-	-	-	-	-	2 e
Greece	-	-	-	-	-	-	-	-	-
Hungary	-	-	-	-	-	-	-	-	24 e
Ireland	-	-	-	-	-	-	-	-	-
Italy	-	-	-	-	-	-	-	-	-
Japan	-	-	-	-	-	-	-	-	-
Korea	-	-	-	-	-	-	-	-	-
Mexico	-	-	-	-	-	-	-	-	-
Netherlands	-	-	-	-	-	-	-	-	-
New Zealand	-	-	-	-	-	-	-	-	-
Norway	-	-	-	-	-	-	-	-	-
Poland	-	-	-	-	-	-	-	-	21 e
Spain	-	-	-	-	-	-	-	-	-
Sweden	-	-	-	-	-	-	-	-	-
Turkey	-	-	-	-	-	-	-	-	-
United Kingdom	-	-	-	-	-	-	-	-	-
United States	-	-	-	-	-	-	-	-	-
Other	-	-	-	-	-	-	-	-	12 e
OECD	**58 e**	**50**	**38 e**	**-**	**-**	**-**	**-**	**-**	**75 e**
Argentina	-	-	-	-	-	-	-	-	-
Brazil	-	-	-	-	-	-	-	-	-
Colombia	-	-	-	-	-	-	-	-	-
Ecuador	-	-	-	-	-	-	-	-	-
Netherlands Antilles	-	-	-	-	-	-	-	-	-
Peru	-	-	-	-	-	-	-	-	-
Trinidad and Tobago	-	-	-	-	-	-	-	-	-
Venezuela	-	-	-	-	-	-	-	-	-
Other	-	-	-	-	-	-	-	-	-
Latin America	**-**	**-**	**-**	**-**	**-**	**-**	**-**	**-**	**-**
Romania	-	-	-	-	-	-	-	-	-
Other	-	-	-	-	-	-	-	-	-
Non-OECD Europe	**-**	**-**	**-**	**-**	**-**	**-**	**-**	**-**	**-**
FSU	**-**	**5 354**	**5 518 e**	**-**	**-**	**-**	**-**	**-**	**1 e**
Iran	-	-	-	-	-	-	-	-	-
Iraq	-	-	-	-	-	-	-	-	-
Kuwait	-	-	-	-	-	-	-	-	-
Oman	-	-	-	-	-	-	-	-	-
Qatar	-	-	-	-	-	-	-	-	-
Saudi Arabia	-	-	-	-	-	-	-	-	-
United Arab Emirates	-	-	-	-	-	-	-	-	-
Other	-	-	-	-	-	-	-	-	-
Middle East	**-**	**-**	**-**	**-**	**-**	**-**	**-**	**-**	**-**
India	-	-	-	-	-	-	-	-	-
Indonesia	-	-	-	-	-	-	-	-	-
Singapore	-	-	-	-	-	-	-	-	-
Other	-	-	-	-	-	-	-	-	-
Asia	**-**	**-**	**-**	**-**	**-**	**-**	**-**	**-**	**-**
China	**-**	**-**	**-**	**-**	**-**	**-**	**-**	**-**	**-**
Algeria	-	-	-	-	-	-	-	-	-
Angola/Cabinada	-	-	-	-	-	-	-	-	-
Cameroon	-	-	-	-	-	-	-	-	-
Egypt	-	-	-	-	-	-	-	-	-
Gabon	-	-	-	-	-	-	-	-	-
Libya	-	-	-	-	-	-	-	-	-
Nigeria	-	-	-	-	-	-	-	-	-
Tunisia	-	-	-	-	-	-	-	-	-
Other	-	-	-	-	-	-	-	-	-
Africa	**-**	**-**	**-**	**-**	**-**	**-**	**-**	**-**	**-**
Non-spec./Other	**5 271 e**	**-**	**-**	**-**	**-**	**-**	**-**	**87**	**-**
Total Imports	**5 329 e**	**5 404**	**5 556 e**	**-**	**-**	**-**	**-**	**87**	**76 e**
Memo: Total OPEC	-	-	-	-	-	-	-	-	-

Source: Annual Oil Statistics.

SLOVAK REPUBLIC
Table 7B: Imports by Country of Origin

000 metric tons

	2000	2001	2002	2000	2001	2002	2000	2001	2002
	Total Products			**Liquefied Petroleum Gases**			**Naphtha**		
Australia	-	-	-	-	-	-	-	-	-
Belgium	-	2	-	-	-	-	-	-	-
Canada	-	-	-	-	-	-	-	-	-
Czech Republic	81 e	54	163 e	20 e	2	1 e	-	-	-
Denmark	-	-	-	-	-	-	-	-	-
Finland	-	-	-	-	-	-	-	-	-
France	-	1	-	-	-	-	-	-	-
Germany	-	7	53 e	-	1	-	-	-	-
Greece	-	-	-	-	-	-	-	-	-
Hungary	73 e	104	165 e	-	-	-	-	-	-
Ireland	-	-	-	-	-	-	-	-	-
Italy	-	-	-	-	-	-	-	-	-
Japan	-	-	-	-	-	-	-	-	-
Korea	-	-	-	-	-	-	-	-	-
Mexico	-	-	-	-	-	-	-	-	-
Netherlands	-	-	-	-	-	-	-	-	-
New Zealand	-	-	-	-	-	-	-	-	-
Norway	-	-	-	-	-	-	-	-	-
Poland	7 e	10	2 e	7 e	-	1 e	-	-	-
Spain	-	-	-	-	-	-	-	-	-
Sweden	-	-	-	-	-	-	-	-	-
Turkey	-	-	-	-	-	-	-	-	-
United Kingdom	-	-	-	-	-	-	-	-	-
United States	-	-	-	-	-	-	-	-	-
Other	99 e	208	288 e	-	-	-	-	-	-
OECD	**260 e**	**386**	**671 e**	**27 e**	**3**	**2 e**	**-**	**-**	**-**
Argentina	-	-	-	-	-	-	-	-	-
Brazil	-	-	-	-	-	-	-	-	-
Colombia	-	-	-	-	-	-	-	-	-
Ecuador	-	-	-	-	-	-	-	-	-
Netherlands Antilles	-	-	-	-	-	-	-	-	-
Peru	-	-	-	-	-	-	-	-	-
Trinidad and Tobago	-	-	-	-	-	-	-	-	-
Venezuela	-	-	-	-	-	-	-	-	-
Other	-	-	3 e	-	-	-	-	-	-
Latin America	**-**	**-**	**3 e**	**-**	**-**	**-**	**-**	**-**	**-**
Romania	-	-	-	-	-	-	-	-	-
Other	-	4	17 e	-	4	7 e	-	-	-
Non-OECD Europe	**-**	**4**	**17 e**	**-**	**4**	**7 e**	**-**	**-**	**-**
FSU	**-**	**-**	**42 e**	**-**	**-**	**24 e**	**-**	**-**	**-**
Iran	-	-	-	-	-	-	-	-	-
Iraq	-	-	-	-	-	-	-	-	-
Kuwait	-	-	-	-	-	-	-	-	-
Oman	-	-	-	-	-	-	-	-	-
Qatar	-	-	-	-	-	-	-	-	-
Saudi Arabia	-	-	-	-	-	-	-	-	-
United Arab Emirates	-	-	-	-	-	-	-	-	-
Other	-	-	-	-	-	-	-	-	-
Middle East	**-**	**-**	**-**	**-**	**-**	**-**	**-**	**-**	**-**
India	-	-	-	-	-	-	-	-	-
Indonesia	-	-	-	-	-	-	-	-	-
Singapore	-	-	-	-	-	-	-	-	-
Other	-	-	-	-	-	-	-	-	-
Asia	**-**	**-**	**-**	**-**	**-**	**-**	**-**	**-**	**-**
China	**-**	**-**	**-**	**-**	**-**	**-**	**-**	**-**	**-**
Algeria	-	-	-	-	-	-	-	-	-
Angola/Cabinada	-	-	-	-	-	-	-	-	-
Cameroon	-	-	-	-	-	-	-	-	-
Egypt	-	-	-	-	-	-	-	-	-
Gabon	-	-	-	-	-	-	-	-	-
Libya	-	-	-	-	-	-	-	-	-
Nigeria	-	-	-	-	-	-	-	-	-
Tunisia	-	-	-	-	-	-	-	-	-
Other	-	-	-	-	-	-	-	-	-
Africa	**-**	**-**	**-**	**-**	**-**	**-**	**-**	**-**	**-**
Non-spec./Other	**10 e**	**162**	**10 e**	**-**	**25**	**-**	**-**	**30**	**-**
Total Imports	**270 e**	**552**	**743 e**	**27 e**	**32**	**33 e**	**-**	**30**	**-**
Memo: Total OPEC	-	-	-	-	-	-	-	-	-

Source: Annual Oil Statistics.

SLOVAK REPUBLIC
Table 7C: Imports by Country of Origin

000 metric tons

	2000	2001	2002	2000	2001	2002	2000	2001	2002
	Total Gasoline			Total Kerosene			Gas/Diesel		
Australia	-	-	-	-	-	-	-	-	-
Belgium	-	-	-	-	-	-	-	-	-
Canada	-	-	-	-	-	-	-	-	-
Czech Republic	6 e	-	64 e	-	-	-	19 e	38	85 e
Denmark	-	-	-	-	-	-	-	-	-
Finland	-	-	-	-	-	-	-	-	-
France	-	-	-	-	-	-	-	-	-
Germany	-	-	-	-	-	-	-	-	-
Greece	-	-	-	-	-	-	-	-	-
Hungary	27 e	84	114 e	-	-	-	46 e	-	46 e
Ireland	-	-	-	-	-	-	-	-	-
Italy	-	-	-	-	-	-	-	-	-
Japan	-	-	-	-	-	-	-	-	-
Korea	-	-	-	-	-	-	-	-	-
Mexico	-	-	-	-	-	-	-	-	-
Netherlands	-	-	-	-	-	-	-	-	-
New Zealand	-	-	-	-	-	-	-	-	-
Norway	-	-	-	-	-	-	-	-	-
Poland	-	-	-	-	-	-	-	-	-
Spain	-	-	-	-	-	-	-	-	-
Sweden	-	-	-	-	-	-	-	-	-
Turkey	-	-	-	-	-	-	-	-	-
United Kingdom	-	-	-	-	-	-	-	-	-
United States	-	-	-	-	-	-	-	-	-
Other	65 e	114	142 e	-	-	-	34 e	72	131 e
OECD	**98 e**	**198**	**320 e**	**-**	**-**	**-**	**99 e**	**110**	**262 e**
Argentina	-	-	-	-	-	-	-	-	-
Brazil	-	-	-	-	-	-	-	-	-
Colombia	-	-	-	-	-	-	-	-	-
Ecuador	-	-	-	-	-	-	-	-	-
Netherlands Antilles	-	-	-	-	-	-	-	-	-
Peru	-	-	-	-	-	-	-	-	-
Trinidad and Tobago	-	-	-	-	-	-	-	-	-
Venezuela	-	-	-	-	-	-	-	-	-
Other	-	-	3 e	-	-	-	-	-	-
Latin America	**-**	**-**	**3 e**	**-**	**-**	**-**	**-**	**-**	**-**
Romania	-	-	-	-	-	-	-	-	-
Other	-	-	-	-	-	-	-	-	-
Non-OECD Europe	**-**	**-**	**-**	**-**	**-**	**-**	**-**	**-**	**-**
FSU	**-**	**-**	**3 e**	**-**	**-**	**1 e**	**-**	**-**	**7 e**
Iran	-	-	-	-	-	-	-	-	-
Iraq	-	-	-	-	-	-	-	-	-
Kuwait	-	-	-	-	-	-	-	-	-
Oman	-	-	-	-	-	-	-	-	-
Qatar	-	-	-	-	-	-	-	-	-
Saudi Arabia	-	-	-	-	-	-	-	-	-
United Arab Emirates	-	-	-	-	-	-	-	-	-
Other	-	-	-	-	-	-	-	-	-
Middle East	**-**	**-**	**-**	**-**	**-**	**-**	**-**	**-**	**-**
India	-	-	-	-	-	-	-	-	-
Indonesia	-	-	-	-	-	-	-	-	-
Singapore	-	-	-	-	-	-	-	-	-
Other	-	-	-	-	-	-	-	-	-
Asia	**-**	**-**	**-**	**-**	**-**	**-**	**-**	**-**	**-**
China	**-**	**-**	**-**	**-**	**-**	**-**	**-**	**-**	**-**
Algeria	-	-	-	-	-	-	-	-	-
Angola/Cabinada	-	-	-	-	-	-	-	-	-
Cameroon	-	-	-	-	-	-	-	-	-
Egypt	-	-	-	-	-	-	-	-	-
Gabon	-	-	-	-	-	-	-	-	-
Libya	-	-	-	-	-	-	-	-	-
Nigeria	-	-	-	-	-	-	-	-	-
Tunisia	-	-	-	-	-	-	-	-	-
Other	-	-	-	-	-	-	-	-	-
Africa	**-**	**-**	**-**	**-**	**-**	**-**	**-**	**-**	**-**
Non-spec./Other	**-**	**23**	**-**	**-**	**3**	**-**	**-**	**8**	**-**
Total Imports	**98 e**	**221**	**326 e**	**-**	**3**	**1 e**	**99 e**	**118**	**269 e**
Memo: Total OPEC	-	-	-	-	-	-	-	-	-

Source: Annual Oil Statistics.

SLOVAK REPUBLIC
Table 7D: Imports by Country of Origin

000 metric tons

	2000	2001	2002	2000	2001	2002	2000	2001	2002
	Fuel Oil (Residual)			**Petroleum Coke**			**Other Products**		
Australia	-	-	-	-	-	-	-	-	-
Belgium	-	-	-	-	-	-	-	2	-
Canada	-	-	-	-	-	-	-	-	-
Czech Republic	36 e	-	5 e	-	-	-	-	14	8 e
Denmark	-	-	-	-	-	-	-	-	-
Finland	-	-	-	-	-	-	-	-	-
France	-	-	-	-	-	-	-	1	-
Germany	-	-	10 e	-	-	42 e	-	6	1 e
Greece	-	-	-	-	-	-	-	-	-
Hungary	-	19	-	-	-	-	-	1	5 e
Ireland	-	-	-	-	-	-	-	-	-
Italy	-	-	-	-	-	-	-	-	-
Japan	-	-	-	-	-	-	-	-	-
Korea	-	-	-	-	-	-	-	-	-
Mexico	-	-	-	-	-	-	-	-	-
Netherlands	-	-	-	-	-	-	-	-	-
New Zealand	-	-	-	-	-	-	-	-	-
Norway	-	-	-	-	-	-	-	-	-
Poland	-	-	-	-	-	-	-	10	1 e
Spain	-	-	-	-	-	-	-	-	-
Sweden	-	-	-	-	-	-	-	-	-
Turkey	-	-	-	-	-	-	-	-	-
United Kingdom	-	-	-	-	-	-	-	-	-
United States	-	-	-	-	-	-	-	-	-
Other	-	-	1 e	-	-	-	-	22	14 e
OECD	**36 e**	**19**	**16 e**	**-**	**-**	**42 e**	**-**	**56**	**29 e**
Argentina	-	-	-	-	-	-	-	-	-
Brazil	-	-	-	-	-	-	-	-	-
Colombia	-	-	-	-	-	-	-	-	-
Ecuador	-	-	-	-	-	-	-	-	-
Netherlands Antilles	-	-	-	-	-	-	-	-	-
Peru	-	-	-	-	-	-	-	-	-
Trinidad and Tobago	-	-	-	-	-	-	-	-	-
Venezuela	-	-	-	-	-	-	-	-	-
Other	-	-	-	-	-	-	-	-	-
Latin America	**-**	**-**	**-**	**-**	**-**	**-**	**-**	**-**	**-**
Romania	-	-	-	-	-	-	-	-	-
Other	-	-	2 e	-	-	8 e	-	-	-
Non-OECD Europe	**-**	**-**	**2 e**	**-**	**-**	**8 e**	**-**	**-**	**-**
FSU	**-**	**-**	**6 e**	**-**	**-**	**-**	**-**	**-**	**1 e**
Iran	-	-	-	-	-	-	-	-	-
Iraq	-	-	-	-	-	-	-	-	-
Kuwait	-	-	-	-	-	-	-	-	-
Oman	-	-	-	-	-	-	-	-	-
Qatar	-	-	-	-	-	-	-	-	-
Saudi Arabia	-	-	-	-	-	-	-	-	-
United Arab Emirates	-	-	-	-	-	-	-	-	-
Other	-	-	-	-	-	-	-	-	-
Middle East	**-**	**-**	**-**	**-**	**-**	**-**	**-**	**-**	**-**
India	-	-	-	-	-	-	-	-	-
Indonesia	-	-	-	-	-	-	-	-	-
Singapore	-	-	-	-	-	-	-	-	-
Other	-	-	-	-	-	-	-	-	-
Asia	**-**	**-**	**-**	**-**	**-**	**-**	**-**	**-**	**-**
China	**-**	**-**	**-**	**-**	**-**	**-**	**-**	**-**	**-**
Algeria	-	-	-	-	-	-	-	-	-
Angola/Cabinada	-	-	-	-	-	-	-	-	-
Cameroon	-	-	-	-	-	-	-	-	-
Egypt	-	-	-	-	-	-	-	-	-
Gabon	-	-	-	-	-	-	-	-	-
Libya	-	-	-	-	-	-	-	-	-
Nigeria	-	-	-	-	-	-	-	-	-
Tunisia	-	-	-	-	-	-	-	-	-
Other	-	-	-	-	-	-	-	-	-
Africa	**-**	**-**	**-**	**-**	**-**	**-**	**-**	**-**	**-**
Non-spec./Other	**-**	**11**	**-**	**-**	**49**	**10 e**	**10 e**	**13**	**-**
Total Imports	**36 e**	**30**	**24 e**	**-**	**49**	**60 e**	**10 e**	**69**	**30 e**
Memo: Total OPEC	-	-	-	-	-	-	-	-	-

Source: Annual Oil Statistics.

SLOVAK REPUBLIC
Table 8A: Exports by Country of Destination

000 metric tons

	2000	2001	2002	2000	2001	2002	2000	2001	2002
	Crude Oil			Natural Gas Liquids			Feedstocks and Other		
Australia	-	-	-	-	-	-	-	-	-
Austria	-	-	12 e	-	-	-	-	-	-
Belgium	-	-	-	-	-	-	-	-	-
Canada	-	-	-	-	-	-	-	-	-
Denmark	-	-	-	-	-	-	-	-	-
Finland	-	-	-	-	-	-	-	-	-
France	-	-	-	-	-	-	-	-	-
Germany	-	-	-	-	-	-	-	-	-
Greece	-	-	-	-	-	-	-	-	-
Ireland	-	-	-	-	-	-	-	-	-
Italy	-	-	-	-	-	-	-	-	-
Japan	-	-	-	-	-	-	-	-	-
Luxembourg	-	-	-	-	-	-	-	-	-
Mexico	-	-	-	-	-	-	-	-	-
Netherlands	-	-	-	-	-	-	-	-	-
Norway	-	-	-	-	-	-	-	-	-
Portugal	-	-	-	-	-	-	-	-	-
Spain	-	-	-	-	-	-	-	-	-
Sweden	-	-	-	-	-	-	-	-	-
Switzerland	-	-	-	-	-	-	-	-	-
Turkey	-	-	-	-	-	-	-	-	-
United Kingdom	-	-	-	-	-	-	-	-	-
United States	-	-	-	-	-	-	-	-	-
Other OECD	15 e	-	3 e	-	-	-	-	-	-
OECD	**15 e**	**-**	**15 e**	**-**	**-**	**-**	**-**	**-**	**-**
Africa	-	-	-	-	-	-	-	-	-
Latin America	-	-	-	-	-	-	-	-	-
Asia	-	-	-	-	-	-	-	-	-
China	-	-	-	-	-	-	-	-	-
Non-OECD Europe	-	-	-	-	-	-	-	-	-
Former Soviet Union	-	-	-	-	-	-	-	-	-
Middle East	-	-	-	-	-	-	-	-	-
Non-specified/Other	-	17	-	-	-	-	-	-	-
Non-OECD	**-**	**17**	**-**	**-**	**-**	**-**	**-**	**-**	**-**
Total Exports	**15**	**17**	**15 e**	**-**	**-**	**-**	**-**	**-**	**-**
	Total Products			Liquefied Petroleum Gas			Naphtha		
Australia	-	-	19 e	-	-	-	-	-	-
Austria	572 e	-	802 e	-	-	-	-	-	-
Belgium	-	-	-	-	-	-	-	-	-
Canada	-	-	-	-	-	-	-	-	-
Denmark	-	-	-	-	-	-	-	-	-
Finland	-	-	-	-	-	-	-	-	-
France	-	-	-	-	-	-	-	-	-
Germany	114 e	-	113 e	-	-	-	-	-	-
Greece	-	-	-	-	-	-	-	-	-
Ireland	-	-	-	-	-	-	-	-	-
Italy	-	-	12 e	-	-	-	-	-	-
Japan	-	-	-	-	-	-	-	-	-
Luxembourg	-	-	-	-	-	-	-	-	-
Mexico	-	-	-	-	-	-	-	-	-
Netherlands	21 e	-	-	-	-	-	-	-	-
Norway	-	-	-	-	-	-	-	-	-
Portugal	-	-	-	-	-	-	-	-	-
Spain	-	-	-	-	-	-	-	-	-
Sweden	-	-	-	-	-	-	-	-	-
Switzerland	-	-	22 e	-	-	-	-	-	-
Turkey	-	-	-	-	-	-	-	-	-
United Kingdom	-	-	-	-	-	-	-	-	-
United States	-	-	-	-	-	-	-	-	-
Other OECD	1 646 e	1 259	2 382 e	-	-	1 e	-	-	3 e
OECD	**2 353 e**	**1 259**	**3 350 e**	**-**	**-**	**1 e**	**-**	**-**	**3 e**
Africa	-	-	-	-	-	-	-	-	-
Latin America	-	-	-	-	-	-	-	-	-
Asia	-	-	-	-	-	-	-	-	-
China	-	-	-	-	-	-	-	-	-
Non-OECD Europe	-	1	7 e	-	-	-	-	-	-
Former Soviet Union	-	2	1 e	-	-	-	-	-	-
Middle East	-	-	-	-	-	-	-	-	-
Non-specified/Other	453 e	1 948	131 e	-	-	-	-	7	-
Non-OECD	**453 e**	**1 951**	**139 e**	**-**	**-**	**-**	**-**	**7**	**-**
Total Exports	**2 806**	**3 210**	**3 489 e**	**-**	**-**	**1 e**	**-**	**7**	**3 e**

Source: Annual Oil Statistics.

SLOVAK REPUBLIC
Table 8B: Exports by Country of Destination

000 metric tons

	2000	2001	2002	2000	2001	2002	2000	2001	2002
	Total Gasoline			**Total Kerosene**			**Gas/Diesel Oil**		
Australia	-	-	19 e	-	-	-	-	-	-
Austria	136 e	-	178 e	-	-	-	318 e	-	525 e
Belgium	-	-	-	-	-	-	-	-	-
Canada	-	-	-	-	-	-	-	-	-
Denmark	-	-	-	-	-	-	-	-	-
Finland	-	-	-	-	-	-	-	-	-
France	-	-	-	-	-	-	-	-	-
Germany	19 e	-	45 e	-	-	-	68 e	-	45 e
Greece	-	-	-	-	-	-	-	-	-
Ireland	-	-	-	-	-	-	-	-	-
Italy	-	-	-	-	-	-	-	-	-
Japan	-	-	-	-	-	-	-	-	-
Luxembourg	-	-	-	-	-	-	-	-	-
Mexico	-	-	-	-	-	-	-	-	-
Netherlands	-	-	-	-	-	-	-	-	-
Norway	-	-	-	-	-	-	-	-	-
Portugal	-	-	-	-	-	-	-	-	-
Spain	-	-	-	-	-	-	-	-	-
Sweden	-	-	-	-	-	-	-	-	-
Switzerland	-	-	-	-	-	-	-	-	-
Turkey	-	-	-	-	-	-	-	-	-
United Kingdom	-	-	-	-	-	-	-	-	-
United States	-	-	-	-	-	-	-	-	-
Other OECD	482 e	452	966 e	13 e	-	9 e	912 e	739	1 069 e
OECD	**637 e**	**452**	**1 208 e**	**13 e**	**-**	**9 e**	**1 298 e**	**739**	**1 639 e**
Africa	-	-	-	-	-	-	-	-	-
Latin America	-	-	-	-	-	-	-	-	-
Asia	-	-	-	-	-	-	-	-	-
China	-	-	-	-	-	-	-	-	-
Non-OECD Europe	-	-	-	-	-	-	-	-	-
Former Soviet Union	-	-	-	-	-	-	-	-	-
Middle East	-	-	-	-	-	-	-	-	-
Non-specified/Other	232 e	573	31 e	-	6	-	124 e	830	-
Non-OECD	**232 e**	**573**	**31 e**	**-**	**6**	**-**	**124 e**	**830**	**-**
Total Exports	**869**	**1 025**	**1 239 e**	**13 e**	**6**	**9 e**	**1 422**	**1 569**	**1 639 e**
	Fuel Oil (Residual)			**Petroleum Coke**			**Other Products**		
Australia	-	-	-	-	-	-	-	-	-
Austria	118 e	-	71 e	-	-	-	-	-	28 e
Belgium	-	-	-	-	-	-	-	-	-
Canada	-	-	-	-	-	-	-	-	-
Denmark	-	-	-	-	-	-	-	-	-
Finland	-	-	-	-	-	-	-	-	-
France	-	-	-	-	-	-	-	-	-
Germany	27 e	-	-	-	-	-	-	-	23 e
Greece	-	-	-	-	-	-	-	-	-
Ireland	-	-	-	-	-	-	-	-	-
Italy	-	-	1 e	-	-	-	-	-	11 e
Japan	-	-	-	-	-	-	-	-	-
Luxembourg	-	-	-	-	-	-	-	-	-
Mexico	-	-	-	-	-	-	-	-	-
Netherlands	21 e	-	-	-	-	-	-	-	-
Norway	-	-	-	-	-	-	-	-	-
Portugal	-	-	-	-	-	-	-	-	-
Spain	-	-	-	-	-	-	-	-	-
Sweden	-	-	-	-	-	-	-	-	-
Switzerland	-	-	-	-	-	-	-	-	22 e
Turkey	-	-	-	-	-	-	-	-	-
United Kingdom	-	-	-	-	-	-	-	-	-
United States	-	-	-	-	-	-	-	-	-
Other OECD	187 e	-	201 e	-	-	-	52 e	68	133 e
OECD	**353 e**	**-**	**273 e**	**-**	**-**	**-**	**52 e**	**68**	**217 e**
Africa	-	-	-	-	-	-	-	-	-
Latin America	-	-	-	-	-	-	-	-	-
Asia	-	-	-	-	-	-	-	-	-
China	-	-	-	-	-	-	-	-	-
Non-OECD Europe	-	-	-	-	-	-	-	1	7 e
Former Soviet Union	-	-	1 e	-	-	-	-	2	-
Middle East	-	-	-	-	-	-	-	-	-
Non-specified/Other	97 e	363	-	-	1	-	-	168	100 e
Non-OECD	**97 e**	**363**	**1 e**	**-**	**1**	**-**	**-**	**171**	**107 e**
Total Exports	**450**	**363**	**274 e**	**-**	**1**	**-**	**52 e**	**239**	**324 e**

Source: Annual Oil Statistics.

SPAIN

Figure 1
FUEL SHARES OF TPES *

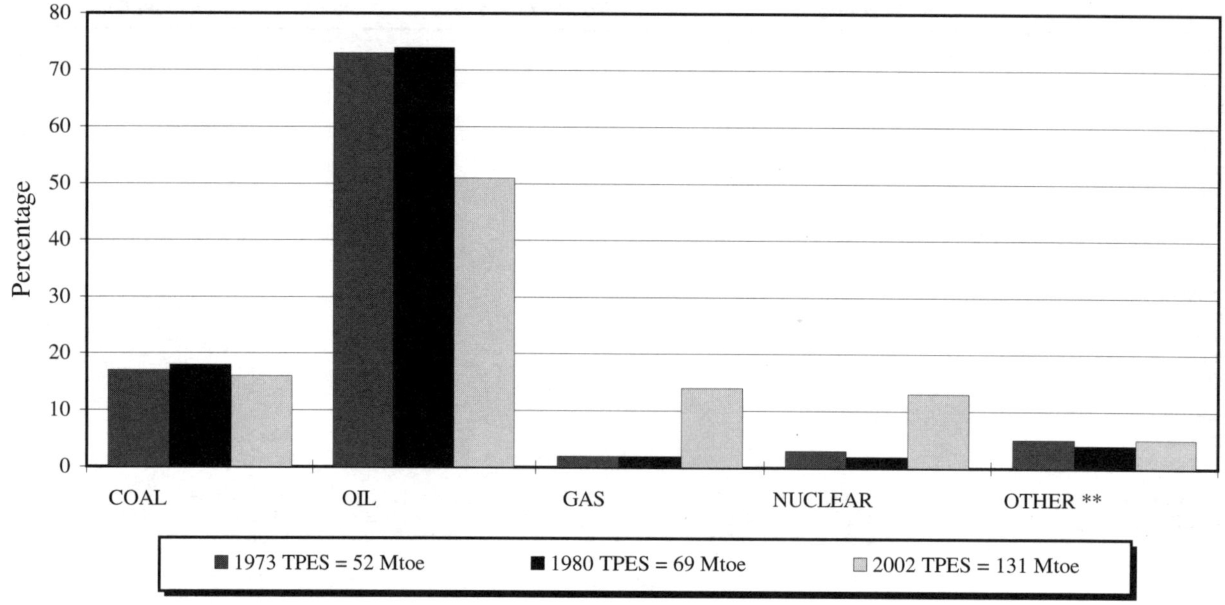

Source: Annual Energy Balances.

* Excluding electricity trade.

** Includes hydro, geothermal, solar, wind, combustible renewables and waste, etc.

Figure 2
OIL PRODUCTION AND CONSUMPTION

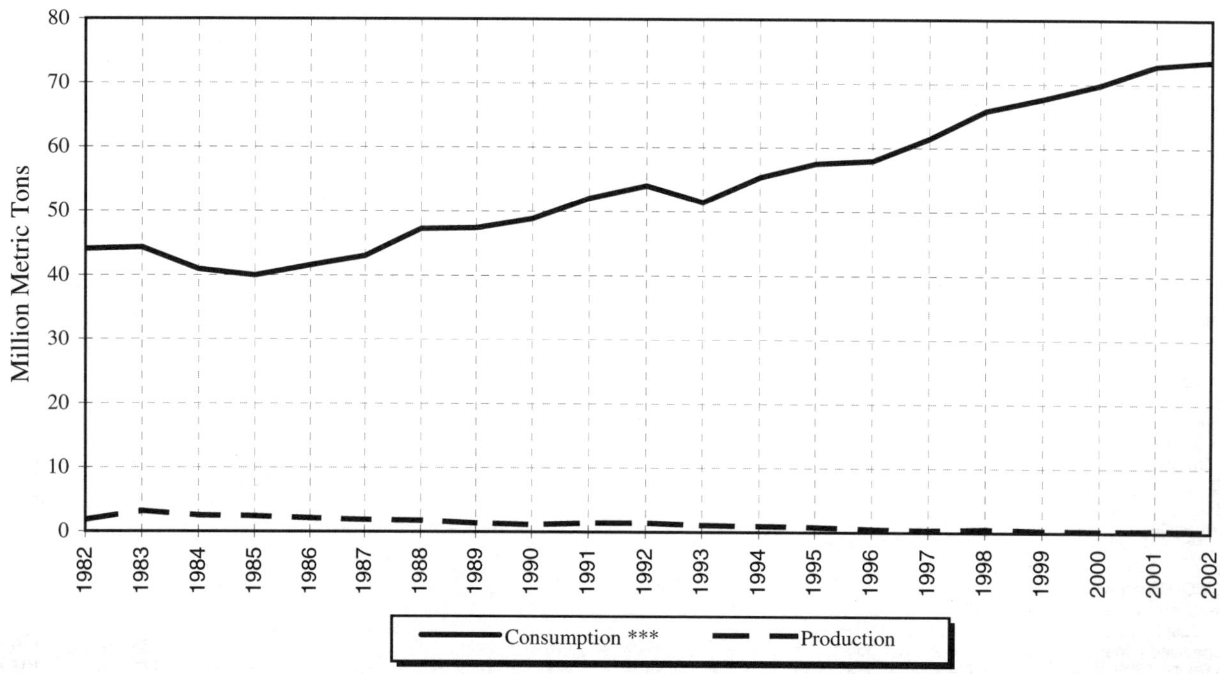

Source: Annual Oil Statistics.

*** Includes refinery fuel and international marine bunkers.

SPAIN

Table 1: Oil Supply and Consumption - Summary Table

000 metric tons

	1973	1978	1980	1990	1999	2000	2001	2002
Indigenous Production	654	1 211	1 755	1 144	299	227	338	316 e
Crude Oil	..	980	1 602	795	299	227	338	316 e
NGL	..	231	153	349	-	-	-	-
Other	..	-	-	-	-	-	-	-
+ Net Imports	40 256	47 397	48 914	48 706	68 976	70 972	72 804	74 410 e
Crude Oil, NGL, Other	42 970	46 758	47 516	53 268	58 911	58 782	57 833	57 694 e
Petroleum Products	-2 714	639	1 398	-4 562	10 065	12 190	14 971	16 716 e
- Intl. Marine Bunkers	1 393	1 425	1 648	3 716	6 033	6 173	6 913	7 059 e
+ Stock Changes	-1 862	1 519	561	- 596	- 440	- 941	167	-1 180 e
TOTAL SUPPLY	**37 655**	**48 702**	**49 582**	**45 538**	**62 802**	**64 085**	**66 396**	**66 487 e**
Refinery Intake	*43 070*	*49 671*	*49 326*	*53 757*	*61 297*	*60 404*	*58 349*	*58 489 e*
Refinery Output	*42 423*	*48 265*	*48 517*	*52 796*	*60 841*	*59 830*	*57 806*	*57 931 e*
LPG + Ethane	*1 439*	*1 035*	*962*	*1 783*	*1 586*	*1 519*	*1 574*	*1 561 e*
Naphtha	*1 789*	*2 816*	*2 375*	*2 196*	*3 102*	*3 358*	*3 076*	*3 000 e*
Kerosenes	*2 167*	*2 426*	*2 236*	*3 890*	*3 925*	*3 969*	*3 923*	*3 830 e*
Gasolines	*4 491*	*5 215*	*5 344*	*9 797*	*9 784*	*9 765*	*9 351*	*8 871 e*
Gas/Diesel Oil	*9 662*	*10 867*	*10 790*	*14 571*	*20 368*	*20 066*	*20 177*	*20 820 e*
Fuel Oil	*20 603*	*22 268*	*22 924*	*14 980*	*14 126*	*13 080*	*11 784*	*12 122 e*
Other Products	*2 272*	*3 638*	*3 886*	*5 579*	*7 950*	*8 073*	*7 921*	*7 727 e*
- Refinery Losses	647	1 406	809	961	456	574	543	558 e
+ Returns	-	-	-	491	30	96	103	97 e
- Statistical Differences	15	1 238	- 429	- 54	727	- 36	69	- 429 e
TOTAL CONSUMPTION	**36 993**	**46 058**	**49 202**	**45 122**	**61 649**	**63 643**	**65 887**	**66 455 e**
Transformation and Energy	**7 270**	**9 175**	**11 747**	**6 181**	**9 510**	**8 960**	**9 436 e**	..
Crude Oil	-	-	-	-	-	-	-	..
LPG + Ethane	4	25	21	29	25	34	25	..
Gas/Diesel Oil	108	90	83	202	430	326	786	..
Fuel Oil	6 417	8 138	10 634	3 975	6 991	6 379	6 365	..
Petroleum Coke	-	-	-	-	150	158	192	..
Other	741	922	1 009	1 975	1 914	2 063	2 068 e	..
Transport	**11 264**	**14 421**	**15 381**	**21 698**	**31 150**	**32 000**	**33 357**	..
Motor Gasoline	4 083	5 454	5 421	8 145	8 928	8 524	8 481	..
Gas/Diesel Oil	4 819	6 089	6 377	10 465	17 500	18 485	19 834	..
Subtotal	*8 902*	*11 543*	*11 798*	*18 610*	*26 428*	*27 009*	*28 315*	..
Aviation Fuels	1 703	2 222	2 014	2 402	4 097	4 378	4 418	..
Other	659	656	1 569	686	625	613	624	..
Industry	**13 773**	**16 888**	**16 320**	**11 513**	**14 084**	**15 162**	**15 717**	..
Naphtha	1 336	2 376	2 364	2 588	4 275	4 691	4 893	..
LPG + Ethane	194	325	399	290	389	405	403	..
Gas/Diesel Oil	144	433	568	442	925	1 136	1 208	..
Fuel Oil	9 853	11 530	10 591	3 769	1 862	1 743	1 649	..
Other	2 246	2 224	2 398	4 424	6 633	7 187	7 564	..
Other Sectors	**4 686**	**5 574**	**5 754**	**5 730**	**6 905**	**7 521**	**7 377**	..
Gas/Diesel Oil	1 690	3 316	3 562	3 149	4 305	5 047	5 087	..
Fuel Oil	1 154	261	198	260	491	465	419	..
Other	1 842	1 997	1 994	2 321	2 109	2 009	1 871	..
Memo: Non-Energy Use	*3 412*	*4 522*	*4 678*	*5 519*	*8 532*	*9 017*	*9 269*	..
Chem/Petchem. Feedst.	*1 336*	*2 376*	*2 364*	*3 101*	*5 008*	*5 533*	*5 837*	..
Other	*2 076*	*2 146*	*2 314*	*2 418*	*3 524*	*3 484*	*3 432*	..

Source: Annual Oil Statistics.

SPAIN
Table 2: Supply of Crude Oil, NGL, Feedstocks and Other Hydrocarbons

000 metric tons

	1973	1978	1980	1990	1999	2000	2001	2002
CRUDE OIL								
Indigenous Production	..	980	1 602	795	299	227	338	316 e
+ Imports	..	46 758	47 430	50 630	58 046	57 475	56 792	56 449 e
- Exports	..	-	-	-	-	-	-	-
- Direct Use	..	61	61	28	12	12	12	12 e
+ Stock Change	..	2 423	56	- 767	463	- 594	- 769	- 287 e
- Statistical Difference	..	429	6	-	-	-	-	-
Refinery Intake	**..**	**49 671**	**49 021**	**50 630**	**58 796**	**57 096**	**56 349**	**56 466 e**
Closing Stock Level	*..*	*3 194*	*2 914*	*3 410*	*3 422*	*4 016*	*4 785*	*5 072 e*
NATURAL GAS LIQUIDS								
Indigenous Production	..	231	153	349	-	-	-	-
+ Imports	..	-	-	-	-	-	-	-
- Exports	..	-	-	-	-	-	-	-
- Direct Use	..	231	153	349	-	-	-	-
+ Stock Change	..	-	-	-	-	-	-	-
- Statistical Difference	..	-	-	-	-	-	-	-
Refinery Intake	**..**	**-**	**-**	**-**	**-**	**-**	**-**	**-**
Closing Stock Level	*..*	*-*	*-*	*-*	*-*	*-*	*-*	*-*
REFINERY FEEDSTOCKS								
Backflows	..	-	-	491	30	96	103	97 e
+ Transfers	..	-	-	- 318	1 540	2 102	632	768 e
+ Imports	..	-	86	2 638	829	1 272	1 024	1 209 e
- Exports	..	-	-	-	-	-	-	-
+ Stock Change	..	-	67	225	62	- 194	221	- 82 e
- Statistical Difference	..	-	- 152	- 91	-	-	-	3 e
Refinery Intake	**..**	**-**	**305**	**3 127**	**2 461**	**3 276**	**1 980**	**1 989 e**
Closing Stock Level	*..*	*-*	*346*	*556*	*1 018*	*1 212*	*991*	*1 073 e*
OTHER HYDROCARBONS								
Indigenous Production	..	-	-	-	-	-	-	-
+ Imports	..	-	-	-	36	35	17	36 e
- Exports	..	-	-	-	-	-	-	-
- Direct Use	..	-	-	-	-	-	-	-
+ Stock Change	..	-	-	-	4	- 3	3	- 5 e
- Statistical Difference	..	-	-	-	-	-	-	- 3 e
Refinery Intake	**..**	**-**	**-**	**-**	**40**	**32**	**20**	**34 e**
Closing Stock Level	*..*	*-*	*-*	*-*	*9*	*12*	*9*	*14 e*
CRUDE OIL, NGL, FEEDSTOCKS AND OTHER								
Indigenous Production	654	1 211	1 755	1 144	299	227	338	316 e
+ Backflows	-	-	-	491	30	96	103	97 e
+ Transfers	-	-	-	- 318	1 540	2 102	632	768 e
+ Imports	42 970	46 758	47 516	53 268	58 911	58 782	57 833	57 694 e
- Exports	-	-	-	-	-	-	-	-
- Direct Use	111	292	214	377	12	12	12	12 e
+ Stock Change	- 554	2 423	123	- 542	529	- 791	- 545	- 374 e
- Statistical Difference	- 111	429	- 146	- 91	-	-	-	-
Refinery Intake	**43 070**	**49 671**	**49 326**	**53 757**	**61 297**	**60 404**	**58 349**	**58 489 e**
Closing Stock Level	*-*	*3 194*	*3 260*	*3 966*	*4 449*	*5 240*	*5 785*	*6 159 e*

Source: Annual Oil Statistics.

SPAIN

Table 3: Motor Gasoline - Supply

000 metric tons

	1973	1978	1980	1990	1999	2000	2001	2002
MOTOR GASOLINE								
Refinery Output	**4 491**	**5 215**	**5 344**	**9 230**	**9 432**	**9 615**	**9 272**	**8 871 e**
+ Imports	32	520	191	597	879	931	927	963 e
- Exports	314	21	3	1 392	1 612	2 374	2 291	2 174 e
+ Transfers	-	-	-	78	112	347	356	540 e
+ Stock Change	- 163	- 260	- 111	- 377	125	- 13	219	- 13 e
- Statistical Difference	- 37	-	-	- 9	8	- 18	2	- 19 e
Total Consumption	**4 083**	**5 454**	**5 421**	**8 145**	**8 928**	**8 524**	**8 481**	**8 206 e**
Closing Stock Level	-	*1 446*	*1 617*	*1 383*	*1 797*	*1 810*	*1 591*	*1 604 e*

Table 4: Aviation Fuels - Supply

000 metric tons

	1973	1978	1980	1990	1999	2000	2001	2002
AVIATION GASOLINE								
Refinery Output	-	-	-	-	**5**	**1**	-	-
+ Imports	31	11	16	12	7	7	6	4 e
- Exports	-	-	-	-	-	-	-	-
+ Transfers	-	-	-	-	- 1	1	5	9 e
+ Stock Change	-	2	- 2	- 1	- 1	1	-	1 e
- Statistical Difference	- 4	-	1	-	-	-	-	1 e
Total Consumption	**35**	**13**	**13**	**11**	**10**	**10**	**11**	**13 e**
Closing Stock Level	-	*4*	*6*	*1*	*4*	*3*	*3*	*2 e*
GASOLINE TYPE JET FUEL								
Refinery Output	-	-	-	**567**	**347**	**149**	**79**	-
+ Imports	-	-	-	-	-	6	-	-
- Exports	-	-	-	255	72	76	31	-
+ Transfers	-	-	-	- 15	- 214	21	4	-
+ Stock Change	-	-	-	- 9	49	1	7	-
- Statistical Difference	-	-	-	10	2	1	-	-
Total Consumption	-	-	-	**272**	**108**	**100**	**59**	-
Closing Stock Level	-	-	-	*20*	*8*	*7*	-	-
KEROSENE TYPE JET FUEL								
Refinery Output	**1 943**	**2 405**	**2 194**	**3 663**	**3 753**	**3 692**	**3 666**	**3 567 e**
+ Imports	-	31	635	55	424	544	815	821 e
- Exports	50	323	682	1 283	175	73	99	240 e
+ Transfers	-	-	-	- 122	48	151	90	- 82 e
+ Stock Change	-	96	- 158	- 189	- 57	- 61	- 138	97 e
- Statistical Difference	225	-	- 12	5	14	- 15	- 14	- 18 e
Total Consumption	**1 668**	**2 209**	**2 001**	**2 119**	**3 979**	**4 268**	**4 348**	**4 181 e**
Closing Stock Level	-	*263*	*479*	*497*	*483*	*544*	*682*	*585 e*

Source: Annual Oil Statistics.

SPAIN
Table 5A: Gas/Diesel Oil - Supply and Consumption

000 metric tons

	1973	1978	1980	1990	1999	2000	2001	2002
Refinery Output	**9 662**	**10 867**	**10 790**	**14 571**	**20 368**	**20 066**	**20 177**	**20 820 e**
+ Recycled Products	-	-	-	-	-	-	-	-
+ Imports	-	350	457	1 231	6 749	7 244	8 055	9 342 e
- Exports	2 228	566	394	401	729	848	572	783 e
- Intl. Marine Bunkers	320	456	395	1 206	1 147	967	998	948 e
+ Transfers	-	-	-	65	-1 482	- 469	131	- 366 e
+ Stock Changes	- 782	- 267	154	47	- 566	- 118	207	- 315 e
- Statistical Differences	- 429	-	22	49	33	- 86	85	- 29 e
Total Consumption	**6 761**	**9 928**	**10 590**	**14 258**	**23 160**	**24 994**	**26 915**	**27 779 e**
Transformation	**106**	**85**	**78**	**143**	**359**	**326**	**786**	**..**
Public Electricity	100	80	70	142	220	236 e	413	..
Autoproducer Electricity	-	-	4	1	-	-	-	..
Public CHP Plants	-	-	-	-	-	-	-	..
Autoproducers of CHP	-	-	-	-	139	90 e	373	..
Public Heat Plants	-	-	-	-	-	-	-	..
Autoproducer Heat Plants	-	-	-	-	-	-	-	..
Petrochemical Industry	-	-	-	-	-	-	-	..
Other Transformation	6	5	4	-	-	-	-	..
Energy Sector	**2**	**5**	**5**	**59**	**71**	**-**	**-**	**..**
Refinery Fuel	-	-	-	13	-	-	-	..
Other Energy Sector	2	5	5	46	71	-	-	..
Final Consumption	**6 653**	**9 838**	**10 507**	**14 056**	**22 730**	**24 668**	**26 129**	**..**
Transport	**4 819**	**6 089**	**6 377**	**10 465**	**17 500**	**18 485**	**19 834**	**..**
Air	-	-	-	-	-	-	-	..
Road	2 896	5 135	4 561	8 995	15 670	16 847	18 165	..
Rail	203	332	192	210	480	484	516	..
Inland Waterways	1 716	622	1 618	1 260	1 350	1 154	1 153	..
Non-Specified	4	-	6	-	-	-	-	..
Industry	**144**	**433**	**568**	**442**	**925**	**1 136**	**1 208**	**..**
Iron and Steel	10	20	36	28	50	72	73	..
Chemical (incl. Pet.-Chem.)	6	60	162	55	85	102	110	..
Non-Ferrous Metals	-	8	26	12	24	32	33	..
Non-Metallic Minerals	11	89	47	25	56	70	76	..
Transport Equipment	14	19	23	20	65	84	87	..
Machinery	4	20	25	20	42	53	51	..
Mining and Quarrying	30	34	38	51	75	94	97	..
Food and Tobacco	13	104	139	145	235	276	295	..
Pulp, Paper and Printing	-	12	11	8	33	40	41	..
Wood and Wood Products	1	6	8	4	15	18	17	..
Construction	24	18	8	12	60	76	90	..
Textiles and Leather	2	20	27	42	80	93	95	..
Non-Specified	29	23	18	20	105	126	143	..
Other Sectors	**1 690**	**3 316**	**3 562**	**3 149**	**4 305**	**5 047**	**5 087**	**..**
Commerce and Public	5	481	680	630	900	1 140	1 290	..
Residential	3	790	836	1 260	1 855	2 056	2 055	..
Agriculture	1 682	2 045	2 046	1 259	1 550	1 851	1 742	..
Non-Specified	-	-	-	-	-	-	-	..
Memo: Non-Energy Use	-	-	-	-	-	-	-	..
Chem/Petchem. Feedst.	-	-	-	-	-	-	-	..
Other	-	-	-	-	-	-	-	..
Closing Stock Level	-	*2 049*	*2 061*	*2 020*	*4 529*	*4 647*	*4 440*	*4 755 e*

Source: Annual Oil Statistics.

SPAIN
Table 5B: Fuel Oil (Residual) - Supply and Consumption

000 metric tons

	1973	1978	1980	1990	1999	2000	2001	2002
Refinery Output	**20 603**	**22 268**	**22 924**	**14 980**	**14 126**	**13 080**	**11 784**	**12 122 e**
+ Recycled Products	-	-	-	-	-	-	-	-
+ Imports	576	718	872	1 107	2 235	2 756	3 297	3 715 e
- Exports	846	745	652	6 231	1 400	1 054	876	328 e
- Intl. Marine Bunkers	1 073	969	1 253	2 510	4 870	5 183	5 886	6 101 e
+ Transfers	-	-	-	958	- 156	- 917	15	414 e
+ Stock Changes	- 565	- 10	558	87	- 372	135	304	- 362 e
- Statistical Differences	634	795	- 428	- 13	- 11	8	- 12	- 23 e
Total Consumption	**18 061**	**20 467**	**22 877**	**8 404**	**9 574**	**8 809**	**8 650**	**9 483 e**
Transformation	**5 344**	**6 491**	**9 095**	**2 126**	**4 834**	**4 305**	**4 480**	..
Public Electricity	5 344	6 242	8 856	1 356	3 304	3 278 e	3 822	..
Autoproducer Electricity	-	249	226	53	40	48	-	..
Public CHP Plants	-	-	-	450	-	-	-	..
Autoproducers of CHP	-	-	-	247	1 490	979 e	658	..
Public Heat Plants	-	-	-	-	-	-	-	..
Autoproducer Heat Plants	-	-	-	-	-	-	-	..
Petrochemical Industry	-	-	-	20	-	-	-	..
Other Transformation	-	-	13	-	-	-	-	..
Energy Sector	**1 073**	**1 647**	**1 539**	**1 849**	**2 157**	**2 074**	**1 885**	..
Refinery Fuel	1 073	1 647	1 539	1 829	2 132	2 074	1 885	..
Other Energy Sector	-	-	-	20	25	-	-	..
Final Consumption	**11 644**	**12 329**	**12 243**	**4 429**	**2 583**	**2 430**	**2 285**	..
Transport	**637**	**538**	**1 454**	**400**	**230**	**222**	**217**	..
Air	-	-	-	-	-	-	-	..
Road	-	-	-	-	-	-	-	..
Rail	182	19	11	-	-	-	-	..
Inland Waterways	455	519	1 443	400	230	222	217	..
Non-Specified	-	-	-	-	-	-	-	..
Industry	**9 853**	**11 530**	**10 591**	**3 769**	**1 862**	**1 743**	**1 649**	..
Iron and Steel	1 385	1 253	1 182	360	125	135	121	..
Chemical (incl. Pet.-Chem.)	1 390	1 909	1 795	1 083	373	334	311	..
Non-Ferrous Metals	-	600	561	180	110	109	101	..
Non-Metallic Minerals	3 730	4 296	3 797	535	201	189	182	..
Transport Equipment	145	173	153	80	55	55	53	..
Machinery	214	183	168	80	65	65	56	..
Mining and Quarrying	182	197	203	60	45	44	41	..
Food and Tobacco	1 233	1 207	1 148	760	320	291	285	..
Pulp, Paper and Printing	745	833	829	360	255	235	219	..
Wood and Wood Products	122	141	139	25	20	19	17	..
Construction	69	50	41	35	40	40	45	..
Textiles and Leather	558	537	537	211	102	91	84	..
Non-Specified	80	151	38	-	151	136	134	..
Other Sectors	**1 154**	**261**	**198**	**260**	**491**	**465**	**419**	..
Commerce and Public	623	202	150	230	338	315	296	..
Residential	490	31	26	15	90	87	87	..
Agriculture	41	28	22	15	63	63	36	..
Non-Specified	-	-	-	-	-	-	-	..
Memo: Non-Energy Use	-	-	-	-	-	-	-	..
Chem/Petchem. Feedst.	-	-	-	-	-	-	-	..
Other	-	-	-	-	-	-	-	..
Closing Stock Level	-	*2 400*	*2 302*	*1 721*	*2 026*	*1 891*	*1 587*	*1 949 e*

Source: Annual Oil Statistics.

SPAIN
Table 5C: Total Products - Supply and Consumption

000 metric tons

	1973	1978	1980	1990	1999	2000	2001	2002
Refinery Output	**42 423**	**48 265**	**48 517**	**52 796**	**60 841**	**59 830**	**57 806**	**57 931 e**
+ Recycled Products	-	-	-	-	-	-	-	-
+ Imports	1 502	3 248	4 600	7 712	16 991	19 557	21 194	22 642 e
- Exports	4 216	2 609	3 202	12 274	6 926	7 367	6 223	5 926 e
- Intl. Marine Bunkers	1 393	1 425	1 648	3 716	6 033	6 173	6 913	7 059 e
+ Transfers	111	292	214	695	-1 528	-2 090	- 620	- 756 e
+ Stock Changes	-1 308	- 904	438	- 54	- 969	- 150	712	- 806 e
- Statistical Differences	126	809	- 283	37	727	- 36	69	- 429 e
Total Consumption	**36 993**	**46 058**	**49 202**	**45 122**	**61 649**	**63 643**	**65 887**	**66 455 e**
Transformation	**5 640**	**6 799**	**9 432**	**2 804**	**5 358**	**4 885**	**5 474**	..
Public Electricity	5 444	6 322	8 926	1 498	3 524	3 514	4 235	..
Autoproducer Electricity	-	249	230	54	40	48	-	..
Public CHP Plants	-	-	-	450	5	7	5	..
Autoproducers of CHP	-	-	-	259	1 629	1 069	1 031	..
Public Heat Plants	-	-	-	-	-	-	-	..
Autoproducer Heat Plants	-	-	-	-	-	-	-	..
Petrochemical Industry	-	-	-	491	30	96	103	..
Other Transformation	196	228	276	52	130	151	100	..
Energy Sector	**1 630**	**2 376**	**2 315**	**3 377**	**4 152**	**4 075**	**3 962 e**	..
Refinery Fuel	1 628	2 371	2 310	3 186	4 001	4 000	3 859	..
Other Energy Sector	2	5	5	191	151	75	103 e	..
Final Consumption	**29 723**	**36 883**	**37 455**	**38 941**	**52 139**	**54 683**	**56 451**	..
Transport	**11 264**	**14 421**	**15 381**	**21 698**	**31 150**	**32 000**	**33 357**	..
Air	1 703	2 222	2 014	2 402	4 097	4 378	4 418	..
Road	7 001	10 707	10 097	17 416	24 948	25 718	27 007	..
Rail	385	351	203	210	480	484	516	..
Inland Waterways	2 171	1 141	3 061	1 670	1 625	1 420	1 416	..
Non-Specified	4	-	6	-	-	-	-	..
Industry	**13 773**	**16 888**	**16 320**	**11 513**	**14 084**	**15 162**	**15 717**	..
Iron and Steel	1 395	1 324	1 273	412	425	538	540	..
Chemical (incl. Pet.-Chem.)	2 917	4 460	4 461	4 449	5 791	6 312	6 619	..
Non-Ferrous Metals	-	616	597	416	694	828	866	..
Non-Metallic Minerals	3 823	4 508	3 975	2 230	2 532	2 903	3 217	..
Transport Equipment	159	208	194	109	132	152	153	..
Machinery	218	256	248	204	262	299	285	..
Mining and Quarrying	212	234	244	112	126	144	144	..
Food and Tobacco	1 253	1 325	1 305	924	587	599	612	..
Pulp, Paper and Printing	745	850	847	379	313	301	285	..
Wood and Wood Products	124	151	151	30	42	44	40	..
Construction	1 009	929	801	1 613	2 100	1 987	1 956	..
Textiles and Leather	561	561	569	260	185	187	182	..
Non-Specified	1 357	1 466	1 655	375	895	868	818	..
Other Sectors	**4 686**	**5 574**	**5 754**	**5 730**	**6 905**	**7 521**	**7 377**	..
Commerce and Public	690	903	990	1 030	1 433	1 664	1 811	..
Residential	2 023	2 545	2 662	3 344	3 760	3 849	3 692	..
Agriculture	1 973	2 126	2 102	1 356	1 712	2 008	1 874	..
Non-Specified	-	-	-	-	-	-	-	..
Memo: Non-Energy Use	*3 412*	*4 522*	*4 678*	*5 519*	*8 532*	*9 017*	*9 269*	..
Chem/Petchem. Feedst.	*1 336*	*2 376*	*2 364*	*3 101*	*5 008*	*5 533*	*5 837*	..
Other	*2 076*	*2 146*	*2 314*	*2 418*	*3 524*	*3 484*	*3 432*	..
Closing Stock Level	-	*7 407*	*7 487*	*6 838*	*10 484*	*10 634*	*9 922*	*10 728 e*

Source: Annual Oil Statistics.

SPAIN
Table 6A: Supply and Consumption

000 metric tons

	1980	2000	2001	2002	1980	2000	2001	2002
	LPG AND ETHANE				**REFINERY GAS**			
Refinery Output	**962**	**1 519**	**1 574**	**1 561 e**	**685**	**1 626**	**1 542**	**1 554 e**
+ Recycled Products	-	-	-	-	-	-	-	-
+ Imports	1 503	1 224	894	1 089 e	-	-	-	-
- Exports	10	116	83	117 e	-	-	-	-
- Intl. Marine Bunkers	-	-	-	-	-	-	-	-
+ Transfers	153	- 148	- 154	- 128 e	95	66	4	- 88 e
+ Stock Change	- 25	- 7	112	- 88 e	-	-	-	-
- Statistical Difference	75	- 18	14	- 17 e	-	-	-	-
Total Consumption	**2 508**	**2 490**	**2 329**	**2 334 e**	**780**	**1 692**	**1 546**	**1 466 e**
of which: Ethane	-	-	-	-	-	-	-	-
Transformation	17	34	25	..	-	-	-	..
Energy	4	-	-	..	760	1 692	1 546	..
Transport	115	75	72	..	-	-	-	..
Industry	399	405	403	..	20	-	-	..
Other	1 973	1 976	1 829	..	-	-	-	..
Memo: Non-Energy Use	-	-	-	..	-	-	-	..
Closing Stock Level	*318*	*347*	*235*	*323 e*	-	-	-	-
	NAPHTHA				**OTHER KEROSENE**			
Refinery Output	**2 375**	**3 358**	**3 076**	**3 000 e**	**42**	**277**	**257**	**263 e**
+ Recycled Products	-	-	-	-	-	-	-	-
+ Imports	196	2 382	2 569	2 336 e	13	-	-	-
- Exports	23	1 344	1 298	1 427 e	2	31	4	7 e
- Intl. Marine Bunkers	-	-	-	-	-	-	-	-
+ Transfers	-	487	651	963 e	-	- 229	- 214	- 211 e
+ Stock Change	60	- 3	- 12	- 3 e	- 29	- 14	- 29	- 37 e
- Statistical Difference	-	- 31	- 90	- 48 e	-	-	-	-
Total Consumption	**2 608**	**4 911**	**5 076**	**4 917 e**	**24**	**3**	**10**	**8 e**
Transformation	242	220	183	..	-	-	-	..
Energy	2	-	-	..	-	-	-	..
Transport	-	-	-	..	-	-	-	..
Industry	2 364	4 691	4 893	..	3	-	-	..
Other	-	-	-	..	21	3	10	..
Memo: Non-Energy Use	*2 364*	*4 691*	*4 893*	..	-	-	-	..
Closing Stock Level	*1*	*4*	*10*	*19 e*	*30*	*120*	*149*	*186 e*
	WHITE SPIRIT + SBP				**LUBRICANTS**			
Refinery Output	**96**	**163**	**142**	**..**	**319**	**370**	**427**	**..**
+ Recycled Products	-	-	-	..	-	-	-	..
+ Imports	9	42	41	..	151	155	148	..
- Exports	13	87	75	..	133	60	58	..
- Intl. Marine Bunkers	-	-	-	..	-	-	-	..
+ Transfers	-	6	3	..	-	- 22	- 46	..
+ Stock Change	- 8	1	- 7	..	- 23	- 14	- 5	..
- Statistical Difference	- 8	- 4	- 4	..	- 14	- 3	- 4	..
Total Consumption	**92**	**129**	**108**	**..**	**328**	**432**	**470**	**..**
Transformation	-	-	-	..	-	-	-	..
Energy	-	-	-	..	-	-	-	..
Transport	-	-	-	..	-	316	335	..
Industry	92	129	108	..	328	96	115	..
Other	-	-	-	..	-	20	20	..
Memo: Non-Energy Use	*92*	*129*	*108*	..	*328*	*432*	*470*	..
Closing Stock Level	*18*	*5*	*12*	..	*221*	*24*	*29*	..

Source: Annual Oil Statistics.

SPAIN
Table 6B: Supply and Consumption

000 metric tons

	1980	2000	2001	2002	1980	2000	2001	2002
	BITUMEN				PARAFFIN WAXES			
Refinery Output	**1 014**	**2 292**	**2 195**	**..**	**140**	**432**	**373**	**..**
+ Recycled Products	-	-	-	..	-	-	-	..
+ Imports	1	101	90	..	2	13	8	..
- Exports	402	469	420	..	9	61	70	..
- Intl. Marine Bunkers	-	-	-	..	-	-	-	..
+ Transfers	-	- 17	- 28	..	-	-	-	..
+ Stock Change	13	- 23	- 7	..	- 3	- 1	1	..
- Statistical Difference	- 126	18	15	..	84	2	- 6	..
Total Consumption	**752**	**1 866**	**1 815**	**..**	**46**	**381**	**318**	**..**
Transformation	-	-	-	..	-	-	-	..
Energy	-	-	-	..	-	-	-	..
Transport	-	-	-	..	-	-	-	..
Industry	752	1 866	1 815	..	46	381	318	..
Other	-	-	-	..	-	-	-	..
Memo: Non-Energy Use	*752*	*1 866*	*1 815*	..	*46*	*381*	*318*	..
Closing Stock Level	*72*	*162*	*169*	..	*9*	*26*	*25*	..
	PETROLEUM COKE				OTHER PRODUCTS			
Refinery Output	**-**	**1 024**	**1 077**	**1 001 e**	**1 632**	**2 166**	**2 165**	**5 172 e**
+ Recycled Products	-	-	-	-	-	-	-	-
+ Imports	494	3 127	3 545	3 171 e	60	1 025	799	1 201 e
- Exports	-	70	98	84 e	879	704	248	766 e
- Intl. Marine Bunkers	-	-	-	-	-	23	29	10 e
+ Transfers	-	- 17	- 5	- 64 e	- 34	-1 350	-1 432	-1 743 e
+ Stock Change	-	- 30	- 5	- 3 e	12	- 4	65	- 83 e
- Statistical Difference	-	5	58	- 31 e	123	105	25	- 245 e
Total Consumption	**494**	**4 029**	**4 456**	**4 052 e**	**668**	**1 005**	**1 295**	**4 016 e**
Transformation	-	-	-	..	-	-	-	..
Energy	-	158	192	..	5	151	339 e	..
Transport	-	-	-	..	-	-	-	..
Industry	494	3 861	4 252	..	663	854	956	..
Other	-	10	12	..	-	-	-	..
Memo: Non-Energy Use	*494*	*676*	*721*	..	*602*	*842*	*944*	..
Closing Stock Level	*43*	*776*	*781*	*784 e*	*310*	*268*	*203*	*521 e*

Source: Annual Oil Statistics.

SPAIN
Table 7A: Imports by Country of Origin

000 metric tons

	Crude Oil			Natural Gas Liquids			Feedstocks and Other		
	2000	2001	2002	2000	2001	2002	2000	2001	2002
Australia	-	-	-	-	-	-	-	-	-
Belgium	-	-	-	-	-	-	-	52	25 e
Canada	-	-	-	-	-	-	-	-	-
Czech Republic	-	-	-	-	-	-	-	-	-
Denmark	-	-	172 e	-	-	-	-	-	-
Finland	-	-	-	-	-	-	-	17	-
France	-	-	-	-	-	-	24	25	66 e
Germany	141	-	-	-	-	-	-	-	26 e
Greece	-	-	-	-	-	-	33	-	-
Hungary	-	-	-	-	-	-	-	-	-
Ireland	-	-	-	-	-	-	-	24	79 e
Italy	104	127	85 e	-	-	-	44	-	27 e
Japan	-	-	-	-	-	-	-	-	-
Korea	-	-	-	-	-	-	-	-	-
Mexico	7 622	7 735	7 786 e	-	-	-	-	-	-
Netherlands	-	-	-	-	-	-	-	29	75 e
New Zealand	-	-	-	-	-	-	-	-	-
Norway	249	420	910 e	-	-	-	19	11	10 e
Poland	-	-	-	-	-	-	-	-	-
Spain	-	-	-	-	-	-	-	-	-
Sweden	-	-	-	-	-	-	26	26	28 e
Turkey	-	-	-	-	-	-	-	21	-
United Kingdom	2 039	1 829	1 052 e	-	-	-	117	-	65 e
United States	-	-	-	-	-	-	-	-	-
Other	-	-	-	-	-	-	10	-	3 e
OECD	**10 155**	**10 111**	**10 005 e**	**-**	**-**	**-**	**273**	**205**	**404 e**
Argentina	-	92	188 e	-	-	-	-	-	-
Brazil	30	30	125 e	-	-	-	-	-	-
Colombia	-	-	332 e	-	-	-	-	-	-
Ecuador	-	-	-	-	-	-	-	-	-
Netherlands Antilles	-	-	-	-	-	-	-	-	-
Peru	-	-	-	-	-	-	-	-	-
Trinidad and Tobago	-	-	-	-	-	-	-	-	-
Venezuela	1 562	2 608	4 455 e	-	-	-	-	-	-
Other	-	-	-	-	-	-	-	-	-
Latin America	**1 592**	**2 730**	**5 100 e**	**-**	**-**	**-**	**-**	**-**	**-**
Romania	-	-	-	-	-	-	-	-	-
Other	-	-	-	-	-	-	-	-	-
Non-OECD Europe	**-**	**-**	**-**	**-**	**-**	**-**	**-**	**-**	**-**
FSU	**5 767**	**6 067**	**9 185 e**	**-**	**-**	**-**	**844**	**628**	**648 e**
Iran	3 880	4 098	3 272 e	-	-	-	-	14	-
Iraq	5 995	2 568	2 352 e	-	-	-	-	-	-
Kuwait	-	-	-	-	-	-	-	-	-
Oman	-	-	-	-	-	-	-	-	-
Qatar	-	-	-	-	-	-	-	-	-
Saudi Arabia	6 628	6 291	6 750 e	-	-	-	-	-	5 e
United Arab Emirates	-	-	-	-	-	-	-	-	-
Other	654	1 291	1 149 e	-	-	-	-	-	10 e
Middle East	**17 157**	**14 248**	**13 523 e**	**-**	**-**	**-**	**-**	**14**	**15 e**
India	-	-	-	-	-	-	-	-	-
Indonesia	-	-	-	-	-	-	-	-	-
Singapore	-	-	-	-	-	-	-	-	-
Other	-	-	-	-	-	-	-	-	-
Asia	**-**	**-**	**-**	**-**	**-**	**-**	**-**	**-**	**-**
China	**-**	**-**	**-**	**-**	**-**	**-**	**-**	**-**	**-**
Algeria	1 476	1 571	1 058 e	-	-	-	18	15	-
Angola/Cabinada	644	651	131 e	-	-	-	-	-	-
Cameroon	381	1 171	1 602 e	-	-	-	-	-	-
Egypt	274	137	-	-	-	-	52	77	-
Gabon	273	186	68 e	-	-	-	-	-	-
Libya	6 901	7 205	6 469 e	-	-	-	-	6	-
Nigeria	9 165	8 678	5 278 e	-	-	-	-	18	98 e
Tunisia	772	711	696 e	-	-	-	-	-	73 e
Other	2 918	3 326	3 334 e	-	-	-	60	-	-
Africa	**22 804**	**23 636**	**18 636 e**	**-**	**-**	**-**	**130**	**116**	**171 e**
Non-spec./Other	**-**	**-**	**-**	**-**	**-**	**-**	**60**	**78**	**7 e**
Total Imports	**57 475**	**56 792**	**56 449 e**	**-**	**-**	**-**	**1 307**	**1 041**	**1 245 e**
Memo: Total OPEC	*35 607*	*33 019*	*29 634 e*	-	-	-	*18*	*53*	*103 e*

Source: Annual Oil Statistics.

SPAIN
Table 7B: Imports by Country of Origin

000 metric tons

	2000	2001	2002	2000	2001	2002	2000	2001	2002
	Total Products			**Liquefied Petroleum Gases**			**Naphtha**		
Australia	-	-	25 e	-	-	-	-	-	-
Belgium	253	289	56 e	-	-	1 e	-	-	-
Canada	25	28	114 e	-	-	-	-	-	-
Czech Republic	-	-	-	-	-	-	-	-	-
Denmark	192	89	143 e	-	-	-	-	-	-
Finland	43	-	-	-	-	-	-	-	-
France	1 104	834	1 179 e	-	-	4 e	13	28	-
Germany	475	683	454 e	-	-	1 e	-	-	-
Greece	161	278	128 e	-	-	-	56	82	30 e
Hungary	-	-	-	-	-	-	-	-	-
Ireland	-	-	-	-	-	-	-	-	-
Italy	3 740	4 558	4 266 e	-	-	-	11	168	40 e
Japan	1	-	5 e	-	-	-	-	-	-
Korea	-	-	-	-	-	-	-	-	-
Mexico	8	7	47 e	-	-	-	-	-	-
Netherlands	473	493	927 e	-	-	-	-	-	-
New Zealand	-	-	-	-	-	-	-	-	-
Norway	193	123	350 e	-	-	73 e	-	-	-
Poland	21	-	5 e	-	-	-	-	-	-
Spain	-	-	-	-	-	-	-	-	-
Sweden	221	167	478 e	-	-	9 e	-	-	-
Turkey	144	78	45 e	-	-	-	63	39	-
United Kingdom	972	996	1 491 e	-	-	114 e	-	-	-
United States	2 671	3 141	3 281 e	-	-	-	-	-	-
Other	300	309	267 e	-	-	2 e	-	-	-
OECD	**10 997**	**12 073**	**13 261 e**	**-**	**-**	**204 e**	**143**	**317**	**70 e**
Argentina	6	10	-	-	-	-	-	-	-
Brazil	-	9	5 e	-	-	-	-	-	-
Colombia	12	-	-	-	-	-	-	-	-
Ecuador	58	-	-	-	-	-	-	-	-
Netherlands Antilles	-	50	-	-	-	-	-	-	-
Peru	-	-	30 e	-	-	-	-	-	-
Trinidad and Tobago	55	-	6 e	-	-	6 e	-	-	-
Venezuela	443	576	243 e	-	-	-	-	-	-
Other	92	183	344 e	-	-	-	-	-	-
Latin America	**666**	**828**	**628 e**	**-**	**-**	**6 e**	**-**	**-**	**-**
Romania	6	28	435 e	-	-	-	-	-	-
Other	105	147	35 e	-	-	-	-	-	-
Non-OECD Europe	**111**	**175**	**470 e**	**-**	**-**	**-**	**-**	**-**	**-**
FSU	**2 517**	**2 898**	**3 298 e**	**-**	**-**	**-**	**-**	**60**	**83 e**
Iran	47	38	30 e	-	-	-	-	-	-
Iraq	-	-	-	-	-	-	-	-	-
Kuwait	77	4	45 e	-	-	-	-	-	-
Oman	-	-	-	-	-	-	-	-	-
Qatar	2	4	-	-	-	-	-	-	-
Saudi Arabia	109	70	149 e	-	-	-	-	-	-
United Arab Emirates	8	42	80 e	-	-	-	-	-	-
Other	84	282	164 e	-	-	-	80	80	-
Middle East	**327**	**440**	**468 e**	**-**	**-**	**-**	**80**	**80**	**-**
India	-	160	-	-	-	-	-	-	-
Indonesia	-	-	-	-	-	-	-	-	-
Singapore	38	-	-	-	-	-	-	-	-
Other	-	10	-	-	-	-	-	-	-
Asia	**38**	**170**	**-**	**-**	**-**	**-**	**-**	**-**	**-**
China	**1**	**-**	**-**	**-**	**-**	**-**	**-**	**-**	**-**
Algeria	3 667	3 652	3 397 e	1 224	894	861 e	1 572	1 614	1 789 e
Angola/Cabinada	3	-	-	-	-	-	-	-	-
Cameroon	7	-	-	-	-	-	-	-	-
Egypt	209	239	226 e	-	-	-	209	50	24 e
Gabon	-	-	-	-	-	-	-	-	-
Libya	813	638	645 e	-	-	-	378	396	342 e
Nigeria	-	10	-	-	-	-	-	-	-
Tunisia	32	52	105 e	-	-	-	-	52	28 e
Other	99	19	144 e	-	-	18 e	-	-	-
Africa	**4 830**	**4 610**	**4 517 e**	**1 224**	**894**	**879 e**	**2 159**	**2 112**	**2 183 e**
Non-spec./Other	**70**	**-**	**-**	**-**	**-**	**-**	**-**	**-**	**-**
Total Imports	**19 557**	**21 194**	**22 642 e**	**1 224**	**894**	**1 089 e**	**2 382**	**2 569**	**2 336 e**
Memo: Total OPEC	*5 166*	*5 034*	*4 589 e*	*1 224*	*894*	*861 e*	*1 950*	*2 010*	*2 131 e*

Source: Annual Oil Statistics.

SPAIN
Table 7C: Imports by Country of Origin

000 metric tons

	2000	2001	2002	2000	2001	2002	2000	2001	2002
	Total Gasoline			**Total Kerosene**			**Gas/Diesel**		
Australia	-	-	-	-	-	-	-	-	-
Belgium	-	-	-	12	-	-	114	210	28 e
Canada	-	-	-	-	-	-	-	28	62 e
Czech Republic	-	-	-	-	-	-	-	-	-
Denmark	1	2	5 e	-	-	-	164	24	67 e
Finland	-	-	-	-	-	-	41	-	-
France	395	354	304 e	51	7	2 e	385	169	290 e
Germany	-	2	-	-	-	-	124	89	205 e
Greece	44	79	31 e	39	63	-	-	29	40 e
Hungary	-	-	-	-	-	-	-	-	-
Ireland	-	-	-	-	-	-	-	-	-
Italy	326	350	289 e	220	225	196 e	2 661	3 307	3 545 e
Japan	-	-	-	-	-	-	-	-	-
Korea	-	-	-	-	-	-	-	-	-
Mexico	-	-	-	-	-	-	-	-	-
Netherlands	59	23	37 e	18	-	-	227	275	717 e
New Zealand	-	-	-	-	-	-	-	-	-
Norway	41	20	5 e	-	-	-	109	100	42 e
Poland	-	-	-	-	-	-	21	-	-
Spain	-	-	-	-	-	-	-	-	-
Sweden	30	-	-	-	-	-	169	115	253 e
Turkey	-	-	25 e	-	-	-	-	-	-
United Kingdom	37	-	80 e	-	-	-	430	406	506 e
United States	-	-	-	-	-	-	82	72	42 e
Other	-	-	-	-	-	-	55	85	29 e
OECD	**933**	**830**	**776 e**	**340**	**295**	**198 e**	**4 582**	**4 909**	**5 826 e**
Argentina	-	-	-	-	-	-	-	-	-
Brazil	-	-	-	-	-	-	-	-	-
Colombia	-	-	-	12	-	-	-	-	-
Ecuador	-	-	-	-	-	-	-	-	-
Netherlands Antilles	-	-	-	-	25	-	-	25	-
Peru	-	-	-	-	-	-	-	-	-
Trinidad and Tobago	-	-	-	-	-	-	-	-	-
Venezuela	-	-	-	24	42	-	-	-	-
Other	-	-	-	-	-	-	-	8	62 e
Latin America	**-**	**-**	**-**	**36**	**67**	**-**	**-**	**33**	**62 e**
Romania	-	-	37 e	-	-	-	-	27	377 e
Other	-	-	-	40	5	-	63	115	35 e
Non-OECD Europe	**-**	**-**	**37 e**	**40**	**5**	**-**	**63**	**142**	**412 e**
FSU	**-**	**-**	**-**	**-**	**-**	**-**	**1 302**	**1 742**	**2 151 e**
Iran	-	-	-	-	-	-	31	-	-
Iraq	-	-	-	-	-	-	-	-	-
Kuwait	-	-	-	-	4	45 e	77	-	-
Oman	-	-	-	-	-	-	-	-	-
Qatar	-	-	-	-	-	-	-	-	-
Saudi Arabia	-	10	42 e	7	10	36 e	68	20	53 e
United Arab Emirates	-	-	17 e	8	-	32 e	-	42	31 e
Other	-	-	-	-	100	115 e	-	92	-
Middle East	**-**	**10**	**59 e**	**15**	**114**	**228 e**	**176**	**154**	**84 e**
India	-	-	-	-	-	-	-	-	-
Indonesia	-	-	-	-	-	-	-	-	-
Singapore	-	-	-	-	-	-	38	-	-
Other	-	-	-	-	-	-	-	-	-
Asia	**-**	**-**	**-**	**-**	**-**	**-**	**38**	**-**	**-**
China	**-**	**-**	**-**	**-**	**-**	**-**	**-**	**-**	**-**
Algeria	9	-	-	81	263	192 e	729	858	548 e
Angola/Cabinada	-	-	-	-	-	-	-	-	-
Cameroon	-	-	-	-	-	-	-	-	-
Egypt	-	93	51 e	-	71	151 e	-	-	-
Gabon	-	-	-	-	-	-	-	-	-
Libya	-	-	-	-	-	44 e	336	217	232 e
Nigeria	-	-	-	-	-	-	-	-	-
Tunisia	-	-	41 e	32	-	-	-	-	27 e
Other	2	-	3 e	-	-	8 e	18	-	-
Africa	**11**	**93**	**95 e**	**113**	**334**	**395 e**	**1 083**	**1 075**	**807 e**
Non-spec./Other	**-**	**-**	**-**	**-**	**-**	**-**	**-**	**-**	**-**
Total Imports	**944**	**933**	**967 e**	**544**	**815**	**821 e**	**7 244**	**8 055**	**9 342 e**
Memo: Total OPEC	*9*	*10*	*59 e*	*120*	*319*	*349 e*	*1 241*	*1 137*	*864 e*

Source: Annual Oil Statistics.

SPAIN
Table 7D: Imports by Country of Origin

000 metric tons

	2000	2001	2002	2000	2001	2002	2000	2001	2002
	Fuel Oil (Residual)			**Petroleum Coke**			**Other Products**		
Australia	-	-	25 e	-	-	-	-	-	-
Belgium	110	45	-	-	-	-	17	34	27 e
Canada	25	-	52 e	-	-	-	-	-	-
Czech Republic	-	-	-	-	-	-	-	-	-
Denmark	27	61	71 e	-	-	-	-	2	-
Finland	-	-	-	-	-	-	2	-	-
France	35	133	471 e	-	-	-	225	143	108 e
Germany	343	580	244 e	8	6	2 e	-	6	2 e
Greece	-	25	27 e	-	-	-	22	-	-
Hungary	-	-	-	-	-	-	-	-	-
Ireland	-	-	-	-	-	-	-	-	-
Italy	383	373	132 e	22	-	-	117	135	64 e
Japan	-	-	-	1	-	5 e	-	-	-
Korea	-	-	-	-	-	-	-	-	-
Mexico	-	-	47 e	-	-	-	8	7	-
Netherlands	114	101	74 e	4	2	1 e	51	92	98 e
New Zealand	-	-	-	-	-	-	-	-	-
Norway	30	-	218 e	3	1	6 e	10	2	6 e
Poland	-	-	5 e	-	-	-	-	-	-
Spain	-	-	-	-	-	-	-	-	-
Sweden	22	52	216 e	-	-	-	-	-	-
Turkey	-	-	-	-	-	-	81	39	20 e
United Kingdom	204	255	500 e	69	176	154 e	232	159	137 e
United States	34	504	596 e	2 538	2 565	2 643 e	17	-	-
Other	72	80	72 e	-	4	-	173	140	164 e
OECD	**1 399**	**2 209**	**2 750 e**	**2 645**	**2 754**	**2 811 e**	**955**	**759**	**626 e**
Argentina	-	-	-	6	10	-	-	-	-
Brazil	-	-	-	-	-	-	-	9	5 e
Colombia	-	-	-	-	-	-	-	-	-
Ecuador	-	-	-	58	-	-	-	-	-
Netherlands Antilles	-	-	-	-	-	-	-	-	-
Peru	-	-	30 e	-	-	-	-	-	-
Trinidad and Tobago	55	-	-	-	-	-	-	-	-
Venezuela	-	-	-	417	534	243 e	2	-	-
Other	78	49	157 e	-	77	117 e	14	49	8 e
Latin America	**133**	**49**	**187 e**	**481**	**621**	**360 e**	**16**	**58**	**13 e**
Romania	-	-	-	-	-	-	6	1	21 e
Other	-	27	-	-	-	-	2	-	-
Non-OECD Europe	**-**	**27**	**-**	**-**	**-**	**-**	**8**	**1**	**21 e**
FSU	**1 118**	**977**	**670 e**	**-**	**-**	**-**	**97**	**119**	**394 e**
Iran	-	-	-	-	-	-	16	38	30 e
Iraq	-	-	-	-	-	-	-	-	-
Kuwait	-	-	-	-	-	-	-	-	-
Oman	-	-	-	-	-	-	-	-	-
Qatar	-	-	-	-	-	-	2	4	-
Saudi Arabia	-	-	-	-	-	-	34	30	18 e
United Arab Emirates	-	-	-	-	-	-	-	-	-
Other	-	-	27 e	-	-	-	4	10	22 e
Middle East	**-**	**-**	**27 e**	**-**	**-**	**-**	**56**	**82**	**70 e**
India	-	-	-	-	160	-	-	-	-
Indonesia	-	-	-	-	-	-	-	-	-
Singapore	-	-	-	-	-	-	-	-	-
Other	-	-	-	-	10	-	-	-	-
Asia	**-**	**-**	**-**	**-**	**170**	**-**	**-**	**-**	**-**
China	**-**	**-**	**-**	**1**	**-**	**-**	**-**	**-**	**-**
Algeria	27	-	-	-	-	-	25	23	7 e
Angola/Cabinada	-	-	-	-	-	-	3	-	-
Cameroon	-	-	-	-	-	-	7	-	-
Egypt	-	25	-	-	-	-	-	-	-
Gabon	-	-	-	-	-	-	-	-	-
Libya	-	-	-	-	-	-	99	25	27 e
Nigeria	-	10	-	-	-	-	-	-	-
Tunisia	-	-	-	-	-	-	-	-	9 e
Other	79	-	81 e	-	-	-	-	19	34 e
Africa	**106**	**35**	**81 e**	**-**	**-**	**-**	**134**	**67**	**77 e**
Non-spec./Other	**-**	**-**	**-**	**-**	**-**	**-**	**70**	**-**	**-**
Total Imports	**2 756**	**3 297**	**3 715 e**	**3 127**	**3 545**	**3 171 e**	**1 336**	**1 086**	**1 201 e**
Memo: Total OPEC	*27*	*10*	*-*	*417*	*534*	*243 e*	*178*	*120*	*82 e*

Source: Annual Oil Statistics.

SPAIN
Table 8A: Exports by Country of Destination

000 metric tons

	2000	2001	2002	2000	2001	2002	2000	2001	2002
	Crude Oil			Natural Gas Liquids			Feedstocks and Other		
Australia	-	-	-	-	-	-	-	-	-
Austria	-	-	-	-	-	-	-	-	-
Belgium	-	-	-	-	-	-	-	-	-
Canada	-	-	-	-	-	-	-	-	-
Denmark	-	-	-	-	-	-	-	-	-
Finland	-	-	-	-	-	-	-	-	-
France	-	-	-	-	-	-	-	-	-
Germany	-	-	-	-	-	-	-	-	-
Greece	-	-	-	-	-	-	-	-	-
Ireland	-	-	-	-	-	-	-	-	-
Italy	-	-	-	-	-	-	-	-	-
Japan	-	-	-	-	-	-	-	-	-
Luxembourg	-	-	-	-	-	-	-	-	-
Mexico	-	-	-	-	-	-	-	-	-
Netherlands	-	-	-	-	-	-	-	-	-
Norway	-	-	-	-	-	-	-	-	-
Portugal	-	-	-	-	-	-	-	-	-
Spain	-	-	-	-	-	-	-	-	-
Sweden	-	-	-	-	-	-	-	-	-
Switzerland	-	-	-	-	-	-	-	-	-
Turkey	-	-	-	-	-	-	-	-	-
United Kingdom	-	-	-	-	-	-	-	-	-
United States	-	-	-	-	-	-	-	-	-
Other OECD	-	-	-	-	-	-	-	-	-
OECD	**-**	**-**	**-**	**-**	**-**	**-**	**-**	**-**	**-**
Africa	-	-	-	-	-	-	-	-	-
Latin America	-	-	-	-	-	-	-	-	-
Asia	-	-	-	-	-	-	-	-	-
China	-	-	-	-	-	-	-	-	-
Non-OECD Europe	-	-	-	-	-	-	-	-	-
Former Soviet Union	-	-	-	-	-	-	-	-	-
Middle East	-	-	-	-	-	-	-	-	-
Non-specified/Other	-	-	-	-	-	-	-	-	-
Non-OECD	**-**	**-**	**-**	**-**	**-**	**-**	**-**	**-**	**-**
Total Exports	**-**	**-**	**-**	**-**	**-**	**-**	**-**	**-**	**-**
	Total Products			Liquefied Petroleum Gas			Naphtha		
Australia	-	-	37 e	-	-	-	-	-	-
Austria	-	-	-	-	-	-	-	-	-
Belgium	164	154	90 e	-	-	-	164	150	89 e
Canada	105	31	123 e	-	-	-	-	-	-
Denmark	-	-	-	-	-	-	-	-	-
Finland	-	4	3 e	-	4	3 e	-	-	-
France	692	621	839 e	12	1	2 e	166	148	286 e
Germany	7	1	5 e	-	-	-	-	-	-
Greece	81	20	21 e	-	-	-	-	-	-
Ireland	-	-	-	-	-	-	-	-	-
Italy	353	385	306 e	-	2	-	123	11	45 e
Japan	-	-	-	-	-	-	-	-	-
Luxembourg	-	-	-	-	-	-	-	-	-
Mexico	117	-	60 e	-	-	3 e	-	-	-
Netherlands	591	688	869 e	3	-	2 e	363	550	757 e
Norway	-	32	1 e	-	-	-	-	32	-
Portugal	967	503	563 e	14	1	4 e	393	177	215 e
Spain	-	-	-	-	-	-	-	-	-
Sweden	63	38	13 e	-	-	-	20	38	13 e
Switzerland	-	-	-	-	-	-	-	-	-
Turkey	75	28	64 e	-	-	-	-	-	-
United Kingdom	289	292	209 e	-	-	2 e	41	63	-
United States	1 799	1 777	1 050 e	6	17	9 e	-	31	-
Other OECD	-	24	-	-	-	-	-	24	-
OECD	**5 303**	**4 598**	**4 253 e**	**35**	**25**	**25 e**	**1 270**	**1 224**	**1 405 e**
Africa	1 068	739	869 e	66	54	88 e	-	10	22 e
Latin America	123	130	47 e	-	-	2 e	-	-	-
Asia	-	160	15 e	-	-	-	-	-	-
China	-	-	-	-	-	-	-	-	-
Non-OECD Europe	695	426	525 e	9	1	-	74	64	-
Former Soviet Union	-	52	2 e	-	-	2 e	-	-	-
Middle East	141	98	214 e	6	-	-	-	-	-
Non-specified/Other	37	20	1 e	-	3	-	-	-	-
Non-OECD	**2 064**	**1 625**	**1 673 e**	**81**	**58**	**92 e**	**74**	**74**	**22 e**
Total Exports	**7 367**	**6 223**	**5 926 e**	**116**	**83**	**117 e**	**1 344**	**1 298**	**1 427 e**

Source: Annual Oil Statistics.

SPAIN
Table 8B: Exports by Country of Destination

000 metric tons

	2000	2001	2002	2000	2001	2002	2000	2001	2002
	Total Gasoline			**Total Kerosene**			**Gas/Diesel Oil**		
Australia	-	-	-	-	-	8 e	-	-	29 e
Austria	-	-	-	-	-	-	-	-	-
Belgium	-	-	-	-	-	-	-	-	-
Canada	53	31	123 e	-	-	-	-	-	-
Denmark	-	-	-	-	-	-	-	-	-
Finland	-	-	-	-	-	-	-	-	-
France	107	118	147 e	24	23	63 e	179	155	180 e
Germany	-	-	-	-	-	-	-	-	-
Greece	79	-	4 e	-	20	6 e	-	-	11 e
Ireland	-	-	-	-	-	-	-	-	-
Italy	71	135	17 e	-	28	84 e	-	14	47 e
Japan	-	-	-	-	-	-	-	-	-
Luxembourg	-	-	-	-	-	-	-	-	-
Mexico	-	-	57 e	-	-	-	-	-	-
Netherlands	124	119	12 e	-	-	-	-	-	14 e
Norway	-	-	1 e	-	-	-	-	-	-
Portugal	33	2	24 e	8	3	9 e	136	69	75 e
Spain	-	-	-	-	-	-	-	-	-
Sweden	-	-	-	-	-	-	-	-	-
Switzerland	-	-	-	-	-	-	-	-	-
Turkey	10	20	24 e	-	-	10 e	-	-	19 e
United Kingdom	105	97	5 e	24	4	44 e	-	28	91 e
United States	959	1 256	810 e	17	-	15 e	125	57	-
Other OECD	-	-	-	-	-	-	-	-	-
OECD	**1 541**	**1 778**	**1 224 e**	**73**	**78**	**239 e**	**440**	**323**	**466 e**
Africa	504	216	389 e	-	-	1 e	284	152	207 e
Latin America	17	119	25 e	20	-	-	-	-	-
Asia	-	-	-	-	-	-	-	-	-
China	-	-	-	-	-	-	-	-	-
Non-OECD Europe	313	121	360 e	8	9	6 e	124	97	110 e
Former Soviet Union	-	27	-	-	-	-	-	-	-
Middle East	41	60	176 e	-	-	-	-	-	-
Non-specified/Other	34	1	-	3	16	1 e	-	-	-
Non-OECD	**909**	**544**	**950 e**	**31**	**25**	**8 e**	**408**	**249**	**317 e**
Total Exports	**2 450**	**2 322**	**2 174 e**	**104**	**103**	**247 e**	**848**	**572**	**783 e**

	2000	2001	2002	2000	2001	2002	2000	2001	2002
	Fuel Oil (Residual)			**Petroleum Coke**			**Other Products**		
Australia	-	-	-	-	-	-	-	-	-
Austria	-	-	-	-	-	-	-	-	-
Belgium	-	-	-	-	-	-	-	4	1 e
Canada	52	-	-	-	-	-	-	-	-
Denmark	-	-	-	-	-	-	-	-	-
Finland	-	-	-	-	-	-	-	-	-
France	29	16	11 e	9	21	4 e	166	139	146 e
Germany	-	-	-	4	1	5 e	3	-	-
Greece	-	-	-	-	-	-	2	-	-
Ireland	-	-	-	-	-	-	-	-	-
Italy	130	165	-	-	10	12 e	29	20	101 e
Japan	-	-	-	-	-	-	-	-	-
Luxembourg	-	-	-	-	-	-	-	-	-
Mexico	117	-	-	-	-	-	-	-	-
Netherlands	30	-	55 e	-	-	-	71	19	29 e
Norway	-	-	-	-	-	-	-	-	-
Portugal	256	95	73 e	26	33	44 e	101	123	119 e
Spain	-	-	-	-	-	-	-	-	-
Sweden	-	-	-	-	-	-	43	-	-
Switzerland	-	-	-	-	-	-	-	-	-
Turkey	58	-	-	-	8	-	7	-	11 e
United Kingdom	70	30	29 e	22	4	2 e	27	66	36 e
United States	-	148	129 e	2	-	-	690	268	87 e
Other OECD	-	-	-	-	-	-	-	-	-
OECD	**742**	**454**	**297 e**	**63**	**77**	**67 e**	**1 139**	**639**	**530 e**
Africa	125	89	1 e	-	6	7 e	89	212	154 e
Latin America	25	-	-	7	9	10 e	54	2	10 e
Asia	-	156	-	-	-	-	-	4	15 e
China	-	-	-	-	-	-	-	-	-
Non-OECD Europe	107	124	30 e	-	-	-	60	10	19 e
Former Soviet Union	-	25	-	-	-	-	-	-	-
Middle East	55	28	-	-	6	-	39	4	38 e
Non-specified/Other	-	-	-	-	-	-	-	-	-
Non-OECD	**312**	**422**	**31 e**	**7**	**21**	**17 e**	**242**	**232**	**236 e**
Total Exports	**1 054**	**876**	**328 e**	**70**	**98**	**84 e**	**1 381**	**871**	**766 e**

Source: Annual Oil Statistics.

SWEDEN

Figure 1
FUEL SHARES OF TPES *

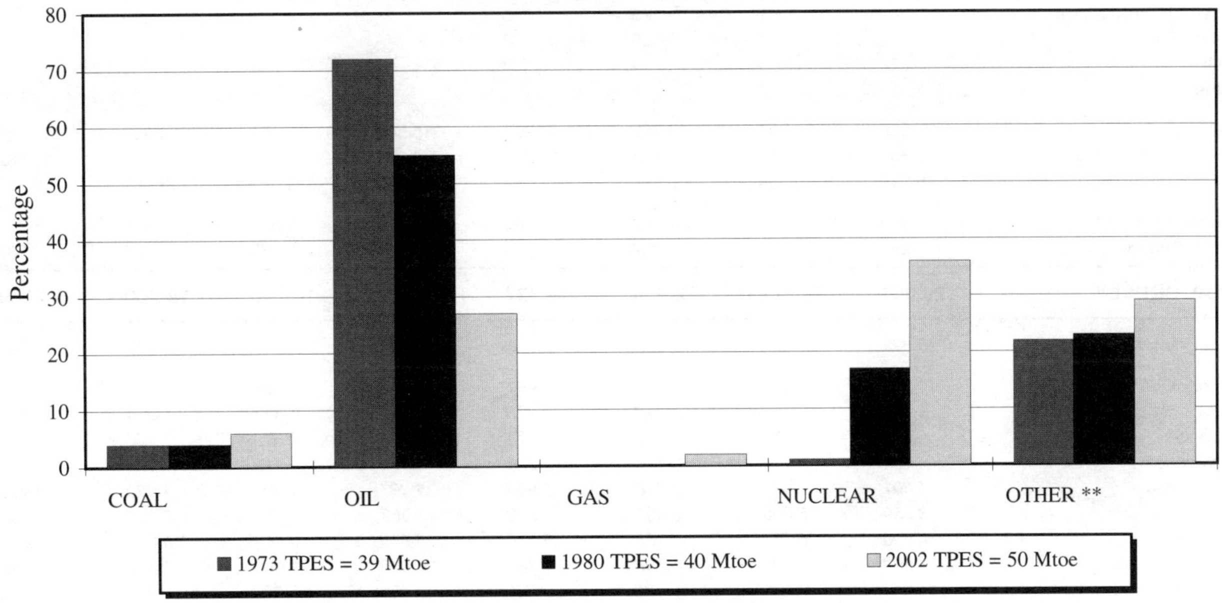

Source: Annual Energy Balances.

* Excluding electricity trade.

** Includes hydro, geothermal, solar, wind, combustible renewables and waste, etc.

Figure 2
OIL PRODUCTION AND CONSUMPTION

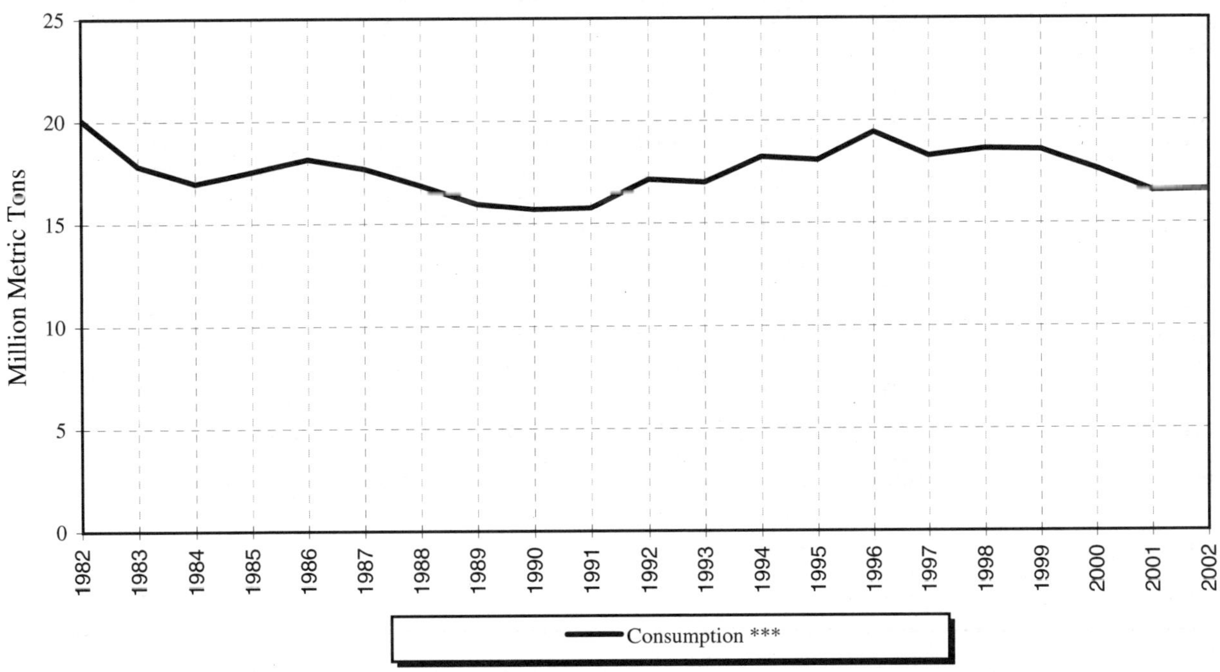

Source: Annual Oil Statistics.

*** Includes refinery fuel and international marine bunkers.

SWEDEN
Table 1: Oil Supply and Consumption - Summary Table

000 metric tons

	1973	1978	1980	1990	1999	2000	2001	2002
Indigenous Production	-	1	25	3	-	-	5	-
Crude Oil	..	1	25	3	-	-	-	-
NGL	..	-	-	-	-	-	-	-
Other	..	-	-	-	-	-	5	-
+ Net Imports	28 625	25 857	25 759	14 954	15 544	15 421	16 053	14 930 e
Crude Oil, NGL, Other	10 606	15 519	17 713	17 016	19 254	20 343	19 504	17 882 e
Petroleum Products	18 019	10 338	8 046	-2 062	-3 710	-4 922	-3 451	-2 952 e
- Intl. Marine Bunkers	1 135	1 111	867	680	1 553	1 396	1 429	1 258 e
+ Stock Changes	564	525	-2 267	50	981	75	101	587 e
TOTAL SUPPLY	**28 054**	**25 272**	**22 650**	**14 327**	**14 972**	**14 100**	**14 730**	**14 259 e**
Refinery Intake	*10 699*	*15 627*	*17 689*	*18 060*	*21 294*	*21 569*	*20 918*	*19 356 e*
Refinery Output	*10 542*	*15 288*	*17 613*	*17 491*	*21 160*	*21 368*	*20 446*	*18 464 e*
LPG + Ethane	*187*	*117*	*110*	*220*	*268*	*311*	*314*	*309 e*
Naphtha	*178*	*65*	*119*	*574*	*1 847*	*1 320*	*1 226 e*	*988 e*
Kerosenes	*134*	*174*	*180*	*284*	*72*	*112*	*143*	*73 e*
Gasolines	*1 301*	*2 437*	*2 745*	*3 993*	*4 303*	*4 394*	*4 020*	*3 933 e*
Gas/Diesel Oil	*3 250*	*5 079*	*6 005*	*6 213*	*7 452*	*7 765*	*7 677*	*6 962 e*
Fuel Oil	*4 500*	*6 777*	*7 758*	*5 070*	*5 673*	*6 403*	*5 919*	*5 053 e*
Other Products	*992*	*639*	*696*	*1 137*	*1 545*	*1 063*	*1 147*	*1 146 e*
- Refinery Losses	157	339	76	569	134	201 e	472 e	892 e
+ Returns	-	197	131	558	1 617	1 344	1 077	807 e
- Statistical Differences	624	- 256	- 919	- 695	- 558	- 993 e	206 e	-1 175 e
TOTAL CONSUMPTION	**27 273**	**25 386**	**23 624**	**15 011**	**17 013**	**16 236 e**	**15 129**	**15 349 e**
Transformation and Energy	**2 714**	**3 657**	**3 348**	**1 527**	**3 093**	**2 434**	**2 298**	..
Crude Oil	-	-	-	-	-	-	-	..
LPG + Ethane	22	33	28	182	244	265	283	..
Gas/Diesel Oil	47	21	28	49	174	130	138	..
Fuel Oil	2 486	3 253	3 069	845	1 275	915	1 011	..
Petroleum Coke	63	92	42	43	40	31	25	..
Other	96	258	181	408	1 360	1 093	841	..
Transport	**5 102**	**5 981**	**5 675**	**6 878**	**7 657**	**7 710**	**7 765**	..
Motor Gasoline	3 108	3 659	3 516	4 166	4 035	3 977	3 961	..
Gas/Diesel Oil	1 333	1 632	1 462	1 830	2 493	2 607	2 614	..
Subtotal	*4 441*	*5 291*	*4 978*	*5 996*	*6 528*	*6 584*	*6 575*	..
Aviation Fuels	561	566	531	744	919	909	925	..
Other	100	124	166	138	210	217	265	..
Industry	**8 328**	**6 670**	**6 231**	**3 376**	**3 490**	**3 160**	**3 161**	..
Naphtha	955	870	650	670	682	689	812	..
LPG + Ethane	141	134	147	526	451	477	549	..
Gas/Diesel Oil	1 145	818	806	437	425	429 e	266	..
Fuel Oil	5 094	4 217	3 882	980	937	891	896	..
Other	993	631	746	763	995	674	638	..
Other Sectors	**11 129**	**9 078**	**8 370**	**3 230**	**2 773**	**2 932**	**1 905**	..
Gas/Diesel Oil	6 942	6 479	5 736	2 784	2 443	2 808 e	1 816 e	..
Fuel Oil	4 030	2 536	2 596	391	289	83	35	..
Other	157	63	38	55	41	41	54	..
Memo: Non-Energy Use	*2 148*	*1 587*	*1 533*	*1 739*	*1 697*	*1 428*	*1 645*	..
Chem/Petchem. Feedst.	*1 070*	*870*	*694*	*903*	*528*	*576*	*817*	..
Other	*1 078*	*717*	*839*	*836*	*1 169*	*852*	*828*	..

Source: Annual Oil Statistics.

SWEDEN
Table 2: Supply of Crude Oil, NGL, Feedstocks and Other Hydrocarbons

000 metric tons

	1973	1978	1980	1990	1999	2000	2001	2002
CRUDE OIL								
Indigenous Production	..	1	25	3	-	-	-	-
+ Imports	..	15 628	17 899	16 865	19 544	20 685	19 891	18 156 e
- Exports	..	-	88	-	-	-	-	-
- Direct Use	..	-	-	-	-	-	-	-
+ Stock Change	..	- 884	-1 117	20	14	- 127	- 192	368 e
- Statistical Difference	..	- 225	- 654	- 459	- 462	- 31	- 470	- 324 e
Refinery Intake	**..**	**14 970**	**17 373**	**17 347**	**20 020**	**20 589**	**20 169**	**18 848 e**
Closing Stock Level	..	*1 293*	*2 340*	*952*	*1 161*	*1 288*	*1 480*	*1 112 e*
NATURAL GAS LIQUIDS								
Indigenous Production	..	-	-	-	-	-	-	-
+ Imports	..	-	-	-	-	-	-	-
- Exports	..	-	-	-	-	-	-	-
- Direct Use	..	-	-	-	-	-	-	-
+ Stock Change	..	-	-	-	-	-	-	-
- Statistical Difference	..	-	-	-	-	-	-	-
Refinery Intake	**..**	**-**	**-**	**-**	**-**	**-**	**-**	**-**
Closing Stock Level	..	-	-	-	-	-	-	-
REFINERY FEEDSTOCKS								
Backflows	..	197	131	558	1 617	1 344	1 077	807 e
+ Transfers	..	636	302	-	-	-	-	-
+ Imports	..	12	-	527	149	204	149	121 e
- Exports	..	121	98	376	439	558	536	398 e
+ Stock Change	..	71	181	- 40	16	15	56	- 13 e
- Statistical Difference	..	138	200	- 44	69	35	3	17 e
Refinery Intake	**..**	**657**	**316**	**713**	**1 274**	**970**	**743**	**500 e**
Closing Stock Level	..	*316*	*162*	*420*	*201*	*186*	*130*	*143 e*
OTHER HYDROCARBONS								
Indigenous Production	..	-	-	-	-	-	5	-
+ Imports	..	-	-	-	-	12	-	3 e
- Exports	..	-	-	-	-	-	-	-
- Direct Use	..	-	-	-	-	-	-	-
+ Stock Change	..	-	-	-	-	- 3	2	-
- Statistical Difference	..	-	-	-	-	- 1	1	- 5 e
Refinery Intake	**..**	**-**	**-**	**-**	**-**	**10**	**6**	**8 e**
Closing Stock Level	..	-	-	-	-	*3*	*1*	*1 e*
CRUDE OIL, NGL, FEEDSTOCKS AND OTHER								
Indigenous Production	-	1	25	3	-	-	5	-
+ Backflows	-	197	131	558	1 617	1 344	1 077	807 e
+ Transfers	-	636	302	-	-	-	-	-
+ Imports	10 606	15 640	17 899	17 392	19 693	20 901	20 040	18 280 e
- Exports	-	121	186	376	439	558	536	398 e
- Direct Use	-	-	-	-	-	-	-	-
+ Stock Change	45	- 813	- 936	- 20	30	- 115	- 134	355 e
- Statistical Difference	- 48	- 87	- 454	- 503	- 393	3	- 466	- 312 e
Refinery Intake	**10 699**	**15 627**	**17 689**	**18 060**	**21 294**	**21 569**	**20 918**	**19 356 e**
Closing Stock Level	-	*1 609*	*2 502*	*1 372*	*1 362*	*1 477*	*1 611*	*1 256 e*

Source: Annual Oil Statistics.

SWEDEN
Table 3: Motor Gasoline - Supply

000 metric tons

	1973	1978	1980	1990	1999	2000	2001	2002
MOTOR GASOLINE								
Refinery Output	**1 301**	**2 437**	**2 745**	**3 741**	**4 303**	**4 394**	**4 020**	**3 933 e**
+ Imports	1 881	1 526	1 511	1 385	1 594	1 586	1 616	1 689 e
- Exports	64	391	526	962	2 062	1 993	1 705	1 585 e
+ Transfers	-	5	1	57	135	- 22	12	59 e
+ Stock Change	- 10	125	- 170	- 25	152	- 32	- 12	3 e
- Statistical Difference	-	43	45	30	87	- 44	- 30	58 e
Total Consumption	**3 108**	**3 659**	**3 516**	**4 166**	**4 035**	**3 977**	**3 961**	**4 041 e**
Closing Stock Level	-	*343*	*231*	*917*	*681*	*713*	*725*	*722 e*

Table 4: Aviation Fuels - Supply

000 metric tons

	1973	1978	1980	1990	1999	2000	2001	2002
AVIATION GASOLINE								
Refinery Output	-	-	-	-	-	-	-	-
+ Imports	31	20	10	11	6	7	3	5 e
- Exports	-	-	-	-	-	-	-	-
+ Transfers	-	- 3	-	-	-	-	-	-
+ Stock Change	-	-	1	- 3	2	- 1	-	- 1 e
- Statistical Difference	-	-	-	4	3	1	- 1	1 e
Total Consumption	**31**	**17**	**11**	**4**	**5**	**5**	**4**	**3 e**
Closing Stock Level	-	*25*	*16*	*5*	*2*	*3*	*3*	*4 e*
GASOLINE TYPE JET FUEL								
Refinery Output	-	-	-	**252**	-	-	-	-
+ Imports	284	215	221	-	-	-	-	-
- Exports	-	-	-	-	-	-	-	-
+ Transfers	-	- 33	9	- 15	-	-	-	-
+ Stock Change	-	25	- 12	21	-	-	-	-
- Statistical Difference	-	- 23	14	49	-	-	-	-
Total Consumption	**284**	**230**	**204**	**209**	-	-	-	-
Closing Stock Level	-	*265*	*290*	*15*	-	-	-	-
KEROSENE TYPE JET FUEL								
Refinery Output	**122**	**173**	**176**	**269**	**72**	**112**	**143**	**73 e**
+ Imports	172	207	256	548	694	754	838	516 e
- Exports	10	4	-	85	22	45	-	17 e
+ Transfers	-	- 39	- 65	- 75	11	41	- 8	-
+ Stock Change	- 38	- 22	- 68	- 21	13	- 12	- 29	27 e
- Statistical Difference	-	- 4	- 17	105	- 146	- 54	23	- 175 e
Total Consumption	**246**	**319**	**316**	**531**	**914**	**904**	**921**	**774 e**
Closing Stock Level	-	*74*	*139*	*182*	*101*	*113*	*142 e*	*115 e*

Source: Annual Oil Statistics.

SWEDEN
Table 5A: Gas/Diesel Oil - Supply and Consumption

000 metric tons

	1973	1978	1980	1990	1999	2000	2001	2002
Refinery Output	**3 250**	**5 079**	**6 005**	**6 213**	**7 452**	**7 765**	**7 677**	**6 962 e**
+ Recycled Products	-	-	-	-	-	-	-	-
+ Imports	6 798	4 385	4 268	1 640	1 548	1 282	1 267	1 956 e
- Exports	444	599	1 112	3 141	3 651	3 859	3 771	3 672 e
- Intl. Marine Bunkers	456	172	138	150	216	166	149	147 e
+ Transfers	-	- 233	- 121	72	119	187 e	265	- 6 e
+ Stock Changes	319	278	-1 116	64	255	- 39	280	93 e
- Statistical Differences	-	- 212	- 246	- 402	- 28	- 804 e	735	34 e
Total Consumption	**9 467**	**8 950**	**8 032**	**5 100**	**5 535**	**5 974 e**	**4 834**	**5 152 e**
Transformation	**47**	**18**	**20**	**48**	**174**	**130**	**138**	**..**
Public Electricity	47	18	20	2	4	12	9	..
Autoproducer Electricity	-	-	-	-	-	-	-	..
Public CHP Plants	-	-	-	12	80	30	47	..
Autoproducers of CHP	-	-	-	7	-	1	2	..
Public Heat Plants	-	-	-	16	75	74	73	..
Autoproducer Heat Plants	-	-	-	-	-	-	-	..
Petrochemical Industry	-	-	-	11	15	13	7	..
Other Transformation	-	-	-	-	-	-	-	..
Energy Sector	**-**	**3**	**8**	**1**	**-**	**-**	**-**	**..**
Refinery Fuel	-	1	-	-	-	-	-	..
Other Energy Sector	-	2	8	1	-	-	-	..
Final Consumption	**9 420**	**8 929**	**8 004**	**5 051**	**5 361**	**5 844 e**	**4 696**	**..**
Transport	**1 333**	**1 632**	**1 462**	**1 830**	**2 493**	**2 607**	**2 614**	**..**
Air	-	-	-	-	-	-	-	..
Road	1 116	1 491	1 388	1 676	2 382	2 467	2 478	..
Rail	87	74	18	39	8	24	24	..
Inland Waterways	130	67	56	86	103	116	112	..
Non-Specified	-	-	-	29	-	-	-	..
Industry	**1 145**	**818**	**806**	**437**	**425**	**429 e**	**266**	**..**
Iron and Steel	107	78	65	32	22	25	18	..
Chemical (incl. Pet.-Chem.)	74	59	41	20	8	7	7	..
Non-Ferrous Metals	23	24	25	13	8	9	7	..
Non-Metallic Minerals	160	92	71	32	38	35	31	..
Transport Equipment	47	51	54	-	23	12	24	..
Machinery	153	162	178	125	78	69	51	..
Mining and Quarrying	10	10	13	8	11	8	13	..
Food and Tobacco	69	75	64	40	60	45	61	..
Pulp, Paper and Printing	20	24	31	26	20	25	9	..
Wood and Wood Products	30	39	38	9	12	7	10	..
Construction	-	-	-	-	-	-	-	..
Textiles and Leather	34	21	25	6	12	-	10	..
Non-Specified	418	175	201	128	133	187 e	25	..
Other Sectors	**6 942**	**6 479**	**5 736**	**2 784**	**2 443**	**2 808 e**	**1 816 e**	**..**
Commerce and Public	-	-	-	1 045	1 118	1 288 e	710 e	..
Residential	6 526	6 121	5 228	1 382	1 003	1 105 e	803	..
Agriculture	347	315	466	357	321	376 e	303	..
Non-Specified	69	43	42	-	1	39	-	..
Memo: Non-Energy Use	-	-	-	-	-	-	-	..
Chem/Petchem. Feedst.	-	-	-	-	-	-	-	..
Other	-	-	-	-	-	-	-	..
Closing Stock Level	-	*2 710*	*2 190*	*1 394*	*1 215*	*1 340*	*1 060*	*967 e*

Source: Annual Oil Statistics.

SWEDEN
Table 5B: Fuel Oil (Residual) - Supply and Consumption

000 metric tons

	1973	1978	1980	1990	1999	2000	2001	2002
Refinery Output	**4 500**	**6 777**	**7 758**	**5 070**	**5 673**	**6 403**	**5 919**	**5 053 e**
+ Recycled Products	-	-	-	-	-	-	-	-
+ Imports	8 616	5 090	5 182	1 017	475	519	715	495 e
- Exports	494	1 552	2 652	3 208	2 752	3 538	3 041	2 601 e
- Intl. Marine Bunkers	679	939	729	530	1 337	1 227	1 277	1 106 e
+ Transfers	-	- 146	- 39	- 3	- 246	- 221 e	- 316	- 56 e
+ Stock Changes	342	896	44	93	623	249 e	115	- 30 e
- Statistical Differences	670	93	- 48	164	- 101	257	130	- 387 e
Total Consumption	**11 615**	**10 033**	**9 612**	**2 275**	**2 537**	**1 928**	**1 985**	**2 142 e**
Transformation	**2 386**	**2 827**	**2 718**	**413**	**738**	**382**	**557**	..
Public Electricity	2 386	2 808	125	17	62 e	10	7	..
Autoproducer Electricity	-	-	-	-	-	-	-	..
Public CHP Plants	-	-	2 102	122	248 e	146	189	..
Autoproducers of CHP	-	-	475	70	200 e	80	154	..
Public Heat Plants	-	-	-	164	152	92	151	..
Autoproducer Heat Plants	-	-	-	-	-	-	-	..
Petrochemical Industry	-	19	16	40	76	54	56	..
Other Transformation	-	-	-	-	-	-	-	..
Energy Sector	**100**	**426**	**351**	**432**	**537**	**533**	**454**	..
Refinery Fuel	96	423	348	431	535	533	454	..
Other Energy Sector	4	3	3	1	2	-	-	..
Final Consumption	**9 129**	**6 780**	**6 543**	**1 430**	**1 262**	**1 013**	**974**	..
Transport	**5**	**27**	**65**	**59**	**36**	**39**	**43**	..
Air	-	-	-	-	-	-	-	..
Road	-	-	-	-	-	-	-	..
Rail	2	11	10	-	-	-	-	..
Inland Waterways	3	16	55	59	36	39	43	..
Non-Specified	-	-	-	-	-	-	-	..
Industry	**5 094**	**4 217**	**3 882**	**980**	**937**	**891**	**896**	..
Iron and Steel	775	571	474	138	128	123	124	..
Chemical (incl. Pet.-Chem.)	425	490	484	69	40	36	59	..
Non-Ferrous Metals	44	49	51	10	8	11	9	..
Non-Metallic Minerals	750	454	412	105	76	99	70	..
Transport Equipment	198	172	184	23	23	10	18	..
Machinery	286	244	202	70	24	27	22	..
Mining and Quarrying	160	108	144	41	53	43	46	..
Food and Tobacco	406	364	360	123	89	72	68	..
Pulp, Paper and Printing	1 796	1 546	1 373	343	446	434	437	..
Wood and Wood Products	120	113	106	34	20	18	16	..
Construction	-	-	-	-	-	-	-	..
Textiles and Leather	131	104	89	23	15	9	15	..
Non-Specified	3	2	3	1	15	9	12	..
Other Sectors	**4 030**	**2 536**	**2 596**	**391**	**289**	**83**	**35**	..
Commerce and Public	-	-	-	247	212 e	58	11	..
Residential	2 055	1 386	1 371	130	31 e	11	16	..
Agriculture	81	114	96	14	42 e	9	8	..
Non-Specified	1 894	1 036	1 129	-	4 e	5	-	..
Memo: Non-Energy Use	*113*	-	*51*	*44*	*38*	*36*	*59*	..
Chem/Petchem. Feedst.	*113*	-	*51*	*44*	*38*	*36*	*59*	..
Other	-	-	-	-	-	-	-	..
Closing Stock Level	-	*5 081*	*3 414*	*1 929*	*1 369*	*1 120 e*	*1 005 e*	*1 035 e*

Source: Annual Oil Statistics.

SWEDEN
Table 5C: Total Products - Supply and Consumption

000 metric tons

	1973	1978	1980	1990	1999	2000	2001	2002
Refinery Output	**10 542**	**15 288**	**17 613**	**17 491**	**21 160**	**21 368**	**20 446**	**18 464 e**
+ Recycled Products	-	-	-	-	-	-	-	-
+ Imports	19 417	13 198	12 851	6 193	5 571	5 453	5 984	6 050 e
- Exports	1 398	2 860	4 805	8 255	9 281	10 375	9 435	9 002 e
- Intl. Marine Bunkers	1 135	1 111	867	680	1 553	1 396	1 429	1 258 e
+ Transfers	-	- 636	- 302	-	-	-	-	-
+ Stock Changes	519	1 338	-1 331	70	951	190	235	232 e
- Statistical Differences	672	- 169	- 465	- 192	- 165	- 996 e	672	- 863 e
Total Consumption	**27 273**	**25 386**	**23 624**	**15 011**	**17 013**	**16 236 e**	**15 129**	**15 349 e**
Transformation	**2 614**	**3 228**	**2 989**	**1 094**	**2 556**	**1 882**	**1 823**	..
Public Electricity	2 433	2 826	145	21	66	22	16	..
Autoproducer Electricity	-	-	-	-	-	-	-	..
Public CHP Plants	-	-	2 102	151	343	182	247	..
Autoproducers of CHP	-	-	475	87	200	81	156	..
Public Heat Plants	-	-	-	202	246	178	240	..
Autoproducer Heat Plants	-	-	-	-	-	-	-	..
Petrochemical Industry	-	197	131	558	1 617	1 344	1 077	..
Other Transformation	181	205	136	75	84	75	87	..
Energy Sector	**100**	**429**	**359**	**433**	**537**	**552**	**475**	..
Refinery Fuel	96	424	348	431	535	552	475	..
Other Energy Sector	4	5	11	2	2	-	-	..
Final Consumption	**24 559**	**21 729**	**20 276**	**13 484**	**13 920**	**13 802**	**12 831**	..
Transport	**5 102**	**5 981**	**5 675**	**6 878**	**7 657**	**7 710**	**7 765**	..
Air	561	566	531	744	919	909	925	..
Road	4 303	5 231	4 987	5 908	6 556	6 586	6 591	..
Rail	89	85	28	39	8	24	24	..
Inland Waterways	149	99	129	158	174	191	193	..
Non-Specified	-	-	-	29	-	-	32	..
Industry	**8 328**	**6 670**	**6 231**	**3 376**	**3 490**	**3 160**	**3 161**	..
Iron and Steel	943	697	595	278	309	301	303	..
Chemical (incl. Pet.-Chem.)	1 456	1 419	1 168	979	839	897	1 122	..
Non-Ferrous Metals	77	87	93	38	28	36	23	..
Non-Metallic Minerals	935	576	514	191	171	174	130	..
Transport Equipment	245	223	238	23	63	41	58	..
Machinery	466	426	400	246	118	111	91	
Mining and Quarrying	170	126	157	47	64	51	61	..
Food and Tobacco	478	443	428	186	180	133	150	..
Pulp, Paper and Printing	1 825	1 583	1 419	406	500	500	485	..
Wood and Wood Products	150	152	144	56	32	26	26	..
Construction	826	477	597	620	813	488	438	..
Textiles and Leather	169	129	118	38	38	18	35	..
Non-Specified	588	332	360	268	335	384	239	..
Other Sectors	**11 129**	**9 078**	**8 370**	**3 230**	**2 773**	**2 932**	**1 905**	..
Commerce and Public	28	13	14	1 341	1 368	1 383	771	..
Residential	8 700	7 557	6 623	1 518	1 037	1 120	823	..
Agriculture	438	429	562	371	363	385	311	..
Non-Specified	1 963	1 079	1 171	-	5	44	-	..
Memo: Non-Energy Use	*2 148*	*1 587*	*1 533*	*1 739*	*1 697*	*1 428*	*1 645*	..
Chem/Petchem. Feedst.	*1 070*	*870*	*694*	*903*	*528*	*576*	*817*	..
Other	*1 078*	*717*	*839*	*836*	*1 169*	*852*	*828*	..
Closing Stock Level	-	*8 747*	*6 456*	*4 806*	*3 823*	*3 719*	*3 488*	*3 256 e*

Source: Annual Oil Statistics.

SWEDEN
Table 6A: Supply and Consumption

000 metric tons

	1980	2000	2001	2002	1980	2000	2001	2002
	LPG AND ETHANE				REFINERY GAS			
Refinery Output	**110**	**311**	**314**	**309 e**	**-**	**-**	**-**	**-**
+ Recycled Products	-	-	-	-	-	-	-	-
+ Imports	131	724	900	784 e	-	-	-	-
- Exports	7	261	226	332 e	-	-	-	-
- Intl. Marine Bunkers	-	-	-	-	-	-	-	-
+ Transfers	-	- 4	- 7	-	-	-	-	-
+ Stock Change	- 20	52	- 105	139 e	-	-	-	-
- Statistical Difference	24	41	- 40	- 58 e	-	-	-	-
Total Consumption	**190**	**781**	**916**	**958 e**	**-**	**-**	**-**	**-**
of which: Ethane	-	-	*32*	*45 e*	-	-	-	-
Transformation	28	265	283	..	-	-	-	..
Energy	-	-	-	..	-	-	-	..
Transport	1	-	32	..	-	-	-	..
Industry	147	477	549	..	-	-	-	..
Other	14	39	52	..	-	-	-	..
Memo: Non-Energy Use	-	*165*	*244*	..	-	-	-	..
Closing Stock Level	*64*	*205*	*314*	*175 e*	-	-	-	-
	NAPHTHA				OTHER KEROSENE			
Refinery Output	**119**	**1 320**	**1 226 e**	**988 e**	**4**	**-**	**-**	**-**
+ Recycled Products	-	-	-	-	-	-	-	-
+ Imports	721	529	571	555 e	23	-	1	-
- Exports	13	358	334	336 e	47	-	-	-
- Intl. Marine Bunkers	-	-	-	-	-	-	-	-
+ Transfers	- 22	14	44	6 e	- 16	-	-	-
+ Stock Change	19	34	37	25 e	3	20	-	- 21 e
- Statistical Difference	- 7	- 243	- 109	- 150 e	- 64	18	- 1	- 21 e
Total Consumption	**831**	**1 782**	**1 653**	**1 388 e**	**31**	**2**	**2**	**-**
Transformation	181	1 074	820	..	-	-	-	..
Energy	-	19	21	..	-	-	-	..
Transport	-	-	-	..	7	-	-	..
Industry	650	689	812	..	-	-	-	..
Other	-	-	-	..	24	2	2	..
Memo: Non-Energy Use	*643*	*375*	*514*	..	-	-	-	..
Closing Stock Level	*28*	*120*	*83*	*58 e*	*- 2*	-	-	*21 e*
	WHITE SPIRIT + SBP				LUBRICANTS			
Refinery Output	**68**	**-**	**-**	**..**	**41**	**335**	**340**	**..**
+ Recycled Products	-	-	-	..	-	-	-	..
+ Imports	61	6	1	..	284	4	13	..
- Exports	248	1	1	..	64	5	4	..
- Intl. Marine Bunkers	-	-	-	..	-	3	3	..
+ Transfers	- 14	-	10	..	- 77	3	3	..
+ Stock Change	- 4	-	-	..	- 9	- 46	8	..
- Statistical Difference	- 177	- 1	1	..	- 17	- 68	- 24	..
Total Consumption	**40**	**6**	**9**	**..**	**192**	**356**	**381**	**..**
Transformation	-	-	-	..	-	-	-	..
Energy	-	-	-	..	-	-	-	..
Transport	-	-	-	..	93	178	190	..
Industry	40	6	9	..	99	178	191	..
Other	-	-	-	..	-	-	-	..
Memo: Non-Energy Use	*40*	*6*	*9*	..	*192*	*356*	*381*	..
Closing Stock Level	*33*	-	-	..	*46*	*46*	*38*	..

Source: Annual Oil Statistics.

SWEDEN
Table 6B: Supply and Consumption

000 metric tons

	1980	2000	2001	2002	1980	2000	2001	2002
	BITUMEN				PARAFFIN WAXES			
Refinery Output	**561**	**727**	**807**	**..**	**-**	**-**	**-**	**..**
+ Recycled Products	-	-	-	..	-	-	-	..
+ Imports	75	16	26	..	11	1	-	..
- Exports	72	315 e	353 e	..	1	-	-	..
- Intl. Marine Bunkers	-	-	-	..	-	-	-	..
+ Transfers	- 1	2	- 3	..	-	-	-	..
+ Stock Change	-	- 40	- 51	..	-	-	-	..
- Statistical Difference	2	- 97	- 12	..	-	-	-	..
Total Consumption	**561**	**487**	**438**	**..**	**10**	**1**	**-**	**..**
Transformation	-	-	-	..	-	-	-	..
Energy	-	-	-	..	-	-	-	..
Transport	-	-	-	..	-	-	-	..
Industry	561	487	438	..	10	1	-	..
Other	-	-	-	..	-	-	-	..
Memo: Non-Energy Use	*561*	*487*	*438*	..	*10*	*1*	-	..
Closing Stock Level	-	*40*	*91*	..	-	-	-	..
	PETROLEUM COKE				OTHER PRODUCTS			
Refinery Output	**-**	**-**	**-**	**-**	**26**	**1**	**-**	**1 146 e**
+ Recycled Products	-	-	-	-	-	-	-	-
+ Imports	42	25	33	-	55	-	-	50 e
- Exports	-	-	-	-	63	-	-	459 e
- Intl. Marine Bunkers	-	-	-	-	-	-	-	5 e
+ Transfers	-	-	-	-	43	-	-	- 3 e
+ Stock Change	-	5	- 8	16 e	1	-	-	- 19 e
- Statistical Difference	-	- 1	-	-	26	- 1	-	- 165 e
Total Consumption	**42**	**31**	**25**	**16 e**	**36**	**2**	**-**	**875 e**
Transformation	42	31	25	..	-	-	-	..
Energy	-	-	-	..	-	-	-	..
Transport	-	-	-	..	-	-	-	..
Industry	-	-	-	..	36	2	-	..
Other	-	-	-	..	-	-	-	..
Memo: Non-Energy Use	-	-	-	..	*36*	*2*	-	..
Closing Stock Level	-	*19*	*27*	*11 e*	*7*	-	-	*148 e*

Source: Annual Oil Statistics.

SWEDEN
Table 7A: Imports by Country of Origin

000 metric tons

	2000	2001	2002	2000	2001	2002	2000	2001	2002
	Crude Oil			Natural Gas Liquids			Feedstocks and Other		
Australia	-	-	-	-	-	-	-	-	-
Belgium	-	-	-	-	-	-	72	72	35 e
Canada	-	-	-	-	-	-	-	-	-
Czech Republic	-	-	-	-	-	-	-	-	-
Denmark	3 396	2 397	2 795 e	-	-	-	-	-	-
Finland	-	-	-	-	-	-	23	9	2 e
France	10	-	8 e	-	-	-	3	-	-
Germany	1 310	2	12 e	-	-	-	5	-	-
Greece	-	-	-	-	-	-	-	-	-
Hungary	-	-	-	-	-	-	-	-	-
Ireland	-	-	-	-	-	-	-	-	-
Italy	-	-	-	-	-	-	-	-	-
Japan	-	-	-	-	-	-	-	-	-
Korea	-	-	-	-	-	-	-	-	-
Mexico	-	-	-	-	-	-	-	-	-
Netherlands	249	70	46 e	-	-	-	2	1	17 e
New Zealand	-	-	-	-	-	-	-	-	-
Norway	9 601	9 143	6 098 e	-	-	-	45	9	11 e
Poland	70	-	-	-	-	-	-	-	-
Spain	23	-	-	-	-	-	5	-	-
Sweden	-	-	-	-	-	-	-	-	-
Turkey	213	-	-	-	-	-	-	-	-
United Kingdom	510	1 808	2 411 e	-	-	-	17	34	48 e
United States	-	-	-	-	-	-	-	-	-
Other	-	-	-	-	-	-	-	-	-
OECD	**15 382**	**13 420**	**11 370 e**	**-**	**-**	**-**	**172**	**125**	**113 e**
Argentina	-	-	-	-	-	-	-	-	-
Brazil	-	-	-	-	-	-	-	-	-
Colombia	-	-	-	-	-	-	-	-	-
Ecuador	-	-	-	-	-	-	-	-	-
Netherlands Antilles	-	-	-	-	-	-	-	-	-
Peru	-	-	-	-	-	-	-	-	-
Trinidad and Tobago	-	-	-	-	-	-	-	-	-
Venezuela	1 030	835	916 e	-	-	-	-	-	-
Other	-	-	-	-	-	-	-	-	-
Latin America	**1 030**	**835**	**916 e**	**-**	**-**	**-**	**-**	**-**	**-**
Romania	-	-	-	-	-	-	-	-	-
Other	-	-	-	-	-	-	-	-	-
Non-OECD Europe	**-**	**-**	**-**	**-**	**-**	**-**	**-**	**-**	**-**
FSU	**1 381**	**1 130**	**3 681 e**	**-**	**-**	**-**	**44**	**24**	**11 e**
Iran	1 441	3 081	1 956 e	-	-	-	-	-	-
Iraq	323	-	-	-	-	-	-	-	-
Kuwait	-	-	-	-	-	-	-	-	-
Oman	-	-	-	-	-	-	-	-	-
Qatar	-	-	-	-	-	-	-	-	-
Saudi Arabia	832	1 146	-	-	-	-	-	-	-
United Arab Emirates	-	-	-	-	-	-	-	-	-
Other	-	-	-	-	-	-	-	-	-
Middle East	**2 596**	**4 227**	**1 956 e**	**-**	**-**	**-**	**-**	**-**	**-**
India	-	-	-	-	-	-	-	-	-
Indonesia	-	-	-	-	-	-	-	-	-
Singapore	-	-	-	-	-	-	-	-	-
Other	-	-	-	-	-	-	-	-	-
Asia	**-**	**-**	**-**	**-**	**-**	**-**	**-**	**-**	**-**
China	**-**	**-**	**-**	**-**	**-**	**-**	**-**	**-**	**-**
Algeria	-	27	-	-	-	-	-	-	-
Angola/Cabinada	-	-	-	-	-	-	-	-	-
Cameroon	-	-	-	-	-	-	-	-	-
Egypt	30	-	-	-	-	-	-	-	-
Gabon	-	-	-	-	-	-	-	-	-
Libya	-	-	42 e	-	-	-	-	-	-
Nigeria	266	252	128 e	-	-	-	-	-	-
Tunisia	-	-	-	-	-	-	-	-	-
Other	-	-	63 e	-	-	-	-	-	-
Africa	**296**	**279**	**233 e**	**-**	**-**	**-**	**-**	**-**	**-**
Non-spec./Other	**-**	**-**	**-**	**-**	**-**	**-**	**-**	**-**	**-**
Total Imports	**20 685**	**19 891**	**18 156 e**	**-**	**-**	**-**	**216**	**149**	**124 e**
Memo: Total OPEC	*3 892*	*5 341*	*3 042 e*	-	-	-	-	-	-

Source: Annual Oil Statistics.

SWEDEN
Table 7B: Imports by Country of Origin

000 metric tons

	2000	2001	2002	2000	2001	2002	2000	2001	2002
	Total Products			**Liquefied Petroleum Gases**			**Naphtha**		
Australia	-	-	-	-	-	-	-	-	-
Belgium	-	3	3 e	-	-	-	-	-	-
Canada	-	-	-	-	-	-	-	-	-
Czech Republic	-	-	-	-	-	-	-	-	-
Denmark	1 111	947	860 e	40	37	22 e	7	11	28 e
Finland	1 550	1 571	1 271 e	3	1	-	-	-	26 e
France	60	75	-	-	-	-	35	75	-
Germany	664	458	226 e	20	17	21 e	131	12	20 e
Greece	-	-	-	-	-	-	-	-	-
Hungary	-	-	-	-	-	-	-	-	-
Ireland	-	25	-	-	-	-	-	-	-
Italy	1	-	-	-	-	-	-	-	-
Japan	-	-	-	-	-	-	-	-	-
Korea	-	-	-	-	-	-	-	-	-
Mexico	-	-	-	-	-	-	-	-	-
Netherlands	76	42	4 e	-	-	-	-	-	-
New Zealand	-	-	-	-	-	-	-	-	-
Norway	442	963	1 805 e	227	549	418 e	88	179	192 e
Poland	110	212	103 e	-	-	-	-	-	-
Spain	39	37	35 e	-	-	-	-	37	35 e
Sweden	-	-	-	-	-	-	-	-	-
Turkey	-	-	-	-	-	-	-	-	-
United Kingdom	556	625	504 e	329	289	278 e	107	17	-
United States	25	55	-	-	-	-	-	-	-
Other	1	1	44 e	-	-	-	-	-	19 e
OECD	**4 635**	**5 014**	**4 855 e**	**619**	**893**	**739 e**	**368**	**331**	**320 e**
Argentina	-	-	-	-	-	-	-	-	-
Brazil	-	-	-	-	-	-	-	-	-
Colombia	-	-	-	-	-	-	-	-	-
Ecuador	-	-	-	-	-	-	-	-	-
Netherlands Antilles	-	-	-	-	-	-	-	-	-
Peru	-	-	-	-	-	-	-	-	-
Trinidad and Tobago	-	-	-	-	-	-	-	-	-
Venezuela	-	-	7 e	-	-	-	-	-	-
Other	6	-	-	-	-	-	-	-	-
Latin America	**6**	**-**	**7 e**	**-**	**-**	**-**	**-**	**-**	**-**
Romania	-	-	-	-	-	-	-	-	-
Other	-	-	-	-	-	-	-	-	-
Non-OECD Europe	**-**	**-**	**-**	**-**	**-**	**-**	**-**	**-**	**-**
FSU	**711**	**903**	**1 080 e**	**4**	**7**	**-**	**161**	**206**	**235 e**
Iran	-	-	-	-	-	-	-	-	-
Iraq	-	-	-	-	-	-	-	-	-
Kuwait	-	33	-	-	-	-	-	-	-
Oman	-	-	-	-	-	-	-	-	-
Qatar	-	-	-	-	-	-	-	-	-
Saudi Arabia	-	-	-	-	-	-	-	-	-
United Arab Emirates	-	-	-	-	-	-	-	-	-
Other	-	-	-	-	-	-	-	-	-
Middle East	**-**	**33**	**-**	**-**	**-**	**-**	**-**	**-**	**-**
India	-	-	-	-	-	-	-	-	-
Indonesia	-	-	-	-	-	-	-	-	-
Singapore	-	-	-	-	-	-	-	-	-
Other	-	-	-	-	-	-	-	-	-
Asia	**-**	**-**	**-**	**-**	**-**	**-**	**-**	**-**	**-**
China	**-**	**-**	**-**	**-**	**-**	**-**	**-**	**-**	**-**
Algeria	101	-	45 e	101	-	45 e	-	-	-
Angola/Cabinada	-	-	-	-	-	-	-	-	-
Cameroon	-	-	-	-	-	-	-	-	-
Egypt	-	-	-	-	-	-	-	-	-
Gabon	-	-	-	-	-	-	-	-	-
Libya	-	-	63 e	-	-	-	-	-	-
Nigeria	-	-	-	-	-	-	-	-	-
Tunisia	-	-	-	-	-	-	-	-	-
Other	-	34	-	-	-	-	-	34	-
Africa	**101**	**34**	**108 e**	**101**	**-**	**45 e**	**-**	**34**	**-**
Non-spec./Other	**-**	**-**	**-**	**-**	**-**	**-**	**-**	**-**	**-**
Total Imports	**5 453**	**5 984**	**6 050 e**	**724**	**900**	**784 e**	**529**	**571**	**555 e**
Memo: Total OPEC	*101*	*33*	*115 e*	*101*	*-*	*45 e*	*-*	*-*	*-*

Source: Annual Oil Statistics.

SWEDEN
Table 7C: Imports by Country of Origin

000 metric tons

	2000	2001	2002	2000	2001	2002	2000	2001	2002
	Total Gasoline			Total Kerosene			Gas/Diesel		
Australia	-	-	-	-	-	-	-	-	-
Belgium	-	-	-	-	1	-	-	-	-
Canada	-	-	-	-	-	-	-	-	-
Czech Republic	-	-	-	-	-	-	-	-	-
Denmark	830	712	613 e	135	120	142 e	70	6	22 e
Finland	628	782	640 e	172	-	21 e	740	758	540 e
France	-	-	-	-	-	-	-	-	-
Germany	2	19	4 e	-	11	-	252	13	15 e
Greece	-	-	-	-	-	-	-	-	-
Hungary	-	-	-	-	-	-	-	-	-
Ireland	-	-	-	-	-	-	-	-	-
Italy	-	-	-	-	-	-	-	-	-
Japan	-	-	-	-	-	-	-	-	-
Korea	-	-	-	-	-	-	-	-	-
Mexico	-	-	-	-	-	-	-	-	-
Netherlands	36	11	4 e	32	6	-	7	23	-
New Zealand	-	-	-	-	-	-	-	-	-
Norway	73	74	313 e	22	31	71 e	20	121	771 e
Poland	11	-	-	96	212	101 e	-	-	-
Spain	-	-	-	-	-	-	-	-	-
Sweden	-	-	-	-	-	-	-	-	-
Turkey	-	-	-	-	-	-	-	-	-
United Kingdom	13	21	116 e	66	80	26 e	6	35	36 e
United States	-	-	-	-	-	-	-	22	-
Other	-	-	4 e	-	-	10 e	-	-	11 e
OECD	**1 593**	**1 619**	**1 694 e**	**523**	**461**	**371 e**	**1 095**	**978**	**1 395 e**
Argentina	-	-	-	-	-	-	-	-	-
Brazil	-	-	-	-	-	-	-	-	-
Colombia	-	-	-	-	-	-	-	-	-
Ecuador	-	-	-	-	-	-	-	-	-
Netherlands Antilles	-	-	-	-	-	-	-	-	-
Peru	-	-	-	-	-	-	-	-	-
Trinidad and Tobago	-	-	-	-	-	-	-	-	-
Venezuela	-	-	-	-	-	7 e	-	-	-
Other	-	-	-	-	-	-	-	-	-
Latin America	**-**	**-**	**-**	**-**	**-**	**7 e**	**-**	**-**	**-**
Romania	-	-	-	-	-	-	-	-	-
Other	-	-	-	-	-	-	-	-	-
Non-OECD Europe	**-**	**-**	**-**	**-**	**-**	**-**	**-**	**-**	**-**
FSU	**-**	**-**	**-**	**231**	**345**	**138 e**	**187**	**289**	**561 e**
Iran	-	-	-	-	-	-	-	-	-
Iraq	-	-	-	-	-	-	-	-	-
Kuwait	-	-	-	-	33	-	-	-	-
Oman	-	-	-	-	-	-	-	-	-
Qatar	-	-	-	-	-	-	-	-	-
Saudi Arabia	-	-	-	-	-	-	-	-	-
United Arab Emirates	-	-	-	-	-	-	-	-	-
Other	-	-	-	-	-	-	-	-	-
Middle East	**-**	**-**	**-**	**-**	**33**	**-**	**-**	**-**	**-**
India	-	-	-	-	-	-	-	-	-
Indonesia	-	-	-	-	-	-	-	-	-
Singapore	-	-	-	-	-	-	-	-	-
Other	-	-	-	-	-	-	-	-	-
Asia	**-**	**-**	**-**	**-**	**-**	**-**	**-**	**-**	**-**
China	**-**	**-**	**-**	**-**	**-**	**-**	**-**	**-**	**-**
Algeria	-	-	-	-	-	-	-	-	-
Angola/Cabinada	-	-	-	-	-	-	-	-	-
Cameroon	-	-	-	-	-	-	-	-	-
Egypt	-	-	-	-	-	-	-	-	-
Gabon	-	-	-	-	-	-	-	-	-
Libya	-	-	-	-	-	-	-	-	-
Nigeria	-	-	-	-	-	-	-	-	-
Tunisia	-	-	-	-	-	-	-	-	-
Other	-	-	-	-	-	-	-	-	-
Africa	**-**	**-**	**-**	**-**	**-**	**-**	**-**	**-**	**-**
Non-spec./Other	**-**	**-**	**-**	**-**	**-**	**-**	**-**	**-**	**-**
Total Imports	**1 593**	**1 619**	**1 694 e**	**754**	**839**	**516 e**	**1 282**	**1 267**	**1 956 e**
Memo: Total OPEC	*-*	*-*	*-*	*-*	*33*	*7 e*	*-*	*-*	*-*

Source: Annual Oil Statistics.

SWEDEN
Table 7D: Imports by Country of Origin

000 metric tons

	2000	2001	2002	2000	2001	2002	2000	2001	2002
	Fuel Oil (Residual)			**Petroleum Coke**			**Other Products**		
Australia	-	-	-	-	-	-	-	-	-
Belgium	-	-	-	-	-	-	-	2	3 e
Canada	-	-	-	-	-	-	-	-	-
Czech Republic	-	-	-	-	-	-	-	-	-
Denmark	29	57	32 e	-	-	-	-	4	1 e
Finland	-	6	-	-	-	-	7	24	44 e
France	25	-	-	-	-	-	-	-	-
Germany	257	380	166 e	-	-	-	2	6	-
Greece	-	-	-	-	-	-	-	-	-
Hungary	-	-	-	-	-	-	-	-	-
Ireland	-	25	-	-	-	-	-	-	-
Italy	-	-	-	-	-	-	1	-	-
Japan	-	-	-	-	-	-	-	-	-
Korea	-	-	-	-	-	-	-	-	-
Mexico	-	-	-	-	-	-	-	-	-
Netherlands	-	-	-	-	-	-	1	2	-
New Zealand	-	-	-	-	-	-	-	-	-
Norway	12	9	40 e	-	-	-	-	-	-
Poland	3	-	-	-	-	-	-	-	2 e
Spain	39	-	-	-	-	-	-	-	-
Sweden	-	-	-	-	-	-	-	-	-
Turkey	-	-	-	-	-	-	-	-	-
United Kingdom	26	182	48 e	-	-	-	9	1	-
United States	-	-	-	25	33	-	-	-	-
Other	-	-	-	-	-	-	1	1	-
OECD	**391**	**659**	**286 e**	**25**	**33**	**-**	**21**	**40**	**50 e**
Argentina	-	-	-	-	-	-	-	-	-
Brazil	-	-	-	-	-	-	-	-	-
Colombia	-	-	-	-	-	-	-	-	-
Ecuador	-	-	-	-	-	-	-	-	-
Netherlands Antilles	-	-	-	-	-	-	-	-	-
Peru	-	-	-	-	-	-	-	-	-
Trinidad and Tobago	-	-	-	-	-	-	-	-	-
Venezuela	-	-	-	-	-	-	-	-	-
Other	6	-	-	-	-	-	-	-	-
Latin America	**6**	**-**	**-**	**-**	**-**	**-**	**-**	**-**	**-**
Romania	-	-	-	-	-	-	-	-	-
Other	-	-	-	-	-	-	-	-	-
Non-OECD Europe	**-**	**-**	**-**	**-**	**-**	**-**	**-**	**-**	**-**
FSU	**122**	**56**	**146 e**	**-**	**-**	**-**	**6**	**-**	**-**
Iran	-	-	-	-	-	-	-	-	-
Iraq	-	-	-	-	-	-	-	-	-
Kuwait	-	-	-	-	-	-	-	-	-
Oman	-	-	-	-	-	-	-	-	-
Qatar	-	-	-	-	-	-	-	-	-
Saudi Arabia	-	-	-	-	-	-	-	-	-
United Arab Emirates	-	-	-	-	-	-	-	-	-
Other	-	-	-	-	-	-	-	-	-
Middle East	**-**	**-**	**-**	**-**	**-**	**-**	**-**	**-**	**-**
India	-	-	-	-	-	-	-	-	-
Indonesia	-	-	-	-	-	-	-	-	-
Singapore	-	-	-	-	-	-	-	-	-
Other	-	-	-	-	-	-	-	-	-
Asia	**-**	**-**	**-**	**-**	**-**	**-**	**-**	**-**	**-**
China	**-**	**-**	**-**	**-**	**-**	**-**	**-**	**-**	**-**
Algeria	-	-	-	-	-	-	-	-	-
Angola/Cabinada	-	-	-	-	-	-	-	-	-
Cameroon	-	-	-	-	-	-	-	-	-
Egypt	-	-	-	-	-	-	-	-	-
Gabon	-	-	-	-	-	-	-	-	-
Libya	-	-	63 e	-	-	-	-	-	-
Nigeria	-	-	-	-	-	-	-	-	-
Tunisia	-	-	-	-	-	-	-	-	-
Other	-	-	-	-	-	-	-	-	-
Africa	**-**	**-**	**63 e**	**-**	**-**	**-**	**-**	**-**	**-**
Non-spec./Other	**-**	**-**	**-**	**-**	**-**	**-**	**-**	**-**	**-**
Total Imports	**519**	**715**	**495 e**	**25**	**33**	**-**	**27**	**40**	**50 e**
Memo: Total OPEC	-	-	*63 e*	-	-	-	-	-	-

Source: Annual Oil Statistics.

SWEDEN
Table 8A: Exports by Country of Destination

000 metric tons

	2000	2001	2002	2000	2001	2002	2000	2001	2002
	Crude Oil			**Natural Gas Liquids**			**Feedstocks and Other**		
Australia	-	-	-	-	-	-	-	-	-
Austria	-	-	-	-	-	-	-	-	-
Belgium	-	-	-	-	-	-	6	51	-
Canada	-	-	-	-	-	-	-	-	-
Denmark	-	-	-	-	-	-	27	8	9 e
Finland	-	-	-	-	-	-	1	4	20 e
France	-	-	-	-	-	-	10	13	6 e
Germany	-	-	-	-	-	-	132	144	60 e
Greece	-	-	-	-	-	-	-	-	-
Ireland	-	-	-	-	-	-	-	-	-
Italy	-	-	-	-	-	-	-	-	-
Japan	-	-	-	-	-	-	-	-	-
Luxembourg	-	-	-	-	-	-	-	-	-
Mexico	-	-	-	-	-	-	-	-	-
Netherlands	-	-	-	-	-	-	339	290	245 e
Norway	-	-	-	-	-	-	17	5	5 e
Portugal	-	-	-	-	-	-	-	-	-
Spain	-	-	-	-	-	-	-	-	-
Sweden	-	-	-	-	-	-	-	-	-
Switzerland	-	-	-	-	-	-	-	-	-
Turkey	-	-	-	-	-	-	-	-	-
United Kingdom	-	-	-	-	-	-	20	21	53 e
United States	-	-	-	-	-	-	6	-	-
Other OECD	-	-	-	-	-	-	-	-	-
OECD	**-**	**-**	**-**	**-**	**-**	**-**	**558**	**536**	**398 e**
Africa	-	-	-	-	-	-	-	-	-
Latin America	-	-	-	-	-	-	-	-	-
Asia	-	-	-	-	-	-	-	-	-
China	-	-	-	-	-	-	-	-	-
Non-OECD Europe	-	-	-	-	-	-	-	-	-
Former Soviet Union	-	-	-	-	-	-	-	-	-
Middle East	-	-	-	-	-	-	-	-	-
Non-specified/Other	-	-	-	-	-	-	-	-	-
Non-OECD	**-**	**-**	**-**	**-**	**-**	**-**	**-**	**-**	**-**
Total Exports	**-**	**-**	**-**	**-**	**-**	**-**	**558**	**536**	**398 e**

	2000	2001	2002	2000	2001	2002	2000	2001	2002
	Total Products			**Liquefied Petroleum Gas**			**Naphtha**		
Australia	-	-	-	-	-	-	-	-	-
Austria	-	-	-	-	-	-	-	-	-
Belgium	91	134	85 e	5	4	-	-	-	-
Canada	34	-	-	-	-	-	-	-	-
Denmark	1 693	1 440	1 460 e	5	3	-	-	-	-
Finland	535	524	477 e	12	-	9 e	-	-	-
France	866	510	490 e	8	33	37 e	-	-	-
Germany	1 271	1 440	1 107 e	12	19	27 e	-	-	-
Greece	-	-	-	-	-	-	-	-	-
Ireland	-	13	29 e	-	-	-	-	-	-
Italy	119	47	26 e	11	-	26 e	-	-	-
Japan	5	-	-	-	-	-	-	-	-
Luxembourg	-	-	-	-	-	-	-	-	-
Mexico	-	-	-	-	-	-	-	-	-
Netherlands	597	499	825 e	38	26	47 e	-	-	-
Norway	958	1 226	1 460 e	63	64	83 e	-	-	-
Portugal	68	51	248 e	7	5	2 e	-	-	-
Spain	151	145	333 e	4	-	11 e	-	-	-
Sweden	-	-	-	-	-	-	-	-	-
Switzerland	43	1	1 e	-	-	-	-	-	-
Turkey	-	-	-	-	-	-	-	-	-
United Kingdom	2 266	2 113	1 518 e	32	31	19 e	322	334	336 e
United States	1 003	980	645 e	13	8	-	36	-	-
Other OECD	336	250	195 e	30	31	47 e	-	-	-
OECD	**10 036**	**9 373**	**8 899 e**	**240**	**224**	**308 e**	**358**	**334**	**336 e**
Africa	-	2	39 e	-	2	5 e	-	-	-
Latin America	2	-	-	2	-	-	-	-	-
Asia	114	-	-	-	-	-	-	-	-
China	-	-	-	-	-	-	-	-	-
Non-OECD Europe	127	44	17 e	19	-	17 e	-	-	-
Former Soviet Union	18	16	43 e	-	-	2 e	-	-	-
Middle East	78	-	-	-	-	-	-	-	-
Non-specified/Other	-	-	4 e	-	-	-	-	-	-
Non-OECD	**339**	**62**	**103 e**	**21**	**2**	**24 e**	**-**	**-**	**-**
Total Exports	**10 375**	**9 435**	**9 002 e**	**261**	**226**	**332 e**	**358**	**334**	**336 e**

Source: Annual Oil Statistics.

SWEDEN
Table 8B: Exports by Country of Destination

000 metric tons

	2000	2001	2002	2000	2001	2002	2000	2001	2002
	Total Gasoline			**Total Kerosene**			**Gas/Diesel Oil**		
Australia	-	-	-	-	-	-	-	-	-
Austria	-	-	-	-	-	-	-	-	-
Belgium	11	-	-	-	-	-	32	102	46 e
Canada	-	-	-	-	-	-	-	-	-
Denmark	615	533	576 e	-	-	-	766	709	592 e
Finland	19	7	7 e	-	-	-	190	170	177 e
France	75	67	-	37	-	-	675	397	453 e
Germany	180	171	54 e	-	-	-	1 057	1 142	966 e
Greece	-	-	-	-	-	-	-	-	-
Ireland	-	-	-	-	-	-	-	3	-
Italy	-	-	-	-	-	-	-	-	-
Japan	-	-	-	-	-	-	-	-	-
Luxembourg	-	-	-	-	-	-	-	-	-
Mexico	-	-	-	-	-	-	-	-	-
Netherlands	194	129	385 e	8	-	-	176	200	194 e
Norway	248	230	259 e	-	-	17 e	217	444	608 e
Portugal	-	-	-	-	-	-	35	-	56 e
Spain	-	-	-	-	-	-	93	69	203 e
Sweden	-	-	-	-	-	-	-	-	-
Switzerland	1	1	-	-	-	-	-	-	1 e
Turkey	-	-	-	-	-	-	-	-	-
United Kingdom	402	473	215 e	-	-	-	418	452	234 e
United States	90	32	58 e	-	-	-	6	-	-
Other OECD	105	62	31 e	-	-	-	194	83	83 e
OECD	**1 940**	**1 705**	**1 585 e**	**45**	**-**	**17 e**	**3 859**	**3 771**	**3 613 e**
Africa	-	-	-	-	-	-	-	-	34 e
Latin America	-	-	-	-	-	-	-	-	-
Asia	-	-	-	-	-	-	-	-	-
China	-	-	-	-	-	-	-	-	-
Non-OECD Europe	29	-	-	-	-	-	-	-	-
Former Soviet Union	-	-	-	-	-	-	-	-	21 e
Middle East	24	-	-	-	-	-	-	-	-
Non-specified/Other	-	-	-	-	-	-	-	-	4 e
Non-OECD	**53**	**-**	**-**	**-**	**-**	**-**	**-**	**-**	**59 e**
Total Exports	**1 993**	**1 705**	**1 585 e**	**45**	**-**	**17 e**	**3 859**	**3 771**	**3 672 e**

	2000	2001	2002	2000	2001	2002	2000	2001	2002
	Fuel Oil (Residual)			**Petroleum Coke**			**Other Products**		
Australia	-	-	-	-	-	-	-	-	-
Austria	-	-	-	-	-	-	-	-	-
Belgium	35	11	-	-	-	-	8	17	39 e
Canada	34	-	-	-	-	-	-	-	-
Denmark	265	161	248 e	-	-	-	42 e	34 e	44 e
Finland	260	303	224 e	-	-	-	54 e	44 e	60 e
France	71	13	-	-	-	-	-	-	-
Germany	22	108	60 e	-	-	-	-	-	-
Greece	-	-	-	-	-	-	-	-	-
Ireland	-	10	29 e	-	-	-	-	-	-
Italy	108	47			-	-	-	-	-
Japan	5	-	-	-	-	-	-	-	-
Luxembourg	-	-	-	-	-	-	-	-	-
Mexico	-	-	-	-	-	-	-	-	-
Netherlands	181	144	199 e	-	-	-	-	-	-
Norway	238	292	242 e	-	-	-	192 e	196 e	251 e
Portugal	26	46	190 e	-	-	-	-	-	-
Spain	54	76	111 e	-	-	-	-	-	8 e
Sweden	-	-	-	-	-	-	-	-	-
Switzerland	42	-	-	-	-	-	-	-	-
Turkey	-	-	-	-	-	-	-	-	-
United Kingdom	1 092	823	711 e	-	-	-	-	-	3 e
United States	858	940	587 e	-	-	-	-	-	-
Other OECD	-	23	-	-	-	-	7	51	34 e
OECD	**3 291**	**2 997**	**2 601 e**	**-**	**-**	**-**	**303 e**	**342 e**	**439 e**
Africa	-	-	-	-	-	-	-	-	-
Latin America	-	-	-	-	-	-	-	-	-
Asia	114	-	-	-	-	-	-	-	-
China	-	-	-	-	-	-	-	-	-
Non-OECD Europe	79	44	-	-	-	-	-	-	-
Former Soviet Union	-	-	-	-	-	-	18 e	16 e	20 e
Middle East	54	-	-	-	-	-	-	-	-
Non-specified/Other	-	-	-	-	-	-	-	-	-
Non-OECD	**247**	**44**	**-**	**-**	**-**	**-**	**18 e**	**16 e**	**20 e**
Total Exports	**3 538**	**3 041**	**2 601 e**	**-**	**-**	**-**	**321**	**358**	**459 e**

Source: Annual Oil Statistics.

SWITZERLAND

Figure 1
FUEL SHARES OF TPES *

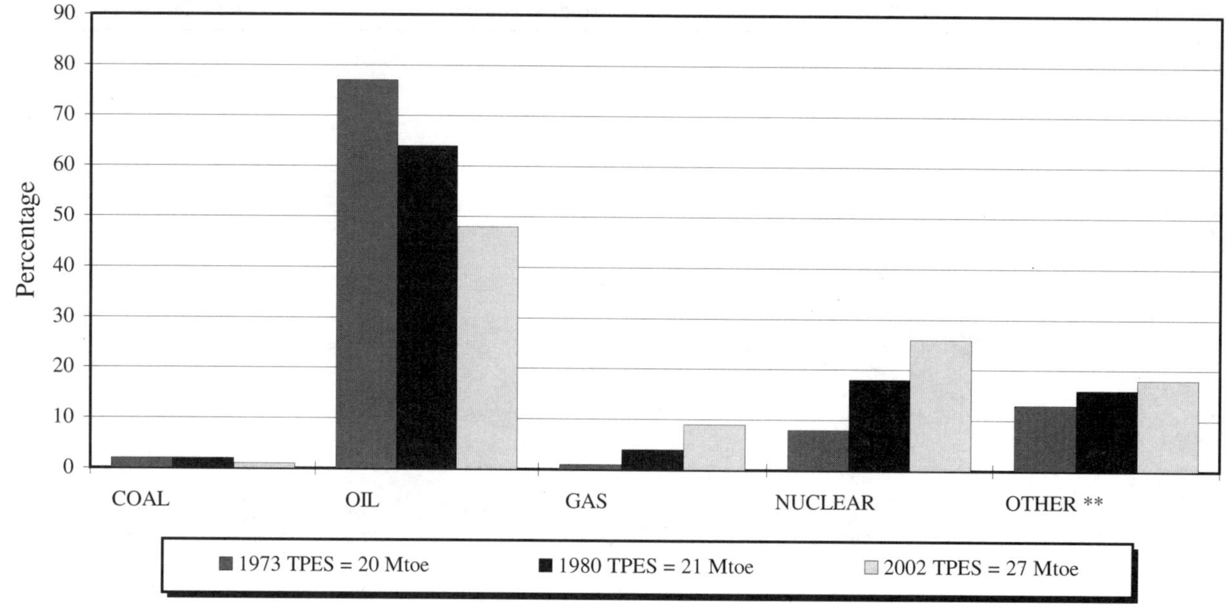

Source: Annual Energy Balances.

* Excluding electricity trade.

** Includes hydro, geothermal, solar, wind, combustible renewables and waste, etc.

Figure 2
OIL PRODUCTION AND CONSUMPTION

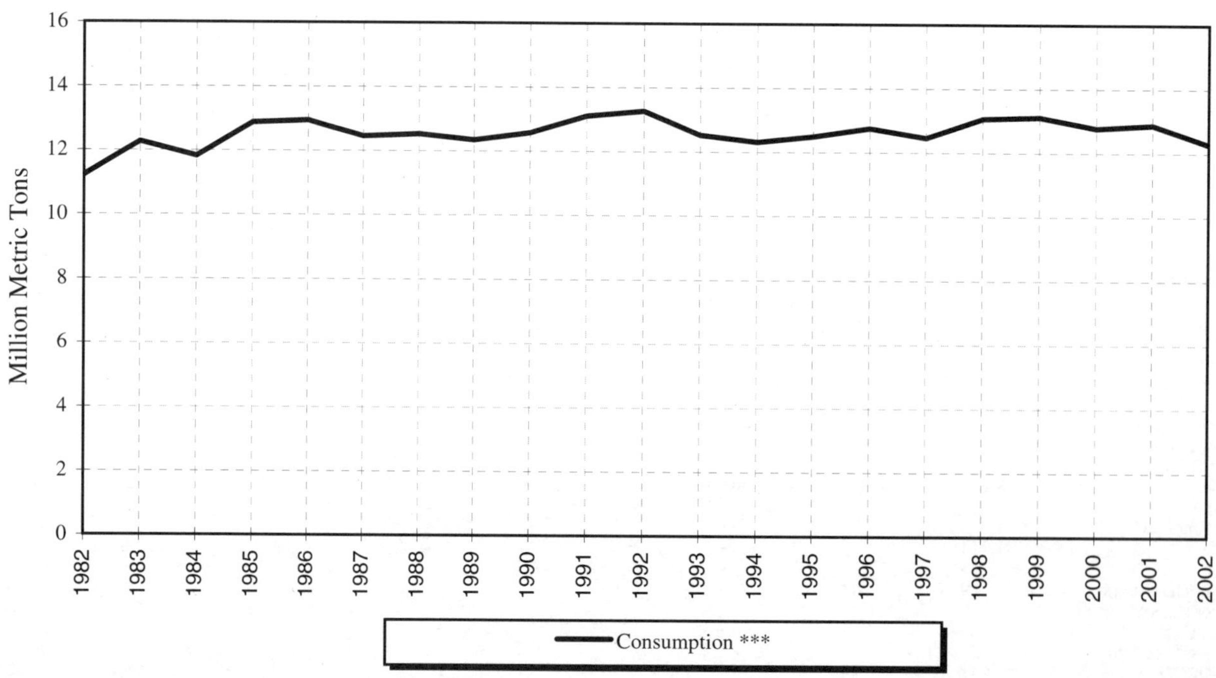

Source: Annual Oil Statistics.

*** Includes refinery fuel and international marine bunkers.

SWITZERLAND
Table 1: Oil Supply and Consumption - Summary Table

000 metric tons

	1973	1978	1980	1990	1999	2000	2001	2002
Indigenous Production	-	-	-	-	-	-	-	-
Crude Oil	..	-	-	-	-	-	-	-
NGL	..	-	-	-	-	-	-	-
Other	..	-	-	-	-	-	-	-
+ Net Imports	14 664	13 666	13 109	12 845	12 100	11 722	13 134	12 469 e
Crude Oil, NGL, Other	6 180	4 246	4 604	3 128	5 093	4 649	4 908	4 876 e
Petroleum Products	8 484	9 420	8 505	9 717	7 007	7 073	8 226	7 593 e
- Intl. Marine Bunkers	-	-	-	18	12	11	12	12 e
+ Stock Changes	104	- 188	- 204	100	569	459	157	- 63 e
TOTAL SUPPLY	**14 768**	**13 478**	**12 905**	**12 927**	**12 657**	**12 170**	**13 279**	**12 394 e**
Refinery Intake	*6 181*	*4 277*	*4 590*	*3 071*	*5 105*	*4 654*	*4 923*	*4 927 e*
Refinery Output	*6 126*	*4 236*	*4 550*	*3 047*	*5 100*	*4 647*	*4 908*	*4 904 e*
LPG + Ethane	*68*	*86*	*94*	*139*	*222*	*191*	*183*	*210 e*
Naphtha	*131*	*22*	*20*	*13*	*- 12*	*16*	*- 5*	*- 2 e*
Kerosenes	*128*	*207*	*227*	*233*	*498*	*454*	*407*	*406 e*
Gasolines	*924*	*886*	*1 096*	*752*	*1 255*	*1 068*	*1 159*	*1 176 e*
Gas/Diesel Oil	*2 376*	*1 815*	*2 018*	*1 167*	*2 010*	*1 847*	*2 025*	*2 011 e*
Fuel Oil	*2 125*	*921*	*802*	*504*	*795*	*754*	*775*	*744 e*
Other Products	*374*	*299*	*293*	*239*	*332*	*317*	*364*	*359 e*
- Refinery Losses	55	41	40	24	5	7	15	23 e
+ Returns	-	-	-	-	-	-	-	-
- Statistical Differences	- 15	-	-	347	- 444	- 598	408	111 e
TOTAL CONSUMPTION	**14 728**	**13 437**	**12 865**	**12 556**	**13 096**	**12 761**	**12 856**	**12 260 e**
Transformation and Energy	**810**	**552**	**339**	**235**	**280**	**229**	**247**	..
Crude Oil	-	-	-	-	-	-	-	..
LPG + Ethane	32	10	11	7	3	2	2	..
Gas/Diesel Oil	19	20	2	28	17	12	13	..
Fuel Oil	489	351	157	105	61	47	47	..
Petroleum Coke	-	-	-	-	-	-	-	..
Other	270	171	169	95	199	168	185	..
Transport	**3 930**	**3 892**	**4 075**	**5 750**	**6 330**	**6 462 e**	**6 277 e**	..
Motor Gasoline	2 462	2 573	2 709	3 701	3 851	3 855 e	3 749 e	..
Gas/Diesel Oil	756	511	556	879	921	981 e	1 000 e	..
Subtotal	*3 218*	*3 084*	*3 265*	*4 580*	*4 772*	*4 836 e*	*4 749 e*	..
Aviation Fuels	650	705	707	1 120	1 517	1 582	1 492	..
Other	54	43	43	50	41	44 e	36 e	..
Industry	**3 707**	**2 897**	**2 771**	**1 299**	**1 758**	**1 669 e**	**1 747 e**	..
Naphtha	104	81	79	43	35	40	45	..
LPG + Ethane	33	36	45	73	100	97 e	94	..
Gas/Diesel Oil	1 075	1 250	1 120	310	912	865 e	923 e	..
Fuel Oil	1 887	1 046	964	403	208	146 e	177	..
Other	608	484	563	470	503	521 e	508 e	..
Other Sectors	**6 281**	**6 096**	**5 680**	**5 272**	**4 728**	**4 401 e**	**4 585 e**	..
Gas/Diesel Oil	5 951	5 793	5 460	5 164	4 586	4 264 e	4 450	..
Fuel Oil	265	240	150	-	-	-	-	..
Other	65	63	70	108	142	137 e	135 e	..
Memo: Non-Energy Use	*781*	*623*	*644*	*577*	*553*	*581 e*	*566 e*	..
Chem/Petchem. Feedst.	*149*	*122*	*135*	*127*	*133*	*137 e*	*138 e*	..
Other	*632*	*501*	*509*	*450*	*420*	*444 e*	*428 e*	..

Source: Annual Oil Statistics.

SWITZERLAND
Table 2: Supply of Crude Oil, NGL, Feedstocks and Other Hydrocarbons

000 metric tons

	1973	1978	1980	1990	1999	2000	2001	2002
CRUDE OIL								
Indigenous Production	..	-	-	-	-	-	-	-
+ Imports	..	3 873	3 832	3 055	5 058	4 613	4 732	4 805 e
- Exports	..	-	-	-	-	-	-	-
- Direct Use	..	-	-	-	-	-	-	-
+ Stock Change	..	33	- 18	- 56	12	6	- 5	3 e
- Statistical Difference	..	-	-	-	-	-	-	-
Refinery Intake	**..**	**3 906**	**3 814**	**2 999**	**5 070**	**4 619**	**4 727**	**4 808 e**
Closing Stock Level	*..*	*89*	*109*	*92*	*75*	*69*	*74*	*71 e*
NATURAL GAS LIQUIDS								
Indigenous Production	..	-	-	-	-	-	-	-
+ Imports	..	-	-	-	-	-	-	-
- Exports	..	-	-	-	-	-	-	-
- Direct Use	..	-	-	-	-	-	-	-
+ Stock Change	..	-	-	-	-	-	-	-
- Statistical Difference	..	-	-	-	-	-	-	-
Refinery Intake	**..**	**-**	**-**	**-**	**-**	**-**	**-**	**-**
Closing Stock Level	*..*	*-*	*-*	*-*	*-*	*-*	*-*	*-*
REFINERY FEEDSTOCKS								
Backflows	..	-	-	-	-	-	-	-
+ Transfers	..	-	-	-	-	-	18	49 e
+ Imports	..	373	772	-	2	-	130	40 e
- Exports	..	-	-	-	-	-	-	-
+ Stock Change	..	- 2	4	5	-	-	-	-
- Statistical Difference	..	-	-	-	-	-	-	1 e
Refinery Intake	**..**	**371**	**776**	**5**	**2**	**-**	**148**	**88 e**
Closing Stock Level	*..*	*18*	*25*	*-*	*-*	*-*	*-*	*-*
OTHER HYDROCARBONS								
Indigenous Production	..	-	-	-	-	-	-	-
+ Imports	..	-	-	73	33	36	46	31 e
- Exports	..	-	-	-	-	-	-	-
- Direct Use	..	-	-	-	-	-	-	-
+ Stock Change	..	-	-	- 6	-	- 1	2	- 1 e
- Statistical Difference	..	-	-	-	-	-	-	- 1 e
Refinery Intake	**..**	**-**	**-**	**67**	**33**	**35**	**48**	**31 e**
Closing Stock Level	*..*	*-*	*-*	*8*	*3*	*4*	*2*	*3 e*
CRUDE OIL, NGL, FEEDSTOCKS AND OTHER								
Indigenous Production	-	-	-	-	-	-	-	-
+ Backflows	-	-	-	-	-	-	-	-
+ Transfers	-	-	-	-	-	-	18	49 e
+ Imports	6 180	4 246	4 604	3 128	5 093	4 649	4 908	4 876 e
- Exports	-	-	-	-	-	-	-	-
- Direct Use	-	-	-	-	-	-	-	-
+ Stock Change	1	31	- 14	- 57	12	5	- 3	2 e
- Statistical Difference	-	-	-	-	-	-	-	-
Refinery Intake	**6 181**	**4 277**	**4 590**	**3 071**	**5 105**	**4 654**	**4 923**	**4 927 e**
Closing Stock Level	*-*	*107*	*134*	*100*	*78*	*73*	*76*	*74 e*

Source: Annual Oil Statistics.

SWITZERLAND
Table 3: Motor Gasoline - Supply

000 metric tons

	1973	1978	1980	1990	1999	2000	2001	2002
MOTOR GASOLINE								
Refinery Output	**924**	**886**	**1 096**	**752**	**1 255**	**1 068**	**1 159**	**1 176 e**
+ Imports	1 598	1 759	1 676	2 985	2 636	2 721	2 696	2 642 e
- Exports	1	6	11	-	-	1	-	-
+ Transfers	-	-	-	18	-	-	-	-
+ Stock Change	- 11	- 31	- 17	- 31	110	164	19	- 41 e
- Statistical Difference	-	-	-	-	22	- 31	1	- 19 e
Total Consumption	**2 510**	**2 608**	**2 744**	**3 724**	**3 979**	**3 983**	**3 873**	**3 796 e**
Closing Stock Level	-	*916*	*966*	*953*	*1 644*	*1 480*	*1 461*	*1 502 e*

Table 4: Aviation Fuels - Supply

000 metric tons

	1973	1978	1980	1990	1999	2000	2001	2002
AVIATION GASOLINE								
Refinery Output	-	-	-	-	-	-	-	-
+ Imports	5	8	8	6	5	5	5	2 e
- Exports	-	-	-	-	-	-	-	-
+ Transfers	-	-	-	-	-	-	-	-
+ Stock Change	1	-	- 1	- 1	-	1	- 1	-
- Statistical Difference	-	-	-	-	-	-	- 1	- 3 e
Total Consumption	**6**	**8**	**7**	**5**	**5**	**6**	**5**	**5 e**
Closing Stock Level	-	*1*	*2*	*2*	*1*	-	*1*	*1 e*
GASOLINE TYPE JET FUEL								
Refinery Output	-	-	-	-	-	-	-	-
+ Imports	-	-	-	-	-	-	-	-
- Exports	-	-	-	-	-	-	-	-
+ Transfers	-	-	-	-	-	-	-	-
+ Stock Change	-	-	-	-	-	-	-	-
- Statistical Difference	-	-	-	-	-	-	-	-
Total Consumption	-	-	-	-	-	-	-	-
Closing Stock Level	-	-	-	-	-	-	-	-
KEROSENE TYPE JET FUEL								
Refinery Output	**124**	**202**	**222**	**231**	**498**	**454**	**407**	**406 e**
+ Imports	525	535	562	891	1 006	1 139	1 108	965 e
- Exports	-	-	-	-	-	-	-	-
+ Transfers	-	-	-	-	- 4	- 3	- 3	-
+ Stock Change	3	20	- 24	- 7	12	- 17	- 22	1 e
- Statistical Difference	-	-	-	-	-	- 3	3	- 6 e
Total Consumption	**652**	**757**	**760**	**1 115**	**1 512**	**1 576**	**1 487**	**1 378 e**
Closing Stock Level	-	*33*	*76*	*142*	*258*	*275*	*297*	*296 e*

Source: Annual Oil Statistics.

SWITZERLAND
Table 5A: Gas/Diesel Oil - Supply and Consumption

000 metric tons

	1973	1978	1980	1990	1999	2000	2001	2002
Refinery Output	**2 376**	**1 815**	**2 018**	**1 167**	**2 010**	**1 847**	**2 025**	**2 011 e**
+ Recycled Products	-	-	-	-	-	-	-	-
+ Imports	5 312	5 798	5 295	5 471	3 534	3 477	4 581	4 239 e
- Exports	3	1	6	8	-	-	-	-
- Intl. Marine Bunkers	-	-	-	18	12	11	12	12 e
+ Transfers	-	-	-	- 9	7	- 6	1	-
+ Stock Changes	102	- 38	- 169	85	442	269	183	- 58 e
- Statistical Differences	- 14	-	-	307	- 455	- 546	392	136 e
Total Consumption	**7 801**	**7 574**	**7 138**	**6 381**	**6 436**	**6 122**	**6 386**	**6 044 e**
Transformation	**19**	**20**	**2**	**28**	**17**	**12**	**13**	**..**
Public Electricity	-	-	-	-	-	-	-	..
Autoproducer Electricity	-	-	-	-	-	-	-	..
Public CHP Plants	19	20	2	27	5	3	3	..
Autoproducers of CHP	-	-	-	1	12	9	10	..
Public Heat Plants	-	-	-	-	-	-	-	..
Autoproducer Heat Plants	-	-	-	-	-	-	-	..
Petrochemical Industry	-	-	-	-	-	-	-	..
Other Transformation	-	-	-	-	-	-	-	..
Energy Sector	**-**	**-**	**-**	**-**	**-**	**-**	**-**	**..**
Refinery Fuel	-	-	-	-	-	-	-	..
Other Energy Sector	-	-	-	-	-	-	-	..
Final Consumption	**7 782**	**7 554**	**7 136**	**6 353**	**6 419**	**6 110**	**6 373**	**..**
Transport	**756**	**511**	**556**	**879**	**921**	**981 e**	**1 000 e**	**..**
Air	-	-	-	-	-	-	-	..
Road	691	493	536	862	902	961 e	980 e	..
Rail	20	10	11	9	10	11 e	11 e	..
Inland Waterways	45	8	9	8	9	9 e	9 e	..
Non-Specified	-	-	-	-	-	-	-	..
Industry	**1 075**	**1 250**	**1 120**	**310**	**912**	**865 e**	**923 e**	**..**
Iron and Steel	-	-	-	-	17	16 e	17 e	..
Chemical (incl. Pet.-Chem.)	-	-	-	67	74	68 e	74 e	..
Non-Ferrous Metals	-	-	-	13	7	7 e	7 e	..
Non-Metallic Minerals	-	-	-	19	47	45 e	48 e	..
Transport Equipment	-	-	-	-	-	-	-	..
Machinery	-	-	-	122	70	65 e	70 e	..
Mining and Quarrying	-	-	-	-	-	-	-	..
Food and Tobacco	-	-	-	33	123	114 e	123 e	..
Pulp, Paper and Printing	-	-	-	8	47	43 e	47 e	..
Wood and Wood Products	-	-	-	-	-	-	-	..
Construction	-	-	-	23	170	170 e	178 e	..
Textiles and Leather	-	-	-	8	54	50 e	54 e	..
Non-Specified	1 075	1 250	1 120	17 e	303	287 e	305 e	..
Other Sectors	**5 951**	**5 793**	**5 460**	**5 164**	**4 586**	**4 264 e**	**4 450**	**..**
Commerce and Public	-	1 746	1 521	1 812	1 375	1 278 e	1 312	..
Residential	-	3 982	3 872	3 267	3 100	2 868 e	2 999	..
Agriculture	55	65	67	85	111	118 e	120 e	..
Non-Specified	5 896	-	-	-	-	-	19	..
Memo: Non-Energy Use	-	-	-	-	-	-	-	..
Chem/Petchem. Feedst.	-	-	-	-	-	-	-	..
Other	-	-	-	-	-	-	-	..
Closing Stock Level	-	*3 377*	*3 883*	*3 598*	*3 027*	*2 758*	*2 575*	*2 633 e*

Source: Annual Oil Statistics.

SWITZERLAND
Table 5B: Fuel Oil (Residual) - Supply and Consumption

000 metric tons

	1973	1978	1980	1990	1999	2000	2001	2002
Refinery Output	**2 125**	**921**	**802**	**504**	**795**	**754**	**775**	**744 e**
+ Recycled Products	-	-	-	-	-	-	-	-
+ Imports	730	915	470	70	-	4	-	-
- Exports	225	21	19	130	506	609	503	554 e
- Intl. Marine Bunkers	-	-	-	-	-	-	-	-
+ Transfers	-	-	-	-	3	9	- 16	- 49 e
+ Stock Changes	10	- 178	18	104	- 31	25	- 17	26 e
- Statistical Differences	- 1	-	-	40	- 8	- 10	15	9 e
Total Consumption	**2 641**	**1 637**	**1 271**	**508**	**269**	**193**	**224**	**158 e**
Transformation	**413**	**316**	**110**	**74**	**20**	**-**	**-**	**..**
Public Electricity	-	-	-	-	-	-	-	..
Autoproducer Electricity	-	-	-	-	-	-	-	..
Public CHP Plants	413	316	110	54	6	-	-	..
Autoproducers of CHP	-	-	-	20	14	-	-	..
Public Heat Plants	-	-	-	-	-	-	-	..
Autoproducer Heat Plants	-	-	-	-	-	-	-	..
Petrochemical Industry	-	-	-	-	-	-	-	..
Other Transformation	-	-	-	-	-	-	-	..
Energy Sector	**76**	**35**	**47**	**31**	**41**	**47**	**47**	**..**
Refinery Fuel	76	35	47	31	41	47	47	..
Other Energy Sector	-	-	-	-	-	-	-	..
Final Consumption	**2 152**	**1 286**	**1 114**	**403**	**208**	**146 e**	**177**	**..**
Transport	**-**	**-**	**-**	**-**	**-**	**-**	**-**	**..**
Air	-	-	-	-	-	-	-	..
Road	-	-	-	-	-	-	-	..
Rail	-	-	-	-	-	-	-	..
Inland Waterways	-	-	-	-	-	-	-	..
Non-Specified	-	-	-	-	-	-	-	..
Industry	**1 887**	**1 046**	**964**	**403**	**208**	**146 e**	**177**	**..**
Iron and Steel	-	-	-	-	-	-	-	..
Chemical (incl. Pet.-Chem.)	-	-	-	50	9	6 e	8 e	..
Non-Ferrous Metals	-	-	-	-	-	-	-	..
Non-Metallic Minerals	-	-	-	124	87	61 e	74 e	..
Transport Equipment	-	-	-	-	-	-	-	..
Machinery	-	-	-	49	1	1 e	1 e	..
Mining and Quarrying	-	-	-	-	-	-	-	..
Food and Tobacco	-	-	-	27	12	8 e	10 e	..
Pulp, Paper and Printing	-	-	-	126	74	52 e	63 e	..
Wood and Wood Products	-	-	-	-	-	-	-	..
Construction	-	-	-	-	4	3 e	3 e	..
Textiles and Leather	-	-	-	26	7	5 e	6 e	..
Non-Specified	1 887	1 046	964	1	14	10 e	12 e	..
Other Sectors	**265**	**240**	**150**	**-**	**-**	**-**	**-**	**..**
Commerce and Public	-	-	-	-	-	-	-	..
Residential	-	-	-	-	-	-	-	..
Agriculture	-	-	-	-	-	-	-	..
Non-Specified	265	240	150	-	-	-	-	..
Memo: Non-Energy Use	-	-	-	-	-	-	-	..
Chem/Petchem. Feedst.	-	-	-	-	-	-	-	..
Other	-	-	-	-	-	-	-	..
Closing Stock Level	-	*530*	*512*	*191*	*177*	*152*	*169*	*143 e*

Source: Annual Oil Statistics.

SWITZERLAND
Table 5C: Total Products - Supply and Consumption

000 metric tons

	1973	1978	1980	1990	1999	2000	2001	2002
Refinery Output	**6 126**	**4 236**	**4 550**	**3 047**	**5 100**	**4 647**	**4 908**	**4 904 e**
+ Recycled Products	-	-	-	-	-	-	-	-
+ Imports	8 720	9 471	8 553	9 880	7 569	7 733	8 801	8 219 e
- Exports	236	51	48	163	562	660	575	626 e
- Intl. Marine Bunkers	-	-	-	18	12	11	12	12 e
+ Transfers	-	-	-	-	-	-	- 18	- 49 e
+ Stock Changes	103	- 219	- 190	157	557	454	160	- 65 e
- Statistical Differences	- 15	-	-	347	- 444	- 598	408	111 e
Total Consumption	**14 728**	**13 437**	**12 865**	**12 556**	**13 096**	**12 761**	**12 856**	**12 260 e**
Transformation	**549**	**366**	**142**	**109**	**40**	**14**	**15**	..
Public Electricity	19	-	-	-	-	-	-	..
Autoproducer Electricity	-	-	-	-	-	-	-	..
Public CHP Plants	432	336	112	81	11	3	3	..
Autoproducers of CHP	-	7	10	21	26	9	10	..
Public Heat Plants	-	-	-	-	-	-	-	..
Autoproducer Heat Plants	-	-	-	-	-	-	-	..
Petrochemical Industry	-	-	-	-	-	-	-	..
Other Transformation	98	23	20	7	3	2	2	..
Energy Sector	**261**	**186**	**197**	**126**	**240**	**215**	**232**	..
Refinery Fuel	261	186	197	126	240	215	232	..
Other Energy Sector	-	-	-	-	-	-	-	..
Final Consumption	**13 918**	**12 885**	**12 526**	**12 321**	**12 816**	**12 532**	**12 609**	..
Transport	**3 930**	**3 892**	**4 075**	**5 750**	**6 330**	**6 462 e**	**6 277 e**	..
Air	658	765	767	1 120	1 517	1 582 e	1 492	..
Road	3 207	3 109	3 288	4 613	4 792	4 859 e	4 764 e	..
Rail	20	10	11	9	10	11 e	11 e	..
Inland Waterways	45	8	9	8	11	10 e	10 e	..
Non-Specified	-	-	-	-	-	-	-	..
Industry	**3 707**	**2 897**	**2 771**	**1 299**	**1 758**	**1 669 e**	**1 747 e**	..
Iron and Steel	-	-	-	-	17	16 e	17 e	..
Chemical (incl. Pet.-Chem.)	149	122	135	247	232	226 e	234 e	..
Non-Ferrous Metals	-	-	-	39	30	31 e	31 e	..
Non-Metallic Minerals	-	-	-	160	144	117 e	133 e	..
Transport Equipment	-	-	-	-	-	-	-	..
Machinery	-	-	-	171	81	76 e	81 e	..
Mining and Quarrying	-	-	-	-	-	-	-	..
Food and Tobacco	-	-	-	60	135	122 e	133 e	..
Pulp, Paper and Printing	-	-	-	134	121	95 e	110 e	..
Wood and Wood Products	-	-	-	-	-	-	-	..
Construction	461	345	351	337	494	499 e	491 e	..
Textiles and Leather	-	-	-	34	61	55 e	60 e	..
Non-Specified	3 097	2 430	2 285	117	443	432 e	457 e	..
Other Sectors	**6 281**	**6 096**	**5 680**	**5 272**	**4 728**	**4 401 e**	**4 585 e**	..
Commerce and Public	-	1 746	1 521	1 812	1 395	1 298 e	1 332	..
Residential	-	3 982	3 872	3 270	3 100	2 868 e	2 999	..
Agriculture	92	90	91	104	132	139 e	140 e	..
Non-Specified	6 189	278	196	86	101	96 e	114	..
Memo: Non-Energy Use	*781*	*623*	*644*	*577*	*553*	*581 e*	*566 e*	..
Chem/Petchem. Feedst.	*149*	*122*	*135*	*127*	*133*	*137 e*	*138 e*	..
Other	*632*	*501*	*509*	*450*	*420*	*444 e*	*428 e*	..
Closing Stock Level	-	*4 954*	*5 544*	*5 005*	*5 195*	*4 741*	*4 581*	*4 646 e*

Source: Annual Oil Statistics.

SWITZERLAND
Table 6A: Supply and Consumption

000 metric tons

	1980	2000	2001	2002	1980	2000	2001	2002
	LPG AND ETHANE				**REFINERY GAS**			
Refinery Output	**94**	**191**	**183**	**210 e**	**160**	**168**	**185**	**203 e**
+ Recycled Products	-	-	-	-	-	-	-	-
+ Imports	11	31	39	39 e	-	-	-	-
- Exports	8	24	29	35 e	-	-	-	-
- Intl. Marine Bunkers	-	-	-	-	-	-	-	-
+ Transfers	-	-	-	-	-	-	-	-
+ Stock Change	1	- 1	-	1 e	-	-	-	-
- Statistical Difference	-	-	-	-	-	-	-	-
Total Consumption	**98**	**197**	**193**	**215 e**	**160**	**168**	**185**	**203 e**
of which: Ethane	-	-	-	-	-	-	-	-
Transformation	11	2	2	..	10	-	-	..
Energy	-	-	-	..	150	168	185	..
Transport	-	4 e	4	..	-	-	-	..
Industry	45	97 e	94	..	-	-	-	..
Other	42	94 e	93	..	-	-	-	..
Memo: Non-Energy Use	*43*	*82 e*	*80 e*	..	-	-	-	..
Closing Stock Level	-	*2*	*2*	*1 e*	-	-	-	-

	1980	2000	2001	2002	1980	2000	2001	2002
	NAPHTHA				**OTHER KEROSENE**			
Refinery Output	**20**	**16**	**- 5**	**- 2 e**	**5**	**-**	**-**	**-**
+ Recycled Products	-	-	-	-	-	-	-	-
+ Imports	67	40	44	18 e	5	6	5	-
- Exports	-	20	-	-	-	-	-	-
- Intl. Marine Bunkers	-	-	-	-	-	-	-	-
+ Transfers	-	-	-	-	-	-	-	-
+ Stock Change	1	- 4	5	3 e	-	-	-	-
- Statistical Difference	-	- 8	- 1	- 4 e	-	-	-	- 2 e
Total Consumption	**88**	**40**	**45**	**23 e**	**10**	**6**	**5**	**2 e**
Transformation	9	-	-	..	-	-	-	..
Energy	-	-	-	..	-	-	-	..
Transport	-	-	-	..	-	1 e	-	..
Industry	79	40	45	..	6	3 e	3 e	..
Other	-	-	-	..	4	2 e	2 e	..
Memo: Non-Energy Use	*79*	*40*	*45*	..	-	*1 e*	-	..
Closing Stock Level	*27*	*34*	*29*	*26 e*	*1*			-

	1980	2000	2001	2002	1980	2000	2001	2002
	WHITE SPIRIT + SBP				**LUBRICANTS**			
Refinery Output	**-**	**-**	**-**	**..**	**-**	**-**	**-**	**..**
+ Recycled Products	-	-	-	..	-	-	-	..
+ Imports	22	6	7	..	83	74	67	..
- Exports	-	-	-	..	4	4	4	..
- Intl. Marine Bunkers	-	-	-	..	-	-	-	..
+ Transfers	-	-	-	..	-	-	-	..
+ Stock Change	-	-	-	..	-	7	3	..
- Statistical Difference	-	-	- 1	..	-	-	-	..
Total Consumption	**22**	**6**	**8**	**..**	**79**	**77**	**66**	**..**
Transformation	-	-	-	..	-	-	-	..
Energy	-	-	-	..	-	-	-	..
Transport	-	-	-	..	43	39 e	32 e	..
Industry	22	6	8	..	36	38 e	34 e	..
Other	-	-	-	..	-	-	-	..
Memo: Non-Energy Use	*22*	*6*	*8*	..	*79*	*77 e*	*66 e*	..
Closing Stock Level	*6*	-	-	..	*34*	*34*	*31*	..

Source: Annual Oil Statistics.

SWITZERLAND
Table 6B: Supply and Consumption

000 metric tons

	1980	2000	2001	2002	1980	2000	2001	2002
	BITUMEN				PARAFFIN WAXES			
Refinery Output	**131**	**119**	**140**	**..**	**-**	**-**	**-**	**..**
+ Recycled Products	-	-	-	..	-	-	-	..
+ Imports	208	177	198	..	13	5	3	..
- Exports	-	2	39	..	-	-	-	..
- Intl. Marine Bunkers	-	-	-	..	-	-	-	..
+ Transfers	-	-	-	..	-	-	-	..
+ Stock Change	1	10	- 10	..	-	-	-	..
- Statistical Difference	-	-	-	..	-	-	-	..
Total Consumption	**340**	**304**	**289**	**..**	**13**	**5**	**3**	**..**
Transformation	-	-	-	..	-	-	-	..
Energy	-	-	-	..	-	-	-	..
Transport	-	-	-	..	-	-	-	..
Industry	340	304	289	..	13	5	3	..
Other	-	-	-	..	-	-	-	..
Memo: Non-Energy Use	*340*	*304*	*289*	*..*	*13*	*5*	*3*	*..*
Closing Stock Level	*24*	*6*	*16*	*..*	*1*	*-*	*-*	*..*
	PETROLEUM COKE				OTHER PRODUCTS			
Refinery Output	**-**	**-**	**-**	**-**	**2**	**30**	**39**	**156 e**
+ Recycled Products	-	-	-	-	-	-	-	-
+ Imports	117	46	46	53 e	16	2	2	261 e
- Exports	-	-	-	-	-	-	-	37 e
- Intl. Marine Bunkers	-	-	-	-	-	-	-	-
+ Transfers	-	-	-	-	-	-	-	-
+ Stock Change	-	-	-	-	-	-	-	3 e
- Statistical Difference	-	-	-	-	-	-	-	-
Total Consumption	**117**	**46**	**46**	**53 e**	**18**	**32**	**41**	**383 e**
Transformation	-	-	-	..	-	-	-	..
Energy	-	-	-	..	-	-	-	..
Transport	-	-	-	..	-	-	-	..
Industry	117	46	46	..	18	32	41	..
Other	-	-	-	..	-	-	-	..
Memo: Non-Energy Use	*50*	*34 e*	*34 e*	*..*	*18*	*32*	*41*	*..*
Closing Stock Level	*10*	*-*	*-*	*-*	*2*	*-*	*-*	*44 e*

Source: Annual Oil Statistics.

SWITZERLAND
Table 7A: Imports by Country of Origin

000 metric tons

	2000	2001	2002	2000	2001	2002	2000	2001	2002
	Crude Oil			Natural Gas Liquids			Feedstocks and Other		
Australia	-	-	-	-	-	-	-	-	-
Belgium	-	-	-	-	-	-	-	-	-
Canada	-	-	-	-	-	-	-	-	-
Czech Republic	-	-	-	-	-	-	-	-	-
Denmark	-	-	-	-	-	-	-	-	-
Finland	-	-	-	-	-	-	-	-	-
France	-	-	-	-	-	-	-	-	-
Germany	-	-	-	-	-	-	1	78	40 e
Greece	-	-	-	-	-	-	-	-	-
Hungary	-	-	-	-	-	-	-	-	-
Ireland	-	-	-	-	-	-	-	-	-
Italy	-	-	-	-	-	-	-	-	-
Japan	-	-	-	-	-	-	-	-	-
Korea	-	-	-	-	-	-	-	-	-
Mexico	-	-	-	-	-	-	-	-	-
Netherlands	-	-	-	-	-	-	35	45	31 e
New Zealand	-	-	-	-	-	-	-	-	-
Norway	-	-	-	-	-	-	-	-	-
Poland	-	-	-	-	-	-	-	-	-
Spain	-	-	-	-	-	-	-	-	-
Sweden	-	-	-	-	-	-	-	-	-
Turkey	-	-	-	-	-	-	-	-	-
United Kingdom	-	-	-	-	-	-	-	-	-
United States	-	-	-	-	-	-	-	-	-
Other	-	-	-	-	-	-	-	-	-
OECD	**-**	**-**	**-**	**-**	**-**	**-**	**36**	**123**	**71 e**
Argentina	-	-	-	-	-	-	-	-	-
Brazil	-	-	-	-	-	-	-	-	-
Colombia	-	-	-	-	-	-	-	-	-
Ecuador	-	-	-	-	-	-	-	-	-
Netherlands Antilles	-	-	-	-	-	-	-	-	-
Peru	-	-	-	-	-	-	-	-	-
Trinidad and Tobago	-	-	-	-	-	-	-	-	-
Venezuela	-	-	-	-	-	-	-	-	-
Other	-	-	-	-	-	-	-	-	-
Latin America	**-**	**-**	**-**	**-**	**-**	**-**	**-**	**-**	**-**
Romania	-	-	-	-	-	-	-	-	-
Other	-	-	-	-	-	-	-	-	-
Non-OECD Europe	**-**	**-**	**-**	**-**	**-**	**-**	**-**	**-**	**-**
FSU	**-**	**-**	**164 e**	**-**	**-**	**-**	**-**	**-**	**-**
Iran	341	408	364 e	-	-	-	-	-	-
Iraq	-	-	-	-	-	-	-	-	-
Kuwait	-	-	-	-	-	-	-	-	-
Oman	-	-	-	-	-	-	-	-	-
Qatar	-	-	-	-	-	-	-	-	-
Saudi Arabia	-	-	-	-	-	-	-	-	-
United Arab Emirates	-	-	-	-	-	-	-	-	-
Other	-	-	-	-	-	-	-	-	-
Middle East	**341**	**408**	**364 e**	**-**	**-**	**-**	**-**	**-**	**-**
India	-	-	-	-	-	-	-	-	-
Indonesia	-	-	-	-	-	-	-	-	-
Singapore	-	-	-	-	-	-	-	-	-
Other	-	-	-	-	-	-	-	-	-
Asia	**-**	**-**	**-**	**-**	**-**	**-**	**-**	**-**	**-**
China	**-**	**-**	**-**	**-**	**-**	**-**	**-**	**-**	**-**
Algeria	749	487	221 e	-	-	-	-	53	-
Angola/Cabinada	-	-	-	-	-	-	-	-	-
Cameroon	-	-	-	-	-	-	-	-	-
Egypt	-	-	-	-	-	-	-	-	-
Gabon	-	-	-	-	-	-	-	-	-
Libya	1 799	2 359	2 168 e	-	-	-	-	-	-
Nigeria	1 712	1 478	1 500 e	-	-	-	-	-	-
Tunisia	-	-	388 e	-	-	-	-	-	-
Other	12	-	-	-	-	-	-	-	-
Africa	**4 272**	**4 324**	**4 277 e**	**-**	**-**	**-**	**-**	**53**	**-**
Non-spec./Other	**-**	**-**	**-**	**-**	**-**	**-**	**-**	**-**	**-**
Total Imports	**4 613**	**4 732**	**4 805 e**	**-**	**-**	**-**	**36**	**176**	**71 e**
Memo: Total OPEC	*4 601*	*4 732*	*4 253 e*	-	-	-	-	*53*	-

Source: Annual Oil Statistics.

SWITZERLAND
Table 7B: Imports by Country of Origin

000 metric tons

	2000	2001	2002	2000	2001	2002	2000	2001	2002
	Total Products			Liquefied Petroleum Gases			Naphtha		
Australia	-	-	-	-	-	-	-	-	-
Belgium	1 184	1 273	1 068 e	-	5	3 e	-	1	-
Canada	-	-	-	-	-	-	-	-	-
Czech Republic	-	1	-	-	-	-	-	-	-
Denmark	-	-	-	-	-	-	-	-	-
Finland	-	-	1 e	-	-	-	-	-	-
France	1 819	1 905	1 443 e	-	-	-	-	-	-
Germany	1 770	1 700	1 969 e	27	30	30 e	-	1	-
Greece	-	-	-	-	-	-	-	-	-
Hungary	2	4	8 e	-	-	-	-	-	-
Ireland	-	-	-	-	-	-	-	-	-
Italy	755	869	911 e	1	2	4 e	39	40	18 e
Japan	-	-	-	-	-	-	-	-	-
Korea	-	-	-	-	-	-	-	-	-
Mexico	-	-	-	-	-	-	-	-	-
Netherlands	2 194	3 042	2 810 e	3	2	2 e	1	2	-
New Zealand	-	-	-	-	-	-	-	-	-
Norway	-	-	-	-	-	-	-	-	-
Poland	-	-	-	-	-	-	-	-	-
Spain	-	-	-	-	-	-	-	-	-
Sweden	1	1	-	-	-	-	-	-	-
Turkey	-	-	-	-	-	-	-	-	-
United Kingdom	1	-	-	-	-	-	-	-	-
United States	-	4	3 e	-	-	-	-	-	-
Other	2	2	-	-	-	-	-	-	-
OECD	**7 728**	**8 801**	**8 213 e**	**31**	**39**	**39 e**	**40**	**44**	**18 e**
Argentina	2	-	-	-	-	-	-	-	-
Brazil	-	-	-	-	-	-	-	-	-
Colombia	-	-	-	-	-	-	-	-	-
Ecuador	-	-	-	-	-	-	-	-	-
Netherlands Antilles	-	-	-	-	-	-	-	-	-
Peru	-	-	-	-	-	-	-	-	-
Trinidad and Tobago	-	-	-	-	-	-	-	-	-
Venezuela	-	-	-	-	-	-	-	-	-
Other	-	-	-	-	-	-	-	-	-
Latin America	**2**	**-**	**-**	**-**	**-**	**-**	**-**	**-**	**-**
Romania	-	-	-	-	-	-	-	-	-
Other	-	-	-	-	-	-	-	-	-
Non-OECD Europe	**-**	**-**	**-**	**-**	**-**	**-**	**-**	**-**	**-**
FSU	**-**	**-**	**-**	**-**	**-**	**-**	**-**	**-**	**-**
Iran	-	-	-	-	-	-	-	-	-
Iraq	-	-	-	-	-	-	-	-	-
Kuwait	-	-	-	-	-	-	-	-	-
Oman	-	-	-	-	-	-	-	-	-
Qatar	-	-	-	-	-	-	-	-	-
Saudi Arabia	-	-	-	-	-	-	-	-	-
United Arab Emirates	-	-	-	-	-	-	-	-	-
Other	-	-	-	-	-	-	-	-	-
Middle East	**-**	**-**	**-**	**-**	**-**	**-**	**-**	**-**	**-**
India	-	-	-	-	-	-	-	-	-
Indonesia	2	-	-	-	-	-	-	-	-
Singapore	-	-	-	-	-	-	-	-	-
Other	-	-	-	-	-	-	-	-	-
Asia	**2**	**-**	**-**	**-**	**-**	**-**	**-**	**-**	**-**
China	**-**	**-**	**-**	**-**	**-**	**-**	**-**	**-**	**-**
Algeria	-	-	-	-	-	-	-	-	-
Angola/Cabinada	-	-	-	-	-	-	-	-	-
Cameroon	-	-	-	-	-	-	-	-	-
Egypt	-	-	-	-	-	-	-	-	-
Gabon	-	-	-	-	-	-	-	-	-
Libya	-	-	-	-	-	-	-	-	-
Nigeria	-	-	-	-	-	-	-	-	-
Tunisia	-	-	-	-	-	-	-	-	-
Other	-	-	-	-	-	-	-	-	-
Africa	**-**	**-**	**-**	**-**	**-**	**-**	**-**	**-**	**-**
Non-spec./Other	**1**	**-**	**6 e**	**-**	**-**	**-**	**-**	**-**	**-**
Total Imports	**7 733**	**8 801**	**8 219 e**	**31**	**39**	**39 e**	**40**	**44**	**18 e**
Memo: Total OPEC	*2*	-	-	-	-	-	-	-	-

Source: Annual Oil Statistics.

SWITZERLAND
Table 7C: Imports by Country of Origin

000 metric tons

	2000	2001	2002	2000	2001	2002	2000	2001	2002
	Total Gasoline			**Total Kerosene**			**Gas/Diesel**		
Australia	-	-	-	-	-	-	-	-	-
Belgium	154	122	236 e	208	242	73 e	810	892	744 e
Canada	-	-	-	-	-	-	-	-	-
Czech Republic	-	-	-	-	-	-	-	-	-
Denmark	-	-	-	-	-	-	-	-	-
Finland	-	-	1 e	-	-	-	-	-	-
France	564	610	716 e	565	487	269 e	610	718	385 e
Germany	998	975	973 e	197	257	513 e	385	268	286 e
Greece	-	-	-	-	-	-	-	-	-
Hungary	-	4	8 e	-	-	-	-	-	-
Ireland	-	-	-	-	-	-	-	-	-
Italy	315	255	249 e	8	9	15 e	355	526	580 e
Japan	-	-	-	-	-	-	-	-	-
Korea	-	-	-	-	-	-	-	-	-
Mexico	-	-	-	-	-	-	-	-	-
Netherlands	695	735	461 e	167	118	95 e	1 317	2 177	2 244 e
New Zealand	-	-	-	-	-	-	-	-	-
Norway	-	-	-	-	-	-	-	-	-
Poland	-	-	-	-	-	-	-	-	-
Spain	-	-	-	-	-	-	-	-	-
Sweden	-	-	-	-	-	-	-	-	-
Turkey	-	-	-	-	-	-	-	-	-
United Kingdom	-	-	-	-	-	-	-	-	-
United States	-	-	-	-	-	-	-	-	-
Other	-	-	-	-	-	-	-	-	-
OECD	**2 726**	**2 701**	**2 644 e**	**1 145**	**1 113**	**965 e**	**3 477**	**4 581**	**4 239 e**
Argentina	-	-	-	-	-	-	-	-	-
Brazil	-	-	-	-	-	-	-	-	-
Colombia	-	-	-	-	-	-	-	-	-
Ecuador	-	-	-	-	-	-	-	-	-
Netherlands Antilles	-	-	-	-	-	-	-	-	-
Peru	-	-	-	-	-	-	-	-	-
Trinidad and Tobago	-	-	-	-	-	-	-	-	-
Venezuela	-	-	-	-	-	-	-	-	-
Other	-	-	-	-	-	-	-	-	-
Latin America	**-**	**-**	**-**	**-**	**-**	**-**	**-**	**-**	**-**
Romania	-	-	-	-	-	-	-	-	-
Other	-	-	-	-	-	-	-	-	-
Non-OECD Europe	**-**	**-**	**-**	**-**	**-**	**-**	**-**	**-**	**-**
FSU	**-**	**-**	**-**	**-**	**-**	**-**	**-**	**-**	**-**
Iran	-	-	-	-	-	-	-	-	-
Iraq	-	-	-	-	-	-	-	-	-
Kuwait	-	-	-	-	-	-	-	-	-
Oman	-	-	-	-	-	-	-	-	-
Qatar	-	-	-	-	-	-	-	-	-
Saudi Arabia	-	-	-	-	-	-	-	-	-
United Arab Emirates	-	-	-	-	-	-	-	-	-
Other	-	-	-	-	-	-	-	-	-
Middle East	**-**	**-**	**-**	**-**	**-**	**-**	**-**	**-**	**-**
India	-	-	-	-	-	-	-	-	-
Indonesia	-	-	-	-	-	-	-	-	-
Singapore	-	-	-	-	-	-	-	-	-
Other	-	-	-	-	-	-	-	-	-
Asia	**-**	**-**	**-**	**-**	**-**	**-**	**-**	**-**	**-**
China	**-**	**-**	**-**	**-**	**-**	**-**	**-**	**-**	**-**
Algeria	-	-	-	-	-	-	-	-	-
Angola/Cabinada	-	-	-	-	-	-	-	-	-
Cameroon	-	-	-	-	-	-	-	-	-
Egypt	-	-	-	-	-	-	-	-	-
Gabon	-	-	-	-	-	-	-	-	-
Libya	-	-	-	-	-	-	-	-	-
Nigeria	-	-	-	-	-	-	-	-	-
Tunisia	-	-	-	-	-	-	-	-	-
Other	-	-	-	-	-	-	-	-	-
Africa	**-**	**-**	**-**	**-**	**-**	**-**	**-**	**-**	**-**
Non-spec./Other	**-**	**-**	**-**	**-**	**-**	**-**	**-**	**-**	**-**
Total Imports	**2 726**	**2 701**	**2 644 e**	**1 145**	**1 113**	**965 e**	**3 477**	**4 581**	**4 239 e**
Memo: Total OPEC	-	-	-	-	-	-	-	-	-

Source: Annual Oil Statistics.

SWITZERLAND
Table 7D: Imports by Country of Origin

000 metric tons

	2000	2001	2002	2000	2001	2002	2000	2001	2002
	Fuel Oil (Residual)			**Petroleum Coke**			**Other Products**		
Australia	-	-	-	-	-	-	-	-	-
Belgium	-	-	-	-	-	-	12	11	12 e
Canada	-	-	-	-	-	-	-	-	-
Czech Republic	-	-	-	-	-	-	-	1	-
Denmark	-	-	-	-	-	-	-	-	-
Finland	-	-	-	-	-	-	-	-	-
France	-	-	-	-	-	-	80	90	73 e
Germany	3	-	-	42	42	46 e	118	127	121 e
Greece	-	-	-	-	-	-	-	-	-
Hungary	-	-	-	-	-	-	2	-	-
Ireland	-	-	-	-	-	-	-	-	-
Italy	-	-	-	-	-	4 e	37	37	41 e
Japan	-	-	-	-	-	-	-	-	-
Korea	-	-	-	-	-	-	-	-	-
Mexico	-	-	-	-	-	-	-	-	-
Netherlands	1	-	-	-	-	-	10	8	8 e
New Zealand	-	-	-	-	-	-	-	-	-
Norway	-	-	-	-	-	-	-	-	-
Poland	-	-	-	-	-	-	-	-	-
Spain	-	-	-	-	-	-	-	-	-
Sweden	-	-	-	-	-	-	1	1	-
Turkey	-	-	-	-	-	-	-	-	-
United Kingdom	-	-	-	-	-	-	1	-	-
United States	-	-	-	-	4	3 e	-	-	-
Other	-	-	-	-	-	-	2	2	-
OECD	**4**	**-**	**-**	**42**	**46**	**53 e**	**263**	**277**	**255 e**
Argentina	-	-	-	2	-	-	-	-	-
Brazil	-	-	-	-	-	-	-	-	-
Colombia	-	-	-	-	-	-	-	-	-
Ecuador	-	-	-	-	-	-	-	-	-
Netherlands Antilles	-	-	-	-	-	-	-	-	-
Peru	-	-	-	-	-	-	-	-	-
Trinidad and Tobago	-	-	-	-	-	-	-	-	-
Venezuela	-	-	-	-	-	-	-	-	-
Other	-	-	-	-	-	-	-	-	-
Latin America	**-**	**-**	**-**	**2**	**-**	**-**	**-**	**-**	**-**
Romania	-	-	-	-	-	-	-	-	-
Other	-	-	-	-	-	-	-	-	-
Non-OECD Europe	**-**	**-**	**-**	**-**	**-**	**-**	**-**	**-**	**-**
FSU	**-**	**-**	**-**	**-**	**-**	**-**	**-**	**-**	**-**
Iran	-	-	-	-	-	-	-	-	-
Iraq	-	-	-	-	-	-	-	-	-
Kuwait	-	-	-	-	-	-	-	-	-
Oman	-	-	-	-	-	-	-	-	-
Qatar	-	-	-	-	-	-	-	-	-
Saudi Arabia	-	-	-	-	-	-	-	-	-
United Arab Emirates	-	-	-	-	-	-	-	-	-
Other	-	-	-	-	-	-	-	-	-
Middle East	**-**	**-**	**-**	**-**	**-**	**-**	**-**	**-**	**-**
India	-	-	-	-	-	-	-	-	-
Indonesia	-	-	-	2	-	-	-	-	-
Singapore	-	-	-	-	-	-	-	-	-
Other	-	-	-	-	-	-	-	-	-
Asia	**-**	**-**	**-**	**2**	**-**	**-**	**-**	**-**	**-**
China	**-**	**-**	**-**	**-**	**-**	**-**	**-**	**-**	**-**
Algeria	-	-	-	-	-	-	-	-	-
Angola/Cabinada	-	-	-	-	-	-	-	-	-
Cameroon	-	-	-	-	-	-	-	-	-
Egypt	-	-	-	-	-	-	-	-	-
Gabon	-	-	-	-	-	-	-	-	-
Libya	-	-	-	-	-	-	-	-	-
Nigeria	-	-	-	-	-	-	-	-	-
Tunisia	-	-	-	-	-	-	-	-	-
Other	-	-	-	-	-	-	-	-	-
Africa	**-**	**-**	**-**	**-**	**-**	**-**	**-**	**-**	**-**
Non-spec./Other	**-**	**-**	**-**	**-**	**-**	**-**	**1**	**-**	**6 e**
Total Imports	**4**	**-**	**-**	**46**	**46**	**53 e**	**264**	**277**	**261 e**
Memo: Total OPEC	*-*	*-*	*-*	*2*	*-*	*-*	*-*	*-*	*-*

Source: Annual Oil Statistics.

SWITZERLAND

Table 8A: Exports by Country of Destination

000 metric tons

	Crude Oil			Natural Gas Liquids			Feedstocks and Other		
	2000	**2001**	**2002**	**2000**	**2001**	**2002**	**2000**	**2001**	**2002**
Australia	-	-	-	-	-	-	-	-	-
Austria	-	-	-	-	-	-	-	-	-
Belgium	-	-	-	-	-	-	-	-	-
Canada	-	-	-	-	-	-	-	-	-
Denmark	-	-	-	-	-	-	-	-	-
Finland	-	-	-	-	-	-	-	-	-
France	-	-	-	-	-	-	-	-	-
Germany	-	-	-	-	-	-	-	-	-
Greece	-	-	-	-	-	-	-	-	-
Ireland	-	-	-	-	-	-	-	-	-
Italy	-	-	-	-	-	-	-	-	-
Japan	-	-	-	-	-	-	-	-	-
Luxembourg	-	-	-	-	-	-	-	-	-
Mexico	-	-	-	-	-	-	-	-	-
Netherlands	-	-	-	-	-	-	-	-	-
Norway	-	-	-	-	-	-	-	-	-
Portugal	-	-	-	-	-	-	-	-	-
Spain	-	-	-	-	-	-	-	-	-
Sweden	-	-	-	-	-	-	-	-	-
Switzerland	-	-	-	-	-	-	-	-	-
Turkey	-	-	-	-	-	-	-	-	-
United Kingdom	-	-	-	-	-	-	-	-	-
United States	-	-	-	-	-	-	-	-	-
Other OECD	-	-	-	-	-	-	-	-	-
OECD	**-**	**-**	**-**	**-**	**-**	**-**	**-**	**-**	**-**
Africa	-	-	-	-	-	-	-	-	-
Latin America	-	-	-	-	-	-	-	-	-
Asia	-	-	-	-	-	-	-	-	-
China	-	-	-	-	-	-	-	-	-
Non-OECD Europe	-	-	-	-	-	-	-	-	-
Former Soviet Union	-	-	-	-	-	-	-	-	-
Middle East	-	-	-	-	-	-	-	-	-
Non-specified/Other	-	-	-	-	-	-	-	-	-
Non-OECD	**-**	**-**	**-**	**-**	**-**	**-**	**-**	**-**	**-**
Total Exports	**-**	**-**	**-**	**-**	**-**	**-**	**-**	**-**	**-**

	Total Products			Liquefied Petroleum Gas			Naphtha		
	2000	**2001**	**2002**	**2000**	**2001**	**2002**	**2000**	**2001**	**2002**
Australia	-	-	-	-	-	-	-	-	-
Austria	-	6	9 e	-	-	-	-	-	-
Belgium	43	107	68 e	-	-	-	-	-	-
Canada	-	-	-	-	-	-	-	-	-
Denmark	-	-	-	-	-	-	-	-	-
Finland	-	-	-	-	-	-	-	-	-
France	109	139	225 e	-	-	-	-	-	-
Germany	162	80	102 e	2	2	10 e	20	-	-
Greece	-	-	-	-	-	-	-	-	-
Ireland	-	-	-	-	-	-	-	-	-
Italy	266	233	207 e	21	25	23 e	-	-	-
Japan	-	-	-	-	-	-	-	-	-
Luxembourg	-	-	-	-	-	-	-	-	-
Mexico	-	-	-	-	-	-	-	-	-
Netherlands	75	5	13 e	-	-	-	-	-	-
Norway	-	-	-	-	-	-	-	-	-
Portugal	-	-	-	-	-	-	-	-	-
Spain	-	-	-	-	-	-	-	-	-
Sweden	1	1	-	-	-	-	-	-	-
Switzerland	-	-	-	-	-	-	-	-	-
Turkey	-	-	-	-	-	-	-	-	-
United Kingdom	-	-	-	-	-	-	-	-	-
United States	-	-	-	-	-	-	-	-	-
Other OECD	2	1	1 e	1	-	1 e	-	-	-
OECD	**658**	**572**	**625 e**	**24**	**27**	**34 e**	**20**	**-**	**-**
Africa	-	-	-	-	-	-	-	-	-
Latin America	-	-	-	-	-	-	-	-	-
Asia	-	-	-	-	-	-	-	-	-
China	-	-	-	-	-	-	-	-	-
Non-OECD Europe	-	1	-	-	1	-	-	-	-
Former Soviet Union	-	-	-	-	-	-	-	-	-
Middle East	-	-	-	-	-	-	-	-	-
Non-specified/Other	2	2	1 e	-	1	1 e	-	-	-
Non-OECD	**2**	**3**	**1 e**	**-**	**2**	**1 e**	**-**	**-**	**-**
Total Exports	**660**	**575**	**626 e**	**24**	**29**	**35 e**	**20**	**-**	**-**

Source: Annual Oil Statistics.

SWITZERLAND
Table 8B: Exports by Country of Destination

000 metric tons

	2000	2001	2002	2000	2001	2002	2000	2001	2002
	Total Gasoline			Total Kerosene			Gas/Diesel Oil		
Australia	-	-	-	-	-	-	-	-	-
Austria	-	-	-	-	-	-	-	-	-
Belgium	-	-	-	-	-	-	-	-	-
Canada	-	-	-	-	-	-	-	-	-
Denmark	-	-	-	-	-	-	-	-	-
Finland	-	-	-	-	-	-	-	-	-
France	1	-	-	-	-	-	-	-	-
Germany	-	-	-	-	-	-	-	-	-
Greece	-	-	-	-	-	-	-	-	-
Ireland	-	-	-	-	-	-	-	-	-
Italy	-	-	-	-	-	-	-	-	-
Japan	-	-	-	-	-	-	-	-	-
Luxembourg	-	-	-	-	-	-	-	-	-
Mexico	-	-	-	-	-	-	-	-	-
Netherlands	-	-	-	-	-	-	-	-	-
Norway	-	-	-	-	-	-	-	-	-
Portugal	-	-	-	-	-	-	-	-	-
Spain	-	-	-	-	-	-	-	-	-
Sweden	-	-	-	-	-	-	-	-	-
Switzerland	-	-	-	-	-	-	-	-	-
Turkey	-	-	-	-	-	-	-	-	-
United Kingdom	-	-	-	-	-	-	-	-	-
United States	-	-	-	-	-	-	-	-	-
Other OECD	-	-	-	-	-	-	-	-	-
OECD	**1**	**-**	**-**	**-**	**-**	**-**	**-**	**-**	**-**
Africa	-	-	-	-	-	-	-	-	-
Latin America	-	-	-	-	-	-	-	-	-
Asia	-	-	-	-	-	-	-	-	-
China	-	-	-	-	-	-	-	-	-
Non-OECD Europe	-	-	-	-	-	-	-	-	-
Former Soviet Union	-	-	-	-	-	-	-	-	-
Middle East	-	-	-	-	-	-	-	-	-
Non-specified/Other	-	-	-	-	-	-	-	-	-
Non-OECD	**-**	**-**	**-**	**-**	**-**	**-**	**-**	**-**	**-**
Total Exports	**1**	**-**	**-**	**-**	**-**	**-**	**-**	**-**	**-**
	Fuel Oil (Residual)			Petroleum Coke			Other Products		
Australia	-	-	-	-	-	-	-	-	-
Austria	-	6	9 e	-	-	-	-	-	-
Belgium	43	107	68 e	-	-	-	-	-	-
Canada	-	-	-	-	-	-	-	-	-
Denmark	-	-	-	-	-	-	-	-	-
Finland	-	-	-	-	-	-	-	-	-
France	106	101	196 e	-	-	-	2	38	29 e
Germany	139	76	84 e	-	-	-	1	2	8 e
Greece	-	-	-	-	-	-	-	-	-
Ireland	-	-	-	-	-	-	-	-	-
Italy	245	208	184 e	-	-	-	-	-	-
Japan	-	-	-	-	-	-	-	-	-
Luxembourg	-	-	-	-	-	-	-	-	-
Mexico	-	-	-	-	-	-	-	-	-
Netherlands	75	5	13 e	-	-	-	-	-	-
Norway	-	-	-	-	-	-	-	-	-
Portugal	-	-	-	-	-	-	-	-	-
Spain	-	-	-	-	-	-	-	-	-
Sweden	-	-	-	-	-	-	1	1	-
Switzerland	-	-	-	-	-	-	-	-	-
Turkey	-	-	-	-	-	-	-	-	-
United Kingdom	-	-	-	-	-	-	-	-	-
United States	-	-	-	-	-	-	-	-	-
Other OECD	-	-	-	-	-	-	1	1	-
OECD	**608**	**503**	**554 e**	**-**	**-**	**-**	**5**	**42**	**37 e**
Africa	-	-	-	-	-	-	-	-	-
Latin America	-	-	-	-	-	-	-	-	-
Asia	-	-	-	-	-	-	-	-	-
China	-	-	-	-	-	-	-	-	-
Non-OECD Europe	-	-	-	-	-	-	-	-	-
Former Soviet Union	-	-	-	-	-	-	-	-	-
Middle East	-	-	-	-	-	-	-	-	-
Non-specified/Other	1	-	-	-	-	-	1	1	-
Non-OECD	**1**	**-**	**-**	**-**	**-**	**-**	**1**	**1**	**-**
Total Exports	**609**	**503**	**554 e**	**-**	**-**	**-**	**6**	**43**	**37 e**

Source: Annual Oil Statistics.

TURKEY

Figure 1
FUEL SHARES OF TPES *

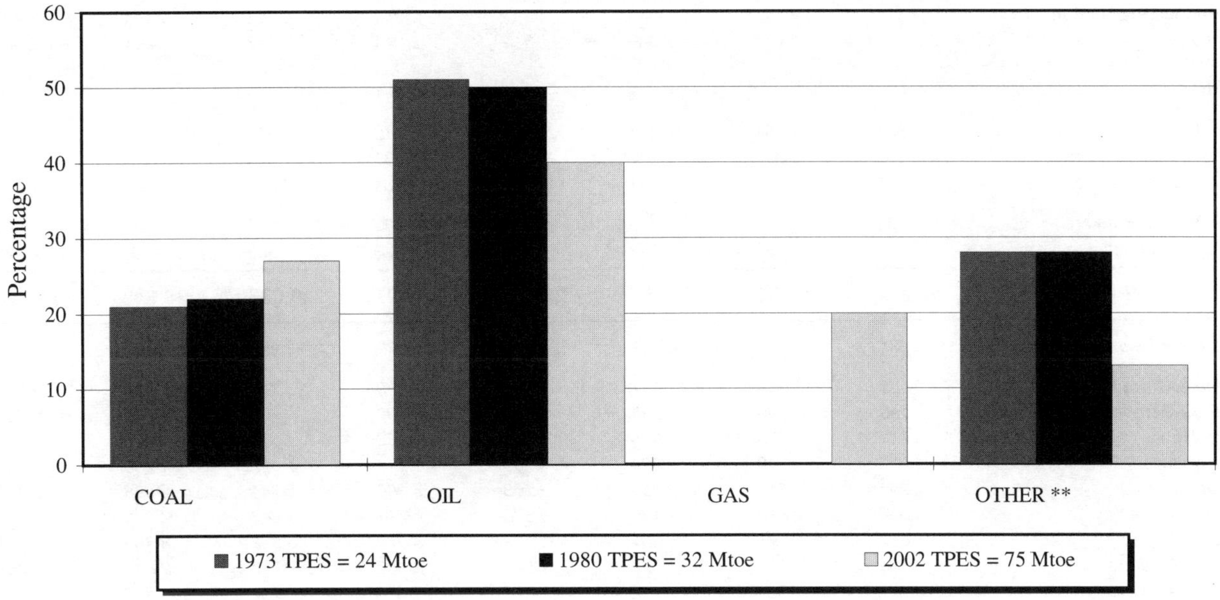

Source: Annual Energy Balances.

* Excluding electricity trade.

** Includes hydro, geothermal, solar, wind, combustible renewables and waste, etc.

Figure 2
OIL PRODUCTION AND CONSUMPTION

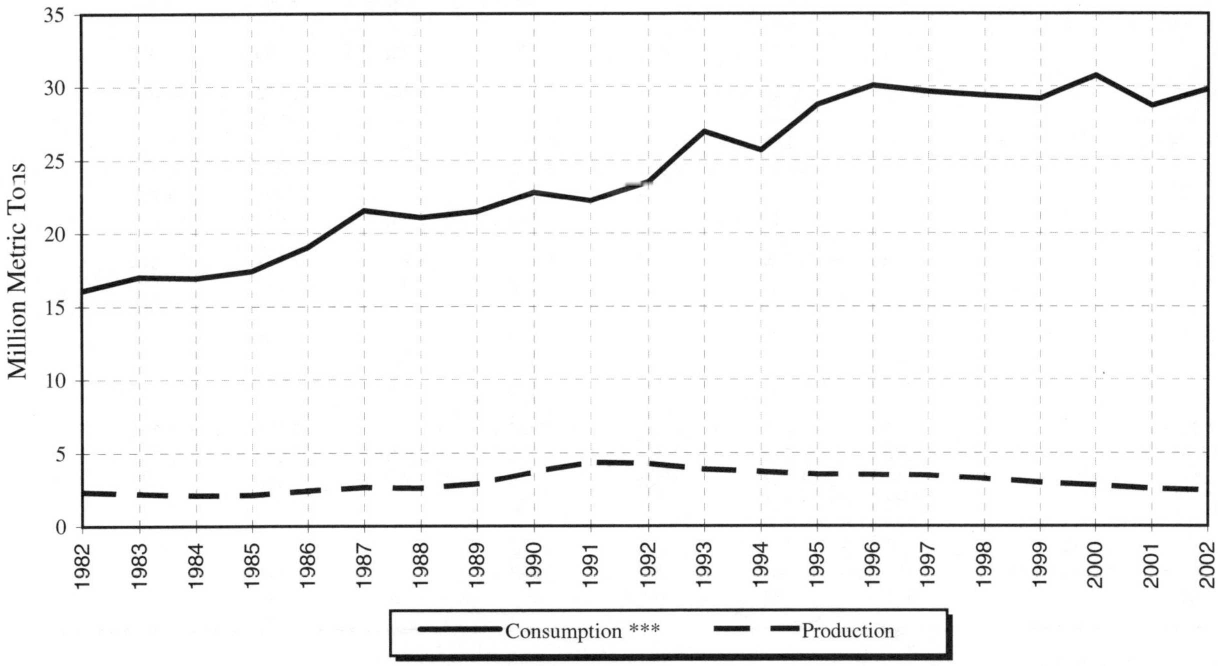

Source: Annual Oil Statistics.

*** Includes refinery fuel and international marine bunkers.

TURKEY
Table 1: Oil Supply and Consumption - Summary Table

000 metric tons

	1973	1978	1980	1990	1999	2000	2001	2002
Indigenous Production	3 511	2 736	2 330	3 712	2 941	2 762	2 520	2 420 e
Crude Oil	..	2 736	2 330	3 712	2 941	2 762	2 520	2 420 e
NGL	..	-	-	-	-	-	-	-
Other	..	-	-	-	-	-	-	-
+ Net Imports	8 661	14 251	13 281	20 270	26 119	28 942	26 463	27 115 e
Crude Oil, NGL, Other	9 306	10 354	10 490	20 062	22 985	21 583	23 242	23 662 e
Petroleum Products	- 645	3 897	2 791	208	3 134	7 359	3 221	3 453 e
- Intl. Marine Bunkers	90	-	-	120	283	405	237	409 e
+ Stock Changes	177	- 31	- 343	-1 137	420	- 663	- 97	448 e
TOTAL SUPPLY	**12 259**	**16 956**	**15 268**	**22 725**	**29 197**	**30 636**	**28 649**	**29 574 e**
Refinery Intake	*12 981*	*13 214*	*12 839*	*23 026*	*25 979*	*23 957*	*25 560*	*26 306 e*
Refinery Output	*12 533*	*13 053*	*12 657*	*22 884*	*25 653*	*23 745*	*25 374*	*26 088 e*
LPG + Ethane	*399*	*390*	*394*	*700*	*737*	*705*	*712*	*751 e*
Naphtha	*241*	*395*	*504*	*1 466*	*1 882*	*1 852*	*2 026*	*1 454 e*
Kerosenes	*807*	*748*	*624*	*786*	*1 144*	*1 050*	*1 230*	*1 652 e*
Gasolines	*1 901*	*2 217*	*2 083*	*3 103*	*3 300*	*2 744*	*2 968*	*3 752 e*
Gas/Diesel Oil	*3 055*	*2 670*	*3 106*	*6 566*	*7 711*	*6 646*	*7 340*	*7 721 e*
Fuel Oil	*5 697*	*5 797*	*5 264*	*8 723*	*8 379*	*8 022*	*8 456*	*7 971 e*
Other Products	*433 e*	*836 e*	*682*	*1 540*	*2 500*	*2 726*	*2 642*	*2 787 e*
- Refinery Losses	448	161	182	142	326	212	186	218 e
+ Returns	-	87	73	62	62	102	104	73 e
- Statistical Differences	289	- 87	3	- 24	37	180	118	2 e
TOTAL CONSUMPTION	**11 522**	**16 969**	**15 156**	**22 669**	**28 896**	**30 346**	**28 449**	**29 427 e**
Transformation and Energy	**1 925**	**2 938**	**2 465**	**2 476**	**3 882**	**4 451**	**4 380**	..
Crude Oil	-	11	6	-	-	-	-	..
LPG + Ethane	-	-	-	-	68	64	27	..
Gas/Diesel Oil	193	352	231	108	279	551	254	..
Fuel Oil	1 543	2 302	1 943	1 891	2 829	3 043	3 283	..
Petroleum Coke	-	-	-	-	-	-	-	..
Other	189	273	285	477	706	793	816	..
Transport	**3 831**	**6 025**	**5 185**	**9 227**	**11 405**	**11 976**	**11 440**	..
Motor Gasoline	1 341	2 431	1 953	3 196	4 257	3 619	2 987	..
Gas/Diesel Oil	1 503	3 206	2 934	5 300	4 941	5 455	5 648	..
Subtotal	*2 844*	*5 637*	*4 887*	*8 496*	*9 198*	*9 074*	*8 635*	..
Aviation Fuels	142	172	133	467	1 151	1 228	1 208	..
Other	845	216	165	264	1 056	1 674	1 597	..
Industry	**2 682**	**4 725**	**4 322**	**6 085**	**7 711**	**7 957**	**7 318**	..
Naphtha	146	383	333	1 470	1 367	1 269	1 371	..
LPG + Ethane	-	11	10	46	85	469	395	..
Gas/Diesel Oil	109	72	67	169	284	260	142	..
Fuel Oil	1 995	3 554	3 406	3 430	3 989	4 048	3 562	..
Other	432	705	506	970	1 986	1 911	1 848 e	..
Other Sectors	**3 084**	**3 281**	**3 184**	**4 881**	**5 898**	**5 962**	**5 311**	..
Gas/Diesel Oil	930	852	897	1 870	2 607	2 607	2 597	..
Fuel Oil	1 166	1 127	1 122	1 275	743	636	596	..
Other	988	1 302	1 165	1 736	2 548	2 719	2 118	..
Memo: Non-Energy Use	*578*	*1 089*	*841*	*2 604*	*3 608*	*3 461*	*3 354*	..
Chem/Petchem. Feedst.	*157*	*400*	*350*	*1 533*	*1 400*	*1 301*	*1 439*	..
Other	*421*	*689*	*491*	*1 071*	*2 208*	*2 160*	*1 915*	..

Source: Annual Oil Statistics.

TURKEY
Table 2: Supply of Crude Oil, NGL, Feedstocks and Other Hydrocarbons

000 metric tons

	1973	1978	1980	1990	1999	2000	2001	2002
CRUDE OIL								
Indigenous Production	..	2 736	2 330	3 712	2 941	2 762	2 520	2 420 e
+ Imports	..	10 354	10 490	20 062	22 985	21 583	23 242	23 662 e
- Exports	..	-	-	-	-	-	-	-
- Direct Use	..	11	6	-	-	-	-	-
+ Stock Change	..	47	- 32	- 777	- 211	- 273	- 125	45 e
- Statistical Difference	..	-	-	- 24	- 2	180	118	2 e
Refinery Intake	**..**	**13 126**	**12 782**	**23 021**	**25 717**	**23 892**	**25 519**	**26 125 e**
Closing Stock Level	*..*	*448*	*308*	*2 240*	*4 592*	*4 865*	*4 990*	*4 945 e*
NATURAL GAS LIQUIDS								
Indigenous Production	..	-	-	-	-	-	-	-
+ Imports	..	-	-	-	-	-	-	-
- Exports	..	-	-	-	-	-	-	-
- Direct Use	..	-	-	-	-	-	-	-
+ Stock Change	..	-	-	-	-	-	-	-
- Statistical Difference	..	-	-	-	-	-	-	-
Refinery Intake	**..**	**-**	**-**	**-**	**-**	**-**	**-**	**-**
Closing Stock Level	*..*	-	-	-	-	-	-	-
REFINERY FEEDSTOCKS								
Backflows	..	87	73	62	62	102	104	73 e
+ Transfers	..	-	-	-	-	-	-	-
+ Imports	..	-	-	-	-	-	-	-
- Exports	..	-	-	-	-	-	-	-
+ Stock Change	..	1	- 16	- 57	200	- 37	- 63	108 e
- Statistical Difference	..	-	-	-	-	-	-	-
Refinery Intake	**..**	**88**	**57**	**5**	**262**	**65**	**41**	**181 e**
Closing Stock Level	*..*	*102*	*140*	*280*	*190*	*227*	*290*	*182 e*
OTHER HYDROCARBONS								
Indigenous Production	..	-	-	-	-	-	-	-
+ Imports	..	-	-	-	-	-	-	-
- Exports	..	-	-	-	-	-	-	-
- Direct Use	..	-	-	-	-	-	-	-
+ Stock Change	..	-	-	-	-	-	-	-
- Statistical Difference	..	-	-	-	-	-	-	-
Refinery Intake	**..**	**-**	**-**	**-**	**-**	**-**	**-**	**-**
Closing Stock Level	*..*	-	-	-	-	-	-	-
CRUDE OIL, NGL, FEEDSTOCKS AND OTHER								
Indigenous Production	3 511	2 736	2 330	3 712	2 941	2 762	2 520	2 420 e
+ Backflows	-	87	73	62	62	102	104	73 e
+ Transfers	-	-	-	-	-	-	-	-
+ Imports	9 306	10 354	10 490	20 062	22 985	21 583	23 242	23 662 e
- Exports	-	-	-	-	-	-	-	-
- Direct Use	-	11	6	-	-	-	-	-
+ Stock Change	164	48	- 48	- 834	- 11	- 310	- 188	153 e
- Statistical Difference	-	-	-	- 24	- 2	180	118	2 e
Refinery Intake	**12 981**	**13 214**	**12 839**	**23 026**	**25 979**	**23 957**	**25 560**	**26 306 e**
Closing Stock Level	-	*550*	*448*	*2 520*	*4 782*	*5 092*	*5 280*	*5 127 e*

Source: Annual Oil Statistics.

TURKEY
Table 3: Motor Gasoline - Supply

000 metric tons

	1973	1978	1980	1990	1999	2000	2001	2002
MOTOR GASOLINE								
Refinery Output	1 901	2 217	2 083	3 103	3 300	2 744	2 968	3 752 e
+ Imports	-	263	9	154	1 145	837	437	365 e
- Exports	368	18	70	48	42	-	174	1 113 e
+ Transfers	-	-	-	-	-	-	-	-
+ Stock Change	-	- 31	- 70	- 13	- 146	38	- 179	137 e
- Statistical Difference	192	-	- 1	-	-	-	-	-
Total Consumption	**1 341**	**2 431**	**1 953**	**3 196**	**4 257**	**3 619**	**3 052**	**3 141 e**
Closing Stock Level	-	*150*	*155*	*257*	*524*	*486*	*665*	*528 e*

Table 4: Aviation Fuels - Supply

000 metric tons

	1973	1978	1980	1990	1999	2000	2001	2002
AVIATION GASOLINE								
Refinery Output	-	-	-	-	-	-	-	-
+ Imports	-	-	-	-	-	-	-	-
- Exports	-	-	-	-	-	-	-	-
+ Transfers	-	-	-	-	-	-	-	-
+ Stock Change	-	-	-	-	-	-	-	-
- Statistical Difference	-	-	-	-	-	-	-	-
Total Consumption	-	-	-	-	-	-	-	-
Closing Stock Level	-	-	-	-	-	-	-	-
GASOLINE TYPE JET FUEL								
Refinery Output	-	-	-	-	-	-	-	-
+ Imports	-	-	-	-	-	-	-	-
- Exports	-	-	-	-	-	-	-	-
+ Transfers	-	-	-	-	-	-	-	-
+ Stock Change	-	-	-	-	-	-	-	-
- Statistical Difference	-	-	-	-	-	-	-	-
Total Consumption	-	-	-	-	-	-	-	-
Closing Stock Level	-	-	-	-	-	-	-	-
KEROSENE TYPE JET FUEL								
Refinery Output	241	171	149	621	1 099	1 024	1 199	1 607 e
+ Imports	-	-	-	-	85	166	-	-
- Exports	94	5	7	150	35	2	5	9 e
+ Transfers	-	-	-	-	-	-	-	-
+ Stock Change	-	6	- 8	- 4	2	40	14	45 e
- Statistical Difference	5	-	1	-	-	-	-	-
Total Consumption	**142**	**172**	**133**	**467**	**1 151**	**1 228**	**1 208**	**1 643 e**
Closing Stock Level	-	*6*	*15*	*28*	*168*	*128*	*114*	*69 e*

Source: Annual Oil Statistics.

TURKEY
Table 5A: Gas/Diesel Oil - Supply and Consumption

000 metric tons

	1973	1978	1980	1990	1999	2000	2001	2002
Refinery Output	**3 055**	**2 670**	**3 106**	**6 566**	**7 711**	**6 646**	**7 340**	**7 721 e**
+ Recycled Products	-	-	-	-	-	-	-	-
+ Imports	-	1 836	1 129	964	1 111	2 470	1 596	2 136 e
- Exports	186	28	27	1	701	26	208	582 e
- Intl. Marine Bunkers	-	-	-	72	145	192	105	219 e
+ Transfers	-	-	-	-	-	-	-	-
+ Stock Changes	-	4	- 79	- 10	135	- 25	18	88 e
- Statistical Differences	134	-	-	-	-	-	-	1 e
Total Consumption	**2 735**	**4 482**	**4 129**	**7 447**	**8 111**	**8 873**	**8 641**	**9 143 e**
Transformation	**180**	**346**	**229**	**108**	**279**	**551**	**253**	**..**
Public Electricity	180	345	229	107	279	551	253	..
Autoproducer Electricity	-	-	-	-	-	-	-	..
Public CHP Plants	-	-	-	-	-	-	-	..
Autoproducers of CHP	-	-	-	-	-	-	-	..
Public Heat Plants	-	-	-	-	-	-	-	..
Autoproducer Heat Plants	-	-	-	-	-	-	-	..
Petrochemical Industry	-	-	-	1	-	-	-	..
Other Transformation	-	1	-	-	-	-	-	..
Energy Sector	**13**	**6**	**2**	**-**	**-**	**-**	**1**	**..**
Refinery Fuel	13	6	2	-	-	-	-	..
Other Energy Sector	-	-	-	-	-	-	1	..
Final Consumption	**2 542**	**4 130**	**3 898**	**7 339**	**7 832**	**8 322**	**8 387**	**..**
Transport	**1 503**	**3 206**	**2 934**	**5 300**	**4 941**	**5 455**	**5 648**	**..**
Air	-	-	-	-	-	-	-	..
Road	1 422	3 015	2 747	5 040	4 636	5 150	5 328	..
Rail	81	114	124	170	200	200	165	..
Inland Waterways	-	77	63	90	105	105	155	..
Non-Specified	-	-	-	-	-	-	-	..
Industry	**109**	**72**	**67**	**169**	**284**	**260**	**142**	**..**
Iron and Steel	-	4	8	14	7	3	6	..
Chemical (incl. Pet.-Chem.)	-	1	2	31	-	-	-	..
Non-Ferrous Metals	-	15	15	-	-	-	-	..
Non-Metallic Minerals	-	-	-	-	2	2	2	..
Transport Equipment	-	-	-	-	-	-	-	..
Machinery	-	2	1	2	-	-	-	..
Mining and Quarrying	-	-	-	-	-	-	-	..
Food and Tobacco	-	-	-	-	-	-	-	..
Pulp, Paper and Printing	-	-	-	-	-	-	-	..
Wood and Wood Products	-	-	-	-	-	-	-	..
Construction	-	-	-	40	-	-	-	..
Textiles and Leather	-	15	18	35	-	-	-	..
Non-Specified	109	35	23	47	275	255	134	..
Other Sectors	**930**	**852**	**897**	**1 870**	**2 607**	**2 607**	**2 597**	**..**
Commerce and Public	154	-	-	-	-	-	-	..
Residential	-	-	-	-	-	-	-	..
Agriculture	776	852	897	1 870	2 607	2 607	2 597	..
Non-Specified	-	-	-	-	-	-	-	..
Memo: Non-Energy Use	-	*1*	*2*	*31*	-	-	-	..
Chem/Petchem. Feedst.	-	*1*	*2*	*31*	-	-	-	..
Other	-	-	-	-	-	-	-	..
Closing Stock Level	-	*90*	*131*	*375*	*640*	*665*	*647*	*559 e*

Source: Annual Oil Statistics.

TURKEY
Table 5B: Fuel Oil (Residual) - Supply and Consumption

000 metric tons

	1973	1978	1980	1990	1999	2000	2001	2002
Refinery Output	**5 697**	**5 797**	**5 264**	**8 723**	**8 379**	**8 022**	**8 456**	**7 971 e**
+ Recycled Products	-	-	-	-	-	-	-	-
+ Imports	20	1 412	1 496	22	170	701	226	782 e
- Exports	78	2	2	1 693	1 109	575	1 007	443 e
- Intl. Marine Bunkers	90	-	-	48	138	213	132	190 e
+ Transfers	-	-	-	-	-	-	-	-
+ Stock Changes	-	- 8	- 120	- 274	320	- 118	- 7	51 e
- Statistical Differences	-	-	2	-	- 40	-	-	-
Total Consumption	**5 549**	**7 199**	**6 636**	**6 730**	**7 662**	**7 817**	**7 536**	**8 171 e**
Transformation	**1 161**	**1 653**	**1 308**	**1 127**	**1 905**	**2 105**	**2 255**	**..**
Public Electricity	1 132	1 264	974	336	1 107	1 342	1 448	..
Autoproducer Electricity	-	-	-	69	85	43	52	..
Public CHP Plants	-	-	-	-	-	-	-	..
Autoproducers of CHP	-	380	323	722	713	720	755	..
Public Heat Plants	-	-	-	-	-	-	-	..
Autoproducer Heat Plants	-	-	-	-	-	-	-	..
Petrochemical Industry	-	-	-	-	-	-	-	..
Other Transformation	29	9	11	-	-	-	-	..
Energy Sector	**382**	**649**	**635**	**764**	**924**	**938**	**1 028**	**..**
Refinery Fuel	382	649	635	764	924	938	1 028	..
Other Energy Sector	-	-	-	-	-	-	-	..
Final Consumption	**4 006**	**4 897**	**4 693**	**4 839**	**4 833**	**4 774**	**4 253**	**..**
Transport	**845**	**216**	**165**	**134**	**101**	**90**	**95**	**..**
Air	-	-	-	-	-	-	-	..
Road	-	-	-	-	-	-	-	..
Rail	287	111	66	44	-	-	-	..
Inland Waterways	558	105	99	90	101	90	95	..
Non-Specified	-	-	-	-	-	-	-	..
Industry	**1 995**	**3 554**	**3 406**	**3 430**	**3 989**	**4 048**	**3 562**	**..**
Iron and Steel	91	351	308	583	429	460	423	..
Chemical (incl. Pet.-Chem.)	181	366	353	680	821	890	790	..
Non-Ferrous Metals	-	138	155	500	277	277	276	..
Non-Metallic Minerals	-	-	-	-	55	63	59	..
Transport Equipment	-	-	-	-	-	-	-	..
Machinery	-	19	10	4	-	-	-	..
Mining and Quarrying	-	-	-	-	-	-	-	..
Food and Tobacco	-	-	-	150	125	140	130	..
Pulp, Paper and Printing	-	233	243	150	200 e	200 e	208	..
Wood and Wood Products	-	-	-	-	-	-	-	..
Construction	-	-	-	-	-	-	-	..
Textiles and Leather	-	240	270	350	290 e	290 e	301	..
Non-Specified	1 723	2 207	2 067	1 013	1 792 e	1 728 e	1 375	..
Other Sectors	**1 166**	**1 127**	**1 122**	**1 275**	**743**	**636**	**596**	**..**
Commerce and Public	-	-	-	-	-	-	-	..
Residential	1 166	1 127	1 122	1 275	743	636	596	..
Agriculture	-	-	-	-	-	-	-	..
Non-Specified	-	-	-	-	-	-	-	..
Memo: Non-Energy Use	-	-	-	-	-	-	-	..
Chem/Petchem. Feedst.	-	-	-	-	-	-	-	..
Other	-	-	-	-	-	-	-	..
Closing Stock Level	-	*179*	*211*	*604*	*305*	*423*	*430*	*379 e*

Source: Annual Oil Statistics.

TURKEY

Table 5C: Total Products - Supply and Consumption

000 metric tons

	1973	1978	1980	1990	1999	2000	2001	2002
Refinery Output	**12 533**	**13 053**	**12 657**	**22 884**	**25 653**	**23 745**	**25 374**	**26 088 e**
+ Recycled Products	-	-	-	-	-	-	-	-
+ Imports	171	3 951	3 019	2 162	5 593	8 669	5 791	6 505 e
- Exports	816	54	228	1 954	2 459	1 310	2 570	3 052 e
- Intl. Marine Bunkers	90	-	-	120	283	405	237	409 e
+ Transfers	-	11	6	-	-	-	-	-
+ Stock Changes	13	- 79	- 295	- 303	431	- 353	91	295 e
- Statistical Differences	289	- 87	3	-	39	-	-	-
Total Consumption	**11 522**	**16 969**	**15 156**	**22 669**	**28 896**	**30 346**	**28 449**	**29 427 e**
Transformation	**1 341**	**2 086**	**1 610**	**1 299**	**2 379**	**2 957**	**2 731**	..
Public Electricity	1 312	1 609	1 203	443	1 386	1 893	1 701	..
Autoproducer Electricity	-	-	-	69	85	43	52	..
Public CHP Plants	-	-	-	-	-	-	-	..
Autoproducers of CHP	-	380	323	722	841	914	874	..
Public Heat Plants	-	-	-	-	-	-	-	..
Autoproducer Heat Plants	-	-	-	-	-	-	-	..
Petrochemical Industry	-	87	73	62	62	102	104	..
Other Transformation	29	10	11	3	5	5	-	..
Energy Sector	**584**	**852**	**855**	**1 177**	**1 503**	**1 494**	**1 649**	..
Refinery Fuel	584	841	849	1 177	1 503	1 494	1 648	..
Other Energy Sector	-	11	6	-	-	-	1	..
Final Consumption	**9 597**	**14 031**	**12 691**	**20 193**	**25 014**	**25 895**	**24 069**	..
Transport	**3 831**	**6 025**	**5 185**	**9 227**	**11 405**	**11 976**	**11 440**	..
Air	142	172	133	467	1 151	1 228	1 208	..
Road	2 763	5 381	4 649	8 286	9 808	10 307	9 782	..
Rail	368	225	190	224	225	228	185	..
Inland Waterways	558	247	213	250	221	213	265	..
Non-Specified	-	-	-	-	-	-	-	..
Industry	**2 682**	**4 725**	**4 322**	**6 085**	**7 711**	**7 957**	**7 318**	..
Iron and Steel	91	356	317	597	436	463	429	..
Chemical (incl. Pet.-Chem.)	338	766	703	2 231	2 222	2 192	2 229	..
Non-Ferrous Metals	-	153	170	500	277	277	276	..
Non-Metallic Minerals	-	-	-	11	60	67	63	..
Transport Equipment	-	-	-	-	-	-	-	..
Machinery	-	21	11	6	-	-	-	..
Mining and Quarrying	-	-	-	-	-	-	-	..
Food and Tobacco	-	-	-	150	130	145	135	..
Pulp, Paper and Printing	-	233	243	150	260	258	268	..
Wood and Wood Products	-	-	-	-	-	-	-	..
Construction	233	431	300	729	1 310	1 255	1 149	..
Textiles and Leather	-	255	288	385	305	305	326	..
Non-Specified	2 020	2 510	2 290	1 326	2 711	2 995	2 443	..
Other Sectors	**3 084**	**3 281**	**3 184**	**4 881**	**5 898**	**5 962**	**5 311**	..
Commerce and Public	154	-	-	-	-	-	-	..
Residential	2 154	2 429	2 287	3 011	3 291	3 355	2 713	..
Agriculture	776	852	897	1 870	2 607	2 607	2 598	..
Non-Specified	-	-	-	-	-	-	-	..
Memo: Non-Energy Use	*578*	*1 089*	*841*	*2 604*	*3 608*	*3 461*	*3 354*	..
Chem/Petchem. Feedst.	*157*	*400*	*350*	*1 533*	*1 400*	*1 301*	*1 439*	..
Other	*421*	*689*	*491*	*1 071*	*2 208*	*2 160*	*1 915*	..
Closing Stock Level	-	*643*	*710*	*1 542*	*1 936*	*2 289*	*2 198*	*1 903 e*

Source: Annual Oil Statistics.

TURKEY
Table 6A: Supply and Consumption

000 metric tons

	1980	2000	2001	2002	1980	2000	2001	2002
	LPG AND ETHANE				**REFINERY GAS**			
Refinery Output	**394**	**705**	**712**	**751 e**	**212**	**556**	**620**	**631 e**
+ Recycled Products	-	-	-	-	-	-	-	-
+ Imports	348	3 878	3 129	2 566 e	-	-	-	-
- Exports	-	-	13	14 e	-	-	-	-
- Intl. Marine Bunkers	-	-	-	-	-	-	-	-
+ Transfers	-	-	-	-	-	-	-	-
+ Stock Change	-	- 60	- 15	- 19 e	-	-	-	-
- Statistical Difference	-	-	-	-	-	-	-	-
Total Consumption	**742**	**4 523**	**3 813**	**3 284 e**	**212**	**556**	**620**	**631 e**
of which: Ethane	-	-	-	-	-	-	-	-
Transformation	-	64	27	..	-	-	-	..
Energy	-	-	-	..	212	556	620	..
Transport	-	1 308	1 302	..	-	-	-	..
Industry	10	469	395	..	-	-	-	..
Other	732	2 682	2 089	..	-	-	-	..
Memo: Non-Energy Use	-	-	-	..	-	-	-	..
Closing Stock Level	*3*	*81*	*96*	*115 e*	-	-	-	-
	NAPHTHA				**OTHER KEROSENE**			
Refinery Output	**504**	**1 852**	**2 026**	**1 454 e**	**475**	**26**	**31**	**45 e**
+ Recycled Products	-	-	-	-	-	-	-	-
+ Imports	-	132	134	237 e	-	-	-	4 e
- Exports	100	411	680	401 e	1	-	-	10 e
- Intl. Marine Bunkers	-	-	-	-	-	-	-	-
+ Transfers	-	-	-	-	-	-	-	-
+ Stock Change	- 2	- 74	82	- 20 e	- 41	11	- 2	- 9 e
- Statistical Difference	-	-	-	-	-	-	-	- 1 e
Total Consumption	**402**	**1 499**	**1 562**	**1 270 e**	**433**	**37**	**29**	**31 e**
Transformation	69	230	191	..	-	-	-	..
Energy	-	-	-	..	-	-	-	..
Transport	-	-	-	..	-	-	-	..
Industry	333	1 269	1 371	..	-	-	-	..
Other	-	-	-	..	433	37	29	..
Memo: Non-Energy Use	*333*	*1 269*	*1 371*	..	-	-	-	..
Closing Stock Level	*12*	*122*	*40*	*60 e*	*58*	*24*	*26*	*35 e*
	WHITE SPIRIT + SBP				**LUBRICANTS**			
Refinery Output	**16**	**3**	**2**	**..**	**158**	**317**	**248**	**..**
+ Recycled Products	-	-	-	..	-	-	-	..
+ Imports	-	117	178	..	30	161	69	..
- Exports	-	-	-	..	21	-	2	..
- Intl. Marine Bunkers	-	-	-	..	-	-	-	..
+ Transfers	-	-	-	..	-	-	-	..
+ Stock Change	3	- 8	-	..	14	- 18	20	..
- Statistical Difference	-	-	-	..	-	-	-	..
Total Consumption	**19**	**112**	**180**	**..**	**181**	**460**	**335**	**..**
Transformation	4	2	5 e	..	-	-	-	..
Energy	-	-	-	..	-	-	-	..
Transport	-	-	-	..	-	276	200 e	..
Industry	15	110	175	..	181	184	135 e	..
Other	-	-	-	..	-	-	-	..
Memo: Non-Energy Use	*15*	*110*	*175*	..	*181*	*460*	*335*	..
Closing Stock Level	*3*	*20*	*20*	..	*104*	*73*	*53*	..

Source: Annual Oil Statistics.

TURKEY
Table 6B: Supply and Consumption

000 metric tons

	1980	2000	2001	2002	1980	2000	2001	2002
	BITUMEN				**PARAFFIN WAXES**			
Refinery Output	**292**	**1 282**	**1 099**	**..**	**-**	**35**	**31**	**..**
+ Recycled Products	-	-	-	..	-	-	-	..
+ Imports	-	-	-	..	5	-	-	..
- Exports	-	-	5	..	-	-	-	..
- Intl. Marine Bunkers	-	-	-	..	-	-	-	..
+ Transfers	-	-	-	..	-	-	-	..
+ Stock Change	9	- 22	55	..	-	- 25	- 3	..
- Statistical Difference	1	-	-	..	-	-	-	..
Total Consumption	**300**	**1 260**	**1 149**	**..**	**5**	**10**	**28**	**..**
Transformation	-	5	-	..	-	-	-	..
Energy	-	-	-	..	-	-	-	..
Transport	-	-	-	..	-	-	-	..
Industry	300	1 255	1 149	..	5	10	28	..
Other	-	-	-	..	-	-	-	..
Memo: Non-Energy Use	*300*	*1 260*	*1 149*	..	*5*	*10*	*28*	..
Closing Stock Level	*16*	*96*	*41*	..	-	*27*	*30*	..
	PETROLEUM COKE				**OTHER PRODUCTS**			
Refinery Output	**-**	**-**	**-**	**-**	**4**	**533**	**642**	**2 156 e**
+ Recycled Products	-	-	-	-	-	-	-	-
+ Imports	-	-	-	-	2	207	22	415 e
- Exports	-	-	-	-	-	296	476	480 e
- Intl. Marine Bunkers	-	-	-	-	-	-	-	-
+ Transfers	-	-	-	-	6	-	-	-
+ Stock Change	-	-	-	-	- 1	- 92	108	22 e
- Statistical Difference	-	-	-	-	-	-	-	-
Total Consumption	**-**	**-**	**-**	**-**	**11**	**352**	**296**	**2 113 e**
Transformation	-	-	-	..	-	-	-	..
Energy	-	-	-	..	6	-	-	..
Transport	-	-	-	..	-	-	-	..
Industry	-	-	-	..	5	352	296	..
Other	-	-	-	..	-	-	-	..
Memo: Non-Energy Use	-	-	-	..	*5*	*352*	*296*	..
Closing Stock Level	-	-	-	-	*2*	*144*	*36*	*158 e*

Source: Annual Oil Statistics.

TURKEY
Table 7A: Imports by Country of Origin

000 metric tons

	2000	2001	2002	2000	2001	2002	2000	2001	2002
	Crude Oil			**Natural Gas Liquids**			**Feedstocks and Other**		
Australia	-	-	-	-	-	-	-	-	-
Belgium	-	-	-	-	-	-	-	-	-
Canada	-	-	-	-	-	-	-	-	-
Czech Republic	-	-	-	-	-	-	-	-	-
Denmark	-	-	-	-	-	-	-	-	-
Finland	-	-	-	-	-	-	-	-	-
France	-	-	-	-	-	-	-	-	-
Germany	-	-	-	-	-	-	-	-	-
Greece	-	-	-	-	-	-	-	-	-
Hungary	-	-	-	-	-	-	-	-	-
Ireland	-	-	-	-	-	-	-	-	-
Italy	-	-	504 e	-	-	-	-	-	-
Japan	-	-	-	-	-	-	-	-	-
Korea	-	-	-	-	-	-	-	-	-
Mexico	-	-	-	-	-	-	-	-	-
Netherlands	-	-	-	-	-	-	-	-	-
New Zealand	-	-	-	-	-	-	-	-	-
Norway	-	-	-	-	-	-	-	-	-
Poland	-	-	-	-	-	-	-	-	-
Spain	-	-	-	-	-	-	-	-	-
Sweden	-	-	-	-	-	-	-	-	-
Turkey	-	-	-	-	-	-	-	-	-
United Kingdom	-	-	-	-	-	-	-	-	-
United States	-	-	-	-	-	-	-	-	-
Other	-	-	-	-	-	-	-	-	-
OECD	**-**	**-**	**504 e**	**-**	**-**	**-**	**-**	**-**	**-**
Argentina	-	-	-	-	-	-	-	-	-
Brazil	-	-	-	-	-	-	-	-	-
Colombia	-	-	-	-	-	-	-	-	-
Ecuador	-	-	-	-	-	-	-	-	-
Netherlands Antilles	-	-	-	-	-	-	-	-	-
Peru	-	-	-	-	-	-	-	-	-
Trinidad and Tobago	-	-	-	-	-	-	-	-	-
Venezuela	-	-	-	-	-	-	-	-	-
Other	-	-	-	-	-	-	-	-	-
Latin America	**-**	**-**	**-**	**-**	**-**	**-**	**-**	**-**	**-**
Romania	-	-	-	-	-	-	-	-	-
Other	-	-	-	-	-	-	-	-	-
Non-OECD Europe	**-**	**-**	**-**	**-**	**-**	**-**	**-**	**-**	**-**
FSU	**2 497**	**4 765**	**3 927 e**	**-**	**-**	**-**	**-**	**-**	**-**
Iran	3 793	4 894	4 541 e	-	-	-	-	-	-
Iraq	4 087	326	-	-	-	-	-	-	-
Kuwait	-	-	-	-	-	-	-	-	-
Oman	-	-	-	-	-	-	-	-	-
Qatar	-	-	-	-	-	-	-	-	-
Saudi Arabia	3 726	3 525	3 863 e	-	-	-	-	-	-
United Arab Emirates	-	-	-	-	-	-	-	-	-
Other	2 218	2 432	2 405 e	-	-	-	-	-	-
Middle East	**13 824**	**11 177**	**10 809 e**	**-**	**-**	**-**	**-**	**-**	**-**
India	-	-	-	-	-	-	-	-	-
Indonesia	-	-	-	-	-	-	-	-	-
Singapore	-	-	-	-	-	-	-	-	-
Other	-	-	-	-	-	-	-	-	-
Asia	**-**	**-**	**-**	**-**	**-**	**-**	**-**	**-**	**-**
China	**-**	**-**	**-**	**-**	**-**	**-**	**-**	**-**	**-**
Algeria	-	80	268 e	-	-	-	-	-	-
Angola/Cabinada	-	-	-	-	-	-	-	-	-
Cameroon	-	-	-	-	-	-	-	-	-
Egypt	-	-	103 e	-	-	-	-	-	-
Gabon	-	-	-	-	-	-	-	-	-
Libya	3 594	4 414	3 900 e	-	-	-	-	-	-
Nigeria	-	-	-	-	-	-	-	-	-
Tunisia	-	71	64 e	-	-	-	-	-	-
Other	-	-	-	-	-	-	-	-	-
Africa	**3 594**	**4 565**	**4 335 e**	**-**	**-**	**-**	**-**	**-**	**-**
Non-spec./Other	**1 668**	**2 735**	**4 087 e**	**-**	**-**	**-**	**-**	**-**	**-**
Total Imports	**21 583**	**23 242**	**23 662 e**	**-**	**-**	**-**	**-**	**-**	**-**
Memo: Total OPEC	*15 200*	*13 239*	*12 572 e*	-	-	-	-	-	-

Source: Annual Oil Statistics.

TURKEY
Table 7B: Imports by Country of Origin

000 metric tons

	2000	2001	2002	2000	2001	2002	2000	2001	2002
	Total Products			**Liquefied Petroleum Gases**			**Naphtha**		
Australia	-	-	-	-	-	-	-	-	-
Belgium	6	-	-	-	-	-	-	-	-
Canada	-	-	-	-	-	-	-	-	-
Czech Republic	-	-	-	-	-	-	-	-	-
Denmark	-	-	-	-	-	-	-	-	-
Finland	-	-	6 e	-	-	6 e	-	-	-
France	382	1	76 e	45	-	-	-	-	-
Germany	15	-	12 e	14	-	3 e	-	-	-
Greece	381	101	47 e	12	-	-	88	25	30 e
Hungary	-	-	-	-	-	-	-	-	-
Ireland	-	-	-	-	-	-	-	-	-
Italy	1 001	185	209 e	40	22	13 e	-	-	-
Japan	-	-	-	-	-	-	-	-	-
Korea	-	-	-	-	-	-	-	-	-
Mexico	-	-	-	-	-	-	-	-	-
Netherlands	68	-	34 e	-	-	34 e	-	-	-
New Zealand	-	-	-	-	-	-	-	-	-
Norway	668	597	814 e	631	597	814 e	-	-	-
Poland	-	-	-	-	-	-	-	-	-
Spain	85	8	-	-	-	-	-	-	-
Sweden	-	-	-	-	-	-	-	-	-
Turkey	-	-	-	-	-	-	-	-	-
United Kingdom	152	85	52 e	147	85	35 e	-	-	-
United States	15	-	9 e	9	-	6 e	-	-	-
Other	-	-	31 e	-	-	31 e	-	-	-
OECD	**2 773**	**977**	**1 290 e**	**898**	**704**	**942 e**	**88**	**25**	**30 e**
Argentina	-	-	11 e	-	-	11 e	-	-	-
Brazil	-	-	-	-	-	-	-	-	-
Colombia	-	-	-	-	-	-	-	-	-
Ecuador	-	-	-	-	-	-	-	-	-
Netherlands Antilles	-	-	-	-	-	-	-	-	-
Peru	-	-	-	-	-	-	-	-	-
Trinidad and Tobago	-	-	-	-	-	-	-	-	-
Venezuela	25	-	46 e	25	-	-	-	-	-
Other	-	-	11 e	-	-	11 e	-	-	-
Latin America	**25**	**-**	**68 e**	**25**	**-**	**22 e**	**-**	**-**	**-**
Romania	214	308	155 e	-	-	-	-	-	-
Other	161	160	284 e	-	-	18 e	-	4	-
Non-OECD Europe	**375**	**468**	**439 e**	**-**	**-**	**18 e**	**-**	**4**	**-**
FSU	**1 811**	**1 526**	**2 328 e**	**131**	**224**	**138 e**	**19**	**22**	**25 e**
Iran	22	-	-	22	-	-	-	-	-
Iraq	-	-	-	-	-	-	-	-	-
Kuwait	426	332	42 e	425	332	42 e	-	-	-
Oman	-	-	-	-	-	-	-	-	-
Qatar	-	-	19 e	-	-	-	-	-	-
Saudi Arabia	183	19	3 e	167	17	3 e	-	-	-
United Arab Emirates	1	32	37 e	-	32	-	-	-	-
Other	194	47	342 e	7	-	-	25	-	95 e
Middle East	**826**	**430**	**443 e**	**621**	**381**	**45 e**	**25**	**-**	**95 e**
India	-	-	-	-	-	-	-	-	-
Indonesia	-	-	-	-	-	-	-	-	-
Singapore	-	-	-	-	-	-	-	-	-
Other	-	-	-	-	-	-	-	-	-
Asia	**-**	**-**	**-**	**-**	**-**	**-**	**-**	**-**	**-**
China	**-**	**-**	**-**	**-**	**-**	**-**	**-**	**-**	**-**
Algeria	2 232	1 781	1 429 e	2 107	1 678	1 359 e	-	83	61 e
Angola/Cabinada	-	-	-	-	-	-	-	-	-
Cameroon	-	-	-	-	-	-	-	-	-
Egypt	5	53	20 e	5	53	20 e	-	-	-
Gabon	-	-	-	-	-	-	-	-	-
Libya	117	265	22 e	67	87	22 e	-	-	-
Nigeria	12	-	-	12	-	-	-	-	-
Tunisia	23	-	26 e	-	-	-	-	-	26 e
Other	-	-	-	-	-	-	-	-	-
Africa	**2 389**	**2 099**	**1 497 e**	**2 191**	**1 818**	**1 401 e**	**-**	**83**	**87 e**
Non-spec./Other	**470**	**291**	**440 e**	**12**	**2**	**-**	**-**	**-**	**-**
Total Imports	**8 669**	**5 791**	**6 505 e**	**3 878**	**3 129**	**2 566 e**	**132**	**134**	**237 e**
Memo: Total OPEC	*3 018*	*2 429*	*1 598 e*	*2 825*	*2 146*	*1 426 e*	*-*	*83*	*61 e*

Source: Annual Oil Statistics.

TURKEY
Table 7C: Imports by Country of Origin

000 metric tons

	2000	2001	2002	2000	2001	2002	2000	2001	2002
	Total Gasoline			**Total Kerosene**			**Gas/Diesel**		
Australia	-	-	-	-	-	-	-	-	-
Belgium	-	-	-	-	-	-	-	-	-
Canada	-	-	-	-	-	-	-	-	-
Czech Republic	-	-	-	-	-	-	-	-	-
Denmark	-	-	-	-	-	-	-	-	-
Finland	-	-	-	-	-	-	-	-	-
France	239	1	24 e	-	-	-	95	-	28 e
Germany	-	-	-	-	-	-	-	-	9 e
Greece	212	75	3 e	-	-	-	35	1	9 e
Hungary	-	-	-	-	-	-	-	-	-
Ireland	-	-	-	-	-	-	-	-	-
Italy	2	38	8 e	28	-	4 e	492	125	-
Japan	-	-	-	-	-	-	-	-	-
Korea	-	-	-	-	-	-	-	-	-
Mexico	-	-	-	-	-	-	-	-	-
Netherlands	-	-	-	-	-	-	63	-	-
New Zealand	-	-	-	-	-	-	-	-	-
Norway	-	-	-	-	-	-	9	-	-
Poland	-	-	-	-	-	-	-	-	-
Spain	38	-	-	-	-	-	-	-	-
Sweden	-	-	-	-	-	-	-	-	-
Turkey	-	-	-	-	-	-	-	-	-
United Kingdom	-	-	-	-	-	-	-	-	-
United States	-	-	-	-	-	-	-	-	3 e
Other	-	-	-	-	-	-	-	-	-
OECD	**491**	**114**	**35 e**	**28**	**-**	**4 e**	**694**	**126**	**49 e**
Argentina	-	-	-	-	-	-	-	-	-
Brazil	-	-	-	-	-	-	-	-	-
Colombia	-	-	-	-	-	-	-	-	-
Ecuador	-	-	-	-	-	-	-	-	-
Netherlands Antilles	-	-	-	-	-	-	-	-	-
Peru	-	-	-	-	-	-	-	-	-
Trinidad and Tobago	-	-	-	-	-	-	-	-	-
Venezuela	-	-	-	-	-	-	-	-	-
Other	-	-	-	-	-	-	-	-	-
Latin America	**-**	**-**	**-**	**-**	**-**	**-**	**-**	**-**	**-**
Romania	38	69	90 e	-	-	-	176	195	35 e
Other	75	96	90 e	-	-	-	29	-	87 e
Non-OECD Europe	**113**	**165**	**180 e**	**-**	**-**	**-**	**205**	**195**	**122 e**
FSU	**-**	**4**	**4 e**	**-**	**-**	**-**	**1 539**	**1 162**	**1 919 e**
Iran	-	-	-	-	-	-	-	-	-
Iraq	-	-	-	-	-	-	-	-	-
Kuwait	-	-	-	-	-	-	1	-	-
Oman	-	-	-	-	-	-	-	-	-
Qatar	-	-	-	-	-	-	-	-	19 e
Saudi Arabia	16	-	-	-	-	-	-	2	-
United Arab Emirates	-	-	-	-	-	-	-	-	-
Other	83	47	136 e	65	-	-	-	-	27 e
Middle East	**99**	**47**	**136 e**	**65**	**-**	**-**	**1**	**2**	**46 e**
India	-	-	-	-	-	-	-	-	-
Indonesia	-	-	-	-	-	-	-	-	-
Singapore	-	-	-	-	-	-	-	-	-
Other	-	-	-	-	-	-	-	-	-
Asia	**-**	**-**	**-**	**-**	**-**	**-**	**-**	**-**	**-**
China	**-**	**-**	**-**	**-**	**-**	**-**	**-**	**-**	**-**
Algeria	125	20	9 e	-	-	-	-	-	-
Angola/Cabinada	-	-	-	-	-	-	-	-	-
Cameroon	-	-	-	-	-	-	-	-	-
Egypt	-	-	-	-	-	-	-	-	-
Gabon	-	-	-	-	-	-	-	-	-
Libya	-	-	-	50	-	-	-	-	-
Nigeria	-	-	-	-	-	-	-	-	-
Tunisia	-	-	-	23	-	-	-	-	-
Other	-	-	-	-	-	-	-	-	-
Africa	**125**	**20**	**9 e**	**73**	**-**	**-**	**-**	**-**	**-**
Non-spec./Other	**9**	**87**	**1 e**	**-**	**-**	**-**	**31**	**111**	**-**
Total Imports	**837**	**437**	**365 e**	**166**	**-**	**4 e**	**2 470**	**1 596**	**2 136 e**
Memo: Total OPEC	*141*	*20*	*9 e*	*50*	*-*	*-*	*1*	*2*	*19 e*

Source: Annual Oil Statistics.

TURKEY
Table 7D: Imports by Country of Origin

000 metric tons

	2000	2001	2002	2000	2001	2002	2000	2001	2002
	Fuel Oil (Residual)			**Petroleum Coke**			**Other Products**		
Australia	-	-	-	-	-	-	-	-	-
Belgium	-	-	-	-	-	-	6	-	-
Canada	-	-	-	-	-	-	-	-	-
Czech Republic	-	-	-	-	-	-	-	-	-
Denmark	-	-	-	-	-	-	-	-	-
Finland	-	-	-	-	-	-	-	-	-
France	-	-	24 e	-	-	-	3	-	-
Germany	-	-	-	-	-	-	1	-	-
Greece	27	-	5 e	-	-	-	7	-	-
Hungary	-	-	-	-	-	-	-	-	-
Ireland	-	-	-	-	-	-	-	-	-
Italy	414	-	184 e	-	-	-	25	-	-
Japan	-	-	-	-	-	-	-	-	-
Korea	-	-	-	-	-	-	-	-	-
Mexico	-	-	-	-	-	-	-	-	-
Netherlands	-	-	-	-	-	-	5	-	-
New Zealand	-	-	-	-	-	-	-	-	-
Norway	28	-	-	-	-	-	-	-	-
Poland	-	-	-	-	-	-	-	-	-
Spain	34	8	-	-	-	-	13	-	-
Sweden	-	-	-	-	-	-	-	-	-
Turkey	-	-	-	-	-	-	-	-	-
United Kingdom	5	-	17 e	-	-	-	-	-	-
United States	-	-	-	-	-	-	6	-	-
Other	-	-	-	-	-	-	-	-	-
OECD	**508**	**8**	**230 e**	**-**	**-**	**-**	**66**	**-**	**-**
Argentina	-	-	-	-	-	-	-	-	-
Brazil	-	-	-	-	-	-	-	-	-
Colombia	-	-	-	-	-	-	-	-	-
Ecuador	-	-	-	-	-	-	-	-	-
Netherlands Antilles	-	-	-	-	-	-	-	-	-
Peru	-	-	-	-	-	-	-	-	-
Trinidad and Tobago	-	-	-	-	-	-	-	-	-
Venezuela	-	-	46 e	-	-	-	-	-	-
Other	-	-	-	-	-	-	-	-	-
Latin America	**-**	**-**	**46 e**	**-**	**-**	**-**	**-**	**-**	**-**
Romania	-	44	30 e	-	-	-	-	-	-
Other	57	60	89 e	-	-	-	-	-	-
Non-OECD Europe	**57**	**104**	**119 e**	**-**	**-**	**-**	**-**	**-**	**-**
FSU	**122**	**92**	**242 e**	**-**	**-**	**-**	**-**	**22**	**-**
Iran	-	-	-	-	-	-	-	-	-
Iraq	-	-	-	-	-	-	-	-	-
Kuwait	-	-	-	-	-	-	-	-	-
Oman	-	-	-	-	-	-	-	-	-
Qatar	-	-	-	-	-	-	-	-	-
Saudi Arabia	-	-	-	-	-	-	-	-	-
United Arab Emirates	-	-	37 e	-	-	-	1	-	-
Other	14	-	80 e	-	-	-	-	-	4 e
Middle East	**14**	**-**	**117 e**	**-**	**-**	**-**	**1**	**-**	**4 e**
India	-	-	-	-	-	-	-	-	-
Indonesia	-	-	-	-	-	-	-	-	-
Singapore	-	-	-	-	-	-	-	-	-
Other	-	-	-	-	-	-	-	-	-
Asia	**-**	**-**	**-**	**-**	**-**	**-**	**-**	**-**	**-**
China	**-**	**-**	**-**	**-**	**-**	**-**	**-**	**-**	**-**
Algeria	-	-	-	-	-	-	-	-	-
Angola/Cabinada	-	-	-	-	-	-	-	-	-
Cameroon	-	-	-	-	-	-	-	-	-
Egypt	-	-	-	-	-	-	-	-	-
Gabon	-	-	-	-	-	-	-	-	-
Libya	-	-	-	-	-	-	-	178	-
Nigeria	-	-	-	-	-	-	-	-	-
Tunisia	-	-	-	-	-	-	-	-	-
Other	-	-	-	-	-	-	-	-	-
Africa	**-**	**-**	**-**	**-**	**-**	**-**	**-**	**178**	**-**
Non-spec./Other	**-**	**22**	**28 e**	**-**	**-**	**-**	**418**	**69**	**411 e**
Total Imports	**701**	**226**	**782 e**	**-**	**-**	**-**	**485**	**269**	**415 e**
Memo: Total OPEC	-	-	*83 e*	-	-	-	*1*	*178*	-

Source: Annual Oil Statistics.

TURKEY
Table 8A: Exports by Country of Destination

000 metric tons

	Crude Oil 2000	Crude Oil 2001	Crude Oil 2002	Natural Gas Liquids 2000	Natural Gas Liquids 2001	Natural Gas Liquids 2002	Feedstocks and Other 2000	Feedstocks and Other 2001	Feedstocks and Other 2002
Australia	-	-	-	-	-	-	-	-	-
Austria	-	-	-	-	-	-	-	-	-
Belgium	-	-	-	-	-	-	-	-	-
Canada	-	-	-	-	-	-	-	-	-
Denmark	-	-	-	-	-	-	-	-	-
Finland	-	-	-	-	-	-	-	-	-
France	-	-	-	-	-	-	-	-	-
Germany	-	-	-	-	-	-	-	-	-
Greece	-	-	-	-	-	-	-	-	-
Ireland	-	-	-	-	-	-	-	-	-
Italy	-	-	-	-	-	-	-	-	-
Japan	-	-	-	-	-	-	-	-	-
Luxembourg	-	-	-	-	-	-	-	-	-
Mexico	-	-	-	-	-	-	-	-	-
Netherlands	-	-	-	-	-	-	-	-	-
Norway	-	-	-	-	-	-	-	-	-
Portugal	-	-	-	-	-	-	-	-	-
Spain	-	-	-	-	-	-	-	-	-
Sweden	-	-	-	-	-	-	-	-	-
Switzerland	-	-	-	-	-	-	-	-	-
Turkey	-	-	-	-	-	-	-	-	-
United Kingdom	-	-	-	-	-	-	-	-	-
United States	-	-	-	-	-	-	-	-	-
Other OECD	-	-	-	-	-	-	-	-	-
OECD	**-**	**-**	**-**	**-**	**-**	**-**	**-**	**-**	**-**
Africa	-	-	-	-	-	-	-	-	-
Latin America	-	-	-	-	-	-	-	-	-
Asia	-	-	-	-	-	-	-	-	-
China	-	-	-	-	-	-	-	-	-
Non-OECD Europe	-	-	-	-	-	-	-	-	-
Former Soviet Union	-	-	-	-	-	-	-	-	-
Middle East	-	-	-	-	-	-	-	-	-
Non-specified/Other	-	-	-	-	-	-	-	-	-
Non-OECD	**-**	**-**	**-**	**-**	**-**	**-**	**-**	**-**	**-**
Total Exports	**-**	**-**	**-**	**-**	**-**	**-**	**-**	**-**	**-**

	Total Products 2000	Total Products 2001	Total Products 2002	Liquefied Petroleum Gas 2000	Liquefied Petroleum Gas 2001	Liquefied Petroleum Gas 2002	Naphtha 2000	Naphtha 2001	Naphtha 2002
Australia	-	-	-	-	-	-	-	-	-
Austria	-	-	-	-	-	-	-	-	-
Belgium	-	-	-	-	-	-	-	-	-
Canada	-	-	-	-	-	-	-	-	-
Denmark	-	-	-	-	-	-	-	-	-
Finland	-	-	-	-	-	-	-	-	-
France	88	120	8 e	-	-	-	48	7	8 e
Germany	20	-	-	-	-	-	20	-	-
Greece	23	-	94 e	-	-	-	20	-	-
Ireland	-	-	-	-	-	-	-	-	-
Italy	543	1 111	1 053 e	-	-	-	60	486	353 e
Japan	-	-	-	-	-	-	-	-	-
Luxembourg	-	-	-	-	-	-	-	-	-
Mexico	-	-	-	-	-	-	-	-	-
Netherlands	66	151	79 e	-	-	-	66	-	-
Norway	-	-	-	-	-	-	-	-	-
Portugal	-	29	-	-	-	-	-	19	-
Spain	20	77	32 e	-	-	-	20	77	-
Sweden	-	-	-	-	-	-	-	-	-
Switzerland	39	101	28 e	-	-	-	39	-	-
Turkey	-	-	-	-	-	-	-	-	-
United Kingdom	-	-	-	-	-	-	-	-	-
United States	266	230	483 e	-	-	-	52	-	20 e
Other OECD	-	-	-	-	-	-	-	-	-
OECD	**1 065**	**1 819**	**1 777 e**	**-**	**-**	**-**	**325**	**589**	**381 e**
Africa	-	50	207 e	-	-	-	-	-	-
Latin America	-	-	67 e	-	-	-	-	-	-
Asia	64	174	-	-	-	-	-	91	-
China	-	-	-	-	-	-	-	-	-
Non-OECD Europe	70	49	231 e	-	-	-	-	-	-
Former Soviet Union	23	9	11 e	-	-	-	-	-	-
Middle East	60	248	445 e	-	-	-	60	-	-
Non-specified/Other	28	221	314 e	-	13	14 e	26	-	20 e
Non-OECD	**245**	**751**	**1 275 e**	**-**	**13**	**14 e**	**86**	**91**	**20 e**
Total Exports	**1 310**	**2 570**	**3 052 e**	**-**	**13**	**14 e**	**411**	**680**	**401 e**

Source: Annual Oil Statistics.

TURKEY
Table 8B: Exports by Country of Destination

000 metric tons

	Total Gasoline			Total Kerosene			Gas/Diesel Oil		
	2000	**2001**	**2002**	**2000**	**2001**	**2002**	**2000**	**2001**	**2002**
Australia	-	-	-	-	-	-	-	-	-
Austria	-	-	-	-	-	-	-	-	-
Belgium	-	-	-	-	-	-	-	-	-
Canada	-	-	-	-	-	-	-	-	-
Denmark	-	-	-	-	-	-	-	-	-
Finland	-	-	-	-	-	-	-	-	-
France	-	-	-	-	-	-	-	-	-
Germany	-	-	-	-	-	-	-	-	-
Greece	-	-	31 e	-	-	-	3	-	24 e
Ireland	-	-	-	-	-	-	-	-	-
Italy	-	40	115 e	-	-	10 e	-	-	83 e
Japan	-	-	-	-	-	-	-	-	-
Luxembourg	-	-	-	-	-	-	-	-	-
Mexico	-	-	-	-	-	-	-	-	-
Netherlands	-	-	79 e	-	-	-	-	-	-
Norway	-	-	-	-	-	-	-	-	-
Portugal	-	-	-	-	-	-	-	-	-
Spain	-	-	32 e	-	-	-	-	-	-
Sweden	-	-	-	-	-	-	-	-	-
Switzerland	-	-	-	-	-	-	-	-	-
Turkey	-	-	-	-	-	-	-	-	-
United Kingdom	-	-	-	-	-	-	-	-	-
United States	-	66	301 e	-	-	-	-	-	-
Other OECD	-	-	-	-	-	-	-	-	-
OECD	**-**	**106**	**558 e**	**-**	**-**	**10 e**	**3**	**-**	**107 e**
Africa	-	20	98 e	-	-	-	-	30	109 e
Latin America	-	-	67 e	-	-	-	-	-	-
Asia	-	-	-	-	-	-	-	-	-
China	-	-	-	-	-	-	-	-	-
Non-OECD Europe	-	2	115 e	-	-	-	-	5	116 e
Former Soviet Union	-	-	-	-	-	-	23	9	3 e
Middle East	-	20	217 e	-	-	-	-	111	183 e
Non-specified/Other	-	26	58 e	2	5	9 e	-	53	64 e
Non-OECD	**-**	**68**	**555 e**	**2**	**5**	**9 e**	**23**	**208**	**475 e**
Total Exports	**-**	**174**	**1 113 e**	**2**	**5**	**19 e**	**26**	**208**	**582 e**

	Fuel Oil (Residual)			Petroleum Coke			Other Products		
	2000	**2001**	**2002**	**2000**	**2001**	**2002**	**2000**	**2001**	**2002**
Australia	-	-	-	-	-	-	-	-	-
Austria	-	-	-	-	-	-	-	-	-
Belgium	-	-	-	-	-	-	-	-	-
Canada	-	-	-	-	-	-	-	-	-
Denmark	-	-	-	-	-	-	-	-	-
Finland	-	-	-	-	-	-	-	-	-
France	-	113	-	-	-	-	40	-	-
Germany	-	-	-	-	-	-	-	-	-
Greece	-	-	-	-	-	-	-	-	39 e
Ireland	-	-	-	-	-	-	-	-	-
Italy	483	465	260 e	-	-	-	-	120	232 e
Japan	-	-	-	-	-	-	-	-	-
Luxembourg	-	-	-	-	-	-	-	-	-
Mexico	-	-	-	-	-	-	-	-	-
Netherlands	-	56	-	-	-	-	-	95	-
Norway	-	-	-	-	-	-	-	-	-
Portugal	-	10	-	-	-	-	-	-	-
Spain	-	-	-	-	-	-	-	-	-
Sweden	-	-	-	-	-	-	-	-	-
Switzerland	-	101	28 e	-	-	-	-	-	-
Turkey	-	-	-	-	-	-	-	-	-
United Kingdom	-	-	-	-	-	-	-	-	-
United States	-	-	-	-	-	-	214	164	162 e
Other OECD	-	-	-	-	-	-	-	-	-
OECD	**483**	**745**	**288 e**	**-**	**-**	**-**	**254**	**379**	**433 e**
Africa	-	-	-	-	-	-	-	-	-
Latin America	-	-	-	-	-	-	-	-	-
Asia	64	83	-	-	-	-	-	-	-
China	-	-	-	-	-	-	-	-	-
Non-OECD Europe	28	20	-	-	-	-	42	22	-
Former Soviet Union	-	-	8 e	-	-	-	-	-	-
Middle East	-	38	-	-	-	-	-	79	45 e
Non-specified/Other	-	121	147 e	-	-	-	-	3	2 e
Non-OECD	**92**	**262**	**155 e**	**-**	**-**	**-**	**42**	**104**	**47 e**
Total Exports	**575**	**1 007**	**443 e**	**-**	**-**	**-**	**296**	**483**	**480 e**

Source: Annual Oil Statistics.

UNITED KINGDOM

Figure 1
FUEL SHARES OF TPES *

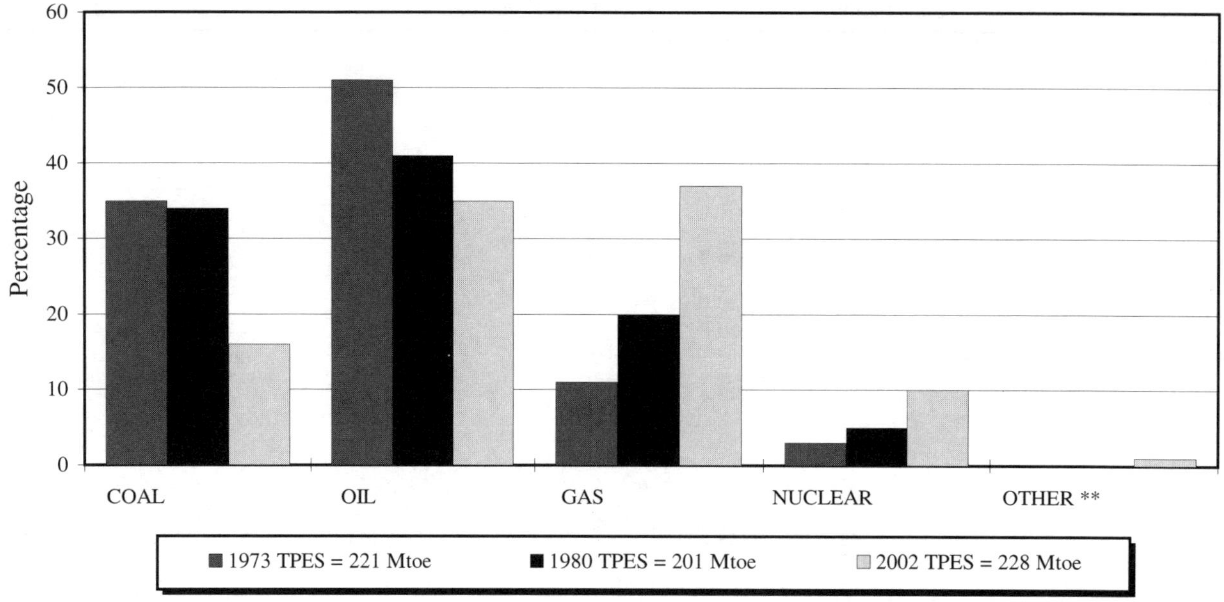

Source: Annual Energy Balances.

* Excluding electricity trade.

** Includes hydro, geothermal, solar, wind, combustible renewables and waste, etc.

Figure 2
OIL PRODUCTION AND CONSUMPTION

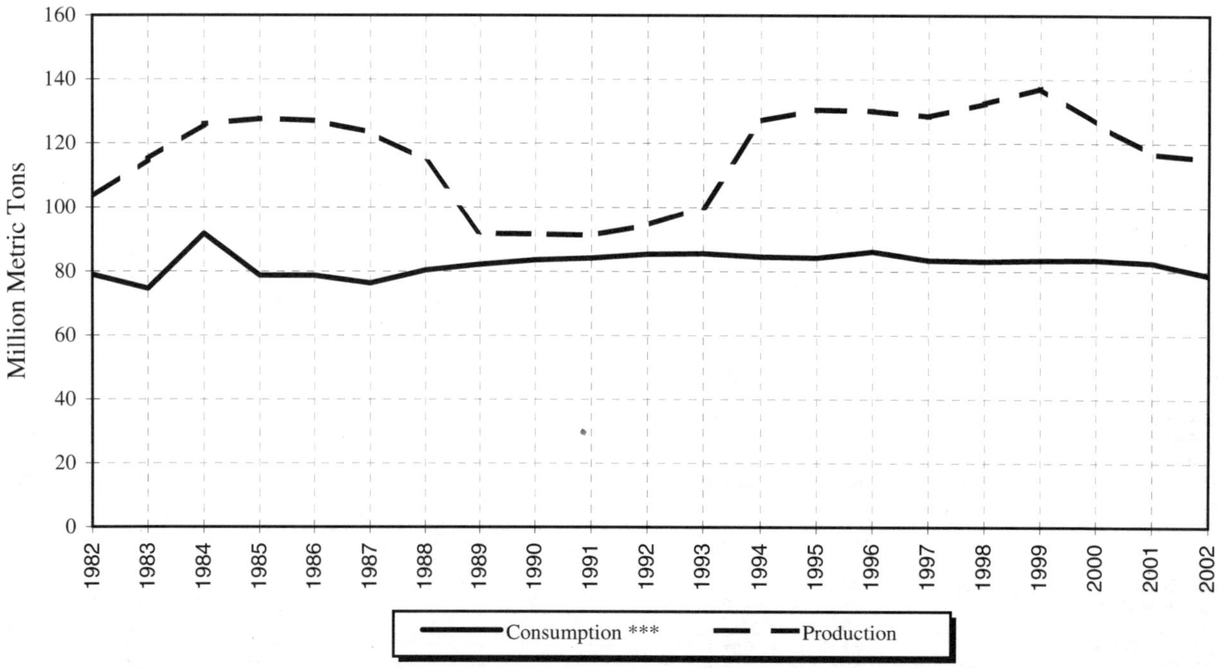

Source: Annual Oil Statistics.

*** Includes refinery fuel and international marine bunkers.

UNITED KINGDOM
Table 1: Oil Supply and Consumption - Summary Table

000 metric tons

	1973	1978	1980	1990	1999	2000	2001	2002
Indigenous Production	547	54 115	80 482	91 602	137 228	126 375	116 805	115 164 e
Crude Oil	..	53 475	78 917	87 990	128 262	117 882	108 387	106 645 e
NGL	..	625	1 550	3 606	8 837	8 363	8 292	8 394 e
Other	..	15	15	6	129	130	126	125 e
+ Net Imports	113 132	41 299	1 672	-10 393	-54 665	-44 996	-35 297	-36 979 e
Crude Oil, NGL, Other	112 237	42 943	6 537	-4 339	-46 906	-38 531	-33 195	-28 347 e
Petroleum Products	895	-1 644	-4 865	-6 054	-7 759	-6 465	-2 102	-8 632 e
- Intl. Marine Bunkers	5 499	2 623	2 457	2 537	2 330	2 079	2 274	2 540 e
+ Stock Changes	511	- 182	180	971	375	754	-1 424	1 543 e
TOTAL SUPPLY	**108 691**	**92 609**	**79 877**	**79 643**	**80 608**	**80 054**	**77 810**	**77 188 e**
Refinery Intake	*114 191*	*96 390*	*86 393*	*88 692*	*88 415*	*88 144*	*84 629*	*85 857 e*
Refinery Output	*113 004*	*95 579*	*85 544*	*88 120*	*86 734*	*86 341*	*82 063*	*84 711 e*
LPG + Ethane	*1 655*	*1 650*	*1 497*	*1 616*	*2 009*	*2 034*	*1 853*	*2 214 e*
Naphtha	*6 607*	*4 809*	*3 666*	*2 165*	*2 451*	*3 099*	*3 463*	*3 251 e*
Kerosenes	*7 267*	*7 397*	*7 232*	*9 850*	*10 802*	*9 562*	*8 998*	*9 000 e*
Gasolines	*15 219*	*16 072*	*16 675*	*26 724*	*25 246*	*23 470*	*21 526 e*	*23 307 e*
Gas/Diesel Oil	*27 853*	*24 064*	*22 172*	*23 404*	*25 869*	*28 298*	*26 796*	*28 270 e*
Fuel Oil	*45 970*	*33 792*	*27 226*	*15 825*	*12 195*	*11 621*	*11 912*	*10 633 e*
Other Products	*8 433*	*7 795*	*7 076*	*8 536*	*8 162*	*8 257*	*7 515*	*8 036 e*
- Refinery Losses	1 187	811	849	572	1 681	1 803	2 566	1 146 e
+ Returns	-	916	2 005	916	2 105	3 493	4 328	1 998 e
- Statistical Differences	649	1 035	1 346	-1 038	- 114	296	- 826	1 801 e
TOTAL CONSUMPTION	**106 855**	**91 679**	**79 687**	**81 025**	**81 146**	**81 448**	**80 398**	**76 239 e**
Transformation and Energy	**33 189**	**23 746**	**18 595**	**15 337**	**9 489**	**11 038**	**11 272**	..
Crude Oil	-	-	-	-	-	-	-	..
LPG + Ethane	261	201	332	365	84	148	133	..
Gas/Diesel Oil	1 406	1 046	811	272	414	520	433	..
Fuel Oil	26 507	18 238	13 106	9 984	4 542	5 900	7 108	..
Petroleum Coke	-	-	467	976	1 165	977	579 e	..
Other	5 015	4 261	3 879	3 740	3 284	3 493	3 019 e	..
Transport	**29 464**	**31 306**	**32 124**	**43 804**	**49 799**	**50 051**	**49 084**	..
Motor Gasoline	16 927	18 348	19 145	24 312	22 597	21 603	20 933	..
Gas/Diesel Oil	7 403	7 813	7 728	12 434	16 857	17 181	17 440	..
Subtotal	*24 330*	*26 161*	*26 873*	*36 746*	*39 454*	*38 784*	*38 373*	..
Aviation Fuels	4 337	4 557	4 728	6 615	9 948	10 886	10 324	..
Other	797	588	523	443	397	381	387	..
Industry	**31 757**	**24 483**	**18 729**	**15 042**	**16 251**	**15 233**	**14 484**	..
Naphtha	6 734	4 884	2 354	2 767	3 102	2 344	1 594	..
LPG + Ethane	1 248	1 126	992	1 941	3 477	3 214	2 733	..
Gas/Diesel Oil	5 718	4 065	3 322	2 582	3 604	3 824	3 787	..
Fuel Oil	13 333	10 038	8 466	3 460	1 044	525	743	..
Other	4 724	4 370	3 595	4 292	5 024	5 326	5 627	..
Other Sectors	**12 445**	**12 144**	**10 239**	**6 842**	**5 607**	**5 126**	**5 558**	..
Gas/Diesel Oil	6 231	6 591	5 783	3 849	2 267	2 114	1 826	..
Fuel Oil	3 262	3 085	2 458	1 023	541	122	109	..
Other	2 952	2 468	1 998	1 970	2 799	2 890	3 623	..
Memo: Non-Energy Use	*11 591*	*9 439*	*7 001*	*9 169*	*10 734*	*10 064*	*8 847*	..
Chem/Petchem. Feedst.	*7 206*	*5 889*	*3 741*	*5 115*	*6 757*	*5 844*	*4 714*	..
Other	*4 385*	*3 550*	*3 260*	*4 054*	*3 977*	*4 220*	*4 133*	..

Source: Annual Oil Statistics.

UNITED KINGDOM
Table 2: Supply of Crude Oil, NGL, Feedstocks and Other Hydrocarbons

000 metric tons

	1973	1978	1980	1990	1999	2000	2001	2002
CRUDE OIL								
Indigenous Production	..	53 475	78 917	87 990	128 262	117 882	108 387	106 645 e
+ Imports	..	65 639	43 262	41 578	39 321	48 868	48 992	53 188 e
- Exports	..	24 411	38 895	54 022	85 052	86 533	80 735	81 418 e
- Direct Use	..	-	-	-	-	-	-	-
+ Stock Change	..	- 578	-1 059	635	- 347	1 171	-1 045	84 e
- Statistical Difference	..	760	1 087	-1 914	636	697	- 311	508 e
Refinery Intake	**..**	**93 365**	**81 138**	**78 095**	**81 548**	**80 691**	**75 910**	**77 991 e**
Closing Stock Level	..	*8 951*	*10 466*	*3 742*	*6 330*	*5 159*	*6 204*	*6 120 e*
NATURAL GAS LIQUIDS								
Indigenous Production	..	625	1 550	3 606	8 837	8 363	8 292	8 394 e
+ Imports	..	-	-	-	-	-	-	-
- Exports	..	342	488	1 197	3 870	3 549	3 522	3 390 e
- Direct Use	..	231	470	1 184	3 646	3 678	3 480	3 280 e
+ Stock Change	..	-	-	- 69	17	- 17	- 17	2 e
- Statistical Difference	..	-	-	- 420	42	- 564	- 314	170 e
Refinery Intake	**..**	**52**	**592**	**1 576**	**1 296**	**1 683**	**1 587**	**1 556 e**
Closing Stock Level	..	-	-	*149*	*81*	*98*	*115*	*113 e*
REFINERY FEEDSTOCKS								
Backflows	..	916	2 005	916	2 105	3 493	4 328	1 998 e
+ Transfers	..	-	-	-	-	-	-	-
+ Imports	..	2 505	3 455	10 922	5 570	5 519	4 559	5 066 e
- Exports	..	448	797	1 776	2 875	2 836	2 489	1 793 e
+ Stock Change	..	-	-	- 112	132	- 56	414	25 e
- Statistical Difference	..	-	-	929	- 510	480	- 194	- 889 e
Refinery Intake	**..**	**2 973**	**4 663**	**9 021**	**5 442**	**5 640**	**7 006**	**6 185 e**
Closing Stock Level	..	-	-	*2 575*	*1 610*	*1 666*	*1 252*	*1 227 e*
OTHER HYDROCARBONS								
Indigenous Production	..	15	15	6	129	130	126	125 e
+ Imports	..	-	-	156	-	-	-	-
- Exports	..	-	-	-	-	-	-	-
- Direct Use	..	15	15	162	-	-	-	-
+ Stock Change	..	-	-	-	-	-	-	-
- Statistical Difference	..	-	-	-	-	-	-	-
Refinery Intake	**..**	**-**	**-**	**-**	**129**	**130**	**126**	**125 e**
Closing Stock Level	..	-	-	-	-	-	-	-
CRUDE OIL, NGL, FEEDSTOCKS AND OTHER								
Indigenous Production	547	54 115	80 482	91 602	137 228	126 375	116 805	115 164 e
+ Backflows	-	916	2 005	916	2 105	3 493	4 328	1 998 e
+ Transfers	-	-	-	-	-	-	-	-
+ Imports	115 472	68 144	46 717	52 656	44 891	54 387	53 551	58 254 e
- Exports	3 235	25 201	40 180	56 995	91 797	92 918	86 746	86 601 e
- Direct Use	316	246	485	1 346	3 646	3 678	3 480	3 280 e
+ Stock Change	437	- 578	-1 059	454	- 198	1 098	- 648	111 e
- Statistical Difference	-1 286	760	1 087	-1 405	168	613	- 819	- 211 e
Refinery Intake	**114 191**	**96 390**	**86 393**	**88 692**	**88 415**	**88 144**	**84 629**	**85 857 e**
Closing Stock Level	-	*8 951*	*10 466*	*6 466*	*8 021*	*6 923*	*7 571*	*7 460 e*

Source: Annual Oil Statistics.

UNITED KINGDOM
Table 3: Motor Gasoline - Supply

000 metric tons

	1973	1978	1980	1990	1999	2000	2001	2002
MOTOR GASOLINE								
Refinery Output	**14 842**	**15 959**	**16 609**	**26 724**	**25 230**	**23 440**	**21 455**	**23 288 e**
+ Imports	3 377	2 404	2 674	1 907	2 492	2 443	3 608	2 301 e
- Exports	1 257	1 049	848	4 265	6 329	4 708	4 447	5 493 e
+ Transfers	-	-	-	-	143	625	943	973 e
+ Stock Change	- 332	111	- 43	39	125	260	- 363	267 e
- Statistical Difference	- 297	- 924	- 753	93	- 936	457	263	1 418 e
Total Consumption	**16 927**	**18 349**	**19 145**	**24 312**	**22 597**	**21 603**	**20 933**	**19 918 e**
Closing Stock Level	-	*1 851*	*1 938*	*1 911*	*1 234*	*974*	*1 337*	*1 070 e*

Table 4: Aviation Fuels - Supply

000 metric tons

	1973	1978	1980	1990	1999	2000	2001	2002
AVIATION GASOLINE								
Refinery Output	**63**	**37**	**60**	**-**	**16**	**30**	**71 e**	**19 e**
+ Imports	59	37	26	31	16	16	10	9 e
- Exports	41	35	49	1	1	-	6	6 e
+ Transfers	-	-	-	-	28	- 11	5	1 e
+ Stock Change	3	13	4	- 8	7	2	3	- 2 e
- Statistical Difference	21	6	1	- 4	22	- 15	24 e	- 31 e
Total Consumption	**63**	**46**	**40**	**26**	**44**	**52**	**59**	**52 e**
Closing Stock Level	-	*29*	*30*	*12*	*5*	*3*	-	*2 e*
GASOLINE TYPE JET FUEL								
Refinery Output	**314**	**76**	**6**	**-**	**-**	**-**	**-**	**-**
+ Imports	31	17	2	-	-	-	-	-
- Exports	302	94	9	-	-	-	-	-
+ Transfers	-	-	-	-	-	-	-	-
+ Stock Change	17	1	5	-	-	-	-	-
- Statistical Difference	- 12	- 5	1	-	-	-	-	-
Total Consumption	**72**	**5**	**3**	**-**	**-**	**-**	**-**	**-**
Closing Stock Level	-	*7*	*1*	-	-	-	-	-
KEROSENE TYPE JET FUEL								
Refinery Output	**4 550**	**4 783**	**5 198**	**7 541**	**7 249**	**6 485**	**5 910**	**5 715 e**
+ Imports	802	415	356	192	2 946	4 675	6 130	5 214 e
- Exports	586	518	622	992	739	487	456	571 e
+ Transfers	-	-	-	-	274	429	- 424	- 348 e
+ Stock Change	- 48	86	99	- 58	173	- 25	- 338	314 e
- Statistical Difference	516	260	346	94	- 36	214	525	93 e
Total Consumption	**4 202**	**4 506**	**4 685**	**6 589**	**9 939**	**10 863**	**10 297**	**10 231 e**
Closing Stock Level	-	*527*	*629*	*587*	*461*	*486*	*824*	*510 e*

Source: Annual Oil Statistics.

UNITED KINGDOM
Table 5A: Gas/Diesel Oil - Supply and Consumption

000 metric tons

	1973	1978	1980	1990	1999	2000	2001	2002
Refinery Output	**27 853**	**24 064**	**22 172**	**23 404**	**25 869**	**28 298**	**26 796**	**28 270 e**
+ Recycled Products	-	-	-	-	-	-	-	-
+ Imports	1 572	1 681	1 417	1 340	5 425	3 815	4 098	3 262 e
- Exports	6 832	5 061	4 737	4 430	6 668	6 416	5 287	6 459 e
- Intl. Marine Bunkers	997	797	758	1 138	1 151	1 141	1 433	1 514 e
+ Transfers	-	-	-	-	- 606	- 643	- 989	- 960 e
+ Stock Changes	- 398	- 171	40	86	232	- 54	31	289 e
- Statistical Differences	440	201	490	125	- 41	220	- 270	- 310 e
Total Consumption	**20 758**	**19 515**	**17 644**	**19 137**	**23 142**	**23 639**	**23 486**	**23 198 e**
Transformation	**894**	**735**	**560**	**124**	**168**	**338**	**249**	..
Public Electricity	853	444	198	44	58	135	9	..
Autoproducer Electricity	-	284	354	-	26	30	36	..
Public CHP Plants	-	-	-	-	-	-	-	..
Autoproducers of CHP	-	-	-	-	-	-	-	..
Public Heat Plants	-	-	-	-	-	-	-	..
Autoproducer Heat Plants	-	-	-	-	-	33	31	..
Petrochemical Industry	-	-	-	67	84	140	173	..
Other Transformation	41	7	8	13	-	-	-	..
Energy Sector	**512**	**311**	**251**	**148**	**246**	**182**	**184**	..
Refinery Fuel	-	40	19	2	40	8	4	..
Other Energy Sector	512	271	232	146	206	174	180	..
Final Consumption	**19 352**	**18 469**	**16 833**	**18 865**	**22 728**	**23 119**	**23 053**	..
Transport	**7 403**	**7 813**	**7 728**	**12 434**	**16 857**	**17 181**	**17 440**	..
Air	-	-	-	-	-	-	-	..
Road	5 658	5 875	5 854	10 652	15 508	15 881	16 418	..
Rail	970	824	790	601	452	429	405	..
Inland Waterways	775	1 114	1 084	1 181	897	871	617	..
Non-Specified	-	-	-	-	-	-	-	..
Industry	**5 718**	**4 065**	**3 322**	**2 582**	**3 604**	**3 824**	**3 787**	..
Iron and Steel	-	-	-	-	13	68	81	..
Chemical (incl. Pet.-Chem.)	288	193	166	561	966	1 071	543	..
Non-Ferrous Metals	172	115	95	26	22	29	41	..
Non-Metallic Minerals	407	270	222	166	155	195	238	..
Transport Equipment	311	246	208	109	98	116	140	..
Machinery	874	721	552	250	168	189	193	..
Mining and Quarrying	478	323	318	259	326	333	382	..
Food and Tobacco	436	389	346	169	130	138	199	..
Pulp, Paper and Printing	124	108	76	25	25	22	37	..
Wood and Wood Products	61	62	49	17	16	15	18	..
Construction	1 152	888	814	641	466	427	469	..
Textiles and Leather	220	157	119	38	39	53	90	..
Non-Specified	1 195	593	357	321	1 180	1 168	1 356	..
Other Sectors	**6 231**	**6 591**	**5 783**	**3 849**	**2 267**	**2 114**	**1 826**	..
Commerce and Public	3 416	3 503	3 022	2 193	1 381	1 272	1 076	..
Residential	888	812	563	286	161	147	193	..
Agriculture	1 245	1 041	813	662	590	559	448	..
Non-Specified	682	1 235	1 385	708	135	136	109	..
Memo: Non-Energy Use	-	-	-	*561*	*844*	*945*	*403*	..
Chem/Petchem. Feedst.	-	-	-	*561*	*844*	*945*	*403*	..
Other	-	-	-	-	-	-	-	..
Closing Stock Level	-	*3 543*	*3 731*	*2 048*	*1 868*	*1 922*	*1 891*	*1 602 e*

Source: Annual Oil Statistics.

UNITED KINGDOM

Table 5B: Fuel Oil (Residual) - Supply and Consumption

000 metric tons

	1973	1978	1980	1990	1999	2000	2001	2002
Refinery Output	**45 970**	**33 792**	**27 226**	**15 825**	**12 195**	**11 621**	**11 912**	**10 633 e**
+ Recycled Products	-	-	-	-	-	-	-	-
+ Imports	7 053	3 845	2 776	4 538	655	596	1 011	861 e
- Exports	5 575	3 707	5 121	4 417	4 928	5 360	5 441	5 075 e
- Intl. Marine Bunkers	4 502	1 826	1 699	1 399	1 179	938	841	1 026 e
+ Transfers	-	-	-	-	64	126	1 137	934 e
+ Stock Changes	1 036	181	1 137	391	65	266	103	- 100 e
- Statistical Differences	589	778	169	390	673	- 278	- 110	526 e
Total Consumption	**43 393**	**31 507**	**24 150**	**14 548**	**6 199**	**6 589**	**7 991**	**5 701 e**
Transformation	**22 561**	**14 964**	**9 580**	**7 964**	**2 604**	**4 424**	**5 016**	..
Public Electricity	18 318	10 772	6 084	6 336	429	272	326 e	..
Autoproducer Electricity	-	1 636	1 522	421 e	-	-	-	..
Public CHP Plants	-	259	291	-	-	-	-	..
Autoproducers of CHP	-	-	-	556 e	158	397	330	..
Public Heat Plants	-	-	-	-	-	-	-	..
Autoproducer Heat Plants	-	-	-	-	-	661	640	..
Petrochemical Industry	-	6	544	105	1 747	2 899	3 592	..
Other Transformation	4 243	2 291	1 139	546	270	195	128	..
Energy Sector	**3 946**	**3 274**	**3 526**	**2 020**	**1 938**	**1 476**	**2 092**	..
Refinery Fuel	3 946	3 274	3 526	2 020	1 920	1 470	2 028	..
Other Energy Sector	-	-	-	-	18	6	64	..
Final Consumption	**16 886**	**13 269**	**11 044**	**4 564**	**1 657**	**689**	**883**	..
Transport	**291**	**146**	**120**	**81**	**72**	**42**	**31**	..
Air	-	-	-	-	-	-	-	..
Road	-	-	-	-	-	-	-	..
Rail	51	57	41	2	-	-	-	..
Inland Waterways	240	89	79	79	72	42	31	..
Non-Specified	-	-	-	-	-	-	-	..
Industry	**13 333**	**10 038**	**8 466**	**3 460**	**1 044**	**525**	**743**	..
Iron and Steel	-	-	-	-	56	-	8	..
Chemical (incl. Pet.-Chem.)	2 757	2 124	2 402	1 357	188	146	140	..
Non-Ferrous Metals	311	208	168	32	16	9	33	..
Non-Metallic Minerals	1 631	965	712	304	57	47	30	..
Transport Equipment	690	519	355	115	37	9	22	..
Machinery	1 512	1 226	930	278	88	36	83	..
Mining and Quarrying	136	62	61	19	3	1	-	..
Food and Tobacco	2 298	1 852	1 574	725	241	67	58	..
Pulp, Paper and Printing	1 467	1 207	884	194	74	19	40	..
Wood and Wood Products	62	64	46	9	12	8	11	..
Construction	106	77	91	79	4	2	2	..
Textiles and Leather	1 409	1 243	881	194	79	89	112	..
Non-Specified	954	491	362	154	189	92	204	..
Other Sectors	**3 262**	**3 085**	**2 458**	**1 023**	**541**	**122**	**109**	..
Commerce and Public	2 480	1 912	1 623	894	431	104	74	..
Residential	67	64	58	17	5	3	6	..
Agriculture	375	316	227	98	93	10	11	..
Non-Specified	340	793	550	14	12	5	18	..
Memo: Non-Energy Use	-	*36*	*923*	*391*	-	-	-	..
Chem/Petchem. Feedst.	-	*36*	*923*	*391*	-	-	-	..
Other	-	-	-	-	-	-	-	..
Closing Stock Level	-	*4 245*	*3 576*	*2 273*	*1 401*	*1 135*	*1 032*	*1 132 e*

Source: Annual Oil Statistics.

UNITED KINGDOM
Table 5C: Total Products - Supply and Consumption

000 metric tons

	1973	1978	1980	1990	1999	2000	2001	2002
Refinery Output	**113 004**	**95 579**	**85 544**	**88 120**	**86 734**	**86 341**	**82 063**	**84 711 e**
+ Recycled Products	-	-	-	-	-	-	-	-
+ Imports	18 299	11 546	9 245	10 849	13 896	14 212	16 978	13 475 e
- Exports	17 404	13 190	14 110	16 903	21 655	20 677	19 080	22 107 e
- Intl. Marine Bunkers	5 499	2 623	2 457	2 537	2 330	2 079	2 274	2 540 e
+ Transfers	316	246	485	1 346	3 646	3 678	3 480	3 280 e
+ Stock Changes	74	396	1 239	517	573	- 344	- 776	1 432 e
- Statistical Differences	1 935	275	259	367	- 282	- 317	- 7	2 012 e
Total Consumption	**106 855**	**91 679**	**79 687**	**81 025**	**81 146**	**81 448**	**80 398**	**76 239 e**
Transformation	**25 624**	**16 979**	**11 858**	**9 025**	**3 046**	**5 233**	**5 828**	..
Public Electricity	19 171	11 216	6 282	6 536	487	407	335	..
Autoproducer Electricity	-	1 920	1 876	421 e	26	30	36	..
Public CHP Plants	-	259	291	-	-	-	-	..
Autoproducers of CHP	-	-	-	556 e	158	397	330	..
Public Heat Plants	-	-	-	-	-	-	-	..
Autoproducer Heat Plants	-	-	-	-	-	711	671	..
Petrochemical Industry	-	916	2 005	916	2 105	3 493	4 328	..
Other Transformation	6 453	2 668	1 404	596	270	195	128	..
Energy Sector	**7 565**	**6 767**	**6 737**	**6 312**	**6 443**	**5 805**	**5 444**	..
Refinery Fuel	7 053	6 423	6 317	5 836	5 846	5 288	5 193	..
Other Energy Sector	512	344	420	476	597	517	251	..
Final Consumption	**73 666**	**67 933**	**61 092**	**65 688**	**71 657**	**70 410**	**69 126**	..
Transport	**29 464**	**31 306**	**32 124**	**43 804**	**49 799**	**50 051**	**49 084**	..
Air	4 341	4 561	4 732	6 618	9 950	10 888	10 327	..
Road	22 996	24 585	25 327	35 261	38 378	37 776	37 662	..
Rail	1 035	895	847	615	464	441	417	..
Inland Waterways	1 092	1 265	1 218	1 310	1 007	946	678	..
Non-Specified	-	-	-	-	-	-	-	..
Industry	**31 757**	**24 483**	**18 729**	**15 042**	**16 251**	**15 233**	**14 484**	..
Iron and Steel	56	60	18	24	90	68	89	..
Chemical (incl. Pet.-Chem.)	10 251	8 170	5 386	6 081	7 067	6 116	4 994	..
Non-Ferrous Metals	483	323	263	58	38	38	74	..
Non-Metallic Minerals	2 038	1 235	934	470	212	242	268	..
Transport Equipment	1 001	765	563	224	170	154	194	..
Machinery	2 386	1 947	1 482	528	256	225	276	..
Mining and Quarrying	614	385	379	278	329	334	382	..
Food and Tobacco	2 734	2 241	1 920	894	371	205	257	..
Pulp, Paper and Printing	1 591	1 315	960	219	99	41	77	..
Wood and Wood Products	123	126	95	26	28	23	29	..
Construction	3 716	2 852	2 731	3 211	2 398	2 404	2 394	..
Textiles and Leather	1 629	1 400	1 000	232	118	142	202	..
Non-Specified	5 135	3 664	2 998	2 797	5 075	5 241	5 248	..
Other Sectors	**12 445**	**12 144**	**10 239**	**6 842**	**5 607**	**5 126**	**5 558**	..
Commerce and Public	5 906	5 428	4 657	3 099	1 824	1 388	1 602	..
Residential	3 783	3 262	2 553	2 230	2 839	2 920	3 164	..
Agriculture	1 696	1 412	1 089	789	710	592	590	..
Non-Specified	1 060	2 042	1 940	724	234	226	202	..
Memo: Non-Energy Use	*11 591*	*9 439*	*7 001*	*9 169*	*10 734*	*10 064*	*8 847*	..
Chem/Petchem. Feedst.	*7 206*	*5 889*	*3 741*	*5 115*	*6 757*	*5 844*	*4 714*	..
Other	*4 385*	*3 550*	*3 260*	*4 054*	*3 977*	*4 220*	*4 133*	..
Closing Stock Level	-	*12 335*	*12 208*	*8 353*	*6 962*	*7 306*	*8 082*	*6 650 e*

Source: Annual Oil Statistics.

UNITED KINGDOM
Table 6A: Supply and Consumption

000 metric tons

	1980	2000	2001	2002	1980	2000	2001	2002
	LPG AND ETHANE				REFINERY GAS			
Refinery Output	**1 497**	**2 034**	**1 853**	**2 214 e**	**2 138**	**2 818**	**2 544**	**2 505 e**
+ Recycled Products	-	-	-	-	-	-	-	-
+ Imports	155	331	423	359 e	-	-	-	-
- Exports	364	710	384	395 e	-	-	-	-
- Intl. Marine Bunkers	-	-	-	-	-	-	-	-
+ Transfers	282	2 663	2 700	2 605 e	-	- 37	- 14	-
+ Stock Change	- 27	- 19	- 89	116 e	2	3	-	-
- Statistical Difference	- 9	635	704	685 e	- 23	- 92	- 161	- 56 e
Total Consumption	**1 552**	**3 664**	**3 799**	**4 214 e**	**2 163**	**2 876**	**2 691**	**2 561 e**
of which: Ethane	-	*1 590*	*1 693*	*1 720 e*	-	-	-	-
Transformation	201	17	-	..	57	-	-	..
Energy	131	131	133	..	2 046	2 710	2 428	..
Transport	-	22	66	..	-	-	-	..
Industry	992	3 214	2 733	..	60	166	263	..
Other	228	280	867	..	-	-	-	..
Memo: Non-Energy Use	*60*	*2 389*	*2 454*	..	*60*	*166*	*263*	..
Closing Stock Level	*95*	*258*	*347*	*231 e*	-	-	-	-
	NAPHTHA				OTHER KEROSENE			
Refinery Output	**3 666**	**3 099**	**3 463**	**3 251 e**	**2 034**	**3 077**	**3 088**	**3 285 e**
+ Recycled Products	-	-	-	-	-	-	-	-
+ Imports	1 283	348	345	70 e	37	86	190	187 e
- Exports	807	973	1 078	1 808 e	113	199	167	362 e
- Intl. Marine Bunkers	-	-	-	-	-	-	-	-
+ Transfers	-	- 114	- 52	-	-	587	294	297 e
+ Stock Change	24	- 58	- 68	92 e	106	- 70	85	24 e
- Statistical Difference	226	- 515	425	- 75 e	- 39	- 357	- 755	- 429 e
Total Consumption	**3 940**	**2 817**	**2 185**	**1 680 e**	**2 103**	**3 838**	**4 245**	**3 860 e**
Transformation	1 460	454	563	..	-	-	-	..
Energy	125	19	28	..	-	-	-	..
Transport	-	-	-	..	16	12	12	..
Industry	2 354	2 344	1 594	..	359	1 312	1 561	..
Other	1	-	-	..	1 728	2 514	2 672	..
Memo: Non-Energy Use	*2 354*	*2 344*	*1 594*	..	-	-	-	..
Closing Stock Level	*635*	*407*	*475*	*383 e*	*568*	*409*	*324*	*300 e*
	WHITE SPIRIT + SBP				LUBRICANTS			
Refinery Output	**73**	**123**	**122**	**..**	**1 253**	**703**	**656**	**..**
+ Recycled Products	-	-	-	..	-	-	-	..
+ Imports	88	38	24	..	331	211	251	..
- Exports	19	9	2	..	677	636	806	..
- Intl. Marine Bunkers	-	-	-	..	-	-	-	..
+ Transfers	-	-	-	..	-	249	170	..
+ Stock Change	- 4	- 13	- 1	..	- 51	- 26	- 3	..
- Statistical Difference	- 32	- 31	- 7	..	- 43	- 303	- 528	..
Total Consumption	**170**	**170**	**150**	**..**	**899**	**804**	**796**	**..**
Transformation	-	-	-	..	-	-	-	..
Energy	-	-	-	..	3	-	-	..
Transport	-	-	-	..	387	305	278	..
Industry	170	85	75	..	468	488	509	..
Other	-	85	75	..	41	11	9	..
Memo: Non-Energy Use	*170*	*170*	*150*	..	*896*	*804*	*796*	..
Closing Stock Level	*57*	*35*	*36*	..	*609*	*195*	*198*	..

Source: Annual Oil Statistics.

UNITED KINGDOM
Table 6B: Supply and Consumption

000 metric tons

	1980	2000	2001	2002	1980	2000	2001	2002
	BITUMEN				PARAFFIN WAXES			
Refinery Output	**1 928**	**1 438**	**1 707**	**..**	**86**	**436**	**416**	**..**
+ Recycled Products	-	-	-	..	-	-	-	..
+ Imports	6	255	238	..	2	23	17	..
- Exports	114	283	269	..	13	51	56	..
- Intl. Marine Bunkers	-	-	-	..	-	-	-	..
+ Transfers	15	319	259	..	-	- 360	- 282	..
+ Stock Change	- 30	25	- 44	..	- 2	35	- 8	..
- Statistical Difference	- 21	- 221	- 32	..	2	51	55	..
Total Consumption	**1 826**	**1 975**	**1 923**	**..**	**71**	**32**	**32**	**..**
Transformation	-	-	-	..	-	-	-	..
Energy	-	-	-	..	-	-	-	..
Transport	-	-	-	..	-	-	-	..
Industry	1 826	1 975	1 923	..	71	32	32	..
Other	-	-	-	..	-	-	-	..
Memo: Non-Energy Use	*1 826*	*1 975*	*1 923*	..	*71*	*32*	*32*	..
Closing Stock Level	*179*	*164*	*208*	..	*25*	*1*	*9*	..
	PETROLEUM COKE				OTHER PRODUCTS			
Refinery Output	**948**	**1 796**	**1 346**	**1 559 e**	**650**	**943**	**724**	**3 972 e**
+ Recycled Products	-	-	-	-	-	-	-	-
+ Imports	-	657	633	659 e	92	718	-	553 e
- Exports	-	502	460	548 e	617	343	221	1 390 e
- Intl. Marine Bunkers	-	-	-	-	-	-	-	-
+ Transfers	-	1	-	-	188	- 156	- 267	- 222 e
+ Stock Change	- 11	36	- 27	16 e	- 10	- 706	- 57	416 e
- Statistical Difference	370	235	153	310 e	- 426	- 317	- 293	- 119 e
Total Consumption	**567**	**1 753**	**1 339 e**	**1 376 e**	**729**	**773**	**472**	**3 448 e**
Transformation	-	-	-	..	-	-	-	..
Energy	467	977	579 e	..	188	310	-	..
Transport	-	-	-	..	-	-	-	..
Industry	100	776	760	..	541	463	472	..
Other	-	-	-	..	-	-	-	..
Memo: Non-Energy Use	*100*	*776*	*760*	..	*541*	*463*	*472*	..
Closing Stock Level	*36*	*291*	*318*	*302 e*	*99*	*1 026*	*1 083*	*1 118 e*

Source: Annual Oil Statistics.

UNITED KINGDOM
Table 7A: Imports by Country of Origin

000 metric tons

	2000	2001	2002	2000	2001	2002	2000	2001	2002
	Crude Oil			**Natural Gas Liquids**			**Feedstocks and Other**		
Australia	-	-	-	-	-	-	-	-	-
Belgium	2	2	-	-	-	-	321	217	108 e
Canada	80	-	-	-	-	-	-	-	-
Czech Republic	-	-	-	-	-	-	-	-	-
Denmark	1 345	1 131	351 e	-	-	-	-	15	-
Finland	-	-	-	-	-	-	10	5	-
France	491	372	-	-	-	-	163	256	140 e
Germany	64	66	60 e	-	-	-	213	166	281 e
Greece	-	-	-	-	-	-	-	-	-
Hungary	-	-	-	-	-	-	-	-	-
Ireland	-	-	-	-	-	-	265	527	311 e
Italy	-	-	-	-	-	-	-	26	-
Japan	-	-	-	-	-	-	1	-	-
Korea	-	-	-	-	-	-	-	-	-
Mexico	729	488	628 e	-	-	-	-	-	-
Netherlands	5	22	590 e	-	-	-	467	318	404 e
New Zealand	-	-	-	-	-	-	-	-	-
Norway	32 179	27 972	26 510 e	-	-	-	138	456	402 e
Poland	-	-	88 e	-	-	-	-	-	-
Spain	-	-	-	-	-	-	-	-	16 e
Sweden	89	91	80 e	-	-	-	657	351	359 e
Turkey	-	-	-	-	-	-	-	-	-
United Kingdom	-	-	-	-	-	-	-	-	-
United States	-	-	82 e	-	-	-	37	1	-
Other	84	-	-	-	-	-	56	21	48 e
OECD	**35 068**	**30 144**	**28 389 e**	**-**	**-**	**-**	**2 328**	**2 359**	**2 069 e**
Argentina	-	-	-	-	-	-	-	10	-
Brazil	-	179	126 e	-	-	-	8	10	20 e
Colombia	-	-	-	-	-	-	-	-	-
Ecuador	-	-	-	-	-	-	-	-	-
Netherlands Antilles	-	-	-	-	-	-	-	-	-
Peru	-	-	-	-	-	-	-	-	-
Trinidad and Tobago	-	-	-	-	-	-	-	-	-
Venezuela	1 092	1 137	1 294 e	-	-	-	-	-	-
Other	-	-	-	-	-	-	-	-	-
Latin America	**1 092**	**1 316**	**1 420 e**	**-**	**-**	**-**	**8**	**20**	**20 e**
Romania	-	-	-	-	-	-	-	-	-
Other	-	-	-	-	-	-	-	26	-
Non-OECD Europe	**-**	**-**	**-**	**-**	**-**	**-**	**-**	**26**	**-**
FSU	**2 330**	**2 907**	**3 865 e**	**-**	**-**	**-**	**2 106**	**1 162**	**1 575 e**
Iran	-	-	-	-	-	-	-	-	-
Iraq	-	-	-	-	-	-	-	-	-
Kuwait	-	-	-	-	-	-	-	-	-
Oman	-	-	-	-	-	-	-	-	-
Qatar	-	-	-	-	-	-	-	-	-
Saudi Arabia	2 053	1 531	2 041 e	-	-	-	-	-	144 e
United Arab Emirates	-	-	-	-	-	-	-	-	-
Other	378	491	950 e	-	-	-	-	-	-
Middle East	**2 431**	**2 022**	**2 991 e**	**-**	**-**	**-**	**-**	**-**	**144 e**
India	-	-	-	-	-	-	-	-	-
Indonesia	-	-	-	-	-	-	-	-	-
Singapore	-	-	-	-	-	-	-	-	-
Other	-	-	-	-	-	-	-	-	-
Asia	**-**	**-**	**-**	**-**	**-**	**-**	**-**	**-**	**-**
China	**-**	**-**	**-**	**-**	**-**	**-**	**-**	**9**	**-**
Algeria	564	888	376 e	-	-	-	-	20	29 e
Angola/Cabinada	-	266	-	-	-	-	-	-	-
Cameroon	-	-	-	-	-	-	-	-	-
Egypt	1 412	132	-	-	-	-	-	55	27 e
Gabon	-	-	-	-	-	-	-	-	-
Libya	341	-	129 e	-	-	-	1 021	908	864 e
Nigeria	-	116	-	-	-	-	-	-	-
Tunisia	-	-	-	-	-	-	29	-	-
Other	73	124	-	-	-	-	-	-	-
Africa	**2 390**	**1 526**	**505 e**	-	-	-	**1 050**	**983**	**920 e**
Non-spec./Other	**5 557**	**11 077**	**16 018 e**	**-**	**-**	**-**	**27**	**-**	**338 e**
Total Imports	**48 868**	**48 992**	**53 188 e**	**-**	**-**	**-**	**5 519**	**4 559**	**5 066 e**
Memo: Total OPEC	*4 050*	*3 672*	*3 840 e*	-	-	-	*1 021*	*928*	*1 037 e*

Source: Annual Oil Statistics.

UNITED KINGDOM
Table 7B: Imports by Country of Origin

000 metric tons

	2000	2001	2002	2000	2001	2002	2000	2001	2002
	Total Products			**Liquefied Petroleum Gases**			**Naphtha**		
Australia	-	-	-	-	-	-	-	-	-
Belgium	537	502	312 e	22	39	25 e	-	-	-
Canada	61	40	54 e	-	-	-	-	-	-
Czech Republic	2	-	-	-	-	-	-	-	-
Denmark	181	150	37 e	7	10	6 e	-	-	11 e
Finland	386	126	82 e	11	7	-	-	-	-
France	1 046	930	568 e	33	29	16 e	70	83	-
Germany	970	1 268	960 e	38	5	20 e	5	-	-
Greece	72	-	-	-	-	-	-	-	-
Hungary	-	-	-	-	-	-	-	-	-
Ireland	43	93	33 e	6	2	3 e	-	-	-
Italy	68	56	72 e	-	-	-	-	-	-
Japan	38	-	2 e	-	-	2 e	-	-	-
Korea	1	-	-	-	-	-	-	-	-
Mexico	-	-	-	-	-	-	-	-	-
Netherlands	1 666	1 336	548 e	42	20	41 e	53	-	-
New Zealand	-	-	-	-	-	-	-	-	-
Norway	2 094	565	167 e	120	100	26 e	33	23	12 e
Poland	87	431	129 e	1	-	-	-	-	-
Spain	131	165	85 e	9	-	-	40	162	-
Sweden	958	355	138 e	26	2	7 e	-	-	-
Turkey	81	27	-	-	-	-	-	-	-
United Kingdom	-	-	-	-	-	-	-	-	-
United States	780	54	313 e	-	4	-	-	-	-
Other	176	72	26 e	13	10	14 e	-	17	-
OECD	**9 378**	**6 170**	**3 526 e**	**328**	**228**	**160 e**	**201**	**285**	**23 e**
Argentina	2	-	-	-	-	-	-	-	-
Brazil	-	-	-	-	-	-	-	-	-
Colombia	2	-	26 e	2	-	-	-	-	-
Ecuador	-	-	-	-	-	-	-	-	-
Netherlands Antilles	-	-	-	-	-	-	-	-	-
Peru	-	-	-	-	-	-	-	-	-
Trinidad and Tobago	-	-	-	-	-	-	-	-	-
Venezuela	180	59	366 e	-	-	-	-	-	-
Other	100	105	79 e	-	-	-	-	-	-
Latin America	**284**	**164**	**471 e**	**2**	**-**	**-**	**-**	**-**	**-**
Romania	-	-	-	-	-	-	-	-	-
Other	-	-	-	-	-	-	-	-	-
Non-OECD Europe	**-**	**-**	**-**	**-**	**-**	**-**	**-**	**-**	**-**
FSU	**1 420**	**1 132**	**1 372 e**	**-**	**3**	**-**	**93**	**16**	**11 e**
Iran	-	44	-	-	-	-	-	-	-
Iraq	-	-	-	-	-	-	-	-	-
Kuwait	834	908	715 e	-	-	-	-	-	-
Oman	-	82	-	-	-	-	-	-	-
Qatar	-	-	-	-	-	-	-	-	-
Saudi Arabia	614	978	625 e	-	-	-	-	-	-
United Arab Emirates	695	1 586	566 e	-	-	-	-	-	-
Other	189	161	137 e	-	-	-	-	-	-
Middle East	**2 332**	**3 759**	**2 043 e**	**-**	**-**	**-**	**-**	**-**	**-**
India	-	-	2 e	-	-	-	-	-	-
Indonesia	-	30	-	-	-	-	-	-	-
Singapore	1	-	228 e	-	-	-	-	-	-
Other	-	-	7 e	-	-	-	-	-	-
Asia	**1**	**30**	**237 e**	**-**	**-**	**-**	**-**	**-**	**-**
China	**-**	**3**	**-**	**-**	**3**	**-**	**-**	**-**	**-**
Algeria	411	60	103 e	-	30	2 e	54	-	11 e
Angola/Cabinada	-	-	-	-	-	-	-	-	-
Cameroon	-	-	-	-	-	-	-	-	-
Egypt	-	63	41 e	-	-	3 e	-	-	-
Gabon	-	-	-	-	-	-	-	-	-
Libya	239	619	319 e	-	-	-	-	-	-
Nigeria	28	96	-	-	-	-	-	-	-
Tunisia	19	4	-	-	-	-	-	-	-
Other	73	59	93 e	1	-	-	-	30	-
Africa	**770**	**901**	**556 e**	**1**	**30**	**5 e**	**54**	**30**	**11 e**
Non-spec./Other	**27**	**4 819**	**5 270 e**	**-**	**159**	**194 e**	**-**	**14**	**25 e**
Total Imports	**14 212**	**16 978**	**13 475 e**	**331**	**423**	**359 e**	**348**	**345**	**70 e**
Memo: Total OPEC	*3 001*	*4 380*	*2 694 e*	*-*	*30*	*2 e*	*54*	*-*	*11 e*

Source: Annual Oil Statistics.

UNITED KINGDOM
Table 7C: Imports by Country of Origin

000 metric tons

	2000	2001	2002	2000	2001	2002	2000	2001	2002
	Total Gasoline			**Total Kerosene**			**Gas/Diesel**		
Australia	-	-	-	-	-	-	-	-	-
Belgium	124	288	57 e	140	67	46 e	208	69	173 e
Canada	-	-	-	-	-	54 e	18	40	-
Czech Republic	-	-	-	-	-	-	2	-	-
Denmark	65	-	-	2	33	-	107	107	20 e
Finland	77	10	37 e	-	-	-	287	109	45 e
France	200	107	104 e	104	235	108 e	238	40	11 e
Germany	250	255	35 e	78	25	-	559	967	889 e
Greece	-	-	-	72	-	-	-	-	-
Hungary	-	-	-	-	-	-	-	-	-
Ireland	11	40	-	8	1	11 e	13	46	11 e
Italy	-	-	-	-	-	13 e	2	16	-
Japan	8	-	-	29	-	-	-	-	-
Korea	-	-	-	-	-	-	-	-	-
Mexico	-	-	-	-	-	-	-	-	-
Netherlands	247	516	199 e	622	180	141 e	538	522	86 e
New Zealand	-	-	-	-	-	-	-	-	-
Norway	818	184	60 e	58	-	-	384	169	50 e
Poland	64	296	85 e	11	36	2 e	-	44	42 e
Spain	69	-	-	-	-	85 e	-	-	-
Sweden	402	120	81 e	-	-	-	390	161	50 e
Turkey	-	-	-	81	27	-	-	-	-
United Kingdom	-	-	-	-	-	-	-	-	-
United States	-	7	10 e	72	38	302 e	10	-	-
Other	-	2	-	49	-	-	101	43	-
OECD	**2 335**	**1 825**	**668 e**	**1 326**	**642**	**762 e**	**2 857**	**2 333**	**1 377 e**
Argentina	-	-	-	-	-	-	-	-	-
Brazil	-	-	-	-	-	-	-	-	-
Colombia	-	-	-	-	-	26 e	-	-	-
Ecuador	-	-	-	-	-	-	-	-	-
Netherlands Antilles	-	-	-	-	-	-	-	-	-
Peru	-	-	-	-	-	-	-	-	-
Trinidad and Tobago	-	-	-	-	-	-	-	-	-
Venezuela	-	-	-	180	59	366 e	-	-	-
Other	-	-	-	94	95	79 e	-	-	-
Latin America	**-**	**-**	**-**	**274**	**154**	**471 e**	**-**	**-**	**-**
Romania	-	-	-	-	-	-	-	-	-
Other	-	-	-	-	-	-	-	-	-
Non-OECD Europe	**-**	**-**	**-**	**-**	**-**	**-**	**-**	**-**	**-**
FSU	**98**	**-**	**-**	**204**	**312**	**525 e**	**878**	**765**	**809 e**
Iran	-	-	-	-	44	-	-	-	-
Iraq	-	-	-	-	-	-	-	-	-
Kuwait	-	-	-	834	908	715 e	-	-	-
Oman	-	-	-	-	82	-	-	-	-
Qatar	-	-	-	-	-	-	-	-	-
Saudi Arabia	-	-	-	566	978	625 e	48	-	-
United Arab Emirates	-	-	-	673	1 580	500 e	22	-	-
Other	-	-	-	179	161	137 e	10	-	-
Middle East	**-**	**-**	**-**	**2 252**	**3 759**	**2 043 e**	**80**	**-**	**-**
India	-	-	-	-	-	2 e	-	-	-
Indonesia	-	-	-	-	-	-	-	30	-
Singapore	-	-	-	-	-	228 e	-	-	-
Other	-	-	-	-	-	7 e	-	-	-
Asia	**-**	**-**	**-**	**-**	**-**	**237 e**	**-**	**30**	**-**
China	**-**	**-**	**-**	**-**	**-**	**-**	**-**	**-**	**-**
Algeria	-	-	-	357	30	90 e	-	-	-
Angola/Cabinada	-	-	-	-	-	-	-	-	-
Cameroon	-	-	-	-	-	-	-	-	-
Egypt	-	-	-	-	63	38 e	-	-	-
Gabon	-	-	-	-	-	-	-	-	-
Libya	26	13	-	213	595	319 e	-	11	-
Nigeria	-	-	-	28	96	-	-	-	-
Tunisia	-	-	-	19	4	-	-	-	-
Other	-	-	-	69	29	76 e	-	-	17 e
Africa	**26**	**13**	**-**	**686**	**817**	**523 e**	**-**	**11**	**17 e**
Non-spec./Other	**-**	**1 780**	**1 642 e**	**19**	**636**	**840 e**	**-**	**959**	**1 059 e**
Total Imports	**2 459**	**3 618**	**2 310 e**	**4 761**	**6 320**	**5 401 e**	**3 815**	**4 098**	**3 262 e**
Memo: Total OPEC	*26*	*13*	*-*	*2 851*	*4 296*	*2 681 e*	*70*	*41*	*-*

Source: Annual Oil Statistics.

UNITED KINGDOM
Table 7D: Imports by Country of Origin

000 metric tons

	2000	2001	2002	2000	2001	2002	2000	2001	2002
	Fuel Oil (Residual)			**Petroleum Coke**			**Other Products**		
Australia	-	-	-	-	-	-	-	-	-
Belgium	14	38	11 e	21	-	-	8	1	-
Canada	-	-	-	-	-	-	43	-	-
Czech Republic	-	-	-	-	-	-	-	-	-
Denmark	-	-	-	-	-	-	-	-	-
Finland	11	-	-	-	-	-	-	-	-
France	85	135	45 e	-	-	-	316	301	284 e
Germany	14	-	6 e	-	-	-	26	16	10 e
Greece	-	-	-	-	-	-	-	-	-
Hungary	-	-	-	-	-	-	-	-	-
Ireland	4	4	8 e	-	-	-	1	-	-
Italy	-	-	-	-	-	-	66	40	59 e
Japan	-	-	-	-	-	-	1	-	-
Korea	-	-	-	-	-	-	1	-	-
Mexico	-	-	-	-	-	-	-	-	-
Netherlands	137	77	49 e	-	-	-	27	21	32 e
New Zealand	-	-	-	-	-	-	-	-	-
Norway	18	89	16 e	5	-	-	658	-	3 e
Poland	11	55	-	-	-	-	-	-	-
Spain	-	-	-	1	3	-	12	-	-
Sweden	139	72	-	-	-	-	1	-	-
Turkey	-	-	-	-	-	-	-	-	-
United Kingdom	-	-	-	-	-	-	-	-	-
United States	38	-	-	630	2	-	30	3	1 e
Other	13	-	12 e	-	-	-	-	-	-
OECD	**484**	**470**	**147 e**	**657**	**5**	**-**	**1 190**	**382**	**389 e**
Argentina	-	-	-	-	-	-	2	-	-
Brazil	-	-	-	-	-	-	-	-	-
Colombia	-	-	-	-	-	-	-	-	-
Ecuador	-	-	-	-	-	-	-	-	-
Netherlands Antilles	-	-	-	-	-	-	-	-	-
Peru	-	-	-	-	-	-	-	-	-
Trinidad and Tobago	-	-	-	-	-	-	-	-	-
Venezuela	-	-	-	-	-	-	-	-	-
Other	-	-	-	-	-	-	6	10	-
Latin America	**-**	**-**	**-**	**-**	**-**	**-**	**8**	**10**	**-**
Romania	-	-	-	-	-	-	-	-	-
Other	-	-	-	-	-	-	-	-	-
Non-OECD Europe	**-**	**-**	**-**	**-**	**-**	**-**	**-**	**-**	**-**
FSU	**112**	**36**	**27 e**	**-**	**-**	**-**	**35**	**-**	**-**
Iran	-	-	-	-	-	-	-	-	-
Iraq	-	-	-	-	-	-	-	-	-
Kuwait	-	-	-	-	-	-	-	-	-
Oman	-	-	-	-	-	-	-	-	-
Qatar	-	-	-	-	-	-	-	-	-
Saudi Arabia	-	-	-	-	-	-	-	-	-
United Arab Emirates	-	-	-	-	-	-	-	-	-
Other	-	-	-	-	-	-	-	-	-
Middle East	**-**	**-**	**-**	**-**	**-**	**-**	**-**	**-**	**-**
India	-	-	-	-	-	-	-	-	-
Indonesia	-	-	-	-	-	-	-	-	-
Singapore	-	-	-	-	-	-	1	-	-
Other	-	-	-	-	-	-	-	-	-
Asia	**-**	**-**	**-**	**-**	**-**	**-**	**1**	**-**	**-**
China	**-**	**-**	**-**	**-**	**-**	**-**	**-**	**-**	**-**
Algeria	-	-	-	-	-	-	-	-	-
Angola/Cabinada	-	-	-	-	-	-	-	-	-
Cameroon	-	-	-	-	-	-	-	-	-
Egypt	-	-	-	-	-	-	-	-	-
Gabon	-	-	-	-	-	-	-	-	-
Libya	-	-	-	-	-	-	-	-	-
Nigeria	-	-	-	-	-	-	-	-	-
Tunisia	-	-	-	-	-	-	-	-	-
Other	-	-	-	-	-	-	3	-	-
Africa	**-**	**-**	**-**	**-**	**-**	**-**	**3**	**-**	**-**
Non-spec./Other	**-**	**505**	**687 e**	**-**	**628**	**659 e**	**8**	**138**	**164 e**
Total Imports	**596**	**1 011**	**861 e**	**657**	**633**	**659 e**	**1 245**	**530**	**553 e**
Memo: Total OPEC	-	-	-	-	-	-	-	-	-

Source: Annual Oil Statistics.

UNITED KINGDOM
Table 8A: Exports by Country of Destination

000 metric tons

	2000	2001	2002	2000	2001	2002	2000	2001	2002
	Crude Oil			**Natural Gas Liquids**			**Feedstocks and Other**		
Australia	-	-	-	-	-	-	-	-	-
Austria	-	-	-	-	-	-	-	-	-
Belgium	1 289	249	707 e	274	470	538 e	337	322	262 e
Canada	1 605	3 272	3 006 e	-	-	16 e	-	-	-
Denmark	-	79	64 e	13	23	29 e	-	20	31 e
Finland	729	1 177	653 e	5	26	-	-	-	-
France	15 809	14 188	11 212 e	622	402	533 e	77	117	162 e
Germany	11 545	12 104	9 485 e	156	139	98 e	32	6	6 e
Greece	-	-	134 e	-	4	-	-	-	-
Ireland	-	81	250 e	9	12	13 e	3	-	24 e
Italy	1 299	618	839 e	26	2	4 e	99	56	64 e
Japan	-	-	-	-	26	36 e	-	-	-
Luxembourg	-	-	-	-	-	-	-	-	-
Mexico	-	-	-	-	-	-	-	-	-
Netherlands	18 955	20 306	19 994 e	775	1 016	1 155 e	973	912	489 e
Norway	497	329	225 e	215	41	75 e	14	43	29 e
Portugal	656	631	912 e	124	89	77 e	16	9	21 e
Spain	2 452	2 427	1 359 e	302	205	139 e	164	86	66 e
Sweden	721	1 255	1 624 e	340	279	330 e	9	36	166 e
Switzerland	-	-	-	-	-	-	-	-	-
Turkey	-	-	-	108	29	30 e	52	66	52 e
United Kingdom	-	-	-	-	-	-	-	-	-
United States	27 688	21 702	25 794 e	68	225	32 e	709	582	210 e
Other OECD	436	-	-	12	22	13 e	4	29	17 e
OECD	**83 681**	**78 418**	**76 258 e**	**3 049**	**3 010**	**3 118 e**	**2 489**	**2 284**	**1 599 e**
Africa	163	-	264 e	16	25	10 e	153	48	-
Latin America	84	-	178 e	157	105	-	24	25	-
Asia	1 638	-	-	-	12	73 e	-	-	-
China	519	262	1 634 e	-	110	53 e	-	-	-
Non-OECD Europe	66	534	155 e	-	-	-	33	115	14 e
Former Soviet Union	251	-	-	-	-	-	7	-	-
Middle East	-	-	-	-	-	-	-	-	-
Non-specified/Other	131	1 521	2 929 e	327	260	136 e	130	17	180 e
Non-OECD	**2 852**	**2 317**	**5 160 e**	**500**	**512**	**272 e**	**347**	**205**	**194 e**
Total Exports	**86 533**	**80 735**	**81 418 e**	**3 549**	**3 522**	**3 390 e**	**2 836**	**2 489**	**1 793 e**
	Total Products			**Liquefied Petroleum Gas**			**Naphtha**		
Australia	-	-	-	-	-	-	-	-	-
Austria	13	15	13 e	-	-	-	-	-	-
Belgium	1 125	1 144	1 334 e	125	75	114 e	457	318	522 e
Canada	162	189	145 e	3	-	-	-	-	-
Denmark	149	218	99 e	14	-	-	-	-	6 e
Finland	11	52	136 e	-	-	-	-	-	-
France	2 588	2 289	2 255 e	-	60	53 e	92	149	187 e
Germany	1 411	1 355	1 187 e	-	-	-	13	-	-
Greece	16	1	11 e	-	-	-	-	-	-
Ireland	3 870	3 636	3 993 e	81	37	49 e	-	3	-
Italy	1 360	1 223	892 e	22	6	13 e	-	-	-
Japan	4	1	-	-	-	-	-	-	-
Luxembourg	-	-	-	-	-	-	-	-	-
Mexico	35	-	36 e	-	-	-	-	-	-
Netherlands	2 427	2 282	3 099 e	113	1	17 e	353	500	916 e
Norway	215	504	696 e	6	-	-	-	12	-
Portugal	293	409	648 e	68	50	54 e	-	-	-
Spain	975	667	1 201 e	78	6	2 e	-	-	-
Sweden	198	412	402 e	-	30	59 e	58	10	-
Switzerland	4	11	2 e	-	9	-	-	-	-
Turkey	55	64	91 e	7	-	-	-	-	-
United Kingdom	-	-	-	-	-	-	-	-	-
United States	3 172	3 146	3 089 e	34	7	15 e	-	50	19 e
Other OECD	120	59	58 e	16	4	10 e	-	-	-
OECD	**18 203**	**17 677**	**19 387 e**	**567**	**285**	**386 e**	**973**	**1 042**	**1 650 e**
Africa	388	128	336 e	103	22	9 e	-	-	12 e
Latin America	153	266	337 e	28	76	-	-	-	6 e
Asia	119	25	33 e	-	-	-	-	-	-
China	13	43	-	12	-	-	-	-	-
Non-OECD Europe	1 075	594	288 e	-	1	-	-	-	-
Former Soviet Union	35	5	30 e	-	-	-	-	-	24 e
Middle East	232	272	153 e	-	-	-	-	11	-
Non-specified/Other	459	70	1 543 e	-	-	-	-	25	116 e
Non-OECD	**2 474**	**1 403**	**2 720 e**	**143**	**99**	**9 e**	**-**	**36**	**158 e**
Total Exports	**20 677**	**19 080**	**22 107 e**	**710**	**384**	**395 e**	**973**	**1 078**	**1 808 e**

Source: Annual Oil Statistics.

UNITED KINGDOM
Table 8B: Exports by Country of Destination

000 metric tons

	2000	2001	2002	2000	2001	2002	2000	2001	2002
	Total Gasoline			**Total Kerosene**			**Gas/Diesel Oil**		
Australia	-	-	-	-	-	-	-	-	-
Austria	-	-	-	-	-	-	-	-	-
Belgium	16	63	49 e	11	-	-	342	327	332 e
Canada	159	186	109 e	-	-	-	-	-	-
Denmark	50	103	51 e	6	-	-	75	86	35 e
Finland	-	-	-	-	-	-	-	45	93 e
France	56	273	262 e	39	13	152 e	2 092	1 526	1 396 e
Germany	646	366	199 e	1	-	-	304	519	571 e
Greece	5	-	11 e	-	-	-	10	-	-
Ireland	686	735	870 e	583	484	670 e	1 532	1 329	1 575 e
Italy	49	27	-	-	12	-	131	61	28 e
Japan	-	-	-	-	-	-	-	-	-
Luxembourg	-	-	-	-	-	-	-	-	-
Mexico	-	-	36 e	-	-	-	35	-	-
Netherlands	665	723	825 e	16	77	36 e	809	568	757 e
Norway	11	63	70 e	-	-	-	-	-	20 e
Portugal	-	-	9 e	-	-	-	-	20	64 e
Spain	46	-	53 e	-	-	-	498	330	458 e
Sweden	12	17	71 e	5	37	29 e	63	213	139 e
Switzerland	-	-	-	-	-	-	-	-	-
Turkey	-	-	-	-	-	-	14	42	91 e
United Kingdom	-	-	-	-	-	-	-	-	-
United States	1 869	1 712	2 058 e	-	-	-	172	150	453 e
Other OECD	25	1	-	-	-	-	11	-	-
OECD	**4 295**	**4 269**	**4 673 e**	**661**	**623**	**887 e**	**6 088**	**5 216**	**6 012 e**
Africa	67	23	148 e	-	-	-	82	-	56 e
Latin America	13	98	313 e	-	-	-	-	32	-
Asia	-	-	-	-	-	-	-	-	-
China	-	-	-	-	-	-	-	-	-
Non-OECD Europe	180	62	-	25	-	-	144	39	-
Former Soviet Union	15	-	-	-	-	-	5	-	-
Middle East	138	-	-	-	-	-	38	-	-
Non-specified/Other	-	1	365 e	-	-	46 e	59	-	391 e
Non-OECD	**413**	**184**	**826 e**	**25**	**-**	**46 e**	**328**	**71**	**447 e**
Total Exports	**4 708**	**4 453**	**5 499 e**	**686**	**623**	**933 e**	**6 416**	**5 287**	**6 459 e**

	2000	2001	2002	2000	2001	2002	2000	2001	2002
	Fuel Oil (Residual)			**Petroleum Coke**			**Other Products**		
Australia	-	-	-	-	-	-	-	-	-
Austria	-	-	-	13	11	13 e	-	4	-
Belgium	29	160	90 e	4	-	1 e	141	201	226 e
Canada	-	-	36 e	-	-	-	-	3	-
Denmark	-	26	-	-	-	-	4	3	7 e
Finland	-	-	40 e	-	-	-	11	7	3 e
France	155	149	53 e	33	58	102 e	121	61	50 e
Germany	296	291	304 e	38	68	47 e	113	111	66 e
Greece	-	-	-	-	-	-	1	1	-
Ireland	760	857	645 e	-	-	-	228	191	184 e
Italy	1 120	837	825 e	38	26	5 e	-	254	21 e
Japan	-	-	-	1	1	-	3	-	-
Luxembourg	-	-	-	-	-	-	-	-	-
Mexico	-	-	-	-	-	-	-	-	-
Netherlands	313	220	426 e	-	8	4 e	158	185	118 e
Norway	111	320	92 e	60	88	105 e	27	21	409 e
Portugal	207	324	495 e	-	-	-	18	15	26 e
Spain	178	180	512 e	158	101	156 e	17	50	20 e
Sweden	-	34	27 e	35	23	32 e	25	48	45 e
Switzerland	-	-	-	-	2	2 e	4	-	-
Turkey	-	22	-	-	-	-	34	-	-
United Kingdom	-	-	-	-	-	-	-	-	-
United States	1 069	1 191	518 e	13	-	-	15	36	26 e
Other OECD	-	-	-	19	16	18 e	49	38	30 e
OECD	**4 238**	**4 611**	**4 063 e**	**412**	**402**	**485 e**	**969**	**1 229**	**1 231 e**
Africa	35	36	68 e	26	14	7 e	75	33	36 e
Latin America	99	40	-	1	1	4 e	12	19	14 e
Asia	87	-	-	22	15	21 e	10	10	12 e
China	-	-	-	1	1	-	-	42	-
Non-OECD Europe	693	470	264 e	33	22	20 e	-	-	4 e
Former Soviet Union	-	-	-	7	5	6 e	8	-	-
Middle East	44	246	110 e	-	-	-	12	15	43 e
Non-specified/Other	164	38	570 e	-	-	5 e	236	6	50 e
Non-OECD	**1 122**	**830**	**1 012 e**	**90**	**58**	**63 e**	**353**	**125**	**159 e**
Total Exports	**5 360**	**5 441**	**5 075 e**	**502**	**460**	**548 e**	**1 322**	**1 354**	**1 390 e**

Source: Annual Oil Statistics.

UNITED STATES

Figure 1
FUEL SHARES OF TPES *

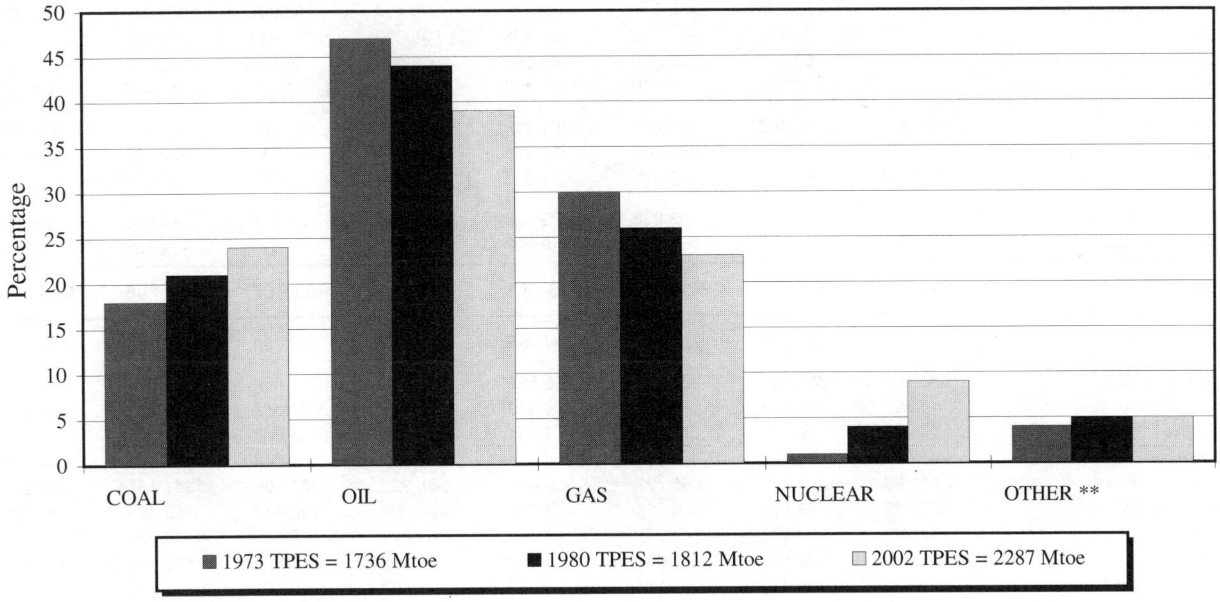

Source: Annual Energy Balances.

* Excluding electricity trade.

** Includes hydro, geothermal, solar, wind, combustible renewables and waste, etc.

Figure 2
OIL PRODUCTION AND CONSUMPTION

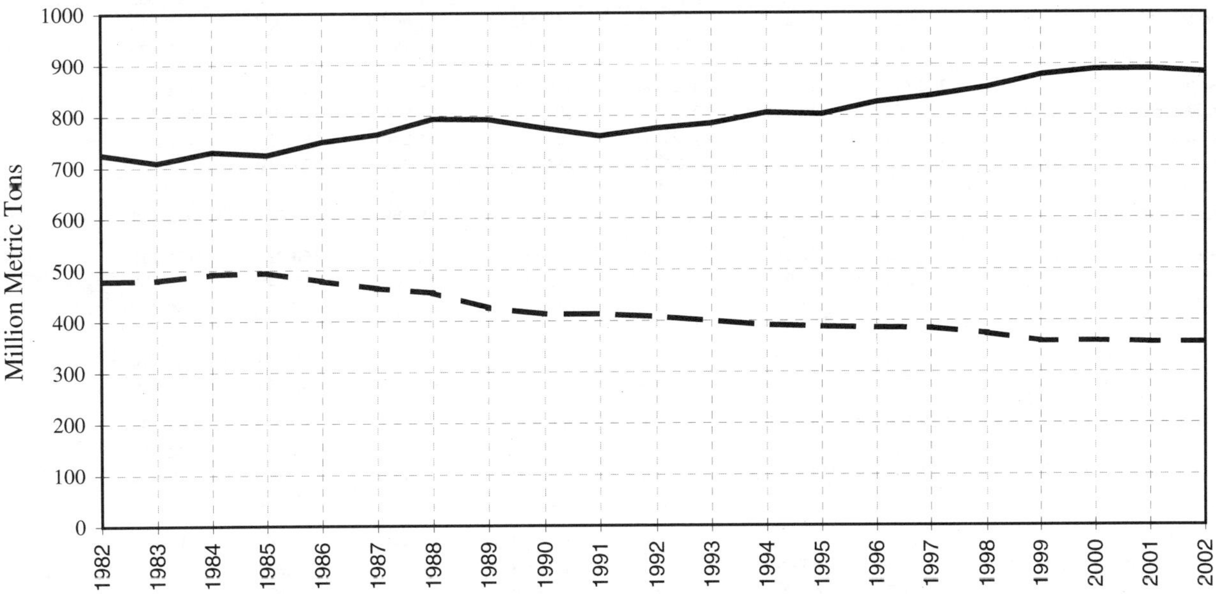

Source: Annual Oil Statistics.

*** Includes refinery fuel and international marine bunkers.

UNITED STATES
Table 1: Oil Supply and Consumption - Summary Table

000 metric tons

	1973	1978	1980	1990	1999	2000	2001	2002
Indigenous Production	513 296	483 076	478 626	413 344	358 436	358 802	356 516	356 379 e
Crude Oil	..	428 447	424 153	362 796	290 099	287 933	286 150	284 407 e
NGL	..	53 462	53 523	49 052	53 491	55 175	53 883	54 134 e
Other	..	1 167	950	1 496	14 846	15 694	16 483	17 838 e
+ Net Imports	298 847	416 188	329 211	361 216	504 554	530 558	552 711	533 231 e
Crude Oil, NGL, Other	171 241	349 711	289 440	339 415	484 660	505 111	522 736	514 443 e
Petroleum Products	127 606	66 477	39 771	21 801	19 894	25 447	29 975	18 788 e
- Intl. Marine Bunkers	9 516	21 239	28 938	29 349	26 604	28 777	19 836	29 423 e
+ Stock Changes	-5 840	3 105	-6 349	-8 228	18 293	3 124	-12 595	4 877 e
TOTAL SUPPLY	**796 787**	**881 130**	**772 550**	**736 983**	**854 679**	**863 707**	**876 796**	**865 064 e**
Refinery Intake	*649 766*	*793 583*	*723 401*	*731 435*	*808 152*	*820 738*	*817 964*	*814 827 e*
Refinery Output	*665 486*	*788 157*	*723 765*	*730 552*	*807 518*	*822 834*	*819 874*	*816 056 e*
LPG + Ethane	*11 595*	*11 494*	*10 785*	*15 869*	*20 424*	*21 192*	*19 989*	*20 269 e*
Naphtha	*21 649*	*10 803*	*13 489*	*7 441*	*8 678*	*9 207*	*7 228*	*10 574 e*
Kerosenes	*49 581*	*44 613*	*46 458*	*64 481*	*77 132*	*79 646*	*75 640*	*74 131 e*
Gasolines	*281 498*	*324 421*	*295 133*	*309 889*	*348 323*	*350 339*	*351 251*	*357 271 e*
Gas/Diesel Oil	*137 982*	*163 154*	*138 323*	*147 719*	*171 296*	*180 877*	*185 755*	*180 334 e*
Fuel Oil	*53 243*	*110 572*	*101 945*	*58 694*	*43 213*	*42 576*	*43 041*	*35 784 e*
Other Products	*109 938*	*123 100*	*117 632*	*126 459*	*138 452*	*138 997*	*136 970*	*137 693 e*
- Refinery Losses	-15 720	5 426	- 364	883	634	-2 096	-1 910	-1 229 e
+ Returns	-	-	-	-	-	-	-	-
- Statistical Differences	-6 761	-3 093	-4 922	-9 367	2 716	6 197	9 611	12 851 e
TOTAL CONSUMPTION	**819 268**	**878 797**	**777 836**	**745 467**	**851 329**	**859 606**	**869 095**	**853 442 e**
Transformation and Energy	**149 917**	**147 329**	**111 402**	**81 195**	**88 205**	**83 036**	**80 986**	..
Crude Oil	800	722	677	1 186	-	-	-	..
LPG + Ethane	1 557	2 246	2 153	1 275	424	496	575	..
Gas/Diesel Oil	6 300	7 539	4 314	2 056	6 055	4 938	4 807	..
Fuel Oil	83 607	95 864	63 832	27 573	26 993	23 003	25 865	..
Petroleum Coke	12 930	12 231	12 757	15 415	21 607	20 997	16 813	..
Other	44 723	28 727	27 669	33 690	33 126	33 602	32 926	..
Transport	**382 757**	**428 136**	**396 894**	**462 306**	**555 552**	**566 066**	**565 306**	..
Motor Gasoline	277 772	312 753	279 252	303 659	359 966	362 736	364 705	..
Gas/Diesel Oil	47 864	57 851	59 366	81 161	111 244	115 981	117 631	..
Subtotal	*325 636*	*370 604*	*338 618*	*384 820*	*471 210*	*478 717*	*482 336*	..
Aviation Fuels	49 002	50 107	52 479	71 918	78 954	82 016	77 859	..
Other	8 119	7 425	5 797	5 568	5 388	5 333	5 111	..
Industry	**154 235**	**179 852**	**184 106**	**141 641**	**152 601**	**150 284**	**161 017**	..
Naphtha	21 334	16 283	17 827	8 881	11 462	13 946	11 135	..
LPG + Ethane	34 521	31 131	39 913	40 499	56 347	56 341	50 432	..
Gas/Diesel Oil	20 826	27 872	24 291	16 598	15 478	16 439	18 230	..
Fuel Oil	30 189	32 435	34 964	10 566	1 184	6 214	7 210	..
Other	47 365	72 131	67 111	65 097	68 130	57 344	74 010	..
Other Sectors	**132 359**	**123 480**	**85 434**	**60 325**	**54 971**	**60 220**	**61 786**	..
Gas/Diesel Oil	74 742	75 017	52 504	40 932	37 198	40 573	42 512	..
Fuel Oil	27 383	23 078	15 067	6 204	419	2 365	2 478	..
Other	30 234	25 385	17 863	13 189	17 354	17 282	16 796	..
Memo: Non-Energy Use	*87 074*	*104 602*	*95 448*	*101 148*	*138 130*	*129 713*	*134 486*	..
Chem/Petchem. Feedst.	*42 071*	*56 908*	*54 665*	*53 440*	*74 010*	*74 609*	*70 402*	..
Other	*45 003*	*47 694*	*40 783*	*47 708*	*64 120*	*55 104*	*64 084*	..

Source: Annual Oil Statistics.

UNITED STATES
Table 2: Supply of Crude Oil, NGL, Feedstocks and Other Hydrocarbons

000 metric tons

	1973	1978	1980	1990	1999	2000	2001	2002
CRUDE OIL								
Indigenous Production	..	428 447	424 153	362 796	290 099	287 933	286 150	284 407 e
+ Imports	..	349 299	285 252	314 165	465 798	484 111	494 431	487 436 e
- Exports	..	3 877	4 121	610	5 971	2 551	1 027	456 e
- Direct Use	..	722	677	1 186	-	-	-	-
+ Stock Change	..	-5 769	-3 821	1 106	3 729	3 264	-3 665	-3 116 e
- Statistical Difference	..	1 802	-1 611	-9 016	1 917	5 626	9 892	12 562 e
Refinery Intake	**..**	**765 576**	**702 397**	**685 287**	**751 738**	**767 131**	**765 997**	**755 709 e**
Closing Stock Level	..	*39 841*	*50 327*	*112 027*	*105 361*	*102 097*	*105 762*	*108 878 e*
NATURAL GAS LIQUIDS								
Indigenous Production	..	53 462	53 523	49 052	53 491	55 175	53 883	54 134 e
+ Imports	..	4 493	7 309	6 199	6 408	7 725	7 831	5 873 e
- Exports	..	569	664	1 302	1 646	2 421	1 400	1 901 e
- Direct Use	..	36 843	41 747	39 206	45 551	46 996	44 627	40 132 e
+ Stock Change	..	514	-1 589	53	497	64	- 270	- 50 e
- Statistical Difference	..	-	-	2	- 1	-	-	3 e
Refinery Intake	**..**	**21 057**	**16 832**	**14 794**	**13 200**	**13 547**	**15 417**	**17 921 e**
Closing Stock Level	..	*9 358*	*10 163*	*1 774*	*1 447*	*1 383*	*1 653*	*1 703 e*
REFINERY FEEDSTOCKS								
Backflows	..	-	-	-	-	-	-	-
+ Transfers	..	-	-	8 061	9 531	9 326	896	1 887 e
+ Imports	..	365	1 664	20 954	18 106	16 090	20 637	21 936 e
- Exports	..	-	-	-	-	-	-	-
+ Stock Change	..	523	-1 076	875	648	- 179	340	1 650 e
- Statistical Difference	..	-4 895	-2 634	-	3	-	-	- 4 e
Refinery Intake	**..**	**5 783**	**3 222**	**29 890**	**28 282**	**25 237**	**21 873**	**25 477 e**
Closing Stock Level	..	*14 950*	*17 304*	*14 774*	*12 786*	*12 965*	*12 625*	*10 975 e*
OTHER HYDROCARBONS								
Indigenous Production	..	1 167	950	1 496	14 846	15 694	16 483	17 838 e
+ Imports	..	-	-	9	3 212	3 485	3 432	2 664 e
- Exports	..	-	-	-	1 247	1 328	1 168	1 109 e
- Direct Use	..	-	-	-	2 596	3 035	3 088	3 029 e
+ Stock Change	..	-	-	- 40	9	174	- 159	80 e
- Statistical Difference	..	-	-	1	- 708	167	823	724 e
Refinery Intake	**..**	**1 167**	**950**	**1 464**	**14 932**	**14 823**	**14 677**	**15 720 e**
Closing Stock Level	..	-	-	*75*	*1 604*	*1 430*	*1 589*	*1 509 e*
CRUDE OIL, NGL, FEEDSTOCKS AND OTHER								
Indigenous Production	513 296	483 076	478 626	413 344	358 436	358 802	356 516	356 379 e
+ Backflows	-	-	-	-	-	-	-	-
+ Transfers	-	-	-	8 061	9 531	9 326	896	1 887 e
+ Imports	171 335	354 157	294 225	341 327	493 524	511 411	526 331	517 909 e
- Exports	94	4 446	4 785	1 912	8 864	6 300	3 595	3 466 e
- Direct Use	38 056	37 565	42 424	40 392	48 147	50 031	47 715	43 161 e
+ Stock Change	-1 538	-4 732	-6 486	1 994	4 883	3 323	-3 754	-1 436 e
- Statistical Difference	-4 823	-3 093	-4 245	-9 013	1 211	5 793	10 715	13 285 e
Refinery Intake	**649 766**	**793 583**	**723 401**	**731 435**	**808 152**	**820 738**	**817 964**	**814 827 e**
Closing Stock Level	-	*64 149*	*77 794*	*128 650*	*121 198*	*117 875*	*121 629*	*123 065 e*

Source: Annual Oil Statistics.

UNITED STATES
Table 3: Motor Gasoline - Supply

000 metric tons

	1973	1978	1980	1990	1999	2000	2001	2002
MOTOR GASOLINE								
Refinery Output	**279 654**	**313 737**	**284 843**	**300 922**	**347 462**	**349 595**	**350 507**	**356 556 e**
+ Imports	5 640	3 103	2 076	12 969	20 904	22 777	28 531	31 143 e
- Exports	172	24	57	2 630	5 168	6 860	6 294	6 852 e
+ Transfers	-	-	592	-	-1 364	2 001	1 072	929 e
+ Stock Change	400	2 237	-2 402	- 463	2 571	- 369	-1 631	287 e
- Statistical Difference	-	-	-	-	-	-	-	-
Total Consumption	**285 522**	**319 053**	**285 052**	**310 798**	**364 405**	**367 144**	**372 185**	**382 063 e**
Closing Stock Level	-	*24 221*	*26 492*	*21 655*	*18 702*	*19 071*	*20 702*	*20 415 e*

Table 4: Aviation Fuels - Supply

000 metric tons

	1973	1978	1980	1990	1999	2000	2001	2002
AVIATION GASOLINE								
Refinery Output	**1 844**	**1 585**	**1 385**	**959**	**837**	**735**	**733**	**712 e**
+ Imports	-	4	1	9	9	38	67	26 e
- Exports	22	-	-	-	-	-	-	-
+ Transfers	-	-	73	-	141	93	117	169 e
+ Stock Change	35	23	-	39	16	26	- 1	10 e
- Statistical Difference	-	-	-	-	-	-	-	1 e
Total Consumption	**1 857**	**1 612**	**1 459**	**1 007**	**1 003**	**892**	**916**	**916 e**
Closing Stock Level	-	*319*	*301*	*201*	*208*	*182*	*183*	*173 e*
GASOLINE TYPE JET FUEL								
Refinery Output	**-**	**9 099**	**8 905**	**8 008**	**24**	**9**	**11**	**3 e**
+ Imports	-	188	656	72	-	-	-	-
- Exports	-	-	-	188	114	5	30	356 e
+ Transfers	-	-	-	-	-	-	-	-
+ Stock Change	-	73	- 114	64	- 3	- 6	3	3 e
- Statistical Difference	-	-	-	1	- 93	- 2	- 16	- 350 e
Total Consumption	**-**	**9 360**	**9 447**	**7 955**	**-**	**-**	**-**	**-**
Closing Stock Level	-	*670*	*810*	*633*	*7*	*13*	*10*	*7 e*
KEROSENE TYPE JET FUEL								
Refinery Output	**39 215**	**36 554**	**39 675**	**62 272**	**73 937**	**76 495**	**72 085**	**71 275 e**
+ Imports	9 299	3 207	2 282	3 817	5 009	6 220	6 177	3 993 e
- Exports	194	28	59	1 808	1 403	1 469	1 340	387 e
+ Transfers	-	-	-	-	-	-	-	-
+ Stock Change	- 387	- 27	- 325	-1 265	410	- 121	21	262 e
- Statistical Difference	2	-	-	- 1	2	1	-	7 e
Total Consumption	**47 931**	**39 706**	**41 573**	**63 017**	**77 951**	**81 124**	**76 943**	**75 136 e**
Closing Stock Level	-	*3 051*	*3 812*	*4 711*	*4 131*	*4 252*	*4 231*	*3 969 e*

Source: Annual Oil Statistics.

UNITED STATES
Table 5A: Gas/Diesel Oil - Supply and Consumption

000 metric tons

	1973	1978	1980	1990	1999	2000	2001	2002
Refinery Output	**137 982**	**163 154**	**138 323**	**147 719**	**171 296**	**180 877**	**185 755**	**180 334 e**
+ Recycled Products	-	-	-	-	-	-	-	-
+ Imports	18 599	2 346	1 495	10 818	9 256	11 170	13 646	10 208 e
- Exports	434	78	164	5 701	8 000	8 588	6 082	5 165 e
- Intl. Marine Bunkers	980	2 030	2 290	7 012	6 589	6 640	6 681	6 730 e
+ Transfers	-	-	22	-1 476	-	-	-	-
+ Stock Changes	-5 649	4 887	3 016	-3 956	3 984	972	-3 816	2 136 e
- Statistical Differences	- 214	-	- 73	- 355	- 28	- 140	- 358	- 732 e
Total Consumption	**149 732**	**168 279**	**140 475**	**140 747**	**169 975**	**177 931**	**183 180**	**181 515 e**
Transformation	**6 300**	**6 350**	**3 900**	**1 969**	**5 862**	**4 792**	**4 572**	..
Public Electricity	6 300	6 350	3 900	1 969	2 744	2 965	2 843	..
Autoproducer Electricity	-	-	-	-	1 336	714	996	..
Public CHP Plants	-	-	-	-	142	88	263	..
Autoproducers of CHP	-	-	-	-	1 640	1 025	470	..
Public Heat Plants	-	-	-	-	-	-	-	..
Autoproducer Heat Plants	-	-	-	-	-	-	-	..
Petrochemical Industry	-	-	-	-	-	-	-	..
Other Transformation	-	-	-	-	-	-	-	..
Energy Sector	**-**	**1 189**	**414**	**87**	**193**	**146**	**235**	..
Refinery Fuel	-	1 189	414	87	193	146	235	..
Other Energy Sector	-	-	-	-	-	-	-	..
Final Consumption	**143 432**	**160 740**	**136 161**	**138 691**	**163 920**	**172 993**	**178 373**	..
Transport	**47 864**	**57 851**	**59 366**	**81 161**	**111 244**	**115 981**	**117 631**	..
Air	-	-	-	-	-	-	-	..
Road	33 549	43 802	45 750	68 361	101 497	105 464	107 006	..
Rail	14 315	14 049	13 616	10 268	8 832	9 775	9 510	..
Inland Waterways	-	-	-	-	-	-	-	..
Non-Specified	-	-	-	2 532	915	742	1 115	..
Industry	**20 826**	**27 872**	**24 291**	**16 598**	**15 478**	**16 439**	**18 230**	..
Iron and Steel	-	-	-	-	78	658	797	..
Chemical (incl. Pet.-Chem.)	-	-	-	-	3 188	4 075	4 765	..
Non-Ferrous Metals	-	-	-	-	1 177	329	314	..
Non-Metallic Minerals	-	-	-	-	2 260	2 038	2 183	..
Transport Equipment	-	-	-	-	665	1 852	1 985	..
Machinery	-	-	-	-	1 130	1 485	1 571	..
Mining and Quarrying	-	-	-	-	-	-	-	..
Food and Tobacco	-	-	-	-	1 811	2 283	2 385	..
Pulp, Paper and Printing	-	-	-	-	1 037	1 153	1 224	..
Wood and Wood Products	-	-	-	-	2 539	1 601	1 838	..
Construction	-	-	-	-	-	-	-	..
Textiles and Leather	-	-	-	-	774	718	849	..
Non-Specified	20 826	27 872	24 291	16 598	819	247	319	..
Other Sectors	**74 742**	**75 017**	**52 504**	**40 932**	**37 198**	**40 573**	**42 512**	..
Commerce and Public	15 920	16 350	12 309	11 253	9 508	10 736	11 292	..
Residential	49 756	47 981	30 364	19 270	18 512	19 751	20 178	..
Agriculture	9 066	10 686	9 831	10 409	9 178	10 086	11 042	..
Non-Specified	-	-	-	-	-	-	-	..
Memo: Non-Energy Use	-	-	-	-	-	-	-	..
Chem/Petchem. Feedst.	-	-	-	-	-	-	-	..
Other	-	-	-	-	-	-	-	..
Closing Stock Level	-	*28 713*	*27 632*	*17 811*	*16 474*	*15 502*	*19 318*	*17 182 e*

Source: Annual Oil Statistics.

UNITED STATES
Table 5B: Fuel Oil (Residual) - Supply and Consumption

000 metric tons

	1973	1978	1980	1990	1999	2000	2001	2002
Refinery Output	**53 243**	**110 572**	**101 945**	**58 694**	**43 213**	**42 576**	**43 041**	**35 784 e**
+ Recycled Products	-	-	-	-	-	-	-	-
+ Imports	100 106	62 041	40 727	24 404	11 426	19 429	17 140	10 195 e
- Exports	1 386	615	913	11 821	7 404	8 068	10 757	11 825 e
- Intl. Marine Bunkers	8 536	19 209	26 648	22 337	20 015	22 137	13 155	22 693 e
+ Transfers	-	-	-	-3 226	-	-	-	-
+ Stock Changes	261	942	- 511	-3 949	2 845	297	-1 206	2 172 e
- Statistical Differences	- 919	-	- 604	-2 920	1 452	414	- 567	586 e
Total Consumption	**144 607**	**153 731**	**115 204**	**44 685**	**28 613**	**31 683**	**35 630**	**13 047 e**
Transformation	**77 050**	**88 350**	**58 750**	**25 666**	**25 799**	**21 948**	**24 803**	..
Public Electricity	77 050	88 350	58 750	25 666	17 852	14 644	15 807	..
Autoproducer Electricity	-	-	-	-	4 196	5 715	7 478	..
Public CHP Plants	-	-	-	-	514	394	311	..
Autoproducers of CHP	-	-	-	-	3 237	1 195	1 207	..
Public Heat Plants	-	-	-	-	-	-	-	..
Autoproducer Heat Plants	-	-	-	-	-	-	-	..
Petrochemical Industry	-	-	-	-	-	-	-	..
Other Transformation	-	-	-	-	-	-	-	..
Energy Sector	**6 557**	**7 514**	**5 082**	**1 907**	**1 194**	**1 055**	**1 062**	..
Refinery Fuel	6 557	7 514	5 082	1 907	1 194	1 055	1 062	..
Other Energy Sector	-	-	-	-	-	-	-	..
Final Consumption	**61 000**	**57 867**	**51 372**	**17 112**	**1 620**	**8 680**	**9 765**	..
Transport	**3 428**	**2 354**	**1 341**	**342**	**17**	**101**	**77**	..
Air	-	-	-	-	-	-	-	..
Road	-	-	-	-	-	-	-	..
Rail	171	28	103	-	-	-	-	..
Inland Waterways	3 257	2 326	1 238	342	17	101	77	..
Non-Specified	-	-	-	-	-	-	-	..
Industry	**30 189**	**32 435**	**34 964**	**10 566**	**1 184**	**6 214**	**7 210**	..
Iron and Steel	4 764	3 183	4 346	-	101	505	538	..
Chemical (incl. Pet.-Chem.)	-	5 195	5 151	-	295	2 225	2 916	..
Non-Ferrous Metals	-	1 715	430	-	4	16	16	..
Non-Metallic Minerals	-	-	915	-	18	68	78	..
Transport Equipment	600	647	710	-	27	87	92	..
Machinery	-	415	368	-	27	87	102	..
Mining and Quarrying	-	-	-	-	-	-	-	..
Food and Tobacco	1 760	2 182	2 045	-	75	279	305	..
Pulp, Paper and Printing	10 925	8 995	9 088	-	417	2 629	2 810	..
Wood and Wood Products	275	299	322	-	7	9	27	..
Construction	-	-	-	-	-	-	-	..
Textiles and Leather	-	-	1 223	-	46	290	314	..
Non-Specified	11 865	9 804	10 366	10 566	167	19	12	..
Other Sectors	**27 383**	**23 078**	**15 067**	**6 204**	**419**	**2 365**	**2 478**	..
Commerce and Public	27 383	23 078	15 067	6 204	419	2 365	2 478	..
Residential	-	-	-	-	-	-	-	..
Agriculture	-	-	-	-	-	-	-	..
Non-Specified	-	-	-	-	-	-	-	..
Memo: Non-Energy Use	-	-	-	-	-	-	-	..
Chem/Petchem. Feedst.	-	-	-	-	-	-	-	..
Other	-	-	-	-	-	-	-	..
Closing Stock Level	-	*29 782*	*32 471*	*18 332*	*10 109*	*9 812*	*11 018*	*8 846 e*

Source: Annual Oil Statistics.

UNITED STATES
Table 5C: Total Products - Supply and Consumption

000 metric tons

	1973	1978	1980	1990	1999	2000	2001	2002
Refinery Output	**665 486**	**788 157**	**723 765**	**730 552**	**807 518**	**822 834**	**819 874**	**816 056 e**
+ Recycled Products	-	-	-	-	-	-	-	-
+ Imports	140 137	77 157	52 824	62 085	60 665	74 404	79 650	69 581 e
- Exports	12 531	10 680	13 053	40 284	40 771	48 957	49 675	50 793 e
- Intl. Marine Bunkers	9 516	21 239	28 938	29 349	26 604	28 777	19 836	29 423 e
+ Transfers	38 056	37 565	42 424	32 331	38 616	40 705	46 819	41 274 e
+ Stock Changes	-4 302	7 837	137	-10 222	13 410	- 199	-8 841	6 313 e
- Statistical Differences	-1 938	-	- 677	- 354	1 505	404	-1 104	- 434 e
Total Consumption	**819 268**	**878 797**	**777 836**	**745 467**	**851 329**	**859 606**	**869 095**	**853 442 e**
Transformation	**85 331**	**96 651**	**64 032**	**28 733**	**37 040**	**31 470**	**30 142**	..
Public Electricity	84 636	95 631	62 812	28 419	22 052	18 700	18 910	..
Autoproducer Electricity	-	-	-	-	5 995	6 773	8 549	..
Public CHP Plants	-	-	-	-	656	520	629	..
Autoproducers of CHP	-	-	-	-	8 337	5 477	2 054	..
Public Heat Plants	-	-	-	-	-	-	-	..
Autoproducer Heat Plants	-	-	-	-	-	-	-	..
Petrochemical Industry	-	-	-	-	-	-	-	..
Other Transformation	695	1 020	1 220	314	-	-	-	..
Energy Sector	**64 586**	**50 678**	**47 370**	**52 462**	**51 165**	**51 566**	**50 844**	..
Refinery Fuel	63 786	49 956	46 693	51 276	51 165	51 566	50 844	..
Other Energy Sector	800	722	677	1 186	-	-	-	..
Final Consumption	**669 351**	**731 468**	**666 434**	**664 272**	**763 124**	**776 570**	**788 109**	..
Transport	**382 757**	**428 136**	**396 894**	**462 306**	**555 552**	**566 066**	**565 306**	..
Air	49 002	50 107	52 479	71 918	78 954	82 016	77 859	..
Road	316 012	361 626	329 458	373 956	463 849	470 353	474 022	..
Rail	14 486	14 077	13 719	10 268	8 832	9 775	9 510	..
Inland Waterways	3 257	2 326	1 238	3 632	3 002	3 180	2 800	..
Non-Specified	-	-	-	2 532	915	742	1 115	..
Industry	**154 235**	**179 852**	**184 106**	**141 641**	**152 601**	**150 284**	**161 017**	..
Iron and Steel	4 764	3 183	4 346	-	244	1 184	1 354	..
Chemical (incl. Pet.-Chem.)	42 071	62 103	59 816	53 440	77 493	80 909	78 083	..
Non-Ferrous Metals	-	1 715	430	-	1 243	396	376	..
Non-Metallic Minerals	-	-	915	-	2 379	2 161	2 310	..
Transport Equipment	600	647	710	-	768	2 020	2 149	..
Machinery	-	415	368	-	1 424	1 797	1 875	..
Mining and Quarrying	-	-	-	-	-	-	-	..
Food and Tobacco	1 760	2 182	2 045	-	2 055	2 687	2 802	..
Pulp, Paper and Printing	10 925	8 995	9 088	-	1 602	3 917	4 155	..
Wood and Wood Products	275	299	322	-	2 694	1 721	1 964	..
Construction	31 425	28 938	24 051	30 138	33 444	32 325	32 667	..
Textiles and Leather	-	-	1 223	-	951	1 066	1 215	..
Non-Specified	62 415	71 375	80 792	58 063	28 304	20 101	32 067	..
Other Sectors	**132 359**	**123 480**	**85 434**	**60 325**	**54 971**	**60 220**	**61 786**	..
Commerce and Public	49 103	45 178	32 002	21 063	13 000	16 268	16 706	..
Residential	68 765	63 841	40 233	25 358	29 169	30 366	30 142	..
Agriculture	14 491	14 461	13 199	13 904	12 802	13 586	14 938	..
Non-Specified	-	-	-	-	-	-	-	..
Memo: Non-Energy Use	*87 074*	*104 602*	*95 448*	*101 148*	*138 130*	*129 713*	*134 486*	..
Chem/Petchem. Feedst.	*42 071*	*56 908*	*54 665*	*53 440*	*74 010*	*74 609*	*70 402*	..
Other	*45 003*	*47 694*	*40 783*	*47 708*	*64 120*	*55 104*	*64 084*	..
Closing Stock Level	-	*98 354*	*101 298*	*77 156*	*62 403*	*62 602*	*71 443*	*65 130 e*

Source: Annual Oil Statistics.

UNITED STATES
Table 6A: Supply and Consumption

000 metric tons

	1980	2000	2001	2002	1980	2000	2001	2002
	LPG AND ETHANE				**REFINERY GAS**			
Refinery Output	**10 785**	**21 192**	**19 989**	**20 269 e**	**29 867**	**33 482**	**33 613**	**33 563 e**
+ Recycled Products	-	-	-	-	-	-	-	-
+ Imports	-	223	176	326 e	-	-	-	-
- Exports	-	-	-	-	-	-	-	-
- Intl. Marine Bunkers	-	-	-	-	-	-	-	-
+ Transfers	40 746	46 996	44 627	40 132 e	-	-	-	-
+ Stock Change	- 34	406	-2 568	1 164 e	-	-	-	-
- Statistical Difference	-	- 1	1	315 e	-	-	-	1 e
Total Consumption	**51 497**	**68 818**	**62 223**	**61 576 e**	**29 867**	**33 482**	**33 613**	**33 562 e**
of which: Ethane	-	*16 820*	*15 008*	*15 425 e*	-	-	-	-
Transformation	1 220	-	-	..	-	-	-	..
Energy	933	496	575	..	27 532	33 089	32 511	..
Transport	431	419	494	..	-	-	-	..
Industry	39 913	56 341	50 432	..	2 335	393	1 102	..
Other	9 000	11 562	10 722	..	-	-	-	..
Memo: Non-Energy Use	*16 235*	*55 465*	*49 647*	..	*2 335*	*393*	*1 102*	..
Closing Stock Level	*558*	*4 556*	*7 124*	*5 960 e*	-	-	-	-
	NAPHTHA				**OTHER KEROSENE**			
Refinery Output	**13 489**	**9 207**	**7 228**	**10 574 e**	**6 783**	**3 151**	**3 555**	**2 856 e**
+ Recycled Products	-	-	-	-	-	-	-	-
+ Imports	4 708	4 809	3 824	2 681 e	-	106	215	244 e
- Exports	224	-	-	-	3	93	99	864 e
- Intl. Marine Bunkers	-	-	-	-	-	-	-	-
+ Transfers	-	-	-	-	5	-	-	-
+ Stock Change	- 146	- 44	86	14 e	528	64	- 171	- 11 e
- Statistical Difference	-	1	-	5 e	-	- 1	1	- 1 e
Total Consumption	**17 827**	**13 971**	**11 138**	**13 264 e**	**7 313**	**3 229**	**3 499**	**2 226 e**
Transformation	-	-	-	..	-	-	-	..
Energy	-	25	3	..	-	-	-	..
Transport	-	-	-	..	-	-	-	..
Industry	17 827	13 946	11 135	..	3 850	289	457	..
Other	-	-	-	..	3 463	2 940	3 042	..
Memo: Non-Energy Use	*17 827*	*13 946*	*11 135*	..	-	-	-	..
Closing Stock Level	*544*	*406*	*320*	*306 e*	*1 401*	*503*	*674*	*685 e*
	WHITE SPIRIT + SBP				**LUBRICANTS**			
Refinery Output	**4 052**	**2 564**	**2 149**	**..**	**9 511**	**9 725**	**9 249**	**..**
+ Recycled Products	-	-	-	..	-	-	-	..
+ Imports	359	428	542	..	138	253	253	..
- Exports	56	819	948	..	1 229	1 369	1 370	..
- Intl. Marine Bunkers	-	-	-	..	-	-	-	..
+ Transfers	165	-	-	..	-	-	-	..
+ Stock Change	- 86	28	13	..	- 146	- 41	- 226	..
- Statistical Difference	-	1	1	..	-	1	- 1	..
Total Consumption	**4 434**	**2 200**	**1 755**	**..**	**8 274**	**8 567**	**7 907**	**..**
Transformation	-	-	-	..	-	-	-	..
Energy	-	-	-	..	-	2	2	..
Transport	-	-	-	..	4 025	4 813	4 540	..
Industry	4 434	2 200	1 755	..	4 249	3 752	3 365	..
Other	-	-	-	..	-	-	-	..
Memo: Non-Energy Use	*4 434*	*2 200*	*1 755*	..	*8 274*	*8 565*	*7 905*	..
Closing Stock Level	*578*	*248*	*235*	..	*1 940*	*1 729*	*1 955*	..

Source: Annual Oil Statistics.

UNITED STATES
Table 6B: Supply and Consumption

000 metric tons

	1980	2000	2001	2002	1980	2000	2001	2002
	BITUMEN				PARAFFIN WAXES			
Refinery Output	**23 775**	**31 744**	**29 239**	**..**	**803**	**907**	**868**	**..**
+ Recycled Products	-	-	-	..	-	-	-	..
+ Imports	331	1 769	1 626	..	21	109	117	..
- Exports	37	354	301	..	125	164	166	..
- Intl. Marine Bunkers	-	-	-	..	-	-	-	..
+ Transfers	-	-	-	..	-	-	-	..
+ Stock Change	- 18	-1 350	726	..	24	- 9	57	..
- Statistical Difference	-	1	-	..	-	-	1	..
Total Consumption	**24 051**	**31 808**	**31 290**	**..**	**723**	**843**	**875**	**..**
Transformation	-	-	-	..	-	-	-	..
Energy	-	8	11	..	-	-	-	..
Transport	-	-	-	..	-	-	-	..
Industry	24 051	31 800	31 279	..	723	843	875	..
Other	-	-	-	..	-	-	-	..
Memo: Non-Energy Use	*24 051*	*31 800*	*31 279*	..	*723*	*843*	*875*	..
Closing Stock Level	*3 152*	*4 132*	*3 406*	..	*69*	*135*	*78*	..
	PETROLEUM COKE				OTHER PRODUCTS			
Refinery Output	**24 749**	**48 313**	**50 831**	**52 234 e**	**24 875**	**12 262**	**11 021**	**51 896 e**
+ Recycled Products	-	-	-	-	-	-	-	-
+ Imports	-	72	366	874 e	30	7 001	6 970	9 891 e
- Exports	8 987	21 160	22 277	22 481 e	1 199	8	11	2 863 e
- Intl. Marine Bunkers	-	-	-	-	-	-	-	-
+ Transfers	-	-	-	-	821	-8 385	1 003	44 e
+ Stock Change	296	- 119	- 135	65 e	55	67	7	211 e
- Statistical Difference	-	129	- 166	- 292 e	-	-	-	26 e
Total Consumption	**16 058**	**26 977**	**28 951**	**30 984 e**	**24 582**	**10 937**	**18 990**	**59 153 e**
Transformation	162	4 730	767	..	-	-	-	..
Energy	12 595	16 267	16 046	..	814	416	340	..
Transport	-	-	-	..	-	-	-	..
Industry	3 301	5 980	12 138	..	23 768	10 521	18 650	..
Other	-	-	-	..	-	-	-	..
Memo: Non-Energy Use	*3 301*	*5 980*	*12 138*	..	*18 268*	*10 521*	*18 650*	..
Closing Stock Level	*815*	*1 681*	*1 816*	*1 751 e*	*723*	*380*	*373*	*5 836 e*

Source: Annual Oil Statistics.

UNITED STATES
Table 7A: Imports by Country of Origin

000 metric tons

	Crude Oil			Natural Gas Liquids			Feedstocks and Other		
	2000	2001	2002	2000	2001	2002	2000	2001	2002
Australia	2 207	1 553	2 302 e	-	-	-	-	-	-
Belgium	-	-	-	-	-	-	1 255	1 654	1 979 e
Canada	68 555	68 770	72 362 e	5 593	4 129	4 176 e	1 052	1 066	709 e
Czech Republic	-	-	-	-	-	-	-	-	-
Denmark	357	-	170 e	-	-	-	-	60	-
Finland	-	-	-	-	-	-	85	136	-
France	-	-	-	-	4	-	568	745	237 e
Germany	-	-	-	-	-	-	458	1 049	1 333 e
Greece	-	-	-	-	-	-	-	6	-
Hungary	-	-	-	-	-	-	-	-	-
Ireland	-	-	-	-	-	-	226	86	-
Italy	-	-	-	-	-	-	293	349	205 e
Japan	-	-	-	-	-	-	-	-	76 e
Korea	-	-	-	-	-	-	22	17	69 e
Mexico	67 161	71 109	76 040 e	68	182	-	184	41	91 e
Netherlands	33	-	-	-	-	-	252	622	397 e
New Zealand	-	-	-	-	-	-	-	-	-
Norway	15 872	15 101	18 303 e	7	301	156 e	713	710	688 e
Poland	-	-	-	-	-	-	-	2	-
Spain	-	-	-	-	7	-	218	193	78 e
Sweden	37	-	-	7	38	-	848	752	620 e
Turkey	-	-	-	-	-	-	230	210	212 e
United Kingdom	14 155	11 867	20 030 e	51	164	21 e	420	854	399 e
United States	-	-	-	-	-	-	-	-	-
Other	-	-	-	-	-	-	17	61	52 e
OECD	**168 377**	**168 400**	**189 207 e**	**5 726**	**4 825**	**4 353 e**	**6 841**	**8 613**	**7 145 e**
Argentina	3 156	3 168	3 717 e	-	-	-	37	151	103 e
Brazil	258	850	2 969 e	-	-	36 e	169	259	299 e
Colombia	16 307	13 182	12 404 e	-	-	-	129	172	298 e
Ecuador	6 465	5 859	5 382 e	-	-	-	164	43	82 e
Netherlands Antilles	-	-	-	-	-	-	2 011	1 766	1 993 e
Peru	217	366	519 e	-	-	-	117	113	92 e
Trinidad and Tobago	2 790	2 548	3 389 e	-	-	14 e	340	208	145 e
Venezuela	76 357	78 510	74 952 e	85	48	97 e	2 953	1 791	2 751 e
Other	1 055	988	1 224 e	-	-	-	17	42	213 e
Latin America	**106 605**	**105 471**	**104 556 e**	**85**	**48**	**147 e**	**5 937**	**4 545**	**5 976 e**
Romania	-	-	-	-	-	-	-	-	-
Other	-	-	-	-	-	-	23	-	20 e
Non-OECD Europe	**-**	**-**	**-**	**-**	**-**	**-**	**23**	**-**	**20 e**
FSU	**350**	**-**	**4 332 e**	**-**	**-**	**-**	**1 477**	**2 104**	**2 485 e**
Iran	-	-	-	-	-	-	-	-	-
Iraq	32 631	42 757	26 790 e	-	-	-	-	-	-
Kuwait	13 394	12 027	10 745 e	-	37	-	14	-	-
Oman	107	974	826 e	-	-	-	-	-	-
Qatar	-	10	444 e	-	-	-	356	335	261 e
Saudi Arabia	77 792	81 181	76 622 e	108	311	-	1 151	899	856 e
United Arab Emirates	124	969	738 e	-	-	-	348	247	-
Other	1 362	1 209	1 495 e	-	-	-	58	162	333 e
Middle East	**125 410**	**139 127**	**117 660 e**	**108**	**348**	**-**	**1 927**	**1 643**	**1 450 e**
India	-	-	-	-	-	-	66	104	19 e
Indonesia	1 797	1 988	2 458 e	-	-	-	209	82	217 e
Singapore	-	-	-	-	-	-	232	481	153 e
Other	3 071	3 386	2 124 e	4	3	-	508	701	632 e
Asia	**4 868**	**5 374**	**4 582 e**	**4**	**3**	**-**	**1 015**	**1 368**	**1 021 e**
China	**1 656**	**639**	**1 038 e**	**-**	**-**	**-**	**-**	**-**	**11 e**
Algeria	27	502	1 378 e	1 747	2 302	1 326 e	1 397	4 641	4 621 e
Angola/Cabinada	15 989	16 819	16 314 e	5	-	-	302	121	315 e
Cameroon	333	213	610 e	-	-	-	-	56	21 e
Egypt	235	-	-	-	-	-	237	114	50 e
Gabon	9 959	9 142	8 909 e	-	-	-	38	-	-
Libya	-	-	-	-	-	-	-	-	-
Nigeria	47 098	43 817	34 919 e	40	294	27 e	133	646	1 189 e
Tunisia	-	-	-	-	-	-	-	27	20 e
Other	3 204	4 927	3 931 e	10	11	20 e	248	191	276 e
Africa	**76 845**	**75 420**	**66 061 e**	**1 802**	**2 607**	**1 373 e**	**2 355**	**5 796**	**6 492 e**
Non-spec./Other	**-**	**-**	**-**	**-**	**-**	**-**	**-**	**-**	**-**
Total Imports	**484 111**	**494 431**	**487 436 e**	**7 725**	**7 831**	**5 873 e**	**19 575**	**24 069**	**24 600 e**
Memo: Total OPEC	*249 220*	*261 761*	*229 046 e*	*1 980*	*2 992*	*1 450 e*	*6 561*	*8 641*	*9 895 e*

Source: Annual Oil Statistics.

UNITED STATES
Table 7B: Imports by Country of Origin

000 metric tons

	2000	2001	2002	2000	2001	2002	2000	2001	2002
	Total Products			**Liquefied Petroleum Gases**			**Naphtha**		
Australia	344	414	303 e	-	-	-	-	-	-
Belgium	1 120	1 539	1 480 e	-	-	9 e	-	42	8 e
Canada	11 588	14 542	16 672 e	220	176	283 e	130	150	161 e
Czech Republic	-	-	-	-	-	-	-	-	-
Denmark	243	191	46 e	-	-	-	-	-	-
Finland	676	421	855 e	-	-	3 e	-	-	-
France	1 267	1 353	816 e	-	-	6 e	116	33	1 e
Germany	927	996	884 e	-	-	-	-	-	-
Greece	186	147	98 e	-	-	-	144	60	-
Hungary	2	1	-	-	-	-	-	-	-
Ireland	32	66	-	-	-	-	-	6	-
Italy	1 122	1 552	1 310 e	-	-	2 e	43	-	10 e
Japan	693	539	280 e	-	-	-	4	-	-
Korea	2 629	2 866	2 270 e	-	-	-	28	51	45 e
Mexico	2 524	1 756	1 822 e	-	-	9 e	1 392	1 253	873 e
Netherlands	1 115	1 493	2 044 e	-	-	9 e	57	45	86 e
New Zealand	96	58	-	-	-	-	32	58	-
Norway	1 321	1 833	1 264 e	-	-	-	190	218	18 e
Poland	-	-	-	-	-	-	-	-	-
Spain	935	1 203	700 e	3	-	-	5	32	-
Sweden	595	467	235 e	-	-	-	11	-	-
Turkey	-	31	383 e	-	-	-	-	23	31 e
United Kingdom	3 227	2 958	2 760 e	-	-	-	23	17	14 e
United States	-	-	-	-	-	-	-	-	-
Other	548	394	406 e	-	-	-	-	-	-
OECD	**31 190**	**34 820**	**34 628 e**	**223**	**176**	**321 e**	**2 175**	**1 988**	**1 247 e**
Argentina	1 317	1 941	2 141 e	-	-	-	6	189	64 e
Brazil	2 467	3 765	2 184 e	-	-	1 e	9	12	16 e
Colombia	1 167	1 725	1 019 e	-	-	-	88	19	173 e
Ecuador	107	320	261 e	-	-	-	19	35	74 e
Netherlands Antilles	2 793	2 463	2 143 e	-	-	-	648	180	148 e
Peru	193	457	298 e	-	-	-	26	70	51 e
Trinidad and Tobago	1 334	1 546	1 344 e	-	-	-	147	47	-
Venezuela	14 389	12 445	7 039 e	-	-	-	508	322	364 e
Other	177	1 005	1 256 e	-	-	-	38	41	31 e
Latin America	**23 944**	**25 667**	**17 685 e**	**-**	**-**	**1 e**	**1 489**	**915**	**921 e**
Romania	-	98	674 e	-	-	-	-	-	-
Other	60	275	58 e	-	-	-	-	-	-
Non-OECD Europe	**60**	**373**	**732 e**	**-**	**-**	**-**	**-**	**-**	**-**
FSU	**4 397**	**3 526**	**4 664 e**	**-**	**-**	**-**	**110**	**64**	**130 e**
Iran	-	-	-	-	-	-	-	-	-
Iraq	-	-	-	-	-	-	-	-	-
Kuwait	380	595	591 e	-	-	-	48	-	-
Oman	-	-	-	-	-	-	-	-	-
Qatar	19	264	-	-	-	-	-	-	-
Saudi Arabia	891	951	621 e	-	-	-	142	130	-
United Arab Emirates	211	608	215 e	-	-	-	81	84	-
Other	273	127	235 e	-	-	-	34	37	-
Middle East	**1 774**	**2 545**	**1 662 e**	**-**	**-**	**-**	**305**	**251**	**-**
India	222	536	924 e	-	-	-	142	-	34 e
Indonesia	438	540	73 e	-	-	4 e	-	-	-
Singapore	1 054	1 000	595 e	-	-	-	8	9	-
Other	481	808	480 e	-	-	-	-	-	-
Asia	**2 195**	**2 884**	**2 072 e**	**-**	**-**	**4 e**	**150**	**9**	**34 e**
China	**472**	**489**	**237 e**	**-**	**-**	**-**	**-**	**-**	**28 e**
Algeria	7 722	5 776	5 873 e	-	-	-	285	350	164 e
Angola/Cabinada	186	403	113 e	-	-	-	-	-	22 e
Cameroon	143	114	52 e	-	-	-	-	-	-
Egypt	171	219	410 e	-	-	-	124	70	53 e
Gabon	-	-	-	-	-	-	-	-	-
Libya	-	-	-	-	-	-	-	-	-
Nigeria	956	1 380	784 e	-	-	-	92	32	49 e
Tunisia	173	213	40 e	-	-	-	-	-	-
Other	1 021	1 241	629 e	-	-	-	79	145	33 e
Africa	**10 372**	**9 346**	**7 901 e**	-	-	-	**580**	**597**	**321 e**
Non-spec./Other	**-**	**-**	**-**	**-**	**-**	**-**	**-**	**-**	**-**
Total Imports	**74 404**	**79 650**	**69 581 e**	**223**	**176**	**326 e**	**4 809**	**3 824**	**2 681 e**
Memo: Total OPEC	*25 006*	*22 559*	*15 196 e*	-	-	*4 e*	*1 156*	*918*	*577 e*

Source: Annual Oil Statistics.

UNITED STATES
Table 7C: Imports by Country of Origin

000 metric tons

	2000	2001	2002	2000	2001	2002	2000	2001	2002
	Total Gasoline			Total Kerosene			Gas/Diesel		
Australia	48	62	-	18	66	-	-	25	-
Belgium	785	1 261	1 371 e	-	-	-	93	13	85 e
Canada	4 039	6 403	7 498 e	161	189	242 e	4 716	4 726	5 391 e
Czech Republic	-	-	-	-	-	-	-	-	-
Denmark	73	1	6 e	-	-	-	-	-	-
Finland	613	421	852 e	-	-	-	-	-	-
France	604	1 050	779 e	-	-	-	15	138	-
Germany	401	326	651 e	-	-	-	81	132	-
Greece	9	62	98 e	-	25	-	33	-	-
Hungary	-	-	-	-	-	-	-	-	-
Ireland	-	1	-	-	-	-	32	44	-
Italy	835	1 211	1 179 e	26	22	-	22	202	92 e
Japan	77	39	-	584	318	193 e	20	109	20 e
Korea	168	489	334 e	1 984	1 738	1 509 e	192	519	227 e
Mexico	276	330	282 e	88	120	94 e	-	14	40 e
Netherlands	643	1 027	1 662 e	-	-	-	99	121	61 e
New Zealand	-	-	-	-	-	-	-	-	-
Norway	349	582	430 e	-	-	10 e	5	-	-
Poland	-	-	-	-	-	-	-	-	-
Spain	747	1 056	575 e	-	-	-	-	34	-
Sweden	103	28	14 e	-	-	-	149	142	28 e
Turkey	-	-	352 e	-	-	-	-	-	-
United Kingdom	1 797	2 065	2 498 e	-	-	1 e	126	180	20 e
United States	-	-	-	-	-	-	-	-	-
Other	307	394	406 e	-	-	-	-	-	-
OECD	**11 874**	**16 808**	**18 987 e**	**2 861**	**2 478**	**2 049 e**	**5 583**	**6 399**	**5 964 e**
Argentina	1 219	1 274	1 562 e	-	-	8 e	7	98	94 e
Brazil	1 444	1 604	1 433 e	-	-	-	-	246	46 e
Colombia	210	460	15 e	107	231	96 e	20	86	101 e
Ecuador	58	21	37 e	-	-	-	-	-	-
Netherlands Antilles	65	268	54 e	630	702	484 e	496	663	667 e
Peru	13	87	-	-	-	-	41	44	-
Trinidad and Tobago	312	269	267 e	28	54	-	28	113	3 e
Venezuela	4 508	3 527	2 409 e	1 404	1 201	595 e	3 108	2 696	1 530 e
Other	71	258	340 e	-	-	-	-	27	193 e
Latin America	**7 900**	**7 768**	**6 117 e**	**2 169**	**2 188**	**1 183 e**	**3 700**	**3 973**	**2 634 e**
Romania	-	33	661 e	-	-	-	-	65	-
Other	60	275	58 e	-	-	-	-	-	-
Non-OECD Europe	**60**	**308**	**719 e**	**-**	**-**	**-**	**-**	**65**	**-**
FSU	**1 533**	**1 584**	**2 461 e**	**-**	**-**	**-**	**1 284**	**1 595**	**1 219 e**
Iran	-	-	-	-	-	-	-	-	-
Iraq	-	-	-	-	-	-	-	-	-
Kuwait	-	-	-	332	412	373 e	-	-	-
Oman	-	-	-	-	-	-	-	-	-
Qatar	5	19	-	-	-	-	14	-	-
Saudi Arabia	396	341	557 e	295	208	58 e	58	92	6 e
United Arab Emirates	-	93	206 e	-	251	9 e	-	80	-
Other	72	28	170 e	34	62	33 e	41	-	32 e
Middle East	**473**	**481**	**933 e**	**661**	**933**	**473 e**	**113**	**172**	**38 e**
India	80	246	565 e	-	82	-	-	208	174 e
Indonesia	1	-	-	-	-	-	26	15	-
Singapore	298	507	389 e	287	246	68 e	136	80	11 e
Other	11	32	19 e	284	353	418 e	139	415	28 e
Asia	**390**	**785**	**973 e**	**571**	**681**	**486 e**	**301**	**718**	**213 e**
China	**439**	**392**	**134 e**	**-**	**28**	**-**	**-**	**-**	**-**
Algeria	-	95	146 e	36	81	33 e	189	133	85 e
Angola/Cabinada	-	-	-	-	-	-	-	101	-
Cameroon	28	-	-	-	-	-	-	73	-
Egypt	-	149	357 e	-	-	-	-	-	-
Gabon	-	-	-	-	-	-	-	-	-
Libya	-	-	-	-	-	-	-	-	-
Nigeria	25	121	308 e	28	3	-	-	95	55 e
Tunisia	-	-	10 e	-	-	-	-	122	-
Other	93	107	24 e	-	-	13 e	-	200	-
Africa	**146**	**472**	**845 e**	**64**	**84**	**46 e**	**189**	**724**	**140 e**
Non-spec./Other	**-**	**-**	**-**	**-**	**-**	**-**	**-**	**-**	**-**
Total Imports	**22 815**	**28 598**	**31 169 e**	**6 326**	**6 392**	**4 237 e**	**11 170**	**13 646**	**10 208 e**
Memo: Total OPEC	*4 935*	*4 196*	*3 626 e*	*2 095*	*2 156*	*1 068 e*	*3 395*	*3 111*	*1 676 e*

Source: Annual Oil Statistics.

UNITED STATES
Table 7D: Imports by Country of Origin

000 metric tons

	2000	2001	2002	2000	2001	2002	2000	2001	2002
	Fuel Oil (Residual)			**Petroleum Coke**			**Other Products**		
Australia	-	-	-	-	-	-	278	261	303 e
Belgium	242	189	-	-	-	-	-	34	7 e
Canada	1 351	1 827	1 683 e	72	13	99 e	899	1 058	1 315 e
Czech Republic	-	-	-	-	-	-	-	-	-
Denmark	170	190	40 e	-	-	-	-	-	-
Finland	63	-	-	-	-	-	-	-	-
France	410	40	-	-	-	-	122	92	30 e
Germany	445	537	207 e	-	-	-	-	1	26 e
Greece	-	-	-	-	-	-	-	-	-
Hungary	-	-	-	-	-	-	2	1	-
Ireland	-	15	-	-	-	-	-	-	-
Italy	141	58	-	-	-	-	55	59	27 e
Japan	-	66	62 e	-	-	-	8	7	5 e
Korea	-	-	-	-	-	-	257	69	155 e
Mexico	491	-	494 e	-	-	-	277	39	30 e
Netherlands	269	277	174 e	-	-	-	47	23	52 e
New Zealand	59	-	-	-	-	-	5	-	-
Norway	259	477	181 e	-	-	-	518	556	625 e
Poland	-	-	-	-	-	-	-	-	-
Spain	27	40	-	-	-	-	153	41	125 e
Sweden	332	297	193 e	-	-	-	-	-	-
Turkey	-	-	-	-	-	-	-	8	-
United Kingdom	1 274	690	199 e	-	-	-	7	6	28 e
United States	-	-	-	-	-	-	-	-	-
Other	241	-	-	-	-	-	-	-	-
OECD	**5 774**	**4 703**	**3 233 e**	**72**	**13**	**99 e**	**2 628**	**2 255**	**2 728 e**
Argentina	85	239	167 e	-	140	246 e	-	1	-
Brazil	918	1 860	586 e	-	5	50 e	96	38	52 e
Colombia	703	918	621 e	-	-	-	39	11	13 e
Ecuador	30	245	128 e	-	-	-	-	19	22 e
Netherlands Antilles	701	613	483 e	-	-	202 e	253	37	105 e
Peru	113	256	247 e	-	-	-	-	-	-
Trinidad and Tobago	676	1 063	1 074 e	-	-	-	143	-	-
Venezuela	3 470	3 537	1 195 e	-	-	-	1 391	1 162	946 e
Other	68	665	685 e	-	-	-	-	14	7 e
Latin America	**6 764**	**9 396**	**5 186 e**	**-**	**145**	**498 e**	**1 922**	**1 282**	**1 145 e**
Romania	-	-	-	-	-	13 e	-	-	-
Other	-	-	-	-	-	-	-	-	-
Non-OECD Europe	**-**	**-**	**-**	**-**	**-**	**13 e**	**-**	**-**	**-**
FSU	**1 285**	**249**	**713 e**	**-**	**-**	**-**	**185**	**34**	**141 e**
Iran	-	-	-	-	-	-	-	-	-
Iraq	-	-	-	-	-	-	-	-	-
Kuwait	-	-	-	-	183	218 e	-	-	-
Oman	-	-	-	-	-	-	-	-	-
Qatar	-	-	-	-	-	-	-	245	-
Saudi Arabia	-	150	-	-	-	-	-	30	-
United Arab Emirates							130	100	
Other	50	-	-	-	-	-	42	-	-
Middle East	**50**	**150**	**-**	**-**	**183**	**218 e**	**172**	**375**	**-**
India	-	-	-	-	-	-	-	-	151 e
Indonesia	411	482	69 e	-	-	-	-	43	-
Singapore	249	158	117 e	-	-	-	76	-	10 e
Other	-	-	-	-	-	-	47	8	15 e
Asia	**660**	**640**	**186 e**	**-**	**-**	**-**	**123**	**51**	**176 e**
China	**-**	**-**	**-**	**-**	**25**	**46 e**	**33**	**44**	**29 e**
Algeria	2 887	61	137 e	-	-	-	4 325	5 056	5 308 e
Angola/Cabinada	150	302	62 e	-	-	-	36	-	29 e
Cameroon	115	41	52 e	-	-	-	-	-	-
Egypt	47	-	-	-	-	-	-	-	-
Gabon	-	-	-	-	-	-	-	-	-
Libya	-	-	-	-	-	-	-	-	-
Nigeria	811	1 032	360 e	-	-	-	-	97	12 e
Tunisia	173	91	30 e	-	-	-	-	-	-
Other	713	475	236 e	-	-	-	136	314	323 e
Africa	**4 896**	**2 002**	**877 e**	**-**	**-**	**-**	**4 497**	**5 467**	**5 672 e**
Non-spec./Other	**-**	**-**	**-**	**-**	**-**	**-**	**-**	**-**	**-**
Total Imports	**19 429**	**17 140**	**10 195 e**	**72**	**366**	**874 e**	**9 560**	**9 508**	**9 891 e**
Memo: Total OPEC	*7 579*	*5 262*	*1 761 e*	*-*	*183*	*218 e*	*5 846*	*6 733*	*6 266 e*

Source: Annual Oil Statistics.

UNITED STATES
Table 8A: Exports by Country of Destination

000 metric tons

	2000	2001	2002	2000	2001	2002	2000	2001	2002
	Crude Oil			Natural Gas Liquids			Feedstocks and Other		
Australia	-	-	-	1	10	11 e	-	-	-
Austria	-	-	-	-	-	-	-	-	-
Belgium	-	-	-	2	-	1 e	17	21	18 e
Canada	1 024	1 027	448 e	587	368	270 e	132	203	188 e
Denmark	-	-	-	-	-	-	-	-	-
Finland	-	-	-	4	-	-	-	-	-
France	-	-	-	10	4	12 e	2	2	3 e
Germany	-	-	-	28	-	-	-	2	-
Greece	-	-	-	-	-	-	-	-	-
Ireland	-	-	-	-	-	-	4	-	-
Italy	-	-	-	-	12	14 e	3	-	-
Japan	954	-	-	-	9	151 e	63	49	69 e
Luxembourg	-	-	-	-	-	-	-	-	-
Mexico	4	-	8 e	1 379	779	1 021 e	479	363	375 e
Netherlands	-	-	-	10	-	-	66	24	8 e
Norway	-	-	-	-	-	-	-	-	-
Portugal	-	-	-	-	-	-	-	-	10 e
Spain	-	-	-	-	-	7 e	-	-	-
Sweden	-	-	-	-	-	-	-	-	-
Switzerland	-	-	-	-	-	8 e	-	-	-
Turkey	-	-	-	-	-	10 e	-	-	-
United Kingdom	-	-	-	2	2	3 e	2	3	2 e
United States	-	-	-	-	-	-	-	-	-
Other OECD	567	-	-	27	-	49 e	23	17	16 e
OECD	**2 549**	**1 027**	**456 e**	**2 050**	**1 184**	**1 557 e**	**791**	**684**	**689 e**
Africa	-	-	-	-	5	12 e	1	1	-
Latin America	-	-	-	308	195	298 e	491	445	338 e
Asia	2	-	-	6	16	25 e	34	34	71 e
China	-	-	-	57	-	1 e	3	-	2 e
Non-OECD Europe	-	-	-	-	-	-	7	-	-
Former Soviet Union	-	-	-	-	-	-	-	-	-
Middle East	-	-	-	-	-	-	1	4	-
Non-specified/Other	-	-	-	-	-	8 e	-	-	9 e
Non-OECD	**2**	**-**	**-**	**371**	**216**	**344 e**	**537**	**484**	**420 e**
Total Exports	**2 551**	**1 027**	**456 e**	**2 421**	**1 400**	**1 901 e**	**1 328**	**1 168**	**1 109 e**

	2000	2001	2002	2000	2001	2002	2000	2001	2002
	Total Products			Liquefied Petroleum Gas			Naphtha		
Australia	582	658	792 e	-	-	-	-	-	-
Austria	1	-	-	-	-	-	-	-	-
Belgium	873	994	1 269 e	-	-	-	-	-	-
Canada	3 647	3 998	4 435 e	-	-	-	-	-	-
Denmark	149	205	243 e	-	-	-	-	-	-
Finland	4	190	193 e	-	-	-	-	-	-
France	599	796	724 e	-	-	-	-	-	-
Germany	186	149	224 e	-	-	-	-	-	-
Greece	259	395	369 e	-	-	-	-	-	-
Ireland	194	126	338 e	-	-	-	-	-	-
Italy	2 399	2 159	1 957 e	-	-	-	-	-	-
Japan	4 192	3 716	3 897 e	-	-	-	-	-	-
Luxembourg	-	-	-	-	-	-	-	-	-
Mexico	14 809	12 108	11 259 e	-	-	-	-	-	-
Netherlands	2 369	2 567	1 382 e	-	-	-	-	-	-
Norway	133	171	223 e	-	-	-	-	-	-
Portugal	347	97	33 e	-	-	-	-	-	-
Spain	2 597	3 273	3 421 e	-	-	-	-	-	-
Sweden	62	197	64 e	-	-	-	-	-	-
Switzerland	64	8	9 e	-	-	-	-	-	-
Turkey	1 150	1 079	1 128 e	-	-	-	-	-	-
United Kingdom	594	803	781 e	-	-	-	-	-	-
United States	-	-	-	-	-	-	-	-	-
Other OECD	535	870	763 e	-	-	-	-	-	-
OECD	**35 745**	**34 559**	**33 504 e**	**-**	**-**	**-**	**-**	**-**	**-**
Africa	1 037	1 282	1 165 e	-	-	-	-	-	-
Latin America	7 858	7 812	7 924 e	-	-	-	-	-	-
Asia	2 754	4 071	4 654 e	-	-	-	-	-	-
China	78	357	853 e	-	-	-	-	-	-
Non-OECD Europe	380	383	510 e	-	-	-	-	-	-
Former Soviet Union	14	51	66 e	-	-	-	-	-	-
Middle East	1 091	1 160	983 e	-	-	-	-	-	-
Non-specified/Other	-	-	1 134 e	-	-	-	-	-	-
Non-OECD	**13 212**	**15 116**	**17 289 e**	**-**	**-**	**-**	**-**	**-**	**-**
Total Exports	**48 957**	**49 675**	**50 793 e**	**-**	**-**	**-**	**-**	**-**	**-**

Source: Annual Oil Statistics.

UNITED STATES
Table 8B: Exports by Country of Destination

000 metric tons

	Total Gasoline 2000	Total Gasoline 2001	Total Gasoline 2002	Total Kerosene 2000	Total Kerosene 2001	Total Kerosene 2002	Gas/Diesel Oil 2000	Gas/Diesel Oil 2001	Gas/Diesel Oil 2002
Australia	1	33	-	-	-	1 e	1	-	-
Austria	-	-	-	-	-	-	-	-	-
Belgium	-	-	1 e	-	-	-	1	39	-
Canada	389	517	429 e	746	524	767 e	466	577	260 e
Denmark	-	-	-	-	-	-	-	-	-
Finland	-	-	-	-	44	21 e	2	145	135 e
France	9	32	1 e	3	-	-	109	69	109 e
Germany	-	-	2 e	-	-	-	6	-	2 e
Greece	-	-	-	-	-	-	-	-	-
Ireland	-	-	-	-	-	-	-	-	-
Italy	-	-	-	-	-	-	91	-	-
Japan	32	-	105 e	5	-	-	57	15	-
Luxembourg	-	-	-	-	-	-	-	-	-
Mexico	5 705	5 028	5 541 e	195	108	238 e	3 663	1 814	913 e
Netherlands	114	30	20 e	-	205	2 e	601	309	390 e
Norway	-	-	-	-	-	-	-	-	-
Portugal	-	-	-	-	-	-	-	-	-
Spain	-	-	-	-	-	-	48	59	202 e
Sweden	-	-	-	-	-	-	4	130	-
Switzerland	-	-	-	-	-	-	8	6	-
Turkey	-	-	-	-	-	-	-	-	-
United Kingdom	1	1	18 e	27	68	73 e	44	89	33 e
United States	-	-	-	-	-	-	-	-	-
Other OECD	31	34	9 e	-	-	-	83	25	75 e
OECD	**6 282**	**5 675**	**6 126 e**	**976**	**949**	**1 102 e**	**5 184**	**3 277**	**2 119 e**
Africa	-	-	20 e	-	28	2 e	1	24	23 e
Latin America	552	483	767 e	234	151	83 e	2 310	1 336	1 406 e
Asia	-	52	1 e	1	9	-	1 076	1 413	1 336 e
China	-	83	-	32	-	-	4	31	218 e
Non-OECD Europe	-	-	-	-	-	-	1	-	50 e
Former Soviet Union	-	-	-	-	-	-	2	1	-
Middle East	31	31	117 e	319	302	64 e	10	-	12 e
Non-specified/Other	-	-	177 e	-	-	-	-	-	1 e
Non-OECD	**583**	**649**	**1 082 e**	**586**	**490**	**149 e**	**3 404**	**2 805**	**3 046 e**
Total Exports	**6 865**	**6 324**	**7 208 e**	**1 562**	**1 439**	**1 251 e**	**8 588**	**6 082**	**5 165 e**

	Fuel Oil (Residual) 2000	Fuel Oil (Residual) 2001	Fuel Oil (Residual) 2002	Petroleum Coke 2000	Petroleum Coke 2001	Petroleum Coke 2002	Other Products 2000	Other Products 2001	Other Products 2002
Australia	-	1	-	563	610	781 e	17	14	10 e
Austria	-	-	-	-	-	-	1	-	-
Belgium	1	3	-	847	924	1 247 e	24	28	21 e
Canada	598	775	1 337 e	811	1 085	982 e	637	520	660 e
Denmark	-	-	-	149	205	243 e	-	-	-
Finland	-	-	27 e	-	-	10 e	2	1	-
France	-	1	-	472	683	597 e	6	11	17 e
Germany	-	-	-	168	134	207 e	12	15	13 e
Greece	-	-	-	257	394	368 e	2	1	1 e
Ireland	-	-	50 e	193	126	288 e	1	-	-
Italy	193	85	132 e	2 099	2 058	1 801 e	16	16	24 e
Japan	110	82	102 e	3 301	3 117	3 241 e	687	502	449 e
Luxembourg	-	-	-	-	-	-	-	-	-
Mexico	3 041	2 625	2 250 e	1 808	2 072	1 630 e	397	461	687 e
Netherlands	271	703	168 e	1 376	1 266	795 e	7	54	7 e
Norway	-	-	-	133	170	223 e	-	1	-
Portugal	-	-	-	347	97	33 e	-	-	-
Spain	84	389	402 e	2 464	2 822	2 811 e	1	3	6 e
Sweden	1	-	-	55	65	64 e	2	2	-
Switzerland	-	1	-	54	-	9 e	2	1	-
Turkey	-	-	-	1 146	1 074	1 124 e	4	5	4 e
United Kingdom	-	47	1 e	496	584	643 e	26	14	13 e
United States	-	-	-	-	-	-	-	-	-
Other OECD	22	97	94 e	270	393	537 e	129	321	48 e
OECD	**4 321**	**4 809**	**4 563 e**	**17 009**	**17 879**	**17 634 e**	**1 973**	**1 970**	**1 960 e**
Africa	33	1	62 e	961	1 178	971 e	42	51	87 e
Latin America	2 409	3 593	3 697 e	1 868	1 711	1 387 e	485	538	584 e
Asia	1 090	2 166	3 106 e	439	244	68 e	148	187	143 e
China	1	132	47 e	3	85	571 e	38	26	17 e
Non-OECD Europe	126	49	66 e	252	333	394 e	1	1	-
Former Soviet Union	-	5	-	8	41	65 e	4	4	1 e
Middle East	88	2	274 e	620	806	472 e	23	19	44 e
Non-specified/Other	-	-	10 e	-	-	919 e	-	-	27 e
Non-OECD	**3 747**	**5 948**	**7 262 e**	**4 151**	**4 398**	**4 847 e**	**741**	**826**	**903 e**
Total Exports	**8 068**	**10 757**	**11 825 e**	**21 160**	**22 277**	**22 481 e**	**2 714**	**2 796**	**2 863 e**

Source: Annual Oil Statistics.

PART IV

OECD HISTORICAL SERIES

QUATRIEME PARTIE

SERIES HISTORIQUES DES PAYS DE L'OCDE

Table 1: Production of Crude Oil
Tableau 1 : Production de pétrole brut

000 metric tons

	1973	1974	1975	1976	1977	1978	1979	1980	1981	1982
Australia	18 140	17 926	19 125	19 554	20 602	20 053	20 183	18 868	18 418	18 126
Austria	2 578	2 238	2 037	1 931	1 762	1 791	1 726	1 448	1 305	1 261
Belgium	-	-	-	-	-	-	-	-	-	-
Canada	88 423	77 484	77 616	71 160	71 278	70 194	78 268	64 128	57 273	56 104
Czech Republic	..	..	85	78	35	66	61	51	50	54
Denmark	68	89	148	194	503	432	432	298	758	1 686
Finland	-	-	-	-	-	-	-	-	-	-
France	1 254	1 080	1 028	1 057	1 037	1 117	1 197	1 415	1 676	1 646
Germany	..	..	..	..	..	5 130	4 799	4 667	4 495	4 292
Greece	-	-	-	-	-	-	-	-	196	986
Hungary	1 989	1 997	2 006	2 118	2 191	2 198	2 027	2 031	2 024	2 027
Iceland	-	-	-	-	-	-	-	-	-	-
Ireland	-	-	-	-	-	-	-	-	-	-
Italy	1 048	1 078	1 071	1 102	1 083	1 422	1 637	1 800	1 436	1 727
Japan	699	671	603	576	589	539	480	430	389	208
Korea	-	-	-	-	-	-	-	-	-	-
Luxembourg	-	-	-	-	-	-	-	-	-	-
Mexico	24 804	31 373	38 586	43 530	53 057	65 457	78 908	107 017	127 620	150 894
Netherlands	1 492	1 461	1 419	1 371	1 382	1 402	1 316	1 280	1 348	1 637
New Zealand	160	170	180	477	662	562	370	320	409	650
Norway	1 595	1 736	9 256	13 828	13 686	17 000	18 223	23 055	22 284	23 267
Poland	392	549	553	455	364	363	331	329	315	241
Portugal	-	-	-	-	-	-	-	-	-	-
Slovak Republic	128	104	94	80	69	61	52	42	39	35
Spain	654	1 892	2 028	1 508	844	980	1 121	1 602	1 226	1 531
Sweden	-	-	-	-	-	1	1	25	6	14
Switzerland	-	-	-	-	-	-	-	-	-	-
Turkey	..	..	3 096	2 595	2 604	2 736	2 823	2 330	2 363	2 333
United Kingdom	231	239	1 568	12 062	37 804	53 475	77 116	78 917	87 946	100 311
United States	452 729	432 109	411 438	401 211	405 671	428 447	420 775	424 153	421 761	425 548
OECD TOTAL	..	..	..	..	..	**673 426**	**711 846**	**734 206**	**753 337**	**794 578**
OECD North America	**565 956**	**540 966**	**527 640**	**515 901**	**530 006**	**564 098**	**577 951**	**595 298**	**606 654**	**632 546**
OECD Pacific	**18 999**	**18 767**	**19 908**	**20 607**	**21 853**	**21 154**	**21 033**	**19 618**	**19 216**	**18 984**
OECD Europe	..	..	..	..	..	**88 174**	**112 862**	**119 290**	**127 467**	**143 048**
IEA Total	..	..	..	..	..	**607 545**	**632 555**	**626 818**	**625 363**	**643 408**
EU15	..	..	..	..	..	**65 750**	**89 345**	**91 452**	**100 392**	**115 091**

Table 1: Production of Crude Oil (continued)
Tableau 1 : Production de pétrole brut (suite)

000 metric tons

1983	1984	1985	1986	1987	1988	1989	1990	1991	1992	
17 655	21 462	24 679	25 282	25 128	25 056	22 540	25 465	25 432	22 311	Australie
1 237	1 175	1 121	1 094	1 047	1 175	1 156	1 149	1 280	1 180	Autriche
-	-	-	-	-	-	-	-	-	-	Belgique
58 265	65 358	64 988	65 120	68 778	71 931	68 571	68 490	67 248	70 012	Canada
56	56	57	56	51	45	47	50	69	82	République Tchèque
2 152	2 314	2 892	3 621	4 602	4 734	5 531	6 037	7 043	7 811	Danemark
-	-	-	-	-	-	-	-	-	-	Finlande
1 661	2 065	2 642	2 942	3 236	3 355	3 244	3 024	2 952	2 866	France
4 145	4 071	4 136	4 068	3 767	3 986	3 837	3 648	3 405	3 279	Allemagne
1 176	1 251	1 250	1 253	1 149	1 050	851	773	789	653	Grèce
2 004	2 007	1 962	1 955	1 861	1 889	1 906	1 915	1 841	1 769	Hongrie
-	-	-	-	-	-	-	-	-	-	Islande
-	-	-	-	-	-	-	-	-	-	Irlande
2 208	2 240	2 352	2 529	3 908	4 811	4 578	4 641	4 307	4 479	Italie
207	209	342	457	435	413	363	343	516	625	Japon
-	-	-	-	-	-	-	-	-	-	Corée
-	-	-	-	-	-	-	-	-	-	Luxembourg
146 070	147 734	144 133	133 643	140 596	139 128	139 187	138 656	145 650	145 395	Mexique
2 589	3 102	3 728	4 628	4 291	3 909	3 391	3 533	3 258	2 845	Pays-Bas
638	815	1 197	1 319	1 259	1 521	1 713	1 767	1 849	1 739	Nouvelle Zélande
29 267	33 780	37 014	40 505	47 585	54 002	72 498	79 663	91 383	104 488	Norvège
210	189	195	167	149	159	157	160	158	200	Pologne
-	-	-	-	-	-	-	-	-	-	Portugal
37	35	66	87	96	98	97	73	71	73	République Slovaque
2 977	2 316	2 183	1 860	1 640	1 480	1 038	795	1 067	1 073	Espagne
23	13	8	4	4	3	3	3	2	1	Suède
-	-	-	-	-	-	-	-	-	-	Suisse
2 203	2 087	2 110	2 394	2 627	2 569	2 876	3 712	4 364	4 276	Turquie
110 839	121 180	122 492	121 238	117 613	109 455	87 407	87 990	86 831	89 179	Royaume-Uni
427 474	439 148	442 507	428 142	411 808	402 585	375 510	362 796	365 816	354 680	Etats-Unis
813 093	**852 607**	**862 054**	**842 364**	**841 630**	**833 354**	**796 501**	**794 683**	**815 331**	**819 016**	**OCDE TOTAL**
631 809	**652 240**	**651 628**	**626 905**	**621 182**	**613 644**	**583 268**	**569 942**	**578 714**	**570 087**	**OCDE Amérique du N.**
18 500	**22 486**	**26 218**	**27 058**	**26 822**	**26 990**	**24 616**	**27 575**	**27 797**	**24 675**	**OCDE Pacifique**
162 784	**177 881**	**184 208**	**188 401**	**193 626**	**192 720**	**188 617**	**197 166**	**208 820**	**224 254**	**OCDE Europe**
666 776	**704 649**	**717 660**	**708 467**	**700 789**	**693 969**	**657 060**	**655 794**	**669 452**	**673 348**	**AIE Total**
129 007	**139 727**	**142 804**	**143 237**	**141 257**	**133 958**	**111 036**	**111 593**	**110 934**	**113 366**	**UE15**

Table 1: Production of Crude Oil (continued)
Tableau 1 : Production de pétrole brut (suite)

000 metric tons

	1993	1994	1995	1996	1997	1998	1999	2000	2001	2002
Australia	21 836	20 262	21 625	19 551	19 508	20 949	15 755	23 874	25 331	23 561 e
Austria	1 154	1 099	1 035	992	972	959	1 002	971	957	947 e
Belgium	-	-	-	-	-	-	-	-	-	-
Canada	73 259	75 657	78 222	80 031	84 589	89 160	82 624	87 937	88 279	90 993 e
Czech Republic	111	128	146	152	163	179	183	175	183	259 e
Denmark	8 265	9 118	9 170	10 122	11 145	11 432	14 465	17 780	16 887	18 183 e
Finland	-	-	-	-	-	-	-	-	-	-
France	2 752	2 769	2 503	2 107	1 780	1 709	1 539	1 417	1 382	1 313 e
Germany	3 064	2 938	2 926	2 874	2 804	2 934	2 746	3 166	3 278	3 510 e
Greece	537	500	435	483	436	293	15	256	171	165 e
Hungary	1 654	1 575	1 668	1 477	1 360	1 260	1 243	1 136	1 065	991 e
Iceland	-	-	-	-	-	-	-	-	-	-
Ireland	-	-	-	-	-	-	-	-	-	-
Italy	4 620	4 877	5 208	5 430	5 926	5 600	4 998	4 555	4 066	5 273 e
Japan	570	543	531	518	498	450	378	330	374	326 e
Korea	-	-	-	-	-	-	-	-	-	-
Luxembourg	-	-	-	-	-	-	-	-	-	-
Mexico	145 815	145 188	141 319	148 454	157 846	160 266	154 081	156 403	162 559	165 963 e
Netherlands	2 672	3 437	2 721	2 221	2 069	1 714	1 594	1 443	1 370	2 232 e
New Zealand	1 842	1 772	1 475	1 950	2 587	2 054	1 811	1 608	1 497	1 395 e
Norway	111 854	125 043	134 617	152 752	152 491	146 166	145 843	157 522	157 551	151 181 e
Poland	235	284	292	317	289	360	434	653	767	672 e
Portugal	-	-	-	-	-	-	-	-	-	-
Slovak Republic	67	67	74	71	64	60	66	59	46	45 e
Spain	874	807	652	512	371	529	299	227	338	316 e
Sweden	-	5	4	-	-	-	-	-	-	-
Switzerland	-	-	-	-	-	-	-	-	-	-
Turkey	3 892	3 696	3 516	3 500	3 449	3 223	2 941	2 762	2 520	2 420 e
United Kingdom	93 950	119 032	121 794	121 930	120 321	124 222	128 262	117 882	108 387	106 645 e
United States	337 707	328 578	323 550	319 732	318 220	308 367	290 099	287 933	286 150	284 407 e
OECD TOTAL	**816 730**	**847 375**	**853 483**	**875 176**	**886 888**	**881 886**	**850 378**	**868 089**	**863 158**	**860 797 e**
OECD North America	**556 781**	**549 423**	**543 091**	**548 217**	**560 655**	**557 793**	**526 804**	**532 273**	**536 988**	**541 363 e**
OECD Pacific	**24 248**	**22 577**	**23 631**	**22 019**	**22 593**	**23 453**	**17 944**	**25 812**	**27 202**	**25 282 e**
OECD Europe	**235 701**	**275 375**	**286 761**	**304 940**	**303 640**	**300 640**	**305 630**	**310 004**	**298 968**	**294 152 e**
IEA Total	**670 613**	**701 836**	**711 798**	**726 334**	**728 689**	**721 200**	**695 797**	**710 974**	**699 786**	**694 117 e**
EU15	**117 888**	**144 582**	**146 448**	**146 671**	**145 824**	**149 392**	**154 920**	**147 697**	**136 836**	**138 584 e**

Table 2: Production of Crude Oil, NGL and Non Crude
Tableau 2 : Production de pétrole brut, de LGN, additives et autres hydrocarbures

000 metric tons

1973	1974	1975	1976	1977	1978	1979	1980	1981	1982	
19 182	18 977	20 332	20 864	22 058	21 891	21 941	20 564	20 064	19 766	Australie
2 596	2 261	2 037	1 931	1 787	1 815	1 751	1 500	1 352	1 306	Autriche
-	-	-	-	-	-	-	-	-	-	Belgique
94 128	88 457	83 029	75 939	76 095	75 080	84 386	81 389	73 723	73 446	Canada
43	45	338	309	250	282	251	236	282	272	République Tchèque
68	89	148	194	503	432	432	298	758	1 686	Danemark
-	-	-	-	-	-	-	-	-	-	Finlande
2 066	1 899	1 845	1 809	1 875	1 943	2 005	2 254	2 436	2 338	France
6 714	6 257	5 809	5 577	5 466	5 130	4 799	6 215	6 119	5 907	Allemagne
-	-	-	-	-	-	-	-	196	1 032	Grèce
2 033	2 076	2 116	2 258	2 357	2 518	2 432	2 517	2 510	2 555	Hongrie
-	-	-	-	-	-	-	-	-	-	Islande
-	-	-	-	-	-	-	-	-	-	Irlande
1 105	1 138	1 071	1 144	1 123	1 489	1 717	1 825	1 487	1 853	Italie
792	758	686	675	689	641	594	542	506	461	Japon
-	-	-	-	-	-	-	-	-	-	Corée
-	-	-	-	-	-	-	-	-	-	Luxembourg
27 182	33 901	41 475	46 545	56 487	69 227	83 723	113 350	135 132	156 889	Mexique
1 556	1 575	1 573	1 546	1 598	1 520	1 581	1 568	1 606	1 894	Pays-Bas
161	172	183	486	672	575	386	338	435	681	Nouvelle Zélande
1 595	1 736	9 256	13 828	13 686	17 000	18 776	24 277	23 376	24 411	Norvège
392	549	553	455	364	363	331	335	319	245	Pologne
-	-	-	-	-	-	-	-	-	-	Portugal
130	107	97	83	72	64	55	44	41	37	République Slovaque
654	1 892	2 261	1 820	1 084	1 211	1 368	1 755	1 445	1 765	Espagne
-	-	-	3	1	1	1	25	6	14	Suède
-	-	-	-	-	-	-	-	-	-	Suisse
3 511	3 309	3 096	2 595	2 604	2 736	2 823	2 330	2 363	2 333	Turquie
547	596	1 772	12 480	38 530	54 115	77 955	80 482	89 489	103 232	Royaume-Uni
513 296	490 286	466 499	456 394	460 937	483 076	475 465	478 626	477 059	478 452	Etats-Unis
677 751	**656 080**	**644 176**	**646 935**	**688 238**	**741 109**	**782 772**	**820 470**	**840 704**	**880 575**	**OCDE TOTAL**
634 600	**612 644**	**591 003**	**578 878**	**593 519**	**627 383**	**643 574**	**673 365**	**685 914**	**708 787**	**OCDE Amérique du N.**
20 135	**19 907**	**21 201**	**22 025**	**20 119**	**23 107**	**22 921**	**21 444**	**21 005**	**20 908**	**OCDE Pacifique**
23 010	**23 529**	**31 972**	**46 032**	**71 300**	**90 619**	**116 277**	**125 601**	**133 785**	**150 880**	**OCDE Europe**
650 047	**621 523**	**602 051**	**599 852**	**631 315**	**671 455**	**698 663**	**706 741**	**705 212**	**723 404**	**AIE Total**
15 306	**15 707**	**16 516**	**26 504**	**51 967**	**67 656**	**91 609**	**95 922**	**104 894**	**121 027**	**UE15**

Table 2: Production of Crude Oil, NGL and Non Crude (continued)
Tableau 2 : Production de pétrole brut, de LGN, additives et autres hydrocarbures (suite)

000 metric tons

	1983	1984	1985	1986	1987	1988	1989	1990	1991	1992
Australia	19 226	23 138	26 764	27 436	27 236	27 171	24 581	27 493	27 513	26 927
Austria	1 281	1 257	1 169	1 134	1 084	1 217	1 197	1 190	1 321	1 220
Belgium	-	-	-	-	-	-	-	-	-	-
Canada	77 272	83 315	84 049	85 448	90 266	93 932	91 821	91 951	92 351	96 719
Czech Republic	290	287	286	289	275	250	243	220	221	237
Denmark	2 152	2 314	2 892	3 621	4 602	4 734	5 531	6 037	7 043	7 811
Finland	-	-	-	-	-	-	-	-	-	-
France	2 317	2 699	3 243	3 446	3 645	3 831	3 740	3 470	3 798	3 568
Germany	6 272	6 455	6 755	6 663	6 390	6 691	6 421	5 520	4 620	4 396
Greece	1 241	1 320	1 322	1 333	1 221	1 118	916	830	836	687
Hungary	2 465	2 512	2 515	2 477	2 416	2 381	2 384	2 299	2 211	2 151
Iceland	-	-	-	-	-	-	-	-	-	-
Ireland	-	-	-	-	-	-	-	-	-	-
Italy	2 341	2 377	2 488	2 637	4 015	4 917	4 682	4 723	4 489	4 705
Japan	481	473	601	700	685	677	636	635	840	951
Korea	-	-	-	-	-	-	-	-	-	-
Luxembourg	-	-	-	-	-	-	-	-	-	-
Mexico	154 527	156 131	153 711	144 492	151 604	150 810	151 631	151 054	157 967	157 765
Netherlands	2 900	3 433	4 069	4 995	4 663	4 271	3 814	3 976	3 719	3 324
New Zealand	672	864	1 333	2 008	1 857	2 252	2 470	2 533	2 500	2 555
Norway	30 611	35 050	38 431	42 375	49 541	56 351	74 872	82 088	93 716	106 862
Poland	210	189	195	170	166	179	185	178	173	218
Portugal	-	-	-	-	-	-	-	-	-	-
Slovak Republic	39	37	68	89	99	100	99	75	120	92
Spain	3 190	2 515	2 388	2 094	1 852	1 718	1 333	1 144	1 422	1 402
Sweden	23	13	8	4	4	3	3	3	2	1
Switzerland	-	-	-	-	-	-	-	-	-	-
Turkey	2 203	2 087	2 110	2 394	2 627	2 569	2 876	3 712	4 364	4 276
United Kingdom	114 937	125 935	127 658	127 071	123 317	114 473	91 824	91 602	91 261	94 395
United States	480 024	491 474	494 350	478 081	463 262	454 909	425 265	413 344	412 898	407 476
OECD TOTAL	**904 674**	**943 875**	**956 405**	**938 957**	**940 827**	**934 554**	**896 524**	**894 077**	**913 385**	**927 738**
OECD North America	**711 823**	**730 920**	**732 110**	**708 021**	**705 132**	**699 651**	**668 717**	**656 349**	**663 216**	**661 960**
OECD Pacific	**20 379**	**24 475**	**28 698**	**30 144**	**29 778**	**30 100**	**27 687**	**30 661**	**30 853**	**30 433**
OECD Europe	**172 472**	**188 480**	**195 597**	**200 792**	**205 917**	**204 803**	**200 120**	**207 067**	**219 316**	**235 345**
IEA Total	**749 898**	**787 518**	**802 431**	**794 206**	**788 958**	**783 465**	**744 609**	**742 770**	**755 125**	**769 663**
EU15	**136 654**	**148 318**	**151 992**	**152 998**	**150 793**	**142 973**	**119 461**	**118 495**	**118 511**	**121 509**

Table 2: Production of Crude Oil, NGL and Non Crude (continued)
Tableau 2 : Production de pétrole brut, de LGN, additives et autres hydrocarbures (suite)

000 metric tons

1993	1994	1995	1996	1997	1998	1999	2000	2001	2002	
26 654	25 067	26 870	26 366	26 686	29 432	23 657	32 076	33 138	32 453 e	Australie
1 194	1 146	1 078	1 045	1 027	1 047	1 063	1 072	1 012	1 003 e	Autriche
-	-	-	-	-	-	-	-	-	-	Belgique
101 658	105 751	110 699	113 931	119 420	125 130	120 298	125 249	127 001	133 121 e	Canada
257	270	276	251	463	433	386	385	366	415 e	République Tchèque
8 265	9 118	9 170	10 122	11 145	11 432	14 465	17 780	16 887	18 183 e	Danemark
-	-	15	13	94	114	114	120	114 e	121 e	Finlande
3 494	3 490	3 218	2 933	2 629	2 410	2 339	2 306	2 294	2 195 e	France
4 174	3 954	3 918	3 818	3 764	3 870	3 850	4 393	4 347	4 641 e	Allemagne
562	531	457	514	465	315	16	279	191	189 e	Grèce
2 041	1 955	2 302	2 119	1 972	1 834	1 788	1 663	1 549	1 499 e	Hongrie
-	-	-	-	-	-	-	-	-	-	Islande
-	-	-	-	-	-	-	-	-	-	Irlande
4 819	5 033	5 454	5 655	6 094	5 819	5 216	4 764	4 223	5 643 e	Italie
875	843	835	820	823	778	722	729	674	560 e	Japon
-	-	-	-	470	391	446	670	572	523 e	Corée
-	-	-	-	-	-	-	-	-	-	Luxembourg
158 176	158 104	153 555	160 443	169 016	171 909	166 882	169 261	175 498	178 261 e	Mexique
3 248	4 323	3 479	3 103	2 930	2 661	2 535	2 350	2 263	3 073 e	Pays-Bas
2 626	2 421	1 958	2 354	2 861	2 289	2 047	1 849	1 749	1 633 e	Nouvelle Zélande
114 471	128 514	138 500	156 804	156 502	149 826	149 350	160 990	162 536	156 381 e	Norvège
256	326	365	394	435	765	894	1 233	1 297	1 211 e	Pologne
-	-	-	-	-	-	-	-	-	-	Portugal
69	70	76	72	65	61	66	59	353	346 e	République Slovaque
1 102	948	783	512	371	529	299	227	338	316 e	Espagne
-	5	4	-	-	-	-	-	5	-	Suède
-	-	-	-	-	-	-	-	-	-	Suisse
3 892	3 696	3 516	3 500	3 449	3 223	2 941	2 762	2 520	2 420 e	Turquie
100 233	127 077	130 456	130 141	128 368	132 498	137 228	126 375	116 805	115 164 e	Royaume-Uni
399 488	390 951	387 408	385 463	384 618	373 946	358 436	358 802	356 516	356 379 e	Etats-Unis
937 554	**973 593**	**984 392**	**1010 373**	**1023 667**	**1020 712**	**995 038**	**1015 394**	**1012 248**	**1015 730 e**	**OCDE TOTAL**
659 322	**654 806**	**651 662**	**659 837**	**673 054**	**670 985**	**645 616**	**653 312**	**659 015**	**667 761 e**	**OCDE Amérique du N.**
30 155	**28 331**	**29 663**	**29 540**	**30 840**	**32 890**	**26 872**	**35 324**	**36 133**	**35 169 e**	**OCDE Pacifique**
248 077	**290 456**	**303 067**	**320 996**	**319 773**	**316 837**	**322 550**	**326 758**	**317 100**	**312 800 e**	**OCDE Europe**
779 053	**815 093**	**830 396**	**849 464**	**854 151**	**847 977**	**827 196**	**844 841**	**835 100**	**835 912 e**	**AIE Total**
127 091	**155 625**	**158 032**	**157 856**	**156 887**	**160 695**	**167 125**	**159 666**	**148 479**	**150 528 e**	**UE15**

Table 3: Net Imports of Total Oil
Tableau 3 : Importations nettes de pétrole

000 metric tons

	1973	1974	1975	1976	1977	1978	1979	1980	1981	1982
Australia	9 052	10 651	8 424	8 929	9 348	10 129	10 758	11 136	10 465	10 394
Austria	9 635	8 768	8 390	9 824	9 415	10 321	11 234	10 894	9 578	8 462
Belgium	30 837	28 645	25 526	25 896	26 562	27 145	28 990	25 856	21 192	21 904
Canada	-13 907	-5 986	1 419	8 920	13 372	10 779	8 098	8 767	9 249	- 682
Czech Republic	8 857	9 251	9 675	9 832	10 860	10 940	11 188	10 900	11 143	10 655
Denmark	18 453	17 628	16 477	16 132	16 860	16 281	15 514	13 158	10 502	9 687
Finland	13 695	13 650	12 691	13 251	13 285	11 765	14 668	13 743	11 657	10 524
France	128 811	125 536	102 631	119 673	111 831	111 782	121 178	112 582	93 700	87 925
Germany	157 949	145 305	135 358	152 099	151 618	154 506	159 266	145 824	123 455	115 493
Greece	11 291	10 728	7 936	11 600	10 423	11 990	12 991	13 191	12 108	10 318
Hungary	6 489	7 296	8 046	8 132	8 410	9 865	9 820	8 300	8 062	7 671
Iceland	678	634	565	560	630	614	649	576	557	522
Ireland	5 409	5 445	5 167	5 320	5 835	6 052	6 351	5 805	5 028	4 470
Italy	105 030	102 582	90 253	97 794	97 179	96 147	101 112	98 051	93 434	90 498
Japan	268 675	263 765	243 383	256 877	265 319	258 345	271 470	246 681	222 093	212 029
Korea	13 005	14 470	15 238	16 650	20 828	22 826	26 595	26 805	26 449	26 075
Luxembourg	1 666	1 479	1 323	1 425	1 410	1 408	1 318	1 083	1 031	1 020
Mexico	5 405	1 330	-2 774	-3 710	-10 153	-17 930	-27 648	-47 117	-64 062	-84 689
Netherlands	40 199	35 840	31 591	37 649	35 564	37 274	40 723	37 525	31 251	27 747
New Zealand	4 246	4 739	4 209	3 995	3 993	3 700	3 890	3 980	3 389	3 103
Norway	6 839	6 438	-1 616	-5 165	-4 061	-7 279	-9 398	-14 620	-15 346	-15 698
Poland	11 624	11 191	14 280	14 443	16 377	16 864	17 590	17 668	15 548	14 855
Portugal	6 082	6 690	6 654	7 393	7 417	7 344	8 962	9 272	8 495	9 510
Slovak Republic	5 298	5 608	6 022	6 428	6 875	7 045	6 902	7 481	6 907	6 231
Spain	40 256	42 388	40 766	47 214	46 616	47 397	48 605	48 914	47 686	42 163
Sweden	28 625	27 873	29 000	28 740	29 050	25 857	28 961	25 759	21 164	19 533
Switzerland	14 664	13 727	12 531	13 254	13 578	13 666	13 343	13 109	11 805	11 169
Turkey	8 661	9 684	9 914	12 702	14 654	14 251	11 536	13 281	13 350	13 754
United Kingdom	113 132	111 322	88 704	80 902	52 600	41 299	18 858	1 672	-18 205	-28 029
United States	298 847	290 294	306 211	365 929	444 214	416 188	410 447	329 211	280 813	226 921
OECD TOTAL	**1359 503**	**1326 971**	**1237 994**	**1372 688**	**1429 909**	**1376 571**	**1383 971**	**1199 487**	**1012 498**	**873 535**
OECD North America	**290 345**	**285 638**	**304 856**	**371 139**	**447 433**	**409 037**	**390 897**	**290 861**	**226 000**	**141 550**
OECD Pacific	**294 978**	**293 625**	**271 254**	**286 451**	**299 488**	**295 000**	**312 713**	**288 602**	**262 396**	**251 601**
OECD Europe	**774 180**	**747 708**	**661 884**	**715 098**	**682 988**	**672 534**	**680 361**	**620 024**	**524 102**	**480 384**
IEA Total	**1336 498**	**1308 208**	**1219 901**	**1354 967**	**1416 180**	**1369 978**	**1386 478**	**1220 879**	**1053 548**	**936 616**
EU15	**711 070**	**683 879**	**602 467**	**654 912**	**615 665**	**606 568**	**618 731**	**563 329**	**472 076**	**431 225**

Table 3: Net Imports of Total Oil (continued)
Tableau 3 : Importations nettes de pétrole (suite)

000 metric tons

1983	1984	1985	1986	1987	1988	1989	1990	1991	1992	
9 078	4 782	833	- 52	2 042	1 957	6 512	5 283	2 782	4 128	Australie
8 241	8 702	8 655	9 313	9 438	8 527	8 953	9 590	10 005	9 924	Autriche
19 984	18 647	19 186	22 909	22 231	22 731	22 859	22 227	24 762	25 370	Belgique
-9 606	-11 028	-16 802	-13 596	-13 104	-19 935	-12 527	-14 075	-19 884	-23 742	Canada
10 048	10 114	10 061	10 016	10 165	9 490	9 781	8 597	7 092	7 779	République Tchèque
8 656	8 294	8 388	7 631	6 088	5 379	3 895	2 699	1 843	1 033	Danemark
10 917	9 239	10 847	11 764	12 659	10 409	11 442	10 423	10 273	9 534	Finlande
82 800	83 773	80 439	80 849	84 096	83 904	85 429	85 602	90 729	86 610	France
113 224	114 719	118 013	128 642	124 247	124 374	116 341	119 579	128 074	131 552	Allemagne
9 872	9 929	10 546	12 193	11 495	12 841	13 525	14 445	14 749	16 380	Grèce
7 886	7 713	7 121	7 201	6 898	6 483	6 158	6 555	4 917	5 512	Hongrie
488	493	537	551	612	579	658	729	619	738	Islande
4 093	4 227	4 045	5 029	4 456	3 893	4 062	4 998	4 874	4 737	Irlande
83 783	84 958	82 691	83 353	88 731	84 815	90 582	90 361	87 069	90 796	Italie
211 182	219 470	207 540	211 631	213 444	227 015	243 146	252 374	253 795	258 628	Japon
26 772	26 924	26 403	31 464	30 930	36 223	41 221	50 428	57 717	72 188	Corée
976	964	1 052	1 142	1 305	1 299	1 441	1 596	1 821	1 922	Luxembourg
-88 487	-86 929	-81 614	-74 083	-75 544	-73 962	-68 463	-68 918	-72 301	-72 235	Mexique
25 604	25 155	24 349	29 350	26 125	30 642	30 349	30 436	32 664	33 697	Pays-Bas
3 160	3 127	2 877	1 963	2 696	2 042	2 353	2 348	2 317	2 559	Nouvelle Zélande
-22 700	-26 807	-30 108	-32 778	-39 909	-47 790	-65 455	-71 352	-84 629	-97 881	Norvège
14 947	15 815	15 844	16 536	16 393	16 825	16 553	14 347	12 679	13 775	Pologne
9 494	9 840	8 523	9 589	9 599	9 552	12 193	12 197	12 384	13 575	Portugal
6 174	5 936	5 925	5 880	5 760	5 621	5 470	4 745	3 786	3 490	République Slovaque
41 974	38 546	38 909	38 466	41 285	45 540	47 859	48 706	49 409	52 168	Espagne
17 647	14 877	16 978	19 339	15 363	15 662	14 545	14 954	14 556	14 608	Suède
12 387	11 957	12 004	13 370	11 927	12 282	12 029	12 845	12 843	12 959	Suisse
14 972	14 908	14 974	16 700	19 398	18 860	18 248	20 270	17 097	19 556	Turquie
-43 023	-37 537	-49 131	-49 738	-48 088	-35 642	-9 367	-10 393	-7 257	-9 624	Royaume-Uni
225 209	244 643	224 404	279 004	301 978	333 675	361 615	361 216	338 450	359 745	Etats-Unis
815 752	**835 451**	**783 489**	**883 638**	**902 716**	**953 291**	**1031 407**	**1042 812**	**1013 235**	**1049 481**	**OCDE TOTAL**
127 116	**146 686**	**125 988**	**191 325**	**213 330**	**239 778**	**280 625**	**278 223**	**246 265**	**263 768**	**OCDE Amérique du N.**
250 192	**254 303**	**237 653**	**245 006**	**249 112**	**267 237**	**293 232**	**310 433**	**316 611**	**337 503**	**OCDE Pacifique**
438 444	**434 462**	**419 848**	**447 307**	**440 274**	**446 276**	**457 550**	**454 156**	**450 359**	**448 210**	**OCDE Europe**
882 630	**900 136**	**842 797**	**934 754**	**955 495**	**1004 228**	**1077 189**	**1091 909**	**1068 452**	**1103 713**	**AIE Total**
394 242	**394 333**	**383 490**	**409 831**	**409 030**	**423 926**	**454 108**	**457 420**	**475 955**	**482 282**	**UE15**

Table 3: Net Imports of Total Oil (continued)
Tableau 3 : Importations nettes de pétrole (suite)

000 metric tons

	1993	1994	1995	1996	1997	1998	1999	2000	2001	2002
Australia	7 503	8 788	7 864	10 229	8 498	5 923	11 366	4 240	921	2 559 e
Austria	9 768	10 116	10 223	11 100	10 834	12 094	11 152	10 755	11 436	12 239 e
Belgium	24 603	25 399	24 441	27 928	28 654	30 067	27 161	28 776	29 335	28 839 e
Canada	-25 486	-27 519	-33 784	-34 953	-34 949	-40 097	-33 989	-36 981	-37 029	-46 903 e
Czech Republic	7 274	7 664	7 867	7 999	7 893	8 233	7 782	7 470	8 121	7 910 e
Denmark	759	1 023	1 493	1 071	- 495	- 174	-4 457	-7 809	-6 433	-8 870 e
Finland	9 395	11 970	8 587	9 844	10 306	10 622	10 168	10 405	9 992	9 599 e
France	85 637	81 173	84 852	89 114	87 846	92 451	89 724	89 953	93 972	93 891 e
Germany	130 408	130 799	129 187	133 869	133 597	137 527	126 050	124 037	129 315	121 315 e
Greece	16 345	14 857	17 274	17 616	18 220	19 475	17 839	19 694	19 699	20 635 e
Hungary	6 108	5 586	5 475	4 910	5 460	6 050	5 414	5 224	4 746	4 777 e
Iceland	723	737	752	803	822	878	928	949	871	888 e
Ireland	5 115	5 495	5 689	6 102	6 674	7 401	8 285	7 932	8 856	8 899 e
Italy	87 962	88 620	90 207	89 493	88 431	90 392	87 753	88 911	84 825	83 402 e
Japan	256 831	271 928	269 336	273 329	273 648	254 002	259 219	263 215	250 842	252 885 e
Korea	81 131	89 703	95 719	104 877	111 997	94 462	102 517	106 846	103 440	103 320 e
Luxembourg	1 875	1 895	1 720	1 823	1 876	2 000	2 051	2 278	2 342	2 443 e
Mexico	-72 762	-66 719	-70 329	-76 270	-81 623	-81 763	-73 714	-76 759	-80 541	-85 233 e
Netherlands	32 513	33 776	31 339	34 040	35 288	36 025	36 245	41 602	41 605	40 119 e
New Zealand	2 598	3 270	3 782	3 756	3 438	4 047	4 490	4 498	4 586	5 068 e
Norway	-104 663	-119 849	-129 396	-146 236	-147 316	-140 804	-139 465	-150 750	-155 565	-149 609 e
Poland	14 689	14 456	15 422	17 322	18 229	18 445	18 634	19 461	18 790	18 704 e
Portugal	12 797	12 699	14 083	13 291	14 351	16 050	16 720	15 827	16 642	16 055 e
Slovak Republic	3 209	3 099	3 699	3 507	3 613	3 610	3 138	2 778	2 816	2 871 e
Spain	49 943	54 125	58 654	57 925	62 417	67 376	68 976	70 972	72 804	74 410 e
Sweden	15 093	16 872	15 801	17 236	16 359	17 407	15 544	15 421	16 053	14 930 e
Switzerland	11 737	12 327	11 669	12 208	12 676	13 108	12 100	11 722	13 134	12 469 e
Turkey	23 591	22 945	26 418	27 637	26 605	26 681	26 119	28 942	26 463	27 115 e
United Kingdom	-15 518	-41 829	-47 565	-45 828	-47 257	-49 614	-54 665	-44 996	-35 297	-36 979 e
United States	393 796	414 689	405 952	435 960	467 893	497 065	504 554	530 558	552 711	533 231 e
OECD TOTAL	**1072 974**	**1088 095**	**1066 431**	**1109 702**	**1143 985**	**1158 939**	**1167 639**	**1195 171**	**1209 452**	**1170 979 e**
OECD North America	**295 548**	**320 451**	**301 839**	**324 737**	**351 321**	**375 205**	**396 851**	**416 818**	**435 141**	**401 095 e**
OECD Pacific	**348 063**	**373 689**	**376 701**	**392 191**	**397 581**	**358 434**	**377 592**	**378 799**	**359 789**	**363 832 e**
OECD Europe	**429 363**	**393 955**	**387 891**	**392 774**	**395 083**	**425 300**	**393 196**	**399 554**	**414 522**	**406 052 e**
IEA Total	**1127 115**	**1136 522**	**1116 887**	**1164 340**	**1202 944**	**1217 769**	**1218 653**	**1248 742**	**1267 516**	**1233 749 e**
EU15	**466 695**	**446 990**	**445 985**	**464 624**	**467 101**	**489 099**	**458 546**	**473 758**	**495 146**	**480 927 e**

Table 4: Net Imports of Crude Oil, NGL and Non Crude
Tableau 4 : Importations nettes de pétrole, de LGN et additives et autres hydrocarbures

000 metric tons

1973	1974	1975	1976	1977	1978	1979	1980	1981	1982	
8 657	8 921	8 294	8 404	8 953	10 244	10 196	10 098	9 841	10 799	Australie
6 633	6 505	6 145	7 348	7 018	8 158	9 061	8 868	7 864	6 691	Autriche
37 320	30 472	29 404	29 226	36 567	33 577	34 237	33 436	28 832	24 830	Belgique
-13 560	-4 235	3 690	9 907	13 896	14 404	11 894	12 369	12 522	1 011	Canada
7 193	7 567	7 280	7 868	8 985	9 068	9 562	9 213	9 488	8 950	République Tchèque
9 760	9 363	7 909	7 634	7 284	7 681	8 466	6 355	5 432	4 189	Danemark
9 524	9 468	9 642	11 136	11 517	10 454	12 716	12 876	10 774	9 700	Finlande
134 920	129 814	106 081	121 143	117 397	115 635	127 156	113 557	95 133	80 807	France
126 630	119 070	106 712	119 276	118 157	115 680	126 837	119 561	99 434	93 114	Allemagne
13 717	11 514	10 428	11 870	9 898	11 620	15 699	14 534	16 207	13 701	Grèce
6 064	6 613	7 598	7 799	7 715	8 497	8 585	7 451	7 220	6 980	Hongrie
-	-	-	-	-	-	-	-	-	-	Islande
2 684	2 684	2 574	1 944	2 276	2 294	2 324	2 034	668	541	Irlande
128 944	119 966	95 859	102 778	106 907	111 490	116 345	94 322	92 952	86 002	Italie
245 094	239 696	224 590	229 423	237 220	230 555	239 096	218 760	195 697	182 603	Japon
14 004	15 292	15 983	17 966	20 970	22 596	25 171	24 812	24 975	24 202	Corée
-	-	-	-	-	-	-	-	-	-	Luxembourg
3 552	73	-5 048	-5 061	-10 825	-19 562	-28 590	-45 601	-60 689	-82 550	Mexique
72 161	64 576	55 219	63 881	58 927	55 689	59 380	49 891	38 370	37 187	Pays-Bas
3 176	3 467	2 912	2 951	2 694	2 417	2 591	2 633	2 303	1 784	Nouvelle Zélande
4 545	4 748	-1 867	-5 498	-4 798	-8 226	-10 207	-15 692	-15 576	-16 226	Norvège
11 140	10 582	13 306	15 095	16 404	16 615	16 608	16 351	13 514	13 227	Pologne
4 348	5 758	5 573	5 971	6 214	6 296	8 485	8 316	7 585	7 970	Portugal
6 684	7 088	7 653	8 139	8 502	8 843	8 887	9 777	8 931	8 026	République Slovaque
42 970	43 660	41 072	48 517	47 563	46 758	47 125	47 516	45 677	43 054	Espagne
10 606	10 026	12 312	13 571	14 288	15 519	15 796	17 713	14 647	13 264	Suède
6 180	5 999	4 702	4 866	4 594	4 246	4 633	4 604	3 989	4 060	Suisse
9 306	9 962	9 634	11 230	12 065	10 354	8 174	10 490	11 598	14 033	Turquie
112 237	111 418	89 842	86 181	53 844	42 943	19 811	6 537	-15 351	-27 916	Royaume-Uni
171 241	179 501	247 211	303 679	368 720	349 711	358 728	289 440	247 619	204 616	Etats-Unis
1195 730	**1169 568**	**1124 710**	**1247 244**	**1292 952**	**1243 556**	**1268 766**	**1090 221**	**919 656**	**774 649**	**OCDE TOTAL**
161 233	**175 339**	**245 853**	**308 525**	**371 791**	**344 553**	**342 032**	**256 208**	**199 452**	**123 077**	**OCDE Amérique du N.**
270 931	**267 376**	**251 779**	**258 744**	**269 837**	**265 812**	**277 054**	**256 303**	**232 816**	**219 388**	**OCDE Pacifique**
763 566	**726 853**	**627 078**	**679 975**	**651 024**	**633 191**	**649 680**	**577 710**	**487 388**	**432 184**	**OCDE Europe**
1174 354	**1151 825**	**1108 799**	**1229 071**	**1278 871**	**1237 660**	**1271 861**	**1109 694**	**957 900**	**835 946**	**AIE Total**
712 454	**674 294**	**578 772**	**630 476**	**597 857**	**583 794**	**603 438**	**535 516**	**448 224**	**393 134**	**UE15**

Table 4: Net Imports of Crude Oil, NGL and Non Crude (continued)
Tableau 4 : Importations nettes de pétrole, de LGN et additives et autres hydrocarbures (suite)

000 metric tons

	1983	1984	1985	1986	1987	1988	1989	1990	1991	1992
Australia	9 914	6 474	1 627	939	1 737	2 819	6 236	4 327	3 337	5 043
Austria	5 942	6 510	7 124	7 309	7 395	6 932	7 384	8 100	8 455	8 671
Belgium	22 901	22 961	20 302	26 172	26 903	25 698	26 759	26 708	30 198	29 772
Canada	-6 638	-9 197	-13 587	-12 950	-12 435	-16 988	-11 702	-10 080	-14 794	-20 316
Czech Republic	8 781	8 654	8 759	8 940	9 075	8 874	8 804	7 392	6 282	6 611
Denmark	4 895	4 909	3 991	4 039	3 109	3 069	2 941	1 988	1 203	999
Finland	10 304	9 343	9 828	9 316	10 685	8 937	8 832	8 713	9 925	8 865
France	74 537	75 134	75 547	71 322	68 216	74 320	72 856	75 274	77 575	76 011
Germany	85 013	86 213	83 527	86 024	83 631	91 141	86 185	88 060	88 752	98 857
Greece	12 079	11 392	10 516	15 492	14 982	14 386	14 276	14 698	13 430	15 784
Hungary	7 589	7 484	6 427	6 783	6 469	6 918	6 321	6 418	5 284	5 702
Iceland	-	-	-	-	-	-	-	-	-	-
Ireland	1 174	1 220	1 247	1 507	1 482	1 377	1 573	2 007	1 738	2 004
Italy	77 425	76 878	74 902	83 013	78 281	77 797	80 393	83 713	84 221	86 306
Japan	176 765	182 611	168 791	165 656	157 944	165 218	178 667	194 623	206 548	214 171
Korea	26 172	27 094	26 908	31 216	29 330	35 425	40 219	41 841	54 180	69 115
Luxembourg	-	-	-	-	-	-	-	-	-	-
Mexico	-85 063	-84 669	-79 373	-71 276	-74 556	-72 460	-70 675	-69 631	-74 849	-74 847
Netherlands	39 989	44 189	38 356	45 704	47 205	50 387	50 089	47 694	51 592	55 180
New Zealand	1 713	1 747	644	879	2 190	2 153	2 602	2 313	2 289	2 302
Norway	-22 854	-27 016	-30 410	-34 250	-39 399	-47 081	-64 225	-66 622	-80 699	-92 011
Poland	13 105	13 636	13 714	14 142	14 172	14 995	14 727	13 129	11 548	13 178
Portugal	7 927	7 662	7 193	8 267	8 042	8 539	10 285	11 189	10 237	11 727
Slovak Republic	8 026	7 924	7 942	8 075	8 159	7 923	7 926	6 169	4 924	4 299
Spain	42 940	41 794	44 238	47 066	45 510	50 093	51 486	53 268	52 931	54 892
Sweden	14 744	13 629	14 135	15 907	15 009	14 649	16 166	17 016	16 238	17 024
Switzerland	4 309	4 224	4 213	4 298	4 183	3 959	3 064	3 128	4 671	4 317
Turkey	14 127	15 589	15 452	16 982	20 155	21 695	18 615	20 062	17 632	19 316
United Kingdom	-39 599	-48 141	-47 404	-46 228	-41 677	-29 058	-2 116	-4 339	2 371	1 343
United States	196 980	205 905	196 213	246 451	270 284	290 774	328 872	339 415	331 264	351 786
OECD TOTAL	**713 197**	**714 153**	**670 822**	**760 795**	**766 081**	**822 491**	**896 560**	**926 573**	**926 483**	**976 101**
OECD North America	**105 279**	**112 039**	**103 253**	**162 225**	**183 293**	**201 326**	**246 495**	**259 704**	**241 621**	**256 623**
OECD Pacific	**214 564**	**217 926**	**197 970**	**198 690**	**191 201**	**205 615**	**227 724**	**243 104**	**266 354**	**290 631**
OECD Europe	**393 354**	**384 188**	**369 599**	**399 880**	**391 587**	**415 550**	**422 341**	**423 765**	**418 508**	**428 847**
IEA Total	**777 129**	**777 262**	**728 539**	**809 854**	**818 306**	**872 033**	**944 582**	**976 906**	**984 860**	**1033 471**
EU15	**360 271**	**353 693**	**343 502**	**374 910**	**368 773**	**398 267**	**427 109**	**434 089**	**448 866**	**467 435**

Table 4: Net Imports of Crude Oil, NGL and Non Crude (continued)

Tableau 4 : Importations nettes de pétrole, de LGN et additives et autres hydrocarbures (suite)

000 metric tons

1993	1994	1995	1996	1997	1998	1999	2000	2001	2002	
7 694	8 622	7 321	10 261	8 871	6 935	11 128	3 507	400	832 e	Australie
8 195	8 674	8 231	8 416	9 113	9 113	8 374	7 812	8 364	8 708 e	Autriche
28 316	28 680	26 411	31 849	33 065	34 580	32 202	33 986	33 234	33 895 e	Belgique
-20 633	-22 661	-28 625	-27 814	-28 272	-34 683	-27 615	-29 971	-28 740	-35 923 e	Canada
6 055	6 712	6 966	7 412	6 911	6 872	5 955	5 614	5 918	6 049 e	République Tchèque
196	- 361	1 616	1 640	-1 181	-2 347	-5 055	-8 205	-7 524	-9 756 e	Danemark
8 468	10 111	8 642	9 999	10 129	11 931	11 378	11 834	11 095	11 522 e	Finlande
77 444	76 443	78 246	83 284	87 049	89 052	80 801	83 914	85 377	80 885 e	France
99 476	105 141	100 113	101 827	96 961	107 990	102 192	100 467	104 141	103 574 e	Allemagne
13 531	14 362	16 871	18 286	18 337	19 206	17 171	20 383	20 081	20 218 e	Grèce
6 026	5 524	5 869	5 228	5 669	6 151	5 784	5 808	5 634	5 038 e	Hongrie
-	-	-	-	-	-	-	-	-	-	Islande
1 889	2 321	2 253	2 167	2 909	3 148	2 848	2 973	3 371	3 190 e	Irlande
85 418	85 070	81 907	81 421	87 255	91 186	87 874	89 882	90 645	88 898 e	Italie
217 617	231 487	228 435	225 355	232 117	217 559	213 872	216 594	208 806	208 031 e	Japon
76 060	79 522	87 351	99 886	121 066	113 702	120 491	123 221	118 793	107 809 e	Corée
-	-	-	-	-	-	-	-	-	-	Luxembourg
-73 360	-71 087	-70 781	-80 191	-89 880	-90 886	-82 601	-91 274	-92 445	-94 354 e	Mexique
54 416	55 218	58 618	60 742	60 120	61 912	58 944	60 369	60 455	54 342 e	Pays-Bas
2 523	2 818	2 890	2 672	2 402	3 332	3 282	3 520	3 441	3 915 e	Nouvelle Zélande
-98 993	-113 419	-124 337	-140 638	-141 801	-135 022	-134 238	-145 268	-150 494	-144 553 e	Norvège
13 940	13 413	14 059	14 984	15 115	15 742	16 627	18 131	17 462	17 362 e	Pologne
11 508	13 824	13 465	12 121	13 480	13 919	13 807	12 133	13 158	12 238 e	Portugal
4 495	4 762	5 390	5 341	5 270	5 421	5 322	5 314	5 474	5 617 e	République Slovaque
52 506	54 776	55 666	54 938	56 636	60 679	58 911	58 782	57 833	57 694 e	Espagne
17 832	18 175	17 807	18 955	20 071	19 847	19 254	20 343	19 504	17 882 e	Suède
4 764	4 876	4 657	5 323	4 984	5 070	5 093	4 649	4 908	4 876 e	Suisse
21 976	21 822	24 694	24 224	23 336	23 734	22 985	21 583	23 242	23 662 e	Turquie
-1 105	-28 882	-34 552	-30 592	-29 223	-36 655	-46 906	-38 531	-33 195	-28 347 e	Royaume-Uni
391 677	407 173	407 636	427 639	462 285	481 670	484 660	505 111	522 736	514 443 e	Etats-Unis
1017 931	**1023 116**	**1006 819**	**1034 735**	**1092 794**	**1109 158**	**1092 540**	**1102 681**	**1111 674**	**1077 747 e**	**OCDE TOTAL**
297 684	**310 426**	**308 230**	**319 634**	**344 133**	**356 101**	**374 444**	**383 866**	**401 551**	**384 166 e**	**OCDE Amérique du N.**
303 894	**322 449**	**325 997**	**338 174**	**304 456**	**341 528**	**348 773**	**346 842**	**331 440**	**320 587 e**	**OCDE Pacifique**
416 353	**387 242**	**372 592**	**376 927**	**384 205**	**411 529**	**369 323**	**371 073**	**378 683**	**372 994 e**	**OCDE Europe**
1072 856	**1076 028**	**1058 151**	**1094 601**	**1162 289**	**1178 881**	**1153 192**	**1170 510**	**1181 183**	**1149 122 e**	**AIE Total**
458 090	**443 552**	**435 294**	**455 053**	**464 721**	**483 561**	**441 795**	**456 142**	**466 539**	**454 943 e**	**UE15**

Table 5: Net Imports of Petroleum Products
Tableau 5 : Importations nettes de produits pétroliers

000 metric tons

	1973	1974	1975	1976	1977	1978	1979	1980	1981	1982
Australia	395	1 730	130	525	395	- 115	562	1 038	624	- 405
Austria	3 002	2 263	2 245	2 476	2 397	2 163	2 173	2 026	1 714	1 771
Belgium	-6 483	-1 827	-3 878	-3 330	-10 005	-6 432	-5 247	-7 580	-7 640	-2 926
Canada	- 347	-1 751	-2 271	- 987	- 524	-3 625	-3 796	-3 602	-3 273	-1 693
Czech Republic	1 664	1 684	2 395	1 964	1 875	1 872	1 626	1 687	1 655	1 705
Denmark	8 693	8 265	8 568	8 498	9 576	8 600	7 048	6 803	5 070	5 498
Finland	4 171	4 182	3 049	2 115	1 768	1 311	1 952	867	883	824
France	-6 109	-4 278	-3 450	-1 470	-5 566	-3 853	-5 978	- 975	-1 433	7 118
Germany	31 319	26 235	28 646	32 823	33 461	38 826	32 429	26 263	24 021	22 379
Greece	-2 426	- 786	-2 492	- 270	525	370	-2 708	-1 343	-4 099	-3 383
Hungary	425	683	448	333	695	1 368	1 235	849	842	691
Iceland	678	634	565	560	630	614	649	576	557	522
Ireland	2 725	2 761	2 593	3 376	3 559	3 758	4 027	3 771	4 360	3 929
Italy	-23 914	-17 384	-5 606	-4 984	-9 728	-15 343	-15 233	3 729	482	4 496
Japan	23 581	24 069	18 793	27 454	28 099	27 790	32 374	27 921	26 396	29 426
Korea	- 999	- 822	- 745	-1 316	- 142	230	1 424	1 993	1 474	1 873
Luxembourg	1 666	1 479	1 323	1 425	1 410	1 408	1 318	1 083	1 031	1 020
Mexico	1 853	1 257	2 274	1 351	672	1 632	942	-1 516	-3 373	-2 139
Netherlands	-31 962	-28 736	-23 628	-26 232	-23 363	-18 415	-18 657	-12 366	-7 119	-9 440
New Zealand	1 070	1 272	1 297	1 044	1 299	1 283	1 299	1 347	1 086	1 319
Norway	2 294	1 690	251	333	737	947	809	1 072	230	528
Poland	484	609	974	- 652	- 27	249	982	1 317	2 034	1 628
Portugal	1 734	932	1 081	1 422	1 203	1 048	477	956	910	1 540
Slovak Republic	-1 386	-1 480	-1 631	-1 711	-1 627	-1 798	-1 985	-2 296	-2 024	-1 795
Spain	-2 714	-1 272	- 306	-1 303	- 947	639	1 480	1 398	2 009	- 891
Sweden	18 019	17 847	16 688	15 169	14 762	10 338	13 165	8 046	6 517	6 269
Switzerland	8 484	7 728	7 829	8 388	8 984	9 420	8 710	8 505	7 816	7 109
Turkey	- 645	- 278	280	1 472	2 589	3 897	3 362	2 791	1 752	- 279
United Kingdom	895	- 96	-1 138	-5 279	-1 244	-1 644	- 953	-4 865	-2 854	- 113
United States	127 606	110 793	59 000	62 250	75 494	66 477	51 719	39 771	33 194	22 305
OECD TOTAL	**163 773**	**157 403**	**113 284**	**125 444**	**136 957**	**133 015**	**115 205**	**109 266**	**92 842**	**98 886**
OECD North America	**129 112**	**110 299**	**59 003**	**62 614**	**75 642**	**64 484**	**48 865**	**34 653**	**26 548**	**18 473**
OECD Pacific	**24 047**	**26 249**	**19 475**	**27 707**	**29 651**	**29 188**	**35 659**	**32 299**	**29 580**	**32 213**
OECD Europe	**10 614**	**20 855**	**34 806**	**35 123**	**31 664**	**39 343**	**30 681**	**42 314**	**36 714**	**48 200**
IEA Total	**162 144**	**156 383**	**111 102**	**125 896**	**137 309**	**132 318**	**114 617**	**111 185**	**95 648**	**100 670**
EU15	**-1 384**	**9 585**	**23 695**	**24 436**	**17 808**	**22 774**	**15 293**	**27 813**	**23 852**	**38 091**

Table 5: Net Imports of Petroleum Products (continued)
Tableau 5 : Importations nettes de produits pétroliers (suite)

000 metric tons

1983	1984	1985	1986	1987	1988	1989	1990	1991	1992	
- 836	-1 692	- 794	- 991	305	- 862	276	956	- 555	- 915	Australie
2 299	2 192	1 531	2 004	2 043	1 595	1 569	1 490	1 550	1 253	Autriche
-2 917	-4 314	-1 116	-3 263	-4 672	-2 967	-3 900	-4 481	-5 436	-4 402	Belgique
-2 968	-1 831	-3 215	- 646	- 669	-2 947	- 825	-3 995	-5 090	-3 426	Canada
1 267	1 460	1 302	1 076	1 090	616	977	1 205	810	1 168	République Tchèque
3 761	3 385	4 397	3 592	2 979	2 310	954	711	640	34	Danemark
613	- 104	1 019	2 448	1 974	1 472	2 610	1 710	348	669	Finlande
8 263	8 639	4 892	9 527	15 880	9 584	12 573	10 328	13 154	10 599	France
28 211	28 506	34 486	42 618	40 616	33 233	30 156	31 519	39 322	32 695	Allemagne
-2 207	-1 463	30	-3 299	-3 487	-1 545	- 751	- 253	1 319	596	Grèce
297	229	694	418	429	- 435	- 163	137	- 367	- 190	Hongrie
488	493	537	551	612	579	658	729	619	738	Islande
2 919	3 007	2 798	3 522	2 974	2 516	2 489	2 991	3 136	2 733	Irlande
6 358	8 080	7 789	340	10 450	7 018	10 189	6 648	2 848	4 490	Italie
34 417	36 859	38 749	45 975	55 500	61 797	64 479	57 751	47 247	44 457	Japon
600	- 170	- 505	248	1 600	798	1 002	8 587	3 537	3 073	Corée
976	964	1 052	1 142	1 305	1 299	1 441	1 596	1 821	1 922	Luxembourg
-3 424	-2 260	-2 241	-2 807	- 988	-1 502	2 212	713	2 548	2 612	Mexique
-14 385	-19 034	-14 007	-16 354	-21 080	-19 745	-19 740	-17 258	-18 928	-21 483	Pays-Bas
1 447	1 380	2 233	1 084	506	- 111	- 249	35	28	257	Nouvelle Zélande
154	209	302	1 472	- 510	- 709	-1 230	-4 730	-3 930	-5 870	Norvège
1 842	2 179	2 130	2 394	2 221	1 830	1 826	1 218	1 131	597	Pologne
1 567	2 178	1 330	1 322	1 557	1 013	1 908	1 008	2 147	1 848	Portugal
-1 852	-1 988	-2 017	-2 195	-2 399	-2 302	-2 456	-1 424	-1 138	- 809	République Slovaque
- 966	-3 248	-5 329	-8 600	-4 225	-4 553	-3 627	-4 562	-3 522	-2 724	Espagne
2 903	1 248	2 843	3 432	354	1 013	-1 621	-2 062	-1 682	-2 416	Suède
8 078	7 733	7 791	9 072	7 744	8 323	8 965	9 717	8 172	8 642	Suisse
845	- 681	- 478	- 282	- 757	-2 835	- 367	208	- 535	240	Turquie
-3 424	10 604	-1 727	-3 510	-6 411	-6 584	-7 251	-6 054	-9 628	-10 967	Royaume-Uni
28 229	38 738	28 191	32 553	31 694	42 901	32 743	21 801	7 186	7 959	Etats-Unis
102 555	**121 298**	**112 667**	**122 843**	**136 635**	**130 800**	**134 847**	**116 239**	**86 752**	**73 380**	**OCDE TOTAL**
21 837	**34 047**	**22 735**	**29 100**	**30 037**	**38 452**	**34 130**	**18 519**	**4 644**	**7 145**	**OCDE Amérique du N.**
35 628	**36 377**	**39 683**	**46 316**	**57 911**	**61 622**	**65 508**	**67 329**	**50 257**	**46 872**	**OCDE Pacifique**
45 090	**50 274**	**50 249**	**47 427**	**48 687**	**30 726**	**35 209**	**30 391**	**31 861**	**19 363**	**OCDE Europe**
105 501	**122 874**	**114 258**	**124 900**	**137 189**	**132 195**	**132 607**	**115 003**	**83 592**	**70 242**	**AIE Total**
33 971	**40 640**	**39 988**	**34 921**	**40 257**	**25 659**	**26 999**	**23 331**	**27 089**	**14 847**	**UE15**

Table 5: Net Imports of Petroleum Products (continued)
Tableau 5 : Importations nettes de produits pétroliers (suite)

000 metric tons

	1993	1994	1995	1996	1997	1998	1999	2000	2001	2002
Australia	- 191	166	543	- 32	- 373	-1 012	238	733	521	1 727 e
Austria	1 573	1 442	1 992	2 684	1 721	2 981	2 778	2 943	3 072	3 531 e
Belgium	-3 713	-3 281	-1 970	-3 921	-4 411	-4 513	-5 041	-5 210	-3 899	-5 056 e
Canada	-4 853	-4 858	-5 159	-7 139	-6 677	-5 414	-6 374	-7 010	-8 289	-10 980 e
Czech Republic	1 219	952	901	587	982	1 361	1 827	1 856	2 203	1 861 e
Denmark	563	1 384	- 123	- 569	686	2 173	598	396	1 091	886 e
Finland	927	1 859	- 55	- 155	177	-1 309	-1 210	-1 429	-1 103	-1 923 e
France	8 193	4 730	6 606	5 830	797	3 399	8 923	6 039	8 595	13 006 e
Germany	30 932	25 658	29 074	32 042	36 636	29 537	23 858	23 570	25 174	17 741 e
Greece	2 814	495	403	- 670	- 117	269	668	- 689	- 382	417 e
Hungary	82	62	- 394	- 318	- 209	- 101	- 370	- 584	- 888	- 261 e
Iceland	723	737	752	803	822	878	928	949	871	888 e
Ireland	3 226	3 174	3 436	3 935	3 765	4 253	5 437	4 959	5 485	5 709 e
Italy	2 544	3 550	8 300	8 072	1 176	- 794	- 121	- 971	-5 820	-5 496 e
Japan	39 214	40 441	40 901	47 974	41 531	36 443	45 347	46 621	42 036	44 854 e
Korea	5 071	10 181	8 368	4 991	-9 069	-19 240	-17 974	-16 375	-15 353	-4 489 e
Luxembourg	1 875	1 895	1 720	1 823	1 876	2 000	2 051	2 278	2 342	2 443 e
Mexico	598	4 368	452	3 921	8 257	9 123	8 887	14 515	11 904	9 121 e
Netherlands	-21 903	-21 442	-27 279	-26 702	-24 832	-25 887	-22 699	-18 767	-18 850	-14 223 e
New Zealand	75	452	892	1 084	1 036	715	1 208	978	1 145	1 153 e
Norway	-5 670	-6 430	-5 059	-5 598	-5 515	-5 782	-5 227	-5 482	-5 071	-5 056 e
Poland	749	1 043	1 363	2 338	3 114	2 703	2 007	1 330	1 328	1 342 e
Portugal	1 289	-1 125	618	1 170	871	2 131	2 913	3 694	3 484	3 817 e
Slovak Republic	-1 286	-1 663	-1 691	-1 834	-1 657	-1 811	-2 184	-2 536	-2 658	-2 746 e
Spain	-2 563	- 651	2 988	2 987	5 781	6 697	10 065	12 190	14 971	16 716 e
Sweden	-2 739	-1 303	-2 006	-1 719	-3 712	-2 440	-3 710	-4 922	-3 451	-2 952 e
Switzerland	6 973	7 451	7 012	6 885	7 692	8 038	7 007	7 073	8 226	7 593 e
Turkey	1 615	1 123	1 724	3 413	3 269	2 947	3 134	7 359	3 221	3 453 e
United Kingdom	-14 413	-12 947	-13 013	-15 236	-18 034	-12 959	-7 759	-6 465	-2 102	-8 632 e
United States	2 119	7 516	-1 684	8 321	5 608	15 395	19 894	25 447	29 975	18 788 e
OECD TOTAL	**55 043**	**64 979**	**59 612**	**74 967**	**51 191**	**49 781**	**75 099**	**92 490**	**97 778**	**93 232 e**
OECD North America	**-2 136**	**7 026**	**-6 391**	**5 103**	**7 188**	**19 104**	**22 407**	**32 952**	**33 590**	**16 929 e**
OECD Pacific	**44 169**	**51 240**	**50 704**	**54 017**	**33 125**	**16 906**	**28 819**	**31 957**	**28 349**	**43 245 e**
OECD Europe	**13 010**	**6 713**	**15 299**	**15 847**	**10 878**	**13 771**	**23 873**	**27 581**	**35 839**	**33 058 e**
IEA Total	**54 259**	**60 494**	**58 736**	**69 739**	**40 655**	**38 888**	**65 461**	**78 232**	**86 333**	**84 627 e**
EU15	**8 605**	**3 438**	**10 691**	**9 571**	**2 380**	**5 538**	**16 751**	**17 616**	**28 607**	**25 984 e**

Table 6: Refinery Input of Crude Oil, NGL and Non Crude

Tableau 6 : Entrées en raffinerie de pétrole brut, de LGN et additives et autres hydrocarbures

000 metric tons

1973	1974	1975	1976	1977	1978	1979	1980	1981	1982	
26 920	26 846	27 418	27 956	29 553	29 441	30 113	29 328	27 672	28 588	Australie
9 226	8 698	8 248	9 319	8 759	9 989	10 718	10 739	9 198	8 122	Autriche
37 248	30 348	29 266	29 267	36 466	33 880	34 004	33 631	29 323	24 968	Belgique
82 996	87 241	84 988	85 108	89 612	88 804	95 306	92 908	84 810	74 547	Canada
7 236	7 612	7 652	8 251	9 191	9 508	9 887	9 686	9 957	9 280	République Tchèque
10 107	9 324	8 086	8 215	8 212	8 605	8 748	6 649	6 229	6 031	Danemark
9 297	9 549	8 789	11 214	11 868	11 283	12 360	12 671	11 315	9 632	Finlande
134 215	127 258	109 050	123 510	120 700	119 450	129 710	116 317	99 311	85 190	France
138 364	131 625	116 577	128 000	127 236	126 680	139 057	138 234	122 532	119 195	Allemagne
12 600	11 163	11 627	11 099	10 552	11 568	15 425	14 268	15 934	15 104	Grèce
7 838	8 575	9 595	9 883	10 207	10 882	10 942	10 084	10 043	9 481	Hongrie
-	-	-	-	-	-	-	-	-	-	Islande
2 920	2 715	2 574	1 944	2 310	2 241	2 313	2 036	735	500	Irlande
132 723	123 705	102 827	108 132	109 733	116 227	119 444	99 247	96 736	89 222	Italie
224 920	220 097	204 063	211 327	217 626	215 755	220 501	203 677	185 591	169 357	Japon
14 565	15 039	16 390	18 792	20 679	22 662	24 819	24 413	24 197	23 480	Corée
-	-	-	-	-	-	-	-	-	-	Luxembourg
28 348	31 593	32 558	36 536	41 205	43 394	46 626	55 212	61 517	60 859	Mexique
73 505	64 688	57 176	65 264	60 958	60 803	65 916	57 537	51 497	53 690	Pays-Bas
3 311	3 561	3 005	3 493	3 256	2 989	3 084	2 913	2 757	2 300	Nouvelle Zélande
6 497	6 362	7 730	8 369	8 463	8 739	8 730	7 884	7 306	7 042	Norvège
11 221	11 636	13 836	15 613	16 601	17 100	16 745	16 315	13 745	13 476	Pologne
4 274	5 780	5 661	5 811	5 880	6 581	8 378	7 685	8 039	8 045	Portugal
6 801	7 155	7 739	8 222	8 574	8 907	9 168	9 726	9 001	7 985	République Slovaque
43 070	44 987	43 899	51 235	47 533	49 671	48 706	49 326	46 719	45 498	Espagne
10 699	10 184	11 597	14 692	15 024	15 627	16 178	17 689	14 032	13 004	Suède
6 181	5 953	4 690	4 911	4 593	4 277	4 620	4 590	4 018	4 057	Suisse
12 981	12 970	13 046	13 487	14 680	13 214	11 210	12 839	13 508	16 530	Turquie
114 191	111 055	94 851	98 505	94 359	96 390	97 854	86 393	78 287	77 130	Royaume-Uni
649 766	631 409	679 899	722 651	787 003	793 583	788 969	723 401	679 605	645 007	Etats-Unis
1822 020	**1767 128**	**1722 837**	**1840 806**	**1920 833**	**1938 250**	**1989 531**	**1855 398**	**1723 614**	**1627 320**	**OCDE TOTAL**
761 110	**750 243**	**797 445**	**844 295**	**917 820**	**925 781**	**930 901**	**871 521**	**825 932**	**780 413**	**OCDE Amérique du N.**
269 716	**265 543**	**250 876**	**261 568**	**271 114**	**270 847**	**278 517**	**260 331**	**240 217**	**223 725**	**OCDE Pacifique**
791 194	**751 342**	**674 516**	**734 943**	**731 899**	**741 622**	**780 113**	**723 546**	**657 465**	**623 182**	**OCDE Europe**
1775 650	**1716 744**	**1668 704**	**1780 435**	**1854 453**	**1868 849**	**1916 992**	**1774 145**	**1639 351**	**1545 000**	**AIE Total**
732 439	**691 079**	**610 228**	**666 207**	**659 590**	**668 995**	**708 811**	**652 422**	**589 887**	**555 331**	**UE15**

Table 6: Refinery Input of Crude Oil, NGL and Non Crude (continued)
Tableau 6 : Entrées en raffinerie de pétrole brut, de LGN et additives et autres hydrocarbures (suite)

000 metric tons

	1983	1984	1985	1986	1987	1988	1989	1990	1991	1992
Australia	26 952	27 344	27 252	26 736	26 741	27 884	29 332	29 571	30 331	31 463
Austria	7 443	7 923	8 555	8 617	8 736	8 685	8 727	9 407	9 945	10 011
Belgium	23 145	23 679	20 489	27 659	28 480	29 531	30 225	29 591	32 754	32 625
Canada	72 321	74 083	73 995	72 224	77 500	80 828	82 082	84 116	80 987	78 731
Czech Republic	9 208	9 150	9 234	9 352	9 468	9 390	9 279	7 950	6 846	7 164
Denmark	6 947	7 342	6 983	7 736	7 629	7 958	8 369	7 939	8 223	8 530
Finland	10 672	10 593	10 498	9 436	10 527	10 428	9 240	10 479	10 799	10 677
France	80 847	79 327	79 852	76 731	73 276	78 318	78 223	78 859	81 740	80 462
Germany	113 990	112 936	111 268	109 285	106 254	112 676	108 264	107 459	105 552	113 121
Greece	14 077	12 693	12 138	16 105	16 451	15 902	16 454	16 549	15 234	16 073
Hungary	9 122	9 428	9 257	9 481	9 162	9 873	9 496	8 751	8 261	8 134
Iceland	-	-	-	-	-	-	-	-	-	-
Ireland	1 187	1 239	1 335	1 544	1 516	1 387	1 495	1 732	1 811	1 977
Italy	83 882	82 479	79 091	89 308	85 630	86 840	88 362	90 955	93 346	93 392
Japan	162 880	168 462	157 185	151 639	145 955	151 266	159 953	175 290	188 051	198 141
Korea	25 867	27 400	27 291	28 457	28 618	34 792	40 261	41 693	56 009	69 283
Luxembourg	-	-	-	-	-	-	-	-	-	-
Mexico	58 823	64 131	65 288	63 614	65 811	65 234	67 492	68 467	70 483	69 954
Netherlands	59 589	62 350	55 808	67 669	67 845	69 505	69 863	68 287	70 733	71 873
New Zealand	2 463	2 516	1 536	2 660	3 685	4 146	4 599	4 759	4 698	4 741
Norway	7 270	7 557	7 742	7 709	9 274	8 401	10 030	13 229	12 522	14 179
Poland	13 726	13 715	14 196	14 541	14 718	15 496	15 778	13 198	12 036	13 734
Portugal	8 280	7 576	7 263	8 584	7 716	8 624	10 706	10 984	10 283	11 795
Slovak Republic	8 045	7 944	7 995	8 104	8 254	8 054	8 237	6 586	5 262	4 625
Spain	46 118	45 531	45 911	50 597	48 739	52 139	52 553	53 757	54 954	57 264
Sweden	14 579	14 383	14 175	16 162	16 404	15 818	17 022	18 060	17 313	18 413
Switzerland	4 317	4 225	4 197	4 300	4 162	4 005	3 071	3 071	4 708	4 293
Turkey	16 251	17 910	17 957	19 410	22 941	24 059	21 839	23 026	22 618	23 225
United Kingdom	76 880	79 117	78 431	80 155	80 263	85 662	87 699	88 692	92 001	92 478
United States	633 027	658 205	653 248	690 453	699 100	717 499	730 108	731 435	722 324	733 214
OECD TOTAL	**1597 908**	**1639 238**	**1608 170**	**1678 268**	**1684 855**	**1744 400**	**1778 759**	**1803 892**	**1829 824**	**1879 572**
OECD North America	**764 171**	**796 419**	**792 531**	**826 291**	**842 411**	**863 561**	**879 682**	**884 018**	**873 794**	**881 899**
OECD Pacific	**218 162**	**225 722**	**213 264**	**209 492**	**204 999**	**218 088**	**234 145**	**251 313**	**279 089**	**303 628**
OECD Europe	**615 575**	**617 097**	**602 375**	**642 485**	**637 445**	**662 751**	**664 932**	**668 561**	**676 941**	**694 045**
IEA Total	**1517 314**	**1553 448**	**1520 691**	**1592 009**	**1596 072**	**1655 616**	**1687 252**	**1715 641**	**1742 043**	**1791 259**
EU15	**547 636**	**547 168**	**531 797**	**569 588**	**559 466**	**583 473**	**587 202**	**592 750**	**604 688**	**618 691**

Table 6: Refinery Input of Crude Oil, NGL and Non Crude (continued)

Tableau 6 : Entrées en raffinerie de pétrole brut, de LGN et additives et autres hydrocarbures (suite)

000 metric tons

1993	1994	1995	1996	1997	1998	1999	2000	2001	2002	
32 639	33 059	33 552	35 159	36 136	37 309	36 373	36 270	34 892	33 640 e	Australie
9 764	9 938	9 326	9 529	10 318	10 071	9 583	8 891	9 510	9 545 e	Autriche
31 649	32 094	29 327	35 562	36 984	38 232	36 119	38 344	40 254	46 374 e	Belgique
81 258	82 979	82 451	87 027	89 162	90 029	91 139	93 309	95 786	97 140 e	Canada
6 594	6 966	7 284	7 859	7 486	7 167	6 314	6 171	6 357	6 425 e	République Tchèque
8 605	8 877	9 810	10 681	8 627	7 829	8 142	8 289	8 143	7 865 e	Danemark
10 214	12 078	11 823	12 588	11 537	13 370	13 153	12 910	12 118	13 402 e	Finlande
82 991	80 546	82 301	86 427	91 215	94 380	86 042	89 536	91 012	84 250 e	France
117 360	121 222	115 881	117 180	112 851	118 994	117 041	118 213	116 365	116 244 e	Allemagne
14 186	16 383	17 765	20 396	20 717	21 118	18 979	22 320	21 595	21 416 e	Grèce
8 476	7 884	8 402	7 630	7 906	7 976	7 687	7 524	7 551	6 907 e	Hongrie
-	-	-	-	-	-	-	-	-	-	Islande
1 895	2 314	2 256	2 150	2 921	3 177	2 817	3 294	3 406	3 221 e	Irlande
93 543	93 558	91 658	90 153	97 124	100 623	95 393	96 363	97 381	97 571 e	Italie
203 253	212 067	214 203	212 922	218 869	215 831	211 996	210 912	205 708	204 259 e	Japon
73 808	79 250	89 722	105 102	123 497	117 015	122 838	125 265	122 148	115 178 e	Corée
-	-	-	-	-	-	-	-	-	-	Luxembourg
71 815	74 028	70 717	67 048	67 210	69 675	70 658	65 371	66 529	67 360 e	Mexique
73 758	75 282	79 021	80 176	80 112	81 018	77 173	81 294	82 598	79 828 e	Pays-Bas
4 898	5 046	4 800	4 649	5 041	5 163	4 942	5 096	5 055	5 226 e	Nouvelle Zélande
14 099	14 706	12 904	14 569	15 231	14 891	15 368	15 076	14 367	13 321 e	Norvège
14 573	14 986	15 001	15 983	16 155	17 442	18 355	19 820	19 330	18 525 e	Pologne
11 366	13 931	13 496	12 462	13 141	14 186	13 594	12 362	13 079	12 733 e	Portugal
4 695	5 024	5 261	5 267	5 399	5 686	5 638	5 810	6 315	6 482 e	République Slovaque
53 936	56 250	56 342	55 743	57 581	62 291	61 297	60 404	58 349	58 489 e	Espagne
19 816	19 613	19 444	20 586	21 683	21 335	21 294	21 569	20 918	19 356 e	Suède
4 784	4 878	4 675	5 288	4 975	5 075	5 105	4 654	4 923	4 927 e	Suisse
25 720	25 043	27 359	26 461	26 673	27 717	25 979	23 957	25 560	26 306 e	Turquie
96 416	93 298	92 876	96 794	97 157	93 928	88 415	88 144	84 629	85 857 e	Royaume-Uni
750 477	756 060	761 518	776 652	795 799	808 264	808 152	820 738	817 964	814 827 e	Etats-Unis
1922 588	**1957 360**	**1969 175**	**2022 043**	**2081 507**	**2109 792**	**2079 586**	**2101 906**	**2091 842**	**2076 674 e**	**OCDE TOTAL**
903 550	**913 007**	**914 686**	**930 727**	**952 171**	**967 968**	**969 949**	**979 418**	**980 279**	**979 327 e**	**OCDE Amérique du N.**
314 598	**329 422**	**342 277**	**357 832**	**383 543**	**376 318**	**376 149**	**377 543**	**367 803**	**358 303 e**	**OCDE Pacifique**
704 440	**714 871**	**712 212**	**733 484**	**745 793**	**766 506**	**733 488**	**744 945**	**743 700**	**739 044 e**	**OCDE Europe**
1831 505	**1863 322**	**1878 196**	**1933 745**	**1992 743**	**2016 989**	**1984 935**	**2010 905**	**1999 668**	**1984 307 e**	**AIE Total**
625 499	**635 384**	**631 326**	**650 427**	**661 968**	**680 552**	**649 042**	**661 933**	**659 357**	**656 151 e**	**UE15**

Table 7: Gross Refinery Output of Petroleum Products
Tableau 7 : Production brute des raffineries en produits pétroliers

000 metric tons

	1973	1974	1975	1976	1977	1978	1979	1980	1981	1982
Australia	25 251	28 439	28 673	29 865	30 230	30 065	30 204	28 933	28 202	28 770
Austria	8 817	8 326	8 125	9 018	8 485	9 874	10 618	10 252	8 912	7 790
Belgium	35 510	28 814	28 300	29 137	36 415	33 545	33 862	33 439	28 991	24 708
Canada	82 322	86 637	83 952	83 600	88 420	88 487	95 569	92 595	86 003	74 753
Czech Republic	7 586	7 785	7 626	8 226	9 133	9 450	9 820	9 660	9 903	9 278
Denmark	9 760	8 930	8 001	8 032	7 996	8 020	8 715	6 616	6 208	5 968
Finland	9 045	9 471	8 608	10 982	11 684	11 104	12 193	12 484	11 115	9 499
France	133 792	126 378	108 180	122 510	119 702	118 583	128 510	115 717	98 478	84 483
Germany	139 508	131 211	118 302	127 304	127 359	126 584	139 461	136 512	120 402	117 460
Greece	12 423	11 067	11 555	11 015	10 458	11 445	15 275	14 134	15 795	14 993
Hungary	7 980	8 646	9 741	9 932	10 202	10 900	11 174	10 265	10 049	9 366
Iceland	-	-	-	-	-	-	-	-	-	-
Ireland	2 675	2 642	2 548	1 928	2 304	2 236	2 302	2 014	730	493
Italy	130 307	121 479	101 176	106 363	107 806	113 861	117 101	97 992	94 739	87 982
Japan	222 333	215 883	204 735	210 272	215 048	213 895	217 169	200 694	183 263	169 739
Korea	14 989	15 372	16 949	19 274	22 245	24 214	26 243	25 601	25 097	24 523
Luxembourg	-	-	-	-	-	-	-	-	-	-
Mexico	25 858	29 487	30 402	33 943	37 890	40 308	42 709	50 432	55 809	56 336
Netherlands	73 081	64 360	56 888	64 944	60 668	60 530	65 457	57 238	51 244	53 335
New Zealand	3 265	3 500	2 956	3 451	3 204	2 954	3 037	2 904	2 749	2 294
Norway	6 112	6 069	7 497	8 213	8 385	8 237	8 442	7 770	7 209	6 937
Poland	10 814	10 955	12 919	14 956	15 489	16 041	15 786	15 426	13 045	12 783
Portugal	4 222	5 702	5 484	5 619	5 752	6 471	8 250	7 609	7 945	7 967
Slovak Republic	6 043	6 293	6 597	6 935	6 982	7 411	7 806	8 099	7 633	6 644
Spain	42 423	44 638	42 593	50 126	48 045	48 265	48 256	48 517	46 294	44 990
Sweden	10 542	10 236	11 486	14 358	14 498	15 288	16 100	17 613	13 755	12 884
Switzerland	6 126	5 900	4 634	4 863	4 552	4 236	4 572	4 550	3 981	4 021
Turkey	12 533	12 793	13 011	13 253	14 433	13 053	11 053	12 657	13 395	16 403
United Kingdom	113 004	109 999	92 678	96 626	92 576	95 579	97 121	85 544	77 500	76 296
United States	665 486	647 718	673 127	724 991	788 327	788 157	783 070	723 765	684 431	652 244
OECD TOTAL	**1821 807**	**1768 730**	**1706 743**	**1829 736**	**1908 288**	**1918 793**	**1969 875**	**1839 032**	**1712 877**	**1622 939**
OECD North America	**773 666**	**763 842**	**787 481**	**842 534**	**914 637**	**916 952**	**921 348**	**866 792**	**826 243**	**783 333**
OECD Pacific	**265 838**	**263 194**	**253 313**	**262 862**	**270 727**	**271 128**	**276 653**	**258 132**	**239 311**	**225 326**
OECD Europe	**782 303**	**741 694**	**665 949**	**724 340**	**722 924**	**730 713**	**771 874**	**714 108**	**647 323**	**614 280**
IEA Total	**1779 092**	**1721 995**	**1656 825**	**1773 902**	**1847 927**	**1855 033**	**1903 574**	**1765 075**	**1636 390**	**1547 176**
EU15	**725 109**	**683 253**	**603 924**	**657 962**	**653 748**	**661 385**	**703 221**	**645 681**	**582 108**	**548 848**

Table 7: Gross Refinery Output of Petroleum Products (continued)
Tableau 7 : Production brute des raffineries en produits pétroliers (suite)

000 metric tons

1983	1984	1985	1986	1987	1988	1989	1990	1991	1992	
27 654	27 824	27 826	26 960	27 160	29 034	29 845	30 758	31 574	31 990	Australie
7 067	7 529	8 097	8 118	7 961	8 167	8 351	8 911	9 369	9 564	Autriche
23 009	23 397	20 384	27 524	28 245	29 392	30 031	29 372	32 494	32 250	Belgique
71 263	73 395	72 153	70 829	76 091	80 466	82 025	84 184	81 598	79 888	Canada
9 184	9 123	9 214	9 338	9 454	9 389	9 237	7 948	6 837	7 151	République Tchèque
6 883	7 284	6 937	7 687	7 556	7 892	8 295	7 846	8 172	8 442	Danemark
10 507	10 454	10 339	9 320	10 429	10 310	9 069	10 344	10 696	10 605	Finlande
80 124	78 736	79 376	76 322	72 987	77 898	77 786	78 333	81 128	79 853	France
111 659	110 456	108 912	107 088	103 943	110 415	106 603	105 957	104 349	111 747	Allemagne
13 946	12 591	12 054	16 035	16 352	15 786	16 349	16 442	15 114	15 976	Grèce
9 048	9 352	9 180	9 335	9 049	9 342	8 978	8 327	7 933	8 146	Hongrie
-	-	-	-	-	-	-	-	-	-	Islande
1 187	1 235	1 296	1 518	1 495	1 337	1 495	1 725	1 801	1 973	Irlande
82 679	81 238	78 151	88 380	85 152	86 211	87 688	90 564	90 906	91 368	Italie
163 452	168 260	156 141	151 189	146 388	152 334	159 091	173 817	184 841	195 022	Japon
27 017	28 545	27 682	28 793	28 908	35 085	40 786	42 365	57 290	70 811	Corée
-	-	-	-	-	-	-	-	-	-	Luxembourg
54 181	57 260	59 140	59 156	60 933	60 640	62 373	65 927	67 812	67 308	Mexique
59 325	62 060	55 592	67 272	67 432	69 064	69 540	67 914	70 332	71 544	Pays-Bas
2 451	2 498	1 529	2 594	3 552	4 082	4 527	4 682	4 645	4 742	Nouvelle Zélande
7 220	7 594	7 723	7 317	9 319	8 554	9 806	13 024	12 550	14 158	Norvège
13 036	13 104	13 367	13 708	14 266	14 816	14 997	12 812	11 568	12 832	Pologne
8 239	7 515	7 166	8 484	7 631	8 534	10 586	11 428	10 191	11 653	Portugal
6 769	6 637	6 614	6 661	6 825	6 585	7 600	6 434	5 184	4 557	République Slovaque
45 587	44 994	45 320	49 786	48 015	51 303	51 664	52 796	54 809	56 740	Espagne
14 223	13 944	13 395	14 596	15 479	14 657	16 549	17 491	17 264	19 354	Suède
4 288	4 189	4 166	4 267	4 147	3 981	3 050	3 047	4 691	4 263	Suisse
16 164	17 775	17 834	19 274	22 731	23 919	21 715	22 884	22 490	23 118	Turquie
76 224	78 503	78 076	79 493	79 874	85 321	87 207	88 120	91 532	92 007	Royaume-Uni
634 639	658 022	653 493	687 249	695 157	713 140	723 405	730 552	724 333	732 835	Etats-Unis
1587 025	**1623 514**	**1591 157**	**1658 293**	**1666 531**	**1727 654**	**1758 648**	**1794 004**	**1821 503**	**1869 897**	**OCDE TOTAL**
760 083	**788 677**	**784 786**	**817 234**	**832 181**	**854 246**	**867 803**	**880 663**	**873 743**	**880 031**	**OCDE Amérique du N.**
220 574	**227 127**	**213 178**	**209 536**	**200 008**	**220 535**	**234 249**	**251 622**	**278 350**	**302 565**	**OCDE Pacifique**
606 368	**607 710**	**593 193**	**631 523**	**628 342**	**652 873**	**656 596**	**661 719**	**669 410**	**687 301**	**OCDE Europe**
1513 039	**1546 513**	**1512 036**	**1578 768**	**1584 507**	**1645 613**	**1673 678**	**1708 831**	**1736 939**	**1785 200**	**AIE Total**
540 659	**539 936**	**525 095**	**561 623**	**552 551**	**576 287**	**581 213**	**587 243**	**598 157**	**613 076**	**UE15**

Table 7: Gross Refinery Output of Petroleum Products (continued)
Tableau 7 : Production brute des raffineries en produits pétroliers (suite)

000 metric tons

	1993	1994	1995	1996	1997	1998	1999	2000	2001	2002
Australia	33 400	33 997	34 894	36 263	36 652	37 144	36 399	36 708	35 403	34 319 e
Austria	9 550	9 737	9 110	9 573	10 199	9 827	9 413	8 768	9 352	9 340 e
Belgium	31 293	31 807	28 993	35 181	36 644	37 944	35 773	38 092	39 863	45 984 e
Canada	82 303	83 756	83 683	87 750	89 907	90 672	91 583	93 712	96 413	98 193 e
Czech Republic	6 587	6 929	7 260	7 830	7 410	7 090	6 247	6 131	6 309	6 376 e
Denmark	8 554	8 827	9 728	10 592	8 590	7 800	8 110	8 231	7 945	7 598 e
Finland	10 069	11 949	11 533	12 451	11 471	13 171	12 893	12 777	11 812	12 998 e
France	82 022	79 923	81 809	85 749	90 968	93 516	85 283	88 576	88 832	82 927 e
Germany	116 045	119 538	114 348	115 435	111 014	117 155	115 017	115 973	114 292	114 085 e
Greece	14 106	16 283	17 663	20 278	20 624	21 026	18 894	22 232	21 483	21 319 e
Hungary	8 557	8 153	8 272	7 645	7 888	7 992	7 659	7 483	7 452	6 893 e
Iceland	-	-	-	-	-	-	-	-	-	-
Ireland	1 889	2 276	2 257	2 151	2 908	3 163	2 831	3 286	3 365	3 191 e
Italy	92 286	92 461	90 473	89 385	96 152	99 694	94 074	94 771	96 409	97 480 e
Japan	200 510	209 870	210 747	209 350	215 326	212 110	208 255	206 162	200 406	198 477 e
Korea	75 757	79 548	89 464	100 163	120 972	113 963	120 349	122 001	119 210	111 998 e
Luxembourg	-	-	-	-	-	-	-	-	-	-
Mexico	69 062	71 718	68 603	63 895	63 874	66 816	67 718	64 176	63 886	66 153 e
Netherlands	73 463	75 001	78 385	79 710	79 412	80 051	76 263	80 235	81 306	78 714 e
New Zealand	4 877	5 079	4 708	4 585	4 952	5 145	4 877	5 060	4 981	5 298 e
Norway	14 130	14 632	13 093	14 470	15 211	14 799	15 335	15 196	13 984	13 104 e
Poland	13 821	13 653	13 676	15 079	15 586	16 682	17 811	18 480	18 411	16 886 e
Portugal	11 266	13 760	13 346	12 341	13 046	14 019	13 482	12 308	13 061	12 881 e
Slovak Republic	4 654	4 950	4 954	5 267	5 387	5 686	5 638	5 810	6 231	6 297 e
Spain	54 617	55 689	55 946	53 785	56 746	61 657	60 841	59 830	57 806	57 931 e
Sweden	20 073	19 089	19 185	20 367	21 405	21 001	21 160	21 368	20 446	18 464 e
Switzerland	4 746	4 837	4 636	5 245	4 954	5 061	5 100	4 647	4 908	4 904 e
Turkey	25 568	24 849	27 140	26 249	26 502	27 346	25 653	23 745	25 374	26 088 e
United Kingdom	96 108	92 902	92 615	96 508	96 756	93 564	86 734	86 341	82 063	84 711 e
United States	749 986	754 572	758 614	776 619	794 543	808 363	807 518	822 834	819 874	816 056 e
OECD TOTAL	**1915 299**	**1945 785**	**1955 135**	**2003 916**	**2065 099**	**2092 457**	**2060 910**	**2084 933**	**2070 877**	**2058 665 e**
OECD North America	**901 351**	**910 046**	**910 900**	**928 264**	**948 324**	**965 851**	**966 819**	**980 722**	**980 173**	**980 402 e**
OECD Pacific	**314 544**	**328 494**	**339 813**	**350 361**	**377 902**	**368 362**	**369 880**	**369 931**	**360 000**	**350 092 e**
OECD Europe	**699 404**	**707 245**	**704 422**	**725 291**	**738 873**	**758 244**	**724 211**	**734 280**	**730 704**	**728 171 e**
IEA Total	**1827 762**	**1855 464**	**1867 902**	**1919 675**	**1980 252**	**2003 273**	**1969 743**	**1996 467**	**1982 349**	**1969 329 e**
EU15	**621 341**	**629 242**	**625 391**	**643 506**	**655 935**	**673 588**	**640 768**	**652 788**	**648 035**	**647 623 e**

Table 8: Gross Consumption of Total Oil *
*Tableau 8 : Consommation brute de pétrole **

000 metric tons

1973	1974	1975	1976	1977	1978	1979	1980	1981	1982	
25 022	28 021	28 048	28 418	29 952	30 660	30 774	30 327	29 129	29 036	Australie
11 605	10 240	10 465	11 090	10 962	11 539	12 094	11 633	10 473	10 091	Autriche
25 705	22 602	21 641	23 662	23 262	24 806	25 124	23 044	20 360	19 463	Belgique
83 889	86 330	82 376	84 051	86 352	87 499	89 605	88 075	82 839	73 347	Canada
9 270	9 490	10 035	10 233	11 079	11 218	11 512	11 294	11 609	11 076	République Tchèque
17 046	15 095	15 400	16 223	16 284	15 723	15 867	13 459	11 655	11 043	Danemark
13 402	12 170	12 098	13 012	12 765	12 653	12 805	12 191	11 577	10 805	Finlande
120 460	113 152	104 400	112 610	107 077	114 736	115 219	107 226	95 135	88 840	France
159 250	144 647	141 816	149 596	149 079	156 718	161 655	147 468	133 577	126 909	Allemagne
9 172	8 322	8 371	9 141	9 525	10 846	11 283	11 453	10 793	10 824	Grèce
8 227	9 021	9 978	10 204	11 078	12 296	11 685	11 145	11 014	10 378	Hongrie
661	618	579	582	609	642	625	583	569	519	Islande
5 096	5 323	5 055	5 293	5 646	5 910	6 305	5 719	5 179	4 402	Irlande
99 002	97 380	91 228	96 126	91 200	93 847	97 659	96 439	91 850	88 866	Italie
250 331	243 000	227 328	239 081	251 741	255 079	256 595	230 966	217 334	205 162	Japon
13 476	14 055	15 633	17 653	21 018	23 906	26 774	26 467	26 101	25 809	Corée
1 643	1 479	1 312	1 421	1 396	1 403	1 293	1 075	1 031	1 020	Luxembourg
30 645	33 753	35 977	39 735	42 205	45 649	48 269	54 793	59 664	61 880	Mexique
29 019	25 582	23 575	27 807	26 089	29 419	35 337	32 994	33 253	25 028	Pays-Bas
4 134	4 299	4 011	4 162	4 053	3 900	3 813	3 846	3 687	3 667	Nouvelle Zélande
7 737	7 267	7 456	7 959	8 500	8 706	9 169	8 740	8 207	7 810	Norvège
10 637	11 366	12 650	13 763	14 861	15 745	15 998	16 360	14 826	14 019	Pologne
5 363	5 573	6 034	6 468	6 391	6 713	7 437	8 027	8 445	8 983	Portugal
4 657	4 813	4 966	5 224	5 355	5 597	5 803	5 883	5 693	4 845	République Slovaque
36 993	40 964	41 788	47 817	44 521	46 058	48 198	49 202	46 085	42 389	Espagne
27 273	25 070	25 637	28 105	27 139	25 386	26 307	23 624	21 442	19 516	Suède
14 728	13 138	12 450	12 978	13 079	13 437	12 882	12 865	11 842	11 220	Suisse
11 522	11 894	13 203	14 776	16 747	16 969	14 708	15 156	15 051	16 044	Turquie
106 855	100 371	90 131	88 583	89 679	91 679	91 866	79 687	74 422	76 342	Royaume-Uni
819 268	783 052	753 006	811 060	867 031	878 797	850 488	777 836	738 521	704 157	Etats-Unis
1962 088	**1888 087**	**1816 647**	**1936 833**	**2004 675**	**2057 536**	**2057 149**	**1917 577**	**1811 363**	**1723 490**	**OCDE TOTAL**
933 802	**903 135**	**871 359**	**934 846**	**995 588**	**1011 945**	**988 362**	**920 704**	**881 024**	**839 384**	**OCDE Amérique du N.**
292 963	**289 375**	**275 020**	**289 314**	**306 764**	**313 545**	**317 956**	**291 606**	**276 251**	**263 674**	**OCDE Pacifique**
735 323	**695 577**	**670 268**	**712 673**	**702 323**	**732 046**	**750 831**	**705 267**	**654 088**	**620 432**	**OCDE Europe**
1915 488	**1837 537**	**1762 475**	**1877 529**	**1941 645**	**1989 903**	**1986 454**	**1839 958**	**1730 611**	**1642 227**	**AIE Total**
667 884	**627 970**	**598 951**	**636 954**	**621 015**	**647 436**	**668 449**	**623 241**	**575 277**	**544 521**	**UE15**

* Gross consumption includes consumption of fuels by refineries and excludes international marine bunkers. / *La consommation brute comprend la consommation d'énergie des raffineries mais non les soutages maritimes internationaux.*

Table 8: Gross Consumption of Total Oil * (continued)
*Tableau 8 : Consommation brute de pétrole * (suite)*

000 metric tons

	1983	1984	1985	1986	1987	1988	1989	1990	1991	1992
Australia	27 213	28 129	28 232	28 394	28 729	30 160	31 588	32 394	31 882	31 796
Austria	9 900	9 543	9 662	10 015	10 214	9 918	9 896	10 172	10 914	10 738
Belgium	17 446	16 685	17 261	19 275	18 865	19 200	18 533	18 359	20 035	20 651
Canada	69 276	69 837	69 496	70 160	73 044	76 408	79 775	78 375	73 990	76 194
Czech Republic	10 364	10 652	10 504	10 406	10 557	9 960	10 346	9 085	7 562	8 156
Denmark	10 383	10 355	10 645	10 403	9 781	9 109	8 668	7 956	8 224	8 021
Finland	10 142	9 774	9 638	10 497	10 880	10 850	10 638	10 474	10 444	10 310
France	86 965	83 399	81 720	82 154	83 347	84 108	86 090	85 082	90 272	90 312
Germany	124 691	124 524	127 664	133 289	130 732	131 499	123 922	128 479	135 257	136 622
Greece	10 574	10 727	10 867	10 501	11 406	11 820	12 688	12 998	13 292	13 320
Hungary	9 728	10 012	10 263	9 669	9 905	9 006	9 020	8 671	8 119	8 241
Iceland	503	522	517	517	556	582	607	671	646	697
Ireland	4 037	4 062	4 024	4 855	4 349	3 816	3 950	4 449	4 793	4 927
Italy	89 218	85 702	83 436	85 593	90 171	91 200	94 531	93 543	92 735	94 384
Japan	207 214	214 934	202 696	206 435	210 128	226 100	234 699	245 848	249 414	255 421
Korea	27 089	27 060	26 451	27 704	28 706	34 388	39 468	48 582	58 019	69 785
Luxembourg	970	971	1 032	1 126	1 280	1 315	1 445	1 583	1 843	1 895
Mexico	59 377	62 686	65 719	67 188	70 144	70 813	75 782	78 843	81 651	82 840
Netherlands	24 055	22 987	22 657	28 222	25 071	23 217	23 029	23 157	24 127	24 718
New Zealand	3 582	3 654	3 572	3 704	3 972	4 127	4 333	4 516	4 394	4 708
Norway	7 751	8 206	8 469	8 782	9 017	8 674	8 500	8 608	8 155	8 042
Poland	14 611	14 981	15 083	15 707	16 067	16 212	16 127	13 417	12 897	13 152
Portugal	8 903	8 944	8 323	8 998	8 678	9 156	11 473	11 382	11 467	12 851
Slovak Republic	4 899	4 727	4 662	4 498	4 440	4 293	5 134	4 971	4 358	3 925
Spain	42 200	39 092	37 798	38 091	39 280	43 992	44 129	45 122	48 079	50 015
Sweden	17 259	16 444	16 966	17 479	16 812	16 186	15 265	15 011	14 931	16 213
Switzerland	12 267	11 818	12 871	12 935	12 429	12 493	12 316	12 556	13 092	13 239
Turkey	17 035	16 952	17 337	18 814	21 362	20 883	21 295	22 669	22 087	23 391
United Kingdom	72 520	89 434	76 499	76 520	74 581	78 530	79 747	81 025	81 697	82 743
United States	690 241	711 980	706 010	732 307	746 826	775 082	772 713	745 467	729 127	742 462
OECD TOTAL	**1690 413**	**1728 793**	**1700 074**	**1754 238**	**1781 329**	**1843 097**	**1865 707**	**1863 465**	**1873 503**	**1919 769**
OECD North America	**818 894**	**844 503**	**841 225**	**869 655**	**890 014**	**922 303**	**928 270**	**902 685**	**884 768**	**901 496**
OECD Pacific	**265 098**	**273 777**	**260 951**	**266 237**	**271 535**	**294 775**	**310 088**	**331 340**	**343 709**	**361 710**
OECD Europe	**606 421**	**610 513**	**597 898**	**618 346**	**619 780**	**626 019**	**627 349**	**629 440**	**645 026**	**656 563**
IEA Total	**1611 023**	**1645 877**	**1614 093**	**1666 328**	**1690 122**	**1751 197**	**1768 057**	**1765 563**	**1773 951**	**1819 155**
EU15	**529 263**	**532 643**	**518 192**	**537 018**	**535 447**	**543 916**	**544 004**	**548 792**	**568 110**	**577 720**

* Gross consumption includes consumption of fuels by refineries and excludes international marine bunkers. / *La consommation brute comprend la consommation d'énergie des raffineries mais non les soutages maritimes internationaux.*

Table 8: Gross Consumption of Total Oil * (continued)
*Tableau 8 : Consommation brute de pétrole * (suite)*

000 metric tons

1993	1994	1995	1996	1997	1998	1999	2000	2001	2002	
32 645	33 511	34 796	36 177	36 368	37 019	37 467	38 559	38 591	38 932 e	Australie
11 192	11 164	11 152	12 260	12 160	12 744	12 077	11 847	12 724	12 899 e	Autriche
20 017	20 625	20 205	22 804	23 311	23 620	23 024	22 978	23 414	21 825 e	Belgique
77 172	78 743	80 288	83 180	86 417	87 061	89 264	89 715	90 155	92 198 e	Canada
7 683	7 845	8 115	8 373	7 969	8 282	8 186	7 949	8 326	8 128 e	République Tchèque
8 123	8 531	8 972	9 912	9 596	9 272	9 284	8 936	8 613	8 171 e	Danemark
9 777	10 396	9 649	9 614	9 533	9 790	9 797	9 340	9 537	9 683 e	Finlande
87 664	87 412	89 762	91 213	91 853	95 413	94 777	93 717	96 131	92 734 e	France
138 478	136 987	136 491	138 887	138 325	138 943	134 924	133 383	134 867	130 445 e	Allemagne
13 298	13 500	13 994	15 016	15 286	15 803	15 741	16 140	16 511	17 006 e	Grèce
8 157	8 306	7 778	7 199	7 413	7 696	7 285	7 005	6 771	6 547 e	Hongrie
701	720	707	793	789	820	845	844	828	835 e	Islande
5 022	5 546	5 596	5 804	6 418	7 204	8 107	8 194	8 789	8 551 e	Irlande
93 695	92 931	97 083	95 794	96 137	96 166	91 749	90 237	88 912	89 732 e	Italie
249 829	263 891	265 112	268 909	265 449	254 984	258 576	252 590 e	243 871 e	245 607 e	Japon
76 105	84 503	91 968	95 979	105 672	88 280	95 386	98 018	97 811	99 984 e	Corée
1 890	1 885	1 741	1 807	1 885	1 962	2 094	2 227	2 368	2 421 e	Luxembourg
81 678	86 568	80 559	80 013	83 121	87 728	86 351	89 601	87 431	85 431 e	Mexique
24 247	24 783	24 687	24 489	25 020	25 302	26 212	26 870	27 504	27 809 e	Pays-Bas
4 695	5 104	5 327	5 380	5 540	5 514	5 720	5 877	5 824	6 058 e	Nouvelle Zélande
8 217	8 463	8 662	9 231	9 087	9 197	9 327	8 687	8 929	8 333 e	Norvège
13 931	14 413	14 914	16 890	18 139	18 685	19 518	19 006	18 864	18 090 e	Pologne
12 134	12 264	13 380	12 766	13 641	15 297	15 636	15 291	15 652	16 089 e	Portugal
3 322	3 362	3 272	3 452	3 689	3 785	3 431	3 364	3 382	3 444 e	République Slovaque
47 929	52 233	54 274	53 203	55 617	59 590	61 649	63 643	65 887	66 455 e	Espagne
16 066	17 131	16 997	18 295	16 918	16 999	17 013	16 236 e	15 129	15 349 e	Suède
12 498	12 290	12 469	12 735	12 449	13 045	13 096	12 761	12 856	12 260 e	Suisse
26 840	25 546	28 587	29 958	29 499	29 240	28 896	30 346	28 449	29 427 e	Turquie
83 058	82 239	81 774	83 463	80 437	80 050	81 146	81 448	80 398	76 239 e	Royaume-Uni
755 240	776 995	772 286	797 806	813 728	831 061	851 329	859 606	869 095	853 442 e	Etats-Unis
1931 303	**1987 887**	**2000 597**	**2051 402**	**2081 466**	**2090 552**	**2117 907**	**2124 415**	**2127 619**	**2104 124 e**	**OCDE TOTAL**
914 090	**942 306**	**933 133**	**960 999**	**983 266**	**1005 850**	**1026 944**	**1038 922**	**1046 681**	**1031 071 e**	**OCDE Amérique du N.**
363 274	**387 009**	**397 203**	**400 415**	**413 029**	**385 797**	**397 149**	**395 044**	**386 097**	**390 581 e**	**OCDE Pacifique**
653 939	**658 572**	**670 261**	**683 958**	**685 171**	**698 905**	**693 814**	**690 449**	**694 841**	**682 472 e**	**OCDE Europe**
1831 671	**1882 824**	**1901 145**	**1950 254**	**1975 728**	**1979 534**	**2007 762**	**2011 600**	**2017 114**	**1996 324 e**	**AIE Total**
572 590	**577 627**	**585 757**	**595 327**	**596 137**	**608 155**	**603 230**	**600 487**	**606 436**	**595 408 e**	**UE15**

* Gross consumption includes consumption of fuels by refineries and excludes international marine bunkers. / *La consommation brute comprend la consommation d'énergie des raffineries mais non les soutages maritimes internationaux.*

Table 9: Gross Consumption of LPG *
Tableau 9 : Consommation brute de GPL *

000 metric tons

	1973	1974	1975	1976	1977	1978	1979	1980	1981	1982
Australia	361	408	429	477	478	474	499	562	603	638
Austria	110	112	124	141	168	151	202	166	157	129
Belgium	541	514	562	549	549	581	641	550	510	475
Canada	1 651	2 053	1 918	1 780	2 159	1 931	1 767	1 948	2 007	1 672
Czech Republic	91	96	97	102	135	132	148	131	89	100
Denmark	303	235	183	200	201	207	229	239	248	258
Finland	90	97	87	93	95	100	110	120	135	144
France	2 696	2 327	2 704	2 796	2 834	3 004	3 452	3 539	3 482	3 305
Germany	3 325	3 369	2 864	3 367	3 036	3 079	3 411	2 575	2 475	2 527
Greece	165	153	146	157	159	164	165	189	193	188
Hungary	203	212	224	238	246	256	263	277	286	297
Iceland	1	-	1	1	1	-	-	1	1	1
Ireland	92	99	138	164	168	147	168	169	159	171
Italy	1 997	2 138	2 118	2 215	2 362	2 303	2 324	2 488	2 544	2 538
Japan	11 588	11 566	11 650	12 683	13 052	14 424	15 262	14 942	15 469	16 542
Korea	83	111	146	138	146	202	287	388	428	623
Luxembourg	21	21	22	20	22	22	24	25	25	24
Mexico	2 820	2 725	2 763	2 819	2 822	2 962	3 157	4 059	4 464	5 088
Netherlands	475	485	541	739	843	1 218	1 983	2 071	2 675	2 359
New Zealand	3	3	3	5	6	8	11	14	21	26
Norway	28	27	28	32	85	329	704	885	913	966
Poland	142	142	149	159	173	187	189	200	201	195
Portugal	363	379	415	453	453	459	477	500	508	521
Slovak Republic	14	19	17	21	15	17	25	18	18	18
Spain	1 821	1 910	2 002	2 153	2 263	2 418	2 578	2 508	2 455	2 293
Sweden	191	173	187	171	170	180	185	190	172	189
Switzerland	83	77	75	56	70	79	86	98	97	126
Turkey	440	478	541	580	640	748	699	742	765	781
United Kingdom	1 600	1 414	1 276	1 330	1 320	1 481	1 686	1 552	1 489	1 835
United States	53 040	49 537	47 861	46 049	49 732	48 367	53 979	51 497	51 214	52 443
OECD TOTAL	**84 338**	**80 880**	**79 271**	**79 688**	**84 403**	**85 630**	**94 711**	**92 643**	**93 803**	**96 472**
OECD North America	**57 511**	**54 315**	**52 542**	**50 648**	**54 713**	**53 260**	**58 903**	**57 504**	**57 685**	**59 203**
OECD Pacific	**12 035**	**12 088**	**12 228**	**13 303**	**13 682**	**15 108**	**16 059**	**15 906**	**16 521**	**17 829**
OECD Europe	**14 792**	**14 477**	**14 501**	**15 737**	**16 008**	**17 262**	**19 749**	**19 233**	**19 597**	**19 440**
IEA Total	**81 361**	**77 994**	**76 341**	**76 688**	**81 392**	**82 464**	**91 340**	**88 365**	**89 119**	**91 170**
EU15	**13 790**	**13 426**	**13 369**	**14 548**	**14 643**	**15 514**	**17 635**	**16 881**	**17 227**	**16 956**

* Gross consumption includes consumption of fuels by refineries and excludes international marine bunkers. / *La consommation brute comprend la consommation d'énergie des raffineries mais non les soutages maritimes internationaux.*

Table 9: Gross Consumption of LPG * (continued)
*Tableau 9 : Consommation brute de GPL * (suite)*

000 metric tons

1983	1984	1985	1986	1987	1988	1989	1990	1991	1992	
650	772	991	1 121	1 162	1 307	1 384	1 571	1 585	1 793	Australie
113	95	117	112	116	160	117	133	166	163	Autriche
424	500	482	429	518	477	489	426	540	655	Belgique
2 248	2 824	3 394	3 235	4 259	4 933	4 959	2 485	2 950	3 667	Canada
94	83	100	95	89	89	77	131	141	150	République Tchèque
255	240	209	181	149	133	114	99	89	82	Danemark
151	155	160	163	205	166	148	151	133	132	Finlande
3 256	3 338	3 444	3 324	3 587	3 498	3 681	3 264	3 768	3 667	France
2 334	2 414	2 706	2 752	3 028	2 691	2 770	2 663	2 961	2 966	Allemagne
186	181	182	191	205	215	250	269	301	335	Grèce
305	327	327	309	315	309	312	332	303	293	Hongrie
-	-	-	-	1	1	1	1	1	1	Islande
151	144	146	146	138	134	143	134	136	133	Irlande
2 724	2 739	2 894	3 097	3 376	3 361	3 441	3 409	3 465	3 417	Italie
16 364	18 145	17 910	18 238	18 658	17 016	18 208	18 815	19 422	19 766	Japon
856	1 052	1 228	1 516	1 829	2 233	2 615	3 079	3 718	4 648	Corée
24	25	24	23	22	21	20	20	20	18	Luxembourg
5 383	5 875	6 181	6 438	6 538	6 685	7 164	7 352	7 799	8 489	Mexique
2 426	2 990	3 035	3 113	3 110	2 498	2 427	2 794	2 761	2 555	Pays-Bas
30	47	84	99	103	104	106	104	111	117	Nouvelle Zélande
940	1 089	1 128	859	1 210	1 124	925	846	834	722	Norvège
189	177	181	191	292	294	328	219	157	159	Pologne
527	525	532	557	605	674	699	783	831	892	Portugal
18	19	18	17	21	17	16	15	77	45	République Slovaque
2 312	2 339	2 460	2 382	2 512	2 659	2 416	2 591	2 904	2 688	Espagne
329	469	421	500	653	585	668	759	800	990	Suède
130	138	161	154	168	181	161	162	184	161	Suisse
932	922	933	1 024	1 181	1 325	1 457	1 600	1 710	1 859	Turquie
2 367	2 618	2 247	2 874	2 901	2 825	2 640	1 986	2 290	1 906	Royaume-Uni
49 080	54 723	54 229	51 392	55 398	56 272	56 521	37 866	40 420	42 920	Etats-Unis
94 798	**104 965**	**105 924**	**104 532**	**112 349**	**111 987**	**114 257**	**94 059**	**100 577**	**105 389**	**OCDE TOTAL**
56 711	**63 422**	**63 001**	**61 065**	**66 195**	**67 890**	**68 644**	**47 703**	**51 169**	**55 076**	**OCDE Amérique du N.**
17 900	**20 016**	**20 213**	**20 974**	**21 752**	**20 660**	**22 313**	**23 569**	**24 836**	**26 324**	**OCDE Pacifique**
20 187	**21 527**	**21 907**	**22 493**	**24 402**	**23 437**	**23 300**	**22 787**	**24 572**	**23 989**	**OCDE Europe**
89 208	**98 894**	**99 544**	**97 886**	**105 497**	**104 990**	**106 748**	**86 472**	**92 543**	**96 695**	**AIE Total**
17 579	**18 772**	**19 059**	**19 844**	**21 125**	**20 097**	**20 023**	**19 481**	**21 165**	**20 599**	**UE15**

* Gross consumption includes consumption of fuels by refineries and excludes international marine bunkers. / *La consommation brute comprend la consommation d'énergie des raffineries mais non les soutages maritimes internationaux.*

Table 9: Gross Consumption of LPG * (continued)
*Tableau 9 : Consommation brute de GPL * (suite)*

000 metric tons

	1993	1994	1995	1996	1997	1998	1999	2000	2001	2002
Australia	1 981	2 152	2 315	2 333	2 174	2 275	2 358	2 387	2 449	2 611 e
Austria	170	174	186	158	132	145	151	170	142	170 e
Belgium	560	621	573	585	672	667	528	570	585	503 e
Canada	3 777	3 657	3 610	3 582	3 798	2 653	3 179	2 792	2 936	2 754 e
Czech Republic	104	122	183	191	223	234	245	264	274	241 e
Denmark	83	83	83	89	78	83	78	77	71	68 e
Finland	129	152	155	166	238	276	236	239	242	238 e
France	3 371	3 350	3 103	3 433	3 321	3 637	3 836	3 759	3 785	3 594 e
Germany	3 172	3 695	3 683	3 584	3 330	3 155	2 930	3 001	3 099	2 876 e
Greece	357	361	404	443	466	495	457	442	450	402 e
Hungary	292	278	301	346	335	352	331	335	296	282 e
Iceland	1	1	1	1	2	2	2	2	2	1 e
Ireland	135	139	135	128	127	132	136	142	146	146 e
Italy	3 706	3 672	3 917	3 915	3 787	3 739	4 204	4 181	3 997	3 918 e
Japan	19 963	19 425	19 702	20 212	19 882	18 917	19 017	17 442	17 544	17 262 e
Korea	5 124	5 496	5 744	6 014	6 187	5 874	6 648	7 309	7 281	7 881 e
Luxembourg	15	15	14	15	19	24	24	24	28	20 e
Mexico	8 929	9 384	9 333	9 088	8 842	9 206	9 185	10 600	10 439	10 696 e
Netherlands	2 323	1 938	1 950	1 786	2 080	2 142	2 034	2 081	1 903	2 029 e
New Zealand	117	118	119	122	121	108	104	114	130	129 e
Norway	845	868	901	910	936	895	851	925	1 276	876 e
Poland	265	398	609	760	911	982	1 063	1 107	1 331	1 449 e
Portugal	956	1 027	1 075	1 173	1 163	1 091	1 094	1 041	1 023	1 002 e
Slovak Republic	37	38	34	33	35	32	32	30	243	200 e
Spain	2 520	2 718	2 511	2 669	2 511	2 576	2 559	2 490	2 329	2 334 e
Sweden	884	975	847	840	861	954	734	781	884	913 e
Switzerland	179	173	162	183	205	220	203	197	193	215 e
Turkey	2 101	2 062	2 362	2 518	2 860	3 191	3 357	4 523	3 813	3 284 e
United Kingdom	2 004	2 492	2 509	2 512	2 436	2 371	2 202	2 074	2 106	2 494 e
United States	41 670	46 297	46 450	48 298	48 568	46 239	52 250	51 998	47 215	46 151 e
OECD TOTAL	**105 770**	**111 881**	**112 971**	**116 087**	**116 300**	**112 667**	**120 028**	**121 097**	**116 212**	**114 739 e**
OECD North America	**54 376**	**59 338**	**59 393**	**60 968**	**61 208**	**58 098**	**64 614**	**65 390**	**60 590**	**59 601 e**
OECD Pacific	**27 185**	**27 191**	**27 880**	**28 681**	**28 364**	**27 174**	**28 127**	**27 252**	**27 404**	**27 883 e**
OECD Europe	**24 209**	**25 352**	**25 698**	**26 438**	**26 728**	**27 395**	**27 287**	**28 455**	**28 218**	**27 255 e**
IEA Total	**96 538**	**102 060**	**102 994**	**106 205**	**106 510**	**102 445**	**109 746**	**109 358**	**104 197**	**102 393 e**
EU15	**20 385**	**21 412**	**21 145**	**21 496**	**21 221**	**21 487**	**21 203**	**21 072**	**20 790**	**20 707 e**

* Gross consumption includes consumption of fuels by refineries and excludes international marine bunkers. / *La consommation brute comprend la consommation d'énergie des raffineries mais non les soutages maritimes internationaux.*

Table 10: Gross Consumption of Motor Gasoline *
*Tableau 10 : Consommation brute d'essence moteur **

000 metric tons

1973	1974	1975	1976	1977	1978	1979	1980	1981	1982	
8 894	9 097	9 455	9 822	10 296	10 674	10 995	10 915	10 964	11 265	Australie
2 218	2 030	2 140	2 160	2 290	2 285	2 415	2 457	2 430	2 406	Autriche
2 573	2 511	2 769	2 869	2 978	3 108	3 132	2 948	2 719	2 669	Belgique
23 570	24 444	25 386	26 163	26 499	27 236	28 153	28 290	27 368	25 197	Canada
1 137	1 030	1 136	1 231	1 315	1 241	1 266	1 003	1 260	1 005	République Tchèque
1 635	1 488	1 577	1 662	1 693	1 767	1 668	1 534	1 441	1 409	Danemark
1 251	1 181	1 331	1 328	1 333	1 353	1 409	1 340	1 343	1 376	Finlande
15 772	15 230	15 932	16 758	16 977	17 587	17 706	17 746	18 111	18 133	France
20 780	20 467	22 205	22 165	23 528	25 829	26 132	26 531	25 104	25 507	Allemagne
951	847	932	1 056	1 212	1 363	1 375	1 377	1 447	1 513	Grèce
918	1 086	1 151	1 127	1 289	1 398	1 388	1 405	1 357	1 340	Hongrie
80	72	86	78	78	91	88	90	92	96	Islande
796	798	796	822	873	976	981	1 018	1 020	986	Irlande
11 257	10 275	10 958	10 609	10 350	11 008	11 971	12 106	11 985	11 928	Italie
19 599	19 398	20 721	21 946	22 957	24 371	25 344	25 432	25 828	25 947	Japon
726	470	439	501	712	881	1 012	823	704	541	Corée
158	153	180	208	228	244	279	286	311	310	Luxembourg
8 405	8 571	8 778	9 747	9 884	10 743	12 529	14 279	16 136	16 224	Mexique
3 556	3 202	3 473	3 659	3 800	3 954	3 979	3 855	3 686	3 632	Pays-Bas
1 522	1 607	1 660	1 639	1 668	1 692	1 649	1 645	1 650	1 685	Nouvelle Zélande
1 095	1 013	1 142	1 227	1 316	1 348	1 411	1 391	1 380	1 405	Norvège
2 422	2 375	2 597	2 722	3 025	3 220	3 289	3 269	3 244	3 020	Pologne
714	683	804	781	760	753	748	751	779	811	Portugal
532	484	501	501	541	550	566	674	694	648	République Slovaque
4 083	4 148	4 535	4 842	5 064	5 454	5 743	5 421	5 655	5 746	Espagne
3 108	2 900	3 248	3 425	3 559	3 659	3 636	3 516	3 461	3 486	Suède
2 510	2 457	2 444	2 439	2 582	2 608	2 594	2 744	2 851	2 889	Suisse
1 341	1 406	1 926	2 049	2 505	2 431	2 063	1 953	1 890	2 058	Turquie
16 927	16 484	16 125	16 879	17 336	18 349	18 686	19 145	18 718	19 248	Royaume-Uni
285 522	279 739	286 639	301 239	309 055	319 053	302 672	285 052	284 846	285 023	Etats-Unis
444 052	**435 646**	**451 066**	**471 654**	**485 703**	**505 226**	**494 879**	**478 996**	**478 474**	**477 503**	**OCDE TOTAL**
317 497	**312 754**	**320 803**	**337 149**	**345 438**	**357 032**	**343 354**	**327 621**	**328 350**	**326 444**	**OCDE Amérique du N.**
30 741	**30 572**	**32 275**	**33 908**	**35 633**	**37 618**	**39 000**	**38 815**	**39 146**	**39 438**	**OCDE Pacifique**
95 814	**92 320**	**97 988**	**100 597**	**104 632**	**110 576**	**112 525**	**112 560**	**110 978**	**111 621**	**OCDE Europe**
432 613	**424 144**	**439 104**	**458 606**	**472 175**	**490 622**	**478 407**	**460 684**	**458 308**	**457 515**	**AIE Total**
85 779	**82 397**	**87 005**	**89 223**	**91 981**	**97 689**	**99 860**	**100 031**	**98 210**	**99 160**	**UE15**

* Gross consumption includes consumption of fuels by refineries and excludes international marine bunkers. / *La consommation brute comprend la consommation d'énergie des raffineries mais non les soutages maritimes internationaux.*

Table 10: Gross Consumption of Motor Gasoline * (continued)
*Tableau 10 : Consommation brute d'essence moteur * (suite)*

000 metric tons

	1983	1984	1985	1986	1987	1988	1989	1990	1991	1992
Australia	11 098	11 354	11 540	11 755	11 742	12 157	12 534	12 943	12 389	12 451
Austria	2 491	2 473	2 430	2 463	2 504	2 560	2 596	2 554	2 796	2 676
Belgium	2 579	2 592	2 502	2 720	2 837	2 934	2 877	2 727	2 738	2 903
Canada	24 420	24 211	24 078	24 174	24 423	25 090	25 581	24 950	24 131	24 465
Czech Republic	1 001	1 104	1 069	1 043	1 061	1 115	1 250	1 161	1 085	1 412
Denmark	1 439	1 493	1 530	1 527	1 529	1 552	1 525	1 582	1 681	1 765
Finland	1 421	1 457	1 521	1 648	1 736	1 818	1 942	1 986	1 984	1 992
France	18 241	18 325	18 006	18 484	18 537	18 829	18 490	18 231	17 876	17 597
Germany	25 903	26 630	26 212	27 322	28 245	29 366	29 548	31 274	31 376	31 430
Greece	1 569	1 672	1 795	1 868	1 998	2 137	2 325	2 423	2 501	2 582
Hungary	1 261	1 296	1 336	1 407	1 506	1 579	1 750	1 790	1 567	1 463
Iceland	93	95	99	105	118	126	125	134	139	135
Ireland	931	886	841	867	833	857	861	885	905	971
Italy	11 936	12 022	12 093	12 603	12 718	13 041	13 678	14 502	15 440	17 085
Japan	26 266	26 604	26 905	27 568	28 225	29 021	30 971	32 735	33 700	34 711
Korea	542	632	812	1 003	1 214	1 592	2 145	2 778	3 366	4 132
Luxembourg	296	293	303	308	327	328	377	412	483	523
Mexico	14 559	14 354	14 917	15 366	15 938	16 220	18 099	21 392	23 437	23 044
Netherlands	3 593	3 613	3 395	3 482	3 407	3 355	3 417	3 445	3 456	3 591
New Zealand	1 676	1 704	1 624	1 680	1 712	1 793	1 872	1 929	1 915	1 969
Norway	1 441	1 495	1 591	1 700	1 758	1 778	1 783	1 788	1 730	1 694
Poland	3 182	3 158	2 954	3 108	2 940	3 029	3 349	3 088	3 594	3 802
Portugal	814	810	855	940	1 048	1 160	1 263	1 370	1 507	1 692
Slovak Republic	386	421	408	436	452	493	489	469	396	453
Spain	5 727	5 801	5 894	6 392	6 841	7 604	7 763	8 145	8 534	9 194
Sweden	3 577	3 719	3 750	3 935	4 094	4 247	4 402	4 166	4 251	4 350
Switzerland	3 014	3 086	3 058	3 237	3 334	3 459	3 524	3 724	3 857	3 998
Turkey	2 086	2 043	1 916	2 171	2 502	2 665	2 729	3 196	3 102	3 406
United Kingdom	19 566	20 227	20 403	21 470	22 184	23 249	23 924	24 312	24 021	24 044
United States	287 000	288 512	289 922	296 996	307 129	315 690	315 901	310 798	307 691	316 099
OECD TOTAL	**478 108**	**482 082**	**483 759**	**497 778**	**512 892**	**528 844**	**537 090**	**540 889**	**541 648**	**555 629**
OECD North America	**325 979**	**327 077**	**328 917**	**336 536**	**347 490**	**357 000**	**359 581**	**357 140**	**355 259**	**363 608**
OECD Pacific	**39 582**	**40 294**	**40 881**	**42 006**	**42 893**	**44 563**	**47 522**	**50 385**	**51 370**	**53 263**
OECD Europe	**112 547**	**114 711**	**113 961**	**119 236**	**122 509**	**127 281**	**129 987**	**133 364**	**135 019**	**138 758**
IEA Total	**459 888**	**464 054**	**465 381**	**478 763**	**493 444**	**508 976**	**515 028**	**515 806**	**514 082**	**528 195**
EU15	**100 083**	**102 013**	**101 530**	**106 029**	**108 838**	**113 037**	**114 988**	**118 014**	**119 549**	**122 395**

* Gross consumption includes consumption of fuels by refineries and excludes international marine bunkers. / *La consommation brute comprend la consommation d'énergie des raffineries mais non les soutages maritimes internationaux.*

Table 10: Gross Consumption of Motor Gasoline * (continued)
*Tableau 10 : Consommation brute d'essence moteur * (suite)*

000 metric tons

1993	1994	1995	1996	1997	1998	1999	2000	2001	2002	
12 552	12 776	12 957	13 054	13 069	13 102	13 306	13 618	13 023	13 394 e	Australie
2 568	2 513	2 402	2 218	2 105	2 204	2 047	1 980	1 994	2 141 e	Autriche
2 839	2 839	2 833	2 739	2 535	2 512	2 394	2 245	2 186	2 082 e	Belgique
25 019	25 717	25 766	26 093	26 730	27 466	28 168	28 180	28 527	29 096 e	Canada
1 348	1 568	1 637	1 843	1 841	1 816	1 924	1 858	1 901	1 926 e	République Tchèque
1 811	1 874	1 894	1 913	1 963	2 001	1 995	1 968	1 939	1 953 e	Danemark
1 876	1 918	1 897	1 834	1 881	1 858	1 850	1 785	1 823	1 840 e	Finlande
17 069	16 489	15 651	14 998	14 633	14 554	14 457	13 896	13 685	13 202 e	France
31 528	29 801	30 134	30 276	30 226	30 281	30 250	28 806	27 949	27 194 e	Allemagne
2 644	2 695	2 774	2 940	3 035	3 156	3 215	3 280	3 385	3 543 e	Grèce
1 488	1 445	1 427	1 345	1 353	1 386	1 402	1 336	1 391	1 434 e	Hongrie
134	137	136	156	135	142	145	143	144	148 e	Islande
953	984	1 037	1 099	1 175	1 306	1 414	1 493	1 551	1 600 e	Irlande
17 557	18 344	19 189	19 236	19 279	19 252	18 061	17 235	17 369	16 647 e	Italie
35 261	36 983	37 584	38 791	39 868	40 796	41 828	43 008	43 958	44 617 e	Japon
4 983	6 056	7 013	8 037	8 432	7 206	7 548	7 314	7 352	7 549 e	Corée
526	545	516	521	541	545	563	582	571	556 e	Luxembourg
23 055	23 523	22 500	21 641	21 578	23 029	22 437	23 118	23 134	24 144 e	Mexique
3 789	3 908	4 023	4 078	4 041	4 085	4 140	4 031	4 121	4 172 e	Pays-Bas
1 957	2 071	2 096	2 075	2 130	2 151	2 183	2 162	2 175	2 066 e	Nouvelle Zélande
1 681	1 686	1 667	1 691	1 666	1 683	1 671	1 620	1 681	1 669 e	Norvège
3 910	4 316	4 384	4 568	4 946	4 998	5 755	5 005	4 629	4 282 e	Pologne
1 795	1 851	1 890	1 965	1 949	2 027	2 079	2 123	1 995	2 023 e	Portugal
472	496	535	485	644	667	643	602	612	643 e	République Slovaque
8 851	9 160	8 534	9 093	8 970	9 007	8 928	8 524	8 481	8 206 e	Espagne
4 119	4 158	4 251	4 205	4 126	4 017	4 035	3 977	3 961	4 041 e	Suède
3 705	3 703	3 590	3 682	3 824	3 851	3 979	3 983	3 873	3 796 e	Suisse
3 955	3 974	4 330	4 570	4 580	4 543	4 257	3 619	3 052	3 141 e	Turquie
23 766	22 843	21 953	22 411	22 252	21 848	22 597	21 603	20 933	19 918 e	Royaume-Uni
322 659	328 768	336 755	341 085	346 431	356 799	364 405	367 144	372 185	382 063 e	Etats-Unis
500 870	**573 141**	**581 355**	**588 642**	**595 938**	**608 288**	**617 676**	**616 238**	**619 580**	**629 086 e**	**OCDE TOTAL**
370 733	**378 008**	**385 021**	**388 819**	**394 739**	**407 294**	**415 010**	**418 442**	**423 846**	**435 303 e**	**OCDE Amérique du N.**
54 753	**57 886**	**59 650**	**61 957**	**63 499**	**63 255**	**64 865**	**66 102**	**66 508**	**67 626 e**	**OCDE Pacifique**
138 384	**137 247**	**136 684**	**137 866**	**137 700**	**137 739**	**137 801**	**131 694**	**129 226**	**126 157 e**	**OCDE Europe**
536 299	**544 669**	**553 800**	**561 792**	**568 635**	**579 452**	**588 696**	**587 370**	**591 061**	**599 869 e**	**AIE Total**
121 691	**119 922**	**118 978**	**119 526**	**118 711**	**118 653**	**118 025**	**113 528**	**111 943**	**109 118 e**	**UE15**

* Gross consumption includes consumption of fuels by refineries and excludes international marine bunkers. / *La consommation brute comprend la consommation d'énergie des raffineries mais non les soutages maritimes internationaux.*

Table 11: Gross Consumption of Gas/Diesel Oil *
*Tableau 11 : Consommation brute de gazole / carburant diesel **

000 metric tons

	1973	1974	1975	1976	1977	1978	1979	1980	1981	1982
Australia	5 185	5 053	5 533	5 933	6 394	6 686	6 473	6 814	6 782	6 966
Austria	2 374	2 182	2 321	2 484	2 527	2 819	2 966	2 713	2 415	2 439
Belgium	9 275	8 385	8 287	8 887	8 913	9 529	9 818	8 532	7 808	7 176
Canada	23 093	23 640	22 527	23 792	23 153	24 266	24 732	24 649	22 480	20 641
Czech Republic	2 165	2 219	3 415	3 600	3 893	3 844	3 946	3 893	3 819	3 687
Denmark	6 473	5 407	5 804	6 437	6 359	6 354	6 654	5 587	4 916	4 746
Finland	4 531	4 045	4 245	4 800	4 694	4 774	4 810	4 506	4 119	4 012
France	45 643	39 914	37 863	38 348	37 733	42 570	41 359	39 395	35 779	33 338
Germany	67 103	59 150	59 590	63 922	63 219	67 063	68 143	58 555	53 962	50 629
Greece	2 527	2 180	2 244	2 503	2 600	3 074	3 116	3 239	3 001	3 312
Hungary	3 274	3 547	3 644	3 781	3 996	4 474	4 454	4 345	4 507	4 289
Iceland	341	346	344	310	319	321	295	242	228	201
Ireland	1 090	1 038	1 033	1 060	1 217	1 334	1 418	1 298	1 272	1 208
Italy	18 645	19 524	19 889	22 156	21 670	23 791	24 936	24 319	23 919	24 345
Japan	29 974	30 095	29 158	30 721	32 209	35 167	36 902	36 052	35 695	34 817
Korea	2 347 e	2 409 e	2 798 e	3 424 e	4 087 e	5 032 e	5 470 e	5 439 e	5 527 e	5 671 e
Luxembourg	605	534	509	527	515	560	573	528	494	488
Mexico	6 081	7 446	8 357	9 168	9 457	10 407	11 224	12 624	13 549	12 839
Netherlands	7 329	5 964	6 186	7 170	6 812	7 452	7 448	7 423	7 692	4 748
New Zealand	867	884	886	900	942	921	891	988	894	920
Norway	2 979	2 576	2 962	3 055	3 307	3 351	3 474	3 199	3 028	2 893
Poland	3 490	3 756	4 198	4 477	4 865	5 250	5 307	5 184	4 709	4 532
Portugal	1 035	1 042	1 126	1 244	1 247	1 400	1 534	1 742	1 796	1 887
Slovak Republic	1 205	1 235	1 040	1 065	1 055	1 168	1 237	1 329	1 371	1 214
Spain	6 761	6 753	7 163	8 281	9 280	9 928	10 722	10 590	9 777	10 156
Sweden	9 467	7 883	8 321	9 199	8 772	8 950	9 278	8 032	7 614	6 701
Switzerland	7 801	6 736	6 741	7 365	7 149	7 574	7 079	7 138	6 319	5 796
Turkey	2 735	2 895	3 262	3 779	4 626	4 482	3 936	4 129	4 207	4 686
United Kingdom	20 758	19 099	18 464	18 578	19 625	19 515	19 893	17 644	17 122	16 771
United States	149 732	142 712	139 450	153 339	164 230	168 279	161 854	140 475	131 697	122 072
OECD TOTAL	**444 885**	**418 649**	**417 360**	**450 305**	**464 865**	**490 335**	**489 942**	**450 603**	**426 498**	**403 180**
OECD North America	**178 906**	**173 798**	**170 334**	**186 299**	**196 840**	**202 952**	**197 810**	**177 748**	**167 726**	**155 552**
OECD Pacific	**38 373**	**38 441**	**38 375**	**40 978**	**43 632**	**47 806**	**49 736**	**49 293**	**48 898**	**48 374**
OECD Europe	**227 606**	**206 410**	**208 651**	**223 028**	**224 393**	**239 577**	**242 396**	**223 562**	**209 874**	**199 254**
IEA Total	**433 768**	**405 866**	**403 421**	**435 285**	**449 169**	**473 189**	**471 879**	**431 224**	**406 641**	**384 394**
EU15	**203 616**	**183 100**	**183 045**	**195 596**	**195 183**	**209 113**	**212 668**	**194 103**	**181 686**	**171 956**

* Gross consumption includes consumption of fuels by refineries and excludes international marine bunkers. / *La consommation brute comprend la consommation d'énergie des raffineries mais non les soutages maritimes internationaux.*

Table 11: Gross Consumption of Gas/Diesel Oil * (continued)
*Tableau 11 : Consommation brute de gazole / carburant diesel * (suite)*

000 metric tons

1983	1984	1985	1986	1987	1988	1989	1990	1991	1992	
6 518	6 861	7 046	7 156	7 449	7 732	8 289	8 551	8 303	8 467	Australie
2 508	2 346	2 567	2 752	2 847	3 018	3 159	3 341	3 849	3 868	Autriche
6 817	6 945	7 643	8 573	8 389	8 434	8 254	8 416	9 003	9 175	Belgique
19 439	19 625	19 487	19 063	19 092	20 087	20 691	19 792	18 406	18 451	Canada
3 605	3 825	3 772	3 885	3 821	3 606	3 798	3 369	3 081	2 909	République Tchèque
4 612	4 654	5 075	5 070	4 986	4 643	4 380	3 877	4 050	3 790	Danemark
3 775	3 793	3 839	3 952	4 156	4 101	4 014	4 075	3 945	3 893	Finlande
33 436	33 166	33 383	34 368	34 022	33 862	34 884	35 748	38 234	39 280	France
51 146	51 741	55 440	60 111	58 421	57 286	50 347	53 655	60 582	60 917	Allemagne
3 386	3 601	3 685	3 617	4 112	4 352	4 663	4 724	4 938	4 808	Grèce
3 836	3 776	3 944	3 524	3 657	3 336	3 289	3 233	2 798	2 833	Hongrie
203	210	236	245	269	273	274	278	266	281	Islande
1 200	1 214	1 292	1 496	1 423	1 455	1 550	1 679	1 761	1 852	Irlande
24 821	27 137	26 512	26 756	27 917	28 090	28 067	27 359	27 396	26 698	Italie
35 872	38 174	38 537	40 408	42 892	47 535	50 820	54 080	56 537	57 704	Japon
6 243 e	6 919 e	7 249 e	7 934 e	8 842 e	10 312 e	11 560 e	13 393 e	15 696 e	17 450 e	Corée
489	485	523	569	594	602	652	756	930	967	Luxembourg
10 531	11 380	11 909	11 359	11 365	10 430	11 081	11 691	13 020	13 532	Mexique
5 411	5 036	5 400	7 896	5 652	5 324	5 646	5 281	5 951	6 044	Pays-Bas
900	938	923	918	895	895	928	1 007	1 012	1 173	Nouvelle Zélande
2 824	2 860	3 001	3 173	3 264	3 085	3 134	3 045	2 919	2 870	Norvège
4 917	5 206	5 368	5 598	5 651	5 930	5 716	4 956	4 603	4 736	Pologne
1 874	1 828	1 832	1 890	2 039	2 205	2 296	2 442	2 564	2 707	Portugal
1 220	1 164	1 171	1 169	1 199	1 225	1 208	1 124	878	842	République Slovaque
10 446	10 794	11 033	11 136	11 469	13 316	13 617	14 258	15 055	15 573	Espagne
5 957	5 606	5 919	5 981	5 692	5 678	5 254	5 100	4 859	5 120	Suède
6 735	6 427	7 472	7 100	6 631	6 484	6 284	6 381	6 837	6 723	Suisse
5 201	5 315	5 561	6 110	7 022	6 977	6 972	7 447	7 048	6 915	Turquie
16 617	17 256	17 366	17 719	17 580	18 452	18 989	19 137	19 324	19 689	Royaume-Uni
126 124	133 032	132 941	133 820	137 184	146 315	144 814	140 747	136 107	140 709	Etats-Unis
406 663	**421 314**	**430 126**	**443 348**	**448 532**	**465 040**	**464 630**	**468 942**	**479 952**	**489 976**	**OCDE TOTAL**
156 094	**164 037**	**164 337**	**164 242**	**167 641**	**176 832**	**176 586**	**172 230**	**167 533**	**172 692**	**OCDE Amérique du N.**
49 533	**52 892**	**53 755**	**56 416**	**60 078**	**66 474**	**71 597**	**77 031**	**81 548**	**84 794**	**OCDE Pacifique**
201 036	**204 385**	**212 034**	**222 690**	**220 813**	**221 734**	**216 447**	**219 681**	**200 871**	**232 490**	**OCDE Europe**
389 792	**403 354**	**411 442**	**424 977**	**430 048**	**447 182**	**446 351**	**450 893**	**461 185**	**470 585**	**AIE Total**
172 495	**175 602**	**181 509**	**191 886**	**189 299**	**190 818**	**185 772**	**189 848**	**202 441**	**204 381**	**UE15**

* Gross consumption includes consumption of fuels by refineries and excludes international marine bunkers. / *La consommation brute comprend la consommation d'énergie des raffineries mais non les soutages maritimes internationaux.*

Table 11: Gross Consumption of Gas/Diesel Oil * (continued)
*Tableau 11 : Consommation brute de gazole / carburant diesel * (suite)*

000 metric tons

	1993	1994	1995	1996	1997	1998	1999	2000	2001	2002
Australia	8 748	9 045	9 384	10 035	10 352	10 569	10 833	11 182	11 328	11 772 e
Austria	4 244	4 248	4 526	5 538	5 302	5 835	5 752	5 861	6 632	6 942 e
Belgium	9 147	9 465	9 686	10 717	11 005	11 029	10 769	10 541	11 183	10 672 e
Canada	19 298	20 515	20 981	22 143	22 982	22 360	23 197	24 346	23 564	23 660 e
Czech Republic	2 801	2 896	2 651	2 609	2 444	2 701	2 641	2 755	2 935	2 942 e
Denmark	3 906	3 781	3 836	3 960	3 815	3 713	3 718	3 533	3 417	3 299 e
Finland	3 729	3 881	3 797	3 859	3 879	3 945	4 015	4 004	4 145	4 099 e
France	40 162	39 034	41 076	42 385	42 266	45 112	44 589	44 805	47 554	46 863 e
Germany	63 177	61 624	61 049	64 466	62 549	61 806	58 494	56 886	60 511	57 255 e
Greece	4 784	4 823	5 044	5 736	5 838	6 225	6 168	6 332	6 666	6 978 e
Hungary	2 303	2 275	1 812	1 723	1 784	1 897	1 898	2 060	2 103	2 361 e
Iceland	294	295	293	339	329	329	347	343	311	329 e
Ireland	1 862	2 070	2 099	2 226	2 309	2 603	2 867	3 030	3 132	3 111 e
Italy	26 082	24 065	24 841	24 613	24 670	25 356	26 353	26 958	27 831	28 388 e
Japan	58 338	60 568	62 027	63 464	62 583	60 722	61 379	60 403	60 338	59 482 e
Korea	18 886 e	20 554 e	22 700 e	23 709	22 887	16 530	17 344	17 754	18 085	19 111 e
Luxembourg	957	942	903	966	1 021	1 091	1 158	1 290	1 413	1 464 e
Mexico	12 641	12 435	12 538	12 417	13 148	14 065	13 051	13 712	13 330	13 240 e
Netherlands	5 969	5 957	5 408	5 676	5 667	5 975	6 364	6 666	6 667	6 880 e
New Zealand	1 200	1 329	1 514	1 567	1 651	1 810	1 887	2 026	2 006	2 278 e
Norway	3 063	3 123	3 215	3 611	3 516	3 651	3 858	3 420	3 554	3 558 e
Poland	4 891	4 897	5 122	6 475	6 997	7 163	7 375	7 271	7 601	6 879 e
Portugal	2 777	2 931	3 070	3 310	3 580	4 035	4 228	4 760	5 183	5 221 e
Slovak Republic	701	735	852	806	828	855	845	825	887	984 e
Spain	15 898	16 670	17 869	19 170	19 470	21 491	23 160	24 994	26 915	27 779 e
Sweden	5 015	5 354	5 339	5 688	5 420	5 562	5 535	5 974 e	4 834	5 152 e
Switzerland	6 391	6 076	6 345	6 467	6 112	6 508	6 436	6 122	6 386	6 044 e
Turkey	8 344	7 867	8 689	9 042	7 812	6 688	8 111	8 873	8 641	9 143 e
United Kingdom	20 257	21 081	21 395	22 593	23 150	23 267	23 142	23 639	23 486	23 198 e
United States	144 581	150 587	153 459	160 371	163 480	165 411	169 975	177 931	183 180	181 515 e
OECD TOTAL	**500 446**	**509 123**	**521 520**	**545 681**	**546 846**	**548 304**	**555 489**	**568 296**	**583 818**	**580 599 e**
OECD North America	**176 520**	**183 537**	**186 978**	**194 931**	**199 610**	**201 836**	**206 223**	**215 989**	**220 074**	**218 415 e**
OECD Pacific	**87 172**	**91 496**	**95 625**	**98 775**	**97 473**	**89 631**	**91 443**	**91 365**	**91 757**	**92 643 e**
OECD Europe	**236 754**	**234 090**	**238 917**	**251 975**	**249 763**	**256 837**	**257 823**	**260 942**	**271 987**	**269 541 e**
IEA Total	**481 919**	**490 761**	**502 715**	**525 644**	**525 544**	**525 892**	**533 871**	**546 145**	**561 689**	**559 167 e**
EU15	**207 966**	**205 926**	**209 938**	**220 903**	**219 941**	**227 045**	**226 312**	**229 273**	**239 569**	**237 301 e**

* Gross consumption includes consumption of fuels by refineries and excludes international marine bunkers. / *La consommation brute comprend la consommation d'énergie des raffineries mais non les soutages maritimes internationaux.*

Table 12: Gross Consumption of Residual Fuel Oil *
*Tableau 12 : Consommation brute de fioul lourd **

000 metric tons

1973	1974	1975	1976	1977	1978	1979	1980	1981	1982	
5 750	6 338	5 434	5 140	5 465	5 444	5 723	5 186	4 459	3 963	Australie
5 487	4 603	4 419	4 985	4 688	4 882	4 840	4 653	3 942	3 711	Autriche
9 664	7 784	7 029	7 478	6 993	8 163	7 684	7 373	5 779	5 868	Belgique
20 164	20 685	17 400	17 519	18 031	15 837	15 386	14 761	13 040	10 606	Canada
4 406	4 708	4 117	4 059	4 453	4 512	4 532	4 634	4 699	4 275	République Tchèque
6 873	6 399	6 166	6 156	6 203	5 678	5 592	4 554	3 563	3 046	Danemark
5 705	4 641	4 455	4 729	4 534	4 369	4 262	4 119	3 886	3 383	Finlande
37 536	38 241	32 489	37 346	32 064	32 594	32 602	28 531	21 484	17 764	France
41 680	36 600	34 685	36 499	34 741	35 072	35 036	30 402	24 375	20 773	Allemagne
4 101	4 079	3 836	4 196	4 256	4 695	4 893	4 945	4 407	4 036	Grèce
2 808	3 144	3 469	3 375	3 759	4 189	3 607	3 150	2 914	2 675	Hongrie
106	110	81	102	106	133	152	180	170	144	Islande
2 310	2 577	2 316	2 416	2 546	2 640	2 893	2 577	2 105	1 488	Irlande
47 648	47 566	42 148	44 254	41 184	42 628	44 095	42 813	39 324	35 686	Italie
105 595	99 585	87 669	91 972	94 368	90 926	90 525	78 930	67 000	58 718	Japon
8 195 e	8 905 e	9 795 e	10 794 e	12 564 e	13 623 e	15 199 e	14 965 e	14 620 e	14 120 e	Corée
776	672	516	573	537	475	313	131	101	106	Luxembourg
8 729	10 089	11 247	12 745	14 399	15 632	14 729	16 705	17 441	18 765	Mexique
6 704	5 284	4 362	5 034	4 493	5 652	8 370	8 395	7 985	5 823	Pays-Bas
1 074	1 087	836	951	749	537	527	470	388	346	Nouvelle Zélande
1 794	1 866	1 536	1 737	1 801	1 614	1 681	1 495	1 279	1 054	Norvège
2 727	3 070	3 546	4 043	4 312	4 471	4 654	4 644	3 997	3 612	Pologne
2 048	2 310	2 555	2 930	2 761	2 835	3 223	3 797	4 111	4 303	Portugal
2 503	2 672	2 997	3 235	3 327	3 437	3 534	3 515	3 261	2 697	République Slovaque
18 061	21 940	21 718	25 804	20 587	20 467	21 132	22 877	20 725	15 913	Espagne
11 615	11 166	11 087	12 539	12 217	10 033	10 377	9 612	8 080	6 870	Suède
2 641	2 227	1 694	1 652	1 739	1 637	1 521	1 271	1 073	893	Suisse
5 549	5 666	5 951	6 701	7 169	7 199	6 155	6 636	6 394	6 749	Turquie
43 393	40 705	34 768	31 410	31 257	31 507	31 549	24 150	20 511	21 535	Royaume-Uni
144 607	134 601	124 886	145 615	158 258	153 731	136 784	115 204	97 477	82 472	Etats-Unis
560 249	**539 320**	**493 207**	**535 989**	**539 561**	**534 612**	**521 570**	**470 675**	**408 590**	**361 394**	**OCDE TOTAL**
173 500	**165 375**	**150 533**	**175 879**	**190 688**	**185 200**	**166 899**	**146 670**	**127 958**	**111 843**	**OCDE Amérique du N.**
120 614	**115 915**	**103 734**	**108 857**	**113 146**	**110 530**	**111 974**	**99 551**	**86 467**	**77 147**	**OCDE Pacifique**
266 135	**258 030**	**235 940**	**251 253**	**235 727**	**238 882**	**242 697**	**224 454**	**194 165**	**172 404**	**OCDE Europe**
546 184	**523 379**	**475 336**	**515 864**	**517 417**	**510 939**	**498 501**	**445 631**	**383 721**	**336 176**	**AIE Total**
243 601	**234 567**	**212 549**	**226 349**	**209 061**	**211 690**	**216 861**	**198 929**	**170 378**	**150 305**	**UE15**

* Gross consumption includes consumption of fuels by refineries and excludes international marine bunkers. / *La consommation brute comprend la consommation d'énergie des raffineries mais non les soutages maritimes internationaux.*

Table 12: Gross Consumption of Residual Fuel Oil * (continued)
*Tableau 12 : Consommation brute de fioul lourd * (suite)*

000 metric tons

	1983	1984	1985	1986	1987	1988	1989	1990	1991	1992
Australia	3 187	3 351	2 709	2 358	2 137	1 897	1 949	1 947	1 758	1 510
Austria	3 139	2 807	2 774	2 837	2 732	2 222	2 006	2 036	1 984	1 789
Belgium	4 211	3 299	2 896	3 551	2 929	3 000	2 531	2 238	2 902	2 999
Canada	8 561	7 531	6 106	6 481	6 721	7 937	9 856	9 402	7 829	7 894
Czech Republic	3 834	3 770	3 760	3 513	3 605	3 163	3 033	2 368	1 892	2 190
Denmark	2 495	2 295	2 212	1 933	1 467	1 207	1 046	870	1 005	991
Finland	2 873	2 617	2 372	2 667	2 538	2 263	2 030	1 912	1 850	1 796
France	14 413	11 247	9 339	7 958	8 173	7 522	8 499	7 791	8 951	7 349
Germany	16 025	14 233	14 285	14 781	13 157	11 978	10 339	10 088	10 461	10 420
Greece	3 670	3 500	3 282	2 867	3 137	2 931	3 202	3 172	3 304	3 253
Hungary	2 530	2 828	2 786	2 514	2 534	1 967	1 807	1 712	2 102	2 364
Iceland	141	139	91	91	83	97	94	98	95	122
Ireland	1 247	1 274	1 187	1 719	1 300	746	689	952	1 143	1 218
Italy	33 759	28 404	27 559	27 282	28 455	29 123	31 183	29 273	28 225	30 116
Japan	55 054	54 151	45 358	41 955	40 490	42 264	42 923	44 889	41 765	42 590
Korea	13 976 e	12 549 e	11 177 e	10 561 e	9 792 e	12 472 e	14 289 e	16 893 e	20 040 e	22 939 e
Luxembourg	75	69	89	127	224	242	269	245	258	240
Mexico	19 782	21 431	22 662	23 760	25 905	26 383	28 033	28 499	27 083	26 546
Netherlands	2 607	1 997	1 671	1 910	1 671	1 642	1 496	1 287	1 366	1 249
New Zealand	281	204	224	187	185	191	157	170	166	228
Norway	879	875	849	1 012	756	748	572	477	373	243
Poland	3 429	3 424	3 531	3 388	3 472	3 273	3 120	2 642	2 335	2 284
Portugal	4 339	4 281	3 452	3 510	2 990	2 668	4 597	4 070	4 073	4 774
Slovak Republic	2 731	2 546	2 537	2 363	2 240	2 052	1 861	1 555	1 453	1 235
Spain	14 127	10 836	9 060	8 487	8 308	8 794	8 641	8 404	9 201	9 894
Sweden	5 084	4 517	4 898	4 731	4 190	3 166	2 638	2 275	2 207	2 234
Switzerland	814	720	658	823	686	686	619	508	455	587
Turkey	6 862	6 615	6 638	6 455	6 915	6 128	6 637	6 730	6 450	7 295
United Kingdom	16 814	31 928	18 988	15 755	12 889	14 369	13 761	14 548	14 326	13 081
United States	64 579	62 476	54 000	65 705	57 751	61 429	60 293	44 685	39 734	38 327
OECD TOTAL	**311 518**	**305 914**	**267 150**	**271 281**	**257 432**	**262 560**	**268 170**	**251 736**	**244 786**	**247 757**
OECD North America	**92 922**	**91 438**	**82 768**	**95 946**	**90 377**	**95 749**	**98 182**	**82 586**	**74 646**	**72 767**
OECD Pacific	**72 498**	**70 255**	**59 468**	**55 061**	**52 604**	**56 824**	**59 318**	**63 899**	**63 729**	**67 267**
OECD Europe	**146 098**	**144 221**	**124 914**	**120 274**	**114 451**	**109 987**	**110 670**	**105 251**	**106 411**	**107 723**
IEA Total	**285 435**	**278 374**	**238 329**	**241 679**	**225 732**	**230 755**	**235 062**	**218 942**	**213 820**	**217 570**
EU15	**124 878**	**123 304**	**104 064**	**100 115**	**94 160**	**91 873**	**92 927**	**89 161**	**91 256**	**91 403**

* Gross consumption includes consumption of fuels by refineries and excludes international marine bunkers. / *La consommation brute comprend la consommation d'énergie des raffineries mais non les soutages maritimes internationaux.*

Table 12: Gross Consumption of Residual Fuel Oil * (continued)
*Tableau 12 : Consommation brute de fioul lourd * (suite)*

000 metric tons

1993	1994	1995	1996	1997	1998	1999	2000	2001	2002	
1 429	1 346	1 461	1 419	1 318	1 208	1 192	1 348 e	1 286	1 324 e	Australie
2 080	1 960	1 888	1 826	1 936	1 963	1 608	1 335	1 351	1 096 e	Autriche
2 699	2 727	2 502	2 564	2 462	2 571	2 343	1 695	1 988	1 254 e	Belgique
7 288	6 227	5 526	5 288	5 934	7 203	6 273	6 511	7 837	7 044 e	Canada
2 121	1 797	1 294	1 409	1 305	1 390	1 218	919	915	533 e	République Tchèque
852	1 183	856	956	667	739	577	463	515	651 e	Danemark
1 664	1 767	1 730	1 708	1 554	1 648	1 627	1 347	1 572	1 631 e	Finlande
6 551	6 041	6 411	6 540	6 198	6 967	6 107	5 417	5 009	4 803 e	France
9 355	9 031	9 556	8 700	7 382	7 640	6 775	7 945	8 665	8 518 e	Allemagne
3 041	3 029	3 217	3 280	3 287	3 327	3 310	3 179	3 115	3 098 e	Grèce
2 607	2 616	2 493	2 262	2 277	2 237	2 057	1 519	1 342	856 e	Hongrie
126	128	117	121	124	100	77	63	85	93 e	Islande
1 203	1 402	1 311	1 269	1 496	1 771	2 086	1 770	1 860	1 542 e	Irlande
29 405	29 720	30 888	29 606	28 910	27 883	25 075	23 986	21 891	24 224 e	Italie
38 628	42 133	38 361	37 646	34 958	33 763	33 212	31 116	27 167	27 757 e	Japon
23 843 e	26 115 e	26 762 e	25 538	26 594	18 659	20 167	21 409	21 265	20 294 e	Corée
250	208	99	81	40	7	5	6	7	5 e	Luxembourg
26 517	29 742	26 412	26 329	29 084	31 171	29 598	30 076	28 858	25 091 e	Mexique
1 232	1 218	1 177	1 224	468	434	402	319	418	277 e	Pays-Bas
150	186	172	198	165	35	31	94	84	147 e	Nouvelle Zélande
234	366	306	399	344	336	299	209	268	188 e	Norvège
2 536	2 633	2 454	2 352	2 294	2 444	2 214	2 054	2 060	1 941 e	Pologne
3 884	3 400	4 363	3 242	3 362	4 291	4 135	3 438	3 459	3 795 e	Portugal
1 040	920	642	689	680	657	424	208	454	543 e	République Slovaque
8 769	9 691	10 708	8 164	7 847	8 452	9 574	8 809	8 650	9 483 e	Espagne
2 529	2 972	2 770	3 659	2 615	2 670	2 537	1 928	1 985	2 142 e	Suède
417	451	420	430	316	386	269	193	224	158 e	Suisse
7 762	7 301	8 101	8 532	8 374	8 082	7 662	7 817	7 536	8 171 e	Turquie
11 982	10 800	9 885	9 178	6 910	6 387	6 199	6 589	7 991	5 701 e	Royaume-Uni
40 536	38 504	26 350	27 396	27 769	33 702	28 613	31 683	35 630	13 047 e	Etats-Unis
240 730	**245 614**	**228 232**	**222 005**	**216 670**	**218 123**	**205 666**	**203 445**	**203 487**	**175 407 e**	**OCDE TOTAL**
74 341	**74 473**	**58 288**	**59 013**	**62 787**	**72 076**	**64 484**	**68 270**	**72 325**	**45 182 e**	**OCDE Amérique du N.**
64 050	**69 780**	**66 756**	**64 801**	**00 035**	**53 665**	**54 602**	**53 967**	**49 802**	**49 522 e**	**OCDE Pacifique**
102 339	**101 361**	**103 188**	**98 191**	**90 848**	**92 382**	**86 580**	**01 208**	**81 360**	**80 703 e**	**OCDE Europe**
210 511	**212 191**	**198 607**	**192 514**	**184 488**	**183 751**	**173 353**	**171 044**	**172 030**	**147 739 e**	**AIE Total**
85 496	**85 149**	**87 361**	**81 997**	**75 134**	**76 750**	**72 360**	**68 226**	**68 476**	**68 220 e**	**UE15**

* Gross consumption includes consumption of fuels by refineries and excludes international marine bunkers. / *La consommation brute comprend la consommation d'énergie des raffineries mais non les soutages maritimes internationaux.*

Table 13: Gross Consumption of Lubricants *
Tableau 13 : Consommation brute de lubrifiants *

000 metric tons

	1973	1974	1975	1976	1977	1978	1979	1980	1981	1982
Australia	410	424	395	389	393	394	416	425	418	432
Austria	263	182	174	196	193	206	218	193	188	187
Belgium	220	220	190	196	199	195	214	197	194	196
Canada	874	886	850	873	951	909	941	933	910	775
Czech Republic	228	226	186	173	190	184	197	186	178	178
Denmark	106	105	71	87	86	84	83	74	70	74
Finland	105	102	97	98	91	95	112	108	104	112
France	1 051	1 009	932	1 007	1 020	932	961	940	893	879
Germany	1 166	1 057	1 031	1 087	1 126	1 150	1 212	1 561	1 496	1 448
Greece	53	60	61	105	83	91	99	100	103	92
Hungary	133	146	163	172	183	202	189	173	192	148
Iceland	7	13	6	6	8	7	8	6	7	6
Ireland	40	44	37	27	30	50	58	40	49	42
Italy	395	458	527	545	441	489	630	657	736	688
Japan	2 492	2 133	1 969	1 806	1 825	1 850	1 973	1 896	1 886	1 736
Korea	-	-	-	-	-	-	-	-	-	-
Luxembourg	12	12	9	9	9	10	10	10	9	9
Mexico	-	-	-	-	-	-	-	-	-	-
Netherlands	224	217	198	230	184	185	209	191	184	141
New Zealand	72	60	57	62	63	66	44	48	65	65
Norway	84	73	76	74	72	63	86	62	59	60
Poland	346	359	391	455	504	502	506	618	544	563
Portugal	70	74	68	75	87	83	89	100	99	95
Slovak Republic	-	-	-	-	-	-	-	-	-	-
Spain	373	356	368	393	431	403	352	328	320	357
Sweden	177	197	191	197	186	180	201	192	185	180
Switzerland	101	88	71	80	74	81	89	79	79	73
Turkey	176	137	177	211	230	247	208	181	178	206
United Kingdom	1 185	1 045	992	1 010	1 029	1 024	1 035	899	841	831
United States	8 362	8 026	6 972	7 947	8 310	8 914	9 327	8 274	7 989	7 175
OECD TOTAL	**18 725**	**17 709**	**16 259**	**17 510**	**17 998**	**18 596**	**19 467**	**18 471**	**17 976**	**16 748**
OECD North America	**9 236**	**8 912**	**7 822**	**8 820**	**9 261**	**9 823**	**10 268**	**9 207**	**8 899**	**7 950**
OECD Pacific	**2 974**	**2 617**	**2 421**	**2 257**	**2 281**	**2 310**	**2 433**	**2 369**	**2 369**	**2 233**
OECD Europe	**6 515**	**6 180**	**6 016**	**6 433**	**6 456**	**6 463**	**6 766**	**6 895**	**6 708**	**6 565**
IEA Total	**18 372**	**17 337**	**15 862**	**17 049**	**17 486**	**18 087**	**18 953**	**17 847**	**17 425**	**16 179**
EU15	**5 440**	**5 138**	**4 946**	**5 262**	**5 195**	**5 177**	**5 483**	**5 590**	**5 471**	**5 331**

* Gross consumption includes consumption of fuels by refineries and excludes international marine bunkers. / *La consommation brute comprend la consommation d'énergie des raffineries mais non les soutages maritimes internationaux.*

Table 13: Gross Consumption of Lubricants * (continued)
*Tableau 13 : Consommation brute de lubrifiants * (suite)*

000 metric tons

1983	1984	1985	1986	1987	1988	1989	1990	1991	1992	
390	401	402	410	413	433	456	462	427	426	Australie
272	243	170	187	205	171	128	164	144	120	Autriche
191	202	206	203	205	213	222	234	180	198	Belgique
778	818	811	848	905	816	821	818	765	759	Canada
143	187	186	192	195	206	203	197	177	206	République Tchèque
76	77	77	82	79	79	76	75	68	67	Danemark
103	107	108	112	117	116	119	118	95	94	Finlande
836	837	832	847	873	906	928	933	898	876	France
1 444	1 444	1 433	1 512	1 454	1 517	1 562	1 467	1 230	1 186	Allemagne
93	99	108	104	112	122	142	132	86	87	Grèce
167	162	164	182	175	168	260	133	73	26	Hongrie
6	6	6	6	7	7	6	7	7	7	Islande
30	52	48	47	46	43	42	57	40	38	Irlande
701	633	526	572	785	699	704	689	646	764	Italie
1 767	1 854	1 917	1 874	1 866	1 958	2 040	2 170	2 139	2 105	Japon
-	-	-	-	-	-	-	-	-	-	Corée
8	9	9	10	10	11	11	10	10	10	Luxembourg
-	-	-	-	-	-	-	-	-	-	Mexique
152	157	157	143	147	120	167	178	177	187	Pays-Bas
47	56	60	39	46	37	54	29	42	32	Nouvelle Zélande
65	79	98	98	100	99	106	110	104	97	Norvège
556	565	594	569	566	524	486	343	236	225	Pologne
93	94	98	101	103	116	121	125	125	121	Portugal
-	-	-	-	-	-	-	-	-	-	République Slovaque
367	336	400	244	470	362	273	380	296	276	Espagne
195	220	218	216	222	223	230	161	127	110	Suède
72	67	66	63	77	77	78	73	70	69	Suisse
207	247	215	272	252	283	278	297	310	270	Turquie
821	821	816	803	828	849	839	822	759	786	Royaume-Uni
7 477	8 051	7 486	7 407	8 340	8 074	8 267	8 487	7 583	7 756	Etats-Unis
17 057	**17 824**	**17 211**	**17 143**	**18 598**	**18 229**	**18 619**	**18 671**	**16 814**	**16 898**	**OCDE TOTAL**
8 255	**8 869**	**8 297**	**8 265**	**9 245**	**8 890**	**9 088**	**9 305**	**8 348**	**8 515**	**OCDE Amérique du N.**
2 204	**2 311**	**2 379**	**2 323**	**2 325**	**2 428**	**2 550**	**2 661**	**2 608**	**2 563**	**OCDE Pacifique**
6 598	**6 644**	**6 535**	**6 565**	**7 028**	**6 911**	**6 981**	**6 705**	**5 858**	**5 820**	**OCDE Europe**
16 495	**17 253**	**16 611**	**16 568**	**18 025**	**17 698**	**18 127**	**18 321**	**16 571**	**16 666**	**AIE Total**
5 382	**5 331**	**5 206**	**5 183**	**5 656**	**5 547**	**5 564**	**5 545**	**4 881**	**4 920**	**UE15**

* Gross consumption includes consumption of fuels by refineries and excludes international marine bunkers. / *La consommation brute comprend la consommation d'énergie des raffineries mais non les soutages maritimes internationaux.*

Table 13: Gross Consumption of Lubricants * (continued)
*Tableau 13 : Consommation brute de lubrifiants * (suite)*

000 metric tons

	1993	1994	1995	1996	1997	1998	1999	2000	2001	2002
Australia	440	448	444	461	462	462	449	471	497	..
Austria	92	89	87	105	108	106	103	109	108	..
Belgium	173	157	167	180	182	176	202	174	104	..
Canada	798	870	886	886	965	925	939	965	903	..
Czech Republic	152	187	168	144	128	199	191	228	183	..
Denmark	67	66	66	62	60	54	56	52	50	..
Finland	97	104	103	93	97	94	92	88	87	..
France	840	856	866	840	857	889	875	889	860	..
Germany	1 132	1 151	1 171	1 129	1 168	1 147	1 160	1 122	1 058	..
Greece	85	86	74	67	80	52	64	56	79	..
Hungary	82	165	339	145	175	235	169	247	141	..
Iceland	7	7	8	7	7	7	7	7	6	..
Ireland	39	38	18	46	36	33	34	34	33	..
Italy	830	686	703	753	700	735	615	826	764	..
Japan	2 031	2 081	2 103	2 126	2 159	2 078	2 027	2 002	1 949	..
Korea	-	-	-	-	527	398	447	535	504	..
Luxembourg	10	10	10	10	9	9	8	7	7	..
Mexico	-	-	-	-	-	-	-	-	-	..
Netherlands	178	176	188	188	180	191	192	184	176	..
New Zealand	36	27	20	10	7	14	24	17	14	..
Norway	83	74	86	87	90	82	82	89	72	..
Poland	221	234	232	200	216	311	297	350	377	..
Portugal	109	109	121	119	107	118	117	111	104	..
Slovak Republic	32	50	55	55	55	79	93	55	78	..
Spain	183	273	264	302	375	395	437	432	470	..
Sweden	117	182	133	179	251	261	347	356	381	..
Switzerland	68	64	71	67	73	74	71	77	66	..
Turkey	287	290	339	371	406	340	420	460	335	..
United Kingdom	806	795	895	864	872	813	790	804	796	..
United States	7 948	8 263	8 114	7 793	8 271	8 657	8 835	8 567	7 907	..
OECD TOTAL	**16 943**	**17 538**	**17 731**	**17 289**	**18 623**	**18 934**	**19 143**	**19 314**	**18 109**	**..**
OECD North America	**8 746**	**9 133**	**9 000**	**8 679**	**9 236**	**9 582**	**9 774**	**9 532**	**8 810**	**..**
OECD Pacific	**2 507**	**2 556**	**2 567**	**2 597**	**3 155**	**2 952**	**2 947**	**3 025**	**2 964**	**..**
OECD Europe	**5 690**	**5 849**	**6 164**	**6 013**	**6 232**	**6 400**	**6 422**	**6 757**	**6 335**	**..**
IEA Total	**16 683**	**17 247**	**17 436**	**17 027**	**18 345**	**18 537**	**18 746**	**18 902**	**17 648**	**..**
EU15	**4 758**	**4 778**	**4 866**	**4 937**	**5 082**	**5 073**	**5 092**	**5 244**	**5 077**	**..**

* Gross consumption includes consumption of fuels by refineries and excludes international marine bunkers. / *La consommation brute comprend la consommation d'énergie des raffineries mais non les soutages maritimes internationaux.*

Table 14: Gross Consumption of Bitumen *
Tableau 14 : Consommation brute de bitume *

000 metric tons

1973	1974	1975	1976	1977	1978	1979	1980	1981	1982	
474	479	464	509	538	568	594	584	525	528	Australie
613	657	608	615	625	607	628	579	532	496	Autriche
538	476	513	555	493	463	516	459	396	391	Belgique
3 226	3 039	2 919	2 896	3 223	3 227	3 266	3 269	3 005	2 531	Canada
623	615	506	510	460	512	513	482	453	470	République Tchèque
373	294	271	315	376	333	319	304	244	247	Danemark
352	354	337	306	315	318	336	316	292	312	Finlande
3 395	3 154	2 984	3 130	2 846	2 867	2 913	2 773	2 643	2 356	France
4 656	4 648	4 241	3 721	3 794	3 891	3 936	3 861	3 448	3 345	Allemagne
194	85	152	152	125	107	118	120	152	158	Grèce
476	511	573	562	591	612	571	559	562	534	Hongrie
15	8	12	6	10	5	11	11	14	12	Islande
115	117	113	83	87	119	108	85	99	84	Irlande
2 057	1 819	1 865	1 800	1 738	1 742	1 923	1 990	1 715	1 899	Italie
5 265	4 631	4 065	4 110	4 589	5 155	5 140	4 759	4 566	4 535	Japon
219	132	190	170	240	317	324	293	313	329	Corée
16	20	23	23	25	29	24	29	30	28	Luxembourg
-	-	-	-	-	-	-	-	-	-	Mexique
765	694	699	729	626	662	685	586	503	415	Pays-Bas
121	125	79	97	113	126	93	104	103	98	Nouvelle Zélande
258	246	262	280	285	285	260	257	249	273	Norvège
824	966	1 037	1 078	1 207	1 315	1 215	1 320	1 022	1 042	Pologne
80	64	70	76	112	134	152	158	179	188	Portugal
-	-	-	-	-	-	-	-	-	-	République Slovaque
916	863	886	904	853	861	790	752	735	846	Espagne
781	873	847	661	568	422	662	561	518	546	Suède
449	428	343	351	323	335	338	340	325	322	Suisse
233	236	287	284	243	431	330	300	345	350	Turquie
2 458	2 240	2 089	1 867	1 847	1 887	1 977	1 826	1 666	1 956	Royaume-Uni
31 425	28 979	25 257	24 862	26 314	28 938	28 757	24 051	20 632	20 658	Etats-Unis
60 917	**56 753**	**51 692**	**50 652**	**52 566**	**56 268**	**56 499**	**50 728**	**45 266**	**44 949**	**OCDE TOTAL**
34 651	**32 018**	**28 176**	**27 758**	**29 537**	**32 165**	**32 023**	**27 320**	**23 637**	**23 189**	**OCDE Amérique du N.**
6 079	**5 367**	**4 798**	**4 886**	**5 480**	**6 166**	**6 151**	**5 740**	**5 507**	**5 490**	**OCDE Pacifique**
20 187	**19 368**	**18 718**	**18 008**	**17 549**	**17 937**	**18 325**	**17 668**	**16 122**	**16 270**	**OCDE Europe**
60 078	**55 779**	**50 643**	**49 568**	**51 349**	**54 948**	**55 273**	**49 397**	**44 230**	**43 895**	**AIE Total**
17 309	**16 358**	**15 698**	**14 937**	**14 430**	**14 442**	**15 087**	**14 399**	**13 152**	**13 267**	**UE15**

* Gross consumption includes consumption of fuels by refineries and excludes international marine bunkers. / *La consommation brute comprend la consommation d'énergie des raffineries mais non les soutages maritimes internationaux.*

Table 14: Gross Consumption of Bitumen * (continued)
*Tableau 14 : Consommation brute de bitume * (suite)*

000 metric tons

	1983	1984	1985	1986	1987	1988	1989	1990	1991	1992
Australia	504	571	594	600	578	569	602	656	634	610
Austria	502	545	513	549	532	502	481	538	492	441
Belgium	334	288	274	294	301	352	336	355	378	356
Canada	2 637	2 464	2 803	2 862	3 112	2 821	3 015	3 131	2 849	2 901
Czech Republic	483	478	500	523	522	521	532	502	332	367
Denmark	252	284	270	274	277	267	284	216	211	218
Finland	341	336	343	410	402	458	490	475	418	411
France	2 139	2 021	2 138	2 380	2 539	2 788	2 989	3 012	3 076	3 079
Germany	3 169	3 358	3 061	3 124	2 970	2 961	3 011	2 929	3 349	3 688
Greece	211	247	289	207	170	186	233	204	223	235
Hungary	480	435	460	465	479	466	428	394	334	223
Iceland	7	12	14	19	17	15	13	19	17	18
Ireland	63	101	86	92	93	85	109	135	166	151
Italy	1 863	1 815	1 985	2 095	2 366	2 510	2 447	2 414	2 242	2 280
Japan	4 905	5 149	4 957	5 584	5 738	5 963	6 015	6 220	6 043	6 450
Korea	326	420	470	479	462	566	569	841	1 154	1 640
Luxembourg	25	20	8	6	1	3	5	9	6	5
Mexico	-	-	-	-	-	-	-	-	-	-
Netherlands	449	510	500	540	554	528	488	474	404	413
New Zealand	104	137	116	128	132	139	129	141	101	135
Norway	305	327	318	365	368	342	365	369	388	351
Poland	1 196	1 141	1 100	1 310	1 229	1 215	1 078	609	524	552
Portugal	150	140	151	183	224	250	279	278	363	372
Slovak Republic	-	-	-	-	-	-	1	1	-	1
Spain	821	735	870	908	880	1 045	1 405	1 566	1 948	1 995
Sweden	515	476	460	530	537	571	505	601	448	483
Switzerland	290	286	286	316	292	283	307	310	320	287
Turkey	415	461	500	631	840	577	478	692	874	841
United Kingdom	1 987	1 900	1 887	2 019	2 162	2 342	2 470	2 647	2 932	3 841
United States	22 293	24 801	25 679	27 095	28 207	28 444	27 374	29 249	26 943	27 599
OECD TOTAL	**46 766**	**49 458**	**50 632**	**53 988**	**55 984**	**56 769**	**56 438**	**58 987**	**57 169**	**59 943**
OECD North America	**24 930**	**27 265**	**28 482**	**29 957**	**31 319**	**31 265**	**30 389**	**32 380**	**29 792**	**30 500**
OECD Pacific	**5 839**	**6 277**	**6 137**	**6 791**	**6 910**	**7 237**	**7 315**	**7 858**	**7 932**	**8 835**
OECD Europe	**15 997**	**15 916**	**16 013**	**17 240**	**17 755**	**18 267**	**18 734**	**18 749**	**19 445**	**20 608**
IEA Total	**45 563**	**48 305**	**49 518**	**52 659**	**54 738**	**55 539**	**55 346**	**58 358**	**56 628**	**59 372**
EU15	**12 821**	**12 776**	**12 835**	**13 611**	**14 008**	**14 848**	**15 532**	**15 853**	**16 656**	**17 968**

* Gross consumption includes consumption of fuels by refineries and excludes international marine bunkers. / *La consommation brute comprend la consommation d'énergie des raffineries mais non les soutages maritimes internationaux.*

Table 14: Gross Consumption of Bitumen * (continued)
*Tableau 14 : Consommation brute de bitume * (suite)*

000 metric tons

1993	1994	1995	1996	1997	1998	1999	2000	2001	2002	
674	729	711	714	747	736	738	804	664	..	Australie
410	446	440	495	542	576	560	587	618	..	Autriche
340	401	362	334	365	342	399	395	335	..	Belgique
2 821	3 100	3 505	3 652	4 124	4 207	4 261	4 038	3 242	..	Canada
274	315	357	250	184	368	364	378	373	..	République Tchèque
226	227	938	1 561	1 710	1 391	1 472	1 459	1 286	..	Danemark
382	401	378	336	353	336	316	303	258	..	Finlande
2 893	3 209	3 103	2 948	3 272	3 248	3 369	3 375	3 410	..	France
3 385	3 934	3 576	3 397	3 514	3 381	3 625	3 348	3 135	..	Allemagne
249	253	299	303	307	345	354	406	414	..	Grèce
232	273	293	314	233	335	227	228	243	..	Hongrie
21	22	23	24	20	30	25	30	32	..	Islande
122	119	106	142	160	161	138	133	163	..	Irlande
2 060	2 058	2 194	2 205	2 350	2 493	2 522	2 428	2 564	..	Italie
6 533	6 629	6 538	6 409	6 359	5 941	5 644	7 158	6 803	..	Japon
1 555	1 550	1 567	1 764	1 987	1 620	1 693	1 692	1 815	..	Corée
2	1	13	12	8	9	12	5	4	..	Luxembourg
-	-	-	-	-	-	-	-	-	..	Mexique
412	419	399	358	338	299	358	287	381	..	Pays-Bas
145	172	164	166	165	114	168	154	124	..	Nouvelle Zélande
356	316	378	381	379	389	361	325	297	..	Norvège
523	435	285	628	665	610	712	892	725	..	Pologne
453	397	475	425	630	745	575	552	800	..	Portugal
162	204	318	318	318	334	393	432	100	..	République Slovaque
1 778	2 032	2 050	1 339	1 805	2 092	1 995	1 866	1 815	..	Espagne
566	513	491	500	493	516	810	487	438	..	Suède
265	280	284	265	265	277	298	304	289	..	Suisse
1 267	871	970	1 120	1 325	1 702	1 315	1 260	1 149	..	Turquie
3 939	3 823	3 696	3 018	2 199	1 967	1 928	1 975	1 923	..	Royaume-Uni
28 704	29 379	29 517	29 403	30 458	31 424	32 969	31 808	31 290	..	Etats-Unis
60 749	**62 508**	**63 430**	**62 781**	**65 275**	**65 988**	**67 601**	**67 109**	**64 690**	..	**OCDE TOTAL**
31 525	**32 479**	**33 022**	**33 056**	**34 582**	**35 631**	**37 230**	**35 846**	**34 532**	..	**OCDE Amérique du N.**
8 907	**9 080**	**8 980**	**9 053**	**9 258**	**8 411**	**8 243**	**9 808**	**9 406**	..	**OCDE Pacifique**
20 317	**20 949**	**21 428**	**20 673**	**21 435**	**21 946**	**22 128**	**21 455**	**20 752**	..	**OCDE Europe**
60 043	**61 847**	**62 804**	**61 811**	**64 272**	**65 014**	**66 471**	**65 755**	**63 833**	..	**AIE Total**
17 217	**18 233**	**18 520**	**17 373**	**18 046**	**17 901**	**18 433**	**17 606**	**17 544**	..	**UE15**

* Gross consumption includes consumption of fuels by refineries and excludes international marine bunkers. / *La consommation brute comprend la consommation d'énergie des raffineries mais non les soutages maritimes internationaux.*

Table 15: Total Products Consumption in the Transformation Sector *
*Tableau 15 : Consommation de produits pétroliers dans le secteur transformation **

000 metric tons

	1973	1974	1975	1976	1977	1978	1979	1980	1981	1982
Australia	674	1 527	1 446	1 472	1 403	1 407	1 442	1 263	1 223	1 317
Austria	1 304	875	1 025	1 372	1 144	1 303	1 294	1 225	941	874
Belgium	5 108	4 890	3 601	3 802	3 300	3 783	3 876	4 023	3 132	2 651
Canada	2 615	2 750	2 820	3 268	3 477	3 443	3 453	3 290	2 424	2 242
Czech Republic	1 201	1 469	1 677	1 818	1 849	1 886	1 830	1 584	1 940	1 045
Denmark	3 060	3 253	3 004	2 716	2 806	2 377	2 210	1 360	806	1 589
Finland	1 554	1 638	1 644	2 334	1 951	1 700	1 583	1 572	1 475	1 073
France	16 845	15 694	12 643	17 939	12 953	14 488	14 998	12 763	9 529	8 328
Germany	14 944	11 710	11 041	13 089	12 237	14 062	14 775	14 878	13 173	12 447
Greece	1 810	1 702	1 583	1 478	1 470	1 910	1 972	2 218	1 785	1 681
Hungary	1 403	1 675	2 060	1 907	2 284	2 798	2 083	1 641	1 609	1 615
Iceland	19	18	19	14	14	12	12	10	12	5
Ireland	1 256	1 411	1 377	1 507	1 551	1 790	1 843	1 602	1 173	789
Italy	23 374	24 486	23 567	24 640	22 258	24 255	24 610	26 184	24 971	24 102
Japan	72 001	69 916	66 552	67 934	79 149	78 447	76 563	68 172	63 483	53 696
Korea	2 972	3 468	4 439	4 911	5 723	6 501	6 861	6 704	7 282	7 450
Luxembourg	118	107	82	76	57	55	44	24	18	28
Mexico	4 086	5 012	5 945	6 696	7 143	8 587	8 487	10 108	9 865	10 897
Netherlands	1 623	997	943	874	1 047	4 487	9 610	9 835	12 290	5 410
New Zealand	438	495	237	344	208	69	33	23	22	42
Norway	46	19	33	43	66	55	105	110	108	147
Poland	1 306	1 437	1 575	1 828	2 137	2 417	2 700	2 698	2 380	2 155
Portugal	557	609	959	1 217	821	837	1 239	1 625	1 931	1 942
Slovak Republic	441	472	529	571	588	607	624	672	623	515
Spain	5 640	7 464	8 136	11 086	6 608	6 799	6 661	9 432	9 141	7 336
Sweden	2 614	3 067	1 993	2 899	2 930	3 228	3 620	2 989	2 007	1 889
Switzerland	549	429	454	480	437	366	349	142	110	126
Turkey	1 341	1 340	1 427	1 639	1 899	2 086	1 533	1 610	1 722	1 827
United Kingdom	25 624	23 873	17 924	15 586	15 638	16 979	15 807	11 858	10 579	12 666
United States	85 331	81 468	76 874	84 313	94 908	96 651	79 905	64 032	53 418	38 045
OECD TOTAL	**279 854**	**273 271**	**255 609**	**277 853**	**288 056**	**303 385**	**290 122**	**263 647**	**239 172**	**203 929**
OECD North America	**92 032**	**89 230**	**85 639**	**94 277**	**105 528**	**108 681**	**91 845**	**77 430**	**65 707**	**51 184**
OECD Pacific	**76 085**	**75 406**	**72 674**	**74 661**	**86 483**	**86 424**	**84 899**	**76 162**	**72 010**	**62 505**
OECD Europe	**111 737**	**108 635**	**97 296**	**108 915**	**96 045**	**108 280**	**113 378**	**110 055**	**101 455**	**90 240**
IEA Total	**274 002**	**266 332**	**247 541**	**268 744**	**278 174**	**291 762**	**278 299**	**250 159**	**226 292**	**190 357**
EU15	**105 431**	**101 776**	**89 522**	**100 615**	**86 771**	**98 053**	**104 142**	**101 588**	**92 951**	**82 805**

* Consumption in the transformation sector includes backflows from the petrochemical industry. / *La consommation du secteur transformation comprend les retours de l'industrie pétrochimique.*

Table 15: Total Products Consumption in the Transformation Sector * (continued)
*Tableau 15 : Consommation de produits pétroliers dans le secteur transformation * (suite)*

000 metric tons

1983	1984	1985	1986	1987	1988	1989	1990	1991	1992	
1 151	1 123	978	841	638	511	680	911	935	651	Australie
728	623	531	575	544	418	425	382	465	364	Autriche
1 625	987	834	529	469	335	338	265	322	281	Belgique
1 752	1 564	1 658	1 398	2 241	2 634	4 070	3 689	3 125	3 418	Canada
1 022	1 234	1 284	1 420	1 458	1 448	1 245	1 079	1 007	1 029	République Tchèque
1 245	1 039	1 004	795	602	480	481	334	378	357	Danemark
802	691	917	988	943	782	638	663	543	547	Finlande
6 036	4 145	3 586	3 209	3 340	3 388	4 562	3 608	4 776	3 929	France
11 296	10 186	10 466	10 990	11 461	10 829	10 304	10 095	10 124	9 968	Allemagne
1 630	1 714	1 732	1 403	1 525	1 473	1 619	1 788	1 941	1 922	Grèce
1 642	1 887	2 213	1 943	1 951	1 363	1 215	1 118	1 638	1 874	Hongrie
3	2	2	2	2	1	1	1	1	1	Islande
584	537	569	1 259	657	256	185	352	576	592	Irlande
23 628	20 077	19 297	20 795	23 708	23 765	26 145	26 216	26 540	28 449	Italie
53 066	54 305	44 112	44 431	42 881	48 976	52 503	57 734	56 373	57 828	Japon
7 302	5 690	4 264	3 045	1 132	2 645	3 235	4 434	6 187	7 717	Corée
14	7	5	21	5	18	6	3	8	18	Luxembourg
11 818	12 764	13 084	13 928	15 343	15 968	16 902	16 765	16 961	16 616	Mexique
4 552	3 139	3 303	6 966	3 605	801	745	696	756	719	Pays-Bas
64	22	38	13	6	5	2	3	6	40	Nouvelle Zélande
142	166	143	122	164	152	147	101	101	74	Norvège
2 130	2 147	2 151	2 057	2 277	2 074	2 024	1 911	1 679	1 535	Pologne
2 171	2 084	1 605	1 926	1 303	913	2 764	2 268	2 383	3 248	Portugal
523	487	486	451	428	392	359	317	466	401	République Slovaque
6 281	3 322	2 304	2 101	2 404	2 468	2 897	2 804	3 670	4 484	Espagne
1 305	1 574	2 144	1 795	1 588	1 062	805	1 094	1 479	2 236	Suède
122	120	135	148	147	126	119	109	171	203	Suisse
1 943	2 013	1 986	2 109	1 771	1 176	1 340	1 299	1 005	1 270	Turquie
9 557	25 443	12 916	9 302	7 184	7 496	8 260	9 025	8 105	8 612	Royaume-Uni
37 598	32 480	26 336	35 622	30 388	37 847	40 570	28 733	28 454	23 087	Etats-Unis
191 732	**191 572**	**160 083**	**170 184**	**160 165**	**169 802**	**184 586**	**177 797**	**180 175**	**181 470**	**OCDE TOTAL**
51 168	**40 008**	**41 078**	**50 948**	**47 972**	**56 449**	**61 542**	**49 187**	**48 540**	**43 121**	**OCDE Amérique du N.**
61 583	**61 140**	**49 392**	**48 330**	**44 657**	**52 137**	**56 420**	**63 082**	**63 501**	**66 236**	**OCDE Pacifique**
78 981	**83 624**	**69 613**	**70 906**	**67 536**	**61 216**	**66 624**	**65 528**	**60 104**	**72 113**	**OCDE Europe**
177 258	**176 172**	**144 360**	**153 746**	**142 115**	**151 367**	**165 300**	**158 803**	**161 068**	**162 917**	**AIE Total**
71 454	**75 568**	**61 213**	**62 654**	**59 338**	**54 484**	**60 174**	**59 593**	**62 066**	**65 726**	**UE15**

* Consumption in the transformation sector includes backflows from the petrochemical industry. / *La consommation du secteur transformation comprend les retours de l'industrie pétrochimique.*

Table 15: Total Products Consumption in the Transformation Sector * (continued)
*Tableau 15 : Consommation de produits pétroliers dans le secteur transformation * (suite)*

000 metric tons

	1993	1994	1995	1996	1997	1998	1999	2000	2001	2002
Australia	651	659	752	745	587	594	509	525	612	..
Austria	719	662	585	543	647	734	644	361	487	..
Belgium	262	372	232	229	268	401	293	178	278	..
Canada	2 560	2 132	2 383	1 889	2 701	3 934	3 251	3 003	3 655	..
Czech Republic	660	685	695	742	604	468	404	366	360	..
Denmark	354	780	1 124	1 833	1 764	1 561	1 520	1 420	1 366	..
Finland	584	699	652	891	850	877	941	798	819	..
France	3 070	3 040	3 317	3 345	3 498	4 236	3 908	3 531	3 120	..
Germany	9 397	9 089	8 287	8 182	8 288	8 530	8 215	8 839	8 602	..
Greece	1 972	1 916	2 085	2 045	1 967	1 884	1 945	2 043	1 934	..
Hungary	2 325	2 349	2 003	1 674	1 944	1 923	1 613	1 331	1 326	..
Iceland	1	1	3	2	2	7	11	1	1	..
Ireland	582	669	639	655	846	1 146	1 435	1 071	1 275	..
Italy	28 043	28 548	30 562	29 284	28 334	26 951	22 161	22 266	19 673	..
Japan	52 061	59 765	54 226	53 683	49 339	42 383	42 608	33 009	26 533	..
Korea	7 282	9 350	11 154	10 361	15 187	8 853	9 109	11 001	11 675	..
Luxembourg	12	5	2	7	4	-	-	-	-	..
Mexico	16 831	20 064	17 571	18 107	20 786	22 932	22 487	23 861	22 762	..
Netherlands	687	735	863	852	777	739	1 370	616	647	..
New Zealand	1	1	1	-	-	-	-	-	-	..
Norway	74	82	49	85	79	90	117	67	71	..
Poland	1 471	1 497	758	785	787	545	524	456	479	..
Portugal	2 461	1 770	2 482	1 408	1 503	2 313	2 338	1 758	2 007	..
Slovak Republic	894	690	577	617	460	468	262	111	372	..
Spain	3 341	4 182	4 014	3 266	3 053	3 681	5 358	4 885	5 474	..
Sweden	2 389	2 655	2 447	3 347	2 348	2 610	2 556	1 882	1 823	..
Switzerland	85	60	62	83	43	123	40	14	15	..
Turkey	1 778	2 046	1 917	1 995	2 136	2 455	2 379	2 957	2 731	..
United Kingdom	6 951	5 093	5 308	4 893	2 936	2 632	3 046	5 233	5 828	..
United States	25 378	23 413	15 776	17 597	28 772	39 780	37 040	31 470	30 142	..
OECD TOTAL	**172 876**	**183 009**	**170 526**	**169 145**	**180 510**	**182 850**	**176 084**	**163 053**	**154 067**	**..**
OECD North America	**44 769**	**45 609**	**35 730**	**37 593**	**52 259**	**66 646**	**62 778**	**58 334**	**56 559**	**..**
OECD Pacific	**59 995**	**69 775**	**66 133**	**64 789**	**65 113**	**51 830**	**52 226**	**44 535**	**38 820**	**..**
OECD Europe	**68 112**	**67 625**	**68 663**	**66 763**	**63 138**	**64 374**	**61 080**	**60 184**	**58 688**	**..**
IEA Total	**153 679**	**160 757**	**151 617**	**149 634**	**158 475**	**158 898**	**152 800**	**138 624**	**130 453**	**..**
EU15	**60 824**	**60 215**	**62 599**	**60 780**	**57 083**	**58 295**	**55 730**	**54 881**	**53 333**	**..**

* Consumption in the transformation sector includes backflows from the petrochemical industry. / *La consommation du secteur transformation comprend les retours de l'industrie pétrochimique.*

Table 16: Total Products Consumption in the Energy Sector *
*Tableau 16 : Consommation de produits pétroliers dans le secteur de l'énergie **

000 metric tons

1973	1974	1975	1976	1977	1978	1979	1980	1981	1982	
405	2 591	2 541	2 635	2 815	2 905	2 632	2 356	2 228	2 144	Australie
183	151	262	345	331	341	421	514	503	447	Autriche
-	-	530	1 452	1 948	1 800	1 796	2 006	1 729	1 470	Belgique
5 726	5 740	5 752	5 592	5 960	6 669	6 788	6 650	6 178	5 566	Canada
25	25	119	147	152	173	205	190	196	163	République Tchèque
-	-	361	315	314	309	357	290	311	264	Danemark
465	515	457	460	531	503	542	515	520	456	Finlande
6 637	6 579	6 069	6 579	6 620	6 516	6 484	6 211	5 613	5 098	France
9 179	9 043	8 750	8 898	8 901	8 810	9 427	9 125	7 383	7 024	Allemagne
350	457	391	380	387	391	485	442	502	472	Grèce
356	388	408	421	424	467	534	491	510	442	Hongrie
-	-	-	-	-	-	-	-	-	-	Islande
57	53	104	80	98	76	70	58	23	22	Irlande
4 656	4 866	4 836	5 162	5 262	5 278	4 888	5 071	4 953	4 894	Italie
10 405	10 439	9 810	10 773	10 722	10 380	10 439	9 870	9 092	8 819	Japon
737	765	833	923	1 059	1 139	1 213	1 201	1 153	1 106	Corée
-	-	-	-	-	-	-	-	-	-	Luxembourg
4 684	4 413	2 812	3 092	4 593	3 849	3 723	5 244	6 025	6 165	Mexique
3 645	3 391	2 897	3 335	3 256	2 994	3 238	3 001	2 725	2 899	Pays-Bas
130	144	122	138	121	133	127	153	158	123	Nouvelle Zélande
84	183	267	357	421	429	376	370	374	333	Norvège
280	295	311	448	498	520	545	567	578	565	Pologne
341	357	346	339	320	372	396	378	484	415	Portugal
339	359	385	410	429	446	458	131	122	109	République Slovaque
1 630	2 118	2 281	2 506	2 545	2 376	2 308	2 315	2 142	2 106	Espagne
100	120	251	400	442	429	334	359	329	375	Suède
261	267	218	210	199	186	192	197	184	173	Suisse
584	785	777	797	805	852	789	855	808	912	Turquie
7 565	7 375	6 348	6 565	6 482	6 767	7 009	6 737	5 905	6 129	Royaume-Uni
64 586	61 499	45 592	46 841	49 508	50 678	49 825	47 370	44 633	45 358	Etats-Unis
123 410	**122 918**	**103 830**	**109 600**	**115 143**	**115 788**	**115 601**	**112 667**	**105 361**	**104 049**	**OCDE TOTAL**
74 996	**71 652**	**54 156**	**55 525**	**60 061**	**61 196**	**60 336**	**59 264**	**56 836**	**57 089**	**OCDE Amérique du N.**
11 677	**13 939**	**13 306**	**14 469**	**14 717**	**14 557**	**14 411**	**13 580**	**12 631**	**12 192**	**OCDE Pacifique**
36 737	**37 327**	**36 368**	**39 606**	**40 365**	**40 035**	**40 854**	**39 823**	**35 894**	**34 768**	**OCDE Europe**
118 107	**117 851**	**100 322**	**105 650**	**109 623**	**110 973**	**110 875**	**106 725**	**98 636**	**97 210**	**AIE Total**
34 808	**35 025**	**33 883**	**36 816**	**37 437**	**36 962**	**37 755**	**37 022**	**33 122**	**32 071**	**UE15**

* Consumption in the energy sector includes refinery fuel and distribution losses. / *La consommation du secteur de l'énergie comprend la consommation propre des raffineries et les pertes de distribution.*

Table 16: Total Products Consumption in the Energy Sector * (continued)
*Tableau 16 : Consommation de produits pétroliers dans le secteur de l'énergie * (suite)*

000 metric tons

	1983	1984	1985	1986	1987	1988	1989	1990	1991	1992
Australia	2 021	1 860	1 881	1 784	2 164	2 180	2 273	2 243	2 403	2 037
Austria	394	353	363	468	476	502	444	528	486	760
Belgium	1 404	1 319	1 165	1 408	1 435	1 447	1 367	1 335	1 573	1 491
Canada	5 030	5 379	5 206	5 547	5 800	6 166	6 375	6 914	6 811	7 228
Czech Republic	157	162	160	152	173	162	149	119	59	71
Denmark	287	304	290	321	308	300	320	301	328	371
Finland	495	459	480	434	476	465	411	415	425	423
France	5 241	4 445	4 523	4 351	4 319	4 629	4 698	4 830	4 975	5 083
Germany	6 473	6 129	5 947	5 441	5 292	5 538	5 491	5 510	5 727	6 028
Greece	387	352	331	543	617	704	798	789	804	780
Hungary	482	521	600	594	613	543	556	572	547	507
Iceland	-	-	-	-	-	-	-	-	-	-
Ireland	40	43	28	35	53	50	57	59	58	61
Italy	5 094	4 780	4 670	5 181	5 010	4 911	5 238	4 906	4 333	4 586
Japan	8 619	8 243	8 419	8 095	8 092	8 436	8 691	9 246	9 701	10 191
Korea	1 192	1 204	1 088	1 106	1 095	1 342	1 575	1 580	2 225	2 717
Luxembourg	-	-	-	-	-	-	-	-	-	-
Mexico	5 066	4 725	5 150	6 620	6 208	5 667	6 231	8 111	8 852	8 116
Netherlands	1 511	1 516	1 638	2 036	2 351	3 119	3 376	3 672	3 247	3 610
New Zealand	124	117	78	128	260	243	261	265	262	264
Norway	370	424	440	406	391	425	451	797	674	789
Poland	540	559	581	566	639	656	630	522	503	663
Portugal	522	408	331	410	373	436	544	484	404	458
Slovak Republic	123	123	123	118	118	110	138	109	122	118
Spain	2 272	2 359	2 534	2 937	2 939	3 266	3 307	3 377	3 525	3 778
Sweden	314	255	296	349	392	322	360	433	408	571
Switzerland	175	169	164	172	173	162	128	126	169	171
Turkey	939	1 021	1 035	1 067	1 259	1 300	1 261	1 177	1 129	1 210
United Kingdom	5 875	5 980	5 809	6 066	5 818	5 972	6 266	6 312	6 558	6 529
United States	46 276	47 223	47 123	49 657	51 043	52 977	53 244	52 462	50 785	51 793
OECD TOTAL	**101 423**	**100 432**	**100 453**	**105 992**	**107 887**	**112 030**	**114 640**	**117 194**	**117 093**	**120 404**
OECD North America	**56 372**	**57 327**	**57 479**	**61 824**	**63 051**	**64 810**	**65 850**	**67 487**	**66 448**	**67 137**
OECD Pacific	**11 956**	**11 424**	**11 466**	**11 113**	**11 611**	**12 201**	**12 800**	**13 334**	**14 591**	**15 209**
OECD Europe	**33 095**	**31 681**	**31 508**	**33 055**	**33 225**	**35 019**	**35 990**	**36 373**	**36 054**	**38 058**
IEA Total	**95 694**	**95 025**	**94 599**	**98 688**	**100 922**	**105 597**	**107 641**	**108 452**	**107 616**	**111 507**
EU15	**30 309**	**28 702**	**28 405**	**29 980**	**29 859**	**31 661**	**32 677**	**32 951**	**32 851**	**34 529**

* Consumption in the energy sector includes refinery fuel and distribution losses. / *La consommation du secteur de l'énergie comprend la consommation propre des raffineries et les pertes de distribution.*

Table 16: Total Products Consumption in the Energy Sector * (continued)
*Tableau 16 : Consommation de produits pétroliers dans le secteur de l'énergie * (suite)*

000 metric tons

1993	1994	1995	1996	1997	1998	1999	2000	2001	2002	
2 029	2 152	2 131	2 278	2 139	2 400	2 471	2 431	2 437	..	Australie
841	896	846	924	927	915	810	746	746	..	Autriche
1 429	1 541	1 403	1 679	1 672	1 609	1 547	1 531	1 559 e	..	Belgique
7 299	7 042	7 445	7 810	8 068	8 202	8 158	8 242	8 545	..	Canada
78	163	80	233	241	273	317	282	350	..	République Tchèque
377	378	424	442	365	319	329	330	235	..	Danemark
422	509	496	530	470	493	502	523	528	..	Finlande
5 123	5 221	5 295	5 433	5 588	5 624	5 359	5 417	5 254	..	France
6 413	6 672	6 491	6 532	6 413	6 699	6 692	6 765	6 523	..	Allemagne
777	833	843	924	942	968	875	1 050	1 068	..	Grèce
535	498	512	553	543	470	533	377	373	..	Hongrie
-	-	-	-	-	-	-	-	-	..	Islande
61	63	64	63	77	88	90	105	113	..	Irlande
4 871	4 854	5 085	5 007	5 266	5 653	5 113	4 922	5 167	..	Italie
10 554	10 963	11 175	11 393	11 877	11 994	11 985	9 033 e	8 794 e	..	Japon
2 881	3 062	3 313	3 660	4 494	4 514	4 563	4 632	4 627	..	Corée
-	-	-	-	-	-	-	-	-	..	Luxembourg
7 037	7 406	7 051	7 490	6 922	8 029	8 126	7 621	6 814	..	Mexique
3 736	3 784	4 035	4 044	3 690	3 519	3 081	3 051	3 158	..	Pays-Bas
285	281	251	255	286	292	279	282	283	..	Nouvelle Zélande
779	806	726	831	880	926	900	903	683	..	Norvège
649	614	1 184	1 323	1 138	1 250	1 124	1 197	1 330	..	Pologne
538	805	885	874	934	814	943	841	913	..	Portugal
94	27	31	53	42	72	16	13	407	..	République Slovaque
3 572	3 792	3 884	4 055	4 103	4 118	4 152	4 075	3 962 e	..	Espagne
554	527	524	523	447	492	537	552	475	..	Suède
196	226	219	234	227	233	240	215	232	..	Suisse
1 406	1 343	1 476	1 491	1 569	1 615	1 503	1 494	1 649	..	Turquie
6 851	7 017	7 169	7 455	7 224	7 130	6 443	5 805	5 444	..	Royaume-Uni
51 123	51 706	50 371	51 928	51 231	51 338	51 165	51 566	50 844	..	Etats-Unis
120 510	**123 181**	**123 409**	**128 017**	**127 775**	**130 049**	**127 853**	**124 001**	**122 513**	**..**	**OCDE TOTAL**
65 459	**66 154**	**64 867**	**67 228**	**66 221**	**67 569**	**67 449**	**67 429**	**66 203**	**..**	**OCDE Amérique du N.**
15 749	**16 458**	**16 870**	**17 586**	**18 790**	**19 200**	**19 298**	**16 378**	**16 141**	**..**	**OCDE Pacifique**
39 302	**40 569**	**41 672**	**43 203**	**42 758**	**43 280**	**41 106**	**40 194**	**40 169**	**..**	**OCDE Europe**
112 730	**115 134**	**115 143**	**119 151**	**119 673**	**120 698**	**118 587**	**115 170**	**113 962**	**..**	**AIE Total**
35 565	**36 892**	**37 444**	**38 485**	**38 118**	**38 441**	**36 473**	**35 713**	**35 145**	**..**	**UE15**

* Consumption in the energy sector includes refinery fuel and distribution losses. / *La consommation du secteur de l'énergie comprend la consommation propre des raffineries et les pertes de distribution.*

Table 17: Total Products Final Consumption
Tableau 17 : Consommation finale de produits pétroliers

000 metric tons

	1973	1974	1975	1976	1977	1978	1979	1980	1981	1982
Australia	23 943	23 903	24 061	24 311	25 734	26 348	26 700	26 708	25 678	25 575
Austria	10 118	9 214	9 178	9 373	9 487	9 895	10 379	9 894	9 029	8 770
Belgium	20 597	17 712	17 510	18 408	18 014	19 223	19 452	17 015	15 499	15 342
Canada	75 548	77 840	73 804	75 191	76 915	77 387	79 364	78 135	74 237	65 539
Czech Republic	8 044	7 996	8 239	8 268	9 078	9 159	9 477	9 520	9 473	9 868
Denmark	13 986	11 842	12 035	13 192	13 164	13 037	13 300	11 809	10 538	9 190
Finland	11 383	10 017	9 997	10 218	10 283	10 450	10 680	10 104	9 582	9 276
France	96 978	90 879	85 688	88 092	87 504	93 732	93 737	88 252	79 993	75 414
Germany	135 127	123 894	122 025	127 609	127 941	133 846	137 453	123 465	113 021	107 438
Greece	7 012	6 163	6 397	7 283	7 668	8 545	8 826	8 793	8 506	8 671
Hungary	6 468	6 958	7 510	7 876	8 370	9 031	9 068	9 013	8 895	8 321
Iceland	642	600	560	568	595	630	613	573	557	514
Ireland	3 783	3 859	3 574	3 706	3 997	4 044	4 392	4 059	3 983	3 591
Italy	70 972	68 028	62 825	66 324	63 680	64 314	68 161	65 184	61 926	59 870
Japan	167 925	162 645	150 966	160 374	161 870	166 252	169 593	152 924	144 759	142 647
Korea	9 767	9 822	10 361	11 819	14 236	16 266	18 700	18 562	17 666	17 253
Luxembourg	1 525	1 372	1 230	1 345	1 339	1 348	1 249	1 051	1 013	992
Mexico	21 875	24 328	27 220	29 947	30 469	33 213	36 059	39 441	43 774	44 818
Netherlands	23 751	21 194	19 735	23 598	21 786	21 938	22 489	20 158	18 238	16 719
New Zealand	3 566	3 660	3 652	3 680	3 724	3 698	3 653	3 670	3 507	3 502
Norway	7 607	7 065	7 156	7 559	8 013	8 222	8 688	8 260	7 725	7 330
Poland	9 051	9 634	10 764	11 487	12 226	12 808	12 753	13 095	11 868	11 299
Portugal	4 465	4 607	4 729	4 912	5 250	5 504	5 802	6 024	6 030	6 626
Slovak Republic	3 877	3 982	4 052	4 243	4 338	4 544	4 721	5 080	4 948	4 221
Spain	29 723	31 382	31 371	34 225	35 368	36 883	39 229	37 455	34 802	32 947
Sweden	24 559	21 883	23 393	24 806	23 767	21 729	22 353	20 276	19 106	17 252
Switzerland	13 918	12 442	11 778	12 288	12 443	12 885	12 341	12 526	11 548	10 921
Turkey	9 597	9 769	10 999	12 340	14 043	14 031	12 386	12 691	12 521	13 305
United Kingdom	73 666	69 123	65 859	66 432	67 559	67 933	69 050	61 092	57 938	57 547
United States	669 351	640 085	630 540	679 906	722 615	731 468	720 758	666 434	640 470	620 754
OECD TOTAL	**1558 824**	**1491 898**	**1457 208**	**1549 380**	**1601 476**	**1638 363**	**1651 426**	**1541 263**	**1466 830**	**1415 512**
OECD North America	**766 774**	**742 253**	**731 564**	**785 044**	**829 999**	**842 068**	**836 181**	**784 010**	**758 481**	**731 111**
OECD Pacific	**205 201**	**200 030**	**189 040**	**200 184**	**205 564**	**212 564**	**218 646**	**201 864**	**191 610**	**188 977**
OECD Europe	**586 849**	**549 615**	**536 604**	**564 152**	**565 913**	**583 731**	**596 599**	**555 389**	**516 739**	**495 424**
IEA Total	**1523 379**	**1453 354**	**1414 612**	**1503 135**	**1553 848**	**1587 168**	**1597 280**	**1483 074**	**1405 683**	**1354 660**
EU15	**527 645**	**491 169**	**475 546**	**499 523**	**496 807**	**512 421**	**526 552**	**484 631**	**449 204**	**429 645**

Table 17: Total Products Final Consumption (continued)
Tableau 17 : Consommation finale de produits pétroliers (suite)

000 metric tons

1983	1984	1985	1986	1987	1988	1989	1990	1991	1992	
24 041	25 146	25 373	25 769	25 927	27 469	28 635	29 240	28 544	29 108	Australie
8 778	8 567	8 768	8 972	9 194	8 998	9 027	9 262	9 963	9 614	Autriche
14 417	14 379	15 262	17 338	16 961	17 418	16 828	16 759	18 140	18 879	Belgique
62 494	62 894	62 632	63 215	65 003	67 608	69 330	67 772	64 054	65 548	Canada
9 185	9 256	9 060	8 834	8 926	8 350	8 952	7 887	6 496	7 056	République Tchèque
8 851	9 012	9 351	9 287	8 871	8 329	7 867	7 321	7 518	7 293	Danemark
8 845	8 624	8 241	9 075	9 461	9 603	9 589	9 396	9 476	9 340	Finlande
75 688	74 809	73 611	74 594	75 688	76 091	76 830	76 644	80 521	81 300	France
106 922	108 209	111 251	116 858	113 979	115 132	108 127	112 874	119 406	120 626	Allemagne
8 557	8 661	8 804	8 555	9 264	9 643	10 271	10 421	10 547	10 618	Grèce
7 604	7 604	7 450	7 132	7 341	7 100	7 249	6 981	5 934	5 860	Hongrie
500	520	515	515	554	581	606	670	645	696	Islande
3 413	3 482	3 427	3 561	3 639	3 510	3 708	4 038	4 159	4 274	Irlande
60 496	60 845	59 469	59 617	61 453	62 524	63 148	62 421	61 862	61 349	Italie
145 529	152 386	150 165	153 909	159 155	168 688	173 505	178 868	183 340	187 402	Japon
18 595	20 166	21 099	23 553	26 479	30 401	34 658	42 568	49 607	59 351	Corée
956	964	1 027	1 105	1 275	1 297	1 439	1 580	1 835	1 877	Luxembourg
42 493	45 197	47 485	46 640	48 593	49 178	52 649	53 967	55 838	58 108	Mexique
17 992	18 332	17 716	19 220	19 115	19 297	18 908	18 789	20 124	20 389	Pays-Bas
3 394	3 515	3 456	3 563	3 706	3 879	4 070	4 248	4 126	4 404	Nouvelle Zélande
7 239	7 616	7 886	8 254	8 462	8 097	7 902	7 710	7 380	7 179	Norvège
11 941	12 275	12 351	13 084	13 151	13 482	13 473	10 984	10 715	10 954	Pologne
6 210	6 452	6 387	6 662	7 002	7 807	8 165	8 630	8 680	9 145	Portugal
4 253	4 117	4 053	3 929	3 894	3 791	4 637	4 545	3 770	3 406	République Slovaque
33 647	33 411	32 960	33 053	33 937	38 258	37 925	38 941	40 884	41 753	Espagne
15 640	14 615	14 526	15 335	14 832	14 802	14 100	13 484	13 044	13 406	Suède
11 970	11 529	12 572	12 615	12 109	12 205	12 069	12 321	12 752	12 865	Suisse
14 153	13 918	14 316	15 638	18 332	18 407	18 694	20 193	19 953	20 911	Turquie
57 088	58 011	57 774	61 152	61 579	65 062	65 221	65 688	67 034	67 602	Royaume-Uni
606 367	632 277	632 551	647 028	665 395	684 258	678 899	664 272	649 888	667 582	Etats-Unis
1397 258	**1436 789**	**1439 538**	**1470 062**	**1513 277**	**1561 265**	**1566 481**	**1568 474**	**1576 235**	**1617 895**	**OCDE TOTAL**
711 354	**740 368**	**742 668**	**756 883**	**778 991**	**801 014**	**800 878**	**786 011**	**769 780**	**791 238**	**OCDE Amérique du N.**
191 559	**201 213**	**200 093**	**206 794**	**215 267**	**230 437**	**240 868**	**254 921**	**265 617**	**280 265**	**OCDE Pacifique**
494 345	**495 208**	**496 777**	**514 385**	**519 019**	**529 784**	**524 735**	**527 539**	**540 838**	**546 392**	**OCDE Europe**
1338 071	**1374 680**	**1375 134**	**1413 894**	**1447 085**	**1494 233**	**1495 116**	**1498 308**	**1505 267**	**1544 731**	**AIE Total**
427 500	**428 373**	**428 574**	**444 384**	**446 250**	**457 771**	**451 153**	**456 248**	**473 193**	**477 465**	**UE15**

Table 17: Total Products Final Consumption (continued)
Tableau 17 : Consommation finale de produits pétroliers (suite)

000 metric tons

	1993	1994	1995	1996	1997	1998	1999	2000	2001	2002
Australia	29 965	30 700	31 913	33 154	33 642	34 025	34 487	35 603	35 542	..
Austria	9 632	9 606	9 721	10 793	10 586	11 095	10 623	10 740	11 491	..
Belgium	18 326	18 712	18 570	20 896	21 371	21 610	21 184	21 269	21 577	..
Canada	67 313	69 569	70 460	73 481	75 648	74 925	77 855	78 470	77 955	..
Czech Republic	6 945	6 997	7 340	7 398	7 124	7 541	7 465	7 301	7 616	..
Denmark	7 392	7 373	7 424	7 637	7 467	7 392	7 435	7 186	7 012	..
Finland	8 771	9 188	8 501	8 193	8 213	8 420	8 354	8 019	8 190	..
France	79 471	79 151	81 150	82 435	82 767	85 553	85 510	84 769	87 757	..
Germany	122 668	121 226	121 713	124 173	123 624	123 714	120 017	117 779	119 742	..
Greece	10 549	10 751	11 066	12 047	12 377	12 951	12 921	13 047	13 509	..
Hungary	5 297	5 459	5 263	4 972	4 926	5 303	5 139	5 297	5 072	..
Iceland	700	719	704	791	787	813	834	843	827	..
Ireland	4 379	4 814	4 893	5 086	5 495	5 970	6 582	7 018	7 401	..
Italy	60 781	59 529	61 436	61 503	62 537	63 562	64 475	63 049	64 072	..
Japan	187 214	193 163	199 711	203 833	204 233	200 607	203 983	210 548 e	208 544 e	..
Korea	65 942	72 091	77 501	81 958	85 991	74 913	81 714	82 385	81 509	..
Luxembourg	1 878	1 880	1 739	1 800	1 881	1 962	2 094	2 227	2 368	..
Mexico	57 810	59 098	55 937	54 416	55 413	56 767	55 738	58 119	57 855	..
Netherlands	19 824	20 264	19 789	19 593	20 553	21 044	21 761	23 203	23 699	..
New Zealand	4 409	4 822	5 075	5 125	5 254	5 222	5 441	5 595	5 541	..
Norway	7 364	7 575	7 887	8 315	8 128	8 181	8 310	7 717	8 175	..
Poland	11 811	12 302	12 972	14 782	16 214	16 890	17 870	17 353	17 055	..
Portugal	9 135	9 689	10 013	10 484	11 204	12 170	12 355	12 692	12 732	..
Slovak Republic	2 334	2 645	2 664	2 782	3 187	3 245	3 153	3 240	2 603	..
Spain	41 016	44 259	46 376	45 882	48 461	51 791	52 139	54 683	56 451	..
Sweden	13 123	13 949	14 026	14 425	14 123	13 897	13 920	13 802	12 831	..
Switzerland	12 217	12 004	12 188	12 418	12 179	12 689	12 816	12 532	12 609	..
Turkey	23 656	22 157	25 194	26 472	25 794	25 170	25 014	25 895	24 069	..
United Kingdom	69 256	70 129	69 297	71 115	70 277	70 288	71 657	70 410	69 126	..
United States	678 739	701 876	706 139	728 281	733 725	739 943	763 124	776 570	788 109	..
OECD TOTAL	**1637 917**	**1681 697**	**1706 662**	**1754 240**	**1773 181**	**1777 653**	**1813 970**	**1837 361**	**1851 039**	**..**
OECD North America	**803 862**	**830 543**	**832 536**	**856 178**	**864 786**	**871 635**	**896 717**	**913 159**	**923 919**	**..**
OECD Pacific	**287 530**	**300 776**	**314 200**	**324 070**	**329 120**	**314 767**	**325 625**	**334 131**	**331 136**	**..**
OECD Europe	**546 525**	**550 378**	**559 926**	**573 992**	**579 275**	**591 251**	**591 628**	**590 071**	**595 984**	**..**
IEA Total	**1565 262**	**1606 933**	**1634 385**	**1681 469**	**1697 580**	**1699 938**	**1736 375**	**1757 806**	**1772 699**	**..**
EU15	**476 201**	**480 520**	**485 714**	**496 062**	**500 936**	**511 419**	**511 027**	**509 893**	**517 958**	**..**

Table 18: Total Products Consumption in the Transport Sector
Tableau 18 : Consommation de produits pétroliers dans le secteur du transport

000 metric tons

1973	1974	1975	1976	1977	1978	1979	1980	1981	1982	
12 703	13 148	13 651	14 077	14 978	15 835	16 203	16 655	16 878	17 343	Australie
3 814	3 468	3 541	3 664	3 806	4 054	4 295	4 014	3 813	3 774	Autriche
4 687	4 411	4 691	4 956	5 205	5 322	5 633	5 630	5 518	5 545	Belgique
33 024	34 349	34 858	36 477	37 053	37 794	39 768	40 622	40 192	35 260	Canada
2 274	2 161	2 328	2 450	2 553	2 598	2 654	2 398	2 587	2 227	République Tchèque
3 314	3 012	3 315	3 383	3 484	3 614	3 607	3 434	3 330	3 366	Danemark
2 403	2 348	2 565	2 587	2 601	2 638	2 844	2 837	2 865	2 930	Finlande
25 402	25 226	26 017	27 950	28 305	30 226	30 537	30 850	31 301	31 463	France
35 535	34 684	36 282	37 249	39 418	42 556	43 903	44 508	43 252	43 267	Allemagne
2 538	2 060	2 085	2 760	3 145	3 538	3 689	3 819	3 948	4 055	Grèce
1 826	2 044	2 160	2 185	2 438	2 630	2 655	2 687	2 646	2 632	Hongrie
229	177	168	193	201	216	200	180	215	196	Islande
1 363	1 390	1 392	1 448	1 576	1 811	1 856	1 714	1 707	1 635	Irlande
18 824	17 634	18 377	19 406	20 630	21 704	23 803	24 217	24 150	25 080	Italie
39 577	39 693	41 735	43 923	45 951	48 995	51 046	51 774	50 702	49 565	Japon
2 473	2 323	2 374	2 636	3 919	4 580	5 651	4 859	3 900	4 406	Corée
273	275	327	351	359	409	462	479	524	521	Luxembourg
12 309	13 544	14 429	15 725	16 691	18 074	20 495	22 983	25 563	25 460	Mexique
6 985	6 533	6 971	7 721	8 114	8 020	7 994	8 180	8 286	8 034	Pays-Bas
2 026	2 134	2 224	2 242	2 256	2 318	2 344	2 342	2 308	2 322	Nouvelle Zélande
2 442	2 372	2 742	2 639	2 896	2 927	3 224	2 964	2 902	2 907	Norvège
5 039	5 224	5 741	6 054	6 613	7 045	7 139	6 966	6 623	6 231	Pologne
1 847	1 837	2 005	2 006	2 129	2 202	2 370	2 523	2 600	2 727	Portugal
1 576	1 554	1 352	1 359	1 374	1 490	1 578	1 170	1 205	1 098	République Slovaque
11 264	12 045	12 607	13 720	13 596	14 421	16 063	15 381	14 571	14 886	Espagne
5 102	5 050	5 215	5 594	5 812	5 981	5 910	5 675	5 577	5 614	Suède
3 930	3 721	3 567	3 583	3 831	3 892	3 883	4 075	4 147	4 225	Suisse
3 831	4 001	4 750	5 203	6 149	6 025	5 410	5 185	5 193	5 573	Turquie
29 464	28 366	28 006	29 029	29 796	31 306	32 014	32 124	30 977	31 665	Royaume-Uni
382 757	374 318	380 807	398 944	412 983	428 136	418 209	396 894	394 772	392 341	Etats-Unis
658 831	**649 102**	**666 282**	**699 514**	**727 862**	**760 357**	**765 439**	**747 139**	**742 252**	**736 348**	**OCDE TOTAL**
428 090	**422 211**	**400 094**	**451 146**	**466 727**	**484 004**	**478 472**	**460 499**	**460 527**	**453 061**	**OCDE Amérique du N.**
56 779	**57 298**	**59 984**	**62 878**	**67 104**	**71 728**	**75 244**	**75 630**	**73 788**	**73 636**	**OCDE Pacifique**
173 962	**169 593**	**176 204**	**185 490**	**194 031**	**204 625**	**211 723**	**211 010**	**207 937**	**209 651**	**OCDE Europe**
639 678	**628 603**	**644 592**	**676 183**	**702 983**	**733 532**	**736 027**	**715 840**	**708 646**	**703 363**	**AIE Total**
152 815	**148 339**	**153 396**	**161 824**	**167 976**	**177 802**	**184 980**	**185 385**	**182 419**	**184 562**	**UE15**

Table 18: Total Products Consumption in the Transport Sector (continued)
Tableau 18 : Consommation de produits pétroliers dans le secteur du transport (suite)

000 metric tons

	1983	1984	1985	1986	1987	1988	1989	1990	1991	1992
Australia	17 131	17 755	18 352	18 840	19 190	20 030	20 778	21 177	20 694	21 234
Austria	4 001	3 933	3 946	4 065	4 155	4 386	4 546	4 603	5 140	5 157
Belgium	5 594	5 679	5 904	6 409	6 589	7 164	7 363	7 462	7 590	8 052
Canada	34 574	35 495	35 671	35 924	37 346	39 438	40 103	38 904	37 380	38 107
Czech Republic	2 205	2 373	2 331	2 293	2 351	2 377	2 551	2 652	2 321	2 879
Denmark	3 345	3 564	4 023	4 177	4 168	4 222	4 293	3 914	4 075	4 083
Finland	2 992	3 076	3 222	3 415	3 670	3 769	4 007	4 102	3 977	3 927
France	31 806	32 376	32 531	34 100	35 272	37 553	39 020	40 362	41 240	42 434
Germany	43 984	45 333	45 634	47 970	49 934	51 796	53 235	56 149	56 744	57 834
Greece	4 272	4 439	4 628	4 616	4 781	5 125	5 334	5 757	5 878	6 040
Hungary	2 502	2 582	2 629	2 766	2 875	2 892	3 009	2 882	2 508	2 436
Iceland	191	209	221	210	235	244	276	275	269	270
Ireland	1 547	1 535	1 673	1 752	1 701	1 786	1 881	1 962	2 000	2 098
Italy	25 221	26 255	27 197	28 256	29 878	30 965	32 188	32 972	33 901	35 341
Japan	51 079	52 929	53 625	55 615	57 888	61 156	65 706	68 962	71 699	74 292
Korea	5 576	6 112	6 657	7 753	9 231	10 603	12 215	14 178	15 949	17 789
Luxembourg	512	537	577	605	684	718	822	977	1 149	1 238
Mexico	22 822	23 953	24 199	23 946	24 552	24 839	27 432	29 549	31 693	32 354
Netherlands	8 267	8 375	8 494	8 850	8 970	9 344	9 762	9 918	10 102	10 737
New Zealand	2 309	2 432	2 442	2 586	2 780	2 950	3 133	3 279	3 234	3 398
Norway	3 012	3 136	3 295	3 580	3 744	3 791	3 859	3 940	3 670	3 773
Poland	6 749	6 872	6 813	7 342	7 649	7 701	7 925	6 653	6 910	7 053
Portugal	2 667	2 565	2 606	2 768	2 989	3 255	3 445	3 669	3 911	4 236
Slovak Republic	868	875	867	890	918	965	954	933	784	887
Spain	14 960	15 531	15 328	15 711	16 657	19 697	20 726	21 698	22 675	24 059
Sweden	5 714	6 004	6 174	6 598	6 747	7 137	7 354	6 878	6 772	7 033
Switzerland	4 419	4 555	4 650	4 881	5 016	5 284	5 448	5 750	5 837	6 025
Turkey	5 857	5 971	6 234	7 198	8 170	8 391	8 539	9 227	8 878	9 071
United Kingdom	32 596	34 178	34 807	37 025	38 588	40 972	42 892	43 804	43 183	44 121
United States	399 534	411 514	415 069	427 230	441 926	460 429	464 342	462 306	454 540	465 219
OECD TOTAL	**746 306**	**770 143**	**779 799**	**807 371**	**838 654**	**878 979**	**903 138**	**914 894**	**914 703**	**941 177**
OECD North America	**456 930**	**470 962**	**474 939**	**487 100**	**503 824**	**524 706**	**531 877**	**530 759**	**523 613**	**535 680**
OECD Pacific	**76 095**	**79 228**	**81 076**	**84 794**	**89 089**	**94 739**	**101 832**	**107 596**	**111 576**	**116 713**
OECD Europe	**213 281**	**219 953**	**223 784**	**235 477**	**245 741**	**259 534**	**269 429**	**276 539**	**279 514**	**288 784**
IEA Total	**715 676**	**738 234**	**747 699**	**774 983**	**805 300**	**845 230**	**866 551**	**877 484**	**875 047**	**900 613**
EU15	**187 478**	**193 380**	**196 744**	**206 317**	**214 783**	**227 889**	**236 868**	**244 227**	**248 337**	**256 390**

Table 18: Total Products Consumption in the Transport Sector (continued)
Tableau 18 : Consommation de produits pétroliers dans le secteur du transport (suite)

000 metric tons

1993	1994	1995	1996	1997	1998	1999	2000	2001	2002	
21 692	22 260	23 263	24 189	24 674	24 892	25 320	26 022	26 026	..	Australie
5 030	5 055	5 134	5 637	5 379	5 917	5 817	6 125	6 486	..	Autriche
8 094	8 208	8 247	8 683	8 966	9 316	9 331	9 325	9 238	..	Belgique
38 464	40 280	41 075	42 147	43 503	44 546	45 781	45 804	45 492	..	Canada
2 824	3 124	2 700	3 511	3 511	3 677	4 499	4 588	4 823	..	République Tchèque
4 165	4 360	4 413	4 488	4 544	4 604	4 663	4 620	4 415	..	Danemark
3 834	3 988	3 941	3 872	4 075	4 129	4 263	4 223	4 308	..	Finlande
42 942	43 466	43 990	44 208	45 173	47 657	48 189	49 460	49 996	..	France
59 570	59 165	59 802	60 277	60 698	61 822	63 843	62 856	61 360	..	Allemagne
6 152	6 323	6 297	6 403	6 579	7 123	7 291	7 007	7 192	..	Grèce
2 420	2 423	2 479	2 486	2 611	2 898	3 078	3 080	3 211	..	Hongrie
267	281	263	303	281	306	318	333	318	..	Islande
2 228	2 258	2 283	2 597	2 778	3 223	3 598	3 917	4 177	..	Irlande
36 141	36 099	36 968	37 244	37 888	39 016	39 488	39 613	40 112	..	Italie
74 757	78 321	80 828	83 819	85 733	86 315	87 302	88 775	89 785	..	Japon
20 091	22 721	25 729	27 936	28 598	24 605	26 818	29 007	29 934	..	Corée
1 247	1 298	1 265	1 311	1 422	1 507	1 664	1 823	1 929	..	Luxembourg
33 103	33 917	32 642	31 661	32 422	33 541	33 592	35 208	35 403	..	Mexique
11 104	11 343	11 879	12 306	12 590	13 108	13 567	13 704	13 763	..	Pays-Bas
3 500	3 762	4 033	4 120	4 261	4 323	4 529	4 656	4 708	..	Nouvelle Zélande
3 943	3 959	4 069	4 252	4 323	4 452	4 582	4 226	4 296	..	Norvège
6 952	7 338	7 628	8 640	9 131	9 128	10 135	8 729	8 728	..	Pologne
4 378	4 575	4 755	5 002	5 150	5 596	5 918	6 381	6 414	..	Portugal
848	1 051	1 193	1 115	1 306	1 326	1 338	1 294	1 367	..	République Slovaque
23 783	24 787	25 263	27 026	27 183	29 701	31 150	32 000	33 357	..	Espagne
6 907	7 180	7 234	7 184	7 304	7 418	7 657	7 710	7 765	..	Suède
5 748	5 934	5 900	5 963	6 169	6 271	6 330	6 462 e	6 277 e	..	Suisse
10 804	10 467	11 738	12 403	11 740	10 884	11 405	11 976	11 440	..	Turquie
44 921	45 182	45 121	46 927	47 578	48 091	49 799	50 051	49 084	..	Royaume-Uni
475 860	492 555	503 147	514 639	524 109	539 315	555 552	566 066	565 306	..	Etats-Unis
961 769	**991 680**	**1013 279**	**1040 349**	**1050 679**	**1084 707**	**1116 817**	**1135 041**	**1136 710**	**..**	**OCDE TOTAL**
547 427	**566 752**	**576 864**	**588 447**	**600 034**	**617 402**	**634 925**	**647 078**	**646 201**	**..**	**OCDE Amérique du N.**
120 040	**127 064**	**133 853**	**140 064**	**143 266**	**140 135**	**143 969**	**148 460**	**150 453**	**..**	**OCDE Pacifique**
294 302	**297 864**	**302 562**	**311 838**	**316 379**	**327 170**	**337 923**	**339 503**	**340 056**	**..**	**OCDE Europe**
920 599	**949 093**	**971 553**	**998 630**	**1016 539**	**1040 406**	**1071 434**	**1089 477**	**1090 894**	**..**	**AIE Total**
260 496	**263 287**	**266 592**	**273 165**	**277 307**	**288 228**	**296 238**	**298 815**	**299 596**	**..**	**UE15**

Table 19: Total Products Consumption in the Industry Sector
Tableau 19 : Consommation de produits pétroliers dans le secteur de l'industrie

000 metric tons

	1973	1974	1975	1976	1977	1978	1979	1980	1981	1982
Australia	7 852	8 352	7 884	7 746	8 184	8 045	8 051	7 874	6 864	6 298
Austria	3 192	3 061	2 916	2 716	2 763	2 737	2 784	1 975	1 765	1 628
Belgium	7 969	6 625	5 298	6 407	5 779	6 239	5 738	4 521	4 038	4 614
Canada	21 261	21 631	19 027	17 932	19 894	20 982	21 753	20 603	19 272	15 405
Czech Republic	5 192	5 243	4 940	4 823	5 518	5 249	5 702	6 030	5 661	6 762
Denmark	3 437	2 600	2 531	2 846	2 856	2 743	2 720	2 583	2 131	1 984
Finland	5 138	4 616	3 945	4 106	3 882	3 885	3 956	3 791	3 713	3 597
France	35 097	34 951	30 504	30 642	30 250	31 051	33 737	30 625	23 211	20 974
Germany	47 083	44 086	40 296	40 201	39 908	40 604	42 117	36 452	31 831	29 509
Greece	2 455	2 245	2 369	2 649	2 787	2 964	3 194	3 155	2 860	2 603
Hungary	2 288	2 368	2 782	3 031	3 139	3 269	3 288	3 299	3 202	2 900
Iceland	129	136	111	120	130	150	159	156	110	103
Ireland	1 654	1 790	1 530	1 606	1 696	1 513	1 747	1 633	1 604	1 318
Italy	30 092	29 379	23 768	26 219	24 957	22 678	24 726	22 940	20 906	18 653
Japan	93 518	89 152	75 581	81 122	79 880	79 040	80 936	65 831	57 716	60 532
Korea	6 297	6 520	6 805	7 747	7 779	9 179	10 217	9 875	8 861	8 100
Luxembourg	828	725	550	601	566	511	360	207	187	167
Mexico	5 385	6 412	7 562	8 738	8 185	9 215	9 282	9 163	10 615	11 173
Netherlands	9 925	9 780	8 154	10 555	9 360	9 989	10 725	9 008	7 440	6 222
New Zealand	989	961	887	903	926	881	886	832	787	711
Norway	3 130	3 172	3 009	3 194	3 335	3 523	3 567	3 616	3 337	3 122
Poland	2 999	3 422	3 912	4 205	4 340	4 421	4 243	4 729	3 956	3 763
Portugal	1 786	2 000	1 919	2 136	2 280	2 377	2 548	2 609	2 590	3 020
Slovak Republic	1 799	1 958	2 195	2 348	2 436	2 515	2 556	2 979	2 831	2 333
Spain	13 773	14 969	14 091	15 618	16 547	16 888	17 222	16 320	14 867	12 292
Sweden	8 328	7 794	7 709	7 699	6 986	6 670	7 016	6 231	5 356	4 813
Switzerland	3 707	3 288	2 899	2 755	2 819	2 897	2 737	2 771	2 669	2 260
Turkey	2 682	2 686	3 003	3 619	3 919	4 725	4 052	4 322	4 409	4 546
United Kingdom	31 757	29 654	26 231	25 911	25 730	24 483	24 830	18 729	17 301	16 766
United States	154 235	145 504	134 288	152 074	182 265	179 852	199 935	184 106	170 744	159 049
OECD TOTAL	**513 977**	**495 080**	**446 696**	**480 269**	**509 096**	**509 275**	**540 784**	**486 965**	**440 834**	**415 217**
OECD North America	**180 881**	**173 547**	**160 877**	**178 744**	**210 344**	**210 049**	**230 970**	**213 872**	**200 631**	**185 627**
OECD Pacific	**108 656**	**104 985**	**91 157**	**97 518**	**96 769**	**97 145**	**100 090**	**84 412**	**74 228**	**75 641**
OECD Europe	**224 440**	**216 548**	**194 662**	**204 007**	**201 983**	**202 081**	**209 724**	**188 681**	**165 975**	**153 949**
IEA Total	**503 665**	**483 152**	**432 916**	**464 858**	**494 005**	**492 974**	**524 544**	**469 938**	**423 322**	**397 845**
EU15	**202 514**	**194 275**	**171 811**	**179 912**	**176 347**	**175 332**	**183 420**	**160 779**	**139 800**	**128 160**

Table 19: Total Products Consumption in the Industry Sector (continued)
Tableau 19 : Consommation de produits pétroliers dans le secteur de l'industrie (suite)

000 metric tons

1983	1984	1985	1986	1987	1988	1989	1990	1991	1992	
5 212	5 640	5 324	5 288	5 058	5 814	6 166	6 390	6 176	6 131	Australie
1 656	1 596	1 609	1 783	1 905	1 913	1 966	2 052	2 076	1 866	Autriche
4 174	4 027	4 156	4 982	4 542	4 671	4 321	4 183	5 038	5 007	Belgique
14 318	15 177	15 540	16 086	16 842	17 297	17 288	17 696	16 136	16 723	Canada
6 102	6 036	5 787	5 572	5 609	4 995	5 446	3 969	2 941	3 221	République Tchèque
1 920	2 040	1 913	1 930	1 626	1 491	1 309	1 220	1 264	1 259	Danemark
3 287	3 216	2 571	3 082	3 055	3 003	2 958	2 608	2 842	2 769	Finlande
21 340	20 179	19 479	18 043	19 336	19 025	18 597	17 451	19 464	19 463	France
29 168	26 991	27 401	27 276	26 733	27 741	26 901	26 269	27 027	27 556	Allemagne
2 361	2 195	1 997	2 008	2 148	2 196	2 324	2 131	2 054	2 074	Grèce
2 681	2 628	2 401	2 310	2 331	2 261	2 340	1 957	1 683	1 574	Hongrie
99	120	115	113	111	118	108	166	138	172	Islande
1 242	1 329	1 196	1 073	1 120	940	1 004	867	900	908	Irlande
19 059	17 666	16 713	16 450	16 732	17 667	17 463	17 148	15 876	14 922	Italie
60 354	62 631	61 098	61 462	64 058	66 561	68 294	68 615	68 205	67 969	Japon
8 416	9 076	9 284	10 103	11 035	12 474	13 602	17 271	21 051	26 700	Corée
144	144	153	173	278	280	324	301	313	290	Luxembourg
11 908	13 402	15 143	14 756	15 625	15 821	16 932	15 623	15 045	15 968	Mexique
7 275	7 570	6 925	8 168	8 364	8 031	7 338	7 575	8 143	7 914	Pays-Bas
649	658	608	587	591	606	607	615	552	613	Nouvelle Zélande
3 053	3 382	3 402	3 407	3 411	3 099	2 978	2 788	2 493	2 270	Norvège
3 850	4 077	4 104	4 300	4 402	4 375	4 073	2 958	2 515	2 424	Pologne
2 650	2 921	2 805	2 902	2 886	3 432	3 611	3 791	3 572	3 651	Portugal
2 601	2 496	2 440	2 313	2 247	2 117	3 006	2 973	2 550	2 126	République Slovaque
12 835	11 921	11 519	11 232	11 065	12 494	11 794	11 513	12 011	11 426	Espagne
4 480	4 252	4 085	4 162	3 998	3 861	3 424	3 376	3 143	3 184	Suède
2 730	2 099	2 170	2 001	1 646	1 481	1 379	1 299	1 290	1 342	Suisse
4 883	4 675	4 543	4 837	6 007	5 776	5 806	6 085	6 226	6 746	Turquie
15 759	15 119	14 640	16 087	15 727	16 772	15 486	15 042	16 408	15 950	Royaume-Uni
137 285	147 766	143 672	146 884	149 281	148 018	142 101	141 641	137 378	141 959	Etats-Unis
391 491	**401 029**	**392 793**	**399 370**	**407 769**	**414 330**	**408 946**	**405 573**	**404 510**	**414 177**	**OCDE TOTAL**
163 511	**176 345**	**174 355**	**177 726**	**181 748**	**181 136**	**176 321**	**174 960**	**168 559**	**174 650**	**OCDE Amérique du N.**
74 631	**78 005**	**76 314**	**77 440**	**80 742**	**85 455**	**88 669**	**92 891**	**95 984**	**101 413**	**OCDE Pacifique**
153 349	**146 679**	**142 124**	**144 204**	**145 279**	**147 739**	**143 056**	**137 722**	**139 967**	**138 114**	**OCDE Europe**
373 033	**380 934**	**370 991**	**377 888**	**385 384**	**391 899**	**384 827**	**383 853**	**384 262**	**393 487**	**AIE Total**
127 350	**121 166**	**117 162**	**119 351**	**119 515**	**123 517**	**118 820**	**115 527**	**120 131**	**118 239**	**UE15**

Table 19: Total Products Consumption in the Industry Sector (continued)
Tableau 19 : Consommation de produits pétroliers dans le secteur de l'industrie (suite)

000 metric tons

	1993	1994	1995	1996	1997	1998	1999	2000	2001	2002
Australia	6 482	6 666	6 841	7 158	7 110	7 253	7 274	7 650	7 591	..
Austria	2 016	2 173	2 005	2 246	2 614	2 565	2 226	2 288	2 316	..
Belgium	4 734	4 766	4 244	5 134	5 770	5 897	5 896	6 754	7 101	..
Canada	17 648	17 870	17 842	19 097	19 469	18 916	20 370	19 692	19 482	..
Czech Republic	3 137	2 997	3 670	3 284	3 150	3 367	2 833	2 535	2 638	..
Denmark	1 171	1 155	1 177	1 210	1 140	1 087	1 108	1 068	1 085	..
Finland	2 597	2 857	2 172	1 963	1 900	1 993	2 028	1 797	1 705	..
France	18 098	18 598	19 959	19 947	20 351	19 344	18 882	18 845	19 333	..
Germany	26 988	27 826	28 067	26 628	27 429	27 839	26 831	27 069	26 553	..
Greece	1 939	1 951	2 268	2 545	2 617	2 537	2 384	2 610	2 621	..
Hungary	1 427	1 499	1 632	1 533	1 502	1 666	1 349	1 523	1 249	..
Iceland	168	171	176	200	228	241	257	267	294	..
Ireland	869	1 024	1 014	926	1 056	1 035	1 086	1 208	1 241	..
Italy	14 065	14 538	14 761	14 646	15 549	15 319	15 075	14 173	13 547	..
Japan	68 971	70 929	73 408	73 989	74 224	70 695	74 371	69 867 e	65 743 e	..
Korea	28 502	30 113	30 084	31 492	34 855	33 984	35 721	36 303	35 665	..
Luxembourg	296	262	169	149	117	95	96	83	77	..
Mexico	14 595	14 804	12 797	12 469	12 616	12 444	12 757	12 560	12 318	..
Netherlands	6 892	7 246	6 960	6 220	7 019	6 862	7 193	8 596	9 049	..
New Zealand	603	670	687	679	664	549	573	630	536	..
Norway	2 322	2 464	2 681	2 720	2 570	2 503	2 487	2 451	2 703	..
Poland	2 622	2 364	2 683	3 199	3 468	3 743	3 526	3 796	3 379	..
Portugal	3 447	3 788	3 857	3 748	4 390	4 948	4 809	4 633	4 490	..
Slovak Republic	1 126	1 289	1 206	1 411	1 595	1 689	1 609	1 770	990	..
Spain	11 096	12 883	14 452	12 014	14 559	15 582	14 084	15 162	15 717	..
Sweden	3 240	3 643	3 693	3 900	3 791	3 560	3 490	3 160	3 161	..
Switzerland	1 130	1 272	1 277	1 270	1 259	1 255	1 758	1 669 e	1 747 e	..
Turkey	7 184	6 268	7 256	7 678	8 003	8 580	7 711	7 957	7 318	..
United Kingdom	16 859	17 468	16 853	17 265	16 402	15 993	16 251	15 233	14 484	..
United States	141 970	149 786	144 446	152 577	150 622	146 636	152 601	150 284	161 017	..
OECD TOTAL	**412 194**	**429 340**	**428 337**	**437 297**	**446 039**	**438 177**	**446 636**	**441 633**	**445 150**	**..**
OECD North America	**174 213**	**182 460**	**175 085**	**184 143**	**182 707**	**177 996**	**185 728**	**182 536**	**192 817**	**..**
OECD Pacific	**104 558**	**108 378**	**111 020**	**113 318**	**116 853**	**112 481**	**117 939**	**114 450**	**109 535**	**..**
OECD Europe	**133 423**	**138 502**	**142 232**	**139 836**	**146 479**	**147 700**	**142 969**	**144 647**	**142 798**	**..**
IEA Total	**393 683**	**410 712**	**411 475**	**420 018**	**428 132**	**420 060**	**428 487**	**423 240**	**428 169**	**..**
EU15	**114 307**	**120 178**	**121 651**	**118 541**	**124 704**	**124 656**	**121 439**	**122 679**	**122 480**	**..**

Table 20: Total Products Consumption in the Other Sectors
Tableau 20 : Consommation de produits pétroliers dans les autres secteurs

000 metric tons

1973	1974	1975	1976	1977	1978	1979	1980	1981	1982	
3 388	2 403	2 526	2 488	2 572	2 468	2 446	2 179	1 936	1 934	Australie
3 112	2 685	2 721	2 993	2 918	3 104	3 300	3 905	3 451	3 368	Autriche
7 941	6 676	7 521	7 045	7 030	7 662	8 081	6 864	5 943	5 183	Belgique
21 263	21 860	19 919	20 782	19 968	18 611	17 843	16 910	14 773	14 874	Canada
578	592	971	995	1 007	1 312	1 121	1 092	1 225	879	République Tchèque
7 235	6 230	6 189	6 963	6 824	6 680	6 973	5 792	5 077	3 840	Danemark
3 842	3 053	3 487	3 525	3 800	3 927	3 880	3 476	3 004	2 749	Finlande
36 479	30 702	29 167	29 500	28 949	32 455	29 463	26 777	25 481	22 977	France
52 509	45 124	45 447	50 159	48 615	50 686	51 433	42 505	37 938	34 662	Allemagne
2 019	1 858	1 943	1 874	1 736	2 043	1 943	1 819	1 698	2 013	Grèce
2 354	2 546	2 568	2 660	2 793	3 132	3 125	3 027	3 047	2 789	Hongrie
284	287	281	255	264	264	254	237	232	215	Islande
766	679	652	652	725	720	789	712	672	638	Irlande
22 056	21 015	20 680	20 699	18 093	19 932	19 632	18 027	16 870	16 137	Italie
34 830	33 800	33 650	35 329	36 039	38 217	37 611	35 319	36 341	32 550	Japon
997	979	1 182	1 436	2 538	2 507	2 832	3 828	4 905	4 747	Corée
424	372	353	393	414	428	427	365	302	304	Luxembourg
4 181	4 372	5 229	5 484	5 593	5 924	6 282	7 295	7 596	8 185	Mexique
6 841	4 881	4 610	5 322	4 312	3 929	3 770	2 970	2 512	2 463	Pays-Bas
551	565	541	535	542	499	423	496	412	469	Nouvelle Zélande
2 035	1 521	1 405	1 726	1 782	1 772	1 897	1 680	1 486	1 301	Norvège
1 013	988	1 111	1 228	1 273	1 342	1 371	1 400	1 289	1 305	Pologne
832	770	805	770	841	925	884	892	840	879	Portugal
502	470	505	536	528	539	587	931	912	790	République Slovaque
4 686	4 368	4 673	4 887	5 225	5 574	5 944	5 754	5 364	5 769	Espagne
11 129	9 039	10 469	11 513	10 969	9 078	9 427	8 370	8 173	6 825	Suède
6 281	5 433	5 312	5 950	5 793	6 096	5 721	5 680	4 732	4 436	Suisse
3 084	3 082	3 246	3 518	3 975	3 281	2 924	3 184	2 919	3 186	Turquie
12 445	11 103	11 622	11 492	12 033	12 144	12 206	10 239	9 660	9 116	Royaume-Uni
132 359	120 263	115 445	128 888	127 367	123 480	102 614	85 434	74 954	69 364	Etats-Unis
386 016	**347 716**	**344 230**	**369 597**	**364 518**	**368 731**	**345 203**	**307 159**	**283 744**	**263 947**	**OCDE TOTAL**
157 803	**146 495**	**140 593**	**155 154**	**152 928**	**148 016**	**126 739**	**109 639**	**97 323**	**92 423**	**OCDE Amérique du N.**
39 766	**37 747**	**37 899**	**39 788**	**41 691**	**43 691**	**43 312**	**41 822**	**43 594**	**39 700**	**OCDE Pacifique**
188 447	**163 474**	**165 738**	**174 655**	**169 899**	**177 025**	**175 152**	**155 698**	**142 827**	**131 824**	**OCDE Europe**
380 036	**341 599**	**337 104**	**362 094**	**356 860**	**360 662**	**336 709**	**297 296**	**273 715**	**253 452**	**AIE Total**
172 316	**148 555**	**150 339**	**157 787**	**152 484**	**159 287**	**158 152**	**138 467**	**126 985**	**116 923**	**UE15**

Table 20: Total Products Consumption in the Other Sectors (continued)
Tableau 20 : Consommation de produits pétroliers dans les autres secteurs (suite)

000 metric tons

	1983	1984	1985	1986	1987	1988	1989	1990	1991	1992
Australia	1 698	1 751	1 697	1 641	1 679	1 625	1 691	1 673	1 674	1 743
Austria	3 121	3 038	3 213	3 124	3 134	2 699	2 515	2 607	2 747	2 591
Belgium	4 649	4 673	5 202	5 947	5 830	5 583	5 144	5 114	5 512	5 820
Canada	13 602	12 222	11 421	11 205	10 815	10 873	11 939	11 172	10 538	10 718
Czech Republic	878	847	942	969	966	978	955	1 266	1 234	956
Denmark	3 586	3 408	3 415	3 180	3 077	2 616	2 265	2 187	2 179	1 951
Finland	2 566	2 332	2 448	2 578	2 736	2 831	2 624	2 686	2 657	2 644
France	22 542	22 254	21 601	22 451	21 080	19 513	19 213	18 831	19 817	19 403
Germany	33 770	35 885	38 216	41 612	37 312	35 595	27 991	30 456	35 635	35 236
Greece	1 924	2 027	2 179	1 931	2 335	2 322	2 613	2 533	2 615	2 504
Hungary	2 421	2 394	2 420	2 056	2 135	1 947	1 900	2 142	1 743	1 850
Iceland	210	191	179	192	208	219	222	229	238	254
Ireland	624	618	558	736	818	784	823	1 209	1 259	1 268
Italy	16 216	16 924	15 559	14 911	14 843	13 892	13 497	12 301	12 085	11 086
Japan	34 096	36 826	35 442	36 832	37 209	40 971	39 505	41 291	43 436	45 141
Korea	4 603	4 978	5 158	5 697	6 213	7 324	8 841	11 119	12 607	14 862
Luxembourg	300	283	297	327	313	299	293	302	373	349
Mexico	7 763	7 842	8 143	7 938	8 416	8 518	8 285	8 795	9 100	9 786
Netherlands	2 450	2 387	2 297	2 202	1 781	1 922	1 808	1 296	1 879	1 738
New Zealand	436	425	406	390	335	323	330	354	340	393
Norway	1 174	1 098	1 189	1 267	1 307	1 207	1 065	982	1 217	1 136
Poland	1 342	1 326	1 434	1 442	1 100	1 406	1 475	1 373	1 290	1 477
Portugal	893	966	976	992	1 127	1 120	1 109	1 170	1 197	1 258
Slovak Republic	784	746	746	726	729	709	677	639	436	393
Spain	5 852	5 959	6 113	6 110	6 215	6 067	5 405	5 730	6 198	6 268
Sweden	5 446	4 359	4 267	4 575	4 087	3 804	3 322	3 230	3 129	3 189
Switzerland	4 821	4 875	5 752	5 733	5 447	5 440	5 242	5 272	5 625	5 498
Turkey	3 413	3 272	3 539	3 603	4 155	4 240	4 349	4 881	4 849	5 094
United Kingdom	8 733	8 714	8 327	8 040	7 264	7 318	6 843	6 842	7 443	7 531
United States	69 548	72 997	73 810	72 914	74 188	75 811	72 456	60 325	57 970	60 404
OECD TOTAL	**259 461**	**265 617**	**266 946**	**271 321**	**266 854**	**267 956**	**254 397**	**248 007**	**257 022**	**262 541**
OECD North America	**90 913**	**93 061**	**93 374**	**92 057**	**93 419**	**95 202**	**92 680**	**80 292**	**77 608**	**80 908**
OECD Pacific	**40 833**	**43 980**	**42 703**	**44 560**	**45 436**	**50 243**	**50 367**	**54 437**	**58 057**	**62 139**
OECD Europe	**127 715**	**128 576**	**130 869**	**134 704**	**127 999**	**122 511**	**111 350**	**113 278**	**121 357**	**119 494**
IEA Total	**249 362**	**255 512**	**256 444**	**261 023**	**256 401**	**257 104**	**243 738**	**236 971**	**245 958**	**250 631**
EU15	**112 672**	**113 827**	**114 668**	**118 716**	**111 952**	**106 365**	**95 465**	**96 494**	**104 725**	**102 836**

Table 20: Total Products Consumption in the Other Sectors (continued)
Tableau 20 : Consommation de produits pétroliers dans les autres secteurs (suite)

000 metric tons

1993	1994	1995	1996	1997	1998	1999	2000	2001	2002	
1 791	1 774	1 809	1 807	1 858	1 880	1 893	1 931	1 925	..	Australie
2 586	2 378	2 582	2 910	2 593	2 613	2 580	2 327	2 689	..	Autriche
5 498	5 738	6 079	7 079	6 635	6 397	5 957	5 190	5 238	..	Belgique
11 201	11 419	11 543	12 237	12 676	11 463	11 704	12 974	12 981	..	Canada
984	876	970	603	463	497	133	178	155	..	République Tchèque
2 056	1 858	1 834	1 939	1 783	1 701	1 664	1 498	1 512	..	Danemark
2 340	2 343	2 388	2 358	2 238	2 298	2 063	1 999	2 177	..	Finlande
18 431	17 087	17 201	18 280	17 243	18 552	18 439	16 464	18 428	..	France
36 110	34 235	33 844	37 268	35 497	34 053	29 343	27 854	31 829	..	Allemagne
2 458	2 477	2 501	3 099	3 181	3 291	3 246	3 430	3 696	..	Grèce
1 450	1 537	1 152	953	813	739	712	694	612	..	Hongrie
265	267	265	288	278	266	259	243	215	..	Islande
1 282	1 532	1 596	1 563	1 661	1 712	1 898	1 893	1 983	..	Irlande
10 575	8 892	9 707	9 613	9 100	9 227	9 912	9 263	10 413	..	Italie
43 486	43 913	45 475	46 025	44 276	43 597	42 310	51 906	53 016	..	Japon
17 349	19 257	21 688	22 530	22 538	16 324	19 175	17 075	15 910	..	Corée
335	320	305	340	342	360	334	321	362	..	Luxembourg
10 112	10 377	10 498	10 286	10 375	10 782	9 389	10 351	10 134	..	Mexique
1 828	1 675	950	1 067	944	1 074	1 001	903	887	..	Pays-Bas
306	390	355	326	329	350	339	309	297	..	Nouvelle Zélande
1 099	1 152	1 137	1 343	1 235	1 226	1 241	1 040	1 176	..	Norvège
2 237	2 600	2 661	2 943	3 615	4 019	4 209	4 828	4 948	..	Pologne
1 310	1 326	1 401	1 734	1 664	1 626	1 628	1 678	1 828	..	Portugal
360	305	265	256	286	230	206	176	246	..	République Slovaque
6 137	6 589	6 661	6 842	6 719	6 508	6 905	7 521	7 377	..	Espagne
2 976	3 126	3 099	3 341	3 028	2 919	2 773	2 932	1 905	..	Suède
5 339	4 798	5 011	5 185	4 751	5 163	4 728	4 401 e	4 585 e	..	Suisse
5 668	5 422	6 200	6 391	6 051	5 706	5 898	5 962	5 311	..	Turquie
7 476	7 479	7 323	6 923	6 297	6 204	5 607	5 126	5 558	..	Royaume-Uni
60 909	59 535	58 546	61 065	58 994	53 992	54 971	60 220	61 786	..	Etats-Unis
263 954	**260 677**	**265 046**	**276 594**	**267 463**	**254 769**	**250 517**	**260 687**	**269 179**	..	**OCDE TOTAL**
[illegible]	**81 331**	**80 587**	**83 588**	**82 045**	**76 237**	**76 064**	**83 545**	**84 901**	..	**OCDE Amérique du N.**
62 932	**65 334**	**69 327**	**70 000**	**69 001**	**62 151**	**63 717**	**71 221**	**71 148**	..	**OCDE Pacifique**
118 800	**114 012**	**115 132**	**122 318**	**116 417**	**116 381**	**110 736**	**106 921**	**113 130**	..	**OCDE Europe**
250 980	**247 128**	**251 357**	**262 821**	**252 909**	**239 472**	**236 454**	**245 089**	**253 636**	..	**AIE Total**
101 398	**97 055**	**97 471**	**104 356**	**98 925**	**98 535**	**93 350**	**88 399**	**95 882**	..	**UE15**

INTERNATIONAL ENERGY AGENCY
ENERGY STATISTICS DIVISION
POSSIBLE STAFF VACANCIES

The Division is responsible for statistical support and advice to the policy and operational Divisions of the International Energy Agency. It also produces a wide range of annual and quarterly publications complemented by CD-ROMs and on-line data services. For these purposes, the Division maintains extensive international databases covering most aspects of energy supply and use.

Vacancies for statisticians occur from time to time. Typically their work includes:

- Gathering and vetting data from questionnaires and publications, discussions on data issues with respondents to questionnaires in national administrations and fuel companies.
- Managing energy databases in order to maintain accuracy and timeliness of output.
- Creating and maintaining computer procedures for the production of tables, reports and analyses.
- Preparing studies on an ad-hoc basis as required by other Divisions of the International Energy Agency.

Nationals of any OECD Member country are eligible for appointment. Basic salaries range from 2 590 to 3 430 Euros per month, depending on qualifications. The possibilities for advancement are good for candidates with appropriate qualifications and experience. Tentative enquiries about future vacancies are welcomed from men and women with relevant qualifications and experience. Applications in French or English, accompanied by a curriculum vitae, should be sent to:

Ms. Beth Hunter
Head of Administrative Unit
IEA, 9 rue de la Fédération
75739 Paris Cedex 15, FRANCE
beth.hunter@iea.org

AGENCE INTERNATIONALE DE L'ENERGIE
DIVISION DES STATISTIQUES DE L'ENERGIE
VACANCES D'EMPLOI EVENTUELLES

La Division est chargée de fournir une aide et des conseils dans le domaine statistique aux Divisions administratives et opérationnelles de l'Agence internationale de l'énergie. En outre, elle diffuse une large gamme de publications annuelles et trimestrielles complétées par CD-ROMs ainsi que des services de données sur internet. A cet effet, la Division tient à jour de vastes bases de données internationales portant sur la plupart des aspects de l'offre et de la consommation d'énergie.

Des postes de statisticien sont susceptibles de se libérer de temps à autre. Les fonctions dévolues aux titulaires de ces postes sont notamment les suivantes :

- Rassembler et valider les données tirées de questionnaires et de publications, ainsi que d'échanges de vues sur les données avec les personnes des Administrations nationales ou des entreprises du secteur de l'énergie qui répondent aux questionnaires.
- Gérer des bases de données relatives à l'énergie en vue de s'assurer de l'exactitude et de l'actualisation des données de sortie.
- Gérer et maintenir des procédures informatiques pour la réalisation de tableaux, rapports et analyses.
- Effectuer des études en fonction des besoins des autres Divisions de l'Agence internationale de l'énergie.

Ces postes sont ouverts aux ressortissants des pays Membres de l'OCDE. Les traitements de base sont compris entre 2 590 et 3 430 euros par mois, suivant les qualifications. Les candidats possédant les qualifications et l'expérience appropriées se verront offrir des perspectives de promotion. Les demandes de renseignements sur les postes susceptibles de se libérer qui émanent de personnes dotées des qualifications et de l'expérience voulues seront les bienvenues. Les candidatures, rédigées en français ou en anglais et accompagnées d'un curriculum vitae, doivent être envoyées à l'adresse suivante :

Ms. Beth Hunter
Head of Administrative Unit
IEA, 9 rue de la Fédération
75739 Paris Cedex 15, FRANCE
beth.hunter@iea.org

New!!

On-Line Data Services

Users can now instantly access not only all the data published in this book but also all the time-series used for preparing this publication and all the other statistics publications of the IEA. The data are available on-line, either through annual subscription or pay-per-view access. More information on this new service can be found on our website: http://data.iea.org/.

TEN ANNUAL PUBLICATIONS

Renewables Information 2003

This annual publication of comprehensive information on the use of renewables and waste in the OECD region was first published in 2002. The report addresses a need for development of reliable statistics on this energy form. Support for increasing renewable energy's role in the energy sector has never been greater, and this publication seeks to increase understanding of the current market and trends over recent years. The report contains analysis of renewables and waste energy supply, electricity production and installed electricity generating capacity in OECD countries. The analysis is supported by detailed statistical tables for eight regional aggregates and for each of the thirty OECD countries. *Published September 2003.*

Coal Information 2003

Issued annually since 1983, this publication provides comprehensive information on current world coal market trends and long-term prospects. Compiled in co-operation with the Coal Industry Advisory Board, it contains thorough analysis and current country-specific statistics for OECD Member countries and selected non-OECD countries on coal prices, demand, trade, production, productive capacity, emissions standards for coal-fired boilers, coal ports, coal-fired power stations and coal data for non-OECD countries. This publication is a key reference tool for all sectors of the coal industry as well as for OECD Member country governments. *Published August 2003.*

Electricity Information 2003

This publication brings together in one volume the IEA's data on electricity and heat supply and demand in the OECD. The report presents a comprehensive picture of electricity capacity and production, consumption, trade and prices for the OECD regions and individual countries in over 20 separate tables for each OECD country. Detailed data on the fuels used for electricity and heat production are also presented. *Published September 2003.*

Natural Gas Information 2003

A detailed reference work on gas supply and demand, covering not only the OECD countries but also the rest of the world. Contains essential information on LNG and pipeline trade, gas reserves, storage capacity and prices. The main part of the book, however, concentrates on OECD countries, showing a detailed gas supply and demand balance for each individual country and for the three OECD regions: North America, Europe and Asia-Pacific, as well as a breakdown of gas consumption by end-user. Import and export data are reported by source and destination. *Published September 2003.*

Oil Information 2003

A comprehensive reference book on current developments in oil supply and demand. The first part of this publication contains key data on world production, trade, prices and consumption of major oil product groups, with time series back to the early 1970s. The second part gives a more detailed and comprehensive picture of oil supply, demand, trade, production and consumption by end-user for each OECD country individually and for the OECD regions. Trade data are reported extensively by origin and destination. *Published August 2003.*

Energy Statistics of OECD Countries 2000-2001

No other publication offers such in-depth statistical coverage. It is intended for anyone involved in analytical or policy work related to energy issues. It contains data on energy supply and consumption in original units for coal, oil, natural gas, combustible renewables/wastes and products derived from these primary fuels, as well as for electricity and heat. Data are presented for the two most recent years available in detailed supply and consumption tables. Historical tables summarise data on production, trade and final consumption. Each issue includes definitions of products and flows and explanatory notes on the individual country data. *Published June 2003.*

Energy Balances of OECD Countries 2000-2001

A companion volume to *Energy Statistics of OECD Countries*, this publication presents standardised energy balances expressed in million tonnes of oil equivalent. Energy supply and consumption data are divided by main fuel: coal, oil, gas, nuclear, hydro, geothermal/solar, combustible renewables/wastes, electricity and heat. This allows for easy comparison of the contributions each fuel makes to the economy and their interrelationships through the conversion of one fuel to another. All of this is essential for estimating total energy supply, forecasting, energy conservation, and analysing the potential for interfuel substitution. Complete energy balances are presented for the two most recent years available. Historical tables summarise key energy and economic indicators as well as data on production, trade and final consumption. Each issue includes definitions of products and flows and explanatory notes on the individual country data as well as conversion factors from original units to tonnes of oil equivalent. *Published June 2003.*

Energy Statistics of Non-OECD Countries 2000-2001

This publication offers the same in-depth statistical coverage as the homonymous publication covering OECD countries. It includes data in original units for over 100 individual countries and nine main regions. The consistency of OECD and non-OECD countries' detailed statistics provides an accurate picture of the global energy situation. For a description of the content, please see *Energy Statistics of OECD Countries* above. *Published September 2003.*

Energy Balances of Non-OECD Countries 2000-2001

A companion volume to the publication *Energy Statistics of Non-OECD Countries*, this publication presents energy balances in million tonnes of oil equivalent and key economic and energy indicators for over 100 individual countries and nine main regions. It offers the same statistical coverage as the homonymous publication covering OECD Countries, and thus provides an accurate picture of the global energy situation. For a description of the content, please see *Energy Balances of OECD Countries* above. *Published September 2003.*

CO_2 Emissions from Fuel Combustion - 2003 Edition

In order for nations to tackle the problem of climate change, they need accurate greenhouse gas emissions data. This publication provides a new basis for comparative analysis of CO_2 emissions from fossil fuel combustion, a major source of anthropogenic emissions. The data in this book are designed to assist in understanding the evolution of these emissions from 1971 to 2001 on a country, regional and worldwide basis. They should help in the preparation and the follow-up to the Ninth Conference of the Parties (COP-9) meeting under the U.N. Climate Convention in Milan, Italy in December 2003. Emissions were calculated using IEA energy databases and the default methods and emissions factors from the *Revised 1996 IPCC Guidelines for National Greenhouse Gas Inventories*. *Published October 2003.*

TWO QUARTERLIES

Oil, Gas, Coal and Electricity, Quarterly Statistics

Oil statistics cover OECD production, trade (by origin and destination), refinery intake and output, stock changes and consumption for crude oil, NGL and nine selected oil product groups. Statistics for natural gas show OECD supply, consumption and trade (by origin and destination). Coal data cover the main OECD and world-wide producers of hard and brown coal and major exporters and importers of steam and coking coal. Trade data for the main OECD countries are reported by origin and destination. Electricity statistics cover production (by major fuel category), consumption and trade for 29 OECD countries. Quarterly data on world oil and coal production are included, as well as world steam and coking coal trade.

Energy Prices and Taxes

This publication responds to the needs of the energy industry and OECD governments for up-to-date information on prices and taxes in national and international energy markets. It contains for OECD countries and certain non-OECD countries prices at all market levels: import prices, industry prices and consumer prices. The statistics cover the main petroleum products, gas, coal and electricity, giving for imported products an average price both for importing country and country of origin. Every issue includes full notes on sources and methods and a description of price mechanisms in each country.

CD-ROMs

To complement its publications, the Energy Statistics Division produces CD-ROMs containing the complete databases which are used for preparing the statistics publications. State-of-the-art software allows you to access and manipulate all these data in a very user-friendly manner and includes graphic and mapping facilities.

Annual CD-ROMs

- Energy Statistics of OECD Countries, 1960-2001
- Energy Balances of OECD Countries, 1960-2001
- Energy Statistics of Non-OECD Countries, 1971-2001
- Energy Balances of Non-OECD Countries, 1971-2001
- CO_2 Emissions from Fuel Combustion 1960/1971-2001
- Coal Information 2003
- Electricity Information 2003
- Natural Gas Information 2003
- Oil Information 2003

Quarterly CD-ROMs

- Energy Prices and Taxes

ON-LINE DATA SERVICES

All the databases available on CD-ROM are now also accessible over the Internet at http://data.iea.org/. For additional information on this new service see the note at the beginning of these pages.

Furthermore, the IEA *Monthly Oil Data Service* and a *Monthly Gas Data Service* (see boxes below) can also be accessed over the Internet.

The IEA Monthly Oil Data Service

The IEA Monthly Oil Data Service provides the detailed databases of historical and projected information which is used in preparing the IEA's monthly Oil Market Report (OMR). The IEA Monthly Oil Data Service comprises three packages:

- ***Supply, Demand, Balances and Stocks;***
- ***Trade;***
- ***Field-by-Field Supply;***

available separately or combined as a subscriber service on the Internet. The data are available two days after the official release of the Oil Market Report.

A full description of this service is available on our web site (http://www.iea.org.)

The IEA Monthly Gas Data Service

The Monthly Gas Data Service provides for OECD countries historical and current data on natural gas supply and demand, as well as detailed information on trade origins and destinations. Data can be obtained in different ways from the Internet.

A description of the service is available on our web site (http://data.iea.org).

For more information on any of the IEA statistics products, please feel free to contact Ms. Sharon Michel in the Energy Statistics Division, E-mail: sharon.michel@iea.org; Tel: (+33 1) 40 57 66 25; Fax: (+33 1) 40 57 66 49.

ORDER FORM

IEA PRODUCTS

Fax: +33 (0)1 40 57 67 75
E-mail: books@iea.org
www.iea.org/books

INTERNATIONAL ENERGY AGENCY

9, rue de la Fédération
F-75739 Paris Cedex 15

PUBLICATIONS

I would like to order the following publications

ANNUAL PUBLICATIONS - 2003 Edition	PUBLICATION DATE	QTY	PRICE* PAPER COPY	PRICE* PDF	TOTAL
☐ Energy Statistics of OECD Countries 2000-2001	June 2003		€110	€88	
☐ Energy Balances of OECD Countries 2000-2001	June 2003		€110	€88	
☐ Energy Statistics of Non-OECD Countries 2000-2001	September 2003		€110	€88	
☐ Energy Balances of Non-OECD Countries 2000-2001	September 2003		€110	€88	
☐ Coal Information 2003	August 2003		€200	€160	
☐ Electricity Information 2003	September 2003		€130	€104	
☐ Natural Gas Information 2003	September 2003		€150	€120	
☐ Oil Information 2003	August 2003		€150	€120	
☐ Renewables Information 2003	November 2003		Free	Free	
☐ CO_2 Emissions from Fuel Combustion 1971-2001	November 2003		€150	€120	

Please enter my subscription as indicated below

QUARTERLY PUBLICATIONS	QTY	SINGLE COPY* PAPER COPY	SINGLE COPY* PDF	ANNUAL* PAPER COPY	ANNUAL* PDF	TOTAL
☐ Energy Prices and Taxes		€110	€88	€350	€280	
☐ Oil, Gas, Coal and Electricity Statistics		€110	NA	€350	NA	

*Please cercle the appropriate choice.

DELIVERY DETAILS

Name　　　　Organisation

Address

Country　　　　Postcode

Telephone　　　　Fax or e-mail

PAYMENT DETAILS

☐ I enclose a cheque payable to IEA Publications for the sum of € ____________

☐ Please debit my credit card (tick choice).　☐ Mastercard　☐ VISA　☐ American Express

Card no:

Expiry date:

Signature:

IEA PARIS
Tel: (+33-1) 40 57 66 90
Fax: (+52-5) 40 57 67 75
E-mail: books@iea.org

OECD PARIS CENTRE
Tel: (+33-01) 45 24 81 67
Fax: (+33-01) 49 10 42 76
E-mail: distribution@oecd.org

OECD BONN CENTRE
Tel: (+49-228) 959 12 15
Fax: (+49-228) 959 12 18
E-mail: bonn.contact@oecd.org

You can also send your order to your nearest OECD sales point or through the IEA online services: www.iea.org/books

OECD MEXICO CENTRE
Tel: (+52-5) 280 12 09
Fax: (+52-5) 280 04 80
E-mail: mexico.contact@oecd.org

OECD TOKYO CENTRE
Tel: (+81-3) 3586 2016
Fax: (+81-3) 3584 7929
E-mail: center@oecdtokyo.org

OECD WASHINGTON CENTER
Tel: (+1-202) 785-6323
Toll-free number for orders:
(+1-800) 456-6323
Fax: (+1-202) 785-0350
E-mail: washington.contact@oecd.org

ORDER FORM

IEA PRODUCTS

Fax: +33 (0)1 40 57 67 75
E-mail: books@iea.org
www.iea.org/books

INTERNATIONAL ENERGY AGENCY

9, rue de la Fédération
F-75739 Paris Cedex 15

CD-ROMs

*I would like to order the following CD-ROMs**

ANNUAL CD-ROMs* - 2003 Edition	PUBLICATION DATE	QTY	PRICE	TOTAL
☐ Energy Statistics of OECD Countries 1960-2001	June 2003		€500	
☐ Energy Balances of OECD Countries 1960-2001	June 2003		€500	
☐ Energy Statistics of Non-OECD Countries 1971-2001	September 2003		€500	
☐ Energy Balances of Non-OECD Countries 1971-2001	September 2003		€500	
☐ *Combined subscription of the above four series*	-		*€1200*	
☐ Coal Information 2003	August 2003		€500	
☐ Electricity Information 2003	September 2003		€500	
☐ Natural Gas Information 2003	September 2003		€500	
☐ Oil Information 2003	August 2003		€500	
☐ CO_2 Emissions from Fuel Combustion 1971-2001	November 2003		€500	

Please enter my subscription as indicated below

QUARTERLY CD-ROMs*	QTY	PRICE	TOTAL
☐ Energy Prices and Taxes (four quarters)		€800	

*Prices are for single user licence. Please contact us for pricing information on multi-user licences.

DELIVERY DETAILS

Name ____ Organisation ____

Address ____

Country ____ Postcode ____

Telephone ____ Fax or e-mail ____

PAYMENT DETAILS

☐ I enclose a cheque payable to IEA Publications for the sum of € ____

☐ Please debit my credit card (tick choice). ☐ Mastercard ☐ VISA ☐ American Express

Card no: ____

Expiry date: ____

Signature: ____

IEA PARIS
Tel: (+33-1) 40 57 66 90
Fax: (+52-5) 40 57 67 75
E-mail: books@iea.org

OECD PARIS CENTRE
Tel: (+33-01) 45 24 81 67
Fax: (+33-01) 49 10 42 76
E-mail: distribution@oecd.org

OECD BONN CENTRE
Tel: (+49-228) 959 12 15
Fax: (+49-228) 959 12 18
E-mail: bonn.contact@oecd.org

You can also send your order to your nearest OECD sales point or through the IEA online services: www.iea.org/books

OECD MEXICO CENTRE
Tel: (+52-5) 280 12 09
Fax: (+52-5) 280 04 80
E-mail: mexico.contact@oecd.org

OECD TOKYO CENTRE
Tel: (+81-3) 3586 2016
Fax: (+81-3) 3584 7929
E-mail: center@oecdtokyo.org

OECD WASHINGTON CENTER
Tel: (+1-202) 785-6323
Toll-free number for orders: (+1-800) 456-6323
Fax: (+1-202) 785-0350
E-mail: washington.contact@oecd.org

ORDER FORM

IEA PRODUCTS

Fax: +33 (0)1 40 57 66 49
E-mail: mos@iea.org

INTERNATIONAL ENERGY AGENCY

9, rue de la Fédération
F-75739 Paris Cedex 15

Monthly Oil Data Service
Annual Subscription Form

Please tick the package you require:

DATABASE Annual Subscription for 12 months	PRICE FOR SINGLE USER LICENCE	PRICE FOR MULTI-USER LICENCE*
☐ Supply, Demand, Balances and Stocks	€5 500	€11 000
☐ Trade	€1 650	€3 300
☐ Field-by-Field Supply	€2 750	€5 500
☐ COMPLETE SERVICE (Supply, Demand, Balances and Stocks, Trade and Field-by-Field Supply)	€8 250	€16 500
TOTAL COST		

*Multi-user licences must be purchased for 2 users and above.

Monthly Gas Data Service

Please tick the package you require:

DATABASE	PRICE FOR SINGLE USER LICENCE	PRICE FOR MULTI-USER LICENCE*
☐ **Natural Gas Balances & Trade - *Historical plus 12 Monthly updates*** Annual Subscription for 12 months	€440	€880
☐ **Natural Gas Balances & Trade - *Historical*** Historical database only	€330	€660

*Multi-user licences must be purchased for 2 users and above.

DELIVERY DETAILS

Name ______ Organisation ______

Address ______

______

Country ______ Postcode ______

Telephone ______ Fax or e-mail ______

PAYMENT DETAILS

☐ I enclose a cheque payable to IEA Publications for the sum of € ______

☐ Please debit my credit card (tick choice). ☐ Mastercard ☐ VISA ☐ American Express

Card no: ______ Expiry date: ______	Signature: